Conversion between USCS and SI

To convert from USCS to SI: To convert the value of a variable from USCS units to equivalent SI units, ***multiply*** the value to be converted by the right-hand side of the corresponding equivalency statement in the Table of Equivalencies.

Example: Convert a length $L = 3.25$ in to its equivalent value in millimeters.

Solution: The corresponding equivalency statement is: 1.0 in $= 25.4$ mm

$$L = 3.25 \text{ in} \times (25.4 \text{ mm/in}) = \textbf{82.55 mm}$$

To convert from SI to USCS: To convert the value of a variable from SI units to equivalent USCS units, ***divide*** the value to be converted by the right-hand side of the corresponding equivalency statement in the Table of Equivalencies.

Example: Convert an area $A = 1000$ mm^2 to its equivalent in square inches.

Solution: The corresponding equivalency statement is: 1.0 in$^2 = 645.16$ mm^2

$$A = 1000 \text{ mm}^2/(645.16 \text{ mm}^2/\text{in}^2) = \textbf{1.55 in}^2$$

FUNDAMENTALS OF MODERN MANUFACTURING

Materials, Processes, and Systems

Fourth Edition

Mikell P. Groover

Professor of Industrial and
Systems Engineering
Lehigh University

The author and publisher gratefully acknowledge the contributions of
Dr. Gregory L. Tonkay, Associate Professor of Industrial and
Systems Engineering, Lehigh University.

WILEY

JOHN WILEY & SONS, INC.

ACQUISITIONS EDITOR	Michael McDonald
EDITORIAL ASSISTANT	Renata Marchione
SENIOR PRODUCTION EDITOR	Anna Melhorn
MARKETING MANAGER	Christopher Ruel
SENIOR DESIGNER	James O'Shea
MEDIA EDITOR	Lauren Sapira
OUTSIDE PRODUCTION MANAGMENT	Thomson Digital
COVER PHOTO	Courtesy of Kennametal, Inc.

This book was set in Times New Roman by Thomson Digital and printed and bound by World Color. The cover was printed by World Color.

This book is printed on acid-free paper. ∞

Groover, Mikell P.
Fundamentals of modern manufacturing: materials, processes and systems, 4th ed.

ISBN 978-0470-467002

Printed in the United States of America

10 9 8 7 6 5 4 3 2 1

PREFACE

Fundamentals of Modern Manufacturing: Materials, Processes, and Systems is designed for a first course or two-course sequence in manufacturing at the junior level in mechanical, industrial, and manufacturing engineering curricula. Given its coverage of engineering materials, it is also suitable for materials science and engineering courses that emphasize materials processing. Finally, it may be appropriate for technology programs related to the preceding engineering disciplines. Most of the book's content is concerned with manufacturing processes (about 65% of the text), but it also provides significant coverage of engineering materials and production systems. Materials, processes, and systems are the basic building blocks of modern manufacturing and the three broad subject areas covered in the book.

APPROACH

The author's objective in this edition and its predecessors is to provide a treatment of manufacturing that is ***modern*** and ***quantitative***. Its claim to be "modern" is based on (1) its balanced coverage of the basic engineering materials (metals, ceramics, polymers, and composite materials), (2) its inclusion of recently developed manufacturing processes in addition to the traditional processes that have been used and refined over many years, and (3) its comprehensive coverage of electronics manufacturing technologies. Competing textbooks tend to emphasize metals and their processing at the expense of the other engineering materials, whose applications and methods of processing have grown significantly in the last several decades. Also, most competing books provide minimum coverage of electronics manufacturing. Yet the commercial importance of electronics products and their associated industries have increased substantially during recent decades.

The book's claim to be more "quantitative" is based on its emphasis on manufacturing science and its greater use of mathematical models and quantitative (end-of-chapter) problems than other manufacturing textbooks. In the case of some processes, it was the first manufacturing processes book to ever provide a quantitative engineering coverage of the topic.

NEW TO THIS EDITION

This fourth edition is an updated version of the third edition. The publisher's instructions to the author were to increase content but reduce page count. As this preface is being written, it is too early to tell whether the page count is reduced, but the content has definitely been increased. Additions and changes in the fourth edition include the following:

> The chapter count has been reduced from 45 to 42 through consolidation of several chapters.

> Selected end-of-chapter problems have been revised to make use of PC spread sheet calculations.

> A new section on trends in manufacturing has been added in Chapter 1.

> Chapter 5 on dimensions, tolerances, and surfaces has been modified to include measuring and gauging techniques used for these part features.
> A new section on specialty steels has been added to Chapter 6 on metals.
> Sections on polymer recycling and biodegradable plastics have been added in Chapter 8 on polymers.
> Several new casting processes are discussed in Chapter 11.
> Sections on thread cutting and gear cutting have been added in Chapter 22 on machining operations and machine tools.
> Several additional hole-making tools have been included in Chapter 23 on cutting tool technology.
> Former Chapters 28 and 29 on industrial cleaning and coating processes have been consolidated into a single chapter.
> A new section on friction-stir welding has been added to Chapter 30 on welding processes.
> Chapter 37 on nanotechnology has been reorganized with several new topics and processes added.
> The three previous Chapters 39, 40, and 41 on manufacturing systems have been consolidated into two chapters: Chapter 38 titled Automation for Manufacturing Systems and Chapter 39 on Integrated Manufacturing Systems. New topics covered in these chapters include automation components and material handling technologies.
> Former Chapters 44 on Quality Control and 45 on Measurement and Inspection have been consolidated into a single chapter, Chapter 42 titled Quality Control and Inspection. New sections have been added on Total Quality Management, Six Sigma, and ISO 9000. The text on conventional measuring techniques has been moved to Chapter 5.

OTHER KEY FEATURES

Additional features of the book continued from the third edition include the following:

> A DVD showing action videos of many of the manufacturing processes is included with the book.
> A large number of end-of-chapter problems, review questions, and multiple choice questions are available to instructors to use for homework exercises and quizzes.
> Sections on *Guide to Processing* are included in each of the chapters on engineering materials.
> Sections on *Product Design Considerations* are provided in many of the manufacturing process chapters.
> *Historical Notes* on many of the technologies are included throughout the book.
> The principal engineering units are System International (metric), but both metric and U.S. Customary Units are used throughout the text.

SUPPORT MATERIAL FOR INSTRUCTORS

For instructors who adopt the book for their courses, the following support materials are available:

➤ A *Solutions Manual* (in digital format) covering all problems, review questions, and multiple-choice quizzes.
➤ A complete set of PowerPoint slides for all chapters.

These support materials may be found at the website www.wiley.com/college/groover. Evidence that the book has been adopted as the main textbook for the course must be verified. Individual questions or comments may be directed to the author personally at Mikell.Groover@Lehigh.edu.

ACKNOWLEDGEMENTS

I would like to express my appreciation to the following people who served as technical reviewers of individual sets of chapters for the first edition: Iftikhar Ahmad (George Mason University), J. T. Black (Auburn University), David Bourell (University of Texas at Austin), Paul Cotnoir (Worcester Polytechnic Institute), Robert E. Eppich (American Foundryman's Society), Osama Eyeda (Virginia Polytechnic Institute and State University), Wolter Fabricky (Virginia Polytechnic Institute and State University), Keith Gardiner (Lehigh University), R. Heikes (Georgia Institute of Technology), Jay R. Geddes (San Jose State University), Ralph Jaccodine (Lehigh University), Steven Liang (Georgia Institute of Technology), Harlan MacDowell (Michigan State University), Joe Mize (Oklahoma State University), Colin Moodie (Purdue University), Michael Philpott (University of Illinois at Urbana-Champaign), Corrado Poli (University of Massachusetts at Amherst), Chell Roberts (Arizona State University), Anil Saigal (Tufts University), G. Sathyanarayanan (Lehigh University), Malur Srinivasan (Texas A&M University), A. Brent Strong (Brigham Young University), Yonglai Tian (George Mason University), Gregory L. Tonkay (Lehigh University), Chester VanTyne (Colorado School of Mines), Robert Voigt (Pennsylvania State University), and Charles White (GMI Engineering and Management Institute).

For their reviews of certain chapters in the second edition, I would like to thank John T. Berry (Mississippi State University), Rajiv Shivpuri (The Ohio State University), James B. Taylor (North Carolina State University), Joel Troxler (Montana State University), and Ampere A. Tseng (Arizona State University).

For their advice and encouragement on the third edition, I would like to thank several of my colleagues at Lehigh, including John Coulter, Keith Gardiner, Andrew Herzing, Wojciech Misiolek, Nicholas Odrey, Gregory Tonkay, and Marvin White. I am especially grateful to Andrew Herzing in the Materials Science and Engineering Department at Lehigh for his review of the new nanofabrication chapter and to Greg Tonkay in my own department for developing many of the new and revised problems and questions in this new edition. For their reviews of the third edition, I would like to thank Mica Grujicic (Clemson University), Wayne Nguyen Hung (Texas A&M University), Patrick Kwon (Michigan State University), Yuan-Shin Lee (North Carolina State University), T. Warren Liao (Louisiana State University), Fuewen Frank Liou (Missouri University of Science and Technology), Val Marinov (North Dakota State University), William J. Riffe (Kettering University), John E. Wyatt (Mississippi State University), Y. Lawrence Yao (Columbia University), Allen Yi (The Ohio State University), and Henry Daniel Young (Wright State University).

For their advice on this fourth edition, I would like to thank the following people: Barbara Mizdail (The Pennsylvania State University – Berks campus) and Jack Feng (formerly of Bradley University and now at Caterpillar, Inc.) for conveying questions and feedback from their students, Larry Smith (St. Clair College, Windsor, Ontario) for his advice on using the ASME standards for hole drilling, Richard Budihas (Voltaic LLC) for his contributed research on nanotechnology and integrated circuit processing, and colleague Marvin White at Lehigh for his insights on integrated circuit technology.

In addition, it seems appropriate to acknowledge my colleagues at Wiley, Senior Acquisition Editor Michael McDonald and Production Editor Anna Melhorn. Last but certainly not least, I appreciate the kind efforts of editor Sumit Shridhar of Thomson Digital.

ABOUT THE AUTHOR

Mikell P. Groover is Professor of Industrial and Systems Engineering at Lehigh University, where he also serves as faculty member in the Manufacturing Systems Engineering Program. He received his B.A. in Arts and Science (1961), B.S. in Mechanical Engineering (1962), M.S. in Industrial Engineering (1966), and Ph.D. (1969), all from Lehigh. He is a Registered Professional Engineer in Pennsylvania. His industrial experience includes several years as a manufacturing engineer with Eastman Kodak Company. Since joining Lehigh, he has done consulting, research, and project work for a number of industrial companies.

His teaching and research areas include manufacturing processes, production systems, automation, material handling, facilities planning, and work systems. He has received a number of teaching awards at Lehigh University, as well as the *Albert G. Holzman Outstanding Educator Award* from the Institute of Industrial Engineers (1995) and the *SME Education Award* from the Society of Manufacturing Engineers (2001). His publications include over 75 technical articles and ten books (listed below). His books are used throughout the world and have been translated into French, German, Spanish, Portuguese, Russian, Japanese, Korean, and Chinese. The first edition of the current book *Fundamentals of Modern Manufacturing* received the *IIE Joint Publishers Award* (1996) and the *M. Eugene Merchant Manufacturing Textbook Award* from the Society of Manufacturing Engineers (1996).

Dr. Groover is a member of the Institute of Industrial Engineers, American Society of Mechanical Engineers (ASME), the Society of Manufacturing Engineers (SME), the North American Manufacturing Research Institute (NAMRI), and ASM International. He is a Fellow of IIE (1987) and SME (1996).

PREVIOUS BOOKS BY THE AUTHOR

Automation, Production Systems, and Computer-Aided Manufacturing, Prentice Hall, 1980.

CAD/CAM: Computer-Aided Design and Manufacturing, Prentice Hall, 1984 (co-authored with E. W. Zimmers, Jr.).

Industrial Robotics: Technology, Programming, and Applications, McGraw-Hill Book Company, 1986 (co-authored with M. Weiss, R. Nagel, and N. Odrey).

Automation, Production Systems, and Computer Integrated Manufacturing, Prentice Hall, 1987.

Fundamentals of Modern Manufacturing: Materials, Processes, and Systems, originally published by Prentice Hall in 1996, and subsequently published by John Wiley & Sons, Inc., 1999.

Automation, Production Systems, and Computer Integrated Manufacturing, Second Edition, Prentice Hall, 2001.

Fundamentals of Modern Manufacturing: Materials, Processes, and Systems, Second Edition, John Wiley & Sons, Inc., 2002.

Work Systems and the Methods, Measurement, and Management of Work, Pearson Prentice Hall, 2007.

Fundamentals of Modern Manufacturing: Materials, Processes, and Systems, Third Edition, John Wiley & Sons, Inc., 2007.

Automation, Production Systems, and Computer Integrated Manufacturing, Third Edition, Pearson Prentice Hall, 2008.

CONTENTS

1 INTRODUCTION AND OVERVIEW OF MANUFACTURING

Chapter Contents

Making things has been an essential activity of human civilizations since before recorded history. Today, the term *manufacturing* is used for this activity. For technological and economic reasons, manufacturing is important to the welfare of the United States and most other developed and developing nations. *Technology* can be defined as the application of science to provide society and its members with those things that are needed or desired. Technology affects our daily lives, directly and indirectly, in many ways. Consider the list of products in Table 1.1. They represent various technologies that help society and its members to live better. What do all these products have in common? They are all manufactured. These technological wonders would not be available to society if they could not be manufactured. Manufacturing is the critical factor that makes technology possible.

Economically, manufacturing is an important means by which a nation creates material wealth. In the United States, the manufacturing industries account for about 15% of gross domestic product (GDP). A country's natural resources, such as agricultural lands, mineral deposits, and oil reserves, also create wealth. In the U.S., agriculture, mining, and similar industries account for less than 5% of GDP (agriculture alone is only about 1%). Construction and public utilities make up around 5%. The rest is service industries, which include retail, transportation, banking, communication, education, and government. The service sector accounts for more than 75% of U.S. GDP. Government alone accounts for about as much of GDP as the manufacturing sector; however, government services do not create wealth. In the modern global economy, a nation must have a strong manufacturing base (or it must have significant natural resources) if it is to provide a strong economy and a high standard of living for its people.

In this opening chapter, we consider some general topics about manufacturing. What is manufacturing? How is it organized in industry? What are the materials, processes, and systems by which it is accomplished?

1

TABLE 1.1 Products representing various technologies, most of which affect nearly everyone.		
Athletic shoes	Fax machine	One-piece molded plastic patio chair
Automatic teller machine	Flat-screen high-definition television	Optical scanner
Automatic dishwasher	Hand-held electronic calculator	Personal computer (PC)
Ballpoint pen	High density PC diskette	Photocopying machine
Cell phone	Home security system	Pull-tab beverage cans
Compact disc (CD)	Hybrid gas-electric automobile	Quartz crystal wrist watch
Compact disc player	Industrial robot	Self-propelled mulching lawnmower
Compact fluorescent light bulb	Ink-jet color printer	Supersonic aircraft
Contact lenses	Integrated circuit	Tennis racket of composite materials
Digital camera	Magnetic resonance imaging	Video games
Digital video disc (DVD)	(MRI) machine for medical diagnosis	Washing machine and dryer
Digital video disc player	Microwave oven	

1.1 WHAT IS MANUFACTURING?

The word *manufacture* is derived from two Latin words, *manus* (hand) and *factus* (make); the combination means made by hand. The English word *manufacture* is several centuries old, and "made by hand" accurately described the manual methods used when the word was first coined.[1] Most modern manufacturing is accomplished by automated and computer-controlled machinery (Historical Note 1.1).

Historical Note 1.1 *History of manufacturing*

The history of manufacturing can be separated into two subjects: (1) human's discovery and invention of materials and processes to make things, and (2) development of the systems of production. The materials and processes to make things predate the systems by several millennia. Some of the processes—casting, hammering (forging), and grinding—date back 6000 years or more. The early fabrication of implements and weapons was accomplished more as crafts and trades than manufacturing as it is known today. The ancient Romans had what might be called factories to produce weapons, scrolls, pottery and glassware, and other products of the time, but the procedures were largely based on handicraft.

The systems aspects of manufacturing are examined here, and the materials and processes are postponed until Historical Note 1.2. *Systems of manufacturing* refer to the ways of organizing people and equipment so that production can be performed more efficiently. Several historical events and discoveries stand out as having had

a major impact on the development of modern manufacturing systems.

Certainly one significant discovery was the principle of *division of labor*—dividing the total work into tasks and having individual workers each become a specialist at performing only one task. This principle had been practiced for centuries, but the economist Adam Smith (1723–1790) is credited with first explaining its economic significance in *The Wealth of Nations.*

The *Industrial Revolution* (circa 1760–1830) had a major impact on production in several ways. It marked the change from an economy based on agriculture and handicraft to one based on industry and manufacturing. The change began in England, where a series of machines were invented and steam power replaced water, wind, and animal power. These advances gave British industry significant advantages over other nations, and England attempted to restrict export of the new technologies. However, the revolution eventually spread to other European countries and the United States.

---

[1] As a noun, the word *manufacture* first appeared in English around 1567 AD. As a verb, it first appeared around 1683 AD.

Several inventions of the Industrial Revolution greatly contributed to the development of manufacturing: (1) *Watt's steam engine,* a new power-generating technology for industry; (2) *machine tools,* starting with John Wilkinson's boring machine around 1775 (Historical Note 22.1); (3) the *spinning jenny, power loom,* and other machinery for the textile industry that permitted significant increases in productivity; and (4) the *factory system,* a new way of organizing large numbers of production workers based on division of labor.

While England was leading the industrial revolution, an important concept was being introduced in the United States: *interchangeable parts* manufacture. Much credit for this concept is given to Eli Whitney (1765–1825), although its importance had been recognized by others [9]. In 1797, Whitney negotiated a contract to produce 10,000 muskets for the U.S. government. The traditional way of making guns at the time was to custom fabricate each part for a particular gun and then hand-fit the parts together by filing. Each musket was unique, and the time to make it was considerable. Whitney believed that the components could be made accurately enough to permit parts assembly without fitting. After several years of development in his Connecticut factory, he traveled to Washington in 1801 to demonstrate the principle. He laid out components for 10 muskets before government officials, including Thomas Jefferson, and proceeded to select parts randomly to assemble the guns. No special filing or fitting was required, and all of the guns worked perfectly. The secret behind his achievement was the collection of special machines, fixtures, and gages that he had developed in his factory. Interchangeable parts manufacture required many years of development before becoming a practical reality, but it revolutionized methods of manufacturing. It is a prerequisite for mass production. Because its origins were in the United States, interchangeable parts production came to be known as the *American System* of manufacture.

The mid- and late 1800s witnessed the expansion of railroads, steam-powered ships, and other machines that created a growing need for iron and steel. New steel production methods were developed to meet this demand (Historical Note 6.1). Also during this period, several consumer products were developed, including the sewing machine, bicycle, and automobile. To meet the mass demand for these products, more efficient production methods were required. Some historians identify developments during this period as the *Second Industrial Revolution,* characterized in terms of its effects on manufacturing systems by: (1) mass production, (2) scientific management movement, (3) assembly lines, and (4) electrification of factories.

In the late 1800s, the *scientific management* movement was developing in the United States in response to the need to plan and control the activities of growing numbers of production workers. The movement's leaders included Frederick W. Taylor (1856–1915), Frank Gilbreth (1868–1924), and his wife Lilian (1878–1972). Scientific management included several features [2]: (1) *motion study,* aimed at finding the best method to perform a given task; (2) *time study,* to establish work standards for a job; (3) extensive use of *standards* in industry; (4) the *piece rate system* and similar labor incentive plans; and (5) use of data collection, record keeping, and cost accounting in factory operations.

Henry Ford (1863–1947) introduced the *assembly line* in 1913 at his Highland Park, MI plant. The assembly line made possible the mass production of complex consumer products. Use of assembly line methods permitted Ford to sell a Model T automobile for as little as $500, thus making ownership of cars feasible for a large segment of the U.S. population.

In 1881, the first electric power generating station had been built in New York City, and soon electric motors were being used as a power source to operate factory machinery. This was a far more convenient power delivery system than steam engines, which required overhead belts to distribute power to the machines. By 1920, electricity had overtaken steam as the principal power source in U.S. factories. The twentieth century was a time of more technological advances than in all other centuries combined. Many of these developments resulted in the *automation* of manufacturing.

1.1.1 MANUFACTURING DEFINED

As a field of study in the modern context, manufacturing can be defined two ways, one technologic and the other economic. Technologically, *manufacturing* is the application of physical and chemical processes to alter the geometry, properties, and/or appearance of a given starting material to make parts or products; manufacturing also includes assembly of multiple parts to make products. The processes to accomplish manufacturing involve a combination of machinery, tools, power, and labor, as depicted in Figure 1.1(a).

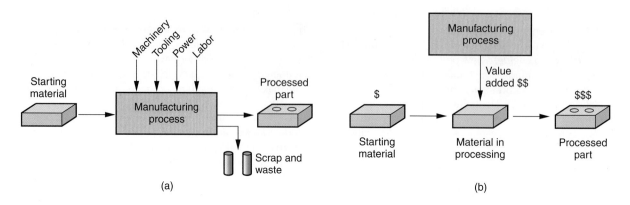

FIGURE 1.1   Two ways to define manufacturing: (a) as a technical process, and (b) as an economic process.

Manufacturing is almost always carried out as a sequence of operations. Each operation brings the material closer to the desired final state.

Economically, ***manufacturing*** is the transformation of materials into items of greater value by means of one or more processing and/or assembly operations, as depicted in Figure 1.1(b). The key point is that manufacturing ***adds value*** to the material by changing its shape or properties, or by combining it with other materials that have been similarly altered. The material has been made more valuable through the manufacturing operations performed on it. When iron ore is converted into steel, value is added. When sand is transformed into glass, value is added. When petroleum is refined into plastic, value is added. And when plastic is molded into the complex geometry of a patio chair, it is made even more valuable.

The words ***manufacturing*** and ***production*** are often used interchangeably. The author's view is that production has a broader meaning than manufacturing. To illustrate, one might speak of "crude oil production," but the phrase "crude oil manufacturing" seems out of place. Yet when used in the context of products such as metal parts or automobiles, either word seems okay.

1.1.2 MANUFACTURING INDUSTRIES AND PRODUCTS

Manufacturing is an important commercial activity performed by companies that sell products to customers. The type of manufacturing done by a company depends on the kind of product it makes. Let us explore this relationship by examining the types of industries in manufacturing and identifying the products they make.

Manufacturing Industries   Industry consists of enterprises and organizations that produce or supply goods and services. Industries can be classified as primary, secondary, or tertiary. ***Primary industries*** cultivate and exploit natural resources, such as agriculture and mining. ***Secondary industries*** take the outputs of the primary industries and convert them into consumer and capital goods. Manufacturing is the principal activity in this category, but construction and power utilities are also included. ***Tertiary industries*** constitute the service sector of the economy. A list of specific industries in these categories is presented in Table 1.2.

This book is concerned with the secondary industries in Table 1.2, which include the companies engaged in manufacturing. However, the International Standard Industrial Classification (ISIC) used to compile Table 1.2 includes several industries whose production technologies are not covered in this text; for example, beverages, chemicals, and food processing. In this book, manufacturing means production of ***hardware,*** which ranges from nuts and bolts to digital computers and military weapons. Plastic and ceramic

TABLE 1.2   Specific industries in the primary, secondary, and tertiary categories.

Primary	Secondary		Tertiary (Service)	
Agriculture	Aerospace	Food processing	Banking	Insurance
Forestry	Apparel	Glass, ceramics	Communications	Legal
Fishing	Automotive	Heavy machinery	Education	Real estate
Livestock	Basic metals	Paper	Entertainment	Repair and
Quarries	Beverages	Petroleum refining	Financial services	maintenance
Mining	Building materials	Pharmaceuticals	Government	Restaurant
Petroleum	Chemicals	Plastics (shaping)	Health and	Retail trade
	Computers	Power utilities	medical	Tourism
	Construction	Publishing	Hotel	Transportation
	Consumer	Textiles	Information	Wholesale trade
	appliances	Tire and rubber		
	Electronics	Wood and furniture		
	Equipment			
	Fabricated metals			

products are included, but apparel, paper, pharmaceuticals, power utilities, publishing, and wood products are excluded.

Manufactured Products   Final products made by the manufacturing industries can be divided into two major classes: consumer goods and capital goods. ***Consumer goods*** are products purchased directly by consumers, such as cars, personal computers, TVs, tires, and tennis rackets. ***Capital goods*** are those purchased by companies to produce goods and/or provide services. Examples of capital goods include aircraft, computers, communication equipment, medical apparatus, trucks and buses, railroad locomotives, machine tools, and construction equipment. Most of these capital goods are purchased by the service industries. It was noted in the Introduction that manufacturing accounts for about 15% of GDP and services about 75% of GDP in the United States. Yet the manufactured capital goods purchased by the service sector are the enablers of that sector. Without the capital goods, the service industries could not function.

In addition to final products, other manufactured items include the ***materials, components,*** and ***supplies*** used by the companies that make the final products. Examples of these items include sheet steel, bar stock, metal stampings, machined parts, plastic moldings and extrusions, cutting tools, dies, molds, and lubricants. Thus, the manufacturing industries consist of a complex infrastructure with various categories and layers of intermediate suppliers with whom the final consumer never deals.

This book is generally concerned with ***discrete items*** —individual parts and assembled products, rather than items produced by ***continuous processes.*** A metal stamping is a discrete item, but the sheet-metal coil from which it is made is continuous (almost). Many discrete parts start out as continuous or semicontinuous products, such as extrusions and electrical wire. Long sections made in almost continuous lengths are cut to the desired size. An oil refinery is a better example of a continuous process.

Production Quantity and Product Variety   The quantity of products made by a factory has an important influence on the way its people, facilities, and procedures are organized. Annual production quantities can be classified into three ranges: (1) ***low*** production, quantities in the range 1 to 100 units per year; (2) ***medium*** production, from 100 to 10,000 units annually; and (3) ***high*** production, 10,000 to millions of units. The boundaries

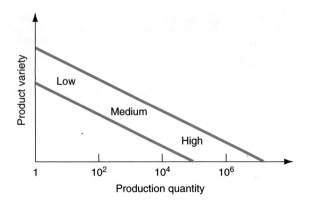

FIGURE 1.2   Relationship between product variety and production quantity in discrete product manufacturing.

between the three ranges are somewhat arbitrary (in the author's judgment). Depending on the kinds of products, these boundaries may shift by an order of magnitude or so.

Production quantity refers to the number of units produced annually of a particular product type. Some plants produce a variety of different product types, each type being made in low or medium quantities. Other plants specialize in high production of only one product type. It is instructive to identify product variety as a parameter distinct from production quantity. *Product variety* refers to different product designs or types that are produced in the plant. Different products have different shapes and sizes; they perform different functions; they are intended for different markets; some have more components than others; and so forth. The number of different product types made each year can be counted. When the number of product types made in the factory is high, this indicates high product variety.

There is an inverse correlation between product variety and production quantity in terms of factory operations. If a factory's product variety is high, then its production quantity is likely to be low; but if production quantity is high, then product variety will be low, as depicted in Figure 1.2. Manufacturing plants tend to specialize in a combination of production quantity and product variety that lies somewhere inside the diagonal band in Figure 1.2.

Although product variety has been identified as a quantitative parameter (the number of different product types made by the plant or company), this parameter is much less exact than production quantity, because details on how much the designs differ are not captured simply by the number of different designs. Differences between an automobile and an air conditioner are far greater than between an air conditioner and a heat pump. Within each product type, there are differences among specific models.

The extent of the product differences may be small or great, as illustrated in the automotive industry. Each of the U.S. automotive companies produces cars with two or three different nameplates in the same assembly plant, although the body styles and other design features are virtually the same. In different plants, the company builds heavy trucks. The terms "soft" and "hard" might be used to describe these differences in product variety. *Soft product variety* occurs when there are only small differences among products, such as the differences among car models made on the same production line. In an assembled product, soft variety is characterized by a high proportion of common parts among the models. *Hard product variety* occurs when the products differ substantially, and there are few common parts, if any. The difference between a car and a truck exemplifies hard variety.

1.1.3   MANUFACTURING CAPABILITY

A manufacturing plant consists of a set of *processes* and *systems* (and people, of course) designed to transform a certain limited range of *materials* into products of increased value. These three building blocks—materials, processes, and systems—constitute the subject of modern manufacturing. There is a strong interdependence among these factors. A company engaged in manufacturing cannot do everything. It must do only certain things, and it must do those things well. ***Manufacturing capability*** refers to the technical and physical limitations of a manufacturing firm and each of its plants. Several dimensions of this capability can be identified: (1) technological processing capability, (2) physical size and weight of product, and (3) production capacity.

Technological Processing Capability   The technological processing capability of a plant (or company) is its available set of manufacturing processes. Certain plants perform machining operations, others roll steel billets into sheet stock, and others build automobiles. A machine shop cannot roll steel, and a rolling mill cannot build cars. The underlying feature that distinguishes these plants is the processes they can perform. Technological processing capability is closely related to material type. Certain manufacturing processes are suited to certain materials, whereas other processes are suited to other materials. By specializing in a certain process or group of processes, the plant is simultaneously specializing in certain material types. Technological processing capability includes not only the physical processes, but also the expertise possessed by plant personnel in these processing technologies. Companies must concentrate on the design and manufacture of products that are compatible with their technological processing capability.

Physical Product Limitations   A second aspect of manufacturing capability is imposed by the physical product. A plant with a given set of processes is limited in terms of the size and weight of the products that can be accommodated. Large, heavy products are difficult to move. To move these products about, the plant must be equipped with cranes of the required load capacity. Smaller parts and products made in large quantities can be moved by conveyor or other means. The limitation on product size and weight extends to the physical capacity of the manufacturing equipment as well. Production machines come in different sizes. Larger machines must be used to process larger parts. The production and material handling equipment must be planned for products that lie within a certain size and weight range.

Production Capacity   A third limitation on a plant's manufacturing capability is the production quantity that can be produced in a given time period (e.g., month or year). This quantity limitation is commonly called ***plant capacity,*** or ***production capacity,*** defined as the maximum rate of production that a plant can achieve under assumed operating conditions. The operating conditions refer to number of shifts per week, hours per shift, direct labor manning levels in the plant, and so on. These factors represent inputs to the manufacturing plant. Given these inputs, how much output can the factory produce?

Plant capacity is usually measured in terms of output units, such as annual tons of steel produced by a steel mill, or number of cars produced by a final assembly plant. In these cases, the outputs are homogeneous. In cases in which the output units are not homogeneous, other factors may be more appropriate measures, such as available labor hours of productive capacity in a machine shop that produces a variety of parts.

Materials, processes, and systems are the basic building blocks of manufacturing and the three broad subject areas of this book. This introductory chapter provides an overview of these three subjects before embarking on detailed coverage in the remaining chapters.

1.2   MATERIALS IN MANUFACTURING

Most engineering materials can be classified into one of three basic categories:(1) metals, (2) ceramics, and (3) polymers.Their chemistries are different, their mechanical and physical properties are different, and these differences affect the manufacturing processes that can be used to produce products from them. In addition to the three basic categories, there are

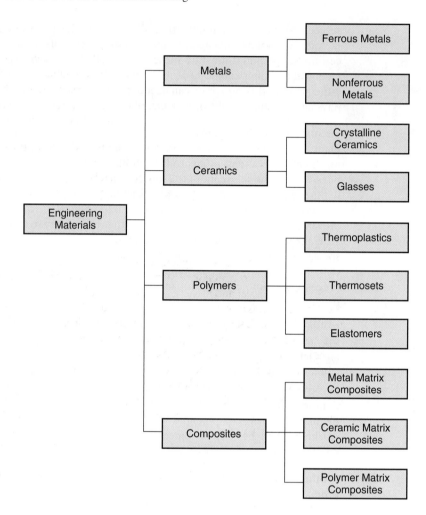

FIGURE 1.3
Classification of the four
engineering materials.

(4) *composites*—nonhomogeneous mixtures of the other three basic types rather than a unique category. The classification of the four groups is pictured in Figure 1.3. This section surveys these materials. Chapters 6 through 9 cover the four material types in more detail.

1.2.1 METALS

Metals used in manufacturing are usually ***alloys,*** which are composed of two or more elements, with at least one being a metallic element. Metals and alloys can be divided into two basic groups: (1) ferrous and (2) nonferrous.

Ferrous Metals   Ferrous metals are based on iron; the group includes steel and cast iron. These metals constitute the most important group commercially, more than three fourths of the metal tonnage throughout the world. Pure iron has limited commercial use, but when alloyed with carbon, iron has more uses and greater commercial value than any other metal. Alloys of iron and carbon form steel and cast iron.

Steel can be defined as an iron–carbon alloy containing 0.02% to 2.11% carbon. It is the most important category within the ferrous metal group. Its composition often includes other alloying elements as well, such as manganese, chromium, nickel, and molybdenum, to enhance the properties of the metal. Applications of steel include construction (bridges, I-beams, and nails), transportation (trucks, rails, and rolling stock for railroads), and consumer products (automobiles and appliances).

Cast iron is an alloy of iron and carbon (2% to 4%) used in casting (primarily sand casting). Silicon is also present in the alloy (in amounts from 0.5% to 3%), and other elements are often added also, to obtain desirable properties in the cast part. Cast iron is available in several different forms, of which gray cast iron is the most common; its applications include blocks and heads for internal combustion engines.

Nonferrous Metals   Nonferrous metals include the other metallic elements and their alloys. In almost all cases, the alloys are more important commercially than the pure metals. The nonferrous metals include the pure metals and alloys of aluminum, copper, gold, magnesium, nickel, silver, tin, titanium, zinc, and other metals.

1.2.2 CERAMICS

A *ceramic* is defined as a compound containing metallic (or semimetallic) and nonmetallic elements. Typical nonmetallic elements are oxygen, nitrogen, and carbon. Ceramics include a variety of traditional and modern materials. Traditional ceramics, some of which have been used for thousands of years, include: *clay* (abundantly available, consisting of fine particles of hydrous aluminum silicates and other minerals used in making brick, tile, and pottery); *silica* (the basis for nearly all glass products); and *alumina* and *silicon carbide* (two abrasive materials used in grinding). Modern ceramics include some of the preceding materials, such as alumina, whose properties are enhanced in various ways through modern processing methods. Newer ceramics include: *carbides*—metal carbides such as tungsten carbide and titanium carbide, which are widely used as cutting tool materials; and *nitrides*—metal and semimetal nitrides such as titanium nitride and boron nitride, used as cutting tools and grinding abrasives.

For processing purposes, ceramics can be divided into crystalline ceramics and glasses. Different methods of manufacturing are required for the two types. Crystalline ceramics are formed in various ways from powders and then fired (heated to a temperature below the melting point to achieve bonding between the powders). The glass ceramics (namely, glass) can be melted and cast, and then formed in processes such as traditional glass blowing.

1.2.3 POLYMERS

A *polymer* is a compound formed of repeating structural units called *mers,* whose atoms share electrons to form very large molecules. Polymers usually consist of carbon plus one or more other elements, such as hydrogen, nitrogen, oxygen, and chlorine. Polymers are divided into three categories: (1) thermoplastic polymers, (2) thermosetting polymers, and (3) elastomers.

Thermoplastic polymers can be subjected to multiple heating and cooling cycles without substantially altering the molecular structure of the polymer. Common thermoplastics include polyethylene, polystyrene, polyvinylchloride, and nylon. *Thermosetting polymers* chemically transform (cure) into a rigid structure on cooling from a heated plastic condition; hence the name thermosetting. Members of this type include phenolics, amino resins, and epoxies. Although the name thermosetting is used, some of these polymers cure by mechanisms other than heating. *Elastomers* are polymers that exhibit significant elastic behavior; hence the name elastomer. They include natural rubber, neoprene, silicone, and polyurethane.

1.2.4 COMPOSITES

Composites do not really constitute a separate category of materials; they are mixtures of the other three types. A *composite* is a material consisting of two or more phases that are processed separately and then bonded together to achieve properties superior to those of its constituents. The term *phase* refers to a homogeneous mass of material, such as an aggregation of grains of identical unit cell structure in a solid metal. The usual structure of a composite consists of particles or fibers of one phase mixed in a second phase, called the *matrix.*

Composites are found in nature (e.g., wood), and they can be produced synthetically. The synthesized type is of greater interest here, and it includes glass fibers in a polymer matrix, such as fiber-reinforced plastic; polymer fibers of one type in a matrix of a second polymer, such as an epoxy-Kevlar composite; and ceramic in a metal matrix, such as a tungsten carbide in a cobalt binder to form a cemented carbide cutting tool.

Properties of a composite depend on its components, the physical shapes of the components, and the way they are combined to form the final material. Some composites combine high strength with light weight and are suited to applications such as aircraft components, car bodies, boat hulls, tennis rackets, and fishing rods. Other composites are strong, hard, and capable of maintaining these properties at elevated temperatures, for example, cemented carbide cutting tools.

1.3   MANUFACTURING PROCESSES

A *manufacturing process* is a designed procedure that results in physical and/or chemical changes to a starting work material with the intention of increasing the value of that material. A manufacturing process is usually carried out as a *unit operation*, which means that it is a single step in the sequence of steps required to transform the starting material into a final product. Manufacturing operations can be divided into two basic types: (1) processing operations and (2) assembly operations. A *processing operation* transforms a work material from one state of completion to a more advanced state that is closer to the final desired product. It adds value by changing the geometry, properties, or appearance of the starting material. In general, processing operations are performed on discrete work-parts, but certain processing operations are also applicable to assembled items (e.g., painting a spot-welded car body). An *assembly operation* joins two or more components to create a new entity, called an assembly, subassembly, or some other term that refers to the joining process (e.g., a welded assembly is called a *weldment*). A classification of manufacturing processes is presented in Figure 1.4. Many of the manufacturing processes covered in this text can be viewed on the DVD that comes with this book. Alerts are provided on these video clips throughout the text. Some of the basic processes used in modern manufacturing date from antiquity (Historical Note 1.2).

1.3.1   PROCESSING OPERATIONS

A processing operation uses energy to alter a workpart's shape, physical properties, or appearance to add value to the material. The forms of energy include mechanical, thermal, electrical, and chemical. The energy is applied in a controlled way by means of machinery and tooling. Human energy may also be required, but the human workers are generally employed to control the machines, oversee the operations, and load and unload parts before and after each cycle of operation. A general model of a processing operation is illustrated in Figure 1.1(a). Material is fed into the process, energy is applied by the machinery and tooling to transform the material, and the completed workpart exits the process. Most production operations produce waste or scrap, either as a natural aspect of the process (e.g., removing material, as in machining) or in the form of occasional defective pieces. It is an important objective in manufacturing to reduce waste in either of these forms.

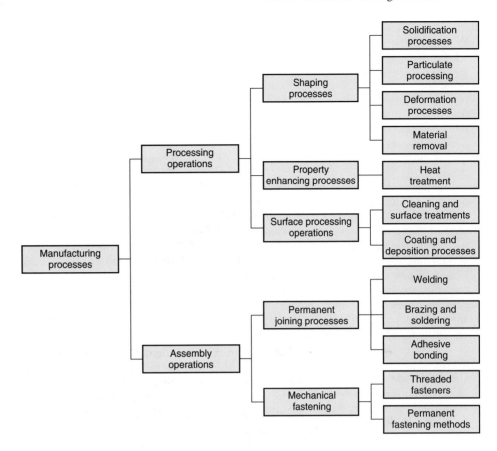

FIGURE 1.4
Classification of manufacturing processes.

Historical Note 1.2   *Manufacturing materials and processes*

Although most of the historical developments that form the modern practice of manufacturing have occurred only during the last few centuries (Historical Note 1.1), several of the basic fabrication processes date as far back as the Neolithic period (circa 8000–3000 BCE.). It was during this period that processes such as the following were developed: carving and other *woodworking,* hand forming and *firing* of clay pottery, *grinding* and *polishing* of stone, *spinning* and *weaving* of textiles, and *dyeing* of cloth.

Metallurgy and metalworking also began during the Neolithic period, in Mesopotamia and other areas around the Mediterranean. It either spread to, or developed independently in, regions of Europe and Asia. Gold was found by early humans in relatively pure form in nature; it could be *hammered* into shape. Copper was probably the first metal to be extracted from ores, thus requiring *smelting* as a processing technique. Copper could not be hammered readily because it strain hardened; instead, it was shaped by *casting* (Historical Note 10.1). Other metals used during this period were silver and tin. It was discovered that copper alloyed with tin produced a more workable metal than copper alone (casting and hammering could both be used). This heralded the important period known as the *Bronze Age* (circa 3500–1500 BCE.).

Iron was also first smelted during the Bronze Age. Meteorites may have been one source of the metal, but iron ore was also mined. Temperatures required to reduce iron ore to metal are significantly higher than for copper, which made furnace operations more difficult. Other processing methods were also more difficult for the same reason. Early blacksmiths learned that when certain irons (those containing small amounts of carbon) were sufficiently *heated* and then *quenched,* they became very hard. This permitted grinding a very sharp cutting edge on knives and weapons, but it also made the metal brittle. Toughness could be increased by reheating at a lower temperature, a process known as *tempering.*

What we have described is, of course, the **heat treatment** of steel. The superior properties of steel caused it to succeed bronze in many applications (weaponry, agriculture, and mechanical devices). The period of its use has subsequently been named the **Iron Age** (starting around 1000 bce.). It was not until much later, well into the nineteenth century, that the demand for steel grew significantly and more modern steelmaking techniques were developed (Historical Note 6.1).

The beginnings of machine tool technology occurred during the Industrial Revolution. During the period 1770–1850, machine tools were developed for most of the conventional **material removal processes,** such as **boring, turning, drilling, milling, shaping,** and **planing** (Historical Note 22.1). Many of the individual processes predate the machine tools by centuries; for example, drilling and sawing (of wood) date from ancient times, and turning (of wood) from around the time of Christ.

Assembly methods were used in ancient cultures to make ships, weapons, tools, farm implements, machinery, chariots and carts, furniture, and garments. The earliest processes included **binding** with twine and rope, **riveting** and **nailing,** and **soldering.** Around 2000 years ago, **forge welding** and **adhesive bonding** were developed. Widespread use of screws, bolts, and nuts as

fasteners—so common in today's assembly—required the development of machine tools that could accurately cut the required helical shapes (e.g., Maudsley's screw cutting lathe, 1800). It was not until around 1900 that **fusion welding** processes started to be developed as assembly techniques (Historical Note 29.1).

Natural rubber was the first polymer to be used in manufacturing (if we overlook wood, which is a polymer composite). The **vulcanization** process, discovered by Charles Goodyear in 1839, made rubber a useful engineering material (Historical Note 8.2). Subsequent developments included plastics such as cellulose nitrate in 1870, Bakelite in 1900, polyvinylchloride in 1927, polyethylene in 1932, and nylon in the late 1930s (Historical Note 8.1). Processing requirements for plastics led to the development of **injection molding** (based on die casting, one of the metal casting processes) and other polymer-shaping techniques.

Electronics products have imposed unusual demands on manufacturing in terms of miniaturization. The evolution of the technology has been to package more and more devices into smaller and smaller areas—in some cases millions of transistors onto a flat piece of semiconductor material that is only 12 mm (0.50 in.) on a side. The history of electronics processing and packaging dates from only a few decades (Historical Notes 34.1, 35.1, and 35.2).

More than one processing operation is usually required to transform the starting material into final form. The operations are performed in the particular sequence required to achieve the geometry and condition defined by the design specification.

Three categories of processing operations are distinguished: (1) shaping operations, (2) property-enhancing operations, and (3) surface processing operations. **Shaping operations** alter the geometry of the starting work material by various methods. Common shaping processes include casting, forging, and machining. **Property-enhancing operations** add value to the material by improving its physical properties without changing its shape. Heat treatment is the most common example. **Surface processing operations** are performed to clean, treat, coat, or deposit material onto the exterior surface of the work. Common examples of coating are plating and painting. Shaping processes are covered in Parts III through VI, corresponding to the four main categories of shaping processes in Figure 1.4. Property-enhancing processes and surface processing operations are covered in Part VII.

Shaping Processes   Most shape processing operations apply heat, mechanical force, or a combination of these to effect a change in geometry of the work material. There are various ways to classify the shaping processes. The classification used in this book is based on the state of the starting material, by which we have four categories: (1) **solidification processes,** in which the starting material is a heated **liquid** or **semifluid** that cools and solidifies to form the part geometry; (2) **particulate processing,** in which the starting material is a **powder,** and the powders are formed and heated into the desired geometry; (3) **deformation processes,** in which the starting material is a **ductile solid** (commonly metal) that is deformed to shape the part; and (4) **material removal processes,** in which

FIGURE 1.5 Casting and molding processes start with a work material heated to a fluid or semifluid state. The process consists of: (1) pouring the fluid into a mold cavity and (2) allowing the fluid to solidify, after which the solid part is removed from the mold.

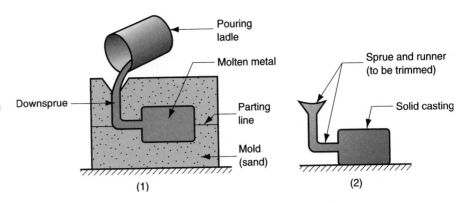

the starting material is a ***solid*** (ductile or brittle), from which material is removed so that the resulting part has the desired geometry.

In the first category, the starting material is heated sufficiently to transform it into a liquid or highly plastic (semifluid) state. Nearly all materials can be processed in this way. Metals, ceramic glasses, and plastics can all be heated to sufficiently high temperatures to convert them into liquids. With the material in a liquid or semifluid form, it can be poured or otherwise forced to flow into a mold cavity and allowed to solidify, thus taking a solid shape that is the same as the cavity. Most processes that operate this way are called casting or molding. ***Casting*** is the name used for metals, and ***molding*** is the common term used for plastics. This category of shaping process is depicted in Figure 1.5.

In ***particulate processing,*** the starting materials are powders of metals or ceramics. Although these two materials are quite different, the processes to shape them in particulate processing are quite similar. The common technique involves pressing and sintering, illustrated in Figure 1.6, in which the powders are first squeezed into a die cavity under high pressure and then heated to bond the individual particles together.

In ***deformation processes,*** the starting workpart is shaped by the application of forces that exceed the yield strength of the material. For the material to be formed in this way, it must be sufficiently ductile to avoid fracture during deformation. To increase ductility (and for other reasons), the work material is often heated before forming to a temperature below the melting point. Deformation processes are associated most closely with metalworking and include operations such as ***forging*** and ***extrusion,*** shown in Figure 1.7.

FIGURE 1.6 Particulate processing: (1) the starting material is powder; the usual process consists of (2) pressing and (3) sintering.

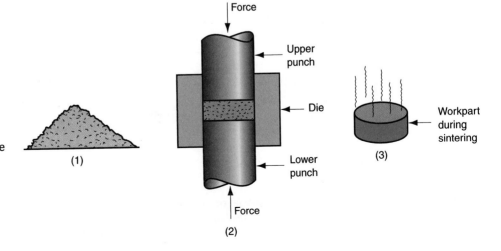

FIGURE 1.7 Some common deformation processes: (a) **forging**, in which two halves of a die squeeze the workpart, causing it to assume the shape of the die cavity; and (b) **extrusion**, in which a billet is forced to flow through a die orifice, thus taking the cross-sectional shape of the orifice.

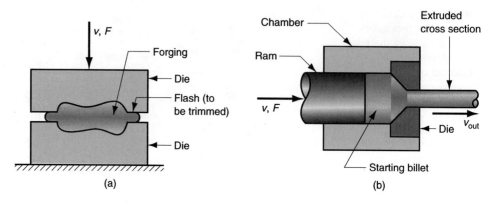

Material removal processes are operations that remove excess material from the starting workpiece so that the resulting shape is the desired geometry. The most important processes in this category are **machining** operations such as **turning, drilling,** and **milling,** shown in Figure 1.8. These cutting operations are most commonly applied to solid metals, performed using cutting tools that are harder and stronger than the work metal. **Grinding** is another common process in this category. Other material removal processes are known as **nontraditional processes** because they use lasers, electron beams, chemical erosion, electric discharges, and electrochemical energy to remove material rather than cutting or grinding tools.

It is desirable to minimize waste and scrap in converting a starting workpart into its subsequent geometry. Certain shaping processes are more efficient than others in terms of material conservation. Material removal processes (e.g., machining) tend to be wasteful of material, simply by the way they work. The material removed from the starting shape is waste, at least in terms of the unit operation. Other processes, such as certain casting and molding operations, often convert close to 100% of the starting material into final product. Manufacturing processes that transform nearly all of the starting material into product and require no subsequent machining to achieve final part geometry are called *net shape processes.* Other processes require minimum machining to produce the final shape and are called *near net shape processes.*

Property-Enhancing Processes The second major type of part processing is performed to improve mechanical or physical properties of the work material. These processes do not alter the shape of the part, except unintentionally in some cases. The most important property-enhancing processes involve **heat treatments,** which include various annealing

FIGURE 1.8 Common machining operations: (a) **turning,** in which a single-point cutting tool removes metal from a rotating workpiece to reduce its diameter; (b) **drilling,** in which a rotating drill bit is fed into the work to create a round hole; and (c) **milling,** in which a workpart is fed past a rotating cutter with multiple edges.

and strengthening processes for metals and glasses. ***Sintering*** of powdered metals and ceramics is also a heat treatment that strengthens a pressed powder metal workpart.

Surface Processing   Surface processing operations include (1) cleaning, (2) surface treatments, and (3) coating and thin film deposition processes. ***Cleaning*** includes both chemical and mechanical processes to remove dirt, oil, and other contaminants from the surface. ***Surface treatments*** include mechanical working such as shot peening and sand blasting, and physical processes such as diffusion and ion implantation. ***Coating*** and ***thin film deposition*** processes apply a coating of material to the exterior surface of the workpart. Common coating processes include ***electroplating, anodizing*** of aluminum, organic ***coating*** (call it ***painting***), and porcelain enameling. Thin film deposition processes include ***physical vapor deposition*** and ***chemical vapor deposition*** to form extremely thin coatings of various substances.

Several surface-processing operations have been adapted to fabricate semiconductor materials into integrated circuits for microelectronics. These processes include chemical vapor deposition, physical vapor deposition, and oxidation. They are applied to very localized areas on the surface of a thin wafer of silicon (or other semiconductor material) to create the microscopic circuit.

1.3.2 ASSEMBLY OPERATIONS

The second basic type of manufacturing operation is ***assembly,*** in which two or more separate parts are joined to form a new entity. Components of the new entity are connected either permanently or semipermanently. Permanent joining processes include ***welding, brazing, soldering,*** and ***adhesive bonding.*** They form a joint between components that cannot be easily disconnected. Certain ***mechanical assembly*** methods are available to fasten two (or more) parts together in a joint that can be conveniently disassembled. The use of screws, bolts, and other ***threaded fasteners*** are important traditional methods in this category. Other mechanical assembly techniques form a more permanent connection; these include ***rivets, press fitting,*** and ***expansion fits.*** Special joining and fastening methods are used in the assembly of electronic products. Some of the methods are identical to or are adaptations of the preceding processes, for example, soldering. Electronics assembly is concerned primarily with the assembly of components such as integrated circuit packages to printed circuit boards to produce the complex circuits used in so many of today's products. Joining and assembly processes are discussed in Part VIII, and the specialized assembly techniques for electronics are described in Part IX.

1.3.3 PRODUCTION MACHINES AND TOOLING

Manufacturing operations are accomplished using machinery and tooling (and people). The extensive use of machinery in manufacturing began with the Industrial Revolution. It was at that time that metal cutting machines started to be developed and widely used. These were called ***machine tools***—power-driven machines used to operate cutting tools previously operated by hand. Modern machine tools are described by the same basic definition, except that the power is electrical rather than water or steam, and the level of precision and automation is much greater today. Machine tools are among the most versatile of all production machines. They are used to make not only parts for consumer products, but also components for other production machines. Both in a historic and a reproductive sense, the machine tool is the mother of all machinery.

Other production machines include ***presses*** for stamping operations, ***forge hammers*** for forging, ***rolling mills*** for rolling sheet metal, ***welding machines*** for welding, and ***insertion machines*** for inserting electronic components into printed circuit boards. The name of the equipment usually follows from the name of the process.

TABLE 1.3 Production equipment and tooling used for various manufacturing processes.

Process	Equipment	Special Tooling (Function)
Casting	[a]	Mold (cavity for molten metal)
Molding	Molding machine	Mold (cavity for hot polymer)
Rolling	Rolling mill	Roll (reduce work thickness)
Forging	Forge hammer or press	Die (squeeze work to shape)
Extrusion	Press	Extrusion die (reduce cross-section)
Stamping	Press	Die (shearing, forming sheet metal)
Machining	Machine tool	Cutting tool (material removal)
		Fixture (hold workpart)
		Jig (hold part and guide tool)
Grinding	Grinding machine	Grinding wheel (material removal)
Welding	Welding machine	Electrode (fusion of work metal)
		Fixture (hold parts during welding)

[a]Various types of casting setups and equipment (Chapter 11).

Production equipment can be general purpose or special purpose. *General purpose equipment* is more flexible and adaptable to a variety of jobs. It is commercially available for any manufacturing company to invest in. *Special purpose equipment* is usually designed to produce a specific part or product in very large quantities. The economics of mass production justify large investments in special purpose machinery to achieve high efficiencies and short cycle times. This is not the only reason for special purpose equipment, but it is the dominant one. Another reason may be because the process is unique and commercial equipment is not available. Some companies with unique processing requirements develop their own special purpose equipment.

Production machinery usually requires *tooling* that customizes the equipment for the particular part or product. In many cases, the tooling must be designed specifically for the part or product configuration. When used with general purpose equipment, it is designed to be exchanged. For each workpart type, the tooling is fastened to the machine and the production run is made. When the run is completed, the tooling is changed for the next workpart type. When used with special purpose machines, the tooling is often designed as an integral part of the machine. Because the special purpose machine is likely being used for mass production, the tooling may never need changing except for replacement of worn components or for repair of worn surfaces.

The type of tooling depends on the type of manufacturing process. Table 1.3 lists examples of special tooling used in various operations. Details are provided in the chapters that discuss these processes.

1.4 PRODUCTION SYSTEMS

To operate effectively, a manufacturing firm must have systems that allow it to efficiently accomplish its type of production. Production systems consist of people, equipment, and procedures designed for the combination of materials and processes that constitute a firm's manufacturing operations. Production systems can be divided into two categories: (1) production facilities and (2) manufacturing support systems, as shown in Figure 1.10. *Production facilities* refer to the physical equipment and the arrangement of equipment in the factory. *Manufacturing support systems* are the procedures used by the company to manage production and solve the technical and logistics problems encountered in ordering materials, moving work through the factory, and ensuring that products meet quality standards. Both categories include people. People make these systems work. In general, direct labor workers are responsible for operating the manufacturing equipment; and professional staff workers are responsible for manufacturing support.

1.4.1 PRODUCTION FACILITIES

Production facilities consist of the factory and the production, material handling, and other equipment in the factory. The equipment comes in direct physical contact with the parts and/or assemblies as they are being made. The facilities "touch" the product. Facilities also include the way the equipment is arranged in the factory—the ***plant layout.*** The equipment is usually organized into logical groupings; which can be called ***manufacturing systems,*** such as an automated production line, or a machine cell consisting of an industrial robot and two machine tools.

A manufacturing company attempts to design its manufacturing systems and organize its factories to serve the particular mission of each plant in the most efficient way. Over the years, certain types of production facilities have come to be recognized as the most appropriate way to organize for a given combination of product variety and production quantity, as discussed in Section 1.1.2. Different types of facilities are required for each of the three ranges of annual production quantities.

Low-Quantity Production   In the low-quantity range (1–100 units/year), the term ***job shop*** is often used to describe the type of production facility. A job shop makes low quantities of specialized and customized products. The products are typically complex, such as space capsules, prototype aircraft, and special machinery. The equipment in a job shop is general purpose, and the labor force is highly skilled.

A job shop must be designed for maximum flexibility to deal with the wide product variations encountered (hard product variety). If the product is large and heavy, and therefore difficult to move, it typically remains in a single location during its fabrication or assembly. Workers and processing equipment are brought to the product, rather than moving the product to the equipment. This type of layout is referred to as a ***fixed-position layout,*** shown in Figure 1.9(a). In a pure situation, the product remains in a single location during its entire production. Examples of such products include ships, aircraft, locomotives, and heavy machinery. In actual practice, these items are usually built in large modules at single locations, and then the completed modules are brought together for final assembly using large-capacity cranes.

The individual components of these large products are often made in factories in which the equipment is arranged according to function or type. This arrangement is called a ***process layout.*** The lathes are in one department, the milling machines are in another department, and so on, as in Figure 1.9(b). Different parts, each requiring a different operation sequence, are routed through the departments in the particular order needed for their processing, usually in batches. The process layout is noted for its flexibility; it can accommodate a great variety of operation sequences for different part configurations. Its disadvantage is that the machinery and methods to produce a part are not designed for high efficiency.

Medium Quantity Production   In the medium-quantity range (100–10,000 units annually), two different types of facility are distinguished, depending on product variety. When product variety is hard, the usual approach is ***batch production,*** in which a batch of one product is made, after which the manufacturing equipment is changed over to produce a batch of the next product, and so on. The production rate of the equipment is greater than the demand rate for any single product type, and so the same equipment can be shared among multiple products. The changeover between production runs takes time—time to change tooling and set up the machinery. This setup time is lost production time, and this is a disadvantage of batch manufacturing. Batch production is commonly used for make-to-stock situations, in which

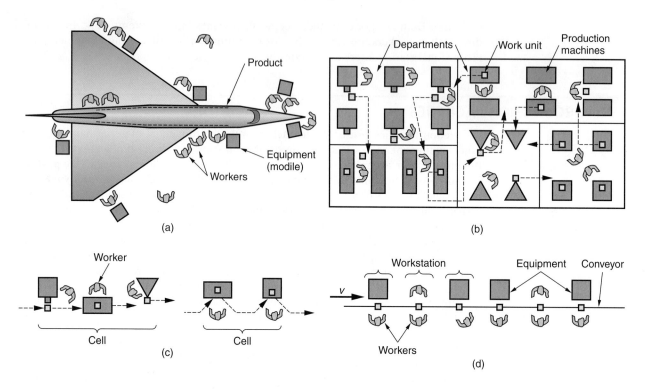

FIGURE 1.9 Various types of plant layout: (a) fixed-position layout, (b) process layout, (c) cellular layout, and (d) product layout.

items are manufactured to replenish inventory that has been gradually depleted by demand. The equipment is usually arranged in a process layout, as in Figure 1.9(b).

An alternative approach to medium-range production is possible if product variety is soft. In this case, extensive changeovers between one product style and the next may not be necessary. It is often possible to configure the manufacturing system so that groups of similar products can be made on the same equipment without significant lost time because of setup. The processing or assembly of different parts or products is accomplished in cells consisting of several workstations or machines. The term *cellular manufacturing* is often associated with this type of production. Each cell is designed to produce a limited variety of part configurations; that is, the cell specializes in the production of a given set of similar parts, according to the principles of *group technology* (Section 39.5). The layout is called a *cellular layout*, depicted in Figure 1.9(c).

High Production The high-quantity range (10,000 to millions of units per year) is referred to as *mass production.* The situation is characterized by a high demand rate for the product, and the manufacturing system is dedicated to the production of that single item. Two categories of mass production can be distinguished: quantity production and flow line production. *Quantity production* involves the mass production of single parts on single pieces of equipment. It typically involves standard machines (e.g., stamping presses) equipped with special tooling (e.g., dies and material handling devices), in effect dedicating the equipment to the production of one part type. Typical layouts used in quantity production are the process layout and cellular layout.

Flow line production involves multiple pieces of equipment or workstations arranged in sequence, and the work units are physically moved through the sequence to complete the product. The workstations and equipment are designed specifically for the product to maximize efficiency. The layout is called a *product layout,* and the workstations are arranged into one long line, as in Figure 1.9(d), or into a series of connected line segments. The work is usually moved between stations by mechanized conveyor. At each station, a small amount of the total work is completed on each unit of product.

The most familiar example of flow line production is the assembly line, associated with products such as cars and household appliances. The pure case of flow line production occurs when there is no variation in the products made on the line. Every product is identical, and the line is referred to as a ***single model production line.*** To successfully market a given product, it is often beneficial to introduce feature and model variations so that individual customers can choose the exact merchandise that appeals to them. From a production viewpoint, the feature differences represent a case of soft product variety. The term ***mixed-model production line*** applies to situations in which there is soft variety in the products made on the line. Modern automobile assembly is an example. Cars coming off the assembly line have variations in options and trim representing different models and in many cases different nameplates of the same basic car design.

1.4.2 MANUFACTURING SUPPORT SYSTEMS

To operate its facilities efficiently, a company must organize itself to design the processes and equipment, plan and control the production orders, and satisfy product quality requirements. These functions are accomplished by manufacturing support systems—people and procedures by which a company manages its production operations. Most of these support systems do not directly contact the product, but they plan and control its progress through the factory. Manufacturing support functions are often carried out in the firm by people organized into departments such as the following:

➤ ***Manufacturing engineering.*** The manufacturing engineering department is responsible for planning the manufacturing processes—deciding what processes should be used to make the parts and assemble the products. This department is also involved in designing and ordering the machine tools and other equipment used by the operating departments to accomplish processing and assembly.

➤ ***Production planning and control.*** This department is responsible for solving the logistics problem in manufacturing—ordering materials and purchased parts, scheduling production, and making sure that the operating departments have the necessary capacity to meet the production schedules.

➤ ***Quality control.*** Producing high-quality products should be a top priority of any manufacturing firm in today's competitive environment. It means designing and

FIGURE 1.10   Overview of major topics covered in the book.

building products that conform to specifications and satisfy or exceed customer expectations. Much of this effort is the responsibility of the QC department.

1.5   TRENDS IN MANUFACTURING

This section considers several trends that are affecting the materials, processes, and systems used in manufacturing. These trends are motivated by technological and economic factors occurring throughout the world. Their effects are not limited to manufacturing; they impact society as a whole. The discussion is organized into the following topic areas: (1) lean production and Six Sigma, (2) globalization, (3) environmentally conscious manufacturing, and (4) microfabrication and nanotechnology.

1.5.1   LEAN PRODUCTION AND SIX SIGMA

These are two programs aimed at improving efficiency and quality in manufacturing. They address the demands by customers for the products they buy to be both low in cost and high in quality. The reason why lean and Six Sigma are trends is because they are being so widely adopted by companies, especially in the United States.

Lean production is based on the Toyota Production System developed by Toyota Motors in Japan. Its origins date from the 1950s, when Toyota began using unconventional methods to improve quality, reduce inventories, and increase flexibility in its operations. *Lean production* can be defined simply as "doing more work with fewer resources."[2] It means that fewer workers and less equipment are used to accomplish more production in less time, and yet achieve higher quality in the final product. The underlying objective of lean production is the elimination of waste. In the Toyota Production System, the seven forms of waste in production are (1) production of defective parts, (2) production of more parts than required, (3) excessive inventories, (4) unnecessary processing steps, (5) unnecessary movement of workers, (6) unnecessary movement and handling of materials, and (7) workers waiting. The methods used by Toyota to reduce waste include techniques for preventing errors, stopping a process when something goes wrong, improved equipment maintenance, involving workers in process improvements (so-called continuous improvement), and standardized work procedures. Probably the most important development was the just-in-time delivery system, which is described in Section 41.4 in the chapter on production and inventory control.

Six Sigma was started in the 1980s at Motorola Corporation in the United States. The objective was to reduce variability in the company's processes and products to increase customer satisfaction. Today, *Six Sigma* can be defined as "a quality-focused program that utilizes worker teams to accomplish projects aimed at improving an organization's operational performance."[3] Six Sigma is discussed in more detail in Section 42.4.2.

1.5.2   GLOBALIZATION AND OUTSOURCING

The world is becoming more and more integrated, creating an international economy in which barriers once established by national boundaries have been reduced or eliminated. This has enabled a freer flow of goods and services, capital, technology, and people among regions and countries. *Globalization* is the term that describes this trend, which was recognized in the late 1980s and is now a dominant economic reality. Of interest here is that once underdeveloped

---

[2]M. P. Groover, *Work Systems and the Methods, Measurement, and Management of Work* [7], p. 514. The term *lean production* was coined by researchers at the Massachusetts Institute of Technology who studied the production operations at Toyota and other automobile companies in the 1980s.
[3]Ibid, p. 541.

nations such as China, India, and Mexico have developed their manufacturing infrastructures and technologies to a point where they are now important producers in the global economy. The advantages of these three countries in particular are their large populations (therefore, large workforce pool) and low labor costs. Hourly wages are currently an order of magnitude or more higher in the United States than in these countries, making it difficult for domestic U.S. companies to compete in many products requiring a high labor content. Examples include garments, furniture, many types of toys, and electronic gear. The result has been a loss of manufacturing jobs in the United States and a gain of related work to these countries.

Globalization is closely related to outsourcing. In manufacturing, **outsourcing** refers to the use of outside contractors to perform work that was traditionally accomplished in-house. Outsourcing can be done in several ways, including the use of local suppliers. In this case the jobs remain in the United States. Alternatively, U.S. companies can outsource to foreign countries, so that parts and products once made in the United States are now made outside the country. In this case U.S. jobs are displaced. Two possibilities can be distinguished: (1) **offshore outsourcing,** which refers to production in China or other overseas locations and transporting the items by cargo ship to the United States, and (2) **near-shore outsourcing,** which means the items are made in Canada, Mexico, or Central America and shipped by rail or truck into the United States.

China is a country of particular interest in this discussion of globalization because of its fast-growing economy, the importance of manufacturing in that economy, and the extent to which U.S. companies have outsourced work to China. To take advantage of the low labor rates, U.S. companies have outsourced much of their production to China (and other east Asian countries). Despite the logistics problems and costs of shipping the goods back into the United States, the result has been lower costs and higher profits for the outsourcing companies, as well as lower prices and a wider variety of available products for U.S. consumers. The downside has been the loss of well-paying manufacturing jobs in the United States. Another consequence of U.S. outsourcing to China has been a reduction in the relative contribution of the manufacturing sector to GDP. In the 1990s, the manufacturing industries accounted for about 20% of GDP in the United States. Today that contribution is less than 15%. At the same time, the manufacturing sector in China has grown (along with the rest of its economy), now accounting for almost 35% of Chinese GDP. Because the U.S. GDP is roughly three times China's, the United States' manufacturing sector is still larger. However, China is the world leader in several industries. Its tonnage output of steel is greater than the combined outputs of the next six largest steel producing nations (in order, Japan, United States, Russia, India, South Korea, and Germany).[4] China is also the largest producer of metal castings, accounting for more tonnage than the next three largest producers (in order, United States, Japan, and India) [5].

Steel production and casting are considered "dirty" industries, and environmental pollution is an issue not only in China, but in many places throughout the World. This issue is addressed in the next trend.

1.5.3 ENVIRONMENTALLY CONSCIOUS MANUFACTURING

An inherent feature of virtually all manufacturing processes is waste (Section 1.3.1). The most obvious examples are material removal processes, in which chips are removed from a starting workpiece to create the desired part geometry. Waste in one form or another is a by-product of nearly all production operations. Another unavoidable aspect of manufacturing is that power is required to accomplish any given process. Generating that power requires fossil fuels (at least in the United States and China), the burning of which results in pollution of the environment. At the end of the manufacturing sequence, a product is created that is sold to a

---

[4]Source: World Steel Association, 2008 data.

customer. Ultimately, the product wears out and is disposed of, perhaps in some landfill, with the associated environmental degradation. More and more attention is being paid by society to the environmental impact of human activities throughout the world and how modern civilization is using our natural resources at an unsustainable rate. Global warming is presently a major concern. The manufacturing industries contribute to these problems.

Environmentally conscious manufacturing refers to programs that seek to determine the most efficient use of materials and natural resources in production, and minimize the negative consequences on the environment. Other associated terms for these programs include *green manufacturing, cleaner production,* and *sustainable manufacturing*. They all boil down to two basic approaches: (1) design products that minimize their environmental impact, and (2) design processes that are environmentally friendly.

Product design is the logical starting point in environmentally conscious manufacturing. The term *design for environment* (DFE) is sometimes used for the techniques that attempt to consider environmental impact during product design prior to production. Considerations in DFE include the following: (1) select materials that require minimum energy to produce, (2) select processes that minimize waste of materials and energy, (3) design parts that can be recycled or reused, (4) design products that can be readily disassembled to recover the parts, (5) design products that minimize the use of hazardous and toxic materials, and (6) give attention to how the product will be disposed of at the end of its useful life.

To a great degree, decisions made during design dictate the materials and processes that are used to make the product. These decisions limit the options available to the manufacturing departments to achieve sustainability. However, various approaches can be applied to make plant operations more environmentally friendly. They include the following: (1) adopt good housekeeping practices—keep the factory clean, (2) prevent pollutants from escaping into the environment (rivers and atmosphere), (3) minimize waste of materials in unit operations, (4) recycle rather than discard waste materials, (5) use net shape processes, (6) use renewable energy sources when feasible, (7) provide maintenance to production equipment so that it operates at maximum efficiency, and (8) invest in equipment that minimizes power requirements.

Various topics related to environmentally conscious manufacturing are discussed in the text. The topics of polymer recycling and biodegradable plastics are covered in Section 8.5. Cutting fluid filtration and dry machining, which reduce the adverse effects of contaminated cutting fluids, are considered in Section 23.4.2.

1.5.4 MICROFABRICATION AND NANOTECHNOLOGY

Another trend in manufacturing is the emergence of materials and products whose dimensions are sometimes so small that they cannot be seen by the naked eye. In extreme cases, the items cannot even be seen under an optical microscope. Products that are so miniaturized require special fabrication technologies. *Microfabrication* refers to the processes needed to make parts and products whose features sizes are in the micrometer range $(1\ \mu m = 10^{-3}\ mm = 10^{-6}\ m)$. Examples include ink-jet printing heads, compact discs (CDs and DVDs), and microsensors used in automotive applications (e.g., air-bag deployment sensors). *Nanotechnology* refers to materials and products whose feature sizes are in the nanometer scale $(1\ nm = 10^{-3}\ \mu m = 10^{-6}\ mm = 10^{-9}\ m)$, a scale that approaches the size of atoms and molecules. Ultra-thin coatings for catalytic converters, flat screen TV monitors, and cancer drugs are examples of products based on nanotechnology. Microscopic and nanoscopic materials and products are expected to increase in importance in the future, both technologically and economically, and processes are needed to produce them commercially. The purpose here is to make the reader aware of this trend toward miniaturization. Chapters 36 and 37 are devoted to these technologies.

1.6   ORGANIZATION OF THE BOOK

The preceding sections provide an overview of the book. The remaining 41 chapters are organized into 11 parts. The block diagram in previous Figure 1.10 summarizes the major topics that are covered. It shows the production system (outlined in dashed lines) with engineering materials entering from the left and finished products exiting at the right. Part I, Material Properties and Product Attributes, consists of four chapters that describe the important characteristics and specifications of materials and the products made from them. Part II discusses the four basic engineering materials: metals, ceramics, polymers, and composites.

The largest block in Figure 1.10 is labeled "Manufacturing processes and assembly operations." The processes and operations included in the text are those identified in Figure 1.4. Part III begins the coverage of the four categories of shaping processes. Part III consists of six chapters on the solidification processes that include casting of metals, glassworking, and polymer shaping. In Part IV, the particulate processing of metals and ceramics is covered in two chapters. Part V deals with metal deformation processes such as rolling, forging, extrusion, and sheet metalworking. Finally, Part VI discusses the material removal processes. Four chapters are devoted to machining, and two chapters cover grinding (and related abrasive processes) and the nontraditional material removal technologies.

The other types of processing operations, property enhancing and surface processing, are covered in two chapters in Part VII. Property enhancing is accomplished by heat treatment, and surface processing includes operations such as cleaning, electroplating, and coating (painting).

Joining and assembly processes are considered in Part VIII, which is organized into four chapters on welding, brazing, soldering, adhesive bonding, and mechanical assembly.

Several unique processes that do not neatly fit into the classification scheme of Figure 1.4 are covered in Part IX, Special Processing and Assembly Technologies. Its five chapters cover rapid prototyping, processing of integrated circuits, electronics, micro-fabrication, and nanofabrication.

The remaining blocks in Figure 1.10 deal with the systems of production. Part X, "Manufacturing Systems," covers the major systems technologies and equipment groupings located in the factory: numerical control, industrial robotics, group technology, cellular manufacturing, flexible manufacturing systems, and production lines. Finally, Part XI deals with manufacturing support systems: manufacturing engineering, production planning and control, and quality control and inspection.

REFERENCES

[1] Black, J., and Kohser, R. *DeGarmo's Materials and Processes in Manufacturing,* 10th ed. John Wiley & Sons, Hoboken, New Jersey, 2008.

[2] Emerson, H. P., and Naehring, D. C. E. *Origins of Industrial Engineering.* Industrial Engineering & Management Press, Institute of Industrial Engineers, Norcross, Georgia, 1988.

[3] Flinn, R. A., and Trojan, P. K. *Engineering Materials and Their Applications,* 5th ed. John Wiley & Sons, New York, 1995.

[4] Garrison, E. *A History of Engineering and Technology.* CRC Taylor & Francis, Boca Raton, Florida, 1991.

[5] Gray, A. "Global Automotive Metal Casting," *Advanced Materials & Processes,* April 2009, pp. 33–35.

[6] Groover, M. P. *Automation, Production Systems, and Computer Integrated Manufacturing,* 3rd ed. Pearson Prentice-Hall, Upper Saddle River, New Jersey, 2008.

[7] Groover, M. P. *Work Systems and the Methods, Measurement, and Management of Work,* Pearson Prentice-Hall, Upper Saddle River, New Jersey, 2007.

[8] Hornyak, G. L., Moore, J. J., Tibbals, H. F., and Dutta, J., *Fundamentals of Nanotechnology,* CRC Taylor & Francis, Boca Raton, Florida, 2009.

[9] Hounshell, D. A. *From the American System to Mass Production, 1800–1932.* The Johns Hopkins University Press, Baltimore, Maryland, 1984.

[10] Kalpakjian, S., and Schmid S. R. *Manufacturing Processes for Engineering Materials,* 6th ed.

Pearson Prentice Hall, Upper Saddle River, New Jersey, 2010.

[11] wikipedia.org/wiki/globalization

[12] www.bsdglobal.com/tools

REVIEW QUESTIONS

1.1. What are the differences among primary, secondary, and tertiary industries? Give an example of each category.

1.2. What is a capital good? Provide an example.

1.3. How are product variety and production quantity related when comparing typical factories?

1.4. Define manufacturing capability.

1.5. Name the three basic categories of materials.

1.6. How does a shaping process differ from a surface processing operation?

1.7. What are two subclasses of assembly processes? Provide an example process for each subclass.

1.8. Define batch production and describe why it is often used for medium-quantity production products.

1.9. What is the difference between a process layout and a product layout in a production facility?

1.10. Name two departments that are typically classified as manufacturing support departments.

MULTIPLE CHOICE QUIZ

There are 18 correct answers in the following multiple choice questions (some questions have multiple answers that are correct). To attain a perfect score on the quiz, all correct answers must be given. Each correct answer is worth 1 point. Each omitted answer or wrong answer reduces the score by 1 point, and each additional answer beyond the correct number of answers reduces the score by 1 point. Percentage score on the quiz is based on the total number of correct answers.

1.1. Which of the following industries are classified as secondary industries (three correct answers): (a) beverages (b) financial services, (c) fishing, (d) mining, (e) power utilities, (f) publishing, and, (g) transportation?

1.2. Mining is classified in which one of the following industry categories: (a) agricultural industry, (b) manufacturing industry, (c) primary industry, (d) secondary industry, (e) service industry, or, (f) tertiary industry?

1.3. Inventions of the Industrial Revolution include which one of the following: (a) automobile, (b) cannon, (c) printing press, (d) steam engine, or, (e) sword?

1.4. Ferrous metals include which of the following (two correct answers): (a) aluminum, (b) cast iron, (c) copper, (d) gold, and, (e) steel?

1.5. Which one of the following engineering materials is defined as a compound containing metallic and nonmetallic elements: (a) ceramic, (b) composite, (c) metal, or, (d) polymer?

1.6. Which of the following processes start with a material that is in a fluid or semifluid state and solidifies the material in a cavity (two best answers): (a) casting, (b) forging, (c) machining, (d) molding, (e) pressing, and, (f) turning?

1.7. Particulate processing of metals and ceramics involves which of the following steps (two best answers): (a) adhesive bonding, (b) deformation, (c) forging, (d) material removal, (e) melting, (f) pressing, and, (g) sintering?

1.8. Deformation processes include which of the following (two correct answers): (a) casting, (b) drilling, (c) extrusion, (d) forging, (e) milling, (f) painting, and, (g) sintering?

1.9. Which one of the following is a machine used to perform extrusion: (a) forge hammer, (b) milling machine, (c) rolling mill, (d) press, (e) torch?

1.10. High-volume production of assembled products is most closely associated with which one of the following layout types: (a) cellular layout, (b) fixed position layout, (c) process layout, or, (d) product layout?

1.11. A production planning and control department accomplishes which of the following functions in its role of providing manufacturing support (two best answers): (a) designs and orders machine tools, (b) develops corporate strategic plans, (c) orders materials and purchased parts, (d) performs quality inspections, and, (e) schedules the order of products on a machine?

Part I Material Properties and Product Attributes

2

THE NATURE OF MATERIALS

Chapter Contents

An understanding of materials is fundamental in the study of manufacturing processes. In Chapter 1, manufacturing was defined as a transformation process. It is the material that is transformed; and it is the behavior of the material when subjected to the particular forces, temperatures, and other physical parameters of the process that determines the success of the operation. Certain materials respond well to certain types of manufacturing processes, and poorly or not at all to others. What are the characteristics and properties of materials that determine their capacity to be transformed by the different processes?

Part I of this book consists of four chapters that address this question. The current chapter considers the atomic structure of matter and the bonding between atoms and molecules. It also shows how atoms and molecules in engineering materials organize themselves into two structural forms: crystalline and noncrystalline. It turns out that the basic engineering materials—metals, ceramics, and polymers—can exist in either form, although a preference for a particular form is usually exhibited by a given material. For example, metals almost always exist as crystals in their solid state. Glass (e.g., window glass), a ceramic, assumes a noncrystalline form. Some polymers are mixtures of crystalline and amorphous structures.

Chapters 3 and 4 discuss the mechanical and physical properties that are relevant in manufacturing. Of course, these properties are also important in product design. Chapter 5 is concerned with several part and product attributes that are specified during product design and must be achieved in manufacturing: dimensions, tolerances, and surface finish. Chapter 5 also describes how these attributes are measured.

2.1 ATOMIC STRUCTURE AND THE ELEMENTS

The basic structural unit of matter is the atom. Each atom is composed of a positively charged nucleus, surrounded by a sufficient number of negatively charged electrons so that the charges are balanced. The number of electrons identifies the atomic number and the element of the atom. There are slightly more than 100 elements (not counting a few extras that have been artificially synthesized), and these elements are the chemical building blocks of all matter.

Just as there are differences among the elements, there are also similarities. The elements can be grouped into families and relationships established between and within the families by means of the Periodic Table, shown in Figure 2.1. In the horizontal direction there is a certain repetition, or periodicity, in the arrangement of elements. Metallic elements occupy the left and center portions of the chart, and nonmetals are located to the right. Between them, along a diagonal, is a transition zone containing elements called *metalloids* or *semimetals.* In principle, each of the elements can exist as a solid, liquid, or gas, depending on temperature and pressure. At room temperature and atmospheric pressure, they each have a natural phase; e.g., iron (Fe) is a solid, mercury (Hg) is a liquid, and nitrogen (N) is a gas.

In the table, the elements are arranged into vertical columns and horizontal rows in such a way that similarities exist among elements in the same columns. For example, in the extreme right column are the *noble gases* (helium, neon, argon, krypton, xenon, and radon), all of which exhibit great chemical stability and low reaction rates. The *halogens* (fluorine, chlorine, bromine, iodine, and astatine) in column VIIA share similar properties (hydrogen is not included among the halogens). The *noble metals* (copper, silver, and gold) in column IB have similar properties. Generally there are correlations in properties among elements within a given column, whereas differences exist among elements in different columns.

FIGURE 2.1   Periodic Table of Elements. The atomic number and symbol are listed for the 103 elements.

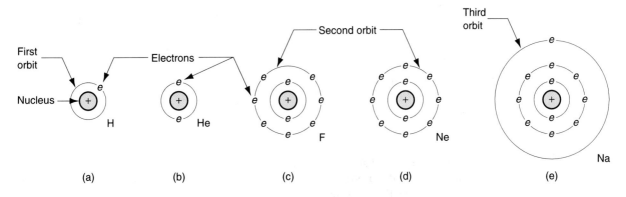

FIGURE 2.2   Simple model of atomic structure for several elements: (a) hydrogen, (b) helium, (c) fluorine, (d) neon, and (e) sodium.

Many of the similarities and differences among the elements can be explained by their respective atomic structures. The simplest model of atomic structure, called the planetary model, shows the electrons of the atom orbiting around the nucleus at certain fixed distances, called shells, as shown in Figure 2.2. The hydrogen atom (atomic number 1) has one electron in the orbit closest to the nucleus. Helium (atomic number 2) has two. Also shown in the figure are the atomic structures for fluorine (atomic number 9), neon (atomic number 10), and sodium (atomic number 11). One might infer from these models that there is a maximum number of electrons that can be contained in a given orbit. This turns out to be correct, and the maximum is defined by

$$\text{Maximum number of electrons in an orbit} = 2n^2 \qquad (2.1)$$

where n identifies the orbit, with $n = 1$ closest to the nucleus.

The number of electrons in the outermost shell, relative to the maximum number allowed, determines to a large extent the atom's chemical affinity for other atoms. These outer-shell electrons are called **_valence electrons._** For example, because a hydrogen atom has only one electron in its single orbit, it readily combines with another hydrogen atom to form a hydrogen molecule H_2. For the same reason, hydrogen also reacts readily with various other elements (e.g., to form H_2O). In the helium atom, the two electrons in its only orbit are the maximum allowed $(2n^2 = 2(1)^2 = 2)$, and so helium is very stable. Neon is stable for the same reason: Its outermost orbit $(n = 2)$ has eight electrons (the maximum allowed), so neon is an inert gas.

In contrast to neon, fluorine has one fewer electron in its outer shell $(n = 2)$ than the maximum allowed and is readily attracted to other elements that might share an electron to make a more stable set. The sodium atom seems divinely made for the situation, with one electron in its outermost orbit. It reacts strongly with fluorine to form the compound sodium fluoride, as pictured in Figure 2.3.

FIGURE 2.3   The sodium fluoride molecule, formed by the transfer of the "extra" electron of the sodium atom to complete the outer orbit of the fluorine atom.

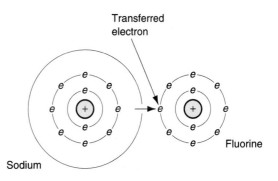

At the low atomic numbers considered here, the prediction of the number of electrons in the outer orbit is straightforward. As the atomic number increases to higher levels, the allocation of electrons to the different orbits becomes somewhat more complicated. There are rules and guidelines, based on quantum mechanics, that can be used to predict the positions of the electrons among the various orbits and explain their characteristics. A discussion of these rules is somewhat beyond the scope of the coverage of materials for manufacturing.

2.2   BONDING BETWEEN ATOMS AND MOLECULES

Atoms are held together in molecules by various types of bonds that depend on the valence electrons. By comparison, molecules are attracted to each other by weaker bonds, which generally result from the electron configuration in the individual molecules. Thus, we have two types of bonding: (1) primary bonds, generally associated with the formation of molecules; and (2) secondary bonds, generally associated with attraction between molecules. Primary bonds are much stronger than secondary bonds.

Primary Bonds   Primary bonds are characterized by strong atom-to-atom attractions that involve the exchange of valence electrons. Primary bonds include the following forms: (a) ionic, (b) covalent, and (c) metallic, as illustrated in Figure 2.4. Ionic and covalent bonds are called *intra*molecular bonds because they involve attractive forces between atoms within the molecule.

In the *ionic bond,* the atoms of one element give up their outer electron(s), which are in turn attracted to the atoms of some other element to increase their electron count in the outermost shell to eight. In general, eight electrons in the outer shell is the most stable atomic configuration (except for the very light atoms), and nature provides a very strong bond between atoms that achieves this configuration. The previous example of the reaction of sodium and fluorine to form sodium fluoride (Figure 2.3) illustrates this form of atomic bond. Sodium chloride (table salt) is a more common example. Because of the transfer of electrons between the atoms, sodium and fluorine (or sodium and chlorine) *ions* are formed, from which this bonding derives its name. Properties of solid materials with ionic bonding include low electrical conductivity and poor ductility.

The *covalent bond* is one in which electrons are shared (as opposed to transferred) between atoms in their outermost shells to achieve a stable set of eight. Fluorine and diamond are two examples of covalent bonds. In fluorine, one electron from each of two atoms is shared to form F_2 gas, as shown in Figure 2.5(a). In the case of diamond, which is carbon (atomic number 6), each atom has four neighbors with which it shares electrons. This produces a very rigid three-dimensional structure, not adequately represented in Figure 2.5(b), and accounts for the extreme high hardness of this material. Other forms of

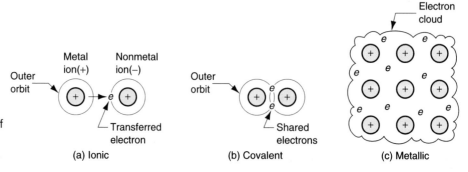

FIGURE 2.4   Three forms of primary bonding: (a) ionic, (b) covalent, and (c) metallic.

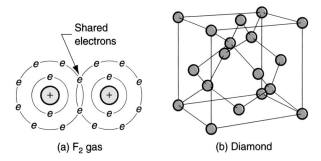

FIGURE 2.5   Two examples of covalent bonding: (a) fluorine gas F_2, and (b) diamond.

(a) F_2 gas   (b) Diamond

carbon (e.g., graphite) do not exhibit this rigid atomic structure. Solids with covalent bonding generally possess high hardness and low electrical conductivity.

The metallic bond is, of course, the atomic bonding mechanism in pure metals and metal alloys. Atoms of the metallic elements generally possess too few electrons in their outermost orbits to complete the outer shells for all of the atoms in, say, a given block of metal. Accordingly, instead of sharing on an atom-to-atom basis, **metallic bonding** involves the sharing of outer-shell electrons by all atoms to form a general electron cloud that permeates the entire block. This cloud provides the attractive forces to hold the atoms together and forms a strong, rigid structure in most cases. Because of the general sharing of electrons, and their freedom to move within the metal, metallic bonding provides for good electrical conductivity. Other typical properties of materials characterized by metallic bonding include good conduction of heat and good ductility. (Although some of these terms are yet to be defined, the text relies on the reader's general understanding of material properties.)

Secondary Bonds   Whereas primary bonds involve atom-to-atom attractive forces, secondary bonds involve attraction forces between molecules, or **inter**molecular forces. There is no transfer or sharing of electrons in secondary bonding, and these bonds are therefore weaker than primary bonds. There are three forms of secondary bonding: (a) dipole forces, (b) London forces, and (c) hydrogen bonding, illustrated in Figure 2.6. Types (a) and (b) are often referred to as **van der Waals** forces, after the scientist who first studied and quantified them.

Dipole forces arise in a molecule comprised of two atoms that have equal and opposite electrical charges. Each molecule therefore forms a dipole, as shown in Figure 2.6(a) for hydrogen chloride. Although the material is electrically neutral in its aggregate form, on a molecular scale the individual dipoles attract each other, given the proper orientation of positive and negative ends of the molecules. These dipole forces provide a net intermolecular bonding within the material.

London forces involve attractive forces between nonpolar molecules; that is, the atoms in the molecule do not form dipoles in the sense of the preceding paragraph. However, owing to the rapid motion of the electrons in orbit around the molecule, temporary dipoles form when more electrons happen to be on one side of the molecule than the other, as suggested by

(a) Dipole forces   (b) London forces   (c) Hydrogen bonding

FIGURE 2.6   Three types of secondary bonding: (a) dipole forces, (b) London forces, and (c) hydrogen bonding.

Figure 2.6(b). These instantaneous dipoles provide a force of attraction between molecules in the material.

Finally, *hydrogen bonding* occurs in molecules containing hydrogen atoms that are covalently bonded to another atom (e.g., oxygen in H_2O). Because the electrons needed to complete the shell of the hydrogen atom are aligned on one side of its nucleus, the opposite side has a net positive charge that attracts the electrons of atoms in neighboring molecules. Hydrogen bonding is illustrated in Figure 2.6(c) for water, and is generally a stronger intermolecular bonding mechanism than the other two forms of secondary bonding. It is important in the formation of many polymers.

2.3   CRYSTALLINE STRUCTURES

Atoms and molecules are used as building blocks for the more macroscopic structure of matter that is considered here and in the following section. When materials solidify from the molten state, they tend to close ranks and pack tightly, in many cases arranging themselves into a very orderly structure, and in other cases, not quite so orderly. Two fundamentally different material structures can be distinguished:(1) crystalline and (2) noncrystalline. Crystalline structures are examined in this section, and noncrystalline in the next. The video clip on heat treatment shows how metals naturally form into crystal structures.

VIDEO CLIP _____

Heat treatment: View the segment titled "metal and alloy structures."

_____

Many materials form into crystals on solidification from the molten or liquid state. It is characteristic of virtually all metals, as well as many ceramics and polymers. A *crystalline structure* is one in which the atoms are located at regular and recurring positions in three dimensions. The pattern may be replicated millions of times within a given crystal. The structure can be viewed in the form of a *unit cell,* which is the basic geometric grouping of atoms that is repeated. To illustrate, consider the unit cell for the body-centered cubic (BCC) crystal structure shown in Figure 2.7, one of the common structures found in metals. The simplest model of the BCC unit cell is illustrated in Figure 2.7(a). Although this model clearly depicts the locations of the atoms within the cell, it does not indicate the close packing of the atoms that occurs in the real crystal, as in Figure 2.7(b). Figure 2.7(c) shows the repeating nature of the unit cell within the crystal.

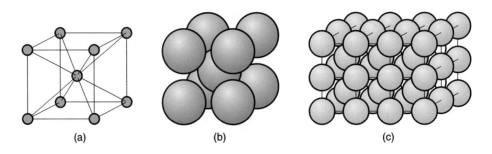

(a)   (b)   (c)

FIGURE 2.7   Body-centered cubic (BCC) crystal structure: (a) unit cell, with atoms indicated as point locations in a three-dimensional axis system; (b) unit cell model showing closely packed atoms (sometimes called the hard-ball model); and (c) repeated pattern of the BCC structure.

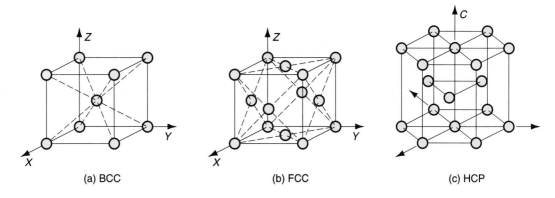

FIGURE 2.8   Three types of crystal structures in metals: (a) body-centered cubic, (b) face-centered cubic, and (c) hexagonal close-packed.

2.3.1 TYPES OF CRYSTAL STRUCTURES

In metals, three lattice structures are common: (1) body-centered cubic (BCC), (2) face-centered cubic (FCC), and (3) hexagonal close-packed (HCP), illustrated in Figure 2.8. Crystal structures for the common metals are presented in Table 2.1. Note that some metals undergo a change of structure at different temperatures. Iron, for example, is BCC at room temperature; it changes to FCC above 912°C (1674°F) and back to BCC at temperatures above 1400°C (2550°F). When a metal (or other material) changes structure like this, it is referred to as being ***allotropic.***

2.3.2 IMPERFECTIONS IN CRYSTALS

Thus far, crystal structures have been discussed as if they were perfect—the unit cell repeated in the material over and over in all directions. A perfect crystal is sometimes desirable to satisfy aesthetic or engineering purposes. For instance, a perfect diamond (contains no flaws) is more valuable than one containing imperfections. In the production of integrated circuit chips, large single crystals of silicon possess desirable processing characteristics for forming the microscopic details of the circuit pattern.

However, there are various reasons why a crystal's lattice structure may not be perfect. The imperfections often arise naturally because of the inability of the solidifying material to continue the replication of the unit cell indefinitely without interruption. Grain boundaries in metals are an example. In other cases, the imperfections are introduced purposely during the

TABLE 2.1 Crystal structures for the common metals (at room temperature).		
Body-Centered Cubic (BCC)	**Face-Centered Cubic (FCC)**	**Hexagonal Close-Packed (HCP)**
Chromium (Cr)	Aluminum (Al)	Magnesium (Mg)
Iron (Fe)	Copper (Cu)	Titanium (Ti)
Molybdenum (Mo)	Gold (Au)	Zinc (Zn)
Tantalum (Ta)	Lead (Pb)	
Tungsten (W)	Silver (Ag)	
	Nickel (Ni)	

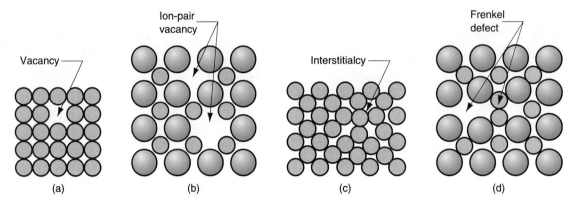

FIGURE 2.9   Point defects: (a) vacancy, (b) ion-pair vacancy, (c) interstitialcy, and (d) displaced ion.

manufacturing process; for example, the addition of an alloying ingredient in a metal to increase its strength.

The various imperfections in crystalline solids are also called defects. Either term, *imperfection* or *defect,* refers to deviations in the regular pattern of the crystalline lattice structure. They can be catalogued as (1) point defects, (2) line defects, and (3) surface defects.

Point defects are imperfections in the crystal structure involving either a single atom or a few atoms. The defects can take various forms including, as shown in Figure 2.9: (a) *vacancy,* the simplest defect, involving a missing atom within the lattice structure; (b) *ion-pair vacancy,* also called a *Schottky defect,* which involves a missing pair of ions of opposite charge in a compound that has an overall charge balance; (c) *interstitialcy,* a lattice distortion produced by the presence of an extra atom in the structure; and (d) *displaced ion,* known as a *Frenkel defect,* which occurs when an ion becomes removed from a regular position in the lattice structure and inserted into an interstitial position not normally occupied by such an ion.

A *line defect* is a connected group of point defects that forms a line in the lattice structure. The most important line defect is the *dislocation,* which can take two forms: (a) edge dislocation and (b) screw dislocation. An *edge dislocation* is the edge of an extra plane of atoms that exists in the lattice, as illustrated in Figure 2.10(a). A *screw dislocation,* Figure 2.10(b), is a spiral within the lattice structure wrapped around an imperfection line, like a screw is wrapped around its axis. Both types of dislocations can arise in the crystal structure during solidification (e.g., casting), or they can be initiated during a

FIGURE 2.10   Line defects: (a) edge dislocation and (b) screw dislocation.

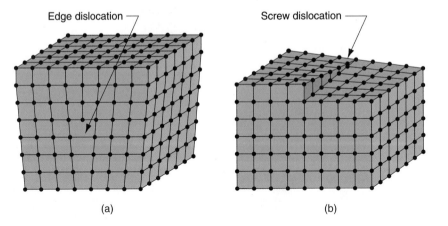

deformation process (e.g., metal forming) performed on the solid material. Dislocations are useful in explaining certain aspects of mechanical behavior in metals.

Surface defects are imperfections that extend in two directions to form a boundary. The most obvious example is the external surface of a crystalline object that defines its shape. The surface is an interruption in the lattice structure. Surface boundaries can also lie inside the material. Grain boundaries are the best example of these internal surface interruptions. Metallic grains are discussed in a moment, but first consider how deformation occurs in a crystal lattice, and how the process is aided by the presence of dislocations.

2.3.3 DEFORMATION IN METALLIC CRYSTALS

When a crystal is subjected to a gradually increasing mechanical stress, its initial response is to deform *elastically.* This can be likened to a tilting of the lattice structure without any changes of position among the atoms in the lattice, in the manner depicted in Figure 2.11(a) and (b). If the force is removed, the lattice structure (and therefore the crystal) returns to its original shape. If the stress reaches a high value relative to the electrostatic forces holding the atoms in their lattice positions, a permanent shape change occurs, called *plastic deformation.* What has happened is that the atoms in the lattice have permanently moved from their previous locations, and a new equilibrium lattice has been formed, as suggested by Figure 2.11(c).

The lattice deformation shown in (c) of the figure is one possible mechanism, called slip, by which plastic deformation can occur in a crystalline structure. The other is called twinning, discussed later.

Slip involves the relative movement of atoms on opposite sides of a plane in the lattice, called the *slip plane.* The slip plane must be somehow aligned with the lattice structure (as indicated in the sketch), and so there are certain preferred directions along which slip is more likely to occur. The number of these *slip directions* depends on the lattice type. The three common metal crystal structures are somewhat more complicated, especially in three dimensions, than the square lattice depicted in Figure 2.11. It turns out that HCP has the fewest slip directions, BCC the most, and FCC falls in between. HCP metals show poor ductility and are generally difficult to deform at room temperature. Metals with BCC structure would figure to have the highest ductility, if the number of slip directions were the only criterion. However, nature is not so simple. These metals are generally stronger than the others, which complicates the issue; and the BCC metals usually require higher stresses to cause slip. In fact, some of the BCC metals exhibit poor ductility. Low carbon steel is a notable exception; although relatively strong, it is widely used with great commercial success in sheet-metal-forming operations, in which it exhibits good ductility. The FCC metals are generally the most ductile of the three crystal structures, combining a good number of slip directions with (usually) relatively low to moderate strength. All three of these metal structures become more ductile at elevated temperatures, and this fact is often exploited in shaping them.

Dislocations play an important role in facilitating slip in metals. When a lattice structure containing an edge dislocation is subjected to a shear stress, the material deforms

FIGURE 2.11 Deformation of a crystal structure: (a) original lattice; (b) elastic deformation, with no permanent change in positions of atoms; and (c) plastic deformation, in which atoms in the lattice are forced to move to new "homes."

(a)

Shear stress

(b)

Shear stress

Slip plane

(c)

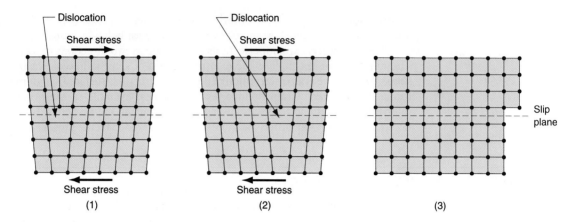

FIGURE 2.12 Effect of dislocations in the lattice structure under stress. In the series of diagrams, the movement of the dislocation allows deformation to occur under a lower stress than in a perfect lattice.

much more readily than in a perfect structure. This is explained by the fact that the dislocation is put into motion within the crystal lattice in the presence of the stress, as shown in the series of sketches in Figure 2.12. Why is it easier to move a dislocation through the lattice than it is to deform the lattice itself? The answer is that the atoms at the edge dislocation require a smaller displacement within the distorted lattice structure to reach a new equilibrium position. Thus, a lower energy level is needed to realign the atoms into the new positions than if the lattice were missing the dislocation. A lower stress level is therefore required to effect the deformation. Because the new position manifests a similar distorted lattice, movement of atoms at the dislocation continues at the lower stress level.

The slip phenomenon and the influence of dislocations have been explained here on a very microscopic basis. On a larger scale, slip occurs many times over throughout the metal when subjected to a deforming load, thus causing it to exhibit the familiar macroscopic behavior. Dislocations represent a good-news–bad-news situation. Because of dislocations, the metal is more ductile and yields more readily to plastic deformation (forming) during manufacturing. However, from a product design viewpoint, the metal is not nearly as strong as it would be in the absence of dislocations.

Twinning is a second way in which metal crystals plastically deform. *Twinning* can be defined as a mechanism of plastic deformation in which atoms on one side of a plane (called the twinning plane) are shifted to form a mirror image of the other side of the plane. It is illustrated in Figure 2.13. The mechanism is important in HCP metals (e.g., magnesium, zinc)

FIGURE 2.13 Twinning involves the formation of an atomic mirror image (i.e., a "twin") on the opposite side of the twinning plane: (a) before, and (b) after twinning.

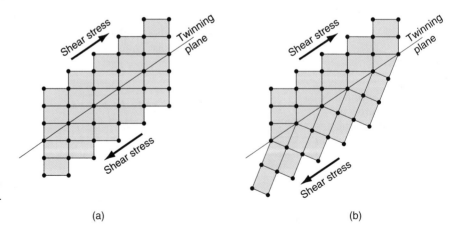

(a)   (b)

because they do not slip readily. Besides structure, another factor in twinning is the rate of deformation. The slip mechanism requires more time than twinning, which can occur almost instantaneously. Thus, in situations in which the deformation rate is very high, metals twin that would otherwise slip. Low carbon steel is an example that illustrates this rate sensitivity; when subjected to high strain rates it twins, whereas at moderate rates it deforms by slip.

2.3.4 GRAINS AND GRAIN BOUNDARIES IN METALS

A given block of metal may contain millions of individual crystals, called *grains.* Each grain has its own unique lattice orientation; but collectively, the grains are randomly oriented within the block. Such a structure is referred to as *polycrystalline.* It is easy to understand how such a structure is the natural state of the material. When the block is cooled from the molten state and begins to solidify, nucleation of individual crystals occurs at random positions and orientations throughout the liquid. As these crystals grow they finally interfere with each other, forming at their interface a surface defect—a *grain boundary.* The grain boundary consists of a transition zone, perhaps only a few atoms thick, in which the atoms are not aligned with either grain.

The size of the grains in the metal block is determined by the number of nucleation sites in the molten material and the cooling rate of the mass, among other factors. In a casting process, the nucleation sites are often created by the relatively cold walls of the mold, which motivate a somewhat preferred grain orientation at these walls.

Grain size is inversely related to cooling rate: Faster cooling promotes smaller grain size, whereas slower cooling has the opposite effect. Grain size is important in metals because it affects mechanical properties. Smaller grain size is generally preferable from a design viewpoint because it means higher strength and hardness. It is also desirable in certain manufacturing operations (e.g., metal forming), because it means higher ductility during deformation and a better surface on the finished product.

Another factor influencing mechanical properties is the presence of grain boundaries in the metal. They represent imperfections in the crystalline structure that interrupt the continued movement of dislocations. This helps to explain why smaller grain size—therefore more grains and more grain boundaries—increases the strength of the metal. By interfering with dislocation movement, grain boundaries also contribute to the characteristic property of a metal to become stronger as it is deformed. The property is called *strain hardening,* and it is examined more closely in the discussion of mechanical properties in Chapter 3.

2.4 NONCRYSTALLINE (AMORPHOUS) STRUCTURES

Many important materials are noncrystalline—liquids and gases, for example. Water and air have noncrystalline structures. A metal loses its crystalline structure when it is melted. Mercury is a liquid metal at room temperature, with its melting point of $-38°C$ ($-37°F$). Important classes of engineering materials have a noncrystalline form in their solid state; the term *amorphous* is often used to describe these materials. Glass, many plastics, and rubber fall into this category. Many important plastics are mixtures of crystalline and noncrystalline forms. Even metals can be amorphous rather than crystalline, given that the cooling rate during transformation from liquid to solid is fast enough to inhibit the atoms from arranging themselves into their preferred regular patterns. This can happen, for instance, if the molten metal is poured between cold, closely spaced, rotating rolls.

Two closely related features distinguish noncrystalline from crystalline materials: (1) absence of a long-range order in the molecular structure, and (2) differences in melting and thermal expansion characteristics.

FIGURE 2.14 Illustration of difference in structure between: (a) crystalline and (b) noncrystalline materials. The crystal structure is regular, repeating, and denser, whereas the noncrystalline structure is more loosely packed and random.

(a)   (b)

The difference in molecular structure can be visualized with reference to Figure 2.14. The closely packed and repeating pattern of the crystal structure is shown on the left; and the less dense and random arrangement of atoms in the noncrystalline material on the right. The difference is demonstrated by a metal when it melts. The more loosely packed atoms in the molten metal show an increase in volume (reduction in density) compared with the material's solid crystalline state. This effect is characteristic of most materials when melted. (Ice is a notable exception; liquid water is denser than solid ice.) It is a general characteristic of liquids and solid amorphous materials that they are absent of long-range order as on the right in our figure.

The melting phenomenon will now be examined in more detail, and in doing so, the second important difference between crystalline and noncrystalline structures will be defined. As indicated, a metal experiences an increase in volume when it melts from the solid to the liquid state. For a pure metal, this volumetric change occurs rather abruptly, at a constant temperature (i.e., the melting temperature T_m), as indicated in Figure 2.15. The change represents a discontinuity from the slopes on either side in the plot. The gradual slopes characterize the metal's **thermal expansion**—the change in volume as a function of temperature, which is usually different in the solid and liquid states. Associated with the sudden volume increase as the metal transforms from solid to liquid at the melting point is the addition of a certain quantity of heat, called the **heat of fusion,** which causes the atoms to lose the dense, regular arrangement of the crystalline structure. The process is reversible; it operates in both directions. If the molten metal is cooled through its melting temperature, the same abrupt change in volume occurs (except that it is a decrease), and the same quantity of heat is given off by the metal.

An amorphous material exhibits quite different behavior than that of a pure metal when it changes from solid to liquid, as shown in Figure 2.15. The process is again reversible, but observe the behavior of the amorphous material during cooling from the liquid state, rather

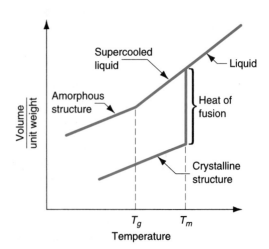

FIGURE 2.15 Characteristic change in volume for a pure metal (a crystalline structure), compared to the same volumetric changes in glass (a noncrystalline structure).

than during melting from the solid, as before. Glass (silica, SiO_2) is used to illustrate. At high temperatures, glass is a true liquid, and the molecules are free to move about as in the usual definition of a liquid. As the glass cools, it gradually transforms into the solid state, going through a transition phase, called a ***supercooled liquid,*** before finally becoming rigid. It does not show the sudden volumetric change that is characteristic of crystalline materials; instead, it passes through its melting temperature T_m without a change in its thermal expansion slope. In this supercooled liquid region, the material becomes increasingly viscous as the temperature continues to decrease. As it cools further, a point is finally reached at which the supercooled liquid converts to a solid. This is called the ***glass-transition temperature*** T_g. At this point, there is a change in the thermal expansion slope. (It might be more precise to refer to it as the thermal contraction slope; however, the slope is the same for expansion and contraction.) The rate of thermal expansion is lower for the solid material than for the supercooled liquid.

The difference in behavior between crystalline and noncrystalline materials can be traced to the response of their respective atomic structures to changes in temperature. When a pure metal solidifies from the molten state, the atoms arrange themselves into a regular and recurring structure. This crystal structure is much more compact than the random and loosely packed liquid from which it formed. Thus, the process of solidification produces the abrupt volumetric contraction observed in Figure 2.15 for the crystalline material. By contrast, amorphous materials do not achieve this repeating and closely packed structure at low temperatures. The atomic structure is the same random arrangement as in the liquid state; thus, there is no abrupt volumetric change as these materials transition from liquid to solid.

2.5 ENGINEERING MATERIALS

Let us summarize how atomic structure, bonding, and crystal structure (or absence thereof) are related to the type of engineering material—metals, ceramics, and polymer.

Metals   Metals have crystalline structures in the solid state, almost without exception. The unit cells of these crystal structures are almost always BCC, FCC, or HCP. The atoms of the metals are held together by metallic bonding, which means that their valence electrons can move about with relative freedom (compared with the other types of atomic and molecular bonding). These structures and bonding generally make the metals strong and hard. Many of the metals are quite ductile (capable of being deformed, which is useful in manufacturing), especially the FCC metals. Other general properties of metals related to structure and bonding include: high electrical and thermal conductivity, opaqueness (impervious to light rays), and reflectivity (capacity to reflect light rays).

Ceramics   Ceramic molecules are characterized by ionic or covalent bonding, or both. The metallic atoms release or share their outermost electrons to the nonmetallic atoms, and a strong attractive force exists within the molecules. The general properties that result from these bonding mechanisms include: high hardness and stiffness (even at elevated temperatures), brittleness (no ductility), electrical insulation (nonconducting) properties, refractoriness (being thermally resistant), and chemical inertness.

Ceramics possess either a crystalline or noncrystalline structure. Most ceramics have a crystal structure, whereas glasses based on silica (SiO_2) are amorphous. In certain cases, either structure can exist in the same ceramic material. For example, silica occurs in nature as crystalline quartz. When this mineral is melted and then cooled, it solidifies to form fused silica, which has a noncrystalline structure.

Polymers   A polymer molecule consists of many repeating ***mers*** to form very large molecules held together by covalent bonding. Elements in polymers are usually carbon plus one or more other elements such as hydrogen, nitrogen, oxygen, and chlorine. Secondary bonding (van der Waals) holds the molecules together within the aggregate material (intermolecular bonding). Polymers have either a glassy structure or mixture of glassy and crystalline. There are differences among the three polymer types. In ***thermoplastic polymers,*** the molecules consist of long chains of mers in a linear structure. These materials can be heated and cooled without substantially altering their linear structure. In ***thermosetting polymers,*** the molecules transform into a rigid, three-dimensional structure on cooling from a heated plastic condition. If thermosetting polymers are reheated, they degrade chemically rather than soften. ***Elastomers*** have large molecules with coiled structures. The uncoiling and recoiling of the molecules when subjected to stress cycles motivate the aggregate material to exhibit its characteristic elastic behavior.

The molecular structure and bonding of polymers provide them with the following typical properties: low density, high electrical resistivity (some polymers are used as insulating materials), and low thermal conductivity. Strength and stiffness of polymers vary widely. Some are strong and rigid (although not matching the strength and stiffness of metals or ceramics), whereas others exhibit highly elastic behavior.

REFERENCES

[1] Callister, W. D., Jr., ***Materials Science and Engineering: An Introduction,*** 7th ed. John Wiley & Sons, Hoboken, New Jersey, 2007.

[2] Dieter, G. E. ***Mechanical Metallurgy,*** 3rd ed. McGraw-Hill, New York, 1986.

[3] Flinn, R. A., and Trojan, P. K. ***Engineering Materials and Their Applications,*** 5th ed. John Wiley & Sons, New York, 1995.

[4] Guy, A. G., and Hren, J. J. ***Elements of Physical Metallurgy,*** 3rd ed. Addison-Wesley, Reading, Massachusetts, 1974.

[5] Van Vlack, L. H. ***Elements of Materials Science and Engineering,*** 6th ed. Addison-Wesley, Reading, Massachusetts, 1989.

REVIEW QUESTIONS

2.1. The elements listed in the Periodic Table can be divided into three categories. What are these categories? Give an example of each.

2.2. Which elements are the noble metals?

2.3. What is the difference between primary and secondary bonding in the structure of materials?

2.4. Describe how ionic bonding works.

2.5. What is the difference between crystalline and noncrystalline structures in materials?

2.6. What are some common point defects in a crystal lattice structure?

2.7. Define the difference between elastic and plastic deformation in terms of the effect on the crystal lattice structure.

2.8. How do grain boundaries contribute to the strain hardening phenomenon in metals?

2.9. Identify some materials that have a crystalline structure.

2.10. Identify some materials that possess a noncrystalline structure.

2.11. What is the basic difference in the solidification (or melting) process between crystalline and noncrystalline structures?

MULTIPLE CHOICE QUIZ

There are 20 correct answers in the following multiple choice questions (some questions have multiple answers that are correct). To attain a perfect score on the quiz, all correct answers must be given. Each correct answer is worth 1 point. Each

omitted answer or wrong answer reduces the score by 1 point, and each additional answer beyond the correct number of answers reduces the score by 1 point. Percentage score on the quiz is based on the total number of correct answers.

2.1. The basic structural unit of matter is which one of the following: (a) atom, (b) electron, (c) element, (d) molecule, or (e) nucleus?

2.2. Approximately how many different elements have been identified (one best answer): (a) 10, (b) 50, (c) 100, (d) 200, or (e) 500?

2.3. In the Periodic Table, the elements can be divided into which of the following categories (three best answers): (a) ceramics, (b) gases, (c) liquids, (d) metals, (e) nonmetals, (f) polymers, (g) semi-metals, and (h) solids?

2.4. The element with the lowest density and smallest atomic weight is which one of the following: (a) aluminum, (b) argon, (c) helium, (d) hydrogen, or (e) magnesium?

2.5. Which of the following bond types are classified as primary bonds (three correct answers): (a) covalent bonding, (b) hydrogen bonding, (c) ionic bonding, (d) metallic bonding, and (e) van der Waals forces?

2.6. How many atoms are there in the face-centered cubic (FCC) unit cell (one correct answer): (a) 8, (b) 9, (c) 10, (d) 12, or (e) 14?

2.7. Which of the following are not point defects in a crystal lattice structure (three correct answers): (a) edge dislocation, (b) grain boundaries, (c) interstitialcy, (d) Schottky defect, (e) screw dislocation, or (f) vacancy?

2.8. Which one of the following crystal structures has the fewest slip directions, thus making the metals with this structure generally more difficult to deform at room temperature: (a) BCC, (b) FCC, or (c) HCP?

2.9. Grain boundaries are an example of which one of the following types of crystal structure defects: (a) dislocation, (b) Frenkel defect, (c) line defects, (d) point defects, or (e) surface defects?

2.10. Twinning is which of the following (three best answers): (a) elastic deformation, (b) mechanism of plastic deformation, (c) more likely at high deformation rates, (d) more likely in metals with HCP structure, (e) slip mechanism, and (f) type of dislocation?

2.11. Polymers are characterized by which of the following bonding types (two correct answers): (a) adhesive, (b) covalent, (c) hydrogen, (d) ionic, (e) metallic, and (f) van der Waals?

3 MECHANICAL PROPERTIES OF MATERIALS

Chapter Contents

Mechanical properties of a material determine its behavior when subjected to mechanical stresses. These properties include elastic modulus, ductility, hardness, and various measures of strength. Mechanical properties are important in design because the function and performance of a product depend on its capacity to resist deformation under the stresses encountered in service. In design, the usual objective is for the product and its components to withstand these stresses without significant change in geometry. This capability depends on properties such as elastic modulus and yield strength. In manufacturing, the objective is just the opposite. Here, stresses that exceed the yield strength of the material must be applied to alter its shape. Mechanical processes such as forming and machining succeed by developing forces that exceed the material's resistance to deformation. Thus, there is the following dilemma: Mechanical properties that are desirable to the designer, such as high strength, usually make the manufacture of the product more difficult. It is helpful for the manufacturing engineer to appreciate the design viewpoint and for the designer to be aware of the manufacturing viewpoint.

This chapter examines the mechanical properties of materials that are most relevant in manufacturing.

3.1 STRESS–STRAIN RELATIONSHIPS

There are three types of static stresses to which materials can be subjected: tensile, compressive, and shear. Tensile stresses tend to stretch the material, compressive stresses tend to squeeze it, and shear involves stresses that tend to cause adjacent portions of the material to slide against each other. The stress–strain curve is the basic relationship that describes the mechanical properties of materials for all three types.

FIGURE 3.1   Tensile test: (a) tensile force applied in (1) and (2) resulting elongation of material; (b) typical test specimen; and (c) setup of the tensile test.

3.1.1 TENSILE PROPERTIES

The tensile test is the most common procedure for studying the stress–strain relationship, particularly for metals. In the test, a force is applied that pulls the material, tending to elongate it and reduce its diameter, as shown in Figure 3.1(a). Standards by ASTM (American Society for Testing and Materials) specify the preparation of the test specimen and the conduct of the test itself. The typical specimen and general setup of the tensile test is illustrated in Figure 3.1(b) and (c), respectively.

The starting test specimen has an original length L_o and area A_o. The length is measured as the distance between the gage marks, and the area is measured as the (usually round) cross section of the specimen. During the testing of a metal, the specimen stretches, then necks, and finally fractures, as shown in Figure 3.2. The load and the change in length of the specimen are recorded as testing proceeds, to provide the data required to determine

FIGURE 3.2   Typical progress of a tensile test: (1) beginning of test, no load; (2) uniform elongation and reduction of cross-sectional area; (3) continued elongation, maximum load reached; (4) necking begins, load begins to decrease; and (5) fracture. If pieces are put back together as in, (6) final length can be measured.

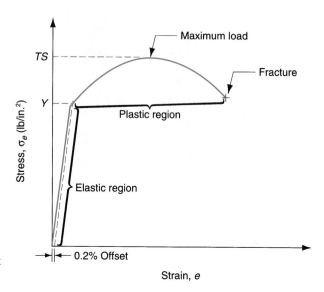

FIGURE 3.3 Typical engineering stress–strain plot in a tensile test of a metal.

the stress–strain relationship. There are two different types of stress–strain curves: (1) engineering stress–strain and (2) true stress–strain. The first is more important in design, and the second is more important in manufacturing.

Engineering Stress–Strain   The engineering stress and strain in a tensile test are defined relative to the original area and length of the test specimen. These values are of interest in design because the designer expects that the strains experienced by any component of the product will not significantly change its shape. The components are designed to withstand the anticipated stresses encountered in service.

A typical engineering stress–strain curve from a tensile test of a metallic specimen is illustrated in Figure 3.3. The *engineering stress* at any point on the curve is defined as the force divided by the original area:

$$s = \frac{F}{A_o} \tag{3.1}$$

where s = engineering stress, MPa (lb/in.²), F = applied force in the test, N (lb), and A_o = original area of the test specimen, mm² (in.²).

The *engineering strain* at any point in the test is given by

$$e = \frac{L - L_o}{L_o} \tag{3.2}$$

where e = engineering strain, mm/mm (in/in); L = length at any point during the elongation, mm (in); and L_o = original gage length, mm (in).

The units of engineering strain are given as mm/mm (in/in), but think of it as representing elongation per unit length, without units.

The stress–strain relationship in Figure 3.3 has two regions, indicating two distinct forms of behavior: (1) elastic and (2) plastic. In the elastic region, the relationship between stress and strain is linear, and the material exhibits elastic behavior by returning to its original length when the load (stress) is released. The relationship is defined by *Hooke's law:*

$$s = Ee \tag{3.3}$$

where E = *modulus of elasticity,* MPa (lb/in²), a measure of the inherent stiffness of a material.

TABLE 3.1   Elastic modulus for selected materials.

Metals	Modulus of Elasticity		Ceramics and Polymers	Modulus of Elasticity	
	MPa	lb/in^2		MPa	lb/in^2
Aluminum and alloys	69×10^3	10×10^6	Alumina	345×10^3	50×10^6
Cast iron	138×10^3	20×10^6	Diamond[a]	1035×10^3	150×10^6
Copper and alloys	110×10^3	16×10^6	Plate glass	69×10^3	10×10^6
Iron	209×10^3	30×10^6	Silicon carbide	448×10^3	65×10^6
Lead	21×10^3	3×10^6	Tungsten carbide	552×10^3	80×10^6
Magnesium	48×10^3	7×10^6	Nylon	3.0×10^3	0.40×10^6
Nickel	209×10^3	30×10^6	Phenol formaldehyde	7.0×10^3	1.00×10^6
Steel	209×10^3	30×10^6	Polyethylene (low density)	0.2×10^3	0.03×10^6
Titanium	117×10^3	17×10^6	Polyethylene (high density)	0.7×10^3	0.10×10^6
Tungsten	407×10^3	59×10^6	Polystyrene	3.0×10^3	0.40×10^6

[a]Compiled from [8], [10], [11], [15], [16], and other sources.
Although diamond is not a ceramic, it is often compared with the ceramic materials.

It is a constant of proportionality whose value is different for different materials. Table 3.1 presents typical values for several materials, metals and nonmetals.

As stress increases, some point in the linear relationship is finally reached at which the material begins to yield. This **yield point Y** of the material can be identified in the figure by the change in slope at the end of the linear region. Because the start of yielding is usually difficult to see in a plot of test data (it does not usually occur as an abrupt change in slope), Y is typically defined as the stress at which a strain offset of 0.2% from the straight line has occurred. More specifically, it is the point where the stress–strain curve for the material intersects a line that is parallel to the straight portion of the curve but offset from it by a strain of 0.2%. The yield point is a strength characteristic of the material, and is therefore often referred to as the **yield strength** (other names include **yield stress** and **elastic limit**).

The yield point marks the transition to the plastic region and the start of plastic deformation of the material. The relationship between stress and strain is no longer guided by Hooke's law. As the load is increased beyond the yield point, elongation of the specimen proceeds, but at a much faster rate than before, causing the slope of the curve to change dramatically, as shown in Figure 3.3. Elongation is accompanied by a uniform reduction in cross-sectional area, consistent with maintaining constant volume. Finally, the applied load F reaches a maximum value, and the engineering stress calculated at this point is called the **tensile strength** or **ultimate tensile strength** of the material. It is denoted as TS where $TS = F_{\max}/A_o$. TS and Y are important strength properties in design calculations. (They are also used in manufacturing calculations.) Some typical values of yield strength and tensile strength are listed in Table 3.2 for selected metals. Conventional tensile testing of ceramics is difficult, and an alternative test is used to measure the strength of these brittle materials (Section 3.1.3). Polymers differ in their strength properties from metals and ceramics because of viscoelasticity (Section 3.5).

To the right of the tensile strength on the stress–strain curve, the load begins to decline, and the test specimen typically begins a process of localized elongation known as **necking**. Instead of continuing to strain uniformly throughout its length, straining becomes concentrated in one small section of the specimen. The area of that section narrows down (necks) significantly until failure occurs. The stress calculated immediately before failure is known as the **fracture stress**.

The amount of strain that the material can endure before failure is also a mechanical property of interest in many manufacturing processes. The common measure of this property is **ductility**, the ability of a material to plastically strain without fracture. This

TABLE 3.2   Yield strength and tensile strength for selected metals.

Metal	Yield Strength		Tensile Strength		Metal	Yield Strength		Tensile Strength	
	MPa	lb/in^2	MPa	lb/in^2		MPa	lb/in^2	MPa	lb/in^2
Aluminum, annealed	28	4,000	69	10,000	Nickel, annealed	150	22,000	450	65,000
Aluminum, CW[a]	105	15,000	125	18,000	Steel, low C[a]	175	25,000	300	45,000
Aluminum alloys[a]	175	25,000	350	50,000	Steel, high C[a]	400	60,000	600	90,000
Cast iron[a]	275	40,000	275	40,000	Steel, alloy[a]	500	75,000	700	100,000
Copper, annealed	70	10,000	205	30,000	Steel, stainless[a]	275	40,000	650	95,000
Copper alloys[a]	205	30,000	410	60,000	Titanium, pure	350	50,000	515	75,000
Magnesium alloys[a]	175	25,000	275	40,000	Titanium alloy	800	120,000	900	130,000

Compiled from [8], [10], [11], [16], and other sources.
[a]Values given are typical. For alloys, there is a wide range in strength values depending on composition and treatment (e.g., heat treatment, work hardening).

measure can be taken as either elongation or area reduction. Elongation is defined as

$$EL = \frac{L_f - L_o}{L_o} \tag{3.4}$$

where EL = elongation, often expressed as a percent; L_f = specimen length at fracture, mm (in), measured as the distance between gage marks after the two parts of the specimen have been put back together; and L_o = original specimen length, mm (in).

Area reduction is defined as

$$AR = \frac{A_o - A_f}{A_o} \tag{3.5}$$

where AR = area reduction, often expressed as a percent; A_f = area of the cross section at the point of fracture, mm^2(in^2); and A_o = original area, mm^2 (in^2).

There are problems with both of these ductility measures because of necking that occurs in metallic test specimens and the associated nonuniform effect on elongation and area reduction. Despite these difficulties, percent elongation and percent area reduction are the most commonly used measures of ductility in engineering practice. Some typical values of percent elongation for various materials (mostly metals) are listed in Table 3.3.

True Stress–Strain   Thoughtful readers may be troubled by the use of the original area of the test specimen to calculate engineering stress, rather than the actual (instantaneous) area that becomes increasingly smaller as the test proceeds. If the actual area were used, the calculated stress value would be higher. The stress value obtained by dividing the instantaneous value of area into the applied load is defined as the **true stress:**

$$\sigma = \frac{F}{A} \tag{3.6}$$

where σ = true stress, MPa (lb/in^2); F = force, N (lb); and A = actual (instantaneous) area resisting the load, mm^2 (in^2).

Similarly, **true strain** provides a more realistic assessment of the "instantaneous" elongation per unit length of the material. The value of true strain in a tensile test can be estimated by dividing the total elongation into small increments, calculating the engineering strain for each increment on the basis of its starting length, and then adding up the strain values. In the limit, true strain is defined as

$$\epsilon = \int_{L_o}^{L} \frac{dL}{L} = \ln \frac{L}{L_o} \tag{3.7}$$

TABLE 3.3   Ductility as a percent of elongation (typical values) for various selected materials.

Material	Elongation	Material	Elongation
Metals		*Metals, continued*	
Aluminum, annealed	40%	Steel, low C[a]	30%
Aluminum, cold worked	8%	Steel, high C[a]	10%
Aluminum alloys, annealed[a]	20%	Steel, alloy[a]	20%
Aluminum alloys, heat treated[a]	8%	Steel, stainless, austenitic[a]	55%
Aluminum alloys, cast[a]	4%	Titanium, nearly pure	20%
Cast iron, gray[a]	0.6%	Zinc alloy	10%
Copper, annealed	45%	*Ceramics*	0[b]
Copper, cold worked	10%	*Polymers*	
Copper alloy: brass, annealed	60%	Thermoplastic polymers	100%
Magnesium alloys[a]	10%	Thermosetting polymers	1%
Nickel, annealed	45%	Elastomers (e.g., rubber)	1%[c]

Compiled from [8], [10], [11], [16], and other sources.
[a]Values given are typical. For alloys, there is a range of ductility that depends on composition and treatment (e.g., heat treatment, degree of work hardening).
[b]Ceramic materials are brittle; they withstand elastic strain but virtually no plastic strain.
[c]Elastomers endure significant elastic strain, but their plastic strain is very limited, only around 1% being typical.

where $L =$ instantaneous length at any moment during elongation.

At the end of the test (or other deformation), the final strain value can be calculated using $L = L_f$.

When the engineering stress–strain data in Figure 3.3 are plotted using the true stress and strain values, the resulting curve would appear as in Figure 3.4. In the elastic region, the plot is virtually the same as before. Strain values are small, and true strain is nearly equal to engineering strain for most metals of interest. The respective stress values are also very close to each other. The reason for these near equalities is that the cross-sectional area of the test specimen is not significantly reduced in the elastic region. Thus, Hooke's law can be used to relate true stress to true strain: $\sigma = E\epsilon$.

The difference between the true stress–strain curve and its engineering counterpart occurs in the plastic region. The stress values are higher in the plastic region because the

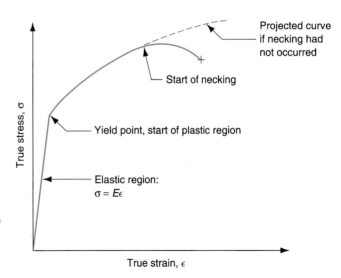

FIGURE 3.4 True stress–strain curve for the previous engineering stress–strain plot in Figure 3.3.

instantaneous cross-sectional area of the specimen, which has been continuously reduced during elongation, is now used in the computation. As in the previous curve, a downturn finally occurs as a result of necking. A dashed line is used in the figure to indicate the projected continuation of the true stress–strain plot if necking had not occurred.

As strain becomes significant in the plastic region, the values of true strain and engineering strain diverge. True strain can be related to the corresponding engineering strain by

$$\epsilon = \ln(1 + e) \tag{3.8}$$

Similarly, true stress and engineering stress can be related by the expression

$$\sigma = s(1 + e) \tag{3.9}$$

In Figure 3.4, note that stress increases continuously in the plastic region until necking begins. When this happened in the engineering stress–strain curve, its significance was lost because an admittedly erroneous area value was used to calculate stress. Now when the true stress also increases, it cannot be dismissed so lightly. What it means is that the metal is becoming stronger as strain increases. This is the property called **strain hardening** that was mentioned in the previous chapter in the discussion of metallic crystal structures, and it is a property that most metals exhibit to a greater or lesser degree.

Strain hardening, or **work hardening** as it is often called, is an important factor in certain manufacturing processes, particularly metal forming. Consider the behavior of a metal as it is affected by this property. If the portion of the true stress–strain curve representing the plastic region were plotted on a log–log scale, the result would be a linear relationship, as shown in Figure 3.5. Because it is a straight line in this transformation of the data, the relationship between true stress and true strain in the plastic region can be expressed as

$$\sigma = K\epsilon^n \tag{3.10}$$

This equation is called the **flow curve,** and it provides a good approximation of the behavior of metals in the plastic region, including their capacity for strain hardening. The constant K is called the **strength coefficient,** MPa (lb/in²), and it equals the value of true stress at a true strain value equal to one. The parameter n is called the **strain hardening exponent,** and it is the slope of the line in Figure 3.5. Its value is directly related to a metal's tendency to work harden. Typical values of K and n for selected metals are given in Table 3.4.

Necking in a tensile test and metal-forming operations that stretch the workpart is closely related to strain hardening. As the test specimen is elongated during the initial part of the test (before necking begins), uniform straining occurs throughout the length because if any element in the specimen becomes strained more than the surrounding metal, its strength increases because of work hardening, thus making it more resistant to additional strain until

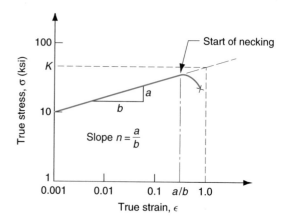

FIGURE 3.5   True stress–strain curve plotted on log–log scale.

TABLE 3.4   Typical values of strength coefficient *K* and strain hardening exponent *n* for selected metals.

Material	Strength Coefficient, K		Strain Hardening Exponent, n
	MPa	**lb/in^2**	
Aluminum, pure, annealed	175	25,000	0.20
Aluminum alloy, annealed[a]	240	35,000	0.15
Aluminum alloy, heat treated	400	60,000	0.10
Copper, pure, annealed	300	45,000	0.50
Copper alloy: brass[a]	700	100,000	0.35
Steel, low C, annealed[a]	500	75,000	0.25
Steel, high C, annealed[a]	850	125,000	0.15
Steel, alloy, annealed[a]	700	100,000	0.15
Steel, stainless, austenitic, annealed	1200	175,000	0.40

Compiled from [9], [10], [11], and other sources.
[a]Values of K and n vary according to composition, heat treatment, and work hardening.

the surrounding metal has been strained an equal amount. Finally, the strain becomes so large that uniform straining cannot be sustained. A weak point in the length develops (because of buildup of dislocations at grain boundaries, impurities in the metal, or other factors), and necking is initiated, leading to failure. Empirical evidence reveals that necking begins for a particular metal when the true strain reaches a value equal to the strain-hardening exponent n. Therefore, a higher n value means that the metal can be strained further before the onset of necking during tensile loading.

Types of Stress–Strain Relationships   Much information about elastic–plastic behavior is provided by the true stress–strain curve. As indicated, Hooke's law ($\sigma = E\epsilon$) governs the metal's behavior in the elastic region, and the flow curve ($\sigma = K\epsilon^n$) determines the behavior in the plastic region. Three basic forms of stress–strain relationship describe the behavior of nearly all types of solid materials, shown in Figure 3.6:

1. ***Perfectly elastic.*** The behavior of this material is defined completely by its stiffness, indicated by the modulus of elasticity E. It fractures rather than yielding to plastic flow. Brittle materials such as ceramics, many cast irons, and thermosetting polymers possess stress–strain curves that fall into this category. These materials are not good candidates for forming operations.

2. ***Elastic and perfectly plastic.*** This material has a stiffness defined by E. Once the yield strength Y is reached, the material deforms plastically at the same stress level. The flow curve is given by $K = Y$ and $n = 0$. Metals behave in this fashion when they have been

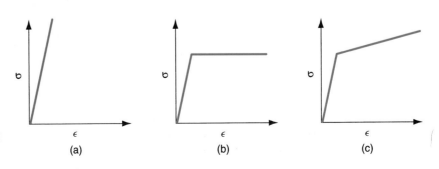

FIGURE 3.6   Three categories of stress–strain relationship: (a) perfectly elastic, (b) elastic and perfectly plastic, and (c) elastic and strain hardening.

heated to sufficiently high temperatures that they recrystallize rather than strain harden during deformation. Lead exhibits this behavior at room temperature because room temperature is above the recrystallization point for lead.

3. ***Elastic and strain hardening.*** This material obeys Hooke's law in the elastic region. It begins to flow at its yield strength Y. Continued deformation requires an ever-increasing stress, given by a flow curve whose strength coefficient K is greater than Y and whose strain-hardening exponent n is greater than zero. The flow curve is generally represented as a linear function on a natural logarithmic plot. Most ductile metals behave this way when cold worked.

Manufacturing processes that deform materials through the application of tensile stresses include wire and bar drawing (Section 19.6) and stretch forming (Section 20.6.1).

3.1.2 COMPRESSION PROPERTIES

A compression test applies a load that squeezes a cylindrical specimen between two platens, as illustrated in Figure 3.7. As the specimen is compressed, its height is reduced and its cross-sectional area is increased. Engineering stress is defined as

$$s = \frac{F}{A_o} \tag{3.11}$$

where A_o = original area of the specimen.

This is the same definition of engineering stress used in the tensile test. The engineering strain is defined as

$$e = \frac{h - h_o}{h_o} \tag{3.12}$$

where h = height of the specimen at a particular moment into the test, mm (in); and h_o = starting height, mm (in).

Because the height is decreased during compression, the value of e will be negative. The negative sign is usually ignored when expressing values of compression strain.

When engineering stress is plotted against engineering strain in a compression test, the results appear as in Figure 3.8. The curve is divided into elastic and plastic regions, as before,

FIGURE 3.7
Compression test:
(a) compression force applied to test piece in (1), and (2) resulting change in height; and (b) setup for the test, with size of test specimen exaggerated.

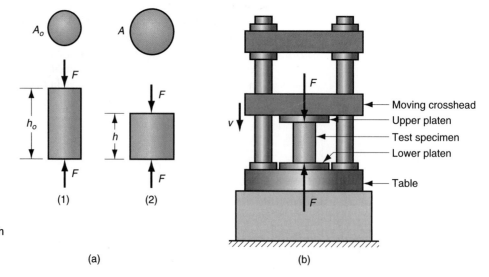

Moving crosshead
Upper platen
Test specimen
Lower platen

Table

(a)   (b)

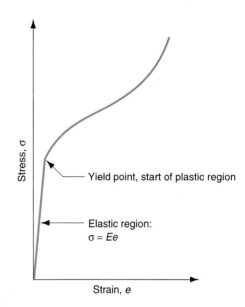

FIGURE 3.8   Typical engineering stress–strain curve for a compression test.

but the shape of the plastic portion of the curve is different from its tensile test complement. Because compression causes the cross section to increase (rather than decrease as in the tensile test), the load increases more rapidly than previously. This results in a higher value of calculated engineering stress.

Something else happens in the compression test that contributes to the increase in stress. As the cylindrical specimen is squeezed, friction at the surfaces in contact with the platens tends to prevent the ends of the cylinder from spreading. Additional energy is consumed by this friction during the test, and this results in a higher applied force. It also shows up as an increase in the computed engineering stress. Hence, owing to the increase in cross-sectional area and friction between the specimen and the platens, the characteristic engineering stress–strain curve is obtained in a compression test as seen in the figure.

Another consequence of the friction between the surfaces is that the material near the middle of the specimen is permitted to increase in area much more than at the ends. This results in the characteristic ***barreling*** of the specimen, as seen in Figure 3.9.

Although differences exist between the engineering stress–strain curves in tension and compression, when the respective data are plotted as true stress–strain, the relationships are nearly identical (for almost all materials). Because tensile test results are more abundant in the literature, values of the flow curve parameters (K and n) can be derived from tensile test data

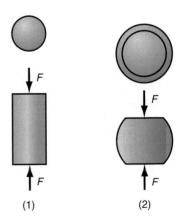

FIGURE 3.9   Barreling effect in a compression test: (1) start of test; and (2) after considerable compression has occurred.

(1)   (2)

and applied with equal validity to a compression operation. What must be done in using the tensile test results for a compression operation is to ignore the effect of necking, a phenomenon that is peculiar to straining induced by tensile stresses. In compression, there is no corresponding collapse of the work. In previous plots of tensile stress–strain curves, the data were extended beyond the point of necking by means of the dashed lines. The dashed lines better represent the behavior of the material in compression than the actual tensile test data.

Compression operations in metal forming are much more common than stretching operations. Important compression processes in industry include rolling, forging, and extrusion (Chapter 19).

3.1.3 BENDING AND TESTING OF BRITTLE MATERIALS

Bending operations are used to form metal plates and sheets. As shown in Figure 3.10, the process of bending a rectangular cross section subjects the material to tensile stresses (and strains) in the outer half of the bent section and compressive stresses (and strains) in the inner half. If the material does not fracture, it becomes permanently (plastically) bent as shown in (3.1) of Figure 3.10.

Hard, brittle materials (e.g., ceramics), which possess elasticity but little or no plasticity, are often tested by a method that subjects the specimen to a bending load. These materials do not respond well to traditional tensile testing because of problems in preparing the test specimens and possible misalignment of the press jaws that hold the specimen. The **bending test** (also known as the **flexure test**) is used to test the strength of these materials, using a setup illustrated in the first diagram in Figure 3.10. In this procedure, a specimen of rectangular cross section is positioned between two supports, and a load is applied at its center. In this configuration, the test is called a three-point bending test. A four-point configuration is also sometimes used. These brittle materials do not flex to the exaggerated extent shown in Figure 3.10; instead they deform elastically until immediately before fracture. Failure usually occurs because the ultimate tensile strength of the outer fibers of the specimen has been exceeded. This results in **cleavage,** a failure mode associated with ceramics and metals operating at low service temperatures, in which separation rather than slip occurs along certain crystallographic planes. The strength value derived from this test is called the **transverse rupture strength,** calculated from the formula

$$TRS = \frac{1.5\,FL}{bt^2} \qquad\qquad (3.13)$$

(1) (2) (3)

FIGURE 3.10   Bending of a rectangular cross section results in both tensile and compressive stresses in the material: (1) initial loading; (2) highly stressed and strained specimen; and (3) bent part.

where TRS = transverse rupture strength, MPa (lb/in^2); F = applied load at fracture, N (lb); L = length of the specimen between supports, mm (in); and b and t are the dimensions of the cross section of the specimen as shown in the figure, mm (in).

The flexure test is also used for certain nonbrittle materials such as thermoplastic polymers. In this case, because the material is likely to deform rather than fracture, TRS cannot be determined based on failure of the specimen. Instead, either of two measures is used: (1) the load recorded at a given level of deflection, or (2) the deflection observed at a given load.

3.1.4 SHEAR PROPERTIES

Shear involves application of stresses in opposite directions on either side of a thin element to deflect it, as shown in Figure 3.11. The shear stress is defined as

$$\tau = \frac{F}{A} \tag{3.14}$$

where τ = shear stress, lb/in^2 (MPa); F = applied force, N (lb); and A = area over which the force is applied, in^2 (mm^2).

Shear strain can be defined as

$$\gamma = \frac{\delta}{b} \tag{3.15}$$

where γ = shear strain, mm/mm (in/in); δ = the deflection of the element, mm (in); and b = the orthogonal distance over which deflection occurs, mm (in).

Shear stress and strain are commonly tested in a **torsion test,** in which a thin-walled tubular specimen is subjected to a torque as shown in Figure 3.12. As torque is increased, the tube deflects by twisting, which is a shear strain for this geometry.

The shear stress can be determined in the test by the equation

$$\tau = \frac{T}{2\pi R^2 t} \tag{3.16}$$

FIGURE 3.11 Shear (a) stress and (b) strain.

FIGURE 3.12 Torsion test setup.

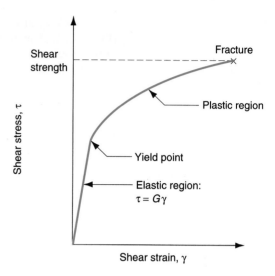

FIGURE 3.13 Typical shear stress–strain curve from a torsion test.

where $T =$ applied torque, N-mm (lb-in); $R =$ radius of the tube measured to the neutral axis of the wall, mm (in); and $t =$ wall thickness, mm (in).

The shear strain can be determined by measuring the amount of angular deflection of the tube, converting this into a distance deflected, and dividing by the gauge length L. Reducing this to a simple expression

$$\gamma = \frac{R\alpha}{L} \qquad (3.17)$$

where $\alpha =$ the angular deflection (radians).

A typical shear stress–strain curve is shown in Figure 3.13. In the elastic region, the relationship is defined by

$$\tau = G\gamma \qquad (3.18)$$

where $G =$ the **shear modulus,** or **shear modulus of elasticity,** MPa (lb/in^2). For most materials, the shear modulus can be approximated by $G = 0.4E$, where E is the conventional elastic modulus.

In the plastic region of the shear stress–strain curve, the material strain hardens to cause the applied torque to continue to increase until fracture finally occurs. The relationship in this region is similar to the flow curve. The shear stress at fracture can be calculated and this is used as the **shear strength S** of the material. Shear strength can be estimated from tensile strength data by the approximation: $S = 0.7(TS)$.

Because the cross-sectional area of the test specimen in the torsion test does not change as it does in the tensile and compression tests, the engineering stress–strain curve for shear derived from the torsion test is virtually the same as the true stress–strain curve.

Shear processes are common in industry. Shearing action is used to cut sheet metal in blanking, punching, and other cutting operations (Section 20.1). In machining, the material is removed by the mechanism of shear deformation (Section 21.2).

3.2 HARDNESS

The hardness of a material is defined as its resistance to permanent indentation. Good hardness generally means that the material is resistant to scratching and wear. For many engineering applications, including most of the tooling used in manufacturing, scratch and wear resistance are important characteristics. As the reader shall see later in this section, there is a strong correlation between hardness and strength.

3.2.1 HARDNESS TESTS

Hardness tests are commonly used for assessing material properties because they are quick and convenient. However, a variety of testing methods are appropriate because of differences in hardness among different materials. The best-known hardness tests are Brinell and Rockwell.

Brinell Hardness Test   The Brinell hardness test is widely used for testing metals and nonmetals of low to medium hardness. It is named after the Swedish engineer who developed it around 1900. In the test, a hardened steel (or cemented carbide) ball of 10-mm diameter is pressed into the surface of a specimen using a load of 500, 1500, or 3000 kg. The load is then divided into the indentation area to obtain the Brinell Hardness Number (BHN). In equation form

$$HB = \frac{2F}{\pi D_b \left(D_b - \sqrt{D_b^2 - D_i^2} \right)} \tag{3.19}$$

where HB = Brinell Hardness Number (BHN); F = indentation load, kg; D_b = diameter of the ball, mm; and D_i = diameter of the indentation on the surface, mm.

These dimensions are indicated in Figure 3.14(a). The resulting BHN has units of kg/mm², but the units are usually omitted in expressing the number. For harder materials (above 500 BHN), the cemented carbide ball is used because the steel ball experiences elastic deformation that compromises the accuracy of the reading. Also, higher loads (1500 and 3000 kg) are typically used for harder materials. Because of differences in results under different loads, it is considered good practice to indicate the load used in the test when reporting HB readings.

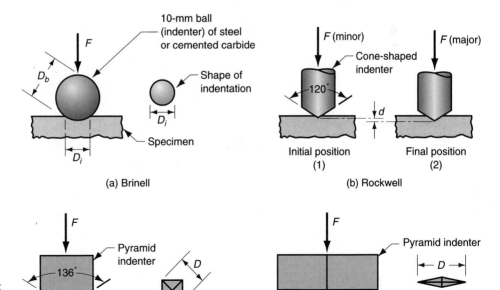

FIGURE 3.14
Hardness testing methods:
(a) Brinell; (b) Rockwell:
(1) initial minor load
and (2) major load,
(c) Vickers, and
(d) Knoop.

TABLE 3.5	Common Rockwell hardness scales.			
Rockwell Scale	**Hardness Symbol**	**Indenter**	**Load (kg)**	**Typical Materials Tested**
A	HRA	Cone	60	Carbides, ceramics
B	HRB	1.6 mm ball	100	Nonferrous metals
C	HRC	Cone	150	Ferrous metals, tool steels

Rockwell Hardness Test   This is another widely used test, named after the metallurgist who developed it in the early 1920s. It is convenient to use, and several enhancements over the years have made the test adaptable to a variety of materials.

In the Rockwell Hardness Test, a cone-shaped indenter or small-diameter ball, with diameter = 1.6 or 3.2 mm (1/16 or 1/8 in) is pressed into the specimen using a minor load of 10 kg, thus seating the indenter in the material. Then, a major load of 150 kg (or other value) is applied, causing the indenter to penetrate into the specimen a certain distance beyond its initial position. This additional penetration distance d is converted into a Rockwell hardness reading by the testing machine. The sequence is depicted in Figure 3.14(b). Differences in load and indenter geometry provide various Rockwell scales for different materials. The most common scales are indicated in Table 3.5.

Vickers Hardness Test   This test, also developed in the early 1920s, uses a pyramid-shaped indenter made of diamond. It is based on the principle that impressions made by this indenter are geometrically similar regardless of load. Accordingly, loads of various size are applied, depending on the hardness of the material to be measured. The Vickers Hardness (HV) is then determined from the formula

$$HV = \frac{1.854\,F}{D^2} \tag{3.20}$$

where F = applied load, kg, and D = the diagonal of the impression made by the indenter, mm, as indicated in Figure 3.14(c).

The Vickers test can be used for all metals and has one of the widest scales among hardness tests.

Knoop Hardness Test   The Knoop test, developed in 1939, uses a pyramid-shaped diamond indenter, but the pyramid has a length-to-width ratio of about 7:1, as indicated in Figure 3.14(d), and the applied loads are generally lighter than in the Vickers test. It is a microhardness test, meaning that it is suitable for measuring small, thin specimens or hard materials that might fracture if a heavier load were applied. The indenter shape facilitates reading of the impression under the lighter loads used in this test. The Knoop hardness value (HK) is determined according to the formula

$$HK = 14.2\frac{F}{D^2} \tag{3.21}$$

where F = load, kg; and D = the long diagonal of the indentor, mm.

Because the impression made in this test is generally very small, considerable care must be taken in preparing the surface to be measured.

Scleroscope   The previous tests base their hardness measurements either on the ratio of applied load divided by the resulting impression area (Brinell, Vickers, and Knoop) or by the depth of the impression (Rockwell). The Scleroscope is an instrument that measures the rebound height of a "hammer" dropped from a certain distance above the surface of the material to be tested. The hammer consists of a weight with diamond indenter attached to it.

The Scleroscope therefore measures the mechanical energy absorbed by the material when the indenter strikes the surface. The energy absorbed gives an indication of resistance to penetration, which matches the definition of hardness given here. If more energy is absorbed, the rebound will be less, meaning a softer material. If less energy is absorbed, the rebound will be higher—thus a harder material. The primary use of the Scleroscope seems to be in measuring the hardness of large parts of steel and other ferrous metals.

Durometer   The previous tests are all based on resistance to permanent or plastic deformation (indentation). The durometer is a device that measures the elastic deformation of rubber and similar flexible materials by pressing an indenter into the surface of the object. The resistance to penetration is an indication of hardness, as the term is applied to these types of materials.

3.2.2 HARDNESS OF VARIOUS MATERIALS

This section compares the hardness values of some common materials in the three engineering material classes: metals, ceramics, and polymers.

Metals   The Brinell and Rockwell hardness tests were developed at a time when metals were the principal engineering materials. A significant amount of data has been collected using these tests on metals. Table 3.6 lists hardness values for selected metals.

For most metals, hardness is closely related to strength. Because the method of testing for hardness is usually based on resistance to indentation, which is a form of compression, one would expect a good correlation between hardness and strength properties determined in a compression test. However, strength properties in a compression test are nearly the same as those from a tension test, after allowances for changes in cross-sectional area of the respective test specimens; so the correlation with tensile properties should also be good.

Brinell hardness (HB) exhibits a close correlation with the ultimate tensile strength TS of steels, leading to the relationship [9, 15]:

$$TS = K_h(HB) \tag{3.22}$$

where K_h is a constant of proportionality. If TS is expressed in MPa, then $K_h = 3.45$; and if TS is in lb/in^2, then $K_h = 500$.

TABLE 3.6   Typical hardness of selected metals.

Metal	Brinell Hardness, HB	Rockwell Hardness, HR[a]	Metal	Brinell Hardness, HB	Rockwell Hardness, HR[a]
Aluminum, annealed	20		Magnesium alloys, hardened[b]	70	35B
Aluminum, cold worked	35		Nickel, annealed	75	40B
Aluminum alloys, annealed[b]	40		Steel, low C, hot rolled[b]	100	60B
Aluminum alloys, hardened[b]	90	52B	Steel, high C, hot rolled[b]	200	95B, 15C
Aluminum alloys, cast[b]	80	44B	Steel, alloy, annealed[b]	175	90B, 10C
Cast iron, gray, as cast[b]	175	10C	Steel, alloy, heat treated[b]	300	33C
Copper, annealed	45		Steel, stainless, austenitic[b]	150	85B
Copper alloy: brass, annealed	100	60B	Titanium, nearly pure	200	95B
Lead	4		Zinc	30	

Compiled from [10], [11], [16], and other sources.
[a]HR values are given in the B or C scale as indicated by the letter designation. Missing values indicate that the hardness is too low for Rockwell scales.
[b]HB values given are typical. Hardness values will vary according to composition, heat treatment, and degree of work hardening.

TABLE 3.7   Hardness of selected ceramics and other hard materials, arranged in ascending order of hardness.

Material	Vickers Hardness, HV	Knoop Hardness, HK	Material	Vickers Hardness, HV	Knoop Hardness, HK
Hardened tool steel[a]	800	850	Titanium nitride, TiN	3000	2300
Cemented carbide (WC – Co)[a]	2000	1400	Titanium carbide, TiC	3200	2500
Alumina, Al_2O_3	2200	1500	Cubic boron nitride, BN	6000	4000
Tungsten carbide, WC	2600	1900	Diamond, sintered polycrystal	7000	5000
Silicon carbide, SiC	2600	1900	Diamond, natural	10,000	8000

Compiled from [14], [16], and other sources.
[a]Hardened tool steel and cemented carbide are the two materials commonly used in the Brinell hardness test.

TABLE 3.8   Hardness of selected polymers.

Polymer	Brinell Hardness, HB	Polymer	Brinell Hardness, HB
Nylon	12	Polypropylene	7
Phenol formaldehyde	50	Polystyrene	20
Polyethylene, low density	2	Polyvinyl-chloride	10
Polyethylene, high density	4		

Compiled from [5], [8], and other sources.

Ceramics   The Brinell hardness test is not appropriate for ceramics because the materials being tested are often harder than the indenter ball. The Vickers and Knoop hardness tests are used to test these hard materials. Table 3.7 lists hardness values for several ceramics and hard materials. For comparison, the Rockwell C hardness for hardened tool steel is 65 HRC. The HRC scale does not extend high enough to be used for the harder materials.

Polymers   Polymers have the lowest hardness among the three types of engineering materials. Table 3.8 lists several of the polymers on the Brinell hardness scale, although this testing method is not normally used for these materials. It does, however, allow comparison with the hardness of metals.

3.3   EFFECT OF TEMPERATURE ON PROPERTIES

Temperature has a significant effect on nearly all properties of a material. It is important for the designer to know the material properties at the operating temperatures of the product when in service. It is also important to know how temperature affects mechanical properties in manufacturing. At elevated temperatures, materials are lower in strength and higher in ductility. The general relationships for metals are depicted in Figure 3.15. Thus, most metals can be formed more easily at elevated temperatures than when they are cold.

Hot Hardness   A property often used to characterize strength and hardness at elevated temperatures is hot hardness. *Hot hardness* is simply the ability of a material to retain hardness at elevated temperatures; it is usually presented as either a listing of hardness values at different temperatures or as a plot of hardness versus temperature, as in Figure 3.16. Steels can be alloyed to achieve significant improvements in hot hardness, as shown in the figure.

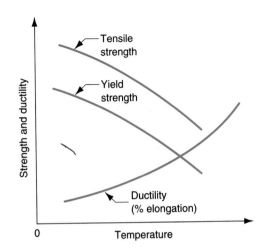

FIGURE 3.15   General effect of temperature on strength and ductility.

Ceramics exhibit superior properties at elevated temperatures. These materials are often selected for high temperature applications, such as turbine parts, cutting tools, and refractory applications. The outside skin of a shuttle spacecraft is lined with ceramic tiles to withstand the friction heat of high-speed re-entry into the atmosphere.

Good hot hardness is also desirable in the tooling materials used in many manufacturing operations. Significant amounts of heat energy are generated in most metalworking processes, and the tools must be capable of withstanding the high temperatures involved.

Recrystallization Temperature   Most metals behave at room temperature according to the flow curve in the plastic region. As the metal is strained, it increases in strength because of strain hardening (the strain-hardening exponent $n > 0$). However, if the metal is heated to a sufficiently elevated temperature and then deformed, strain hardening does not occur. Instead, new grains are formed that are free of strain, and the metal behaves as a perfectly plastic material; that is, with a strain-hardening exponent $n = 0$. The formation of new strain-free grains is a process called ***recrystallization,*** and the temperature at which it occurs is about one-half the melting point ($0.5\ T_m$), as measured on an absolute scale (R or K). This is called the ***recrystallization temperature***. Recrystallization takes time. The recrystallization temperature for a particular metal is usually specified as the temperature at which complete formation of new grains requires about 1 hour.

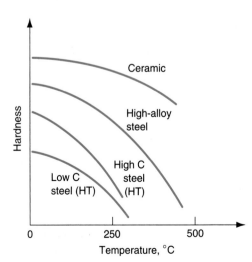

FIGURE 3.16   Hot hardness—typical hardness as a function of temperature for several materials.

Recrystallization is a temperature-dependent characteristic of metals that can be exploited in manufacturing. By heating the metal to the recrystallization temperature before deformation, the amount of straining that the metal can endure is substantially increased, and the forces and power required to carry out the process are significantly reduced. Forming metals at temperatures above the recrystallization temperature is called *hot working* (Section 18.3).

3.4 FLUID PROPERTIES

Fluids behave quite differently than solids. A fluid flows; it takes the shape of the container that holds it. A solid does not flow; it possesses a geometric form that is independent of its surroundings. Fluids include liquids and gases; the interest in this section is on the former. Many manufacturing processes are accomplished on materials that have been converted from solid to liquid state by heating. Metals are cast in the molten state; glass is formed in a heated and highly fluid state; and polymers are almost always shaped as thick fluids.

Viscosity Although flow is a defining characteristic of fluids, the tendency to flow varies for different fluids. Viscosity is the property that determines fluid flow. Roughly, *viscosity* can be defined as the resistance to flow that is characteristic of a fluid. It is a measure of the internal friction that arises when velocity gradients are present in the fluid—the more viscous the fluid is, the higher the internal friction and the greater the resistance to flow. The reciprocal of viscosity is *fluidity*—the ease with which a fluid flows.

Viscosity is defined more precisely with respect to the setup in Figure 3.17, in which two parallel plates are separated by a distance d. One of the plates is stationary while the other is moving at a velocity v, and the space between the plates is occupied by a fluid. Orienting these parameters relative to an axis system, d is in the y-axis direction and v is in the x-axis direction. The motion of the upper plate is resisted by force F that results from the shear viscous action of the fluid. This force can be reduced to a shear stress by dividing F by the plate area A

$$\tau = \frac{F}{A} \tag{3.23}$$

where τ = shear stress, N/m^2 or Pa (lb/in^2).

This shear stress is related to the rate of shear, which is defined as the change in velocity dv relative to dy. That is

$$\dot{\gamma} = \frac{dv}{dy} \tag{3.24}$$

where $\dot{\gamma}$ = shear rate, 1/s; dv = incremental change in velocity, m/s (in/sec); and dy = incremental change in distance y, m (in).

FIGURE 3.17 Fluid flow between two parallel plates, one stationary and the other moving at velocity v.

The shear viscosity is the fluid property that defines the relationship between F/A and dv/dy; that is

$$\frac{F}{A} = \eta \frac{dv}{dy} \quad \text{or} \quad \tau = \eta \dot{\gamma} \tag{3.25}$$

where η = a constant of proportionality called the coefficient of viscosity, Pa-s (lb-sec/in^2). Rearranging Equation 3.25, the coefficient of viscosity can be expressed as follows

$$\eta = \frac{\tau}{\dot{\gamma}} \tag{3.26}$$

Thus, the viscosity of a fluid can be defined as the ratio of shear stress to shear rate during flow, where shear stress is the frictional force exerted by the fluid per unit area, and shear rate is the velocity gradient perpendicular to the flow direction. The viscous characteristics of fluids defined by Equation 3.26 were first stated by Newton. He observed that viscosity was a constant property of a given fluid, and such a fluid is referred to as a ***Newtonian fluid***.

The units of coefficient of viscosity require explanation. In the International System of units (SI), because shear stress is expressed in N/m^2 or Pascals and shear rate in 1/s, it follows that η has units of N-s/m^2 or Pascal-seconds, abbreviated Pa-s. In the U.S. customary units, the corresponding units are lb/in^2 and 1/sec, so that the units for coefficient of viscosity are lb-sec/in^2. Other units sometimes given for viscosity are poise, which = dyne-sec/cm^2 (10 poise = 1 Pa-s and 6895 Pa-s = 1 lb-sec/in^2). Some typical values of coefficient of viscosity for various fluids are given in Table 3.9. One can observe in several of the materials listed that viscosity varies with temperature.

Viscosity in Manufacturing Processes   For many metals, the viscosity in the molten state compares with that of water at room temperature. Certain manufacturing processes, notably casting and welding, are performed on metals in their molten state, and success in these operations requires low viscosity so that the molten metal fills the mold cavity or weld seam before solidifying. In other operations, such as metal forming and machining, lubricants and coolants are used in the process, and again the success of these fluids depends to some extent on their viscosities.

Glass ceramics exhibit a gradual transition from solid to liquid states as temperature is increased; they do not suddenly melt as pure metals do. The effect is illustrated by the viscosity values for glass at different temperatures in Table 3.9. At room temperature, glass is solid and brittle, exhibiting no tendency to flow; for all practical purposes, its viscosity is infinite. As glass is heated, it gradually softens, becoming less and less viscous (more and more fluid), until it can finally be formed by blowing or molding at around 1100°C (2000°F).

TABLE 3.9   Viscosity values for selected fluids.

Material	Coefficient of Viscosity		Material	Coefficient of Viscosity	
	Pa-s	**lb-sec/in²**		**Pa-s**	**lb-sec/in²**
Glass[b], 540 C (1000 F)	10^{12}	10^8	Pancake syrup (room temp)	50	73×10^{-4}
Glass[b], 815 C (1500 F)	10^5	14	Polymer,[a] 151 C (300 F)	115	167×10^{-4}
Glass[b], 1095 C (2000 F)	10^3	0.14	Polymer,[a] 205 C (400 F)	55	80×10^{-4}
Glass[b], 1370 C (2500 F)	15	22×10^{-4}	Polymer,[a] 260 C (500 F)	28	41×10^{-4}
Mercury, 20 C (70 F)	0.0016	0.23×10^{-6}	Water, 20 C (70 F)	0.001	0.15×10^{-6}
Machine oil (room temp.)	0.1	0.14×10^{-4}	Water, 100 C (212 F)	0.0003	0.04×10^{-6}

Compiled from various sources.
[a]Low-density polyethylene is used as the polymer example here; most other polymers have slightly higher viscosities.
[b]Glass composition is mostly SiO_2; compositions and viscosities vary; values given are representative.

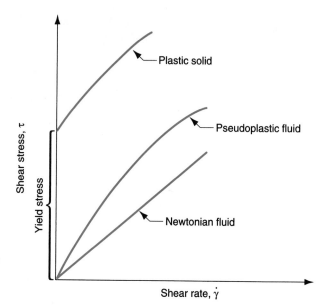

FIGURE 3.18 Viscous behaviors of Newtonian and pseudoplastic fluids. Polymer melts exhibit pseudoplastic behavior. For comparison, the behavior of a plastic solid material is shown.

Most polymer-shaping processes are performed at elevated temperatures, at which the material is in a liquid or highly plastic condition. Thermoplastic polymers represent the most straightforward case, and they are also the most common polymers. At low temperatures, thermoplastic polymers are solid; as temperature is increased, they typically transform first into a soft rubbery material, and then into a thick fluid. As temperature continues to rise, viscosity decreases gradually, as in Table 3.9 for polyethylene, the most widely used thermoplastic polymer. However, with polymers the relationship is complicated by other factors. For example, viscosity is affected by flow rate. The viscosity of a thermoplastic polymer is not a constant. A polymer melt does not behave in a Newtonian fashion. Its relationship between shear stress and shear rate can be seen in Figure 3.18. A fluid that exhibits this decreasing viscosity with increasing shear rate is called *pseudoplastic*. This behavior complicates the analysis of polymer shaping.

3.5 VISCOELASTIC BEHAVIOR OF POLYMERS

Another property that is characteristic of polymers is viscoelasticity. *Viscoelasticity* is the property of a material that determines the strain it experiences when subjected to combinations of stress and temperature over time. As the name suggests, it is a combination of viscosity and elasticity. Viscoelasticity can be explained with reference to Figure 3.19. The two parts of the figure show the typical response of two materials to an applied stress below the yield point during some time period. The material in (a) exhibits perfect elasticity; when the stress is removed, the material returns to its original shape. By contrast, the material in (b) shows viscoelastic behavior. The amount of strain gradually increases over time under the applied stress. When stress is removed, the material does not immediately return to its original shape; instead, the strain decays gradually. If the stress had been applied and then immediately removed, the material would have returned immediately to its starting shape. However, time has entered the picture and played a role in affecting the behavior of the material.

A simple model of viscoelasticity can be developed using the definition of elasticity as a starting point. Elasticity is concisely expressed by Hooke's law, $\sigma = E\epsilon$, which simply relates stress to strain through a constant of proportionality. In a viscoelastic solid, the

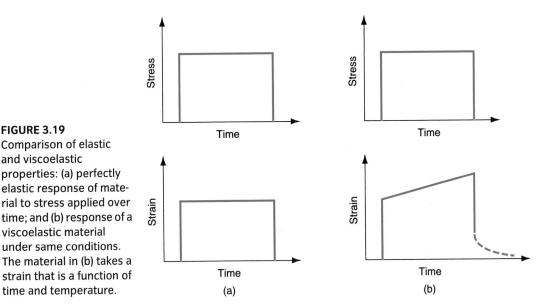

FIGURE 3.19
Comparison of elastic and viscoelastic properties: (a) perfectly elastic response of material to stress applied over time; and (b) response of a viscoelastic material under same conditions. The material in (b) takes a strain that is a function of time and temperature.

relationship between stress and strain is time dependent; it can be expressed as

$$\sigma(t) = f(t)\epsilon \qquad (3.27)$$

The time function $f(t)$ can be conceptualized as a modulus of elasticity that depends on time. It might be written $E(t)$ and referred to as a viscoelastic modulus. The form of this time function can be complex, sometimes including strain as a factor. Without getting into the mathematical expressions for it, nevertheless the effect of the time dependency can be explored. One common effect can be seen in Figure 3.20, which shows the stress–strain behavior of a thermoplastic polymer under different strain rates. At low strain rate, the material exhibits significant viscous flow. At high strain rate, it behaves in a much more brittle fashion.

Temperature is a factor in viscoelasticity. As temperature increases, the viscous behavior becomes more and more prominent relative to elastic behavior. The material becomes more like a fluid. Figure 3.21 illustrates this temperature dependence for a thermoplastic polymer. At low temperatures, the polymer shows elastic behavior. As T increases above the glass transition temperature T_g, the polymer becomes viscoelastic. As temperature increases further, it becomes soft and rubbery. At still higher temperatures, it exhibits viscous characteristics. The temperatures at which these modes of behavior are observed vary, depending on the plastic. Also, the shapes of the modulus versus temperature curve differ according to the

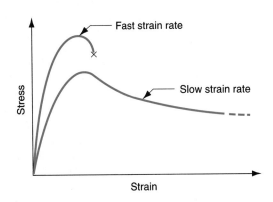

FIGURE 3.20   Stress–strain curve of a viscoelastic material (thermoplastic polymer) at high and low strain rates.

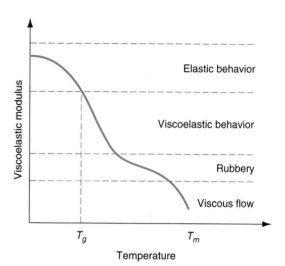

FIGURE 3.21   Viscoelastic modulus as a function of temperature for a thermoplastic polymer.

proportions of crystalline and amorphous structures in the thermoplastic. Thermosetting polymers and elastomers behave differently than shown in the figure; after curing, these polymers do not soften as thermoplastics do at elevated temperatures. Instead, they degrade (char) at high temperatures.

Viscoelastic behavior manifests itself in polymer melts in the form of shape memory. As the thick polymer melt is transformed during processing from one shape to another, it "remembers" its previous shape and attempts to return to that geometry. For example, a common problem in extrusion of polymers is die swell, in which the profile of the extruded material grows in size, reflecting its tendency to return to its larger cross section in the extruder barrel immediately before being squeezed through the smaller die opening. The properties of viscosity and viscoelasticity are examined in more detail in the discussion of plastic shaping (Chapter 13).

REFERENCES

[1] Avallone, E. A., and Baumeister III, T. (eds.). *Mark's Standard Handbook for Mechanical Engineers,* 11th ed. McGraw-Hill, New York, 2006.

[2] Beer, F. P., Russell, J. E., Eisenberg, E., and Mazurek, D., *Vector Mechanics for Engineers: Statics,* 9th ed. McGraw-Hill, New York, 2009.

[3] Black, J. T., and Kohser, R. A. *DeGarmo's Materials and Processes in Manufacturing,* 10th ed. John Wiley & Sons, Hoboken, New Jersey, 2008.

[4] Budynas, R. G. *Advanced Strength and Applied Stress Analysis,* 2nd ed. McGraw-Hill, New York, 1998.

[5] Chandra, M., and Roy, S. K. *Plastics Technology Handbook,* 4th ed. CRC Press, Inc., Boca Raton, Florida, 2006.

[6] Dieter, G. E. *Mechanical Metallurgy,* 3rd ed. McGraw-Hill, New York, 1986.

[7] *Engineering Plastics.* Engineered Materials Handbook, Vol. 2. ASM International, Metals Park, Ohio, 1987.

[8] Flinn, R. A., and Trojan, P. K. *Engineering Materials and Their Applications,* 5th ed. John Wiley & Sons, Hoboken, New Jersey, 1995.

[9] Kalpakjian, S., and Schmid S. R. *Manufacturing Processes for Engineering Materials,* 5th ed. Prentice Hall, Upper Saddle River, New Jersey, 2007.

[10] *Metals Handbook,* Vol. 1, Properties and Selection: Iron, Steels, and High Performance Alloys. ASM International, Metals Park, Ohio, 1990.

[11] *Metals Handbook,* Vol. 2, Properties and Selection: Nonferrous Alloys and Special Purpose Materials, ASM International, Metals Park, Ohio, 1991.

[12] *Metals Handbook,* Vol. 8, Mechanical Testing and Evaluation, ASM International, Metals Park, Ohio, 2000.

[13] Morton-Jones, D. H. *Polymer Processing.* Chapman and Hall, London, 2008.

[14] Schey, J. A. *Introduction to Manufacturing Processes.* 3rd ed. McGraw-Hill, New York, 2000.

[15] Van Vlack, L. H. *Elements of Materials Science and Engineering,* 6th ed. Addison-Wesley, Reading, Massachusetts, 1991.

[16] Wick, C., and Veilleux, R. F. (eds.). *Tool and Manufacturing Engineers Handbook,* 4th ed. Vol. 3, Materials, Finishing, and Coating. Society of Manufacturing Engineers, Dearborn, Michigan, 1985.

REVIEW QUESTIONS

3.1. What is the dilemma between design and manufacturing in terms of mechanical properties?

3.2. What are the three types of static stresses to which materials are subjected?

3.3. State Hooke's law.

3.4. What is the difference between engineering stress and true stress in a tensile test?

3.5. Define tensile strength of a material.

3.6. Define yield strength of a material.

3.7. Why cannot a direct conversion be made between the ductility measures of elongation and reduction in area using the assumption of constant volume?

3.8. What is work hardening?

3.9. In what case does the strength coefficient have the same value as the yield strength?

3.10. How does the change in cross-sectional area of a test specimen in a compression test differ from its counterpart in a tensile test specimen?

3.11. What is the complicating factor that occurs in a compression test?

3.12. Tensile testing is not appropriate for hard brittle materials such as ceramics. What is the test commonly used to determine the strength properties of such materials?

3.13. How is the shear modulus of elasticity G related to the tensile modulus of elasticity E, on average?

3.14. How is shear strength S related to tensile strength TS, on average?

3.15. What is hardness, and how is it generally tested?

3.16. Why are different hardness tests and scales required?

3.17. Define the recrystallization temperature for a metal.

3.18. Define viscosity of a fluid.

3.19. What is the defining characteristic of a Newtonian fluid?

3.20. What is viscoelasticity, as a material property?

MULTIPLE CHOICE QUIZ

There are 15 correct answers in the following multiple choice questions (some questions have multiple answers that are correct). To attain a perfect score on the quiz, all correct answers must be given. Each correct answer is worth 1 point. Each omitted answer or wrong answer reduces the score by 1 point, and each additional answer beyond the correct number of answers reduces the score by 1 point. Percentage score on the quiz is based on the total number of correct answers.

3.1. Which of the following are the three basic types of static stresses to which a material can be subjected (three correct answers): (a) compression, (b) hardness, (c) reduction in area, (d) shear, (e) tensile, (f) true stress, and (g) yield?

3.2. Which one of the following is the correct definition of ultimate tensile strength, as derived from the results of a tensile test on a metal specimen: (a) the stress encountered when the stress–strain curve transforms from elastic to plastic behavior, (b) the maximum load divided by the final area of the specimen, (c) the maximum load divided by the original area of the specimen, or (d) the stress observed when the specimen finally fails?

3.3. If stress values were measured during a tensile test, which of the following would have the higher value: (a) engineering stress or (b) true stress?

3.4. If strain measurements were made during a tensile-test, which of the following would have the higher value: (a) engineering strain, or (b) true strain?

3.5. The plastic region of the stress–strain curve for a metal is characterized by a proportional relationship between stress and strain: (a) true or (b) false?

3.6. Which one of the following types of stress–strain relationship best describes the behavior of brittle materials such as ceramics and thermosetting plastics: (a) elastic and perfectly plastic, (b) elastic and strain hardening, (c) perfectly elastic, or (d) none of the above?

3.7. Which one of the following types of stress–strain relationship best describes the behavior of most metals at room temperature: (a) elastic and perfectly plastic, (b) elastic and strain hardening, (c) perfectly elastic, or (d) none of the above?

3.8. Which one of the following types of stress–strain relationship best describes the behavior of metals at temperatures above their respective re-crystallization points: (a) elastic and perfectly plastic, (b) elastic and strain hardening, (c) perfectly elastic, or (d) none of the above?

3.9. Which one of the following materials has the highest modulus of elasticity: (a) aluminum, (b) diamond, (c) steel, (d) titanium, or (e) tungsten?

3.10. The shear strength of a metal is usually (a) greater than or (b) less than its tensile strength?

3.11. Most hardness tests involve pressing a hard object into the surface of a test specimen and measuring the indentation (or its effect) that results: (a) true or (b) false?

3.12. Which one of the following materials has the highest hardness: (a) alumina ceramic, (b) gray cast iron, (c) hardened tool steel, (d) high carbon steel, or (e) polystyrene?

3.13. Viscosity can be defined as the ease with which a fluid flows: (a) true or (b) false?

PROBLEMS

Strength and Ductility in Tension

3.1. A tensile test uses a test specimen that has a gage length of 50 mm and an area $= 200$ mm^2. During the test the specimen yields under a load of 98,000 N. The corresponding gage length $= 50.23$ mm. This is the 0.2% yield point. The maximum load of 168,000 N is reached at a gage length $= 64.2$ mm. Determine (a) yield strength, (b) modulus of elasticity, and (c) tensile strength. (d) If fracture occurs at a gage length of 67.3 mm, determine the percent elongation. (e) If the specimen necked to an area $= 92$ mm^2, determine the percent reduction in area.

3.2. A test specimen in a tensile test has a gage length of 2.0 in and an area $= 0.5$ in^2. During the test the specimen yields under a load of 32,000 lb. The corresponding gage length $= 2.0083$ in. This is the 0.2 percent yield point. The maximum load of 60,000 lb is reached at a gage length $= 2.60$ in. Determine (a) yield strength, (b) modulus of elasticity, and (c) tensile strength. (d) If fracture occurs at a gage length of 2.92 in, determine the percent elongation. (e) If the specimen necked to an area $= 0.25$ in^2, determine the percent reduction in area.

3.3. During a tensile test in which the starting gage length $= 125.0$ mm and the cross-sectional area $= 62.5$ mm^2, the following force and gage length data are collected (1) 17,793 N at 125.23 mm, (2) 23,042 N at 131.25 mm, (3) 27,579 N at 140.05 mm, (4) 28,913 N at 147.01 mm, (5) 27,578 N at 153.00 mm, and (6) 20,462 N at 160.10 mm. The maximum load is 28,913 N and the final data point occurred immediately before failure. (a) Plot the engineering stress strain curve. Determine (b) yield strength, (c) modulus of elasticity, and (d) tensile strength.

Flow Curve

3.4. In Problem 3.3, determine the strength coefficient and the strain-hardening exponent in the flow curve equation. Be sure not to use data after the point at which necking occurred.

3.5. In a tensile test on a metal specimen, true strain $= 0.08$ at a stress $= 265$ MPa. When true stress $= 325$ MPa, true strain $= 0.27$. Determine the strength coefficient and the strain-hardening exponent in the flow curve equation.

3.6. During a tensile test, a metal has a true strain $= 0.10$ at a true stress $= 37,000$ lb/in^2. Later, at a true stress $= 55,000$ lb/in^2, true strain $= 0.25$. Determine the strength coefficient and strain-hardening exponent in the flow curve equation.

3.7. In a tensile test a metal begins to neck at a true strain $= 0.28$ with a corresponding true stress $= 345.0$ MPa. Without knowing any more about the test, can you estimate the strength coefficient and the strain-hardening exponent in the flow curve equation?

3.8. A tensile test for a certain metal provides flow curve parameters: strain-hardening exponent is 0.3 and strength coefficient is 600 MPa. Determine (a) the flow stress at a true strain $= 1.0$ and (b) true strain at a flow stress $= 600$ MPa.

3.9. The flow curve for a certain metal has a strain-hardening exponent of 0.22 and strength coefficient of 54,000 lb/in^2. Determine (a) the flow stress at a true strain $= 0.45$ and (b) the true strain at a flow stress $= 40,000$ lb/in^2.

3.10. A metal is deformed in a tension test into its plastic region. The starting specimen had a gage length $= 2.0$ in and an area $= 0.50$ in^2. At one point in the tensile test, the gage length $= 2.5$ in, and the corresponding engineering stress $= 24,000$ lb/in^2; at another point in the test before necking, the gage length $= 3.2$ in, and the corresponding engineering stress $= 28,000$ lb/in^2. Determine the strength coefficient and the strain-hardening exponent for this metal.

3.11. A tensile test specimen has a starting gage length = 75.0 mm. It is elongated during the test to a length = 110.0 mm before necking occurs. Determine (a) the engineering strain and (b) the true strain. (c) Compute and sum the engineering strains as the specimen elongates from: (1) 75.0 to 80.0 mm, (2) 80.0 to 85.0 mm, (3) 85.0 to 90.0 mm, (4) 90.0 to 95.0 mm, (5) 95.0 to 100.0 mm, (6) 100.0 to 105.0 mm, and (7) 105.0 to 110.0 mm. (d) Is the result closer to the answer to part (a) or part (b)? Does this help to show what is meant by the term true strain?

3.12. A tensile specimen is elongated to twice its original length. Determine the engineering strain and true strain for this test. If the metal had been strained in compression, determine the final compressed length of the specimen such that (a) the engineering strain is equal to the same value as in tension (it will be negative value because of compression), and (b) the true strain would be equal to the same value as in tension (again, it will be negative value because

of compression). Note that the answer to part (a) is an impossible result. True strain is therefore a better measure of strain during plastic deformation.

3.13. Derive an expression for true strain as a function of D and D_o for a tensile test specimen of round cross section, where D = the instantaneous diameter of the specimen and D_o is its original diameter.

3.14. Show that true strain = $\ln(1 + e)$, where e = engineering strain.

3.15. Based on results of a tensile test, the flow curve strain-hardening exponent = 0.40 and strength coefficient = 551.6 MPa. Based on this information, calculate the (engineering) tensile strength for the metal.

3.16. A copper wire of diameter 0.80 mm fails at an engineering stress = 248.2 MPa. Its ductility is measured as 75% reduction of area. Determine the true stress and true strain at failure.

3.17. A steel tensile specimen with starting gage length = 2.0 in and cross-sectional area = 0.5 in² reaches a maximum load of 37,000 lb. Its elongation at this point is 24%. Determine the true stress and true strain at this maximum load.

Compression

3.18. A metal alloy has been tested in a tensile test with the following results for the flow curve parameters: strength coefficient = 620.5 MPa and strain-hardening exponent = 0.26. The same metal is now tested in a compression test in which the starting height of the specimen = 62.5 mm and its diameter = 25 mm. Assuming that the cross section increases uniformly, determine the load required to compress the specimen to a height of (a) 50 mm and (b) 37.5 mm.

3.19. The flow curve parameters for a certain stainless steel are strength coefficient = 1100 MPa and strain-hardening exponent = 0.35. A cylindrical specimen of starting cross-sectional area = 1000

mm² and height = 75 mm is compressed to a height of 58 mm. Determine the force required to achieve this compression, assuming that the cross section increases uniformly.

3.20. A steel test specimen (modulus of elasticity = 30 × 10⁶ lb/in²) in a compression test has a starting height = 2.0 in and diameter = 1.5 in. The metal yields (0.2% offset) at a load = 140,000 lb. At a load of 260,000 lb, the height has been reduced to 1.6 in. Determine (a) yield strength and (b) flow curve parameters (strength coefficient and strain-hardening exponent). Assume that the cross-sectional area increases uniformly during the test.

Bending and Shear

3.21. A bend test is used for a certain hard material. If the transverse rupture strength of the material is known to be 1000 MPa, what is the anticipated load at which the specimen is likely to fail, given that its width = 15 mm, thickness = 10 mm, and length = 60 mm?

3.22. A special ceramic specimen is tested in a bend test. Its width = 0.50 in and thickness = 0.25 in. The length of the specimen between supports = 2.0 in. Determine the transverse rupture strength if failure occurs at a load = 1700 lb.

3.23. A torsion test specimen has a radius = 25 mm, wall thickness = 3 mm, and gage length = 50 mm. In testing, a torque of 900 N-m results in an angular deflection = 0.3° Determine (a) the shear stress, (b)

shear strain, and (c) shear modulus, assuming the specimen had not yet yielded. (d) If failure of the specimen occurs at a torque = 1200 N-m and a corresponding angular deflection = 10°, what is the shear strength of the metal?

3.24. In a torsion test, a torque of 5000 ft-lb is applied which causes an angular deflection = 1° on a thin-walled tubular specimen whose radius = 1.5 in, wall thickness = 0.10 in, and gage length = 2.0 in. Determine (a) the shear stress, (b) shear strain, and (c) shear modulus, assuming the specimen had not yet yielded. (d) If the specimen fails at a torque = 8000 ft-lb and an angular deflection = 23°, calculate the shear strength of the metal.

Hardness

3.25. In a Brinell hardness test, a 1500-kg load is pressed into a specimen using a 10-mm-diameter hardened steel ball. The resulting indentation has a diameter = 3.2 mm. (a) Determine the Brinell hardness number for the metal. (b) If the specimen is steel, estimate the tensile strength of the steel.

3.26. One of the inspectors in the quality control department has frequently used the Brinell and Rockwell hardness tests, for which equipment is available in the company. He claims that all hardness tests are based on the same principle as the Brinell test, which is that hardness is always measured as the applied load divided by the area of the impressions made by an indentor. (a) Is he correct? (b) If not, what are some of the other principles involved in hardness testing, and what are the associated tests?

3.27. A batch of annealed steel has just been received from the vendor. It is supposed to have a tensile strength in the range 60,000 to 70,000 lb/in^2. A Brinell hardness test in the receiving department yields a value of $HB = 118$. (a) Does the steel meet the specification on tensile strength? (b) Estimate the yield strength of the material.

Viscosity of Fluids

3.28. Two flat plates, separated by a space of 4 mm, are moving relative to each other at a velocity of 5 m/sec. The space between them is occupied by a fluid of unknown viscosity. The motion of the plates is resisted by a shear stress of 10 Pa because of the viscosity of the fluid. Assuming that the velocity gradient of the fluid is constant, determine the coefficient of viscosity of the fluid.

3.29. Two parallel surfaces, separated by a space of 0.5 in that is occupied by a fluid, are moving relative to each other at a velocity of 25 in/sec. The motion is resisted by a shear stress of 0.3 lb/in^2 because of the viscosity of the fluid. If the velocity gradient in the space between the surfaces is constant, determine the viscosity of the fluid.

3.30. A 125.0-mm-diameter shaft rotates inside a stationary bushing whose inside diameter = 125.6 mm and length = 50.0 mm. In the clearance between the shaft and the bushing is a lubricating oil whose viscosity = 0.14 Pa-s. The shaft rotates at a velocity of 400 rev/min; this speed and the action of the oil are sufficient to keep the shaft centered inside the bushing. Determine the magnitude of the torque due to viscosity that acts to resist the rotation of the shaft.

4 PHYSICAL PROPERTIES OF MATERIALS

Chapter Contents

Physical properties, as the term is used here, defines the behavior of materials in response to physical forces other than mechanical. They include volumetric, thermal, electrical, and electrochemical properties. Components in a product must do more than simply withstand mechanical stresses. They must conduct electricity (or prevent its conduction), allow heat to be transferred (or allow it to escape), transmit light (or block its transmission), and satisfy myriad other functions.

Physical properties are important in manufacturing because they often influence the performance of the process. For example, thermal properties of the work material in machining determine the cutting temperature, which affects how long the tool can be used before it fails. In microelectronics, electrical properties of silicon and the way in which these properties can be altered by various chemical and physical processes comprise the basis of semiconductor manufacturing.

This chapter discusses the physical properties that are most important in manufacturing—properties that will be encountered in subsequent chapters of the book. They are divided into major categories such as volumetric, thermal, electrical, and so on. We also relate these properties to manufacturing, as we did in the previous chapter on mechanical properties.

4.1 VOLUMETRIC AND MELTING PROPERTIES

These properties are related to the volume of solids and how they are affected by temperature. The properties include density, thermal expansion, and melting point. They are explained in the following, and a listing of typical values for selected engineering materials is presented in Table 4.1.

4.1.1 DENSITY

In engineering, the density of a material is its weight per unit volume. Its symbol is ρ, and typical units are g/cm^3 (lb/in^3).

TABLE 4.1 Volumetric properties in U.S. customary units for selected engineering materials.

Material	Density, ρ		Coefficient of Thermal Expansion, α		Melting Point, T_m	
	g/cm³	lb/in³	°C⁻¹ × 10⁻⁶	°F⁻¹ × 10⁻⁶	°C	°F
Metals						
Aluminum	2.70	0.098	24	13.3	660	1220
Copper	8.97	0.324	17	9.4	1083	1981
Iron	7.87	0.284	12.1	6.7	1539	2802
Lead	11.35	0.410	29	16.1	327	621
Magnesium	1.74	0.063	26	14.4	650	1202
Nickel	8.92	0.322	13.3	7.4	1455	2651
Steel	7.87	0.284	12	6.7	a	a
Tin	7.31	0.264	23	12.7	232	449
Tungsten	19.30	0.697	4.0	2.2	3410	6170
Zinc	7.15	0.258	40	22.2	420	787
Ceramics						
Glass	2.5	0.090	1.8–9.0	1.0–5.0	b	b
Alumina	3.8	0.137	9.0	5.0	NA	NA
Silica	2.66	0.096	NA	NA	b	b
Polymers						
Phenol resins	1.3	0.047	60	33	c	c
Nylon	1.16	0.042	100	55	b	b
Teflon	2.2	0.079	100	55	b	b
Natural rubber	1.2	0.043	80	45	b	b
Polyethylene (low density)	0.92	0.033	180	100	b	b
Polystyrene	1.05	0.038	60	33	b	b

Compiled from, [2], [3], [4], and other sources.
[a]Melting characteristics of steel depend on composition.
[b]Softens at elevated temperatures and does not have a well-defined melting point.
[c]Chemically degrades at high temperatures. NA = not available; value of property for this material could not be obtained.

The density of an element is determined by its atomic number and other factors, such as atomic radius and atomic packing. The term *specific gravity* expresses the density of a material relative to the density of water and is therefore a ratio with no units.

Density is an important consideration in the selection of a material for a given application, but it is generally not the only property of interest. Strength is also important, and the two properties are often related in a *strength-to-weight ratio,* which is the tensile strength of the material divided by its density. The ratio is useful in comparing materials for structural applications in aircraft, automobiles, and other products in which weight and energy are of concern.

4.1.2 THERMAL EXPANSION

The density of a material is a function of temperature. The general relationship is that density decreases with increasing temperature. Put another way, the volume per unit weight increases with temperature. Thermal expansion is the name given to this effect that temperature has on density. It is usually expressed as the *coefficient of thermal expansion,* which measures the change in length per degree of temperature, as mm/mm/°C (in/in/°F). It is a length ratio rather than a volume ratio because this is easier to measure and apply. It is consistent with the usual design situation in which dimensional changes are of greater interest than volumetric changes. The change in length corresponding to a given temperature change is given by

$$L_2 - L_1 = \alpha L_1 (T_2 - T_1) \tag{4.1}$$

where α = coefficient of thermal expansion, $°C^{-1}(°F^{-1})$; and L_1 and L_2 are lengths, mm (in), corresponding, respectively, to temperatures T_1 and T_2, $°C$ ($°F$).

Values of coefficient of thermal expansion given in Table 4.1 suggest that it has a linear relationship with temperature. This is only an approximation. Not only is length affected by temperature, but the thermal expansion coefficient itself is also affected. For some materials it increases with temperature; for other materials it decreases. These changes are usually not significant enough to be of much concern, and values like those in the table are quite useful in design calculations for the range of temperatures contemplated in service. Changes in the coefficient are more substantial when the metal undergoes a phase transformation, such as from solid to liquid, or from one crystal structure to another.

In manufacturing operations, thermal expansion is put to good use in shrink fit and expansion fit assemblies (Section 32.3) in which a part is heated to increase its size or cooled to decrease its size to permit insertion into some other part. When the part returns to ambient temperature, a tightly fitted assembly is obtained. Thermal expansion can be a problem in heat treatment (Chapter 27) and welding (Section 30.6) because of thermal stresses that develop in the material during these processes.

4.1.3 MELTING CHARACTERISTICS

For a pure element, the **melting point** T_m is the temperature at which the material transforms from solid to liquid state. The reverse transformation, from liquid to solid, occurs at the same temperature and is called the **freezing point**. For crystalline elements, such as metals, the melting and freezing temperatures are the same. A certain amount of heat energy, called the **heat of fusion,** is required at this temperature to accomplish the transformation from solid to liquid.

Melting of a metal element at a specific temperature, as it has been described, assumes equilibrium conditions. Exceptions occur in nature; for example, when a molten metal is cooled, it may remain in the liquid state below its freezing point if nucleation of crystals does not initiate immediately. When this happens, the liquid is said to be **supercooled**.

There are other variations in the melting process—differences in the way melting occurs in different materials. For example, unlike pure metals, most metal alloys do not have a single melting point. Instead, melting begins at a certain temperature, called the **solidus,** and continues as the temperature increases until finally converting completely to the liquid state at a temperature called the **liquidus**. Between the two temperatures, the alloy is a mixture of solid and molten metals, the amounts of each being inversely proportional to their relative distances from the liquidus and solidus. Although most alloys behave in this way, exceptions are eutectic alloys that melt (and freeze) at a single temperature. These issues are examined in the discussion of phase diagrams in Chapter 6.

Another difference in melting occurs with noncrystalline materials (glasses). In these materials, there is a gradual transition from solid to liquid states. The solid material gradually softens as temperature increases, finally becoming liquid at the melting point. During softening, the material has a consistency of increasing plasticity (increasingly like a fluid) as it gets closer to the melting point.

These differences in melting characteristics among pure metals, alloys, and glass are portrayed in Figure 4.1. The plots show changes in density as a function of temperature for three hypothetical materials: a pure metal, an alloy, and glass. Plotted in the figure is the volumetric change, which is the reciprocal of density.

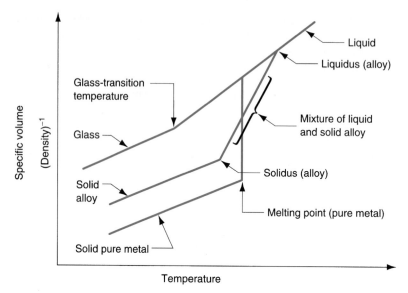

FIGURE 4.1 Changes in volume per unit weight (1/density) as a function of temperature for a hypothetical pure metal, alloy, and glass; all exhibiting similar thermal expansion and melting characteristics.

The importance of melting in manufacturing is obvious. In metal casting (Chapters 10 and 11), the metal is melted and then poured into a mold cavity. Metals with lower melting points are generally easier to cast, but if the melting temperature is too low, the metal loses its applicability as an engineering material. Melting characteristics of polymers are important in plastic molding and other polymer shaping processes (Chapter 13). Sintering of powdered metals and ceramics requires knowledge of melting points. Sintering does not melt the materials, but the temperatures used in the process must approach the melting point to achieve the required bonding of the powders.

4.2 THERMAL PROPERTIES

Much of the previous section is concerned with the effects of temperature on volumetric properties of materials. Certainly, thermal expansion, melting, and heat of fusion are thermal properties because temperature determines the thermal energy level of the atoms, leading to the changes in the materials. The current section examines several additional thermal properties—ones that relate to the storage and flow of heat within a substance. The usual properties of interest are specific heat and thermal conductivity, values of which are compiled for selected materials in Table 4.2.

4.2.1 SPECIFIC HEAT AND THERMAL CONDUCTIVITY

The specific heat C of a material is defined as the quantity of heat energy required to increase the temperature of a unit mass of the material by one degree. Some typical values are listed in Table 4.2. To determine the amount of energy needed to heat a certain weight of a metal in a furnace to a given elevated temperature, the following equation can be used

$$H = CW(T_2 - T_1) \tag{4.2}$$

where H = amount of heat energy, J (Btu); C = specific heat of the material, J/kg °C (Btu/lb °F); W = its weight, kg (lb); and $(T_2 - T_1)$ = change in temperature, °C (°F).

TABLE 4.2   Values of common thermal properties for selected materials. Values are at room temperature, and these values change for different temperatures.

Material	Specific Heat Cal/g °C[a] or Btu/lbm °F	Thermal Conductivity J/s mm °C	Thermal Conductivity Btu/hr in °F	Material	Specific Heat Cal/g °C[a] or Btu/lbm °F	Thermal Conductivity J/s mm °C	Thermal Conductivity Btu/hr in °F
Metals				*Ceramics*			
Aluminum	0.21	0.22	9.75	Alumina	0.18	0.029	1.4
Cast iron	0.11	0.06	2.7	Concrete	0.2	0.012	0.6
Copper	0.092	0.40	18.7	*Polymers*			
Iron	0.11	0.072	2.98	Phenolics	0.4	0.00016	0.0077
Lead	0.031	0.033	1.68	Polyethylene	0.5	0.00034	0.016
Magnesium	0.25	0.16	7.58	Teflon	0.25	0.00020	0.0096
Nickel	0.105	0.070	2.88	Natural rubber	0.48	0.00012	0.006
Steel	0.11	0.046	2.20	*Other*			
Stainless steel[b]	0.11	0.014	0.67	Water (liquid)	1.00	0.0006	0.029
Tin	0.054	0.062	3.0	Ice	0.46	0.0023	0.11
Zinc	0.091	0.112	5.41				

Compiled from [2], [3], [6], and other sources.
[a]Specific heat has the same numerical value in Btu/lbm-F or Cal/g-C. 1.0 Calory = 4.186 Joule.
[b]Austenitic (18-8) stainless steel.

The volumetric heat storage capacity of a material is often of interest. This is simply density multiplied by specific heat ρC. Thus, *volumetric specific heat* is the heat energy required to raise the temperature of a unit volume of material by one degree, J/mm^3 °C (Btu/in^3 °F).

Conduction is a fundamental heat-transfer process. It involves transfer of thermal energy within a material from molecule to molecule by purely thermal motions; no transfer of mass occurs. The thermal conductivity of a substance is therefore its capability to transfer heat through itself by this physical mechanism. It is measured by the *coefficient of thermal conductivity* k, which has typical units of J/s mm °C (Btu/in hr °F). The coefficient of thermal conductivity is generally high in metals, low in ceramics and plastics.

The ratio of thermal conductivity to volumetric specific heat is frequently encountered in heat transfer analysis. It is called the *thermal diffusivity* K and is determined as

$$K = \frac{k}{\rho C} \tag{4.3}$$

It can be used to calculate cutting temperatures in machining (Section 21.5.1).

4.2.2   THERMAL PROPERTIES IN MANUFACTURING

Thermal properties play an important role in manufacturing because heat generation is common in so many processes. In some operations heat is the energy that accomplishes the process; in others heat is generated as a consequence of the process.

Specific heat is of interest for several reasons. In processes that require heating of the material (e.g., casting, heat treating, and hot metal forming), specific heat determines the amount of heat energy needed to raise the temperature to a desired level, according to Eq. (4.2).

In many processes carried out at ambient temperature, the mechanical energy to perform the operation is converted to heat, which raises the temperature of the workpart. This is common in machining and cold forming of metals. The temperature rise is a function of the metal's specific heat. Coolants are often used in machining to reduce these temperatures, and here the fluid's heat capacity is critical. Water is almost always employed as the base for these fluids because of its high heat-carrying capacity.

Thermal conductivity functions to dissipate heat in manufacturing processes, sometimes beneficially, sometimes not. In mechanical processes such as metal forming and machining, much of the power required to operate the process is converted to heat. The ability of the work material and tooling to conduct heat away from its source is highly desirable in these processes.

On the other hand, high thermal conductivity of the work metal is undesirable in fusion welding processes such as arc welding. In these operations, the heat input must be concentrated at the joint location so that the metal can be melted. For example, copper is generally difficult to weld because its high thermal conductivity allows heat to be conducted from the energy source into the work too rapidly, inhibiting heat buildup for melting at the joint.

4.3 MASS DIFFUSION

In addition to heat transfer in a material, there is also mass transfer. *Mass diffusion* involves movement of atoms or molecules within a material or across a boundary between two materials in contact. It is perhaps more appealing to one's intuition that such a phenomenon occurs in liquids and gases, but it also occurs in solids. It occurs in pure metals, in alloys, and between materials that share a common interface. Because of thermal agitation of the atoms in a material (solid, liquid, or gas), atoms are continuously moving about. In liquids and gases, where the level of thermal agitation is high, it is a free-roaming movement. In solids (metals in particular), the atomic motion is facilitated by vacancies and other imperfections in the crystal structure.

Diffusion can be illustrated by the series of sketches in Figure 4.2 for the case of two metals suddenly brought into intimate contact with each other. At the start, both metals have their own atomic structure; but with time there is an exchange of atoms, not only across the boundary, but within the separate pieces. Given enough time, the assembly of two pieces will finally reach a uniform composition throughout.

Temperature is an important factor in diffusion. At higher temperatures, thermal agitation is greater and the atoms can move about more freely. Another factor is the concentration gradient dc/dx, which indicates the concentration of the two types of atoms in a direction of interest defined by x. The concentration gradient is plotted in Figure 4.2(b) to correspond to the instantaneous distribution of atoms in the assembly. The relationship often used to describe mass diffusion is *Fick's first law:*

$$dm = -D\left(\frac{dc}{dt}\right)A\,dt \tag{4.4}$$

where dm = small amount of material transferred, D = diffusion coefficient of the metal, which increases rapidly with temperature, dc/dx = concentration gradient, A = area of the boundary, and dt represents a small time increment. An alternative expression of Eq. (4.4) gives the mass diffusion rate:

$$\frac{dm}{dt} = -D\left(\frac{dc}{dt}\right)A \tag{4.5}$$

Although these equations are difficult to use in calculations because of the problem of assessing D, they are helpful in understanding diffusion and the variables on which D depends.

Mass diffusion is used in several processes. A number of surface-hardening treatments are based on diffusion (Section 27.4), including carburizing and nitriding. Among the welding processes, diffusion welding (Section 30.5.2) is used to join two components by pressing them together and allowing diffusion to occur across the boundary to create a permanent bond. Diffusion is also used in electronics manufacturing to alter the surface chemistry of a semiconductor chip in very localized regions to create circuit details (Section 34.4.3).

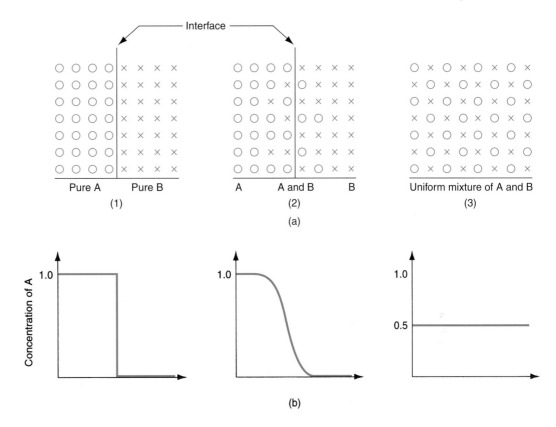

FIGURE 4.2   Mass diffusion: (a) model of atoms in two solid blocks in contact: (1) at the start when two pieces are brought together, they each have their individual compositions; (2) after some time, an exchange of atoms has occurred; and (3) eventually, a condition of uniform concentration occurs. The concentration gradient dc/dx for metal A is plotted in (b) of the figure.

4.4 ELECTRICAL PROPERTIES

Engineering materials exhibit a great variation in their capacity to conduct electricity. This section defines the physical properties by which this capacity is measured.

4.4.1 RESISTIVITY AND CONDUCTIVITY

The flow of electrical current involves movement of **charge carriers**—infinitesimally small particles possessing an electrical charge. In solids, these charge carriers are electrons. In a liquid solution, charge carriers are positive and negative ions. The movement of charge carriers is driven by the presence of an electric voltage and resisted by the inherent characteristics of the material, such as atomic structure and bonding between atoms and molecules. This is the familiar relationship defined by Ohm's law

$$I = \frac{E}{R} \tag{4.6}$$

where I = current, A; E = voltage, V; and R = electrical resistance, Ω.

The resistance in a uniform section of material (e.g., a wire) depends on its length L, cross-sectional area A, and the resistivity of the material r; thus,

$$R = r\frac{L}{A} \quad \text{or} \quad r = R\frac{A}{L} \qquad (4.7)$$

where resistivity has units of $\Omega\text{-m}^2/\text{m}$ or $\Omega\text{-m}$ (Ω-in).

Resistivity is the basic property that defines a material's capability to resist current flow. Table 4.3 lists values of resistivity for selected materials. Resistivity is not a constant; instead it varies, as do so many other properties, with temperature. For metals, it increases with temperature.

It is often more convenient to consider a material as conducting electrical current rather than resisting its flow. The *conductivity* of a material is simply the reciprocal of resistivity:

$$\text{Electrical conductivity} = \frac{1}{r} \qquad (4.8)$$

where conductivity has units of $(\Omega\text{-m})^{-1}$ $((\Omega\text{-in})^{-1})$.

4.4.2 CLASSES OF MATERIALS BY ELECTRICAL PROPERTIES

Metals are the best *conductors* of electricity, because of their metallic bonding. They have the lowest resistivity (Table 4.3). Most ceramics and polymers, whose electrons are tightly bound by covalent and/or ionic bonding, are poor conductors. Many of these materials are used as *insulators* because they possess high resistivities.

An insulator is sometimes referred to as a dielectric, because the term *dielectric* means nonconductor of direct current. It is a material that can be placed between two electrodes without conducting current between them. However, if the voltage is high enough, the current will suddenly pass through the material; for example, in the form of an arc. The *dielectric strength* of an insulating material, then, is the electrical potential required to break down the insulator per unit thickness. Appropriate units are volts/m (volts/in).

In addition to conductors and insulators (or dielectrics), there are also superconductors and semiconductors. A *superconductor* is a material that exhibits zero resistivity. It is a phenomenon that has been observed in certain materials at low temperatures

TABLE 4.3 Resistivity of selected materials.

Material	Resistivity Ω-m	Ω-in	Material	Resistivity Ω-m	Ω-in
Conductors	$10^{-6}-10^{-8}$	$10^{-4}-10^{-7}$	*Conductors, continued*		
Aluminum	2.8×10^{-8}	1.1×10^{-6}	Steel, low C	17.0×10^{-8}	6.7×10^{-6}
Aluminum alloys	4.0×10^{-8a}	1.6×10^{-6a}	Steel, stainless	70.0×10^{-8a}	27.6×10^{-6}
Cast iron	65.0×10^{-8a}	25.6×10^{-6a}	Tin	11.5×10^{-8}	4.5×10^{-6}
Copper	1.7×10^{-8}	0.67×10^{-6}	Zinc	6.0×10^{-8}	2.4×10^{-6}
Gold	2.4×10^{-8}	0.95×10^{-6}	Carbon	5000×10^{-8b}	2000×10^{-6b}
Iron	9.5×10^{-8}	3.7×10^{-6}	*Semiconductors*	10^1-10^5	10^2-10^7
Lead	20.6×10^{-8}	8.1×10^{-6}	Silicon	1.0×10^3	
Magnesium	4.5×10^{-8}	1.8×10^{-6}	*Insulators*	$10^{12}-10^{15}$	$10^{13}-10^{17}$
Nickel	6.8×10^{-8}	2.7×10^{-6}	Natural rubber	1.0×10^{12b}	0.4×10^{14b}
Silver	1.6×10^{-8}	0.63×10^{-6}	Polyethylene	100×10^{12b}	40×10^{14b}

Compiled from various standard sources.
[a]Value varies with alloy composition.
[b]Value is approximate.

approaching absolute zero. One might expect the existence of this phenomenon, because of the significant effect that temperature has on resistivity. That these superconducting materials exist is of great scientific interest. If materials could be developed that exhibit this property at more normal temperatures, there would be significant practical implications in power transmission, electronic switching speeds, and magnetic field applications.

Semiconductors have already proved their practical worth: Their applications range from mainframe computers to household appliances and automotive engine controllers. As one would guess, a **semiconductor** is a material whose resistivity lies between insulators and conductors. The typical range is shown in Table 4.3. The most commonly used semiconductor material today is silicon (Section 7.5.2), largely because of its abundance in nature, relative low cost, and ease of processing. What makes semiconductors unique is the capacity to significantly alter conductivities in their surface chemistries in very localized areas to fabricate integrated circuits (Chapter 34).

Electrical properties play an important role in various manufacturing processes. Some of the nontraditional processes use electrical energy to remove material. Electric discharge machining (Section 26.3.1) uses the heat generated by electrical energy in the form of sparks to remove material from metals. Most of the important welding processes use electrical energy to melt the joint metal. Finally, the capacity to alter the electrical properties of semiconductor materials is the basis for microelectronics manufacturing.

4.5   ELECTROCHEMICAL PROCESSES

Electrochemistry is a field of science concerned with the relationship between electricity and chemical changes, and with the conversion of electrical and chemical energy.

In a water solution, the molecules of an acid, base, or salt are dissociated into positively and negatively charged ions. These ions are the charge carriers in the solution—they allow electric current to be conducted, playing the same role that electrons play in metallic conduction. The ionized solution is called an **electrolyte;** and electrolytic conduction requires that current enter and leave the solution at **electrodes**. The positive electrode is called the **anode,** and the negative electrode is the **cathode**. The whole arrangement is called an **electrolytic cell**. At each electrode, some chemical reaction occurs, such as the deposition or dissolution of material, or the decomposition of gas from the solution. **Electrolysis** is the name given to these chemical changes occurring in the solution.

Consider a specific case of electrolysis: decomposition of water, illustrated in Figure 4.3. To accelerate the process, dilute sulfuric acid (H_2SO_4) is used as the electrolyte, and platinum and carbon (both chemically inert) are used as electrodes. The electrolyte dissociates in the ions H^+ and $SO_4^=$. The H^+ ions are attracted to the negatively charged cathode; upon

FIGURE 4.3   Example of electrolysis: decomposition of water.

reaching it they acquire an electron and combine into molecules of hydrogen gas:

$$2H^+ + 2e \rightarrow H_2 \text{ (gas)} \tag{4.9a}$$

The $SO_4^=$ ions are attracted to the anode, transferring electrons to it to form additional sulfuric acid and liberate oxygen:

$$2SO_4^= - 4e + 2H_2O \rightarrow 2H_2SO_4 + O_2 \text{ (gas)} \tag{4.9b}$$

The product H_2SO_4 is dissociated into ions of H^+ and $SO_4^=$ again and so the process continues.

In addition to the production of hydrogen and oxygen gases, as illustrated by the example, electrolysis is also used in several other industrial processes. Two examples are (1) *electroplating* (Section 28.3.1), an operation that adds a thin coating of one metal (e.g., chromium) to the surface of a second metal (e.g., steel) for decorative or other purposes; and (2) *electrochemical machining* (Section 26.2), a process in which material is removed from the surface of a metal part. Both these operations rely on electrolysis to either add or remove material from the surface of a metal part. In electroplating, the workpart is set up in the electrolytic circuit as the cathode, so that the positive ions of the coating metal are attracted to the negatively charged part. In electrochemical machining, the workpart is the anode, and a tool with the desired shape is the cathode. The action of electrolysis in this setup is to remove metal from the part surface in regions determined by the shape of the tool as it slowly feeds into the work.

The two physical laws that determine the amount of material deposited or removed from a metallic surface were first stated by the British scientist Michael Faraday:

1. The mass of a substance liberated in an electrolytic cell is proportional to the quantity of electricity passing through the cell.
2. When the same quantity of electricity is passed through different electrolytic cells, the masses of the substances liberated are proportional to their chemical equivalents.

Faraday's laws are used in the subsequent coverage of electroplating and electrochemical machining.

REFERENCES

[1] Guy, A. G., and Hren, J. J. *Elements of Physical Metallurgy,* 3rd ed. Addison-Wesley Publishing Company, Reading, Massachusetts, 1974.

[2] Flinn, R. A., and Trojan, P. K. *Engineering Materials and Their Applications,* 5th ed. John Wiley & Sons, New York, 1995.

[3] Kreith, F., and Bohn, M. S., *Principles of Heat Transfer,* 6th ed. CL-Engineering, New York, 2000.

[4] *Metals Handbook,* 10th ed., Vol. 1, Properties and Selection: Iron, Steel, and High Performance Alloys. ASM International, Metals Park, Ohio, 1990.

[5] *Metals Handbook,* 10th ed., Vol. 2, Properties and Selection: Nonferrous Alloys and Special Purpose Materials. ASM International, Metals Park, Ohio, 1990.

[6] Van Vlack, L. H. *Elements of Materials Science and Engineering,* 6th ed. Addison-Wesley, Reading, Massachusetts, 1989.

REVIEW QUESTIONS

4.1. Define density as a material property.
4.2. What is the difference in melting characteristics between a pure metal element and an alloy metal?
4.3. Describe the melting characteristics of a non-crystalline material such as glass.
4.4. Define specific heat as a material property.
4.5. What is thermal conductivity as a material property?
4.6. Define thermal diffusivity.
4.7. What are the important variables that affect mass diffusion?
4.8. Define resistivity as a material property.
4.9. Why are metals better conductors of electricity than ceramics and polymers?
4.10. What is dielectric strength as a material property?
4.11. What is an electrolyte?

MULTIPLE CHOICE QUIZ

There are 12 correct answers in the following multiple choice questions (some questions have multiple answers that are correct). To attain a perfect score on the quiz, all correct answers must be given. Each correct answer is worth 1 point. Each omitted answer or wrong answer reduces the score by 1 point, and each additional answer beyond the correct number of answers reduces the score by 1 point. Percentage score on the quiz is based on the total number of correct answers.

4.1. Which one of the following metals has the lowest density: (a) aluminum, (b) copper, (c) magnesium, or (d) tin?

4.2. The thermal expansion properties of polymers are generally (a) greater than, (b) less than, or (c) the same as those of metals?

4.3. In the heating of most metal alloys, melting begins at a certain temperature and concludes at a higher temperature. In these cases, which of the following temperatures marks the beginning of melting: (a) liquidus or (b) solidus?

4.4. Which one of the following materials has the highest specific heat: (a) aluminum, (b) concrete, (c) polyethylene, or (d) water?

4.5. Copper is generally considered easy to weld because of its high thermal conductivity: (a) true or (b) false?

4.6. The mass diffusion rate dm/dt across a boundary between two different metals is a function of which of the following variables (four best answers): (a) concentration gradient dc/dx, (b) contact area, (c) density, (d) melting point, (e) thermal expansion, (f) temperature, and (g) time?

4.7. Which of the following pure metals is the best conductor of electricity: (a) aluminum, (b) copper, (c) gold, or (d) silver?

4.8. A superconductor is characterized by which of the following (one best answer): (a) high conductivity, (b) resistivity properties between those of conductors and semiconductors, (c) very low resistivity, or (d) zero resistivity?

4.9. In an electrolytic cell, the anode is the electrode that is (a) positive or (b) negative.

PROBLEMS

4.1. The starting diameter of a shaft is 25.00 mm. This shaft is to be inserted into a hole in an expansion fit assembly operation. To be readily inserted, the shaft must be reduced in diameter by cooling. Determine the temperature to which the shaft must be reduced from room temperature (20°C) in order to reduce its diameter to 24.98 mm. Refer to Table 4.1.

4.2. A bridge built with steel girders is 500 m in length and 12 m in width. Expansion joints are provided to compensate for the change in length in the support girders as the temperature fluctuates. Each expansion joint can compensate for a maximum of 40 mm of change in length. From historical records it is estimated that the minimum and maximum temperatures in the region will be −35°C and 38°C, respectively. What is the minimum number of expansion joints required?

4.3. Aluminum has a density of 2.70 g/cm³ at room temperature (20°C). Determine its density at 650°C, using data in Table 4.1 as a reference.

4.4. With reference to Table 4.1, determine the increase in length of a steel bar whose length = 10.0 in, if the bar is heated from room temperature of 70°F to 500°F.

4.5. With reference to Table 4.2, determine the quantity of heat required to increase the temperature of an aluminum block that is 10 cm × 10 cm × 10 cm from room temperature (21°C) to 300°C.

4.6. What is the resistance R of a length of copper wire whose length = 10 m and whose diameter = 0.10 mm? Use Table 4.3 as a reference.

4.7. A 16-gage nickel wire (0.0508-in diameter) connects a solenoid to a control circuit that is 32.8 ft away. (a) What is the resistance of the wire? Use Table 4.3 as a reference. (b) If a current was passed through the wire, it would heat up. How does this affect the resistance?

4.8. Aluminum wiring was used in many homes in the 1960s because of the high cost of copper at the time. Aluminum wire that was 12 gauge (a measure of cross-sectional area) was rated at 15 A of current. If copper wire of the same gauge were used to replace the aluminum wire, what current should the wire be capable of carrying if all factors except resistivity are considered equal? Assume that the resistance of the wire is the primary factor that determines the current it can carry and the cross-sectional area and length are the same for the aluminum and copper wires.

5 DIMENSIONS, SURFACES, AND THEIR MEASUREMENT

Chapter Contents

In addition to mechanical and physical properties of materials, other factors that determine the performance of a manufactured product include the dimensions and surfaces of its components. *Dimensions* are the linear or angular sizes of a component specified on the part drawing. Dimensions are important because they determine how well the components of a product fit together during assembly. When fabricating a given component, it is nearly impossible and very costly to make the part to the exact dimension given on the drawing. Instead a limited variation is allowed from the dimension, and that allowable variation is called a *tolerance*.

The surfaces of a component are also important. They affect product performance, assembly fit, and aesthetic appeal that a potential customer might have for the product. A *surface* is the exterior boundary of an object with its surroundings, which may be another object, a fluid, or space, or combinations of these. The surface encloses the object's bulk mechanical and physical properties.

This chapter discusses dimensions, tolerances, and surfaces—three attributes specified by the product designer and determined by the manufacturing processes used to make the parts and products. It also considers how these attributes are assessed using measuring and gaging devices. A closely related topic is inspection, covered in Chapter 42.

5.1 DIMENSIONS, TOLERANCES AND RELATED ATTRIBUTES

The basic parameters used by design engineers to specify sizes of geometric features on a part drawing are defined in this section. The parameters include dimensions and tolerances, flatness, roundness, and angularity.

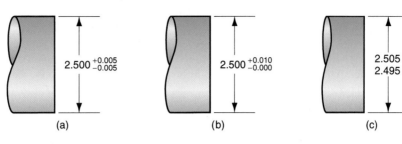

FIGURE 5.1   Three ways to specify tolerance limits for a nominal dimension of 2.500: (a) bilateral, (b) unilateral, and (c) limit dimensions.

5.1.1 DIMENSIONS AND TOLERANCES

ANSI [3] defines a ***dimension*** as "a numerical value expressed in appropriate units of measure and indicated on a drawing and in other documents along with lines, symbols, and notes to define the size or geometric characteristic, or both, of a part or part feature." Dimensions on part drawings represent nominal or basic sizes of the part and its features. These are the values that the designer would like the part size to be, if the part could be made to an exact size with no errors or variations in the fabrication process. However, there are variations in the manufacturing process, which are manifested as variations in the part size. Tolerances are used to define the limits of the allowed variation. Quoting again from the ANSI standard [3], a ***tolerance*** is "the total amount by which a specific dimension is permitted to vary. The tolerance is the difference between the maximum and minimum limits."

Tolerances can be specified in several ways, illustrated in Figure 5.1. Probably most common is the ***bilateral tolerance***, in which the variation is permitted in both positive and negative directions from the nominal dimension. For example, in Figure 5.1(a), the nominal dimension = 2.500 linear units (e.g., mm, in), with an allowable variation of 0.005 units in either direction. Parts outside these limits are unacceptable. It is possible for a bilateral tolerance to be unbalanced; for example, 2.500 +0.010, –0.005 dimensional units. A ***unilateral tolerance*** is one in which the variation from the specified dimension is permitted in only one direction, either positive or negative, as in Figure 5.1(b). ***Limit dimensions*** are an alternative method to specify the permissible variation in a part feature size; they consist of the maximum and minimum dimensions allowed, as in Figure 5.1(c).

5.1.2 OTHER GEOMETRIC ATTRIBUTES

Dimensions and tolerances are normally expressed as linear (length) values. There are other geometric attributes of parts that are also important, such as flatness of a surface, roundness of a shaft or hole, parallelism between two surfaces, and so on. Definitions of these terms are listed in Table 5.1.

5.2 CONVENTIONAL MEASURING INSTRUMENTS AND GAGES

Measurement is a procedure in which an unknown quantity is compared with a known standard, using an accepted and consistent system of units. Two systems of units have evolved in the world: (1) the U.S. customary system (U.S.C.S.), and (2) the International System of Units (or SI, for Systeme Internationale d'Unites), more popularly known as the metric system. Both systems are used in parallel throughout this book. The metric system is widely accepted in nearly every part of the industrialized world except the United States, which has stubbornly clung to its U.S.C.S. Gradually, the United States is adopting SI.

Measurement provides a numerical value of the quantity of interest, within certain limits of accuracy and precision. ***Accuracy*** is the degree to which the measured value agrees with the true value of the quantity of interest. A measurement procedure is accurate when it is

TABLE 5.1	Definitions of geometric attributes of parts.

Angularity—The extent to which a part feature such as a surface or axis is at a specified angle relative to a reference surface. If the angle = 90°, then the attribute is called perpendicularity or squareness.

Circularity—For a surface of revolution such as a cylinder, circular hole, or cone, circularity is the degree to which all points on the intersection of the surface and a plane perpendicular to the axis of revolution are equidistant from the axis. For a sphere, circularity is the degree to which all points on the intersection of the surface and a plane passing through the center are equidistant from the center.

Concentricity—The degree to which any two (or more) part features such as a cylindrical surface and a circular hole have a common axis.

Cylindricity—The degree to which all points on a surface of revolution such as a cylinder are equidistant from the axis of revolution.

Flatness—The extent to which all points on a surface lie in a single plane.

Parallelism—The degree to which all points on a part feature such as a surface, line, or axis are equidistant from a reference plane or line or axis.

Perpendicularity—The degree to which all points on a part feature such as a surface, line, or axis are 90 from a reference plane or line or axis.

Roundness—Same as circularity.

Squareness—Same as perpendicularity.

Straightness—The degree to which a part feature such as a line or axis is a straight line.

absent of systematic errors, which are positive or negative deviations from the true value that are consistent from one measurement to the next. **Precision** is the degree of repeatability in the measurement process. Good precision means that random errors in the measurement procedure are minimized. Random errors are usually associated with human participation in the measurement process. Examples include variations in the setup, imprecise reading of the scale, round-off approximations, and so on. Nonhuman contributors to random error include temperature changes, gradual wear and/or misalignment in the working elements of the device, and other variations.

Closely related to measurement is gaging. **Gaging** (also spelled **gauging**) determines simply whether the part characteristic meets or does not meet the design specification. It is usually faster than measuring, but scant information is provided about the actual value of the characteristic of interest. The video clip on measurement and gaging illustrates some of the topics discussed in this chapter.

VIDEO CLIP _____

Measurement and Gaging. This clip contains three segments: (1) precision, resolution, and accuracy, (2) how to read a vernier caliper, and (3) how to read a micrometer.

_____

This section considers the variety of manually operated measuring instruments and gages used to evaluate dimensions such as length and diameter, as well as features such as angles, straightness, and roundness. This type of equipment is found in metrology labs, inspection departments, and tool rooms. The logical starting topic is precision gage blocks.

5.2.1 PRECISION GAGE BLOCKS

Precision gage blocks are the standards against which other dimensional measuring instruments and gages are compared. Gage blocks are usually square or rectangular. The measuring surfaces are finished to be dimensionally accurate and parallel to within several millionths of an inch and are polished to a mirror finish. Several grades of precision gage blocks are available, with closer tolerances for higher precision grades. The highest grade—the **master laboratory standard**—is made to a tolerance of ±0.000,03 mm (±0.000,001 in). Depending on degree of hardness desired and price the user is willing to pay, gage blocks can be made out of any of several hard materials, including tool steel, chrome-plated steel, chromium carbide, or tungsten carbide.

Precision gage blocks are available in certain standard sizes or in sets, the latter containing a variety of different-sized blocks. The sizes in a set are systematically determined so they can be stacked to achieve virtually any dimension desired to within 0.0025 mm (0.0001 in).

For best results, gage blocks must be used on a flat reference surface, such as a surface plate. A *surface plate* is a large solid block whose top surface is finished to a flat plane. Most surface plates today are made of granite. Granite has the advantage of being hard, non-rusting, nonmagnetic, long wearing, thermally stable, and easy to maintain.

Gage blocks and other high-precision measuring instruments must be used under standard conditions of temperature and other factors that might adversely affect the measurement. By international agreement, 20°C (68°F) has been established as the standard temperature. Metrology labs operate at this standard. If gage blocks or other measuring instruments are used in a factory environment in which the temperature differs from this standard, corrections for thermal expansion or contraction may be required. Also, working gage blocks used for inspection in the shop are subject to wear and must be calibrated periodically against more precise laboratory gage blocks.

5.2.2 MEASURING INSTRUMENTS FOR LINEAR DIMENSIONS

Measuring instruments can be divided into two types: graduated and nongraduated. *Graduated measuring devices* include a set of markings (called *graduations*) on a linear or angular scale to which the object's feature of interest can be compared for measurement. *Nongraduated measuring devices* possess no such scale and are used to make comparisons between dimensions or to transfer a dimension for measurement by a graduated device.

The most basic of the graduated measuring devices is the *rule* (made of steel, and often called a *steel rule*), used to measure linear dimensions. Rules are available in various lengths. Metric rule lengths include 150, 300, 600, and 1000 mm, with graduations of 1 or 0.5 mm. Common U.S. sizes are 6, 12, and 24 in, with graduations of 1/32, 1/64, or 1/100 in.

Calipers are available in either nongraduated or graduated styles. A nongraduated caliper (referred to simply as a *caliper*) consists of two legs joined by a hinge mechanism, as in Figure 5.2. The ends of the legs are made to contact the surfaces of the object being measured,

FIGURE 5.2   Two sizes of outside calipers. (Courtesy of L.S. Starrett Co.)

FIGURE 5.3   Slide caliper, opposite sides of instrument shown. (Courtesy of L.S. Starrett Co.)

and the hinge is designed to hold the legs in position during use. The contacts point either inward or outward. When they point inward, as in Figure 5.2, the instrument is an *outside caliper* and is used for measuring outside dimensions such as a diameter. When the contacts point outward, it is an *inside caliper*, which is used to measure the distance between two internal surfaces. An instrument similar in configuration to the caliper is a *divider*, except that both legs are straight and terminate in hard, sharply pointed contacts. Dividers are used for scaling distances between two points or lines on a surface, and for scribing circles or arcs onto a surface.

A variety of graduated calipers are available for various measurement purposes. The simplest is the *slide caliper*, which consists of a steel rule to which two jaws are added, one fixed at the end of the rule and the other movable, shown in Figure 5.3. Slide calipers can be used for inside or outside measurements, depending on whether the inside or outside jaw faces are used. In use, the jaws are forced into contact with the part surfaces to be measured, and the location of the movable jaw indicates the dimension of interest. Slide calipers permit more accurate and precise measurements than simple rules. A refinement of the slide caliper is the *vernier caliper*, shown in Figure 5.4. In this device, the movable jaw includes a vernier scale, named after P. Vernier (1580–1637), a French mathematician who invented it. The vernier provides graduations of 0.01 mm in the SI (and 0.001 inch in the U.S. customary scale), much more precise than the slide caliper.

The *micrometer* is a widely used and very accurate measuring device, the most common form of which consists of a spindle and a C-shaped anvil, as in Figure 5.5. The spindle is moved relative to the fixed anvil by means of an accurate screw thread. On a typical U.S. micrometer, each rotation of the spindle provides 0.025 in of linear travel. Attached to the spindle is a thimble graduated with 25 marks around its circumference, each mark corresponding to 0.001 in. The micrometer sleeve is usually equipped with a vernier,

FIGURE 5.4   Vernier caliper. (Courtesy of L.S. Starrett Co.)

allowing resolutions as close as 0.0001 in. On a micrometer with metric scale, graduations are 0.01 mm. Modern micrometers (and graduated calipers) are available with electronic devices that display a digital readout of the measurement (as in the figure). These instruments are easier to read and eliminate much of the human error associated with reading conventional graduated devices.

The most common micrometer types are (1) *external micrometer,* Figure 5.5, also called an *outside micrometer,* which comes in a variety of standard anvil sizes; (2) *internal micrometer,* or *inside micrometer,* which consists of a head assembly and a set of rods of different lengths to measure various inside dimensions that might be encountered; and (3) *depth micrometer,* similar to an inside micrometer but adapted to measure hole depths.

FIGURE 5.5   External micrometer, standard 1-in size with digital readout. (Courtesy of L. S. Starrett Co.)

FIGURE 5.6 Dial indicator: top view shows dial and graduated face; bottom view shows rear of instrument with cover plate removed. (Courtesy of Federal Products Co., Providence, RI.)

5.2.3 COMPARATIVE INSTRUMENTS

Comparative instruments are used to make dimensional comparisons between two objects, such as a workpart and a reference surface. They are usually not capable of providing an absolute measurement of the quantity of interest; instead, they measure the magnitude and direction of the deviation between two objects. Instruments in this category include mechanical and electronic gages.

Mechanical Gages: Dial Indicators   *Mechanical gages* are designed to mechanically magnify the deviation to permit observation. The most common instrument in this category is the **dial indicator** (Figure 5.6), which converts and amplifies the linear movement of a contact pointer into rotation of a dial needle. The dial is graduated in small units such as 0.01 mm (or 0.001 in). Dial indicators are used in many applications to measure straightness, flatness, parallelism, squareness, roundness, and runout. A typical setup for measuring runout is illustrated in Figure 5.7.

Electronic Gages   Electronic gages are a family of measuring and gaging instruments based on transducers capable of converting a linear displacement into an electrical signal. The electrical signal is then amplified and transformed into a suitable data format such as a digital readout, as in Figure 5.5. Applications of electronic gages have grown rapidly in recent years, driven by advances in microprocessor technology. They are gradually replacing many of the conventional measuring and gaging devices. Advantages of electronic gages include (1) good sensitivity, accuracy, precision, repeatability, and speed of response; (2) ability to sense very small dimensions—down to 0.025 μm (1 μ-in.); (3) ease of operation; (4) reduced

FIGURE 5.7 Dial indicator setup to measure runout; as part is rotated about its center, variations in outside surface relative to center are indicated on the dial.

human error; (5) electrical signal that can be displayed in various formats; and (6) capability to be interfaced with computer systems for data processing.

5.2.4 FIXED GAGES

A fixed gage is a physical replica of the part dimension to be assessed. There are two basic categories: master gage and limit gage. A *master gage* is fabricated to be a direct replica of the nominal size of the part dimension. It is generally used for setting up a comparative measuring instrument, such as a dial indicator; or for calibrating a measuring device.

A *limit gage* is fabricated to be a reverse replica of the part dimension and is designed to check the dimension at one or more of its tolerance limits. A limit gage often consists of two gages in one piece, the first for checking the lower limit of the tolerance on the part dimension, and the other for checking the upper limit. These gages are popularly known as *GO/NO-GO gages,* because one gage limit allows the part to be inserted, whereas the other limit does not. The *GO limit* is used to check the dimension at its maximum material condition; this is the minimum size for an internal feature such as a hole, and it is the maximum size for an external feature such as an outside diameter. The *NO-GO limit* is used to inspect the minimum material condition of the dimension in question.

Common limit gages are snap gages and ring gages for checking outside part dimensions, and plug gages for checking inside dimensions. A *snap gage* consists of a C-shaped frame with gaging surfaces located in the jaws of the frame, as in Figure 5.8. It has two gage buttons, the first being the GO gage, and the second being the NO-GO gage. Snap gages are used for checking outside dimensions such as diameter, width, thickness, and similar surfaces.

Ring gages are used for checking cylindrical diameters. For a given application, a pair of gages is usually required, one GO and the other NO-GO. Each gage is a ring whose opening is machined to one of the tolerance limits of the part diameter. For ease of handling, the outside of the ring is knurled. The two gages are distinguished by the presence of a groove around the outside of the NO-GO ring.

The most common limit gage for checking hole diameter is the *plug gage*. The typical gage consists of a handle to which are attached two accurately ground cylindrical pieces (plugs) of hardened steel, as in Figure 5.9. The cylindrical plugs serve as the GO and NO-GO

FIGURE 5.8   Snap gage for measuring diameter of a part; difference in height of GO and NO-GO gage buttons is exaggerated.

FIGURE 5.9   Plug gage; difference in diameters of GO and NO-GO plugs is exaggerated.

gages. Other gages similar to the plug gage include *taper gages,* consisting of a tapered plug for checking tapered holes; and *thread gages,* in which the plug is threaded for checking internal threads on parts.

Fixed gages are easy to use, and the time required to complete an inspection is almost always less than when a measuring instrument is employed. Fixed gages were a fundamental element in the development of interchangeable parts manufacturing (Historical Note 1.1). They provided the means by which parts could be made to tolerances that were sufficiently close for assembly without filing and fitting. Their disadvantage is that they provide little if any information on the actual part size; they only indicate whether the size is within tolerance. Today, with the availability of high-speed electronic measuring instruments, and with the need for statistical process control of part sizes, use of gages is gradually giving way to instruments that provide actual measurements of the dimension of interest.

5.2.5 ANGULAR MEASUREMENTS

Angles can be measured using any of several styles of *protractor.* A *simple protractor* consists of a blade that pivots relative to a semicircular head that is graduated in angular units (e.g., degrees, radians). To use, the blade is rotated to a position corresponding to some part angle to be measured, and the angle is read off the angular scale. A *bevel protractor* (Figure 5.10) consists of two straight blades that pivot relative to each other. The pivot assembly has a protractor scale that permits the angle formed by the blades to be read. When equipped with a vernier, the bevel protractor can be read to about 5 min; without a vernier the resolution is only about 1 degree.

High precision in angular measurements can be made using a *sine bar,* illustrated in Figure 5.11. One possible setup consists of a flat steel straight edge (the sine bar), and two precision rolls set a known distance apart on the bar. The straight edge is aligned with the part angle to be measured, and gage blocks or other accurate linear measurements are made to determine height. The procedure is carried out on a surface plate to achieve most accurate results. This height H and the length L of the sine bar between rolls are used to calculate the angle A using

$$\sin A = \frac{H}{L} \tag{5.1}$$

FIGURE 5.10   Bevel protractor with vernier scale. (Courtesy of L.S. Starrett Co.)

FIGURE 5.11 Setup for using a sine bar.

5.3 SURFACES

A surface is what one touches when holding an object, such as a manufactured part. The designer specifies the part dimensions, relating the various surfaces to each other. These *nominal surfaces,* representing the intended surface contour of the part, are defined by lines in the engineering drawing. The nominal surfaces appear as absolutely straight lines, ideal circles, round holes, and other edges and surfaces that are geometrically perfect. The actual surfaces of a manufactured part are determined by the processes used to make it. The variety of processes available in manufacturing result in wide variations in surface characteristics, and it is important for engineers to understand the technology of surfaces.

Surfaces are commercially and technologically important for a number of reasons, different reasons for different applications: (1) Aesthetic reasons—surfaces that are smooth and free of scratches and blemishes are more likely to give a favorable impression to the customer. (2) Surfaces affect safety. (3) Friction and wear depend on surface characteristics. (4) Surfaces affect mechanical and physical properties; for example, surface flaws can be points of stress concentration. (5) Assembly of parts is affected by their surfaces; for example, the strength of adhesively bonded joints (Section 31.3) is increased when the surfaces are slightly rough. (6) Smooth surfaces make better electrical contacts.

Surface technology is concerned with (1) defining the characteristics of a surface, (2) surface texture, (3) surface integrity, and (4) the relationship between manufacturing processes and the characteristics of the resulting surface. The first three topics are covered in this section; the final topic is presented in Section 5.5.

5.3.1 CHARACTERISTICS OF SURFACES

A microscopic view of a part's surface reveals its irregularities and imperfections. The features of a typical surface are illustrated in the highly magnified cross section of the surface of a metal part in Figure 5.12. Although the discussion here is focused on metallic surfaces,

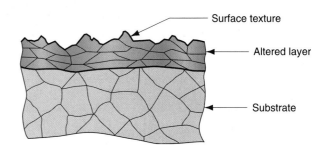

FIGURE 5.12 A magnified cross section of a typical metallic part surface.

these comments apply to ceramics and polymers, with modifications owing to differences in structure of these materials. The bulk of the part, referred to as the ***substrate,*** has a grain structure that depends on previous processing of the metal; for example, the metal's substrate structure is affected by its chemical composition, the casting process originally used on the metal, and any deformation operations and heat treatments performed on the casting.

The exterior of the part is a surface whose topography is anything but straight and smooth. In this highly magnified cross section, the surface has roughness, waviness, and flaws. Although not shown here, it also possesses a pattern and/or direction resulting from the mechanical process that produced it. All of these geometric features are included in the term ***surface texture***.

Just below the surface is a layer of metal whose structure differs from that of the substrate. This is called the ***altered layer,*** and it is a manifestation of the actions that have been visited upon the surface during its creation and afterward. Manufacturing processes involve energy, usually in large amounts, which operates on the part against its surface. The altered layer may result from work hardening (mechanical energy), heating (thermal energy), chemical treatment, or even electrical energy. The metal in this layer is affected by the application of energy, and its microstructure is altered accordingly. This altered layer falls within the scope of ***surface integrity,*** which is concerned with the definition, specification, and control of the surface layers of a material (most commonly metals) in manufacturing and subsequent performance in service. The scope of surface integrity is usually interpreted to include surface texture as well as the altered layer beneath.

In addition, most metal surfaces are coated with an ***oxide film,*** given sufficient time after processing for the film to form. Aluminum forms a hard, dense, thin film of Al_2O_3 on its surface (which serves to protect the substrate from corrosion), and iron forms oxides of several chemistries on its surface (rust, which provides virtually no protection at all). There is also likely to be moisture, dirt, oil, adsorbed gases, and other contaminants on the part's surface.

5.3.2 SURFACE TEXTURE

Surface texture consists of the repetitive and/or random deviations from the nominal surface of an object; it is defined by four features: roughness, waviness, lay, and flaws, shown in Figure 5.13. ***Roughness*** refers to the small, finely spaced deviations from the nominal surface that are determined by the material characteristics and the process that formed the surface. ***Waviness*** is defined as the deviations of much larger spacing; they occur because of work

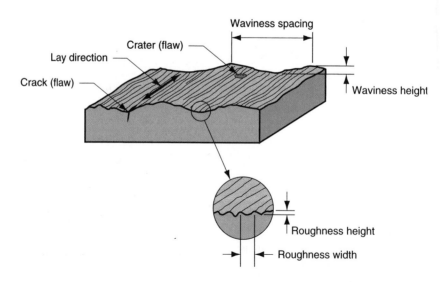

FIGURE 5.13 Surface texture features.

Lay symbol	Surface pattern	Description	Lay symbol	Surface pattern	Description
=		Lay is parallel to line representing surface to which symbol is applied.	C		Lay is circular relative to center of surface to which symbol is applied.
⊥		Lay is perpendicular to line representing surface to which symbol is applied.	R		Lay is approximately radial relative to the center of the surface to which symbol is applied.
X		Lay is angular in both directions to line representing surface to which symbol is applied.	P		Lay is particulate, nondirectional, or protuberant.

FIGURE 5.14   Possible lays of a surface. (Source: [1]).

deflection, vibration, heat treatment, and similar factors. Roughness is superimposed on waviness. *Lay* is the predominant direction or pattern of the surface texture. It is determined by the manufacturing method used to create the surface, usually from the action of a cutting tool. Figure 5.14 presents most of the possible lays a surface can take, together with the symbol used by a designer to specify them. Finally, *flaws* are irregularities that occur occasionally on the surface; these include cracks, scratches, inclusions, and similar defects in the surface. Although some of the flaws relate to surface texture, they also affect surface integrity (Section 5.2.3).

Surface Roughness and Surface Finish   Surface roughness is a measurable characteristic based on the roughness deviations as defined in the preceding. *Surface finish* is a more subjective term denoting smoothness and general quality of a surface. In popular usage, surface finish is often used as a synonym for surface roughness.

The most commonly used measure of surface texture is surface roughness. With respect to Figure 5.15, *surface roughness* can be defined as the average of the vertical deviations from the nominal surface over a specified surface length. An arithmetic average (AA) is generally used, based on the absolute values of the deviations, and this roughness value is referred to by the name *average roughness.* In equation form

$$R_a = \int_0^{L_m} \frac{|y|}{L_m} dx \qquad (5.2)$$

where R_a = arithmetic mean value of roughness, m (in); y = the vertical deviation from nominal surface (converted to absolute value), m (in); and L_m = the specified distance over which the surface deviations are measured.

FIGURE 5.15
Deviations from nominal surface used in the two definitions of surface roughness.

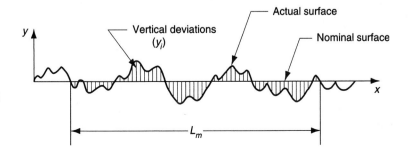

An approximation of Eq. (5.2), perhaps easier to comprehend, is given by

$$R_a = \sum_{i=1}^{n} \frac{|y_i|}{n} \qquad (5.3)$$

where R_a has the same meaning as above; y_i = vertical deviations converted to absolute value and identified by the subscript i, m (in); and n = the number of deviations included in L_m. The units in these equations are meters and inches.

In fact, the scale of the deviations is very small, so more appropriate units are μm (μm = m $\times 10^{-6}$ = mm $\times 10^{-3}$) or μ-in (μ-in = inch $\times 10^{-6}$). These are the units commonly used to express surface roughness.

The AA method is the most widely used averaging method for surface roughness today. An alternative, sometimes used in the United States, is the ***root-mean-square*** (RMS) average, which is the square root of the mean of the squared deviations over the measuring length. RMS surface roughness values will almost always be greater than the AA values because the larger deviations will figure more prominently in the calculation of the RMS value.

Surface roughness suffers the same kinds of deficiencies of any single measure used to assess a complex physical attribute. For example, it fails to account for the lay of the surface pattern; thus, surface roughness may vary significantly, depending on the direction in which it is measured.

Another deficiency is that waviness can be included in the R_a computation. To deal with this problem, a parameter called the ***cutoff length*** is used as a filter that separates the waviness in a measured surface from the roughness deviations. In effect, the cutoff length is a sampling distance along the surface. A sampling distance shorter than the waviness width will eliminate the vertical deviations associated with waviness and only include those associated with roughness. The most common cutoff length used in practice is 0.8 mm (0.030 in). The measuring length L_m is normally set at about five times the cutoff length.

The limitations of surface roughness have motivated the development of additional measures that more completely describe the topography of a given surface. These measures include three-dimensional graphical renderings of the surface, as described in [17].

Symbols for Surface Texture   Designers specify surface texture on an engineering drawing by means of symbols as in Figure 5.16. The symbol designating surface texture parameters is a check mark (looks like a square root sign), with entries as indicated for average roughness, waviness, cutoff, lay, and maximum roughness spacing. The symbols for lay are from Figure 5.14.

FIGURE 5.16   Surface texture symbols in engineering drawings: (a) the symbol, and (b) symbol with identification labels. Values of R_a are given in microinches; units for other measures are given in inches. Designers do not always specify all of the parameters on engineering drawings.

5.3.3 SURFACE INTEGRITY

Surface texture alone does not completely describe a surface. There may be metallurgical or other changes in the material immediately beneath the surface that can have a significant effect on its mechanical properties. ***Surface integrity*** is the study and control of this subsurface layer and any changes in it because of processing that may influence the performance of the finished part or product. This subsurface layer was previously referred to as the altered layer when its structure differs from the substrate, as in Figure 5.12.

The possible alterations and injuries to the subsurface layer that can occur in manufacturing are listed in Table 5.2. The surface changes are caused by the application of various forms of energy during processing—mechanical, thermal, chemical, and electrical. Mechanical energy is the most common form used in manufacturing; it is applied against the work material in operations such as metal forming (e.g., forging, extrusion), pressworking, and machining. Although its primary function in these processes is to change the geometry of the workpart, mechanical energy can also cause residual stresses, work hardening, and cracks

---

TABLE 5.2   Surface and subsurface alterations that define surface integrity.[a]

Absorption are impurities that are absorbed and retained in surface layers of the base material, possibly leading to embrittlement or other property changes.

Alloy depletion occurs when critical alloying elements are lost from the surface layers, with possible loss of properties in the metal.

Cracks are narrow ruptures or separations either at or below the surface that alter the continuity of the material. Cracks are characterized by sharp edges and length-to-width ratios of 4:1 or more. They are classified as macroscopic (can be observed with magnification of 10× or less) and microscopic (requires magnification of more than 10×).

Craters are rough surface depressions left in the surface by short circuit discharges; associated with electrical processing methods such as electric discharge machining and electrochemical machining (Chapter 26).

Hardness changes refer to hardness differences at or near the surface.

Heat affected zone are regions of the metal that are affected by the application of thermal energy; the regions are not melted but are sufficiently heated that they undergo metallurgical changes that affect properties. Abbreviated HAZ, the effect is most prominent in fusion welding operations (Chapter 31).

Inclusions are small particles of material incorporated into the surface layers during processing; they are a discontinuity in the base material. Their composition usually differs from the base material.

Intergranular attack refers to various forms of chemical reactions at the surface, including intergranular corrosion and oxidation.

Laps, folds, seams are irregularities and defects in the surface caused by plastic working of overlapping surfaces.

Pits are shallow depressions with rounded edges formed by any of several mechanisms, including selective etching or corrosion; removal of surface inclusions; mechanically formed dents; or electrochemical action.

Plastic deformation refers to microstructural changes from deforming the metal at the surface; it results in strain hardening.

Recrystallization involves the formation of new grains in strain hardened metals; associated with heating of metal parts that have been deformed.

Redeposited metal is metal that is removed from the surface in the molten state and then reattached prior to solidification.

Resolidified metal is a portion of the surface that is melted during processing and then solidified without detaching from the surface. The name ***remelted metal*** is also used for resolidified metal. ***Recast metal*** is a term that includes both redeposited and resolidified metal.

Residual stresses are stresses remaining in the material after processing.

Selective etch is a form of chemical attack that concentrates on certain components in the base material.

---

[a]Compiled from [2].

TABLE 5.3 Forms of energy applied in manufacturing and the resulting possible surface and subsurface alterations that can occur.[a]

Mechanical	Thermal	Chemical	Electrical
Residual stresses in subsurface layer	Metallurgical changes (recrystallization, grain size changes, phase changes at surface)	Intergranular attack	Changes in conductivity and/or magnetism
Cracks—microscopic and macroscopic	Redeposited or resolidified material	Chemical contamination	Craters resulting from short circuits during certain electrical processing techniques
Plastic deformation	Heat-affected zone	Absorption of elements such as H and Cl	
Laps, folds, or seams	Hardness changes	Corrosion, pitting, and etching	
Voids or inclusions		Dissolving of microconstituents	
Hardness variations (e.g., work hardening)		Alloy depletion	

[a]Based on [2].

in the surface layers. Table 5.3 indicates the various types of surface and subsurface alterations that are attributable to the different forms of energy applied in manufacturing. Most of the alterations in the table refer to metals, for which surface integrity has been most intensively studied.

5.4   MEASUREMENT OF SURFACES

Surfaces are described as consisting of two parameters: (1) surface texture and (2) surface integrity. This section is concerned with the measurement of these two parameters.

5.4.1   MEASUREMENT OF SURFACE ROUGHNESS

Various methods are used to assess surface roughness. They can be divided into three categories: (1) subjective comparison with standard test surfaces, (2) stylus electronic instruments, and (3) optical techniques.

Standard Test Surfaces   Sets of standard surface finish blocks are available, produced to specified roughness values.[1] To estimate the roughness of a given test specimen, the surface is compared with the standard both visually and by the "fingernail test." In this test, the user gently scratches the surfaces of the specimen and the standards, judging which standard is closest to the specimen. Standard test surfaces are a convenient way for a machine operator to obtain an estimate of surface roughness. They are also useful for design engineers in judging what value of surface roughness to specify on a part drawing.

Stylus Instruments   The disadvantage of the fingernail test is its subjectivity. Several stylus-type instruments are commercially available to measure surface roughness—similar to

---

[1]In the U.S.C.S., these blocks have surfaces with roughness values of 2, 4, 8, 16, 32, 64, or 128 microinches.

FIGURE 5.17   Stylus-type instrument for measuring surface roughness. (Courtesy of Giddings & Lewis, Measurement Systems Division.)

the fingernail test, but more scientific. An example is the Profilometer, shown in Figure 5.17. In these electronic devices, a cone-shaped diamond stylus with point radius of about 0.005 mm (0.0002 in) and 90° tip angle is traversed across the test surface at a constant slow speed. The operation is depicted in Figure 5.18. As the stylus head is traversed horizontally, it also moves vertically to follow the surface deviations. The vertical movement is converted into an electronic signal that represents the topography of the surface. This can be displayed as either a profile of the actual surface or an average roughness value. ***Profiling devices*** use a separate flat plane as the nominal reference against which deviations are measured. The output is a plot of the surface contour along the line traversed by the stylus. This type of system can identify both roughness and waviness in the test surface. ***Averaging devices*** reduce the roughness deviations to a single value R_a. They use skids riding on the actual surface to establish the nominal reference plane. The skids act as a mechanical filter to reduce the effect of waviness in the surface; in effect, these averaging devices electronically perform the computations in Eq. (5.1).

Optical Techniques   Most other surface-measuring instruments employ optical techniques to assess roughness. These techniques are based on light reflectance from the surface, light scatter or diffusion, and laser technology. They are useful in applications where stylus contact with the surface is undesirable. Some of the techniques permit very-high-speed operation, thus making 100% inspection feasible. However, the optical techniques yield values that do not always correlate well with roughness measurements made by stylus-type instruments.

FIGURE 5.18   Sketch illustrating the operation of stylus-type instrument. Stylus head traverses horizontally across surface, while stylus moves vertically to follow surface profile. Vertical movement is converted into either (1) a profile of the surface or (2) the average roughness value.

Traversing direction

Stylus head

Vertical motion of stylus →

Stylus

Workpart

5.4.2 EVALUATION OF SURFACE INTEGRITY

Surface integrity is more difficult to assess than surface roughness. Some of the techniques to inspect for subsurface changes are destructive to the material specimen. Evaluation techniques for surface integrity include the following:

➤ *Surface texture.* Surface roughness, designation of lay, and other measures provide superficial data on surface integrity. This type of testing is relatively simple to perform and is always included in the evaluation of surface integrity.

➤ *Visual examination.* Visual examination can reveal various surface flaws such as cracks, craters, laps, and seams. This type of assessment is often augmented by fluorescent and photographic techniques.

➤ *Microstructural examination.* This involves standard metallographic techniques for preparing cross sections and obtaining photomicrographs for examination of microstructure in the surface layers compared with the substrate.

➤ *Microhardness profile.* Hardness differences near the surface can be detected using microhardness measurement techniques such as Knoop and Vickers (Section 3.2.1). The part is sectioned, and hardness is plotted against distance below the surface to obtain a hardness profile of the cross section.

➤ *Residual stress profile.* X-ray diffraction techniques can be employed to measure residual stresses in the surface layers of a part.

5.5 EFFECT OF MANUFACTURING PROCESSES

The ability to achieve a certain tolerance or surface is a function of the manufacturing process. This section describes the general capabilities of various processes in terms of tolerance and surface roughness and surface integrity.

Some manufacturing processes are inherently more accurate than others. Most machining processes are quite accurate, capable of tolerances of ± 0.05 mm (± 0.002 in) or better. By contrast, sand castings are generally inaccurate, and tolerances of 10 to 20 times those used for machined parts should be specified. Table 5.4 lists a variety of manufacturing processes and indicates the typical tolerances for each process. Tolerances are

TABLE 5.4   Typical tolerance limits, based on process capability (Section 42.2), for various manufacturing processes.[b]

Process	Typical Tolerance, mm (in)	Process	Typical Tolerance, mm (in)
Sand casting		Abrasive	
Cast iron	± 1.3 (± 0.050)	Grinding	± 0.008 (± 0.0003)
Steel	± 1.5 (± 0.060)	Lapping	± 0.005 (± 0.0002)
Aluminum	± 0.5 (± 0.020)	Honing	± 0.005 (± 0.0002)
Die casting	± 0.12 (± 0.005)	Nontraditional and thermal	
Plastic molding:		Chemical machining	± 0.08 (± 0.003)
Polyethylene	± 0.3 (± 0.010)	Electric discharge	± 0.025 (± 0.001)
Polystyrene	± 0.15 (± 0.006)	Electrochem. grind	± 0.025 (± 0.001)
Machining:		Electrochem. machine	± 0.05 (± 0.002)
Drilling, 6 mm (0.25 in)	$\pm 0.08\pm 0.03$ ($+0.003/-0.001$)	Electron beam cutting	± 0.08 (± 0.003)
Milling	± 0.08 (± 0.003)	Laser beam cutting	± 0.08 (± 0.003)
Turning	± 0.05 (± 0.002)	Plasma arc cutting	± 1.3 (± 0.050)

[b]Compiled from [4], [5], and other sources. For each process category, tolerances vary depending on process parameters. Also, tolerances increase with part size.

TABLE 5.5 Surface roughness values produced by the various manufacturing processes.[a]

Process	Typical Finish	Roughness Range[b]	Process	Typical Finish	Roughness Range[b]
Casting:			*Abrasive:*		
Die casting	Good	1–2 (30–65)	Grinding	Very good	0.1–2 (5–75)
Investment	Good	1.5–3 (50–100)	Honing	Very good	0.1–1 (4–30)
Sand casting	Poor	12–25 (500–1000)	Lapping	Excellent	0.05–0.5 (2–15)
Metal forming:			Polishing	Excellent	0.1–0.5 (5–15)
Cold rolling	Good	1–3 (25–125)	Superfinish	Excellent	0.02–0.3 (1–10)
Sheet metal draw	Good	1–3 (25–125)	*Nontraditional:*		
Cold extrusion	Good	1–4 (30–150)	Chemical milling	Medium	1.5–5 (50–200)
Hot rolling	Poor	12–25 (500–1000)	Electrochemical	Good	0.2–2 (10–100)
Machining:	Good	0.5–6 (15–250)	Electric discharge	Medium	1.5–15 (50–500)
Boring			Electron beam	Medium	1.5–15 (50–500)
Drilling	Medium	1.5–6 (60–250)	Laser beam	Medium	1.5–15 (50–500)
Milling	Good	1–6 (30–250)	*Thermal:*		
Reaming	Good	1–3 (30–125)	Arc welding	Poor	5–25 (250–1000)
Shaping and planing	Medium	1.5–12 (60–500)	Flame cutting	Poor	12–25 (500–1000)
Sawing	Poor	3–25 (100–1000)	Plasma arc cutting	Poor	12–25 (500–1000)
Turning	Good	0.5–6 (15–250)			

[a]Compiled from [1], [2], and other sources.
[b]Roughness range values are given, μm (μ-in). Roughness can vary significantly for a given process, depending on process parameters.

based on the process capability for the particular manufacturing operation, as defined in Section 42.2. The tolerance that should be specified is a function of part size; larger parts require more generous tolerances. The table lists tolerance for moderately sized parts in each processing category.

The manufacturing process determines surface finish and surface integrity. Some processes are capable of producing better surfaces than others. In general, processing cost increases with improvement in surface finish. This is because additional operations and more time are usually required to obtain increasingly better surfaces. Processes noted for providing superior finishes include honing, lapping, polishing, and superfinishing (Chapter 25). Table 5.5 indicates the usual surface roughness that can be expected from various manufacturing processes.

REFERENCES

[1] American National Standards Institute, Inc. *Surface Texture,* ANSI B46.1-1978. American Society of Mechanical Engineers, New York, 1978.

[2] American National Standards Institute, Inc. *Surface Integrity,* ANSI B211.1-1986. Society of Manufacturing Engineers, Dearborn, Michigan, 1986.

[3] American National Standards Institute, Inc. *Dimensioning and Tolerancing,* ANSI Y14.5M-1982. American Society of Mechanical Engineers, New York, 1982.

[4] Bakerjian, R. and Mitchell, P. *Tool and Manufacturing Engineers Handbook,* 4th ed., Vol. VI, *Design for Manufacturability.* Society of Manufacturing Engineers, Dearborn, Michigan, 1992.

[5] Brown & Sharpe. *Handbook of Metrology.* North Kingston, Rhode Island, 1992.

[6] Curtis, M., *Handbook of Dimensional Measurement,* 4th ed. Industrial Press, New York, 2007.

[7] Drozda, T. J. and Wick, C. *Tool and Manufacturing Engineers Handbook,* 4th ed., Vol. I, Machining. Society of Manufacturing Engineers, Dearborn, Michigan, 1983.

[8] Farago, F. T. ***Handbook of Dimensional Measurement,*** 3rd ed. Industrial Press Inc., New York, 1994.

[9] ***Machining Data Handbook,*** 3rd ed., Vol. II. Machinability Data Center, Cincinnati, Ohio, 1980, Ch. 18.

[10] Mummery, L. ***Surface Texture Analysis—The Handbook.*** Hommelwerke Gmbh, Germany, 1990.

[11] Oberg, E., Jones, F. D., Horton, H. L., and Ryffel, H. ***Machinery's Handbook,*** 26th ed. Industrial Press, New York, 2000.

[12] Schaffer, G. H. "The Many Faces of Surface Texture," Special Report 801, ***American Machinist and Automated Manufacturing,*** June 1988, pp. 61–68.

[13] Sheffield Measurement, a Cross & Trecker Company, ***Surface Texture and Roundness Measurement Handbook, Dayton,*** Ohio, 1991.

[14] Spitler, D., Lantrip, J., Nee, J., and Smith, D. A. ***Fundamentals of Tool Design,*** 5th ed. Society of Manufacturing Engineers, Dearborn, Michigan, 2003.

[15] S. Starrett Company. ***Tools and Rules.*** Athol, Massachusetts, 1992.

[16] Wick, C., and Veilleux, R. F. ***Tool and Manufacturing Engineers Handbook,*** 4th ed., Vol. IV, Quality Control and Assembly. Society of Manufacturing Engineers, Dearborn, Michigan, 1987, Section 1.

[17] Zecchino, M. "Why Average Roughness Is Not Enough," ***Advanced Materials & Processes,*** March 2003, pp. 25–28.

REVIEW QUESTIONS

5.1. What is a tolerance?

5.2. What is the difference between a bilateral tolerance and a unilateral tolerance?

5.3. What is accuracy in measurement?

5.4. What is precision in measurement?

5.5. What is meant by the term graduated measuring device?

5.6. What are some of the reasons why surfaces are important?

5.7. Define nominal surface.

5.8. Define surface texture.

5.9. How is surface texture distinguished from surface integrity?

5.10. Within the scope of surface texture, how is roughness distinguished from waviness?

5.11. Surface roughness is a measurable aspect of surface texture; what does **surface roughness** mean?

5.12. Indicate some of the limitations of using surface roughness as a measure of surface texture.

5.13. Identify some of the changes and injuries that can occur at or immediately below the surface of a metal.

5.14. What causes the various types of changes that occur in the altered layer just beneath the surface?

5.15. What are the common methods for assessing surface roughness?

5.16. Name some manufacturing processes that produce very poor surface finishes.

5.17. Name some manufacturing processes that produce very good or excellent surface finishes.

5.18. (**Video**) Based on the video about vernier calipers, are the markings on the vernier plate (moveable scale) the same spacing, slightly closer, or slightly further apart compared to the stationary bar?

5.19. (**Video**) Based on the video about vernier calipers, explain how to read the scale on a vernier caliper.

5.20. (**Video**) Based on the video about micrometers, explain the primary factor that makes an English micrometer different from a metric micrometer.

MULTIPLE CHOICE QUIZ

There are 19 correct answers in the following multiple choice questions (some questions have multiple answers that are correct). To attain a perfect score on the quiz, all correct answers must be given. Each correct answer is worth 1 point. Each omitted answer or wrong answer reduces the score by 1 point, and each additional answer beyond the correct number of answers reduces the score by 1 point. Percentage score on the quiz is based on the total number of correct answers.

5.1. A tolerance is which one of the following: (a) clearance between a shaft and a mating hole, (b) measurement error, (c) total permissible variation from a specified dimension, or (d) variation in manufacturing?

5.2. Which of the following two geometric terms have the same meaning: (a) circularity, (b) concentricity, (c) cylindricity, and (d) roundness?

5.3. A surface plate is most typically made of which one of the following materials: (a) aluminum oxide ceramic, (b) cast iron, (c) granite, (d) hard polymers, or (e) stainless steel?

5.4. An outside micrometer would be appropriate for measuring which of the following (two correct answers): (a) hole depth, (b) hole diameter, (c) part length, (d) shaft diameter, and (e) surface roughness?

5.5. In a GO/NO-GO gage, which one of the following best describes the function of the GO gage: (a) checks limit of maximum tolerance, (b) checks maximum material condition, (c) checks maximum size, (d) checks minimum material condition, or (e) checks minimum size?

5.6. Which of the following are likely to be GO/NO-GO gages (three correct answers): (a) gage blocks, (b) limit gage, (c) master gage, (d) plug gage, and (e) snap gage?

5.7. Surface texture includes which of the following characteristics of a surface (three correct answers): (a) deviations from the nominal surface, (b) feed marks of the tool that produced the surface, (c)

hardness variations, (d) oil films, and (e) surface cracks?

5.8. Surface texture is included within the scope of surface integrity: (a) true or (b) false?

5.9. Thermal energy is normally associated with which of the following changes in the altered layer (three best answers): (a) cracks, (b) hardness variations, (c) heat affected zone, (d) plastic deformation, (e) recrystallization, or (f) voids?

5.10. Which one of the following manufacturing processes will likely result in the best surface finish: (a) arc welding, (b) grinding, (c) machining, (d) sand casting, or (e) sawing?

5.11. Which one of the following manufacturing processes will likely result in the worst surface finish: (a) cold rolling, (b) grinding, (c) machining, (d) sand casting, or (e) sawing?

PROBLEMS

5.1. Design the nominal sizes of a GO/NO-GO plug gage to inspect a 1.500 ± 0.030 in diameter hole. There is a wear allowance applied only to the GO side of the gage. The wear allowance is 2% of the entire tolerance band for the inspected feature. Determine (a) the nominal size of the GO gage including the wear allowance and (b) the nominal size of the NO-GO gage.

5.2. Design the nominal sizes of a GO/NO-GO snap gage to inspect the diameter of a shaft that is 1.500 ± 0.030. A wear allowance of 2% of the entire tolerance band is applied to the GO side. Determine (a) the nominal size of the GO gage including the wear allowance and (b) the nominal size of the NO-GO gage.

5.3. Design the nominal sizes of a GO/NO-GO plug gage to inspect a 30.00 ± 0.18 mm diameter hole. There is a wear allowance applied only to the GO side of the gage. The wear allowance is 3% of the entire tolerance band for the inspected feature. Determine (a) the nominal size of the GO gage including the wear allowance and (b) the nominal size of the NO-GO gage.

5.4. Design the nominal sizes of a GO/NO-GO snap gage to inspect the diameter of a shaft that is 30.00 ±

0.18 mm. A wear allowance of 3% of the entire tolerance band is applied to the GO side. Determine (a) the nominal size of the GO gage including the wear allowance and (b) the nominal size of the NO-GO gage.

5.5. A sine bar is used to determine the angle of a part feature. The length of the sine bar is 6.000 in. The rolls have a diameter of 1.000 in. All inspection is performed on a surface plate. In order for the sine bar to match the angle of the part, the following gage blocks must be stacked: 2.0000, 0.5000, 0.3550. Determine the angle of the part feature.

5.6. A 200.00 mm sine bar is used to inspect an angle on a part. The angle has a dimension of 35.0 ± 1.8. The sine bar rolls have a diameter of 30.0 mm. A set of gage blocks is available that can form any height from 10.0000 to 199.9975 mm in increments of 0.0025 mm. Determine (a) the height of the gage block stack to inspect the minimum angle, (b) height of the gage block stack to inspect the maximum angle, and (c) smallest increment of angle that can be setup at the nominal angle size. All inspection is performed on a surface plate.

Part II Engineering Materials

6

METALS

Chapter Contents

Part II discusses the four types of engineering materials: (1) metals, (2) ceramics, (3) polymers, and (4) composites. Metals are the most important engineering materials and the topic of this chapter. A *metal* is a category of materials generally characterized by properties of ductility, malleability, luster, and high electrical and thermal conductivity. The category includes both metallic elements and their alloys. Metals have properties that satisfy a wide variety of design requirements. The manufacturing processes by which they are shaped into products have been developed and refined over many years; indeed, some of the processes date from ancient times (Historical Note 1.2). In addition, the properties of metals can be enhanced through heat treatment (covered in Chapter 27).

The technological and commercial importance of metals results from the following general properties possessed by virtually all of the common metals:

➢ *High stiffness and strength.* Metals can be alloyed for high rigidity, strength, and hardness; thus, they are used to provide the structural framework for most engineered products.

➢ *Toughness.* Metals have the capacity to absorb energy better than other classes of materials.

➢ *Good electrical conductivity.* Metals are conductors because of their metallic bonding that permits the free movement of electrons as charge carriers.

➢ *Good thermal conductivity.* Metallic bonding also explains why metals generally conduct heat better than ceramics or polymers.

In addition, certain metals have specific properties that make them attractive for specialized applications. Many common metals are available at relatively low cost per unit weight and are often the material of choice simply because of their low cost.

Metals are converted into parts and products using a variety of manufacturing processes. The starting form of the metal differs, depending on the process. The major categories are (1) *cast metal,* in which the initial form is a casting; (2) *wrought metal,* in which the metal has been worked or can be worked (e.g., rolled or otherwise formed) after casting; better mechanical properties are generally associated with wrought metals compared with cast metals; and (3) *powdered metal,* in which the metal is purchased in the form of very small powders for conversion into parts using powder metallurgy techniques. Most metals are available in all three forms. The discussion in this chapter focuses on categories (1) and (2), which are of greatest commercial and engineering interest. Powder metallurgy techniques are examined in Chapter 16.

Metals are classified into two major groups: (1) *ferrous*—those based on iron; and (2) *nonferrous*—all other metals. The ferrous group can be further subdivided into steels and cast irons. Most of the discussion in the present chapter is organized around this classification, but first the general topic of alloys and phase diagrams is examined.

6.1 ALLOYS AND PHASE DIAGRAMS

Although some metals are important as pure elements (e.g., gold, silver, copper), most engineering applications require the improved properties obtained by alloying. Through alloying, it is possible to enhance strength, hardness, and other properties compared with pure metals. This section defines and classifies alloys; it then discusses phase diagrams, which indicate the phases of an alloy system as a function of composition and temperature.

6.1.1 ALLOYS

An alloy is a metal composed of two or more elements, at least one of which is metallic. The two main categories of alloys are (1) solid solutions and (2) intermediate phases.

Solid Solutions A solid solution is an alloy in which one element is dissolved in another to form a single-phase structure. The term *phase* describes any homogeneous mass of material, such as a metal in which the grains all have the same crystal lattice structure. In a solid solution, the solvent or base element is metallic, and the dissolved element can be either metallic or nonmetallic. Solid solutions come in two forms, shown in Figure 6.1. The first is a *substitutional solid solution,* in which atoms of the solvent element are replaced in its unit cell by the dissolved element. Brass is an example, in which zinc is dissolved in copper. To make the substitution, several rules must be satisfied [3], [6], [7]: (1) the atomic radii of the two elements must be similar, usually within 15%; (2) their lattice types must be the

FIGURE 6.1 Two forms of solid solutions: (a) substitutional solid solution, and (b) interstitial solid solution.

(a) (b)

same; (3) if the elements have different valences, the lower valence metal is more likely to be the solvent; and (4) if the elements have high chemical affinity for each other, they are less likely to form a solid solution and more likely to form a compound.

The second type of solid solution is an *interstitial solid solution,* in which atoms of the dissolving element fit into the vacant spaces between base metal atoms in the lattice structure. It follows that the atoms fitting into these interstices must be small compared with those of the solvent metal. The most important example of this second type is carbon dissolved in iron to form steel.

In both forms of solid solution, the alloy structure is generally stronger and harder than either of the component elements.

Intermediate Phases There are usually limits to the solubility of one element in another. When the amount of the dissolving element in the alloy exceeds the solid solubility limit of the base metal, a second phase forms in the alloy. The term *intermediate phase* is used to describe it because its chemical composition is intermediate between the two pure elements. Its crystalline structure is also different from those of the pure metals. Depending on composition, and recognizing that many alloys consist of more than two elements, these intermediate phases can be of several types, including (1) metallic compounds consisting of a metal and nonmetal such as Fe_3C; and (2) intermetallic compounds—two metals that form a compound, such as Mg_2Pb. The composition of the alloy is often such that the intermediate phase is mixed with the primary solid solution to form a two-phase structure, one phase dispersed throughout the second. These two-phase alloys are important because they can be formulated and heat treated for significantly higher strength than solid solutions.

6.1.2 PHASE DIAGRAMS

As the term is used in this text, a phase diagram is a graphical means of representing the phases of a metal alloy system as a function of composition and temperature. This discussion of the diagram will be limited to alloy systems consisting of two elements at atmospheric pressures. This type of diagram is called a *binary phase diagram.* Other forms of phase diagrams are discussed in texts on materials science, such as [6].

The Copper–Nickel Alloy System The best way to introduce the phase diagram is by example. Figure 6.2 presents one of the simplest cases, the Cu–Ni alloy system. Composition is plotted on the horizontal axis and temperature on the vertical axis. Thus, any point in the diagram indicates the overall composition and the phase or phases present at the given temperature. Pure copper melts at 1083°C (1981°F), and pure nickel at 1455°C (2651°F). Alloy compositions between these extremes exhibit gradual melting that commences at the solidus and concludes at the liquidus as temperature is increased.

The copper–nickel system is a solid solution alloy throughout its entire range of compositions. Anywhere in the region below the solidus line, the alloy is a solid solution; there are no intermediate solid phases in this system. However, there is a mixture of phases in the region bounded by the solidus and liquidus. Recall from Chapter 4 that the solidus is the temperature at which the solid metal begins to melt as temperature is increased, and the liquidus is the temperature at which melting is completed. It can now be seen from the phase diagram that these temperatures vary with composition. Between the solidus and liquidus, the metal is a solid–liquid mix.

Determining Chemical Compositions of Phases Although the overall composition of the alloy is given by its position along the horizontal axis, the compositions of the liquid

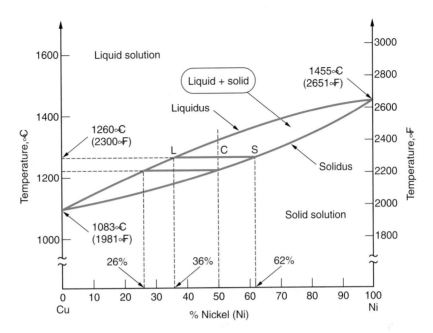

FIGURE 6.2 Phase diagram for the copper–nickel alloy system.

and solid phases are not the same. It is possible to determine these compositions from the phase diagram by drawing a horizontal line at the temperature of interest. The points of intersection between the horizontal line and the solidus and liquidus indicate the compositions of the solid and liquid phases present, respectively. Simply construct the vertical projections from the intersection points to the *x*-axis and read the corresponding compositions.

Example 6.1 Determining Compositions from the Phase Diagram

To illustrate the procedure, suppose one wants to analyze the compositions of the liquid and solid phases present in the copper-nickel system at an aggregate composition of 50% nickel and a temperature of 1260°C (2300°F).

Solution: A horizontal line is drawn at the given temperature level as shown in Figure 6.2. The line intersects the solidus at a composition of 62% nickel, thus indicating the composition of the solid phase. The intersection with the liquidus occurs at a composition of 36% Ni, corresponding to the analysis of the liquid phase.   ■

As the temperature of the 50–50 Cu–Ni alloy is reduced, the solidus line is reached at about 1221°C (2230°F). Applying the same procedure used in the example, the composition of the solid metal is 50% nickel, and the composition of the last remaining liquid to freeze is about 26% nickel. How is it, the reader might ask, that the last ounce of molten metal has a composition so different from the solid metal into which it freezes? The answer is that the phase diagram assumes equilibrium conditions are allowed to prevail. In fact, the binary phase diagram is sometimes called an equilibrium diagram because of this assumption. What it means is that enough time is permitted for the solid metal to gradually change its composition by diffusion to achieve the composition indicated by the intersection point along the liquidus. In practice, when an alloy freezes (e.g., a casting), *segregation* occurs in the solid mass because of nonequilibrium conditions. The first liquid to solidify has a composition that is rich in the metal element with the higher melting point. Then as additional metal solidifies, its composition is different from that of the first metal to freeze. As the nucleation sites grow into a solid mass, compositions are distributed within the mass, depending on the temperature and time in the process at which freezing occurred. The overall composition is the average of the distribution.

Determining Amounts of Each Phase The amounts of each phase present at a given temperature from the phase diagram can also be determined. This is done by the ***inverse lever rule:*** (1) using the same horizontal line as before that indicates the overall composition at a given temperature, measure the distances between the aggregate composition and the intersection points with the liquidus and solidus, identifying the distances as CL and CS, respectively (refer back to Figure 6.2); (2) the proportion of liquid phase present is given by

$$L \text{ phase proportion} = \frac{CS}{(CS + CL)} \qquad (6.1)$$

(3) the proportion of solid phase present is given by

$$S \text{ phase proportion} = \frac{CL}{(CS + CL)} \qquad (6.2)$$

**Example 6.2
Determining
Proportions of
Each Phase**

Determine the proportions of liquid and solid phases for the 50% nickel composition of the copper–nickel system at the temperature of 1260°C (2300°F).

Solution: Using the same horizontal line in Figure 6.2 as in previous Example 6.1, the distances CS and CL are measured as 10 mm and 12 mm, respectively. Thus the proportion of the liquid phase is $10/22 = 0.45$ (45%), and the proportion of solid phase is $12/22 = 0.55$ (55%). ∎

    The proportions given by Eqs. (6.1) and (6.2) are by weight, same as the phase diagram percentages. Note that the proportions are based on the distance on the opposite side of the phase of interest; hence the name inverse lever rule. One can see the logic in this by taking the extreme case when, say, $CS = 0$; at that point, the proportion of the liquid phase is zero because the solidus has been reached and the alloy is therefore completely solidified.
    The methods for determining chemical compositions of phases and the amounts of each phase are applicable to the solid region of the phase diagram as well as the liquidus–solidus region. Wherever there are regions in the phase diagram in which two phases are present, these methods can be used. When only one phase is present (in Figure 6.2, this is the entire solid region), the composition of the phase is its aggregate composition under equilibrium conditions; and the inverse lever rule does not apply because there is only one phase.

The Tin–Lead Alloy System A more complicated phase diagram is the Sn–Pb system, shown in Figure 6.3. Tin–lead alloys have traditionally been used as solders for making electrical and mechanical connections (Section 31.2).[1] The phase diagram exhibits several features not included in the previous Cu–Ni system. One feature is the presence of two solid phases, alpha (α) and beta (β). The α phase is a solid solution of tin in lead at the left side of the diagram, and the β phase is a solid solution of lead in tin that occurs only at elevated temperatures around 200°C (375°F) at the right side of the diagram. Between these solid solutions lies a mixture of the two solid phases, $\alpha + \beta$.
    Another feature of interest in the tin–lead system is how melting differs for different compositions. Pure tin melts at 232°C (449°F), and pure lead melts at 327°C (621°F). Alloys of these elements melt at lower temperatures. The diagram shows two liquidus lines that begin at the melting points of the pure metals and meet at a composition of 61.9% Sn. This is the eutectic composition for the tin–lead system. In general, a ***eutectic alloy*** is a particular composition in an alloy system for which the solidus and liquidus are at the same temperature. The corresponding ***eutectic temperature,*** the melting point of the eutectic

---

[1]Because lead is a poisonous substance, alternative alloying elements have been substituted for lead in many commercial solders. These are called lead-free solders.

FIGURE 6.3 Phase diagram for the tin–lead alloy system.

composition, is 183°C (362°F) in the present case. The eutectic temperature is always the lowest melting point for an alloy system (eutectic is derived from the Greek word *eutektos,* meaning easily melted).

Methods for determining the chemical analysis of the phases and the proportions of phases present can be readily applied to the Sn–Pb system just as it was used in the Cu–Ni system. In fact, these methods are applicable in any region containing two phases, including two solid phases. Most alloy systems are characterized by the existence of multiple solid phases and eutectic compositions, and so the phase diagrams of these systems are often similar to the tin–lead diagram. Of course, many alloy systems are considerably more complex. One of these is the alloy system of iron and carbon.

6.2 FERROUS METALS

The ferrous metals are based on iron, one of the oldest metals known to humans (Historical Note 6.1). The properties and other data relating to iron are listed in Table 6.1(a). The ferrous metals of engineering importance are alloys of iron and carbon. These alloys divide into two major groups: steel and cast iron. Together, they constitute approximately 85% of the metal tonnage in the United States [6]. This discussion of the ferrous metals begins with the iron–carbon phase diagram.

TABLE 6.1   Basic data on the metallic elements: (a) Iron.

Symbol: Fe	Principal ore: **Hematite** (Fe_2O_3)
Atomic number: 26	Alloying elements: Carbon; also chromium, manganese, nickel, molybdenum, vanadium, and silicon
Specific gravity: 7.87	
Crystal structure: BCC	
Melting temperature: 1539°C (2802°F)	Typical applications: Construction, machinery, automotive, railway tracks and equipment
Elastic modulus: 209,000 MPa (30×10^6 lb/in^2)	

Compiled from [6], [11], [12], and other references.

Historical Note 6.1   *Iron and steel*

Iron was discovered sometime during the Bronze Age. It was probably uncovered from ashes of fires built near iron ore deposits. Use of the metal grew, finally surpassing bronze in importance. The Iron Age is usually dated from about 1200 BCE, although artifacts made of iron have been found in the Great Pyramid of Giza in Egypt, which dates to 2900 BCE. Iron-smelting furnaces have been discovered in Israel dating to 1300 BCE. Iron chariots, swords, and tools were made in ancient Assyria (northern Iraq) around 1000 BCE. The Romans inherited ironworking from their provinces, mainly Greece, and they developed the technology to new heights, spreading it throughout Europe. The ancient civilizations learned that iron was harder than bronze and that it took a sharper, stronger edge.

During the Middle Ages in Europe, the invention of the cannon created the first real demand for iron; only then did it finally exceed copper and bronze in usage. Also, the cast iron stove, the appliance of the seventeenth and eighteenth centuries, significantly increased demand for iron (Historical Note 11.3).

In the nineteenth century, industries such as railroads, shipbuilding, construction, machinery, and the military created a dramatic growth in the demand for iron and steel in Europe and America. Although large quantities of (crude) *pig iron* could be produced by **blast furnaces,** the subsequent processes for producing wrought iron and steel were slow. The necessity to improve productivity of these vital metals

was the "mother of invention." Henry Bessemer in England developed the process of blowing air up through the molten iron that led to the **Bessemer converter** (patented in 1856). Pierre and Emile Martin in France built the first **open hearth furnace** in 1864. These methods permitted up to 15 tons of steel to be produced in a single batch (heat), a substantial increase from previous methods.

In the United States, expansion of the railroads after the Civil War created a huge demand for steel. In the 1880s and 1890s, steel beams were first used in significant quantities in construction. Skyscrapers came to rely on these steel frames.

When electricity became available in abundance in the late 1800s, this energy source was used for steelmaking. The first commercial **electric furnace** for production of steel was operated in France in 1899. By 1920, this had become the principal process for making alloy steels.

The use of pure oxygen in steelmaking was initiated just before World War II in several European countries and the United States. Work in Austria after the war culminated in the development of the **basic oxygen furnace** (BOF). This has become the leading modern technology for producing steel, surpassing the open hearth method around 1970. The Bessemer converter had been surpassed by the open hearth method around 1920 and ceased to be a commercial steelmaking process in 1971.

6.2.1   THE IRON–CARBON PHASE DIAGRAM

The iron–carbon phase diagram is shown in Figure 6.4. Pure iron melts at 1539°C (2802°F). During the rise in temperature from ambient, it undergoes several solid phase transformations, as indicated in the diagram. Starting at room temperature the phase is alpha (α), also called *ferrite.* At 912°C (1674°F), ferrite transforms to gamma (γ), called *austenite.* This, in turn, transforms at 1394°C (2541°F) to delta (δ), which remains until melting occurs. The three phases are distinct; alpha and delta have BCC lattice structures (Section 2.3.1), and between them, gamma is FCC. The video clip on heat treatment describes the iron–carbon phase diagram and how it is used to strengthen steel.

*VIDEO CLIP*_____

Heat Treatment: View the segment on the iron–carbon phase diagram.

Iron as a commercial product is available at various levels of purity. *Electrolytic iron* is the most pure, at about 99.99%, for research and other purposes where the pure metal is required. *Ingot iron,* containing about 0.1% impurities (including about 0.01% carbon), is

FIGURE 6.4 Phase diagram for iron–carbon system, up to about 6% carbon.

used in applications in which high ductility or corrosion resistance are needed. **Wrought iron** contains about 3% slag but very little carbon, and is easily shaped in hot forming operations such as forging.

Solubility limits of carbon in iron are low in the ferrite phase—only about 0.022% at 723°C (1333°F). Austenite can dissolve up to about 2.1% carbon at a temperature of 1130°C (2066°F). This difference in solubility between alpha and gamma leads to opportunities for strengthening by heat treatment (but leave that for Chapter 27). Even without heat treatment, the strength of iron increases dramatically as carbon content increases, and the metal is called steel. More precisely, **steel** is defined as an iron–carbon alloy containing from 0.02% to 2.11% carbon.[2] Of course, steels can also contain other alloying elements as well.

A eutectic composition at 4.3% carbon can be seen in the diagram. There is a similar feature in the solid region of the diagram at 0.77% carbon and 723°C (1333°F). This is called the **eutectoid composition.** Steels below this carbon level are known as **hypoeutectoid steels,** and above this carbon level, from 0.77% to 2.1%, they are called **hypereutectoid steels.**

In addition to the phases mentioned, one other phase is prominent in the iron–carbon alloy system. This is Fe_3C, also known as **cementite,** an intermediate phase. It is a metallic compound of iron and carbon that is hard and brittle. At room temperature under equilibrium conditions, iron–carbon alloys form a two-phase system at carbon levels even slightly above zero. The carbon content in steel ranges between these very low levels and about 2.1% C. Above 2.1% C, up to about 4% or 5%, the alloy is defined as **cast iron.**

6.2.2 IRON AND STEEL PRODUCTION

Coverage of iron and steel production begins with the iron ores and other raw materials required. Ironmaking is then discussed, in which iron is reduced from the ores, and

---

[2]This is the conventional definition of steel, but exceptions exist. A recently developed steel for sheet-metal forming, called **interstitial-free steel,** has a carbon content of only 0.005%. It is discussed in Section 6.2.3.

steelmaking, in which the iron is refined to obtain the desired purity and composition (alloying). The casting processes that are accomplished at the steel mill are then considered.

Iron Ores and Other Raw Materials   The principal ore used in the production of iron and steel is **hematite** (Fe_2O_3). Other iron ores include **magnetite** (Fe_3O_4), **siderite** ($FeCO_3$), and **limonite** (Fe_2O_3-xH_2O, in which x is typically around 1.5). Iron ores contain from 50% to around 70% iron, depending on grade (hematite is almost 70% iron). In addition, scrap iron and steel are widely used today as raw materials in iron- and steelmaking.

Other raw materials needed to reduce iron from the ores are coke and limestone. **Coke** is a high carbon fuel produced by heating bituminous coal in a limited oxygen atmosphere for several hours, followed by water spraying in special quenching towers. Coke serves two functions in the reduction process: (1) it is a fuel that supplies heat for the chemical reactions; and (2) it produces carbon monoxide (CO) to reduce the iron ore. **Limestone** is a rock containing high proportions of calcium carbonate ($CaCO_3$). The limestone is used in the process as a flux to react with and remove impurities in the molten iron as slag.

Ironmaking   To produce iron, a charge of ore, coke, and limestone are dropped into the top of a blast furnace. A **blast furnace** is a refractory-lined chamber with a diameter of about 9 to 11 m (30–35 ft) at its widest and a height of 40 m (125 ft), in which hot gases are forced into the lower part of the chamber at high rates to accomplish combustion and reduction of the iron. A typical blast furnace and some of its technical details are illustrated in Figures 6.5 and 6.6. The charge slowly descends from the top of the furnace toward the

FIGURE 6.5   Cross section of ironmaking blast furnace showing major components.

Iron ore, coke, and limestone

Gas to cleaning and reheating

Typical temperature profile

200°C (400°F)

800°C (1500°F)

1100°C (2000°F)

1400°C (2500°F)

1650°C (3000°F)

Direction of motion of charge material

Direction of motion of hot gases

Hot blast air

Slag

Molten pig iron

FIGURE 6.6   Schematic diagram indicating details of the blast furnace operation.

base and is heated to temperatures around 1650°C (3000°F). Burning of the coke is accomplished by the hot gases (CO, H_2, CO_2, H_2O, N_2, O_2, and fuels) as they pass upward through the layers of charge material. The carbon monoxide is supplied as hot gas, and it is also formed from combustion of coke. The CO gas has a reducing effect on the iron ore; the reaction (simplified) can be written as follows (using hematite as the starting ore)

$$Fe_2O_3 + CO \rightarrow 2FeO + CO_2 \qquad (6.3a)$$

Carbon dioxide reacts with coke to form more carbon monoxide

$$CO_2 + C(coke) \rightarrow 2CO \qquad (6.3b)$$

which then accomplishes the final reduction of FeO to iron

$$FeO + CO \rightarrow Fe + CO_2 \qquad (6.3c)$$

The molten iron drips downward, collecting at the base of the blast furnace. This is periodically tapped into hot iron ladle cars for transfer to subsequent steelmaking operations.

The role played by limestone can be summarized as follows. First the limestone is reduced to lime (CaO) by heating, as follows

$$CaCO_3 \rightarrow CaO + CO_2 \qquad (6.4)$$

The lime combines with impurities such as silica (SiO_2), sulfur (S), and alumina (Al_2O_3) in reactions that produce a molten slag that floats on top of the iron.

It is instructive to note that approximately 7 tons of raw materials are required to produce 1 ton of iron. The ingredients are proportioned about as follows: 2.0 tons of iron ore, 1.0 ton of coke, 0.5 ton of limestone, and (here's the amazing statistic) 3.5 tons of gases. A significant proportion of the byproducts are recycled.

The iron tapped from the base of the blast furnace (called ***pig iron***) contains more than 4% C, plus other impurities: 0.3–1.3% Si, 0.5–2.0% Mn, 0.1–1.0% P, and 0.02–0.08% S [11]. Further refinement of the metal is required for both cast iron and steel. A furnace called a ***cupola*** (Section 11.4.1) is commonly used for converting pig iron into gray cast iron. For steel, compositions must be more closely controlled and impurities brought to much lower levels.

FIGURE 6.7 Basic oxygen furnace showing BOF vessel during processing of a heat.

Steelmaking   Since the mid-1800s, a number of processes have been developed for refining pig iron into steel. Today, the two most important processes are the basic oxygen furnace (BOF) and the electric furnace. Both are used to produce carbon and alloy steels.

The ***basic oxygen furnace*** accounts for about 70% of U.S. steel production. The BOF is an adaptation of the Bessemer converter. Whereas the Bessemer process used air blown up through the molten pig iron to burn off impurities, the basic oxygen process uses pure oxygen. A diagram of the conventional BOF during the middle of a heat is illustrated in Figure 6.7. The typical BOF vessel is about 5 m (16 ft) inside diameter and can process 150 to 200 tons in a heat.

The BOF steelmaking sequence is shown in Figure 6.8. Integrated steel mills transfer the molten pig iron from the blast furnace to the BOF in railway cars called hot-iron ladle cars. In modern practice, steel scrap is added to the pig iron, accounting for about 30% of a typical BOF charge. Lime (CaO) is also added. After charging, the lance is inserted into the vessel so that its tip is about 1.5 m (5 ft) above the surface of the molten iron. Pure O_2 is blown at high velocity through the lance, causing combustion and heating at the surface of the molten pool. Carbon dissolved in the iron and other impurities such as silicon, manganese, and phosphorus are oxidized. The reactions are

$$2C + O_2 \rightarrow 2CO \text{ (}CO_2 \text{ is also produced)} \tag{6.5a}$$

$$Si + O_2 \rightarrow SiO_2 \tag{6.5b}$$

$$2Mn + O_2 \rightarrow 2MnO \tag{6.5c}$$

$$4P + 5O_2 \rightarrow 2P_2O_5 \tag{6.5d}$$

The CO and CO_2 gases produced in the first reaction escape through the mouth of the BOF vessel and are collected by the fume hood; the products of the other three reactions are removed as slag, using the lime as a fluxing agent. The C content in the iron decreases almost linearly with time during the process, thus permitting fairly predictable control over carbon levels in the steel. After refining to the desired level, the molten steel is tapped; alloying ingredients and other additives are poured into the heat; then the slag is

FIGURE 6.8   BOF sequence during processing cycle: (1) charging of scrap and (2) pig iron; (3) blowing (Figure 6.7); (4) tapping the molten steel; and (5) pouring off the slag.

poured. A 200-ton heat of steel can be processed in about 20 min, although the entire cycle time (tap-to-tap time) takes about 45 min.

Recent advances in the technology of the basic oxygen process include the use of nozzles in the bottom of the vessel through which oxygen is injected into the molten iron. This allows better mixing than the conventional BOF lance, resulting in shorter processing times (a reduction of about 3 min), lower carbon contents, and higher yields.

The *electric arc furnace* accounts for about 30% of U.S. steel production. Although pig iron was originally used as the charge in this type of furnace, scrap iron and scrap steel are the primary raw materials today. Electric arc furnaces are available in several designs; the direct arc type shown in Figure 6.9 is currently the most economical type. These furnaces have removable roofs for charging from above; tapping is accomplished by tilting the entire furnace. Scrap iron and steel selected for their compositions, together with alloying ingredients and limestone (flux), are charged into the furnace and heated by an electric arc that flows between large electrodes and the charge metal. Complete melting requires about 2 hours; tap-to-tap time is 4 hours. Capacities of electric furnaces commonly range between 25 and 100 tons per heat. Electric arc furnaces are noted for better-quality steel but higher cost per ton, compared with the BOF. The electric arc furnace is generally associated with production of alloy steels, tool steels, and stainless steels.

Casting of Ingots   Steels produced by BOF or electric furnace are solidified for subsequent processing either as cast ingots or by continuous casting. Steel *ingots* are large discrete castings weighing from less than 1 ton up to around 300 tons (the weight of an entire heat). Ingot molds are made of high carbon iron and are tapered at the top or bottom for removal of the solid casting. A *big-end-down mold* is illustrated in Figure 6.10. The cross

FIGURE 6.9 Electric arc furnace for steelmaking.

section may be square, rectangular, or round, and the perimeter is usually corrugated to increase surface area for faster cooling. The mold is placed on a platform called a ***stool;*** after solidification the mold is lifted, leaving the casting on the stool.

The solidification process for ingots as well as other castings is described in the chapter on casting principles (Chapter 10). Because ingots are such large castings, the time required for solidification and the associated shrinkage are significant. Porosity caused by the reaction of carbon and oxygen to form CO during cooling and solidification is a problem that must be addressed in ingot casting. These gases are liberated from the molten steel because of their reduced solubility with decreasing temperature. Cast steels are often treated to limit or prevent CO gas evolution during solidification. The treatment involves adding elements such as Si and Al that react with the oxygen dissolved in the molten steel, so it is not available for CO reaction. The structure of the solid steel is thus free of pores and other defects caused by gas formation.

Continuous Casting   Continuous casting is widely applied in aluminum and copper production, but its most noteworthy application is in steelmaking. The process is replacing ingot casting because it dramatically increases productivity. Ingot casting is a discrete process. Because the molds are relatively large, solidification time is significant. For a large

FIGURE 6.10   A big-end-down ingot mold typical of type used in steelmaking.

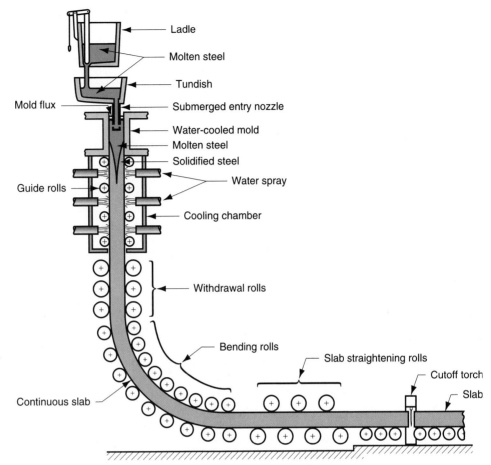

FIGURE 6.11
Continuous casting; steel is poured into tundish and distributed to a water-cooled continuous casting mold; it solidifies as it travels down through the mold. The slab thickness is exaggerated for clarity.

Labels in figure: Ladle, Molten steel, Tundish, Submerged entry nozzle, Mold flux, Water-cooled mold, Molten steel, Solidified steel, Water spray, Guide rolls, Cooling chamber, Withdrawal rolls, Bending rolls, Slab straightening rolls, Cutoff torch, Slab, Continuous slab

steel ingot, it may take 10 to 12 hours for the casting to solidify. The use of continuous casting reduces solidification time by an order of magnitude.

The continuous casting process, also called *strand casting,* is illustrated in Figure 6.11. Molten steel is poured from a ladle into a temporary container called a *tundish,* which dispenses the metal to one or more continuous casting molds. The steel begins to solidify at the outer regions as it travels down through the water-cooled mold. Water sprays accelerate the cooling process. While still hot and plastic, the metal is bent from vertical to horizontal orientation. It is then cut into sections or fed continuously into a rolling mill (Section 19.1) in which it is formed into plate or sheet stock or other cross sections.

6.2.3 STEELS

As defined earlier, *Steel* is an alloy of iron that contains carbon ranging by weight between 0.02% and 2.11% (most steels range between 0.05% and 1.1%C). It often includes other alloying ingredients, such as manganese, chromium, nickel, and/or molybdenum (see Table 6.2); but it is the carbon content that turns iron into steel. Hundreds of compositions of steel are available commercially. For purposes of organization here, the vast majority of commercially important steels can be grouped into the following categories: (1) plain carbon steels, (2) low alloy steels, (3) stainless steels, (4) tool steels, and (5) specialty steels.

Plain Carbon Steels   These steels contain carbon as the principal alloying element, with only small amounts of other elements (about 0.4% manganese plus lesser amounts of

TABLE 6.2   AISI-SAE designations of steels.

Code	Name of Steel	Nominal Chemical Analysis, %							
		Cr	Mn	Mo	Ni	V	P	S	Si
10XX	Plain carbon		0.4				0.04	0.05	
11XX	Resulfurized		0.9				0.01	0.12	0.01
12XX	Resulfurized, rephosphorized		0.9				0.10	0.22	0.01
13XX	Manganese		1.7				0.04	0.04	0.3
20XX	Nickel steels		0.5		0.6		0.04	0.04	0.2
31XX	Nickel–chrome	0.6			1.2		0.04	0.04	0.3
40XX	Molybdenum		0.8	0.25			0.04	0.04	0.2
41XX	Chrome–molybdenum	1.0	0.8	0.2			0.04	0.04	0.3
43XX	Ni–Cr–Mo	0.8	0.7	0.25	1.8		0.04	0.04	0.2
46XX	Nickel–molybdenum		0.6	0.25	1.8		0.04	0.04	0.3
47XX	Ni–Cr–Mo	0.4	0.6	0.2	1.0		0.04	0.04	0.3
48XX	Nickel–molybdenum		0.6	0.25	3.5		0.04	0.04	0.3
50XX	Chromium	0.5	0.4				0.04	0.04	0.3
52XX	Chromium	1.4	0.4				0.02	0.02	0.3
61XX	Cr–Vanadium	0.8	0.8			0.1	0.04	0.04	0.3
81XX	Ni–Cr–Mo	0.4	0.8	0.1	0.3		0.04	0.04	0.3
86XX	Ni–Cr–Mo	0.5	0.8	0.2	0.5		0.04	0.04	0.3
88XX	Ni–Cr–Mo	0.5	0.8	0.35	0.5		0.04	0.04	0.3
92XX	Silicon–Manganese		0.8				0.04	0.04	2.0
93XX	Ni–Cr–Mo	1.2	0.6	0.1	3.2		0.02	0.02	0.3
98XX	Ni–Cr–Mo	0.8	0.8	0.25	1.0		0.04	0.04	0.3

silicon, phosphorus, and sulfur). The strength of plain carbon steels increases with carbon content. A typical plot of the relationship is illustrated in Figure 6.12. As seen in the phase diagram for iron and carbon (Figure 6.4), steel at room temperature is a mixture of ferrite (α) and cementite (Fe_3C). The cementite particles distributed throughout the ferrite act as

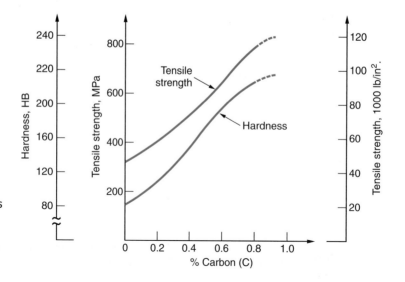

FIGURE 6.12 Tensile strength and hardness as a function of carbon content in plain carbon steel (hot-rolled, unheat-treated).

obstacles to the movement of dislocations during slip (Section 2.3.3); more carbon leads to more barriers, and more barriers mean stronger and harder steel.

According to a designation scheme developed by the American Iron and Steel Institute (AISI) and the Society of Automotive Engineers (SAE), plain carbon steels are specified by a four-digit number system: 10XX, in which 10 indicates that the steel is plain carbon, and XX indicates the percent of carbon in hundredths of percentage points. For example, 1020 steel contains 0.20% C. The plain carbon steels are typically classified into three groups according to their carbon content:

1. *Low carbon steels* contain less than 0.20% C and are by far the most widely used steels. Typical applications are automobile sheet-metal parts, plate steel for fabrication, and railroad rails. These steels are relatively easy to form, which accounts for their popularity where high strength is not required. Steel castings usually fall into this carbon range, also.

2. *Medium carbon steels* range in carbon between 0.20% and 0.50% and are specified for applications requiring higher strength than the low-C steels. Applications include machinery components and engine parts such as crankshafts and connecting rods.

3. *High carbon steels* contain carbon in amounts greater than 0.50%. They are specified for still higher strength applications and where stiffness and hardness are needed. Springs, cutting tools and blades, and wear-resistant parts are examples.

Increasing carbon content strengthens and hardens the steel, but its ductility is reduced. Also, high carbon steels can be heat treated to form martensite, making the steel very hard and strong (Section 27.2).

Low Alloy Steels   Low alloy steels are iron–carbon alloys that contain additional alloying elements in amounts totaling less than about 5% by weight. Owing to these additions, low alloy steels have mechanical properties that are superior to those of the plain carbon steels for given applications. Superior properties usually mean higher strength, hardness, hot hardness, wear resistance, toughness, and more desirable combinations of these properties. Heat treatment is often required to achieve these improved properties.

Common alloying elements added to steel are chromium, manganese, molybdenum, nickel, and vanadium, sometimes individually but usually in combinations. These elements typically form solid solutions with iron and metallic compounds with carbon (carbides), assuming sufficient carbon is present to support a reaction. The effects of the principal alloying ingredients can be summarized as follows:

➢ *Chromium* (Cr) improves strength, hardness, wear resistance, and hot hardness. It is one of the most effective alloying ingredients for increasing hardenability (Section 27.2.3). In significant proportions, Cr improves corrosion resistance.

➢ *Manganese* (Mn) improves the strength and hardness of steel. When the steel is heat treated, hardenability is improved with increased manganese. Because of these benefits, manganese is a widely used alloying ingredient in steel.

➢ *Molybdenum* (Mo) increases toughness and hot hardness. It also improves hardenability and forms carbides for wear resistance.

➢ *Nickel* (Ni) improves strength and toughness. It increases hardenability but not as much as some of the other alloying elements in steel. In significant amounts it improves corrosion resistance and is the other major ingredient (besides chromium) in certain types of stainless steel.

TABLE 6.3   Treatments and mechanical properties of selected steels.

Code	Treatment[a]	Tensile Strength		Elongation, %
		MPa	lb/in^2	
1010	HR	304	44,000	47
1010	CD	366	53,000	12
1020	HR	380	55,000	28
1020	CD	421	61,000	15
1040	HR	517	75,000	20
1040	CD	587	85,000	10
1055	HT	897	130,000	16
1315	None	545	79,000	34
2030	None	566	82,000	32
3130	HT	697	101,000	28
4130	HT	890	129,000	17
4140	HT	918	133,000	16
4340	HT	1279	185,000	12
4815	HT	635	92,000	27
9260	HT	994	144,000	18
HSLA	None	586	85,000	20

Compiled from [6], [11], and other sources.
[a]HR = hot-rolled; CD = cold-drawn; HT = heat treatment involving heating and quenching, followed by tempering to produce tempered martensite (Section 27.2).

➤ *Vanadium* (V) inhibits grain growth during elevated temperature processing and heat treatment, which enhances strength and toughness of steel. It also forms carbides that increase wear resistance.

The AISI-SAE designations of many of the low alloy steels are presented in Table 6.2, which indicates nominal chemical analysis. As before, carbon content is specified by XX in 1/100% of carbon. For completeness, plain carbon steels (10XX) have been included. To obtain an idea of the properties possessed by some of these steels, Table 6.3 was compiled, which lists the treatment to which the steel is subjected for strengthening and its strength and ductility.

Low alloy steels are not easily welded, especially at medium and high carbon levels. Since the 1960s, research has been directed at developing low carbon, low alloy steels that have better strength-to-weight ratios than plain carbon steels but are more weldable than low alloy steels. The products developed out of these efforts are called *high-strength low-alloy* (HSLA) steels. They generally have low carbon contents (in the range 0.10%–0.30% C) plus relatively small amounts of alloying ingredients (usually only about 3% total of elements such as Mn, Cu, Ni, and Cr). HSLA steels are hot-rolled under controlled conditions designed to provide improved strength compared with plain C steels, yet with no sacrifice in formability or weldability. Strengthening is by solid solution alloying; heat treatment is not feasible because of low carbon content. Table 6.3 lists one HSLA steel, together with properties (chemistry is: 0.12 C, 0.60 Mn, 1.1 Ni, 1.1 Cr, 0.35 Mo, and 0.4 Si).

Stainless Steels   Stainless steels are a group of highly alloyed steels designed to provide high corrosion resistance. The principal alloying element in stainless steel is chromium, usually above 15%. The chromium in the alloy forms a thin, impervious oxide film in an oxidizing atmosphere, which protects the surface from corrosion. Nickel is another alloying ingredient used in certain stainless steels to increase corrosion protection. Carbon is used to strengthen and harden the metal; however, increasing the carbon content has the effect of reducing corrosion protection because chromium carbide forms to reduce the amount of free Cr available in the alloy.

In addition to corrosion resistance, stainless steels are noted for their combination of strength and ductility. Although these properties are desirable in many applications, they generally make these alloys difficult to work in manufacturing. Also, stainless steels are significantly more expensive than plain C or low alloy steels.

Stainless steels are traditionally divided into three groups, named for the predominant phase present in the alloy at ambient temperature.

1. *Austenitic stainless* have a typical composition of around 18% Cr and 8% Ni and are the most corrosion resistant of the three groups. Owing to this composition, they are sometimes identified as 18-8 stainless. They are nonmagnetic and very ductile; but they show significant work hardening. The nickel has the effect of enlarging the austenite region in the iron–carbon phase diagram, making it stable at room temperature. Austenitic stainless steels are used to fabricate chemical and food processing equipment, as well as machinery parts requiring high corrosion resistance.

2. *Ferritic stainless* have around 15% to 20% chromium, low carbon, and no nickel. This provides a ferrite phase at room temperature. Ferritic stainless steels are magnetic and are less ductile and corrosion resistant than the austenitics. Parts made of ferritic stainless range from kitchen utensils to jet engine components.

3. *Martensitic stainless* have a higher carbon content than ferritic stainlesses, thus permitting them to be strengthened by heat treatment (Section 27.2). They have as much as 18% Cr but no Ni. They are strong, hard, and fatigue resistant, but not generally as corrosion resistant as the other two groups. Typical products include cutlery and surgical instruments.

Most stainless steels are designated by a three-digit AISI numbering scheme. The first digit indicates the general type, and the last two digits give the specific grade within the type. Table 6.4 lists the common stainless steels with typical compositions and mechanical properties. The traditional stainless steels were developed in the early 1900s. Since then, several additional high alloy steels have been developed that have good corrosion resistance and other desirable properties. These are also classified as stainless steels. Continuing the list:

4. *Precipitation hardening stainless,* which have a typical composition of 17% Cr and 7% Ni, with additional small amounts of alloying elements such as aluminum, copper, titanium, and molybdenum. Their distinguishing feature among stainlesses is that they can be strengthened by precipitation hardening (Section 27.3). Strength and corrosion resistance are maintained at elevated temperatures, which suits these alloys to aerospace applications.

5. *Duplex stainless* possess a structure that is a mixture of austenite and ferrite in roughly equal amounts. Their corrosion resistance is similar to the austenitic grades, and they show improved resistance to stress-corrosion cracking. Applications include heat exchangers, pumps, and wastewater treatment plants.

Tool Steels   Tool steels are a class of (usually) highly alloyed steels designed for use as industrial cutting tools, dies, and molds. To perform in these applications, they must possess high strength, hardness, hot hardness, wear resistance, and toughness under impact. To obtain these properties, tool steels are heat treated. Principal reasons for the high levels of alloying elements are (1) improved hardenability, (2) reduced distortion during heat

TABLE 6.4 Compositions and mechanical properties of selected stainless steels.

Type	Chemical Analysis, %						Tensile Strength		Elongation, %
	Fe	Cr	Ni	C	Mn	Other[a]	MPa	lb/in^2	
Austenitic									
301	73	17	7	0.15	2		620	90,000	40
302	71	18	8	0.15	2		515	75,000	40
304	69	19	9	0.08	2		515	75,000	40
309	61	23	13	0.20	2		515	75,000	40
316	65	17	12	0.08	2	2.5 Mo	515	75,000	40
Ferritic									
405	85	13	—	0.08	1		415	60,000	20
430	81	17	—	0.12	1		415	60,000	20
Martensitic									
403	86	12	—	0.15	1		485	70,000	20
403[b]	86	12	—	0.15	1		825	120,000	12
416	85	13	—	0.15	1		485	70,000	20
416[b]	85	13	—	0.15	1		965	140,000	10
440	81	17	—	0.65	1		725	105,000	20
440[b]	81	17	—	0.65	1		1790	260,000	5

Compiled from [11].
[a]All of the grades in the table contain about 1% (or less) Si plus small amounts (well below 1%) of phosphorus, sulfur, and other elements such as aluminum.
[b]Heat treated.

treatment, (3) hot hardness, (4) formation of hard metallic carbides for abrasion resistance, and (5) enhanced toughness.

The tool steels divide into major types, according to application and composition. The AISI uses a classification scheme that includes a prefix letter to identify the tool steel. In the following list of tool steel types, the prefix and some typical compositions are presented in Table 6.5:

TABLE 6.5 Tool steels by AISI prefix identification, with examples of composition and typical hardness values.

AISI	Example	Chemical Analysis, %[a]							Hardness, HRC
		C	Cr	Mn	Mo	Ni	V	W	
T	T1	0.7	4.0				1.0	18.0	65
M	M2	0.8	4.0		5.0		2.0	6.0	65
H	H11	0.4	5.0		1.5		0.4		55
D	D1	1.0	12.0		1.0				60
A	A2	1.0	5.0		1.0				60
O	O1	0.9	0.5	1.0				0.5	61
W	W1	1.0							63
S	S1	0.5	1.5					2.5	50
P	P20	0.4	1.7		0.4				40[b]
L	L6	0.7	0.8		0.2	1.5			45[b]

[a]Percent composition rounded to nearest tenth.
[b]Hardness estimated.

T, M   ***High-speed tool steels*** are used as cutting tools in machining processes (Section 23.2.1). They are formulated for high wear resistance and hot hardness. The original high-speed steels (HSS) were developed around 1900. They permitted dramatic increases in cutting speed compared to previously used tools; hence their name. The two AISI designations indicate the principal alloying element: T for tungsten and M for molybdenum.

H   ***Hot-working tool steels*** are intended for hot-working dies in forging, extrusion, and die-casting.

D   ***Cold-work tool steels*** are die steels used for cold working operations such as sheetmetal pressworking, cold extrusion, and certain forging operations. The designation D stands for die. Closely related AISI designations are A and O. A and O stand for air- and oil-hardening. They all provide good wear resistance and low distortion.

W   ***Water-hardening tool steels*** have high carbon with little or no other alloying elements. They can only be hardened by fast quenching in water. They are widely used because of low cost, but they are limited to low temperature applications. Cold heading dies are a typical application.

S   ***Shock-resistant tool steels*** are intended for use in applications where high toughness is required, as in many sheetmetal shearing, punching, and bending operations.

P   ***Mold steels*** are used to make molds for molding plastics and rubber.

L   ***Low-alloy tool steels*** are generally reserved for special applications.

Tool steels are not the only tool materials. Plain carbon, low alloy, and stainless steels are used for many tool and die applications. Cast irons and certain nonferrous alloys are also suitable for certain tooling applications. In addition, several ceramic materials (e.g., Al_2O_3) are used as high-speed cutting inserts, abrasives, and other tools.

Specialty Steels   To complete this survey, several specialty steels are mentioned that are not included in the previous coverage. One of the reasons why these steels are special is that they possess unique processing characteristics.

Maraging steels are low carbon alloys containing high amounts of nickel (15% to 25%) and lesser proportions of cobalt, molybdenum, and titanium. Chromium is also sometimes added for corrosion resistance. Maraging steels are strengthened by precipitation hardening (Section 27.3), but in the unhardened condition, they are quite processable by forming and/or machining. They can also be readily welded. Heat treatment results in very high strength together with good toughness. Tensile strengths of 2000 MPa (290,000 lb/in^2) and 10% elongation are not unusual. Applications include parts for missiles, machinery, dies, and other situations where these properties are required and justify the high cost of the alloy.

Free-machining steels are carbon steels formulated to improve machinability (Section 24.1). Alloying elements include sulfur, lead, tin, bismuth, selenium, tellurium, and/or phosphorus. Lead is less-frequently used today because of environmental and health concerns. Added in small amounts, these elements act to lubricate the cutting operation, reduce friction, and break up chips for easier disposal. Although more expensive than non-free-machining steels, they often pay for themselves in higher production rates and longer tool lives.

Because of their good ductility, low-carbon sheet steels are widely used in sheet-metal forming operations. Further improvements in formability have been achieved using a new class of sheet steel product called ***interstitial-free steels.*** These steels have extremely low carbon levels (0.005% C), which result from the use of alloying elements such as niobium and titanium that combine with C and leave the steel virtually free of interstitial atoms. The result

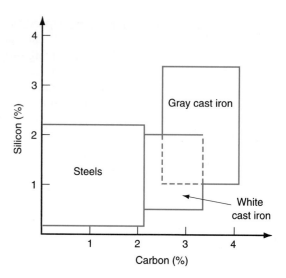

FIGURE 6.13 Carbon and silicon compositions for cast irons, with comparison to steels (most steels have relatively low silicon contents—cast steels have the higher Si content). Ductile iron is formed by special melting and pouring treatment of gray cast iron, and malleable iron is formed by heat treatment of white cast iron.

is excellent ductility, even greater than low-C steels. Applications include deep-drawing operations in the automotive industry.

6.2.4 CAST IRONS

Cast iron is an iron alloy containing from 2.1% to about 4% carbon and from 1% to 3% silicon. Its composition makes it highly suitable as a casting metal. In fact, the tonnage of cast iron castings is several times that of all other cast metal parts combined (excluding cast ingots made during steelmaking, which are subsequently rolled into bars, plates, and similar stock). The overall tonnage of cast iron is second only to steel among metals.

There are several types of cast iron, the most important being gray cast iron. Other types include ductile iron, white cast iron, malleable iron, and various alloy cast irons. Typical chemical compositions of gray and white cast irons are shown in Figure 6.13, indicating their relationship with cast steel. Ductile and malleable irons possess chemistries similar to the gray and white cast irons, respectively, but result from special treatments to be described in the following. Table 6.6 presents a listing of chemistries for the principal types together with mechanical properties.

Gray Cast Iron   Gray cast iron accounts for the largest tonnage among the cast irons. It has a composition in the range 2.5% to 4% carbon and 1% to 3% silicon. This chemistry results in the formation of graphite (carbon) flakes distributed throughout the cast product upon solidification. The structure causes the surface of the metal to have a gray color when fractured; hence the name gray cast iron. The dispersion of graphite flakes accounts for two attractive properties: (1) good vibration damping, which is desirable in engines and other machinery; and (2) internal lubricating qualities, which makes the cast metal machinable.

The strength of gray cast iron spans a significant range. The American Society for Testing of Materials (ASTM) uses a classification method for gray cast iron that is intended to provide a minimum tensile strength (*TS*) specification for the various classes: Class 20 gray cast iron has a *TS* of 20,000 lb/in^2, Class 30 has a *TS* of 30,000 lb/in^2, and so forth, up to around 70,000 lb/in^2 (see Table 6.6 for equivalent *TS* in metric units). The compressive strength of gray cast iron is significantly greater than its tensile strength. Properties of the casting can be controlled to some extent by heat treatment. Ductility of gray cast iron is very low; it is a relatively brittle material. Products made from gray cast iron include automotive engine blocks and heads, motor housings, and machine tool bases.

TABLE 6.6   Compositions and mechanical properties of selected cast irons.

| Type | Typical Composition, % | | | | | Tensile Strength | | |
	Fe	C	Si	Mn	Other[a]	MPa	lb/in²	Elongation, %
Gray cast irons								
ASTM Class 20	93.0	3.5	2.5	0.65		138	20,000	0.6
ASTM Class 30	93.6	3.2	2.1	0.75		207	30,000	0.6
ASTM Class 40	93.8	3.1	1.9	0.85		276	40,000	0.6
ASTM Class 50	93.5	3.0	1.6	1.0	0.67 Mo	345	50,000	0.6
Ductile irons								
ASTM A395	94.4	3.0	2.5		414	60,000	18	
ASTM A476	93.8	3.0	3.0		552	80,000	3	
White cast iron								
Low-C	92.5	2.5	1.3	0.4	1.5Ni, 1Cr, 0.5Mo	276	40,000	0
Malleable irons								
Ferritic	95.3	2.6	1.4	0.4		345	50,000	10
Pearlitic	95.1	2.4	1.4	0.8		414	60,000	10

Compiled from [11]. Cast irons are identified by various systems. This table attempts to indicate the particular cast iron grade using the most common identification for each type.

[a]Cast irons also contain phosphorus and sulfur usually totaling less than 0.3%.

Ductile Iron   This is an iron with the composition of gray iron in which the molten metal is chemically treated before pouring to cause the formation of graphite spheroids rather than flakes. This results in a stronger and more ductile iron, hence its name. Applications include machinery components requiring high strength and good wear resistance.

White Cast Iron   This cast iron has less carbon and silicon than gray cast iron. It is formed by more rapid cooling of the molten metal after pouring, thus causing the carbon to remain chemically combined with iron in the form of cementite (Fe_3C), rather than precipitating out of solution in the form of flakes. When fractured, the surface has a white crystalline appearance that gives the iron its name. Owing to the cementite, white cast iron is hard and brittle, and its wear resistance is excellent. Strength is good, with *TS* of 276 MPa (40,000 lb/in²) being typical. These properties make white cast iron suitable for applications in which wear resistance is required. Railway brake shoes are an example.

Malleable Iron   When castings of white cast iron are heat treated to separate the carbon out of solution and form graphite aggregates, the resulting metal is called malleable iron. The new microstructure can possess substantial ductility (up to 20% elongation)—a significant difference from the metal out of which it was transformed. Typical products made of malleable cast iron include pipe fittings and flanges, certain machine components, and railroad equipment parts.

Alloy Cast Irons   Cast irons can be alloyed for special properties and applications. These alloy cast irons are classified as follows: (1) heat-treatable types that can be hardened by martensite formation; (2) corrosion-resistant types, whose alloying elements include nickel and chromium; and (3) heat-resistant types containing high proportions of nickel for hot hardness and resistance to high temperature oxidation.

6.3   NONFERROUS METALS

The nonferrous metals include metal elements and alloys not based on iron. The most important engineering metals in the nonferrous group are aluminum, copper, magnesium, nickel, titanium, and zinc, and their alloys.

Although the nonferrous metals as a group cannot match the strength of the steels, certain nonferrous alloys have corrosion resistance and/or strength-to-weight ratios that make them competitive with steels in moderate-to-high stress applications. In addition, many of the nonferrous metals have properties other than mechanical that make them ideal for applications in which steel would be quite unsuitable. For example, copper has one of the lowest electrical resistivities among metals and is widely used for electrical wire. Aluminum is an excellent thermal conductor, and its applications include heat exchangers and cooking pans. It is also one of the most readily formed metals, and is valued for that reason also. Zinc has a relatively low melting point, so zinc is widely used in die casting operations. The common nonferrous metals have their own combination of properties that make them attractive in a variety of applications. The following nine sections discuss the nonferrous metals that are the most commercially and technologically important.

6.3.1   ALUMINUM AND ITS ALLOYS

Aluminum and magnesium are light metals, and they are often specified in engineering applications for this feature. Both elements are abundant on Earth, aluminum on land and magnesium in the sea, although neither is easily extracted from their natural states.

Properties and other data on aluminum are listed in Table 6.1(b). Among the major metals, it is a relative newcomer, dating only to the late 1800s (Historical Note 6.2). The coverage in this section includes (1) a brief description of how aluminum is produced and (2) a discussion of the properties and the designation system for the metal and its alloys.

Aluminum Production   The principal aluminum ore is **bauxite**, which consists largely of hydrated aluminum oxide (Al_2O_3-H_2O) and other oxides. Extraction of the aluminum from bauxite can be summarized in three steps: (1) washing and crushing the ore into fine powders; (2) the Bayer process, in which the bauxite is converted to pure alumina (Al_2O_3); and (3) electrolysis, in which the alumina is separated into aluminum and

---

TABLE 6.1   (continued): (b) Aluminum.

Symbol: Al	Principal ore: Bauxite (impure mix of Al_2O_3 and $Al(OH)_3$)
Atomic number: 13	
Specific gravity: 2.7	Alloying elements: Copper, magnesium, manganese, silicon, and zinc
Crystal structure: FCC	
Melting temperature: 660°C (1220°F)	Typical applications: Containers (aluminum cans), wrapping foil, electrical conductors, pots and pans, parts for construction, aerospace, automotive, and other uses in which light weight is important
Elastic modulus: 69,000 MPa (10×10^6 lb/in^2)	

Historical Note 6.2   *Aluminum*

In 1807, the English chemist Humphrey Davy, believing that the mineral *alumina* (Al_2O_3) had a metallic base, attempted to extract the metal. He did not succeed, but was sufficiently convinced that he proceeded to name the metal anyway: *alumium,* later changing the name to *aluminum.* In 1825, the Danish physicist/chemist Hans Orsted finally succeeded in separating the metal. He noted that it "resembles tin." In 1845, the German physicist Friedrich Wohler was the first to determine the specific gravity, ductility, and various other properties of aluminum.

The modern electrolytic process for producing aluminum was based on the concurrent but independent work of Charles Hall in the United States and Paul Heroult in France around 1886. In 1888, Hall and a group of businessmen started the Pittsburgh Reduction Co. The first ingot of aluminum was produced by the electrolytic smelting process that same year. Demand for aluminum grew. The need for large amounts of electricity in the production process led the company to relocate in Niagara Falls in 1895, where hydroelectric power was becoming available at very low cost. In 1907, the company changed its name to the Aluminum Company of America (Alcoa). It was the sole producer of aluminum in the United States until World War II.

---

oxygen gas (O_2). The ***Bayer process,*** named after the German chemist who developed it, involves solution of bauxite powders in aqueous caustic soda (NaOH) under pressure, followed by precipitation of pure Al_2O_3 from solution. Alumina is commercially important in its own right as an engineering ceramic (Chapter 7).

Electrolysis to separate Al_2O_3 into its constituent elements requires dissolving the precipitate in a molten bath of cryolite (Na_3AlF_6) and subjecting the solution to direct current between the plates of an electrolytic furnace. The electrolyte dissociates to form aluminum at the cathode and oxygen gas at the anode.

Properties and Designation Scheme   Aluminum has high electrical and thermal conductivity, and its resistance to corrosion is excellent because of the formation of a hard, thin oxide surface film. It is a very ductile metal and is noted for its formability. Pure aluminum is relatively low in strength, but it can be alloyed and heat treated to compete with some steels, especially when weight is an important consideration.

The designation system for aluminum alloys is a four-digit code number. The system has two parts, one for wrought aluminums and the other for cast aluminums. The difference is that a decimal point is used after the third digit for cast aluminums. The designations are presented in Table 6.7(a).

TABLE 6.7(a)   Designations of wrought and cast aluminum alloys.

Alloy Group	Wrought Code	Cast Code
Aluminum, 99.0% or higher purity	1XXX	1XX.X
Aluminum alloys, by major element(s):		
Copper	2XXX	2XX.X
Manganese	3XXX	
Silicon + copper and/or magnesium		3XX.X
Silicon	4XXX	4XX.X
Magnesium	5XXX	5XX.X
Magnesium and silicon	6XXX	
Zinc	7XXX	7XX.X
Tin		8XX.X
Other	8XXX	9XX.X

TABLE 6.7(b)	Temper designations for aluminum alloys.
Temper	**Description**
F	As fabricated—no special treatment.
H	Strain hardened (wrought aluminums). H is followed by two digits, the first indicating a heat treatment, if any; and the second indicating the degree of work hardening remaining; for example: H1X No heat treatment after strain hardening, and X = 1 to 9, indicating degree of work hardening. H2X Partially annealed, and X = degree of work hardening remaining in product. H3X Stabilized, and X = degree of work hardening remaining. ***Stabilized*** means heating to slightly above service temperature anticipated.
O	Annealed to relieve strain hardening and improve ductility; reduces strength to lowest level.
T	Thermal treatment to produce stable tempers other than F, H, or O. It is followed by a digit to indicate specific treatments; for example: T1 = cooled from elevated temperature, naturally aged. T2 = cooled from elevated temperature, cold worked, naturally aged. T3 = solution heat treated, cold worked, naturally aged. T4 = solution heat treated and naturally aged. T5 = cooled from elevated temperature, artificially aged. T6 = solution heat treated and artificially aged. T7 = solution heat treated and overaged or stabilized. T8 = solution heat treated, cold worked, artificially aged. T9 = solution heat treated, artificially aged, and cold worked. T10 = cooled from elevated temperature, cold worked, and artificially aged.
W	Solution heat treatment, applied to alloys that age harden in service; it is an unstable temper.

Because properties of aluminum alloys are so influenced by work hardening and heat treatment, the temper (strengthening treatment, if any) must be designated in addition to the composition code. The principal temper designations are presented in Table 6.7(b). This designation is attached to the preceding four-digit number, separated from it by a hyphen, to indicate the treatment or absence thereof; for example, 2024-T3. Of course, temper treatments that specify strain hardening do not apply to the cast alloys. Some examples of the remarkable differences in the mechanical properties of aluminum alloys that result from the different treatments are presented in Table 6.8.

6.3.2 MAGNESIUM AND ITS ALLOYS

Magnesium (Mg) is the lightest of the structural metals. Its specific gravity and other basic data are presented in Table 6.1(c). Magnesium and its alloys are available in both wrought and cast forms. It is relatively easy to machine. However, in all processing of magnesium, small

TABLE 6.1	(continued): (c) Magnesium.		
Symbol:	Mg	Extracted from:	$MgCl_2$ in sea water by electrolysis
Atomic number:	12	Alloying elements:	See Table 6.9
Specific gravity:	1.74	Typical applications:	Aerospace, missiles, bicycles, chain
Crystal structure:	HCP		saw housings, luggage, and other
Melting temperature:	650°C (1202°F)		applications in which light weight is
Elastic modulus:	48,000 MPa (7×10^6 lb/in^2)		a primary requirement

TABLE 6.8   Compositions and mechanical properties of selected aluminum alloys.

Code	Typical Composition, %[a]						Temper	Tensile Strength		Elongation
	Al	Cu	Fe	Mg	Mn	Si		MPa	lb/in^2	
1050	99.5		0.4			0.3	O	76	11,000	39
							H18	159	23,000	7
1100	99.0		0.6			0.3	O	90	13,000	40
							H18	165	24,000	10
2024	93.5	4.4	0.5	1.5	0.6	0.5	O	185	27,000	20
							T3	485	70,000	18
3004	96.5	0.3	0.7	1.0	1.2	0.3	O	180	26,000	22
							H36	260	38,000	7
4043	93.5	0.3	0.8			5.2	O	130	19,000	25
							H18	285	41,000	1
5050	96.9	0.2	0.7	1.4	0.1	0.4	O	125	18,000	18
							H38	200	29,000	3
6063	98.5		0.3	0.7		0.4	O	90	13,000	25
							T4	172	25,000	20

Compiled from [12].
[a]In addition to elements listed, alloy may contain trace amounts of other elements such as copper, magnesium, manganese, vanadium, and zinc.

particles of the metal (such as small metal cutting chips) oxidize rapidly, and care must be taken to avoid fire hazards.

Magnesium Production   Sea water contains about 0.13% $MgCl_2$, and this is the source of most commercially produced magnesium. To extract Mg, a batch of sea water is mixed with milk of lime–calcium hydroxide $(Ca(OH)_2)$. The resulting reaction precipitates magnesium hydroxide $(Mg(OH)_2)$ that settles and is removed as a slurry. The slurry is then filtered to increase $Mg(OH)_2$ content and then mixed with hydrochloric acid (HCl), which reacts with the hydroxide to form concentrated $MgCl_2$—much more concentrated than the original sea water. Electrolysis is used to decompose the salt into magnesium (Mg) and chlorine gas (Cl_2). The magnesium is then cast into ingots for subsequent processing. The chlorine is recycled to form more $MgCl_2$.

Properties and Designation Scheme   As a pure metal, magnesium is relatively soft and lacks sufficient strength for most engineering applications. However, it can be alloyed and heat treated to achieve strengths comparable to aluminum alloys. In particular, its strength-to-weight ratio is an advantage in aircraft and missile components.

The designation scheme for magnesium alloys uses a three-to-five character alphanumeric code. The first two characters are letters that identify the principal alloying elements (up to two elements can be specified in the code, in order of decreasing percentages, or alphabetically if equal percentages). These code letters are listed in Table 6.9. The letters are followed by a two-digit number that indicates, respectively, the amounts of the two alloying ingredients to the nearest percent. Finally, the last symbol is a letter that indicates some variation in composition, or simply the chronological order in which it was standardized for commercial availability. Magnesium alloys also require specification of a temper, and the same basic scheme presented in Table 6.7(b) for aluminum is used for magnesium alloys.

Some examples of magnesium alloys, illustrating the designation scheme and indicating tensile strength and ductility of these alloys, are presented in Table 6.10.

TABLE 6.9 Code letters used to identify alloying elements in magnesium alloys.

| A | Aluminum (Al) | H | Thorium (Th) | M | Manganese (Mn) | Q | Silver (Ag) | T | Tin (Sn) |
| E | Rate earth metals | K | Zirconium (Zr) | P | Lead (Pb) | S | Silicon (Si) | Z | Zinc (Zn) |

TABLE 6.10 Compositions and mechanical properties of selected magnesium alloys.

| Code | Typical Composition, % | | | | | | Process | Tensile Strength | | Elongation |
	Mg	Al	Mn	Si	Zn	Other		MPa	lb/in^2	
AZ10A	98.0	1.3	0.2	0.1	0.4		Wrought	240	35,000	10
AZ80A	91.0	8.5			0.5		Forged	330	48,000	11
HM31A	95.8		1.2			3.0 Th	Wrought	283	41,000	10
ZK21A	97.1				2.3	6 Zr	Wrought	260	38,000	4
AM60	92.8	6.0	0.1	0.5	0.2	0.3 Cu	Cast	220	32,000	6
AZ63A	91.0	6.0			3.0		Cast	200	29,000	6

Compiled from [12].

6.3.3 COPPER AND ITS ALLOYS

Copper (Cu) is one of the oldest metals known (Historical Note 6.3). Basic data on the element copper are presented in Table 6.1(d).

Copper Production   In ancient times, copper was available in nature as a free element. Today these natural deposits are more difficult to find, and copper is now extracted from ores that are mostly sulfides, such as *chalcopyrite* ($CuFeS_2$). The ore is crushed (Section 17.1.1), concentrated by flotation, and then *smelted* (melted or fused, often with an associated chemical reaction to separate a metal from its ore). The resulting copper is called *blister copper,* which is between 98% and 99% pure. Electrolysis is used to obtain higher purity levels suitable for commercial use.

Properties and Designation Scheme   Pure copper has a distinctive reddish-pink color, but its most distinguishing engineering property is its low electrical resistivity—one of the lowest

---

Historical Note 6.3   *Copper*

Copper was one of the first metals used by human cultures (gold was the other). Discovery of the metal was probably around 6000 BCE. At that time, copper was found in the free metallic state. Ancient peoples fashioned implements and weapons out of it by hitting the metal (cold forging). Pounding copper made it harder (strain hardening); this and its attractive reddish color made it valuable in early civilizations.

Around 4000 BCE, it was discovered that copper could be melted and cast into useful shapes. It was later found that copper mixed with tin could be more readily cast and worked than the pure metal. This led to the widespread use of bronze and the subsequent naming of the Bronze Age, dated from about 2000 BCE to the time of Christ.

To the ancient Romans, the island of Cyprus was almost the only source of copper. They called the metal *aes cyprium* (ore of Cyprus). This was shortened to *Cyprium* and subsequently renamed *Cuprium.* From this derives the chemical symbol Cu.

TABLE 6.1   (continued): (d) Copper.

Symbol: Cu	Ore extracted from: Several: e.g., chalcopyrite ($CuFeS_2$).
Atomic number: 29	Alloying elements: Tin (bronze), zinc (brass),
Specific gravity: 8.96	aluminum, silicon, nickel, and
Crystal structure: FCC	beryllium.
Melting temperature: 1083°C (1981°F)	Typical applications: Electrical conductors and
Elastic modulus: 110,000 MPa (16×10^6 lb/in^2)	components, ammunition (brass), pots and pans, jewelry, plumbing, marine applications, heat exchangers, springs (Be-Cu).

of all elements. Because of this property, and its relative abundance in nature, commercially pure copper is widely used as an electrical conductor. (Note that the conductivity of copper decreases significantly as alloying elements are added.) Cu is also an excellent thermal conductor. Copper is one of the noble metals (gold and silver are also noble metals), so it is corrosion resistant. All of these properties combine to make copper one of the most important metals.

On the downside, the strength and hardness of copper are relatively low, especially when weight is taken into account. Accordingly, to improve strength (as well as for other reasons), copper is frequently alloyed. ***Bronze*** is an alloy of copper and tin (typically about 90% Cu and 10% Sn), still widely used today despite its ancient ancestry. Additional bronze alloys have been developed, based on other elements than tin; these include aluminum bronzes, and silicon bronzes. ***Brass*** is another familiar copper alloy, composed of copper and zinc (typically around 65% Cu and 35% Zn). The highest strength alloy of copper is beryllium-copper (only about 2% Be). It can be heat treated to tensile strengths of 1035 MPa (150,000 lb/in^2). Be-Cu alloys are used for springs.

The designation of copper alloys is based on the Unified Numbering System for Metals and Alloys (UNS), which uses a five-digit number preceded by the letter C (C for copper). The alloys are processed in wrought and cast forms, and the designation system includes both. Some copper alloys with compositions and mechanical properties are presented in Table 6.11.

6.3.4 NICKEL AND ITS ALLOYS

Nickel (Ni) is similar to iron in many respects. It is magnetic, and its modulus of elasticity is virtually the same as that of iron and steel. However, it is much more corrosion resistant, and the high temperature properties of its alloys are generally superior. Because of its corrosion-resistant characteristics, it is widely used as an alloying element in steel, such as stainless steel, and as a plating metal on other metals such as plain carbon steel.

TABLE 6.1   (continued): (e) Nickel.

Symbol: Ni	Ore extracted from: Pentlandite (($Fe, Ni)_9S_8$)
Atomic number: 28	Alloying elements: Copper, chromium, iron, aluminum.
Specific gravity: 8.90	Typical applications: Stainless steel alloying ingredient,
Crystal structure: FCC	plating metal for steel, applications
Melting temperature: 1453°C (2647°F)	requiring high temperature and
Elastic Modulus: 209,000 MPa (30×10^6 lb/in^2)	corrosion resistance.

TABLE 6.11   Compositions and mechanical properties of selected copper alloys.

Code	Typical Composition, %					Tensile Strength		Elongation, %
	Cu	Be	Ni	Sn	Zn	MPa	lb/in^2	
C10100	99.99					235	34,000	45
C11000	99.95					220	32,000	45
C17000	98.0	1.7	a			500	70,000	45
C24000	80.0				20.0	290	42,000	52
C26000	70.0				30.0	300	44,000	68
C52100	92.0			8.0		380	55,000	70
C71500	70.0		30.0			380	55,000	45
C71500[b]	70.0		30.0			580	84,000	3

Compiled from [12].
[a]Small amounts of Ni and Fe + 0.3 Co.
[b]Heat treated for high strength.

Nickel Production   The most important ore of nickel is ***pentlandite*** ((Ni, Fe)$_9$S$_8$). To extract the nickel, the ore is first crushed and ground with water. Flotation techniques are used to separate the sulfides from other minerals mixed with the ore. The nickel sulfide is then heated to burn off some of the sulfur, followed by smelting to remove iron and silicon. Further refinement is accomplished in a Bessemer-style converter to yield high-concentration nickel sulfide (NiS). Electrolysis is then used to recover high-purity nickel from the compound. Ores of nickel are sometimes mixed with copper ores, and the recovery technique described here also yields copper in these cases.

Nickel Alloys   Alloys of nickel are commercially important in their own right and are noted for corrosion resistance and high temperature performance. Composition, tensile strength, and ductility of some of the nickel alloys are given in Table 6.12. In addition, a number of superalloys are based on nickel (Section 6.4).

6.3.5   TITANIUM AND ITS ALLOYS

Titanium (Ti) is fairly abundant in nature, constituting about 1% of Earth's crust (aluminum, the most abundant, is about 8%). The density of Ti is between aluminum and iron; these and other data are presented in Table 6.1(f). Its importance has grown in recent decades due to

TABLE 6.12   Compositions and mechanical properties of selected nickel alloys.

Code	Typical Composition, %							Tensile Strength		Elongation, %
	Ni	Cr	Cu	Fe	Mn	Si	Other	MPa	lb/in^2	
270	99.9		a	a				345	50,000	50
200	99.0		0.2	0.3	0.2	0.2	C, S	462	67,000	47
400	66.8		30.0	2.5	0.2	0.5	C	550	80,000	40
600	74.0	16.0	0.5	8.0	1.0	0.5		655	95,000	40
230	52.8	22.0		3.0	0.4	0.4	b	860	125,000	47

Compiled from [12].
[a]Trace amounts.
[b]Other alloying ingredients in Grade 230: 5% Co, 2% Mo, 14% W, 0.3% Al, 0.1% C.

TABLE 6.1   (continued): (f) Titanium.

Symbol: Ti Atomic number: 22 Specific gravity: 4.51 Crystal structure: HCP Melting temperature: 1668°C (3034°F) Elastic modulus: 117,000 MPa (17×10^6 lb/in²)	Ores extracted from: Rutile (TiO_2) and Ilmenite ($FeTiO_3$) Alloying elements: Aluminum, tin, vanadium, copper, and magnesium Typical applications: Jet engine components, other aerospace applications, prosthetic implants

its aerospace applications, in which its light weight and good strength-to-weight ratio are exploited.

Titanium Production   The principal ores of titanium are *rutile,* which is 98% to 99% TiO_2, and *ilmenite,* which is a combination of FeO and TiO_2. Rutile is preferred as an ore because of its higher Ti content. In recovery of the metal from its ores, the TiO_2 is converted to titanium tetrachloride ($TiCl_4$) by reacting the compound with chlorine gas. This is followed by a sequence of distillation steps to remove impurities. The highly concentrated $TiCl_4$ is then reduced to metallic titanium by reaction with magnesium; this is known as the **Kroll process.** Sodium can also be used as a reducing agent. In either case, an inert atmosphere must be maintained to prevent O_2, N_2, or H_2 from contaminating the Ti, owing to its chemical affinity for these gases. The resulting metal is used to cast ingots of titanium and its alloys.

Properties of Titanium   Ti's coefficient of thermal expansion is relatively low among metals. It is stiffer and stronger than aluminum, and it retains good strength at elevated temperatures. Pure titanium is reactive, which presents problems in processing, especially in the molten state. However, at room temperature it forms a thin adherent oxide coating (TiO_2) that provides excellent corrosion resistance.

These properties give rise to two principal application areas for titanium: (1) in the commercially pure state, Ti is used for corrosion resistant components, such as marine components and prosthetic implants; and (2) titanium alloys are used as high-strength components in temperatures ranging from ambient to above 550°C (1000°F), especially where its excellent strength-to-weight ratio is exploited. These latter applications include aircraft and missile components. Some of the alloying elements used with titanium include aluminum, manganese, tin, and vanadium. Some compositions and mechanical properties for several alloys are presented in Table 6.13.

TABLE 6.13   Compositions and mechanical properties of selected titanium alloys.

Code[a]	Typical Composition, %						Tensile Strength		Elongation, %
	Ti	Al	Cu	Fe	V	Other	MPa	lb/in²	
R50250	99.8			0.2			240	35,000	24
R56400	89.6	6.0		0.3	4.0	[b]	1000	145,000	12
R54810	90.0	8.0			1.0	1 Mo[b]	985	143,000	15
R56620	84.3	6.0	0.8	0.8	6.0	2 Sn[b]	1030	150,000	14

Compiled from [1] and [12].
[a]United Numbering System (UNS).
[b]Traces of C, H, O.

TABLE 6.1   (continued): (g) Zinc.

Symbol: Zn	Elastic modulus: 90,000 MPa (13×10^6 lb/in^2)a
Atomic number: 30	Ore extracted from: Sphalerite (ZnS)
Specific gravity: 7.13	Alloying elements: Aluminum, magnesium, copper
Crystal structure: HCP	Typical applications: Galvanized steel and iron, die
Melting temperature: 419°C (786°F)	castings, alloying element in brass

aZinc creeps, which makes it difficult to measure modulus of elasticity; some tables of properties omit E for zinc for this reason.

6.3.6   ZINC AND ITS ALLOYS

Table 6.1(g) lists basic data on zinc. Its low melting point makes it attractive as a casting metal. It also provides corrosion protection when coated onto steel or iron; **galvanized steel** is steel that has been coated with zinc.

Production of Zinc   Zinc blende or **sphalerite** is the principal ore of zinc; it contains zinc sulfide (ZnS). Other important ores include **smithsonite,** which is zinc carbonate (ZnCO$_3$), and **hemimorphate,** which is hydrous zinc silicate (Zn$_4$Si$_2$O$_7$OH·H$_2$O).

Sphalerite must be concentrated (**beneficiated,** as it is called) because of the small fraction of zinc sulfide present in the ore. This is accomplished by first crushing the ore, then grinding with water in a ball mill (Section 17.1.1) to create a slurry. In the presence of a frothing agent, the slurry is agitated so that the mineral particles float to the top and can be skimmed off (separated from the lower-grade minerals). The concentrated zinc sulfide is then roasted at around 1260°C (2300°F), so that zinc oxide (ZnO) is formed from the reaction.

There are various thermochemical processes for recovering zinc from this oxide, all of which reduce zinc oxide by means of carbon. The carbon combines with oxygen in ZnO to form CO and/or CO$_2$, thus freeing Zn in the form of vapor that is condensed to yield the desired metal.

An electrolytic process is also widely used, accounting for about half the world's production of zinc. This process also begins with the preparation of ZnO, which is mixed with dilute sulfuric acid (H$_2$SO$_4$), followed by electrolysis to separate the resulting zinc sulfate (ZnSO$_4$) solution to yield the pure metal.

Zinc Alloys and Applications   Several alloys of zinc are listed in Table 6.14, with data on composition, tensile strength, and applications. Zinc alloys are widely used in die casting to mass produce components for the automotive and appliance industries. Another major application of zinc is in galvanized steel. As the name suggests, a galvanic cell is created in

TABLE 6.14   Compositions, tensile strength, and applications of selected zinc alloys.

	Typical Composition, %					Tensile Strength		
Code	Zn	Al	Cu	Mg	Fe	MPa	lb/in^2	Application
Z33520	95.6	4.0	0.25	0.04	0.1	283	41,000	Die casting
Z35540	93.4	4.0	2.5	0.04	0.1	359	52,000	Die casting
Z35635	91.0	8.0	1.0	0.02	0.06	374	54,000	Foundry alloy
Z35840	70.9	27.0	2.0	0.02	0.07	425	62,000	Foundry alloy
Z45330	98.9		1.0	0.01		227	33,000	Rolled alloy

Compiled from [12].
[a] UNS, Unified Numbering System for metals.

TABLE 6.1 (continued): (h) Lead and tin

	Lead	Tin
Symbol:	Pb	Sn
Atomic number:	82	50
Specific gravity:	11.35	7.30
Crystal structure:	FCC	HCP
Melting temperature:	327°C (621°F)	232°C (449°F)
Modulus of elasticity:	21,000 MPa (3×10^6 lb/in^2)	42,000 MPa (6×10^6 lb/in^2)
Ore from which extracted:	Galena (PbS)	Cassiterite (SnO$_2$)
Typical alloying elements:	Tin, antimony	Lead, copper
Typical applications:	See text	Bronze, solder, tin cans

galvanized steel (Zn is the anode and steel is the cathode) that protects the steel from corrosive attack. A third important use of zinc is in brass. As previously indicated in the discussion of copper, this alloy consists of copper and zinc, in the ratio of about 2/3 Cu to 1/3 Zn. Finally, readers may be interested to know that the U.S. one cent coin is mostly zinc. The penny is coined out of zinc and then electroplated with copper, so that the final proportions are 97.5% Zn and 2.5% Cu. It costs the U.S. Mint about 1.5 cents to produce each penny.

6.3.7 LEAD AND TIN

Lead (Pb) and tin (Sn) are often considered together because of their low melting temperatures, and because they are used in soldering alloys to make electrical connections. The phase diagram for the tin–lead alloy system is depicted in Figure 6.3. Basic data for lead and tin are presented in Table 6.1(h).

Lead is a dense metal with a low melting point; other properties include low strength, low hardness (the word "soft" is appropriate), high ductility, and good corrosion resistance. In addition to its use in solder, applications of lead and its alloys include ammunition, type metals, x-ray shielding, storage batteries, bearings, and vibration damping. It has also been widely used in chemicals and paints. Principal alloying elements with lead are tin and antimony.

Tin has an even lower melting point than lead; other properties include low strength, low hardness, and good ductility. The earliest use of tin was in bronze, the alloy consisting of copper and tin developed around 3000 BCE in Mesopotamia and Egypt. Bronze is still an important commercial alloy (although its relative importance has declined during 5000 years). Other uses of tin include tin-coated sheet steel containers ("tin cans") for storing food and, of course, solder metal.

6.3.8 REFRACTORY METALS

The refractory metals are metals capable of enduring high temperatures. The most important metals in this group are molybdenum and tungsten; see Table 6.1(i). Other refractory metals are columbium (Cb) and tantalum (Ta). In general, these metals and their alloys are capable of maintaining high strength and hardness at elevated temperatures.

Molybdenum has a high melting point and is relatively dense, stiff, and strong. It is used both as a pure metal (99.9+% Mo) and as an alloy. The principal alloy is TZM, which contains small amounts of titanium and zirconium (less than 1% total). Mo and its alloys possess good high temperature strength, and this accounts for many of its applications, which include heat shields, heating elements, electrodes for resistance welding, dies for high

TABLE 6.1   (continued): (i) Refractory metals.

	Molybdenum	Tungsten
Symbol:	Mo	W
Atomic number:	42	74
Specific gravity:	10.2	19.3
Crystal structure:	BCC	BCC
Melting point:	2619°C (4730°F)	3400°C (6150°F)
Elastic modulus:	324,000 MPa (47×10^6 lb/in^2)	407,000 MPa (59×10^6 lb/in^3)
Principal ores:	Molybdenite (MoS_2)	Scheelite ($CaWO_4$), Wolframite (($Fe,Mn)WO_4$) [a]
Alloying elements:	See text	
Applications:	See text	Light filaments, rocket engine parts, WC tools.

[a]Tungsten is used as a pure metal and as an alloying ingredient, but few alloys are based on W.

temperature work (e.g., die casting molds), and parts for rocket and jet engines. In addition to these applications, molybdenum is also widely used as an alloying ingredient in other metals, such as steels and superalloys.

Tungsten (W) has the highest melting point among metals and is one of the densest. It is also the stiffest and hardest of all pure metals. Its most familiar application is filament wire in incandescent light bulbs. Applications of tungsten are typically characterized by high operating temperatures, such as parts for rocket and jet engines and electrodes for arc welding. W is also widely used as an element in tool steels, heat resistant alloys, and tungsten carbide (Section 7.3.2).

A major disadvantage of both Mo and W is their propensity to oxidize at high temperatures, above about 600°C (1000°F), thus detracting from their high temperature properties. To overcome this deficiency, either protective coatings must be used on these metals in high temperature applications or the metal parts must operate in a vacuum. For example, the tungsten filament must be energized in a vacuum inside the glass light bulb.

6.3.9   PRECIOUS METALS

The precious metals, also called the ***noble metals*** because they are chemically inactive, include silver, gold, and platinum. They are attractive metals, available in limited supply, and have been used throughout civilized history for coinage and to underwrite paper

TABLE 6.1   (continued): (j) The precious metals.

	Gold	Platinum	Silver
Symbol:	Au	Pt	Ag
Atomic number:	79	78	47
Specific gravity:	19.3	21.5	10.5
Crystal structure:	FCC	FCC	FCC
Melting temperature:	1063°C (1945°F)	1769°C (3216°F)	961°C (1762°F)
Principal ores:	[a]	[a]	[a]
Applications:	See text	See text	See text

[a]All three precious metals are mined from deposits in which the pure metal is mixed with other ores and metals. Silver is also mined from the ore ***Argentite*** (Ag_2S).

currency. They are also widely used in jewelry and similar applications that exploit their high value. As a group, these precious metals possess high density, good ductility, high electrical conductivity, and good corrosion resistance; see Table 6.1(j).

Silver (Ag) is less expensive per unit weight than gold or platinum. Nevertheless, its attractive "silvery" luster makes it a highly valued metal in coins, jewelry, and tableware (which even assumes the name of the metal: "silverware"). It is also used for fillings in dental work. Silver has the highest electrical conductivity of any metal, which makes it useful for contacts in electronics applications. Finally, it should be mentioned that light-sensitive silver chloride and other silver halides are the basis for photography.

Gold (Au) is one of the heaviest metals; it is soft and easily formed, and possesses a distinctive yellow color that adds to its value. In addition to currency and jewelry, its applications include electrical contacts (owing to its good electrical conductivity and corrosion resistance), dental work, and plating onto other metals for decorative purposes.

Platinum (Pt) is also used in jewelry and is in fact more expensive than gold. It is the most important of six precious metals known as the platinum group metals, which consists of Ruthenium (Ru), Rhodium (Rh), Palladium (Pd), Osmium (Os), and Iridium (Ir), in addition to Pt. They are clustered in a rectangle in the periodic table (Figure 2.1). Osmium, Iridium, and Platinum are all denser than gold (Ir is the densest material known, at 22.65 g/cm^3). Because the platinum group metals are all scarce and very expensive, their applications are generally limited to situations in which only small amounts are needed and their unique properties are required (e.g., high melting temperatures, corrosion resistance, and catalytic characteristics). The applications include thermocouples, electrical contacts, spark plugs, corrosion resistant devices, and catalytic pollution control equipment for automobiles.

6.4 SUPERALLOYS

Superalloys constitute a category that straddles the ferrous and nonferrous metals. Some of them are based on iron, whereas others are based on nickel and cobalt. In fact, many of the superalloys contain substantial amounts of three or more metals, rather than consisting of one base metal plus alloying elements. Although the tonnage of these metals is not significant compared with most of the other metals discussed in this chapter, they are nevertheless commercially important because they are very expensive; and they are technologically important because of what they can do.

The *superalloys* are a group of high-performance alloys designed to meet very demanding requirements for strength and resistance to surface degradation (corrosion and oxidation) at high service temperatures. Conventional room temperature strength is usually not the important criterion for these metals, and most of them possess room temperature strength properties that are good but not outstanding. Their high temperature performance is what distinguishes them; tensile strength, hot hardness, creep resistance, and corrosion resistance at very elevated temperatures are the mechanical properties of interest. Operating temperatures are often in the vicinity of 1100°C (2000°F). These metals are widely used in gas turbines—jet and rocket engines, steam turbines, and nuclear power plants—systems in which operating efficiency increases with higher temperatures.

The superalloys are usually divided into three groups, according to their principal constituent: iron, nickel, or cobalt:

➢ *Iron-based alloys* have iron as the main ingredient, although in some cases the iron is less than 50% of the total composition.

➢ *Nickel-based alloys* generally have better high temperature strength than alloy steels. Nickel is the base metal. The principal alloying elements are chromium and

TABLE 6.15 Some typical superalloy compositions together with strength properties at room temperature and elevated temperature.

Superalloy	Chemical Analysis, % [a]							Tensile Strength at Room Temperature		Tensile Strength at 870°C (1600°F)	
	Fe	Ni	Co	Cr	Mo	W	Other[b]	MPa	lb/in²	MPa	lb/in²
Iron-based											
Incoloy 802	46	32		21			<2	690	100,000	195	28,000
Haynes 556	29	20	20	22	3		6	815	118,000	330	48,000
Nickel-based											
Incoloy 718	18	53		19	3		6	1435	208,000	340	49,000
Rene 41		55	11	19	1		5	1420	206,000	620	90,000
Hastelloy S	1	67		16	15		1	845	130,000	340	50,000
Nimonic 75	3	76		20			<2	745	108,000	150	22,000
Cobalt-based											
Stellite 6B	3	3	53	30	2	5	4	1010	146,000	385	56,000
Haynes 188	3	22	39	22		14		960	139,000	420	61,000
L-605		10	53	20		15	2	1005	146,000	325	47,000

Compiled from [11] and [12].
[a]Compositions to nearest percent.
[b]Other elements include carbon, niobium, titanium, tungsten, manganese, and silicon.

cobalt; lesser elements include aluminum, titanium, molybdenum, niobium (Nb), and iron. Some familiar names in this group include Inconel, Hastelloy, and Rene 41.

➢ **Cobalt-based alloys** consist of cobalt (around 40%) and chromium (perhaps 20%) as their main components. Other alloying elements include nickel, molybdenum, and tungsten.

In virtually all of the superalloys, including those based on iron, strengthening is accomplished by precipitation hardening. The iron-based superalloys do not use martensite formation for strengthening. Typical compositions and strength properties at room temperature and elevated temperature for some of the alloys are presented in Table 6.15.

6.5 GUIDE TO THE PROCESSING OF METALS

A wide variety of manufacturing processes are available to shape metals, enhance their properties, assemble them, and finish them for appearance and protection.

Shaping, Assembly, and Finishing Processes Metals are shaped by all of the basic processes, including casting, powder metallurgy, deformation processes, and material removal. In addition, metal parts are joined to form assemblies by welding, brazing, soldering, and mechanical fastening; and finishing processes are commonly used to improve the appearance of metal parts and/or to provide corrosion protection. These finishing operations include electroplating and painting.

Enhancement of Mechanical Properties in Metals Mechanical properties of metals can be altered by a number of techniques. Some of these techniques have been referred to in the discussion of the various metals. Methods for enhancing mechanical properties of metals can be grouped into three categories: (1) alloying, (2) cold working, and (3) heat treatment. ***Alloying*** has been discussed throughout the present chapter and is an important technique for strengthening metals. ***Cold working*** has previously been referred to as strain hardening; its effect is to increase strength and reduce ductility. The degree to which these mechanical properties are affected depends on the amount of strain and the strain hardening exponent in the flow curve, Eq. (3.10). Cold working can be used on both pure metals and alloys. It is accomplished during deformation of the workpart by one of the shape forming processes, such as rolling, forging, or extrusion. Strengthening of the metal therefore occurs as a by-product of the shaping operation.

Heat treatment refers to several types of heating and cooling cycles performed on a metal to beneficially change its properties. They operate by altering the basic microstructure of the metal, which in turn determines mechanical properties. Some heat treatment operations are applicable only to certain types of metals; for example, the heat treatment of steel to form martensite is somewhat specialized because martensite is unique to steel. Heat treatments for steels and other metals are discussed in Chapter 27.

REFERENCES

[1] Bauccio. M. (ed.). ***ASM Metals Reference Book,*** 3rd ed. ASM International, Materials Park, Ohio, 1993.

[2] Black, J, and Kohser, R. ***DeGarmo's Materials and Processes in Manufacturing,*** 10th ed., John Wiley & Sons, Hoboken, New Jersey, 2008.

[3] Brick, R. M., Pense, A. W., and Gordon, R. B. ***Structure and Properties of Engineering Materials,*** 4th ed. McGraw-Hill, New York, 1977.

[4] Carnes, R., and Maddock, G., "Tool Steel Selection," *Advanced Materials & Processes,* June 2004, pp. 37–40.

[5] ***Encyclopaedia Britannica,*** Vol. 21, *Macropaedia.* Encyclopaedia Britannica, Chicago, 1990, under section: Industries, Extraction and Processing.

[6] Flinn, R. A., and Trojan, P. K. ***Engineering Materials and Their Applications,*** 5th ed. John Wiley & Sons, New York, 1995.

[7] Guy, A. G., and Hren, J. J. ***Elements of Physical Metallurgy,*** 3rd ed. Addison-Wesley, Reading, Massachusetts, 1974.

[8] Hume-Rothery, W., Smallman, R. E., and Haworth, C. W. ***The Structure of Metals and Alloys.*** Institute of Materials, London, 1988.

[9] Keefe, J. "A Brief Introduction to Precious Metals," ***The AMMTIAC Quarterly,*** Vol. 2, No. 1, 2007.

[10] Lankford, W. T., Jr., Samways, N. L., Craven, R. F., and McGannon, H. E. ***The Making, Shaping, and Treating of Steel,*** 10th ed. United States Steel Co., Pittsburgh, 1985.

[11] ***Metals Handbook,*** Vol. 1, ***Properties and Selection: Iron, Steels, and High Performance Alloys.*** ASM International, Metals Park, Ohio, 1990.

[12] ***Metals Handbook,*** Vol. 2, ***Properties and Selection: Nonferrous Alloys and Special Purpose Materials,*** ASM International, Metals Park, Ohio, 1990.

[13] Moore, C., and Marshall, R. I. ***Steelmaking.*** The Institute for Metals, The Bourne Press, Ltd., Bournemouth, U.K., 1991.

[14] Wick, C., and Veilleux, R. F. (eds.). ***Tool and Manufacturing Engineers Handbook,*** 4, Vol. 3, ***Materials, Finishing, and Coating.*** Society of Manufacturing Engineers, Dearborn, Michigan, 1985.

REVIEW QUESTIONS

6.1. What are some of the general properties that distinguish metals from ceramics and polymers?

6.2. What are the two major groups of metals? Define them.

6.3. What is an alloy?

6.4. What is a solid solution in the context of alloys?

6.5. Distinguish between a substitutional solid solution and an interstitial solid solution.

6.6. What is an intermediate phase in the context of alloys?

6.7. The copper-nickel system is a simple alloy system, as indicated by its phase diagram. Why is it so simple?

6.8. What is the range of carbon percentages that defines an iron–carbon alloy as a steel?

6.9. What is the range of carbon percentages that defines an iron–carbon alloy as cast iron?

6.10. Identify some of the common alloying elements other than carbon in low alloy steels.

6.11. What are some of the mechanisms by which the alloying elements other than carbon strengthen steel?

6.12. What is the predominant alloying element in all of the stainless steels?

6.13. Why is austenitic stainless steel called by that name?

6.14. Besides high carbon content, what other alloying element is characteristic of the cast irons?

6.15. Identify some of the properties for which aluminum is noted.

6.16. What are some of the noteworthy properties of magnesium?

6.17. What is the most important engineering property of copper that determines most of its applications?

6.18. What elements are traditionally alloyed with copper to form (a) bronze and (b) brass?

6.19. What are some of the important applications of nickel?

6.20. What are the noteworthy properties of titanium?

6.21. Identify some of the important applications of zinc.

6.22. What important alloy is formed from lead and tin?

6.23. (a) Name the important refractory metals. (b) What does the term *refractory* mean?

6.24. (a) Name the four principal noble metals. (b) Why are they called noble metals?

6.25. The superalloys divide into three basic groups, according to the base metal used in the alloy. Name the three groups.

6.26. What is so special about the superalloys? What distinguishes them from other alloys?

6.27. What are the three basic methods by which metals can be strengthened?

MULTIPLE CHOICE QUIZ

There are 20 correct answers in the following multiple choice questions (some questions have multiple answers that are correct). To attain a perfect score on the quiz, all correct answers must be given. Each correct answer is worth 1 point. Each omitted answer or wrong answer reduces the score by 1 point, and each additional answer beyond the correct number of answers reduces the score by 1 point. Percentage score on the quiz is based on the total number of correct answers.

6.1. Which of the following properties or characteristics are inconsistent with the metals (two correct answers): (a) good thermal conductivity, (b) high strength, (c) high electrical resistivity, (d) high stiffness, and (e) ionic bonding?

6.2. Which one of the metallic elements is the most abundant on the earth: (a) aluminum, (b) copper, (c) iron, (d) magnesium, or (e) silicon?

6.3. The predominant phase in the iron–carbon alloy system for a composition with 99% Fe at room temperature is which one of the following: (a) austenite, (b) cementite, (c) delta, (d) ferrite, or (e) gamma?

6.4. A steel with 1.0% carbon is known as which one of the following: (a) eutectoid, (b) hypoeutectoid, (c) hypereutectoid, or (d) wrought iron?

6.5. The strength and hardness of steel increases as carbon content (a) increases or (b) decreases?

6.6. Plain carbon steels are designated in the AISI code system by which of the following: (a) 01XX, (b) 10XX, (c) 11XX, (d) 12XX, or (e) 30XX?

6.7. Which one of the following elements is the most important alloying ingredient in steel: (a) carbon, (b) chromium, (c) nickel, (d) molybdenum, or (e) vanadium?

6.8. Which one of the following is not a common alloying ingredient in steel: (a) chromium, (b) manganese, (c) nickel, (d) vanadium, (e) zinc?

6.9. Solid solution alloying is the principal strengthening mechanism in high-strength low-alloy (HSLA) steels: (a) true or (b) false?

6.10. Which of the following alloying elements are most commonly associated with stainless steel (two best answers): (a) chromium, (b) manganese, (c) molybdenum, (d) nickel, and (e) tungsten?

6.11. Which of the following is the most important cast iron commercially: (a) ductile cast iron, (b) gray cast iron, (c) malleable iron, or (d) white cast iron?

6.12. Which one of the following metals has the lowest density: (a) aluminum, (b) magnesium, (c) tin, or (d) titanium?

6.13. Which of the following metals has the highest density: (a) gold, (b) lead, (c) platinum, (d) silver, or (e) tungsten?

6.14. From which of the following ores is aluminum derived: (a) alumina, (b) bauxite, (c) cementite, (d) hematite, or (e) scheelite?

6.15. Which of the following metals is noted for its good electrical conductivity (one best answer): (a) copper, (b) gold, (c) iron, (d) nickel, or (e) tungsten?

6.16. Traditional brass is an alloy of which of the following metallic elements (two correct answers):

(a) aluminum, (b) copper, (c) gold, (d) tin, and (e) zinc?

6.17. Which one of the following metals has the lowest melting point: (a) aluminum, (b) lead, (c) magnesium, (d) tin, or (e) zinc?

PROBLEMS

6.1. For the copper-nickel phase diagram in Figure 6.2, find the compositions of the liquid and solid phases for a nominal composition of 70% Ni and 30% Cu at 1371°C (2500°F).

6.2. For the preceding problem, use the inverse lever rule to determine the proportions of liquid and solid phases present in the alloy.

6.3. Using the lead–tin phase diagram in Figure 6.3, determine the liquid and solid phase compositions for a nominal composition of 40% Sn and 60% Pb at 204°C (400°F).

6.4. For the preceding problem, use the inverse lever rule to determine the proportions of liquid and solid phases present in the alloy.

6.5. Using the lead–tin phase diagram in Figure 6.3, determine the liquid and solid phase compositions for a nominal composition of 90% Sn and 10% Pb at 204°C (400°F).

6.6. For the preceding problem, use the inverse lever rule to determine the proportions of liquid and solid phases present in the alloy.

6.7. In the iron–iron carbide phase diagram of Figure 6.4, identify the phase or phases present at the following temperatures and nominal compositions: (a) 650°C (1200°F) and 2% Fe_3C, (b) 760°C (1400°F) and 2% Fe_3C, and (c) 1095°C (2000°F) and 1% Fe_3C.

CERAMICS

Chapter Contents

We usually consider metals to be the most important class of engineering materials. However, it is of interest to note that ceramic materials are actually more abundant and widely used. Included in this category are clay products (e.g., bricks and pottery), glass, cement, and more modern ceramic materials such as tungsten carbide and cubic boron nitride. This is the class of materials discussed in this chapter. We also include coverage of several elements related to ceramics because they are sometimes used in similar applications. These elements are carbon, silicon, and boron.

The importance of ceramics as engineering materials derives from their abundance in nature and their mechanical and physical properties, which are quite different from those of metals. A *ceramic* material is an inorganic compound consisting of a metal (or semimetal) and one or more nonmetals. The word *ceramic* traces from the Greek *keramos* meaning potter's clay or wares made from fired clay. Important examples of ceramic materials are *silica*, or silicon dioxide (SiO_2), the main ingredient in most glass products; *alumina*, or aluminum oxide (Al_2O_3), used in applications ranging from abrasives to artificial bones; and more complex compounds such as hydrous aluminum silicate ($Al_2Si_2O_5(OH)_4$), known as *kaolinite*, the principal ingredient in most clay products. The elements in these compounds are the most common in Earth's crust; see Table 7.1. The group includes many additional compounds, some of which occur naturally while others are manufactured.

The general properties that make ceramics useful in engineered products are high hardness, good electrical and thermal insulating characteristics, chemical stability, and high melting temperatures. Some ceramics are translucent—window glass being the clearest example. They are also brittle and possess virtually no ductility, which can cause problems in both processing and performance of ceramic products.

The commercial and technological importance of ceramics is best demonstrated by the variety of products and applications that are based on this class of material. The list includes:

TABLE 7.1 Most common elements in the Earth's crust, with approximate percentages.

Oxygen	Silicon	Aluminum	Iron	Calcium	Sodium	Potassium	Magnesium
50%	26%	7.6%	4.7%	3.5%	2.7%	2.6%	2.0%

Compiled from [6].

> ➤ *Clay construction products*, such as bricks, clay pipe, and building tile
> ➤ *Refractory ceramics*, which are capable of high temperature applications such as furnace walls, crucibles, and molds
> ➤ *Cement used in concrete*, used for construction and roads (concrete is a composite material, but its components are ceramics)
> ➤ *Whiteware products*, including pottery, stoneware, fine china, porcelain, and other tableware, based on mixtures of clay and other minerals
> ➤ *Glass* used in bottles, glasses, lenses, window panes, and light bulbs
> ➤ *Glass fibers* for thermal insulating wool, reinforced plastics (fiberglass), and fiber optics communications lines
> ➤ *Abrasives*, such as aluminum oxide and silicon carbide
> ➤ *Cutting tool materials*, including tungsten carbide, aluminum oxide, and cubic boron nitride
> ➤ *Ceramic insulators*, which are used in applications such as electrical transmission components, spark plugs, and microelectronic chip substrates
> ➤ *Magnetic ceramics*, for example, in computer memories
> ➤ *Nuclear fuels* based on uranium oxide (UO_2)
> ➤ *Bioceramics*, which include materials used in artificial teeth and bones

For purposes of organization, we classify ceramic materials into three basic types: (1) *traditional ceramics*—silicates used for clay products such as pottery and bricks, common abrasives, and cement; (2) *new ceramics*—more recently developed ceramics based on nonsilicates such as oxides and carbides, and generally possessing mechanical or physical properties that are superior or unique compared to traditional ceramics; and (3) *glasses*—based primarily on silica and distinguished from the other ceramics by their noncrystalline structure. In addition to the three basic types, we have *glass ceramics*—glasses that have been transformed into a largely crystalline structure by heat treatment.

7.1 STRUCTURE AND PROPERTIES OF CERAMICS

Ceramic compounds are characterized by covalent and ionic bonding. These bonds are stronger than metallic bonding in metals, which accounts for the high hardness and stiffness but low ductility of ceramic materials. Just as the presence of free electrons in the metallic bond explains why metals are good conductors of heat and electricity, the presence of tightly held electrons in ceramic molecules explains why these materials are poor conductors. The strong bonding also provides these materials with high melting temperatures, although some ceramics decompose, rather than melt, at elevated temperatures.

Most ceramics take a crystalline structure. The structures are generally more complex than those of most metals. There are several reasons for this. First, ceramic molecules usually consist of atoms that are significantly different in size. Second, the ion charges are often different, as in many of the common ceramics such as SiO_2 and Al_2O_3. Both of these factors tend to force a more complicated physical arrangement of the atoms in the molecule and in the resulting crystal structure. In addition, many ceramic materials consist of more than two elements, such as (Al$_2$Si$_2$O$_5$(OH)$_4$), also leading to further complexity in the molecular structure. Crystalline ceramics can be single crystals or polycrystalline substances. In the more common second form, mechanical and physical properties are affected by grain size; higher strength and toughness are achieved in the finer-grained materials.

Some ceramic materials tend to assume an amorphous structure or *glassy* phase, rather than a crystalline form. The most familiar example is, of course, glass. Chemically, most glasses consist of fused silica. Variations in properties and colors are obtained by adding other glassy ceramic materials such as oxides of aluminum, boron, calcium, and magnesium. In addition to these pure glasses, many ceramics that have a crystal structure use the glassy phase as a binder for their crystalline phase.

7.1.1 MECHANICAL PROPERTIES

Basic mechanical properties of ceramics are presented in Chapter 3. Ceramic materials are rigid and brittle, exhibiting a stress-strain behavior best characterized as perfectly elastic (see Figure 3.6). As seen in Table 7.2, hardness and elastic modulus for many of the new ceramics are greater than those of metals (see Tables 3.1, 3.6, and 3.7). Stiffness and hardness of traditional ceramics and glasses are significantly less than for new ceramics.

Theoretically, the strength of ceramics should be higher than that of metals because of their atomic bonding. The covalent and ionic bonding types are stronger than metallic bonding. However, metallic bonding has the advantage that it allows for slip, the basic mechanism by which metals deform plastically when subjected to high stresses. Bonding in ceramics is more rigid and does not permit slip under stress. The inability to slip makes it much more difficult for ceramics to absorb stresses. Yet ceramics contain the same imperfections in their crystal structure as metals—vacancies, interstitialcies, displaced atoms, and microscopic cracks. These internal flaws tend to concentrate the stresses, especially when a tensile, bending, or impact loading is involved. As a result of these factors, ceramics fail by brittle fracture under applied stress much more readily than metals. Their

TABLE 7.2   Selected mechanical and physical properties of ceramic materials.

Material	Hardness (Vickers)	Elastic modulus, E		Specific Gravity	Melting Temperature	
		Gpa	(lb/in^2)		°C	°F
Traditional ceramics						
Brick-fireclay	NA	95	14×10^6	2.3	NA	NA
Cement, Portland	NA	50	7×10^6	2.4	NA	NA
Silicon carbide (SiC)	2600 HV	460	68×10^6	3.2	27,007[a]	48,927[a]
New ceramics						
Alumina (Al$_2$O$_3$)	2200 HV	345	50×10^6	3.8	2054	3729
Cubic boron nitride (cBN)	6000 HV	NA	NA	2.3	30,007[a]	54,307[a]
Titanium carbide (TiC)	3200 HV	300	45×10^6	4.9	3250	5880
Tungsten carbide (WC)	2600 HV	700	100×10^6	15.6	2870	5198
Glass						
Silica glass (SiO$_2$)	500 HV	69	10×10^6	2.2	7[b]	7[b]

NA = Not available or not applicable.
[a]The ceramic material chemically dissociates or, in the case of diamond and graphite, sublimes (vaporizes), rather than melts.
[b]Glass, being noncrystalline, does not melt at a specific melting point. Instead, it gradually exhibits fluid properties with increasing temperature. It becomes liquid at around 1400°C (2550°F).
Compiled from [3], [4], [5], [6], [9], [10], and other sources.

tensile strength and toughness are relatively low. Also, their performance is much less predictable due to the random nature of the imperfections and the influence of processing variations, especially in products made of traditional ceramics.

The frailties that limit the tensile strength of ceramic materials are not nearly so operative when compressive stresses are applied. Ceramics are substantially stronger in compression than in tension. For engineering and structural applications, designers have learned to use ceramic components so that they are loaded in compression rather than tension or bending.

Various methods have been developed to strengthen ceramics, nearly all of which have as their fundamental approach the minimization of surface and internal flaws and their effects. These methods include [7]: (1) making the starting materials more uniform; (2) decreasing grain size in polycrystalline ceramic products; (3) minimizing porosity; (4) introducing compressive surface stresses, for example, through application of glazes with low thermal expansions, so that the body of the product contracts after firing more than the glaze, thus putting the glaze in compression; (5) using fiber reinforcement; and (6) heat treatments, such as quenching alumina from temperatures in the slightly plastic region to strengthen it.

7.1.2 PHYSICAL PROPERTIES

Several of the physical properties of ceramics are presented in Table 7.2. Most ceramic materials are lighter than metals and heavier than polymers (see Table 4.1). Melting temperatures are higher than for most metals, some ceramics preferring to decompose rather than melt.

Electrical and thermal conductivities of most ceramics are lower than for metals; but the range of values is greater, permitting some ceramics to be used as insulators while others are electrical conductors. Thermal expansion coefficients are somewhat less than for the metals, but the effects are more damaging in ceramics because of their brittleness. Ceramic materials with relatively high thermal expansions and low thermal conductivities are especially susceptible to failures of this type, which result from significant temperature gradients and associated volumetric changes in different regions of the same part. The terms *thermal shock* and *thermal cracking* are used in connection with such failures. Certain glasses (for example, those containing high proportions of SiO_2) and glass ceramics are noted for their low thermal expansion and are particularly resistant to these thermal failures (*Pyrex* is a familiar example).

7.2 TRADITIONAL CERAMICS

These materials are based on mineral silicates, silica, and mineral oxides. The primary products are fired clay (pottery, tableware, brick, and tile), cement, and natural abrasives such as alumina. These products, and the processes used to make them, date back thousands of years (see Historical Note 7.1). Glass is also a silicate ceramic material and is often included within the traditional ceramics group [5], [6]. We cover glass in a later section because it is distinguished from the above crystalline materials by its amorphous or vitreous structure (the term *vitreous* means glassy, or possessing the characteristics of glass).

7.2.1 RAW MATERIALS

Mineral silicates, such as clays of various compositions, and silica, such as quartz, are among the most abundant substances in nature and constitute the principal raw materials for

---

Historical Note 7.1   *Ancient pottery ceramics*

Making pottery has been an art since the earliest civilizations. Archeologists examine ancient pottery and similar artifacts to study the cultures of the ancient world. Ceramic pottery does not corrode or disintegrate with age nearly as rapidly as artifacts made of wood, metal, or cloth.

Somehow, early tribes discovered that clay is transformed into a hard solid when placed near an open fire. Burnt clay articles have been found in the Middle East that date back nearly 10,000 years. Earthenware pots and similar products became an established commercial trade in Egypt by around 4000 BCE.

The greatest advances in pottery making were made in China, where fine white stoneware was first crafted as early as 1400 BCE. By the ninth century, the Chinese were making articles of porcelain, which was fired at higher temperatures than earthenware or stoneware to partially vitrify the more complex mixture of raw materials and produce translucency in the final product. Dinnerware made of Chinese porcelain was highly valued in Europe; it was called "china." It contributed significantly to trade between China and Europe and influenced the development of European culture.

---

traditional ceramics. These solid crystalline compounds have been formed and mixed in the Earth's crust over billions of years by complex geological processes.

The clays are the raw materials used most widely in ceramics. They consist of fine particles of hydrous aluminum silicate that become a plastic substance that is formable and moldable when mixed with water. The most common clays are based on the mineral **kaolinite** ($Al_2Si_2O_5(OH)_4$). Other clay minerals vary in composition, both in terms of proportions of the basic ingredients and through additions of other elements such as magnesium, sodium, and potassium.

Besides its plasticity when mixed with water, a second characteristic of clay that makes it so useful is that it fuses into a dense, strong material when heated to a sufficiently elevated temperature. The heat treatment is known as **firing**. Suitable firing temperatures depend on clay composition. Thus, clay can be shaped while wet and soft, and then fired to obtain the final hard ceramic product.

Silica (SiO_2) is another major raw material for the traditional ceramics. It is the principal component in glass, and an important ingredient in other ceramic products including whiteware, refractories, and abrasives. Silica is available naturally in various forms, the most important of which is **quartz**. The main source of quartz is **sandstone**. The abundance of sandstone and its relative ease of processing means that silica is low in cost; it is also hard and chemically stable. These features account for its widespread use in ceramic products. It is generally mixed in various proportions with clay and other minerals to achieve the appropriate characteristics in the final product. Feldspar is one of the other minerals often used. **Feldspar** refers to any of several crystalline minerals that consist of aluminum silicate combined with either potassium, sodium, calcium, or barium. The potassium blend, for example, has the chemical composition $KAlSi_3O_8$. Mixtures of clay, silica, and feldspar are used to make stoneware, china, and other tableware.

Still another important raw material for traditional ceramics is **alumina**. Most alumina is processed from the mineral **bauxite**, which is an impure mixture of hydrous aluminum oxide and aluminum hydroxide plus similar compounds of iron or manganese. Bauxite is also the principal ore in the production of aluminum metal. A purer but less common form of Al_2O_3 is the mineral **corundum**, which contains alumina in massive amounts. Slightly impure forms of corundum crystals are the colored gemstones sapphire and ruby. Alumina ceramic is used as an abrasive in grinding wheels and as a refractory brick in furnaces.

Silicon carbide, also used as an abrasive, does not occur as a mineral. Instead, it is produced by heating mixtures of sand (source of silicon) and coke (carbon) to a temperature of around 2200°C (4000°F), so that the resulting chemical reaction forms SiC and carbon monoxide.

7.2.2 TRADITIONAL CERAMIC PRODUCTS

The minerals discussed above are the ingredients for a variety of ceramic products. We organize our coverage here by major categories of traditional ceramic products. A summary of these products, and the raw materials and ceramics out of which they are made, is presented in Table 7.3. We limit our coverage to materials commonly in with manufactured products, thus omitting certain commercially important ceramics such as cement.

Pottery and Tableware   This category is one of the oldest, dating back thousands of years; yet it is still one of the most important. It includes tableware products that we all use: earthenware, stoneware, and china. The raw materials for these products are clay usually combined with other minerals such as silica and feldspar. The wetted mixture is shaped and then fired to produce the finished piece.

Earthenware is the least refined of the group; it includes pottery and similar articles made in ancient times. Earthenware is relatively porous and is often glazed. *Glazing* involves application of a surface coating, usually a mixture of oxides such as silica and alumina, to make the product less pervious to moisture and more attractive to the eye. *Stoneware* has lower porosity than earthenware, resulting from closer control of ingredients and higher firing temperatures. *China* is fired at even higher temperatures, which produces the translucence in the finished pieces that characterize their fine quality. The reason for this is that much of the ceramic material has been converted to the glassy (vitrified) phase, which is relatively transparent compared to the polycrystalline form. Modern *porcelain* is nearly the same as china and is produced by firing the components, mainly clay, silica, and feldspar, at still higher temperatures to achieve a very hard, dense, glassy material. Porcelain is used in a variety of products ranging from electrical insulation to bathtub coatings.

Brick and Tile   Building brick, clay pipe, unglazed roof tile, and drain tile are made from various low-cost clays containing silica and gritty matter widely available in natural deposits. These products are shaped by pressing (molding) and firing at relatively low temperatures.

Refractories   Refractory ceramics, often in the form of bricks, are critical in many industrial processes that require furnaces and crucibles to heat and/or melt materials. The useful properties of refractory materials are high temperature resistance, thermal insulation, and resistance to chemical reaction with the materials (usually molten metals) being heated. As we have mentioned, alumina is often used as a refractory ceramic, together with silica. Other refractory materials include magnesium oxide (MgO) and calcium oxide (CaO). The refractory lining often contains two layers, the outside layer being more porous because this increases the insulation properties.

Abrasives   Traditional ceramics used for abrasive products, such as grinding wheels and sandpaper, are *alumina* and *silicon carbide*. Although SiC is the harder material (hardness of SiC is 2600 HV vs. 2200 HV for alumina), the majority of grinding wheels are based on

TABLE 7.3 Summary of traditional ceramic products.		
Product	**Principal Chemistry**	**Minerals and Raw Materials**
Pottery, tableware	$Al_2Si_2O_5(OH)_4$, SiO_2, $KAlSi_3O_8$	Clay + silica + feldspar
Porcelain	$Al_2Si_2O_5(OH)_4$, SiO_2, $KAlSi_3O_8$	Clay + silica + feldspar
Brick, tile	$Al_2Si_2O_5(OH)_4$, SiO_2 plus fine stones	Clay + silica + other
Refractory	Al_2O_3, SiO_2 Others: MgO, CaO	Alumina and silica
Abrasive: silicon carbide	SiC	Silica + coke
Abrasive: aluminum oxide	Al_2O_3	Bauxite or alumina

Al_2O_3 because it gives better results when grinding steel, the most widely used metal. The abrasive particles (grains of ceramic) are distributed throughout the wheel using a bonding material such as shellac, polymer resin, or rubber. The use of abrasives in industry involves material removal, and the technology of grinding wheels and other abrasive methods to remove material is presented in Chapter 25.

7.3   NEW CERAMICS

The term *new ceramics* refers to ceramic materials that have been developed synthetically over the last several decades and to improvements in processing techniques that have provided greater control over the structures and properties of ceramic materials. In general, new ceramics are based on compounds other than variations of aluminum silicate (which form the bulk of the traditional ceramic materials). New ceramics are usually simpler chemically than traditional ceramics; for example, oxides, carbides, nitrides, and borides. The dividing line between traditional and new ceramics is sometimes fuzzy, because aluminum oxide and silicon carbide are included among the traditional ceramics. The distinction in these cases is based more on methods of processing than chemical composition.

We organize the new ceramics into chemical compound categories: oxides, carbides, and nitrides, discussed in the following sections. More complete coverage of new ceramics is presented in references [3], [5], and [8].

7.3.1   OXIDE CERAMICS

The most important oxide new ceramic is *alumina*. Although also discussed in the context of traditional ceramics, alumina is today produced synthetically from bauxite, using an electric furnace method. Through control of particle size and impurities, refinements in processing methods, and blending with small amounts of other ceramic ingredients, strength and toughness of alumina have been improved substantially compared to its natural counterpart. Alumina also has good hot hardness, low thermal conductivity, and good corrosion resistance. This is a combination of properties that promote a wide variety of applications, including [13]: abrasives (grinding wheel grit), bioceramics (artificial bones and teeth), electrical insulators, electronic components, alloying ingredients in glass, refractory brick, cutting tool inserts (Section 23.2.4), spark plug barrels, and engineering components (see Figure 7.1).

FIGURE 7.1   Alumina ceramic components. (Photo courtesy of Insaco Inc.)

7.3.2 CARBIDES

The carbide ceramics include silicon carbide (SiC), tungsten carbide (WC), titanium carbide (TiC), tantalum carbide (TaC), and chromium carbide (Cr_3C_2). Silicon carbide was discussed previously. Although it is a man-made ceramic, the methods for its production were developed a century ago, and therefore it is generally included in the traditional ceramics group. In addition to its use as an abrasive, other SiC applications include resistance heating elements and additives in steelmaking.

WC, TiC, and TaC are valued for their hardness and wear resistance in cutting tools and other applications requiring these properties. *Tungsten carbide* was the first to be developed (Historical Note 7.2) and is the most important and widely used material in the group. WC is typically produced by carburizing tungsten powders that have been reduced from tungsten ores such as *wolframite* ($FeMnWO_4$) and scheelite ($CaWO_4$). *Titanium carbide* is produced by carburizing the minerals *rutile* (TiO_2) or *ilmenite* ($FeTiO_3$). And *tantalum carbide* is made by carburizing either pure tantalum powders or tantalum pentoxide (Ta_2O_5) [11]. *Chromium carbide* is more suited to applications where chemical stability and oxidation resistance are important. Cr_3C_2 is prepared by carburizing chromium oxide (Cr_2O_3) as the starting compound. Carbon black is the usual source of carbon in all of these reactions.

Historical Note 7.2   *Tungsten carbide*

The compound WC does not occur in nature. It was first fabricated in the late 1890s by the Frenchman Henri Moissan. However, the technological and commercial importance of the development was not recognized for two decades.

Tungsten became an important metal for incandescent lamp filaments in the early 1900s. Wire drawing was required to produce the filaments. The traditional tool steel draw dies of the period were unsatisfactory for drawing tungsten wire due to excessive wear. There was a need for a much harder material. The compound WC was known to possess such hardness. In 1914 in Germany, H. Voigtlander and H. Lohmann developed a fabrication process for hard carbide draw dies by sintering parts pressed from powders of tungsten carbide and/or molybdenum carbide. Lohmann is credited with the first commercial production of sintered carbides.

The breakthrough leading to the modern technology of cemented carbides is linked to the work of K. Schroter in Germany in the early and mid-1920s. He used WC powders mixed with about 10% of a metal from the iron group, finally settling on cobalt as the best binder, and sintering the mixture at a temperature close to the melting point of the metal. The hard material was first marketed in Germany as "Widia" in 1926. The Schroter patents were assigned to the General Electric Company under the trade name "Carboloy"—first produced in the United States around 1928.

Widia and Carboloy were used as cutting tool materials, with cobalt content in the range 4% to 13%. They were effective in the machining of cast iron and many nonferrous metals, but not in the cutting of steel. When steel was machined, the tools would wear rapidly by cratering. In the early 1930s, carbide cutting tool grades with WC and TiC were developed for steel cutting. In 1931, the German firm Krupp started production of Widia X, which had a composition 84% WC, 10% TiC, and 6% cobalt (Co). And Carboloy Grade 831 was introduced in the United States in 1932; it contained 69% WC, 21% TiC, and 10% Co.

Except for SiC, each carbide discussed here must be combined with a metallic binder such as cobalt or nickel in order to fabricate a useful solid product. In effect, the carbide powders bonded in a metal framework creates what is known as a *cemented carbide*—a composite material, specifically a *cermet* (reduced from *cer*amic and *met*al). We examine cemented carbides and other cermets in Section 9.2.1. The carbides have little engineering value except as constituents in a composite system.

7.3.3 NITRIDES

The important nitride ceramics are silicon nitride (Si_3N_4), boron nitride (BN), and titanium nitride (TiN). As a group, the nitride ceramics are hard and brittle, and they melt at high temperatures (but not generally as high as the carbides). They are usually electrically insulating, except for TiN.

Silicon nitride shows promise in high temperature structural applications. Si_3N_4 oxidizes at about $1200°C$ ($2200°F$) and chemically decomposes at around $1900°C$ ($3400°F$). It has low thermal expansion, good resistance to thermal shock and creep, and resists corrosion by molten nonferrous metals. These properties have provided applications for this ceramic in gas turbines, rocket engines, and melting crucibles.

Boron nitride exists in several structures, similar to carbon. The important forms of BN are (1) hexagonal, similar to graphite; and (2) cubic, same as diamond; in fact, its hardness is comparable to that of diamond. This latter structure goes by the names *cubic boron nitride* and *borazon*, symbolized cBN, and is produced by heating hexagonal BN under very high pressures. Owing to its extreme hardness, the principal applications of cBN are in cutting tools (Section 23.2.5) and abrasive wheels (Section 25.1.1). Interestingly, it does not compete with diamond cutting tools and grinding wheels. Diamond is suited to nonsteel machining and grinding, while cBN is appropriate for steel.

Titanium nitride has properties similar to those of other nitrides in this group, except for its electrical conductivity; it is a conductor. TiN has high hardness, good wear resistance, and a low coefficient of friction with the ferrous metals. This combination of properties makes TiN an ideal material as a surface coating on cutting tools. The coating is only around 0.006 mm (0.00024 in) thick, so the amounts of material used in this application are low.

A new ceramic material related to the nitride group, and also to the oxides, is the oxynitride ceramic called *sialon*. It consists of the elements silicon, aluminum, oxygen, and nitrogen; and its name derives from these ingredients: Si-Al-O-N. Its chemical composition is variable, a typical composition being $Si_4Al_2O_2N_6$. Properties of sialon are similar to those of silicon nitride, but it has better resistance to oxidation at high temperatures than Si_3N_4. Its principal application is for cutting tools, but its properties may make it suitable for other high temperature applications in the future.

7.4 GLASS

The term glass is somewhat confusing because it describes a state of matter as well as a type of ceramic. As a state of matter, the term refers to an amorphous, or noncrystalline, structure of a solid material. The glassy state occurs in a material when insufficient time is allowed during cooling from the molten condition for the crystalline structure to form. It turns out that all three categories of engineering materials (metals, ceramics, and polymers) can assume the glassy state, although the circumstances for metals to do so are quite rare.

As a type of ceramic, *glass* is an inorganic, nonmetallic compound (or mixture of compounds) that cools to a rigid condition without crystallizing; it is a ceramic that is in the glassy state as a solid material. This is the material we shall discuss in this section—a material that dates back 4500 years (Historical Note 7.3).

7.4.1 CHEMISTRY AND PROPERTIES OF GLASS

The principal ingredient in virtually all glasses is *silica*, most commonly found as the mineral quartz in sandstone and silica sand. Quartz occurs naturally as a crystalline substance; but when melted and then cooled, it forms vitreous silica. Silica glass has a very low thermal expansion coefficient and is therefore quite resistant to thermal shock. These properties are

Historical Note 7.3   *History of glass*

The oldest glass specimens, dating from around 2500 BCE, are glass beads and other simple shapes found in Mesopotamia and ancient Egypt. These were made by painstakingly sculpturing glass solids, rather than by molding or shaping molten glass. It was a thousand years before the ancient cultures exploited the fluid properties of hot glass, by pouring it in successive layers over a sand core until sufficient thickness and rigidity had been attained in the product, a cup-shaped vessel. This pouring technique was used until around 200 BCE, when a simple tool was developed that revolutionized glassworking—the blowpipe.

Glassblowing was probably first accomplished in Babylon and later by the Romans. It was performed using an iron tube several feet long, with a mouthpiece on one

end and a fixture for holding the molten glass on the other. A blob of hot glass in the required initial shape and viscosity was attached to the end of the iron tube, and then blown into shape by an artisan either freely in air or into a mold cavity. Other simple tools were utilized to add the stem and/or base to the object.

The ancient Romans showed great skill in their use of various metallic oxides to color glass. Their technology is evident in the stained glass windows of cathedrals and churches of the Middle Ages in Italy and the rest of Europe. The art of glassblowing is still practiced today for certain consumer glassware; and automated versions of glassblowing are used for mass-produced glass products such as bottles and light bulbs (Chapter 12).

ideal for elevated temperature applications; accordingly, Pyrex and chemical glassware designed for heating are made with high proportions of silica glass.

In order to reduce the melting point of glass for easier processing, and to control properties, the composition of most commercial glasses includes other oxides as well as silica. Silica remains as the main component in these glass products, usually comprising 50% to 75% of total chemistry. The reason SiO_2 is used so widely in these compositions is because it is the best ***glass former***. It naturally transforms into a glassy state upon cooling from the liquid, whereas most ceramics crystallize upon solidification. Table 7.4 lists typical

TABLE 7.4   Typical compositions of selected glass products.

Product	Chemical Composition (by weight to nearest %)								
	SiO_2	Na_2O	CaO	Al_2O_3	MgO	K_2O	PbO	B_2O_3	Other
Soda-lime glass	71	14	13	2					
Window glass	72	15	8	1	4				
Container glass	72	13	10	2[a]	2	1			
Light bulb glass	73	17	5	1	4				
Laboratory glass									
Vycor	96			1				3	
Pyrex	81	4		2				13	
E-glass (fibers)	54	1	17	15	4			9	
S-glass (fibers)	64			26	10				
Optical glasses									
Crown glass	67	8				12		12	ZnO
Flint glass	46	3				6	45		

Compiled from [4], [5] and [10], and other sources.
[a]May include Fe_2O_3 with Al_2O_3

chemistries for some common glasses. The additional ingredients are contained in a solid solution with SiO_2, and each has a function: (1) acting as flux (promoting fusion) during heating; (2) increasing fluidity in the molten glass for processing; (3) retarding ***devitrification***—the tendency to crystallize from the glassy state; (4) reducing thermal expansion in the final product; (5) improving the chemical resistance against attack by acids, basic substances, or water; (6) adding color to the glass; and (7) altering the index of refraction for optical applications (e.g., lenses).

7.4.2 GLASS PRODUCTS

Following is a list of the major categories of glass products. We examine the roles played by the different ingredients in Table 7.4 as we discuss these products.

Window Glass   This glass is represented by two chemistries in Table 7.4: (1) soda-lime glass and (2) window glass. The soda-lime formula dates back to the glass-blowing industry of the 1800s and earlier. It was (and is) made by mixing soda (Na_2O) and lime (CaO) with silica (SiO_2) as the major ingredient. The blending of ingredients has evolved empirically to achieve a balance between avoiding crystallization during cooling and achieving chemical durability of the final product. Modern window glass and the techniques for making it have required slight adjustments in composition and closer control over its variation. Magnesia (MgO) has been added to help reduce devitrification.

Containers   In previous times, the same basic soda-lime composition was used for manual glass-blowing to make bottles and other containers. Modern processes for shaping glass containers cool the glass more rapidly than older methods. Also, the importance of chemical stability in container glass is better understood today. Resulting changes in composition have attempted to optimize the proportions of lime (CaO) and soda (Na_2O_3). Lime promotes fluidity. It also increases devitrification, but since cooling is more rapid, this effect is not as important as in prior processing techniques with slower cooling rates. Soda reduces chemical instability and solubility of the container glass.

Light Bulb Glass   Glass used in light bulbs and other thin glass items (e.g., drinking glasses, Christmas ornaments) is high in soda and low in lime; it also contains small amounts of magnesia and alumina. The chemistry is dictated largely by the economics of large volumes involved in light bulb manufacture. The raw materials are inexpensive and suited to the continuous melting furnaces used today.

Laboratory Glassware   These products include containers for chemicals (e.g., flasks, beakers, glass tubing). The glass must be resistant to chemical attack and thermal shock. Glass that is high in silica is suitable because of its low thermal expansion. The trade name "Vicor" is used for this high-silica glass. This product is very insoluble in water and acids. Additions of boric oxide also produce a glass with low coefficient of thermal expansion, so some glass for laboratory ware contains B_2O_3 in amounts of around 13%. The trade name "Pyrex" is used for the borosilicate glass developed by the Corning Glass Works. Both Vicor and Pyrex are included in our listing as examples of this product category.

Glass Fibers   Glass fibers are manufactured for a number of important applications, including fiberglass reinforced plastics, insulation wool, and fiber optics. The compositions vary according to function. The most commonly used glass reinforcing fibers in plastics are E-glass. It is high in CaO and Al_2O_3 content, it is economical, and it possesses good tensile strength in fiber form. Another glass fiber material is S-glass, which has higher strength but is not as economical as E-glass. Compositions are indicated in our table.

Insulating fiberglass wool can be manufactured from regular soda-lime-silica glasses. The glass product for fiber optics consists of a long, continuous core of glass with high refractive index surrounded by a sheath of lower refractive glass. The inside glass must have a very high transmittance for light in order to accomplish long distance communication.

Optical Glasses   Applications for these glasses include lenses for eyeglasses and optical instruments such as cameras, microscopes, and telescopes. To achieve their function, the glasses must have different refractive indices, but each lens must be homogenous in composition. Optical glasses are generally divided into: crowns and flints. ***Crown glass*** has a low index of refraction, while ***flint glass*** contains lead oxide (PbO) that gives it a high index of refraction.

7.4.3 GLASS-CERAMICS

Glass-ceramics are a class of ceramic material produced by conversion of glass into a polycrystalline structure through heat treatment. The proportion of crystalline phase in the final product typically ranges between 90% and 98%, with the remainder being unconverted vitreous material. Grain size is usually between 0.1 and 1.0 μm (4 and 40 μ-in), significantly smaller than the grain size of conventional ceramics. This fine crystal microstructure makes glass-ceramics much stronger than the glasses from which they are derived. Also, due to their crystal structure, glass-ceramics are opaque (usually gray or white) rather than clear.

The processing sequence for glass-ceramics is as follows: (1) The first step involves heating and forming operations used in glassworking (Section 12.2) to create the desired product geometry. Glass shaping methods are generally more economical than pressing and sintering to shape traditional and new ceramics made from powders. (2) The product is cooled. (3) The glass is reheated to a temperature sufficient to cause a dense network of crystal nuclei to form throughout the material. It is the high density of nucleation sites that inhibits grain growth of individual crystals, thus leading ultimately to the fine grain size in the glass-ceramic material. The key to the propensity for nucleation is the presence of small amounts of nucleating agents in the glass composition. Common nucleating agents are TiO_2, P_2O_5, and ZrO_2. (4) Once nucleation is initiated, the heat treatment is continued at a higher temperature to cause growth of the crystalline phases.

Several examples of glass-ceramic systems and typical compositions are listed in Table 7.5. The Li_2O-Al_2O_3-SiO_2 system is the most important commercially; it includes Corning Ware (Pyroceram), the familiar product of the Corning Glass Works.

The significant advantages of glass-ceramics include (1) efficiency of processing in the glassy state, (2) close dimensional control over the final product shape, and (3) good mechanical and physical properties. Properties include high strength (stronger than glass), absence of porosity, low coefficient of thermal expansion, and high resistance to thermal

TABLE 7.5   Several glass-ceramic systems.

Glass-Ceramic System	Typical Composition (to nearest %)						
	Li_2O	MgO	Na_2O	BaO	Al_2O_3	SiO_2	TiO_2
Li_2O–Al_2O_3–SiO_2	3				18	70	5
MgO–Al_2O_3–SiO_2		13			30	47	10
Na_2O–BaO–Al_2O_3–SiO_2			13	9	29	41	7

Compiled from [5], [6], and [10].

shock. These properties have resulted in applications in cooking ware, heat exchangers, and missile radomes. Certain systems (e.g., MgO-Al_2O_3-SiO_2 system) are also characterized by high electrical resistance, suitable for electrical and electronics applications.

7.5 SOME IMPORTANT ELEMENTS RELATED TO CERAMICS

In this section, several elements of engineering importance are discussed: carbon, silicon, and boron. We encounter these materials on occasion in subsequent chapters. Although they are not ceramic materials according to our definition, they sometimes compete for applications with ceramics. And they have important applications of their own. Basic data on these elements are presented in Table 7.6.

7.5.1 CARBON

Carbon occurs in two alternative forms of engineering and commercial importance: graphite and diamond. They compete with ceramics in various applications: graphite in situations where its refractory properties are important, and diamond in industrial applications where hardness is the critical factor (such as cutting and grinding tools).

Graphite   Graphite has a high content of crystalline carbon in the form of layers. Bonding between atoms in the layers is covalent and therefore strong, but the parallel layers are bonded to each other by weak van der Waals forces. This structure makes graphite quite anisotropic; strength and other properties vary significantly with direction. It explains why graphite can be used both as a lubricant and as a fiber in advanced composite materials. In powder form, graphite possesses low frictional characteristics due to the ease with which it shears between the layers; in this form, graphite is valued as a lubricant. In fiber form, graphite is oriented in the hexagonal planar direction to produce a filament material of very high strength and elastic modulus. These graphite fibers are used in structural composites ranging from tennis rackets to fighter aircraft components.

Graphite exhibits certain high temperature properties that are both useful and unusual. It is resistant to thermal shock, and its strength actually increases with temperature. Tensile strength at room temperature is about 100 MPa (14,500 lb/in^2), but increases to about twice this value at 2500°C (4500°F) [5]. Theoretical density of carbon is 2.22 g/cm^3, but apparent density of bulk graphite is lower due to porosity (around 1.7 g/cm^3). This is

TABLE 7.6   Some basic data and properties of carbon, silicon, and boron.

	Carbon	Silicon	Boron
Symbol	C	Si	B
Atomic number	6	14	5
Specific gravity	2.25	2.42	2.34
Melting temperature	3727°C[a] (6740°F)	1410°C (2570°F)	2030°C (3686°F)
Elastic modulus, GPa (lb/in^2)	240[b] (35×10^6)[c] 10357[c] (150×10^6)[c]	NA	393 (57×10^6)
Hardness (Mohs scale)	1[b], 10[c]	7	9.3

NA = not available.
[a]Carbon sublimes (vaporizes) rather than melt.
[b]Carbon in the form of graphite (typical value given).
[c]Carbon in the form of diamond.

increased through compacting and heating. It is electrically conductive, but its conductivity is not as high as most metals. A disadvantage of graphite is that it oxidizes in air above around 500°C (932°F). In a reducing atmosphere it can be used up to around 3000°C (5400°F), not far below its sublimation point of 3727°C (6740°F).

The traditional form of graphite is polycrystalline with a certain amount of amorphous carbon in the mixture. Graphite crystals are often oriented (to a limited degree) in the commercial production process to enhance properties in a preferred direction for the application. Also, strength is improved by reducing grain size (similar to ceramics). Graphite in this form is used for crucibles and other refractory applications, electrodes, resistance heating elements, antifriction materials, and fibers in composite materials. Thus, graphite is a very versatile material. As a powder it is a lubricant. In traditional solid form it is a refractory. And when formed into graphite fibers, it is a high-strength structural material.

Diamond   Diamond is carbon that possesses a cubic crystalline structure with covalent bonding between atoms, as shown in Figure 2.5(b). This structure is three-dimensional, rather than layered as in graphite carbon, and this accounts for the very high hardness of diamond. Single crystal natural diamonds (mined in South Africa) have a hardness of 10,000 HV, while the hardness of an industrial diamond (polycrystalline) is around 7000 HV. The high hardness accounts for most of the applications of industrial diamond. It is used in cutting tools and grinding wheels for machining hard, brittle materials, or materials that are very abrasive. For example, diamond tools and wheels are used to cut ceramics, fiberglass, and hardened metals other than steels. Diamond is also used in dressing tools to sharpen grinding wheels that consist of other abrasives such as alumina and silicon carbide. Similar to graphite, diamond has a propensity to oxidize (decompose) in air at temperatures above about 650°C (1200°F).

Industrial or synthetic diamonds date back to the 1950s and are fabricated by heating graphite to around 3000°C (5400°F) under very high pressures (Figure 7.2). This process approximates the geological conditions by which natural diamonds were formed millions of years ago.

FIGURE 7.2
Synthetically produced diamond powders. (Photo courtesy of GE Superabrasives, General Electric Company.)

7.5.2 SILICON

Silicon is a semimetallic element in the same group in the periodic table as carbon (Figure 2.1). Silicon is one of the most abundant elements in the Earth's crust, comprising about 26% by weight (Table 7.1). It occurs naturally only as a chemical compound—in rocks, sand, clay, and soil—either as silicon dioxide or as more complex silicate compounds. As an element it has the same crystalline structure as diamond, but its hardness is lower. It is hard but brittle, lightweight, chemically inactive at room temperature, and is classified as a semiconductor.

The greatest amounts of silicon in manufacturing are in ceramic compounds (SiO_2 in glass and silicates in clays) and alloying elements in steel, aluminum, and copper alloys. It is also used as a reducing agent in certain metallurgical processes. Of significant technological importance is pure silicon as the base material in semiconductor manufacturing in electronics. The vast majority of integrated circuits produced today are made from silicon (Chapter 34).

7.5.3 BORON

Boron is a semimetallic element in the same periodic group as aluminum. It is only about 0.001% of the Earth's crust by weight, commonly occurring as the minerals **borax** ($Na_2B_4O_7–10H_2O$) and **kernite** ($Na_2B_4O_7–4H_2O$). Boron is lightweight and very stiff (high modulus of elasticity) in fiber form. In terms of electrical properties, it is classified as a semiconductor (its conductivity varies with temperature; it is an insulator at low temperatures but a conductor at high temperatures).

As a material of industrial significance, boron is usually found in compound form. As such, it is used as a solution in nickel electroplating operations, an ingredient (B_2O_3) in certain glass compositions, a catalyst in organic chemical reactions, and as a nitride (cubic boron nitride) for cutting tools. In nearly pure form it is used as a fiber in composite materials (Sections 9.4.1 and 15.1.2).

7.6 GUIDE TO PROCESSING CERAMICS

The processing of ceramics can be divided into two basic categories: molten ceramics and particulate ceramics. The major category of molten ceramics is glassworking (Chapter 12). Particulate ceramics include traditional and new ceramics; their processing methods constitute most of the rest of the shaping technologies for ceramics (Chapter 17). Cermets, such as cemented carbides, are a special case because they are metal matrix composites (Section 17.3). Table 7.7 provides a guide to the processing of ceramic materials and the elements carbon, silicon, and boron.

TABLE 7.7 Guide to the processing of ceramic materials and the elements carbon, silicon, and boron.

Material	Chapter or Section	Material	Chapter or Section
Glass	Chapter 12	Synthetic diamonds	Section 23.2.6
Glass fibers	Section 12.2.3	Silicon	Section 35.2
Particulate ceramics	Chapter 17	Carbon fibers	Section 15.1.2
Cermets	Section 17.3	Boron fibers	Section 15.1.2

REFERENCES

[1] Carter, C. B., and Norton, M. G. *Ceramic Materials: Science and Engineering.* Springer, New York, 2007.

[2] Chiang, Y-M., Birnie, III, D. P., and Kingery, W. D. *Physical Ceramics.* John Wiley & Sons, Inc., New York, 1997.

[3] *Engineered Materials Handbook,* Vol. 4, *Ceramics and Glasses.* ASM International, Materials Park, Ohio, 1991.

[4] Flinn, R. A., and Trojan, P. K. *Engineering Materials and Their Applications,* 5th ed. John Wiley & Sons, Inc., New York, 1995.

[5] Hlavac, J. *The Technology of Glass and Ceramics.* Elsevier Scientific Publishing Company, New York, 1983.

[6] Kingery, W. D., Bowen, H. K., and Uhlmann, D. R. *Introduction to Ceramics,* 2nd ed. John Wiley & Sons, Inc., New York, 1995.

[7] Kirchner, H. P. *Strengthening of Ceramics.* Marcel Dekker, Inc., New York, 1979.

[8] Richerson, D. W. *Ceramics—Applications in Manufacturing.* Society of Manufacturing Engineers, Dearborn, Michigan, 1989.

[9] Richerson, D. W. *Modern Ceramic Engineering: Properties, Processing, and Use in Design,* 3rd ed. CRC Taylor & Francis, Boca Raton, Florida, 2006.

[10] Scholes, S. R., and Greene, C. H. *Modern Glass Practice,* 7th ed. CBI Publishing Company, Boston, 1993.

[11] Schwarzkopf, P., and Kieffer, R. *Cemented Carbides.* The Macmillan Company, New York, 1960.

[12] Singer, F., and Singer, S. S. *Industrial Ceramics.* Chemical Publishing Company, New York, 1963.

[13] Somiya, S. (ed.). *Advanced Technical Ceramics.* Academic Press, San Diego, California, 1989.

REVIEW QUESTIONS

7.1. What is a ceramic?
7.2. What are the four most common elements in the Earth's crust?
7.3. What is the difference between the traditional ceramics and the new ceramics?
7.4. What is the feature that distinguishes glass from the traditional and new ceramics?
7.5. What are the general mechanical properties of ceramic materials?
7.6. What are the general physical properties of ceramic materials?
7.7. What type of atomic bonding characterizes the ceramics?
7.8. What do bauxite and corundum have in common?
7.9. What is clay, as used in making ceramic products?
7.10. What is glazing, as applied to ceramics?
7.11. What does the term refractory mean?
7.12. What are some of the principal applications of cemented carbides, such as WC–Co?
7.13. What is one of the important applications of titanium nitride, as mentioned in the text?
7.14. What are the elements in the ceramic material Sialon?
7.15. Define glass.
7.16. What is the primary mineral in glass products?
7.17. What are some of the functions of the ingredients that are added to glass in addition to silica? Name at least three.
7.18. What does the term devitrification mean?
7.19. What is graphite?

MULTIPLE CHOICE QUIZ

There are 17 correct answers in the following multiple choice questions (some questions have multiple answers that are correct). To attain a perfect score on the quiz, all correct answers must be given. Each correct answer is worth 1 point. Each omitted answer or wrong answer reduces the score by 1 point, and each additional answer beyond the correct number of answers reduces the score by 1 point. Percentage score on the quiz is based on the total number of correct answers.

7.1. Which one of the following is the most common element in the Earth's crust: (a) aluminum, (b) calcium, (c) iron, (d) oxygen, or (e) silicon?
7.2. Glass products are based primarily on which one of the following minerals: (a) alumina, (b) corundum, (c) feldspar, (d) kaolinite, or (e) silica?
7.3. Which of the following contains significant amounts of aluminum oxide (three correct answers): (a) alumina, (b) bauxite, (c) corundum, (d) feldspar, (e) kaolinite, (f) quartz, (g) sandstone, and (h) silica?
7.4. Which of the following ceramics are commonly used as abrasives in grinding wheels (two best answers):

(a) aluminum oxide, (b) calcium oxide, (c) carbon monoxide, (d) silicon carbide, and (e) silicon dioxide?

7.5. Which one of the following is generally the most porous of the clay-based pottery ware: (a) china, (b) earthenware, (c) porcelain, or (d) stoneware?

7.6. Which one of the following is fired at the highest temperatures: (a) china, (b) earthenware, (c) porcelain, or (d) stoneware?

7.7. Which one of the following comes closest to expressing the chemical composition of clay: (a) Al_2O_3, (b) $Al_2(Si_2O_5)(OH)_4$, (c) $3AL_2O_3-2SiO_2$, (d) MgO, or (e) SiO_2?

7.8. Glass ceramics are polycrystalline ceramic structures that have been transformed into the glassy state: (a) true or (b) false?

7.9. Which one of the following materials is closest to diamond in hardness: (a) aluminum oxide, (b) carbon dioxide, (c) cubic boron nitride, (d) silicon dioxide, or (e) tungsten carbide?

7.10. Which of the following best characterizes the structure of glass-ceramics: (a) 95% polycrystalline, (b) 95% vitreous, or (c) 50% polycrystalline?

7.11. Properties and characteristics of the glass-ceramics include which of the following (two best answers): (a) efficiency in processing, (b) electrical conductor, (c) high-thermal expansion, and (d) strong, relative to other glasses?

7.12. Diamond is the hardest material known: (a) true or (b) false?

7.13. Synthetic diamonds date to (a) ancient times, (b) 1800s, (c) 1950s, or (d) 1980.

POLYMERS

Chapter Contents

Of the three basic types of materials, polymers are the newest and at the same time the oldest known to man. Polymers form the living organisms and vital processes of all life on Earth. To ancient man, biological polymers were the source of food, shelter, and many of his implements. However, our interest in this chapter is in polymers other than biological. With the exception of natural rubber, nearly all of the polymeric materials used in engineering today are synthetic. The materials themselves are made by chemical processing, and most of the products are made by solidification processes.

A *polymer* is a compound consisting of long-chain molecules, each molecule made up of repeating units connected together. There may be thousands, even millions of units in a single polymer molecule. The word is derived from the Greek words *poly,* meaning many, and *meros* (reduced to *mer*), meaning part. Most polymers are based on carbon and are therefore considered organic chemicals.

Polymers can be separated into *plastics* and *rubbers*. As engineering materials, they are relatively new compared to metals and ceramics, dating only from around the mid-1800s (Historical Note 8.1). For our purposes in covering polymers as a technical subject, it is appropriate to divide them into the following three categories, where (1) and (2) are plastics and (3) is the rubber category:

1. *Thermoplastic polymers,* also called *thermoplastics* (TP), are solid materials at room temperature, but they become viscous liquids when heated to temperatures of only a few hundred degrees. This characteristic allows them to be easily and economically shaped into products. They can be subjected to this heating and cooling cycle repeatedly without significant degradation of the polymer.

2. *Thermosetting polymers,* or *thermosets* (TS), cannot tolerate repeated heating cycles as thermoplastics can. When initially heated, they soften and flow for molding, but the elevated temperatures also produce a chemical reaction

---

Historical Note 8.1 *History of polymers*

Certainly one of the milestones in the history of polymers was Charles Goodyear's discovery of vulcanization of rubber in 1839 (Historical Note 8.2). In 1851, his brother Nelson patented hard rubber, called *ebonite*, which in reality is a thermosetting polymer. It was used for many years for combs, battery cases, and dental prostheses.

At the 1862 International Exhibition in London, an English chemist Alexander Parkes demonstrated the possibilities of the first thermoplastic, a form of *cellulose nitrate* (cellulose is a natural polymer in wood and cotton). He called it *Parkesine* and described it as a replacement for ivory and tortoiseshell. The material became commercially important due to the efforts of American John W. Hyatt, Jr., who combined cellulose nitrate and camphor (which acts as a plasticizer) together with heat and pressure to form the product he called *Celluloid*. His patent was issued in 1870. Celluloid plastic was transparent, and the applications subsequently developed for it included photographic and motion picture film and windshields for carriages and early motorcars.

Several additional products based on cellulose were developed around the turn of the last century. Cellulose fibers, called *Rayon*, were first produced around 1890. Packaging film, called *Cellophane*, was first marketed around 1910. *Cellulose acetate* was adopted as the base for photographic film around the same time. This material was to become an important thermoplastic for injection molding during the next several decades.

The first synthetic plastic was developed in the early 1900s by the Belgian-born American chemist L. H. Baekeland. It involved the reaction and polymerization of phenol and formaldehyde to form what its inventor called *Bakelite*. This thermosetting resin is still commercially important today. It was followed by other similar polymers: urea-formaldehyde in 1918 and melamineformaldehyde in 1939.

The late 1920s and 1930s saw the development of a number of thermoplastics of major importance today. A Russian I. Ostromislensky had patented *polyvinyl-chloride* in 1912, but it was first commercialized in 1927 as a wall covering. Around the same time, *polystyrene* was first produced in Germany. In England, fundamental research was started in 1932 that led to the synthesis of *polyethylene*; the first production plant came on line just before the outbreak of World War II. This was low density polyethylene. Finally, a major research program initiated in 1928 under the direction of W. Carothers at DuPont in the United States led to the synthesis of the polyamide *nylon*; it was commercialized in the late 1930s. Its initial use was in ladies' hosiery; subsequent applications during the war included low-friction bearings and wire insulation. Similar efforts in Germany provided an alternative form of nylon in 1939.

Several important special-purpose polymers were developed in the 1940s: *fluorocarbons (Teflon)*, *silicones*, and *polyurethanes* in 1943; *epoxy* resins in 1947, and *acrylonitrile-butadiene-styrene* copolymer (ABS) in 1948. During the 1950s: *polyester* fibers in 1950; and *polypropylene*, *polycarbonate*, and *high-density polyethylene* in 1957. *Thermoplastic elastomers* were first developed in the 1960s. The ensuing years have witnessed a tremendous growth in the use of plastics.

---

that hardens the material into an infusible solid. If reheated, thermosetting polymers degrade and char rather than soften.

3. *Elastomers* are the rubbers. Elastomers (E) are polymers that exhibit extreme elastic extensibility when subjected to relatively low mechanical stress. Some elastomers can be stretched by a factor of 10 and yet completely recover to their original shape. Although their properties are quite different from thermosets, they have a similar molecular structure that is different from the thermoplastics.

Thermoplastics are commercially the most important of the three types, constituting around 70% of the tonnage of all synthetic polymers produced. Thermosets and elastomers share the remaining 30% about evenly, with a slight edge for the former. Common TP polymers include polyethylene, polyvinylchloride, polypropylene, polystyrene, and nylon. Examples of TS polymers are phenolics, epoxies, and certain polyesters. The most common example given for elastomers is natural (vulcanized) rubber; however, synthetic rubbers exceed the tonnage of natural rubber.

Although the classification of polymers into the TP, TS, and E categories suits our purposes for organizing the topic in this chapter, we should note that the three types sometimes overlap. Certain polymers that are normally thermoplastic can be made into thermosets. Some polymers can be either thermosets or elastomers (we indicated that their molecular structures are similar). And some elastomers are thermoplastic. However, these are exceptions to the general classification scheme.

The growth in applications of synthetic polymers is truly impressive. On a volumetric basis, current annual usage of polymers exceeds that of metals. There are several reasons for the commercial and technological importance of polymers:

➤ Plastics can be formed by molding into intricate part geometries, usually with no further processing required. They are very compatible with *net shape* processing.

➤ Plastics possess an attractive list of properties for many engineering applications where strength is not a factor: (1) low density relative to metals and ceramics; (2) good strength-to-weight ratios for certain (but not all) polymers; (3) high corrosion resistance; and (4) low electrical and thermal conductivity.

➤ On a volumetric basis, polymers are cost-competitive with metals.

➤ On a volumetric basis, polymers generally require less energy to produce than metals. This is generally true because the temperatures for working these materials are much lower than for metals.

➤ Certain plastics are translucent and/or transparent, which makes them competitive with glass in some applications.

➤ Polymers are widely used in composite materials (Chapter 9).

On the negative side, polymers in general have the following limitations: (1) strength is low relative to metals and ceramics; (2) modulus of elasticity or stiffness is also low—in the case of elastomers, of course, this may be a desirable characteristic; (3) service temperatures are limited to only a few hundred degrees because of the softening of thermoplastic polymers or degradation of thermosetting polymers and elastomers; (4) some polymers degrade when subjected to sunlight and other forms of radiation; and (5) plastics exhibit viscoelastic properties (Section 3.5), which can be a distinct limitation in load bearing applications.

In this chapter we examine the technology of polymeric materials. The first section is devoted to an introductory discussion of polymer science and technology. Subsequent sections survey the three basic categories of polymers: thermoplastics, thermosets, and elastomers.

8.1 FUNDAMENTALS OF POLYMER SCIENCE AND TECHNOLOGY

Polymers are synthesized by joining many small molecules together to form very large molecules, called *macromolecules,* that possess a chain-like structure. The small units, called *monomers,* are generally simple unsaturated organic molecules such as ethylene (C_2H_4). The atoms in these molecules are held together by covalent bonds; and when joined to form the polymer, the same covalent bonding holds the links of the chain together. Thus, each large molecule is characterized by strong primary bonding. Synthesis of the polyethylene molecule is depicted in Figure 8.1. As we have described its structure here, polyethylene is a linear polymer; its mers form one long chain.

A mass of polymer material consists of many macromolecules; the analogy of a bowl of just-cooked spaghetti (without sauce) is sometimes used to visualize the relationship of the individual molecules to the bulk material. Entanglement among the long strands helps

FIGURE 8.1 Synthesis of polyethylene from ethylene monomers: (1) *n* ethylene monomers yields (2a) polyethylene of chain length *n*; (2b) concise notation for depicting the polymer structure of chain length *n*.

to hold the mass together, but atomic bonding is more significant. The bonding between macromolecules in the mass is due to van der Waals and other secondary bonding types. Thus, the aggregate polymer material is held together by forces that are substantially weaker than the primary bonds holding the molecules together. This explains why plastics in general are not nearly as stiff and strong as metals or ceramics.

When a thermoplastic polymer is heated, it softens. The heat energy causes the macromolecules to become thermally agitated, exciting them to move relative to each other within the polymer mass (here, the wet spaghetti analogy loses its appeal). The material begins to behave like a viscous liquid, viscosity decreasing (fluidity increasing) with rising temperature.

Let us expand on these opening remarks, tracing how polymers are synthesized and examining the characteristics of the materials that result from the synthesis.

8.1.1 POLYMERIZATION

As a chemical process, the synthesis of polymers can occur by either of two methods: (1) addition polymerization and (2) step polymerization. Production of a given polymer is generally associated with one method or the other.

Addition Polymerization In this process, exemplified by polyethylene, the double bonds between carbon atoms in the ethylene monomers are induced to open so that they join with other monomer molecules. The connections occur on both ends of the expanding macromolecule, developing long chains of repeating mers. Because of the way the molecules are formed, the process is also known as ***chain polymerization***. It is initiated using a chemical catalyst (called an ***initiator***) to open the carbon double bond in some of the monomers. These monomers, which are now highly reactive because of their unpaired electrons, then capture other monomers to begin forming chains that are reactive. The chains propagate by capturing still other monomers, one at a time, until large molecules have been produced and the reaction is terminated. The process proceeds as indicated in Figure 8.2. The entire polymerization reaction takes only seconds for any given macromolecule. However, in the industrial process, it may take many minutes or even hours to complete the polymerization of a given batch, since all of the chain reactions do not occur simultaneously in the mixture.

FIGURE 8.2 Model of addition (chain) polymerization: (1) initiation, (2) rapid addition of monomers, and (3) resulting long-chain polymer molecule with *n* mers at termination of reaction.

Polymer	Monomer	Repeating mer	Chemical formula
Polypropylene	H H │ │ C = C │ │ H CH$_3$	⎡ H H ⎤ ─C ─ C─ │ │ ⎣ H CH$_3$ ⎦$_n$	$(C_3H_6)_n$
Polyvinyl chloride	H H │ │ C = C │ │ H Cl	⎡ H H ⎤ ─C ─ C─ │ │ ⎣ H Cl ⎦$_n$	$(C_2H_3Cl)_n$
Polystyrene	H H │ │ C = C │ │ H C$_6$H$_5$	⎡ H H ⎤ ─C ─ C─ │ │ ⎣ H C$_6$H$_5$ ⎦$_n$	$(C_8H_8)_n$
Polytetrafluoroethylene (Teflon)	F F │ │ C = C │ │ F F	⎡ F F ⎤ ─C ─ C─ │ │ ⎣ F F ⎦$_n$	$(C_2F_4)_n$
Polyisoprene (natural rubber)	H H H │ │ │ C ─ C = C ─ C │ │ │ H CH$_3$ H	⎡ H H H ⎤ ─C ─ C = C ─ C─ │ │ │ ⎣ H CH$_3$ H ⎦$_n$	$(C_5H_8)_n$

FIGURE 8.3 Some typical polymers formed by addition (chain) polymerization.

Other polymers typically formed by addition polymerization are presented in Figure 8.3, along with the starting monomer and the repeating mer. Note that the chemical formula for the monomer is the same as that of the mer in the polymer. This is a characteristic of this method of polymerization. Note also that many of the common polymers involve substitution of some alternative atom or molecule in place of one of the H atoms in polyethylene. Polypropylene, polyvinylchloride, and polystyrene are examples of this substitution. Polytetrafluoroethylene replaces all four H atoms in the structure with atoms of fluorine (F). Most addition polymers are thermoplastics. The exception in Figure 8.3 is polyisoprene, the polymer of natural rubber. Although formed by addition polymerization, it is an elastomer.

Step Polymerization   In this form of polymerization, two reacting monomers are brought together to form a new molecule of the desired compound. In most (but not all) step polymerization processes, a byproduct of the reaction is also produced. The byproduct is typically water, which condenses; hence, the term *condensation polymerization* is often used for processes that yield the condensate. As the reaction continues, more molecules of the reactants combine with the molecules first synthesized to form polymers of length $n = 2$, then polymers of length $n = 3$, and so on. Polymers of increasing n are created in a slow, stepwise fashion. In addition to this gradual elongation of the molecules, intermediate polymers of length n_1 and n_2 also combine to form molecules of length $n = n_1 + n_2$, so that two types of reactions are proceeding simultaneously once the process is under way, as illustrated in Figure 8.4. Accordingly, at any point in the process, the batch contains polymers of various lengths. Only after sufficient time has elapsed are molecules of adequate length formed.

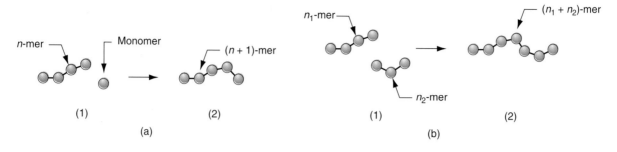

FIGURE 8.4   Model of step polymerization showing the two types of reactions occurring: (a) *n*-mer attaching a single monomer to form a (*n* + 1) -mer; and (b) n_1-mer combining with n_2-mer to form a ($n_1 + n_2$) -mer. Sequence is shown by (1) and (2).

It should be noted that water is not always the byproduct of the reaction; for example, ammonia (NH_3) is another simple compound produced in some reactions. Nevertheless, the term condensation polymerization is still used. It should also be noted that although most step polymerization processes involve condensation of a byproduct, some do not. Examples of commercial polymers produced by step (condensation) polymerization are given in Figure 8.5. Both thermoplastic and thermosetting polymers are synthesized by this method; nylon-6,6 and polycarbonate are TP polymers, while phenol formaldehyde and urea formaldehyde are TS polymers.

Degree of Polymerization and Molecular Weight   A macromolecule produced by polymerization consists of *n* repeating mers. Since molecules in a given batch of polymerized

Polymer	Repeating unit	Chemical formula	Condensate
Nylon-6, 6	$\left[\begin{pmatrix} H \\ \| \\ C \\ \| \\ H \end{pmatrix}_6 - N - \overset{O}{\overset{\|}{C}} - \begin{pmatrix} H \\ \| \\ C \\ \| \\ H \end{pmatrix}_4 - \overset{O}{\overset{\|}{C}} - N \right]_n$	$[(CH_2)_6 (CONH)_2 (CH_2)_4]_n$	H_2O
Polycarbonate	$+(C_6H_4) - \overset{CH_3}{\underset{CH_3}{\overset{\|}{\underset{\|}{C}}}} - (C_6H_4) - O - \overset{O}{\overset{\|}{C}} - O \, \}_n$	$(C_3H_6 (C_6H_4)_2 CO_3)_n$	HCl
Phenol formaldehyde	$+ C_6H_4 - \overset{H}{\underset{OH}{\overset{\|}{\underset{\|}{C}}}} \overset{H}{\underset{}{}} \}_n$	$[(C_6H_4)CH_2OH]_n$	H_2O
Urea formaldehyde	$+ \overset{NH}{\underset{NH}{\overset{\|}{\underset{\|}{C}}}} = O - \overset{H}{\underset{H}{\overset{\|}{\underset{\|}{C}}}} \}_n$	$(CO(NH)_2 CH_2)_n$	H_2O

FIGURE 8.5   Some typical polymers formed by step (condensation) polymerization (simplified expression of structure and formula; ends of polymer chain are not shown).

TABLE 8.1   Typical values of degree of polymerization and molecular weight for selected thermoplastic polymers.

Polymer	Degree of Polymerization (n)	Molecular Weight
Polyethylene	10,000	300,000
Polystyrene	3,000	300,000
Polyvinylchloride	1,500	100,000
Nylon	120	15,000
Polycarbonate	200	40,000

Compiled from [7].

material vary in length, n for the batch is an average; its statistical distribution is normal. The mean value of n is called the ***degree of polymerization*** (DP) for the batch. The degree of polymerization affects the properties of the polymer: higher DP increases mechanical strength but also increases viscosity in the fluid state, which makes processing more difficult.

The ***molecular weight*** (MW) of a polymer is the sum of the molecular weights of the mers in the molecule; it is n times the molecular weight of each repeating unit. Since n varies for different molecules in a batch, the molecule weight must be interpreted as an average. Typical values of DP and MW for selected polymers are presented in Table 8.1.

8.1.2 POLYMER STRUCTURES AND COPOLYMERS

There are structural differences among polymer molecules, even molecules of the same polymer. In this section we examine three aspects of molecular structure: (1) stereo-regularity, (2) branching and cross-linking, and (3) copolymers.

Stereoregularity   Stereoregularity is concerned with the spatial arrangement of the atoms and groups of atoms in the repeating units of the polymer molecule. An important aspect of stereoregularity is the way the atom groups are located along the chain for a polymer that has one of the H atoms in its mers replaced by some other atom or atom group. Polypropylene is an example; it is similar to polyethylene except that CH_3 is substituted for one of the four H atoms in the mer. Three tactic arrangements are possible, illustrated in Figure 8.6: (a) ***isotactic,*** in which the odd atom groups are all on the same side; (b) ***syndiotactic,*** in which the atom groups alternate on opposite sides; and (c) ***atactic,*** in which the groups are randomly along either side.

The tactic structure is important in determining the properties of the polymer. It also influences the tendency of a polymer to crystallize (Section 8.1.3). Continuing with

FIGURE 8.6 Possible arrangement of atom groups in polypropylene: (a) isotactic, (b) syndiotactic, and (c) atactic.

FIGURE 8.7   Various structures of polymer molecules: (a) linear, characteristic of thermoplastics; (b) branched; (c) loosely cross-linked as in an elastomer; and (d) tightly cross-linked or networked structure as in a thermoset.

our polypropylene example, this polymer can be synthesized in any of the three tactic structures. In its isotactic form, it is strong and melts at 175°C (347°F); the syndiotactic structure is also strong, but melts at 131°C (268°F); but atactic polypropylene is soft and melts at around 75°C (167°F) and has little commercial use [6], [9].

Linear, Branched, and Cross-Linked Polymers   We have described the polymerization process as yielding macromolecules of a chain-like structure, called a *linear polymer*. This is the characteristic structure of a thermoplastic polymer. Other structures are possible, as portrayed in Figure 8.7. One possibility is for side branches to form along the chain, resulting in the *branched polymer* shown in Figure 8.7(b). In polyethylene, this occurs because hydrogen atoms are replaced by carbon atoms at random points along the chain, initiating the growth of a branch chain at each location. For certain polymers, primary bonding occurs between branches and other molecules at certain connection points to form *cross-linked polymers* as pictured in Figure 8.7(c) and (d). Cross-linking occurs because a certain proportion of the monomers used to form the polymer are capable of bonding to adjacent monomers on more than two sides, thus allowing branches from other molecules to attach. Lightly cross-linked structures are characteristic of elastomers. When the polymer is highly cross-linked we refer to it as having a *network structure,* as in (d); in effect, the entire mass is one gigantic macromolecule. Thermosetting plastics take this structure after curing.

The presence of branching and cross-linking in polymers has a significant effect on properties. It is the basis of the difference between the three categories of polymers: TP, TS, and E. Thermoplastic polymers always possess linear or branched structures, or a mixture of the two. Branching increases entanglement among the molecules, usually making the polymer stronger in the solid state and more viscous at a given temperature in the plastic or liquid state.

Thermosetting plastics and elastomers are cross-linked polymers. Cross-linking causes the polymer to become chemically set; the reaction cannot be reversed. The effect

FIGURE 8.8 Various structures of copolymers: (a) alternating, (b) random, (c) block, and (d) graft.

is to permanently change the structure of the polymer; upon heating, it degrades or burns rather than melts. Thermosets possess a high degree of cross-linking, while elastomers possess a low degree of cross-linking. Thermosets are hard and brittle, while elastomers are elastic and resilient.

Copolymers   Polyethylene is a ***homopolymer;*** so are polypropylene, polystyrene, and many other common plastics; their molecules consist of repeating mers that are all the same type. ***Copolymers*** are polymers whose molecules are made of repeating units of two different types. An example is the copolymer synthesized from ethylene and propylene to produce a copolymer with elastomeric properties. The ethylene-propylene copolymer can be represented as follows:

$$-(C_2H_4)_n(C_3H_6)_m-$$

where n and m range between 10 and 20, and the proportions of the two constituents are around 50% each. We find in Section 8.4.3 that the combination of polyethylene and polypropylene with small amounts of diene is an important synthetic rubber.

Copolymers can possess different arrangements of their constituent mers. The possibilities are shown in Figure 8.8: (a) ***alternating copolymer,*** in which the mers repeat every other place; (b) ***random,*** in which the mers are in random order, the frequency depending on the relative proportions of the starting monomers; (c) ***block,*** in which mers of the same type tend to group themselves into long segments along the chain; and (d) ***graft,*** in which mers of one type are attached as branches to a main backbone of mers of the other type. The ethylene–propylene diene rubber, mentioned previously, is a block type.

Synthesis of copolymers is analogous to alloying of metals to form solid solutions. As with metallic alloys, differences in the ingredients and structure of copolymers can have a substantial effect on properties. An example is the polyethylene–polypropylene mixture we have been discussing. Each of these polymers alone is fairly stiff; yet a 50–50 mixture forms a copolymer of random structure that is rubbery.

It is also possible to synthesize ***ternary polymers,*** or ***terpolymers,*** which consist of mers of three different types. An example is the plastic ABS (acrylonitrile–butadiene–styrene—no wonder they call it ABS).

8.1.3 CRYSTALLINITY

Both amorphous and crystalline structures are possible with polymers, although the tendency to crystallize is much less than for metals or nonglass ceramics. Not all polymers can form crystals. For those that can, the ***degree of crystallinity*** (the proportion of

TABLE 8.2	Comparison of low-density polyethylene and high-density polyethylene.	
Polyethylene Type	**Low Density**	**High Density**
Degree of crystallinity	55%	92%
Specific gravity	0.92	0.96
Modulus of elasticity	140 MPa (20,305 lb/in^2)	700 MPa (101,530 lb/in^2)
Melting temperature	115°C (239°F)	135°C (275°F)

Compiled from [6]. Values given are typical.

crystallized material in the mass) is always less than 100%. As crystallinity is increased in a polymer, so are (1) density, (2) stiffness, strength, and toughness, and (3) heat resistance. In addition, (4) if the polymer is transparent in the amorphous state, it becomes opaque when partially crystallized. Many polymers are transparent, but only in the amorphous (glassy) state. Some of these effects can be illustrated by the differences between low-density and high-density polyethylene, presented in Table 8.2. The underlying reason for the property differences between these materials is the degree of crystallinity.

Linear polymers consist of long molecules with thousands of repeated mers. Crystallization in these polymers involves the folding back and forth of the long chains upon themselves to achieve a very regular arrangement of the mers, as pictured in Figure 8.9(a). The crystallized regions are called **_crystallites_**. Owing to the tremendous length of a single molecule (on an atomic scale), it may participate in more than one crystallite. Also, more than one molecule may be combined in a single crystal region. The crystallites take the form of lamellae, as pictured in Figure 8.9(b), that are randomly mixed in with the amorphous material. Thus, a polymer that crystallizes is a two-phase system—crystallites interspersed throughout an amorphous matrix.

A number of factors determine the capacity and/or tendency of a polymer to form crystalline regions within the material. The factors can be summarized as follows: (1) as a general rule, only linear polymers can form crystals; (2) stereoregularity of the molecule is critical [15]: isotactic polymers always form crystals; syndiotactic polymers sometimes form

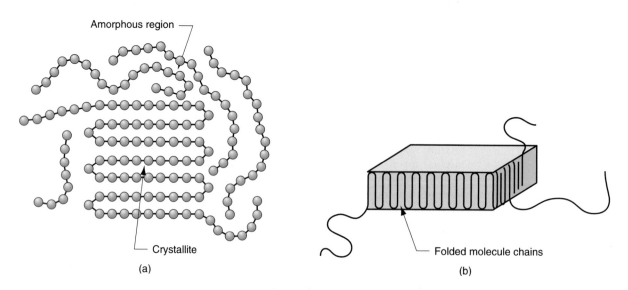

FIGURE 8.9 Crystallized regions in a polymer: (a) long molecules forming crystals randomly mixed in with the amorphous material; and (b) folded chain lamella, the typical form of a crystallized region.

crystals; atactic polymers never form crystals; (3) copolymers, due to their molecular irregularity, rarely form crystals; (4) slower cooling promotes crystal formation and growth, as it does in metals and ceramics; (5) mechanical deformation, as in the stretching of a heated thermoplastic, tends to align the structure and increase crystallization; and (6) plasticizers (chemicals added to a polymer to soften it) reduce the degree of crystallinity.

8.1.4 THERMAL BEHAVIOR OF POLYMERS

The thermal behavior of polymers with crystalline structures is different from that of amorphous polymers (Section 2.4). The effect of structure can be observed on a plot of specific volume (reciprocal of density) as a function of temperature, as plotted in Figure 8.10. A highly crystalline polymer has a melting point T_m at which its volume undergoes an abrupt change. Also, at temperatures above T_m, the thermal expansion of the molten material is greater than for the solid material below T_m. An amorphous polymer does not undergo the same abrupt changes at T_m. As it is cooled from the liquid, its coefficient of thermal expansion continues to decline along the same trajectory as when it was molten, and it becomes increasingly viscous with decreasing temperature. During cooling below T_m, the polymer changes from liquid to rubbery. As temperature continues to drop, a point is finally reached at which the thermal expansion of the amorphous polymer suddenly becomes lower. This is the **_glass-transition temperature,_** T_g (Section 3.5), seen as the change in slope. Below T_g, the material is hard and brittle.

A partially crystallized polymer lies between these two extremes, as indicated in Figure 8.10. It is an average of the amorphous and crystalline states, the average depending on the degree of crystallinity. Above T_m it exhibits the viscous characteristics of a liquid; between T_m and T_g it has viscoelastic properties; and below T_g it has the conventional elastic properties of a solid.

What we have described in this section applies to thermoplastic materials, which can move up and down the curve of Figure 8.10 multiple times. The manner in which they are heated and cooled may change the path that is followed. For example, fast cooling rates may inhibit crystal formation and increase the glass-transition temperature. Thermosets and elastomers cooled from the liquid state behave like an amorphous polymer until cross-linking occurs. Their molecular structure restricts the formation of crystals. And once their molecules are cross-linked, they cannot be reheated to the molten state.

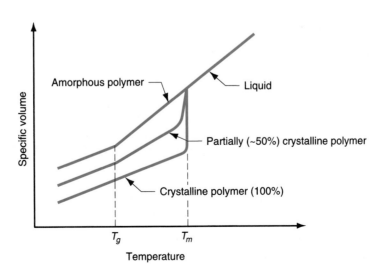

FIGURE 8.10 Behavior of polymers as a function of temperature.

8.1.5 ADDITIVES

The properties of a polymer can often be beneficially changed by combining them with additives. Additives either alter the molecular structure of the polymer or add a second phase to the plastic, in effect transforming a polymer into a composite material. Additives can be classified by function as (1) fillers, (2) plasticizers, (3) colorants, (4) lubricants, (5) flame retardants, (6) cross-linking agents, (7) ultraviolet light absorbers, and (8) antioxidants.

Filler   *Fillers* are solid materials added to a polymer usually in particulate or fibrous form to alter its mechanical properties or to simply reduce material cost. Other reasons for using fillers are to improve dimensional and thermal stability. Examples of fillers used in polymers include cellulosic fibers and powders (e.g., cotton fibers and wood flour, respectively); powders of silica (SiO_2), calcium carbonate ($CaCO_3$), and clay (hydrous aluminum silicate); and fibers of glass, metal, carbon, or other polymers. Fillers that improve mechanical properties are called *reinforcing agents,* and composites thus created are referred to as *reinforced plastics*; they have higher stiffness, strength, hardness, and toughness than the original polymer. Fibers provide the greatest strengthening effect.

Plasticizers   *Plasticizers* are chemicals added to a polymer to make it softer and more flexible, and to improve its flow characteristics during forming. The plasticizer works by reducing the glass transition temperature to below room temperature. Whereas the polymer is hard and brittle below T_g, it is soft and tough above it. Addition of a plasticizer[1] to polyvinylchloride (PVC) is a good example; depending on the proportion of plasticizer in the mix, PVC can be obtained in a range of properties, from rigid and brittle to flexible and rubbery.

Colorants   An advantage of many polymers over metals or ceramics is that the material itself can be obtained in most any color. This eliminates the need for secondary coating operations. Colorants for polymers are of two types: pigments and dies. *Pigments* are finely powdered materials that are insoluble in and must be uniformly distributed throughout the polymer in very low concentrations, usually less than 1%. They often add opacity as well as color to the plastic. *Dies* are chemicals, usually supplied in liquid form, that are generally soluble in the polymer. They are normally used to color transparent plastics such as styrene and acrylics.

Other Additives   *Lubricants* are sometimes added to the polymer to reduce friction and promote flow at the mold interface. Lubricants are also helpful in releasing the part from the mold in injection molding. Mold-release agents, sprayed onto the mold surface, are often used for the same purpose.

Nearly all polymers burn if the required heat and oxygen are supplied. Some polymers are more combustible than others. *Flame retardants* are chemicals added to polymers to reduce flammability by any or a combination of the following mechanisms: (1) interfering with flame propagation, (2) producing large amounts of incombustible gases, and/or (3) increasing the combustion temperature of the material. The chemicals may also function to (4) reduce the emission of noxious or toxic gases generated during combustion.

We should include among the additives those that cause cross-linking to occur in thermosetting polymers and elastomers. The term *cross-linking agent* refers to a variety of ingredients that cause a cross-linking reaction or act as a catalyst to promote such a reaction. Important commercial examples are (1) sulfur in vulcanization of natural rubber, (2) formaldehyde for phenolics to form phenolic thermosetting plastics, and (3) peroxides for polyesters.

---

[1]The common plasticizer in PVC is dioctyl phthalate, a phthalate ester.

Many polymers are susceptible to degradation by ultraviolet light (e.g., from sunlight) and oxidation. The degradation manifests itself as the breaking of links in the long chain molecules. Polyethylene, for example, is vulnerable to both types of degradation, which lead to a loss of mechanical strength. ***Ultraviolet light absorbers*** and ***antioxidants*** are additives that reduce the susceptibility of the polymer to these forms of attack.

8.2 THERMOPLASTIC POLYMERS

In this section, we discuss the properties of the thermoplastic polymer group and then survey its important members.

8.2.1 PROPERTIES OF THERMOPLASTIC POLYMERS

The defining property of a thermoplastic polymer is that it can be heated from a solid state to a viscous liquid state and then cooled back down to solid, and that this heating and cooling cycle can be applied multiple times without degrading the polymer. The reason for this property is that TP polymers consist of linear (and/or branched) macromolecules that do not cross-link when heated. By contrast, thermosets and elastomers undergo a chemical change when heated, which cross-links their molecules and permanently sets these polymers.

In truth, thermoplastics do deteriorate chemically with repeated heating and cooling. In plastic molding, a distinction is made between new or ***virgin*** material, and plastic that has been previously molded (e.g., sprues, defective parts) and therefore has experienced thermal cycling. For some applications, only virgin material is acceptable. Thermoplastic polymers also degrade gradually when subjected to continuous elevated temperatures below T_m. This long-term effect is called ***thermal aging*** and involves slow chemical deterioration. Some TP polymers are more susceptible to thermal aging than others, and for a given material the rate of deterioration depends on temperature.

Mechanical Properties   In our discussion of mechanical properties in Chapter 3, we compared polymers to metals and ceramics. The typical thermoplastic at room temperature is characterized by the following: (1) much lower stiffness, the modulus of elasticity being two (in some cases, three) orders of magnitude lower than metals and ceramics; (2) lower tensile strength, about 10% of the metals; (3) much lower hardness; and (4) greater ductility on average, but there is a tremendous range of values, from 1% elongation for polystyrene to 500% or more for polypropylene.

Mechanical properties of thermoplastics depend on temperature. The functional relationships must be discussed in the context of amorphous and crystalline structures. Amorphous thermoplastics are rigid and glass-like below their glass transition temperature T_g and flexible or rubber-like just above it. As temperature increases above T_g, the polymer becomes increasingly soft, finally becoming a viscous fluid (it never becomes a thin liquid due to its high molecular weight). The effect on mechanical behavior can be portrayed as in Figure 8.11, in which mechanical behavior is defined as deformation resistance. This is analogous to modulus of elasticity but it allows us to observe the effect of temperature on the amorphous polymer as it transitions from solid to liquid. Below T_g, the material is elastic and strong. At T_g, a rather sudden drop in deformation resistance is observed as the material transforms into its rubbery phase; its behavior is viscoelastic in this region. As temperature increases, it gradually becomes more fluid-like.

A theoretical thermoplastic with 100% crystallinity would have a distinct melting point T_m at which it transforms from solid to liquid, but would show no perceptible T_g point. Of course, real polymers have less than 100% crystallinity. For partially crystallized polymers, the resistance to deformation is characterized by the curve that lies between

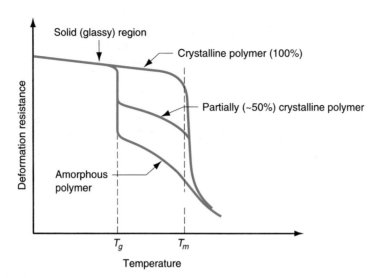

FIGURE 8.11
Relationship of mechanical properties, portrayed as deformation resistance, as a function of temperature for an amorphous thermoplastic, a 100% crystalline (theoretical) thermoplastic, and a partially crystallized thermoplastic.

the two extremes, its position determined by the relative proportions of the two phases. The partially crystallized polymer exhibits features of both amorphous and fully crystallized plastics. Below T_g, it is elastic with deformation resistance sloping downward with rising temperatures. Above T_g, the amorphous portions of the polymer soften, while the crystalline portions remain intact. The bulk material exhibits properties that are generally viscoelastic. As T_m is reached, the crystals now melt, giving the polymer a liquid consistency; resistance to deformation is now due to the fluid's viscous properties. The degree to which the polymer assumes liquid characteristics at and above T_m depends on molecular weight and degree of polymerization. Higher DP and MW reduce flow of the polymer, making it more difficult to process by molding and similar shaping methods. This is a dilemma faced by those who select these materials because higher MW and DP mean higher strength.

Physical Properties   Physical properties of materials are discussed in Chapter 4. In general, thermoplastic polymers have the following characteristics: (1) lower densities than metals or ceramics—typical specific gravities for polymers are around 1.2, for ceramics around 2.5, and for metals around 7.0; (2) much higher coefficient of thermal expansion—roughly 5 times the value for metals and 10 times the value for ceramics; (3) much lower melting temperatures; (4) specific heats that are 2 to 4 times those of metals and ceramics; (5) thermal conductivities that are about three orders of magnitude lower than those of metals; and (6) insulating electrical properties.

8.2.2   IMPORTANT COMMERCIAL THERMOPLASTICS

Thermoplastic products include molded and extruded items, fibers, films, sheets, packaging materials, paints, and varnishes. The starting raw materials for these products are normally supplied to the fabricator in the form of powders or pellets in bags, drums, or larger loads by truck or rail car. The most important TP polymers are discussed in alphabetical order in this section. For each plastic, Table 8.3 lists the chemical formula and selected properties. Approximate market share is given relative to all plastics (thermoplastic and thermosetting).

Acetals   *Acetal* is the popular name given to ***polyoxymethylene***, an engineering polymer prepared from formaldehyde (CH_2O) with high stiffness, strength, toughness, and wear resistance. In addition, it has a high melting point, low moisture absorption, and is insoluble

TABLE 8.3   Important commercial thermoplastic polymers: (a) acetal.

Polymer:	Polyoxymethylene, also known as polyacetal $(OCH_2)_n$		
Symbol:	POM	Elongation:	25%–75%
Polymerization method:	Step (condensation)	Specific gravity:	1.42
Degree of crystallinity:	75% typical	Glass transition temperature:	$-80°C$ $(-112°F)$
Modulus of elasticity:	3500 MPa (507,630 lb/in^2)	Melting temperature:	180°C (356°F)
Tensile strength:	70 MPa (10,150 lb/in^2)	Approximate market share:	Much less than 1%

Table 8.3 is compiled from [2], [4], [6], [7], [9], [16], and other sources.

TABLE 8.3   (continued): (b) acrylics (thermoplastic).

Representative polymer:	Polymethylmethacrylate $(C_5H_8O_2)_n$		
Symbol:	PMMA	Elongation:	5
Polymerization method:	Addition	Specific gravity:	1.2
Degree of crystallinity:	None (amorphous)	Glass transition temperature:	105°C (221°F)
Modulus of elasticity:	2800 MPa (406,110 lb/in^2)	Melting temperature:	200°C (392°F)
Tensile strength:	55 MPa (7975 lb/in^2)	Approximate market share:	About 1%

in common solvents at ambient temperatures. Because of this combination of properties, acetal resins are competitive with certain metals (e.g., brass and zinc) in automotive components such as door handles, pump housings, and similar parts; appliance hardware; and machinery components.

Acrylics   The acrylics are polymers derived from acrylic acid $(C_3H_4O_2)$ and compounds originating from it. The most important thermoplastic in the acrylics group is ***polymethylmethacrylate*** (PMMA) or Plexiglas (Rohm & Haas's trade name for PMMA). Data on PMMA are listed in Table 8.3(b). It is an amorphous linear polymer. Its outstanding property is excellent transparency, which makes it competitive with glass in optical applications. Examples include automotive tail-light lenses, optical instruments, and aircraft windows. Its limitation when compared with glass is a much lower scratch resistance. Other uses of PMMA include floor waxes and emulsion latex paints. Another important use of acrylics is in fibers for textiles; polyacrylonitrile (PAN) is an example that goes by the more familiar trade names Orlon (DuPont) and Acrilan (Monsanto).

Acrylonitrile–Butadiene–Styrene   ABS is called an engineering plastic due to its excellent combination of mechanical properties, some of which are listed in Table 8.3(c). ABS is a two-phase terpolymer, one phase being the hard copolymer styrene–acrylonitrile, while the other phase is styrene-butadiene copolymer that is rubbery. The name of the plastic is derived from the three starting monomers, which may be mixed in various proportions. Typical applications include components for automotive, appliances, business machines; and pipes and fittings.

Cellulosics   *Cellulose* $(C_6H_{10}O_5)$ is a carbohydrate polymer commonly occurring in nature. Wood and cotton fibers, the chief industrial sources of cellulose, contain about 50% and 95%

TABLE 8.3   (continued): (c) acrylonitrile–butadiene–styrene.

Polymer:	Terpolymer of acrylonitrile (C_3H_3N), butadiene (C_4H_6), and styrene (C_8H_8)		
Symbol:	ABS	Tensile strength:	50 MPa (7250 lb/in^2)
Polymerization method:	Addition	Elongation:	10%–30%
Degree of crystallinity:	None (amorphous)	Specific gravity:	1.06
Modulus of elasticity:	2100 MPa (304,580 lb/in^2)	Approximate market share:	About 3%

TABLE 8.3   (continued): (d) cellulosics.

Representative polymer:	Cellulose acetate $(C_6H_9O_5-COCH_3)_n$	Elongation:	10%–50%
Symbol:	CA	Specific gravity:	1.3
Polymerization method:	Step (condensation)	Glass transition temperature:	105°C (221°F)
Degree of crystallinity:	Amorphous	Melting temperature:	306°C (583°F)
Modulus of elasticity:	2800 MPa (406,110 lb/in^2)	Approximate market share:	Less than 1%
Tensile strength:	30 MPa (4350 lb/in^2)		

of the polymer, respectively. When cellulose is dissolved and reprecipitated during chemical processing, the resulting polymer is called ***regenerated cellulose.*** When this is produced as a fiber for apparel it is known as ***rayon*** (of course, cotton itself is a widely used fiber for apparel). When it is produced as a thin film, it is ***cellophane,*** a common packaging material. Cellulose itself cannot be used as a thermoplastic because it decomposes before melting when its temperature is increased. However, it can be combined with various compounds to form several plastics of commercial importance; examples are ***cellulose acetate*** (CA) and ***cellulose acetate–butyrate*** (CAB). CA, data for which are given in Table 8.3(d), is produced in the form of sheets (for wrapping), film (for photography), and molded parts. CAB is a better molding material than CA and has greater impact strength, lower moisture absorption, and better compatibility with plasticizers. The cellulosic thermoplastics share about 1% of the market.

Fluoropolymers   ***Polytetrafluorethylene*** (PTFE), commonly known as ***Teflon,*** accounts for about 85% of the family of polymers called ***fluoropolymers,*** in which F atoms replace H atoms in the hydrocarbon chain. PTFE is extremely resistant to chemical and environmental attack, is unaffected by water, good heat resistance, and very low coefficient of friction. These latter two properties have promoted its use in nonstick household cookware. Other applications that rely on the same property include nonlubricating bearings and similar components. PTFE also finds applications in chemical equipment and food processing.

Polyamides   An important polymer family that forms characteristic amide linkages (CO-NH) during polymerization is the polyamides (PA). The most important members of the PA family are ***nylons,*** of which the two principal grades are nylon-6 and nylon-6,6 (the numbers are codes that indicate the number of carbon atoms in the monomer). The data given in Table 8.3(f) are for nylon-6,6, which was developed at DuPont in the 1930s. Properties of nylon-6, developed in Germany are similar. Nylon is strong, highly elastic, tough, abrasion resistant, and self-lubricating. It retains good mechanical properties at temperatures up to about 125°C (257°F). One shortcoming is that it absorbs water with an accompanying degradation in properties. The majority of applications of nylon (about 90%) are in fibers for carpets, apparel, and tire cord. The remainder (10%) are in engineering components; nylon is commonly a good substitute for metals in bearings, gears, and similar parts where strength and low friction are needed.

A second group of polyamides is the ***aramids*** (aromatic polyamides) of which ***Kevlar*** (DuPont trade name) is gaining in importance as a fiber in reinforced plastics. The reason for the interest in Kevlar is that its strength is the same as steel at 20% of the weight.

TABLE 8.3   (continued): (e) fluoropolymers.

Representative polymer:	Polytetrafluorethylene $(C_2F_4)_n$	Elongation:	100%–300%
Symbol:	PTFE	Specific gravity:	2.2
Polymerization method:	Addition	Glass transition temperature:	127°C (260°F)
Degree of crystallinity:	About 95% crystalline	Melting temperature:	327°C (620°F)
Modulus of elasticity:	425 MPa (61,640 lb/in^2)	Approximate market share:	Less than 1%
Tensile strength:	20 MPa (2900 lb/in^2)		

TABLE 8.3   (continued): (f) polyamides.

Representative polymer:	Nylon-6,6 $((CH_2)_6(CONH)_2(CH_2)_4)_n$			
Symbol:	PA-6,6	Elongation:	300%
Polymerization method:	Step (condensation)	Specific gravity:	1.14
Degree of crystallinity:	Highly crystalline	Glass transition temperature:	50°C (122°F)
Modulus of elasticity:	700 MPa (101,500 lb/in^2)	Melting temperature:	260°C (500°F)
Tensile strength:	70 MPa (10,150 lb/in^2)	Approximate market share:	1% for all polyamides

Polycarbonate   Polycarbonate (PC) is noted for its generally excellent mechanical properties, which include high toughness and good creep resistance. It is one of the best thermoplastics for heat resistance—it can be used to temperatures around 125°C (257°F). In addition, it is transparent and fire resistant. Applications include molded machinery parts, housings for business machines, pump impellers, safety helmets, and compact disks (e.g., audio, video, and computer). It is also widely used in glazing (window and windshield) applications.

Polyesters   The polyesters form a family of polymers made up of the characteristic ester linkages (CO–O). They can be either thermoplastic or thermosetting, depending on whether cross-linking occurs. Of the thermoplastic polyesters, a representative example is *polyethylene terephthalate* (PET), data for which are compiled in the table. It can be either amorphous or partially crystallized (up to about 30%), depending on how it is cooled after shaping. Fast cooling favors the amorphous state, which is highly transparent. Significant applications include blow-molded beverage containers, photographic films, and magnetic recording tape. In addition, PET fibers are widely used in apparel. Polyester fibers have low moisture absorption and good deformation recovery, both of which make them ideal for "wash and wear" garments that resist wrinkling. The PET fibers are almost always blended with cotton or wool. Familiar trade names for polyester fibers include Dacron (DuPont), Fortrel (Celanese), and Kodel (Eastman Kodak).

Polyethylene   Polyethylene (PE) was first synthesized in the 1930s, and today it accounts for the largest volume of all plastics. The features that make PE attractive as an engineering material are low cost, chemical inertness, and easy processing. Polyethylene is available in

TABLE 8.3   (continued): (g) polycarbonate.

Polymer:	Polycarbonate $(C_3H_6(C_6H_4)_2CO_3)_n$			
Symbol:	PC	Elongation:	110%
Polymerization method:	Step (condensation)	Specific gravity:	1.2
Degree of crystallinity:	Amorphous	Glass transition temperature:	150°C (302°F)
Modulus of elasticity:	2500 MPa (362,590 lb/in^2)	Melting temperature:	230°C (446°F)
Tensile strength:	65 MPa (9425 lb/in^2)	Approximate market share:	Less than 1%

TABLE 8.3   (continued): (h) polyesters (thermoplastic).

Representative polymer:	Polyethylene terephthalate $(C_2H_4–C_8H_4O_4)_n$			
Symbol:	PET	Elongation:	200%
Polymerization method:	Step (condensation)	Specific gravity:	1.3
Degree of crystallinity:	Amorphous to 30% crystalline	Glass transition temperature:	70°C (158°F)
Modulus of elasticity:	2300 MPa (333,590 lb/in^2)	Melting temperature:	265°C (509°F)
Tensile strength:	55 MPa (7975 lb/in^2)	Approximate market share:	About 2%

TABLE 8.3   (continued): (i) polyethylene.

Polyethylene:	$(C_2H_4)_n$ (low density)	$(C_2H_4)_n$ (high density)
Symbol:	LDPE	HDPE
Polymerization method:	Addition	Addition
Degree of crystallinity:	55% typical	92% typical
Modulus of elasticity:	140 MPa (20,305 lb/in^2)	700 MPa (101,500 lb/in^2)
Tensile strength:	15 MPa (2175 lb/in^2)	30 MPa (4350 lb/in^2)
Elongation:	100%–500%	20%–100%
Specific gravity:	0.92	0.96
Glass transition temperature:	$-100°C$ $(-148°F)$	$-115°C$ $(-175°F)$
Melting temperature:	115°C (239°F)	135°C (275°F)
Approximate market share:	About 20%	About 15%

several grades, the most common of which are ***low-density polyethylene*** (LDPE) and ***high-density polyethylene*** (HDPE). The low-density grade is a highly branched polymer with lower crystallinity and density. Applications include squeezable bottles, frozen food bags, sheets, film, and wire insulation. HDPE has a more linear structure, with higher crystallinity and density. These differences make HDPE stiffer and stronger and give it a higher melting temperature. HDPE is used to produce bottles, pipes, and housewares. Both grades can be processed by most polymer shaping methods (Chapter 13). Properties for the two grades are given in Table 8.3(i).

Polypropylene   Polypropylene (PP) has become a major plastic, especially for injection molding, since its introduction in the late 1950s. PP can be synthesized in isotactic, syndiotactic, or atactic structures, the first of these being the most important and for which the characteristics are given in the table. It is the lightest of the plastics, and its strength-to-weight ratio is high. PP is frequently compared with HDPE because its cost and many of its properties are similar. However, the high melting point of polypropylene allows certain applications that preclude use of polyethylene—for example, components that must be sterilized. Other applications are injection molded parts for automotive and houseware, and fiber products for carpeting. A special application suited to polypropylene is one-piece hinges that can be subjected to a high number of flexing cycles without failure.

Polystyrene   There are several polymers, copolymers, and terpolymers based on the monomer styrene (C_8H_8), of which polystyrene (PS) is used in the highest volume. It is a linear homopolymer with amorphous structure that is generally noted for its brittleness. PS is transparent, easily colored, and readily molded, but degrades at elevated temperatures and dissolves in various solvents. Because of its brittleness, some PS grades contain 5% to 15% rubber and the term ***high-impact polystyrene*** (HIPS) is used for these types. They have higher toughness, but transparency and tensile strength are reduced. In addition to injection molding applications (e.g., molded toys, housewares), polystyrene also finds uses in packaging in the form of PS foams.

TABLE 8.3   (continued): (j) polypropylene.

Polymer:	Polypropylene $(C_3H_6)_n$		
Symbol:	PP		
Polymerization method:	Addition	Elongation:	10%–500%a
Degree of crystallinity:	High, varies with processing	Specific gravity:	0.90
Modulus of elasticity:	1400 MPa (203,050 lb/in^2)	Glass transition temperature:	$-20°C$ $(-4°F)$
Tensile strength:	35 MPa (5075 lb/in^2)	Melting temperature:	176°C (348°F)
		Approximate market share:	About 13%

[a]Elongation depends on additives.

TABLE 8.3   (continued): (k) polystyrene.

Polymer:	Polystyrene $(C_8H_8)_n$	Elongation:	1%
Symbol:	PS	Specific gravity:	1.05
Polymerization method:	Addition	Glass transition temperature:	100°C (212°F)
Degree of crystallinity:	None (amorphous)	Melting temperature:	240°C (464°F)
Modulus of elasticity:	3200 MPa (464,120 lb/in^2)	Approximate market share:	About 10%
Tensile strength:	50 MPa (7250 lb/in^2)		

TABLE 8.3   (continued): (l) polyvinylchloride.

Polymer:	Polyvinylchloride $(C_2H_3Cl)_n$	Elongation:	2% with no plasticizer
Symbol:	PVC	Specific gravity:	1.40
Polymerization method:	Addition	Glass transition temperature:	81°C (178°F)[b]
Degree of crystallinity:	None (amorphous structure)	Melting temperature:	212°C (414°F)
Modulus of elasticity:	2800 MPa (406,110 lb/in^2)[a]	Approximate market share:	About 16%
Tensile strength:	40 MPa (5800 lb/in^2)		

[b]With no plasticizer.

Polyvinylchloride   Polyvinylchloride (PVC) is a widely used plastic whose properties can be varied by combining additives with the polymer. In particular, plasticizers are used to achieve thermoplastics ranging from rigid PVC (no plasticizers) to flexible PVC (high proportions of plasticizer). The range of properties makes PVC a versatile polymer, with applications that include rigid pipe (used in construction, water and sewer systems, irrigation), fittings, wire and cable insulation, film, sheets, food packaging, flooring, and toys. PVC by itself is relatively unstable to heat and light, and stabilizers must be added to improve its resistance to these environmental conditions. Care must be taken in the production and handling of the vinyl chloride monomer used to polymerize PVC, due to its carcinogenic nature.

8.3   THERMOSETTING POLYMERS

Thermosetting (TS) polymers are distinguished by their highly cross-linked structure. In effect, the formed part (e.g., the pot handle or electrical switch cover) becomes one large macromolecule. Thermosets are always amorphous and exhibit no glass transition temperature. In this section, we examine the general characteristics of the TS plastics and identify the important materials in this category.

8.3.1   GENERAL PROPERTIES AND CHARACTERISTICS

Owing to differences in chemistry and molecular structure, properties of thermosetting plastics are different from those of thermoplastics. In general, thermosets are (1) more rigid—modulus of elasticity is 2 to 3 times greater; (2) brittle—they possess virtually no ductility; (3) less soluble in common solvents; (4) capable of higher service temperatures; and (5) not capable of being remelted—instead they degrade or burn.

The differences in properties of the TS plastics are attributable to cross-linking, which forms a thermally stable, three-dimensional, covalently bonded structure within the molecule. Cross-linking is accomplished in three ways [7]:

1. *Temperature-activated systems*—In the most common systems, the changes are caused by heat supplied during the part-shaping operation (e.g., molding). The starting

material is a linear polymer in granular form supplied by the chemical plant. As heat is added, the material softens for molding; continued heating results in cross-linking of the polymer. The term ***thermosetting*** is most aptly applied to these polymers.

2. ***Catalyst-activated systems***—Cross-linking in these systems occurs when small amounts of a catalyst are added to the polymer, which is in liquid form. Without the catalyst, the polymer remains stable; once combined with the catalyst, it changes into solid form.

3. ***Mixing-activated systems***—Most epoxies are examples of these systems. The mixing of two chemicals results in a reaction that forms a cross-linked solid polymer. Elevated temperatures are sometimes used to accelerate the reactions.

The chemical reactions associated with cross-linking are called ***curing*** or ***setting***. Curing is done at the fabrication plants that shape the parts rather than the chemical plants that supply the starting materials to the fabricator.

8.3.2 IMPORTANT THERMOSETTING POLYMERS

Thermosetting plastics are not as widely used as the thermoplastics, perhaps because of the added processing complications involved in curing the TS polymers. The largest volume thermosets are phenolic resins, whose annual volume is about 6% of the total plastics market. This is significantly less than polyethylene, the leading thermoplastic, whose volume is about 35% of the total. Technical data for these materials are given in Table 8.4. Market share data refer to total plastics (TP plus TS).

Amino Resins   Amino plastics, characterized by the amino group (NH_2), consist of two thermosetting polymers, urea-formaldehyde and melamine-formaldehyde, which are produced by the reaction of formaldehyde (CH_2O) with either urea ($CO(NH_2)_2$) or melamine ($C_3H_6N_6$), respectively. In commercial importance, the amino resins rank just below the other formaldehyde resin, phenol-formaldehyde, discussed below. ***Urea–formaldehyde*** is competitive with the phenols in certain applications, particularly as a plywood and particle-board adhesive. The resins are also used as a molding compound. It is slightly more expensive than the phenol material. ***Melamine–formaldehyde*** plastic is water resistant and is used for dishware and as a coating in laminated table and counter tops (Formica, trade name of Cyanamid Co.). When used as molding materials, amino plastics usually contain significant proportions of fillers, such as cellulose.

Epoxies   Epoxy resins are based on a chemical group called the ***epoxides***. The simplest formulation of epoxide is ethylene oxide (C_2H_3O). Epichlorohydrin (C_3H_5OCl) is a much more widely used epoxide for producing epoxy resins. Uncured, epoxies have a low degree of polymerization. To increase molecular weight and to cross-link the epoxide, a curing agent

TABLE 8.4 Important commercial thermosetting polymers: (a) amino resins.		
Representative polymer: Melamine-formaldehyde		
Monomers: Melamine ($C_3H_6N_6$) and formaldehyde (CH_2O)		
Polymerization method: Step (condensation)		
Modulus of elasticity: 9000 MPa (1,305,000 lb/in^2)	Elongation:	Less than 1%
Tensile strength: 50 MPa (7250 lb/in^2)	Specific gravity:	1.5
	Approximate market share:	About 4% for urea-formaldehyde and melamine-formaldehyde.

Table 8.4 is compiled from [2], [4], [6], [7], [9], [16], and other sources.

TABLE 8.4   (continued): (b) epoxy.

Example chemistry:	Epichlorohydrin (C_3H_5OCl) plus curing agent such as triethylamine (C_6H_5–CH_2N–$(CH_3)_2$)		
Polymerization method:	Condensation	Elongation:	0%
Modulus of elasticity:	7000 MPa (1,015,000 lb/in^2)	Specific gravity:	1.1
Tensile strength:	70 MPa (10,150 lb/in^2)	Approximate market share:	About 1%

TABLE 8.4   (continued): (c) phenol formaldehyde.

Monomer ingredients:	Phenol (C_6H_5OH) and formaldehyde (CH_2O)		
Polymerization method:	Step (condensation)	Elongation:	Less than 1%
Modulus of elasticity:	7000 MPa (1,015,000 lb/in^2)	Specific gravity:	1.4
Tensile strength:	70 MPa (10,150 lb/in^2)	Approximate market share:	6%

must be used. Possible curing agents include polyamines and acid anhydrides. Cured epoxies are noted for strength, adhesion, and heat and chemical resistance. Applications include surface coatings, industrial flooring, glass fiber-reinforced composites, and adhesives. Insulating properties of epoxy thermosets make them useful in various electronic applications, such as encapsulation of integrated circuits and lamination of printed circuit boards.

Phenolics   Phenol (C_6H_5OH) is an acidic compound that can be reacted with aldehydes (dehydrogenated alcohols), formaldehyde (CH_2O) being the most reactive. ***Phenol-formaldehyde*** is the most important of the phenolic polymers; it was first commercialized around 1900 under the trade name ***Bakelite***. It is almost always combined with fillers such as wood flour, cellulose fibers, and minerals when used as a molding material. It is brittle, possesses good thermal, chemical, and dimensional stability. Its capacity to accept colorants is limited—it is available only in dark colors. Molded products constitute only about 10% of total phenolics use. Other applications include adhesives for plywood, printed circuit boards, counter tops, and bonding material for brake linings and abrasive wheels.

Polyesters   Polyesters, which contain the characteristic ester linkages (CO–O), can be thermosetting as well as thermoplastic (Section 8.2). Thermosetting polyesters are used largely in reinforced plastics (composites) to fabricate large items such as pipes, tanks, boat hulls, auto body parts, and construction panels. They can also be used in various molding processes to produce smaller parts. Synthesis of the starting polymer involves reaction of an acid or anhydride such as maleic anhydride ($C_4H_2O_3$) with a glycol such as ethylene glycol ($C_2H_6O_2$). This produces an ***unsaturated polyester*** of relatively low molecular weight (MW = 1000 to 3000). This ingredient is mixed with a monomer capable of polymerizing and cross-linking with the polyester. Styrene (C_8H_8) is commonly used for this purpose, in proportions of 30% to 50%. A third component, called an inhibitor, is added to prevent premature cross-linking. This mixture forms the polyester resin system that is supplied to the fabricator. Polyesters are cured either by heat (temperature-activated systems), or by means of a catalyst

TABLE 8.4   (continued): (d) unsaturated polyester.

Example chemistry:	Maleic anhydride ($C_4H_2O_3$) and ethylene glycol ($C_2H_6O_2$) plus styrene (C_8H_8)		
Polymerization method:	Step (condensation)	Elongation:	0%
Modulus of elasticity:	7000 MPa (1,015,000 lb/in^2)	Specific gravity:	1.1
Tensile strength:	30 MPa (4350 lb/in^2)	Approximate market share:	3%

TABLE 8.4   (continued): (e) polyimides.

Starting monomers:	Pyromellitic dianhydride ($C_6H_2(C_2O_3)_2$), 4,4'-oxydianiline ($O(C_6H_4NH_2)_2$)		
Polymerization method:	Condensation	Elongation:	5%
Modulus of elasticity:	3200 MPa (464,120 lb/in^2)	Specific gravity:	1.43
Tensile strength:	80 MPa (11,600 lb/in^2)	Approximate market share:	Less than 1%

added to the polyester resin (catalyst-activated systems). Curing is done at the time of fabrication (molding or other forming process) and results in cross-linking of the polymer.

An important class of polyesters are the *alkyd* resins (the name derived by abbreviating and combining the words *alcohol* and *acid* and changing a few letters). They are used primarily as bases for paints, varnishes, and lacquers. Alkyd molding compounds are also available, but their applications are limited.

Polyimides   These plastics are available as both thermoplastics and thermosets, but the TS types are more important commercially. They are available under brand names such as Kapton (Dupont) and Kaptrex (Professional Plastics) in several forms including tapes, films, coatings, and molding resins. TS polyimides (PI) are noted for chemical resistance, high tensile strength and stiffness, and stability at elevated temperatures. They are called high-temperature polymers due to their excellent heat resistance. Applications that exploit these properties include insulating films, molded parts used in elevated temperature service, flexible cables in laptop computers, medical tubing, and fibers for protective clothing.

Polyurethanes   This includes a large family of polymers, all characterized by the urethane group (NHCOO) in their structure. The chemistry of the polyurethanes is complex, and there are many chemical varieties in the family. The characteristic feature is the reaction of a *polyol,* whose molecules contain hydroxyl (OH) groups, such as butylene ether glycol ($C_4H_{10}O_2$); and an *isocyanate,* such as diphenylmethane diisocyanate ($C_{15}H_{10}O_2N_2$). Through variations in chemistry, cross-linking, and processing, polyurethanes can be thermoplastic, thermoset-ting, or elastomeric materials, the latter two being the most important commercially. The largest application of polyurethane is in foams. These can range between elastomeric and rigid, the latter being more highly cross-linked. Rigid foams are used as a filler material in hollow construction panels and refrigerator walls. In these types of applications, the material provides excellent thermal insulation, adds rigidity to the structure, and does not absorb water in significant amounts. Many paints, varnishes, and similar coating materials are based on urethane systems. We discuss polyurethane elastomers in Section 8.4.

Silicones   Silicones are inorganic and semi-inorganic polymers, distinguished by the presence of the repeating siloxane link (–Si–O–) in their molecular structure. A typical formulation combines the methyl radical (CH_3) with (SiO) in various proportions to obtain

TABLE 8.4   (continued): (f) polyurethane.

Polymer:	Polyurethane is formed by the reaction of a polyol and an isocyanate. Chemistry varies significantly		
Polymerization method:	Step (condensation)	Elongation:	Depends on cross-linking
Modulus of elasticity:	Depends on chemistry and processing	Specific gravity:	1.2
Tensile strength:	30 MPa (4350 lb/in^2)[a]	Approximate market share:	About 4%, including elastomers

[a]Typical for highly cross-linked polyurethane.

TABLE 8.4   (continued): (g) silicone thermosetting resins.

Example chemistry:	$((CH_3)_6 -SiO)_n$		Elongation:	0%
Polymerization method:	Step (condensation), usually		Specific gravity:	1.65
Tensile strength:	30 MPa (4350 lb/in^2)		Approximate market share:	Less than 1%

the repeating unit $-((CH_3)_m - SiO)-$, where ***m*** establishes the proportionality. By variations in composition and processing, polysiloxanes can be produced in three forms: (1) fluids, (2) elastomers, and (3) thermosetting resins. Fluids (1) are low molecular weight polymers used for lubricants, polishes, waxes, and other liquids—not really polymers in the sense of this chapter, but important commercial products nevertheless. Silicone elastomers (2), covered in Section 8.4, and thermosetting silicones (3), treated here, are cross-linked. When highly cross-linked, polysiloxanes form rigid resin systems used for paints, varnishes, and other coatings; and laminates such as printed circuit boards. They are also used as molding materials for electrical parts. Curing is accomplished by heating or by allowing the solvents containing the polymers to evaporate. Silicones are noted for their good heat resistance and water repellence, but their mechanical strength is not as great as other cross-linked polymers. Data in Table 8.4(g) are for a typical silicone thermosetting polymer.

8.4   ELASTOMERS

Elastomers are polymers capable of large elastic deformation when subjected to relatively low stresses. Some elastomers can withstand extensions of 500% or more and still return to their original shape. The more popular term for elastomer is, of course, rubber. We can divide rubbers into two categories: (1) natural rubber, derived from certain biological plants; and (2) synthetic elastomers, produced by polymerization processes similar to those used for thermoplastic and thermosetting polymers. Before discussing natural and synthetic rubbers, let us consider the general characteristics of elastomers.

8.4.1   CHARACTERISTICS OF ELASTOMERS

Elastomers consist of long-chain molecules that are cross-linked. They owe their impressive elastic properties to the combination of two features: (1) the long molecules are tightly kinked when unstretched, and (2) the degree of cross-linking is substantially below that of the thermosets. These features are illustrated in the model of Figure 8.12(a), which shows a tightly kinked cross-linked molecule under no stress.

When the material is stretched, the molecules are forced to uncoil and straighten as shown in Figure 8.12(b). The molecules' natural resistance to uncoiling provides the initial elastic modulus of the aggregate material. As further strain is experienced, the covalent bonds

FIGURE 8.12   Model of long elastomer molecules, with low degree of cross-linking: (a) unstretched, and (b) under tensile stress.

(a)   (b)

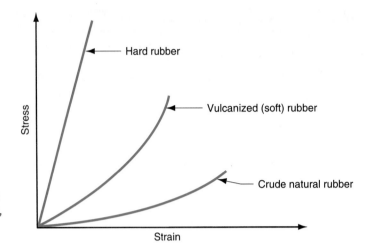

FIGURE 8.13 Increase in stiffness as a function of strain for three grades of rubber: natural rubber, vulcanized rubber, and hard rubber.

of the cross-linked molecules begin to play an increasing role in the modulus, and the stiffness increases as illustrated in Figure 8.13. With greater cross-linking, the elastomer becomes stiffer and its modulus of elasticity is more linear. These characteristics are shown in the figure by the stress–strain curves for three grades of rubber: natural crude rubber, whose cross-linking is very low; cured (vulcanized) rubber with low-to-medium cross-linking; and hard rubber (ebonite), whose high degree of cross-linking transforms it into a thermosetting plastic.

For a polymer to exhibit elastomeric properties, it must be amorphous in the unstretched condition, and its temperature must be above T_g. If below the glass transition temperature, the material is hard and brittle; above T_g the polymer is in the "rubbery" state. Any amorphous thermoplastic polymer will exhibit elastomeric properties above T_g for a short time, because its linear molecules are always coiled to some extent, thus allowing for elastic extension. It is the absence of cross-linking in TP polymers that prevents them from being truly elastic; instead they exhibit viscoelastic behavior.

Curing is required to effect cross-linking in most of the common elastomers today. The term for curing used in the context of natural rubber (and certain synthetic rubbers) is **vulcanization,** which involves the formation of chemical cross-links between the polymer chains. Typical cross-linking in rubber is 1 to 10 links per 100 carbon atoms in the linear polymer chain, depending on the degree of stiffness desired in the material. This is considerably less than the degree of cross-linking in thermosets.

An alternative method of curing involves the use of starting chemicals that react when mixed (sometimes requiring a catalyst or heat) to form elastomers with relatively infrequent cross-links between molecules. These synthetic rubbers are known as **reactive system elastomers**. Certain polymers that cure by this means, such as urethanes and silicones, can be classified as either thermosets or elastomers, depending on the degree of cross-linking achieved during the reaction.

A relatively new class of elastomers, called **thermoplastic elastomers,** possesses elastomeric properties that result from the mixture of two phases, both thermoplastic. One is above its T_g at room temperature while the other is below its T_g. Thus, we have a polymer that includes soft rubbery regions intermixed with hard particles that act as cross-links. The composite material is elastic in its mechanical behavior, although not as extensible as most other elastomers. Because both phases are thermoplastic, the aggregate material can be heated above its T_m for forming, using processes that are generally more economical than those used for rubber.

We discuss the elastomers in the following two sections. The first deals with natural rubber and how it is vulcanized to create a useful commercial material; the second examines the synthetic rubbers.

8.4.2 NATURAL RUBBER

Natural rubber (NR) consists primarily of polyisoprene, a high-molecular-weight polymer of isoprene (C_5H_8). It is derived from latex, a milky substance produced by various plants, the most important of which is the rubber tree (*Hevea brasiliensis*) that grows in tropical climates (Historical Note 8.2). Latex is a water emulsion of polyisoprene (about one-third by weight), plus various other ingredients. Rubber is extracted from the latex by various methods (e.g., coagulation, drying, and spraying) that remove the water.

Historical Note 8.2   *Natural rubber*

The first use of natural rubber seems to have been in the form of rubber balls used for sport by the natives of Central and South America at least 500 hundred years ago. Columbus noted this during his second voyage to the New World in 1493–1496. The balls were made from the dried gum of a rubber tree. The first white men in South America called the tree *caoutchouc*, which was their way of pronouncing the Indian name for it. The name *rubber* came from the English chemist Joseph Priestley, who discovered (around 1770) that gum rubber would ''rub'' away pencil marks.

Early rubber goods were less than satisfactory; they melted in summer heat and hardened in winter cold. One of those in the business of making and selling rubber goods was American Charles Goodyear. Recognizing the deficiencies of the natural material, he experimented with ways to improve its properties and discovered that rubber could be cured by heating it with sulfur. This was

in 1839, and the process, later called ***vulcanization***, was patented by him in 1844.

Vulcanization and the emerging demand for rubber products led to tremendous growth in rubber production and the industry that supported it. In 1876, Henry Wickham collected thousands of rubber tree seeds from the Brazilian jungle and planted them in England; the sprouts were later transplanted to Ceylon and Malaya (then British colonies) to form rubber plantations. Soon, other countries in the region followed the British example. Southeast Asia became the base of the rubber industry.

In 1888, a British veterinary surgeon named John Dunlop patented pneumatic tires for bicycles. By the twentieth century, the motorcar industry was developing in the United States and Europe. Together, the automobile and rubber industries grew to occupy positions of unimagined importance.

Natural crude rubber (without vulcanization) is sticky in hot weather, but stiff and brittle in cold weather. To form an elastomer with useful properties, natural rubber must be vulcanized. Traditionally, vulcanization has been accomplished by mixing small amounts of sulfur and other chemicals with the crude rubber and heating. The chemical effect of vulcanization is cross-linking; the mechanical result is increased strength and stiffness, yet maintenance of extensibility. The dramatic change in properties caused by vulcanization can be seen in the stress–strain curves of Figure 8.13.

Sulfur alone can cause cross-linking, but the process is slow, taking hours to complete. Other chemicals are added to sulfur during vulcanization to accelerate the process and serve other beneficial functions. Also, rubber can be vulcanized using chemicals other than sulfur. Today, curing times have been reduced significantly compared to the original sulfur curing of years ago.

As an engineering material, vulcanized rubber is noted among elastomers for its high tensile strength, tear strength, resilience (capacity to recover shape after deformation), and resistance to wear and fatigue. Its weaknesses are that it degrades when subjected to heat, sunlight, oxygen, ozone, and oil. Some of these limitations can be reduced through the use of additives. Typical properties and other data for vulcanized natural rubber are listed in Table 8.5. Market share is relative to total annual rubber volume, natural plus synthetic. Rubber volume is about 15% of total polymer market.

TABLE 8.5 Characteristics and typical properties of vulcanized rubber.

Polymer:	Polyisoprene $(C_5H_8)_n$		
Symbol:	NR	Specific gravity:	0.93
Modulus of elasticity:	18 MPa (2610 lb/in^2) at 300% elongation	High temperature limit:	80°C (176°F)
Tensile strength:	25 MPa (3625 lb/in^2)	Low temperature limit:	−50°C (−58°F)
Elongation:	700% at failure	Approximate market share:	22%

Compiled from [2], [6], [9], and other sources.

The largest single market for natural rubber is automotive tires. In tires, carbon black is an important additive; it reinforces the rubber, serving to increase tensile strength and resistance to tearing and abrasion. Other products made of rubber include shoe soles, bushings, seals, and shock-absorbing components. In each case, the rubber is compounded to achieve the specific properties required in the application. Besides carbon black, other additives used in rubber and some of the synthetic elastomers include clay, kaolin, silica, talc, and calcium carbonate, as well as chemicals that accelerate and promote vulcanization.

8.4.3 SYNTHETIC RUBBERS

Today, the tonnage of synthetic rubbers is more than three times that of natural rubber. Development of these synthetic materials was motivated largely by the world wars when NR was difficult to obtain (Historical Note 8.3). The most important of the synthetics is styrene–butadiene rubber (SBR), a copolymer of butadiene (C_4H_6) and styrene (C_8H_8). As with most other polymers, the predominant raw material for the synthetic rubbers is petroleum. Only the synthetic rubbers of greatest commercial importance are discussed here. Technical data are presented in Table 8.6. Market share data are for total volume of natural and synthetic

---

Historical Note 8.3 *Synthetic rubbers*

In 1826, Faraday recognized the formula of natural rubber to be C_5H_8. Subsequent attempts at reproducing this molecule over many years were generally unsuccessful. Regrettably, it was the world wars that created the necessity which became the mother of invention for synthetic rubber. In World War I, the Germans, denied access to natural rubber, developed a methyl-based substitute. This material was not very successful, but it marks the first large-scale production of synthetic rubber.

After World War I, the price of natural rubber was so low that many attempts at fabricating synthetics were abandoned. However, the Germans, perhaps anticipating a future conflict, renewed their development efforts. The firm I.G. Farben developed two synthetic rubbers, starting in the early 1930s, called Buna-S and Buna-N. **Buna** is derived from **bu**tadiene (C_4H_6), which has become the critical ingredient in many modern synthetic rubbers, and **Na**, the symbol for sodium, used to accelerate or catalyze the polymerization process

(*Natrium* is the German word for sodium). The symbol S in Buna-S stands for styrene. Buna-S is the copolymer we know today as **styrene–butadiene rubber**, or SBR. The *N* in Buna-N stands for acrylo*N*itrile, and the synthetic rubber is called **nitrile rubber** in current usage.

Other efforts included the work at the DuPont Company in the United States, which led to the development of polychloroprene, first marketed in 1932 under the name Duprene, later changed to **Neoprene**, its current name.

During World War II, the Japanese cut off the supply of natural rubber from Southeast Asia to the United States. Production of Buna-S synthetic rubber was begun on a large scale in America. The federal government preferred to use the name **GR-S** (Government Rubber-Styrene) rather than Buna-S (the German name). By 1944, the United States was outproducing Germany in SBR 10-to-1. Since the early 1960s, worldwide production of synthetic rubbers has exceeded that of natural rubbers.

TABLE 8.6   Characteristics and typical properties of synthetic rubbers: (a) butadiene rubber.

Polymer:	Polybutadiene $(C_4H_6)_n$		
Symbol:	BR	Specific gravity:	0.93
Tensile strength:	15 MPa (2175 lb/in^2)	High temperature limit:	100°C (212°F)
Elongation:	500% at failure	Low temperature limit:	−50°C (−58°F)
		Approx. market share:	12%

Table 8.6 is compiled from [2], [4], [6], [9], [11], and other sources.

rubbers. About 10% of total volume of rubber production is reclaimed; thus, total tonnages in Tables 8.5 and 8.6 do not sum to 100%.

Butadiene Rubber   *Polybutadiene* (BR) is important mainly in combination with other rubbers. It is compounded with natural rubber and with styrene (styrene–butadiene rubber is discussed later) in the production of automotive tires. Without compounding, the tear resistance, tensile strength, and ease of processing of polybutadiene are less than desirable.

Butyl Rubber   Butyl rubber is a copolymer of polyisobutylene (98%–99%) and polyisoprene (1%–2%). It can be vulcanized to provide a rubber with very low air permeability, which has led to applications in inflatable products such as inner tubes, liners in tubeless tires, and sporting goods.

Chloroprene Rubber   Polychloroprene was one of the first synthetic rubbers to be developed (early 1930s). Commonly known today as *Neoprene,* it is an important special-purpose rubber. It crystallizes when strained to provide good mechanical properties. Chloroprene rubber (CR) is more resistant to oils, weather, ozone, heat, and flame (chlorine makes this rubber self-extinguishing) than NR, but somewhat more expensive. Its applications include fuel hoses (and other automotive parts), conveyor belts, and gaskets, but not tires.

Ethylene–Propylene Rubber   Polymerization of ethylene and propylene with small proportions (3%–8%) of a diene monomer produces the terpolymer ethylene–propylene–diene (EPDM), a useful synthetic rubber. Applications are for parts mostly in the automotive industry other than tires. Other uses are wire and cable insulation.

TABLE 8.6   (continued): (b) butyl rubber.

Polymer:	Copolymer of isobutylene $(C_4H_8)_n$ and isoprene $(C_5H_8)_n$		
Symbol:	PIB	Specific gravity:	0.92
Modulus of elasticity:	7 MPa (1015 lb/in^2) at 300% elongation	High temperature limit:	110°C (230°F)
Tensile strength:	20 MPa (2900 lb/in^2)	Low temperature limit:	−50°C (−58°F)
Elongation:	700%	Approximate market share:	About 3%

TABLE 8.6   (continued): (c) chloroprene rubber (neoprene).

Polymer:	Polychloroprene $(C_4H_5Cl)_n$		
Symbol:	CR	Specific gravity:	1.23
Modulus of elasticity:	7 MPa (1015 lb/in^2) at 300% elongation	High temperature limit:	120°C (248°F)
Tensile strength:	25 MPa (3625 lb/in^2)	Low temperature limit:	−20°C (−4°F)
Elongation:	500% at failure	Approximate market share:	2%

TABLE 8.6 (continued): (d) ethylene–propylene–diene rubber.

Representative polymer:	Terpolymer of ethylene (C_2H_4), propylene (C_3H_6), and a diene monomer (3%–8%) for cross-linking		
Symbol:	EPDM	Specific gravity:	0.86
Tensile strength:	15 MPa (2175 lb/in^2)	High temperature limit:	150°C (302°F)
Elongation:	300% at failure	Low temperature limit:	−50°C (−58°F)
		Approximate market share:	5%

TABLE 8.6 (continued): (e) isoprene rubber (synthetic).

Polymer:	Polyisoprene (C_5H_8)$_n$		
Symbol:	IR	Specific gravity:	0.93
Modulus of elasticity:	17 MPa (2465 lb/in^2) at 300% elongation	High temperature limit:	80°C (176°F)
Tensile strength:	25 MPa (3625 lb/in^2)	Low temperature limit:	−50°C (−58°F)
Elongation:	500% at failure	Approximate market share:	2%

Isoprene Rubber   Isoprene can be polymerized to synthesize a chemical equivalent of natural rubber. Synthetic (unvulcanized) *polyisoprene* is softer and more easily molded than raw natural rubber. Applications of the synthetic material are similar to those of its natural counterpart, car tires being the largest single market. It is also used for footwear, conveyor belts, and caulking compound. Cost per unit weight is about 35% higher than for NR.

Nitrile Rubber   This is a vulcanizable copolymer of butadiene (50%–75%) and acrylonitrile (25%–50%). Its more technical name is ***butadiene-acrylonitrile rubber***. It has good strength and resistance to abrasion, oil, gasoline, and water. These properties make it ideal for applications such as gasoline hoses and seals, and also for footwear.

Polyurethanes   Thermosetting polyurethanes (Section 8.3.2) with minimum cross-linking are elastomers, most commonly produced as flexible foams. In this form, they are widely used as cushion materials for furniture and automobile seats. Unfoamed polyurethane can

TABLE 8.6 (continued): (f) nitrile rubber.

Polymer:	Copolymer of butadiene (C_4H_6) and acrylonitrile (C_3H_3N)		
Symbol:	NBR	Specific gravity:	1.00 (without fillers)
Modulus of elasticity:	10 MPa (1450 lb/in^2) at 300% elongation	High temperature limit:	120°C (248°F)
Tensile strength:	30 MPa (4350 lb/in^2)	Low temperature limit:	−50°C (−58°F)
Elongation:	500% at failure	Approximate market share:	2%

TABLE 8.6 (continued): (g) polyurethane.

Polymer:	Polyurethane (chemistry varies)		
Symbol:	PUR	Specific gravity:	1.25
Modulus of elasticity:	10 MPa (1450 lb/in^2) at 300% elongation	High temperature limit:	100°C (212°F)
Tensile strength:	60 MPa (8700 lb/in^2)	Low temperature limit:	−50°C (−58°F)
Elongation:	700% at failure	Approximate market share:	Listed under thermosets, Table 8.4(e)

TABLE 8.6 (continued): (h) silicone rubber.		
Representative polymer:	Polydimethylsiloxane $(SiO(CH_3)_2)_n$	
Symbol:	VMQ	
Tensile strength:	10 MPa (1450 lb/in^2)	Specific gravity: 0.98
Elongation:	700% at failure	High temperature limit: 230°C (446°F)
		Low temperature limit: −50°C (−58°F)
		Approximate market share: Less than 1%

be molded into products ranging from shoe soles to car bumpers, with cross-linking adjusted to achieve the desired properties for the application. With no cross-linking, the material is a thermoplastic elastomer that can be injection molded. As an elastomer or thermoset, reaction injection molding and other shaping methods are used.

Silicones   Like the polyurethanes, silicones can be elastomeric or thermosetting, depending on the degree of cross-linking. Silicone elastomers are noted for the wide temperature range over which they can be used. Their resistance to oils is poor. The silicones possess various chemistries, the most common being *polydimethylsiloxane* (Table 8.6(h)). To obtain acceptable mechanical properties, silicone elastomers must be reinforced, usually with fine silica powders. Owing to their high cost, they are considered special-purpose rubbers for applications such as gaskets, seals, wire and cable insulation, prosthetic devices, and bases for caulking materials.

Styrene–Butadiene Rubber   SBR is a random copolymer of styrene (about 25%) and butadiene (about 75%). It was originally developed in Germany as Buna-S rubber before World War II. Today, it is the largest tonnage elastomer, totaling about 40% of all rubbers produced (natural rubber is second in tonnage). Its attractive features are low cost, resistance to abrasion, and better uniformity than NR. When reinforced with carbon black and vulcanized, its characteristics and applications are very similar to those of natural rubber. Cost is also similar. A close comparison of properties reveals that most of its mechanical properties except wear resistance are inferior to NR, but its resistance to heat aging, ozone, weather, and oils is superior. Applications include automotive tires, footwear, and wire and cable insulation. A material chemically related to SBR is styrene–butadiene–styrene block copolymer, a thermoplastic elastomer discussed below.

Thermoplastic Elastomers   As previously described, a thermoplastic elastomer (TPE) is a thermoplastic that behaves like an elastomer. It constitutes a family of polymers that is a fast-growing segment of the elastomer market. TPEs derive their elastomeric properties not from chemical cross-links, but from physical connections between soft and hard phases that make up the material. Thermoplastic elastomers include *styrene–butadiene–styrene* (SBS), a block copolymer as opposed to styrene–butadiene rubber (SBR) which is a random copolymer (Section 8.1.2); *thermoplastic polyurethanes*;

TABLE 8.6 (continued): (i) styrene–butadiene rubber.		
Polymer:	Copolymer of styrene (C_8H_8) and butadiene (C_4H_6)	
Symbol:	SBR	Elongation: 700% at failure
Modulus of elasticity:	17 MPa (2465 lb/in^2) at 300% elongation	Specific gravity: 0.94
Tensile strength:	20 MPa (2900 lb/in^2) reinforced	High temperature limit: 110°C (230°F)
		Low temperature limit: −50°C (−58°F)
		Approximate market share: Slightly less than 30%

TABLE 8.6	(continued): (j) thermoplastic elastomers (TPE).		
Representative polymer:	Styrene–butadiene–styrene block copolymer		
Symbol:	SBS (also YSBR)	Specific gravity:	1.0
Tensile strength:	14 MPa (2030 lb/in^2)	High temperature limit:	65°C (149°F)
Elongation:	400%	Low temperature limit:	−50°C (−58°F)
		Approximate market share:	12%

thermoplastic polyester copolymers; and other copolymers and polymer blends. Table 8.6 (j) gives data on SBS. The chemistry and structure of these materials are generally complex, involving two materials that are incompatible so that they form distinct phases whose room temperature properties are different. Owing to their thermoplasticity, the TPEs cannot match conventional cross-linked elastomers in elevated temperature strength and creep resistance. Typical applications include footwear, rubber bands, extruded tubing, wire coating, and molded parts for automotive and other uses in which elastomeric properties are required. TPEs are not suitable for tires.

8.5 POLYMER RECYCLING AND BIODEGRADABILITY

It is estimated that since the 1950s, 1 billion tons of plastic have been discarded as garbage.[2] This plastic trash could be around for centuries, because the primary bonds that make plastics so durable also make them resistant to degradation by the environmental and biological processes of nature. In this section, we consider two polymer topics related to environmental concerns: (1) recycling of polymer products and (2) biodegradable plastics.

8.5.1 POLYMER RECYCLING

Approximately 200 million tons of plastic products are made annually throughout the world, more than one-eighth of which are produced in the United States.[3] Only about 6% of the U.S. tonnage is recycled as plastic waste; the rest either remains in products and/or ends up in garbage landfills. *Recycling* means recovering the discarded plastic items and reprocessing them into new products, in some cases products that are quite different from the original discarded items.

In general the recycling of plastics is more difficult that recycling of glass and metal products. There are several reasons for this: (1) compared to plastic parts, many recycled metal items are much larger and heavier (e.g., structural steel from buildings and bridges, steel car body frames), so the economics of recycling are more favorable for recycling metals; most plastic items are lightweight; (2) compared to plastics, which come in a variety of chemical compositions that do not mix well, glass products are all based on silicon dioxide; and (3) many plastic products contain fillers, dyes, and other additives that cannot be readily separated from the polymer itself. Of course, a common problem in all recycling efforts is the fluctuation in prices of recycled materials.

To cope with the problem of mixing different types of plastics and to promote recycling of plastics, the Plastic Identification Code (PIC) was developed by the Society

---

[2]en.wikipedia.org/wiki/Plastic.
[3]According to the Society of Plastics Engineers, as reported in en.wikipedia.org/wiki/Biodegradable_plastic.

of the Plastics Industry. The code is a symbol consisting of a triangle formed by three bent arrows enclosing a number. It is printed or molded on the plastic item. The number identifies the plastic for recycling purposes. The seven plastics (all thermoplastics) used in the PIC recycling program are (1) polyethylene terephthalate, used in 2-liter beverage containers; (2) high-density polyethylene, used in milk jugs and shopping bags; (3) polyvinyl chloride, used in juice bottles and PVC pipes; (4) low-density polyethylene, used in squeezable bottles and flexible container lids; (5) polypropylene, used in yogurt and margarine containers; (6) polystyrene, used in egg cartons, disposable plates, cups, and utensils, and as foamed packing materials; and (7) other, such as polycarbonate or ABS. The PIC facilitates the separation of items made from the different types of plastics for reprocessing. Nevertheless, sorting the plastics is a labor-intensive activity.

Once separated, the thermoplastic items can be readily reprocessed into new products by remelting. This is not the case with thermosets and rubbers because of the cross-linking in these polymers. Thus, these materials must be recycled and reprocessed by different means. Recycled thermosets are typically ground up into particulate matter and used as fillers, for example, in molded plastic parts. Most recycled rubber comes from used tires. While some of these tires are retreaded, others are ground up into granules in forms such as chunks and nuggets that can be used for landscape mulch, playgrounds, and similar purposes.

8.5.2 BIODEGRADABLE POLYMERS

Another approach that addresses the environmental concerns about plastics involves the development of biodegradable plastics, which are defined as plastics that are decomposed by the actions of microorganisms occurring in nature, such as bacteria and fungi. Conventional plastic products usually consist of a combination of a petroleum-based polymer and a filler (Section 8.1.5). In effect, the material is a polymer-matrix composite (Section 9.4). The purpose of the filler is to improve mechanical properties and/or reduce material cost. In many cases, neither the polymer nor the filler are biodegradable. Distinguished from these non-biodegradable plastics are two forms of biodegradable plastics: (1) partially degradable and (2) completely degradable.

Partially biodegradable plastics consist of a conventional polymer and a natural filler. The polymer matrix is petroleum-based, which is non-biodegradable, but the natural filler can be consumed by microorganisms (e.g., in a landfill), thus converting the polymer into a sponge-like structure and possibly leading to its degradation over time.

The plastics of greatest interest from an environmental viewpoint are the ***completely biodegradable plastics*** (aka ***bioplastics***) consisting of a polymer and filler that are both derived from natural and renewable sources. Various agricultural products are used as the raw materials for biodegradable plastics. A common polymeric starting material is starch, which is a major component in corn, wheat, rice, and potatoes. It consists of the two polymers amylose and amylopectin. Starch can be used to synthesize several thermoplastic materials that are processable by conventional plastic shaping methods, such as extrusion and injection molding (Chapter 13). Another starting point for biodegradable plastics involves fermentation of either corn starch or sugar cane to produce lactic acid, which can be polymerized to form polylactide, another thermoplastic material. A common filler used in bioplastics is cellulose, often in the form of reinforcing fibers in the polymer-matrix composite. Cellulose is grown as flax or hemp. It is inexpensive and possesses good mechanical strength.

Applications of biodegradable plastics are inhibited by the fact that these materials are more expensive than petroleum-based polymers. That may change in the future due to technological advances and economies of scale. Biopolymers are most attractive in situations where degradability is a higher priority than cost savings. At the top of the list are packaging materials that are quickly discarded as waste in landfills. It is estimated that approximately 40% of all plastics are used in packaging, mostly for food products [12]. Thus, biodegradable plastics are being used increasingly as substitutes for conventional plastics in packaging applications. Other applications include disposable food service items, coatings for paper and cardboard, waste bags, and mulches for agricultural crops. Medical applications include sutures, catheter bags, and sanitary laundry bags in hospitals.

8.6 GUIDE TO THE PROCESSING OF POLYMERS

Polymers are nearly always shaped in a heated, highly plastic consistency. Common operations are extrusion and molding. The molding of thermosets is generally more complicated because they require curing (cross-linking). Thermoplastics are easier to mold, and a greater variety of molding operations are available to process them (Chapter 13). Although plastics readily lend themselves to net shape processing, machining is sometimes required (Chapter 22); and plastic parts can be assembled into products by permanent joining techniques such as welding (Chapter 29), adhesive bonding (Section 31.3), or mechanical assembly (Chapter 32).

Rubber processing has a longer history than plastics, and the industries associated with these polymer materials have traditionally been separated, even though their processing is similar in many ways. We cover rubber processing technology in Chapter 14.

REFERENCES

[1] Alliger, G., and Sjothum, I. J. (eds.). *Vulcanization of Elastomers.* Krieger Publishing Company, New York, 1978.

[2] Billmeyer, F. W., Jr. *Textbook of Polymer Science*, 3rd ed. John Wiley & Sons, Inc., New York, 1984.

[3] Blow, C. M., and Hepburn, C. *Rubber Technology and Manufacture,* 2nd ed. Butterworth Scientific, London, 1982.

[4] Brandrup, J., and Immergut, E. E. (eds.). *Polymer Handbook,* 4th ed. John Wiley & Sons, Inc., New York, 2004.

[5] Brydson, J. A. *Plastics Materials,* 4th ed. Butterworths & Co., Ltd., London, 1999.

[6] Chanda, M., and Roy, S. K. *Plastics Technology Handbook,* 4th ed. CRC Taylor & Francis, Boca Raton, Florida, 2006.

[7] Charrier, J-M. *Polymeric Materials and Processing.* Oxford University Press, New York, 1991.

[8] *Engineering Materials Handbook,* Vol. 2, *Engineering Plastics.* ASM International, Materials Park, Ohio, 2000.

[9] Flinn, R. A., and Trojan, P. K. *Engineering Materials and Their Applications,* 5th ed. John Wiley & Sons, Inc., New York, 1995.

[10] Hall, C. *Polymer Materials,* 2nd ed. John Wiley & Sons, New York, 1989.

[11] Hofmann, W. *Rubber Technology Handbook.* Hanser Publishers, Munich, Germany, 1988.

[12] Kolybaba, M., Tabil, L. G., Panigrahi, S., Crerar, W. J., Powell, T., and Wang, B. "Biodegradable Polymers: Past Present, and Future," Paper Number RRV03-0007, American Society of Agricultural Engineers, October 2003.

[13] Margolis, J. M. *Engineering Plastics Handbook.* McGraw-Hill, New York, 2006.

[14] Mark, J. E., and Erman, B. (eds.). *Science and Technology of Rubber,* 3rd ed. Academic Press, Orlando, Florida, 2005.

[15] McCrum, N. G., Buckley, C. P., and Bucknall, C. B. *Principles of Polymer Engineering,* 2nd ed. Oxford University Press, Oxford, UK, 1997.

[16] *Modern Plastics Encyclopedia.* Modern Plastics, McGraw-Hill, Inc., New York, 1990.

[17] Reisinger, T. J. G. "Polymers of Tomorrow," *Advanced Materials & Processes,* March 2004, pp. 43–45.

[18] Rudin, A. *The Elements of Polymer Science and Engineering,* 2nd ed. Academic Press, Inc., Orlando, Florida, 1998.

[19] Seymour, R. B., and Carraher, C. E. *Seymour/Carraher's Polymer Chemistry,* 5th ed. Marcel Dekker, Inc., New York, 2000.

[20] Seymour, R. B. *Engineering Polymer Sourcebook.* McGraw-Hill Book Company, New York, 1990.

[21] Wikipedia. "Plastic recycling." Available at: http://en.wikipedia.org/wiki/Plastic_recycling. "Biodegradable plastic." Available at: http://en.wikipedia.org/wiki/

Biodegradable_plastic. "Plastic." Available at: http://en.wikipedia.org/wiki/Plastic.

[22] Green Plastics. Available at: http://www.greenplastics.com/reference.

[23] Young, R. J., and Lovell, P. *Introduction to Polymers,* 3rd ed. CRC Taylor and Francis, Boca Raton, Florida, 2008.

REVIEW QUESTIONS

8.1. What is a polymer?

8.2. What are the three basic categories of polymers?

8.3. How do the properties of polymers compare with those of metals?

8.4. What does the degree of polymerization indicate?

8.5. What is cross-linking in a polymer, and what is its significance?

8.6. What is a copolymer?

8.7. Copolymers can possess four different arrangements of their constituent mers. Name and briefly describe the four arrangements.

8.8. What is a terpolymer?

8.9. How are a polymer's properties affected when it takes on a crystalline structure?

8.10. Does any polymer ever become 100% crystalline?

8.11. What are some of the factors that influence a polymer's tendency to crystallize?

8.12. Why are fillers added to a polymer?

8.13. What is a plasticizer?

8.14. In addition to fillers and plasticizers, what are some other additives used with polymers?

8.15. Describe the difference in mechanical properties as a function of temperature between a highly crystalline thermoplastic and an amorphous thermoplastic.

8.16. What is unique about the polymer cellulose?

8.17. The nylons are members of which polymer group?

8.18. What is the chemical formula of ethylene, the monomer for polyethylene?

8.19. What is the basic difference between low-density and high-density polyethylene?

8.20. How do the properties of thermosetting polymers differ from those of thermoplastics?

8.21. Cross-linking (curing) of thermosetting plastics is accomplished by one of three ways. Name the three ways.

8.22. Elastomers and thermosetting polymers are both cross-linked. Why are their properties so different?

8.23. What happens to an elastomer when it is below its glass transition temperature?

8.24. What is the primary polymer ingredient in natural rubber?

8.25. How do thermoplastic elastomers differ from conventional rubbers?

MULTIPLE CHOICE QUIZ

There are 20 correct answers in the following multiple choice questions (some questions have multiple answers that are correct). To attain a perfect score on the quiz, all correct answers must be given. Each correct answer is worth 1 point. Each omitted answer or wrong answer reduces the score by 1 point, and each additional answer beyond the correct number of answers reduces the score by 1 point. Percentage score on the quiz is based on the total number of correct answers.

8.1. Of the three polymer types, which one is the most important commercially: (a) thermoplastics, (b) thermosets, or (c) elastomers?

8.2. Which one of the three polymer types is not normally considered to be a plastic: (a) thermoplastics, (b) thermosets, or (c) elastomers?

8.3. Which one of the three polymer types does not involve cross-linking: (a) thermoplastics, (b) thermosets, or (c) elastomers?

8.4. As the degree of crystallinity in a given polymer increases, the polymer becomes denser and stiffer, and its melting temperature decreases: (a) true or (b) false?

8.5. Which one of the following is the chemical formula for the repeating unit in polyethylene: (a) CH_2, (b) C_2H_4, (c) C_3H_6, (d) C_5H_8, or (e) C_8H_8?

8.6. Degree of polymerization is which one of the following: (a) average number of mers in the molecule chain; (b) proportion of the monomer that has been polymerized; (c) sum of the molecule weights of the mers in the molecule; or (d) none of the above?

8.7. A branched molecular structure is stronger in the solid state and more viscous in the molten state than a linear structure for the same polymer: (a) true or (b) false?

8.8. A copolymer is a mixture of the macromolecules of two different homopolymers: (a) true or (b) false?

8.9. As the temperature of a polymer increases, its density (a) increases, (b) decreases, or (c) remains fairly constant?

8.10. Which of the following plastics has the highest market share: (a) phenolics, (b) polyethylene,

(c) polypropylene, (d) polystyrene, or (e) polyvinylchloride?

8.11. Which of the following polymers are normally thermoplastic (four best answers): (a) acrylics, (b) cellulose acetate, (c) nylon, (d) phenolics, (e) polychloroprene, (f) polyesters, (g) polyethylene, (h) polyisoprene, and (i) polyurethane?

8.12. Polystyrene (without plasticizers) is amorphous, transparent, and brittle: (a) true or (b) false?

8.13. The fiber rayon used in textiles is based on which one of the following polymers: (a) cellulose, (b) nylon, (c) polyester, (d) polyethylene, or (e) polypropylene?

8.14. The basic difference between low-density polyethylene and high-density polyethylene is that the latter has a much higher degree of crystallinity: (a) true or (b) false?

8.15. Among the thermosetting polymers, the most widely used commercially is which one of the following: (a) epoxies, (b) phenolics, (c) silicones, or (d) urethanes?

8.16. The chemical formula for polyisoprene in natural rubber is which of the following: (a) CH_2, (b) C_2H_4, (c) C_3H_6, (d) C_5H_8, or (e) C_8H_8?

8.17. The leading commercial synthetic rubber is which one of the following: (a) butyl rubber, (b) isoprene rubber, (c) polybutadiene, (d) polyurethane, (e) styrene-butadiene rubber, or (f) thermoplastic elastomers?

9 COMPOSITE MATERIALS

Chapter Contents

In addition to metals, ceramics, and polymers, a fourth material category can be distinguished: composites. A ***composite material*** is a material system composed of two or more physically distinct phases whose combination produces aggregate properties that are different from those of its constituents. In certain respects, composites are the most interesting of the engineering materials because their structure is more complex than the other three types.

The technological and commercial interest in composite materials derives from the fact that their properties are not just different from their components but are often far superior. Some of the possibilities include:

➢ Composites can be designed that are very strong and stiff, yet very light in weight, giving them strength-to-weight and stiffness-to-weight ratios several times greater than steel or aluminum. These properties are highly desirable in applications ranging from commercial aircraft to sports equipment.

➢ Fatigue properties are generally better than for the common engineering metals. Toughness is often greater, too.

➢ Composites can be designed that do not corrode like steel; this is important in automotive and other applications.

➢ With composite materials, it is possible to achieve combinations of properties not attainable with metals, ceramics, or polymers alone.

➢ Better appearance and control of surface smoothness are possible with certain composite materials.

Along with the advantages, there are disadvantages and limitations associated with composite materials. These include: (1) properties of many important composites are anisotropic, which means the properties differ depending on the direction in which they are measured; (2) many of the polymer-based composites are subject to attack by chemicals or solvents, just as the polymers themselves are susceptible to attack; (3) composite materials are generally expensive, although prices may drop as volume increases; and (4) certain of the manufacturing methods for shaping composite materials are slow and costly.

We have already encountered several composite materials in our coverage of the three other material types. Examples include cemented carbides (tungsten carbide with cobalt binder), plastic molding compounds that contain fillers (e.g., cellulose fibers, wood flour), and rubber mixed with carbon black. We did not always identify these materials as composites; however, technically, they fit the above definition. It could even be argued that a two-phase metal alloy (e.g., $Fe + Fe_3C$) is a composite material, although it is not classified as such. Perhaps the most important composite material of all is wood.

In our presentation of composite materials, we first examine their technology and classification. There are many different materials and structures that can be used to form composites; we survey the various categories, devoting the most time to fiber-reinforced plastics, which are commercially the most important type. In the final section, we provide a guide to the manufacturing processes for composites.

9.1 TECHNOLOGY AND CLASSIFICATION OF COMPOSITE MATERIALS

As noted in our definition, a composite material consists of two or more distinct phases. The term *phase* indicates a homogeneous material, such as a metal or ceramic in which all of the grains have the same crystal structure, or a polymer with no fillers. By combining the phases, using methods yet to be described, a new material is created with aggregate performance exceeding that of its parts. The effect is synergistic.

Composite materials can be classified in various ways. One possible classification distinguishes between (1) traditional and (2) synthetic composites. *Traditional composites* are those that occur in nature or have been produced by civilizations for many years. Wood is a naturally occurring composite material, while concrete (Portland cement plus sand or gravel) and asphalt mixed with gravel are traditional composites used in construction. *Synthetic composites* are modern material systems normally associated with the manufacturing industries, in which the components are first produced separately and then combined in a controlled way to achieve the desired structure, properties, and part geometry. These synthetic materials are the composites normally thought of in the context of engineered products. Our attention in this chapter is focused on these materials.

9.1.1 COMPONENTS IN A COMPOSITE MATERIAL

In the simplest manifestation of our definition, a composite material consists of two phases: a primary phase and a secondary phase. The primary phase forms the *matrix* within which the secondary phase is imbedded. The imbedded phase is sometimes referred to as a *reinforcing agent* (or similar term), because it usually serves to strengthen the composite. The reinforcing phase may be in the form of fibers, particles, or various other geometries, as we shall see. The phases are generally insoluble in each other, but strong adhesion must exist at their interface(s).

The matrix phase can be any of three basic material types: polymers, metals, or ceramics. The secondary phase may also be one of the three basic materials, or it may be an element such as carbon or boron. Possible combinations in a two-component composite material can be organized as a 3×4 chart, as in Table 9.1. We see that certain combinations are not feasible, such as a polymer in a ceramic matrix. We also see that the possibilities

TABLE 9.1   Possible combinations of two-component composite materials.

Secondary phase (reinforcement)	Primary Phase (matrix)		
	Metal	**Ceramic**	**Polymer**
Metal	Powder metal parts infiltrated with a second metal	NA	Plastic molding compounds Steel-belted radial tires
Ceramic	Cermets[a] Fiber-reinforced metals	SiC whisker-reinforced Al_2O_3	Plastic molding compounds Fiberglass-reinforced plastic
Polymer	Powder metal parts impregnated with polymer	NA	Plastic molding compounds Kevlar-reinforced epoxy
Elements (C, B)	Fiber-reinforced metals	NA	Rubber with carbon black B or C fiber-reinforced plastic

NA = Not applicable currently.
[a]Cermets include cemented carbides.

include two-phase structures consisting of components of the same material type, such as fibers of Kevlar (polymer) in a plastic (polymer) matrix. In other composites the imbedded material is an element such as carbon or boron.

The classification system for composite materials used in this book is based on the matrix phase. We list the classes here and discuss them in Sections 9.2 through 9.4:

1. ***Metal Matrix Composites*** (MMCs) include mixtures of ceramics and metals, such as cemented carbides and other cermets, as well as aluminum or magnesium reinforced by strong, high stiffness fibers.
2. ***Ceramic Matrix Composites*** (CMCs) are the least common category. Aluminum oxide and silicon carbide are materials that can be imbedded with fibers for improved properties, especially in high temperature applications.
3. ***Polymer Matrix Composites*** (PMCs). Thermosetting resins are the most widely used polymers in PMCs. Epoxy and polyester are commonly mixed with fiber reinforcement, and phenolic is mixed with powders. Thermoplastic molding compounds are often reinforced, usually with powders (Section 8.1.5).

The classification can be applied to traditional composites as well as synthetics. Concrete is a ceramic matrix composite, while asphalt and wood are polymer matrix composites.

The matrix material serves several functions in the composite. First, it provides the bulk form of the part or product made of the composite material. Second, it holds the imbedded phase in place, usually enclosing and often concealing it. Third, when a load is applied, the matrix shares the load with the secondary phase, in some cases deforming so that the stress is essentially born by the reinforcing agent.

9.1.2   THE REINFORCING PHASE

It is important to understand that the role played by the secondary phase is to reinforce the primary phase. The imbedded phase is most commonly one of the shapes illustrated in Figure 9.1: fibers, particles, or flakes. In addition, the secondary phase can take the form of an infiltrated phase in a skeletal or porous matrix.

Fibers   ***Fibers*** are filaments of reinforcing material, generally circular in cross-section, although alternative shapes are sometimes used (e.g., tubular, rectangular, hexagonal).

(a)   (b)   (c)

Diameters range from less than 0.0025 mm (0.0001 in) to about 0.13 mm (0.005 in), depending on material.

Fiber reinforcement provides the greatest opportunity for strength enhancement of composite structures. In fiber-reinforced composites, the fiber is often considered to be the principal constituent since it bears the major share of the load. Fibers are of interest as reinforcing agents because the filament form of most materials is significantly stronger than the bulk form. The effect of fiber diameter on tensile strength can be seen in Figure 9.2. As diameter is reduced, the material becomes oriented in the direction of the fiber axis and the probability of defects in the structure decreases significantly. As a result, tensile strength increases dramatically.

Fibers used in composites can be either continuous or discontinuous. ***Continuous fibers*** are very long; in theory, they offer a continuous path by which a load can be carried by the composite part. In reality, this is difficult to achieve due to variations in the fibrous material and processing. ***Discontinuous fibers*** (chopped sections of continuous fibers) are short lengths ($L/D \approx 100$). An important type of discontinuous fiber are ***whiskers***—hair-like single crystals with diameters down to about 0.001 mm (0.00004 in) and very high strength.

Fiber orientation is another factor in composite parts. We can distinguish three cases, illustrated in Figure 9.3: (a) one-dimensional reinforcement, in which maximum strength and stiffness are obtained in the direction of the fiber; (b) planar reinforcement,

FIGURE 9.2   Relationship between tensile strength and diameter for a carbon fiber. (Source: [1]). Other filament materials show similar relationships.

FIGURE 9.3   Fiber orientation in composite materials: (a) one-dimensional, continuous fibers; (b) planar, continuous fibers in the form of a woven fabric; and (c) random, discontinuous fibers.

TABLE 9.2   Typical properties of fiber materials used as reinforcement in composites.

Fiber Material	Diameter		Tensile Strength		Elastic Modulus	
	mm	mils[a]	MPa	lb/in^2	GPa	lb/in^2
Metal: Steel	0.13	5.0	1000	150,000	206	30×10^6
Metal: Tungsten	0.013	0.5	4000	580,000	407	59×10^6
Ceramic: Al$_2$O$_3$	0.02	0.8	1900	275,000	380	55×10^6
Ceramic: SiC	0.13	5.0	3275	475,000	400	58×10^6
Ceramic: E-glass	0.01	0.4	3450	500,000	73	10×10^6
Ceramic: S-glass	0.01	0.4	4480	650,000	86	12×10^6
Polymer: Kevlar	0.013	0.5	3450	500,000	130	19×10^6
Element: Carbon	0.01	0.4	2750	400,000	240	35×10^6
Element: Boron	0.14	5.5	3100	450,000	393	57×10^6

[a]1 mil = 0.001 in.
Compiled from [3], [7], [11], and other sources. Note that strength depends on fiber diameter (Figure 9.2); the properties in this table must be interpreted accordingly.

in some cases in the form of a two-dimensional woven fabric; and (c) random or three-dimensional in which the composite material tends to possess isotropic properties.

Various materials are used as fibers in fiber-reinforced composites: metals, ceramics, polymers, carbon, and boron. The most important commercial use of fibers is in polymer composites. However, use of fiber-reinforced metals and ceramics is growing. Following is a survey of the important types of fiber materials, with properties listed in Table 9.2:

➢ *Glass*—The most widely used fiber in polymers, the term ***fiberglass*** is applied to denote glass fiber-reinforced plastic (GFRP). The two common glass fibers are E-glass and S-glass (compositions listed in Table 7.4). E-glass is strong and low cost, but its modulus is less than other fibers. S-glass is stiffer, and its tensile strength is one of the highest of all fiber materials; however, it is more expensive than E-glass.

➢ *Carbon*—Carbon (Section 7.5.1) can be made into high-modulus fibers. Besides stiffness, other attractive properties include low-density and low-thermal expansion. C-fibers are generally a combination of graphite and amorphous carbon.

➢ *Boron*—Boron (Section 7.5.3) has a very high elastic modulus, but its high cost limits applications to aerospace components in which this property (and others) are critical.

➢ *Kevlar 49*—This is the most important polymer fiber; it is a highly crystalline aramid, a member of the polyamide family (Section 8.2.2). Its specific gravity is low, giving it one of the highest strength-to-weight ratios of all fibers.

➢ *Ceramics*—Silicon carbide (SiC) and aluminum oxide (Al$_2$O$_3$) are the main fiber materials among ceramics. Both have high elastic moduli and can be used to strengthen low-density, low-modulus metals such as aluminum and magnesium.

➢ *Metal*—Steel filaments, both continuous and discontinuous, are used as reinforcing fibers in plastics. Other metals are currently less common as reinforcing fibers.

Particles and Flakes   A second common shape of the imbedded phase is ***particulate***, ranging in size from microscopic to macroscopic. Particles are an important material form for metals and ceramics; we discuss the characterization and production of engineering powders in Chapters 16 and 17.

The distribution of particles in the composite matrix is random, and therefore strength and other properties of the composite material are usually isotropic. The strengthening mechanism depends on particle size. The microscopic size is represented by very fine powders (around 1 μm) distributed in the matrix in concentrations of 15% or less. The presence of these powders results in dispersion-hardening of the matrix, in which dislocation movement in the matrix material is restricted by the microscopic particles. In effect, the matrix itself is strengthened, and no significant portion of the applied load is carried by the particles.

As particle size increases to the macroscopic range, and the proportion of imbedded material increases to 25% and more, the strengthening mechanism changes. In this case, the applied load is shared between the matrix and the imbedded phase. Strengthening occurs due to the load-carrying ability of the particles and the bonding of particles in the matrix. This form of composite strengthening occurs in cemented carbides, in which tungsten carbide is held in a cobalt binder. The proportion of tungsten carbide (WC) in the cobalt (Co) matrix is typically 80% or more.

Flakes are basically two-dimensional particles—small flat platelets. Two examples of this shape are the minerals mica (silicate of K and Al) and talc ($Mg_3Si_4O_{10}(OH)_2$), used as reinforcing agents in plastics. They are generally lower cost materials than polymers, and they add strength and stiffness to plastic molding compounds. Platelet sizes are usually in the range 0.01– to 1 mm (0.0004–0.040 in) across the flake, with a thickness of 0.001– to 0.005 mm (0.00004–0.00020 in).

Infiltrated Phase   The fourth form of imbedded phase occurs when the matrix has the form of a porous skeleton (like a sponge), and the second phase is simply a *filler*. In this case, the imbedded phase assumes the shape of the pores in the matrix. Metallic fillers are sometimes used to infiltrate the open porous structure of parts made by powder metallurgy techniques (Section 16.3.4), in effect creating a composite material. Oil-impregnated sintered PM components, such as bearings and gears, might be considered another example of this category.

The Interface   There is always an *interface* between constituent phases in a composite material. For the composite to operate effectively, the phases must bond where they join. In some cases, there is a direct bonding between the two ingredients, as suggested by Figure 9.4

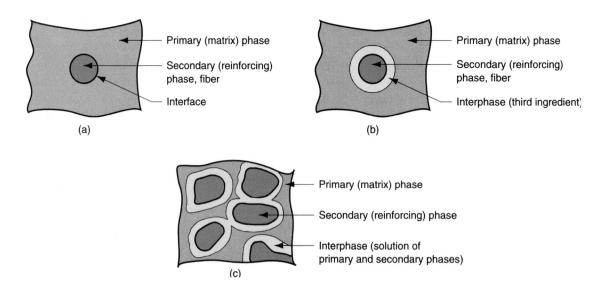

FIGURE 9.4   Interfaces and interphases between phases in a composite material: (a) direct bonding between primary and secondary phases; (b) addition of a third ingredient to bond the primary and secondary phases and form an interphase; and (c) formation of an interphase by solution of the primary and secondary phases at their boundary.

(a). In other cases, a third ingredient is added to promote bonding of the two primary phases. Called an ***interphase,*** this third ingredient can be thought of as an adhesive. An important example is the coating of glass fibers to achieve adhesion with thermosetting resin in fiberglass-reinforced plastics. As illustrated in Figure 9.4(b), this case results in two interfaces, one on either boundary of the interphase. Finally, a third form of interface occurs when the two primary components are not completely insoluble in each other; in this case, the interphase is formed consisting of a solution of the phases, as in Figure 9.4(c). An example occurs in cemented carbides (Section 9.2.1); at the high sintering temperatures used on these materials, some solubility results at the boundaries to create an interphase.

9.1.3 PROPERTIES OF COMPOSITE MATERIALS

In the selection of a composite material, an optimum combination of properties is usually being sought, rather than one particular property. For example, the fuselage and wings of an aircraft must be lightweight as well as strong, stiff, and tough. Finding a monolithic material that satisfies these requirements is difficult. Several fiber-reinforced polymers possess this combination of properties.

Another example is rubber. Natural rubber is a relatively weak material. In the early 1900s, it was discovered that by adding significant amounts of carbon black (almost pure carbon) to natural rubber, its strength is increased dramatically. The two ingredients interact to provide a composite material that is significantly stronger than either one alone. Rubber, of course, must also be vulcanized to achieve full strength.

Rubber itself is a useful additive in polystyrene. One of the distinctive and disadvantageous properties of polystyrene is its brittleness. Although most other polymers have considerable ductility, polystyrene has virtually none. Rubber (natural or synthetic) can be added in modest amounts (5%–15%) to produce high-impact polystyrene, which has much superior toughness and impact strength.

Properties of a composite material are determined by three factors: (1) the materials used as component phases in the composite, (2) the geometric shapes of the constituents and resulting structure of the composite system, and (3) the manner in which the phases interact with one another.

Rule of Mixtures   The properties of a composite material are a function of the starting materials. Certain properties of a composite material can be computed by means of a ***rule of mixtures***, which involves calculating a weighted average of the constituent material properties. Density is an example of this averaging rule. The mass of a composite material is the sum of the masses of the matrix and reinforcing phases:

$$m_c = m_m + m_r \tag{9.1}$$

where m = mass, kg (lb); and the subscripts c, m, and r indicate composite, matrix, and reinforcing phases, respectively. Similarly, the volume of the composite is the sum of its constituents:

$$V_c = V_m + V_r + V_v \tag{9.2}$$

where V = volume, cm^3 (in^3). V_v is the volume of any voids in the composite (e.g., pores). The density of the composite is the mass divided by the volume.

$$\rho_c = \frac{m_c}{V_c} = \frac{m_m + m_r}{V_c} \tag{9.3}$$

Because the masses of the matrix and reinforcing phase are their respective densities multiplied by their volumes,

$$m_m = \rho_m V_m \quad \text{and} \quad m_r = \rho_r V_r$$

we can substitute these terms into Eq. (9.3) and conclude that

$$\rho_c = f_m \rho_m + f_r \rho_r \tag{9.4}$$

where $f_m = V_m/V_c$ and $f_r = V_r/V_c$ are simply the volume fractions of the matrix and reinforcing phases.

Fiber-Reinforced Composites   Determining mechanical properties of composites from constituent properties is usually more involved. The rule of mixtures can sometimes be used to estimate the modulus of elasticity of a fiber-reinforced composite made of continuous fibers where E_c is measured in the longitudinal direction. The situation is depicted in Figure 9.5(a); we assume that the fiber material is much stiffer than the matrix and that the bonding between the two phases is secure. Under this model, the modulus of the composite can be predicted as follows:

$$E_c = f_m E_m + f_r E_r \tag{9.5}$$

where E_c, E_m, and E_r are the elastic moduli of the composite and its constituents, MPa (lb/in^2); and f_m and f_r are again the volume fractions of the matrix and reinforcing phase. The effect of Eq. (9.5) is seen in Figure 9.5(b).

Perpendicular to the longitudinal direction, the fibers contribute little to the overall stiffness except for their filling effect. The composite modulus can be estimated in this

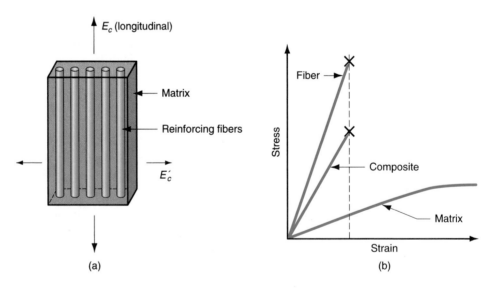

(a)   (b)

FIGURE 9.5   (a) Model of a fiber-reinforced composite material showing direction in which elastic modulus is being estimated by the rule of mixtures. (b) Stress–strain relationships for the composite material and its constituents. The fiber is stiff but brittle, while the matrix (commonly a polymer) is soft but ductile. The composite's modulus is a weighted average of its components' moduli. But when the reinforcing fibers fail, the composite does likewise.

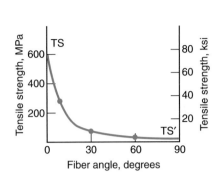

FIGURE 9.6 Variation in elastic modulus and tensile strength as a function of direction of measurement relative to longitudinal axis of carbon fiber-reinforced epoxy composite. (Source: [7]).

direction using the following:

$$E'_c = \frac{E_m E_r}{f_m E_r + f_r E_m} \qquad (9.6)$$

where E'_c = elastic modulus perpendicular to the fiber direction, MPa (lb/in^2). Our two equations for E_c demonstrate the significant anisotropy of fiber-reinforced composites. This directional effect can be seen in Figure 9.6 for a fiber-reinforced polymer composite, in which both elastic modulus and tensile strength are measured relative to fiber direction.

Fibers illustrate the importance of geometric shape. Most materials have tensile strengths several times greater in a fibrous form than in bulk. However, applications of fibers are limited by surface flaws, buckling when subjected to compression, and the inconvenience of the filament geometry when a solid component is needed. By imbedding the fibers in a polymer matrix, a composite material is obtained that avoids the problems of fibers but utilizes their strengths. The matrix provides the bulk shape to protect the fiber surfaces and resist buckling; and the fibers lend their high strength to the composite. When a load is applied, the low-strength matrix deforms and distributes the stress to the high-strength fibers, which then carry the load. If individual fibers break, the load is redistributed through the matrix to other fibers.

9.1.4 OTHER COMPOSITE STRUCTURES

Our model of a composite material described above is one in which a reinforcing phase is imbedded in a matrix phase, the combination having properties that are superior in certain respects to either of the constituents alone. However, composites can take alternative forms that do not fit this model, some of which are of considerable commercial and technological importance.

A *laminar composite structure* consists of two or more layers bonded together to form an integral piece, as in Figure 9.7(a). The layers are usually thick enough that this composite can be readily identified—not always the case with other composites. The layers are often of different materials, but not necessarily. Plywood is such an example; the layers are of the same wood, but the grains are oriented differently to increase overall strength of the laminated piece. A laminar composite often uses different materials in its layers to gain the advantage of combining the particular properties of each. In some cases, the layers themselves may be composite materials. We have mentioned that wood is a composite material; therefore, plywood is a laminar composite structure in which the layers themselves are composite materials. A list of examples of laminar composites is compiled in Table 9.3.

FIGURE 9.7   Laminar composite structures: (a) conventional laminar structure; (b) sandwich structure using a foam core, and (c) honeycomb sandwich structure.

TABLE 9.3   Examples of laminar composite structures.

Laminar Composite	Description (reference in text if applicable)
Automotive tires	A tire consists of multiple layers bonded together; the layers are composite materials (rubber reinforced with carbon black), and the plies consist of rubber-impregnated fabrics (Chapter 14).
Honeycomb sandwich	A lightweight honeycomb structure is bonded on either face to thin sheets, as in Figure 9.7(c).
Fiber-reinforced polymers	Multilayered fiber-reinforced plastic panels are used for aircraft, automobile body panels, and boat hulls (Chapter 15).
Plywood	Alternating sheets of wood are bonded together at different orientations for improved strength.
Printed circuit boards	Layers of copper and reinforced plastic are used for electrical conductivity and insulation, respectively (Section 36.2).
Snow skis	Skis are laminar composite structures consisting of multiple layers of metals, particle board, and phenolic plastic.
Windshield glass	Two layers of glass on either side of a sheet of tough plastic (Section 12.3.1).

The **sandwich structure** is sometimes distinguished as a special case of the laminar composite structure. It consists of a relatively thick core of low-density material bonded on both faces to thin sheets of a different material. The low-density core may be a **foamed material**, as in Figure 9.7(b), or a **honeycomb**, as in (c). The reason for using a sandwich structure is to obtain a material with high strength-to-weight and stiffness-to-weight ratios.

9.2   METAL MATRIX COMPOSITES

Metal matrix composites (MMCs) consist of a metal matrix reinforced by a second phase. Common reinforcing phases include (1) particles of ceramic and (2) fibers of various materials, including other metals, ceramics, carbon, and boron. MMCs of the first type are commonly called cermets.

9.2.1   CERMETS

A **cermet**[1] is a composite material in which a ceramic is contained in a metallic matrix. The ceramic often dominates the mixture, sometimes ranging up to 96% by volume.

[1]The word "cermet" was first used in the English language around 1948.

FIGURE 9.8
Photomicrograph (~1500x) of cemented carbide with 85% WC and 15% Co. (Photo courtesy of Kennametal Inc.)

Bonding can be enhanced by slight solubility between phases at the elevated temperatures used in processing these composites. Cermets can be subdivided into (1) cemented carbides and (2) oxide-based cermets.

Cemented Carbides   *Cemented carbides* are composed of one or more carbide compounds bonded in a metallic matrix. The term *cermet* is not used for all of these materials, even though it is technically correct. The common cemented carbides are based on tungsten carbide (WC), titanium carbide (TiC), and chromium carbide (Cr_3C_2). Tantalum carbide (TaC) and others are also used but less commonly. The principal metallic binders are cobalt and nickel. We have previously discussed the carbide ceramics (Section 7.3.2); they constitute the principal ingredient in cemented carbides, typically ranging in content from 80% to 95% of total weight.

Cemented carbide parts are produced by particulate processing techniques (Section 17.3). Cobalt is the binder used for WC (see Figure 9.8), and nickel is a common binder for TiC and Cr_3C_2. Even though the binder constitutes only about 5% to 15%, its effect on mechanical properties is significant in the composite material. Using WC–Co as an example, as the percentage of Co is increased, hardness is decreased and transverse rupture strength (TRS) is increased, as shown in Figure 9.9. TRS correlates with toughness of the WC–Co composite.

Cutting tools are the most common application of cemented carbides based on *tungsten carbide.* Other applications of WC–Co cemented carbides include wire drawing dies, rock-drilling bits and other mining tools, dies for powder metallurgy, indenters for hardness testers, and other applications where hardness and wear resistance are critical requirements.

Titanium carbide cermets are used principally for high temperature applications. Nickel is the preferred binder; its oxidation resistance at high temperatures is superior to that of cobalt. Applications include gas-turbine nozzle vanes, valve seats, thermocouple protection tubes, torch tips, and hot-working spinning tools [11]. TiC–Ni is also used as a cutting tool material in machining operations.

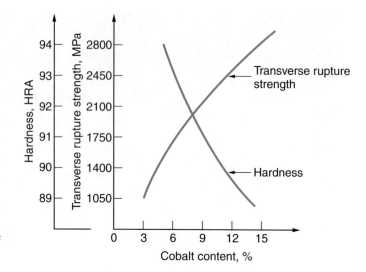

FIGURE 9.9 Typical plot of hardness and transverse rupture strength as a function of cobalt content.

Compared with WC–Co cemented carbides, nickel-bonded **chromium carbides** are more brittle, but have excellent chemical stability and corrosion resistance. This combination, together with good wear resistance, makes it suitable for applications such as gage blocks, valve liners, spray nozzles, and bearing seal rings [11].

Oxide-based Cermets   Most of these composites utilize Al_2O_3 as the particulate phase; MgO is another oxide sometimes used. A common metal matrix is chromium, although other metals can also be used as binders. Relative proportions of the two phases vary significantly, with the possibility for the metal binder to be the major ingredient. Applications include cutting tools, mechanical seals, and thermocouple shields.

9.2.2 FIBER-REINFORCED METAL MATRIX COMPOSITES

These MMCs are of interest because they combine the high tensile strength and modulus of elasticity of a fiber with metals of low density, thus achieving good strength-to-weight and modulus-to-weight ratios in the resulting composite material. Typical metals used as the low-density matrix are aluminum, magnesium, and titanium. Some of the important fiber materials used in the composite include Al_2O_3, boron, carbon, and SiC.

Properties of fiber-reinforced MMCs are anisotropic, as expected. Maximum tensile strength in the preferred direction is obtained by using continuous fibers bonded strongly to the matrix metal. Elastic modulus and tensile strength of the composite material increase with increasing fiber volume. MMCs with fiber reinforcement have good high-temperature strength properties; and they are good electrical and thermal conductors. Applications have largely been components in aircraft and turbine machinery, where these properties can be exploited.

9.3   CERAMIC MATRIX COMPOSITES

Ceramics have certain attractive properties: high stiffness, hardness, hot hardness, and compressive strength; and relatively low density. Ceramics also have several faults: low toughness and bulk tensile strength, and susceptibility to thermal cracking. Ceramic matrix composites (CMCs) represent an attempt to retain the desirable properties of ceramics

FIGURE 9.10 Highly magnified electron microscopy photograph (~3000x) showing fracture surface of SiC whisker reinforced ceramic (Al_2O_3) used as cutting tool material. (Courtesy of Greenleaf Corporation, Saegertown, Pennsylvania.)

while compensating for their weaknesses. CMCs consist of a ceramic primary phase imbedded with a secondary phase. To date, most development work has focused on the use of fibers as the secondary phase. Success has been elusive. Technical difficulties include thermal and chemical compatibility of the constituents in CMCs during processing. Also, as with any ceramic material, limitations on part geometry must be considered.

Ceramic materials used as matrices include alumina (Al_2O_3), boron carbide (B_4C), boron nitride (BN), silicon carbide (SiC), silicon nitride (Si_3N_4), titanium carbide (TiC), and several types of glass [10]. Some of these materials are still in the development stage as CMC matrices. Fiber materials in CMCs include carbon, SiC, and Al_2O_3.

The reinforcing phase in current CMC technology consists of either short fibers, such as whiskers, or long fibers. Products with short fibers have been successfully fabricated using particulate processing methods (Chapter 17), the fibers being treated as a form of powder in these materials. Although there are performance advantages in using long fibers as reinforcement in ceramic matrix composites, development of economical processing techniques for these materials has been difficult. One promising commercial application of CMCs is in metal-cutting tools as a competitor of cemented carbides, as illustrated in Figure 9.10. The composite tool material has whiskers of SiC in a matrix of Al_2O_3. Other potential applications are in elevated temperatures and environments that are chemically corrosive to other materials.

9.4 POLYMER MATRIX COMPOSITES

A *polymer matrix composite* (PMC) consists of a polymer primary phase in which a secondary phase is imbedded in the form of fibers, particles, or flakes. Commercially, PMCs are the most important of the three classes of synthetic composites. They include most plastic molding compounds, rubber reinforced with carbon black, and fiber-reinforced polymers (FRPs). Of the three, FRPs are most closely identified with the term composite. If one mentions "composite material" to a design engineer, FRP is usually the composite that comes to mind. Our video clip on composite materials and manufacturing provides an overview of fiber-reinforced polymer composites.

VIDEO CLIP ———————————————————————————

View the segment titled Composite Materials and Manufacturing.

9.4.1 FIBER-REINFORCED POLYMERS

A *fiber-reinforced polymer* is a composite material consisting of a polymer matrix imbedded with high-strength fibers. The polymer matrix is usually a thermosetting plastic such as unsaturated polyester or epoxy, but thermoplastic polymers, such as nylons (polyamides), polycarbonate, polystyrene, and polyvinylchloride, are also used. In addition, elastomers are also reinforced by fibers for rubber products such as tires and conveyor belts.

Fibers in PMCs come in various forms: discontinuous (chopped), continuous, or woven as a fabric. Principal fiber materials in FRPs are glass, carbon, and Kevlar 49. Less common fibers include boron, SiC, and Al_2O_3, and steel. Glass (in particular E-glass) is the most common fiber material in today's FRPs; its use to reinforce plastics dates from around 1920.

The term *advanced composites* is sometimes used in connection with FRPs developed since the late 1960s that use boron, carbon, or Kevlar, as the reinforcing fibers [13]. Epoxy is the common matrix polymer. These composites generally have high fiber content (>50% by volume) and possess high strength and modulus of elasticity. When two or more fiber materials are combined in the FRP composite, it is called a *hybrid composite*. Advantages cited for hybrids over conventional or advanced FRPs include balanced strength and stiffness, improved toughness and impact resistance, and reduced weight [11]. Advanced and hybrid composites are used in aerospace applications.

The most widely used form of the FRP itself is a laminar structure, made by stacking and bonding thin layers of fiber and polymer until the desired thickness is obtained. By varying the fiber orientation among the layers, a specified level of anisotropy in properties can be achieved in the laminate. This method is used to form parts of thin cross section, such as aircraft wing and fuselage sections, automobile and truck body panels, and boat hulls.

Properties There are a number of attractive features that distinguish fiber-reinforced plastics as engineering materials. Most notable are (1) high strength-to-weight ratio, (2) high modulus-to-weight ratio, and (3) low specific gravity. A typical FRP weighs only about one-fifth as much as steel; yet strength and modulus are comparable in the fiber direction. Table 9.4 compares these properties for several FRPs, steels, and an aluminum alloy. Properties listed in Table 9.4 depend on the proportion of fibers in the composite. Both tensile strength and elastic modulus increase as the fiber content is increased, by Eq. (9.5). Other properties and characteristics of fiber-reinforced plastics include (4) good fatigue strength; (5) good corrosion resistance, although polymers are soluble in various chemicals; (6) low thermal expansion for many FRPs, leading to good dimensional stability; and (7) significant anisotropy in

TABLE 9.4   Comparison of typical properties of fiber-reinforced plastics and representative metal alloys.

Material	Specific Gravity (SG)	Tensile Strength (TS)		Elastic Modulus (E)		Index[a]	
		MPa	lb/in^2	GPa	lb/in^2	TS/SG	E/SG
Low-C steel	7.87	345	50,000	207	30×10^6	1.0	1.0
Alloy steel (heat treated)	7.87	3450	500,000	207	30×10^6	10.0	1.0
Aluminum alloy (heat treated)	2.70	415	60,000	69	10×10^6	3.5	1.0
FRP: fiberglass in polyester	1.50	205	30,000	69	10×10^6	3.1	1.7
FRP: Carbon in epoxy[b]	1.55	1500	220,000	140	20×10^6	22.3	3.4
FRP: Carbon in epoxy[c]	1.65	1200	175,000	214	31×10^6	16.7	4.9
FRP: Kevlar in epoxy matrix	1.40	1380	200,000	76	11×10^6	22.5	2.1

[a]Indices are relative tensile strength-to-weight (*TS/SG*) and elastic modulus-to-weight (*E/SG*) ratios compared to low-C steel as the base (index = 1.0 for the base).
[b]High tensile-strength carbon fibers used in FRP.
[c]High modulus carbon fibers used in FRP.
Compiled from [3], [7], and other sources. Properties are measured in the fiber direction.

properties. With regard to this last feature, the mechanical properties of the FRPs given in Table 9.4 are in the direction of the fiber. As previously noted, their values are significantly less when measured in a different direction.

Applications   During the last three decades there has been a steady growth in the application of fiber-reinforced polymers in products requiring high strength and low weight, often as substitutions for metals. The aerospace industry is one of the biggest users of advanced composites. Designers are continually striving to reduce aircraft weight to increase fuel efficiency and payload capacity. Applications of advanced composites in both military and commercial aircraft have increased steadily. Much of the structural weight of today's airplanes and helicopters consists of FRPs. The new Boeing 787 Dreamliner features 50% (by weight) composite (carbon fiber-reinforced plastic). That's about 80% of the volume of the aircraft. Composites are used for the fuselage, wings, tail, doors, and interior. By comparison, Boeing's 777 has only about 12% composites (by weight).

The automotive industry is another important user of FRPs. The most obvious applications are FRP body panels for cars and truck cabs. A notable example is the Chevrolet Corvette that has been produced with FRP bodies for decades. Less apparent applications are in certain chassis and engine parts. Automotive applications differ from those in aerospace in two significant respects. First, the requirement for high strength-to-weight ratio is less demanding than for aircraft. Car and truck applications can use conventional fiberglass reinforced plastics rather than advanced composites. Second, production quantities are much higher in automotive applications, requiring more economical methods of fabrication. Continued use of low-carbon sheet steel in automobiles in the face of FRP's advantages is evidence of the low cost and processability of steel.

FRPs have been widely adopted for sports and recreational equipment. Fiberglass reinforced plastic has been used for boat hulls since the 1940s. Fishing rods were another early application. Today, FRPs are represented in a wide assortment of sports products, including tennis rackets, golf club shafts, football helmets, bows and arrows, skis, and bicycle wheels.

9.4.2   OTHER POLYMER MATRIX COMPOSITES

In addition to FRPs, other PMCs contain particles, flakes, and short fibers. Ingredients of the secondary phase are called *fillers* when used in polymer molding compounds (Section 8.1.5). Fillers divide into two categories: (1) reinforcements and (2) extenders. ***Reinforcing fillers*** serve to strengthen or otherwise improve mechanical properties of the polymer. Common examples include: wood flour and powdered mica in phenolic and amino resins to increase strength, abrasion resistance, and dimensional stability; and carbon black in rubber to improve strength, wear, and tear resistance. ***Extenders*** simply increase the bulk and reduce the cost-per-unit weight of the polymer, but have little or no effect on mechanical properties. Extenders may be formulated to improve molding characteristics of the resin.

Foamed polymers (Section 13.11) are a form of composite in which gas bubbles are imbedded in a polymer matrix. Styrofoam and polyurethane foam are the most common examples. The combination of near-zero density of the gas and relatively low density of the matrix makes these materials extremely light weight. The gas mixture also lends very low thermal conductivity for applications in which heat insulation is required.

9.5   GUIDE TO PROCESSING COMPOSITE MATERIALS

Composite materials are formed into shapes by many different processing technologies. The two phases are typically produced separately before being combined into the composite part geometry. The matrix phases are generally processed by the technologies described in Chapters 6, 7, and 8 for metals, ceramics, and polymers.

Processing methods for the imbedded phase depend on geometry. Fiber production is described in Section 12.2.3 for glass and Section 13.4 for polymers. Fiber production methods for carbon, boron, and other materials are summarized in Table 15.1. Powder production for metals is described in Section 16.2 and for ceramics in Section 17.1.1. Processing techniques to fabricate MMC and CMC components, are similar to those used for powdered metals and ceramics (Chapters 16 and 17). We deal with the processing of cermets specifically in Section 17.3.

Molding processes are commonly performed on PMCs, both particle and chopped fiber types. Molding processes for these composites are the same as those used for polymers (Chapter 13). Other more specialized processes for polymer matrix composites, fiber-reinforced polymers in particular, are described in Chapter 15. Many laminated composite and honeycomb structures are assembled by adhesive bonding (Section 31.3).

REFERENCES

[1] Chawla, K. K. *Composite Materials: Science and Engineering,* 3rd ed. Springer-Verlag, New York, 2008.

[2] Delmonte, J. *Metal-Polymer Composites.* Van Nostrand Reinhold, New York, 1990.

[3] *Engineering Materials Handbook,* Vol. 1, *Composites.* ASM International, Metals Park, Ohio, 1987.

[4] Flinn, R. A., and Trojan, P. K. *Engineering Materials and Their Applications,* 5th ed. John Wiley & Sons, New York, 1995.

[5] Greenleaf Corporation. *WG-300 — Whisker Reinforced Ceramic/Ceramic Composites* [marketing literature]. Saegertown, Pennsylvania, YEAR??.

[6] Hunt, W. H., Jr., and Herling, S. R. "Aluminum Metal-Matrix Composites," *Advanced Materials & Processes,* February 2004, pp. 39–42.

[7] Mallick, P. K. *Fiber-Reinforced Composites: Materials, Manufacturing, and Designs,* 3rd ed. CRC Taylor & Francis, Boca Raton, Florida, 2007.

[8] McCrum, N. G., Buckley, C. P., and Bucknall, C. B. *Principles of Polymer Engineering,* 2nd ed. Oxford University Press, Oxford, UK, 1997.

[9] Morton-Jones, D. H. *Polymer Processing.* Chapman and Hall, London, 1989.

[10] Naslain, R., and Harris, B. (eds.). *Ceramic Matrix Composites.* Elsevier Applied Science, London and New York, 1990.

[11] Schwartz, M. M. *Composite Materials Handbook,* 2nd ed. McGraw-Hill Book Company, New York, 1992.

[12] Tadmor, Z., and Gogos, C. G. *Principles of Polymer Processing.* Wiley-Interscience, Hoboken, New Jersey, 2006.

[13] Wick, C., and Veilleux R. F. (eds.). *Tool and Manufacturing Engineers Handbook,* 4th ed, Volume III—, *Materials, Finishing, and Coating*, Chapter 8. Society of Manufacturing Engineers, Dearborn, Michigan, 1985.

[14] Wikipedia. "Boeing 787." Available at: wikipedia. org/wiki/Boeing_787.

[15] Zweben, C., Hahn, H. T., and Chou, T-W. *Delaware Composites Design Encyclopedia,* Vol. 1, *Mechanical Behavior and Properties of Composite Materials.* Technomic Publishing, Lancaster, Pennsylvania, 1989.

REVIEW QUESTIONS

9.1. What is a composite material?

9.2. Identify some of the characteristic properties of composite materials.

9.3. What does the term anisotropic mean?

9.4. How are traditional composites distinguished from synthetic composites?

9.5. Name the three basic categories of composite materials.

9.6. What are the common forms of the reinforcing phase in composite materials?

9.7. What is a whisker?

9.8. What are the two forms of sandwich structure among laminar composite structures? Briefly describe each.

9.9. Give some examples of commercial products which are laminar composite structures.

9.10. What are the three general factors that determine the properties of a composite material?

9.11. What is the rule of mixtures?

9.12. What is a cermet?

9.13. Cemented carbides are what class of composites?

9.14. What are some of the weaknesses of ceramics that might be corrected in fiber-reinforced ceramic matrix composites?

9.15. What is the most common fiber material in fiber-reinforced plastics?

9.16. What does the term advanced composites mean?

9.17. What is a hybrid composite?

9.18. Identify some of the important properties of fiber-reinforced plastic composite materials.

9.19. Name some of the important applications of FRPs.

9.20. What is meant by the term interface in the context of composite materials?

MULTIPLE CHOICE QUIZ

There are 19 correct answers in the following multiple-choice questions (some questions have multiple answers that are correct). To attain a perfect score on the quiz, all correct answers must be given. Each correct answer is worth 1 point. Each omitted answer or wrong answer reduces the score by 1 point, and each additional answer beyond the correct number of answers reduces the score by 1 point. Percentage score on the quiz is based on the total number of correct answers.

9.1. Anisotropic means which one of the following: (a) composite materials with composition consisting of more than two materials, (b) properties are the same in every direction, (c) properties vary depending on the direction in which they are measured, or (d) strength and other properties are a function of curing temperature?

9.2. The reinforcing phase is the matrix within which the secondary phase is imbedded: (a) true or (b) false?

9.3. Which one of the following reinforcing geometries offers the greatest potential for strength and stiffness improvement in the resulting composite material: (a) fibers, (b) flakes, (c) particles, or (d) infiltrated phase?

9.4. Wood is which one of the following composite types: (a) CMC, (b) MMC, or (c) PMC?

9.5. Which of the following materials are used as fibers in fiber-reinforced plastics (four best answers): (a) aluminum oxide, (b) boron, (c) cast iron, (d) E-glass, (e) epoxy, (f) Kevlar 49, (g) polyester, and (h) silicon?

9.6. Which of the following metals are used as the matrix material in fiber-reinforced MMCs (two best answers): (a) aluminum, (b) copper, (c) iron, (d) magnesium, and (e) zinc?

9.7. Which of the following metals are used as the matrix metals in nearly all WC cemented carbides and TiC cermets (two correct answers): (a) aluminum, (b) chromium, (c) cobalt, (d) lead, (e) nickel, (f) tungsten, and (g) tungsten carbide?

9.8. Ceramic matrix composites are designed to overcome which of the following weaknesses of ceramics (two best answers): (a) compressive strength, (b) hardness, (c) hot hardness, (d) modulus of elasticity, (e) tensile strength, and (f) toughness?

9.9. Which one of the following polymer types are most commonly used in polymer matrix composites: (a) elastomers, (b) thermoplastics, or (c) thermosets?

9.10. Which of the following materials are not composites (two correct answers): (a) cemented carbide, (b) phenolic molding compound, (c) plywood, (d) Portland cement, (e) rubber in automobile tires, (f) wood, and (g) 1020 steel?

9.11. In the Boeing 787 Dreamliner, what percentage of the aircraft consist of composite materials (two correct answers): (a) 12% by volume, (b) 20% by volume, (c) 50% by volume, (d) 80% by volume, (e) 12% by weight, (f) 20% by weight, (g) 50% by weight, and (h) 80% by weight?

PROBLEMS

9.1. A fiberglass composite is composed of a matrix of vinyl ester and reinforcing fibers of E-glass. The volume fraction of E-glass is 35%. The remainder is vinyl ester. The density of the vinyl ester is 0.882 g/cm^3, and its modulus of elasticity is 3.60 GPa. The density of E-glass is 2.60 g/cm^3, and its modulus of elasticity is 76.0 GPa. A section of composite 1.00 cm × 50.00 cm × 200.00 cm is fabricated with the E-glass fibers running longitudinal along the 200-cm direction. Assume there are no voids in the composite. Determine the (a) mass of vinyl ester in the section, (b) mass of E-glass fibers in the section, and (c) the density of the composite.

9.2. For problem 9.1, determine the modulus of elasticity in (a) the longitudinal direction of the glass fibers and (b) the perpendicular direction to the glass fibers.

9.3. A composite sample of carbon reinforced epoxy has dimensions of 12 in $\times$ 12 in $\times$ 0.25 in and mass of 1.8 lb. The carbon fibers have a modulus of elasticity of $50(10^6)$ lb/in^2 and a density of 0.069 lb/in^3. The epoxy matrix has modulus of elasticity of $0.61(10^6)$ lb/in^2 and a density of 0.042 lb/in^3. What is the volume fraction of (a) the carbon fibers and (b) the epoxy matrix in the sample? Assume there are no voids in the sample.

9.4. In problem 9.3, what is the predicted value for the modulus of elasticity (a) in the longitudinal direction and (b) the perpendicular to the carbon fibers?

9.5. A composite has a matrix of polyester with Kevlar-49 fibers. The volume fractions of polyester and Kevlar are 60% and 40%, respectively. The Kevlar fibers have a modulus of elasticity of 60 GPa in the longitudinal direction and 3 GPa in the transverse direction. The polyester matrix has a modulus of elasticity of 5.6 GPa in both directions. (a) Determine the modulus of elasticity for the composite in the longitudinal direction. (b) Determine the modulus of elasticity in the transverse direction.

Part III Solidification Processes

10 FUNDAMENTALS OF METAL CASTING

In this part of the book, we consider those manufacturing processes in which the starting work material is either a liquid or is in a highly plastic condition, and a part is created through solidification of the material. Casting and molding processes dominate this category of shaping operations. With reference to Figure 10.1, the solidification processes can be classified according to the engineering material that is processed: (1) metals, (2) ceramics, specifically glasses,[1] and (3) polymers and polymer matrix composites (PMCs). Casting of metals is covered in this and the following chapter. Glassworking is covered in Chapter 12, and polymer and PMC processing is treated in Chapters 13, 14, and 15.

Casting is a process in which molten metal flows by gravity or other force into a mold where it solidifies in the shape of the mold cavity. The term *casting* is also applied to the part that is made by this process. It is one of the oldest shaping processes, dating back 6000 years (Historical Note 10.1). The principle of casting seems simple: melt the metal, pour it into a mold, and let it cool and solidify; yet there are many factors and variables that must be considered in order to accomplish a successful casting operation.

Casting includes both the casting of ingots and the casting of shapes. The term *ingot* is usually associated with the primary metals industries; it describes a large casting that is simple in shape and intended for subsequent reshaping by

---

[1]Among the ceramics, only glass is processed by solidification; traditional and new ceramics are shaped using particulate processes (Chapter 17).

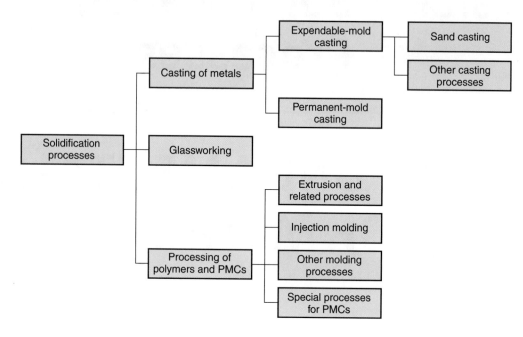

FIGURE 10.1 Classification of solidification processes.

Historical Note 10.1 *Origins of casting*

Casting of metals can be traced back to around 4000 BCE. Gold was the first metal to be discovered and used by the early civilizations; it was malleable and could be readily hammered into shape at room temperature. There seemed to be no need for other ways to shape gold. It was the subsequent discovery of copper that gave rise to the need for casting. Although copper could be forged to shape, the process was more difficult (due to strain hardening) and limited to relatively simple forms. Historians believe that hundreds of years elapsed before the process of casting copper was first performed, probably by accident during the reduction of copper ore in preparation for hammering the metal into some useful form. Thus, through serendipity, the art of casting was born. It is likely that the discovery occurred in Mesopotamia, and the "technology" quickly spread throughout the rest of the ancient world.

It was an innovation of significant importance in the history of mankind. Shapes much more intricate could be formed by casting than by hammering. More sophisticated tools and weapons could be fabricated. More detailed implements and ornaments could be fashioned. Fine gold jewelry could be made more beautiful and valuable than by previous methods. Alloys were first used for casting when it was discovered that mixtures of copper and tin (the alloy thus formed was bronze) yielded much better castings than copper alone. Casting permitted the creation of wealth to those nations that could perform it best. Egypt ruled the Western civilized world during the Bronze Age (nearly 2000 years) largely due to its ability to perform the casting process.

Religion provided an important influence during the Dark Ages (circa 400 to 1400) for perpetuating the foundryman's skills. Construction of cathedrals and churches required the casting of bells that were used in these structures. Indeed, the time and effort needed to cast the large bronze bells of the period helped to move the casting process from the realm of art toward the regimen of technology. Advances in melting and mold-making techniques were made. Pit molding, in which the molds were formed in a deep pit located in front of the furnace to simplify the pouring process, was improved as a casting procedure. In addition, the bellfounder learned the relationships between the tone of the bell, which was the important measure of product quality, and its size, shape, thickness, and metal composition.

Another important product associated with the development of casting was the cannon. Chronologically, it followed the bell, and therefore many of the casting techniques developed for bellfounding were applied to cannon making. The first cast cannon was made in Ghent, Belgium, in the year 1313—by a religious monk, of all people. It was made of bronze, and the bore was formed by means of a core during casting. Because of the rough bore surface created by the casting process, these early guns were not accurate and had to be fired at relatively close range to be effective. It was soon realized that accuracy and range could be improved if the bore were made smooth by machining the surface. Quite appropriately, this machining process was called **boring** (Section 22.1.5).

processes such as rolling or forging. Ingot casting was discussed in Chapter 6. **Shape casting** involves the production of more complex geometries that are much closer to the final desired shape of the part or product. It is with the casting of shapes rather than ingots that this chapter and the next are concerned.

A variety of shape casting methods are available, thus making it one of the most versatile of all manufacturing processes. Among its capabilities and advantages are the following:

➤ Casting can be used to create complex part geometries, including both external and internal shapes.

➤ Some casting processes are capable of producing parts to **net shape.** No further manufacturing operations are required to achieve the required geometry and dimensions of the parts. Other casting processes are **near net shape,** for which some additional shape processing is required (usually machining) in order to achieve accurate dimensions and details.

➤ Casting can be used to produce very large parts. Castings weighing more than 100 tons have been made.

➤ The casting process can be performed on any metal that can be heated to the liquid state.

➤ Some casting methods are quite suited to mass production.

There are also disadvantages associated with casting—different disadvantages for different casting methods. These include limitations on mechanical properties, porosity, poor dimensional accuracy and surface finish for some casting processes, safety hazards to humans when processing hot molten metals, and environmental problems.

Parts made by casting processes range in size from small components weighing only a few ounces up to very large products weighing tons. The list of parts includes dental crowns, jewelry, statues, wood-burning stoves, engine blocks and heads for automotive vehicles, machine frames, railway wheels, frying pans, pipes, and pump housings. All varieties of metals can be cast, ferrous and nonferrous.

Casting can also be used on other materials such as polymers and ceramics; however, the details are sufficiently different that we postpone discussion of the casting processes for these materials until later chapters. This chapter and the next deal exclusively with metal casting. Here we discuss the fundamentals that apply to virtually all casting operations. In the following chapter, the individual casting processes are described, along with some of the product design issues that must be considered when making parts out of castings.

10.1   OVERVIEW OF CASTING TECHNOLOGY

As a production process, casting is usually carried out in a foundry. A **foundry** is a factory equipped for making molds, melting and handling metal in molten form, performing the casting process, and cleaning the finished casting. The workers who perform the casting operations in these factories are called **foundrymen.**

10.1.1   CASTING PROCESSES

Discussion of casting logically begins with the mold. The *mold* contains a cavity whose geometry determines the shape of the cast part. The actual size and shape of the cavity must be slightly oversized to allow for shrinkage that occurs in the metal during solidification and cooling. Different metals undergo different amounts of shrinkage, so the mold cavity must be designed for the particular metal to be cast if dimensional accuracy is critical. Molds are made of a variety of materials, including sand, plaster, ceramic, and metal. The various casting processes are often classified according to these different types of molds.

To accomplish a casting operation, the metal is first heated to a temperature high enough to completely transform it into a liquid state. It is then poured, or otherwise directed, into the cavity of the mold. In an *open mold,* Figure 10.2(a), the liquid metal is simply poured until it fills the open cavity. In a *closed mold,* Figure 10.2(b), a passageway, called the gating system, is provided to permit the molten metal to flow from outside the mold into the cavity. The closed mold is by far the more important category in production casting operations.

As soon as the molten metal is in the mold, it begins to cool. When the temperature drops sufficiently (e.g., to the freezing point for a pure metal), solidification begins. Solidification involves a change of phase of the metal. Time is required to complete the phase change, and considerable heat is given up in the process. It is during this step in the process that the metal assumes the solid shape of the mold cavity and many of the properties and characteristics of the casting are established.

Once the casting has cooled sufficiently, it is removed from the mold. Depending on the casting method and metal used, further processing may be required. This may include trimming the excess metal from the actual cast part, cleaning the surface, inspecting the product, and heat treatment to enhance properties. In addition, machining (Chapter 22) may be required to achieve closer tolerances on certain part features and to remove the cast surface.

Casting processes divide into two broad categories, according to type of mold used: expendable-mold casting and permanent-mold casting. An *expendable mold* means that the mold in which the molten metal solidifies must be destroyed in order to remove the casting. These molds are made out of sand, plaster, or similar materials, whose form is maintained by using binders of various kinds. Sand casting is the most prominent example of the expendable-mold processes. In sand casting, the liquid metal is poured into a mold

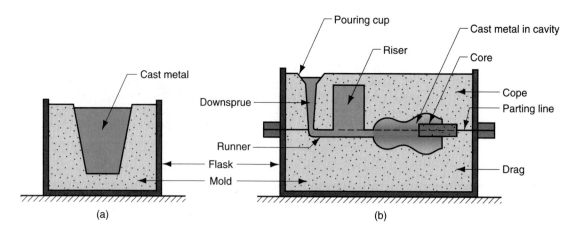

FIGURE 10.2   Two forms of mold: (a) open mold, simply a container in the shape of the desired part; and (b) closed mold, in which the mold geometry is more complex and requires a gating system (passageway) leading into the cavity.

made of sand. After the metal hardens, the mold must be sacrificed in order to recover the casting.

A *permanent mold* is one that can be used over and over to produce many castings. It is made of metal (or, less commonly, a ceramic refractory material) that can withstand the high temperatures of the casting operation. In permanent-mold casting, the mold consists of two (or more) sections that can be opened to permit removal of the finished part. Die casting is the most familiar process in this group.

More intricate casting geometries are generally possible with the expendable-mold processes. Part shapes in the permanent-mold processes are limited by the need to open the mold. On the other hand, some of the permanent mold processes have certain economic advantages in high production operations. We discuss the expendable-mold and permanent-mold casting processes in Chapter 11.

10.1.2 SAND-CASTING MOLDS

Sand casting is by far the most important casting process. A sand-casting mold will be used to describe the basic features of a mold. Many of these features and terms are common to the molds used in other casting processes. Figure 10.2(b) shows the cross-sectional view of a typical sand-casting mold, indicating some of the terminology. The mold consists of two halves: cope and drag. The *cope* is the upper half of the mold, and the *drag* is the bottom half. These two mold parts are contained in a box, called a *flask,* which is also divided into two halves, one for the cope and the other for the drag. The two halves of the mold separate at the *parting line.*

In sand casting (and other expendable-mold processes) the mold cavity is formed by means of a *pattern,* which is made of wood, metal, plastic, or other material and has the shape of the part to be cast. The cavity is formed by packing sand around the pattern, about half each in the cope and drag, so that when the pattern is removed, the remaining void has the desired shape of the cast part. The pattern is usually made oversized to allow for shrinkage of the metal as it solidifies and cools. The sand for the mold is moist and contains a binder to maintain its shape.

The cavity in the mold provides the external surfaces of the cast part. In addition, a casting may have internal surfaces. These surfaces are determined by means of a *core,* a form placed inside the mold cavity to define the interior geometry of the part. In sand casting, cores are generally made of sand, although other materials can be used, such as metals, plaster, and ceramics.

The *gating system* in a casting mold is the channel, or network of channels, by which molten metal flows into the cavity from outside the mold. As shown in the figure, the gating system typically consists of a *downsprue* (also called simply the *sprue*), through which the metal enters a *runner* that leads into the main cavity. At the top of the downsprue, a *pouring cup* is often used to minimize splash and turbulence as the metal flows into the downsprue. It is shown in our diagram as a simple cone-shaped funnel. Some pouring cups are designed in the shape of a bowl, with an open channel leading to the downsprue.

In addition to the gating system, any casting in which shrinkage is significant requires a riser connected to the main cavity. The *riser* is a reservoir in the mold that serves as a source of liquid metal for the casting to compensate for shrinkage during solidification. The riser must be designed to freeze after the main casting in order to satisfy its function.

As the metal flows into the mold, the air that previously occupied the cavity, as well as hot gases formed by reactions of the molten metal, must be evacuated so that the metal will completely fill the empty space. In sand casting, for example, the natural porosity of the sand mold permits the air and gases to escape through the walls of the cavity. In permanent-metal molds, small vent holes are drilled into the mold or machined into the parting line to permit removal of air and gases.

10.2 HEATING AND POURING

To perform a casting operation, the metal must be heated to a temperature somewhat above its melting point and then poured into the mold cavity to solidify. In this section, we consider several aspects of these two steps in casting.

10.2.1 HEATING THE METAL

Heating furnaces of various kinds (Section 11.4.1) are used to heat the metal to a molten temperature sufficient for casting. The heat energy required is the sum of (1) the heat to raise the temperature to the melting point, (2) the heat of fusion to convert it from solid to liquid, and (3) the heat to raise the molten metal to the desired temperature for pouring. This can be expressed:

$$H = \rho V \{ C_s(T_m - T_o) + H_f + C_l(T_p - T_m) \} \qquad (10.1)$$

where H = total heat required to raise the temperature of the metal to the pouring temperature, J (Btu); ρ = density, g/cm^3 (lbm/in^3); C_s = weight specific heat for the solid metal, J/g-C (Btu/lbm-F); T_m = melting temperature of the metal, °C (°F); T_o = starting temperature—usually ambient, °C (°F); H_f = heat of fusion, J/g (Btu/lbm); C_l = weight specific heat of the liquid metal, J/g-C (Btu/lbm-F); T_p = pouring temperature, °C (°F); and V = volume of metal being heated, cm^3 (in^3).

Example 10.1 Heating Metal for Casting

One cubic meter of a certain eutectic alloy is heated in a crucible from room temperature to 100°C above its melting point for casting. The alloy's density = 7.5 g/cm^3, melting point = 800°C, specific heat = 0.33 J/g°C in the solid state and 0.29 J/g°C in the liquid state; and heat of fusion = 160 J/g. How much heat energy must be added to accomplish the heating, assuming no losses?

Solution: We assume ambient temperature in the foundry = 25°C and that the density of the liquid and solid states of the metal are the same. Noting that one m^3 = 10^6 cm^3, and substituting the property values into Eq. (10.1), we have

$$H = (7.5)(10^6) \{ 0.33(800 - 25) + 160 + 0.29(100) \} = 3335(10^6) \text{ J} \qquad \blacksquare$$

The above equation is of conceptual value, but its computational value is limited, notwithstanding our example calculation. Use of Eq. (10.1) is complicated by the following factors: (1) Specific heat and other thermal properties of a solid metal vary with temperature, especially if the metal undergoes a change of phase during heating. (2) A metal's specific heat may be different in the solid and liquid states. (3) Most casting metals are alloys, and most alloys melt over a temperature range between a solidus and liquidus rather than at a single melting point; thus, the heat of fusion cannot be applied so simply as indicated above. (4) The property values required in the equation for a particular alloy are not readily available in most cases. (5) There are significant heat losses to the environment during heating.

10.2.2 POURING THE MOLTEN METAL

After heating, the metal is ready for pouring. Introduction of molten metal into the mold, including its flow through the gating system and into the cavity, is a critical step in the casting process. For this step to be successful, the metal must flow into all regions of the mold before solidifying. Factors affecting the pouring operation include pouring temperature, pouring rate, and turbulence.

The *pouring temperature* is the temperature of the molten metal as it is introduced into the mold. What is important here is the difference between the temperature at pouring and the temperature at which freezing begins (the melting point for a pure metal or the liquidus temperature for an alloy). This temperature difference is sometimes referred to as the *superheat.* This term is also used for the amount of heat that must be removed from the molten metal between pouring and when solidification commences [7].

Pouring rate refers to the volumetric rate at which the molten metal is poured into the mold. If the rate is too slow, the metal will chill and freeze before filling the cavity. If the pouring rate is excessive, turbulence can become a serious problem. *Turbulence* in fluid flow is characterized by erratic variations in the magnitude and direction of the velocity throughout the fluid. The flow is agitated and irregular rather than smooth and streamlined, as in laminar flow. Turbulent flow should be avoided during pouring for several reasons. It tends to accelerate the formation of metal oxides that can become entrapped during solidification, thus degrading the quality of the casting. Turbulence also aggravates *mold erosion,* the gradual wearing away of the mold surfaces due to impact of the flowing molten metal. The densities of most molten metals are much higher than water and other fluids we normally deal with. These molten metals are also much more chemically reactive than at room temperature. Consequently, the wear caused by the flow of these metals in the mold is significant, especially under turbulent conditions. Erosion is especially serious when it occurs in the main cavity because the geometry of the cast part is affected.

10.2.3 ENGINEERING ANALYSIS OF POURING

There are several relationships that govern the flow of liquid metal through the gating system and into the mold. An important relationship is *Bernoulli's theorem,* which states that the sum of the energies (head, pressure, kinetic, and friction) at any two points in a flowing liquid are equal. This can be written in the following form:

$$h_1 + \frac{p_1}{\rho} + \frac{v_1^2}{2g} + F_1 = h_2 + \frac{p_2}{\rho} + \frac{v_2^2}{2g} + F_2 \qquad (10.2)$$

where h = head, cm (in), p = pressure on the liquid, N/cm^2 (lb/in^2); ρ = density, g/cm^3 (lbm/in^3); v = flow velocity, cm/s (in/sec); g = gravitational acceleration constant, 981 cm/s/s ($32.2 \times 12 = 386$ in/sec/sec); and F = head losses due to friction, cm (in). Subscripts 1 and 2 indicate any two locations in the liquid flow.

Bernoulli's equation can be simplified in several ways. If we ignore friction losses (to be sure, friction will affect the liquid flow through a sand mold), and assume that the system remains at atmospheric pressure throughout, then the equation can be reduced to

$$h_1 + \frac{v_1^2}{2g} = h_2 + \frac{v_2^2}{2g} \qquad (10.3)$$

This can be used to determine the velocity of the molten metal at the base of the sprue. Let us define point 1 at the top of the sprue and point 2 at its base. If point 2 is used as the reference plane, then the head at that point is zero ($h_2 = 0$) and h_1 is the height (length) of the sprue. When the metal is poured into the pouring cup and overflows down the sprue, its initial velocity at the top is zero ($v_1 = 0$). Hence, Eq. (10.3) further simplifies to

$$h_1 = \frac{v_2^2}{2g}$$

which can be solved for the flow velocity:

$$v = \sqrt{2gh} \qquad (10.4)$$

where v = the velocity of the liquid metal at the base of the sprue, cm/s (in/sec); g = 981 cm/s/s (386 in/sec/sec); and h = the height of the sprue, cm (in).

Another relationship of importance during pouring is the **continuity law,** which states that the volume rate of flow remains constant throughout the liquid. The volume flow rate is equal to the velocity multiplied by the cross-sectional area of the flowing liquid. The continuity law can be expressed:

$$Q = v_1 A_1 = v_2 A_2 \tag{10.5}$$

where Q = volumetric flow rate, cm³/s (in³/sec); v = velocity as before; A = cross-sectional area of the liquid, cm² (in²); and the subscripts refer to any two points in the flow system. Thus, an increase in area results in a decrease in velocity, and vice versa.

Equations (10.4) and (10.5) indicate that the sprue should be tapered. As the metal accelerates during its descent into the sprue opening, the cross-sectional area of the channel must be reduced; otherwise, as the velocity of the flowing metal increases toward the base of the sprue, air can be aspirated into the liquid and conducted into the mold cavity. To prevent this condition, the sprue is designed with a taper, so that the volume flow rate vA is the same at the top and bottom of the sprue.

Assuming that the runner from the sprue base to the mold cavity is horizontal (and therefore the head h is the same as at the sprue base), then the volume rate of flow through the gate and into the mold cavity remains equal to vA at the base. Accordingly, we can estimate the time required to fill a mold cavity of volume V as

$$T_{MF} = \frac{V}{Q} \tag{10.6}$$

where T_{MF} = mold filling time, s (sec); V = volume of mold cavity, cm³ (in³); and Q = volume flow rate, as before. The mold filling time computed by Eq. (10.6) must be considered a minimum time. This is because the analysis ignores friction losses and possible constriction of flow in the gating system; thus, the mold filling time will be longer than what is given by Eq. (10.6).

Example 10.2
Pouring
Calculations

A mold sprue is 20 cm long, and the cross-sectional area at its base is 2.5 cm². The sprue feeds a horizontal runner leading into a mold cavity whose volume is 1560 cm³. Determine: (a) velocity of the molten metal at the base of the sprue, (b) volume rate of flow, and (c) time to fill the mold.

Solution: (a) The velocity of the flowing metal at the base of the sprue is given by Eq. (10.4):

$$v = \sqrt{2(981)(20)} = 198.1 \text{ cm/s}$$

(b) The volumetric flow rate is

$$Q = (2.5 \text{ cm}^2)(198.1 \text{ cm/s}) = 495 \text{ cm}^2/\text{s}$$

(c) Time required to fill a mold cavity of 100 in³ at this flow rate is

$$T_{MF} = 1560/495 = 3.2\text{s}$$

10.2.4 FLUIDITY

The molten metal flow characteristics are often described by the term *fluidity,* a measure of the capability of a metal to flow into and fill the mold before freezing. Fluidity is the inverse of viscosity (Section 3.4); as viscosity increases, fluidity decreases. Standard

FIGURE 10.3 Spiral mold test for fluidity, in which fluidity is measured as the length of the spiral channel that is filled by the molten metal prior to solidification.

testing methods are available to assess fluidity, including the spiral mold test shown in Figure 10.3, in which fluidity is indicated by the length of the solidified metal in the spiral channel. A longer cast spiral means greater fluidity of the molten metal.

Factors affecting fluidity include pouring temperature relative to melting point, metal composition, viscosity of the liquid metal, and heat transfer to the surroundings. A higher pouring temperature relative to the freezing point of the metal increases the time it remains in the liquid state, allowing it to flow further before freezing. This tends to aggravate certain casting problems such as oxide formation, gas porosity, and penetration of liquid metal into the interstitial spaces between the grains of sand forming the mold. This last problem causes the surface of the casting to contain imbedded sand particles, thus making it rougher and more abrasive than normal.

Composition also affects fluidity, particularly with respect to the metal's solidification mechanism. The best fluidity is obtained by metals that freeze at a constant temperature (e.g., pure metals and eutectic alloys). When solidification occurs over a temperature range (most alloys are in this category), the partially solidified portion interferes with the flow of the liquid portion, thereby reducing fluidity. In addition to the freezing mechanism, metal composition also determines *heat of fusion*—the amount of heat required to solidify the metal from the liquid state. A higher heat of fusion tends to increase the measured fluidity in casting.

10.3   SOLIDIFICATION AND COOLING

After pouring into the mold, the molten metal cools and solidifies. In this section we examine the physical mechanism of solidification that occurs during casting. Issues associated with solidification include the time for a metal to freeze, shrinkage, directional solidification, and riser design.

10.3.1   SOLIDIFICATION OF METALS

Solidification involves the transformation of the molten metal back into the solid state. The solidification process differs depending on whether the metal is a pure element or an alloy.

Pure Metals   A pure metal solidifies at a constant temperature equal to its freezing point, which is the same as its melting point. The melting points of pure metals are well known and documented (Table 4.1). The process occurs over time as shown in the plot of Figure 10.4, called a cooling curve. The actual freezing takes time, called the *local solidification time* in casting, during which the metal's latent heat of fusion is released into the surrounding mold. The *total solidification time* is the time taken between pouring and complete solidification.

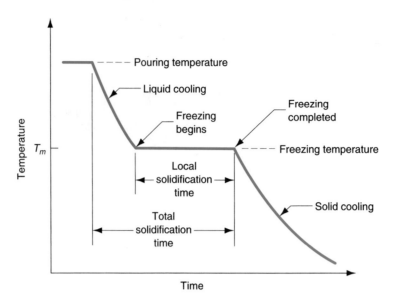

FIGURE 10.4 Cooling curve for a pure metal during casting.

After the casting has completely solidified, cooling continues at a rate indicated by the downward slope of the cooling curve.

Because of the chilling action of the mold wall, a thin skin of solid metal is initially formed at the interface immediately after pouring. Thickness of the skin increases to form a shell around the molten metal as solidification progresses inward toward the center of the cavity. The rate at which freezing proceeds depends on heat transfer into the mold, as well as the thermal properties of the metal.

It is of interest to examine the metallic grain formation and growth during this solidification process. The metal which forms the initial skin has been rapidly cooled by the extraction of heat through the mold wall. This cooling action causes the grains in the skin to be fine and randomly oriented. As cooling continues, further grain formation and growth occur in a direction away from the heat transfer. Since the heat transfer is through the skin and mold wall, the grains grow inwardly as needles or spines of solid metal. As these spines enlarge, lateral branches form, and as these branches grow, further branches form at right angles to the first branches. This type of grain growth is referred to as ***dendritic growth,*** and it occurs not only in the freezing of pure metals but alloys as well. These treelike structures are gradually filled-in during freezing, as additional metal is continually deposited onto the dendrites until complete solidification has occurred. The grains resulting from this dendritic growth take on a preferred orientation, tending to be coarse, columnar grains aligned toward the center of the casting. The resulting grain formation is illustrated in Figure 10.5.

Most Alloys   Most alloys freeze over a temperature range rather than at a single tempera-ture. The exact range depends on the alloy system and the particular composition.

FIGURE 10.5   Characteristic grain structure in a casting of a pure metal, showing randomly oriented grains of small size near the mold wall, and large columnar grains oriented toward the center of the casting.

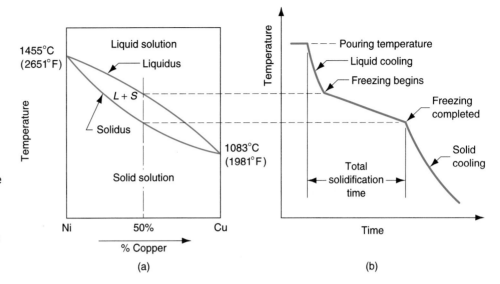

FIGURE 10.6 (a) Phase diagram for a copper–nickel alloy system and (b) associated cooling curve for a 50%Ni–50%Cu composition during casting.

Solidification of an alloy can be explained with reference to Figure 10.6, which shows the phase diagram for a particular alloy system (Section 6.1.2) and the cooling curve for a given composition. As temperature drops, freezing begins at the temperature indicated by the *liquidus* and is completed when the *solidus* is reached. The start of freezing is similar to that of the pure metal. A thin skin is formed at the mold wall due to the large temperature gradient at this surface. Freezing then progresses as before through the formation of dendrites that grow away from the walls. However, owing to the temperature spread between the liquidus and solidus, the nature of the dendritic growth is such that an advancing zone is formed in which both liquid and solid metal coexist. The solid portions are the dendrite structures that have formed sufficiently to trap small islands of liquid metal in the matrix. This solid–liquid region has a soft consistency that has motivated its name as the *mushy zone.* Depending on the conditions of freezing, the mushy zone can be relatively narrow, or it can exist throughout most of the casting. The latter condition is promoted by factors such as slow heat transfer out of the hot metal and a wide difference between liquidus and solidus temperatures. Gradually, the liquid islands in the dendrite matrix solidify as the temperature of the casting drops to the solidus for the given alloy composition.

   Another factor complicating solidification of alloys is that the composition of the dendrites as they start to form favors the metal with the higher melting point. As freezing continues and the dendrites grow, there develops an imbalance in composition between the metal that has solidified and the remaining molten metal. This composition imbalance is finally manifested in the completed casting in the form of segregation of the elements. The segregation is of two types, microscopic and macroscopic. At the microscopic level, the chemical composition varies throughout each individual grain. This is due to the fact that the beginning spine of each dendrite has a higher proportion of one of the elements in the alloy. As the dendrite grows in its local vicinity, it must expand using the remaining liquid metal that has been partially depleted of the first component. Finally, the last metal to freeze in each grain is that which has been trapped by the branches of the dendrite, and its composition is even further out of balance. Thus, we have a variation in chemical composition within single grains of the casting.

   At the macroscopic level, the chemical composition varies throughout the entire casting. Since the regions of the casting that freeze first (at the outside near the mold walls) are richer in one component than the other, the remaining molten alloy is deprived of that component by the time freezing occurs at the interior. Thus, there is a general segregation

FIGURE 10.7   Characteristic grain structure in an alloy casting, showing segregation of alloying components in the center of casting.

through the cross-section of the casting, sometimes called ***ingot segregation,*** as illustrated in Figure 10.7.

Eutectic Alloys   Eutectic alloys constitute an exception to the general process by which alloys solidify. A ***eutectic alloy*** is a particular composition in an alloy system for which the solidus and liquidus are at the same temperature. Hence, solidification occurs at a constant temperature rather than over a temperature range, as described above. The effect can be seen in the phase diagram of the lead–tin system shown in Figure 6.3. Pure lead has a melting point of 327°C (621°F), while pure tin melts at 232°C (450°F). Although most lead–tin alloys exhibit the typical solidus–liquidus temperature range, the particular composition of 61.9% tin and 38.1% lead has a melting (freezing) point of 183°C (362°F). This composition is the ***eutectic composition*** of the lead–tin alloy system, and 183°C is its ***eutectic temperature.*** Lead–tin alloys are not commonly used in casting, but Pb–Sn compositions near the eutectic are used for electrical soldering, where the low melting point is an advantage. Examples of eutectic alloys encountered in casting include aluminum–silicon (11.6% Si) and cast iron (4.3% C).

10.3.2 SOLIDIFICATION TIME

Whether the casting is pure metal or alloy, solidification takes time. The total solidification time is the time required for the casting to solidify after pouring. This time is dependent on the size and shape of the casting by an empirical relationship known as ***Chvorinov's rule,*** which states:

$$T_{TS} = C_m \left(\frac{V}{A}\right)^n \tag{10.7}$$

where T_{TS} = total solidification time, min; V = volume of the casting, cm^3 (in^3); A = surface area of the casting, cm^2 (in^2); n is an exponent usually taken to have a value = 2; and C_m is the ***mold constant.*** Given that $n = 2$, the units of C_m are min/cm^2 (min/in^2), and its value depends on the particular conditions of the casting operation, including mold material (e.g., specific heat, thermal conductivity), thermal properties of the cast metal (e.g., heat of fusion, specific heat, thermal conductivity), and pouring temperature relative to the melting point of the metal. The value of C_m for a given casting operation can be based on experimental data from previous operations carried out using the same mold material, metal, and pouring temperature, even though the shape of the part may be quite different.

Chvorinov's rule indicates that a casting with a higher volume-to-surface area ratio will cool and solidify more slowly than one with a lower ratio. This principle is put to good use in designing the riser in a mold. To perform its function of feeding molten metal to the main cavity, the metal in the riser must remain in the liquid phase longer than the casting. In other words, the T_{TS} for the riser must exceed the T_{TS} for the main casting. Since the mold conditions for both riser and casting are the same, their mold constants will be equal. By designing the riser to have a larger volume-to-area ratio, we can be fairly sure that the main casting solidifies first and that the effects of shrinkage are minimized. Before considering how the riser might be designed using Chvorinov's rule, let us consider the topic of shrinkage, which is the reason why risers are needed.

10.3.3 SHRINKAGE

Our discussion of solidification has neglected the impact of shrinkage that occurs during cooling and freezing. Shrinkage occurs in three steps: (1) liquid contraction during cooling prior to solidification; (2) contraction during the phase change from liquid to solid, called **solidification shrinkage;** and (3) thermal contraction of the solidified casting during cooling to room temperature. The three steps can be explained with reference to a cylindrical casting made in an open mold, as shown in Figure 10.8. The molten metal immediately after pouring is shown in part (0) of the series. Contraction of the liquid metal during cooling from pouring temperature to freezing temperature causes the height of the liquid to be reduced from its starting level as in (1) of the figure. The amount of this liquid contraction is usually around 0.5%. Solidification shrinkage, seen in part (2), has two effects. First, contraction causes a further reduction in the height of the casting. Second, the amount of liquid metal available to feed the top center portion of the casting becomes restricted. This is usually the last region to freeze, and the absence of metal creates a void in the casting at this location. This shrinkage cavity is called a **pipe** by foundrymen. Once

FIGURE 10.8 Shrinkage of a cylindrical casting during solidification and cooling:
(0) starting level of molten metal immediately after pouring; (1) reduction in level caused by liquid contraction during cooling; (2) reduction in height and formation of shrinkage cavity caused by solidification shrinkage; and (3) further reduction in height and diameter due to thermal contraction during cooling of the solid metal. For clarity, dimensional reductions are exaggerated in our sketches.

TABLE 10.1   Typical linear shrinkage values for different casting metals due to solid thermal contraction.

Metal	Linear shrinkage	Metal	Linear shrinkage	Metal	Linear shrinkage
Aluminum alloys	1.3%	Magnesium	2.1%	Steel, chrome	2.1%
Brass, yellow	1.3%–1.6%	Magnesium alloy	1.6%	Tin	2.1%
Cast iron, gray	0.8%–1.3%	Nickel	2.1%	Zinc	2.6%
Cast iron, white	2.1%	Steel, carbon	1.6%–2.1%		

Compiled from [10].

solidified, the casting experiences further contraction in height and diameter while cooling, as in (3). This shrinkage is determined by the solid metal's coefficient of thermal expansion, which in this case is applied in reverse to determine contraction.

Solidification shrinkage occurs in nearly all metals because the solid phase has a higher density than the liquid phase. The phase transformation that accompanies solidification causes a reduction in the volume per unit weight of metal. The exception is cast iron containing high carbon content, whose solidification during the final stages of freezing is complicated by a period of graphitization, which results in expansion that tends to counteract the volumetric decrease associated with the phase change [7]. Compensation for solidification shrinkage is achieved in several ways depending on the casting operation. In sand casting, liquid metal is supplied to the cavity by means of risers (Section 10.3.5). In die casting (Section 11.3.3), the molten metal is applied under pressure.

Pattern-makers account for thermal contraction by making the mold cavities oversized. The amount by which the mold must be made larger relative to the final casting size is called the *pattern shrinkage allowance*. Although the shrinkage is volumetric, the dimensions of the casting are expressed linearly, so the allowances must be applied accordingly. Special "shrink rules" with slightly elongated scales are used to make the patterns and molds larger than the desired casting by the appropriate amount. Table 10.1 lists typical values of linear shrinkage for various cast metals; these values can be used to determine shrink rule scales.

10.3.4 DIRECTIONAL SOLIDIFICATION

In order to minimize the damaging effects of shrinkage, it is desirable for the regions of the casting most distant from the liquid metal supply to freeze first and for solidification to progress from these remote regions toward the riser(s). In this way, molten metal will continually be available from the risers to prevent shrinkage voids during freezing. The term *directional solidification* is used to describe this aspect of the freezing process and the methods by which it is controlled. The desired directional solidification is achieved by observing Chvorinov's rule in the design of the casting itself, its orientation within the mold, and the design of the riser system that feeds it. For example, by locating sections of the casting with lower V/A ratios away from the riser, freezing will occur first in these regions and the supply of liquid metal for the rest of the casting will remain open until these bulkier sections solidify.

Another way to encourage directional solidification is to use *chills*—internal or external heat sinks that cause rapid freezing in certain regions of the casting. *Internal chills* are small metal parts placed inside the cavity before pouring so that the molten metal will solidify first around these parts. The internal chill should have a chemical composition similar to the metal being poured, most readily achieved by making the chill out of the same metal as the casting itself.

External chills are metal inserts in the walls of the mold cavity that can remove heat from the molten metal more rapidly than the surrounding sand in order to promote solidification. They are often used effectively in sections of the casting that are difficult to

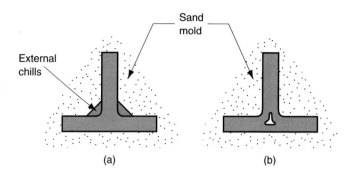

FIGURE 10.9
(a) External chill to encourage rapid freezing of the molten metal in a thin section of the casting; and (b) the likely result if the external chill were not used.

(a)   (b)

feed with liquid metal, thus encouraging rapid freezing in these sections while the connection to liquid metal is still open. Figure 10.9 illustrates a possible application of external chills and the likely result in the casting if the chill were not used.

As important as it is to initiate freezing in the appropriate regions of the cavity, it is also important to avoid premature solidification in sections of the mold nearest the riser. Of particular concern is the passageway between the riser and the main cavity. This connection must be designed in such a way that it does not freeze before the casting, which would isolate the casting from the molten metal in the riser. Although it is generally desirable to minimize the volume in the connection (to reduce wasted metal), the cross-sectional area must be sufficient to delay the onset of freezing. This goal is usually aided by making the passageway short in length, so that it absorbs heat from the molten metal in the riser and the casting.

10.3.5 RISER DESIGN

As described earlier, a riser, Figure 10.2(b), is used in a sand-casting mold to feed liquid metal to the casting during freezing in order to compensate for solidification shrinkage. To function, the riser must remain molten until after the casting solidifies. Chvorinov's rule can be used to compute the size of a riser that will satisfy this requirement. The following example illustrates the calculation.

**Example 10.3
Riser Design
Using
Chvorinov's Rule**

A cylindrical riser must be designed for a sand-casting mold. The casting itself is a steel rectangular plate with dimensions 7.5 cm × 12.5 cm × 2.0 cm. Previous observations have indicated that the total solidification time (T_{TS}) for this casting = 1.6 min. The cylinder for the riser will have a diameter-to-height ratio = 1.0. Determine the dimensions of the riser so that its T_{TS} = 2.0 min.

Solution: First determine the V/A ratio for the plate. Its volume $V = 7.5 \times 12.5 \times 2.0 = 187.5 \text{ cm}^3$, and its surface area $A = 2(7.5 \times 12.5 + 7.5 \times 2.0 + 12.5 \times 2.0) = 267.5 \text{ cm}^2$. Given that $T_{TS} = 1.6$ min, we can determine the mold constant C_m from Eq. (10.7), using a value of $n = 2$ in the equation.

$$C_m = \frac{T_{TS}}{(V/A)^2} = \frac{1.6}{(187.5/267.5)^2} = 3.26 \text{ min/cm}^2$$

Next we must design the riser so that its total solidification time is 2.0 min, using the same value of mold constant. The volume of the riser is given by

$$V = \frac{\pi D^2 h}{4}$$

and the surface area is given by

$$A = \pi Dh + \frac{2\pi D^2}{4}$$

Since we are using a D/H ratio $= 1.0$, then $D = H$. Substituting D for H in the volume and area formulas, we get

$$V = \pi D^3/4$$

and

$$A = \pi D^2 + 2\pi D^2/4 = 1.5\pi D^2$$

Thus the V/A ratio $= D/6$. Using this ratio in Chvorinov's equation, we have

$$T_{TS} = 2.0 = 3.26\left(\frac{D}{6}\right)^2 = 0.09056\, D^2$$

$$D^2 = 2.0/0.09056 = 22.086\,\text{cm}^2$$

$$D = 4.7\,\text{cm}$$

Since $H = D$, then $H = 4.7\,\text{cm}$ also.

The riser represents waste metal that will be separated from the cast part and remelted to make subsequent castings. It is desirable for the volume of metal in the riser to be a minimum. Since the geometry of the riser is normally selected to maximize the V/A ratio, this tends to reduce the riser volume as much as possible. Note that the volume of the riser in our example problem is $V = \pi(4.7)^3/4 = 81.5\,\text{cm}^3$, only 44% of the volume of the plate (casting), even though its total solidification time is 25% longer.

Risers can be designed in different forms. The design shown in Figure 10.2(b) is a **side riser.** It is attached to the side of the casting by means of a small channel. A **top riser** is one that is connected to the top surface of the casting. Risers can be open or blind. An **open riser** is exposed to the outside at the top surface of the cope. This has the disadvantage of allowing more heat to escape, promoting faster solidification. A **blind riser** is entirely enclosed within the mold, as in Figure 10.1(b). ■

REFERENCES

[1] Amstead, B. H., Ostwald, P. F., and Begeman, M. L. *Manufacturing Processes.* John Wiley & Sons, Inc., New York, 1987.

[2] Beeley, P. R. *Foundry Technology.* Butterworths-Heinemann, Oxford, UK, 2001.

[3] Black, J, and Kohser, R. *DeGarmo's Materials and Processes in Manufacturing,* 10th ed. John Wiley & Sons, Inc., Hoboken, New Jersey, 2008.

[4] Datsko, J. *Material Properties and Manufacturing Processes.* John Wiley & Sons, Inc., New York, 1966.

[5] Edwards, L., and Endean, M. *Manufacturing with Materials.* Open University, Milton Keynes, and Butterworth Scientific Ltd., London, 1990.

[6] Flinn, R. A. *Fundamentals of Metal Casting.* American Foundrymen's Society, Inc., Des Plaines, Illinois, 1987.

[7] Heine, R. W., Loper, Jr., C. R., and Rosenthal, C. *Principles of Metal Casting,* 2nd ed. McGraw-Hill Book Co., New York, 1967.

[8] Kotzin, E. L. (ed.). *Metalcasting and Molding Processes.* American Foundrymen's Society, Inc., Des Plaines, Illinois, 1981.

[9] Lessiter, M. J., and K. Kirgin. "Trends in the Casting Industry," *Advanced Materials & Processes,* January 2002, pp. 42–43.

[10] *Metals Handbook,* Vol. 15, *Casting.* ASM International, Materials Park, Ohio, 2008.

[11] Mikelonis, P. J. (ed.). *Foundry Technology.* American Society for Metals, Metals Park, Ohio, 1982.

[12] Niebel, B. W., Draper, A. B., Wysk, R. A. *Modern Manufacturing Process Engineering.* McGraw-Hill Book Co., New York, 1989.

[13] Simpson, B. L. *History of the Metalcasting Industry.* American Foundrymen's Society, Inc., Des Plaines, Illinois, 1997.

[14] Taylor, H. F., Flemings, M. C., and Wulff, J. *Foundry Engineering,* 2nd ed. American Foundrymen's Society, Inc., Des Plaines, Illinois, 1987.

[15] Wick, C., Benedict, J. T., and Veilleux, R. F. *Tool and Manufacturing Engineers Handbook,* 4th ed., Vol. II, *Forming.* Society of Manufacturing Engineers, Dearborn, Michigan, 1984.

REVIEW QUESTIONS

10.1. Identify some of the important advantages of shape-casting processes.

10.2. What are some of the limitations and disadvantages of casting?

10.3. What is a factory that performs casting operations usually called?

10.4. What is the difference between an open mold and a closed mold?

10.5. Name the two basic mold types that distinguish casting processes.

10.6. Which casting process is the most important commercially?

10.7. What is the difference between a pattern and a core in sand molding?

10.8. What is meant by the term superheat?

10.9. Why should turbulent flow of molten metal into the mold be avoided?

10.10. What is the continuity law as it applies to the flow of molten metal in casting?

10.11. What are some of the factors that affect the fluidity of a molten metal during pouring into a mold cavity?

10.12. What does heat of fusion mean in casting?

10.13. How does solidification of alloys differ from solidification of pure metals?

10.14. What is a eutectic alloy?

10.15. What is the relationship known as Chvorinov's rule in casting?

10.16. Identify the three sources of contraction in a metal casting after pouring.

10.17. What is a chill in casting?

MULTIPLE CHOICE QUIZ

There are 15 correct answers in the following multiple choice questions (some questions have multiple answers that are correct). To attain a perfect score on the quiz, all correct answers must be given. Each correct answer is worth 1 point. Each omitted answer or wrong answer reduces the score by 1 point, and each additional answer beyond the correct number of answers reduces the score by 1 point. Percentage score on the quiz is based on the total number of correct answers.

10.1. Sand casting is which of the following types: (a) expendable mold or (b) permanent mold?

10.2. The upper half of a sand-casting mold is called which of the following: (a) cope or (b) drag?

10.3. In casting, a flask is which one of the following: (a) beverage bottle for foundrymen, (b) box which holds the cope and drag, (c) container for holding liquid metal, or (d) metal which extrudes between the mold halves?

10.4. In foundry work, a runner is which one of the following: (a) channel in the mold leading from the downsprue to the main mold cavity, (b) foundryman who moves the molten metal to the mold, or (c) vertical channel into which molten metal is poured into the mold?

10.5. Turbulence during pouring of the molten metal is undesirable for which of the following reasons (two best answers): (a) it causes discoloration of the mold surfaces, (b) it dissolves the binder used to hold together the sand mold, (c) it increases erosion of the mold surfaces, (d) it increases the formation of metallic oxides that can become entrapped during solidification, (e) it increases the mold filling time, and (f) it increases total solidification time?

10.6. Total solidification time is defined as which one of the following: (a) time between pouring and complete solidification, (b) time between pouring and cooling to room temperature, (c) time between solidification and cooling to room temperature, or (d) time to give up the heat of fusion?

10.7. During solidification of an alloy when a mixture of solid and liquid metals is present, the solid-liquid mixture is referred to as which one of the following: (a) eutectic composition, (b) ingot segregation, (c) liquidus, (d) mushy zone, or (e) solidus?

10.8. Chvorinov's rule states that total solidification time is proportional to which one of the following quantities: (a) $(A/V)^n$, (b) H_f, (c) T_m, (d) V, (e) V/A, or (f) $(V/A)^2$; where A = surface area of casting, H_f = heat of fusion, T_m = melting temperature, and V = volume of casting?

10.9. A riser in casting is described by which of the following (three correct answers): (a) an insert in the casting that inhibits buoyancy of the core, (b) gating system in which the sprue feeds directly into the cavity, (c) metal that is not part of the casting, (d) source of molten metal to feed the casting and compensate for shrinkage during solidification, and (e) waste metal that is usually recycled?

10.10. In a sand-casting mold, the V/A ratio of the riser should be (a) equal to, (b) greater than, or (c) smaller than the V/A ratio of the casting itself?

10.11. Which of the following riser types are completely enclosed within the sand mold and connected to the main cavity by a channel to feed the molten metal (two correct answers): (a) blind riser, (b) open riser, (c) side riser, and (d) top riser?

PROBLEMS

Heating and Pouring

10.1. A disk 40 cm in diameter and 5 cm thick is to be cast of pure aluminum in an open-mold casting operation. The melting temperature of aluminum = 660°C, and the pouring temperature will be 800°C. Assume that the amount of aluminum heated will be 5% more than what is needed to fill the mold cavity. Compute the amount of heat that must be added to the metal to heat it to the pouring temperature, starting from a room temperature of 25°C. The heat of fusion of aluminum = 389.3 J/g. Other properties can be obtained from Tables 4.1 and 4.2 in the text. Assume the specific heat has the same value for solid and molten aluminum.

10.2. A sufficient amount of pure copper is to be heated for casting a large plate in an open mold. The plate has dimensions: length = 20 in, width = 10 in, and thickness = 3 in. Compute the amount of heat that must be added to the metal to heat it to a temperature of 2150°F for pouring. Assume that the amount of metal heated will be 10% more than what is needed to fill the mold cavity. Properties of the metal are: density = 0.324 lbm/in^3, melting point = 1981°F, specific heat of the metal = 0.093 Btu/lbm-F in the solid state and 0.090 Btu/lbm-F in the liquid state, and heat of fusion = 80 Btu/lbm.

10.3. The downsprue leading into the runner of a certain mold has a length = 175 mm. The cross-sectional area at the base of the sprue is 400 mm^2. The mold cavity has a volume = 0.001 m^3. Determine (a) the velocity of the molten metal flowing through the base of the downsprue, (b) the volume rate of flow, and (c) the time required to fill the mold cavity.

10.4. A mold has a downsprue of length = 6.0 in. The cross-sectional area at the bottom of the sprue is 0.5 in^2. The sprue leads into a horizontal runner which feeds the mold cavity, whose volume = 75 in^3. Determine (a) the velocity of the molten metal flowing through the base of the downsprue, (b) the volume rate of flow, and (c) the time required to fill the mold cavity.

10.5. The flow rate of liquid metal into the downsprue of a mold = 1 L/s. The cross-sectional area at the top of the sprue = 800 mm^2, and its length = 175 mm. What area should be used at the base of the sprue to avoid aspiration of the molten metal?

10.6. The volume rate of flow of molten metal into the downsprue from the pouring cup is 50 in^3/sec. At the top where the pouring cup leads into the downsprue, the cross-sectional area = 1.0 in^2. Determine what the area should be at the bottom of the sprue if its length = 8.0 in. It is desired to maintain a constant flow rate, top and bottom, in order to avoid aspiration of the liquid metal.

10.7. Molten metal can be poured into the pouring cup of a sand mold at a steady rate of 1000 cm^3/s. The molten metal overflows the pouring cup and flows into the downsprue. The cross-section of the sprue is round, with a diameter at the top = 3.4 cm. If the sprue is 25 cm long, determine the proper diameter at its base so as to maintain the same volume flow rate.

10.8. During pouring into a sand mold, the molten metal can be poured into the downsprue at a constant flow rate during the time it takes to fill the mold. At the end of pouring the sprue is filled and there is negligible metal in the pouring cup. The downsprue is 6.0 in long. Its cross-sectional area at the top = 0.8 in^2 and at the base = 0.6 in^2. The cross-sectional area of the runner leading from the sprue also = 0.6 in^2, and it is 8.0 in long before leading into the mold cavity, whose volume = 65 in^3. The volume of the riser located along the runner near the mold cavity = 25 in^3. It takes a total of 3.0

sec to fill the entire mold (including cavity, riser, runner, and sprue). This is more than the theoretical time required, indicating a loss of velocity due to friction in the sprue and runner. Find (a) the theoretical velocity and flow rate at the base of the

downsprue; (b) the total volume of the mold; (c) the actual velocity and flow rate at the base of the sprue; and (d) the loss of head in the gating system due to friction.

Shrinkage

10.9. Determine the shrink rule to be used by pattern makers for white cast iron. Using the shrinkage value in Table 10.1, express your answer in terms of decimal fraction inches of elongation per foot of length compared to a standard 1-foot scale.

10.10. Determine the shrink rule to be used by mold makers for die casting of zinc. Using the shrinkage value in Table 10.1, express your answer in terms of decimal mm of elongation per 300 mm of length compared to a standard 300-mm scale.

10.11. A flat plate is to be cast in an open mold whose bottom has a square shape that is 200 mm × 200 mm. The mold is 40 mm deep. A total of 1,000,000 mm^3 of molten aluminum is poured into the mold. Solidification shrinkage is known to be 6.0%. Table 10.1 lists the linear shrinkage due to thermal contraction after solidification to be 1.3%. If the availability of molten metal in the mold allows the square shape of the cast plate to maintain its 200 mm × 200 mm dimensions until solidification is completed, determine the final dimensions of the plate.

Solidification Time and Riser Design

10.12. In the casting of steel under certain mold conditions, the mold constant in Chvorinov's rule is known to be 4.0 min/cm^2, based on previous experience. The casting is a flat plate whose length = 30 cm, width = 10 cm, and thickness = 20 mm. Determine how long it will take for the casting to solidify.

10.13. Solve for total solidification time in the previous problem only using an exponent value of 1.9 in Chvorinov's rule instead of 2.0. What adjustment must be made in the units of the mold constant?

10.14. A disk-shaped part is to be cast out of aluminum. The diameter of the disk = 500 mm and its thickness = 20 mm. If the mold constant = 2.0 s/mm^2 in Chvorinov's rule, how long will it take the casting to solidify?

10.15. In casting experiments performed using a certain alloy and type of sand mold, it took 155 s for a cube-shaped casting to solidify. The cube was 50 mm on a side. (a) Determine the value of the mold constant in Chvorinov's rule. (b) If the same alloy and mold type were used, find the total solidification time for a cylindrical casting in which the diameter = 30 mm and length = 50 mm.

10.16. A steel casting has a cylindrical geometry with 4.0 in diameter and weighs 20 lb. This casting takes 6.0 min to completely solidify. Another cylindrical-shaped casting with the same diameter-to-length ratio weighs 12 lb. This casting is made of the same steel, and the same conditions of mold and pouring were used. Determine: (a) the mold constant in Chvorinov's rule, (b) the dimensions, and (c) the total solidification time of the lighter casting. The density of steel = 490 lb/ft^3.

10.17. The total solidification times of three casting shapes are to be compared: (1) a sphere with diameter = 10 cm, (2) a cylinder with diameter and length both = 10 cm, and (3) a cube with each side = 10 cm. The same casting alloy is used in the three cases. (a) Determine the relative solidification times for each geometry. (b) Based on the results of part (a), which geometric element would make the best riser? (c) If the mold constant = 3.5 min/cm^2 in Chvorinov's rule, compute the total solidification time for each casting.

10.18. The total solidification times of three casting shapes are to be compared: (1) a sphere, (2) a cylinder, in which the length-to-diameter ratio = 1.0, and (3) a cube. For all three geometries, the volume = 1000 cm^3. The same casting alloy is used in the three cases. (a) Determine the relative solidification times for each geometry. (b) Based on the results of part (a), which geometric element would make the best riser? (c) If the mold constant = 3.5 min/cm^2 in Chvorinov's rule, compute the total solidification time for each casting.

10.19. A cylindrical riser is to be used for a sand-casting mold. For a given cylinder volume, determine the diameter-to-length ratio that will maximize the time to solidify.

10.20. A riser in the shape of a sphere is to be designed for a sand casting mold. The casting is a rectangular plate, with length = 200 mm, width = 100 mm, and thickness = 18 mm. If the total solidification time of the casting itself is known to be 3.5 min, determine the diameter of the riser so that it will take 25% longer for the riser to solidify.

10.21. A cylindrical riser is to be designed for a sand casting mold. The length of the cylinder is to be 1.25 times its diameter. The casting is a square plate, each side = 10 in and thickness = 0.75 in. If the metal is cast iron, and the mold constant = 16.0 min/in^2 in Chvorinov's rule, determine the dimensions of the riser so that it will take 30% longer for the riser to solidify.

10.22. A cylindrical riser with diameter-to-length ratio = 1.0 is to be designed for a sand casting mold. The casting geometry is illustrated in Figure P10.22, in which the units are inches. If the mold constant in Chvorinov's rule = 19.5 min/in^2, determine the dimensions of the riser so that the riser will take 0.5 min longer to freeze than the casting itself.

FIGURE P10.22 Casting geometry for Problem 10.22 (units are in inches).

METAL CASTING PROCESSES

Chapter Contents

Metal casting processes divide into two categories, based on mold type: (1) expendable mold and (2) permanent mold. In expendable mold casting operations, the mold is sacrificed in order to remove the cast part. Since a new mold is required for each new casting, production rates in expendable-mold processes are often limited by the time required to make the mold rather than the time to make the casting itself. However, for certain part geometries, sand molds can be produced and castings made at rates of 400 parts per hour and higher. In permanent-mold casting processes, the mold is fabricated out of metal (or other durable material) and can be used many times to make many castings. Accordingly, these processes possess a natural advantage in terms of higher production rates.

Our discussion of casting processes in this chapter is organized as follows: (1) sand casting, (2) other expendable-mold casting processes, and (3) permanent-mold casting processes. The chapter also includes casting equipment and practices used in foundries. Another section deals with inspection and quality issues. Product design guidelines are presented in the final section.

11.1 SAND CASTING

Sand casting is the most widely used casting process, accounting for a significant majority of the total tonnage cast. Nearly all casting alloys can be sand cast; indeed, it is one of the few processes that can be used for metals with high melting temperatures, such as steels, nickels, and titaniums. Its versatility permits the casting of parts ranging in size from small to very large (Figure 11.1) and in production quantities from one to millions.

Sand casting, also known as **sand-mold casting,** consists of pouring molten metal into a sand mold, allowing the metal to solidify, and then breaking up the mold to remove the casting. The casting must then be cleaned and

FIGURE 11.1   A large sand casting weighing more than 680 kg (1500 lb) for an air compressor frame. (Courtesy of Elkhart Foundry, photo by Paragon Inc., Elkhart, Indiana.)

inspected, and heat treatment is sometimes required to improve metallurgical properties. The cavity in the sand mold is formed by packing sand around a pattern (an approximate duplicate of the part to be cast), and then removing the pattern by separating the mold into two halves. The mold also contains the gating and riser system. In addition, if the casting is to have internal surfaces (e.g., hollow parts or parts with holes), a core must be included in the mold. Since the mold is sacrificed to remove the

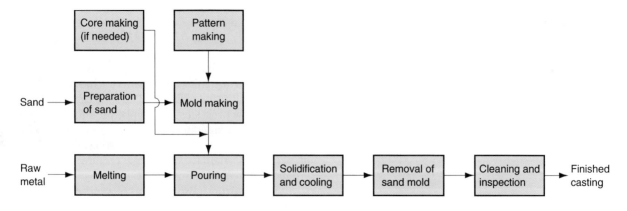

FIGURE 11.2   Steps in the production sequence in sand casting. The steps include not only the casting operation but also pattern making and mold making.

casting, a new sand mold must be made for each part that is produced. From this brief description, sand casting is seen to include not only the casting operation itself, but also the fabrication of the pattern and the making of the mold. The production sequence is outlined in Figure 11.2 Our video clip on casting contains a segment on sand casting.

VIDEO CLIP _____

Casting: View the segment titled Sand-Mold Casting.

_____

11.1.1 PATTERNS AND CORES

Sand casting requires a ***pattern***—a full-sized model of the part, enlarged to account for shrinkage and machining allowances in the final casting. Materials used to make patterns include wood, plastics, and metals. Wood is a common pattern material because it is easily shaped. Its disadvantages are that it tends to warp, and it is abraded by the sand being compacted around it, thus limiting the number of times it can be reused. Metal patterns are more expensive to make, but they last much longer. Plastics represent a compromise between wood and metal. Selection of the appropriate pattern material depends to a large extent on the total quantity of castings to be made.

There are various types of patterns, as illustrated in Figure 11.3. The simplest is made of one piece, called a ***solid pattern***—same geometry as the casting, adjusted in size for shrinkage and machining. Although it is the easiest pattern to fabricate, it is not the easiest to use in making the sand mold. Determining the location of the parting line between the two halves of the mold for a solid pattern can be a problem, and incorporating the gating system and sprue into the mold is left to the judgment and skill of the foundry worker. Consequently, solid patterns are generally limited to very low production quantities.

Split patterns consist of two pieces, dividing the part along a plane coinciding with the parting line of the mold. Split patterns are appropriate for complex part geometries and moderate production quantities. The parting line of the mold is predetermined by the two pattern halves, rather than by operator judgment.

For higher production quantities, match-plate patterns or cope-and-drag patterns are used. In ***match-plate*** patterns, the two pieces of the split pattern are attached to opposite sides of a wood or metal plate. Holes in the plate allow the top and bottom (cope

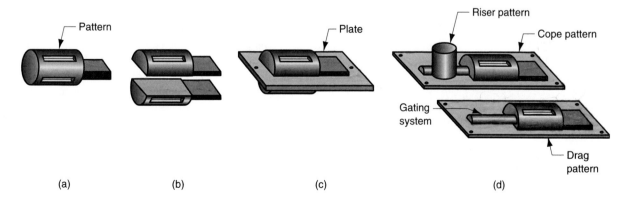

(a)   (b)   (c)   (d)

FIGURE 11.3   Types of patterns used in sand casting: (a) solid pattern, (b) split pattern, (c) match-plate pattern, and (d) cope-and-drag pattern.

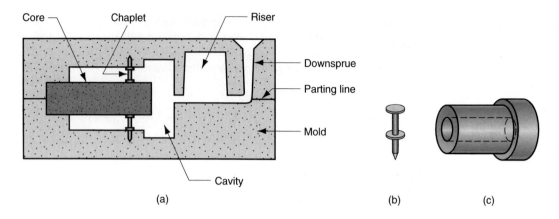

FIGURE 11.4 (a) Core held in place in the mold cavity by chaplets, (b) possible chaplet design, and (c) casting with internal cavity.

and drag) sections of the mold to be aligned accurately. ***Cope-and-drag patterns*** are similar to match-plate patterns except that split pattern halves are attached to separate plates, so that the cope and drag sections of the mold can be fabricated independently, instead of using the same tooling for both. Part (d) of the figure includes the gating and riser system in the cope-and-drag patterns.

Patterns define the external shape of the cast part. If the casting is to have internal surfaces, a core is required. A ***core*** is a full-scale model of the interior surfaces of the part. It is inserted into the mold cavity prior to pouring, so that the molten metal will flow and solidify between the mold cavity and the core to form the casting's external and internal surfaces. The core is usually made of sand, compacted into the desired shape. As with the pattern, the actual size of the core must include allowances for shrinkage and machining. Depending on the geometry of the part, the core may or may not require supports to hold it in position in the mold cavity during pouring. These supports, called ***chaplets,*** are made of a metal with a higher melting temperature than the casting metal. For example, steel chaplets would be used for cast iron castings. On pouring and solidification, the chaplets become bonded into the casting. A possible arrangement of a core in a mold using chaplets is sketched in Figure 11.4. The portion of the chaplet protruding from the casting is subsequently cut off.

11.1.2 MOLDS AND MOLD MAKING

Foundry sands are silica (SiO_2) or silica mixed with other minerals. The sand should possess good refractory properties—capacity to stand up under high temperatures without melting or otherwise degrading. Other important features of the sand include grain size, distribution of grain size in the mixture, and shape of the individual grains (Section 16.1). Small grain size provides a better surface finish on the cast part, but large grain size is more permeable (to allow escape of gases during pouring). Molds made from grains of irregular shape tend to be stronger than molds of round grains because of interlocking, yet interlocking tends to restrict permeability.

In making the mold, the grains of sand are held together by a mixture of water and bonding clay. A typical mixture (by volume) is 90% sand, 3% water, and 7% clay. Other bonding agents can be used in place of clay, including organic resins (e.g., phenolic resins) and inorganic binders (e.g., sodium silicate and phosphate). Besides sand and binder, additives are sometimes combined with the mixture to enhance properties such as strength and/or permeability of the mold.

To form the mold cavity, the traditional method is to compact the molding sand around the pattern for both cope and drag in a container called a *flask.* The packing process is performed by various methods. The simplest is hand ramming, accomplished manually by a foundry worker. In addition, various machines have been developed to mechanize the packing procedure. These machines operate by any of several mechanisms, including (1) squeezing the sand around the pattern by pneumatic pressure; (2) a jolting action in which the sand, contained in the flask with the pattern, is dropped repeatedly in order to pack it into place; and (3) a slinging action, in which the sand grains are impacted against the pattern at high speed.

An alternative to traditional flasks for each sand mold is *flaskless molding,* which refers to the use of one master flask in a mechanized system of mold production. Each sand mold is produced using the same master flask. Mold production rates up to 600 per hour are claimed for this more automated method [8].

Several indicators are used to determine the quality of the sand mold [7]: (1) *strength*—the mold's ability to maintain its shape and resist erosion caused by the flow of molten metal; it depends on grain shape, adhesive qualities of the binder, and other factors; (2) *permeability*—capacity of the mold to allow hot air and gases from the casting operation to pass through the voids in the sand; (3) *thermal stability*—ability of the sand at the surface of the mold cavity to resist cracking and buckling upon contact with the molten metal; (4) *collapsibility*—ability of the mold to give way and allow the casting to shrink without cracking the casting; it also refers to the ability to remove the sand from the casting during cleaning; and (5) *reusability*—can the sand from the broken mold be reused to make other molds? These measures are sometimes incompatible; for example, a mold with greater strength is less collapsible.

Sand molds are often classified as green-sand, dry-sand, or skin-dried molds. *Green-sand molds* are made of a mixture of sand, clay, and water, the word *green* referring to the fact that the mold contains moisture at the time of pouring. Green-sand molds possess sufficient strength for most applications, good collapsibility, good permeability, good reusability, and are the least expensive of the molds. They are the most widely used mold type, but they are not without problems. Moisture in the sand can cause defects in some castings, depending on the metal and geometry of the part. A *dry-sand mold* is made using organic binders rather than clay, and the mold is baked in a large oven at temperatures ranging from 200°C to 320°C (392°F to 608°F) [8]. Oven baking strengthens the mold and hardens the cavity surface. A dry-sand mold provides better dimensional control in the cast product, compared to green-sand molding. However, dry-sand molding is more expensive, and production rate is reduced because of drying time. Applications are generally limited to medium and large castings in low to medium production rates. In a *skin-dried mold,* the advantages of a dry-sand mold is partially achieved by drying the surface of a green-sand mold to a depth of 10 to 25 mm (0.4–1 in) at the mold cavity surface, using torches, heating lamps, or other means. Special bonding materials must be added to the sand mixture to strengthen the cavity surface.

The preceding mold classifications refer to the use of conventional binders consisting of either clay-and-water or ones that require heating to cure. In addition to these classifications, chemically bonded molds have been developed that are not based on either of these traditional binder ingredients. Some of the binder materials used in these "no-bake" systems include furan resins (consisting of furfural alcohol, urea, and formaldehyde), phenolics, and alkyd oils. No-bake molds are growing in popularity due to their good dimensional control in high production applications.

11.1.3 THE CASTING OPERATION

After the core is positioned (if one is used) and the two halves of the mold are clamped together, then casting is performed. Casting consists of pouring, solidification, and cooling

TABLE 11.1   Densities of selected casting alloys.

Metal	Density g/cm³	lb/in³	Metal	Density g/cm³	lb/in³
Aluminum (99% = pure)	2.70	0.098	Cast iron, gray[a]	7.16	0.260
Aluminum-silicon alloy	2.65	0.096	Copper (99% = pure)	8.73	0.317
Aluminum-copper (92% Al)	2.81	0.102	Lead (pure)	11.30	0.410
Brass[a]	8.62	0.313	Steel	7.82	0.284

Source: [7].
[a]Density depends on composition of alloy; value given is typical.

of the cast part (Sections 10.2 and 10.3). The gating and riser system in the mold must be designed to deliver liquid metal into the cavity and provide for a sufficient reservoir of molten metal during solidification shrinkage. Air and gases must be allowed to escape.

One of the hazards during pouring is that the buoyancy of the molten metal will displace the core. Buoyancy results from the weight of molten metal being displaced by the core, according to Archimedes' principle. The force tending to lift the core is equal to the weight of the displaced liquid less the weight of the core itself. Expressing the situation in equation form,

$$F_b = W_m - W_c \qquad (11.1)$$

where F_b = buoyancy force, N (lb); W_m = weight of molten metal displaced, N (lb); and % W_c = weight of the core, N (lb). Weights are determined as the volume of the core multiplied by the respective densities of the core material (typically sand) and the metal being cast. The density of a sand core is approximately $1.6 \, \text{g/cm}^3$ ($0.058 \, \text{lb/in}^3$). Densities of several common casting alloys are given in Table 11.1.

Example 11.1 Buoyancy in Sand Casting

A sand core has a volume = $1875 \, \text{cm}^3$ and is located inside a sand mold cavity. Determine the buoyancy force tending to lift the core during pouring of molten lead into the mold.

Solution: Density of the sand core is $1.6 \, \text{g/cm}^3$. Weight of the core is $1875(1.6) = 3000 \, \text{g} = 3.0 \, \text{kg}$. Density of lead, based on Table 11.1, is $11.3 \, \text{g/cm}^3$. The weight of lead displaced by the core is $1875(11.3) = 21,188 \, \text{g} = 21.19 \, \text{kg}$. The difference $= 21.19 - 3.0 = 18.19 \, \text{kg}$. Given that 1 kg = 9.81 N, the buoyancy force is therefore $F_b = 9.81(18.19) = \textbf{178.4 N.}$

Following solidification and cooling, the sand mold is broken away from the casting to retrieve the part. The part is then cleaned—gating and riser system are separated, and sand is removed. The casting is then inspected (Section 11.5). ∎

11.2   OTHER EXPENDABLE-MOLD CASTING PROCESSES

As versatile as sand casting is, there are other casting processes that have been developed to meet special needs. The differences between these methods are in the composition of the mold material, or the manner in which the mold is made, or in the way the pattern is made.

11.2.1   SHELL MOLDING

Shell molding is a casting process in which the mold is a thin shell (typically 9 mm or 3/8 in) made of sand held together by a thermosetting resin binder. Developed in Germany during the early 1940s, the process is described and illustrated in Figure 11.5.

FIGURE 11.5   Steps in shell molding: (1) a match-plate or cope-and-drag metal pattern is heated and placed over a box containing sand mixed with thermosetting resin; (2) box is inverted so that sand and resin fall onto the hot pattern, causing a layer of the mixture to partially cure on the surface to form a hard shell; (3) box is repositioned so that loose, uncured particles drop away; (4) sand shell is heated in oven for several minutes to complete curing; (5) shell mold is stripped from the pattern; (6) two halves of the shell mold are assembled, supported by sand or metal shot in a box, and pouring is accomplished. The finished casting with sprue removed is shown in (7).

There are many advantages to the shell-molding process. The surface of the shell-mold cavity is smoother than a conventional green-sand mold, and this smoothness permits easier flow of molten metal during pouring and better surface finish on the final casting. Finishes of 2.5 μm (100 μ-in) can be obtained. Good dimensional accuracy is also achieved, with tolerances of $\pm$0.25 mm ($\pm$0.010 in) possible on small-to-medium-sized parts. The good finish and accuracy often precludes the need for further machining. Collapsibility of the mold is generally sufficient to avoid tearing and cracking of the casting.

Disadvantages of shell molding include a more expensive metal pattern than the corresponding pattern for green-sand molding. This makes shell molding difficult to justify for small quantities of parts. Shell molding can be mechanized for mass production and is very economical for large quantities. It seems particularly suited to steel castings of less than 20 lb. Examples of parts made using shell molding include gears, valve bodies, bushings, and camshafts.

11.2.2 VACUUM MOLDING

Vacuum molding, also called the **V-process,** was developed in Japan around 1970. It uses a sand mold held together by vacuum pressure rather than by a chemical binder. Accordingly, the term **vacuum** in this process refers to the making of the mold rather than the casting operation itself. The steps of the process are explained in Figure 11.6.

Because no binders are used, the sand is readily recovered in vacuum molding. Also, the sand does not require extensive mechanical reconditioning normally done when

FIGURE 11.6 Steps in vacuum molding: (1) a thin sheet of preheated plastic is drawn over a match-plate or cope-and-drag pattern by vacuum—the pattern has small vent holes to facilitate vacuum forming; (2) a specially designed flask is placed over the pattern plate and filled with sand, and a sprue and pouring cup are formed in the sand; (3) another thin plastic sheet is placed over the flask, and a vacuum is drawn that causes the sand grains to be held together, forming a rigid mold; (4) the vacuum on the mold pattern is released to permit the pattern to be stripped from the mold; (5) this mold is assembled with its matching half to form the cope and drag, and with vacuum maintained on both halves, pouring is accomplished. The plastic sheet quickly burns away on contacting the molten metal. After solidification, nearly all of the sand can be recovered for reuse.

binders are used in the molding sand. Since no water is mixed with the sand, moisture-related defects are absent from the product. Disadvantages of the V-process are that it is relatively slow and not readily adaptable to mechanization.

11.2.3 EXPANDED POLYSTYRENE PROCESS

The expanded polystyrene casting process uses a mold of sand packed around a polystyrene foam pattern that vaporizes when the molten metal is poured into the mold. The process and variations of it are known by other names, including *lost-foam process, lost-pattern process, evaporative-foam process,* and *full-mold process* (the last being a trade name). The foam pattern includes the sprue, risers, and gating system, and it may also contain internal cores (if needed), thus eliminating the need for a separate core in the mold. Also, since the foam pattern itself becomes the cavity in the mold, considerations of draft and parting lines can be ignored. The mold does not have to be opened into cope and drag sections. The sequence in this casting process is illustrated and described in Figure 11.7. Various methods for making the pattern can be used, depending on the quantities of castings to be produced. For one-of-a-kind castings, the foam is manually cut from large strips and assembled to form the pattern. For large production runs, an

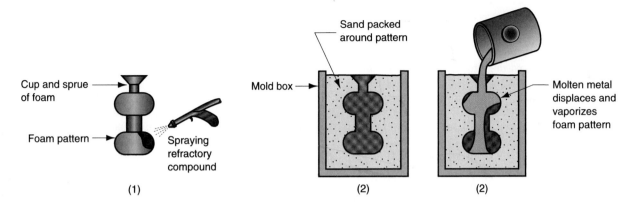

FIGURE 11.7   Expanded polystyrene casting process: (1) pattern of polystyrene is coated with refractory compound; (2) foam pattern is placed in mold box, and sand is compacted around the pattern; and (3) molten metal is poured into the portion of the pattern that forms the pouring cup and sprue. As the metal enters the mold, the polystyrene foam is vaporized ahead of the advancing liquid, thus allowing the resulting mold cavity to be filled.

automated molding operation can be set up to mold the patterns prior to making the molds for casting. The pattern is normally coated with a refractory compound to provide a smoother surface on the pattern and to improve its high temperature resistance. Molding sands usually include bonding agents. However, dry sand is used in certain processes in this group, which aids recovery and reuse. The video clip on casting features a segment titled Evaporative-Foam Casting.

VIDEO CLIP _____

Casting: View the segment titled Evaporative-Foam Casting.

_____

A significant advantage for this process is that the pattern need not be removed from the mold. This simplifies and expedites mold making. In a conventional green-sand mold, two halves are required with proper parting lines, draft allowances must be provided in the mold design, cores must be inserted, and the gating and riser system must be added. With the expanded polystyrene process, these steps are built into the pattern itself. A new pattern is needed for every casting, so the economics of the expanded polystyrene casting process depend largely on the cost of producing the patterns. The process has been applied to mass produce castings for automobiles engines. Automated production systems are installed to mold the polystyrene foam patterns for these applications.

11.2.4 INVESTMENT CASTING

In investment casting, a pattern made of wax is coated with a refractory material to make the mold, after which the wax is melted away prior to pouring the molten metal. The term ***investment*** comes from one of the less familiar definitions of the word ***invest,*** which is "to cover completely," this referring to the coating of the refractory material around the wax pattern. It is a precision casting process, because it is capable of making castings of high accuracy and intricate detail. The process dates back to ancient Egypt (Historical Note 11.1) and is also known as the ***lost-wax process,*** because the wax pattern is lost from the mold prior to casting.

Historical Note 11.1 *Investment casting*

The lost wax casting process was developed by the ancient Egyptians some 3500 years ago. Although written records do not identify when the invention occurred or the artisan responsible, historians speculate that the process resulted from the close association between pottery and molding in early times. It was the potter who crafted the molds that were used for casting. The idea for the lost wax process must have originated with a potter who was familiar with the casting process. As he was working one day on a ceramic piece—perhaps an ornate vase or bowl—it occurred to him that the article might be more attractive and durable if made of metal. So he fashioned a core in the general shape of the piece, but smaller than the desired final dimensions, and coated it with wax to establish the size. The wax proved to be an easy material to form, and intricate designs and shapes could be created by the craftsman. On the wax surface, he carefully plastered several layers of clay and devised a means of holding the resulting components together. He then baked the mold in a kiln, so that the clay hardened and the wax melted and drained out to form a cavity. At last, he poured molten bronze into the cavity and, after the casting had solidified and cooled, broke away the mold to recover the part. Considering the education and experience of this early pottery maker and the tools he had to work with, development of the lost wax casting process demonstrated great innovation and insight. "No other process can be named by archeologists so crowded with deduction, engineering ability and ingenuity" [14].

FIGURE 11.8 Steps in investment casting: (1) wax patterns are produced; (2) several patterns are attached to a sprue to form a pattern tree; (3) the pattern tree is coated with a thin layer of refractory material; (4) the full mold is formed by covering the coated tree with sufficient refractory material to make it rigid; (5) the mold is held in an inverted position and heated to melt the wax and permit it to drip out of the cavity; (6) the mold is preheated to a high temperature, which ensures that all contaminants are eliminated from the mold; it also permits the liquid metal to flow more easily into the detailed cavity; the molten metal is poured; it solidifies; and (7) the mold is broken away from the finished casting. Parts are separated from the sprue.

Steps in investment casting are described in Figure 11.8. Since the wax pattern is melted off after the refractory mold is made, a separate pattern must be made for every casting. Pattern production is usually accomplished by a molding operation—pouring or injecting the hot wax into a ***master die*** that has been designed with proper allowances for shrinkage of both wax and subsequent metal casting. In cases where the part geometry is complicated, several separate wax pieces must be joined to make the pattern. In high-production operations, several patterns are attached to a sprue, also made of wax, to form a ***pattern tree;*** this is the geometry that will be cast out of metal. The video clip on casting contains a segment on investment casting.

VIDEO CLIP

Casting: View the segment titled Investment Casting.

Coating with refractory (step 3) is usually accomplished by dipping the pattern tree into a slurry of very fine grained silica or other refractory (almost in powder form) mixed with plaster to bond the mold into shape. The small grain size of the refractory material provides a smooth surface and captures the intricate details of the wax pattern. The final mold (step 4) is accomplished by repeatedly dipping the tree into the refractory slurry or by gently packing the refractory around the tree in a container. The mold is allowed to air dry for about 8 hours to harden the binder.

Advantages of investment casting include: (1) parts of great complexity and intricacy can be cast; (2) close dimensional control—tolerances of ± 0.075 mm (± 0.003 in) are possible; (3) good surface finish is possible; (4) the wax can usually be recovered for reuse; and (5) additional machining is not normally required—this is a net shape process. Because many steps are involved in this casting operation, it is a relatively expensive process. Investment castings are normally small in size, although parts with complex geometries weighing up to 75 lb have been successfully cast. All types of metals, including steels, stainless steels, and other high temperature alloys, can be investment cast. Examples of parts include complex machinery parts, blades, and other components for turbine engines, jewelry, and dental fixtures. Shown in Figure 11.9 is a part illustrating the intricate features possible with investment casting.

11.2.5 PLASTER-MOLD AND CERAMIC-MOLD CASTING

Plaster-mold casting is similar to sand casting except that the mold is made of plaster of Paris (gypsum, $CaSO_4$–$2H_2O$) instead of sand. Additives such as talc and silica flour are mixed with the plaster to control contraction and setting time, reduce cracking, and increase strength. To make the mold, the plaster mixture combined with water is poured over a plastic or metal pattern in a flask and allowed to set. Wood patterns are generally unsatisfactory due to the extended contact with water in the plaster. The fluid consistency permits the plaster mixture to readily flow around the pattern, capturing its details and surface finish. Thus, the cast product in plaster molding is noted for these attributes.

Curing of the plaster mold is one of the disadvantages of this process, at least in high production. The mold must set for about 20 minutes before the pattern is stripped. The mold is then baked for several hours to remove moisture. Even with the baking, not all of the moisture content is removed from the plaster. The dilemma faced by foundrymen is that mold strength is lost when the plaster becomes too dehydrated, and yet moisture content can cause casting defects in the product. A balance must be achieved between these undesirable alternatives. Another disadvantage with the plaster mold is that it is not permeable, thus limiting escape of gases from the mold cavity. This problem can be solved in

FIGURE 11.9   A one-piece compressor stator with 108 separate airfoils made by investment casting. (Courtesy of Howmet Corp.).

a number of ways: (1) evacuating air from the mold cavity before pouring; (2) aerating the plaster slurry prior to mold making so that the resulting hard plaster contains finely dispersed voids; and (3) using a special mold composition and treatment known as the **Antioch process.** This process involves using about 50% sand mixed with the plaster, heating the mold in an autoclave (an oven that uses superheated steam under pressure), and then drying. The resulting mold has considerably greater permeability than a conventional plaster mold.

Plaster molds cannot withstand the same high temperatures as sand molds. They are therefore limited to the casting of lower-melting-point alloys, such as aluminum, magnesium, and some copper-base alloys. Applications include metal molds for plastic and rubber molding, pump and turbine impellers, and other parts of relatively intricate geometry. Casting sizes range from about 20 g (less than 1 oz) to more than 100 kg (more than 220 lb). Parts weighing less than about 10 kg (22 lb) are most common. Advantages of plaster molding for these applications are good surface finish and dimensional accuracy and the capability to make thin cross-sections in the casting.

Ceramic-mold casting is similar to plaster-mold casting, except that the mold is made of refractory ceramic materials that can withstand higher temperatures than plaster. Thus, ceramic molding can be used to cast steels, cast irons, and other high-temperature alloys. Its applications (relatively intricate parts) are similar to those of plaster-mold casting except for the metals cast. Its advantages (good accuracy and finish) are also similar.

11.3 PERMANENT-MOLD CASTING PROCESSES

The economic disadvantage of any of the expendable-mold processes is that a new mold is required for every casting. In permanent-mold casting, the mold is reused many times. In this section, we treat permanent-mold casting as the basic process in the group of casting processes that all use reusable metal molds. Other members of the group include die casting and centrifugal casting.

11.3.1 THE BASIC PERMANENT-MOLD PROCESS

Permanent-mold casting uses a metal mold constructed of two sections that are designed for easy, precise opening and closing. These molds are commonly made of steel or cast iron. The cavity, with gating system included, is machined into the two halves to provide accurate dimensions and good surface finish. Metals commonly cast in permanent molds include aluminum, magnesium, copper-base alloys, and cast iron. However, cast iron requires a high pouring temperature, $1250°C$ to $1500°C$ ($2282°F–2732°F$), which takes a heavy toll on mold life. The very high pouring temperatures of steel make permanent molds unsuitable for this metal, unless the mold is made of refractory material.

Cores can be used in permanent molds to form interior surfaces in the cast product. The cores can be made of metal, but either their shape must allow for removal from the casting or they must be mechanically collapsible to permit removal. If withdrawal of a metal core would be difficult or impossible, sand cores can be used, in which case the casting process is often referred to as *semipermanent-mold casting.*

Steps in the basic permanent-mold casting process are described in Figure 11.10. In preparation for casting, the mold is first preheated and one or more coatings are sprayed on the cavity. Preheating facilitates metal flow through the gating system and into the cavity. The coatings aid heat dissipation and lubricate the mold surfaces for easier separation of the cast product. After pouring, as soon as the metal solidifies, the mold is opened and the casting is removed. Unlike expendable molds, permanent molds do not collapse, so the mold must be opened before appreciable cooling contraction occurs in order to prevent cracks from developing in the casting.

Advantages of permanent-mold casting include good surface finish and close dimensional control, as previously indicated. In addition, more rapid solidification caused by the metal mold results in a finer grain structure, so stronger castings are produced. The process is generally limited to metals of lower melting points. Other limitations include simple part geometries compared to sand casting (because of the need to open the mold), and the expense of the mold. Because mold cost is substantial, the process is best suited to high-volume production and can be automated accordingly. Typical parts include automotive pistons, pump bodies, and certain castings for aircraft and missiles.

11.3.2 VARIATIONS OF PERMANENT-MOLD CASTING

Several casting processes are quite similar to the basic permanent-mold method. These include slush casting, low-pressure casting, and vacuum permanent-mold casting.

Slush Casting Slush casting is a permanent-mold process in which a hollow casting is formed by inverting the mold after partial freezing at the surface to drain out the liquid metal in the center. Solidification begins at the mold walls because they are relatively cool, and it progresses over time toward the middle of the casting (Section 10.3.1). Thickness of

FIGURE 11.10 Steps in permanent-mold casting: (1) mold is preheated and coated; (2) cores (if used) are inserted, and mold is closed; (3) molten metal is poured into the mold; and (4) mold is opened. Finished part is shown in (5).

the shell is controlled by the length of time allowed before draining. Slush casting is used to make statues, lamp pedestals, and toys out of low-melting-point metals such as zinc and tin. In these items, the exterior appearance is important, but the strength and interior geometry of the casting are minor considerations.

Low-Pressure Casting In the basic permanent-mold casting process and in slush casting, the flow of metal into the mold cavity is caused by gravity. In low-pressure casting, the liquid metal is forced into the cavity under low pressure—approximately 0.1 MPa (14.5 lb/in^2)—from beneath so that the flow is upward, as illustrated in Figure 11.11. The advantage of this approach over traditional pouring is that clean molten metal from the center of the ladle is introduced into the mold, rather than metal that has been exposed to air. Gas porosity and oxidation defects are thereby minimized, and mechanical properties are improved.

Vacuum Permanent-Mold Casting Not to be confused with vacuum molding (Section 11.2.2), this process is a variation of low-pressure casting in which a vacuum is used to draw the molten metal into the mold cavity. The general configuration of the vacuum permanent-mold casting process is similar to the low-pressure casting operation. The difference is that reduced air pressure from the vacuum in the mold is used to draw the liquid metal into the

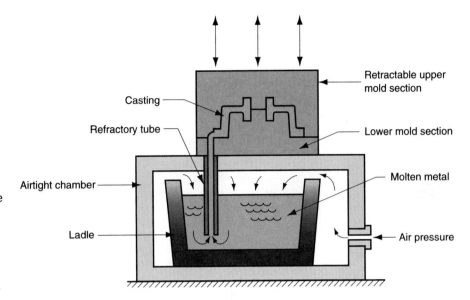

FIGURE 11.11 Low-pressure casting. The diagram shows how air pressure is used to force the molten metal in the ladle upward into the mold cavity. Pressure is maintained until the casting has solidified.

cavity, rather than forcing it by positive air pressure from below. There are several benefits of the vacuum technique relative to low-pressure casting: air porosity and related defects are reduced, and greater strength is given to the cast product.

11.3.3 DIE CASTING

Die casting is a permanent-mold casting process in which the molten metal is injected into the mold cavity under high pressure. Typical pressures are 7 to 350 MPa (1015–50,763 lb/in^2). The pressure is maintained during solidification, after which the mold is opened and the part is removed. Molds in this casting operation are called dies; hence the name die casting. The use of high pressure to force the metal into the die cavity is the most notable feature that distinguishes this process from others in the permanent-mold category. The reader can see the various forms of this process in the video clip on die casting.

VIDEO CLIP _____

Die Casting. This clip contains two segments: (1) die casting machines and (2) die casting tooling.

_____

Die casting operations are carried out in special die casting machines (Historical Note 11.2), which are designed to hold and accurately close the two halves of the mold, and keep them closed while the liquid metal is forced into the cavity. The general configuration is shown in Figure 11.12. There are two main types of die casting machines: (1) hot-chamber and (2) cold-chamber, differentiated by how the molten metal is injected into the cavity.

In *hot-chamber machines,* the metal is melted in a container attached to the machine, and a piston is used to inject the liquid metal under high pressure into the die. Typical injection pressures are 7 to 35 MPa (1015–5076 lb/in^2). The casting cycle is summarized in Figure 11.13. Production rates up to 500 parts per hour are not uncommon. Hot-chamber die casting imposes a special hardship on the injection system because much of it is submerged in the molten metal. The process is therefore limited in its applications to low-melting-point metals that do not chemically attack the plunger and other mechanical components. The metals include zinc, tin, lead, and sometimes magnesium.

Historical Note 11.2   *Die casting machines*

The modern die casting machine has its origins in the printing industry and the need in the mid to late 1800s to satisfy an increasingly literate population with a growing appetite for reading. The linotype, invented and developed by O. Mergenthaler in the late 1800s, is a machine that produces printing type. It is a casting machine because it casts a line of type characters out of lead to be used in preparing printing plates. The name *linotype* derives from the fact that the machine produces a line of type characters during each cycle of operation.

The machine was first used successfully on a commercial basis in New York City by *The Tribune* in 1886.

The linotype proved the feasibility of mechanized casting machines. The first die casting machine was patented by H. Doehler in 1905 (this machine is displayed in the Smithsonian Institute in Washington, DC). In 1907, E. Wagner developed the first die casting machine to utilize the hot-chamber design. It was first used during World War I to cast parts for binoculars and gas masks.

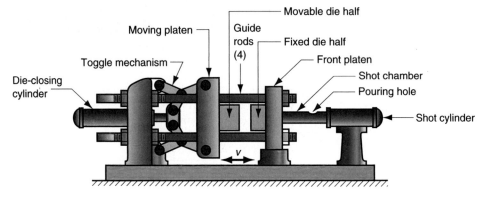

FIGURE 11.12   General configuration of a (cold-chamber) die casting machine.

FIGURE 11.13   Cycle in hot-chamber casting: (1) with die closed and plunger withdrawn, molten metal flows into the chamber; (2) plunger forces metal in chamber to flow into die, maintaining pressure during cooling and solidification; and (3) plunger is withdrawn, die is opened, and solidified part is ejected. Finished part is shown in (4).

In *cold-chamber die casting machines,* molten metal is poured into an unheated chamber from an external melting container, and a piston is used to inject the metal under high pressure into the die cavity. Injection pressures used in these machines are typically 14 to 140 MPa (2031–20,305 lb/in^2). The production cycle is explained in Figure 11.14. Compared to hot-chamber machines, cycle rates are not usually as fast because of the need to ladle the liquid metal into the chamber from an external source. Nevertheless, this casting process is a high production operation. Cold-chamber machines are typically used for casting aluminum, brass, and magnesium alloys. Low-melting-point alloys (zinc, tin, lead) can also be cast on cold-chamber machines, but the advantages of the hot-chamber process usually favor its use on these metals.

Molds used in die casting operations are usually made of tool steel, mold steel, or maraging steel. Tungsten and molybdenum with good refractory qualities are also being used, especially in attempts to die cast steel and cast iron. Dies can be single-cavity or multiple-cavity. Single-cavity dies are shown in Figures 11.13 and 11.14. Ejector pins are required to remove the part from the die when it opens, as in our diagrams. These pins push the part away from the mold surface so that it can be removed. Lubricants must also be sprayed into the cavities to prevent sticking.

Because the die materials have no natural porosity and the molten metal rapidly flows into the die during injection, venting holes and passageways must be built into the dies at the parting line to evacuate the air and gases in the cavity. The vents are quite small; yet they fill with metal during injection. This metal must later be trimmed from the part. Also, formation of *flash* is common in die casting, in which the liquid metal under high pressure squeezes into the small space between the die halves at the parting line or into the clearances around the cores and ejector pins. This flash must be trimmed from the casting, along with the sprue and gating system.

Advantages of die casting include (1) high production rates possible; (2) economical for large production quantities; (3) close tolerances possible, on the order of ±0.076 mm (±0.003

FIGURE 11.14 Cycle in cold-chamber casting: (1) with die closed and ram withdrawn, molten metal is poured into the chamber; (2) ram forces metal to flow into die, maintaining pressure during cooling and solidification; and (3) ram is withdrawn, die is opened, and part is ejected. (Gating system is simplified.)

in) for small parts; (4) good surface finish; (5) thin sections are possible, down to about 0.5 mm (0.020 in); and (6) rapid cooling provides small grain size and good strength to the casting. The limitation of this process, in addition to the metals cast, is the shape restriction. The part geometry must allow for removal from the die cavity.

11.3.4 SQUEEZE CASTING AND SEMISOLID METAL CASTING

These are two processes that are often associated with die casting. *Squeeze casting* is a combination of casting and forging (Section 19.3) in which a molten metal is poured into a pre-heated lower die, and the upper die is closed to create the mold cavity after solidification begins. This differs from the usual permanent-mold casting process in which the die halves are closed prior to pouring or injection. Owing to the hybrid nature of the process, it is also known as *liquid–metal forging.* The pressure applied by the upper die in squeeze casting causes the metal to completely fill the cavity, resulting in good surface finish and low shrinkage. The required pressures are significantly less than in forging of a solid metal billet and much finer surface detail can be imparted by the die than in forging. Squeeze casting can be used for both ferrous and non-ferrous alloys, but aluminum and magnesium alloys are the most common due to their lower melting temperatures. Automotive parts are a common application.

Semi-solid metal casting is a family of net-shape and near net-shape processes performed on metal alloys at temperatures between the liquidus and solidus (Section 10.3.1). Thus the alloy is a mixture of solid and molten metals during casting, like a slurry; it is in the mushy state. In order to flow properly, the mixture must consist of solid metal globules in a liquid rather than the more typical dendritic solid shapes that form during freezing of a molten metal. This is achieved by forcefully stirring the slurry to prevent dendrite formation and instead encourage the spherical shapes, which in turn reduces the viscosity of the work metal. Advantages of semisolid metal casting include the following [16]: (1) complex part geometries, (2) thin walls in parts, (3) close tolerances, (4) zero or low porosity, resulting in high strength of the casting.

There are several forms of semisolid metal casting. When applied to aluminum, the terms thixocasting and rheocasting are used. The prefix in thixocasting is derived from the word *thixotropy,* which refers to the decrease in viscosity of some fluid-like materials when agitated. The prefix in rheocasting comes from *rheology,* the science that relates deformation and flow of materials. In *thixocasting,* the starting work material is a pre-cast billet that has a nondendritic microstructure; this is heated into the semisolid temperature range and injected into a mold cavity using die casting equipment. In *rheocasting,* a semisolid slurry is injected into the mold cavity by a die casting machine, very much like conventional die casting. The difference is that the starting metal in rheocasting is at a temperature between the solidus and liquidus rather than above the liquidus. And the mushy mixture is agitated to prevent dendrite formation.

When applied to magnesium, the term is *thixomolding,* which utilizes equipment similar to an injection-molding machine (Section 13.6.3). Magnesium alloy granules are fed into a barrel and propelled forward by a rotating screw as they are heated into the semisolid temperature range. The required globular form of the solid phase is accomplished by the mixing action of the rotating screw. The slurry is then injected into the mold cavity by a linear forward movement of the screw.

11.3.5 CENTRIFUGAL CASTING

Centrifugal casting refers to several casting methods in which the mold is rotated at high speed so that centrifugal force distributes the molten metal to the outer regions of the die cavity. The group includes (1) true centrifugal casting, (2) semicentrifugal casting, and (3) centrifuge casting.

FIGURE 11.15 Setup for true centrifugal casting.

True Centrifugal Casting In true centrifugal casting, molten metal is poured into a rotating mold to produce a tubular part. Examples of parts made by this process include pipes, tubes, bushings, and rings. One possible setup is illustrated in Figure 11.15. Molten metal is poured into a horizontal rotating mold at one end. In some operations, mold rotation commences after pouring has occurred rather than beforehand. The high-speed rotation results in centrifugal forces that cause the metal to take the shape of the mold cavity. Thus, the outside shape of the casting can be round, octagonal, hexagonal, and so on. However, the inside shape of the casting is (theoretically) perfectly round, due to the radially symmetric forces at work.

Orientation of the axis of mold rotation can be either horizontal or vertical, the former being more common. Let us consider how fast the mold must rotate in *horizontal centrifugal casting* for the process to work successfully. Centrifugal force is defined by this physics equation:

$$F = \frac{mv^2}{R} \tag{11.2}$$

where F = force, N (lb); m = mass, kg (lbm); v = velocity, m/s (ft/sec); and R = inside radius of the mold, m (ft). The force of gravity is its weight $W = mg$, where W is given in kg (lb), and g = acceleration of gravity, 9.8 m/s^2 (32.2 ft/sec^2). The so-called G-factor GF is the ratio of centrifugal force divided by weight:

$$GF = \frac{mv^2}{R} = \frac{mv^2}{Rmg} = \frac{v^2}{Rg} \tag{11.3}$$

Velocity v can be expressed as $2\pi RN/60 = \pi RN/30$, where N = rotational speed, rev/min. Substituting this expression into Eq. (11.3), we obtain

$$GF = \frac{\left(R\frac{\pi N}{30}\right)^2}{g} \tag{11.4}$$

Rearranging this to solve for rotational speed N, and using diameter D rather than radius in the resulting equation, we have

$$N = \frac{30}{\pi}\sqrt{\frac{2gGF}{D}} \tag{11.5}$$

where D = inside diameter of the mold, m (ft). If the G-factor is too low in centrifugal casting, the liquid metal will not remain forced against the mold wall during the upper half of the circular path but will "rain" inside the cavity. Slipping occurs between the molten metal and the mold wall, which means that the rotational speed of the metal is less than that of the mold. On an empirical basis, values of GF = 60 to 80 are found to be appropriate for horizontal centrifugal casting [2], although this depends to some extent on the metal being cast.

**Example 11.2
Rotation Speed in
True Centrifugal
Casting**

A true centrifugal casting operation is to be performed horizontally to make copper tube sections with $OD = 25$ cm and $ID = 22.5$ cm. What rotational speed is required if a G-factor of 65 is used to cast the tubing?

Solution: The inside diameter of the mold $D = OD$ of the casting $= 25$ cm $= 0.25$ m. We can compute the required rotational speed from Eq. (11.5) as follows:

$$N = \frac{30}{\pi} \sqrt{\frac{2(9.8)(26)}{0.25}} = \mathbf{61.7\ rev/min}.$$ ∎

In *vertical centrifugal casting,* the effect of gravity acting on the liquid metal causes the casting wall to be thicker at the base than at the top. The inside profile of the casting wall takes on a parabolic shape. The difference in inside radius between top and bottom is related to speed of rotation as follows:

$$N = \frac{30}{\pi} \sqrt{\frac{2gL}{R_t^2 - R_b^2}} \qquad (11.6)$$

where L = vertical length of the casting, m (ft); R_t = inside radius at the top of the casting, m (ft); and R_b = inside radius at the bottom of the casting, m (ft). Equation (11.6) can be used to determine the required rotational speed for vertical centrifugal casting, given specifications on the inside radii at top and bottom. One can see from the formula that for R_t to equal R_b, the speed of rotation N would have to be infinite, which is impossible of course. As a practical matter, part lengths made by vertical centrifugal casting are usually no more than about twice their diameters. This is quite satisfactory for bushings and other parts that have large diameters relatively to their lengths, especially if machining will be used to accurately size the inside diameter.

Castings made by true centrifugal casting are characterized by high density, especially in the outer regions of the part where F is greatest. Solidification shrinkage at the exterior of the cast tube is not a factor, because the centrifugal force continually reallocates molten metal toward the mold wall during freezing. Any impurities in the casting tend to be on the inner wall and can be removed by machining if necessary.

Semicentrifugal Casting In this method, centrifugal force is used to produce solid castings, as in Figure 11.16, rather than tubular parts. The rotation speed in semicentrifugal casting is usually set so that G-factors of around 15 are obtained [2], and the molds are

FIGURE 11.16 Semicentrifugal casting.

FIGURE 11.17   (a) Centrifuge casting—centrifugal force causes metal to flow to the mold cavities away from the axis of rotation; and (b) the casting.

designed with risers at the center to supply feed metal. Density of metal in the final casting is greater in the outer sections than at the center of rotation. The process is often used on parts in which the center of the casting is machined away, thus eliminating the portion of the casting where the quality is lowest. Wheels and pulleys are examples of castings that can be made by this process. Expendable molds are often used in semicentrifugal casting, as suggested by our illustration of the process.

Centrifuge Casting   In centrifuge casting, Figure 11.17, the mold is designed with part cavities located away from the axis of rotation, so that the molten metal poured into the mold is distributed to these cavities by centrifugal force. The process is used for smaller parts, and radial symmetry of the part is not a requirement as it is for the other two centrifugal casting methods.

11.4 FOUNDRY PRACTICE

In all casting processes, the metal must be heated to the molten state to be poured or otherwise forced into the mold. Heating and melting are accomplished in a furnace. This section covers the various types of furnaces used in foundries and the pouring practices for delivering the molten metal from furnace to mold.

11.4.1 FURNACES

The types of furnaces most commonly used in foundries are (1) cupolas, (2) direct fuel-fired furnaces, (3) crucible furnaces, (4) electric-arc furnaces, and (5) induction furnaces. Selection of the most appropriate furnace type depends on factors such as the casting alloy; its melting and pouring temperatures; capacity requirements of the furnace; costs of investment, operation, and maintenance; and environmental pollution considerations.

Cupolas   A cupola is a vertical cylindrical furnace equipped with a tapping spout near its base. Cupolas are used only for melting cast irons, and although other furnaces are also used, the largest tonnage of cast iron is melted in cupolas. General construction and operating features of the cupola are illustrated in Figure 11.18. It consists of a large shell

Interior — — Exterior

Charging door
Charging floor
Refractory lining
Charge — — Steel shell
Blower
Wind box
Slag — Molten metal ready to be tapped
Slag spout — Bott (plug)
Sand bottom
Supports — Tapping spout

FIGURE 11.18 Cupola used for melting cast iron. Furnace shown is typical for a small foundry and omits details of emissions control system required in a modern cupola.

of steel plate lined with refractory. The "charge," consisting of iron, coke, flux, and possible alloying elements, is loaded through a charging door located less than halfway up the height of the cupola. The iron is usually a mixture of pig iron and scrap (including risers, runners, and sprues left over from previous castings). Coke is the fuel used to heat the furnace. Forced air is introduced through openings near the bottom of the shell for combustion of the coke. The flux is a basic compound such as limestone that reacts with coke ash and other impurities to form slag. The slag serves to cover the melt, protecting it from reaction with the environment inside the cupola and reducing heat loss. As the mixture is heated and melting of the iron occurs, the furnace is periodically tapped to provide liquid metal for the pour.

Direct Fuel-Fired Furnaces   A direct fuel-fired furnace contains a small open-hearth, in which the metal charge is heated by fuel burners located on the side of the furnace. The roof of the furnace assists the heating action by reflecting the flame down against the charge. Typical fuel is natural gas, and the combustion products exit the furnace through a stack. At the bottom of the hearth is a tap hole to release the molten metal. Direct fuel-fired furnaces are generally used in casting for melting nonferrous metals such as copper-base alloys and aluminum.

Crucible Furnaces   These furnaces melt the metal without direct contact with a burning fuel mixture. For this reason, they are sometimes called ***indirect fuel-fired furnaces.*** Three types of crucible furnaces are used in foundries: (a) lift-out type, (b) stationary, and

FIGURE 11.19   Three types of crucible furnaces: (a) lift-out crucible, (b) stationary pot, and (c) tilting-pot furnace.

(c) tilting, illustrated in Figure 11.19. They all utilize a container (the crucible) made out of a suitable refractory material (e.g., a clay–graphite mixture) or high-temperature steel alloy to hold the charge. In the *lift-out crucible furnace,* the crucible is placed in a furnace and heated sufficiently to melt the metal charge. Oil, gas, or powdered coal are typical fuels for these furnaces. When the metal is melted, the crucible is lifted out of the furnace and used as a pouring ladle. The other two types, sometimes referred to as *pot furnaces,* have the heating furnace and container as one integral unit. In the *stationary pot furnace,* the furnace is stationary and the molten metal is ladled out of the container. In the *tilting-pot furnace,* the entire assembly can be tilted for pouring. Crucible furnaces are used for nonferrous metals such as bronze, brass, and alloys of zinc and aluminum. Furnace capacities are generally limited to several hundred pounds.

Electric-Arc Furnaces   In this furnace type, the charge is melted by heat generated from an electric arc. Various configurations are available, with two or three electrodes (see Figure 6.9). Power consumption is high, but electric-arc furnaces can be designed for high melting capacity (23,000–45,000 kg/hr or 25–50 tons/hr), and they are used primarily for casting steel.

Induction Furnaces   An induction furnace uses alternating current passing through a coil to develop a magnetic field in the metal, and the resulting induced current causes rapid heating and melting of the metal. Features of an induction furnace for foundry operations are illustrated in Figure 11.20. The electromagnetic force field causes a mixing

FIGURE 11.20
Induction furnace.

action to occur in the liquid metal. Also, since the metal does not come in direct contact with the heating elements, the environment in which melting takes place can be closely controlled. All of this results in molten metals of high quality and purity, and induction furnaces are used for nearly any casting alloy when these requirements are important. Melting steel, cast iron, and aluminum alloys are common applications in foundry work.

11.4.2 POURING, CLEANING, AND HEAT TREATMENT

Moving the molten metal from the melting furnace to the mold is sometimes done using crucibles. More often, the transfer is accomplished by *ladles* of various kinds. These ladles receive the metal from the furnace and allow for convenient pouring into the molds. Two common ladles are illustrated in Figure 11.21, one for handling large volumes of molten metal using an overhead crane, and the other a "two-man ladle" for manually moving and pouring smaller amounts.

One of the problems in pouring is that oxidized molten metal can be introduced into the mold. Metal oxides reduce product quality, perhaps rendering the casting defective, so measures are taken to minimize the entry of these oxides into the mold during pouring. Filters are sometimes used to catch the oxides and other impurities as the metal is poured from the spout, and fluxes are used to cover the molten metal to retard oxidation. In addition, ladles have been devised to pour the liquid metal from the bottom, since the top surface is where the oxides accumulate.

After the casting has solidified and been removed from the mold, a number of additional steps are usually required. These operations include (1) trimming, (2) removing the core, (3) surface cleaning, (4) inspection, (5) repair, if required, and (6) heat treatment. Steps (1) through (5) are collectively referred to in foundry work as "cleaning." The extent to which these additional operations are required varies with casting processes and metals. When required, they are usually labor intensive and costly.

Trimming involves removal of sprues, runners, risers, parting-line flash, fins, chaplets, and any other excess metal from the cast part. In the case of brittle casting alloys and when the cross sections are relatively small, these appendages on the casting can be broken off. Otherwise, hammering, shearing, hack-sawing, band-sawing, abrasive wheel cutting, or various torch cutting methods are used.

If cores have been used to cast the part, they must be removed. Most cores are chemically bonded or oil-bonded sand, and they often fall out of the casting as the binder deteriorates. In some cases, they are removed by shaking the casting, either manually or mechanically. In rare instances, cores are removed by chemically dissolving the bonding agent used in the sand core. Solid cores must be hammered or pressed out.

FIGURE 11.21 Two common types of ladles: (a) crane ladle and (b) two-man ladle.

Surface cleaning is most important in the case of sand casting. In many of the other casting methods, especially the permanent-mold processes, this step can be avoided. ***Surface cleaning*** involves removal of sand from the surface of the casting and otherwise enhancing the appearance of the surface. Methods used to clean the surface include tumbling, air-blasting with coarse sand grit or metal shot, wire brushing, buffing, and chemical pickling (Chapter 28).

Defects are possible in casting, and inspection is needed to detect their presence. We consider these quality issues in the following section.

Castings are often heat treated to enhance their properties, either for subsequent processing operations such as machining or to bring out the desired properties for application of the part.

11.5   CASTING QUALITY

There are numerous opportunities for things to go wrong in a casting operation, resulting in quality defects in the cast product. In this section, we compile a list of the common defects that occur in casting, and we indicate the inspection procedures to detect them.

Casting Defects   Some defects are common to any and all casting processes. These defects are illustrated in Figure 11.22 and briefly described in the following:

(a) ***Misruns,*** which are castings that solidify before completely filling the mold cavity. Typical causes include (1) fluidity of the molten metal is insufficient, (2) pouring temperature is too low, (3) pouring is done too slowly, and/or (4) cross-section of the mold cavity is too thin.

(b) ***Cold Shuts,*** which occur when two portions of the metal flow together but there is a lack of fusion between them due to premature freezing. Its causes are similar to those of a misrun.

(c) ***Cold shots,*** which result from splattering during pouring, causing the formation of solid globules of metal that become entrapped in the casting. Pouring procedures and gating system designs that avoid splattering can prevent this defect.

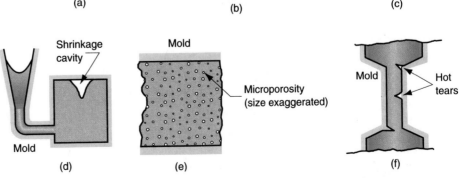

FIGURE 11.22   Some common defects in castings: (a) misrun, (b) cold shut, (c) cold shot, (d) shrinkage cavity, (e) microporosity, and (f) hot tearing.

(d) *Shrinkage cavity* is a depression in the surface or an internal void in the casting, caused by solidification shrinkage that restricts the amount of molten metal available in the last region to freeze. It often occurs near the top of the casting, in which case it is referred to as a "pipe." See Figure 10.8(3). The problem can often be solved by proper riser design.

(e) *Microporosity* consists of a network of small voids distributed throughout the casting caused by localized solidification shrinkage of the final molten metal in the dendritic structure. The defect is usually associated with alloys, because of the protracted manner in which freezing occurs in these metals.

(f) *Hot tearing,* also called *hot cracking,* occurs when the casting is restrained from contraction by an unyielding mold during the final stages of solidification or early stages of cooling after solidification. The defect is manifested as a separation of the metal (hence, the terms *tearing* and *cracking*) at a point of high tensile stress caused by the metal's inability to shrink naturally. In sand casting and other expendable-mold processes, it is prevented by compounding the mold to be collapsible. In permanent-mold processes, hot tearing is reduced by removing the part from the mold immediately after solidification.

Some defects are related to the use of sand molds, and therefore they occur only in sand castings. To a lesser degree, other expendable-mold processes are also susceptible to these problems. Defects found primarily in sand castings are shown in Figure 11.23 and described here:

(a) *Sand blow* is a defect consisting of a balloon-shaped gas cavity caused by release of mold gases during pouring. It occurs at or below the casting surface near the top of the casting. Low permeability, poor venting, and high moisture content of the sand mold are the usual causes.

(b) *Pinholes,* also caused by release of gases during pouring, consist of many small gas cavities formed at or slightly below the surface of the casting.

(c) *Sand wash,* which is an irregularity in the surface of the casting that results from erosion of the sand mold during pouring, and the contour of the erosion is formed in the surface of the final cast part.

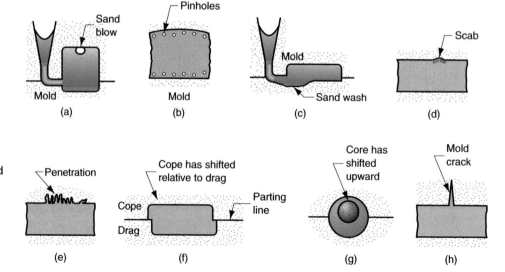

FIGURE 11.23
Common defects in sand castings: (a) sand blow, (b) pin holes, (c) sand wash, (d) scabs, (e) penetration, (f) mold shift, (g) core shift, and (h) mold crack.

(d) *Scabs* are rough areas on the surface of the casting due to encrustations of sand and metal. It is caused by portions of the mold surface flaking off during solidification and becoming imbedded in the casting surface.

(e) *Penetration* refers to a surface defect that occurs when the fluidity of the liquid metal is high, and it penetrates into the sand mold or sand core. Upon freezing, the casting surface consists of a mixture of sand grains and metal. Harder packing of the sand mold helps to alleviate this condition.

(f) *Mold shift* refers to a defect caused by a sidewise displacement of the mold cope relative to the drag, the result of which is a step in the cast product at the parting line.

(g) *Core shift* is similar to mold shift, but it is the core that is displaced, and the displacement is usually vertical. Core shift and mold shift are caused by buoyancy of the molten metal (Section 11.1.3).

(h) *Mold crack* occurs when mold strength is insufficient, and a crack develops, into which liquid metal can seep to form a "fin" on the final casting.

Inspection Methods Foundry inspection procedures include (1) visual inspection to detect obvious defects such as misruns, cold shuts, and severe surface flaws; (2) dimensional measurements to ensure that tolerances have been met; and (3) metallurgical, chemical, physical, and other tests concerned with the inherent quality of the cast metal [7]. Tests in category (3) include: (a) pressure testing—to locate leaks in the casting; (b) radiographic methods, magnetic particle tests, the use of fluorescent penetrants, and supersonic testing—to detect either surface or internal defects in the casting; and (c) mechanical testing to determine properties such as tensile strength and hardness. If defects are discovered but are not too serious, it is often possible to save the casting by welding, grinding, or other salvage methods to which the customer has agreed.

11.6 METALS FOR CASTING

Most commercial castings are made of alloys rather than pure metals. Alloys are generally easier to cast, and properties of the resulting product are better. Casting alloys can be classified as ferrous or nonferrous. The ferrous category is subdivided into cast iron and cast steel.

Ferrous Casting Alloys: Cast Iron Cast iron is the most important of all casting alloys (Historical Note 11.3). The tonnage of cast iron castings is several times that of all other metals combined. There are several types of cast iron: (1) gray cast iron, (2) nodular iron, (3) white cast iron, (4) malleable iron, and (5) alloy cast irons (Section 6.2.4). Typical pouring temperatures for cast iron are around 1400°C (2552°F), depending on composition.

Ferrous Casting Alloys: Steel The mechanical properties of steel make it an attractive engineering material (Section 6.2.3), and the capability to create complex geometries makes casting an appealing process. However, great difficulties are faced by the foundry specializing in steel. First, the melting point of steel is considerably higher than for most other metals that are commonly cast. The solidification range for low carbon steels (Figure 6.4) begins at just under 1540°C (2804°F). This means that the pouring temperature required for steel is very high—about 1650°C (3002°F). At these high temperatures, steel is chemically very reactive. It readily oxidizes, so special procedures must be used during melting and pouring to isolate the molten metal from air. Also, molten steel has relatively poor fluidity, and this limits the design of thin sections in components cast out of steel.

Historical Note 11.3 *Early cast iron products*

In the early centuries of casting, bronze and brass were preferred over cast iron as foundry metals. Iron was more difficult to cast, due to its higher melting temperatures and lack of knowledge about its metallurgy. Also, there was little demand for cast iron products. This all changed starting in the sixteenth and seventeenth centuries.

The art of sand-casting iron entered Europe from China, where iron was cast in sand molds more than 2500 years ago. In 1550 the first cannons were cast from iron in Europe. Cannon balls for these guns were made of cast iron starting around 1568. Guns and their projectiles created a large demand for cast iron. But these items were for military rather than civilian use. Two cast iron products that became significant to the general public in the sixteenth and seventeenth centuries were the cast iron stove and cast iron water pipe.

As unspectacular a product as it may seem today, the cast iron stove brought comfort, health, and improved living conditions to many people in Europe and America. During the 1700s, the manufacture of cast iron stoves was one of the largest and most profitable industries on these two continents. The commercial success of stove making was due to the large demand for the product and the art and technology of casting iron that had been developed to produce it.

Cast iron water pipe was another product that spurred the growth of the iron casting industry. Until the advent of cast iron pipes, a variety of methods had been tried to supply water directly to homes and shops, including hollow wooden pipes (which quickly rotted), lead pipes (too expensive), and open trenches (susceptible to pollution). Development of the iron casting process provided the capability to fabricate water pipe sections at relatively low cost. Cast iron water pipes were used in France starting in 1664, and later in other parts of Europe. By the early 1800s, cast iron pipe lines were being widely installed in England for water and gas delivery. The first significant water pipe installation in the United States was in Philadelphia in 1817, using pipe imported from England.

Several characteristics of steel castings make it worth the effort to solve these problems. Tensile strength is higher than for most other casting metals, ranging upward from about 410 MPa (59,465 lb/in^2) [9]. Steel castings have better toughness than most other casting alloys. The properties of steel castings are isotropic; strength is virtually the same in all directions. By contrast, mechanically formed parts (e.g., rolling, forging) exhibit directionality in their properties. Depending on the requirements of the product, isotropic behavior of the material may be desirable. Another advantage of steel castings is ease of welding. They can be readily welded without significant loss of strength, to repair the casting, or to fabricate structures with other steel components.

Nonferrous Casting Alloys Nonferrous casting metals include alloys of aluminum, magnesium, copper, tin, zinc, nickel, and titanium (Section 6.3). *Aluminum alloys* are generally considered to be very castable. The melting point of pure aluminum is 660°C (1112°F), so pouring temperatures for aluminum casting alloys are low compared to cast iron and steel. Their properties make them attractive for castings: light weight, wide range of strength properties attainable through heat treatment, and ease of machining. *Magnesium alloys* are the lightest of all casting metals. Other properties include corrosion resistance, as well as high strength-to-weight and stiffness-to-weight ratios.

Copper alloys include bronze, brass, and aluminum bronze. Properties that make them attractive include corrosion resistance, attractive appearance, and good bearing qualities. The high cost of copper is a limitation on the use of its alloys. Applications include pipe fittings, marine propeller blades, pump components, and ornamental jewelry.

Tin has the lowest melting point of the casting metals. *Tin-based alloys* are generally easy to cast. They have good corrosion resistant but poor mechanical strength, which limits their applications to pewter mugs and similar products not requiring high strength. *Zinc alloys* are commonly used in die casting. Zinc has a low melting point and good fluidity, making it highly castable. Its major weakness is low creep strength, so its castings cannot be subjected to prolonged high stresses.

Nickel alloys have good hot strength and corrosion resistance, which make them suited to high-temperature applications such as jet engine and rocket components, heat shields, and similar components. Nickel alloys also have a high melting point and are not easy to cast. *Titanium alloys* for casting are corrosion resistant and possess high strength-to-weight ratios. However, titanium has a high melting point, low fluidity, and a propensity to oxidize at high temperatures. These properties make it and its alloys difficult to cast.

11.7 PRODUCT DESIGN CONSIDERATIONS

If casting is selected by the product designer as the primary manufacturing process for a particular component, then certain guidelines should be followed to facilitate production of the part and avoid many of the defects enumerated in Section 11.5. Some of the important guidelines and considerations for casting are presented here.

- ➤ *Geometric simplicity.* Although casting is a process that can be used to produce complex part geometries, simplifying the part design will improve its castability. Avoiding unnecessary complexities simplifies mold making, reduces the need for cores, and improves the strength of the casting.

- ➤ *Corners.* Sharp corners and angles should be avoided, because they are sources of stress concentrations and may cause hot tearing and cracks in the casting. Generous fillets should be designed on inside corners, and sharp edges should be blended.

- ➤ *Section thicknesses.* Section thicknesses should be uniform in order to avoid shrinkage cavities. Thicker sections create *hot spots* in the casting, because greater volume requires more time for solidification and cooling. These are likely locations of shrinkage cavities. Figure 11.24 illustrates the problem and offers some possible solutions.

- ➤ *Draft.* Part sections that project into the mold should have a draft or taper, as defined in Figure 11.25. In expendable-mold casting, the purpose of this draft is to facilitate removal of the pattern from the mold. In permanent-mold casting, its purpose is to aid in removal of the part from the mold. Similar tapers should be allowed if solid cores are used in the casting process. The required draft need only be about 1° for sand casting and 2° to 3° for permanent-mold processes.

- ➤ *Use of cores.* Minor changes in part design can reduce the need for coring, as shown in Figure 11.25.

- ➤ *Dimensional tolerances.* There are significant differences in the dimensional accuracies that can be achieved in castings, depending on which process is used. Table 11.2 provides a compilation of typical part tolerances for various casting processes and metals.

- ➤ *Surface finish.* Typical surface roughness achieved in sand casting is around 6 μm (250 μ-in). Similarly poor finishes are obtained in shell molding, while plaster-mold and investment casting produce much better roughness values: 0.75 μm (30 μ-in).

FIGURE 11.24 (a) Thick section at intersection can result in a shrinkage cavity. Remedies include (b) redesign to reduce thickness and (c) use of a core.

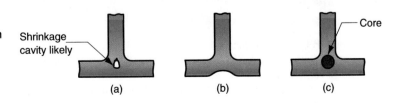

FIGURE 11.25 Design change to eliminate the need for using a core: (a) original design and (b) redesign.

TABLE 11.2 Typical dimensional tolerances for various casting processes and metals.

Casting Process	Part Size	Tolerance mm	Tolerance in	Casting Process	Part Size	Tolerance mm	Tolerance in
Sand casting				Permanent mold			
Aluminum[a]	Small	±0.5	±0.020	Aluminum[a]	Small	±0.25	±0.010
Cast iron	Small	±1.0	±0.040	Cast iron	Small	±0.8	±0.030
	Large	±1.5	±0.060	Copper alloys	Small	±0.4	±0.015
Copper alloys	Small	±0.4	±0.015	Steel	Small	±0.5	±0.020
Steel	Small	±1.3	±0.050	Die casting			
	Large	±2.0	±0.080	Aluminum[a]	Small	±0.12	±0.005
Shell molding				Copper alloys	Small	±0.12	±0.005
Aluminum[a]	Small	±0.25	±0.010	Investment			
Cast iron	Small	±0.5	±0.020	Aluminum[a]	Small	±0.12	± 0.005
Copper alloys	Small	±0.4	±0.015	Cast iron	Small	±0.25	±0.010
Steel	Small	±0.8	±0.030	Copper alloys	Small	±0.12	±0.005
Plaster mold	Small	±0.12	±0.005	Steel	Small	±0.25	±0.010
	Large	±0.4	±0.015				

Compiled from [7], [15], and other sources.
[a]Values for aluminum also apply to magnesium.

Among the permanent-mold processes, die casting is noted for good surface finishes at around 1 μm (40 μ-in).

➤ *Machining allowances.* Tolerances achievable in many casting processes are insufficient to meet functional needs in many applications. Sand casting is the most prominent example of this deficiency. In these cases, portions of the casting must be machined to the required dimensions. Almost all sand castings must be machined to some extent in order for the part to be made functional. Therefore, additional material, called the *machining allowance,* is left on the casting for machining those surfaces where necessary. Typical machining allowances for sand castings range between 1.5 mm and 3 mm (0.06 in and 0.12 in).

REFERENCES

[1] Amstead, B. H., Ostwald, P. F., and Begeman, M. L. *Manufacturing Processes.* John Wiley & Sons, Inc., New York, 1987.

[2] Beeley, P. R. *Foundry Technology.* Newnes-Butterworths, London, 1972.

[3] Black, J, and Kohser, R. *DeGarmo's Materials and Processes in Manufacturing,* 10th ed. John Wiley & Sons, Inc., Hoboken, New Jersey, 2008.

[4] Datsko, J. *Material Properties and Manufacturing Processes.* John Wiley & Sons, Inc., New York, 1966.

[5] Decker, R. F., D. M. Walukas, S. E. LeBeau, R. E. Vining, and N. D. Prewitt."Advances in Semi-Solid Molding," *Advanced Materials & Processes,* April 2004, pp 41–42.

[6] Flinn, R. A. *Fundamentals of Metal Casting.* American Foundrymen's Society, Inc., Des Plaines, Illinois, 1987.

[7] Heine, R. W., Loper, Jr., C. R., and Rosenthal, C. *Principles of Metal Casting,* 2nd ed. McGraw-Hill Book Co., New York, 1967.

[8] Kotzin, E. L. *Metalcasting & Molding Processes.* American Foundrymen's Society, Inc., Des Plaines, Illinois, 1981.

[9] *Metals Handbook,* Vol. 15, *Casting.* ASM International, Materials Park, Ohio, 2008.

[10] Mikelonis, P. J. (ed.). *Foundry Technology.* American Society for Metals, Metals Park, Ohio, 1982.

[11] Mueller, B. "Investment Casting Trends," *Advanced Materials & Processes,* March 2005, pp. 30–32.

[12] Niebel, B. W., Draper, A. B., Wysk, R. A. *Modern Manufacturing Process Engineering.* McGraw-Hill Book Co., New York, 1989.

[13] Perry, M. C. "Investment Casting," *Advanced Materials & Processes,* June 2008, pp. 31–33.

[14] Simpson, B. L. *History of the Metalcasting Industry.* American Foundrymen's Society, Inc., Des Plaines, Illinois, 1997.

[15] Wick, C., Benedict, J. T., and Veilleux, R. F. *Tool and Manufacturing Engineers Handbook,* 4th ed. Vol. II, Forming, Ch. 16. Society of Manufacturing Engineers, Dearborn, Michigan, 1984.

[16] Wikipedia. "Semi-solid metal casting." Available at: http://en.wikipedia.org/wiki/Semi-solid_metal_casting.

REVIEW QUESTIONS

11.1. Name the two basic categories of casting processes.

11.2. There are various types of patterns used in sand casting. What is the difference between a split pattern and a match-plate pattern?

11.3. What is a chaplet?

11.4. What properties determine the quality of a sand mold for sand casting?

11.5. What is the Antioch process?

11.6. What is the difference between vacuum permanent-mold casting and vacuum molding?

11.7. What are the most common metals used in die casting?

11.8. Which die casting machines usually have a higher production rate, cold-chamber or hot-chamber, and why?

11.9. What is flash in die casting?

11.10. What is the difference between true centrifugal casting and semicentrifugal casting?

11.11. What is a cupola?

11.12. What are some of the operations required in sand casting after the casting is removed from the mold?

11.13. What are some of the general defects encountered in casting processes? Name and briefly describe three.

11.14. (**Video**) What is the composition of green sand in the green-sand molding process?

11.15. (**Video**) What are the advantages and disadvantages of sand casting over investment casting?

11.16. (**Video**) Explain the difference between horizontal and vertical die casting machines. Which is more popular?

11.17. (**Video**) Why are aluminum and copper alloys unsuitable for use in hot-chamber die casting?

11.18. (**Video**) According to the die casting video, what materials are most common for die casting dies?

MULTIPLE CHOICE QUIZ

There are 27 correct answers in the following multiple choice questions (some questions have multiple answers that are correct). To attain a perfect score on the quiz, all correct answers must be given. Each correct answer is worth 1 point. Each omitted answer or wrong answer reduces the score by 1 point, and each additional answer beyond the correct number of answers reduces the score by 1 point. Percentage score on the quiz is based on the total number of correct answers.

11.1. Which one of the following casting processes is the most widely used: (a) centrifugal casting, (b) die casting, (c) investment casting, (d) sand casting, or (e) shell casting?

11.2. In sand casting, the volumetric size of the pattern is (a) bigger than, (b) same size as, or (c) smaller than the cast part?

11.3. Silica sand has which one of the following compositions: (a) Al_2O_3, (b) SiO, (c) SiO_2, or (d) $SiSO_4$?

11.4. For which one of the following reasons is a green mold named: (a) green is the color of the mold, (b) moisture is contained in the mold, (c) mold is cured, or (d) mold is dry?

11.5. Given that W_m = weight of the molten metal displaced by a core and W_c = weight of the core, the buoyancy force is which one of the following: (a) downward force = $W_m + W_c$, (b) downward force = $W_m - W_c$, (c) upward force = $W_m + W_c$, or (d) upward force = $W_m - W_c$?

11.6. Which of the following casting processes are expendable-mold operations (four correct answers): (a) centrifugal casting, (b) die casting, (c) investment casting, (d) low pressure casting, (e) sand casting, (f) shell molding, (g) slush casting, and (h) vacuum molding?

11.7. Shell molding is best described by which one of the following: (a) casting operation in which the molten metal has been poured out after a thin shell has been solidified in the mold, (b) casting process in which the mold is a thin shell of sand bonded by a thermosetting resin, (c) sand casting operation in which the pattern is a shell rather than a solid form, or (d) casting operation used to make artificial sea shells?

11.8. Investment casting is also known by which one of the following names: (a) fast-payback molding, (b) full-mold process, (c) lost-foam process, (d) lost-pattern process, or (e) lost-wax process?

11.9. In plaster-mold casting, the mold is made of which one of the following materials: (a) Al_2O_3, (b) $CaSO_4$-H_2O, (c) SiC, or (d) SiO_2?

11.10. Which of the following qualifies as a precision-casting process (two correct answers): (a) ingot casting, (b) investment casting, (c) plaster-mold casting, (d) sand casting, and (e) shell molding?

11.11. Which of the following casting processes are permanent-mold operations (three correct answers): (a) centrifugal casting, (b) die casting, (c) expanded polystyrene process, (d) sand casting, (e) shell molding, (f) slush casting, and (g) vacuum molding.

11.12. Which of the following metals would typically be used in die casting (three best answers): (a) aluminum, (b) cast iron, (c) steel, (d) tin, (e) tungsten, and (f) zinc?

11.13. Which of the following are advantages of die casting over sand casting (four best answers): (a) better surface finish, (b) closer tolerances, (c) higher melting temperature metals, (d) higher production rates, (e) larger parts can be cast, and (f) mold can be reused?

11.14. Cupolas are furnaces used to melt which of the following metals (one best answer): (a) aluminum, (b) cast iron, (c) steel, or (d) zinc?

11.15. A misrun is which one of the following defects in casting: (a) globules of metal becoming entrapped in the casting, (b) metal is not properly poured into the downsprue, (c) metal solidifies before filling the cavity, (d) microporosity, and (e) "pipe" formation?

11.16. Which one of the following casting metals is most important commercially: (a) aluminum and its alloys, (b) bronze, (c) cast iron, (d) cast steel, or (e) zinc alloys?

PROBLEMS

Buoyancy Force

11.1. An 92% aluminum-8% copper alloy casting is made in a sand mold using a sand core that weighs 20 kg. Determine the buoyancy force in Newtons tending to lift the core during pouring.

11.2. A sand core located inside a mold cavity has a volume of 157.0 in^3. It is used in the casting of a cast iron pump housing. Determine the buoyancy force that will tend to lift the core during pouring.

11.3. Caplets are used to support a sand core inside a sand mold cavity. The design of the caplets and the manner in which they are placed in the mold cavity surface allows each caplet to sustain a force of 10 lb. Several caplets are located beneath the core to support it before pouring; and several other caplets are placed above the core to resist the buoyancy force during pouring. If the volume of the core = 325 in^3, and the metal poured is brass, determine the minimum number of caplets that should be placed (a) beneath the core, and (b) above the core.

11.4. A sand core used to form the internal surfaces of a steel casting experiences a buoyancy force of 23 kg. The volume of the mold cavity forming the outside surface of the casting = 5000 cm^3. What is the weight of the final casting? Ignore considerations of shrinkage.

Centrifugal Casting

11.5. A horizontal true centrifugal casting operation will be used to make copper tubing. The lengths will be 1.5 m with outside diameter = 15.0 cm, and inside diameter = 12.5 cm. If the rotational speed of the pipe = 1000 rev/min, determine the G-factor.

11.6. A true centrifugal casting operation is to be performed in a horizontal configuration to make cast iron pipe sections. The sections will have a length = 42.0 in, outside diameter = 8.0 in, and wall thickness = 0.50 in. If the rotational speed of the pipe = 500 rev/min, determine the G-factor. Is the operation likely to be successful?

11.7. A horizontal true centrifugal casting process is used to make brass bushings with the following dimensions: length = 10 cm, outside diameter = 15 cm, and inside diameter = 12 cm. (a) Determine the required rotational speed in order to obtain a G-factor of 70. (b) When operating at this speed, what is the centrifugal force per square meter (Pa) imposed by the molten metal on the inside wall of the mold?

11.8. True centrifugal casting is performed horizontally to make large diameter copper tube sections. The tubes have a length = 1.0 m, diameter = 0.25 m, and wall thickness = 15 mm. (a) If the rotational speed of the pipe = 700 rev/min, determine the G-factor on the molten metal. (b) Is the rotational speed sufficient to avoid "rain?" (c) What volume of molten metal must be poured into the mold to make the casting if solidification shrinkage and contraction after solidification are considered? Solidification shrinkage for copper = 4.5%, and solid thermal contraction = 7.5%.

11.9. If a true centrifugal casting operation were to be performed in a space station circling the Earth, how would weightlessness affect the process?

11.10. A horizontal true centrifugal casting process is used to make aluminum rings with the following dimensions: length = 5 cm, outside diameter = 65 cm, and inside diameter = 60 cm. (a) Determine the rotational speed that will provide a G-factor = 60. (b) Suppose that the ring were made out of steel instead of aluminum. If the rotational speed computed in part (a) were used in the steel casting operation, determine the G-factor and (c) centrifugal force per square meter (Pa) on the mold wall. (d) Would this rotational speed result in a successful operation?

11.11. For the steel ring of preceding Problem 11.10(b), determine the volume of molten metal that must be poured into the mold, given that the liquid shrinkage is 0.5%, solidification shrinkage = 3%, and solid contraction after freezing = 7.2%.

11.12. A horizontal, true centrifugal casting process is used to make lead pipe for chemical plants. The pipe has length = 0.5 m, outside diameter = 70 mm, and wall thickness = 6.0 mm. Determine the rotational speed that will provide a G-factor = 60.

11.13. A vertical, true centrifugal casting process is used to make tube sections with length = 10.0 in and outside diameter = 6.0 in. The inside diameter of the tube = 5.5 in at the top and 5.0 in at the bottom. At what speed must the tube be rotated during the operation in order to achieve these specifications?

11.14. A vertical, true centrifugal casting process is used to produce bushings that are 200 mm long and 200 mm in outside diameter. If the rotational speed during solidification is 500 rev/min, determine the inside diameter at the top of the bushing if the inside diameter at the bottom is 150 mm.

11.15. A vertical, true centrifugal casting process is used to cast brass tubing that is 15.0 in long and whose outside diameter = 8.0 in. If the speed of rotation during solidification is 1000 rev/min, determine the inside diameters at the top and bottom of the tubing if the total weight of the final casting = 75.0 lbs.

Defects and Design Considerations

11.16. The housing for a certain machinery product is made of two components, both aluminum castings. The larger component has the shape of a dish sink, and the second component is a flat cover that is attached to the first component to create an enclosed space for the machinery parts. Sand casting is used to produce the two castings, both of which are plagued by defects in the form of misruns and cold shuts. The foreman complains that the parts are too thin, and that is the reason for the defects. However, it is known that the same components are cast successfully in other foundries. What other explanation can be given for the defects?

11.17. A large, steel sand casting shows the characteristic signs of penetration defect: a surface consisting of a mixture of sand and metal. (a) What steps can be taken to correct the defect? (b) What other possible defects might result from taking each of these steps?

GLASSWORKING

Chapter Contents

Glass products are commercially manufactured in an almost unlimited variety of shapes. Many are produced in very large quantities, such as light bulbs, beverage bottles, and window glass. Others, such as giant telescope lenses, are made individually.

Glass is one of three basic types of ceramics (Chapter 7). It is distinguished by its noncrystalline (vitreous) structure, whereas the other ceramic materials have a crystalline structure. The methods by which glass is shaped into useful products are quite different from those used for the other types. In glassworking, the principal starting material is silica (SiO_2); this is usually combined with other oxide ceramics, which form glasses. The starting material is heated to transform it from a hard solid into a viscous liquid; it is then shaped into the desired geometry while in this highly plastic or fluid condition. When cooled and hard, the material remains in the glassy state rather than crystallizing.

The typical manufacturing sequence in glassworking consists of the steps pictured in Figure 12.1. Shaping is accomplished by various processes, including casting, pressing-and-blowing (to produce bottles and other containers), and rolling (to make plate glass). A finishing step is required for certain products.

12.1 RAW MATERIALS PREPARATION AND MELTING

The main component in nearly all glasses is silica, the primary source of which is natural quartz in sand. The sand must be washed and classified. Washing removes impurities such as clay and certain minerals that would cause undesirable coloring of the glass. *Classifying* the sand means grouping the grains according to size. The most desirable particle size for glassmaking is in the range of 0.1 to 0.6 mm (0.004 to

FIGURE 12.1 The typical process sequence in glassworking: (1) preparation of raw materials and melting, (2) shaping, and (3) heat treatment.

0.025 in) [3]. The various other components, such as soda ash (source of Na_2O), limestone (source of CaO), aluminum oxide, potash (source of K_2O), and other minerals are added in the proper proportions to achieve the desired composition. The mixing is usually done in batches, in amounts that are compatible with the capacities of available melting furnaces.

Recycled glass is usually added to the mixture in modern practice. In addition to preserving the environment, recycled glass facilitates melting. Depending on the amount of waste glass available and the specifications of the final composition, the proportion of recycled glass may be up to 100%.

The batch of starting materials to be melted is referred to as a **charge**, and the procedure of loading it into the melting furnace is called **charging** the furnace. Glass-melting furnaces can be divided into the following types [3]: (1) **pot furnaces**—ceramic pots of limited capacity in which melting occurs by heating the walls of the pot; (2) **day tanks**—larger capacity vessels for batch production in which heating is done by burning fuels above the charge; (3) **continuous tank furnaces**—long tank furnaces in which raw materials are fed in one end, and melted as they move to the other end where molten glass is drawn out for high production; and (4) **electric furnaces** of various designs for a wide range of production rates.

Glass melting is generally carried out at temperatures around 1500°C to 1600°C (2700°F to 2900°F). The melting cycle for a typical charge takes 24 to 48 hours. This is the time required for all of the sand grains to become a clear liquid and the molten glass to be refined and cooled to the appropriate temperature for working. Molten glass is a viscous liquid, the viscosity being inversely related to temperature. Because the shaping operation immediately follows the melting cycle, the temperature at which the glass is tapped from the furnace depends on the viscosity required for the subsequent process.

12.2   SHAPING PROCESSES IN GLASSWORKING

The major categories of glass products were identified in Section 7.4.2 as window glass, containers, light bulbs, laboratory glassware, glass fibers, and optical glass. Despite the variety represented by this list, the shaping processes to fabricate these products can be grouped into only three categories: (1) discrete processes for piece ware, which includes bottles, light bulbs, and other individual items; (2) continuous processes for making flat glass (sheet and plate glass for windows) and tubing (for laboratory ware and fluorescent lights); and (3) fiber-making processes to produce fibers for insulation, fiberglass composite materials, and fiber optics.

12.2.1   SHAPING OF PIECE WARE

The ancient methods of hand-working glass, such as glass blowing, were briefly described in Historical Note 7.3. Handicraft methods are still employed today for making glassware items of high value in small quantities. Most of the processes discussed in this section are

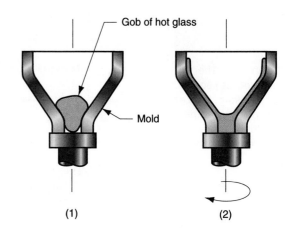

FIGURE 12.2 Spinning of funnel-shaped glass parts: (1) gob of glass dropped into mold; and (2) rotation of mold to cause spreading of molten glass on mold surface.

highly mechanized technologies for producing discrete pieces such as jars, bottles, and light bulbs in high quantities.

Spinning Glass spinning is similar to *centrifugal casting* of metals, and is also known by that name in glassworking. It is used to produce funnel-shaped components such as the back sections of cathode ray tubes for televisions and computer monitors. The setup is pictured in Figure 12.2. A gob of molten glass is dropped into a conical mold made of steel. The mold is rotated so that centrifugal force causes the glass to flow upward and spread itself on the mold surface. The faceplate (i.e., the front viewing screen) is later assembled to the funnel using a sealing glass of low melting point.

Pressing This is a widely used process for mass producing glass pieces such as dishes, bake ware, headlight lenses, TV tube faceplates, and similar items that are relatively flat. The process is illustrated and described in Figure 12.3. The large quantities of most pressed products justify a high level of automation in this production sequence.

Blowing Several shaping sequences include blowing as one or more of the steps. Instead of a manual operation, blowing is performed on highly automated equipment. The two sequences we describe here are the press-and-blow and blow-and-blow methods.

As the name indicates, the *press-and-blow* method is a pressing operation followed by a blowing operation, as portrayed in Figure 12.4. The process is suited to the production of wide-mouth containers. A split mold is used in the blowing operation for part removal.

The *blow-and-blow* method is used to produce smaller-mouthed bottles. The sequence is similar to the preceding, except that two (or more) blowing operations

FIGURE 12.3 Pressing of a flat glass piece: (1) a gob of glass fed into mold from the furnace; (2) pressing into shape by plunger; and (3) plunger is retracted and the finished product is removed. Symbols v and F indicate motion (v = velocity) and applied force, respectively.

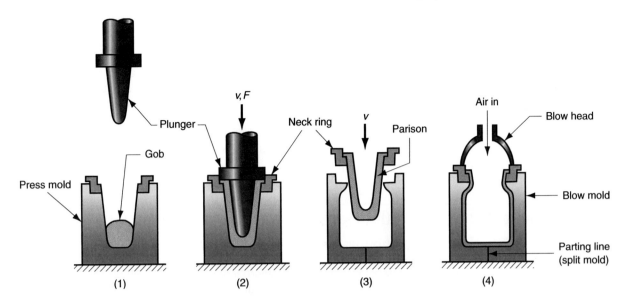

FIGURE 12.4 Press-and-blow forming sequence: (1) molten gob is fed into mold cavity; (2) pressing to form a *parison*; (3) the partially formed parison, held in a neck ring, is transferred to the blow mold; and (4) blown into final shape. Symbols *v* and *F* indicate motion (*v* = velocity) and applied force, respectively.

are used rather than pressing and blowing. There are variations to the process, depending on the geometry of the product, with one possible sequence shown in Figure 12.5. Reheating is sometimes required between blowing steps. Duplicate and triplicate molds are sometimes used along with matching gob feeders to increase production rates. Press-and-blow and blow-and-blow methods are used to make jars, beverage bottles, incandescent light bulb enclosures, and similar geometries.

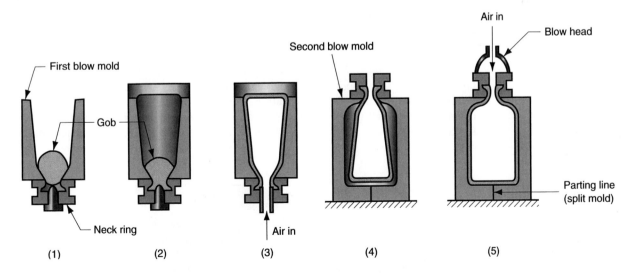

FIGURE 12.5 Blow-and-blow forming sequence: (1) gob is fed into inverted mold cavity; (2) mold is covered; (3) first blowing step; (4) partially formed piece is reoriented and transferred to second blow mold; and (5) blown to final shape.

Casting If the molten glass is sufficiently fluid, it can be poured into a mold. Relatively massive objects, such as astronomical lenses and mirrors, are made by this method. These pieces must be cooled very slowly to avoid internal stresses and possible cracking owing to temperature gradients that would otherwise be set up in the glass. After cooling and solidifying, the piece must be finished by lapping and polishing. Casting is not much used in glassworking except for these kinds of special jobs. Not only is cooling and cracking a problem, but also molten glass is relatively viscous at normal working temperatures, and does not flow through small orifices or into small sections as well as molten metals or heated thermoplastics. Smaller lenses are usually made by pressing, discussed in the preceding.

12.2.2 SHAPING OF FLAT AND TUBULAR GLASS

Here we describe two methods for making plate glass and one method for producing tube stock. They are continuous processes, in which long sections of flat window glass or glass tubing are made and later cut into appropriate sizes and lengths. They are modern technologies in contrast to the ancient method described in Historical Note 12.1.

Historical Note 12.1   *Ancient methods of making flat glass [7]*

Glass windows have been used in buildings for many centuries. The oldest process for making flat window glass was by manual glass blowing. The procedure consisted of the following: (1) a glass globe was blown on a blowpipe; (2) a portion of the globe was made to stick to the end of a "punty," a metal rod used by glassblowers, and then detached from the blowpipe; and (3) after reheating the glass, the punty was rotated with sufficient speed for centrifugal force to shape the open globe into a flat disk. The disk, whose maximum possible size was only about 1 m (3 ft), was later cut into small panes for windows.

At the center of the disk, where the glass was attached to the punty during the third step in the process, a lump would tend to form that had the appearance of a crown. The name "crown glass" was derived from this resemblance. Lenses for spectacles were ground from glass made by this method. Today, the name crown glass is still used for certain types of optical and ophthalmic glass, even though the ancient method has been replaced by modern production technology.

Rolling of Flat Plate Flat plate glass can be produced by rolling, as illustrated in Figure 12.6. The starting glass, in a suitably plastic condition from the furnace, is squeezed through opposing rolls whose separation determines the thickness of the sheet. The rolling operation is usually set up so that the flat glass is moved directly into an annealing furnace. The rolled glass sheet must later be ground and polished for parallelism and smoothness.

FIGURE 12.6 Rolling of flat glass.

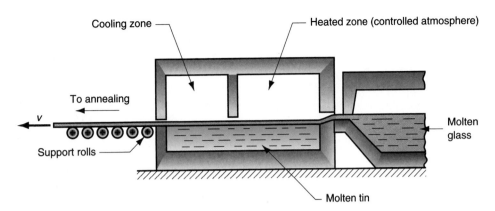

FIGURE 12.7 The float process for producing sheet glass.

Float Process This process was developed in the late 1950s. Its advantage over other methods such as rolling is that it obtains smooth surfaces that need no subsequent finishing. In the *float process*, illustrated in Figure 12.7, the glass flows directly from its melting furnace onto the surface of a molten tin bath. The highly fluid glass spreads evenly across the molten tin surface, achieving a uniform thickness and smoothness. After moving into a cooler region of the bath, the glass hardens and travels through an annealing furnace, after which it is cut to size.

Drawing of Glass Tubes Glass tubing is manufactured by a drawing process known as the *Danner process*, illustrated in Figure 12.8. Molten glass flows around a rotating hollow mandrel through which air is blown while the glass is being drawn. The air temperature and its volumetric flow rate, as well as the drawing velocity, determine the diameter and wall thickness of the tubular cross section. During hardening, the glass tube is supported by a series of rollers extending about 30 m (100 ft) beyond the mandrel. The continuous tubing is then cut into standard lengths. Tubular glass products include laboratory glassware, fluorescent light tubes, and thermometers.

12.2.3 FORMING OF GLASS FIBERS

Glass fibers are used in applications ranging from insulation wool to fiber optics communications lines (Section 7.4.2). Glass fiber products can be divided into two categories [6]: (1) fibrous glass for thermal insulation, acoustical insulation, and air filtration, in which the fibers are in a random, wool-like condition; and (2) long, continuous filaments suitable for fiber-reinforced plastics, yarns and fabrics, and fiber optics. Different production methods are used for the two categories; we describe two methods in the following, representing each of the product categories, respectively.

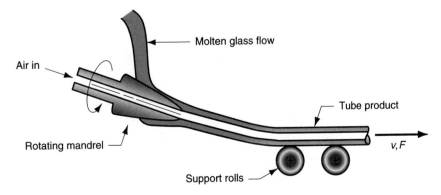

FIGURE 12.8 Drawing of glass tubes by the Danner process. Symbols *v* and *F* indicate motion (*v* = velocity) and applied force, respectively.

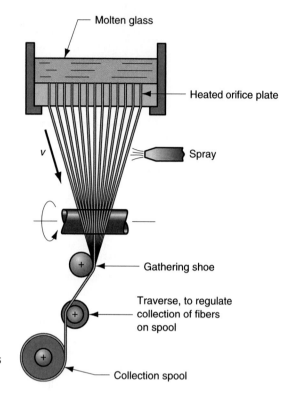

FIGURE 12.9   Drawing of continuous glass fibers.

Centrifugal Spraying   In a typical process for making glass wool, molten glass flows into a rotating bowl with many small orifices around its periphery. Centrifugal force causes the glass to flow through the holes to become a fibrous mass suitable for thermal and acoustical insulation.

Drawing of Continuous Filaments   In this process, illustrated in Figure 12.9, continuous glass fibers of small diameter (the lower size limit is around 0.0025 mm [0.0001 in]) are produced by drawing strands of molten glass through small orifices in a heated plate made of a platinum alloy. The plate may have several hundred holes, each making one fiber. The individual fibers are collected into a strand by reeling them onto a spool. Before spooling, the fibers are coated with various chemicals to lubricate and protect them. Drawing speeds of around 50 m/s (10,000 ft/min) or more are not unusual.

12.3   HEAT TREATMENT AND FINISHING

Heat treatment of the glass product is the third step in the glassworking sequence. For some products, additional finishing operations are performed.

12.3.1   HEAT TREATMENT

We discussed glass-ceramics in Section 7.4.3 This unique material is made by a special heat treatment that transforms most of the vitreous state into a polycrystalline ceramic. Other heat treatments performed on glass cause changes that are less dramatic technologically but perhaps more important commercially; examples include annealing and tempering.

Annealing   Glass products usually have undesirable internal stresses after forming, which reduce their strength. Annealing is done to relieve these stresses; the treatment therefore has the same function in glassworking as it does in metalworking. ***Annealing*** involves heating the glass to an elevated temperature and holding it for a certain period to eliminate stresses and temperature gradients, then slowly cooling the glass to suppress stress formation, followed by more rapid cooling to room temperature. Common annealing temperatures are around 500°C (900°F). The length of time the product is held at the temperature, as well as the heating and cooling rates during the cycle, depend on thickness of the glass, the usual rule being that the required annealing time varies with the square of thickness.

Annealing in modern glass factories is performed in tunnel-like furnaces, called ***lehrs***, in which the products flow slowly through the hot chamber on conveyors. Burners are located only at the front end of the chamber, so that the glass experiences the required heating and cooling cycle.

Tempered Glass and Related Products   A beneficial internal stress pattern can be developed in glass products by a heat treatment known as ***tempering***, and the resulting material is called ***tempered glass***. As in the treatment of hardened steel, tempering increases the toughness of glass. The process involves heating the glass to a temperature somewhat above its annealing temperature and into the plastic range, followed by quenching of the surfaces, usually with air jets. When the surfaces cool, they contract and harden while the interior is still plastic and compliant. As the internal glass slowly cools, it contracts, thus putting the hard surfaces in compression. Like other ceramics, glass is much stronger when subjected to compressive stresses than tensile stresses. Accordingly, tempered glass is much more resistant to scratching and breaking because of the compressive stresses on its surfaces. Applications include windows for tall buildings, all-glass doors, safety glasses, and other products requiring toughened glass.

When tempered glass fails, it does so by shattering into numerous small fragments that are less likely to cut someone than conventional (annealed) window glass. Interestingly, automobile windshields are not made of tempered glass, because of the danger posed to the driver by this fragmentation. Instead, conventional glass is used; however, it is fabricated by sandwiching two pieces of glass on either side of a tough polymer sheet. Should this ***laminated glass*** fracture, the glass splinters are retained by the polymer sheet and the windshield remains relatively transparent.

12.3.2 FINISHING

Finishing operations are sometimes required for glassware products. These secondary operations include grinding, polishing, and cutting. When glass sheets are produced by drawing and rolling, the opposite sides are not necessarily parallel, and the surfaces contain defects and scratch marks caused by the use of hard tooling on soft glass. The glass sheets must be ground and polished for most commercial applications. In pressing and blowing operations when split dies are used, polishing is often required to remove the seam marks from the container product.

In continuous glassworking processes, such as plate and tube production, the continuous sections must be cut into smaller pieces. This is accomplished by first scoring the glass with a glass-cutting wheel or cutting diamond and then breaking the section along the score line. Cutting is generally done as the glass exits the annealing lehr.

Decorative and surface processes are performed on certain glassware products. These processes include mechanical cutting and polishing operations; sandblasting; chemical etching (with hydrofluoric acid, often in combination with other chemicals); and coating (for example, coating of plate glass with aluminum or silver to produce mirrors).

12.4 PRODUCT DESIGN CONSIDERATIONS

Glass possesses special properties that make it desirable in certain applications. The following design recommendations are compiled from Bralla [1] and other sources.

➢ Glass is transparent and has certain optical properties that are unusual if not unique among engineering materials. For applications requiring transparency, light transmittance, magnification, and similar optical properties, glass is likely to be the material of choice. Certain polymers are transparent and may be competitive, depending on design requirements.

➢ Glass is several times stronger in compression than in tension; components should be designed so that they are subjected to compressive stresses, not tensile stresses.

➢ Ceramics, including glass, are brittle. Glass parts should not be used in applications that involve impact loading or high stresses, which might cause fracture.

➢ Certain glass compositions have very low thermal expansion coefficients and are therefore tolerant of thermal shock. These glasses can be selected for applications in which this characteristic is important.

➢ Outside edges and corners on glass parts should have large radii or chamfers; likewise, inside corners should have large radii. Both outside and inside corners are potential points of stress concentration.

➢ Unlike parts made of traditional and new ceramics, threads may be included in the design of glass parts; they are technically feasible with the press-and-blow shaping processes. However, the threads should be coarse.

REFERENCES

[1] Bralla, J. G. (editor). *Design for Manufacturability Handbook.* 2nd ed. McGraw-Hill, New York, 1998.

[2] Flinn, R. A., and Trojan, P. K. *Engineering Materials and Their Applications*, 5th ed. John Wiley & Sons, New York, 1995.

[3] Hlavac, J. *The Technology of Glass and Ceramics.* Elsevier Scientific Publishing, New York, 1983.

[4] McColm, I. J. *Ceramic Science for Materials Technologists.* Chapman and Hall, New York, 1983.

[5] McLellan, G., and Shand, E. B. *Glass Engineering Handbook.* 3rd ed. McGraw-Hill, New York, 1984.

[6] Mohr, J. G., and Rowe, W. P. *Fiber Glass.* Krieger, New York, 1990.

[7] Scholes, S. R., and Greene, C. H. *Modern Glass Practice.* 7th ed. TechBooks, Marietta, Georgia, 1993.

REVIEW QUESTIONS

12.1. Glass is classified as a ceramic material; yet glass is different from the traditional and new ceramics. What is the difference?

12.2. What is the predominant chemical compound in almost all glass products?

12.3. What are the three basic steps in the glassworking sequence?

12.4. Melting furnaces for glassworking can be divided into four types. Name three of the four types.

12.5. Describe the spinning process in glassworking.

12.6. What is the main difference between the press-and-blow and the blow-and-blow shaping processes in glassworking?

12.7. There are several ways of shaping plate or sheet glass. Name and briefly describe one of them.

12.8. Describe the Danner process.

12.9. Two processes for forming glass fibers are discussed in the text. Name and briefly describe one of them.

12.10. What is the purpose of annealing in glassworking?

12.11. Describe how a piece of glass is heat treated to produce tempered glass.

12.12. Describe the type of material that is commonly used to make windshields for automobiles.

12.13. What are some of the design recommendations for glass parts?

MULTIPLE CHOICE QUIZ

There are 10 correct answers in the following multiple choice questions (some questions have multiple answers that are correct). To attain a perfect score on the quiz, all correct answers must be given. Each correct answer is worth 1 point. Each omitted answer or wrong answer reduces the score by 1 point, and each additional answer beyond the correct number of answers reduces the score by 1 point. Percentage score on the quiz is based on the total number of correct answers.

12.1. Which one of the following terms refers to the glassy state of a material: (a) crystalline, (b) devitrified, (c) polycrystalline, (d) vitiated, or (e) vitreous?

12.2. Besides helping to preserve the environment, the use of recycled glass as an ingredient of the starting material in glassmaking serves what other useful purpose (one answer): (a) adds coloring variations to the glass for aesthetic value, (b) makes the glass easier to melt, (c) makes the glass stronger, or (d) reduces odors in the plant?

12.3. The charge in glassworking is which one of the following: (a) the duration of the melting cycle, (b) the electric energy required to melt the glass, (c) the name given to the melting furnace, or (d) the starting materials in melting?

12.4. Typical glass melting temperatures are in which of the following ranges: (a) 400°C to 500°C, (b) 900°C to 1000°C, (c) 1500°C to 1600°C, or (d) 2000°C to 2200°C?

12.5. Casting is a glassworking process used for (a) high production, (b) low production, or (c) medium production?

12.6. Which one of the following processes or processing steps is not applicable in glassworking: (a) annealing, (b) pressing, (c) quenching, (d) sintering, and (e) spinning?

12.7. The press-and-blow process is best suited to the production of (narrow-necked) beverage bottles, whereas the blow-and-blow process is more appropriate for producing (wide-mouthed) jars: (a) true, or (b) false?

12.8. Which one of the following processes is used to produce glass tubing: (a) Danner process, (b) pressing, (c) rolling, or (d) spinning?

12.9. If a glass part with a wall thickness of 5 mm (0.20 in) takes 10 minutes to anneal, how much time would a glass part of similar geometry but with a wall thickness of 7.5 mm (0.30 in) take to anneal (choose the one closest answer): (a) 10 minutes, (b) 15 minutes, (c) 20 minutes, or (d) 30 minutes?

12.10. A lehr is which of the following: (a) a lion's den, (b) a melting furnace, (c) a sintering furnace, (d) an annealing furnace, or (e) none of the above?

13 SHAPING PROCESSES FOR PLASTICS

Chapter Contents

Plastics can be shaped into a wide variety of products, such as molded parts, extruded sections, films and sheets, insulation coatings on electrical wires, and fibers for textiles. In addition, plastics are often the principal ingredient in other materials, such as paints and varnishes; adhesives; and various polymer matrix composites. In this chapter we consider the technologies by which these products are shaped, postponing paints and varnishes, adhesives, and composites until later chapters. Many plastic-shaping processes can be adapted to rubbers (Chapter 14) and polymer matrix composites (Chapter 15).

The commercial and technological importance of these shaping processes derives from the growing importance of the materials being processed. Applications of plastics have increased at a much faster rate than either metals or ceramics during the last 50 years. Indeed, many parts previously made of metals are today being made of plastics and plastic composites. The same is true of glass; plastic containers have been largely substituted for glass bottles and jars in product packaging. The total volume of polymers (plastics and rubbers) now exceeds that of metals. We can identify several reasons why the plastic-shaping processes are important:

➢ The variety of shaping processes, and the ease with which polymers can be processed, allows an almost unlimited variety of part geometries to be formed.

➢ Many plastic parts are formed by molding, which is a *net shape* process. Further shaping is generally not needed.

➢ Although heating is usually required to form plastics, *less energy* is required than for metals because the processing temperatures are much lower.

➢ Because lower temperatures are used in processing, handling of the product is simplified during production. Because many plastic processing methods are one-step operations (e.g., molding), the amount of product handling required is substantially reduced compared with metals.

➤ Finishing by painting or plating is not required (except in unusual circumstances) for plastics.

As discussed in Chapter 8, the two types of plastics are ***thermoplastics*** and ***thermosets.*** The difference is that thermosets undergo a curing process during heating and shaping, which causes a permanent chemical change (***cross-linking***) in their molecular structure. Once they have been cured, they cannot be melted through reheating. By contrast, thermoplastics do not cure, and their chemical structure remains basically unchanged upon reheating even though they transform from solid to fluid. Of the two types, thermoplastics are by far the more important type commercially, comprising more than 80% of the total plastics tonnage.

Plastic-shaping processes can be classified as follows according to the resulting product geometry: (1) continuous extruded products with constant cross section other than sheets, films, and filaments; (2) continuous sheets and films; (3) continuous filaments (fibers); (4) molded parts that are mostly solid; (5) hollow molded parts with relatively thin walls; (6) discrete parts made of formed sheets and films; (7) castings; and (8) foamed products. This chapter examines each of these categories. The most important processes commercially are those associated with thermoplastics; the two processes of greatest significance are extrusion and injection molding. A brief history of plastic-shaping processes is presented in Historical Note 13.1.

Coverage of the plastic-shaping processes begins by examining the properties of polymer melts, because nearly all of the thermoplastic shaping processes share the common step of heating the plastic so that it flows.

Historical Note 13.1   *Plastic shaping processes*

Equipment for shaping plastics evolved largely from rubber processing technology. Noteworthy among the early contributors was Edwin Chaffee, an American who developed a two-roll steam-heated mill for mixing additives into rubber around 1835 (Section 14.1.3). He was also responsible for a similar device called a calender, which consists of a series of heated rolls for coating rubber onto cloth (Section 13.3). Both machines are still used today for plastics as well as rubbers.

The first extruders, dating from around 1845 in the United Kingdom, were ram-driven machines for extruding rubber and coating rubber onto electrical wire. The trouble with ram-type extruders is that they operate in an intermittent fashion. An extruder that could operate continuously, especially for wire and cable coating, was highly desirable. Although several individuals worked with varying degrees of success on a screw-type extruder (Section 13.2.1), Mathew Gray in the United Kingdom is

credited with the invention; his patent is dated 1879. As thermoplastics were subsequently developed, these screw extruders, originally designed for rubber, were adapted. An extrusion machine specifically designed for thermoplastics was introduced in 1935.

Injection molding machines for plastics were adaptations of equipment designed for metal die casting (Historical Note 11.2). Around 1872, John Hyatt, an important figure in the development of plastics (Historical Note 8.1), patented a molding machine specifically for plastics. It was a plunger-type machine (Section 13.6.3). The injection molding machine in its modern form was introduced in 1921, with semiautomatic controls added in 1937. Ram-type machines were the standard in the plastic molding industry for many decades, until the superiority of the reciprocating screw machine, developed by William Willert in the United States in 1952, became obvious.

13.1 PROPERTIES OF POLYMER MELTS

To shape a thermoplastic polymer it must be heated so that it softens to the consistency of a liquid. In this form, it is called a ***polymer melt.*** Polymer melts exhibit several unique properties and characteristics, considered in this section.

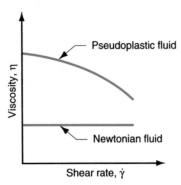

FIGURE 13.1 Viscosity relationships for Newtonian fluid and typical polymer melt.

Viscosity Because of its high molecular weight, a polymer melt is a thick fluid with high viscosity. As we defined the term in Section 3.4, *viscosity* is a fluid property that relates the shear stress experienced during flow of the fluid to the rate of shear. Viscosity is important in polymer processing because most of the shaping methods involve flow of the polymer melt through small channels or die openings. The flow rates are often large, thus leading to high rates of shear; and the shear stresses increase with shear rate, so that significant pressures are required to accomplish the processes.

Figure 13.1 shows viscosity as a function of shear rate for two types of fluids. For a **Newtonian fluid** (which includes most simple fluids such as water and oil), viscosity is a constant at a given temperature; it does not change with shear rate. The relationship between shear stress and shear strain is proportional, with viscosity as the constant of proportionality:

$$\tau = \eta \dot{\gamma} \quad \text{or} \quad \eta = \tau \dot{\gamma} \tag{13.1}$$

where τ = shear stress, Pa (lb/in^2); η = coefficient of shear viscosity, Ns/m^2, or Pa-s (lb-sec/in^2); and $\dot{\gamma}$ = shear rate, 1/s (1/sec).

However, for a polymer melt, viscosity decreases with shear rate, indicating that the fluid becomes thinner at higher rates of shear. This behavior is called **pseudoplasticity** and can be modeled to a reasonable approximation by the expression

$$\tau = k(\dot{\gamma})^n \tag{13.2}$$

where k = a constant corresponding to the viscosity coefficient and n = flow behavior index.

For $n = 1$, the equation reduces to the previous Eq. (13.1) for a Newtonian fluid, and k becomes η. For a polymer melt, values of n are less than 1.

In addition to the effect of shear rate (fluid flow rate), viscosity of a polymer melt is also affected by temperature. Like most fluids, the value decreases with increasing temperature. This is shown in Figure 13.2 for several common polymers at a shear rate of 10^3 s^{-1}, which is approximately the same as the rates encountered in injection molding and high speed extrusion. Thus we see that the viscosity of a polymer melt decreases with increasing values of shear rate and temperature. Equation (13.2) can be applied, except that k depends on temperature as shown in Figure 13.2.

Viscoelasticity Another property possessed by polymer melts is **viscoelasticity.** We discussed this property in the context of solid polymers in Section 3.5. However, liquid polymers exhibit it also. A good example is **die swell** in extrusion, in which the hot plastic expands when exiting the die opening. The phenomenon, illustrated in Figure 13.3, can be explained by noting that the polymer was contained in a much larger cross section before entering the narrow die channel. In effect, the extruded material "remembers" its former shape and attempts to return to it after leaving the die orifice. More technically, the compressive

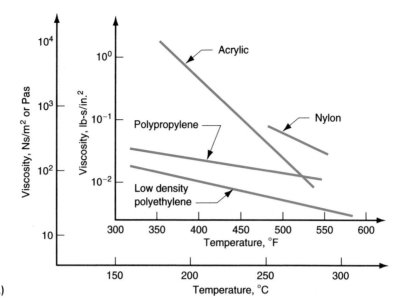

FIGURE 13.2 Viscosity as a function of temperatures for selected polymers at a shear rate of 10^3 s^{-1}. (Data compiled from [12].)

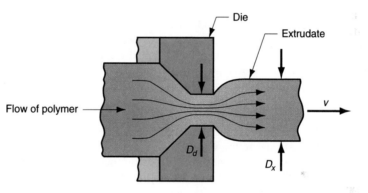

FIGURE 13.3 Die swell, a manifestation of viscoelasticity in polymer melts, as depicted here on exiting an extrusion die.

stresses acting on the material as it enters the small die opening do not relax immediately. When the material subsequently exits the orifice and the restriction is removed, the unrelaxed stresses cause the cross section to expand.

Die swell can be most easily measured for a circular cross section by means of the **swell ratio**, defined as

$$r_s = \frac{D_x}{D_d} \tag{13.3}$$

where r_s = swell ratio; D_x = diameter of the extruded cross section, mm (in); and D_d = diameter of the die orifice, mm (in).

The amount of die swell depends on the time the polymer melt spends in the die channel. Increasing the time in the channel, by means of a longer channel, reduces die swell.

13.2 EXTRUSION

Extrusion is one of the fundamental shaping processes, for metals and ceramics as well as polymers. **Extrusion** is a compression process in which material is forced to flow through a die orifice to provide long continuous product whose cross-sectional shape is determined by the shape of the orifice. As a polymer shaping process, it is widely used for thermoplastics and elastomers (but rarely for thermosets) to mass produce items such as tubing, pipes, hose, structural shapes (such as window and door molding), sheet and film, continuous filaments, and coated electrical wire and cable. For these types of products, extrusion is carried out as a continuous process; the *extrudate* (extruded product) is subsequently cut into desired lengths. This section covers the basic extrusion process, and several subsequent sections examine processes based on extrusion.

13.2.1 PROCESS AND EQUIPMENT

In polymer extrusion, feedstock in pellet or powder form is fed into an extrusion barrel where it is heated and melted and forced to flow through a die opening by means of a rotating screw, as illustrated in Figure 13.4. The two main components of the extruder are the barrel and the screw. The die is not a component of the extruder; it is a special tool that must be fabricated for the particular profile to be produced.

The internal diameter of the extruder barrel typically ranges from 25 to 150 mm (1.0 to 6.0 in). The barrel is long relative to its diameter, with L/D ratios usually between 10 and 30. The L/D ratio is reduced in Figure 13.4 for clarity of drawing. The higher ratios are used for thermoplastic materials, whereas lower L/D values are for elastomers. A hopper containing the feedstock is located at the end of the barrel opposite the die. The pellets are fed by gravity onto the rotating screw whose turning moves the material along the barrel. Electric heaters are used to initially melt the solid pellets; subsequent mixing and mechanical working of the material generate additional heat, which maintains the melt. In some cases, enough heat is supplied through the mixing and shearing action that external heating is not required. Indeed, in some cases the barrel must be externally cooled to prevent overheating of the polymer.

The material is conveyed through the barrel toward the die opening by the action of the extruder screw, which rotates at about 60 rev/min. The screw serves several functions and is divided into sections that correspond to these functions. The sections and functions are the (1) *feed section,* in which the stock is moved from the hopper port and preheated; (2) *compression section,* where the polymer is transformed into liquid consistency, air entrapped amongst the pellets is extracted from the melt, and the material is compressed; and (3) *metering section,* in which the melt is homogenized and sufficient pressure is developed to pump it through the die opening.

FIGURE 13.4 Components and features of a (single-screw) extruder for plastics and elastomers.

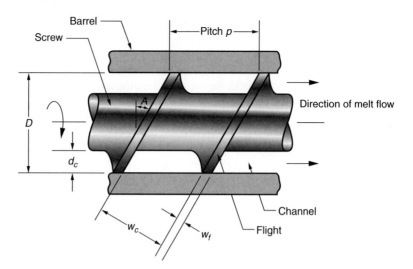

FIGURE 13.5 Details of an extruder screw inside the barrel.

The operation of the screw is determined by its geometry and speed of rotation. Typical extruder screw geometry is depicted in Figure 13.5. The screw consists of spiraled "flights" (threads) with channels between them through which the polymer melt is moved. The channel has a width w_c and depth d_c. As the screw rotates, the flights push the material forward through the channel from the hopper end of the barrel toward the die. Although not discernible in the diagram, the flight diameter is smaller than the barrel diameter D by a very small clearance—around 0.05 mm (0.002 in). The function of the clearance is to limit leakage of the melt backward to the trailing channel. The flight land has a width w_f and is made of hardened steel to resist wear as it turns and rubs against the inside of the barrel. The screw has a pitch whose value is usually close to the diameter D. The flight angle A is the helix angle of the screw and can be determined from the relation

$$\tan A = \frac{p}{\pi D} \tag{13.4}$$

where p = pitch of the screw[1].

The increase in pressure applied to the polymer melt in the three sections of the barrel is determined largely by the channel depth d_c. In Figure 13.4, d_c is relatively large in the feed section to allow large amounts of granular polymer to be admitted into the barrel. In the compression section, d_c is gradually reduced, thus applying increased pressure on the polymer as it melts. In the metering section, d_c is small and pressure reaches a maximum as flow is restrained by the screen pack and backer plate. The three sections of the screw are shown as being about equal in length in Figure 13.4; this is appropriate for a polymer that melts gradually, such as low-density polyethylene. For other polymers, the optimal section lengths are different. For crystalline polymers such as nylon, melting occurs rather abruptly at a specific melting point; therefore, a short compression section is appropriate. Amorphous polymers such as polyvinylchloride melt more slowly than LDPE, and the compression zone for these materials must take almost the entire length of the screw. Although the optimal screw design for each material type is different, it is common practice to use general-purpose screws. These designs represent a compromise among the different materials, and they avoid the need to make frequent screw changes, which result in costly equipment downtime.

---

[1]Unfortunately, p is the natural symbol to use for two variables in this chapter. It represents the screw pitch here and in several other chapters. We use the same symbol p for pressure later in the chapter.

Progress of the polymer along the barrel leads ultimately to the die zone. Before reaching the die, the melt passes through a screen pack—a series of wire meshes supported by a stiff plate (called a **breaker plate**) containing small axial holes. The screen pack assembly functions to (1) filter contaminants and hard lumps from the melt; (2) build pressure in the metering section; and (3) straighten the flow of the polymer melt and remove its "memory" of the circular motion imposed by the screw. This last function is concerned with the polymer's viscoelastic property; if the flow were left unstraightened, the polymer would play back its history of turning inside the extrusion chamber, tending to twist and distort the extrudate.

13.2.2 ANALYSIS OF EXTRUSION

In this section, we develop mathematical models to describe, in a simplified way, several aspects of polymer extrusion.

Melt Flow in the Extruder   As the screw rotates inside the barrel, the polymer melt is forced to move forward toward the die; the system operates much like an Archimedian screw. The principal transport mechanism is **drag flow**, resulting from friction between the viscous liquid and two opposing surfaces moving relative to each other: (1) the stationary barrel and (2) the channel of the turning screw. The arrangement can be likened to the fluid flow that occurs between a stationary plate and a moving plate separated by a viscous liquid, as illustrated in Figure 3.17. Given that the moving plate has a velocity v, it can be reasoned that the average velocity of the fluid is $v/2$, resulting in a volume flow rate of

$$Q_d = 0.5\,v\,d\,w \tag{13.5}$$

where Q_d = volume drag flow rate, m³/s (in³/sec.); v = velocity of the moving plate, m/s (in/sec.); d = distance separating the two plates, m (in); and w = the width of the plates perpendicular to velocity direction, m (in).

These parameters can be compared with those in the channel defined by the rotating extrusion screw and the stationary barrel surface.

$$v = \pi DN \cos A \tag{13.6}$$

$$d = d_c \tag{13.7}$$

and

$$w = w_c = (\pi D \tan A - w_f)\cos A \tag{13.8}$$

where D = screw flight diameter, m (in); N = screw rotational speed, rev/s; d_c = screw channel depth, m (in); w_c = screw channel width, m (in); A = flight angle; and w_f = flight land width, m (in).

If we assume that the flight land width is negligibly small, then the last of these equations reduces to

$$w_c = \pi D \tan A \cos A = \pi D \sin A \tag{13.9}$$

Substituting Eqs. (13.6), (13.7), and (13.9) into Eq. (13.5), and using several trigonometric identities, we get

$$Q_d = 0.5\,\pi^2\,D^2\,N\,d_c \sin A \cos A \tag{13.10}$$

If no forces were present to resist the forward motion of the fluid, this equation would provide a reasonable description of the melt flow rate inside the extruder. However, compressing the polymer melt through the downstream die creates a **back**

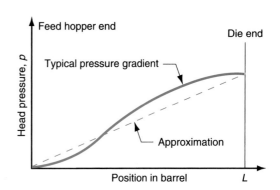

FIGURE 13.6 Typical pressure gradient in an extruder; dashed line indicates a straight line approximation to facilitate computations.

pressure in the barrel that reduces the material moved by drag flow in Eq. (13.10). This flow reduction, called the **back pressure flow,** depends on the screw dimensions, viscosity of the polymer melt, and pressure gradient along the barrel. These dependencies can be summarized in this equation [12]:

$$Q_b = \frac{\pi D d_c^3 \sin^2 A}{12\eta} \left(\frac{dp}{dl}\right) \tag{13.11}$$

where Q_b = back pressure flow, m^3/s (in^3/sec); η = viscosity, N-s/m^2 (lb-sec/in^2); dp/dl = the pressure gradient, MPa/m (lb/in^2/in); and the other terms were previously defined.

The actual pressure gradient in the barrel is a function of the shape of the screw over its length; a typical pressure profile is given in Figure 13.6. If we assume as an approximation that the profile is a straight line, indicated by the dashed line in the figure, then the pressure gradient becomes a constant p/L, and the previous equation reduces to

$$Q_b = \frac{p\pi D d_c^3 \sin^2 A}{12\eta L} \tag{13.12}$$

where p = head pressure in the barrel, MPa (lb/in^2); and L = length of the barrel, m (in).

Recall that this back pressure flow is really not an actual flow by itself; it is a reduction in the drag flow. Thus, we can compute the magnitude of the melt flow in an extruder as the difference between the drag flow and back pressure flow:

$$Q_x = Q_d - Q_b$$
$$Q_x = 0.5\,\pi^2\,D^2\,N\,d_c \sin A \cos A - \frac{p\pi D d_c^3 \sin^2 A}{12\eta L} \tag{13.13}$$

where Q_x = the resulting flow rate of polymer melt in the extruder.

Equation (13.13) assumes that there is minimal **leak flow** through the clearance between flights and barrel. Leak flow of melt will be small compared with drag and back pressure flow except in badly worn extruders.

Equation (13.13) contains many parameters, which can be divided into two types: (1) design parameters, and (2) operating parameters. The design parameters are those that define the geometry of the screw and barrel: diameter D, length L, channel depth d_c, and helix angle A. For a given extruder operation, these factors cannot be changed during the process. The operating parameters are those that can be changed during the process to affect output flow; they include rotational speed N, head pressure p, and melt viscosity η. Of course, melt viscosity is controllable only to the extent to which temperature and shear rate can be manipulated to affect this property. Let us see how the parameters play out their roles in the following example.

**Example 13.1
Extrusion Flow
Rates**

An extruder barrel has a diameter $D = 75$ mm. The screw rotates at $N = 1$ rev/s. Channel depth $d_c = 6.0$ mm and flight angle $A = 20°$. Head pressure at the end of the barrel $p = 7.0 \times 10^6$ Pa, length of the barrel $L = 1.9$ m, and viscosity of the polymer melt is assumed to be $\eta = 100$ Pa-s. Determine the volume flow rate of the plastic in the barrel Q_x.

Solution: Using Eq. (13.13) we can compute the drag flow and opposing back pressure flow in the barrel.

$$Q_d = 0.5\,\pi^2\,(75 \times 10^{-3})^2\,(1.0)\,(6 \times 10^{-3})(\sin 20)(\cos 20) = 53,525(10^{-9})\text{m}^3/\text{s}$$

$$Q_b = \frac{\pi(7 \times 10^6)(75 \times 10^{-3})(6 \times 10^{-3})^3(\sin 20)^2}{12(100)(1.9)} = 18.276(10^{-6}) = 18,276(10^{-9})\text{m}^3/\text{s}$$

$$Q_x = Q_d - Q_b = (53,525 - 18,276)(10^{-9}) = \mathbf{35,249(10^{-9})\text{m}^3/\text{s}}$$

Extruder and Die Characteristics If back pressure is zero, so that melt flow is unrestrained in the extruder, then the flow would equal drag flow Q_d given by Eq. (13.10). Given the design and operating parameters (D, A, N, etc.), this is the maximum possible flow capacity of the extruder. Denote it as Q_{max}:

$$Q_{max} = 0.5\pi^2\,D^2\,N\,d_c\,\sin A\,\cos A \qquad (13.14)$$

On the other hand, if back pressure were so great as to cause zero flow, then back pressure flow would equal drag flow; that is

$$Q_x = Q_d - Q_b = 0, \text{ so } Q_d = Q_b$$

Using the expressions for Q_d and Q_b in Eq. (13.13), we can solve for p to determine what this maximum head pressure p_{max} would have to be to cause no flow in the extruder:

$$p_{max} = \frac{6\pi DNL\eta \cot A}{d_c^2} \qquad (13.15)$$

The two values Q_{max} and p_{max} are points along the axes of a diagram known as the **extruder characteristic** (or **screw characteristic**), as in Figure 13.7. It defines the relationship between head pressure and flow rate in an extrusion machine with given operating parameters.

With a die in the machine and the extrusion process underway, the actual values of Q_x and p will lie somewhere between the extreme values, the location determined by the characteristics of the die. Flow rate through the die depends on the size and shape of the opening and the pressure applied to force the melt through it. This can be expressed as

$$Q_x = K_s p \qquad (13.16)$$

FIGURE 13.7   Extruder characteristic (also called the screw characteristic) and die characteristic. The extruder operating point is at intersection of the two lines.

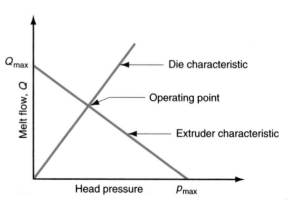

where Q_x = flow rate, m³/s (in³/sec.); p = head pressure, Pa (lb/in²); and K_s = shape factor for the die, m⁵/Ns (in⁵/lb-sec).

For a circular die opening of a given channel length, the shape factor can be computed [12] as

$$K_s = \frac{\pi D_d^4}{128\eta L_d} \qquad (13.17)$$

where D_d = die opening diameter, m (in) η = melt viscosity, N-s/m² (lb-sec/in²); and L_d = die opening length, m (in).

For shapes other than round, the die shape factor is less than for a round of the same cross-sectional area, meaning that greater pressure is required to achieve the same flow rate.

The relationship between Q_x and p in Eq. (13.16) is called the ***die characteristic.*** In Figure 13.7, this is drawn as a straight line that intersects with the previous extruder characteristic. The intersection point identifies the values of Q_x and p that are known as the ***operating point*** for the extrusion process.

Example 13.2
Extruder and Die
Characteristics

Consider the extruder from Example 13.1, in which D = 75 mm, L = 1.9 m, N = 1 rev/s, d_c = 6 mm, and A = 20°. The plastic melt has a shear viscosity η = 100 Pa-s. Determine (a) Q_{max} and p_{max}, (b) shape factor K_s for a circular die opening in which D_d = 6.5 mm and L_d = 20 mm, and (c) values of Q_x and p at the operating point.

Solution: (a) Q_{max} is given by Eq. (13.14).

$$Q_{max} = 0.5\pi^2 D^2 N d_c \sin A \cos A = 0.5\,\pi^2 (75 \times 10^{-3})^2 (1.0)(6 \times 10^{-3})(\sin 20)(\cos 20)$$
$$= \mathbf{53{,}525(10^{-9})\,m^3/s}$$

p_{max} is given by Eq. (13.15).

$$p_{max} = \frac{6\pi DNL\eta \cot A}{d_c^2} = \frac{6\pi(75 \times 10^{-3})(1.9)(1.0)(100) \cot 20}{(6 \times 10^{-3})^2} = \mathbf{20{,}499{,}874\,Pa}$$

These two values define the intersection with the ordinate and abscissa for the extruder characteristic.

(b) The shape factor for a circular die opening with D_d = 6.5 mm and L_d = 20 mm can be determined from Eq. (13.17).

$$K_s = \frac{\pi(6.5 \times 10^{-3})^4}{128(100)(20 \times 10^{-3})} = \mathbf{21.9(10^{-12})\,m^5/Ns}$$

This shape factor defines the slope of the die characteristic.

(c) The operating point is defined by the values of Q_x and p at which the screw characteristic intersects with the die characteristic. The screw characteristic can be expressed as the equation of the straight line between Q_{max} and p_{max}, which is

$$Q_x = Q_{max} - (Q_{max}/p_{max})p$$
$$= 53,525(10^{-9}) - (53,525(10^{-9})/20,499,874)p = 53,525(10^{-9}) - 2.611(10^{-12})p$$
$$(13.18)$$

The die characteristic is given by Eq. (13.16) using the value of K_s computed in part (b).

$$Q_x = 21.9(10^{-12})p$$

Setting the two equations equal, we have

$$53,525(10^{-9}) - 2.611(10^{-12})p = 21.9(10^{-12})p$$
$$p = \mathbf{2.184(10^{-6})\,Pa}$$

Solving for Q_x using one of the starting equations, we obtain

$$Q_x = 53.525(10^{-6}) - 2.611(10^{-12})(2.184)(10^6) = \mathbf{47.822(10^{-6})\,m^3/s}$$

Checking this with the other equation for verification,

$$Q_x = 21.9(10^{-12})(2.184)(10^6) = \mathbf{47.82(10^{-6})\,m^3/s}$$ ∎

13.2.3 DIE CONFIGURATIONS AND EXTRUDED PRODUCTS

The shape of the die orifice determines the cross-sectional shape of the extrudate. We can enumerate the common die profiles and corresponding extruded shapes as follows: (1) solid profiles; (2) hollow profiles, such as tubes; (3) wire and cable coating; (4) sheet and film; and (5) filaments. The first three categories are covered in the present section. Methods for producing sheet and film are examined in Section 13.3; and filament production is discussed in Section 13.4. These latter shapes sometimes involve forming processes other than extrusion.

Solid Profiles Solid profiles include regular shapes such as rounds and squares and irregular cross sections such as structural shapes, door and window moldings, automobile trim, and house siding. The side view cross section of a die for these solid shapes is illustrated in Figure 13.8. Just beyond the end of the screw and before the die, the polymer melt passes through the screen pack and breaker plate to straighten the flow lines. Then it flows into a (usually) converging die entrance, the shape designed to maintain laminar flow and avoid dead spots in the corners that would otherwise be present near the orifice. The melt then flows through the die opening itself.

When the material exits the die, it is still soft. Polymers with high melt viscosities are the best candidates for extrusion, because they hold shape better during cooling. Cooling is accomplished by air blowing, water spray, or passing the extrudate through a water trough.

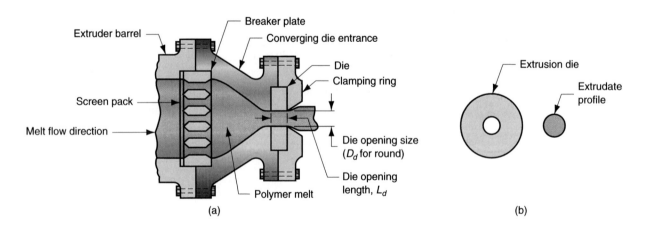

FIGURE 13.8 (a) Side view cross section of an extrusion die for solid regular shapes, such as round stock; (b) front view of die, with profile of extrudate. Die swell is evident in both views. (Some die construction details are simplified or omitted for clarity.)

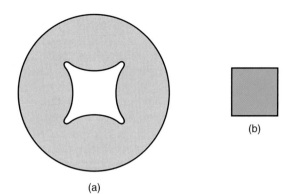

FIGURE 13.9 (a) Die cross section showing required orifice profile to obtain (b) a square extruded profile.

(a)

(b)

To compensate for die swell, the die opening is made long enough to remove some of the memory in the polymer melt. In addition, the extrudate is often drawn (stretched) to offset expansion from die swell.

For shapes other than round, the die opening is designed with a cross section that is slightly different from the desired profile, so that the effect of die swell is to provide shape correction. This correction is illustrated in Figure 13.9 for a square cross section. Because different polymers exhibit varying degrees of die swell, the shape of the die profile depends on the material to be extruded. Considerable skill and judgment are required by the die designer for complex cross sections.

Hollow Profiles   Extrusion of hollow profiles, such as tubes, pipes, hoses, and other cross sections containing holes, requires a mandrel to form the hollow shape. A typical die configuration is shown in Figure 13.10. The mandrel is held in place using a spider, seen in Section A-A of the figure. The polymer melt flows around the legs supporting the mandrel to reunite into a monolithic tube wall. The mandrel often includes an air channel through which air is blown to maintain the hollow form of the extrudate during

FIGURE 13.10   Side view cross section of extrusion die for shaping hollow cross sections such as tubes and pipes; Section A-A is a front view cross section showing how the mandrel is held in place; Section B-B shows the tubular cross section just prior to exiting the die; die swell causes an enlargement of the diameter. (Some die construction details are simplified.)

FIGURE 13.11 Side view cross section of die for coating of electrical wire by extrusion. (Some die construction details are simplified.)

hardening. Pipes and tubes are cooled using open water troughs or by pulling the soft extrudate through a water-filled tank with sizing sleeves that limit the OD of the tube while air pressure is maintained on the inside.

Wire and Cable Coating   The coating of wire and cable for insulation is one of the most important polymer extrusion processes. As shown in Figure 13.11 for wire coating, the polymer melt is applied to the bare wire as it is pulled at high speed through a die. A slight vacuum is drawn between the wire and the polymer to promote adhesion of the coating. The taught wire provides rigidity during cooling, which is usually aided by passing the coated wire through a water trough. The product is wound onto large spools at speeds of up to 50 m/s (10,000 ft/min).

13.2.4   DEFECTS IN EXTRUSION

A number of defects can afflict extruded products. One of the worst is *melt fracture,* in which the stresses acting on the melt immediately before and during its flow through the die are so high as to cause failure, manifested in the form of a highly irregular surface on the extrudate. As suggested by Figure 13.12, melt fracture can be caused by a sharp reduction at the die entrance, causing turbulent flow that breaks up the melt. This contrasts with the streamlined, laminar flow in the gradually converging die in Figure 13.8.

FIGURE 13.12 Melt fracture, caused by turbulent flow of the melt through a sharply reduced die entrance.

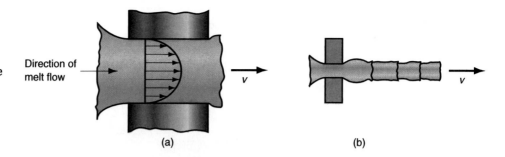

FIGURE 13.13
(a) Velocity profile of the melt as it flows through the die opening, which can lead to defects called sharkskin and (b) bambooing.

A more common defect in extrusion is **sharkskin,** in which the surface of the product becomes roughened upon exiting the die. As the melt flows through the die opening, friction at the interface results in a velocity profile across the cross section, Figure 13.13. Tensile stresses develop at the surface as this material is stretched to keep up with the faster moving center core. These stresses cause minor ruptures that roughen the surface. If the velocity gradient becomes extreme, prominent marks occur on the surface, giving it the appearance of a bamboo pole; hence, the name **bambooing** for this more severe defect.

13.3 PRODUCTION OF SHEET AND FILM

Thermoplastic sheet and film are produced by a number of processes, most important of which are two methods based on extrusion. The term **sheet** refers to stock with a thickness ranging from 0.5 mm (0.020 in) to about 12.5 mm (0.5 in) and used for products such as flat window glazing and stock for thermoforming (Section 13.9). **Film** refers to thicknesses below 0.5 mm (0.020 in). Thin films are used for packaging (product wrapping material, grocery bags, and garbage bags); thicker film applications include covers and liners (pool covers and liners for irrigation ditches).

All of the processes covered in this section are continuous, high-production operations. More than half of the films produced today are polyethylene, mostly low-density PE. The principal other materials are polypropylene, polyvinylchloride, and regenerated cellulose (cellophane). These are all thermoplastic polymers.

Slit-Die Extrusion of Sheet and Film   Sheet and film of various thicknesses are produced by conventional extrusion, using a narrow slit as the die opening. The slit may be up to 3 m (10 ft) wide and as narrow as around 0.4 mm (0.015 in). One possible die configuration is illustrated in Figure 13.14. The die includes a manifold that spreads the polymer melt laterally before it flows through the slit (die orifice). One of the difficulties in this extrusion method is uniformity of thickness throughout the width of the stock. This is caused by the drastic shape change experienced by the polymer melt during its flow through the die and also to temperature and pressure variations in the die. Usually, the edges of the film must be trimmed because of thickening at the edges.

To achieve high production rates, an efficient method of cooling and collecting the film must be integrated with the extrusion process. This is usually done by immediately directing the extrudate into a quenching bath of water or onto chill rolls, as shown in Figure 13.15. The chill roll method seems to be the more important commercially. Contact with the cold rolls quickly quenches and solidifies the extrudate; in effect, the extruder serves as a feeding device for the chill rolls that actually form the film. The process is noted for very high production speeds—5 m/s (1000 ft/min). In addition, close tolerances on film thickness can be achieved. Owing to the cooling method used in this process, it is known as **chill-roll extrusion.**

FIGURE 13.14 One of several die configurations for extruding sheet and film.

Blown-Film Extrusion Process This is the other widely used process for making thin polyethylene film for packaging. It is a complex process, combining extrusion and blowing to produce a tube of thin film; it is best explained with reference to the diagram in Figure 13.16. The process begins with the extrusion of a tube that is immediately drawn upward while still molten and simultaneously expanded in size by air inflated into it through the die mandrel. A "frost line" marks the position along the upward moving bubble where solidification of the polymer occurs. Air pressure in the bubble must be kept constant to maintain uniform film thickness and tube diameter. The air is contained in the tube by pinch rolls that squeeze the tube back together after it has cooled. Guide rolls and collapsing rolls are also used to restrain the blown tube and direct it into the pinch rolls. The flat tube is then collected onto a windup reel.

The effect of air inflation is to stretch the film in both directions as it cools from the molten state. This results in isotropic strength properties, which is an advantage over other processes in which the material is stretched primarily in one direction. Other advantages include the ease with which extrusion rate and air pressure can be changed to control stock width and gage. Comparing this process with slit-die extrusion, the blown-film method produces stronger film (so that a thinner film can be used to package a product), but thickness control and production rates are lower. The final blown film can

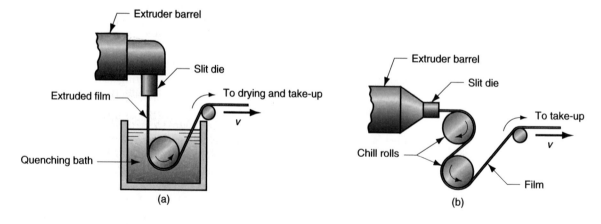

FIGURE 13.15 Use of (a) water quenching bath or (b) chill rolls to achieve fast solidification of the molten film after extrusion.

FIGURE 13.16 Blown-film process for high production of thin tubular film.

be left in tubular form (e.g., for garbage bags), or it can be subsequently cut at the edges to provide two parallel thin films.

Calendering   Calendering is a process for producing sheet and film stock out of rubber (Section 14.1.4) or rubbery thermoplastics such as plasticized PVC. In the process, the initial feedstock is passed through a series of rolls to work the material and reduce its thickness to the desired gage. A typical setup is illustrated in Figure 13.17. The equipment is expensive, but production rate is high; speeds approaching 2.5 m/s (500 ft/min) are possible. Close control is required over roll temperatures, pressures, and rotational speed. The process is noted for its good surface finish and high gage accuracy in the film. Plastic products made by the calendering process include PVC floor covering, shower curtains, vinyl table cloths, pool liners, and inflatable boats and toys.

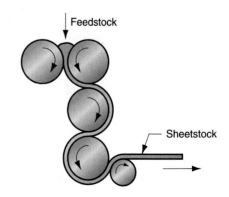

FIGURE 13.17   A typical roll configuration in calendering.

13.4 FIBER AND FILAMENT PRODUCTION (SPINNING)

The most important application of polymer fibers and filaments is in textiles. Their use as reinforcing materials in plastics (composites) is a growing application, but still small compared with textiles. A *fiber* can be defined as a long, thin strand of material whose length is at least 100 times its cross-sectional dimension. A *filament* is a fiber of continuous length.

Fibers can be natural or synthetic. Synthetic fibers constitute about 75% of the total fiber market today, polyester being the most important, followed by nylon, acrylics, and rayon. Natural fibers are about 25% of the total produced, with cotton by far the most important staple (wool production is significantly less than cotton).

The term *spinning* is a holdover from the methods used to draw and twist natural fibers into yarn or thread. In the production of synthetic fibers, the term refers to the process of extruding a polymer melt or solution through a *spinneret* (a die with multiple small holes) to make filaments that are then drawn and wound onto a *bobbin.* There are three principal variations in the spinning of synthetic fibers, depending on the polymer being processed: (1) melt spinning, (2) dry spinning, and (3) wet spinning.

Melt spinning is used when the starting polymer can best be processed by heating to the molten state and pumping through the spinneret, much in the manner of conventional extrusion. A typical spinneret is 6 mm (0.25 in) thick and contains approximately 50 holes of diameter 0.25 mm (0.010 in); the holes are countersunk, so that the resulting bore has an *L/D* ratio of only 5/1 or less. The filaments that emanate from the die are drawn and simultaneously air cooled before being collected together and spooled onto the bobbin, as shown in Figure 13.18. Significant extension and thinning of the filaments occur while the polymer is still molten, so that the final diameter wound onto the bobbin may be only 1/10

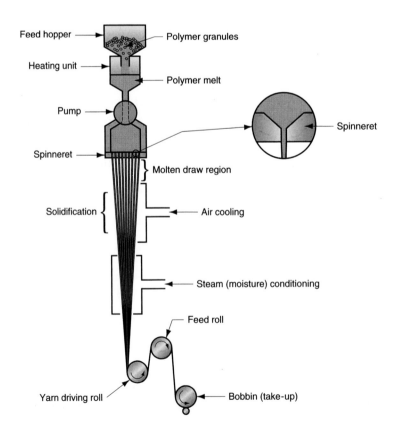

FIGURE 13.18 Melt spinning of continuous filaments.

of the extruded size. Melt spinning is used for polyesters and nylons; because these are the most important synthetic fibers, melt spinning is the most important of the three processes for synthetic fibers.

In *dry spinning,* the starting polymer is in solution and the solvent can be separated by evaporation. The extrudate is pulled through a heated chamber that removes the solvent; otherwise the sequence is similar to the previous. Fibers of cellulose acetate and acrylic are produced by this process. In *wet spinning,* the polymer is also in solution—only the solvent is nonvolatile. To separate the polymer, the extrudate must be passed through a liquid chemical that coagulates or precipitates the polymer into coherent strands that are then collected onto bobbins. This method is used to produce rayon (regenerated cellulose fibers).

Filaments produced by any of the three processes are usually subjected to further cold drawing to align the crystal structure along the direction of the filament axis. Extensions of 2 to 8 are typical [13]. This has the effect of significantly increasing the tensile strength of the fibers. Drawing is accomplished by pulling the thread between two spools, where the winding spool is driven at a faster speed than the unwinding spool.

13.5 COATING PROCESSES

Plastic (or rubber) coating involves application of a layer of the given polymer onto a substrate material. Three categories are distinguished [6]: (1) wire and cable coating; (2) planar coating, which involves the coating of a flat film; and (3) contour coating—the coating of a three-dimensional object. We have already examined wire and cable coating (Section 13.2.3); it is basically an extrusion process. The other two categories are surveyed in the following paragraphs. In addition, there is the technology of applying paints, varnishes, lacquers, and other similar coatings (Section 28.6).

Planar coating is used to coat fabrics, paper, cardboard, and metal foil; these items are major products for some plastics. The important polymers include polyethylene and polypropylene, with lesser applications for nylon, PVC, and polyester. In most cases, the coating is only 0.01 to 0.05 mm (0.0005–0.002 in) thick. The two major planar coating techniques are illustrated in Figure 13.19. In the *roll method,* the polymer coating material is squeezed against the substrate by means of opposing rolls. In the *doctor blade method,* a sharp knife edge controls the amount of polymer melt that is coated onto the

FIGURE 13.19   Planar coating processes: (a) roll method, and (b) doctor-blade method.

substrate. In both cases, the coating material is supplied either by a slit-die extrusion process or by calendering.

Contour coating of three-dimensional objects can be accomplished by dipping or spraying. *Dipping* involves submersion of the object into a suitable bath of polymer melt or solution, followed by cooling or drying. *Spraying* (such as spray painting) is an alternative method for applying a polymer coating to a solid object.

13.6 INJECTION MOLDING

Injection molding is a process in which a polymer is heated to a highly plastic state and forced to flow under high pressure into a mold cavity, where it solidifies. The molded part, called a *molding,* is then removed from the cavity. The process produces discrete components that are almost always net shape. The production cycle time is typically in the range of 10 to 30 sec, although cycles of 1 min or longer are not uncommon for large parts. Also, the mold may contain more than one cavity, so that multiple moldings are produced each cycle. Many aspects of injection molding are illustrated in the video clip.

VIDEO CLIP _____

Plastic Injection Molding. This clip contains three segments: (1) plastic materials and molding, (2) injection molding machines, and (3) injection molds.

Complex and intricate shapes are possible with injection molding. The challenge in these cases is to fabricate a mold whose cavity is the same geometry as the part and that also allows for part removal. Part size can range from about 50 g (2 oz) up to about 25 kg (more than 50 lb), the upper limit represented by components such as refrigerator doors and automobile bumpers. The mold determines the part shape and size and is the special tooling in injection molding. For large, complex parts, the mold can cost hundreds of thousands of dollars. For small parts, the mold can be built to contain multiple cavities, also making the mold expensive. Thus, injection molding is economical only for large production quantities.

Injection molding is the most widely used molding process for thermoplastics. Some thermosets and elastomers are injection molded, with modifications in equipment and operating parameters to allow for cross-linking of these materials. We discuss these and other variations of injection molding in Section 13.6.6.

13.6.1 PROCESS AND EQUIPMENT

Equipment for injection molding evolved from metal die casting (Historical Note 13.1). A large injection molding machine is shown in Figure 13.20. As illustrated in the schematic in Figure 13.21, an injection molding machine consists of two principal components: (1) the plastic injection unit and (2) the mold clamping unit. The *injection unit* is much like an extruder. It consists of a barrel that is fed from one end by a hopper containing a supply of plastic pellets. Inside the barrel is a screw whose operation surpasses that of an extruder screw in the following respect: in addition to turning for mixing and heating the polymer, it also acts as a ram that rapidly moves forward to inject molten plastic into the mold. A nonreturn valve mounted near the tip of the screw prevents the melt from flowing backward along the screw threads. Later in the molding cycle the ram retracts to its former position. Because of its dual action, it is called a *reciprocating screw,* a name that also identifies the machine type. Older injection

FIGURE 13.20   A large (3000-ton capacity) injection molding machine. (Courtesy of Cincinnati Milacron.)

molding machines used a simple ram (without screw flights), but the superiority of the reciprocating screw design has led to its widespread adoption in today's molding plants. To summarize, the functions of the injection unit are to melt and homogenize the polymer, and then inject it into the mold cavity.

The *clamping unit* is concerned with the operation of the mold. Its functions are to (1) hold the two halves of the mold in proper alignment with each other; (2) keep the mold closed during injection by applying a clamping force sufficient to resist the injection force; and (3) open and close the mold at the appropriate times in the molding cycle. The clamping unit consists of two platens, a fixed platen and a moveable platen, and a mechanism for translating the latter. The mechanism is basically a power press that is operated by hydraulic piston or mechanical toggle devices of various types. Clamping forces of several thousand tons are available on large machines.

The cycle for injection molding of a thermoplastic polymer proceeds in the following sequence, illustrated in Figure 13.22. Let us pick up the action with the mold open and the machine ready to start a new molding: (1) The mold is closed and clamped. (2) A *shot* of melt, which has been brought to the right temperature and viscosity by heating and the mechanical working of the screw, is injected under high pressure into the mold cavity. The plastic cools and begins to solidify when it encounters the cold surface of the mold. Ram pressure is maintained to pack additional melt into the cavity to

FIGURE 13.21   Diagram of an injection molding machine, reciprocating screw type (some mechanical details are simplified).

FIGURE 13.22 Typical molding cycle: (1) mold is closed, (2) melt is injected into cavity, (3) screw is retracted, and (4) mold opens, and part is ejected.

compensate for contraction during cooling. (3) The screw is rotated and retracted with the nonreturn valve open to permit fresh polymer to flow into the forward portion of the barrel. Meanwhile, the polymer in the mold has completely solidified. (4) The mold is opened, and the part is ejected and removed.

13.6.2 THE MOLD

The mold is the special tool in injection molding; it is custom designed and fabricated for the given part to be produced. When the production run for that part is finished, the mold is replaced with a new mold for the next part. In this section we examine several types of mold for injection molding.

Two-Plate Mold The conventional *two-plate mold,* illustrated in Figure 13.23, consists of two halves fastened to the two platens of the molding machine's clamping unit. When the clamping unit is opened, the two mold halves open, as shown in (b). The most obvious feature of the mold is the *cavity,* which is usually formed by removing metal from the mating surfaces of the two halves. Molds can contain a single cavity or multiple cavities to produce more than one part in a single shot. The figure shows a mold with two cavities. The *parting surfaces* (or *parting line* in a cross-sectional view of the mold) are where the mold opens to remove the part(s).

In addition to the cavity, other features of the mold serve indispensable functions during the molding cycle. A mold must have a distribution channel through which the polymer melt flows from the nozzle of the injection barrel into the mold cavity. The distribution channel consists of (1) a *sprue,* which leads from the nozzle into the mold; (2) *runners,* which lead from the sprue to the cavity (or cavities); and (3) *gates* that constrict

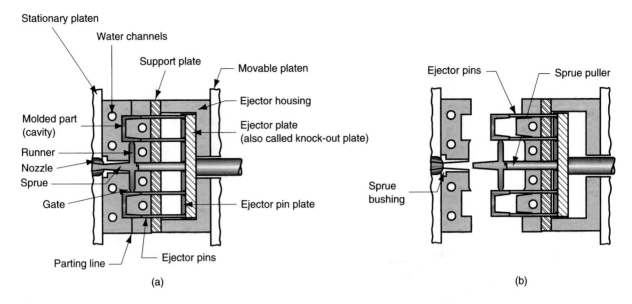

FIGURE 13.23   Details of a two-plate mold for thermoplastic injection molding: (a) closed and (b) open. Mold has two cavities to produce two cup-shaped parts (cross section shown) with each injection shot.

the flow of plastic into the cavity. The constriction increases the shear rate, thereby reducing the viscosity of the polymer melt. There are one or more gates for each cavity in the mold.

An *ejection system* is needed to eject the molded part from the cavity at the end of the molding cycle. *Ejector pins* built into the moving half of the mold usually accomplish this function. The cavity is divided between the two mold halves in such a way that the natural shrinkage of the molding causes the part to stick to the moving half. When the mold opens, the ejector pins push the part out of the mold cavity.

A *cooling system* is required for the mold. This consists of an external pump connected to passageways in the mold, through which water is circulated to remove heat from the hot plastic. Air must be evacuated from the mold cavity as the polymer rushes in. Much of the air passes through the small ejector pin clearances in the mold. In addition, narrow *air vents* are often machined into the parting surface; only about 0.03 mm (0.001 in) deep and 12 to 25 mm (0.5 to 1.0 in) wide, these channels permit air to escape to the outside but are too small for the viscous polymer melt to flow through.

To summarize, a mold consists of (1) one or more cavities that determine part geometry, (2) distribution channels through which the polymer melt flows to the cavities, (3) an ejection system for part removal, (4) a cooling system, and (5) vents to permit evacuation of air from the cavities.

Other Mold Types   The two-plate mold is the most common mold in injection molding. An alternative is a *three-plate mold,* shown in Figure 13.24, for the same part geometry as before. There are advantages to this mold design. First, the flow of molten plastic is through a gate located at the base of the cup-shaped part, rather than at the side. This allows more even distribution of melt into the sides of the cup. In the side gate design in the two-plate mold of Figure 13.23, the plastic must flow around the core and join on the opposite side, possibly creating a weakness at the weld line. Second, the three-plate mold allows more automatic operation of the molding machine. As the mold opens, it divides into three plates with two openings between them. This action separates the runner from the parts, which drop by gravity into containers beneath the mold.

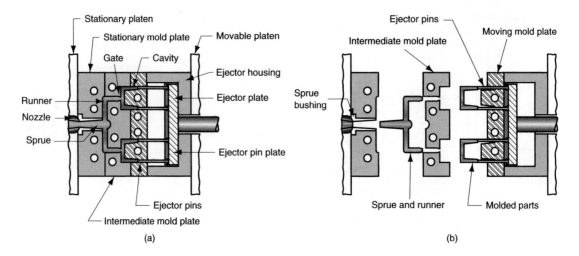

FIGURE 13.24 Three-plate mold: (a) closed, and (b) open.

The sprue and runner in a conventional two- or three-plate mold represent waste material. In many instances they can be ground and reused; however, in some cases the product must be made of "virgin" plastic (plastic that has not been previously molded). The **hot-runner mold** eliminates the solidification of the sprue and runner by locating heaters around the corresponding runner channels. Although the plastic in the mold cavity solidifies, the material in the sprue and runner channels remains molten, ready to be injected into the cavity in the next cycle.

13.6.3 INJECTION MOLDING MACHINES

Injection molding machines differ in both injection unit and clamping unit. This section discusses the important types of machines available today. The name of the injection molding machine is generally based on the type of injection unit used.

Injection Units   Two types of injection units are widely used today. The **reciprocating-screw machine** (Section 13.6.1, Figures 13.21 and 13.22) is the most common. This design uses the same barrel for melting and injection of plastic. The alternative unit involves the use of separate barrels for plasticizing and injecting the polymer, as shown in Figure 13.25 (a). This type is called a **screw-preplasticizer machine** or **two-stage machine.** Plastic pellets are fed from a hopper into the first stage, which uses a screw to drive the polymer forward and melt it. This barrel feeds a second barrel, which uses a plunger to inject the melt into the mold. Older machines used one plunger-driven barrel to melt and inject the plastic. These machines are referred to as **plunger-type injection molding machines** (Figure 13.25(b)).

Clamping Units   Clamping designs are of three types [11]: toggle, hydraulic, and hydro-mechanical. **Toggle clamps** include various designs, one of which is illustrated in Figure 13.26(a). An actuator moves the crosshead forward, extending the toggle links to push the moving platen toward a closed position. At the beginning of the movement, mechanical advantage is low and speed is high; but near the end of the stroke, the reverse is true. Thus, toggle clamps provide both high speed and high force at different points in the cycle when they are desirable. They are actuated either by hydraulic cylinders or ball screws driven by electric motors. Toggle-clamp units seem most suited to relatively low tonnage machines.

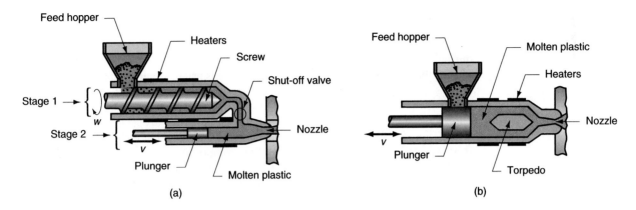

FIGURE 13.25   Two alternative injection systems to the reciprocating screw shown in Figure 13.21: (a) screw preplasticizer, and (b) plunger type.

Hydraulic clamps, shown in Figure 13.26(b), are used on higher-tonnage injection molding machines, typically in the range 1300 to 8900 kN (150 to 1000 tons). These units are also more flexible than toggle clamps in terms of setting the tonnage at given positions during the stroke. *Hydromechanical clamps* are designed for large tonnages, usually above 8900 kN (1000 tons). They operate by (1) using hydraulic cylinders to rapidly move the mold toward closing position, (2) locking the position by mechanical means, and (3) using high-pressure hydraulic cylinders to finally close the mold and build tonnage.

FIGURE 13.26   Two clamping designs: (a) one possible toggle clamp design: (1) open and (2) closed; and (b) hydraulic clamping: (1) open, and (2) closed. Tie rods used to guide moving platens not shown.

13.6.4 SHRINKAGE AND DEFECTS IN INJECTION MOLDING

Polymers have high thermal expansion coefficients, and significant shrinkage can occur during cooling of the plastic in the mold. Contraction of crystalline plastics tends to be greater than for amorphous polymers. Shrinkage is usually expressed as the reduction in linear size that occurs during cooling to room temperature from the molding temperature for the given polymer. Appropriate units are therefore mm/mm (in/in) of the dimension under consideration. Typical values for selected polymers are given in Table 13.1.

Fillers in the plastic tend to reduce shrinkage. In commercial molding practice, shrinkage values for the specific molding compound should be obtained from the producer before making the mold. To compensate for shrinkage, the dimensions of the mold cavity must be made larger than the specified part dimensions. The following formula can be used [14]:

$$D_c = D_p + D_p S + D_p S^2 \qquad (13.19)$$

where D_c = dimension of cavity, mm (in); D_p = molded part dimension, mm (in), and S = shrinkage values obtained from Table 13.1.

The third term on the right-hand side corrects for shrinkage that occurs in the shrinkage.

Example 13.3
Shrinkage in Injection Molding

The nominal length of a part made of polyethylene is to be 80 mm. Determine the corresponding dimension of the mold cavity that will compensate for shrinkage.

Solution: From Table 13.1, the shrinkage for polyethylene is $S = 0.025$. Using Eq. (13.19), the mold cavity diameter should be:

$$D_c = 80.0 + 80.0(0.025) + 80.0(0.025)^2$$
$$= 80.0 + 2.0 + 0.05 = \mathbf{82.05\ mm}$$

∎

Because of differences in shrinkage among plastics, mold dimensions must be determined for the particular polymer to be molded. The same mold will produce different part sizes for different polymer types.

Values in Table 13.1 represent a gross simplification of the shrinkage issue. In reality, shrinkage is affected by a number of factors, any of which can alter the amount of contraction experienced by a given polymer. The most important factors are injection pressure, compaction time, molding temperature, and part thickness. As injection pressure is increased, forcing more material into the mold cavity, shrinkage is reduced. Increasing compaction time has a similar effect, assuming the polymer in the gate does not solidify and seal off the cavity; maintaining pressure forces more material into the cavity while shrinkage is taking place. Net shrinkage is thereby reduced.

Molding temperature refers to the temperature of the polymer in the cylinder immediately before injection. One might expect that a higher polymer temperature would increase shrinkage, on the reasoning that the difference between molding and

TABLE 13.1   Typical values of shrinkage for moldings of selected thermoplastics.

Plastic	Shrinkage, mm/mm (in/in)	Plastic	Shrinkage, mm/mm (in/in)
ABS	0.006	Polyethylene	0.025
Nylon-6,6	0.020	Polystyrene	0.004
Polycarbonate	0.007	PVC	0.005

Compiled from [14].

room temperatures is greater. However, shrinkage is actually lower at higher molding temperatures. The explanation is that higher temperatures significantly lower the viscosity of the polymer melt, allowing more material to be packed into the mold; the effect is the same as higher injection pressures. Thus, the effect on viscosity more than compensates for the larger temperature difference.

Finally, thicker parts show greater shrinkage. A molding solidifies from the outside; the polymer in contact with the mold surface forms a skin that grows toward the center of the part. At some point during solidification, the gate solidifies, isolating the material in the cavity from the runner system and compaction pressure. When this happens, the molten polymer inside the skin accounts for most of the remaining shrinkage that occurs in the part. A thicker part section experiences greater shrinkage because it contains a higher proportion of molten material.

In addition to the shrinkage issue, other things can also go wrong. Here are some of the common defects in injection molded parts:

> *Short shots.* As in casting, a short shot is a molding that has solidified before completely filling the cavity. The defect can be corrected by increasing temperature and/or pressure. The defect may also result from use of a machine with insufficient shot capacity, in which case a larger machine is needed.

> *Flashing.* Flashing occurs when the polymer melt is squeezed into the parting surface between mold plates; it can also occur around ejection pins. The defect is usually caused by (1) vents and clearances in the mold that are too large; (2) injection pressure too high compared with clamping force; (3) melt temperature too high; or (4) excessive shot size.

> *Sink marks and voids.* These are defects usually related to thick molded sections. A *sink mark* occurs when the outer surface on the molding solidifies, but contraction of the internal material causes the skin to be depressed below its intended profile. A *void* is caused by the same basic phenomenon; however, the surface material retains its form and the shrinkage manifests itself as an internal void because of high tensile stresses on the still-molten polymer. These defects can be addressed by increasing the packing pressure after injection. A better solution is to design the part to have uniform section thicknesses and use thinner sections.

> *Weld lines.* Weld lines occur when polymer melt flows around a core or other convex detail in the mold cavity and meets from opposite directions; the boundary thus formed is called a weld line, and it may have mechanical properties that are inferior to those in the rest of the part. Higher melt temperatures, higher injection pressures, alternative gating locations on the part, and better venting are ways of dealing with this defect.

13.6.5   OTHER INJECTION MOLDING PROCESSES

The vast majority of injection molding applications involve thermoplastics. Several variants of the process are described in this section.

Thermoplastic Foam Injection Molding   Plastic foams have a variety of applications, and we discuss these materials and their processing in Section 13.11. One of the processes, sometimes called *structural foam molding,* is appropriate to discuss here because it is injection molding. It involves the molding of thermoplastic parts that possess a dense outer skin surrounding a lightweight foam center. Such parts have high stiffness-to-weight ratios suitable for structural applications.

A structural foam part can be produced either by introducing a gas into the molten plastic in the injection unit or by mixing a gas-producing ingredient with the starting pellets. During injection, an insufficient amount of melt is forced into the mold cavity, where it expands (foams) to fill the mold. The foam cells in contact with the cold mold surface collapse to form a dense skin, while the material in the core retains its cellular structure. Items made of structural foam include electronic cases, business machine housings, furniture components, and washing machine tanks. Advantages cited for structural foam molding include lower injection pressures and clamping forces, and thus the capability to produce large components, as suggested by the preceding list. A disadvantage of the process is that the resulting part surfaces tend to be rough, with occasional voids. If good surface finish is needed for the application, then additional processing is required, such as sanding, painting, and adhesion of a veneer.

Multi-Injection Molding Processes Unusual effects can be achieved by multiple injection of different polymers to mold a part. The polymers are injected either simultaneously or sequentially, and there may be more than one mold cavity involved. Several processes fall under this heading, all characterized by two or more injection units—thus, the equipment for these processes is expensive.

   Sandwich molding involves injection of two separate polymers—one is the outer skin of the part and the other is the inner core, which is typically a polymer foam. A specially designed nozzle controls the flow sequence of the two polymers into the mold. The sequence is designed so that the core polymer is completely surrounded by the skin material inside the mold cavity. The final structure is similar to that of a structural foam molding. However, the molding possesses a smooth surface, thus overcoming one of the major shortcomings of the previous process. In addition, it consists of two distinct plastics, each with its own characteristics suited to the application.

   Another multi-injection molding process involves sequential injection of two polymers into a two-position mold. With the mold in the first position, the first polymer is injected into the cavity. Then the mold opens to the second position, and the second melt is injected into the enlarged cavity. The resulting part consists of two integrally connected plastics. *Bi-injection molding* is used to combine plastics of two different colors (e.g., automobile tail light covers) or to achieve different properties in different sections of the same part.

Injection Molding of Thermosets Injection molding is used for thermosetting (TS) plastics, with certain modifications in equipment and operating procedure to allow for cross-linking. The machines for thermoset injection molding are similar to those used for thermoplastics. They use a reciprocating-screw injection unit, but the barrel length is shorter to avoid premature curing and solidification of the TS polymer. For the same reason, temperatures in the barrel are kept at relatively low levels, usually 50°C to 125°C (120°F to 260°F), depending on the polymer. The plastic, usually in the form of pellets or granules, is fed into the barrel through a hopper. Plasticizing occurs by the action of the rotating screw as the material is moved forward toward the nozzle. When sufficient melt has accumulated ahead of the screw, it is injected into a mold that is heated to 150°C to 230°C (300°F to 450°F), where cross-linking occurs to harden the plastic. The mold is then opened, and the part is ejected and removed. Molding cycle times typically range from 20 sec to 2 min, depending on polymer type and part size.

   Curing is the most time-consuming step in the cycle. In many cases, the part can be removed from the mold before curing is completed, so that final hardening occurs because of retained heat within a minute or two after removal. An alternative approach is to use a multiple-mold machine, in which two or more molds are attached to an indexing head served by a single injection unit.

   The principal thermosets for injection molding are phenolics, unsaturated polyesters, melamines, epoxies, and urea-formaldehyde. Elastomers are also injected molded (Section 14.1.4). More than 50% of the phenolic moldings currently produced in the United States are made by this process [11], representing a shift away from compression and transfer

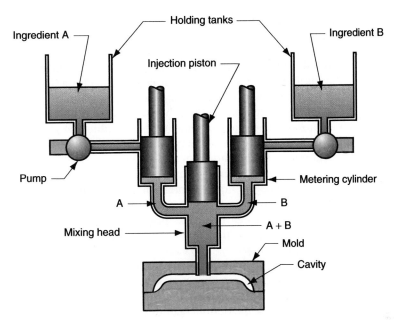

FIGURE 13.27 Reaction injection molding (RIM) system, shown immediately after ingredients A and B have been pumped into the mixing head prior to injection into the mold cavity (some details of processing equipment omitted).

molding, the traditional processes used for thermosets (Section 13.7). Most of the TS molding materials contain large proportions of fillers (up to 70% by weight), including glass fibers, clay, wood fibers, and carbon black. In effect, these are composite materials that are being injection molded.

Reaction Injection Molding   Reaction injection molding (RIM) involves the mixing of two highly reactive liquid ingredients and immediately injecting the mixture into a mold cavity, where chemical reactions leading to solidification occur. The two ingredients form the components used in catalyst-activated or mixing-activated thermoset systems (Section 8.3.1). Urethanes, epoxies, and urea-formaldehyde are examples of these systems. RIM was developed with polyurethane to produce large automotive components such as bumpers, spoilers, and fenders. These kinds of parts still constitute the major application of the process. RIM-molded polyurethane parts typically possess a foam internal structure surrounded by a dense outer skin.

As shown in Figure 13.27, liquid ingredients are pumped in precisely measured amounts from separate holding tanks into a mixing head. The ingredients are rapidly mixed and then injected into the mold cavity at relatively low pressure where polymerization and curing occur. A typical cycle time is around 2 min. For relatively large cavities the molds for RIM are much less costly than corresponding molds for conventional injection molding. This is because of the low clamping forces required in RIM and the opportunity to use lightweight components in the molds. Other advantages of RIM include (1) low energy is required in the process; (2) equipment costs are less than injection molding; (3) a variety of chemical systems are available that enable specific properties to be obtained in the molded product; and (4) the production equipment is reliable, and the chemical systems and machine relationships are well understood [17].

13.7 COMPRESSION AND TRANSFER MOLDING

Discussed in this section are two molding techniques widely used for thermosetting polymers and elastomers. For thermoplastics, these techniques cannot match the efficiency of injection molding, except for very special applications.

13.7.1 COMPRESSION MOLDING

Compression molding is an old and widely used molding process for thermosetting plastics. Its applications also include rubber tires and various polymer matrix composite parts. The process, illustrated in Figure 13.28 for a TS plastic, consists of (1) loading a precise amount of molding compound, called the *charge,* into the bottom half of a heated mold; (2) bringing the mold halves together to compress the charge, forcing it to flow and conform to the shape of the cavity; (3) heating the charge by means of the hot mold to polymerize and cure the material into a solidified part; and (4) opening the mold halves and removing the part from the cavity.

The initial charge of molding compound can be any of several forms, including powders or pellets, liquid, or preform. The amount of polymer must be precisely controlled to obtain repeatable consistency in the molded product. It has become common practice to preheat the charge before its placement into the mold; this softens the polymer and shortens the production cycle time. Preheating methods include infrared heaters, convection heating in an oven, and use of a heated rotating screw in a barrel. The latter technique (borrowed from injection molding) is also used to meter the amount of the charge.

Compression molding presses are oriented vertically and contain two platens to which the mold halves are fastened. The presses involve either of two types of actuation: (1) upstroke of the bottom platen or (2) downstroke of the top platen, the former being the more common machine configuration. They are generally powered by a hydraulic cylinder that can be designed to provide clamping capacities up to several hundred tons.

Molds for compression molding are generally simpler than their injection mold counterparts. There is no sprue and runner system in a compression mold, and the process itself is generally limited to simpler part geometries because of the lower flow capabilities of the starting thermosetting materials. However, provision must be made for heating the mold, usually accomplished by electric resistance heating, steam, or hot oil circulation. Compression molds can be classified as *hand molds,* used for trial runs; *semiautomatic,* in which the press follows a programmed cycle but the operator manually loads and unloads the press; and *automatic,* which operate under a fully automatic press cycle (including automatic loading and unloading).

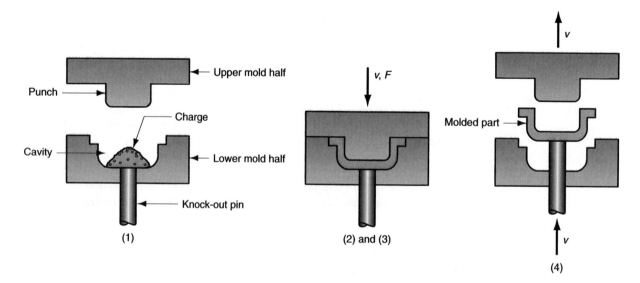

FIGURE 13.28   Compression molding for thermosetting plastics: (1) charge is loaded; (2) and (3) charge is compressed and cured; and (4) part is ejected and removed (some details omitted).

Materials for compression molding include phenolics, melamine, urea-formaldehyde, epoxies, urethanes, and elastomers. Typical moldings include electric plugs and sockets, pot handles, and dinnerware plates. Advantages of compression molding in these applications include (1) molds that are simpler and less expensive, (2) less scrap, and (3) low residual stresses in the molded parts. A typical disadvantage is longer cycle times and therefore lower production rates than injection molding.

13.7.2 TRANSFER MOLDING

In this process, a thermosetting charge is loaded into a chamber immediately ahead of the mold cavity, where it is heated; pressure is then applied to force the softened polymer to flow into the heated mold where curing occurs. There are two variants of the process, illustrated in Figure 13.29: (a) *pot transfer molding,* in which the charge is injected from a

FIGURE 13.29   (a) Pot transfer molding, and (b) plunger transfer molding. Cycle in both processes is: (1) charge is loaded into pot, (2) softened polymer is pressed into mold cavity and cured, and (3) part is ejected.

"pot" through a vertical sprue channel into the cavity; and (b) **plunger transfer molding,** in which the charge is injected by means of a plunger from a heated well through lateral channels into the mold cavity. In both cases, scrap is produced each cycle in the form of the leftover material in the base of the well and lateral channels, called the **cull.** In addition, the sprue in pot transfer is scrap material. Because the polymers are thermosetting, the scrap cannot be recovered.

Transfer molding is closely related to compression molding, because it is used on the same polymer types (thermosets and elastomers). One can also see similarities to injection molding, in the way the charge is preheated in a separate chamber and then injected into the mold. Transfer molding is capable of molding part shapes that are more intricate than compression molding but not as intricate as injection molding. Transfer molding also lends itself to molding with inserts, in which a metal or ceramic insert is placed into the cavity before injection, and the heated plastic bonds to the insert during molding.

13.8 BLOW MOLDING AND ROTATIONAL MOLDING

Both of these processes are used to make hollow, seamless parts out of thermoplastic polymers. Rotational molding can also be used for thermosets. Parts range in size from small plastic bottles of only 5 mL (0.15 oz) to large storage drums of 38,000-L (10,000-gal) capacity. Although the two processes compete in certain cases, generally they have found their own niches. Blow molding is more suited to the mass production of small disposable containers, whereas rotational molding favors large, hollow shapes.

13.8.1 BLOW MOLDING

Blow molding is a molding process in which air pressure is used to inflate soft plastic inside a mold cavity. It is an important industrial process for making one-piece hollow plastic parts with thin walls, such as bottles and similar containers. Because many of these items are used for consumer beverages for mass markets, production is typically organized for very high quantities. The technology is borrowed from the glass industry (Section 12.2.1) with which plastics compete in the disposable and recyclable bottle market.

Blow molding is accomplished in two steps: (1) fabrication of a starting tube of molten plastic, called a **parison** (same as in glass-blowing); and (2) inflation of the tube to the desired final shape. Forming the parison is accomplished by either extrusion or injection molding. The video clip on plastic blow molding illustrates the two categories.

VIDEO CLIP _____

Plastic Blow Molding. This clip contains three segments: (1) blow molding materials and processes, (2) extrusion blow molding, and (3) injection blow molding.

_____

Extrusion Blow Molding   This form of blow molding consists of the cycle illustrated in Figure 13.30. In most cases, the process is organized as a very high production operation for making plastic bottles. The sequence is automated and often integrated with downstream operations such as bottle filling and labeling.

It is usually a requirement that the blown container be rigid, and rigidity depends on wall thickness among other factors. We can relate wall thickness of the blown container to the starting extruded parison [12], assuming a cylindrical shape for the final product. The effect of die swell on the parison is shown in Figure 13.31. The mean diameter of the tube as

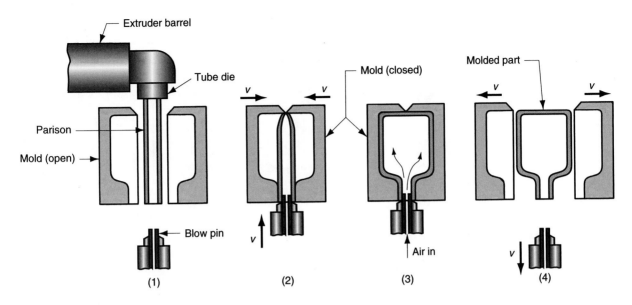

FIGURE 13.30 Extrusion blow molding: (1) extrusion of parison; (2) parison is pinched at the top and sealed at the bottom around a metal blow pin as the two halves of the mold come together; (3) the tube is inflated so that it takes the shape of the mold cavity; and (4) mold is opened to remove the solidified part.

it exits the die is determined by the mean die diameter D_d. Die swell causes expansion to a mean parison diameter D_p. At the same time, wall thickness swells from t_d to t_p. The swell ratio of the parison diameter and wall thickness is given by

$$r_s = \frac{D_p}{D_d} = \frac{t_p}{t_d} \qquad (13.20)$$

When the parison is inflated to the blow mold diameter D_m, there is a corresponding reduction in wall thickness to t_m. Assuming constant volume of cross section, we have

$$\pi D_p t_p = \pi D_m t_m \qquad (13.21)$$

Solving for t_m, we obtain

$$t_m = \frac{D_p t_p}{D_m}$$

FIGURE 13.31
(1) Dimensions of extrusion die, showing parison after die swell; and (2) final blow-molded container in extrusion blow molding.

FIGURE 13.32 Injection blow molding: (1) parison is injected molded around a blowing rod; (2) injection mold is opened and parison is transferred to a blow mold; (3) soft polymer is inflated to conform to the blow mold; and (4) blow mold is opened, and blown product is removed.

Substituting Eq. (13.20) into this equation, we get

$$t_m = \frac{r_s^2 t_d D_d}{D_m} \qquad (13.22)$$

The amount of die swell in the initial extrusion process can be measured by direct observation; and the dimensions of the die are known. Thus, we can determine the wall thickness on the blow-molded container.

Injection Blow Molding   In this process, the starting parison is injection molded rather than extruded. A simplified sequence is outlined in Figure 13.32. Compared to its extrusion-based competitor, injection blow molding usually has the following advantages: (1) higher production rate, (2) greater accuracy in the final dimensions, (3) lower scrap rates, and (4) less wasteful of material. On the other hand, larger containers can be produced with extrusion blow molding because the mold in injection molding is so expensive for large parisons. Also, extrusion blow molding is technically more feasible and economical for double-layer bottles used for storing certain medicines, personal care products, and various chemical compounds.[2]

In a variation of injection blow molding, called ***stretch blow molding*** (Figure 13.33), the blowing rod extends downward into the injection molded parison during step 2, thus stretching the soft plastic and creating a more favorable stressing of the polymer than conventional injection blow molding or extrusion blow molding. The resulting structure is more rigid, with higher transparency and better impact resistance. The most widely used material for stretch blow molding is polyethylene terephthalate (PET), a polyester that has very low permeability and is strengthened by the stretch-blow-molding process. The combination of properties makes it ideal as a container for carbonated beverages (e.g., 2-L soda bottles).

Materials and Products   Blow molding is limited to thermoplastics. Polyethylene is the polymer most commonly used for blow molding—in particular, high density and high molecular weight polyethylene (HDPE and HMWPE). In comparing their properties with those of low density PE given the requirement for stiffness in the final product, it is more

---

[2]The author is indebted to Tom Walko, plant manager at one of Graham Packaging Company's blow molding plants for providing the preceding comparisons between extrusion and injection blow molding.

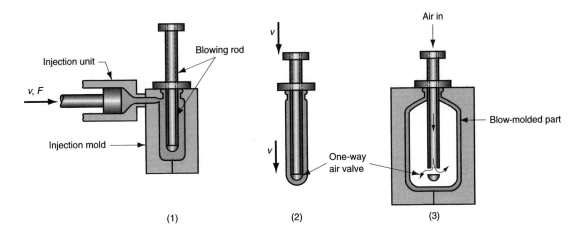

FIGURE 13.33   Stretch blow molding: (1) injection molding of parison, (2) stretching, and (3) blowing.

economical to use these more expensive materials because the container walls can be made thinner. Other blow moldings are made of polypropylene (PP), polyvinylchloride (PVC), and polyethylene terephthalate.

Disposable containers for packaging liquid consumer goods constitute the major share of products made by blow molding; but they are not the only products. Other items include large shipping drums (55-gal) for liquids and powders, large storage tanks (2000-gal), automotive gasoline tanks, toys, and hulls for sail boards and small boats. In the latter case, two boat hulls are made in a single blow molding and subsequently cut into two open hulls.

13.8.2 ROTATIONAL MOLDING

Rotational molding uses gravity inside a rotating mold to achieve a hollow form. Also called *rotomolding,* it is an alternative to blow molding for making large, hollow shapes. It is used principally for thermoplastic polymers, but applications for thermosets and elastomers are becoming more common. Rotomolding tends to favor more complex external geometries, larger parts, and lower production quantities than blow molding. The process consists of the following steps: (1) A predetermined amount of polymer powder is loaded into the cavity of a split mold. (2) The mold is then heated and simultaneously rotated on two perpendicular axes, so that the powder impinges on all internal surfaces of the mold, gradually forming a fused layer of uniform thickness. (3) While still rotating, the mold is cooled so that the plastic skin solidifies. (4) The mold is opened, and the part is unloaded. Rotational speeds used in the process are relatively slow. It is gravity, not centrifugal force, that causes uniform coating of the mold surfaces.

Molds in rotational molding are simple and inexpensive compared with injection molding or blow molding, but the production cycle is much longer, lasting perhaps 10 min or more. To balance these advantages and disadvantages in production, rotational molding is often performed on a multicavity indexing machine, such as the three-station machine shown in Figure 13.34. The machine is designed so that three molds are indexed in sequence through three workstations. Thus, all three molds are working simultaneously. The first workstation is an unload–load station in which the finished part is unloaded from the mold, and the powder for the next part is loaded into the cavity. The second station consists of a heating chamber where hot-air convection heats the mold while it is simultaneously rotated. Temperatures inside the chamber are around 375°C

FIGURE 13.34
Rotational molding cycle performed on a three-station indexing machine: (1) unload–load station; (2) heat and rotate mold; (3) cool the mold.

(700°F), depending on the polymer and the item being molded. The third station cools the mold, using forced cold air or water spray, to cool and solidify the plastic molding inside.

A fascinating variety of articles are made by rotational molding. The list includes hollow toys such as hobby horses and playing balls; boat and canoe hulls, sandboxes, small swimming pools; buoys and other flotation devices; truck body parts, automotive dashboards, fuel tanks; luggage pieces, furniture, garbage cans; fashion mannequins; large industrial barrels, containers, and storage tanks; portable outhouses, and septic tanks. The most popular molding material is polyethylene, especially HDPE. Other plastics include polypropylene, ABS, and high-impact polystyrene.

13.9 THERMOFORMING

Thermoforming is a process in which a flat thermoplastic sheet is heated and deformed into the desired shape. The process is widely used in packaging of consumer products and fabricating large items such as bathtubs, contoured skylights, and internal door liners for refrigerators.

Thermoforming consists of two main steps: heating and forming. Heating is usually accomplished by radiant electric heaters, located on one or both sides of the starting plastic sheet at a distance of roughly 125 mm (5 in). Duration of the heating cycle needed to sufficiently soften the sheet depends on the polymer—its thickness and color. Methods by which forming is accomplished can be classified into three basic categories: (1) vacuum thermoforming, (2) pressure thermoforming, and (3) mechanical thermoforming. In our discussion of these methods, we describe the forming of sheet stock, but in the packaging industry most thermoforming operations are performed on thin films.

Vacuum Thermoforming This was the first thermoforming process (simply called *vacuum forming* when it was developed in the 1950s). Negative pressure is used to draw a preheated sheet into a mold cavity. The process is explained in Figure 13.35 in its most

FIGURE 13.35 Vacuum thermoforming: (1) a flat plastic sheet is softened by heating; (2) the softened sheet is placed over a concave mold cavity; (3) a vacuum draws the sheet into the cavity; and (4) the plastic hardens on contact with the cold mold surface, and the part is removed and subsequently trimmed from the web.

basic form. The holes for drawing the vacuum in the mold are on the order of 0.8 mm (0.031 in) in diameter, so their effect on the plastic surface is minor.

Pressure Thermoforming An alternative to vacuum forming involves positive pressure to force the heated plastic into the mold cavity. This is called *pressure thermoforming* or *blow forming;* its advantage over vacuum forming is that higher pressures can be developed because the latter is limited to a theoretical maximum of 1 atm. Blow-forming pressures of 3 to 4 atm are common. The process sequence is similar to the previous, the difference being that the sheet is pressurized from above into the mold cavity. Vent holes are provided in the mold to exhaust the trapped air. The forming portion of the sequence (steps 2 and 3) is illustrated in Figure 13.36.

At this point it is useful to distinguish between negative and positive molds. The molds shown in Figures 13.35 and 13.36 are *negative molds* because they have concave cavities. A *positive mold* has a convex shape. Both types are used in thermoforming. In the case of the positive mold, the heated sheet is draped over the convex form and negative or positive pressure is used to force the plastic against the mold surface. A positive mold is shown in Figure 13.37 for vacuum thermoforming.

The difference between positive and negative molds may seem unimportant, because the part shapes are the same in the diagrams. However, if the part is drawn into the negative mold, then its exterior surface will have the exact surface contour of the mold cavity. The inside surface will be an approximation of the contour and will possess a finish corresponding to that of the starting sheet. By contrast, if the sheet is draped over a positive mold, then its interior surface will be identical to that of the convex mold; and its outside surface will follow approximately. Depending on the requirements of the product, this distinction might be important.

(2)

(3)

FIGURE 13.36 Pressure thermoforming. The sequence is similar to the previous figure, the difference being: (2) sheet is placed over a mold cavity; and (3) positive pressure forces the sheet into the cavity.

Another difference is in the thinning of the plastic sheet, one of the problems in thermoforming. Unless the contour of the mold is very shallow, there will be significant thinning of the sheet as it is stretched to conform to the mold contour. Positive and negative molds produce a different pattern of thinning in a given part. Consider the tub-shaped part in our figures. In the positive mold, as the sheet is draped over the convex form, the portion making contact with the top surface (corresponding to the base of the tub) solidifies quickly and experiences virtually no stretching. This results in a thick base but significant thinning in the walls of the tub. By contrast, a negative mold results in a more even distribution of stretching and thinning in the sheet before contact is made with the cold surface.

A way to improve the thinning distribution with a positive mold is to prestretch the sheet before draping it over the convex form. As shown in Figure 13.38, the heated plastic sheet is stretched uniformly by vacuum pressure into a spherical shape before drawing it over the mold.

The first step depicted in frame (1) of Figure 13.38 can be used alone as a method to produce globe-shaped parts such as skylight windows and transparent domes. In the process, closely controlled air pressure is applied to inflate the soft sheet. The pressure is maintained until the blown shape has solidified.

FIGURE 13.37 Use of a positive mold in vacuum thermoforming: (1) the heated plastic sheet is positioned above the convex mold and (2) the clamp is lowered into position, draping the sheet over the mold as a vacuum forces the sheet against the mold surface.

(1)

(2)

FIGURE 13.38
Prestretching the sheet in (1) prior to draping and vacuuming it over a positive mold in (2).

Mechanical Thermoforming   The third method, called mechanical thermoforming, uses matching positive and negative molds that are brought together against the heated plastic sheet, forcing it to assume their shape. In pure mechanical forming, air pressure is not used at all. The process is illustrated in Figure 13.39. Its advantages are better dimensional control and the opportunity for surface detailing on both sides of the part. The disadvantage is that two mold halves are required; therefore, the molds for the other two methods are less costly.

Applications   Thermoforming is a secondary shaping process, the primary process being that which produces the sheet or film (Section 13.3). Only thermoplastics can be thermoformed, because extruded sheets of thermosetting or elastomeric polymers have already been cross-linked and cannot be softened by reheating. Common thermoforming plastics are polystyrene, cellulose acetate and cellulose acetate butyrate, ABS, PVC, acrylic (polymethylmethacrylate), polyethylene, and polypropylene.

FIGURE 13.39
Mechanical thermoforming: (1) heated sheet placed above a negative mold, and (2) mold is closed to shape the sheet.

Mass production thermoforming operations are performed in the packaging industry. The starting sheet or film is rapidly fed through a heating chamber and then mechanically formed into the desired shape. The operations are often designed to produce multiple parts with each stroke of the press using molds with multiple cavities. In some cases, the extrusion machine that produces the sheet or film is located directly upstream from the thermoforming process, thereby eliminating the need to reheat the plastic. For best efficiency, the filling process to put the consumable food item into the container is placed immediately downstream from thermoforming.

Thin film packaging items that are mass produced by thermoforming include blister packs and skin packs. They offer an attractive way to display certain commodity products such as cosmetics, toiletries, small tools, and fasteners (nails, screws, etc.). Thermoforming applications include large parts that can be produced from thicker sheet stock. Examples include covers for business machines, boat hulls, shower stalls, diffusers for lights, advertising displays and signs, bathtubs, and certain toys. Contoured skylights and internal door liners for refrigerators are made, respectively, out of acrylic (because of its transparency) and ABS (because of its ease in forming and resistance to oils and fats found in refrigerators).

13.10 CASTING

In polymer shaping, casting involves pouring of a liquid resin into a mold, using gravity to fill the cavity, and allowing the polymer to harden. Both thermoplastics and thermosets are cast. Examples of the former include acrylics, polystyrene, polyamides (nylons), and vinyls (PVC). Conversion of the liquid resin into a hardened thermoplastic can be accomplished in several ways, which include (1) heating the thermoplastic resin to a highly fluid state so that it readily pours and fills the mold cavity, and then permitting it to cool and solidify in the mold; (2) using a low-molecular-weight prepolymer (or monomer) and polymerizing it in the mold to form a high-molecular-weight thermoplastic; and (3) pouring a plastisol (a liquid suspension of fine particles of a thermoplastic resin such as PVC in a plasticizer) into a heated mold so that it gels and solidifies.

Thermosetting polymers shaped by casting include polyurethane, unsaturated polyesters, phenolics, and epoxies. The process involves pouring the liquid ingredients that form the thermoset into a mold so that polymerization and cross-linking occur. Heat and/or catalysts may be required depending on the resin system. The reactions must be sufficiently slow to allow mold pouring to be completed. Fast-reacting thermosetting systems, such as certain polyurethane systems, require alternative shaping processes like reaction injection molding (Section 13.6.5).

Advantages of casting over alternative processes such as injection molding include: (1) the mold is simpler and less costly, (2) the cast item is relatively free of residual stresses and viscoelastic memory, and (3) the process is suited to low production quantities. Focusing on advantage (2), acrylic sheets (Plexiglas, Lucite) are generally cast between two pieces of highly polished plate glass. The casting process permits a high degree of flatness and desirable optical qualities to be achieved in the clear plastic sheets. Such flatness and clarity cannot be obtained by flat sheet extrusion. A disadvantage in some applications is significant shrinkage of the cast part during solidification. For example, acrylic sheets undergo a volumetric contraction of about 20% when cast. This is much more than in injection molding, in which high pressures are used to pack the mold cavity to reduce shrinkage.

Slush casting is an alternative to conventional casting, borrowed from metal casting technology. In **slush casting,** a liquid plastisol is poured into the cavity of a heated split mold, so that a skin forms at the surface of the mold. After a duration that depends on the desired thickness of the skin, the excess liquid is poured out of the mold; the mold is then opened for part removal. The process is also referred to as ***shell casting*** [6].

An important application of casting in electronics is ***encapsulation,*** in which items such as transformers, coils, connectors, and other electrical components are encased in plastic by casting.

13.11   POLYMER FOAM PROCESSING AND FORMING

A polymer foam is a polymer-and-gas mixture, which gives the material a porous or cellular structure. Other terms used for polymer foams include ***cellular polymer, blown polymer,*** and ***expanded polymer.*** The most common polymer foams are polystyrene (Styrofoam) and polyurethane. Other polymers used to make foams include natural rubber ("foamed rubber") and polyvinylchloride (PVC).

The characteristic properties of a foamed polymer include (1) low density, (2) high strength per unit weight, (3) good thermal insulation, and (4) good energy absorbing qualities. The elasticity of the base polymer determines the corresponding property of the foam. Polymer foams can be classified [6] as (1) ***elastomeric,*** in which the matrix polymer is a rubber, capable of large elastic deformation; (2) ***flexible,*** in which the matrix is a highly plasticized polymer such as soft PVC; and (3) ***rigid,*** in which the polymer is a stiff thermoplastic such as polystyrene or a thermosetting plastic such as a phenolic. Depending on chemical formulation and degree of cross-linking, polyurethanes can range over all three categories.

The characteristic properties of polymer foams, and the ability to control their elastic behavior through selection of the base polymer, make these materials highly suitable for certain types of applications, including hot beverage cups, heat insulating structural materials and cores for structural panels, packaging materials, cushion materials for furniture and bedding, padding for automobile dashboards, and products requiring buoyancy.

Common gases used in polymer foams are air, nitrogen, and carbon dioxide. The proportion of gas can range up to 90% or more. The gas is introduced into the polymer by several methods, called foaming processes. These include (1) mixing a liquid resin with air by ***mechanical agitation,*** then hardening the polymer by means of heat or chemical reaction; (2) mixing a ***physical blowing agent*** with the polymer—a gas such as nitrogen (N_2) or pentane (C_5H_{12}), which can be dissolved in the polymer melt under pressure, so that the gas comes out of solution and expands when the pressure is subsequently reduced; and (3) mixing the polymer with chemical compounds, called ***chemical blowing agents,*** that decompose at elevated temperatures to liberate gases such as CO_2 or N_2 within the melt.

The way the gas is distributed throughout the polymer matrix distinguishes two basic foam structures, illustrated in Figure 13.40: (a) ***closed cell,*** in which the gas pores are roughly spherical and completely separated from each other by the polymer matrix; and (b) ***open cell,*** in which the pores are interconnected to some extent, allowing passage of a fluid through the foam. A closed cell structure makes a satisfactory life jacket; an open cell structure would become waterlogged. Other attributes that characterize the structure include the relative proportions of polymer and gas (already mentioned) and the cell density (number of cells per unit volume), which is inversely related to the size of the individual air cells in the foam.

There are many shaping processes for polymer foam products. Because the two most important foams are polystyrene and polyurethane, this discussion is limited to shaping processes for these two materials. Because polystyrene is a thermoplastic and polyurethane can be either a thermoset or an elastomer (it can also be a thermoplastic but

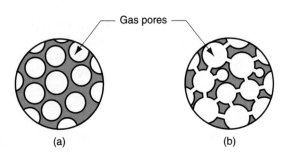

FIGURE 13.40 Two polymer foam structures: (a) closed cell, and (b) open cell.

is less important in this form), the processes covered here for these materials are representative of those used for other polymer foams.

Polystyrene foams are shaped by extrusion and molding. In *extrusion,* a physical or chemical blowing agent is fed into the polymer melt near the die end of the extruder barrel; thus, the extrudate consists of the expanded polymer. Large sheets and boards are made in this way and are subsequently cut to size for heat insulation panels and sections.

Several molding processes are available for polystyrene foam. We previously discussed *structural foam molding* and *sandwich molding* (Section 13.6.5). A more widely used process is *expandable foam molding,* in which the molding material usually consists of prefoamed polystyrene beads. The prefoamed beads are produced from pellets of solid polystyrene that have been impregnated with a physical blowing agent. Prefoaming is performed in a large tank by applying steam heat to partially expand the pellets, simultaneously agitating them to prevent fusion. Then, in the molding process, the prefoamed beads are fed into a mold cavity, where they are further expanded and fused together to form the molded product. Hot beverage cups of polystyrene foam are produced in this way. In some processes, the prefoaming step is omitted, and the impregnated beads are fed directly into the mold cavity, where they are heated, expanded, and fused. In other operations, the expandable foam is first formed into a flat sheet by the *blown-film extrusion process* (Section 13.3) and then shaped by *thermoforming* (Section 13.9) into packaging containers such as egg cartons.

Polyurethane foam products are made in a one-step process in which the two liquid ingredients (polyol and isocyanate) are mixed and immediately fed into a mold or other form, so that the polymer is synthesized and the part geometry is created at the same time. Shaping processes for polyurethane foam can be divided into two basic types [11]: spraying and pouring. *Spraying* involves use of a spray gun into which the two ingredients are continuously fed, mixed, and then sprayed onto a target surface. The reactions leading to polymerization and foaming occur after application on the surface. This method is used to apply rigid insulating foams onto construction panels, railway cars, and similar large items. *Pouring* involves dispensing the ingredients from a mixing head into an open or closed mold in which the reactions occur. An open mold can be a container with the required contour (e. g., for an automobile seat cushion) or a long channel that is slowly moved past the pouring spout to make long, continuous sections of foam. The closed mold is a completely enclosed cavity into which a certain amount of the mixture is dispensed. Expansion of the reactants completely fills the cavity to shape the part. For fast-reacting polyurethanes, the mixture must be rapidly injected into the mold cavity using *reaction injection molding* (Section 13.66). The degree of cross-linking, controlled by the starting ingredients, determines the relative stiffness of the resulting foam.

13.12 PRODUCT DESIGN CONSIDERATIONS

Plastics are an important design material, but the designer must be aware of their limitations. This section lists some design guidelines for plastic components, beginning with those that apply in general, and then ones applicable to extrusion and molding (injection molding, compression molding, and transfer molding).

Several general guidelines apply, irrespective of the shaping process. They are mostly limitations of plastic materials that must be considered by the designer.

➤ **Strength and stiffness.** Plastics are not as strong or stiff as metals. They should not be used in applications in which high stresses will be encountered. Creep resistance is also a limitation. Strength properties vary significantly among plastics, and strength-to-weight ratios for some plastics are competitive with metals in certain applications.

➤ **Impact resistance.** The capacity of plastics to absorb impact is generally good; plastics compare favorably with most metals.

➤ **Service temperatures** of plastics are limited relative to engineering metals and ceramics.

➤ **Thermal expansion** is greater for plastics than metals; so dimensional changes owing to temperature variations are much more significant than for metals.

➤ Many types of plastics are subject to **degradation** from sunlight and certain other forms of radiation. Also, some plastics degrade in oxygen and ozone atmospheres. Finally, plastics are soluble in many common solvents. On the positive side, plastics are resistant to conventional corrosion mechanisms that afflict many metals. The weaknesses of specific plastics must be taken into account by the designer.

Extrusion is one of the most widely used plastic shaping processes. Several design recommendations are presented here for conventional extrusion (compiled mostly from [3]).

➤ **Wall thickness.** Uniform wall thickness is desirable in an extruded cross section. Variations in wall thickness result in nonuniform plastic flow and uneven cooling that tend to warp the extrudate.

➤ **Hollow sections.** Hollow sections complicate die design and plastic flow. It is desirable to use extruded cross sections that are not hollow yet satisfy functional requirements.

➤ **Corners.** Sharp corners, inside and outside, should be avoided in the cross section, because they result in uneven flow during processing and stress concentrations in the final product.

The following guidelines apply to injection molding (the most popular molding process), compression molding, and transfer molding (compiled from Bralla [3], McCrum [10], and other sources).

➤ **Economic production quantities.** Each molded part requires a unique mold, and the mold for any of these processes can be costly, particularly for injection molding. Minimum production quantities for injection molding are usually around 10,000 pieces; for compression molding, minimum quantities are around 1000 parts, because of the simpler mold designs involved. Transfer molding lies between the other two.

➤ **Part complexity.** Although more complex part geometries mean more costly molds, it may nevertheless be economical to design a complex molding if the alternative involves many individual components assembled together. An advantage of plastic molding is that it allows multiple functional features to be combined into one part.

➤ **Wall thickness.** Thick cross sections are generally undesirable; they are wasteful of material, more likely to cause warping caused by shrinkage, and take longer to harden. **Reinforcing ribs** can be used in molded plastic parts to achieve increased stiffness without excessive wall thickness. The ribs should be made thinner than the walls they reinforce, to minimize sink marks on the outside wall.

➤ **Corner radii and fillets.** Sharp corners, both external and internal, are undesirable in molded parts; they interrupt smooth flow of the melt, tend to create surface defects, and cause stress concentrations in the finished part.

TABLE 13.2 Typical tolerances on molded parts for selected plastics.

Plastic	Tolerances for:[a]		Plastic	Tolerances for:[a]	
	50-mm Dimension	10-mm Hole		50-mm Dimension	10-mm Hole
Thermoplastic:			*Thermosetting:*		
ABS	±0.2 mm (±0.007 in)	±0.08 mm (±0.003 in)	Epoxies	±0.15 mm (±0.006 in)	±0.05 mm (±0.002 in)
Polyethylene	±0.3 mm (±0.010 in)	±0.13 mm (±0.005 in)	Phenolics	±0.2 mm (±0.008 in)	±0.08 mm (±0.003 in)
Polystyrene	±0.15 mm (±0.006 in)	±0.1 mm (±0.004 in)			

Values represent typical commercial molding practice. Compiled from [3], [7], [14], and [19].
[a]For smaller sizes, tolerances can be reduced. For larger sizes, more generous tolerances are required.

> ➢ *Holes.* Holes are quite feasible in plastic moldings, but they complicate mold design and part removal. They also cause interruptions in melt flow.

> ➢ *Draft.* A molded part should be designed with a draft on its sides to facilitate removal from the mold. This is especially important on the inside wall of a cup-shaped part because the molded plastic contracts against the positive mold shape. The recommended draft for thermosets is around 1/2° to 1°; for thermoplastics it usually ranges between 1/8° and 1/2°. Suppliers of plastic molding compounds provide recommended draft values for their products.

> ➢ *Tolerances.* Tolerances specify the allowable manufacturing variations for a part. Although shrinkage is predictable under closely controlled conditions, generous tolerances are desirable for injection moldings because of variations in process parameters that affect shrinkage and diversity of part geometries encountered. Table 13.2 lists typical tolerances for molded part dimensions of selected plastics.

REFERENCES

[1] Baird, D. G., and Collias, D. I. *Polymer Processing Principles and Design,* John Wiley & Sons, New York, 1998.

[2] Billmeyer, Fred, W., Jr. *Textbook of Polymer Science,* 3rd ed. John Wiley & Sons, New York, 1984.

[3] Bralla, J. G.(editor in chief). *Design for Manufacturability Handbook,* 2nd ed. McGraw-Hill Book Company, New York, 1998.

[4] Briston, J. H. *Plastic Films,* 3rd ed. Longman Group U.K., Essex, England, 1989.

[5] Chanda, M., and Roy, S. K. *Plastics Technology Handbook,* Marcel Dekker, New York, 1998.

[6] Charrier, J-M. *Polymeric Materials and Processing,* Oxford University Press, New York, 1991.

[7] *Engineering Materials Handbook,* Vol. 2, *Engineering Plastics,* ASM International, Metals Park, Ohio, 1988.

[8] Hall, C. *Polymer Materials,* 2nd ed. John Wiley & Sons. New York, 1989.

[9] Hensen, F. (ed.). *Plastic Extrusion Technology,* Hanser Publishers, Munich, FRG, 1988. (Distributed in United States by Oxford University Press, New York.)

[10] McCrum, N. G., Buckley, C. P., and Bucknall, C. B. *Principles of Polymer Engineering,* 2nd ed., Oxford University Press, Oxford, UK, 1997.

[11] *Modern Plastics Encyclopedia,* Modern Plastics, McGraw-Hill, Hightstown, New Jersey, 1991.

[12] Morton-Jones, D. H. *Polymer Processing,* Chapman and Hall, London, UK, 1989.

[13] Pearson, J. R. A. *Mechanics of Polymer Processing,* Elsevier Applied Science Publishers, London, 1985.

[14] Rubin, I. I. *Injection Molding: Theory and Practice,* John Wiley & Sons, New York, 1973.

[15] Rudin, A. *The Elements of Polymer Science and Engineering,* 2nd ed., Academic Press, Orlando, Florida, 1999.

[16] Strong, A. B. *Plastics: Materials and Processing,* 3rd ed. Pearson Educational, Upper Saddle River, New Jersey, 2006.

[17] Sweeney, F. M. *Reaction Injection Molding Machinery and Processes,* Marcel Dekker, Inc., New York, 1987.

[18] Tadmor, Z., and Gogos, C. G. *Principles of Polymer Processing,* John Wiley & Sons, New York, 1979.

[19] Wick, C., Benedict, J. T., and Veilleux, R. F. *Tool and Manufacturing Engineers Handbook,* 4th ed., Vol. II: *Forming.* Society of Manufacturing Engineers, Dearborn, Michigan, 1984, Chapter 18.

REVIEW QUESTIONS

13.1. What are some of the reasons why plastic shaping processes are important?

13.2. Identify the main categories of plastics shaping processes, as classified by the resulting product geometry.

13.3. Viscosity is an important property of a polymer melt in plastics shaping processes. Upon what parameters does viscosity depend?

13.4. How does the viscosity of a polymer melt differ from most fluids that are Newtonian.

13.5. What does viscoelasticity mean, when applied to a polymer melt?

13.6. Define die swell in extrusion.

13.7. Briefly describe the plastic extrusion process.

13.8. The barrel and screw of an extruder are generally divided into three sections; identify the sections.

13.9. What are the functions of the screen pack and breaker plate at the die end of the extruder barrel?

13.10. What are the various forms of extruded shapes and corresponding dies?

13.11. What is the distinction between plastic sheet and film?

13.12. What is the blown-film process for producing film stock?

13.13. Describe the calendering process.

13.14. Polymer fibers and filaments are used in several applications; what is the most important application commercially?

13.15. Technically, what is the difference between a fiber and a filament?

13.16. Among the synthetic fiber materials, which are the most important?

13.17. Briefly describe the injection molding process.

13.18. An injection-molding machine is divided into two principal components. Name them.

13.19. What are the two basic types of clamping units?

13.20. What is the function of gates in injection molds?

13.21. What are the advantages of a three-plate mold over a two-plate mold in injection molding?

13.22. Discuss some of the defects that can occur in plastic injection molding.

13.23. Describe structural-foam molding.

13.24. What are the significant differences in the equipment and operating procedures between injection molding of thermoplastics and injection molding of thermosets?

13.25. What is reaction injection molding?

13.26. What kinds of products are produced by blow molding?

13.27. What is the form of the starting material in thermoforming?

13.28. What is the difference between a positive mold and a negative mold in thermoforming?

13.29. Why are the molds generally more costly in mechanical thermoforming than in pressure or vacuum thermoforming?

13.30. What are the processes by which polymer foams are produced?

13.31. What are some of the general considerations that product designers must keep in mind when designing components out of plastics?

13.32. (**Video**) According to the injection molding videos, what are the four primary elements that influence the injection molding process?

13.33. (**Video**) According to the injection molding video, name the four types of mold design most common in industry.

13.34. (**Video**) According to the injection molding video, what is the most common type of injection molding machine used in industry?

13.35. (**Video**) According to the blow molding video, what materials are used in blow molding? Name three.

13.36. (**Video**) List the four most common blow-molding processes according to the video on blow molding.

13.37. (**Video**) List the stages of extrusion blow molding according to the video.

13.38. (**Video**) Name the four types of finishing operations performed on plastics, according to the plastics finishing video.

13.39. (**Video**) What are the different processes that can be used to apply decorations to plastic parts according to the plastics finishing video?

MULTIPLE CHOICE QUIZ

There are 29 correct answers in the following multiple choice questions (some questions have multiple answers that are correct). To attain a perfect score on the quiz, all correct answers must be given. Each correct answer is worth 1 point. Each

omitted answer or wrong answer reduces the score by 1 point, and each additional answer beyond the correct number of answers reduces the score by 1 point. Percentage score on the quiz is based on the total number of correct answers.

13.1. The forward movement of polymer melt in an extruder barrel is resisted by drag flow, which is caused by the resistance to flow through the die orifice: (a) true or (b) false?

13.2. Which of the following are sections of a conventional extruder barrel for thermoplastics (three best answers): (a) compression section, (b) die section, (c) feed section, (d) heating section, (e) metering section, and (f) shaping section?

13.3. Which of the following processes are associated with the production of plastic sheet and film (three correct answers): (a) blown-film extrusion process, (b) calendering, (c) chill-roll extrusion, (d) doctor blade method, (e) spinning, (f) thermoforming, and (g) transfer molding?

13.4. The principal components of an injection molding machine are which two of the following: (a) clamping unit, (b) hopper, (c) injection unit, (d) mold, and (e) part ejection unit?

13.5. The parting line in injection molding is which one of the following: (a) the lines formed where polymer melt meets after flowing around a core in the mold, (b) the narrow gate sections where the parts are separated from the runner, (c) where the clamping unit is joined to the injection unit in the molding machine, or (d) where the two mold halves come together?

13.6. The function of the ejection system is which one of the following: (a) move polymer melt into the mold cavity, (b) open the mold halves after the cavity is filled, (c) remove the molded parts from the runner system after molding, or (d) separate the part from the cavity after molding?

13.7. A three-plate mold offers which of the following advantages when compared to a two-plate mold (two best answers): (a) automatic separation of parts from runners, (b) gating is usually at the base of the part to reduce weld lines, (c) sprue does not solidify, and (d) stronger molded parts?

13.8. Which of the following defects or problems is associated with injection molding (three correct answers): (a) bambooing, (b) die swell, (c) drag flow, (d) flash, (e) melt fracture, (f) short shots, or (g) sink marks?

13.9. In rotational molding, centrifugal force is used to force the polymer melt against the surfaces of the mold cavity where solidification occurs: (a) true or (b) false?

13.10. Use of a parison is associated with which one of the following plastic shaping processes: (a) bi-injection molding, (b) blow molding, (c) compression molding, (d) pressure thermoforming, or (e) sandwich molding?

13.11. A thermoforming mold with a convex form is called which one of the following: (a) a die, (b) a negative mold, (c) a positive mold, or (d) a three-plate mold?

13.12. The term encapsulation refers to which one of the following plastics shaping processes: (a) casting, (b) compression molding, (c) extrusion of hollow forms, (d) injection molding in which a metal insert is encased in the molded part, or (e) vacuum thermoforming using a positive mold?

13.13. The two most common polymer foams are which of the following: (a) polyacetal, (b) polyethylene, (c) polystyrene, (d) polyurethane, and (e) polyvinylchloride?

13.14. In which of the following properties do plastic parts often compare favorably with metals (two best answers): (a) impact resistance, (b) resistance to ultraviolet radiation, (c) stiffness, (d) strength, (e) strength-to-weight ratio, and (f) temperature resistance?

13.15. Which of the following processes are generally limited to thermoplastic polymers (two best answers): (a) blow molding, (b) compression molding, (c) reaction injection molding, (d) thermoforming, (e) transfer molding, and (f) wire coating?

13.16. Which of the following processes would be applicable to produce hulls for small boats (three best answers): (a) blow molding, (b) compression molding, (c) injection molding, (d) rotational molding, and (e) vacuum thermoforming?

PROBLEMS

Extrusion

13.1. The diameter of an extruder barrel is 65 mm and its length = 1.75 m. The screw rotates at 55 rev/min. The screw channel depth = 5.0 mm, and the flight angle = 18°. The head pressure at the die end of the barrel is 5.0×10^6 Pa. The viscosity of the polymer melt is given as 100 Pa-s. Find the volume flow rate of the plastic in the barrel.

13.2. An extruder has a diameter of 5.0 in and a length to diameter ratio of 26. The barrel heats the polypropylene melt to 450°F, which provides a melt viscosity of 0.0025 lb-s/in^2. The pitch of the screw is 4.2 in and the channel depth is 0.15 in. In operation the screw rotates at 50 rev/min and a head pressure of 450 lb/in^2 is generated. What is the volume flow rate of polypropylene from the die at the end of the barrel?

13.3. An extruder barrel has a diameter of 110 mm and a length of 3.0 m. The screw channel depth = 7.0 mm, and its pitch = 95 mm. The viscosity of the polymer melt is 105 Pa-s, and the head pressure in the barrel is 4.0 MPa. What rotational speed of the screw is required to achieve a volumetric flow rate of 90 cm^3/s?

13.4. An extruder has a barrel diameter of 2.5 in and a length of 6.0 ft. The screw has a channel depth of 0.25 in, a flight angle of 20°, and rotates at 55 rev/min. The material being extruded is polypropylene. At the present settings, the volumetric flow rate of the polymer melt is 1.50 in^3/sec and the head pressure is 500 lb/in^2. (a) Under these operating characteristics, what is the viscosity of the polypropylene? (b) Using Figure 13.2, approximate the temperature in °F of the polypropylene.

13.5. An extruder has diameter = 80 mm and length = 2.0 m. Its screw has a channel depth = 5 mm, flight angle = 18 degrees, and it rotates at 1 rev/sec. The plastic melt has a shear viscosity = 150 Pa-s. Determine the extruder characteristic by computing Q_{max} and p_{max} and then finding the equation of the straight line between them.

13.6. Determine the helix angle A such that the screw pitch p is equal to the screw diameter D. This is called the "square" angle in plastics extrusion - the angle that provides a flight advance equal to one diameter for each rotation of the screw.

13.7. An extruder barrel has a diameter of 2.5 in. The screw rotates at 60 rev/min; its channel depth = 0.20 in, and its flight angle = 17.5°. The head pressure at the die end of the barrel is 800 lb/in^2 and the length of the barrel is 50 in. The viscosity of the polymer melt is 122 × 10^{-4} lb-sec/in^2. Determine the volume flow rate of the plastic in the barrel.

13.8. An extruder barrel has a diameter of 4.0 in and an L/D ratio of 28. The screw channel depth = 0.25 in, and its pitch = 4.8 in. It rotates at 60 rev/min. The viscosity of the polymer melt is 100 × 10^{-4} lb-sec/in^2. What head pressure is required to obtain a volume flow rate = 150 in^3/min?

13.9. An extrusion operation produces continuous tubing with outside diameter = 2.0 in and inside diameter = 1.7 in. The extruder barrel has a diameter = 4.0 in and length = 10 ft. The screw rotates at 50 rev/min; it has a channel depth = 0.25 in and flight angle = 16°. The head pressure has a value of 350 lb/in^2 and the viscosity of the polymer melt is 80 × 10^{-4} lb-sec/in^2. Under these conditions, what is the production rate in length of tube/min, assuming the extrudate is pulled at a rate that eliminates the effect of die swell (i.e., the tubing has the same OD and ID as the die profile)?

13.10. Continuous tubing is produced in a plastic extrusion operation through a die orifice whose outside diameter = 2.0 in and inside diameter = 1.5 in. The extruder barrel diameter = 5.0 in and length = 12 ft. The screw rotates at 50 rev/min; it has a channel depth = 0.30 in and flight angle = 16°. The head pressure has a value of 350 lb/in^2 and the viscosity of the polymer melt is 90 × 10^{-4} lb-sec/in^2. Under these conditions, what is the production rate in length of tube/min, given that the die swell ratio is 1.25.

13.11. An extruder has barrel diameter and length of 100 mm and 2.8 m, respectively. The screw rotational speed = 50 rev/min, channel depth = 7.5 mm, and flight angle = 17°. The plastic melt has a shear viscosity = 175 Pa-s. Determine: (a) the extruder characteristic, (b) the shape factor K_s for a circular die opening with diameter = 3.0 mm and length = 12.0 mm, and (c) the operating point (Q and p).

13.12. For Problem 01, assume the material is acrylic. (a) Using Figure 13.2, determine the temperature of the polymer melt. (b) If the temperature is lowered 20°C, estimate the resulting viscosity of the polymer melt. (Hint: the y-axis of Figure 13.2 is a log scale, not linear).

13.13. Consider an extruder in which the barrel diameter = 4.5 in and length = 11 ft. The extruder screw rotates at 60 rev/min; it has channel depth = 0.35 in and flight angle = 20°. The plastic melt has a shear viscosity = 125 × 10^{-4} lb-sec/in^2. Determine: (a) Q_{max} and p_{max}; (b) the shape factor K_s for a circular die opening in which $D_d = 0.312$ in and $L_d = 0.75$ in; and (c) the values of Q and p at the operating point.

13.14. An extruder has a barrel diameter = 5.0 in and length = 12 ft. The extruder screw rotates at 50 rev/min; it has channel depth = 0.30 in and flight angle = 17.7°. The plastic melt has a shear viscosity = 100 × 10^{-4} lb-sec/in^2. Find: (a) the extruder characteristic, (b) the values of Q and p at the operating point, given that the die characteristic is $Q_x = 0.00150\ p$.

13.15. Given the data in Problem 13.14, except that the flight angle of the extruder screw is a variable instead of a constant 17.7°. Use a spreadsheet calculator to determine the value of the flight angle that maximizes the volumetric flow rate Q_x. Explore values of flight angle between 10° and

20°. Determine the optimum value to the nearest tenth of a degree.

13.16. An extruder has a barrel diameter of 3.5 in and a length of 5.0 ft. It has a screw channel depth of 0.16 in and a flight angle of 22°. The extruder screw rotates at 75 rev/min. The polymer melt has a shear viscosity $= 65 \times 10^{-4}$ lb-sec/in^2 at the operating temperature of 525°F. The specific gravity of the polymer is 1.2 and its tensile strength is 8000 lb/in^2. A T-shaped cross section is extruded at a rate of 0.11 lb/sec. The density of water is 62.5 lb/ft^3. (a) Find the equation for the extruder characteristic. (b) Find the operating point (Q and p), and (c) the die characteristic that is indicated by the operating point.

Injection Molding

13.17. Compute the percentage volumetric contraction of a polyethylene molded part, based on the value of shrinkage given in Table 13.1.

13.18. The specified dimension $= 225.00$ mm for a certain injection molded part made of ABS. Compute the corresponding dimension to which the mold cavity should be machined, using the value of shrinkage given in Table 13.1.

13.19. The part dimension for a certain injection molded part made of polycarbonate is specified as 3.75 in. Compute the corresponding dimension to which the mold cavity should be machined, using the value of shrinkage given in Table 13.1.

13.20. The foreman in the injection molding department says that a polyethylene part produced in one of the operations has greater shrinkage than the calculations indicate it should have. The important dimension of the part is specified as 112.5 ± 0.25 mm. However, the actual molded part measures 112.02 mm. (a) As a first step, the corresponding mold cavity dimension should be checked. Compute the correct value of the mold dimension, given that the shrinkage value for polyethylene is 0.025 (from Table 13.1). (b) What adjustments in process parameters could be made to reduce the amount of shrinkage?

13.21. An injection molded polyethylene part has a dimension of 2.500 in. A new material, polycarbonate, is used in the same mold. What is the expected corresponding dimension of the polycarbonate molding?

Other Molding Operations and Thermoforming

13.22. The extrusion die for a polyethylene parison used in blow molding has a mean diameter of 18.0 mm. The size of the ring opening in the die is 2.0 mm. The mean diameter of the parison is observed to swell to a size of 21.5 mm after exiting the die orifice. If the diameter of the blow molded container is to be 150 mm, determine (a) the corresponding wall thickness of the container and (b) the wall thickness of the parison.

13.23. A parison is extruded from a die with outside diameter $= 11.5$ mm and inside diameter $= 7.5$ mm. The observed die swell is 1.25. The parison is used to blow mold a beverage container whose outside diameter $= 112$ mm (a standard size 2-L soda bottle). (a) What is the corresponding wall thickness of the container? (b) Obtain an empty 2-L plastic soda bottle and (carefully) cut it across the diameter. Using a micrometer, measure the wall thickness to compare with your answer in (a).

13.24. A blow-molding operation is used to produce a bottle with a diameter of 2.250 in and a wall thickness of 0.045 in. The parison has a thickness of 0.290 in. The observed die swell ratio is 1.30.

(a) What is the required diameter of the parison? (b) What is the diameter of the die?

13.25. An extrusion operation is used to produce a parison whose mean diameter $= 27$ mm. The inside and outside diameters of the die that produced the parison are 18 mm and 22 mm, respectively. If the minimum wall thickness of the blow-molded container is to be 0.40 mm, what is the maximum possible diameter of the blow mold?

13.26. A rotational molding operation is to be used to mold a hollow playing ball out of polypropylene. The ball will be 1.25 ft in diameter and its wall thickness should be 3/32 in. What weight of PP powder should be loaded into the mold to meet these specifications? The specific gravity of the PP grade is 0.90, and the density of water is 62.4 lb/ft^3.

13.27. The problem in a certain thermoforming operation is that there is too much thinning in the walls of the large cup-shaped part. The operation is conventional pressure thermoforming using a positive mold, and the plastic is an ABS sheet with an initial thickness of 3.2 mm. (a) Why is thinning occurring in the walls of the cup? (b) What changes could be made in the operation to correct the problem?

14 RUBBER-PROCESSING TECHNOLOGY

Chapter Contents

Many of the shaping processes used for plastics (Chapter 13) are also applicable to rubbers. However, rubber-processing technology is different in certain respects, and the rubber industry is largely separate from the plastics industry. The rubber industry and goods made of rubber are dominated by one product: tires. Tires are used in large numbers for automobiles, trucks, aircraft, and bicycles. Although pneumatic tires date from the late 1880s, rubber technology can be traced to the discovery in 1839 of vulcanization (Historical Note 8.2), the process by which raw natural rubber is transformed into a usable material through cross–linking of the polymer molecules. During its first century, the rubber industry was concerned only with the processing of natural rubber. Around World War II, synthetic rubbers were developed (Historical Note 8.3); today they account for the majority of rubber production.

14.1 RUBBER PROCESSING AND SHAPING

Production of rubber goods can be divided into two basic steps: (1) production of the rubber itself, and (2) processing of the rubber into finished goods. Production of rubber differs, depending on whether it is natural or synthetic. The difference results from the source of the raw materials. Natural rubber (NR) is produced as an agricultural crop, whereas most synthetic rubbers are made from petroleum.

Production of rubber is followed by processing into final products; this consists of (1) compounding, (2) mixing, (3) shaping, and (4) vulcanizing. Processing techniques for natural and synthetic rubbers are virtually the same, differences being in the chemicals used to effect vulcanization (cross–linking). This sequence does not apply to thermoplastic elastomers, whose shaping techniques are the same as for other thermoplastic polymers.

315

There are several distinct industries involved in the production and processing of rubber. Production of raw natural rubber might be classified as farming because latex, the starting ingredient for natural rubber, is grown on large plantations located in tropical climates. By contrast, synthetic rubbers are produced by the petrochemical industry. Finally, the processing of these materials into tires, shoe soles, and other rubber products occurs at processor (fabricator) plants. The processors are commonly known as the rubber industry. Some of the great names in this industry include Goodyear, B. F. Goodrich, and Michelin. The importance of the tire is reflected in these names.

14.1.1 PRODUCTION OF RUBBER

In this section we briefly survey the production of rubber before it goes to the processor. Our coverage distinguishes natural rubber and synthetic rubber.

Natural Rubber   Natural rubber is tapped from rubber trees (**Hevea brasiliensis**) as latex. The trees are grown on plantations in Southeast Asia and other parts of the world. Latex is a colloidal dispersion of solid particles of the polymer polyisoprene (Section 8.4.2) in water. Polyisoprene is the chemical substance that comprises rubber, and its content in the emulsion is about 30%. The latex is collected in large tanks, thus blending the yield of many trees together.

    The preferred method of recovering rubber from the latex involves coagulation. The latex is first diluted with water to about half its natural concentration. An acid such as formic acid ($HCOOH$) or acetic acid (CH_3COOH) is added to cause the latex to coagulate after about 12 hours. The coagulum, now in the form of soft solid slabs, is then squeezed through a series of rolls that drive out most of the water and reduce the thickness to about 3 mm (1/8 in). The final rolls have grooves that impart a criss-cross pattern to the resulting sheets. The sheets are then draped over wooden frames and dried in smokehouses. The hot smoke contains creosote, which prevents mildew and oxidation of the rubber. Several days are normally required to complete the drying process. The resulting rubber, now in a form called *ribbed smoked sheet,* is folded into large bales for shipment to the processor. This raw rubber has a characteristic dark brown color. In some cases, the sheets are dried in hot air rather than smokehouses, and the term *air-dried sheet* is applied; this is considered to be a better grade of rubber. A still better grade, called *pale crepe* rubber, involves two coagulation steps; the first removes undesirable components of the latex, then the resulting coagulum is subjected to a more involved washing and mechanical working procedure, followed by warm air drying. The color of pale crepe rubber approaches a light tan.

Synthetic Rubber   The various types of synthetic rubber were identified in Section 8.4.3. Most synthetics are produced from petroleum by the same polymerization techniques used to synthesize other polymers (Section 8.1.1). However, unlike thermoplastic and thermo-setting polymers, which are normally supplied to the fabricator as pellets or liquid resins, synthetic rubbers are supplied to rubber processors in the form of large bales. The industry has developed a long tradition of handling natural rubber in these unit loads.

14.1.2 COMPOUNDING

Rubber is always compounded with additives. It is through compounding that the specific rubber is designed to satisfy the given application in terms of properties, cost, and processability. Compounding adds chemicals for vulcanization. Sulfur has traditionally been used for this purpose. The vulcanization process and the chemicals used to accomplish it are discussed in Section 14.1.5.

Additives include fillers that act either to enhance the rubber's mechanical properties (reinforcing fillers) or to extend the rubber to reduce cost (nonreinforcing fillers). The single most important reinforcing filler in rubber is *carbon black,* a colloidal form of carbon, black in color, obtained from the thermal decomposition of hydrocarbons (soot). Its effect is to increase tensile strength and resistance to abrasion and tearing of the final rubber product. Carbon black also provides protection from ultraviolet radiation. These enhancements are especially important in tires. Most rubber parts are black in color because of their carbon black content.

Although carbon black is the most important filler, others are also used. They include china clays—hydrous aluminum silicates ($Al_2Si_2O_5(OH)_4$), which provide less reinforcing than carbon black but are used when the black color is not acceptable; calcium carbonate ($CaCO_3$), which is a nonreinforcing filler; and silica (SiO_2), which can serve reinforcing or nonreinforcing functions, depending on particle size; and other polymers, such as styrene, PVC, and phenolics. Reclaimed (recycled) rubber is also added as a filler in some rubber products, but usually not in proportions exceeding 10%.

Other additives compounded with the rubber include antioxidants, to retard aging by oxidation; fatigue- and ozone-protective chemicals; coloring pigments; plasticizers and softening oils; blowing agents in the production of foamed rubber; and mold-release compounds.

Many products require filament reinforcement to reduce extensibility but retain the other desirable properties of rubber. Tires and conveyor belts are notable examples. Filaments used for this purpose include cellulose, nylon, and polyester. Fiberglass and steel are also used as reinforcements (e.g., steel-belted radial tires). These continuous fiber materials must be added as part of the shaping process; they are not mixed with the other additives.

14.1.3 MIXING

The additives must be thoroughly mixed with the base rubber to achieve uniform dispersion of the ingredients. Uncured rubbers possess high viscosity. Mechanical working experienced by the rubber can increase its temperature up to 150°C (300°F). If vulcanizing agents were present from the start of mixing, premature vulcanization would result—the rubber processor's nightmare [7]. Accordingly, a two-stage mixing process is usually employed. In the first stage, carbon black and other nonvulcanizing additives are combined with the raw rubber. The term *masterbatch* is used for this first-stage mixture. After thorough mixing has been accomplished, and time for cooling has been allowed, the second stage is carried out in which the vulcanizing agents are added.

Equipment for mixing includes the two-roll mill and internal mixers such as the Banbury mixer (Figure 14.1). The *two-roll mill* consists of two parallel rolls, supported in a frame so they can be brought together to obtain a desired "nip" (gap size), and driven to rotate at the same or slightly different speeds. An *internal mixer* has two rotors encased in a jacket, as in Figure 14.1(b) for the Banbury-type internal mixer. The rotors have blades and rotate in opposite directions at different speeds, causing a complex flow pattern in the contained mixture.

14.1.4 SHAPING AND RELATED PROCESSES

Shaping processes for rubber products can be divided into four basic categories: (1) extrusion, (2) calendering, (3) coating, and (4) molding and casting. Most of these

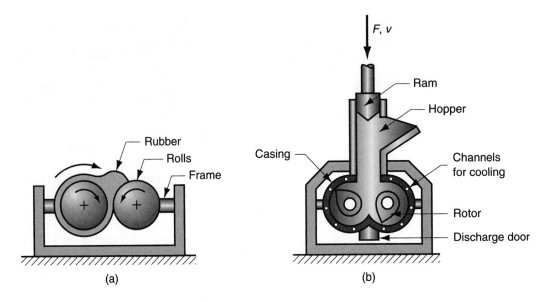

FIGURE 14.1 Mixers used in rubber processing: (a) two-roll mill and (b) Banbury-type internal mixer. These machines can also be used for mastication of natural rubber.

processes are discussed in the previous chapter. We will examine the special issues that arise when they are applied to rubber. Some products require several basic processes plus assembly work in their manufacture, for example, tires.

Extrusion   Extrusion of polymers is discussed in the preceding chapter. Screw extruders are generally used for extrusion of rubber. As with extrusion of thermosetting plastics, the L/D ratio of the extruder barrels is less than for thermoplastics, typically in the range 10 to 15, to reduce the risk of premature cross–linking. Die swell occurs in rubber extrudates, because the polymer is in a highly plastic condition and exhibits the memory property. It has not yet been vulcanized.

Calendering   This process involves passing rubber stock through a series of gaps of decreasing size made by a stand of rotating rolls (Section 13.3). The rubber process must be operated at lower temperatures than for thermoplastic polymers, to avoid premature vulcanization. Also, equipment used in the rubber industry is of heavier construction than that used for thermoplastics, because rubber is more viscous and harder to form. The output of the process is a rubber sheet of thickness determined by the final roll gap; again, swelling occurs in the sheet, causing its thickness to be slightly greater than the gap size. Calendering can also be used to coat or impregnate textile fabrics to produce rubberized fabrics.

There are problems in producing thick sheet by either extrusion or calendering. Thickness control is difficult in the former process, and air entrapment occurs in the latter. These problems are largely solved when extrusion and calendering are combined in the **roller die** process (Figure 14.2). The extruder die is a slit that feeds the calender rolls.

Coating   Coating or impregnating fabrics with rubber is an important process in the rubber industry. These composite materials are used in automobile tires, conveyor belts, inflatable rafts, and waterproof cloth for tarpaulins, tents, and rain coats. The **coating** of rubber onto substrate fabrics includes a variety of processes. We have previously indicated

FIGURE 14.2   Roller die process rubber extrusion followed by rolling.

that calendering is one of the coating methods. Figure 14.3 illustrates one possible way in which the fabric is fed into the calendering rolls to obtain a reinforced rubber sheet.

Alternatives to calendering include skimming, dipping, and spraying. In the *skimming* process, a thick solution of rubber compound in an organic solvent is applied to the fabric as it is unreeled from a supply spool. The coated fabric passes under a doctor blade that skims the solvent to the proper thickness, and then moves into a steam chamber where the solvent is driven off by heat. As its name suggests, *dipping* involves temporary immersion of the fabric into a highly fluid solution of rubber, followed by drying. Likewise, in *spraying,* a spray gun is used to apply the rubber solution.

Molding and Casting   Molded articles include shoe soles and heels, gaskets and seals, suction cups, and bottle stops. Many foamed rubber parts are produced by molding. In addition, molding is an important process in tire production. Principal molding processes for rubber are (1) compression molding, (2) transfer molding, and (3) injection molding. Compression molding is the most important technique because of its use in tire manufacture. Curing (vulcanizing) is accomplished in the mold in all three processes, this representing a departure from the shaping methods already discussed, which require a separate vulcanizing step. With injection molding of rubber, there are risks of premature curing similar to those faced in the same process when applied to thermosetting plastics. Advantages of injection molding over traditional methods for producing rubber parts include better dimensional control, less scrap, and shorter cycle times. In addition to its use in the molding of conventional rubbers, injection molding is also applied for thermoplastic elastomers. Because of high mold costs, large production quantities are required to justify injection molding.

A form of casting, called *dip casting,* is used for producing rubber gloves and overshoes. It involves submersion of a positive mold in a liquid polymer (or a heated form into plastisol) for a certain duration (the process may involve repeated dippings) to form the desired thickness. The coating is then stripped from the form and cured to cross–link the rubber.

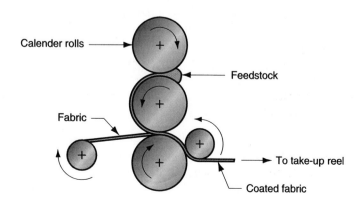

FIGURE 14.3   Coating of fabric with rubber using a calendering process.

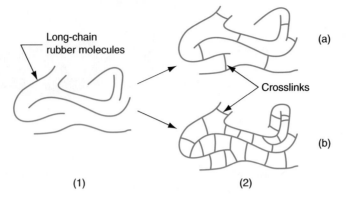

FIGURE 14.4 Effect of vulcanization on the rubber molecules: (1) raw rubber; (2) vulcanized (cross–linked) rubber. Variations of (2) include (a) soft rubber, low degree of cross–linking; and (b) hard rubber, high degree of cross–linking.

14.1.5 VULCANIZATION

Vulcanization is the treatment that accomplishes cross–linking of elastomer molecules, so that the rubber becomes stiffer and stronger but retains extensibility. It is a critical step in the rubber processing sequence. On a submicroscopic scale, the process can be pictured as in Figure 14.4, in which the long-chain molecules of the rubber become joined at certain tie points, the effect of which is to reduce the ability of the elastomer to flow. A typical soft rubber has one or two cross–links per thousand units (mers). As the number of cross–links increases, the polymer becomes stiffer and behaves more like a thermosetting plastic (hard rubber).

Vulcanization, as it was first invented by Goodyear, involved the use of sulfur (about 8 parts by weight of S mixed with 100 parts of natural rubber) at a temperature of 140°C (280°F) for about 5 hours. No other chemicals were included in the process. Vulcanization with sulfur alone is no longer used as a commercial treatment today, because of the long curing times. Various other chemicals, including zinc oxide (ZnO) and stearic acid ($C_{18}H_{36}O_2$), are combined with smaller doses of sulfur to accelerate and strengthen the treatment. The resulting cure time is 15 to 20 minutes for a typical passenger car tire. In addition, various nonsulfur vulcanizing treatments have been developed.

In rubber-molding processes, vulcanization is accomplished in the mold by maintaining the mold temperature at the proper level for curing. In the other forming processes, vulcanization is performed after the part has been shaped. The treatments generally divide between batch processes and continuous processes. Batch methods include the use of an ***autoclave,*** a steam-heated pressure vessel; and ***gas curing,*** in which a heated inert gas such as nitrogen cures the rubber. Many of the basic processes make a continuous product, and if the output is not cut into discrete pieces, continuous vulcanization is appropriate. Continuous methods include ***high-pressure steam,*** suited to the curing of rubber coated wire and cable; ***hot-air tunnel,*** for cellular extrusions and carpet underlays [3]; and ***continuous drum cure,*** in which continuous rubber sheets (e.g., belts and flooring materials) pass through one or more heated rolls to effect vulcanization.

14.2   MANUFACTURE OF TIRES AND OTHER RUBBER PRODUCTS

Tires are the principal product of the rubber industry, accounting for about three fourths of total tonnage. Other important products include footwear, hose, conveyor belts, seals, shock-absorbing components, foamed rubber products, and sports equipment.

14.2.1 TIRES

Pneumatic tires are critical components of the vehicles on which they are used. They are used on automobiles, trucks, buses, farm tractors, earth-moving equipment, military vehicles, bicycles, motorcycles, and aircraft. Tires support the weight of the vehicle and the passengers and cargo on board; they transmit the motor torque to propel the vehicle (except on aircraft); and they absorb vibrations and shock to provide a comfortable ride.

Tire Construction and Production Sequence   A tire is an assembly of many parts, whose manufacture is unexpectedly complex. A passenger car tire consists of about 50 individual pieces; a large earthmover tire may have as many as 175. To begin with, there are three basic tire constructions: (a) diagonal ply, (b) belted bias, and (c) radial ply, pictured in Figure 14.5. In all three cases, the internal structure of the tire, known as the *carcass,* consists of multiple layers of rubber-coated cords, called *plies.* The cords are strands of various materials such as nylon, polyester, fiberglass, and steel, which provide inextensibility to reinforce the rubber in the carcass. The *diagonal ply tire* has the cords running diagonally, but in perpendicular directions in adjacent layers. A typical diagonal ply tire may have four plies. The *belted bias tire* is constructed of diagonal plies with opposite bias but adds several more layers around the outside periphery of the carcass. These *belts* increase the stiffness of the tire in the tread area and limit its diametric expansion during inflation. The cords in the belt also run diagonally, as indicated in the sketch.

A *radial tire* has plies running radially rather than diagonally; it also uses belts around the periphery for support. A *steel-belted radial* is a tire in which the circumferential

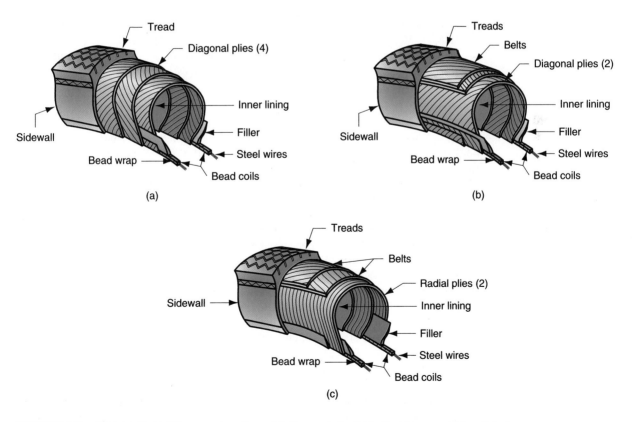

FIGURE 14.5   Three principal tire constructions: (a) diagonal ply, (b) belted bias, and (c) radial ply.

belts have cords made of steel. The radial construction provides a more flexible sidewall, which tends to reduce stress on the belts and treads as they continually deform on contact with the flat road surface during rotation. This effect is accompanied by greater tread life, improved cornering and driving stability, and a better ride at high speeds.

In each construction, the carcass is covered by solid rubber that reaches a maximum thickness in the tread area. The carcass is also lined on the inside with a rubber coating. For tires with inner tubes, the inner liner is a thin coating applied to the innermost ply during its fabrication. For tubeless tires, the inner liner must have low permeability because it holds the air pressure; it is generally a laminated rubber.

Tire production can be summarized in three steps: (1) preforming of components, (2) building the carcass and adding rubber strips to form the sidewalls and treads, and (3) molding and curing the components into one integral piece. The descriptions of these steps that follow are typical; there are variations in processing depending on construction, tire size, and type of vehicle on which the tire will be used.

Preforming of Components   As Figure 14.5 shows, the carcass consists of a number of separate components, most of which are rubber or reinforced rubber. These, as well as the sidewall and tread rubber, are produced by continuous processes and then pre-cut to size and shape for subsequent assembly. The components, labeled in Figure 14.5, and the preforming processes to fabricate them are:

> *Bead coil.* Continuous steel wire is rubber-coated, cut, coiled, and the ends joined.
> *Plies.* Continuous fabric (textile, nylon, fiber glass, steel) is rubber coated in a calendering process and pre-cut to size and shape.
> *Inner lining.* For tube tires, the inner liner is calendered onto the innermost ply. For tubeless tires, the liner is calendered as a two-layered laminate.
> *Belts.* Continuous fabric is rubber coated (similar to plies), but cut at different angles for better reinforcement; then made into a multi-ply belt.
> *Tread.* Extruded as continuous strip; then cut and preassembled to belts.
> *Sidewall.* Extruded as continuous strip; then cut to size and shape.

Building the Carcass   The carcass is traditionally assembled using a machine known as a *building drum,* whose main element is a cylindrical arbor that rotates. Pre-cut strips that form the carcass are built up around this arbor in a step-by-step procedure. The layered plies that form the cross section of the tire are anchored on opposite sides of the rim by two bead coils. The *bead coils* consist of multiple strands of high-strength steel wire. Their function is to provide a rigid support when the finished tire is mounted on the wheel rim. Other components are combined with the plies and bead coils. These include various wrappings and filler pieces to give the tire the proper strength, heat resistance, air retention, and fitting to the wheel rim. After these parts are placed around the arbor and the proper number of plies have been added, the belts are applied. This is followed by the outside rubber that will become the sidewall and tread.[1] At this point in the process, the treads are rubber strips of uniform cross section—the tread design is added later in molding. The building drum is collapsible, so that the unfinished tire can be removed when finished. The form of the tire at this stage is roughly tubular, as portrayed in Figure 14.6.

Molding and Curing   Tire molds are usually two-piece construction (split molds) and contain the tread pattern to be impressed on the tire. The mold is bolted into a press, one half attached to the upper platen (the lid) and the bottom half fastened to the lower

---

[1]Technically, the tread and sidewall are not usually considered to be components of the carcass.

FIGURE 14.6   Tire just before removal from building drum, before molding and curing.

platen (the base). The uncured tire is placed over an expandable diaphragm and inserted between the mold halves, as in Figure 14.7. The press is then closed and the diaphragm expanded, so that the soft rubber is pressed against the cavity of the mold. This causes the tread pattern to be imparted to the rubber. At the same time, the rubber is heated, both from the outside by the mold and from the inside by the diaphragm. Circulating hot water or steam under pressure are used to heat the diaphragm. The duration of this curing step depends on the thickness of the tire wall. A typical passenger tire can be cured in about 15 minutes. Bicycle tires cure in about 4 minutes, whereas tires for large earth-moving equipment take several hours to cure. After curing is completed, the tire is cooled and removed from the press.

14.2.2 OTHER RUBBER PRODUCTS

Most other rubber products are made by less complex processes. ***Rubber belts*** are widely used in conveyors and mechanical power transmission systems. As with tires, rubber is an ideal material for these products, but the belt must have flexibility but little or no extensibility to function. Accordingly, it is reinforced with fibers, commonly polyester or nylon. Fabrics of these polymers are usually coated in calendering operations, assembled together to obtain the required number of plies and thickness, and subsequently vulcanized by continuous or batch heating processes.

   Rubber hose can be either plain or reinforced. Plain hose is extruded tubing. Reinforced tube consists of an inner tube, a reinforcing layer (sometimes called the carcass), and a cover. The internal tubing is extruded of a rubber that has been

FIGURE 14.7   Tire molding (tire is shown in cross-sectional view): (1) the uncured tire is placed over expandable diaphragm; (2) the mold is closed and the diaphragm is expanded to force un-cured rubber against mold cavity, impressing tread pattern into rubber; mold and diaphragm are heated to cure rubber.

compounded for the particular substance that will flow through it. The reinforcement layer is applied to the tube in the form of a fabric, or by spiraling, knitting, braiding, or other application method. The outer layer is compounded to resist environmental conditions. It is applied by extrusion, using rollers, or other techniques.

Footwear components include soles, heels, rubber overshoes, and certain upper parts. Various rubbers are used to make footwear components (Section 8.4). Molded parts are produced by injection molding, compression molding, and certain special molding techniques developed by the shoe industry; the rubbers include both solid and foamed varieties. In some cases, for low volume production, manual methods are used to cut rubber from flat stock.

Rubber is widely used in sports equipment and supplies, including ping pong paddle surfaces, golf club grips, football bladders, and sports balls of various kinds. Tennis balls, for example, are made in significant numbers. Production of these sports products relies on the various shaping processes discussed in Section 14.1.4, as well as special techniques that have been developed for particular items.

14.2.3 PROCESSING OF THERMOPLASTIC ELASTOMERS

A thermoplastic elastomer (TPE) is a thermoplastic polymer that possesses the properties of a rubber (Section 8.4.3); the term *thermoplastic rubber* is also used. TPEs can be processed like thermoplastics, but their applications are those of an elastomer. The most common shaping processes are injection molding and extrusion, which are generally more economical and faster than the traditional processes used for rubbers that must be vulcanized. Molded products include shoe soles, athletic footwear, and automotive components such as fender extensions and corner panels (but not tires—TPEs have been found to be unsatisfactory for that application). Extruded items include insulation coating for electrical wire, tubing for medical applications, conveyor belts, sheet and film stock. Other shaping techniques for TPEs include blow molding and thermoforming (Sections 13.8 and 13.9); these processes cannot be used for vulcanized rubbers.

14.3 PRODUCT DESIGN CONSIDERATIONS

Many of the same guidelines used for plastics apply to rubber products. There are differences, owing to the elastomeric properties of rubber. The following are compiled largely from Bralla [4]; they apply to conventional soft rubber, not hard rubber.

➢ *Economic production quantities.* Rubber parts produced by compression molding (the traditional process) can often be produced in quantities of a thousand or less. The mold cost is relatively low compared with other molding methods. Injection molding, as with plastic parts, requires higher production quantities to justify the more expensive mold.

➢ *Draft.* Draft is usually unnecessary for rubber molded parts. The flexibility of the material allows it to deform for removal from the mold. Shallow undercuts, although undesirable, are possible with rubber-molded parts for the same reason. The low stiffness and high elasticity of the material permits removal from the mold.

➢ *Holes.* Holes are difficult to cut into the rubber after initial forming, due the flexibility of the material. It is generally desirable to mold holes into the rubber during the primary shaping process.

> ➤ **Screw threads.** Screw threads are generally not incorporated into molded rubber parts; the elastic deformability of rubber makes it difficult to assemble parts using the threads, and stripping is a problem once inserted.

REFERENCES

[1] Alliger, G., and Sjothun, I. J. (eds.). *Vulcanization of Elastomers.* Krieger Publishing, New York, 1978.

[2] Billmeyer, Fred, W., Jr. *Textbook of Polymer Science.* 3rd ed. John Wiley & Sons, New York, 1984.

[3] Blow, C. M., and Hepburn, C. *Rubber Technology and Manufacture.* 2nd ed. Butterworth-Heinemann, London, 1982.

[4] Bralla, J. G. (ed.). *Design for Manufacturability Handbook.* 2nd ed. McGraw-Hill, New York, 1999.

[5] Hofmann, W. *Rubber Technology Handbook.* Hanser-Gardner Publications, Cincinnati, Ohio, 1989.

[6] Mark, J. E., and Erman, B. (eds.), *Science and Technology of Rubber,* 3rd ed. Academic Press, Orlando, Florida, 2005.

[7] Morton-Jones, D. H. *Polymer Processing.* Chapman and Hall, London, 1989.

REVIEW QUESTIONS

14.1. How is the rubber industry organized?

14.2. How is raw rubber recovered from the latex that is tapped from a rubber tree?

14.3. What is the sequence of processing steps required to produce finished rubber goods?

14.4. What are some of the additives that are combined with rubber during compounding?

14.5. Name the four basic categories of processes used to shape rubber.

14.6. What does vulcanization do to the rubber?

14.7. Name the three basic tire constructions and briefly identify the differences in their construction.

14.8. What are the three basic steps in the manufacture of a pneumatic tire?

14.9. What is the purpose of the bead coil in a pneumatic tire?

14.10. What is a TPE?

14.11. Many of the design guidelines that are applicable to plastics are also applicable to rubber. However, the extreme flexibility of rubber results in certain differences. What are some examples of these differences?

MULTIPLE CHOICE QUIZ

There are 10 correct answers in the following multiple choice questions (some questions have multiple answers that are correct). To attain a perfect score on the quiz, all correct answers must be given. Each correct answer is worth 1 point. Each omitted answer or wrong answer reduces the score by 1 point, and each additional answer beyond the correct number of answers reduces the score by 1 point. Percentage score on the quiz is based on the total number of correct answers.

14.1. The most important rubber product is which one of the following: (a) footwear, (b) conveyor belts, (c) pneumatic tires, or (d) tennis balls?

14.2. The chemical name of the ingredient recovered from the latex of the rubber tree is which one of the following: (a) polybutadiene, (b) polyisobutylene, (c) polyisoprene, or (d) polystyrene?

14.3. Of the following rubber additives, which one would rank as the single most important: (a) antioxidants, (b) carbon black, (c) clays and other hydrous aluminum silicates, (d) plasticizers and softening oils, or (e) reclaimed rubber?

14.4. Which one of the following molding processes is the most important in the production of products made of conventional rubber: (a) compression molding, (b) injection molding, (c) thermoforming, or (d) transfer molding?

14.5. Which of the following ingredients do not contribute to the vulcanizing process (two correct answers): (a) calcium carbonate, (b) carbon black, (c) stearic acid, (d) sulfur, and (e) zinc oxide?

14.6. How many minutes are required to cure (vulcanize) a modern passenger car tire: (a) 5, (b) 15, (c) 25, or (d) 45?

14.7. When is the tread pattern imprinted onto the circumference of the tire: (a) during preforming, (b) while building the carcass, (c) during molding, or (d) during curing?

14.8. Which of the following are not normally used in the processing of thermoplastic elastomers (two correct answers): (a) blow molding, (b) compression molding, (c) extrusion, (d) injection molding, or (e) vulcanization?

15 SHAPING PROCESSES FOR POLYMER MATRIX COMPOSITES

Chapter Contents

In this chapter we consider manufacturing processes by which polymer matrix composites are shaped into useful components and products. A *polymer matrix composite* (PMC) is a composite material consisting of a polymer embedded with a reinforcing phase such as fibers or powders. The technological and commercial importance of the processes in this chapter derives from the growing use of this class of material, especially *fiber-reinforced polymers* (FRPs). In popular usage, PMC generally refers to fiber-reinforced polymers. FRP composites can be designed with very high strength-to-weight and stiffness-to-weight ratios. These features make them attractive in aircraft, cars, trucks, boats, and sports equipment.

Some of the shaping processes described in this chapter are slow and labor intensive. In general, techniques for shaping composites are less efficient than manufacturing processes for other materials. There are two reasons for this: (1) composite materials are more complex than other materials, consisting as they do of two or more phases and the need to orient the reinforcing phase in the case of fiber-reinforced plastics; and (2) processing technologies for composites have not been the object of improvement and refinement over as many years as processes for other materials.

The variety of shaping methods for fiber-reinforced polymers is sometimes bewildering to students on first reading. Let us provide a road map for the reader entering this new territory. FRP composite shaping processes can be divided into five categories, as organized in Figure 15.1: (1) open mold processes, (2) closed mold processes, (3) filament winding, (4) pultrusion processes, and (5) other. Open mold processes include some of the original manual procedures for laying resins and fibers onto forms. Closed mold processes are much the same as those used in

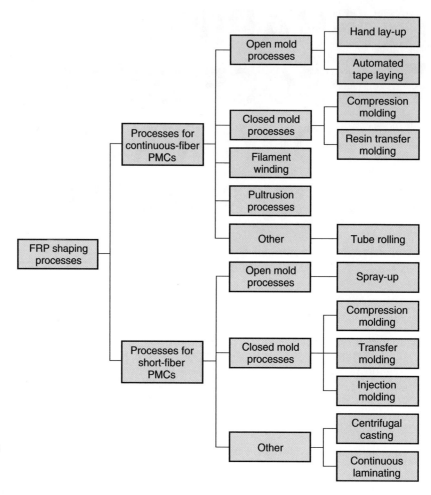

FIGURE 15.1
Classification of manufacturing processes for fiber-reinforced polymer composites.

plastic molding; the reader will recognize the names—compression molding, transfer molding, and injection molding—although the names are sometimes changed and modifications are sometimes made for PMCs. In *filament winding,* continuous filaments that have been dipped in liquid resin are wrapped around a rotating mandrel; when the resin cures, a rigid, hollow, generally cylindrical shape is created. *Pultrusion* is a shaping process for producing long, straight sections of constant cross section; it is similar to extrusion, but adapted to include continuous fiber reinforcement. The "other" category includes several operations that do not fit into the previous categories.

Some of these processes are used to shape composites with continuous fibers, whereas others are used for short fiber PMCs. Figure 15.1 provides an overview of the processes in each division. Let us begin our coverage by exploring how the individual phases in a PMC are produced and how these phases are combined into the starting materials for shaping. For a good overview of the PMC processes, the reader should view the video clip titled Composite Materials and Manufacturing.

VIDEO CLIP

Composite Materials and Manufacturing. This clip contains three segments: (1) composite materials, (2) composites manufacturing processes, and (3) composites overview. Segment (2) is especially relevant to this chapter.

15.1   STARTING MATERIALS FOR PMCs

In a PMC, the starting materials are a polymer and a reinforcing phase. They are processed separately before becoming phases in the composite. This section considers how these materials are produced before being combined, and then how they are combined to make the composite part.

15.1.1   POLYMER MATRIX

All three of the basic polymer types—thermoplastics, thermosets, and elastomers—are used as matrices in PMCs. Thermosetting (TS) polymers are the most common matrix materials. The principal TS polymers are phenolics, unsaturated polyesters, and epoxies. Phenolics are associated with the use of particulate reinforcing phases, whereas polyesters and epoxies are more closely associated with FRPs. Thermoplastic (TP) polymers are also used in PMCs, and in fact, most molding compounds are composite materials that include fillers and/or reinforcing agents. Most elastomers are composite materials because nearly all rubbers are reinforced with carbon black. Shaping processes for rubbers are covered in Chapter 14. In this chapter, coverage is limited to the processing of PMCs that use TS and TP polymers as the matrix. Many of the polymer shaping processes discussed in Chapter 13 are applicable to polymer matrix composites. However, combining the polymer with the reinforcing agent sometimes complicates the operations.

15.1.2   REINFORCING AGENT

The reinforcing phase can be any of several geometries and materials. The geometries include fibers, particles, and flakes, and the materials are ceramics, metals, other polymers, or elements such as carbon or boron. The role of the reinforcing phase and some of its technical features are discussed in Section 9.1.2.

Fibers   Common fiber materials in FRPs are glass, carbon, and the polymer Kevlar. Fibers of these materials are produced by various techniques, some of which we have covered in other chapters. Glass fibers are produced by drawing through small orifices (Section 12.2.3). For carbon, a series of heating treatments is performed to convert a precursor filament containing a carbon compound into a more pure carbon form. The precursor can be any of several substances, including polyacrylonitrile (PAN), pitch (a black carbon resin formed in the distillation of coal tar, wood tar, petroleum, etc.), or rayon (cellulose). Kevlar fibers are produced by extrusion combined with drawing through small orifices in a spinneret (Section 13.4).

Starting as continuous filaments, the fibers are combined with the polymer matrix in any of several forms, depending on the properties desired in the material and the processing method to be used to shape the composite. In some fabrication processes, the filaments are continuous, whereas in others they are chopped into short lengths. In the continuous form, individual filaments are usually available as rovings. A *roving* is a collection of untwisted (parallel) continuous strands; this is a convenient form for handling and processing. Rovings typically contain from 12 to 120 individual strands. By contrast, a *yarn* is a twisted collection of filaments. Continuous rovings are used in several PMC processes, including filament winding and pultrusion.

The most familiar form of continuous fiber is a *cloth*—a fabric of woven yarns. Very similar to a cloth, but distinguished here, is a *woven roving,* a fabric consisting of untwisted filaments rather than yarns. Woven rovings can be produced with unequal numbers of strands in the two directions so that they possess greater strength in one direction than the other. Such unidirectional woven rovings are often preferred in laminated FRP composites.

Fibers can also be prepared in the form of a *mat*—a felt consisting of randomly oriented short fibers held loosely together with a binder, sometimes in a carrier fabric. Mats are commercially available as blankets of various weights, thicknesses, and widths. Mats can be cut and shaped for use as **preforms** in some of the closed mold processes. During molding, the resin impregnates the preform and then cures, thus yielding a fiber-reinforced molding.

Particles and Flakes   Particles and flakes are really in the same class. Flakes are particles whose length and width are large relative to thickness. We discuss these and other issues on characterization of engineering powders in Section 16.1. Production methods for metal powders are discussed in Section 16.2, and techniques for producing ceramic powders are discussed in Section 17.1.1.

15.1.3 COMBINING MATRIX AND REINFORCEMENT

Incorporation of the reinforcing agent into the polymer matrix either occurs during the shaping process or beforehand. In the first case, the starting materials arrive at the fabricating operation as separate entities and are combined into the composite during shaping. Examples of this case are filament winding and pultrusion. The starting reinforcement in these processes consists of continuous fibers. In the second case, the two component materials are combined into some preliminary form that is convenient for use in the shaping process. Nearly all of the thermoplastics and thermosets used in the plastic shaping processes are really polymers combined with fillers (Section 8.1.5). The fillers are either short fibers or particulate (including flakes).

Of greatest interest in this chapter are the starting forms used in processes designed for FRP composites. We might think of the starting forms as prefabricated composites that arrive ready for use at the shaping process. These forms are molding compounds and prepregs.

Molding Compounds   Molding compounds are similar to those used in plastic molding. They are designed for use in molding operations, and so they must be capable of flowing. Most molding compounds for composite processing are thermosetting polymers. Accordingly, they have not been cured before shape processing. Curing is done during and/or after final shaping. FRP composite molding compounds consist of the resin matrix with short, randomly dispersed fibers. They come in several forms.

Sheet molding compound (SMC) is a combination of TS polymer resin, fillers and other additives, and chopped glass fibers (randomly oriented) all rolled into a sheet of typical thickness = 6.5 mm (0.250 in). The most common resin is unsaturated polyester; fillers are usually mineral powders such as talc, silica, limestone; and the glass fibers are typically 12 to 75 mm (0.5 to 3.0 in) long and account for about 30% of the SMC by volume. SMCs are very convenient for handling and cutting to proper size as molding charges. Sheet molding compounds are generally produced between thin layers of polyethylene to limit evaporation of volatiles from the thermosetting resin. The protective coating also improves surface finish on subsequent molded parts. The process for fabricating continuous SMC sheets is depicted in Figure 15.2.

Bulk molding compound (BMC) consists of similar ingredients as those in SMC, but the compounded polymer is in billet form rather than sheet. The fibers in BMC are shorter, typically 2 to 12 mm (0.1 to 0.5 in), because greater fluidity is required in the molding operations for which these materials are designed. Billet diameter is usually

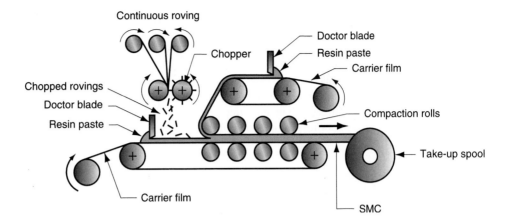

FIGURE 15.2   Process for producing sheet molding compound (SMC).

25 to 50 mm (1 to 2 in). The process for producing BMC is similar to that for SMC, except extrusion is used to obtain the final billet form. BMC is also known as ***dough molding compound*** (DMC), because of its dough-like consistency. Other FRP molding compounds include ***thick molding compound*** (TMC), similar to SMC but thicker—up to 50 mm (2 in); and ***pelletized molding compounds***—basically conventional plastic molding compounds containing short fibers.

Prepregs   Another prefabricated form for FRP shaping operations is ***prepreg,*** which consists of fibers impregnated with partially cured thermosetting resins to facilitate shape processing. Completion of curing must be accomplished during and/or after shaping. Prepregs are available in the form of tapes or cross-plied sheets or fabrics. The advantage of prepregs is that they are fabricated with continuous filaments rather than chopped random fibers, thus increasing strength and modulus of the final product. Prepreg tapes and sheets are associated with advanced composites (reinforced with boron, carbon/graphite, and Kevlar) as well as fiberglass.

15.2   OPEN MOLD PROCESSES

The distinguishing feature of this family of FRP shaping processes is its use of a single positive or negative mold surface (Figure 15.3) to produce laminated FRP structures. Other names for open mold processes include ***contact lamination*** and ***contact molding.*** The starting materials (resins, fibers, mats, and woven rovings) are applied to the mold in layers, building up to the desired thickness. This is followed by curing and part removal. Common resins are unsaturated polyesters and epoxies, using fiberglass as the reinforcement. The moldings are usually large (e.g., boat hulls). The advantage of using an open mold is that the mold costs much less than if two matching molds were used. The disadvantage is that only the part surface in contact with the mold surface is finished; the

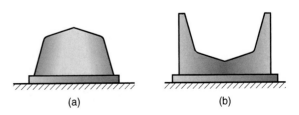

FIGURE 15.3   Types of open mold: (a) positive and (b) negative.

(a)   (b)

other side is rough. For the best possible part surface on the finished side, the mold itself must be very smooth.

There are several important open mold FRP processes. The differences are in the methods of applying the laminations to the mold, alternative curing techniques, and other variations. In this section we describe the family of open mold processes for shaping fiber-reinforced plastics: (1) hand lay-up, (2) spray-up, (3) automated tape-laying machines, and (4) bag molding. We treat hand lay-up as the base process and the others as modifications and refinements.

15.2.1 HAND LAY-UP

Hand lay-up is the oldest open mold method for FRP laminates, dating to the 1940s when it was first used to fabricate boat hulls. It is also the most labor-intensive method. As the name suggests, **hand lay-up** is a shaping method in which successive layers of resin and reinforcement are manually applied to an open mold to build the laminated FRP composite structure. The basic procedure consists of five steps, illustrated in Figure 15.4. The finished molding must usually be trimmed with a power saw to size the outside edges. In general, these same five steps are required for all of the open mold processes; the differences between methods occur in steps 3 and 4.

In step 3, each layer of fiber reinforcement is dry when placed onto the mold. The liquid (uncured) resin is then applied by pouring, brushing, or spraying. Impregnation of resin into the fiber mat or fabric is accomplished by hand rolling. This approach is referred to as **wet lay-up.** An alternative approach is to use **prepregs,** in which the impregnated layers of fiber reinforcement are first prepared outside the mold and then laid onto the mold surface. Advantages cited for the prepregs include closer control over fiber–resin mixture and more efficient methods of adding the laminations [11].

Molds for open mold contact laminating can be made of plaster, metal, glass fiber-reinforced plastic, or other materials. Selection of material depends on economics, surface quality, and other technical factors. For prototype fabrication, in which only one part is produced, plaster molds are usually adequate. For medium quantities, the mold can be

FIGURE 15.4 Hand lay-up procedure: (1) mold is cleaned and treated with a mold release agent; (2) a thin gel coat (resin, possibly pigmented to color) is applied, which will become the outside surface of the molding; (3) when the gel coat has partially set, successive layers of resin and fiber are applied, the fiber being in the form of mat or cloth; each layer is rolled to fully impregnate the fiber with resin and remove air bubbles; (4) the part is cured; and (5) the fully hardened part is removed from the mold.

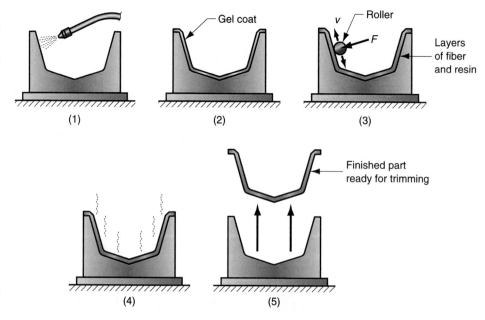

made of fiberglass-reinforced plastic. High production generally requires metal molds. Aluminum, steel, and nickel are used, sometimes with surface hardening on the mold face to resist wear. An advantage of metal, in addition to durability, is its high thermal conductivity that can be used to implement a heat-curing system, or simply to dissipate heat from the laminate while it cures at room temperature.

Products suited to hand lay-up are generally large in size but low in production quantity. In addition to boat hulls, other applications include swimming pools, large container tanks, stage props, radomes, and other formed sheets. Automotive parts have also been made, but the method is not economical for high production. The largest moldings ever made by this process were ship hulls for the British Royal Navy: 85 m (280 ft) long [2].

15.2.2 SPRAY-UP

This represents an attempt to mechanize the application of resin-fiber layers and reduce the time for lay-up. It is an alternative for step 3 in the hand lay-up procedure. In the *spray-up method,* liquid resin and chopped fibers are sprayed onto an open mold to build successive FRP laminations, as in Figure 15.5. The spray gun is equipped with a chopper mechanism that feeds in continuous filament rovings and cuts them into fibers of length 25 to 75 mm (1 to 3 in) that are added to the resin stream as it exits the nozzle. The mixing action results in random orientation of the fibers in the layer—unlike hand lay-up, in which the filaments can be oriented if desired. Another difference is that the fiber content in spray-up is limited to about 35% (compared with a maximum of around 65% in hand lay-up). This is a shortcoming of the spraying and mixing process.

Spraying can be accomplished manually using a portable spray gun or by an automated machine in which the path of the spray gun is preprogrammed and computer controlled. The automated procedure is advantageous for labor efficiency and environmental protection. Some of the volatile emissions from the liquid resins are hazardous, and the path-controlled machines can operate in sealed-off areas without humans present. However, rolling is generally required for each layer, as in hand lay-up.

Products made by the spray-up method include boat hulls, bathtubs, shower stalls, automobile and truck body parts, recreational vehicle components, furniture, large structural panels, and containers. Movie and stage props are sometimes made by this method. Because products made by spray-up have randomly oriented short fibers, they are not as strong as those made by lay-up, in which the fibers are continuous and directed.

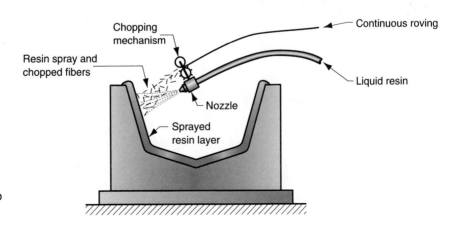

FIGURE 15.5   Spray-up method.

15.2.3 AUTOMATED TAPE-LAYING MACHINES

This is another attempt to automate and accelerate step 3 in the lay-up procedure. Automated tape-laying machines operate by dispensing a prepreg tape onto an open mold following a programmed path. The typical machine consists of an overhead gantry, to which is attached the dispensing head, as shown in Figure 15.6. The gantry permits x-y-z travel of the head, for positioning and following a defined continuous path. The head itself has several rotational axes, plus a shearing device to cut the tape at the end of each path. Prepreg tape widths are commonly 75 mm (3 in), although 300 mm (12 in) widths have been reported [10]; thickness is around 0.13 mm (0.005 in). The tape is stored on the machine in rolls, which are unwound and deposited along the defined path. Each lamination is placed by following a series of back-and-forth passes across the mold surface until the parallel rows of tape complete the layer.

Much of the work to develop automated tape-laying machines has been pioneered by the aircraft industry, which is eager to save labor costs and at the same time achieve the highest possible quality and uniformity in its manufactured components. The disadvantage of this and other computer numerically controlled machines is that it must be programmed, and programming takes time.

15.2.4 CURING

Curing (step 4) is required of all thermosetting resins used in FRP laminated composites. Curing accomplishes cross-linking of the polymer, transforming it from its liquid or highly plastic condition into a hardened product. There are three principal process parameters in curing: time, temperature, and pressure.

Curing normally occurs at room temperature for the TS resins used in hand lay-up and spray-up procedures. Moldings made by these processes are often large (e.g., boat hulls), and heating would be difficult for such parts. In some cases, days are required before room temperature curing is sufficiently complete to remove the part. If feasible, heat is added to speed the curing reaction.

FIGURE 15.6
Automated tape-laying machine. (Courtesy of Cincinnati Milacron.)

Heating is accomplished by several means. Oven curing provides heat at closely controlled temperatures; some curing ovens are equipped to draw a partial vacuum. Infrared heating can be used in applications in which it is impractical or inconvenient to place the molding in an oven.

Curing in an autoclave provides control over both temperature and pressure. An *autoclave* is an enclosed chamber equipped to apply heat and/or pressure at controlled levels. In FRP composites processing, it is usually a large horizontal cylinder with doors at either end. The term *autoclave molding* is sometimes used to refer to the curing of a prepreg laminate in an autoclave. This procedure is used extensively in the aerospace industry to produce advanced composite components of very high quality.

15.3 CLOSED MOLD PROCESSES

These molding operations are performed in molds consisting of two sections that open and close during each molding cycle. The name *matched die molding* is used for some of these processes. One might think that a closed mold is about twice the cost of a comparable open mold. However, tooling cost is even greater owing to the more complex equipment required in these processes. Despite their higher cost, advantages of a closed mold are (1) good finish on all part surfaces, (2) higher production rates,   (3) closer control over tolerances, and (4) more complex three-dimensional shapes are possible.

We divide the closed mold processes into three classes based on their counterparts in conventional plastic molding, even though the terminology is often different when polymer matrix composites are molded: (1) compression molding, (2) transfer molding, and (3) injection molding.

15.3.1 COMPRESSION MOLDING PMC PROCESSES

In compression molding of conventional molding compounds (Section 13.7.1), a charge is placed in the lower mold section, and the sections are brought together under pressure, causing the charge to take the shape of the cavity. The mold halves are heated to cure the thermosetting polymer. When the molding is sufficiently cured, the mold is opened and the part is removed. There are several shaping processes for PMCs based on compression molding; the differences are mostly in the form of the starting materials. The flow of the resin, fibers, and other ingredients during the process is a critical factor in compression molding of FRP composites.

SMC, TMC, and BMC Molding   Several of the FRP molding compounds, namely sheet molding compound (SMC), bulk molding compound (BMC), and thick molding compound (TMC), can be cut to proper size and used as the starting charge in compression molding. Refrigeration is often required to store these materials prior to shape processing. The names of the molding processes are based on the starting molding compound (i.e., *SMC molding* is when the starting charge is precut sheet molding compound; *BMC molding* uses BMC cut to size as the charge; and so on).

Preform Molding   Another form of compression molding, called *preform molding* [11], involves placement of a precut mat into the lower mold section along with a polymer resin charge (e.g., pellets or sheet). The materials are then pressed between heated mold halves, causing the resin to flow and impregnate the fiber mat to produce a fiber reinforced molding. Variations of the process use either thermoplastic or thermosetting polymers.

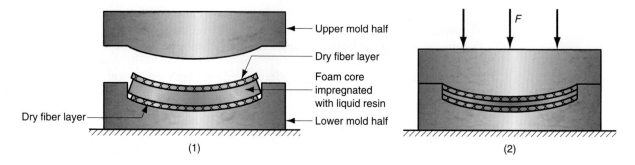

FIGURE 15.7 Elastic reservoir molding: (1) foam is placed into mold between two fiber layers; (2) mold is closed, releasing resin from foam into fiber layers.

Elastic Reservoir Molding The starting charge in elastic reservoir molding (ERM) is a sandwich consisting of a center of polymer foam between two dry fiber layers. The foam core is commonly open-cell polyurethane, impregnated with liquid resin such as epoxy or polyester, and the dry fiber layers can be cloth, woven roving, or other starting fibrous form. As depicted in Figure 15.7, the sandwich is placed in the lower mold section and pressed at moderate pressure—around 0.7 MPa (100 lb/in^2). As the core is compressed, it releases the resin to wet the dry surface layers. Curing produces a lightweight part consisting of a low-density core and thin FRP skins.

15.3.2 TRANSFER MOLDING PMC PROCESSES

In conventional transfer molding (Section 13.7.2), a charge of thermosetting resin is placed in a pot or chamber, heated, and squeezed by ram action into one or more mold cavities. The mold is heated to cure the resin. The name of the process derives from the fact that the fluid polymer is transferred from the pot into the mold. It can be used to mold TS resins in which the fillers include short fibers to produce an FRP composite part. Another form of transfer molding for PMCs is called *resin transfer molding* (RTM) [4], [11]; it refers to a closed mold process in which a preform mat is placed in the lower mold section, the mold is closed, and a thermosetting resin (e.g., polyester resin) is transferred into the cavity under moderate pressure to impregnate the preform. To confuse matters, RTM is sometimes called *resin injection molding* [4], [13]. (The distinction between transfer molding and injection molding is blurry anyway, as the reader may have noted in Chapter 13.) RTM has been used to manufacture such products as bathtubs, swimming pool shells, bench and chair seats, and hulls for small boats.

Several enhancements of the basic RTM process have been developed [5]. One enhancement, called *advanced RTM,* uses high-strength polymers such as epoxy resins and continuous fiber reinforcement instead of mats. Applications include aerospace components, missile fins, and snow skis. Two additional processes are thermal expansion resin transfer molding and ultimately reinforced thermoset resin injection. *Thermal expansion resin transfer molding* (TERTM) is a patented process of TERTM, Inc. that consists of the following steps [5]: (1) A rigid polymer foam (e.g., polyurethane) is shaped into a preform. (2) The preform is enclosed in a fabric reinforcement and placed in a closed mold. (3) A thermosetting resin (e.g., epoxy) is injected into the mold to impregnate the fabric and surround the foam. (4) The mold is heated to expand the foam, fill the mold cavity, and cure the resin. *Ultimately, reinforced thermoset resin injection* (URTRI) is similar to TERTM except that the starting foam core is cast epoxy embedded with miniature hollow glass spheres.

15.3.3 INJECTION MOLDING PMC PROCESSES

Injection molding is noted for low-cost production of plastic parts in large quantities. Although it is most closely associated with thermoplastics, the process can also be adapted to thermosets (Section 13.6.5).

Conventional Injection Molding   In PMC shape processing, injection molding is used for both TP- and TS-type FRPs. In the TP category, virtually all thermoplastic polymers can be reinforced with fibers. Chopped fibers must be used; if continuous fibers were used, they would be reduced anyway by the action of the rotating screw in the barrel. During injection from the chamber into the mold cavity, the fibers tend to become aligned during their journey through the nozzle. Designers can sometimes exploit this feature to optimize directional properties through part design, location of gates, and cavity orientation relative to the gate [8].

Whereas TP molding compounds are heated and then injected into a cold mold, TS polymers are injected into a heated mold for curing. Control of the process with thermosets is trickier because of the risk of premature cross-linking in the injection chamber. Subject to the same risk, injection molding can be applied to fiber-reinforced TS plastics in the form of pelletized molding compound and dough molding compound.

Reinforced Reaction Injection Molding   Some thermosets cure by chemical reaction rather than heat; these resins can be molded by reaction injection molding (Section 13.6.5). In RIM, two reactive ingredients are mixed and immediately injected into a mold cavity where curing and solidification of the chemicals occur rapidly. A closely related process includes reinforcing fibers, typically glass, in the mixture. In this case, the process is called reinforced reaction injection molding (RRIM). Its advantages are similar to those in RIM, with the added benefit of fiber reinforcement. RRIM is used extensively in auto body and truck cab applications for bumpers, fenders, and other body parts.

15.4   FILAMENT WINDING

Filament winding is a process in which resin-impregnated continuous fibers are wrapped around a rotating mandrel that has the internal shape of the desired FRP product. The resin is subsequently cured and the mandrel removed. Hollow axisymmetric components (usually circular in cross section) are produced, as well as some irregular shapes. The most common form of the process is depicted in Figure 15.8. A band of fiber rovings is pulled through a resin bath immediately before being wound in a helical pattern onto a cylindrical mandrel. Continuation of the winding pattern finally completes a surface layer of one filament thickness on the mandrel. The operation is repeated to form additional layers, each having a criss-cross pattern with the previous, until the desired part thickness has been obtained.

There are several methods by which the fibers can be impregnated with resin: (1) *wet winding,* in which the filament is pulled through the liquid resin just before winding, as in the figure; (2) *prepreg winding* (also called *dry winding*), in which filaments preimpregnated with partially cured resin are wrapped around a heated mandrel; and (3) *postimpregnation,* in which filaments are wound onto a mandrel and then impregnated with resin by brushing or other technique.

Two basic winding patterns are used in filament winding: (a) helical and (b) polar (Figure 15.9). In *helical winding,* the filament band is applied in a spiral pattern around the mandrel, at a helix angle θ. If the band is wrapped with a helix angle approaching $90°$, so that the winding advance is one bandwidth per revolution (and the filaments form nearly

FIGURE 15.8   Filament winding.

circular rings around the mandrel), this is referred to as a ***hoop winding;*** it is a special case of helical winding. In ***polar winding,*** the filament is wrapped around the long axis of the mandrel, as in Figure 15.9(b); after each longitudinal revolution, the mandrel is indexed (partially rotated) by one bandwidth, so that a hollow enclosed shape is gradually created. Hoop and polar patterns can be combined in successive windings of the mandrel to produce adjacent layers with filament directions that are approximately perpendicular; this is called a ***bi-axial winding*** [2].

Filament winding machines have motion capabilities similar to those of an engine lathe (Section 22.2.3). The typical machine has a drive motor to rotate the mandrel and a powered feed mechanism to move the carriage. Relative motion between mandrel and carriage must be controlled to accomplish a given winding pattern. In helical winding, the relationship between helix angle and the machine parameters can be expressed as follows:

$$\tan \theta = \frac{v_c}{\pi D N} \tag{15.1}$$

where θ = helix angle of the windings on the mandrel, as in Figure 15.9(a); v_c = speed at which the carriage traverses in the axial direction, m/s (in/sec); D = diameter of the mandrel, m (in); and N = rotational speed, 1/s (rev/sec).

Various types of control are available in filament winding machines. Modern equipment uses ***computer numerical control*** (CNC, Section 38.3), in which mandrel rotation and carriage speed are independently controlled to permit greater adjustment and flexibility in the relative motions. CNC is especially useful in helical winding of contoured shapes, as in Figure 15.10. As indicated in Eq. (15.1), the ratio v_c/DN must remain fixed to maintain a constant helix angle θ. Thus, either v_c and/or N must be adjusted on-line to compensate for changes in D.

The ***mandrel*** is the special tooling that determines the geometry of the filament-wound part. For part removal, mandrels must be capable of collapsing after winding and curing. Various designs are possible, including inflatable/deflatable mandrels, collapsible metal mandrels, and mandrels made of soluble salts or plasters.

FIGURE 15.9   Two basic winding patterns in filament winding: (a) helical and (b) polar.

(a)   (b)

FIGURE 15.10 Filament winding machine. (Courtesy of Cincinnati Milacron.)

Applications of filament winding are often classified as aerospace or commercial [10], the engineering requirements being more demanding in the first category. Aerospace applications include rocket-motor cases, missile bodies, radomes, helicopter blades, and airplane tail sections and stabilizers. These components are made of advanced composites and hybrid composites (Section 9.4.1), with epoxy resins being most common and reinforced with fibers of carbon, boron, Kevlar, and glass. Commercial applications include storage tanks, reinforced pipes and tubing, drive shafts, wind-turbine blades, and lightning rods; these are made of conventional FRPs. Polymers include polyester, epoxy, and phenolic resins; glass is the common reinforcing fiber.

15.5 PULTRUSION PROCESSES

The basic pultrusion process was developed around 1950 for making fishing rods of glass fiber–reinforced polymer (GFRP). The process is similar to extrusion (hence the similarity in name), but it involves pulling of the workpiece (so the prefix "pul-" is used in place of "ex-"). Like extrusion, pultrusion produces continuous, straight sections of constant cross section. A related process, called pulforming, can be used to make parts that are curved and may have variations in cross section throughout their lengths.

15.5.1 PULTRUSION

Pultrusion is a process in which continuous fiber rovings are dipped into a resin bath and pulled through a shaping die where the impregnated resin cures. The setup is sketched in Figure 15.11, which shows the cured product being cut into long, straight sections. The sections are reinforced throughout their length by continuous fibers. Like extrusion, the pieces have a constant cross section, whose profile is determined by the shape of the die opening.

The process consists of five steps (identified in the sketch) performed in a continuous sequence [2]: (1) *filament feeding,* in which the fibers are unreeled from

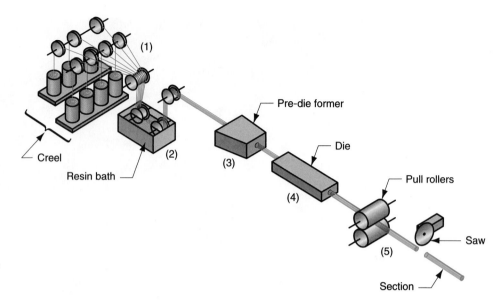

FIGURE 15.11
Pultrusion process (see text for interpretation of sequence numbers).

a creel (shelves with skewers that hold filament bobbins); (2) *resin impregnation,* in which the fibers are dipped in the uncured liquid resin; (3) *pre-die forming*—the collection of filaments is gradually shaped into the approximate cross section desired; (4) *shaping and curing,* in which the impregnated fibers are pulled through the heated die whose length is 1 to 1.5 m (3 to 5 ft) and whose inside surfaces are highly polished; and (5) *pulling and cutting*—pullers are used to draw the cured length through the die, after which it is cut by a cut-off wheel with SiC or diamond grits.

Common resins used in pultrusion are unsaturated polyesters, epoxies, and silicones, all thermosetting polymers. There are difficulties in processing with epoxy polymers because of sticking on the die surface. Thermoplastics have also been studied for possible applications [2]. E-glass is by far the most widely used reinforcing material; proportions range from 30% to 70%. Modulus of elasticity and tensile strength increase with reinforcement content. Products made by pultrusion include solid rods, tubing, long and flat sheets, structural sections (such as channels, angled and flanged beams), tool handles for high-voltage work, and third-rail covers for subways.

15.5.2 PULFORMING

The pultrusion process is limited to straight sections of constant cross section. There is also a need for long parts with continuous fiber reinforcement that are curved rather than straight and whose cross sections may vary throughout the length. The pulforming process is suited to these less-regular shapes. *Pulforming* can be defined as pultrusion with additional steps to form the length into a semicircular contour and alter the cross section at one or more locations along the length. A sketch of the equipment is illustrated in Figure 15.12. After exiting the shaping die, the continuous workpiece is fed into a rotating table with negative molds positioned around its periphery. The work is forced into the mold cavities by a die shoe, which squeezes the cross section at various locations and forms the curvature in the length. The diameter of the table determines the radius of the part. As the work leaves the die table, it is cut to length to provide discrete parts. Resins and fibers similar to those for pultrusion are used in pulforming. An important application of the process is production of automobile leaf springs.

FIGURE 15.12
Pulforming process (not shown in the sketch is the cut-off of the pulformed part).

15.6   OTHER PMC SHAPING PROCESSES

Additional PMC shaping processes worth noting include centrifugal casting, tube rolling, continuous laminating, and cutting. In addition, many of the traditional thermoplastic shaping processes are applicable to (short-fiber) FRPs based on TP polymers; these include blow molding, thermoforming, and extrusion.

Centrifugal Casting   This process is ideal for cylindrical products such as pipes and tanks. The process is the same as its counterpart in metal casting (Section 11.3.5). Chopped fibers combined with liquid resin are poured into a fast-rotating cylindrical mold. Centrifugal force presses the ingredients against the mold wall, where curing takes place. The resulting inner surfaces are quite smooth. Part shrinkage or use of split molds permits part removal.

Tube Rolling   FRP tubes can be fabricated from prepreg sheets by a rolling technique [7], shown in Figure 15.13. Such tubes are used in bicycle frames and space trusses. In the process, a precut prepreg sheet is wrapped around a cylindrical mandrel several times to obtain a tube wall of multiple sheet thicknesses. The rolled sheets are then encased in a heat-shrinking sleeve and oven cured. As the sleeve contracts, entrapped gases are squeezed out the ends of the tube. When curing is complete, the mandrel is removed to yield a rolled FRP tube. The operation is simple, and tooling cost is low. There are variations in the process, such as using different wrapping methods or using a steel mold to enclose the rolled prepreg tube for better dimensional control.

FIGURE 15.13   Tube rolling, showing (a) one possible means of wrapping FRP prepregs around a mandrel, and (b) the completed tube after curing and removal of mandrel.

Continuous Laminating   Fiber-reinforced plastic panels, sometimes translucent and/or corrugated, are used in construction. The process to produce them consists of (1) impregnating layers of glass fiber mat or woven fabric by dipping in liquid resin or by passing beneath a doctor blade; (2) gathering between cover films (cellophane, polyester, or other polymer); and (3) compacting between squeeze rolls and curing. Corrugation (4) is added by formed rollers or mold shoes.

Cutting Methods   FRP laminated composites must be cut in both uncured and cured states. Uncured materials (prepregs, preforms, SMCs, and other starting forms) must be cut to size for lay-up, molding, and so on. Typical cutting tools include knives, scissors, power shears, and steel-rule blanking dies. Also used are nontraditional cutting methods, such as laser beam cutting and water jet cutting (Chapter 26).

Cured FRPs are hard, tough, abrasive, and difficult to cut; but cutting is necessary in many FRP shaping processes to trim excess material, cut holes and outlines, and for other purposes. For fiberglass-reinforced plastics, cemented carbide cutting tools and high-speed steel saw blades must be used. For some advanced composites (e.g., boron-epoxy), diamond cutting tools obtain best results. Water jet cutting is also used with good success on cured FRPs; this process reduces the dust and noise problems associated with conventional sawing methods.

REFERENCES

[1] *ASM Handbook,* Vol. 21: Composites, ASM International, Materials Park, Ohio, 2001.
[2] Bader, M. G., Smith, W., Isham, A. B., Rolston, J. A., and Metzner, A. B. *Delaware Composites Design Encyclopedia.* Vol. 3. *Processing and Fabrication Technology.* Technomic Publishing Co., Lancaster, Pennsylvania, 1990.
[3] Chawla, K. K. *Composite Materials: Science and Engineering,* 3rd ed., Springer-Verlag, New York, 2008.
[4] Charrier, J-M. *Polymeric Materials and Processing.* Oxford University Press, New York, 1991.
[5] Coulter, J. P. "Resin Impregnation During the Manufacture of Composite Materials," *PhD Dissertation.* University of Delaware, 1988.
[6] *Engineering Materials Handbook.* Vol. 1. *Composites.* ASM International, Metals Park, Ohio, 1987.
[7] Mallick, P. K. *Fiber-Reinforced Composites: Materials, Manufacturing, and Design.* 2nd ed. Marcel Dekker, New York, 1993.
[8] McCrum, N. G., Buckley, C. P., and Bucknall, C. B. *Principles of Polymer Engineering.* Oxford University Press, Inc., Oxford, U.K., 1988.
[9] Morton-Jones, D. H. *Polymer Processing.* Chapman and Hall, London, 1989.
[10] Schwartz, M. M. *Composite Materials Handbook.* 2nd ed. McGraw-Hill, New York, 1992.
[11] Strong, A. B. *Fundamentals of Composites Manufacturing: Materials, Methods, and Applications.* 2nd ed. Society of Manufacturing Engineers, Dearborn, Michigan, 2007.
[12] Wick, C., Benedict, J. T., and Veilleux, R. F. (eds.). *Tool and Manufacturing Engineers Handbook.* 4th ed. Vol. II. *Forming,* 1984.
[13] Wick, C., and Veilleux, R. F. (eds.). *Tool and Manufacturing Engineers Handbook.* 4th ed. Vol. III. *Materials, Finishing, and Coating,* 1985.

REVIEW QUESTIONS

15.1. What are the principal polymers used in fiber-reinforced polymers?
15.2. What is the difference between a roving and a yarn?
15.3. In the context of fiber reinforcement, what is a mat?
15.4. Why are particles and flakes members of the same basic class of reinforcing material?
15.5. What is sheet molding compound (SMC)?
15.6. How is a prepreg different from a molding compound?
15.7. Why are laminated FRP products made by the spray-up method not as strong as similar products made by hand lay-up?

15.8. What is the difference between the wet lay-up approach and the prepreg approach in hand lay-up?

15.9. What is an autoclave?

15.10. What are some of the advantages of the closed mold processes for PMCs relative to open mold processes?

15.11. Identify some of the different forms of polymer matrix composite molding compounds.

15.12. What is preform molding?

15.13. Describe reinforced reaction injection molding (RRIM).

15.14. What is filament winding?

15.15. Describe the pultrusion process.

15.16. How does pulforming differ from pultrusion?

15.17. With what kinds of products is tube rolling associated?

15.18. How are FRPs cut?

15.19. (**Video**) According to the video on composites, list the primary purpose of the matrix and the reinforcement in a composite.

15.20. (**Video**) List the primary methods of fiber reinforced thermoset polymer composite production according to the composite video.

15.21. (**Video**) What are the advantages and disadvantages of using prepreg material for lay-up of composites according to the composite video?

MULTIPLE CHOICE QUIZ

There are 14 correct answers in the following multiple choice questions (some questions have multiple answers that are correct). To attain a perfect score on the quiz, all correct answers must be given. Each correct answer is worth 1 point. Each omitted answer or wrong answer reduces the score by 1 point, and each additional answer beyond the correct number of answers reduces the score by 1 point. Percentage score on the quiz is based on the total number of correct answers.

15.1. Which one of the following is the most common polymer type in fiber-reinforced polymer composites: (a) elastomers, (b) thermoplastics, or (c) thermosets?

15.2. Most rubber products are properly classified into which of the following categories (three best answers): (a) elastomer reinforced with carbon black, (b) fiber-reinforced composite, (c) particle-reinforced composite, (d) polymer matrix composite, (e) pure elastomer, and (f) pure polymer?

15.3. Other names for open mold processes include which of the following (two best answers): (a) compression molding, (b) contact lamination, (c) contact molding, (d) filament winding, (e) matched die molding, (f) preform molding, and (g) pultrusion?

15.4. Hand lay-up is classified in which of the following general categories of PMC shaping processes (two best answers): (a) closed mold process, (b) compression molding, (c) contact molding, (d) filament winding, or (e) open mold process?

15.5. A positive mold with a smooth surface will produce a good finish on which surface of the laminated product in the hand lay-up method:(a) inside surface or (b) outside surface?

15.6. A molding operation that uses sheet-molding compound (SMC) is a form of which one of the following: (a) compression molding, (b) contact molding, (c) injection molding, (d) open mold processing, (e) pultrusion, or (f) transfer molding?

15.7. Filament winding involves the use of which one of the following fiber reinforcements: (a) continuous filaments, (b) fabrics, (c) mats, (d) prepregs, (e) short fibers, or (f) woven rovings?

15.8. In filament winding, when the continuous filament is wound around the cylindrical mandrel at a helix angle close to 90°, it is called which of the following (one best answer): (a) bi-axial winding, (b) helical winding, (c) hoop winding, (d) perpendicular winding, (e) polar winding, or (f) radial winding?

15.9. Pultrusion is most similar to which one of the following plastic shaping processes: (a) blow-molding, (b) extrusion, (c) injection molding, or (d) thermoforming?

15.10. Water jet cutting is one of several ways of cutting or trimming uncured or cured FRPs; in the case of cured FRPs, the process is noted for its reduction of dust and noise: (a) true or (b) false?

Part IV Particulate Processing of Metals and Ceramics

16 POWDER METALLURGY

Chapter Contents

This part of the book is concerned with the processing of metals and ceramics that are in the form of powders—very small particulate solids. In the case of traditional ceramics, the powders are produced by crushing and grinding common materials that are found in nature, such as silicate minerals (clay) and quartz. In the case of metals and the new ceramics, the powders are produced by a variety of industrial processes. We cover the powder-making processes as well as the methods used to shape products out of powders in two chapters: Chapter 16 on powder metallurgy and Chapter 17 on particulate processing of ceramics and cermets.

Powder metallurgy (PM) is a metal processing technology in which parts are produced from metallic powders. In the usual PM production sequence, the powders are compressed into the desired shape and then heated to cause bonding of the particles into a hard, rigid mass. Compression, called *pressing,* is accomplished in a press-type machine using tools designed specifically for the part to be manufactured. The tooling, which typically consists of a die and one or more punches, can be expensive, and PM is therefore most appropriate for medium and high production. The heating treatment, called *sintering,* is performed at a temperature below the melting point of the metal. The video clip titled Powder Metallurgy illustrates PM production technology. Considerations that make powder metallurgy an important commercial technology include:

➤ PM parts can be mass produced to *net shape* or *near net shape,* eliminating or reducing the need for subsequent processing.

➤ The PM process itself involves very little waste of material; about 97% of the starting powders are converted to product. This compares favorably with casting processes in which sprues, runners, and risers are wasted material in the production cycle.

➤ Owing to the nature of the starting material in PM, parts having a specified level of porosity can be made. This feature lends itself to the production of porous metal parts such as filters and oil-impregnated bearings and gears.

➤ Certain metals that are difficult to fabricate by other methods can be shaped by powder metallurgy. Tungsten is an example; tungsten filaments used in incandescent lamp bulbs are made using PM technology.

➤ Certain metal alloy combinations and cermets can be formed by PM that cannot be produced by other methods.

➤ PM compares favorably with most casting processes in terms of dimensional control of the product. Tolerances of ±0.13 mm (±0.005 in) are held routinely.

➤ PM production methods can be automated for economical production.

VIDEO CLIP ————————————————————————————

Powder Metallurgy. This clip contains two segments: (1) powder metal parts production and (2) PM overview.

There are limitations and disadvantages associated with PM processing. These include the following: (1) tooling and equipment costs are high, (2) metallic powders are expensive, and (3) there are difficulties with storing and handling metal powders (such as degradation of the metal over time, and fire hazards with particular metals). Also, (4) there are limitations on part geometry because metal powders do not readily flow laterally in the die during pressing, and allowances must be provided for ejection of the part from the die after pressing. In addition, (5) variations in material density throughout the part may be a problem in PM, especially for complex part geometries.

Although parts as large as 22 kg (50 lb) can be produced, most PM components are less than 2.2 kg (5 lb). A collection of typical PM parts is shown in Figure 16.1. The largest tonnage of metals for PM are alloys of iron, steel, and aluminum. Other PM metals include copper, nickel, and refractory metals such as molybdenum and tungsten. Metallic carbides such as tungsten carbide are often included within the scope of powder metallurgy; however, because these materials are ceramics, we defer their consideration until the next chapter.

The development of the modern field of powder metallurgy dates back to the 1800s (Historical Note 16.1). The scope of the modern technology includes not only parts production, but also preparation of the starting powders. Success in powder metallurgy depends to a large degree on the characteristics of the starting powders; we discuss this topic in Section 16.1. Later sections describe powder production, pressing, and sintering. There is a close correlation between PM technology and aspects of ceramics processing (Chapter 17). In ceramics (except glass), the starting material is also powder, so the methods for characterizing the powders are closely related to those in PM. Several of the shape-forming methods are similar, also.

FIGURE 16.1 A collection of powder metallurgy parts. (Courtesy of Dorst America, Inc.)

Historical Note 16.1   *Powder metallurgy*

Powders of metals such as gold and copper, as well as some of the metallic oxides, have been used for decorative purposes since ancient times. The uses included decorations on pottery, bases for paints, and in cosmetics. It is believed that the Egyptians used PM to make tools as far back as 3000 BCE.

The modern field of powder metallurgy dates to the early nineteenth century, when there was a strong interest in the metal platinum. Around 1815, Englishman William Wollaston developed a technique for preparing platinum powders, compacting them under high pressure, and baking (sintering) them at red heat. The Wollaston process marks the beginning of powder metallurgy as it is practiced today.

U.S. patents were issued in 1870 to S. Gwynn that relate to PM self-lubricating bearings. He used a mixture of 99% powdered tin and 1% petroleum, mixing, heating, and finally subjecting the mixture to extreme pressures to form it into the desired shape inside a mold cavity.

By the early 1900s, the incandescent lamp had become an important commercial product. A variety of filament materials had been tried, including carbon, zirconium, vanadium, and osmium; but it was concluded that tungsten was the best filament material. The problem was that tungsten was difficult to process because of its high melting point and unique properties. In 1908, William Coolidge developed a procedure that made production of tungsten incandescent lamp filaments feasible. In his process, fine powders of tungsten oxide (WO_3) were reduced to metallic powders, pressed into compacts, presintered, hot-forged into rounds, sintered, and finally drawn into filament wire. The Coolidge process is still used today to make filaments for incandescent light bulbs.

In the 1920s, cemented carbide tools (WC–Co) were being fabricated by PM techniques (Historical Note 7.2). Self-lubricating bearings were produced in large quantities starting in the 1930s. Powder metal gears and other components were mass produced in the 1960s and 1970s, especially in the automotive industry; and in the 1980s, PM parts for aircraft turbine engines were developed.

16.1 CHARACTERIZATION OF ENGINEERING POWDERS

A *powder* can be defined as a finely divided particulate solid. In this section we characterize metallic powders. However, most of the discussion applies to ceramic powders as well.

16.1.1 GEOMETRIC FEATURES

The geometry of the individual powders can be defined by the following attributes: (1) particle size and distribution, (2) particle shape and internal structure, and (3) surface area.

Particle Size and Distribution   Particle size refers to the dimensions of the individual powders. If the particle shape is spherical, a single dimension is adequate. For other shapes, two or more dimensions are needed. There are various methods available to obtain particle size data. The most common method uses screens of different mesh sizes. The term *mesh count* is used to refer to the number of openings per linear inch of screen. Higher mesh count indicates smaller particle size. A mesh count of 200 means there are 200 openings per linear inch. Because the mesh is square, the count is the same in both directions, and the total number of openings per square inch is $200^2 = 40,000$.

Particles are sorted by passing them through a series of screens of progressively smaller mesh size. The powders are placed on a screen of a certain mesh count and vibrated so that particles small enough to fit through the openings pass through to the next screen below. The second screen empties into a third, and so forth, so that the particles are sorted according to size. A certain powder size might be called size 230 through 200, indicating that the powders have passed through the 200 mesh, but not 230. To make the specification easier, we simply say that the particle size is 200. The procedure of separating the powders by size is called *classification.*

The openings in the screen are less than the reciprocal of the mesh count because of the thickness of the wire in the screen, as illustrated in Figure 16.2. Assuming that the limiting dimension of the particle is equal to the screen opening, we have

$$PS = \frac{1}{MC} - t_w \qquad (16.1)$$

where PS = particle size, in; MC = mesh count, openings per linear inch; and t_w = wire thickness of screen mesh, in.

The figure shows how smaller particles would pass through the openings, whereas larger powders would not. Variations occur in the powder sizes sorted by screening owing to differences in particle shapes, the range of sizes between mesh count steps, and variations in screen openings within a given mesh count. Also, the screening method has a

Particle size that would not pass through mesh

Particle size that would pass through mesh

FIGURE 16.2 Screen mesh for sorting particle sizes.

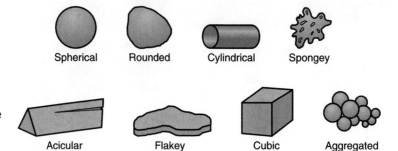

FIGURE 16.3 Several of the possible (ideal) particle shapes in powder metallurgy.

practical upper limit of $MC = 400$ (approximately), because of both the difficulty in making such fine screens and agglomeration of the small powders. Other methods to measure particle size include microscopy and X-ray techniques.

Typical particle sizes used in conventional powder metallurgy (press and sinter) range between 25 and 300 μm (0.001 and 0.012 in).[1] The high end of this range corresponds to a mesh count of around 65. The low end of the range is too small to be measured by the mesh count method.

Particle Shape and Internal Structure Metal powder shapes can be cataloged into various types, several of which are illustrated in Figure 16.3. There will be a variation in the particle shapes in a collection of powders, just as the particle size will vary. A simple and useful measure of shape is the aspect ratio—the ratio of maximum dimension to minimum dimension for a given particle. The aspect ratio for a spherical particle is 1.0, but for an acicular grain the ratio might be 2 to 4. Microscopic techniques are required to determine shape characteristics.

Any volume of loose powders will contain pores between the particles. These are called **open pores** because they are external to the individual particles. Open pores are spaces into which a fluid such as water, oil, or a molten metal can penetrate. In addition, there are **closed pores**—internal voids in the structure of an individual particle. The existence of these internal pores is usually minimal, and their effect when they do exist is minor, but they can influence density measurements, as we shall see later.

Surface Area Assuming that the particle shape is a perfect sphere, its area A and volume V are given by

$$A = \pi D^2 \tag{16.2}$$

$$V = \frac{\pi D^3}{6} \tag{16.3}$$

where D = diameter of the spherical particle, mm (in). The area-to-volume ratio A/V for a sphere is then given by

$$\frac{A}{V} = \frac{6}{D} \tag{16.4}$$

In general, the area-to-volume ratio can be expressed for any particle shape—spherical or nonspherical—as follows:

$$\frac{A}{V} = \frac{K_s}{D} \quad \text{or} \quad K_s = \frac{AD}{V} \tag{16.5}$$

---

[1]These values are provided by Prof. Wojciech Misiolek, my colleague in Lehigh's Department of Materials Science and Engineering. Powder metallurgy is one of his research areas.

where K_s = shape factor; D in the general case = the diameter of a sphere of equivalent volume as the nonspherical particle, mm (in).

Thus, K_s = 6.0 for a sphere. For particle shapes other than spherical, $K_s > 6$.

We can infer the following from these equations. Smaller particle size and higher shape factor (K_s) mean higher surface area for the same total weight of metal powders. This means greater area for surface oxidation to occur. Small powder size also leads to more agglomeration of the particles, which is a problem in automatic feeding of the powders. The reason for using smaller particle sizes is that they provide more uniform shrinkage and better mechanical properties in the final PM product.

16.1.2 OTHER FEATURES

Other features of engineering powders include interparticle friction, flow characteristics, packing, density, porosity, chemistry, and surface films.

Interparticle Friction and Flow Characteristics   Friction between particles affects the ability of a powder to flow readily and pack tightly. A common measure of interparticle friction is the ***angle of repose,*** which is the angle formed by a pile of powders as they are poured from a narrow funnel, as in Figure 16.4. Larger angles indicate greater friction between particles. Smaller particle sizes generally show greater friction and steeper angles. Spherical shapes result in the lowest interpartical friction; as shape deviates more from spherical, friction between particles tends to increase.

Flow characteristics are important in die filling and pressing. Automatic die filling depends on easy and consistent flow of the powders. In pressing, resistance to flow increases density variations in the compacted part; these density gradients are generally undesirable. A common measure of flow is the time required for a certain amount of powder (by weight) to flow through a standard-sized funnel. Smaller flow times indicate easier flow and lower interparticle friction. To reduce interparticle friction and facilitate flow during pressing, lubricants are often added to the powders in small amounts.

Packing, Density, and Porosity   Packing characteristics depend on two density measures. First, ***true density*** is the density of the true volume of the material. This is the density when the powders are melted into a solid mass, values of which are given in Table 4.1. Second, ***bulk density*** is the density of the powders in the loose state after pouring, which includes the effect of pores between particles. Because of the pores, bulk density is less than true density.

The ***packing factor*** is the bulk density divided by the true density. Typical values for loose powders range between 0.5 and 0.7. The packing factor depends on particle shape and

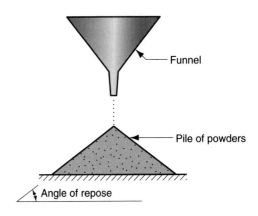

FIGURE 16.4   Interparticle friction as indicated by the angle of repose of a pile of powders poured from a narrow funnel. Larger angles indicate greater interparticle friction.

Funnel

Pile of powders

Angle of repose

the distribution of particle sizes. If powders of various sizes are present, the smaller powders will fit into the interstices of the larger ones that would otherwise be taken up by air, thus resulting in a higher packing factor. Packing can also be increased by vibrating the powders, causing them to settle more tightly. Finally, we should note that external pressure, as applied during compaction, greatly increases packing of powders through rearrangement and deformation of the particles.

Porosity represents an alternative way of considering the packing characteristics of a powder. **Porosity** is defined as the ratio of the volume of the pores (empty spaces) in the powder to the bulk volume. In principle

$$\text{Porosity} + \text{Packing factor} = 1.0 \tag{16.6}$$

The issue is complicated by the possible existence of closed pores in some of the particles. If these internal pore volumes are included in the above porosity, then the equation is exact.

Chemistry and Surface Films   Characterization of the powder would not be complete without an identification of its chemistry. Metallic powders are classified as either elemental, consisting of a pure metal, or pre-alloyed, wherein each particle is an alloy. We discuss these classes and the metals commonly used in PM more thoroughly in Section 16.5.1.

Surface films are a problem in powder metallurgy because of the large area per unit weight of metal when dealing with powders. The possible films include oxides, silica, adsorbed organic materials, and moisture [6]. Generally, these films must be removed before shape processing.

16.2 PRODUCTION OF METALLIC POWDERS

In general, producers of metallic powders are not the same companies as those that make PM parts. The powder producers are the suppliers; the plants that manufacture components out of powder metals are the customers. It is therefore appropriate to separate the discussion of powder production (this section) from the processes used to make PM products (later sections).

Virtually any metal can be made into powder form. There are three principal methods by which metallic powders are commercially produced, each of which involves energy input to increase the surface area of the metal. The methods are (1) atomization, (2) chemical, and (3) electrolytic [13]. In addition, mechanical methods are occasionally used to reduce powder sizes; however, these methods are much more commonly associated with ceramic powder production and we treat them in the next chapter.

16.2.1 ATOMIZATION

This method involves the conversion of molten metal into a spray of droplets that solidify into powders. It is the most versatile and popular method for producing metal powders today, applicable to almost all metals, alloys as well as pure metals. There are multiple ways of creating the molten metal spray, several of which are illustrated in Figure 16.5. Two of the methods shown are based on *gas atomization,* in which a high velocity gas stream (air or inert gas) is utilized to atomize the liquid metal. In Figure 16.5(a), the gas flows through an expansion nozzle, siphoning molten metal from the melt below and spraying it into a container. The droplets solidify into powder form. In a closely related method shown in Figure 16.5(b), molten metal flows by gravity through a nozzle and is immediately atomized

FIGURE 16.5 Several atomization methods for producing metallic powders: (a) and (b) two gas atomization methods; (c) water atomization; and (d) centrifugal atomization by the rotating disk method.

by air jets. The resulting metal powders, which tend to be spherical, are collected in a chamber below.

The approach shown in Figure 16.5(c) is similar to (b), except that a high-velocity water stream is used instead of air. This is known as *water atomization* and is the most common of the atomization methods, particularly suited to metals that melt below 1600°C (2900°F). Cooling is more rapid, and the resulting powder shape is irregular rather than spherical. The disadvantage of using water is oxidation on the particle surface. A recent innovation involves the use of synthetic oil rather than water to reduce oxidation. In both air and water atomization processes, particle size is controlled largely by the velocity of the fluid stream; particle size is inversely related to velocity.

Several methods are based on *centrifugal atomization.* In one approach, the *rotating disk method* shown in Figure 16.5(d), the liquid metal stream pours onto a rapidly rotating disk that sprays the metal in all directions to produce powders.

16.2.2 OTHER PRODUCTION METHODS

Other metal powder production methods include various chemical reduction processes, precipitation methods, and electrolysis.

Chemical reduction includes a variety of chemical reactions by which metallic compounds are reduced to elemental metal powders. A common process involves liberation of metals from their oxides by use of reducing agents such as hydrogen or carbon monoxide. The reducing agent is made to combine with the oxygen in the compound to free the

FIGURE 16.6   Iron powders produced by decomposition of iron pentacarbonyl; particle sizes range from about 0.25 to 3.0 μm (10–125 μ-in). (Photo courtesy of GAF Chemicals Corporation, Advanced Materials Division.)

metallic element. This approach is used to produce powders of iron, tungsten, and copper. Another chemical process for iron powders involves the decomposition of iron pentacarbonyl ($Fe(Co)_5$) to produce spherical particles of high purity. Powders produced by this method are illustrated in the photomicrograph of Figure 16.6. Other chemical processes include *precipitation* of metallic elements from salts dissolved in water. Powders of copper, nickel, and cobalt can be produced by this approach.

In *electrolysis,* an electrolytic cell is set up in which the source of the desired metal is the anode. The anode is slowly dissolved under an applied voltage, transported through the electrolyte, and deposited on the cathode. The deposit is removed, washed, and dried to yield a metallic powder of very high purity. The technique is used for producing powders of beryllium, copper, iron, silver, tantalum, and titanium.

16.3   CONVENTIONAL PRESSING AND SINTERING

After the metallic powders have been produced, the conventional PM sequence consists of three steps: (1) blending and mixing of the powders; (2) compaction, in which the powders are pressed into the desired part shape; and (3) sintering, which involves heating to a temperature below the melting point to cause solid-state bonding of the particles and strengthening of the part. The three steps, sometimes referred to as primary operations in PM, are portrayed in Figure 16.7. In addition, secondary operations are sometimes performed to improve dimensional accuracy, increase density, and for other reasons.

16.3.1   BLENDING AND MIXING OF THE POWDERS

To achieve successful results in compaction and sintering, the metallic powders must be thoroughly homogenized beforehand. The terms blending and mixing are both used in this context. *Blending* refers to when powders of the same chemical composition but possibly different particle sizes are intermingled. Different particle sizes are often blended to reduce porosity. *Mixing* refers to powders of different chemistries being combined. An advantage of PM technology is the opportunity to mix various metals into alloys that would be difficult or impossible to produce by other means. The distinction between blending and mixing is not always precise in industrial practice.

Blending and mixing are accomplished by mechanical means. Four alternatives are illustrated in Figure 16.8: (a) rotation in a drum; (b) rotation in a double-cone container;

FIGURE 16.7 The conventional powder metallurgy production sequence: (1) blending, (2) compacting, and (3) sintering; (a) shows the condition of the particles, whereas (b) shows the operation and/or workpart during the sequence.

(c) agitation in a screw mixer; and (d) stirring in a blade mixer. There is more science to these devices than one would suspect. Best results seem to occur when the container is between 20% and 40% full. The containers are usually designed with internal baffles or other ways of preventing free-fall during blending of powders of different sizes, because variations in settling rates between sizes result in segregation—just the opposite of what is wanted in blending. Vibration of the powder is undesirable, because it also causes segregation.

Other ingredients are usually added to the metallic powders during the blending and/ or mixing step. These additives include (1) *lubricants,* such as stearates of zinc and aluminum, in small amounts to reduce friction between particles and at the die wall during compaction; (2) *binders,* which are required in some cases to achieve adequate strength in the pressed but unsintered parts; and (3) *deflocculants,* which inhibit agglomeration of powders for better flow characteristics during subsequent processing.

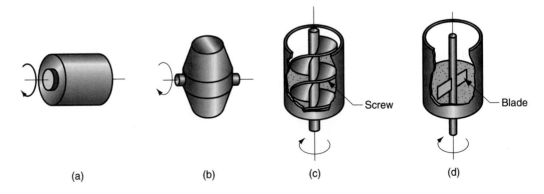

FIGURE 16.8 Several blending and mixing devices: (a) rotating drum, (b) rotating double-cone, (c) screw mixer, and (d) blade mixer.

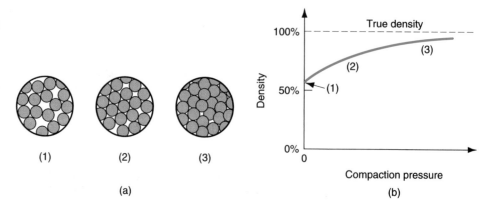

FIGURE 16.9 Pressing, the conventional method of compacting metal powders in PM: (1) filling the die cavity with powder, done by automatic feed in production, (2) initial, and (3) final positions of upper and lower punches during compaction, and (4) ejection of part.

16.3.2 COMPACTION

In compaction, high pressure is applied to the powders to form them into the required shape. The conventional compaction method is *pressing,* in which opposing punches squeeze the powders contained in a die. The steps in the pressing cycle are shown in Figure 16.9. The workpart after pressing is called a *green compact,* the word *green* meaning not yet fully processed. As a result of pressing, the density of the part, called the *green density,* is much greater than the starting bulk density. The *green strength* of the part when pressed is adequate for handling but far less than that achieved after sintering.

The applied pressure in compaction results initially in repacking of the powders into a more efficient arrangement, eliminating "bridges" formed during filling, reducing pore space, and increasing the number of contacting points between particles. As pressure increases, the particles are plastically deformed, causing interparticle contact area to increase and additional particles to make contact. This is accompanied by a further reduction in pore volume. The progression is illustrated in three views in Figure 16.10

FIGURE 16.10 (a) Effect of applied pressure during compaction: (1) initial loose powders after filling, (2) repacking, and (3) deformation of particles; and (b) density of the powders as a function of pressure. The sequence here corresponds to steps 1, 2, and 3 in Figure 16.9.

FIGURE 16.11   A 450-kN (50-ton) hydraulic press for compaction of powder metallurgy components. (Photo courtesy of Dorst America, Inc.)

for starting particles of spherical shape. Also shown is the associated density represented by the three views as a function of applied pressure.

Presses used in conventional PM compaction are mechanical, hydraulic, or a combination of the two. A 450 kN (50 ton) hydraulic unit is shown in Figure 16.11. Because of differences in part complexity and associated pressing requirements, presses can be distinguished as (1) pressing from one direction, referred to as single-action presses; or (2) pressing from two directions, any of several types including opposed ram, double-action, and multiple action. Current available press technology can provide up to 10 separate action controls to produce parts of significant geometric complexity. We examine part complexity and other design issues in Section 16.6.

The capacity of a press for PM production is generally given in tons or kN or MN. The required force for pressing depends on the projected area of the PM part (area in the horizontal plane for a vertical press) multiplied by the pressure needed to compact the given metal powders. Reducing this to equation form

$$F = A_p p_c \qquad (16.7)$$

where F = required force, N (lb); A_p = projected area of the part, mm^2 (in^2); and p_c = compaction pressure required for the given powder material, MPa (lb/in^2).

Compaction pressures typically range from 70 MPa (10,000 lb/in^2) for aluminum powders to 700 MPa (100,000 lb/in^2) for iron and steel powders.

16.3.3 SINTERING

After pressing, the green compact lacks strength and hardness; it is easily crumbled under low stresses. *Sintering* is a heat treatment operation performed on the compact to bond

FIGURE 16.12 Sintering on a microscopic scale: (1) particle bonding is initiated at contact points; (2) contact points grow into "necks"; (3) the pores between particles are reduced in size; and (4) grain boundaries develop between particles in place of the necked regions.

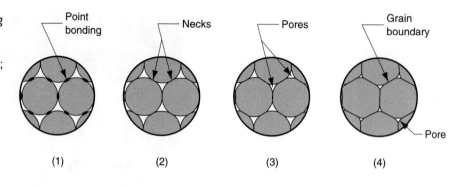

its metallic particles, thereby increasing strength and hardness. The treatment is usually carried out at temperatures between 0.7 and 0.9 of the metal's melting point (absolute scale). The terms *solid-state sintering* or *solid-phase sintering* are sometimes used for this conventional sintering because the metal remains unmelted at these treatment temperatures.

It is generally agreed among researchers that the primary driving force for sintering is reduction of surface energy [6], [16]. The green compact consists of many distinct particles, each with its own individual surface, and so the total surface area contained in the compact is very high. Under the influence of heat, the surface area is reduced through the formation and growth of bonds between the particles, with associated reduction in surface energy. The finer the initial powder size, the higher the total surface area, and the greater the driving force behind the process.

The series of sketches in Figure 16.12 shows on a microscopic scale the changes that occur during sintering of metallic powders. Sintering involves mass transport to create the necks and transform them into grain boundaries. The principal mechanism by which this occurs is diffusion; other possible mechanisms include plastic flow. Shrinkage occurs during sintering as a result of pore size reduction. This depends to a large extent on the density of the green compact, which depends on the pressure during compaction. Shrinkage is generally predictable when processing conditions are closely controlled.

Because PM applications usually involve medium-to-high production, most sintering furnaces are designed with mechanized flow-through capability for the workparts. The heat treatment consists of three steps, accomplished in three chambers in these continuous furnaces: (1) preheat, in which lubricants and binders are burned off; (2) sinter; and (3) cool down. The treatment is illustrated in Figure 16.13. Typical sintering temperatures and times are given for selected metals in Table 16.1.

In modern sintering practice, the atmosphere in the furnace is controlled. The purposes of a controlled atmosphere include (1) protection from oxidation, (2) providing a reducing atmosphere to remove existing oxides, (3) providing a carburizing atmosphere, and (4) assisting in removing lubricants and binders used in pressing. Common sintering furnace atmospheres are inert gas, nitrogen-based, dissociated ammonia, hydrogen, and natural gas [6]. Vacuum atmospheres are used for certain metals, such as stainless steel and tungsten.

16.3.4 SECONDARY OPERATIONS

PM secondary operations include densification, sizing, impregnation, infiltration, heat treatment, and finishing.

FIGURE 16.13
(a) Typical heat treatment cycle in sintering; and (b) schematic cross section of a continuous sintering furnace.

Densification and Sizing A number of secondary operations are performed to increase density, improve accuracy, or accomplish additional shaping of the sintered part. *Repressing* is a pressing operation in which the part is squeezed in a closed die to increase density and improve physical properties. *Sizing* is the pressing of a sintered part to improve dimensional accuracy. *Coining* is a pressworking operation on a sintered part to press details into its surface.

Some PM parts require *machining* after sintering. Machining is rarely done to size the part, but rather to create geometric features that cannot be achieved by pressing, such as internal and external threads, side holes, and other details.

Impregnation and Infiltration Porosity is a unique and inherent characteristic of powder metallurgy technology. It can be exploited to create special products by filling the available pore space with oils, polymers, or metals that have lower melting temperatures than the base powder metal.

Impregnation is the term used when oil or other fluid is permeated into the pores of a sintered PM part. The most common products of this process are oil-impregnated bearings,

TABLE 16.1 Typical sintering temperatures and times for selected powder metals.

Metal	Sintering Temperatures		Typical Time
	°C	°F	
Brass	850	1600	25 min
Bronze	820	1500	15 min
Copper	850	1600	25 min
Iron	1100	2000	30 min
Stainless steel	1200	2200	45 min
Tungsten	2300	4200	480 min

Compiled from [10] and [17].

gears, and similar machinery components. Self-lubricating bearings, usually made of bronze or iron with 10% to 30% oil by volume, are widely used in the automotive industry. The treatment is accomplished by immersing the sintered parts in a bath of hot oil.

An alternative application of impregnation involves PM parts that must be made pressure tight or impervious to fluids. In this case, the parts are impregnated with various types of polymer resins that seep into the pore spaces in liquid form and then solidify. In some cases, resin impregnation is used to facilitate subsequent processing, for example, to permit the use of processing solutions (such as plating chemicals) that would otherwise soak into the pores and degrade the product, or to improve machinability of the PM workpart.

Infiltration is an operation in which the pores of the PM part are filled with a molten metal. The melting point of the filler metal must be below that of the PM part. The process involves heating the filler metal in contact with the sintered component so that capillary action draws the filler into the pores. The resulting structure is relatively nonporous, and the infiltrated part has a more uniform density, as well as improved toughness and strength. An application of the process is copper infiltration of iron PM parts.

16.3.5 HEAT TREATMENT AND FINISHING

Powder metal components can be heat treated (Chapter 27) and finished (electroplated or painted, Chapter 28) by most of the same processes used on parts produced by casting and other metalworking processes. Special care must be exercised in heat treatment because of porosity; for example, salt baths are not used for heating PM parts. Plating and coating operations are applied to sintered parts for appearance purposes and corrosion resistance. Again, precautions must be taken to avoid entrapment of chemical solutions in the pores; impregnation and infiltration are frequently used for this purpose. Common platings for PM parts include copper, nickel, chromium, zinc, and cadmium.

16.4   ALTERNATIVE PRESSING AND SINTERING TECHNIQUES

The conventional press and sinter sequence is the most widely used shaping technology in powder metallurgy. Additional methods for processing PM parts are discussed in this section.

16.4.1 ISOSTATIC PRESSING

A feature of conventional pressing is that pressure is applied uniaxially. This imposes limitations on part geometry, because metallic powders do not readily flow in directions perpendicular to the applied pressure. Uniaxial pressing also leads to density variations in the compact after pressing. In *isostatic pressing*, pressure is applied from all directions against the powders that are contained in a flexible mold; hydraulic pressure is used to achieve compaction. Isostatic pressing takes two alternative forms: (1) cold isostatic pressing and (2) hot isostatic pressing.

Cold isostatic pressing (CIP) involves compaction performed at room temperature. The mold, made of rubber or other elastomer material, is oversized to compensate for shrinkage. Water or oil is used to provide the hydrostatic pressure against the mold inside the chamber. Figure 16.14 illustrates the processing sequence in cold isostatic pressing. Advantages of CIP include more uniform density, less expensive tooling, and greater applicability to shorter production runs. Good dimensional accuracy is difficult to

FIGURE 16.14   Cold isostatic pressing: (1) powders are placed in the flexible mold; (2) hydrostatic pressure is applied against the mold to compact the powders; and (3) pressure is reduced and the part is removed.

achieve in isostatic pressing because of the flexible mold. Consequently, subsequent finish shaping operations are often required to obtain the required dimensions, either before or after sintering.

Hot isostatic pressing (HIP) is carried out at high temperatures and pressures, using a gas such as argon or helium as the compression medium. The mold in which the powders are contained is made of sheet metal to withstand the high temperatures. HIP accomplishes pressing and sintering in one step. Despite this apparent advantage, it is a relatively expensive process and its applications seem to be concentrated in the aerospace industry. PM parts made by HIP are characterized by high density (porosity near zero), thorough interparticle bonding, and good mechanical strength.

16.4.2   POWDER INJECTION MOLDING

Injection molding is closely associated with the plastics industry (Section 13.6). The same basic process can be applied to form parts of metal or ceramic powders, the difference being that the starting polymer contains a high content of particulate matter, typically from 50% to 85% by volume. When used in powder metallurgy, the term *metal injection molding* (MIM) is used. The more general process is *powder injection molding* (PIM), which includes both metal and ceramic powders. The steps in MIM proceed as follows [7]: (1) Metallic powders are mixed with an appropriate binder. (2) Granular pellets are formed from the mixture. (3) The pellets are heated to molding temperature, injected into a mold cavity, and the part is cooled and removed from the mold. (4) The part is processed to remove the binder using any of several thermal or solvent techniques. (5) The part is sintered. (6) Secondary operations are performed as appropriate.

The binder in powder injection molding acts as a carrier for the particles. Its functions are to provide proper flow characteristics during molding and hold the powders in the molded shape until sintering. The five basic types of binders in PIM are: (1) thermosetting polymers, such as phenolics; (2) thermoplastic polymers, such as polyethylene; (3) water; (4) gels; and (5) inorganic materials [7]. Polymers are the most frequently used.

Powder injection molding is suited to part geometries similar to those in plastic injection molding. It is not cost competitive for simple axisymmetric parts, because the

FIGURE 16.15 Powder rolling: (1) powders are fed through compaction rolls to form a green strip; (2) sintering; (3) cold rolling; and (4) resintering.

conventional press-and-sinter process is quite adequate for these cases. PIM seems most economical for small, complex parts of high value. Dimensional accuracy is limited by the shrinkage that accompanies densification during sintering.

16.4.3 POWDER ROLLING, EXTRUSION, AND FORGING

Rolling, extrusion, and forging are familiar bulk metal forming processes (Chapter 19). We describe them here in the context of powder metallurgy.

Powder Rolling Powders can be compressed in a rolling mill operation to form metal strip stock. The process is usually set up to run continuously or semicontinuously, as shown in Figure 16.15. The metallic powders are compacted between rolls into a green strip that is fed directly into a sintering furnace. It is then cold rolled and resintered.

Powder Extrusion Extrusion is one of the basic manufacturing processes (Section 1.3.1). In PM extrusion, the starting powders can be in different forms. In the most popular method, powders are placed in a vacuum-tight sheet metal can, heated, and extruded with the container. In another variation, billets are preformed by a conventional press and sinter process, and then the billet is hot extruded. These methods achieve a high degree of densification in the PM product.

Powder Forging Forging is an important metal forming process (Section 1.3.1). In powder forging, the starting workpart is a powder metallurgy part preformed to proper size by pressing and sintering. Advantages of this approach are: (1) densification of the PM part, (2) lower tooling costs and fewer forging "hits" (and therefore higher production rate) because the starting workpart is preformed, and (3) reduced material waste.

16.4.4 COMBINED PRESSING AND SINTERING

Hot isostatic pressing (Section 16.4.1) accomplishes compaction and sintering in one step. Other techniques that combine the two steps are hot pressing and spark sintering.

Hot Pressing The setup in uniaxial hot pressing is very similar to conventional PM pressing, except that heat is applied during compaction. The resulting product is generally dense, strong, hard, and dimensionally accurate. Despite these advantages, the process presents certain technical problems that limit its adoption. Principal among these are (1) selecting a suitable mold material that can withstand the high sintering temperatures; (2) longer production cycle required to accomplish sintering; and (3) heating and maintaining atmospheric control in the process [2]. Hot pressing has found some application in the production of sintered carbide products using graphite molds.

Spark Sintering   An alternative approach that combines pressing and sintering but overcomes some of the problems in hot pressing is spark sintering. The process consists of two basic steps [2], [17]: (1) powder or a green compacted preform is placed in a die; and (2) upper and lower punches, which also serve as electrodes, compress the part and simultaneously apply a high-energy electrical current that burns off surface contaminants and sinters the powders, forming a dense, solid part in about 15 seconds. The process has been applied to a variety of metals.

16.4.5 LIQUID PHASE SINTERING

Conventional sintering (Section 16.3.3) is solid-state sintering; the metal is sintered at a temperature below its melting point. In systems involving a mixture of two powder metals, in which there is a difference in melting temperature between the metals, an alternative type of sintering is used, called liquid phase sintering. In this process, the two powders are initially mixed, and then heated to a temperature that is high enough to melt the lower-melting-point metal but not the other. The melted metal thoroughly wets the solid particles, creating a dense structure with strong bonding between the metals upon solidification. Depending on the metals involved, prolonged heating may result in alloying of the metals by gradually dissolving the solid particles into the liquid melt and/or diffusion of the liquid metal into the solid. In either case, the resulting product is fully densified (no pores) and strong. Examples of systems that involve liquid phase sintering include Fe–Cu, W–Cu, and Cu–Co [6].

16.5   MATERIALS AND PRODUCTS FOR POWDER METALLURGY

The raw materials for PM processing are more expensive than for other metalworking because of the additional energy required to reduce the metal to powder form. Accordingly, PM is competitive only in a certain range of applications. In this section we identify the materials and products that seem most suited to powder metallurgy.

Powder Metallurgy Materials   From a chemistry standpoint, metal powders can be classified as either elemental or pre-alloyed. *Elemental* powders consist of a pure metal and are used in applications in which high purity is important. For example, pure iron might be used where its magnetic properties are important. The most common elemental powders are those of iron, aluminum, and copper.

Elemental powders are also mixed with other metal powders to produce special alloys that are difficult to formulate using conventional processing methods. Tool steels are an example; PM permits blending of ingredients that is difficult or impossible by traditional alloying techniques. Using mixtures of elemental powders to form an alloy provides a processing benefit, even where special alloys are not involved. Because the powders are pure metals, they are not as strong as pre-alloyed metals. Therefore, they deform more readily during pressing, so that density and green strength are higher than with pre-alloyed compacts.

In ***pre-alloyed*** powders, each particle is an alloy composed of the desired chemical composition. Pre-alloyed powders are used for alloys that cannot be formulated by mixing elemental powders; stainless steel is an important example. The most common pre-alloyed powders are certain copper alloys, stainless steel, and high-speed steel.

The commonly used elemental and pre-alloyed powdered metals, in approximate order of tonnage usage, are: (1) iron, by far the most widely used PM metal, frequently mixed with graphite to make steel parts, (2) aluminum, (3) copper and its alloys, (4) nickel, (5) stainless steel, (6) high-speed steel, and (7) other PM materials such as tungsten, molybdenum, titanium, tin, and precious metals.

Powder Metallurgy Products   A substantial advantage offered by PM technology is that parts can be made to near net shape or net shape; they require little or no additional shaping after PM processing. Some of the components commonly manufactured by powder metallurgy are gears, bearings, sprockets, fasteners, electrical contacts, cutting tools, and various machinery parts. When produced in large quantities, metal gears and bearings are particularly well suited to PM for two reasons: (1) the geometry is defined principally in two dimensions, so the part has a top surface of a certain shape, but there are no features along the sides; and (2) there is a need for porosity in the material to serve as a reservoir for lubricant. More complex parts with true three-dimensional geometries are also feasible in powder metallurgy, by adding secondary operations such as machining to complete the shape of the pressed and sintered part, and by observing certain design guidelines such as those outlined in the following section.

16.6   DESIGN CONSIDERATIONS IN POWDER METALLURGY

Use of PM techniques is generally suited to a certain class of production situations and part designs. In this section we attempt to define the characteristics of this class of applications for which powder metallurgy is most appropriate. We first present a classification system for PM parts, and then offer some guidelines on component design.

The Metal Powder Industries Federation (MPIF) defines four classes of powder metallurgy part designs, by level of difficulty in conventional pressing. The system is useful because it indicates some of the limitations on shape that can be achieved with conventional PM processing. The four part classes are illustrated in Figure 16.16.

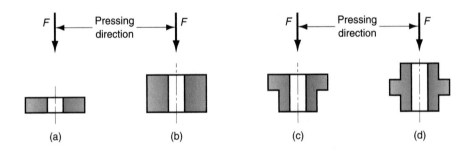

FIGURE 16.16   Four classes of PM parts—side view shown; cross section is circular: (a) Class I—simple thin shapes that can be pressed from one direction; (b) Class II—simple but thicker shapes that require pressing from two directions; (c) Class III—two levels of thickness, pressed from two directions; and (d) Class IV—multiple levels of thickness, pressed from two directions, with separate controls for each level to achieve proper densification throughout the compact.

The MPIF classification system provides some guidance concerning part geome-tries that are suited to conventional PM pressing techniques. Additional advice is offered in the following design guidelines, compiled from [3], [13], and [17].

➤ Economics of PM processing usually require large part quantities to justify the cost of equipment and special tooling required. Minimum quantities of 10,000 units are suggested [17], although exceptions exist.

➤ Powder metallurgy is unique in its capability to fabricate parts with a controlled level of porosity. Porosities up to 50% are possible.

➤ PM can be used to make parts out of unusual metals and alloys — materials that would be difficult if not impossible to fabricate by other means.

➤ The geometry of the part must permit ejection from the die after pressing; this generally means that the part must have vertical or near-vertical sides, although steps in the part are permissible as suggested by the MPIF classification system (Figure 16.16). Design features such as undercuts and holes on the part sides, as shown in Figure 16.17, must be avoided. Vertical undercuts and holes, as in Figure 16.18, are permissible because they do not interfere with ejection. Vertical holes can be of cross-sectional shapes other than round (e.g., squares, keyways) without significant increases in tooling or processing difficulty.

➤ Screw threads cannot be fabricated by PM pressing; if required, they must be machined into the PM component after sintering.

➤ Chamfers and corner radii are possible by PM pressing, as shown in Figure 16.19. Problems are encountered in punch rigidity when angles are too acute.

FIGURE 16.17   Part features to be avoided in PM: (a) side holes and (b) side undercuts. Part ejection is impossible.

(a)   (b)

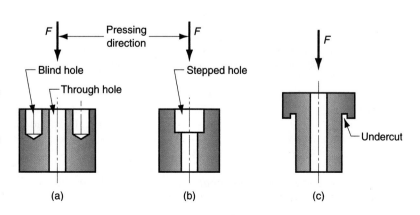

FIGURE 16.18
Permissible part features in PM: (a) vertical hole, blind and through, (b) vertical stepped hole, and (c) undercut in vertical direction. These features allow part ejection.

(a)   (b)   (c)

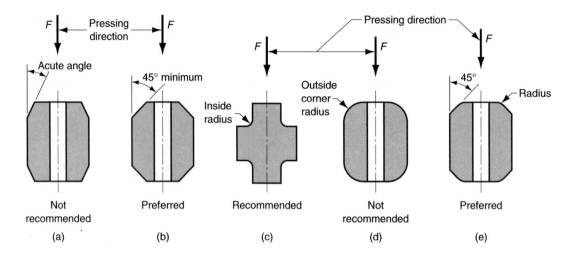

FIGURE 16.19 Chamfers and corner radii are accomplished but certain rules should be observed: (a) avoid acute chamfer angles; (b) larger angles are preferred for punch rigidity; (c) small inside radius is desirable; (d) full outside corner radius is difficult because punch is fragile at corner's edge; (e) outside corner problem can be solved by combining radius and chamfer.

FIGURE 16.20 Minimum recommended wall thickness (a) between holes or (b) between a hole and an outside wall should be 1.5 mm (0.060 in).

> ➢ Wall thickness should be a minimum of 1.5 mm (0.060 in) between holes or a hole and the outside part wall, as indicated in Figure 16.20. Minimum recommended hole diameter is 1.5 mm (0.060 in).

REFERENCES

[1] *ASM Handbook,* Vol. 7: *Powder Metal Technologies and Applications,* ASM International, Materials Park, Ohio, 1998.

[2] Amstead, B. H., Ostwald, P. F., and Begeman, M. L. *Manufacturing Processes.* 8th ed. John Wiley & Sons, New York, 1987.

[3] Bralla, J. G. (ed.). *Design for Manufacturability Handbook.* 2nd ed. McGraw-Hill, New York, 1998.

[4] Bulger, M."Metal Injection Molding," *Advanced Materials & Processes.* March 2005, pp. 39–40.

[5] Dixon, R. H. T., and Clayton, A. *Powder Metallurgy for Engineers.* The Machinery Publishing Co. Ltd., Brighton, U.K., 1971.

[6] German, R. M. *Powder Metallurgy Science.* 2nd ed. Metal Powder Industries Federation, Princeton, New Jersey, 1994.

[7] German, R. M. *Powder Injection Molding.* Metal Powder Industries Federation, Princeton, New Jersey, 1990.

[8] German, R. M. *A-Z of Powder Metallurgy,* Elsevier Science, Amsterdam, Netherlands, 2006.

[9] Johnson, P. K. "P/M Industry Trends in 2005," *Advanced Materials & Processes,* March 2005, pp. 25–28.

[10] *Metals Handbook.* 9th ed. Vol. 7. *Powder Metallurgy.* American Society for Metals, Metals Park, Ohio, 1984.

[11] Pease, L. F. "A Quick Tour of Powder Metallurgy," *Advanced Materials & Processes,* March 2005, pp. 36–38.

[12] Pease, L. F., and West, W. G. *Fundamentals of Powder Metallurgy,* Metal Powder Industries Federation, Princeton, New Jersey, 2002.

[13] *Powder Metallurgy Design Handbook.* Metal Powder Industries Federation, Princeton, New Jersey, 1989.

[14] Schey, J. A. *Introduction to Manufacturing Processes.* 3rd ed. McGraw-Hill, New York, 1999.

[15] Smythe, J. "Superalloy Powders: An Amazing History," *Advanced Materials & Processes,* November 2008, pp. 52–55.

[16] Waldron, M. B., and Daniell, B. L. *Sintering.* Heyden, London, 1978.

[17] Wick, C., Benedict, J. T., and Veilleux, R. F. (eds.). *Tool and Manufacturing Engineers Handbook.* 4th ed. Vol. II, *Forming.* Society of Manufacturing Engineers, Dearborn, Michigan, 1984.

REVIEW QUESTIONS

16.1. Name some of the reasons for the commercial importance of powder metallurgy technology.

16.2. What are some of the disadvantages of PM methods?

16.3. In the screening of powders for sizing, what is meant by the term mesh count?

16.4. What is the difference between open pores and closed pores in metallic powders?

16.5. What is meant by the term aspect ratio for a metallic particle?

16.6. How would one measure the angle of repose for a given amount of metallic powder?

16.7. Define bulk density and true density for metallic powders.

16.8. What are the principal methods used to produce metallic powders?

16.9. What are the three basic steps in the conventional powder metallurgy shaping process?

16.10. What is the technical difference between mixing and blending in powder metallurgy?

16.11. What are some of the ingredients usually added to the metallic powders during blending and/or mixing?

16.12. What is meant by the term green compact?

16.13. Describe what happens to the individual particles during compaction.

16.14. What are the three steps in the sintering cycle in PM?

16.15. What are some of the reasons why a controlled atmosphere furnace is desirable in sintering?

16.16. What are the advantages of infiltration in PM?

16.17. What is the difference between powder injection molding and metal injection molding?

16.18. How is isostatic pressing distinguished from conventional pressing and sintering in PM?

16.19. Describe liquid phase sintering.

16.20. What are the two basic classes of metal powders as far as chemistry is concerned?

16.21. Why is PM technology so well suited to the production of gears and bearings?

16.22. **(Video)** List the most common methods for forming the pressed parts in powder metallurgy according to the powder metallurgy video.

16.23. **(Video)** List the types of environments that can be present during the sintering process according to the powder metallurgy video.

MULTIPLE CHOICE QUIZ

There are 19 correct answers in the following multiple choice questions (some questions have multiple answers that are correct). To attain a perfect score on the quiz, all correct answers must be given. Each correct answer is worth 1 point. Each omitted answer or wrong answer reduces the score by 1 point, and each additional answer beyond the correct number of answers reduces the score by 1 point. Percentage score on the quiz is based on the total number of correct answers.

16.1. The particle size that can pass through a screen is obtained by taking the reciprocal of the mesh count of the screen: (a) true or (b) false?

16.2. For a given weight of metallic powders, the total surface area of the powders is increased by which of the following (two best answers): (a) larger particle size, (b) smaller particle size, (c) higher shape factor, and (d) smaller shape factor?

16.3. As particle size increases, interparticle friction (a) decreases, (b) increases, or (c) remains the same?

16.4. Which of the following powder shapes would tend to have the lowest interparticle friction: (a) acicular, (b) cubic, (c) flakey, (d) spherical, and (e) rounded?

16.5. Which of the following statements is correct in the context of metallic powders (three correct answers): (a) porosity + packing factor = 1.0, (b) packing factor = 1/porosity, (c) packing factor = 1.0 − porosity, (d) packing factor = −porosity, (e) packing factor = bulk density/true density?

16.6. Which of the following most closely typifies the sintering temperatures in PM? (a) 0.5 T_m, (b) 0.8 T_m, (c) T_m, where T_m = melting temperature of the metal?

16.7. Repressing refers to a pressworking operation used to compress a sintered part in a closed die to achieve closer sizing and better surface finish: (a) true or (b) false?

16.8. Impregnation refers to which of the following (two best answers): (a) filling the pores of the PM part with a molten metal, (b) putting polymers into the pores of a PM part, (c) soaking oil by capillary action into the pores of a PM part, and (d) something that should not happen in a factory?

16.9. In cold isostatic pressing, the mold is most typically made of which one of the following: (a) rubber, (b) sheetmetal, (c) textile, (d) thermosetting polymer, or (e) tool steel?

16.10. Which of the following processes combines pressing and sintering of the metal powders (three best answers): (a) hot isostatic pressing, (b) hot pressing, (c) metal injection molding, (d) pressing and sintering, and (e) spark sintering?

16.11. Which of the following design features would be difficult or impossible to achieve by conventional pressing and sintering (three best answers): (a) outside rounded corners, (b) side holes, (c) threaded holes, (d) vertical stepped holes, and (e) vertical wall thickness of 1/8 inch (3 mm)?

PROBLEMS

Characterization of Engineering Powders

16.1. A screen with 325 mesh count has wires with a diameter of 0.001377 in. Determine (a) the maximum particle size that will pass through the wire mesh and (b) the proportion of open space in the screen.

16.2. A screen with 10 mesh count has wires with a diameter of 0.0213 in. Determine (a) the maximum particle size that will pass through the wire mesh and (b) the proportion of open space in the screen.

16.3. What is the aspect ratio of a cubic particle shape?

16.4. Determine the shape factor for metallic particles of the following ideal shapes: (a) sphere, (b) cubic, (c) cylindrical with length-to-diameter ratio of 1:1, (d) cylindrical with length-to-diameter ratio of 2:1, and (e) a disk-shaped flake whose thickness-to-diameter ratio is 1:10.

16.5. A pile of iron powder weighs 2 lb. The particles are spherical in shape and all have the same diameter of 0.002 in. (a) Determine the total surface area of all the particles in the pile. (b) If the packing factor = 0.6, determine the volume taken by the pile. Note: the density of iron = 0.284 lb/in^3.

16.6. Solve Problem 16.5, except that the diameter of the particles is 0.004 in. Assume the same packing factor.

16.7. Suppose in Problem 16.5 that the average particle diameter = 0.002 in; however, the sizes vary, forming a statistical distribution as follows: 25% of the particles by weight are 0.001 in, 50% are 0.002 in, and 25% are 0.003 in. Given this distribution, what is the total surface area of all the particles in the pile?

16.8. A solid cube of copper with each side = 1.0 ft is converted into metallic powders of spherical shape by gas atomization. What is the percentage increase in total surface area if the diameter of each particle is 0.004 in (assume that all particles are the same size)?

16.9. A solid cube of aluminum with each side = 1.0 m is converted into metallic powders of spherical shape by gas atomization. How much total surface area is added by the process if the diameter of each particle is 100 microns (assume that all particles are the same size)?

16.10. Given a large volume of metallic powders, all of which are perfectly spherical and having the same exact diameter, what is the maximum possible packing factor that the powders can take?

Compaction and Design Considerations

16.11. In a certain pressing operation, the metallic powder fed into the open die has a packing factor of 0.5. The pressing operation reduces the powders to two thirds of their starting volume. In the subsequent sintering operation, shrinkage amounts to 10% on a volume basis. Given that these are the only factors that affect the structure of the finished part, determine its final porosity.

16.12. A bearing of simple geometry is to be pressed out of bronze powders, using a compacting pressure of 207 MPa. The outside diameter = 44 mm, the inside diameter = 22 mm, and the length of the bearing = 25 mm. What is the required press tonnage to perform this operation?

16.13. The part shown in Figure P16.13 is to be pressed of iron powders using a compaction pressure of 75,000 lb/in². Dimensions are inches. Determine (a) the most appropriate pressing direction, (b) the required press tonnage to perform this operation, and (c) the final weight of the part if the porosity is 10%. Assume shrinkage during sintering can be neglected.

FIGURE P16.13   Part for Problem 16.13 (dimensions in inches).

16.14. For each of the four part drawings in Figure P16.14, indicate which PM class the parts belong to, whether the part must be pressed from one or two directions, and how many levels of press control will be required? Dimensions are mm.

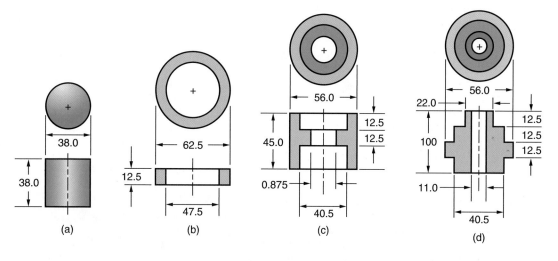

FIGURE P16.14   Parts for Problem 16.14 (dimensions in mm).

17 PROCESSING OF CERAMICS AND CERMETS

Chapter Contents

Ceramic materials divide into three categories (Chapter 7): (1) traditional ceramics, (2) new ceramics, and (3) glasses. The processing of glass involves solidification primarily and is covered in Chapter 12. In the present chapter we consider the particulate processing methods used for traditional and new ceramics. We also consider the processing of metal matrix composites and ceramic matrix composites.

Traditional ceramics are made from minerals occurring in nature. They include pottery, porcelain, bricks, and cement. New ceramics are made from synthetically produced raw materials and cover a wide spectrum of products such as cutting tools, artificial bones, nuclear fuels, and substrates for electronic circuits. The starting material for all of these items is powder. In the case of the traditional ceramics, the powders are usually mixed with water to temporarily bind the particles together and achieve the proper consistency for shaping. For new ceramics, other substances are used as binders during shaping. After shaping, the green parts are sintered. This is often called *firing* in ceramics, but the function is the same as in powder metallurgy: to effect a solid-state reaction that bonds the material into a hard solid mass.

The processing methods discussed in this chapter are commercially and technologically important because virtually all ceramic products are formed by these methods (except, of course, glass products). The manufacturing sequence is similar for traditional and new ceramics because the form of the starting material is the same: powder. However, the processing methods for the two categories are sufficiently different that we discuss them separately.

17.1 PROCESSING OF TRADITIONAL CERAMICS

In this section we describe the production technology used to make traditional ceramic products such as pottery, stoneware and other dinnerware, bricks, tile, and ceramic

FIGURE 17.1   Usual steps in traditional ceramics processing: (1) preparation of raw materials, (2) shaping, (3) drying, and (4) firing. Part (a) shows the workpart during the sequence, whereas (b) shows the condition of the powders.

refractories. Bonded grinding wheels are also produced by the same basic methods. What these products have in common is that their raw materials consist primarily of silicate ceramics—clays. The processing sequence for most of the traditional ceramics consists of the steps depicted in Figure 17.1.

17.1.1   PREPARATION OF THE RAW MATERIAL

The shaping processes for traditional ceramics require that the starting material be in the form of a plastic paste. This paste is made of fine ceramic powders mixed with water, and its consistency determines the ease of forming the material and the quality of the final product. The raw ceramic material usually occurs in nature as rocky lumps, and reduction to powder is the purpose of the preparation step in ceramics processing.

Techniques for reducing particle size in ceramics processing involve mechanical energy in various forms, such as impact, compression, and attrition. The term *comminution* is used for these techniques, which are most effective on brittle materials, including cement, metallic ores, and brittle metals. Two general categories of comminution operations are distinguished: crushing and grinding.

Crushing refers to the reduction of large lumps from the mine to smaller sizes for subsequent further reduction. Several stages may be required (e.g., primary crushing, secondary crushing), the reduction ratio in each stage being in the range 3 to 6. Crushing of minerals is accomplished by compression against rigid surfaces or by impact against surfaces in a rigid constrained motion [1]. Figure 17.2 shows several types of equipment used to perform crushing: (a) jaw crushers, in which a large jaw toggles back and forth to crush lumps against a hard, rigid surface; (b) gyratory crushers, which use a gyrating cone to compress lumps against a rigid surface; (c) roll crushers, in which the ceramic lumps are squeezed between rotating rolls; and (d) hammer mills, which use rotating hammers impacting the material to break up the lumps.

Grinding, in the context here, refers to the operation of reducing the small pieces produced by crushing into a fine powder. Grinding is accomplished by abrasion and impact of the crushed mineral by the free motion of unconnected hard media such as balls, pebbles,

FIGURE 17.2 Crushing operations: (a) jaw crusher, (b) gyratory crusher, (c) roll crusher, and (d) hammer mill.

or rods [1]. Examples of grinding include (a) ball mill, (b) roller mill, and (c) impact grinding, illustrated in Figure 17.3.

In a **ball mill,** hard spheres mixed with the stock to be comminuted are tumbled inside a rotating cylindrical container. The rotation causes the balls and stock to be carried up the container wall, and then pulled back down by gravity to accomplish a grinding action by a combination of impact and attrition. These operations are often carried out with water added to the mixture, so that the ceramic is in the form of a slurry. In a **roller mill,** stock is compressed against a flat horizontal grinding table by rollers riding over the table surface. Although not clearly shown in the sketch, the pressure of the grinding rollers against the table is regulated by mechanical springs or hydraulic-pneumatic means. In **impact grinding,** which seems to be less frequently used, particles of stock are thrown against a hard flat surface, either in a high velocity air stream or a high-speed slurry. The impact fractures the pieces into smaller particles.

The plastic paste required for shaping consists of ceramic powders and water. Clay is usually the main ingredient in the paste because it has ideal forming characteristics. The more

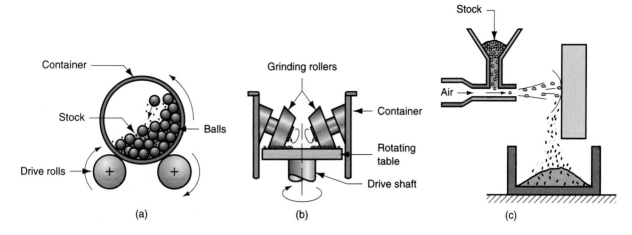

FIGURE 17.3   Mechanical methods of producing ceramic powders: (a) ball mill, (b) roller mill, and (c) impact grinding.

water there is in the mixture, the more plastic and easily formed is the clay paste. However, when the formed part is later dried and fired, shrinkage occurs that can lead to cracking in the product. To address this problem, other ceramic raw materials that do not shrink on drying and firing are usually added to the paste, often in significant amounts. Also, other components can be included to serve special functions. Thus, the ingredients of the ceramic paste can be divided into the following three categories [3]: (1) clay, which provides the consistency and plasticity required for shaping; (2) nonplastic raw materials, such as alumina and silica, which do not shrink in drying and firing but unfortunately reduce plasticity in the mixture during forming; and (3) other ingredients, such as fluxes that melt (vitrify) during firing and promote sintering of the ceramic material, and wetting agents that improve mixing of ingredients.

These ingredients must be thoroughly mixed, either wet or dry. The ball mill often serves this purpose in addition to its grinding function. Also, the proper amounts of powder and water in the paste must be attained, so water must be added or removed, depending on the prior condition of the paste and its desired final consistency.

17.1.2 SHAPING PROCESSES

The optimum proportions of powder and water depend on the shaping process used. Some shaping processes require high fluidity; others act on a composition that contains very low water content. At about 50% water by volume, the mixture is a slurry that flows like a liquid. As the water content is reduced, increased pressure is required on the paste to produce a similar flow. Thus, the shaping processes can be divided according to the consistency of the mixture: (1) slip casting, in which the mixture is a slurry with 25% to 40% water; (2) plastic-forming methods that shape the clay in a plastic condition at 15% to 25% water; (3) semi-dry pressing, in which the clay is moist (10% to 15% water) but has low plasticity; and (4) dry pressing, in which the clay is basically dry, containing less than 5% water. Dry clay has no plasticity. The four categories are represented in the chart of Figure 17.4, which compares the categories with the condition of the clay used as starting material. Each category includes several different shaping processes.

Slip Casting   In slip casting, a suspension of ceramic powders in water, called a *slip,* is poured into a porous plaster of paris ($CaSO_4$–$2H_2O$) mold so that water from the mix is

FIGURE 17.4 Four categories of shaping processes used for traditional ceramics, compared with water content and pressure required to form the clay.

gradually absorbed into the plaster to form a firm layer of clay at the mold surface. The composition of the slip is typically 25% to 40% water, the remainder being clay often mixed with other ingredients. It must be sufficiently fluid to flow into the crevices of the mold cavity, yet lower water content is desirable for faster production rates. Slip casting has two principal variations: drain casting and solid casting. In ***drain casting,*** which is the traditional process, the mold is inverted to drain excess slip after the semi-solid layer has been formed, thus leaving a hollow part in the mold; the mold is then opened and the part removed. The sequence, which is very similar to slush casting of metals, is illustrated in Figure 17.5. It is used to make tea pots, vases, art objects, and other hollow-ware products. In ***solid casting,*** used to produce solid products, adequate time is allowed for the entire body to become firm. The mold must be periodically resupplied with additional slip to account for shrinkage because of absorbed water.

Plastic Forming This category includes a variety of methods, both manual and mechanized. They all require the starting mixture to have a plastic consistency, which is generally achieved with 15% to 25% water. Manual methods generally make use of clay at the upper end of the range because it provides a material that is more easily formed; however, this is accompanied by greater shrinkage in drying. Mechanized methods generally employ a mixture with lower water content so that the starting clay is stiffer.

Although manual forming methods date back thousands of years, they are still used today by skilled artisans, either in production or for artworks. ***Hand modeling*** involves the creation of the ceramic product by manipulating the mass of plastic clay

FIGURE 17.5 Sequence of steps in drain casting, a form of slip casting: (1) slip is poured into mold cavity; (2) water is absorbed into plaster mold to form a firm layer; (3) excess slip is poured out; and (4) part is removed from mold and trimmed.

(1) (2) (3) (4)

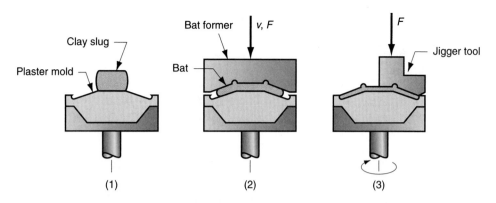

FIGURE 17.6 Sequence in jiggering: (1) wet clay slug is placed on a convex mold; (2) batting; and (3) a jigger tool imparts the final product shape. Symbols *v* and *F* indicate motion (*v* = velocity) and applied force, respectively.

into the desired geometry. In addition to art pieces, patterns for plaster molds in slip casting are often made this way. ***Hand molding*** is a similar method, only a mold or form is used to define portions of the geometry. ***Hand throwing*** on a potter's wheel is another refinement of the handicraft methods. The ***potter's wheel*** is a round table that rotates on a vertical spindle, powered either by motor or foot-operated treadle. Ceramic products of circular cross section can be formed on the rotating table by throwing and shaping the clay, sometimes using a mold to provide the internal shape.

Strictly speaking, use of a motor-driven potter's wheel is a mechanized method. However, most mechanized clay-forming methods are characterized by much less manual participation than the hand-throwing method described above. These more mechanized methods include jiggering, plastic pressing, and extrusion. ***Jiggering*** is an extension of the potter's wheel methods, in which hand throwing is replaced by mechanized techniques. It is used to produce large numbers of identical items such as houseware plates and bowls. Although there are variations in the tools and methods used, reflecting different levels of automation and refinements to the basic process, a typical sequence is as follows, depicted in Figure 17.6: (1) a wet clay slug is placed on a convex mold; (2) a forming tool is pressed into the slug to provide the initial rough shape—the operation is called ***batting*** and the workpiece thus created is called a ***bat;*** and (3) a heated jigger tool is used to impart the final contoured shape to the product by pressing the profile into the surface during rotation of the workpart. The reason for heating the tool is to produce steam from the wet clay that prevents sticking. Closely related to jiggering is ***jolleying,*** in which the basic mold shape is concave rather than convex [8]. In both of these processes, a rolling tool is sometimes used in place of the nonrotating jigger (or jolley) tool; this rolls the clay into shape, avoiding the need to first bat the slug.

Plastic pressing is a forming process in which a plastic clay slug is pressed between upper and lower molds, contained in metal rings. The molds are made of a porous material such as gypsum, so that when a vacuum is drawn on the backs of the mold halves, moisture is removed from the clay. The mold sections are then opened, using positive air pressure to prevent sticking of the part in the mold. Plastic pressing achieves a higher production rate than jiggering and is not limited to radially symmetric parts.

Extrusion is used in ceramics processing to produce long sections of uniform cross section, which are then cut to required piece length. The extrusion equipment utilizes a screw-type action to assist in mixing the clay and pushing the plastic material through the die opening. This production sequence is widely used to make hollow bricks, shaped tiles, drain pipes, tubes, and insulators. It is also used to make the starting clay slugs for other ceramics processing methods such as jiggering and plastic pressing.

Semi-dry Pressing   In semi-dry pressing, the proportion of water in the starting clay is typically 10% to 15%. This results in low plasticity, precluding the use of plastic forming methods that require very plastic clay. Semi-dry pressing uses high pressure to overcome

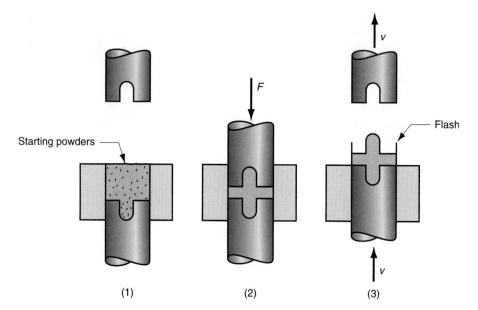

FIGURE 17.7 Semi-dry pressing: (1) depositing moist powder into die cavity, (2) pressing, and (3) opening the die sections and ejection. Symbols *v* and *F* indicate motion (*v* = velocity) and applied force, respectively.

the material's low plasticity and force it to flow into a die cavity, as depicted in Figure 17.7. Flash is often formed from excess clay being squeezed between the die sections.

Dry Pressing The main distinction between semi-dry and dry pressing is the moisture content of the starting mix. The moisture content of the starting clay in dry pressing is typically below 5%. Binders are usually added to the dry powder mix to provide sufficient strength in the pressed part for subsequent handling. Lubricants are also added to prevent die sticking during pressing and ejection. Because dry clay has no plasticity and is very abrasive, there are differences in die design and operating procedures, compared with semi-dry pressing. The dies must be made of hardened tool steel or cemented tungsten carbide to reduce wear. Because dry clay will not flow during pressing, the geometry of the part must be relatively simple, and the amount and distribution of starting powder in the die cavity must be right. No flash is formed in dry pressing, and no drying shrinkage occurs, so drying time is eliminated and good accuracy can be achieved in the dimensions of the final product. The process sequence in dry pressing is similar to semi-dry pressing. Typical products include bathroom tile, electrical insulators, and refractory brick.

17.1.3 DRYING

Water plays an important role in most of the traditional ceramics shaping processes. Thereafter, it serves no purpose and must be removed from the body of the clay piece before firing. Shrinkage is a problem during this step in the processing sequence because water contributes volume to the piece, and when it is removed, the volume is reduced. The effect can be seen in Figure 17.8. As water is initially added to dry clay, it simply replaces the air in the pores between ceramic grains, and there is no volumetric change. Increasing the water content above a certain point causes the grains to become separated and the volume to grow, resulting in wet clay that has plasticity and formability. As more water is added, the mixture eventually becomes a liquid suspension of clay particles in water.

The reverse of this process occurs in drying. As water is removed from the wet clay, the volume of the piece shrinks. The drying process occurs in two stages, as depicted in Figure 17.9. In the first stage, the rate of drying is rapid and constant, as water is evaporated

FIGURE 17.8   Volume of clay as a function of water content. Relationship shown here is typical; it varies for different clay compositions.

from the surface of the clay into the surrounding air and water from the interior migrates by capillary action toward the surface to replace it. It is during this stage that shrinkage occurs, with the associated risk of warping and cracking owing to variations in drying in different sections of the piece. In the second stage of drying, the moisture content has been reduced to where the ceramic grains are in contact, and little or no further shrinkage occurs. The drying process slows, and this is seen in the decreasing rate in the plot.

In production, drying is usually accomplished in drying chambers in which temperature and humidity are controlled to achieve the proper drying schedule. Care must be taken so that water is not removed too rapidly, lest large moisture gradients be set up in the piece, making it more prone to crack. Heating is usually by a combination of convection and radiation, using infrared sources. Typical drying times range between a quarter of an hour for thin sections to several days for very thick sections.

17.1.4 FIRING (SINTERING)

After shaping but before firing, the ceramic piece is said to be ***green*** (the same term as in powder metallurgy), meaning not fully processed or treated. The green piece lacks hardness and strength; it must be fired to fix the part shape and achieve hardness and strength in the finished ware. ***Firing*** is the heat treatment process that sinters the ceramic material; it is performed in a furnace called a ***kiln***. In ***sintering,*** bonds are developed between the ceramic grains, and this is accompanied by densification and reduction of porosity. Therefore, shrinkage occurs in the polycrystalline material in addition to the shrinkage that has already occurred in drying. Sintering in ceramics is basically the same

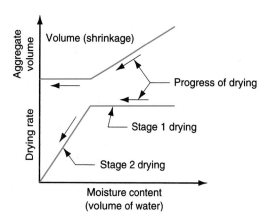

FIGURE 17.9   Typical drying rate curve and associated volume reduction (drying shrinkage) for a ceramic body in drying. Drying rate in the second stage of drying is depicted here as a straight line (constant rate decrease as a function of water content); the function is variously shown as concave or convex in the literature [3], [8].

mechanism as in powder metallurgy. In the firing of traditional ceramics, certain chemical reactions between the components in the mixture may also take place, and a glassy phase also forms among the crystals that acts as a binder. Both of these phenomena depend on the chemical composition of the ceramic material and the firing temperatures used.

Unglazed ceramic ware is fired only once; glazed products are fired twice. *Glazing* refers to the application of a ceramic surface coating to make the piece more impervious to water and to enhance its appearance (Section 7.2.2). The usual processing sequence with glazed ware is (1) fire the ware once before glazing to harden the body of the piece, (2) apply the glaze, and (3) fire the piece a second time to harden the glaze.

17.2   PROCESSING OF NEW CERAMICS

Most of the traditional ceramics are based on clay, which possesses a unique capacity to be plastic when mixed with water but hard when dried and fired. Clay consists of various formulations of hydrous aluminum silicate, usually mixed with other ceramic materials, to form a rather complex chemistry. New ceramics (Section 7.3) are based on simpler chemical compounds, such as oxides, carbides, and nitrides. These materials do not possess the plasticity and formability of traditional clay when mixed with water. Accordingly, other ingredients must be combined with the ceramic powders to achieve plasticity and other desirable properties during forming, so that conventional shaping methods can be used. The new ceramics are generally designed for applications that require higher strength, hardness, and other properties not found in the traditional ceramic materials. These requirements have motivated the introduction of several new processing techniques not previously used for traditional ceramics.

The manufacturing sequence for the new ceramics can be summarized in the following steps: (1) preparation of starting materials, (2) shaping, (3) sintering, and (4) finishing. Although the sequence is nearly the same as for the traditional ceramics, the details are often quite different, as we shall see in the following.

17.2.1   PREPARATION OF STARTING MATERIALS

Because the strength specified for these materials is usually much greater than for traditional ceramics, the starting powders must be more homogeneous in size and composition, and particle size must be smaller (strength of the resulting ceramic product is inversely related to grain size). All of this means that greater control of the starting powders is required. Powder preparation includes mechanical and chemical methods. The mechanical methods consist of the same ball mill grinding operations used for traditional ceramics. The trouble with these methods is that the ceramic particles become contaminated from the materials used in the balls and walls of the mill. This compromises the purity of the ceramic powders and results in microscopic flaws that reduce the strength of the final product.

Two chemical methods are used to achieve greater homogeneity in the powders of new ceramics: freeze drying and precipitation from solution. In *freeze drying,* salts of the appropriate starting chemistry are dissolved in water and the solution is sprayed to form small droplets, which are rapidly frozen. The water is then removed from the droplets in a vacuum chamber, and the resulting freeze-dried salt is decomposed by heating to form the ceramic powders. Freeze drying is not applicable to all ceramics, because in some cases a suitable water-soluble salt cannot be identified as the starting material.

Precipitation from solution is another preparation method used for new ceramics. In the typical process, the desired ceramic compound is dissolved from the starting mineral, thus permitting impurities to be filtered out. An intermediate compound is then precipitated from solution, which is converted into the desired compound by heating. An example of the precipitation method is the **Bayer process** for producing high purity alumina (also used in the production of aluminum). In this process, aluminum oxide is dissolved from the mineral bauxite so that iron compounds and other impurities can be removed. Then, aluminum hydroxide $(Al(OH)_3)$ is precipitated from solution and reduced to Al_2O_3 by heating.

Further preparation of the powders includes classification by size and mixing before shaping. Very fine powders are required for new ceramics applications, and so the grains must be separated and classified according to size. Thorough mixing of the particles, especially when different ceramic powders are combined, is required to avoid segregation.

Various additives are often combined with the starting powders, usually in small amounts. The additives include (1) **plasticizers** to improve plasticity and workability; (2) **binders** to bond the ceramic particles into a solid mass in the final product, (3) **wetting agents** for better mixing; (4) **deflocculants,** which help to prevent clumping and premature bonding of the powders; and (5) **lubricants,** to reduce friction between ceramic grains during forming and to reduce sticking during mold release.

17.2.2 SHAPING

Many of the shaping processes for new ceramics are borrowed from powder metallurgy (PM) and traditional ceramics. The press and sinter methods discussed in Section 16.3 have been adapted to the new ceramic materials. And some of the traditional ceramics-forming techniques (Section 17.1.2) are used to shape the new ceramics, including slip casting, extrusion, and dry pressing. The following processes are not normally associated with the forming of traditional ceramics, although several are associated with PM.

Hot Pressing   Hot pressing is similar to dry pressing (Section 17.1.2), except that the process is carried out at elevated temperatures, so that sintering of the product is accomplished simultaneously with pressing. This eliminates the need for a separate firing step in the sequence. Higher densities and finer grain size are obtained, but die life is reduced by the hot abrasive particles against the die surfaces.

Isostatic Pressing   Isostatic pressing of ceramics is the same process used in powder metallurgy (Section 16.4.1). It uses hydrostatic pressure to compact the ceramic powders from all directions, thus avoiding the problem of nonuniform density in the final product that is often observed in the traditional uniaxial pressing method.

Doctor-Blade Process   This process is used for making thin sheets of ceramic. One common application of the sheets is in the electronics industry as a substrate material for integrated circuits. The process is diagrammed in Figure 17.10. A ceramic slurry is introduced onto a moving carrier film such as cellophane. Thickness of the ceramic on the carrier is determined by a wiper, called a **doctor-blade.** As the slurry moves down the line, it is dried

FIGURE 17.10   The doctor-blade process, used to fabricate thin ceramic sheets. Symbol v indicates motion (v = velocity).

into a flexible green ceramic tape. At the end of the line, a take-up spool reels in the tape for later processing. In its green condition, the tape can be cut or otherwise shaped before firing.

Powder Injection Molding Powder injection molding (PIM) is the same as the PM process (Section 16.4.2), except that the powders are ceramic rather than metallic. Ceramic particles are mixed with a thermoplastic polymer that acts as a carrier and provides the proper flow characteristics at molding temperatures. The mix is then heated and injected into a mold cavity. Upon cooling, which hardens the polymer, the mold is opened and the part is removed. Because the temperatures needed to plasticize the carrier are much lower than those required for sintering the ceramic, the piece is green after molding. Before sintering, the plastic binder must be removed. This is called **_debinding,_** which is usually accomplished by a combination of thermal and solvent treatments.

Applications of ceramic PIM are currently inhibited by difficulties in debinding and sintering. Burning off the polymer is relatively slow, and its removal significantly weakens the green strength of the molded part. Warping and cracking often occur during sintering. Further, ceramic products made by powder injection molding are especially vulnerable to microstructural flaws that limit their strength.

17.2.3 SINTERING

Because the plasticity needed to shape the new ceramics is not normally based on a water mixture, the drying step so commonly required to remove water from the traditional green ceramics can be omitted in the processing of most new ceramic products. The sintering step, however, is still very much required to obtain maximum possible strength and hardness. The functions of sintering are the same as before, to: (1) bond individual grains into a solid mass, (2) increase density, and (3) reduce or eliminate porosity.

Temperatures around 80% to 90% of the melting temperature of the material are commonly used in sintering ceramics. Sintering mechanisms differ somewhat between the new ceramics, which are based predominantly on a single chemical compound (e.g., Al_2O_3), and the clay-based ceramics, which usually consist of several compounds having different melting points. In the case of the new ceramics, the sintering mechanism is mass diffusion across the contacting particle surfaces, probably accompanied by some plastic flow. This mechanism causes the centers of the particles to move closer together, resulting in densification of the final material. In the sintering of traditional ceramics, this mechanism is complicated by the melting of some constituents and the formation of a glassy phase that acts as a binder between the grains.

17.2.4 FINISHING

Parts made of new ceramics sometimes require finishing. In general, these operations have one or more of the following purposes, to: (1) increase dimensional accuracy, (2) improve surface finish, and (3) make minor changes in part geometry. Finishing operations usually involve grinding and other abrasive processes (Chapter 25). Diamond abrasives must be used to cut the hardened ceramic materials.

17.3 PROCESSING OF CERMETS

Many metal matrix composites (MMCs) and ceramic matrix composites (CMCs) are processed by particulate processing methods. The most prominent examples are cemented carbides and other cermets.

17.3.1 CEMENTED CARBIDES

The cemented carbides are a family of composite materials consisting of carbide ceramic particles embedded in a metallic binder. They are classified as metal matrix composites because the metallic binder is the matrix that holds the bulk material together; however, the carbide particles constitute the largest proportion of the composite material, normally ranging between 80% and 96% by volume. Cemented carbides are technically classified as cermets, although they are often distinguished from the other materials in this class.

The most important cemented carbide is tungsten carbide in a cobalt binder (WC–Co). Generally included within this category are certain mixtures of WC, TiC, and TaC in a Co matrix, in which tungsten carbide is the major component. Other cemented carbides include titanium carbide in nickel (TiC–Ni) and chromium carbide in nickel (Cr_3C_2–Ni). These composites are discussed in Section 9.2.1, and the carbide ingredients are described in Section 7.3.2. In our present discussion we are concerned with the particulate processing of cemented carbide.

To provide a strong and pore-free part, the carbide powders must be sintered with a metal binder. Cobalt works best with WC, whereas nickel is better with TiC and Cr_3C_2. The usual proportion of binder metal is from around 4% up to 20%. Powders of carbide and binder metal are thoroughly mixed wet in a ball mill (or other suitable mixing machine) to form a homogeneous sludge. Milling also serves to refine particle size. The sludge is then dried in a vacuum or controlled atmosphere to prevent oxidation in preparation for compaction.

Compaction   Various methods are used to shape the powder mix into a green compact of the desired geometry. The most common process is cold pressing, described earlier and used for high production of cemented carbide parts such as cutting tool inserts. The dies used in cold pressing must be made oversized to account for shrinkage during sintering. Linear shrinkage can be 20% or more. For high production, the dies themselves are made with WC–Co liners to reduce wear, because of the abrasive nature of carbide particles. For smaller quantities, large flat sections are sometimes pressed and then cut into smaller pieces of the specified size.

Other compaction methods used for cemented carbide products include *isostatic pressing* and *hot pressing* for large pieces, such as draw dies and ball mill balls; and *extrusion,* for long sections of circular, rectangular, or other cross section. Each of these processes has been described previously, either in this or the preceding chapter.

Sintering   Although it is possible to sinter WC and TiC without a binder metal, the resulting material is somewhat less than 100% of true density. Use of a binder yields a structure that is virtually free of porosity.

Sintering of WC–Co involves liquid phase sintering (Section 16.4.5). The process can be explained with reference to the binary phase diagram for these constituents in Figure 17.11. The typical composition range for commercial cemented carbide products is identified in the diagram. The usual sintering temperatures for WC–Co are in the range 1370°C to 1425°C (2500°C to 2600°F), which is below cobalt's melting point of 1495°C (2716°F). Thus, the pure binder metal does not melt at the sintering temperature. However, as the phase diagram shows, WC dissolves in Co in the solid state. During the heat treatment, WC is gradually dissolved into the gamma phase, and its melting point is reduced so that melting finally occurs. As the liquid phase forms, it flows and wets the WC particles, further dissolving the solid. The presence of the molten metal also serves to remove gases from the internal regions of the compact. These mechanisms combine to effect a rearrangement of the remaining WC particles into a closer packing, which results in significant densification and shrinkage of the WC–Co mass. Later, during cooling in the sintering cycle, the dissolved carbide is precipitated and deposited onto the existing crystals to form a coherent WC skeleton, throughout which the Co binder is embedded.

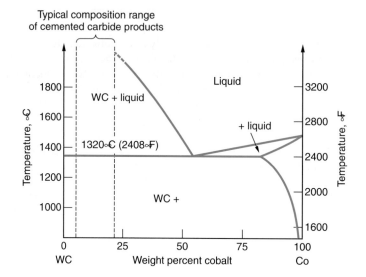

FIGURE 17.11 WC–Co phase diagram. (Source: [7]).

Secondary Operations Subsequent processing is usually required after sintering to achieve adequate dimensional control of cemented carbide parts. Grinding with a diamond abrasive wheel is the most common secondary operation performed for this purpose. Other processes used to shape the hard cemented carbides include electric discharge machining and ultrasonic machining, two nontraditional material removal processes discussed in Chapter 26.

17.3.2 OTHER CERMETS AND CERAMIC MATRIX COMPOSITES

In addition to cemented carbides, other cermets are based on oxide ceramics such as Al_2O_3 and MgO. Chromium is a common metal binder used in these composite materials. The ceramic-to-metal proportions cover a wider range than those of the cemented carbides; in some cases, the metal is the major ingredient. These cermets are formed into useful products by the same basic shaping methods used for cemented carbides.

The current technology of ceramic matrix composites (Section 9.3) includes ceramic materials (e.g., Al_2O_3, BN, Si_3N_4, and glass) reinforced by fibers of carbon, SiC, or Al_2O_3. If the fibers are whiskers (fibers consisting of single crystals), these CMCs can be processed by particulate methods used for new ceramics (Section 17.2).

17.4 PRODUCT DESIGN CONSIDERATIONS

Ceramic materials have special properties that make them attractive to designers if the application is right. The following design recommendations, compiled from Bralla [2] and other sources, apply to both new and traditional ceramic materials, although designers are more likely to find opportunities for new ceramics in engineered products. In general, the same guidelines apply to cemented carbides.

➢ Ceramic materials are several times stronger in compression than in tension; accordingly, ceramic components should be designed to be subjected to compressive stresses, not tensile stresses.

➢ Ceramics are brittle and possess almost no ductility. Ceramic parts should not be used in applications that involve impact loading or high stresses that might cause fracture.

➢ Although many of the ceramic shaping processes allow complex geometries to be formed, it is desirable to keep shapes simple for both economic and technical reasons. Deep holes, channels, and undercuts should be avoided, as should large cantilevered projections.

➢ Outside edges and corners should have radii or chamfers; likewise, inside corners should have radii. This guideline is, of course, violated in cutting tool applications, in which the cutting edge must be sharp to function. The cutting edge is often fabricated with a very small radius or chamfer to protect it from microscopic chipping, which could lead to failure.

➢ Part shrinkage in drying and firing (for traditional ceramics) and sintering (for new ceramics) may be significant and must be taken into account by the designer in dimensioning and tolerancing. This is mostly a problem for manufacturing engineers, who must determine appropriate size allowances so that the final dimensions will be within the tolerances specified.

➢ Screw threads in ceramic parts should be avoided. They are difficult to fabricate and do not have adequate strength in service after fabrication.

REFERENCES

[1] Bhowmick, A. K. Bradley Pulverizer Company, Allentown, Pennsylvania, personal communication, February 1992.

[2] Bralla, J. G. (editor-in-chief). *Design for Manufacturability Handbook.* 2nd ed. McGraw-Hill, New York, 1999.

[3] Hlavac, J. *The Technology of Glass and Ceramics.* Elsevier Scientific Publishing, New York, 1983.

[4] Kingery, W. D., Bowen, H. K., and Uhlmann, D. R. *Introduction to Ceramics.* 2nd ed. John Wiley & Sons, New York, 1995.

[5] Rahaman, M. N. *Ceramic Processing.* CRC Taylor & Francis, Boca Raton, Florida, 2007.

[6] Richerson, D. W. *Modern Ceramic Engineering: Properties, Processing, and Use in Design,* 3rd ed. CRC Taylor & Francis, Boca Raton, Flotida, 2006.

[7] Schwarzkopf, P., and Kieffer, R. *Cemented Carbides.* Macmillan, New York, 1960.

[8] Singer, F., and Singer, S. S. *Industrial Ceramics.* Chemical Publishing Company, New York, 1963.

[9] Somiya, S. (ed.). *Advanced Technical Ceramics.* Academic Press, San Diego, California, 1989.

REVIEW QUESTIONS

17.1. What is the difference between the traditional ceramics and the new ceramics, as far as raw materials are concerned?

17.2. List the basic steps in the traditional ceramics processing sequence.

17.3. What is the technical difference between crushing and grinding in the preparation of traditional ceramic raw materials?

17.4. Describe the slip casting process in traditional ceramics processing.

17.5. List and briefly describe some of the plastic forming methods used to shape traditional ceramic products.

17.6. What is the process of jiggering?

17.7. What is the difference between dry pressing and semi-dry pressing of traditional ceramic parts?

17.8. What happens to a ceramic material when it is sintered?

17.9. What is the name given to the furnace used to fire ceramic ware?

17.10. What is glazing in traditional ceramics processing?

17.11. Why is the drying step, so important in the processing of traditional ceramics, usually not required in processing of new ceramics?

17.12. Why is raw material preparation more important in the processing of new ceramics than for traditional ceramics?

17.13. What is the freeze drying process used to make certain new ceramic powders?

17.14. Describe the doctor-blade process.

17.15. Liquid phase sintering is used for WC–Co compacts, even though the sintering temperatures are below the melting points of either WC or Co. How is this possible?

17.16. What are some design recommendations for ceramic parts?

MULTIPLE CHOICE QUIZ

There are 16 correct answers in the following multiple choice questions (some questions have multiple answers that are correct). To attain a perfect score on the quiz, all correct answers must be given. Each correct answer is worth 1 point. Each omitted answer or wrong answer reduces the score by 1 point, and each additional answer beyond the correct number of answers reduces the score by 1 point. Percentage score on the quiz is based on the total number of correct answers.

17.1. The following equipment is used for crushing and grinding of minerals in the preparation of traditional ceramics raw materials. Which of the pieces listed is used for grinding (two correct answers): (a) ball mill, (b) hammer mill, (c) jaw crusher, (d) roll crusher, and (e) roller mill?

17.2. Which one of the following compounds becomes a plastic and formable material when mixed with suitable proportions of water: (a) aluminum oxide, (b) hydrogen oxide, (c) hydrous aluminum silicate, or (d) silicon dioxide?

17.3. At which one of the following water contents does clay become a suitably plastic material for the traditional ceramics plastic forming processes: (a) 5%, (b) 10%, (c) 20%, or (d) 40%?

17.4. Which of the following processes are not plastic forming methods used in the shaping of traditional ceramics (three correct answers): (a) dry pressing, (b) extrusion, (c) jangling, (d) jiggering, (e) jolleying, (f) slip casting, and (g) spinning?

17.5. The term green piece in ceramics refers to a part that has been shaped but not yet fired: (a) true or (b) false?

17.6. In the final product made of a polycrystalline new ceramic material, strength increases with grain size: (a) true or (b) false?

17.7. Which one of the following processes for the new ceramic materials accomplishes shaping and sintering simultaneously: (a) doctor-blade process, (b) freeze drying, (c) hot pressing, (d) injection molding, or (e) isostatic pressing?

17.8. Which of the following are the purposes of finishing operations used for parts made of the new ceramics (two best answers): (a) apply a surface coating, (b) electroplate the surface, (c) improve surface finish, (d) increase dimensional accuracy, and (e) work harden the surface?

17.9. Which of the following terms describes what a cemented carbide is (one best answer): (a) ceramic, (b) cermet, (c) composite, (d) metal, (e) new ceramic, or (f) traditional ceramic?

17.10. Which of the following geometric features should be avoided if possible in the design of structural components made of new ceramics (three best answers): (a) deep holes, (b) rounded inside corners, (c) rounded outside corners, (d) sharp edges, (e) thick sections, and (f) threads?

Part V Metal Forming and Sheet Metalworking

18

FUNDAMENTALS OF METAL FORMING

Chapter Contents

Metal forming includes a large group of manufacturing processes in which plastic deformation is used to change the shape of metal workpieces. Deformation results from the use of a tool, usually called a *die* in metal forming, which applies stresses that exceed the yield strength of the metal. The metal therefore deforms to take a shape determined by the geometry of the die. Metal forming dominates the class of shaping operations identified in Chapter 1 as the *deformation processes* (Figure 1.4).

Stresses applied to plastically deform the metal are usually compressive. However, some forming processes stretch the metal, while others bend the metal, and still others apply shear stresses to the metal. To be successfully formed, a metal must possess certain properties. Desirable properties include low yield strength and high ductility. These properties are affected by temperature. Ductility is increased and yield strength is reduced when work temperature is raised. The effect of temperature gives rise to distinctions between cold working, warm working, and hot working. Strain rate and friction are additional factors that affect performance in metal forming. We examine all of these issues in this chapter, but first let us provide an overview of the metal forming processes.

18.1 OVERVIEW OF METAL FORMING

Metal forming processes can be classified into two basic categories: bulk deformation processes and sheet metalworking

383

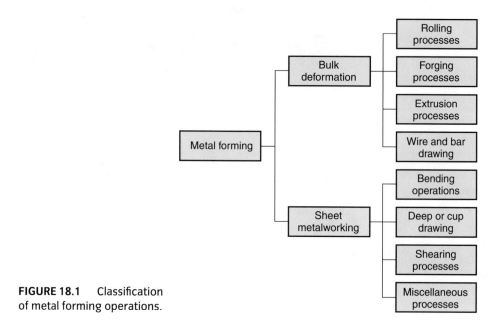

FIGURE 18.1 Classification of metal forming operations.

processes. These two categories are covered in detail in Chapters 19 and 20, respectively. Each category includes several major classes of shaping operations, as indicated in Figure 18.1.

Bulk Deformation Processes Bulk deformation processes are generally characterized by significant deformations and massive shape changes, and the surface area-to-volume of the work is relatively small. The term **bulk** describes the workparts that have this low area-to-volume ratio. Starting work shapes for these processes include cylindrical billets and rectangular bars. Figure 18.2 illustrates the following basic operations in bulk deformation:

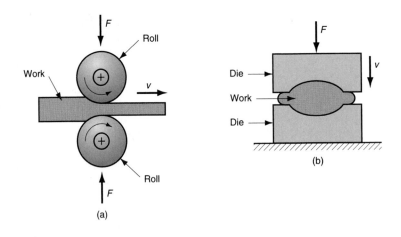

FIGURE 18.2 Basic bulk deformation processes: (a) rolling, (b) forging, (c) extrusion, and (d) drawing. Relative motion in the operations is indicated by *v;* forces are indicated by *F*.

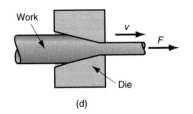

➤ ***Rolling.*** This is a compressive deformation process in which the thickness of a slab or plate is reduced by two opposing cylindrical tools called rolls. The rolls rotate so as to draw the work into the gap between them and squeeze it.

➤ ***Forging.*** In forging, a workpiece is compressed between two opposing dies, so that the die shapes are imparted to the work. Forging is traditionally a hot working process, but many types of forging are performed cold.

➤ ***Extrusion.*** This is a compression process in which the work metal is forced to flow through a die opening, thereby taking the shape of the opening as its own cross section.

➤ ***Drawing.*** In this forming process, the diameter of a round wire or bar is reduced by pulling it through a die opening.

Sheet Metalworking   Sheet metalworking processes are forming and cutting operations performed on metal sheets, strips, and coils. The surface area-to-volume ratio of the starting metal is high; thus, this ratio is a useful means to distinguish bulk deformation from sheet metal processes. ***Pressworking*** is the term often applied to sheet metal operations because the machines used to perform these operations are presses (presses of various types are also used in other manufacturing processes). A part produced in a sheet metal operation is often called a ***stamping.***

Sheet metal operations are always performed as cold working processes and are usually accomplished using a set of tools called a ***punch*** and ***die.*** The punch is the positive portion and the die is the negative portion of the tool set. The basic sheet metal operations are sketched in Figure 18.3 and are defined as follows:

➤ ***Bending.*** Bending involves straining of a metal sheet or plate to take an angle along a (usually) straight axis.

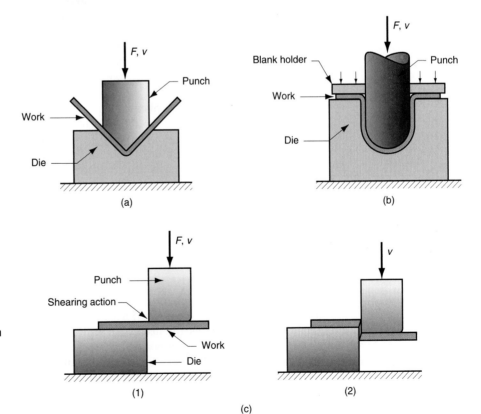

FIGURE 18.3 Basic sheet metalworking operations: (a) bending, (b) drawing, and (c) shearing: (1) as punch first contacts sheet, and (2) after cutting. Force and relative motion in these operations are indicated by *F* and *v*.

➤ ***Drawing.*** In sheet metalworking, drawing refers to the forming of a flat metal sheet into a hollow or concave shape, such as a cup, by stretching the metal. A blankholder is used to hold down the blank while the punch pushes into the sheet metal, as shown in Figure 18.3(b). To distinguish this operation from bar and wire drawing, the terms ***cup drawing*** or ***deep drawing*** are often used.

➤ ***Shearing.*** This process seems somewhat out-of-place in a list of deformation processes, because it involves cutting rather than forming. A shearing operation cuts the work using a punch and die, as in Figure 18.3(c). Although it is not a forming process, it is included here because it is a necessary and very common operation in sheet metalworking.

The miscellaneous processes within the sheet metalworking classification in Figure 18.1 include a variety of related shaping processes that do not use punch and die tooling. Examples of these processes are stretch forming, roll bending, spinning, and bending of tube stock.

18.2 MATERIAL BEHAVIOR IN METAL FORMING

Considerable insight about the behavior of metals during forming can be obtained from the stress–strain curve. The typical stress–strain curve for most metals is divided into an elastic region and a plastic region (Section 3.1.1). In metal forming, the plastic region is of primary interest because the material is plastically and permanently deformed in these processes.

The typical stress–strain relationship for a metal exhibits elasticity below the yield point and strain hardening above it. Figures 3.4 and 3.5 indicate this behavior in linear and logarithmic axes. In the plastic region, the metal's behavior is expressed by the flow curve:

$$\sigma = K\epsilon^n$$

where K = the strength coefficient, MPa (lb/in^2); and n is the strain-hardening exponent. The stress σ and strain ϵ in the flow curve are true stress and true strain. The flow curve is generally valid as a relationship that defines a metal's plastic behavior in cold working. Typical values of K and n for different metals at room temperature are listed in Table 3.4.

Flow Stress The flow curve describes the stress–strain relationship in the region in which metal forming takes place. It indicates the flow stress of the metal—the strength property that determines forces and power required to accomplish a particular forming operation. For most metals at room temperature, the stress–strain plot of Figure 3.5 indicates that as the metal is deformed, its strength increases due to strain hardening. The stress required to continue deformation must be increased to match this increase in strength. ***Flow stress*** is defined as the instantaneous value of stress required to continue deforming the material—to keep the metal "flowing." It is the yield strength of the metal as a function of strain, which can be expressed:

$$Y_f = K\epsilon^n \tag{18.1}$$

where Y_f = flow stress, MPa (lb/in^2).

In the individual forming operations discussed in the following two chapters, the instantaneous flow stress can be used to analyze the process as it is occurring. For example, in certain forging operations, the instantaneous force during compression can be determined from the flow stress value. Maximum force can be calculated based on the flow stress that results from the final strain at the end of the forging stroke.

In other cases, the analysis is based on the average stresses and strains that occur during deformation rather than instantaneous values. Extrusion represents this case, Figure 18.2(c). As the billet is reduced in cross section to pass through the extrusion

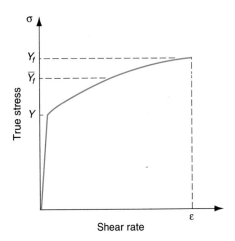

FIGURE 18.4   Stress–strain curve indicating location of average flow stress $\overline{Y}_f$ in relation to yield strength Y and final flow stress Y_f.

die opening, the metal gradually strain hardens to reach a maximum value. Rather than determine a sequence of instantaneous stress–strain values during the reduction, which would be not only difficult but also of limited interest, it is more useful to analyze the process based on the average flow stress during deformation.

Average Flow Stress   The average flow stress (also called the ***mean flow stress***) is the average value of stress over the stress–strain curve from the beginning of strain to the final (maximum) value that occurs during deformation. The value is illustrated in the stress–strain plot of Figure 18.4. The average flow stress is determined by integrating the flow curve equation, Eq. (18.1), between zero and the final strain value defining the range of interest. This yields the equation:

$$\overline{Y}_f = \frac{K\epsilon^n}{1+n} \tag{18.2}$$

where $\overline{Y}_f$ = average flow stress, MPa (lb/in^2); and ϵ = maximum strain value during the deformation process.

   We make extensive use of the average flow stress in our study of the bulk deformation processes in the following chapter. Given values of K and n for the work material, a method of computing final strain will be developed for each process. Based on this strain, Eq. (18.2) can be used to determine the average flow stress to which the metal is subjected during the operation.

18.3 TEMPERATURE IN METAL FORMING

The flow curve is a valid representation of stress–strain behavior of a metal during plastic deformation, particularly for cold working operations. For any metal, the values of K and n depend on temperature. Strength and strain hardening are both reduced at higher temperatures. These property changes are important because they result in lower forces and power during forming. In addition, ductility is increased at higher temperatures, which allows greater plastic deformation of the work metal. We can distinguish three temperature ranges that are used in metal forming: cold, warm, and hot working.

Cold Working   Cold working (also known as ***cold forming***) is metal forming performed at room temperature or slightly above. Significant advantages of cold forming compared to hot working are (1) greater accuracy, meaning closer tolerances can be achieved; (2) better surface finish; (3) higher strength and hardness of the part due to strain hardening; (4) grain flow during deformation provides the opportunity for desirable directional properties to be obtained in the resulting product; and (5) no heating of the work is required, which saves on furnace and fuel costs and permits higher production rates. Owing to this combination of advantages, many cold forming processes have become important mass-production operations. They provide close tolerances and good surfaces, minimizing the amount of machining required so that these operations can be classified as net shape or near net shape processes (Section 1.3.1).

There are certain disadvantages or limitations associated with cold forming operations: (1) higher forces and power are required to perform the operation; (2) care must be taken to ensure that the surfaces of the starting workpiece are free of scale and dirt; and (3) ductility and strain hardening of the work metal limit the amount of forming that can be done to the part. In some operations, the metal must be annealed (Section 27.1) in order to allow further deformation to be accomplished. In other cases, the metal is simply not ductile enough to be cold worked.

To overcome the strain-hardening problem and reduce force and power requirements, many forming operations are performed at elevated temperatures. There are two elevated temperature ranges involved, giving rise to the terms warm working and hot working.

Warm Working   Because plastic deformation properties are normally enhanced by increasing workpiece temperature, forming operations are sometimes performed at temperatures somewhat above room temperature but below the recrystallization temperature. The term ***warm working*** is applied to this second temperature range. The dividing line between cold working and warm working is often expressed in terms of the melting point for the metal. The dividing line is usually taken to be 0.3 T_m, where T_m is the melting point (absolute temperature) for the particular metal.

The lower strength and strain hardening at the intermediate temperatures, as well as higher ductility, provide warm working with the following advantages over cold working: (1) lower forces and power, (2) more intricate work geometries possible, and (3) need for annealing may be reduced or eliminated.

Hot Working   Hot working (also called ***hot forming***) involves deformation at temperatures above the recrystallization temperature (Section 3.3). The recrystallization temperature for a given metal is about one-half of its melting point on the absolute scale. In practice, hot working is usually carried out at temperatures somewhat above $0.5T_m$. The work metal continues to soften as temperature is increased beyond $0.5T_m$, thus enhancing the advantage of hot working above this level. However, the deformation process itself generates heat, which increases work temperatures in localized regions of the part. This can cause melting in these regions, which is highly undesirable. Also, scale on the work surface is accelerated at higher temperatures. Accordingly, hot working temperatures are usually maintained within the range $0.5T_m$ to $0.75T_m$.

The most significant advantage of hot working is the capability to produce substantial plastic deformation of the metal—far more than is possible with cold working or warm working. The principal reason for this is that the flow curve of the hot-worked metal has a strength coefficient that is substantially less than at room temperature, the strain-hardening exponent is zero (at least theoretically), and the ductility of the metal is significantly increased. All of this results in the following advantages relative to cold working: (1) the shape of the workpart can be significantly altered, (2) lower forces and power are required to deform the metal, (3) metals that usually fracture in cold working can be hot formed, (4) strength properties are generally isotropic because of the absence of the oriented grain structure typically created in cold working, and (5) no strengthening of the part occurs from work hardening. This last advantage may seem inconsistent, since strengthening of the metal is often considered an advantage for cold working. However, there are applications in which it is undesirable for the metal to be work hardened because it reduces ductility, for example, if the part is to be subsequently processed by cold forming. Disadvantages of hot working include (1) lower dimensional accuracy, (2) higher total energy required (due to the thermal energy to heat the workpiece), (3) work surface oxidation (scale), (4) poorer surface finish, and (5) shorter tool life.

Recrystallization of the metal in hot working involves atomic diffusion, which is a time-dependent process. Metal forming operations are often performed at high speeds that do not allow sufficient time for complete recrystallization of the grain structure during the deformation cycle itself. However, because of the high temperatures, recrystallization eventually does occur. It may occur immediately following the forming process or later, as the workpiece cools. Even though recrystallization may occur after the actual deformation, its eventual occurrence, and the substantial softening of the metal at high temperatures, are the features that distinguish hot working from warm working or cold working.

Isothermal Forming   Certain metals, such as highly alloyed steels, many titanium alloys, and high-temperature nickel alloys, possess good hot hardness, a property that makes them useful for high-temperature service. However, this very property that makes them attractive in these applications also makes them difficult to form with conventional methods. The problem is that when these metals are heated to their hot working temperatures and then come in contact with the relatively cold forming tools, heat is quickly transferred away from the part surfaces, thus raising the strength in these regions. The variations in temperature and strength in different regions of the workpiece cause irregular flow patterns in the metal during deformation, leading to high residual stresses and possible surface cracking.

Isothermal forming refers to forming operations that are carried out in such a way as to eliminate surface cooling and the resulting thermal gradients in the workpart. It is accomplished by preheating the tools that come in contact with the part to the same temperature as the work metal. This weakens the tools and reduces tool life, but it avoids the problems described above when these difficult metals are formed by conventional methods. In some cases, isothermal forming represents the only way in which these work materials can be formed. The procedure is most closely associated with forging, and we discuss isothermal forging in the following chapter.

18.4   STRAIN RATE SENSITIVITY

Theoretically, a metal in hot working behaves like a perfectly plastic material, with strain-hardening exponent $n = 0$. This means that the metal should continue to flow under the same level of flow stress, once that stress level is reached. However, there is an additional phenomenon that characterizes the behavior of metals during deformation, especially at the elevated temperatures of hot working. That phenomenon is strain rate sensitivity. Let us begin our discussion of this topic by defining strain rate.

The rate at which the metal is strained in a forming process is directly related to the speed of deformation, v. In many forming operations, deformation speed is equal to the velocity of the ram or other moving element of the equipment. It is most easily visualized in a tensile test as the velocity of the testing machine head relative to its fixed base. Given the deformation speed, *strain rate* is defined:

$$\dot{\epsilon} = \frac{v}{h} \tag{18.3}$$

where $\dot{\epsilon}$ = true strain rate, m/s/m (in/sec/in), or simply s^{-1}; and h = instantaneous height of the workpiece being deformed, m (in). If deformation speed v is constant during the operation, strain rate will change as h changes. In most practical forming operations,

FIGURE 18.5 (a) Effect of strain rate on flow stress at an elevated work temperature. (b) Same relationship plotted on log–log coordinates.

valuation of strain rate is complicated by the geometry of the workpart and variations in strain rate in different regions of the part. Strain rate can reach 1000 s^{-1} or more for some metal forming processes such as high-speed rolling and forging.

We have already observed that the flow stress of a metal is a function of temperature. At the temperatures of hot working, flow stress depends on strain rate. The effect of strain rate on strength properties is known as **strain rate sensitivity**. The effect can be seen in Figure 18.5. As strain rate is increased, resistance to deformation increases. This usually plots approximately as a straight line on a log–log graph, thus leading to the relationship:

$$Y_f = C\dot{\epsilon}^m \qquad (18.4)$$

where C is the strength constant (similar but not equal to the strength coefficient in the flow curve equation), and m is the strain rate sensitivity exponent. The value of C is determined at a strain rate of 1.0, and m is the slope of the curve in Figure 18.5(b).

FIGURE 18.6 Effect of temperature on flow stress for a typical metal. The constant C in Eq. (18.4), indicated by the intersection of each plot with the vertical dashed line at strain rate = 1.0, decreases, and m (slope of each plot) increases with increasing temperature.

TABLE 18.1 Typical values of temperature, strain-rate sensitivity, and coefficient of friction in cold, warm, and hot working.

Category	Temperature Range	Strain-Rate Sensitivity Exponent	Coefficient of Friction
Cold working	$\leq 0.3T_m$	$0.000 \leq m \leq 0.05$	0.1
Warm working	$0.3T_m$–$0.5T_m$	$0.05 \leq m \leq 0.1$	0.2
Hot working	$0.5T_m$–$0.75T_m$	$0.05 \leq m \leq 0.4$	0.4–0.5

The effect of temperature on the parameters of Eq. (18.4) is pronounced. Increasing temperature decreases the value of C (consistent with its effect on K in the flow curve equation) and increases the value of m. The general result can be seen in Figure 18.6. At room temperature, the effect of strain rate is almost negligible, indicating that the flow curve is a good representation of the material behavior. As temperature is increased, strain rate plays a more important role in determining flow stress, as indicated by the steeper slopes of the strain rate relationships. This is important in hot working because deformation resistance of the material increases so dramatically as strain rate is increased. To give a sense of the effect, typical values of m for the three temperature ranges of metal working are given in Table 18.1.

Thus we see that even in cold working, strain rate can have an effect, if small, on flow stress. In hot working, the effect can be significant. A more complete expression for flow stress as a function of both strain and strain rate would be the following:

$$Y_f = A\epsilon^n \dot{\epsilon}^m \tag{18.5}$$

where A = a strength coefficient, combining the effects of the previous K and C values. Of course, A, n, and m would all be functions of temperature, and the enormous task of testing and compiling the values of these parameters for different metals and various temperatures would be forbidding.

In our coverage of the various bulk deformation processes in Chapter 19, many of which are performed hot, we neglect the effect of strain rate in analyzing forces and power. For cold working and warm working, and for hot working operations at relatively low deformation speeds, this neglect represents a reasonable assumption.

18.5 FRICTION AND LUBRICATION IN METAL FORMING

Friction in metal forming arises because of the close contact between the tool and work surfaces and the high pressures that drive the surfaces together in these operations. In most metal forming processes, friction is undesirable for the following reasons: (1) metal flow in the work is retarded, causing residual stresses and sometimes defects in the product; (2) forces and power to perform the operation are increased, and (3) tool wear can lead to loss of dimensional accuracy, resulting in defective parts and requiring replacement of the tooling. Since tools in metal forming are generally expensive, tool wear is a major concern. Friction and tool wear are more severe in hot working because of the much harsher environment.

Friction in metal forming is different from that encountered in most mechanical systems, such as gear trains, shafts and bearings, and other components involving relative motion between surfaces. These other cases are generally characterized by low contact pressures, low to moderate temperatures, and ample lubrication to minimize metal-to-metal contact. By contrast, the metal forming environment features high pressures between a hardened tool and a soft workpart, plastic deformation of the softer material, and high temperatures (at least in hot working). These conditions can result in relatively high coefficients of friction in metal working, even in the presence of lubricants. Typical values of coefficient of friction for the three categories of metal forming are listed in Table 18.1.

If the coefficient of friction becomes large enough, a condition known as sticking occurs. *Sticking* in metalworking (also called *sticking friction*) is the tendency for the two surfaces in relative motion to adhere to each other rather than slide. It means that the friction stress between the surfaces exceeds the shear flow stress of the work metal, thus causing the metal to deform by a shear process beneath the surface rather than slip at the surface. Sticking occurs in metal forming operations and is a prominent problem in rolling; we discuss it in that context in the following chapter.

Metalworking lubricants are applied to the tool–work interface in many forming operations to reduce the harmful effects of friction. Benefits include reduced sticking, forces, power, and tool wear; and better surface finish on the product. Lubricants also serve other functions, such as removing heat from the tooling. Considerations in choosing an appropriate metalworking lubricant include (1) type of forming process (rolling, forging, sheet metal drawing, and so on), (2) whether used in hot working or cold working, (3) work material, (4) chemical reactivity with the tool and work metals (it is generally desirable for the lubricant to adhere to the surfaces to be most effective in reducing friction), (5) ease of application, (6) toxicity, (7) flammability, and (8) cost.

Lubricants used for cold working operations include [4], [7] mineral oils, fats and fatty oils, water-based emulsions, soaps, and other coatings. Hot working is sometimes performed dry for certain operations and materials (e.g., hot rolling of steel and extrusion of aluminum). When lubricants are used in hot working, they include mineral oils, graphite, and glass. Molten glass becomes an effective lubricant for hot extrusion of steel alloys. Graphite contained in water or mineral oil is a common lubricant for hot forging of various work materials. More detailed treatments of lubricants in metalworking are found in references [7] and [9].

REFERENCES

[1] Altan, T., Oh, S.-I., and Gegel, H. L. *Metal Forming: Fundamentals and Applications.* ASM International, Materials Park, Ohio, 1983.

[2] Cook, N. H. *Manufacturing Analysis.* Addison-Wesley Publishing Company, Inc., Reading, Massachusetts, 1966.

[3] Hosford, W. F., and Caddell, R. M. *Metal Forming: Mechanics and Metallurgy,* 3rd ed. Cambridge University Press, Cambridge, UK, 2007.

[4] Lange, K. *Handbook of Metal Forming.* Society of Manufacturing Engineers, Dearborn, Michigan, 2006.

[5] Lenard, J. G. *Metal Forming Science and Practice,* Elsevier Science, Amsterdam, The Netherlands, 2002.

[6] Mielnik, E. M. *Metalworking Science and Engineering.* McGraw-Hill, Inc., New York, 1991.

[7] Nachtman, E. S., and Kalpakjian, S. *Lubricants and Lubrication in Metalworking Operations.* Marcel Dekker, Inc., New York, 1985.

[8] Wagoner, R. H., and Chenot, J.-L. *Fundamentals of Metal Forming.* John Wiley & Sons, Inc., New York, 1997.

[9] Wick, C., et al. (eds.). *Tool and Manufacturing Engineers Handbook,* 4th ed. Vol. II, *Forming.* Society of Manufacturing Engineers, Dearborn, Michigan, 1984.

REVIEW QUESTIONS

18.1. What are the differences between bulk deformation processes and sheet metal processes?

18.2. Extrusion is a fundamental shaping process. Describe it.

18.3. Why is the term pressworking often used for sheet metal processes?

18.4. What is the difference between deep drawing and bar drawing?

18.5. Indicate the mathematical equation for the flow curve.

18.6. How does increasing temperature affect the parameters in the flow curve equation?

18.7. Indicate some of the advantages of cold working relative to warm and hot working.

18.8. What is isothermal forming?

18.9. Describe the effect of strain rate in metal forming.

18.10. Why is friction generally undesirable in metal forming operations?

18.11. What is sticking friction in metalworking?

MULTIPLE CHOICE QUIZ

There are 13 correct answers in the following multiple choice questions (some questions have multiple answers that are correct). To attain a perfect score on the quiz, all correct answers must be given. Each correct answer is worth 1 point. Each omitted answer or wrong answer reduces the score by 1 point, and each additional answer beyond the correct number of answers reduces the score by 1 point. Percentage score on the quiz is based on the total number of correct answers.

18.1. Which of the following are bulk deformation processes (three correct answers): (a) bending, (b) deep drawing, (c) extrusion, (d) forging, (e) rolling, and (f) shearing?

18.2. Which of the following is typical of the starting work geometry in sheet metal processes: (a) high volume-to-area ratio or (b) low volume-to-area ratio?

18.3. The flow curve expresses the behavior of a metal in which of the following regions of the stress–strain curve: (a) elastic region or (b) plastic region?

18.4. The average flow stress is the flow stress multiplied by which of the following factors: (a) n, (b) $(1 + n)$, (c) $1/n$, or (d) $1/(1 + n)$, where n is the strain-hardening exponent?

18.5. Hot working of metals refers to which one of the following temperature regions relative to the melting point of the given metal on an absolute temperature scale: (a) room temperature, (b) $0.2T_m$, (c) $0.4T_m$, or (d) $0.6T_m$?

18.6. Which of the following are advantages and characteristics of hot working relative to cold working (four correct answers): (a) fracture of workpart is less likely, (b) friction is reduced, (c) increased strength properties, (d) isotropic mechanical properties, (e) less overall energy is required, (f) lower deformation forces is required, (g) more significant shape changes are possible, and (h) strain-rate sensitivity is reduced?

18.7. Increasing strain rate tends to have which one of the following effects on flow stress during hot forming of metal: (a) decreases flow stress, (b) has no effect, or (c) increases flow stress?

18.8. The coefficient of friction between the part and the tool in cold working tends to be (a) higher, (b) lower, or (c) no different relative to its value in hot working?

PROBLEMS

Flow Curve in Forming

18.1. The strength coefficient = 550 MPa and strain-hardening exponent = 0.22 for a certain metal. During a forming operation, the final true strain that the metal experiences = 0.85. Determine the flow stress at this strain and the average flow stress that the metal experienced during the operation.

18.2. A metal has a flow curve with parameters: strength coefficient = 850 MPa and strain-hardening exponent = 0.30. A tensile specimen of the metal with gage length = 100 mm is stretched to a length = 157 mm. Determine the flow stress at the new length and the average flow stress that the metal has been subjected to during the deformation.

18.3. A particular metal has a flow curve with parameters: strength coefficient = 35,000 lb/in^2 and strain-hardening exponent = 0.26. A tensile specimen of the metal with gage length = 2.0 in is stretched to a length = 3.3 in. Determine the flow stress at this new length and the average flow stress that the metal has been subjected to during deformation.

18.4. The strength coefficient and strain-hardening exponent of a certain test metal are 40,000 lb/in^2 and 0.19, respectively. A cylindrical specimen of the metal with starting diameter = 2.5 in and length = 3.0 in is compressed to a length of 1.5 in. Determine the flow stress at this compressed length and the average flow stress that the metal has experienced during deformation.

18.5. Derive the equation for average flow stress, Eq. (18.2) in the text.

18.6. For a certain metal, the strength coefficient = 700 MPa and strain-hardening exponent = 0.27. Determine the average flow stress that the metal experiences if it is subjected to a stress that is equal to its strength coefficient K.

18.7. Determine the value of the strain-hardening exponent for a metal that will cause the average flow stress to be 3/4 of the final flow stress after deformation.

18.8. The strength coefficient = 35,000 lb/in^2 and strain-hardening exponent = 0.40 for a metal used in a forming operation in which the workpart is reduced in cross-sectional area by stretching. If the average flow stress on the part is 20,000 lb/in^2, determine the amount of reduction in cross-sectional area experienced by the part.

18.9. In a tensile test, two pairs of values of stress and strain were measured for the specimen metal after it had yielded: (1) true stress = 217 MPa and true strain = 0.35, and (2) true stress = 259 MPa and true strain = 0.68. Based on these data points, determine the strength coefficient and strain-hardening exponent.

18.10. The following stress and strain values were measured in the plastic region during a tensile test carried out on a new experimental metal: (1) true stress = 43,608 lb/in^2 and true strain = 0.27 in/in, and (2) true stress = 52,048 lb/in^2 and true strain = 0.85 in/in. Based on these data points, determine the strength coefficient and strain-hardening exponent.

Strain Rate

18.11. The gage length of a tensile test specimen = 150 mm. It is subjected to a tensile test in which the grips holding the end of the test specimen are moved with a relative velocity = 0.1 m/s. Construct a plot of the strain rate as a function of length as the specimen is pulled to a length = 200 mm.

18.12. A specimen with 6.0 in starting gage length is subjected to a tensile test in which the grips holding the end of the test specimen are moved with a relative velocity = 1.0 in/sec. Construct a plot of the strain rate as a function of length as the specimen is pulled to a length = 8.0 in.

18.13. A workpart with starting height h = 100 mm is compressed to a final height of 50 mm. During the deformation, the relative speed of the platens compressing the part = 200 mm/s. Determine the strain rate at (a) h = 100 mm, (b) h = 75 mm, and (c) h = 51 mm.

18.14. A hot working operation is carried out at various speeds. The strength constant = 30,000 lb/in^2 and the strain-rate sensitivity exponent = 0.15. Determine the flow stress if the strain rate is (a) 0.01/sec (b) 1.0/sec, (c) 100/sec.

18.15. A tensile test is performed to determine the parameters strength constant C and strain-rate sensitivity exponent m in Eq. (18.4) for a certain metal. The temperature at which the test is performed = 500°C. At a strain rate = 12/s, the stress is measured at 160 MPa; and at a strain rate = 250/s, the stress = 300 MPa. (a) Determine C and m. (b) If the temperature were 600°C, what changes would you expect in the values of C and m?

18.16. A tensile test is carried out to determine the strength constant C and strain-rate sensitivity exponent m for a certain metal at 1000°F. At a strain rate = 10/sec, the stress is measured at 23,000 lb/in^2; and at a strain rate = 300/sec, the stress = 45,000 lb/in^2. (a) Determine C and m. (b) If the temperature were 900°F, what changes would you expect in the values of C and m?

19 BULK DEFORMATION PROCESSES IN METAL WORKING

Chapter Contents

The deformation processes described in this chapter accomplish significant shape change in metal parts whose initial form is bulk rather than sheet. The starting forms include cylindrical bars and billets, rectangular billets and slabs, and similar elementary geometries. The bulk deformation processes refine the starting shapes, sometimes improving mechanical properties, and always adding commercial value. Deformation processes work by stressing the metal sufficiently to cause it to plastically flow into the desired shape.

Bulk deformation processes are performed as cold, warm, and hot working operations. Cold and warm working is appropriate when the shape change is less severe, and there is a need to improve mechanical properties and achieve good finish on the part. Hot working is generally required when massive deformation of large workparts is involved.

The commercial and technological importance of bulk deformation processes derives from the following:

➤ When performed as hot working operations, they can achieve significant change in the shape of the workpart.
➤ When performed as cold working operations, they can be used not only to shape the product, but also to increase its strength through strain hardening.
➤ These processes produce little or no waste as a by-product of the operation. Some bulk deformation operations are *near net shape* or *net shape* processes; they achieve final product geometry with little or no subsequent machining.

The bulk deformation processes covered in this chapter are (1) rolling, (2) forging, (3) extrusion, and (4) wire and bar drawing. The chapter also documents the variations and related operations of the four basic processes that have been developed over the years.

19.1 ROLLING

Rolling is a deformation process in which the thickness of the work is reduced by compressive forces exerted by two opposing rolls. The rolls rotate as illustrated in Figure 19.1 to pull and simultaneously squeeze the work between them. The basic process shown in our figure is flat rolling, used to reduce the thickness of a rectangular cross section. A closely related process is shape rolling, in which a square cross section is formed into a shape such as an I-beam.

Most rolling processes are very capital intensive, requiring massive pieces of equipment, called rolling mills, to perform them. The high investment cost requires the mills to be used for production in large quantities of standard items such as sheets and plates. Most rolling is carried out by hot working, called **hot rolling**, owing to the large amount of deformation required. Hot-rolled metal is generally free of residual stresses, and its properties are isotropic. Disadvantages of hot rolling are that the product cannot be held to close tolerances, and the surface has a characteristic oxide scale.

Steelmaking provides the most common application of rolling mill operations (Historical Note 19.1). Let us follow the sequence of steps in a steel rolling mill to illustrate the variety of products made. Similar steps occur in other basic metal industries. The work starts out as a cast steel ingot that has just solidified. While it is still hot, the ingot is placed in a furnace where it remains for many hours until it has reached a uniform temperature throughout, so that the metal will flow consistently during rolling. For steel, the desired temperature for rolling is around 1200°C (2200°F). The heating operation is called **soaking**, and the furnaces in which it is carried out are called **soaking pits**.

From soaking, the ingot is moved to the rolling mill, where it is rolled into one of three intermediate shapes called blooms, billets, or slabs. A **bloom** has a square cross section 150 mm × 150 mm (6 in × 6 in) or larger. A **slab** is rolled from an ingot or a bloom and has a rectangular cross section of width 250 mm (10 in) or more and thickness 40 mm (1.5 in) or more. A **billet** is rolled from a bloom and is square with dimensions 40 mm (1.5 in) on a side or larger. These intermediate shapes are subsequently rolled into final product shapes.

Blooms are rolled into structural shapes and rails for railroad tracks. Billets are rolled into bars and rods. These shapes are the raw materials for machining, wire drawing, forging, and other metalworking processes. Slabs are rolled into plates, sheets, and strips. Hot-rolled plates are used in shipbuilding, bridges, boilers, welded structures for various heavy machines, tubes and pipes, and many other products. Figure 19.2 shows some of these rolled steel products. Further flattening of hot-rolled plates and sheets is often accomplished by **cold rolling**, in order to prepare them for subsequent sheet metal operations (Chapter 20). Cold rolling strengthens the metal and permits a tighter tolerance on thickness. In addition, the surface of the cold-rolled sheet is absent of scale and generally superior to the corresponding hot-rolled product. These characteristics make cold-rolled sheets, strips, and coils ideal for stampings, exterior panels, and other parts of products ranging from automobiles to appliances and office furniture.

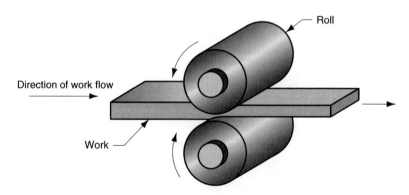

FIGURE 19.1 The rolling process (specifically, flat rolling).

Historical Note 19.1   *Rolling*

Rolling of gold and silver by manual methods dates from the fourteenth century. Leonardo da Vinci designed one of the first rolling mills in 1480, but it is doubtful that his design was ever built. By around 1600, cold rolling of lead and tin was accomplished on manually operated rolling mills. By around 1700, hot rolling of iron was being done in Belgium, England, France, Germany, and Sweden. These mills were used to roll iron bars into sheets. Prior to this time, the only rolls in steelmaking were slitting mills—pairs of opposing rolls with collars (cutting disks) used to slit iron and steel into narrow strips for making nails and similar products. Slitting mills were not intended to reduce thickness.

Modern rolling practice dates from 1783 when a patent was issued in England for using grooved rolls to produce iron bars. The Industrial Revolution created a tremendous demand for iron and steel, stimulating developments in rolling. The first mill for rolling railway rails was started in 1820 in England. The first I-beams were rolled in France in 1849. In addition, the size and capacity of flat rolling mills increased dramatically during this period.

Rolling is a process that requires a very large power source. Water wheels were used to power rolling mills until the eighteenth century. Steam engines increased the capacity of these rolling mills until soon after 1900 when electric motors replaced steam.

19.1.1   FLAT ROLLING AND ITS ANALYSIS

Flat rolling is illustrated in Figures 19.1 and 19.3. It involves the rolling of slabs, strips, sheets, and plates—workparts of rectangular cross section in which the width is greater than the thickness. In flat rolling, the work is squeezed between two rolls so that its thickness is reduced by an amount called the **draft**:

$$d = t_o - t_f \tag{19.1}$$

where d = draft, mm (in); t_o = starting thickness, mm (in); and t_f = final thickness, mm (in). Draft is sometimes expressed as a fraction of the starting stock thickness, called the

FIGURE 19.2   Some of the steel products made in a rolling mill.

reduction:

$$r = \frac{d}{t_o} \tag{19.2}$$

where r = reduction. When a series of rolling operations are used, reduction is taken as the sum of the drafts divided by the original thickness.

In addition to thickness reduction, rolling usually increases work width. This is called **spreading,** and it tends to be most pronounced with low width-to-thickness ratios and low coefficients of friction. Conservation of matter is preserved, so the volume of metal exiting the rolls equals the volume entering

$$t_o w_o L_o = t_f w_f L_f \tag{19.3}$$

where w_o and w_f are the before and after work widths, mm (in); and L_o and L_f are the before and after work lengths, mm (in). Similarly, before and after volume rates of material flow must be the same, so the before and after velocities can be related:

$$t_o w_o v_o = t_f w_f v_f \tag{19.4}$$

where v_o and v_f are the entering and exiting velocities of the work.

The rolls contact the work along an arc defined by the angle θ. Each roll has radius R, and its rotational speed gives it a surface velocity v_r. This velocity is greater than the entering speed of the work v_o and less than its exiting speed v_f. Since the metal flow is continuous, there is a gradual change in velocity of the work between the rolls. However, there is one point along the arc where work velocity equals roll velocity. This is called the **no-slip point,** also known as the **neutral point.** On either side of this point, slipping and friction occur between roll and work. The amount of slip between the rolls and the work can be measured by means of the **forward slip,** a term used in rolling that is defined:

$$s = \frac{v_f - v_r}{v_r} \tag{19.5}$$

where s = forward slip; v_f = final (exiting) work velocity, m/s (ft/sec); and v_r = roll speed, m/s (ft/sec).

The true strain experienced by the work in rolling is based on before and after stock thicknesses. In equation form,

$$\epsilon = \ln \frac{t_o}{t_f} \tag{19.6}$$

The true strain can be used to determine the average flow stress $\overline{Y}_f$ applied to the work material in flat rolling. Recall from the previous chapter, Eq. (18.2), that

$$\overline{Y}_f = \frac{K\epsilon^n}{1+n} \tag{19.7}$$

The average flow stress is used to compute estimates of force and power in rolling.

Friction in rolling occurs with a certain coefficient of friction, and the compression force of the rolls, multiplied by this coefficient of friction, results in a friction force between the rolls and the work. On the entrance side of the no-slip point, friction force is in one direction, and on the other side it is in the opposite direction. However, the two forces are not equal. The friction force on the entrance side is greater, so that the net force pulls the work through the rolls. If this were not the case, rolling would not be possible. There is a limit to the maximum possible draft that can be accomplished in flat rolling with a given coefficient of friction, defined by:

$$d_{\max} = m^2 R \tag{19.8}$$

where d_{max} = maximum draft, mm (in); μ = coefficient of friction; and R = roll radius mm (in). The equation indicates that if friction were zero, draft would be zero, and it would be impossible to accomplish the rolling operation.

Coefficient of friction in rolling depends on lubrication, work material, and working temperature. In cold rolling, the value is around 0.1; in warm working, a typical value is around 0.2; and in hot rolling, μ is around 0.4 [16]. Hot rolling is often characterized by a condition called **sticking,** in which the hot work surface adheres to the rolls over the contact arc. This condition often occurs in the rolling of steels and high-temperature alloys. When sticking occurs, the coefficient of friction can be as high as 0.7. The consequence of sticking is that the surface layers of the work are restricted to move at the same speed as the roll speed v_r; and below the surface, deformation is more severe in order to allow passage of the piece through the roll gap.

Given a coefficient of friction sufficient to perform rolling, roll force F required to maintain separation between the two rolls can be computed by integrating the unit roll pressure (shown as p in Figure 19.3) over the roll-work contact area. This can be expressed:

$$F = w \int_{0}^{L} p\, dL \qquad (19.9)$$

where F = rolling force, N (lb); w = the width of the work being rolled, mm (in); p = roll pressure, MPa (lb/in²); and L = length of contact between rolls and work, mm (in). The integration requires two separate terms, one for either side of the neutral point. Variation in roll pressure along the contact length is significant. A sense of this variation can be obtained from the plot in Figure 19.4. Pressure reaches a maximum at the neutral point, and trails off on either side to the entrance and exit points. As friction increases, maximum pressure increases relative to entrance and exit values. As friction decreases, the neutral point shifts away from the entrance and toward the exit in order to maintain a net pull force in the direction of rolling. Otherwise, with low friction, the work would slip rather than pass between the rolls.

An approximation of the results obtained by Eq. (19.9) can be calculated based on the average flow stress experienced by the work material in the roll gap. That is,

$$F = \overline{Y}_f wL \qquad (19.10)$$

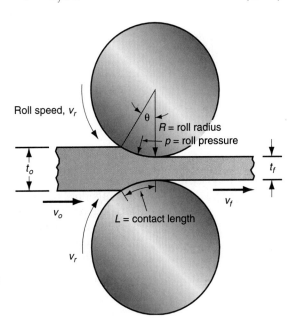

FIGURE 19.3   Side view of flat rolling, indicating before and after thicknesses, work velocities, angle of contact with rolls, and other features.

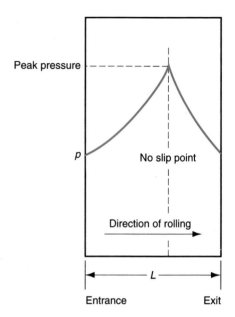

FIGURE 19.4   Typical variation in pressure along the contact length in flat rolling. The peak pressure is located at the neutral point. The area beneath the curve, representing the integration in Eq. (19.9), is the roll force F.

where $\overline{Y}_f$ = average flow stress from Eq. (19.7), MPa (lb/in^2); and the product wL is the roll-work contact area, mm^2 (in^2). Contact length can be approximated by

$$L = \sqrt{R(t_o - t_f)} \qquad (19.11)$$

The torque in rolling can be estimated by assuming that the roll force is centered on the work as it passes between the rolls, and that it acts with a moment arm of one-half the contact length L. Thus, torque for each roll is

$$T = 0.5\,FL \qquad (19.12)$$

The power required to drive each roll is the product of torque and angular velocity. Angular velocity is $2\pi N$, where N = rotational speed of the roll. Thus, the power for each roll is $2\pi NT$. Substituting Eq. (19.12) for torque in this expression for power, and doubling the value to account for the fact that a rolling mill consists of two powered rolls, we get the following expression:

$$P = 2\pi NFL \qquad (19.13)$$

where P = power, J/s or W (in-lb/min); N = rotational speed, 1/s (rev/min); F = rolling force, N (lb); and L = contact length, m (in).

Example 19.1 Flat Rolling

A 300-mm-wide strip 25-mm thick is fed through a rolling mill with two powered rolls each of radius = 250 mm. The work thickness is to be reduced to 22 mm in one pass at a roll speed of 50 rev/min. The work material has a flow curve defined by $K = 275$ MPa and $n = 0.15$, and the coefficient of friction between the rolls and the work is assumed to be 0.12. Determine if the friction is sufficient to permit the rolling operation to be accomplished. If so, calculate the roll force, torque, and horsepower.

Solution: The draft attempted in this rolling operation is

$$d = 25 - 22 = 3\ \text{mm}$$

From Eq. (19.8), the maximum possible draft for the given coefficient of friction is

$$d_{max} = (0.12)^2(250) = 3.6 \, \text{mm}$$

Since the maximum allowable draft exceeds the attempted reduction, the rolling operation is feasible. To compute rolling force, we need the contact length L and the average flow stress $\overline{Y}_f$. The contact length is given by Eq. (19.11):

$$L = \sqrt{250(25 - 22)} = 27.4 \, \text{mm}$$

$\overline{Y}_f$ is determined from the true strain:

$$\epsilon = \ln \frac{25}{22} = 0.128$$

$$\overline{Y}_f = \frac{275(0.128)^{0.15}}{1.15} = 175.7 \, \text{MPa}$$

Rolling force is determined from Eq. (19.10):

$$F = 175.7(300)(27.4) = 1,444,786 \, \text{N}$$

Torque required to drive each roll is given by Eq. (19.12):

$$T = 0.5(1,444,786)(27,4)(10^{-3}) = 19,786 \, \text{N-m}$$

and the power is obtained from Eq. (19.13):

$$P = 2\pi(50)(1,444,786)(27.4)(10^{-3}) = 12,432,086 \, \text{N-m/min} = 207,201 \, \text{N-m/s(W)}$$

For comparison, let us convert this to horsepower (we note that one horsepower = 745.7 W):

$$HP = \frac{207,201}{745.7} = 278 \, \text{hp}$$

It can be seen from this example that large forces and power are required in rolling. Inspection of Eqs. (19.10) and (19.13) indicates that force and/or power to roll a strip of a given width and work material can be reduced by any of the following: (1) using hot rolling rather than cold rolling to reduce strength and strain hardening (K and n) of the work material; (2) reducing the draft in each pass; (3) using a smaller roll radius R to reduce force; and (4) using a lower rolling speed N to reduce power.

19.1.2 SHAPE ROLLING

In shape rolling, the work is deformed into a contoured cross section. Products made by shape rolling include construction shapes such as I-beams, L-beams, and U-channels; rails for railroad tracks; and round and square bars and rods (see Figure 19.2). The process is accomplished by passing the work through rolls that have the reverse of the desired shape.

Most of the principles that apply in flat rolling are also applicable to shape rolling. Shaping rolls are more complicated; and the work, usually starting as a square shape, requires a gradual transformation through several rolls in order to achieve the final cross section. Designing the sequence of intermediate shapes and corresponding rolls is called **roll-pass design.** Its goal is to achieve uniform deformation throughout the cross section in each reduction. Otherwise, certain portions of the work are reduced more than others, causing greater elongation in these sections. The consequence of nonuniform reduction can be warping and cracking of the rolled product. Both horizontal and vertical rolls are utilized to achieve consistent reduction of the work material.

19.1.3 ROLLING MILLS

Various rolling mill configurations are available to deal with the variety of applications and technical problems in the rolling process. The basic rolling mill consists of two opposing rolls and is referred to as a ***two-high*** rolling mill, shown in Figure 19.5(a). The rolls in these mills have diameters in the range of 0.6 to 1.4 m (2.0–4.5 ft). The two-high configuration can be either reversing or nonreversing. In the ***nonreversing mill,*** the rolls always rotate in the same direction, and the work always passes through from the same side. The ***reversing mill*** allows the direction of roll rotation to be reversed, so that the work can be passed through in either direction. This permits a series of reductions to be made through the same set of rolls, simply by passing through the work from opposite directions multiple times. The disadvantage of the reversing configuration is the significant angular momentum possessed by large rotating rolls and the associated technical problems involved in reversing the direction.

Several alternative arrangements are illustrated in Figure 19.5. In the ***three-high*** configuration, Figure 19.5(b), there are three rolls in a vertical column, and the direction of rotation of each roll remains unchanged. To achieve a series of reductions, the work can be passed through from either side by raising or lowering the strip after each pass. The equipment in a three-high rolling mill becomes more complicated, because an elevator mechanism is needed to raise and lower the work.

As several of the previous equations indicate, advantages are gained in reducing roll diameter. Roll-work contact length is reduced with a lower roll radius, and this leads to lower forces, torque, and power. The ***four-high*** rolling mill uses two smaller-diameter rolls to contact the work and two backing rolls behind them, as in Figure 19.5(c). Owing to the high roll forces, these smaller rolls would deflect elastically between their end bearings as the work passes through unless the larger backing rolls were used to support them. Another

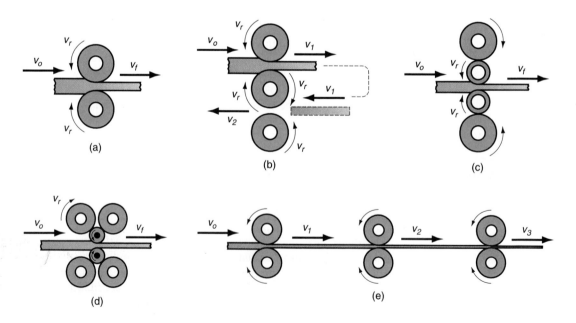

FIGURE 19.5   Various configurations of rolling mills: (a) 2-high, (b) 3-high, (c) 4-high, (d) cluster mill, and (e) tandem rolling mill.

roll configuration that allows smaller working rolls against the work is the ***cluster rolling mill*** (Figure 19.5(d)).

To achieve higher throughput rates in standard products, a ***tandem rolling mill*** is often used. This configuration consists of a series of rolling stands, as represented in Figure 19.5(e). Although only three stands are shown in our sketch, a typical tandem rolling mill may have eight or ten stands, each making a reduction in thickness or a refinement in shape of the work passing through. With each rolling step, work velocity increases, and the problem of synchronizing the roll speeds at each stand is a significant one.

Modern tandem rolling mills are often supplied directly by continuous casting operations (Section 7.2.2). These setups achieve a high degree of integration among the processes required to transform starting raw materials into finished products. Advantages include elimination of soaking pits, reduction in floor space, and shorter manufacturing lead times. These technical advantages translate into economic benefits for a mill that can accomplish continuous casting and rolling.

19.2 OTHER DEFORMATION PROCESSES RELATED TO ROLLING

Several other bulk deformation processes use rolls to form the workpart. The operations include thread rolling, ring rolling, gear rolling, and roll piercing.

Thread Rolling   Thread rolling is used to form threads on cylindrical parts by rolling them between two dies. It is the most important commercial process for mass producing external threaded components (e.g., bolts and screws). The competing process is thread cutting (Section 22.7.1). Most thread rolling operations are performed by cold working in thread rolling machines. These machines are equipped with special dies that determine the size and form of the thread. The dies are of two types: (1) flat dies, which reciprocate relative to each other, as illustrated in Figure 19.6; and (2) round dies, which rotate relative to each other to accomplish the rolling action.

Production rates in thread rolling can be high, ranging up to eight parts per second for small bolts and screws. Not only are these rates significantly higher than thread cutting, but there are other advantages over machining as well: (1) better material utilization, (2) stronger threads due to work hardening, (3) smoother surface, and (4) better fatigue resistance due to compressive stresses introduced by rolling.

Ring Rolling   Ring rolling is a deformation process in which a thick-walled ring of smaller diameter is rolled into a thin-walled ring of larger diameter. The before and after views of the

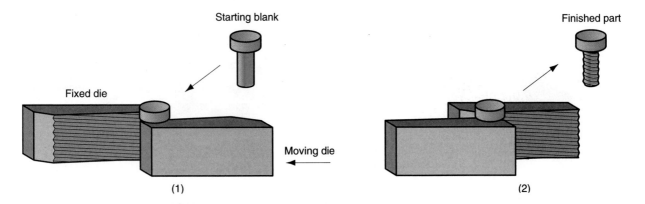

FIGURE 19.6   Thread rolling with flat dies: (1) start of cycle and (2) end of cycle.

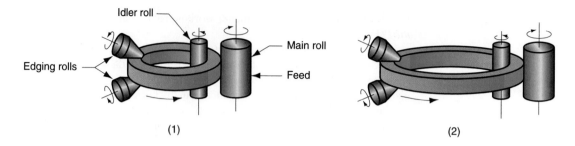

FIGURE 19.7   Ring rolling used to reduce the wall thickness and increase the diameter of a ring: (1) start and (2) completion of process.

process are illustrated in Figure 19.7. As the thick-walled ring is compressed, the deformed material elongates, causing the diameter of the ring to be enlarged. Ring rolling is usually performed as a hot-working process for large rings and as a cold-working process for smaller rings.

Applications of ring rolling include ball and roller bearing races, steel tires for railroad wheels, and rings for pipes, pressure vessels, and rotating machinery. The ring walls are not limited to rectangular cross sections; the process permits rolling of more complex shapes. There are several advantages of ring rolling over alternative methods of making the same parts: raw material savings, ideal grain orientation for the application, and strengthening through cold working.

Gear Rolling   Gear rolling is a cold working process to produce certain gears. The automotive industry is an important user of these products. The setup in gear rolling is similar to thread rolling, except that the deformed features of the cylindrical blank or disk are oriented parallel to its axis (or at an angle in the case of helical gears) rather than spiraled as in thread rolling. Alternative production methods for gears include several machining operations, discussed in Section 22.7.2. Advantages of gear rolling compared to machining are similar to those of thread rolling: higher production rates, better strength and fatigue resistance, and less material waste.

Roll Piercing   Ring rolling is a specialized hot working process for making seamless thick-walled tubes. It utilizes two opposing rolls, and hence it is grouped with the rolling processes. The process is based on the principle that when a solid cylindrical part is compressed on its circumference, as in Figure 19.8(a), high tensile stresses are

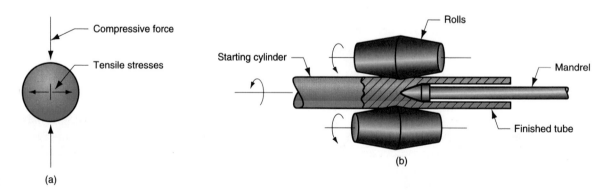

FIGURE 19.8   Roll piercing: (a) formation of internal stresses and cavity by compression of cylindrical part; and (b) setup of Mannesmann roll mill for producing seamless tubing.

developed at its center. If compression is high enough, an internal crack is formed. In roll piercing, this principle is exploited by the setup shown in Figure 19.8(b). Compressive stresses on a solid cylindrical billet are applied by two rolls, whose axes are oriented at slight angles ($\sim6°$) from the axis of the billet, so that their rotation tends to pull the billet through the rolls. A mandrel is used to control the size and finish of the hole created by the action. The terms ***rotary tube piercing*** and ***Mannesmann process*** are also used for this tube-making operation.

19.3   FORGING

Forging is a deformation process in which the work is compressed between two dies, using either impact or gradual pressure to form the part. It is the oldest of the metal forming operations, dating back to perhaps 5000 BCE (Historical Note 19.2). Today, forging is an important industrial process used to make a variety of high-strength components for automotive, aerospace, and other applications. These components include engine crankshafts and connecting rods, gears, aircraft structural components, and jet engine turbine parts. In addition, steel and other basic metals industries use forging to establish the basic form of large components that are subsequently machined to final shape and dimensions.

Historical Note 19.2   *Forging*

The forging process dates from the earliest written records of man, around 7000 years ago. There is evidence that forging was used in ancient Egypt, Greece, Persia, India, China, and Japan to make weapons, jewelry, and a variety of implements. Craftsmen in the art of forging during these times were held in high regard.

Engraved stone platens were used as impression dies in the hammering of gold and silver in ancient Crete around 1600 BCE. This evolved into the fabrication of coins by a similar process around 800 BCE. More complicated impression dies were used in Rome around 200 CE. The blacksmith's trade remained relatively unchanged for many centuries until the drop hammer with guided ram was introduced near the end of the eighteenth century. This development brought forging practice into the Industrial Age.

Forging is carried out in many different ways. One way to classify the operations is by working temperature. Most forging operations are performed hot or warm, owing to the significant deformation demanded by the process and the need to reduce strength and increase ductility of the work metal. However, cold forging is also very common for certain products. The advantage of cold forging is the increased strength that results from strain hardening of the component.

Either impact or gradual pressure is used in forging. The distinction derives more from the type of equipment used than differences in process technology. A forging machine that applies an impact load is called a ***forging hammer,*** while one that applies gradual pressure is called a ***forging press.***

Another difference among forging operations is the degree to which the flow of the work metal is constrained by the dies. By this classification, there are three types of forging operations, shown in Figure 19.9: (a) open-die forging, (b) impression-die forging, and (c) flashless forging. In ***open-die forging,*** the work is compressed between two flat (or almost flat) dies, thus allowing the metal to flow without constraint in a lateral direction relative to the die surfaces. In ***impression-die forging,*** the die surfaces contain a shape or impression that is imparted to the work during compression, thus constraining metal flow to a significant degree. In this type of operation, a portion of the work metal flows beyond the

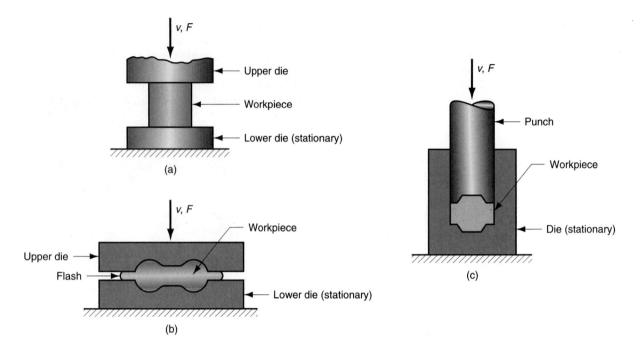

FIGURE 19.9 Three types of forging operation illustrated by cross-sectional sketches: (a) open-die forging, (b) impression-die forging, and (c) flashless forging.

die impression to form *flash,* as shown in the figure. Flash is excess metal that must be trimmed off later. In *flashless forging,* the work is completely constrained within the die and no excess flash is produced. The volume of the starting workpiece must be controlled very closely so that it matches the volume of the die cavity. The reader can obtain a good sense of these operations in our video clip on forging.

VIDEO CLIP _____

Forging. The three segments on this clip are (1) the forging process, (2) open-die forging, and (3) impression-die forging.

19.3.1 OPEN-DIE FORGING

The simplest case of open-die forging involves compression of a workpart of cylindrical cross section between two flat dies, much in the manner of a compression test (Section 3.1.2). This forging operation, known as *upsetting* or *upset forging,* reduces the height of the work and increases its diameter.

Analysis of Open-Die Forging   If open-die forging is carried out under ideal conditions of no friction between work and die surfaces, then homogeneous deformation occurs, and the radial flow of the material is uniform throughout its height, as pictured in Figure 19.10. Under these ideal conditions, the true strain experienced by the work during the process can be determined by

$$\epsilon = \ln \frac{h_o}{h} \tag{19.14}$$

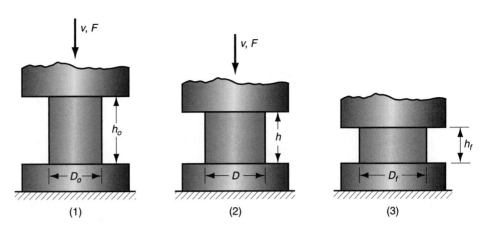

FIGURE 19.10
Homogeneous deformation of a cylindrical workpart under ideal conditions in an open-die forging operation: (1) start of process with workpiece at its original length and diameter, (2) partial compression, and (3) final size.

where h_o = starting height of the work, mm (in); and h = the height at some intermediate point in the process, mm (in). At the end of the compression stroke, h = its final value h_f, and the true strain reaches its maximum value.

Estimates of force to perform upsetting can be calculated. The force required to continue the compression at any given height h during the process can be obtained by multiplying the corresponding cross-sectional area by the flow stress:

$$F = Y_f A \qquad (19.15)$$

where F = force, lb (N); A = cross-sectional area of the part, mm^2 (in^2); and Y_f = flow stress corresponding to the strain given by Eq. (19.14), MPa (lb/in^2). Area A continuously increases during the operation as height is reduced. Flow stress Y_f also increases as a result of work hardening, except when the metal is perfectly plastic (e.g., in hot working). In this case, the strain-hardening exponent $n = 0$, and flow stress Y_f equals the metal's yield strength Y. Force reaches a maximum value at the end of the forging stroke, when both area and flow stress are at their highest values.

An actual upsetting operation does not occur quite as shown in Figure 19.10 because friction opposes the flow of work metal at the die surfaces. This creates the barreling effect shown in Figure 19.11. When performed on a hot workpart with cold dies, the barreling effect is even more pronounced. This results from a higher coefficient of friction typical in hot working and heat transfer at and near the die surfaces, which cools the metal and increases its resistance to deformation. The hotter metal in the middle of the part flows more readily than the cooler metal at the ends. These effects are more significant as the diameter-

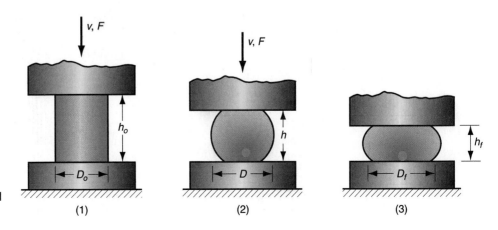

FIGURE 19.11   Actual deformation of a cylindrical workpart in open-die forging, showing pronounced barreling: (1) start of process, (2) partial deformation, and (3) final shape.

to-height ratio of the workpart increases, due to the greater contact area at the work–die interface.

All of these factors cause the actual upsetting force to be greater than what is predicted by Eq. (19.15). As an approximation, we can apply a shape factor to Eq. (19.15) to account for effects of the D/h ratio and friction:

$$F = K_f Y_f A \qquad (19.16)$$

where F, Y_f, and A have the same definitions as in the previous equation; and K_f is the forging shape factor, defined as

$$K_f = 1 + \frac{0.4\,\mu D}{h} \qquad (19.17)$$

where μ = coefficient of friction; D = workpart diameter or other dimension representing contact length with die surface, mm (in); and h = workpart height, mm (in).

Example 19.2 Open-Die Forging

A cylindrical workpiece is subjected to a cold upset forging operation. The starting piece is 75 mm in height and 50 mm in diameter. It is reduced in the operation to a height of 36 mm. The work material has a flow curve defined by $K = 350$ MPa and $n = 0.17$. Assume a coefficient of friction of 0.1. Determine the force as the process begins, at intermediate heights of 62 mm, 49 mm, and at the final height of 36 mm.

Solution: Workpiece volume $V = 75\pi(50^2/4) = 147{,}262$ mm^3. At the moment contact is made by the upper die, $h = 75$ mm and the force $F = 0$. At the start of yielding, h is slightly less than 75 mm, and we assume that strain = 0.002, at which the flow stress is

$$Y_f = K\epsilon^n = 350(0.002)^{0.17} = 121.7 \text{ MPa}$$

The diameter is still approximately $D = 50$ mm and area $A = \pi(50^2/4) = 1963.5$ mm^2. For these conditions, the adjustment factor K_f is computed as

$$K_f = 1 + \frac{0.4(0.1)(50)}{75} = 1.027$$

The forging force is

$$F = 1.027(121.7)(1963.5) = 245{,}410 \text{ MPa}$$

At $h = 62$ mm,

$$\epsilon = \ln\frac{75}{62} = \ln(1.21) = 0.1904$$

$$Y_f = 350(0.1904)^{17} = 264.0 \text{ MPa}$$

Assuming constant volume, and neglecting barreling,

$$A = 147{,}262/62 = 2375.2 \text{ mm}^2 \text{ and } D = \sqrt{\frac{4(2375.2)}{\pi}} = 55.0 \text{ mm}$$

$$K_f = 1 + \frac{0.4(0.1)(55)}{62} = 1.035$$

$$F = 1.035(264)(2375.2) = 649{,}303 \text{ N}$$

Similarly, at $h = 49$ mm, $F = 955{,}642$ N; and at $h = 36$ mm, $F = 1{,}467{,}422$ N. The load-stroke curve in Figure 19.12 was developed from the values in this example.   ∎

Open-Die Forging Practice   Open-die hot forging is an important industrial process. Shapes generated by open-die operations are simple; examples include shafts, disks, and

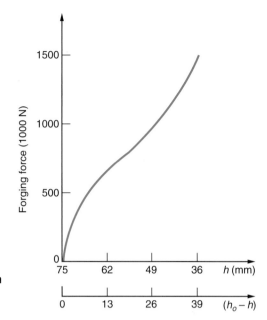

FIGURE 19.12 Upsetting force as a function of height h and height reduction $(h_o - h)$. This plot is sometimes called the load stroke curve.

rings. In some applications, the dies have slightly contoured surfaces that help to shape the work. In addition, the work must often be manipulated (e.g., rotating in steps) to effect the desired shape change. Skill of the human operator is a factor in the success of these operations. An example of open-die forging in the steel industry is the shaping of a large square cast ingot into a round cross section. Open-die forging operations produce rough forms, and subsequent operations are required to refine the parts to final geometry and dimensions. An important contribution of open-die hot-forging is that it creates a favorable grain flow and metallurgical structure in the metal.

Operations classified as open-die forging or related operations include fullering, edging, and cogging, illustrated in Figure 19.13. **Fullering** is a forging operation performed to reduce the cross section and redistribute the metal in a workpart in preparation for subsequent shape forging. It is accomplished by dies with convex surfaces. Fullering die cavities are often designed into multi-cavity impression dies, so that the starting bar can be rough formed before final shaping. **Edging** is similar to fullering, except that the dies have concave surfaces.

A **cogging** operation consists of a sequence of forging compressions along the length of a workpiece to reduce cross section and increase length. It is used in the steel industry to produce blooms and slabs from cast ingots. It is accomplished using open dies with flat or slightly contoured surfaces. The term **incremental forging** is sometimes used for this process.

19.3.2 IMPRESSION-DIE FORGING

Impression-die forging, sometimes called **closed-die forging,** is performed with dies that contain the inverse of the desired shape of the part. The process is illustrated in a three-step sequence in Figure 19.14. The raw workpiece is shown as a cylindrical part similar to that used in the previous open-die operation. As the die closes to its final position, flash is formed by metal that flows beyond the die cavity and into the small gap between the die plates. Although this flash must be cut away from the part in a subsequent trimming operation, it actually serves an important function during impression-die forging. As the flash begins to form in the

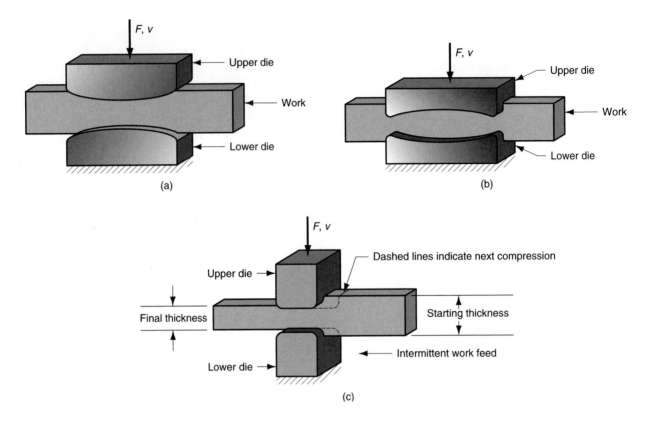

FIGURE 19.13   Several open-die forging operations: (a) fullering, (b) edging, and (c) cogging.

die gap, friction resists continued flow of metal into the gap, thus constraining the bulk of the work material to remain in the die cavity. In hot forging, metal flow is further restricted because the thin flash cools quickly against the die plates, thereby increasing its resistance to deformation. Restricting metal flow in the gap causes the compression pressures on the part to increase significantly, thus forcing the material to fill the sometimes intricate details of the die cavity to ensure a high-quality product.

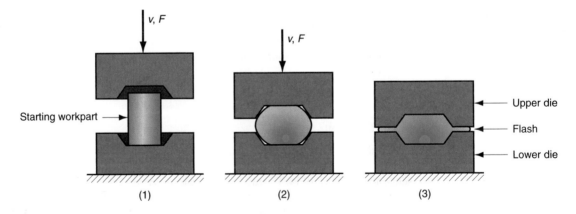

FIGURE 19.14   Sequence in impression-die forging: (1) just prior to initial contact with raw workpiece, (2) partial compression, and (3) final die closure, causing flash to form in gap between die plates.

TABLE 19.1   Typical K_f values for various part shapes in impression-die and flashless forging.

Part Shape	K_f	Part Shape	K_f
Impression-die forging:		Flashless forging:	
Simple shapes with flash	6.0	Coining (top and bottom surfaces)	6.0
Complex shapes with flash	8.0	Complex shapes	8.0
Very complex shapes with flash	10.0		

Several forming steps are often required in impression-die forging to transform the starting blank into the desired final geometry. Separate cavities in the die are needed for each step. The beginning steps are designed to redistribute the metal in the workpart to achieve a uniform deformation and desired metallurgical structure in the subsequent steps. The final steps bring the part to its final geometry. In addition, when drop forging is used, several blows of the hammer may be required for each step. When impression-die drop forging is done manually, as it often is, considerable operator skill is required under adverse conditions to achieve consistent results.

Because of flash formation in impression-die forging and the more complex part shapes made with these dies, forces in this process are significantly greater and more difficult to analyze than in open-die forging. Relatively simple formulas and design factors are often used to estimate forces in impression-die forging. The force formula is the same as previous Eq. (19.16) for open-die forging, but its interpretation is slightly different:

$$F = K_f Y_f A \qquad (19.18)$$

where F = maximum force in the operation, N (lb); A = projected area of the part including flash, mm^2 (in^2); Y_f = flow stress of the material, MPa (lb/in^2); and K_f = forging shape factor. In hot forging, the appropriate value of Y_f is the yield strength of the metal at the elevated temperature. In other cases, selecting the proper value of flow stress is difficult because the strain varies throughout the workpiece for complex shapes. K_f in Eq. (19.18) is a factor intended to account for increases in force required to forge part shapes of various complexities. Table 19.1 indicates the range of values of K_f for different part geometries. Obviously, the problem of specifying the proper K_f value for a given workpart limits the accuracy of the force estimate.

Eq. (19.18) applies to the maximum force during the operation, since this is the load that will determine the required capacity of the press or hammer used in the operation. The maximum force is reached at the end of the forging stroke, when the projected area is greatest and friction is maximum.

Impression-die forging is not capable of close tolerance work, and machining is often required to achieve the accuracies needed. The basic geometry of the part is obtained from the forging process, with machining performed on those portions of the part that require precision finishing (e.g., holes, threads, and surfaces that mate with other components). The advantages of forging, compared to machining the part completely, are higher production rates, conservation of metal, greater strength, and favorable grain orientation of the metal that results from forging. A comparison of the grain flow in forging and machining is illustrated in Figure 19.15.

Improvements in the technology of impression-die forging have resulted in the capability to produce forgings with thinner sections, more complex geometries, drastic reductions in draft requirements on the dies, closer tolerances, and the virtual elimination of machining allowances. Forging processes with these features are known as ***precision forging.*** Common work metals used for precision forging include aluminum and titanium. A comparison of precision and conventional impression-die forging is presented in Figure 19.16. Note that precision forging in this example does not eliminate flash, although it reduces it.

FIGURE 19.15
Comparison of metal grain flow in a part that is: (a) hot forged with finish machining, and (b) machined complete.

(a)

(b)

Some precision forging operations are accomplished without producing flash. Depending on whether machining is required to finish the part geometry, precision forgings are properly classified as *near net shape* or *net shape* processes.

19.3.3 FLASHLESS FORGING

As mentioned above, impression-die forging is sometimes called closed-die forging in industry terminology. However, there is a technical distinction between impression-die forging and true closed-die forging. The distinction is that in closed-die forging, the raw workpiece is completely contained within the die cavity during compression, and no flash is formed. The process sequence is illustrated in Figure 19.17. The term *flashless forging* is appropriate to identify this process.

Flashless forging imposes requirements on process control that are more demanding than impression-die forging. Most important is that the work volume must equal the space in the die cavity within a very close tolerance. If the starting blank is too large, excessive pressures may cause damage to the die or press. If the blank is too small, the cavity will not be filled. Because of the special demands made by flashless forging, the process lends itself best to part geometries that are usually simple and symmetrical, and to work materials such as aluminum and magnesium and their alloys. Flashless forging is often classified as a *precision forging* process [5].

Forces in flashless forging reach values comparable to those in impression-die forging. Estimates of these forces can be computed using the same methods as for impression-die forging: Eq. (19.18) and Table 19.1.

Coining is a special application of closed-die forging in which fine details in the die are impressed into the top and bottom surfaces of the workpart. There is little flow of metal in coining, yet the pressures required to reproduce the surface details in the die cavity are high, as indicated by the value of K_f in Table 19.1. A common application of

FIGURE 19.16
Cross sections of (a) conventional- and (b) precision forgings. Dashed lines in (a) indicate subsequent machining required to make the conventional forging equivalent in geometry to the precision forging. In both cases, flash extensions must be trimmed.

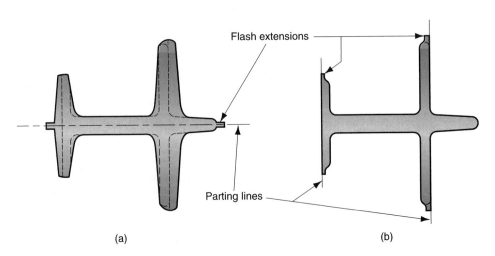

Flash extensions

Parting lines

(a)

(b)

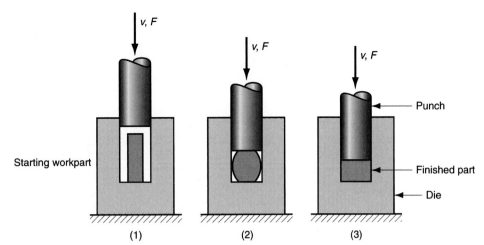

FIGURE 19.17
Flashless forging: (1) just before initial contact with workpiece, (2) partial compression, and (3) final punch and die closure. Symbols *v* and *F* indicate motion (*v* = velocity) and applied force, respectively.

coining is, of course, in the minting of coins, shown in Figure 19.18. The process is also used to provide good surface finish and dimensional accuracy on workparts made by other operations.

19.3.4 FORGING HAMMERS, PRESSES, AND DIES

Equipment used in forging consists of forging machines, classified as hammers or presses, and forging dies, which are the special tooling used in these machines. In addition, auxiliary equipment is needed, such as furnaces to heat the work, mechanical devices to load and unload the work, and trimming stations to cut away the flash in impression-die forging.

Forging Hammers   Forging hammers operate by applying an impact loading against the work. The term ***drop hammer*** is often used for these machines, owing to the means of delivering impact energy (see Figures 19.19 and 19.20). Drop hammers are most frequently

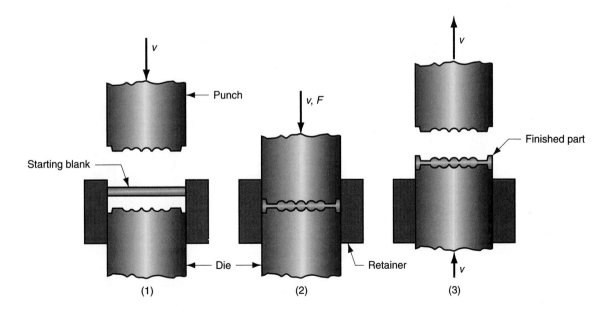

FIGURE 19.18   Coining operation: (1) start of cycle, (2) compression stroke, and (3) ejection of finished part.

FIGURE 19.19 Drop forging hammer, fed by conveyor and heating units at the right of the scene. (Photo courtesy of Chambersburg Engineering Company, Chambersburg, Pennsylvania)

used for impression-die forging. The upper portion of the forging die is attached to the ram, and the lower portion is attached to the anvil. In the operation, the work is placed on the lower die, and the ram is lifted and then dropped. When the upper die strikes the work, the impact energy causes the part to assume the form of the die cavity. Several blows of the hammer are often required to achieve the desired change in shape. Drop hammers can be classified as gravity drop hammers and power drop hammers. *Gravity drop hammers* achieve their energy by the falling weight of a heavy ram. The force of the blow is determined by the height of the drop and the weight of the ram. *Power drop hammers* accelerate the ram by pressurized air or steam. One of the disadvantages of drop hammers is that a large amount of the impact energy is transmitted through the anvil and into the floor of the building.

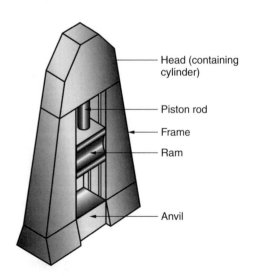

Head (containing cylinder)

Piston rod

Frame

Ram

Anvil

FIGURE 19.20 Diagram showing details of a drop hammer for impression-die forging.

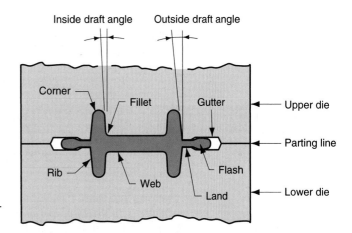

FIGURE 19.21
Terminology for a conventional impression-die in forging.

Forging Presses   Presses apply gradual pressure, rather than sudden impact, to accomplish the forging operation. Forging presses include mechanical presses, hydraulic presses, and screw presses. *Mechanical presses* operate by means of eccentrics, cranks, or knuckle joints, which convert the rotating motion of a drive motor into the translation motion of the ram. These mechanisms are very similar to those used in stamping presses (Section 20.5.2). Mechanical presses typically achieve very high forces at the bottom of the forging stroke. *Hydraulic presses* use a hydraulically driven piston to actuate the ram. *Screw presses* apply force by a screw mechanism that drives the vertical ram. Both screw drive and hydraulic drive operate at relatively low ram speeds and can provide a constant force throughout the stroke. These machines are therefore suitable for forging (and other forming) operations that require a long stroke.

Forging Dies   Proper die design is important in the success of a forging operation. Parts to be forged must be designed based on knowledge of the principles and limitations of this process. Our purpose here is to describe some of the terminology and guidelines used in the design of forgings and forging dies. Design of open dies is generally straightforward because the dies are relatively simple in shape. Our comments apply to impression dies and closed dies. Figure 19.21 defines some of the terminology in an impression die.

We indicate some of the principles and limitations that must be considered in the part design or in the selection of forging as the manufacturing process to make the part in the following discussion of forging die terminology [5]:

➤ *Parting line.* The parting line is the plane that divides the upper die from the lower die. Called the flash line in impression-die forging, it is the plane where the two die halves meet. Its selection by the designer affects grain flow in the part, required load, and flash formation.

➤ *Draft.* Draft is the amount of taper on the sides of the part required to remove it from the die. The term also applies to the taper on the sides of the die cavity. Typical draft angles are 3° on aluminum and magnesium parts and 5° to 7° on steel parts. Draft angles on precision forgings are near zero.

➤ *Webs and ribs.* A web is a thin portion of the forging that is parallel to the parting line, while a rib is a thin portion that is perpendicular to the parting line. These part features cause difficulty in metal flow as they become thinner.

➤ *Fillet and corner radii.* Fillet and corner radii are illustrated in Figure 19.21. Small radii tend to limit metal flow and increase stresses on die surfaces during forging.

➤ *Flash.* Flash formation plays a critical role in impression-die forging by causing pressure buildup inside the die to promote filling of the cavity. This pressure buildup is controlled by designing a flash land and gutter into the die, as pictured in Figure 19.21. The land determines the surface area along which lateral flow of metal occurs, thereby controlling the pressure increase inside the die. The gutter permits excess metal to escape without causing the forging load to reach extreme values.

19.4 OTHER DEFORMATION PROCESSES RELATED TO FORGING

In addition to the conventional forging operations discussed in the preceding sections, other metal forming operations are closely associated with forging.

Upsetting and Heading Upsetting (also called *upset forging*) is a deformation operation in which a cylindrical workpart is increased in diameter and reduced in length. This operation was analyzed in our discussion of open-die forging (Section 19.3.1). However, as an industrial operation, it can also be performed as closed-die forging, as seen in Figure 19.22.

Upsetting is widely used in the fastener industry to form heads on nails, bolts, and similar hardware products. In these applications, the term *heading* is often used to denote the operation. Figure 19.23 illustrates a variety of heading applications, indicating various possible die configurations. Owing to these types of applications, more parts are produced

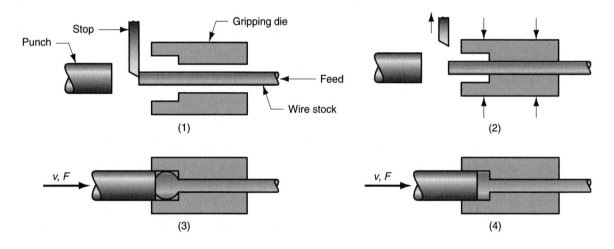

FIGURE 19.22   An upset forging operation to form a head on a bolt or similar hardware item. The cycle is as follows: (1) wire stock is fed to the stop; (2) gripping dies close on the stock and the stop is retracted; (3) punch moves forward; and (4) bottoms to form the head.

FIGURE 19.23
Examples of heading (upset forging) operations: (a) heading a nail using open dies, (b) round head formed by punch, (c) and (d) heads formed by die, and (e) carriage bolt head formed by punch and die.

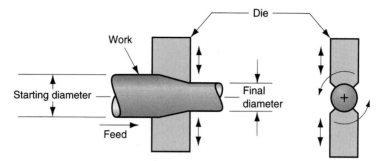

FIGURE 19.24 Swaging process to reduce solid rod stock; the dies rotate as they hammer the work. In radial forging, the workpiece rotates while the dies remain in a fixed orientation as they hammer the work.

by upsetting than by any other forging operation. It is performed as a mass-production operation—cold, warm, or hot—on special upset forging machines, called headers or formers. These machines are usually equipped with horizontal slides, rather than vertical slides as in conventional forging hammers and presses. Long wire or bar stock is fed into the machines, the end of the stock is upset forged, and then the piece is cut to length to make the desired hardware item. For bolts and screws, thread rolling (Section 19.2) is used to form the threads.

There are limits on the amount of deformation that can be achieved in upsetting, usually defined as the maximum length of stock to be forged. The maximum length that can be upset in one blow is three times the diameter of the starting stock. Otherwise, the metal bends or buckles instead of compressing properly to fill the cavity.

Swaging and Radial Forging   Swaging and radial forging are forging processes used to reduce the diameter of a tube or solid rod. Swaging is often performed on the end of a workpiece to create a tapered section. The *swaging* process, shown in Figure 19.24, is accomplished by means of rotating dies that hammer a workpiece radially inward to taper it as the piece is fed into the dies. Figure 19.25 illustrates some of the shapes and products that are made by swaging. A mandrel is sometimes required to control the shape and size of the internal diameter of tubular parts that are swaged. *Radial forging* is similar to swaging in its action against the work and is used to create similar part shapes. The difference is that in radial forging the dies do not rotate around the workpiece; instead, the work is rotated as it feeds into the hammering dies.

Roll Forging   Roll forging is a deformation process used to reduce the cross section of a cylindrical (or rectangular) workpiece by passing it through a set of opposing rolls that have grooves matching the desired shape of the final part. The typical operation is illustrated in Figure 19.26. Roll forging is generally classified as a forging process even though it utilizes rolls. The rolls do not turn continuously in roll forging, but rotate through only a portion of one revolution corresponding to the desired deformation to be accomplished on the part.

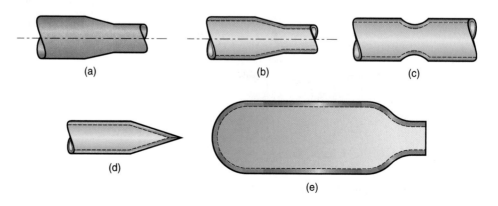

FIGURE 19.25
Examples of parts made by swaging: (a) reduction of solid stock, (b) tapering a tube, (c) swaging to form a groove on a tube, (d) pointing of a tube, and (e) swaging of neck on a gas cylinder.

(a)   (b)   (c)

(d)   (e)

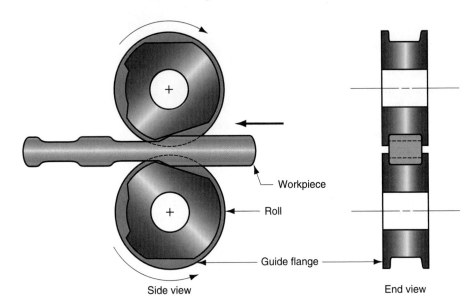

FIGURE 19.26
Roll forging.

Side view   End view

Roll-forged parts are generally stronger and possess favorable grain structure compared to competing processes such as machining that might be used to produce the same part geometry.

Orbital Forging In this process, deformation occurs by means of a cone-shaped upper die that is simultaneously rolled and pressed into the workpart. As illustrated in Figure 19.27, the work is supported on a lower die, which has a cavity into which the work is compressed. Because the axis of the cone is inclined, only a small area of the work surface is compressed at any moment. As the upper die revolves, the area under compression also revolves. These operating characteristics of orbital forging result in a substantial reduction in press load required to accomplish deformation of the work.

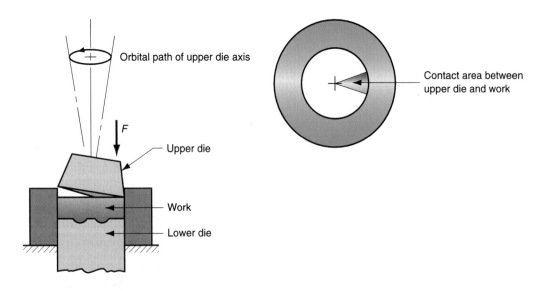

FIGURE 19.27   Orbital forging. At end of deformation cycle, lower die lifts to eject part.

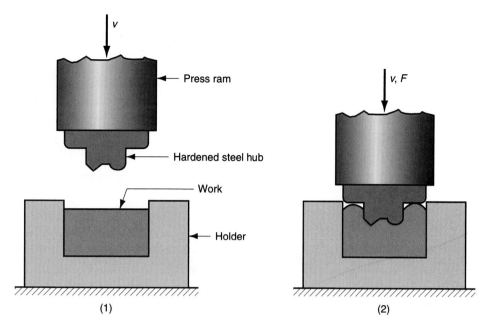

FIGURE 19.28
Hubbing: (1) before deformation, and (2) as the process is completed. Note that the excess material formed by the penetration of the hub must be machined away.

Hubbing   Hubbing is a deformation process in which a hardened steel form is pressed into a soft steel (or other soft metal) block. The process is often used to make mold cavities for plastic molding and die casting, as sketched in Figure 19.28. The hardened steel form, called the *hub,* is machined to the geometry of the part to be molded. Substantial pressures are required to force the hub into the soft block, and this is usually accomplished by a hydraulic press. Complete formation of the die cavity in the block often requires several steps—hubbing followed by annealing to recover the work metal from strain hardening. When significant amounts of material are deformed in the block, as shown in our figure, the excess must be machined away. The advantage of hubbing in this application is that it is generally easier to machine the positive form than the mating negative cavity. This advantage is multiplied in cases where more than one cavity are made in the die block.

Isothermal Forging   Isothermal forging is a term applied to a hot-forging operation in which the workpart is maintained at or near its starting elevated temperature during deformation, usually by heating the forging dies to the same elevated temperature. By avoiding chill of the workpiece on contact with the cold die surfaces as in conventional forging, the metal flows more readily and the force required to perform the process is reduced. Isothermal forging is more expensive than conventional forging and is usually reserved for difficult-to-forge metals, such as titanium and superalloys, and for complex part shapes. The process is sometimes carried out in a vacuum to avoid rapid oxidation of the die material. Similar to isothermal forging is *hot-die forging,* in which the dies are heated to a temperature that is somewhat below that of the work metal.

Trimming   Trimming is an operation used to remove flash on the workpart in impression-die forging. In most cases, trimming is accomplished by shearing, as in Figure 19.29, in which a punch forces the work through a cutting die, the blades for which have the profile of the desired part. Trimming is usually done while the work is still hot, which means that a separate trimming press is included at each forging hammer or press. In cases where the work might be damaged by the cutting process, trimming may be done by alternative methods, such as grinding or sawing.

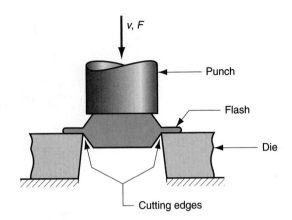

FIGURE 19.29 Trimming operation (shearing process) to remove the flash after impression-die forging.

19.5 EXTRUSION

Extrusion is a compression process in which the work metal is forced to flow through a die opening to produce a desired cross-sectional shape. The process can be likened to squeezing toothpaste out of a toothpaste tube. Extrusion dates from around 1800 (Historical Note 19.3). There are several advantages of the modern process: (1) a variety of shapes are possible, especially with hot extrusion; (2) grain structure and strength properties are enhanced in cold and warm extrusion; (3) fairly close tolerances are possible, especially in cold extrusion; and (4) in some extrusion operations, little or no wasted material is created. However, a limitation is that the cross section of the extruded part must be uniform throughout its length.

Historical Note 19.3   *Extrusion*

Extrusion as an industrial process was invented around 1800 in England, during the Industrial Revolution when that country was leading the world in technological innovations. The invention consisted of the first hydraulic press for extruding lead pipes. An important step forward was made in Germany around 1890, when the first horizontal extrusion press was built for extruding metals with higher melting points than lead. The feature that made this possible was the use of a dummy block that separated the ram from the work billet.

19.5.1 TYPES OF EXTRUSION

Extrusion is carried out in various ways. One important distinction is between direct extrusion and indirect extrusion. Another classification is by working temperature: cold, warm, or hot extrusion. Finally, extrusion is performed as either a continuous process or a discrete process.

Direct versus Indirect Extrusion   Direct extrusion (also called *forward extrusion*) is illustrated in Figure 19.30. A metal billet is loaded into a container, and a ram compresses the material, forcing it to flow through one or more openings in a die at the opposite end of the container. As the ram approaches the die, a small portion of the billet remains that cannot be forced through the die opening. This extra portion, called the *butt,* is separated from the product by cutting it just beyond the exit of the die.

FIGURE 19.30   Direct extrusion.

One of the problems in direct extrusion is the significant friction that exists between the work surface and the walls of the container as the billet is forced to slide toward the die opening. This friction causes a substantial increase in the ram force required in direct extrusion. In hot extrusion, the friction problem is aggravated by the presence of an oxide layer on the surface of the billet. This oxide layer can cause defects in the extruded product. To address these problems, a dummy block is often used between the ram and the work billet. The diameter of the dummy block is slightly smaller than the billet diameter, so that a narrow ring of work metal (mostly the oxide layer) is left in the container, leaving the final product free of oxides.

Hollow sections (e.g., tubes) are possible in direct extrusion by the process setup in Figure 19.31. The starting billet is prepared with a hole parallel to its axis. This allows passage of a mandrel that is attached to the dummy block. As the billet is compressed, the material is forced to flow through the clearance between the mandrel and the die opening. The resulting cross section is tubular. Semi-hollow cross-sectional shapes are usually extruded in the same way.

The starting billet in direct extrusion is usually round in cross section, but the final shape is determined by the shape of the die opening. Obviously, the largest dimension of the die opening must be smaller than the diameter of the billet.

In **indirect extrusion,** also called **backward extrusion** and **reverse extrusion,** Figure 19.32(a), the die is mounted to the ram rather than at the opposite end of the container. As the ram penetrates into the work, the metal is forced to flow through the clearance in a

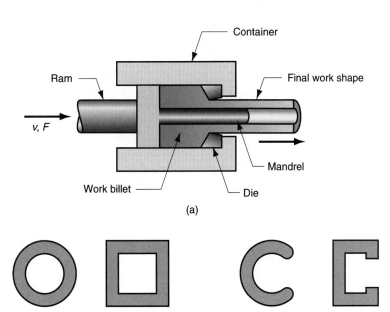

(a)

FIGURE 19.31
(a) Direct extrusion to produce a hollow or semi-hollow cross section; (b) hollow and (c) semi-hollow cross sections.

(b)   (c)

FIGURE 19.32 Indirect extrusion to produce (a) a solid cross section and (b) a hollow cross section.

direction opposite to the motion of the ram. Since the billet is not forced to move relative to the container, there is no friction at the container walls, and the ram force is therefore lower than in direct extrusion. Limitations of indirect extrusion are imposed by the lower rigidity of the hollow ram and the difficulty in supporting the extruded product as it exits the die.

Indirect extrusion can produce hollow (tubular) cross sections, as in Figure 19.32(b). In this method, the ram is pressed into the billet, forcing the material to flow around the ram and take a cup shape. There are practical limitations on the length of the extruded part that can be made by this method. Support of the ram becomes a problem as work length increases.

Hot versus Cold Extrusion   Extrusion can be performed either hot or cold, depending on work metal and amount of strain to which it is subjected during deformation. Metals that are typically extruded hot include aluminum, copper, magnesium, zinc, tin, and their alloys. These same metals are sometimes extruded cold. Steel alloys are usually extruded hot, although the softer, more ductile grades are sometimes cold extruded (e.g., low carbon steels and stainless steel). Aluminum is probably the most ideal metal for extrusion (hot and cold), and many commercial aluminum products are made by this process (structural shapes, door and window frames, etc.).

Hot extrusion involves prior heating of the billet to a temperature above its recrystallization temperature. This reduces strength and increases ductility of the metal, permitting more extreme size reductions and more complex shapes to be achieved in the process. Additional advantages include reduction of ram force, increased ram speed, and reduction of grain flow characteristics in the final product. Cooling of the billet as it contacts the container walls is a problem, and *isothermal extrusion* is sometimes used to overcome this problem. Lubrication is critical in hot extrusion for certain metals (e.g., steels), and special lubricants have been developed that are effective under the harsh conditions in hot extrusion. Glass is sometimes used as a lubricant in hot extrusion; in addition to reducing friction, it also provides effective thermal insulation between the billet and the extrusion container.

Cold extrusion and warm extrusion are generally used to produce discrete parts, often in finished (or near finished) form. The term *impact extrusion* is used to indicate high-speed cold extrusion, and this method is described in more detail in Section 19.5.4. Some important advantages of cold extrusion include increased strength due to strain hardening, close tolerances, improved surface finish, absence of oxide layers, and high production rates. Cold extrusion at room temperature also eliminates the need for heating the starting billet.

Continuous versus Discrete Processing   A true continuous process operates in steady state mode for an indefinite period of time. Some extrusion operations approach this ideal by producing very long sections in one cycle, but these operations are ultimately limited by the size of the starting billet that can be loaded into the extrusion container. These processes are more accurately described as semi-continuous operations. In nearly all cases, the long section is cut into smaller lengths in a subsequent sawing or shearing operation.

In a discrete extrusion operation, a single part is produced in each extrusion cycle. Impact extrusion is an example of the discrete processing case.

19.5.2 ANALYSIS OF EXTRUSION

Let us use Figure 19.33 as a reference in discussing some of the parameters in extrusion. The diagram assumes that both billet and extrudate are round in cross section. One important parameter is the **extrusion ratio**, also called the **reduction ratio.** The ratio is defined:

$$r_x = \frac{A_o}{A_f} \tag{19.19}$$

where r_x = extrusion ratio; A_o = cross-sectional area of the starting billet, mm^2 (in^2); and A_f = final cross-sectional area of the extruded section, mm^2 (in^2). The ratio applies for both direct and indirect extrusion. The value of r_x can be used to determine true strain in extrusion, given that ideal deformation occurs with no friction and no redundant work:

$$\epsilon = \ln r_x = \ln \frac{A_o}{A_f} \tag{19.20}$$

Under the assumption of ideal deformation (no friction and no redundant work), the pressure applied by the ram to compress the billet through the die opening depicted in our figure can be computed as follows:

$$p = \overline{Y}_f \ln r_x \tag{19.21}$$

where $\overline{Y}_f$ = average flow stress during deformation, MPa (lb/in^2). For convenience, we restate Eq. (18.2) from the previous chapter:

$$\overline{Y}_f = \frac{K\epsilon^n}{1+n}$$

In fact, extrusion is not a frictionless process, and the previous equations grossly underestimate the strain and pressure in an extrusion operation. Friction exists between the die and the work as the billet squeezes down and passes through the die opening. In direct extrusion, friction also exists between the container wall and the billet surface. The effect of friction is to increase the strain experienced by the metal. Thus, the actual pressure is greater than that given by Eq. (19.21), which assumes no friction.

FIGURE 19.33   Pressure and other variables in direct extrusion.

Various methods have been suggested to calculate the actual true strain and associated ram pressure in extrusion [1], [3], [6], [11], [12], and [19]. The following empirical equation proposed by Johnson [11] for estimating extrusion strain has gained considerable recognition:

$$\epsilon_x = a + b \ln r_x \tag{19.22}$$

where ϵ_x = extrusion strain; and a and b are empirical constants for a given die angle. Typical values of these constants are: $a = 0.8$ and $b = 1.2$ to 1.5. Values of a and b tend to increase with increasing die angle.

The ram pressure to perform ***indirect extrusion*** can be estimated based on Johnson's extrusion strain formula as follows:

$$p = \overline{Y}_f \epsilon_x \tag{19.23a}$$

where $\overline{Y}_f$ is calculated based on ideal strain from Eq. (19.20), rather than extrusion strain in Eq. (19.22).

In ***direct extrusion,*** the effect of friction between the container walls and the billet causes the ram pressure to be greater than for indirect extrusion. We can write the following expression which isolates the friction force in the direct extrusion container:

$$\frac{p_f \pi D_o^2}{4} = \mu p_c \pi D_o L$$

where p_f = additional pressure required to overcome friction, MPa (lb/in^2); $\pi D_o^2/4$ = billet cross-sectional area, mm^2 (in^2); μ = coefficient of friction at the container wall; p_c = pressure of the billet against the container wall, MPa (lb/in^2); and $\pi D_o L$ = area of the interface between billet and container wall, mm^2 (in^2). The right-hand side of this equation indicates the billet-container friction force, and the left-hand side gives the additional ram force to overcome that friction. In the worst case, sticking occurs at the container wall so that friction stress equals shear yield strength of the work metal:

$$\mu p_s \pi D_o L = Y_s \pi D_o L$$

where Y_s = shear yield strength, MPa (lb/in^2). If we assume that $Y_s = \overline{Y}_f/2$, then p_f reduces to the following:

$$p_f = \overline{Y}_f \frac{2L}{D_o}$$

Based on this reasoning, the following formula can be used to compute ram pressure in direct extrusion:

$$p = \overline{Y}_f \left(\epsilon_x + \frac{2L}{D_o} \right) \tag{19.23b}$$

where the term $2L/D_o$ accounts for the additional pressure due to friction at the container–billet interface. L is the portion of the billet length remaining to be extruded, and D_o is the original diameter of the billet. Note that p is reduced as the remaining billet length decreases during the process. Typical plots of ram pressure as a function of ram stroke for direct and indirect extrusion are presented in Figure 19.34. Eq. (19.23b) probably overestimates ram pressure. With good lubrication, ram pressures would be lower than values calculated by this equation.

Ram force in indirect or direct extrusion is simply pressure p from Eqs. (19.23a) or (19.23b), respectively, multiplied by billet area A_o:

$$F = pA_o \tag{19.24}$$

FIGURE 19.34   Typical plots of ram pressure versus ram stroke (and remaining billet length) for direct and indirect extrusion. The higher values in direct extrusion result from friction at the container wall. The shape of the initial pressure buildup at the beginning of the plot depends on die angle (higher die angles cause steeper pressure buildups). The pressure increase at the end of the stroke is related to formation of the butt.

where F = ram force in extrusion, N (lb). Power required to carry out the extrusion operation is simply

$$P = Fv \qquad (19.25)$$

where P = power, J/s (in-lb/min); F = ram force, N (lb); and v = ram velocity, m/s (in/min).

**Example 19.3
Extrusion
Pressures**

A billet 75 mm long and 25 mm in diameter is to be extruded in a direct extrusion operation with extrusion ratio $r_x = 4.0$. The extrudate has a round cross section. The die angle (half-angle) $= 90°$. The work metal has a strength coefficient $= 415$ MPa, and strain-hardening exponent $= 0.18$. Use the Johnson formula with $a = 0.8$ and $b = 1.5$ to estimate extrusion strain. Determine the pressure applied to the end of the billet as the ram moves forward.

Solution: Let us examine the ram pressure at billet lengths of $L = 75$ mm (starting value), $L = 50$ mm, $L = 25$ mm, and $L = 0$. We compute the ideal true strain, extrusion strain using Johnson's formula, and average flow stress:

$$\epsilon = \ln r_x = \ln 4.0 = 1.3863$$

$$\epsilon_x = 0.8 + 1.5(1.3863) = 2.8795$$

$$\overline{Y}_f = \frac{415(1.3863)^{0.18}}{1.18} = 373 \text{ MPa}$$

$L = 75$ mm: With a die angle of 90°, the billet metal is assumed to be forced through the die opening almost immediately; thus, our calculation assumes that maximum pressure is reached at the billet length of 75 mm. For die angles less than 90°, the pressure would build to a maximum as in Figure 19.34 as the starting billet is squeezed into the cone-shaped portion of the extrusion die. Using Eq. (19.23b),

$$p = 373\left(2.8795 + 2\frac{75}{25}\right) = 3312 \text{ MPa}$$

$$L = 50 \text{ mm}: p = 373\left(2.8795 + 2\frac{50}{25}\right) = 2566 \text{ MPa}$$

$$L = 25 \text{ mm}: p = 373\left(2.8795 + 2\frac{25}{25}\right) = 1820 \text{ MPa}$$

$L = 0$: Zero length is a hypothetical value in direct extrusion. In reality, it is impossible to squeeze all of the metal through the die opening. Instead, a portion of the billet (the "butt") remains unextruded and the pressure begins to increase rapidly as L approaches zero. This increase in pressure at the end of the stroke is seen in the plot of ram pressure versus ram stroke in Figure 19.34. Calculated below is the hypothetical minimum value of ram pressure that would result at $L = 0$.

$$p = 373\left(2.8795 + 2\frac{0}{25}\right) = 1074\,\text{MPa}$$

This is also the value of ram pressure that would be associated with indirect extrusion throughout the length of the billet. ■

19.5.3 EXTRUSION DIES AND PRESSES

Important factors in an extrusion die are die angle and orifice shape. Die angle, more precisely die half-angle, is shown as α in Figure 19.35(a). For low angles, surface area of the die is large, leading to increased friction at the die–billet interface. Higher friction results in larger ram force. On the other hand, a large die angle causes more turbulence in the metal flow during reduction, increasing the ram force required. Thus, the effect of die angle on ram force is a U-shaped function, as in Figure 19.35(b). An optimum die angle exists, as suggested by our hypothetical plot. The optimum angle depends on various factors (e.g., work material, billet temperature, and lubrication) and is therefore difficult to determine for a given extrusion job. Die designers rely on rules of thumb and judgment to decide the appropriate angle.

Our previous equations for ram pressure, Eqs. (19.23a), apply to a circular die orifice. The shape of the die orifice affects the ram pressure required to perform an extrusion operation. A complex cross section, such as the one shown in Figure 19.36, requires a higher pressure and greater force than a circular shape. The effect of the die orifice shape can be assessed by the die **shape factor,** defined as the ratio of the pressure required to extrude a cross section of a given shape relative to the extrusion pressure for a round cross section of the same area. We can express the shape factor as follows:

$$K_x = 0.98 + 0.02\left(\frac{C_x}{C_c}\right)^{2.25} \tag{19.26}$$

where K_x = die shape factor in extrusion; C_x = perimeter of the extruded cross section, mm (in); and C_c = perimeter of a circle of the same area as the extruded shape, mm (in). Eq.

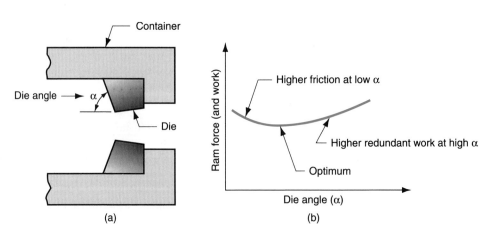

FIGURE 19.35
(a) Definition of die angle in direct extrusion; (b) effect of die angle on ram force.

FIGURE 19.36   A complex extruded cross section for a heat sink. (Photo courtesy of Aluminum Company of America, Pittsburg, Pennsylvania.)

(19.26) is based on empirical data in Altan et al. [1] over a range of C_x/C_c values from 1.0 to about 6.0. The equation may be invalid much beyond the upper limit of this range.

As indicated by Eq. (19.26), the shape factor is a function of the perimeter of the extruded cross section divided by the perimeter of a circular cross section of equal area. A circular shape is the simplest shape, with a value of $K_x = 1.0$. Hollow, thin-walled sections have higher shape factors and are more difficult to extrude. The increase in pressure is not included in our previous pressure equations, Eqs. (19.23a and 19.23b), which apply only to round cross sections. For shapes other than round, the corresponding expression for indirect extrusion is

$$p = K_x \overline{Y}_f \epsilon_x \tag{19.27a}$$

and for direct extrusion,

$$p = K_x \overline{Y}_f \left(\epsilon_x + \frac{2L}{D_o} \right) \tag{19.27b}$$

where p = extrusion pressure, MPa (lb/in^2); K_x = shape factor; and the other terms have the same interpretation as before. Values of pressure given by these equations can be used in Eq. (19.24) to determine ram force.

Die materials used for hot extrusion include tool and alloy steels. Important properties of these die materials include high wear resistance, high hot hardness, and high thermal conductivity to remove heat from the process. Die materials for cold extrusion include tool steels and cemented carbides. Wear resistance and ability to retain shape under high stress are desirable properties. Carbides are used when high production rates, long die life, and good dimensional control are required.

Extrusion presses are either horizontal or vertical, depending on orientation of the work axis. Horizontal types are more common. Extrusion presses are usually hydraulically driven. This drive is especially suited to semi-continuous production of long sections, as in direct extrusion. Mechanical drives are often used for cold extrusion of individual parts, such as in impact extrusion.

19.5.4 OTHER EXTRUSION PROCESSES

Direct and indirect extrusion are the principal methods of extrusion. Various names are given to operations that are special cases of the direct and indirect methods described here. Other extrusion operations are unique. In this section we examine some of these special forms of extrusion and related processes.

Impact Extrusion   Impact extrusion is performed at higher speeds and shorter strokes than conventional extrusion. It is used to make individual components. As the name suggests, the punch impacts the workpart rather than simply applying pressure to it. Impacting can be carried out as forward extrusion, backward extrusion, or combinations of these. Some representative examples are shown in Figure 19.37.

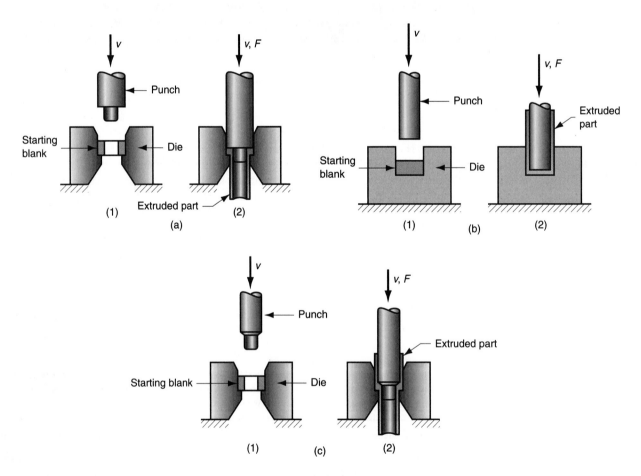

FIGURE 19.37   Several examples of impact extrusion: (a) forward, (b) backward, and (c) combination of forward and backward.

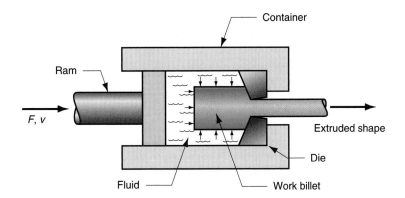

FIGURE 19.38
Hydrostatic extrusion.

Impact extrusion is usually done cold on a variety of metals. Backward impact extrusion is most common. Products made by this process include toothpaste tubes and battery cases. As indicated by these examples, very thin walls are possible on impact extruded parts. The high-speed characteristics of impacting permit large reductions and high production rates, making this an important commercial process.

Hydrostatic Extrusion   One of the problems in direct extrusion is friction along the billet–container interface. This problem can be addressed by surrounding the billet with fluid inside the container and pressurizing the fluid by the forward motion of the ram, as in Figure 19.38. This way, there is no friction inside the container, and friction at the die opening is reduced. Consequently, ram force is significantly lower than in direct extrusion. The fluid pressure acting on all surfaces of the billet gives the process its name. It can be carried out at room temperature or at elevated temperatures. Special fluids and procedures must be used at elevated temperatures. Hydrostatic extrusion is an adaptation of direct extrusion.

Hydrostatic pressure on the work increases the material's ductility. Accordingly, this process can be used on metals that would be too brittle for conventional extrusion operations. Ductile metals can also be hydrostatically extruded, and high reduction ratios are possible on these materials. One of the disadvantages of the process is the required preparation of the starting work billet. The billet must be formed with a taper at one end to fit snugly into the die entry angle. This establishes a seal to prevent fluid from squirting out the die hole when the container is initially pressurized.

19.5.5 DEFECTS IN EXTRUDED PRODUCTS

Owing to the considerable deformation associated with extrusion operations, a number of defects can occur in extruded products. The defects can be classified into the following categories, illustrated in Figure 19.39:

FIGURE 19.39   Some common defects in extrusion: (a) centerburst, (b) piping, and (c) surface cracking.

(a)   (b)   (c)

(a) ***Centerburst.*** This defect is an internal crack that develops as a result of tensile stresses along the centerline of the workpart during extrusion. Although tensile stresses may seem unlikely in a compression process such as extrusion, they tend to occur under conditions that cause large deformation in the regions of the work away from the central axis. The significant material movement in these outer regions stretches the material along the center of the work. If stresses are great enough, bursting occurs. Conditions that promote centerburst are high die angles, low extrusion ratios, and impurities in the work metal that serve as starting points for crack defects. The difficult aspect of centerburst is its detection. It is an internal defect that is usually not noticeable by visual observation. Other names sometimes used for this defect include ***arrowhead fracture, center cracking,*** and ***chevron cracking.***

(b) ***Piping.*** Piping is a defect associated with direct extrusion. As in Figure 19.39(b), it is the formation of a sink hole in the end of the billet. The use of a dummy block whose diameter is slightly less than that of the billet helps to avoid piping. Other names given to this defect include ***tailpipe*** and ***fishtailing.***

(c) ***Surface cracking.*** This defect results from high workpart temperatures that cause cracks to develop at the surface. They often occur when extrusion speed is too high, leading to high strain rates and associated heat generation. Other factors contributing to surface cracking are high friction and surface chilling of high temperature billets in hot extrusion.

19.6 WIRE AND BAR DRAWING

In the context of bulk deformation, drawing is an operation in which the cross section of a bar, rod, or wire is reduced by pulling it through a die opening, as in Figure 19.40. The general features of the process are similar to those of extrusion. The difference is that the work is pulled through the die in drawing, whereas it is pushed through the die in extrusion. Although the presence of tensile stresses is obvious in drawing, compression also plays a significant role because the metal is squeezed down as it passes through the die opening. For this reason, the deformation that occurs in drawing is sometimes referred to as indirect compression. Drawing is a term also used in sheet metalworking (Section 20.3). The term ***wire and bar drawing*** is used to distinguish the drawing process discussed here from the sheet metal process of the same name.

The basic difference between bar drawing and wire drawing is the stock size that is processed. ***Bar drawing*** is the term used for large diameter bar and rod stock, while ***wire drawing*** applies to small diameter stock. Wire sizes down to 0.03 mm (0.001 in) are possible in wire drawing. Although the mechanics of the process are the same for the two cases, the methods, equipment, and even the terminology are somewhat different.

Bar drawing is generally accomplished as a ***single-draft*** operation—the stock is pulled through one die opening. Because the beginning stock has a large diameter, it is in

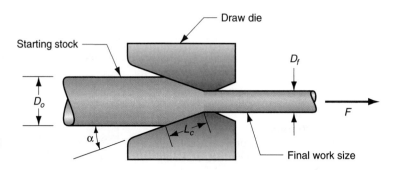

FIGURE 19.40 Drawing of bar, rod, or wire.

the form of a straight cylindrical piece rather than coiled. This limits the length of the work that can be drawn, necessitating a batch type operation. By contrast, wire is drawn from coils consisting of several hundred (or even several thousand) feet of wire and is passed through a series of draw dies. The number of dies varies typically between 4 and 12. The term **continuous drawing** is used to describe this type of operation because of the long production runs that are achieved with the wire coils, which can be butt-welded each to the next to make the operation truly continuous.

In a drawing operation, the change in size of the work is usually given by the area reduction, defined as follows:

$$r = \frac{A_o - A_f}{A_o} \tag{19.28}$$

where r = area reduction in drawing; A_o = original area of work, mm^2 (in^2); and A_f = final area, mm^2 (in^2). Area reduction is often expressed as a percentage.

In bar drawing, rod drawing, and in drawing of large diameter wire for upsetting and heading operations, the term draft is used to denote the before and after difference in size of the processed work. The **draft** is simply the difference between original and final stock diameters:

$$d = D_o - D_f \tag{19.29}$$

where d = draft, mm (in); D_o = original diameter of work, mm (in); and D_f = final work diameter, mm (in).

19.6.1 ANALYSIS OF DRAWING

In this section, we consider the mechanics of wire and bar drawing. How are stresses and forces computed in the process? We also consider how large a reduction is possible in a drawing operation.

Mechanics of Drawing   If no friction or redundant work occurred in drawing, true strain could be determined as follows:

$$\epsilon = \ln \frac{A_o}{A_f} = \ln \frac{1}{1 - r} \tag{19.30}$$

where A_o and A_f are the original and final cross-sectional areas of the work, as previously defined; and r = drawing reduction as given by Eq. (19.28). The stress that results from this ideal deformation is given by

$$\sigma = \overline{Y}_f \epsilon = \overline{Y}_f \ln \frac{A_o}{A_f} \tag{19.31}$$

where $\overline{Y}_f = \frac{K\epsilon^n}{1+n}$ = average flow stress based on the value of strain given by Eq. (19.30).

Because friction is present in drawing and the work metal experiences inhomogeneous deformation, the actual stress is larger than provided by Eq. (19.31). In addition to the ratio A_o/A_f, other variables that influence draw stress are die angle and coefficient of friction at the work–die interface. A number of methods have been proposed for predicting draw stress based on values of these parameters [1], [3], and [19]. We present the equation suggested by Schey [19]:

$$\sigma_d = \overline{Y}_f \left(1 + \frac{\mu}{\tan \alpha}\right) \phi \ln \frac{A_o}{A_f} \tag{19.32}$$

where σ_d = draw stress, MPa (lb/in^2); μ = die-work coefficient of friction; α = die angle (half-angle) as defined in Figure 19.40; and ϕ is a factor that accounts for inhomogeneous deformation which is determined as follows for a round cross section:

$$\phi = 0.88 \pm 0.12 \frac{D}{L_c} \tag{19.33}$$

where D = average diameter of work during drawing, mm (in); and L_c = contact length of the work with the draw die in Figure 19.40, mm (in). Values of D and L_c can be determined from the following:

$$D = \frac{D_o + D_f}{2} \tag{19.34a}$$

$$L_c = \frac{D_o - D_f}{2 \sin \alpha} \tag{19.34b}$$

The corresponding draw force is then the area of the drawn cross section multiplied by the draw stress:

$$F = A_f \sigma_d = A_f \overline{Y}_f \left(1 + \frac{\mu}{\tan \alpha}\right) \phi \ln \frac{A_o}{A_f} \tag{19.35}$$

where F = draw force, N (lb); and the other terms are defined above. The power required in a drawing operation is the draw force multiplied by exit velocity of the work.

**Example 19.4
Stress and Force
in Wire Drawing**

Wire is drawn through a draw die with entrance angle = 15°. Starting diameter is 2.5 mm and final diameter = 2.0 mm. The coefficient of friction at the work–die interface = 0.07. The metal has a strength coefficient K = 205 MPa and a strain-hardening exponent n = 0.20. Determine the draw stress and draw force in this operation.

Solution: The values of D and L_c for Eq. (19.33) can be determined using Eqs. (19.34). D = 2.25 mm and L_c = 0.966 mm. Thus,

$$\phi = 0.88 + 0.12 \frac{2.25}{0.966} = 1.16$$

The areas before and after drawing are computed as A_o = 4.91 mm² and A_f = 3.14 mm². The resulting true strain $\epsilon = \ln(4.91/3.14) = 0.446$, and the average flow stress in the operation is computed:

$$\overline{Y}_f = \frac{205(0.446)^{0.20}}{1.20} = 145.4 \text{ MPa}$$

Draw stress is given by Eq. (19.32):

$$\sigma_d = (145.4)\left(1 + \frac{0.07}{\tan 15}\right)(1.16)(0.446) = 94.1 \text{ MPa}$$

Finally, the draw force is this stress multiplied by the cross-sectional area of the exiting wire:

$$F = 94.1(3.14) = 295.5 \text{ N}$$

Maximum Reduction per Pass   A question that may occur to the reader is: Why is more than one step required to achieve the desired reduction in wire drawing? Why not take the entire reduction in a single pass through one die, as in extrusion? The answer can be explained as follows. From the preceding equations, it is clear that as the reduction increases, draw stress increases. If the reduction is large enough, draw stress will exceed the yield strength of the exiting metal. When that happens, the drawn wire will simply elongate instead of new material being squeezed through the die opening. For wire drawing to be successful, maximum draw stress must be less than the yield strength of the exiting metal.

It is a straightforward matter to determine this maximum draw stress and the resulting maximum possible reduction that can be made in one pass, under certain assumptions. Let us assume a perfectly plastic metal ($n = 0$), no friction, and no redundant work. In this ideal case, the maximum possible draw stress is equal to the yield strength of the work material. Expressing this using the equation for draw stress under conditions of ideal deformation, Eq. (19.31), and setting $\overline{Y}_f = Y$ (because $n = 0$),

$$\sigma_d = \overline{Y}_f \ln \frac{A_o}{A_f} = Y \ln \frac{A_o}{A_f} = Y \ln \frac{1}{1 - r} = Y$$

This means that $\ln(A_o/A_f) = \ln(1/(1 - r)) = 1$. That is, $\epsilon_{max} = 1.0$. In order for ϵ_{max} to be zero, then $A_o/A_f = 1/(1 - r)$ must equal the natural logarithm base e. Accordingly, the maximum possible area ratio is

$$\frac{A_o}{A_f} = e = 2.7183 \tag{19.36}$$

and the maximum possible reduction is

$$r_{max} = \frac{e - 1}{e} = 0.632 \tag{19.37}$$

The value given by Eq. (19.37) is often used as the theoretical maximum reduction possible in a single draw, even though it ignores (1) the effects of friction and redundant work, which would reduce the maximum possible value, and (2) strain hardening, which would increase the maximum possible reduction because the exiting wire would be stronger than the starting metal. In practice, draw reductions per pass are quite below the theoretical limit. Reductions of 0.50 for single-draft bar drawing and 0.30 for multiple-draft wire drawing seem to be the upper limits in industrial operations.

19.6.2 DRAWING PRACTICE

Drawing is usually performed as a cold working operation. It is most frequently used to produce round cross sections, but squares and other shapes are also drawn. Wire drawing is an important industrial process, providing commercial products such as electrical wire and cable; wire stock for fences, coat hangers, and shopping carts; and rod stock to produce nails, screws, rivets, springs, and other hardware items. Bar drawing is used to produce metal bars for machining, forging, and other processes.

Advantages of drawing in these applications include (1) close dimensional control, (2) good surface finish, (3) improved mechanical properties such as strength and hardness, and (4) adaptability to economical batch or mass production. Drawing speeds are as high as 50 m/s (10,000 ft/min) for very fine wire. In the case of bar drawing to provide stock for machining, the operation improves the machinability of the bar (Section 24.1).

Drawing Equipment Bar drawing is accomplished on a machine called a ***draw bench,*** consisting of an entry table, die stand (which contains the draw die), carriage, and exit rack. The arrangement is shown in Figure 19.41. The carriage is used to pull the stock through the draw die. It is powered by hydraulic cylinders or motor-driven chains. The die stand is often designed to hold more than one die, so that several bars can be pulled simultaneously through their respective dies.

FIGURE 19.41
Hydraulically operated draw bench for drawing metal bars.

Wire drawing is done on continuous drawing machines that consist of multiple draw dies, separated by accumulating drums between the dies, as in Figure 19.42. Each drum, called a *capstan,* is motor driven to provide the proper pull force to draw the wire stock through the upstream die. It also maintains a modest tension on the wire as it proceeds to the next draw die in the series. Each die provides a certain amount of reduction in the wire, so that the desired total reduction is achieved by the series. Depending on the metal to be processed and the total reduction, annealing of the wire is sometimes required between groups of dies in the series.

Draw Dies Figure 19.43 identifies the features of a typical draw die. Four regions of the die can be distinguished: (1) entry, (2) approach angle, (3) bearing surface (land), and (4) back relief. The *entry* region is usually a bell-shaped mouth that does not contact the work. Its purpose is to funnel the lubricant into the die and prevent scoring of work and die surfaces. The *approach* is where the drawing process occurs. It is   cone-shaped with an angle (half-angle) normally ranging from about 6° to 20°. The proper angle varies according to work

FIGURE 19.42   Continuous drawing of wire.

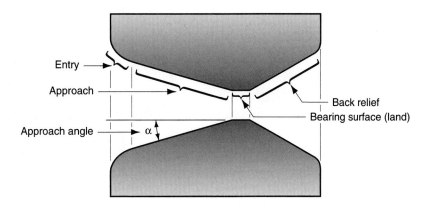

FIGURE 19.43 Draw die for drawing of round rod or wire.

material. The **bearing surface,** or **land,** determines the size of the final drawn stock. Finally, the **back relief** is the exit zone. It is provided with a back relief angle (half-angle) of about 30°. Draw dies are made of tool steels or cemented carbides. Dies for high-speed wire drawing operations frequently use inserts made of diamond (both synthetic and natural) for the wear surfaces.

Preparation of the Work   Prior to drawing, the beginning stock must be properly prepared. This involves three steps: (1) annealing, (2) cleaning, and (3) pointing. The purpose of annealing is to increase the ductility of the stock to accept deformation during drawing. As previously mentioned, annealing is sometimes needed between steps in continuous drawing. Cleaning of the stock is required to prevent damage of the work surface and draw die. It involves removal of surface contaminants (e.g., scale and rust) by means of chemical pickling or shot blasting. In some cases, prelubrication of the work surface is accomplished subsequent to cleaning.

    Pointing involves the reduction in diameter of the starting end of the stock so that it can be inserted through the draw die to start the process. This is usually accomplished by swaging, rolling, or turning. The pointed end of the stock is then gripped by the carriage jaws or other device to initiate the drawing process.

19.6.3 TUBE DRAWING

Drawing can be used to reduce the diameter or wall thickness of seamless tubes and pipes, after the initial tubing has been produced by some other process such as extrusion. Tube drawing can be carried out either with or without a mandrel. The simplest method uses no mandrel and is used for diameter reduction, as in Figure 19.44. The term *tube sinking* is sometimes applied to this operation.

FIGURE 19.44   Tube drawing with no mandrel (tube sinking).

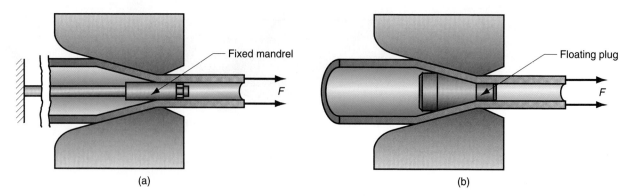

FIGURE 19.45   Tube drawing with mandrels: (a) fixed mandrel, (b) floating plug.

The problem with tube drawing in which no mandrel is used, as in Figure 19.44, is that it lacks control over the inside diameter and wall thickness of the tube. This is why mandrels of various types are used, two of which are illustrated in Figure 19.45. The first, Figure 19.45 (a), uses a *fixed mandrel* attached to a long support bar to establish inside diameter and wall thickness during the operation. Practical limitations on the length of the support bar in this method restrict the length of the tube that can be drawn. The second type, shown in (b), uses a *floating plug* whose shape is designed so that it finds a "natural" position in the reduction zone of the die. This method removes the limitations on work length present with the fixed mandrel.

REFERENCES

[1] Altan, T., Oh, S-I., and Gegel, H. L. *Metal Forming: Fundamentals and Applications.* ASM International, Materials Park, Ohio, 1983.

[2] *ASM Handbook,* Vol. 14A, *Metalworking: Bulk Forming.* ASM International, Materials Park, Ohio, 2005.

[3] Avitzur, B. *Metal Forming: Processes and Analysis.* Robert E. Krieger Publishing Company, Huntington, New York, 1979.

[4] Black, J. T., and Kohser, R. A., *DeGarmo's Materials and Processes in Manufacturing,* 10th ed. John Wiley & Sons, Inc., Hoboken, New Jersey, 2008.

[5] Byrer, T. G.,et al. (eds.). *Forging Handbook.* Forging Industry Association, Cleveland, Ohio; and American Society for Metals, Metals Park, Ohio, 1985.

[6] Cook, N. H. *Manufacturing Analysis.* Addison-Wesley Publishing Company, Inc., Reading, Massachusetts, 1966.

[7] Groover, M. P."An Experimental Study of the Work Components and Extrusion Strain in the Cold Forward Extrusion of Steel," research report. Bethlehem Steel Corporation, Bethlehem, Pennsylvania, 1966.

[8] Harris, J. N. *Mechanical Working of Metals.* Pergamon Press, Oxford, UK, 1983.

[9] Hosford, W. F., and Caddell, R. M. *Metal Forming: Mechanics and Metallurgy,* 3rd ed. Cambridge University Press, Cambridge, UK, 2007.

[10] Jensen, J. E. (ed.). *Forging Industry Handbook.* Forging Industry Association, Cleveland, Ohio, 1970.

[11] Johnson, W."The Pressure for the Cold Extrusion of Lubricated Rod Through Square Dies of Moderate Reduction at Slow Speeds," *Journal of the Institute of Metals,* Vol. 85, 1956.

[12] Kalpakjian, S. *Mechanical Processing of Materials.* D. Van Nostrand Company, Inc., Princeton, New Jersey, 1967, Chapter 5.

[13] Kalpakjian, S., and SchmidS. R. *Manufacturing Processes for Engineering Materials,* 6th ed. Pearson Prentice Hall, Upper Saddle River, New Jersey, 2010.

[14] Lange, K. *Handbook of Metal Forming.* Society of Manufacturing Engineers, Dearborn, Michigan, 2006.

[15] Laue, K., and Stenger, H. *Extrusion: Processes, Machinery, and Tooling.* American Society for Metals, Metals Park, Ohio, 1981.

[16] Mielnik, E. M. *Metalworking Science and Engineering.* McGraw-Hill, Inc., New York, 1991.

[17] Roberts, W. L. *Hot Rolling of Steel.* Marcel Dekker, Inc., New York, 1983.

[18] Roberts, W. L. *Cold Rolling of Steel.* Marcel Dekker, Inc., New York, 1978.

[19] Schey, J. A. *Introduction to Manufacturing Processes,* 3rd ed. McGraw-Hill Book Company, New York, 2000.

[20] Wick, C., et al. (eds.). *Tool and Manufacturing Engineers Handbook,* 4th ed. Vol. II, *Forming.* Society of Manufacturing Engineers, Dearborn, Michigan, 1984.

REVIEW QUESTIONS

19.1. What are the reasons why the bulk deformation processes are important commercially and technologically?

19.2. Name the four basic bulk deformation processes.

19.3. What is rolling in the context of the bulk deformation processes?

19.4. In rolling of steel, what are the differences between a bloom, a slab, and a billet?

19.5. List some of the products produced on a rolling mill.

19.6. What is draft in a rolling operation?

19.7. What is sticking in a hot rolling operation?

19.8. Identify some of the ways in which force in flat rolling can be reduced.

19.9. What is a two-high rolling mill?

19.10. What is a reversing mill in rolling?

19.11. Besides flat rolling and shape rolling, identify some additional bulk forming processes that use rolls to effect the deformation.

19.12. What is forging?

19.13. One way to classify forging operations is by the degree to which the work is constrained in the die. By this classification, name the three basic types.

19.14. Why is flash desirable in impression-die forging?

19.15. What is a trimming operation in the context of impression-die forging?

19.16. What are the two basic types of forging equipment?

19.17. What is isothermal forging?

19.18. What is extrusion?

19.19. Distinguish between direct and indirect extrusion.

19.20. Name some products that are produced by extrusion.

19.21. Why is friction a factor in determining the ram force in direct extrusion but not a factor in indirect extrusion?

19.22. What does the centerburst defect in extrusion have in common with the roll piercing process?

19.23. What is wire drawing and bar drawing?

19.24. Although the workpiece in a wire drawing operation is obviously subjected to tensile stresses, how do compressive stresses also play a role in the process?

19.25. In a wire drawing operation, why must the drawing stress never exceed the yield strength of the work metal?

19.26. (**Video**) According to the video on forming, what is the primary factor that makes the mechanical performance of forged parts better than cast parts in many situations?

19.27. (**Video**) List the accessory tools that can be used during open-die forging according to the video on forging.

19.28. (**Video**) List the performing operations discussed in the forming video.

MULTIPLE CHOICE QUIZ

There are 27 correct answers in the following multiple choice questions (some questions have multiple answers that are correct). To attain a perfect score on the quiz, all correct answers must be given. Each correct answer is worth 1 point. Each omitted answer or wrong answer reduces the score by 1 point, and each additional answer beyond the correct number of answers reduces the score by 1 point. Percentage score on the quiz is based on the total number of correct answers.

19.1. The starting workpiece in steel hot rolling of plate and sheet stock is which of the following (one best answer): (a) bar stock, (b) billet, (c) bloom, (d) slab, or (e) wire stock?

19.2. The maximum possible draft in a rolling operation depends on which of the following parameters (two

correct answers): (a) coefficient of friction between roll and work, (b) roll diameter, (c) roll velocity, (d) stock thickness, (e) strain, and (f) strength coefficient of the work metal?

19.3. Which of the following stress or strength parameters is used in the computation of rolling force (one best answer): (a) average flow stress, (b) compression strength, (c) final flow stress, (d) tensile strength, or (e) yield strength?

19.4. Which of the following rolling mill types are associated with relatively small diameter rolls in contact with the work (two correct answers): (a) cluster mill, (b) continuous rolling mill, (c) four-high mill, (d) reversing mill, and (e) three-high configuration?

19.5. Production of pipes and tubes is associated with which of the following bulk deformation processes (three correct answers): (a) extrusion, (b) hobbing, (c) ring rolling, (d) roll forging, (e) roll piercing, (f) tube sinking, and (g) upsetting?

19.6. Which of the following stress or strength parameters is used in the computation of the maximum force in a forging operation (one best answer): (a) average flow stress, (b) compression strength, (c) final flow stress, (d) tensile strength, or (e) yield strength?

19.7. Which of the following operations are closely related to open-die forging (three best answers): (a) cogging, (b) flashless forging, (c) fullering, (d) impression-die forging, (e) Mannesmann process, (f) precision forging, (g) soaking, and (h) upsetting?

19.8. Flash in impression-die forging serves no useful purpose and is undesirable because it must be trimmed from the part after forming: (a) true or (b) false?

19.9. Which of the following are classified as forging operations (four correct answers): (a) coining, (b) fullering, (c) impact extrusion, (d) roll piercing, (e) swaging, (f) thread rolling, (g) trimming, and (h) upsetting?

19.10. Which of the following are alternative names for indirect extrusion (two correct answers): (a) backward extrusion, (b) direct extrusion, (c) forward extrusion, (d) impact extrusion, and (e) reverse extrusion?

19.11. The production of tubing is possible in indirect extrusion but not in direct extrusion: (a) true or (b) false?

19.12. Which of the following stress or strength parameters is used in the computation of the force in an extrusion operation (one best answer): (a) average flow stress, (b) compression strength, (c) final flow stress, (d) tensile strength, or (e) yield strength?

19.13. In which of the following extrusion operations is friction a factor in determining the extrusion force (one best answer): (a) direct extrusion or (b) indirect extrusion?

19.14. Theoretically, the maximum reduction possible in a wire drawing operation, under the assumptions of a perfectly plastic metal, no friction, and no redundant work, is which of the following (one answer): (a) zero, (b) 0.63, (c) 1.0, or (d) 2.72?

19.15. Which of the following bulk deformation processes are involved in the production of nails for lumber construction (three best answers): (a) bar and wire drawing, (b) extrusion, (c) flashless forging, (d) impression-die forging, (e) rolling, and (f) upsetting?

19.16. Johnson's formula is associated with which one of the four bulk deformation processes: (a) bar and wire drawing, (b) extrusion, (c) forging, and (d) rolling?

PROBLEMS

Rolling

19.1. A 42.0-mm-thick plate made of low carbon steel is to be reduced to 34.0 mm in one pass in a rolling operation. As the thickness is reduced, the plate widens by 4%. The yield strength of the steel plate is 174 MPa and the tensile strength is 290 MPa. The entrance speed of the plate is 15.0 m/min. The roll radius is 325 mm and the rotational speed is 49.0 rev/min. Determine (a) the minimum required coefficient of friction that would make this rolling operation possible, (b) exit velocity of the plate, and (c) forward slip.

19.2. A 2.0-in-thick slab is 10.0 in wide and 12.0 ft long. Thickness is to be reduced in three steps in a hot rolling operation. Each step will reduce the slab to 75% of its previous thickness. It is expected that for this metal and reduction, the slab will widen by 3% in each step. If the entry speed of the slab in the first step is 40 ft/min, and roll speed is the same for the three steps, determine: (a) length and (b) exit velocity of the slab after the final reduction.

19.3. A series of cold rolling operations are to be used to reduce the thickness of a plate from 50 mm down to 25 mm in a reversing two-high mill. Roll diameter = 700 mm and coefficient of friction between rolls and work = 0.15. The specification is that the draft is to be equal on each pass. Determine (a) minimum number of passes required, and (b) draft for each pass?

19.4. In the previous problem, suppose that the percent reduction were specified to be equal for each pass, rather than the draft. (a) What is the minimum number of passes required? (b) What is the draft for each pass?

19.5. A continuous hot rolling mill has two stands. Thickness of the starting plate = 25 mm and width = 300 mm. Final thickness is to be 13 mm. Roll radius at each stand = 250 mm. Rotational speed at the first stand = 20 rev/min. Equal drafts of 6 mm are to be taken at each stand. The plate is wide enough relative to its thickness that no increase in width occurs. Under the assumption that the forward slip is equal at each stand, determine (a) speed v_r at each stand, and (b) forward slip s. (c) Also, determine the exiting speeds at each rolling stand, if the entering speed at the first stand = 26 m/min.

19.6. A continuous hot rolling mill has eight stands. The dimensions of the starting slab are: thickness = 3.0 in, width = 15.0 in, and length = 10 ft. The final thickness is to be 0.3 in. Roll diameter at each stand = 36 in, and rotational speed at stand number 1 = 30 rev/min. It is observed that the speed of the slab entering stand 1 = 240 ft/min. Assume that no widening of the slab occurs during the rolling sequence. Percent reduction in thickness is to be equal at all stands, and it is assumed that the forward slip will be equal at each stand. Determine (a) percentage reduction at each stand, (b) rotational speed of the rolls at stands 2 through 8, and (c) forward slip. (d) What is the draft at stands 1 and 8? (e) What is the length and exit speed of the final strip exiting stand 8?

19.7. A plate that is 250 mm wide and 25 mm thick is to be reduced in a single pass in a two-high rolling mill to a thickness of 20 mm. The roll has a radius = 500 mm, and its speed = 30 m/min. The work material has a strength coefficient = 240 MPa and a strain-hardening exponent = 0.2. Determine (a) roll

force, (b) roll torque, and (c) power required to accomplish this operation.

19.8. Solve Problem 19.7 using a roll radius = 250 mm.

19.9. Solve Problem 19.7, only assume a cluster mill with working rolls of radius = 50 mm. Compare the results with the previous two problems, and note the important effect of roll radius on force, torque and power.

19.10. A 4.50-in-thick slab that is 9 in wide and 24 in long is to be reduced in a single pass in a two-high rolling mill to a thickness of 3.87 in. The roll rotates at a speed of 5.50 rev/min and has a radius of 17.0 in. The work material has a strength coefficient = 30,000 lb/in^2 and a strain-hardening exponent = 0.15. Determine (a) roll force, (b) roll torque, and (c) power required to accomplish this operation.

19.11. A single-pass rolling operation reduces a 20 mm thick plate to 18 mm. The starting plate is 200 mm wide. Roll radius = 250 mm and rotational speed = 12 rev/min. The work material has a strength coefficient = 600 MPa and a strength coefficient = 0.22. Determine (a) roll force, (b) roll torque, and (c) power required for this operation.

19.12. A hot rolling mill has rolls of diameter = 24 in. It can exert a maximum force = 400,000 lb. The mill has a maximum horsepower = 100 hp. It is desired to reduce a 1.5-in thick plate by the maximum possible draft in one pass. The starting plate is 10 in wide. In the heated condition, the work material has a strength coefficient = 20,000 lb/in^2 and a strain-hardening exponent = zero. Determine (a) maximum possible draft, (b) associated true strain, and (c) maximum speed of the rolls for the operation.

19.13. Solve Problem 19.12 except that the operation is warm rolling and the strain-hardening exponent is 0.18. Assume the strength coefficient remains at 20,000 lb/in^2.

Forging

19.14. A cylindrical part is warm upset forged in an open die. The initial diameter is 45 mm and the initial height is 40 mm. The height after forging is 25 mm. The coefficient of friction at the die–work interface is 0.20. The yield strength of the work material is 285 MPa, and its flow curve is defined by a strength coefficient of 600 MPa and a strain-hardening exponent of 0.12. Determine the force in the operation (a) just as the yield point is reached (yield at strain = 0.002), (b) at a height of 35 mm, (c) at a height of 30 mm, and (d) at a height of 25 mm. Use of a spreadsheet calculator is recommended.

19.15. A cylindrical workpart with $D = 2.5$ in and $h = 2.5$ in is upset forged in an open die to a height = 1.5 in. Coefficient of friction at the die–work interface = 0.10. The work material has a flow curve defined by: $K = 40,000$ lb/in^2 and $n = 0.15$. Yield strength = 15,750 lb/in^2. Determine the instantaneous force in the operation (a) just as the yield point is reached (yield at strain = 0.002), (b) at height $h = 2.3$ in, (c) $h = 2.1$ in, (d) $h = 1.9$ in, (e) $h = 1.7$ in, and (f) $h = 1.5$ in. Use of a spreadsheet calculator is recommended.

19.16. A cylindrical workpart has a diameter = 2.5 in and a height = 4.0 in. It is upset forged to a height = 2.75 in.

Coefficient of friction at the die–work interface = 0.10. The work material has a flow curve with strength coefficient = 25,000 lb/in^2 and strain-hardening exponent = 0.22. Determine the plot of force vs. work height. Use of a spreadsheet calculator is recommended.

19.17. A cold heading operation is performed to produce the head on a steel nail. The strength coefficient for this steel is 600 MPa, and the strain-hardening exponent is 0.22. Coefficient of friction at the die–work interface is 0.14. The wire stock out of which the nail is made is 5.00 mm in diameter. The head is to have a diameter of 9.5 mm and a thickness of 1.6 mm. The final length of the nail is 120 mm. (a) What length of stock must project out of the die in order to provide sufficient volume of material for this upsetting operation? (b) Compute the maximum force that the punch must apply to form the head in this open-die operation.

19.18. Obtain a large common nail (flat head). Measure the head diameter and thickness, as well as the diameter of the nail shank. (a) What stock length must project out of the die in order to provide sufficient material to produce the nail? (b) Using appropriate values for strength coefficient and strain-hardening exponent for the metal out of which the nail is made (Table 3.4), compute the maximum force in the heading operation to form the head.

19.19. A hot upset forging operation is performed in an open die. The initial size of the workpart is: $D_o = 25$ mm, and $h_o = 50$ mm. The part is upset to a diameter = 50 mm. The work metal at this elevated temperature yields at 85 MPa ($n = 0$). Coefficient of friction at the die–work interface = 0.40. Determine (a) final height of the part, and (b) maximum force in the operation.

19.20. A hydraulic forging press is capable of exerting a maximum force = 1,000,000 N. A cylindrical workpart is to be cold upset forged. The starting part has diameter = 30 mm and height = 30 mm. The flow curve of the metal is defined by $K = 400$ MPa and $n = 0.2$. Determine the maximum reduction in height to which the part can be compressed with this forging press, if the coefficient of friction = 0.1. Use of a spreadsheet calculator is recommended.

19.21. A part is designed to be hot forged in an impression die. The projected area of the part, including flash, is 16 in^2. After trimming, the part has a projected area of 10 in^2. Part geometry is complex. As heated the work material yields at 10,000 lb/in^2, and has no tendency to strain harden. At room temperature, the material yields at 25,000 lb/in^2 Determine the maximum force required to perform the forging operation.

19.22. A connecting rod is designed to be hot forged in an impression die. The projected area of the part is 6,500 mm^2. The design of the die will cause flash to form during forging, so that the area, including flash, will be 9,000 mm^2. The part geometry is considered to be complex. As heated the work material yields at 75 MPa, and has no tendency to strain harden. Determine the maximum force required to perform the operation.

Extrusion

19.23. A cylindrical billet that is 100 mm long and 50 mm in diameter is reduced by indirect (backward) extrusion to a 20 mm diameter. The die angle is 90°. The Johnson equation has $a = 0.8$ and $b = 1.4$, and the flow curve for the work metal has a strength coefficient of 800 MPa and strain-hardening exponent of 0.13. Determine (a) extrusion ratio, (b) true strain (homogeneous deformation), (c) extrusion strain, (d) ram pressure, and (e) ram force.

19.24. A 3.0-in-long cylindrical billet whose diameter = 1.5 in is reduced by indirect extrusion to a diameter = 0.375 in. Die angle = 90°. In the Johnson equation, $a = 0.8$ and $b = 1.5$. In the flow curve for the work metal, $K = 75,000$ lb/in^2 and $n = 0.25$. Determine (a) extrusion ratio, (b) true strain (homogeneous deformation), (c) extrusion strain, (d) ram pressure, (e) ram force, and (f) power if the ram speed = 20 in/min.

19.25. A billet that is 75 mm long with diameter = 35 mm is direct extruded to a diameter of 20 mm. The extrusion die has a die angle = 75°. For the work metal, $K = 600$ MPa and $n = 0.25$. In the Johnson extrusion strain equation, $a = 0.8$ and $b = 1.4$. Determine (a) extrusion ratio, (b) true strain (homogeneous deformation), (c) extrusion strain, and (d) ram pressure and force at $L = 70, 60, 50, 40, 30, 20$, and 10 mm. Use of a spreadsheet calculator is recommended for part (d).

19.26. A 2.0-in-long billet with diameter = 1.25 in is direct extruded to a diameter of 0.50 in. The extrusion die angle = 90°. For the work metal, $K = 45,000$ lb/in^2, and $n = 0.20$. In the Johnson extrusion strain equation, $a = 0.8$ and $b = 1.5$. Determine (a) extrusion ratio, (b) true strain (homogeneous deformation), (c) extrusion strain, and (d) ram pressure at $L = 2.0, 1.5, 1.0, 0.5$ and 0.0 in. Use of a spreadsheet calculator is recommended for part (d).

19.27. A direct extrusion operation is performed on a cylindrical billet with an initial diameter of 2.0 in and an initial length of 4.0 in. The die angle $= 60°$ and orifice diameter is 0.50 in. In the Johnson extrusion strain equation, $a = 0.8$ and $b = 1.5$. The operation is carried out hot and the hot metal yields at 13,000 lb/in^2 and does not strain harden when hot. (a) What is the extrusion ratio? (b) Determine the ram position at the point when the metal has been compressed into the cone of the die and starts to extrude through the die opening. (c) What is the ram pressure corresponding to this position? (d) Also determine the length of the final part if the ram stops its forward movement at the start of the die cone.

19.28. An indirect extrusion process starts with an aluminum billet with diameter $= 2.0$ in and length $= 3.0$ in. Final cross section after extrusion is a square with 1.0 in on a side. The die angle $= 90°$. The operation is performed cold and the strength coefficient of the metal $K = 26,000$ lb/in^2 and strain-hardening exponent $n = 0.20$. In the Johnson extrusion strain equation, $a = 0.8$ and $b = 1.2$. (a) Compute the extrusion ratio, true strain, and extrusion strain. (b) What is the shape factor of the product? (c) If the butt left in the container at the end of the stroke is 0.5 in thick, what is the length of the extruded section? (d) Determine the ram pressure in the process.

19.29. An L-shaped structural section is direct extruded from an aluminum billet in which $L_o = 500$ mm and $D_o = 100$ mm. Dimensions of the cross section are given in Figure P19.29. Die angle $= 90°$. Determine (a) extrusion ratio, (b) shape factor, and (c) length of the extruded section if the butt remaining in the container at the end of the ram stroke is 25 mm.

19.30. The flow curve parameters for the aluminum alloy of Problem 19.29 are: $K = 240$ MPa and $n = 0.16$. If

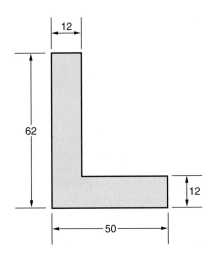

FIGURE P19.29 Part for Problem 19.29 (dimensions are in mm).

the die angle in this operation $= 90°$, and the corresponding Johnson strain equation has constants $a = 0.8$ and $b = 1.5$, compute the maximum force required to drive the ram forward at the start of extrusion.

19.31. A cup-shaped part is backward extruded from an aluminum slug that is 50 mm in diameter. The final dimensions of the cup are: OD $= 50$ mm, ID $= 40$ mm, height $= 100$ mm, and thickness of base $= 5$ mm. Determine (a) extrusion ratio, (b) shape factor, and (c) height of starting slug required to achieve the final dimensions. (d) If the metal has flow curve parameters $K = 400$ MPa and $n = 0.25$, and the constants in the Johnson extrusion strain equation are: $a = 0.8$ and $b = 1.5$, determine the extrusion force.

19.32. Determine the shape factor for each of the extrusion die orifice shapes in Figure P19.32.

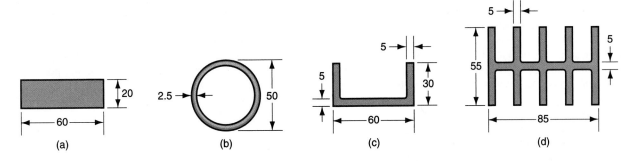

FIGURE P19.32 Cross-sectional shapes for Problem 19.32 (dimensions are in mm): (a) rectangular bar, (b) tube, (c) channel, and (d) cooling fins.

19.33. A direct extrusion operation produces the cross section shown in Figure P19.32(a) from a brass billet whose diameter = 125 mm and length = 350 mm. The flow curve parameters of the brass are $K = 700$ MPa and $n = 0.35$. In the Johnson strain equation, $a = 0.7$ and $b = 1.4$. Determine (a) the extrusion ratio, (b) the shape factor, (c) the force required to drive the ram forward during extrusion at the point in the process when the billet length remaining in the container = 300 mm, and (d) the length of the extruded section at the end of the operation if the volume of the butt left in the container is 600,000 mm^3.

19.34. In a direct extrusion operation the cross section shown in Figure P19.32(b) is produced from a copper billet whose diameter = 100 mm and length = 500 mm. In the flow curve for copper, the strength coefficient = 300 MPa and strain-hardening exponent = 0.50. In the Johnson strain equation, $a = 0.8$ and $b = 1.5$. Determine (a) the extrusion ratio, (b) the shape factor, (c) the force required to drive the ram forward during extrusion at the point in the process when the billet length remaining in the container = 450 mm, and (d) the length of the extruded section at the end of the operation if the volume of the butt left in the container is 350,000 mm^3.

19.35. A direct extrusion operation produces the cross section shown in Figure P19.32(c) from an aluminum billet whose diameter = 150 mm and length = 500 mm. The flow curve parameters for the aluminum are $K = 240$ MPa and $n = 0.16$. In the Johnson strain equation, $a = 0.8$ and $b = 1.2$. Determine (a) the extrusion ratio, (b) the shape factor, (c) the force required to drive the ram forward during extrusion at the point in the process when the billet length remaining in the container = 400 mm, and (d) the length of the extruded section at the end of the operation if the volume of the butt left in the container is 600,000 mm^3.

19.36. A direct extrusion operation produces the cross section shown in Figure P19.32(d) from an aluminum billet whose diameter = 150 mm and length = 900 mm. The flow curve parameters for the aluminum are $K = 240$ MPa and $n = 0.16$. In the Johnson strain equation, $a = 0.8$ and $b = 1.5$. Determine (a) the extrusion ratio, (b) the shape factor, (c) the force required to drive the ram forward during extrusion at the point in the process when the billet length remaining in the container = 850 mm, and (d) the length of the extruded section at the end of the operation if the volume of the butt left in the container is 600,000 mm^3.

Drawing

19.37. A spool of wire has a starting diameter of 2.5 mm. It is drawn through a die with an opening that is to 2.1 mm. The entrance angle of the die is 18°. Coefficient of friction at the work–die interface is 0.08. The work metal has a strength coefficient of 450 MPa and a strain-hardening coefficient of 0.26. The drawing is performed at room temperature. Determine (a) area reduction, (b) draw stress, and (c) draw force required for the operation.

19.38. Rod stock that has an initial diameter of 0.50 in is drawn through a draw die with an entrance angle of 13°. The final diameter of the rod is = 0.375 in. The metal has a strength coefficient of 40,000 lb/in^2 and a strain-hardening exponent of 0.20. Coefficient of friction at the work–die interface = 0.1. Determine (a) area reduction, (b) draw force for the operation, and (c) horsepower to perform the operation if the exit velocity of the stock = 2 ft/sec.

19.39. Bar stock of initial diameter = 90 mm is drawn with a draft = 15 mm. The draw die has an entrance angle = 18°, and the coefficient of friction at the work–die interface = 0.08. The metal behaves as a perfectly plastic material with yield stress = 105 MPa. Determine (a) area reduction, (b) draw stress, (c) draw force required for the operation, and (d) power to perform the operation if exit velocity = 1.0 m/min.

19.40. Wire stock of initial diameter = 0.125 in is drawn through two dies each providing a 0.20 area reduction. The starting metal has a strength coefficient = 40,000 lb/in^2 and a strain-hardening exponent = 0.15. Each die has an entrance angle of 12°, and the coefficient of friction at the work–die interface is estimated to be 0.10. The motors driving the capstans at the die exits can each deliver 1.50 hp at 90% efficiency. Determine the maximum possible speed of the wire as it exits the second die.

20 SHEET METALWORKING

Chapter Contents

Sheet metalworking includes cutting and forming operations performed on relatively thin sheets of metal. Typical sheet-metal thicknesses are between 0.4 mm (1/64 in) and 6 mm (1/4 in). When thickness exceeds about 6 mm, the stock is usually referred to as plate rather than sheet. The sheet or plate stock used in sheet metalworking is produced by flat rolling (Section 19.1). The most commonly used sheet metal is low carbon steel (0.06%–0.15% C typical). Its low cost and good formability, combined with sufficient strength for most product applications, make it ideal as a starting material.

The commercial importance of sheet metalworking is significant. Consider the number of consumer and industrial products that include sheet or plate metal parts: automobile and truck bodies, airplanes, railway cars, locomotives, farm and construction equipment, appliances, office furniture, and more. Although these examples are conspicuous because they have sheet-metal exteriors, many of their internal components are also made of sheet or plate stock. Sheet-metal parts are generally characterized by high strength, good dimensional accuracy, good surface finish, and relatively low cost. For components that must be made in large quantities, economical mass-production operations can be designed to process the parts. Aluminum beverage cans are a prime example.

Sheet-metal processing is usually performed at room temperature (cold working). The exceptions are when the stock is thick, the metal is brittle, or the deformation is significant. These are usually cases of warm working rather than hot working.

Most sheet-metal operations are performed on machine tools called *presses*. The term *stamping press* is used to distinguish these presses from forging and extrusion presses. The tooling that performs sheet metalwork is called a *punch-and-die;* the term *stamping die* is also used. The sheet-metal products are called *stampings.* To facilitate mass production, the sheet metal is often presented to the press as long strips or coils. Various types of punch-and-die tooling and stamping presses are described in Section 20.5. Final sections of the chapter cover various operations that do not utilize conventional punch-and-die tooling, and most of them are not performed on stamping presses. Two video clips on our DVD illustrate many of the topics discussed in this chapter.

VIDEO CLIP _____

Sheet-Metal Shearing and Bending. This clip has two segments on shearing and bending.

_____

VIDEO CLIP _____

Sheet-Metal Stamping Dies and Processes. Two segments are included: (1) sheet metal formability and (2) basic stamping die operations.

_____

   The three major categories of sheet-metal processes are (1) cutting, (2) bending, and (3) drawing. Cutting is used to separate large sheets into smaller pieces, to cut out part perimeters, and to make holes in parts. Bending and drawing are used to form sheet-metal parts into their required shapes.

20.1 CUTTING OPERATIONS

Cutting of sheet metal is accomplished by a shearing action between two sharp cutting edges. The shearing action is depicted in the four stop-action sketches of Figure 20.1, in which the upper cutting edge (the punch) sweeps down past a stationary lower cutting edge (the die). As the punch begins to push into the work, *plastic deformation* occurs in the surfaces of the sheet. As the punch moves downward, *penetration* occurs in which the punch compresses the sheet and cuts into the metal. This penetration zone is generally about one-third the thickness of the sheet. As the punch continues to travel into the work, *fracture* is initiated in the work at the two cutting edges. If the clearance between the punch

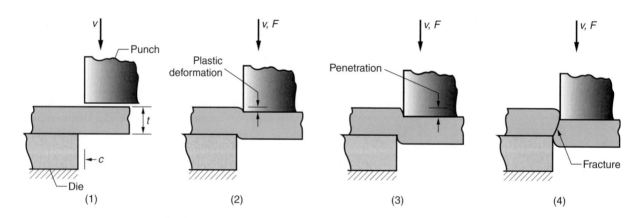

FIGURE 20.1 Shearing of sheet metal between two cutting edges: (1) just before the punch contacts work; (2) punch begins to push into work, causing plastic deformation; (3) punch compresses and penetrates into work causing a smooth cut surface; and (4) fracture is initiated at the opposing cutting edges that separate the sheet. Symbols v and F indicate motion and applied force, respectively, t = stock thickness, c = clearance.

FIGURE 20.2 Characteristic sheared edges of the work.

and die is correct, the two fracture lines meet, resulting in a clean separation of the work into two pieces.

The sheared edges of the sheet have characteristic features as in Figure 20.2. At the top of the cut surface is a region called the ***rollover.*** This corresponds to the depression made by the punch in the work prior to cutting. It is where initial plastic deformation occurred in the work. Just below the rollover is a relatively smooth region called the ***burnish.*** This results from penetration of the punch into the work before fracture began. Beneath the burnish is the ***fractured zone,*** a relatively rough surface of the cut edge where continued downward movement of the punch caused fracture of the metal. Finally, at the bottom of the edge is a ***burr,*** a sharp corner on the edge caused by elongation of the metal during final separation of the two pieces.

20.1.1 SHEARING, BLANKING, AND PUNCHING

The three most important operations in pressworking that cut metal by the shearing mechanism just described are shearing, blanking, and punching.

Shearing is a sheet-metal cutting operation along a straight line between two cutting edges, as shown in Figure 20.3(a). Shearing is typically used to cut large sheets into smaller sections for subsequent pressworking operations. It is performed on a machine called a ***power shears,*** or ***squaring shears.*** The upper blade of the power shears is often inclined, as shown in Figure 20.3(b), to reduce the required cutting force.

Blanking involves cutting of the sheet metal along a closed outline in a single step to separate the piece from the surrounding stock, as in Figure 20.4(a). The part that is cut out is the desired product in the operation and is called the ***blank.*** ***Punching*** is similar to blanking except that it produces a hole, and the separated piece is scrap, called the ***slug.*** The remaining stock is the desired part. The distinction is illustrated in Figure 20.4(b).

FIGURE 20.3 Shearing operation: (a) side view of the shearing operation; (b) front view of power shears equipped with inclined upper cutting blade. Symbol *v* indicates motion.

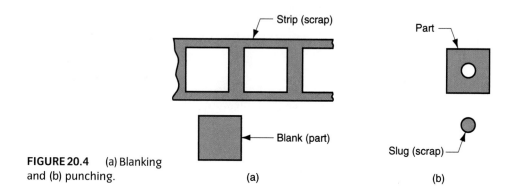

FIGURE 20.4 (a) Blanking and (b) punching.

(a)

(b)

20.1.2 ENGINEERING ANALYSIS OF SHEET-METAL CUTTING

Process parameters in sheet-metal cutting are clearance between punch and die, stock thickness, type of metal and its strength, and length of the cut. Let us define these parameters and some of the relationships among them.

Clearance The clearance c in a shearing operation is the distance between the punch and die, as shown in Figure 20.1(a). Typical clearances in conventional pressworking range between 4% and 8% of the sheet-metal thickness t. The effect of improper clearances is illustrated in Figure 20.5. If the clearance is too small, then the fracture lines tend to pass each other, causing a double burnishing and larger cutting forces. If the clearance is too large, the metal becomes pinched between the cutting edges and an excessive burr results. In special operations requiring very straight edges, such as shaving and fine blanking (Section 20.1.3), clearance is only about 1% of stock thickness.

The correct clearance depends on sheet-metal type and thickness. The recommended clearance can be calculated by the following formula:

$$c = A_c t \tag{20.1}$$

where c = clearance, mm (in); A_c = clearance allowance; and t = stock thickness, mm (in). The clearance allowance is determined according to type of metal. For convenience, metals are classified into three groups given in Table 20.1, with an associated allowance value for each group.

FIGURE 20.5 Effect of clearance: (a) clearance too small causes less-than-optimal fracture and excessive forces; and (b) clearance too large causes oversized burr. Symbols v and F indicate motion and applied force, respectively.

(a)

(b)

TABLE 20.1   Clearance allowance value for three sheet-metal groups.

Metal Group	A_c
1100S and 5052S aluminum alloys, all tempers	0.045
2024ST and 6061ST aluminum alloys; brass, all tempers; soft cold-rolled steel, soft stainless steel	0.060
Cold-rolled steel, half hard; stainless steel, half-hard and full-hard	0.075

Compiled from [3].

These calculated clearance values can be applied to conventional blanking and hole-punching operations to determine the proper punch and die sizes. The die opening must always be larger than the punch size (obviously). Whether to add the clearance value to the die size or subtract it from the punch size depends on whether the part being cut out is a blank or a slug, as illustrated in Figure 20.6 for a circular part. Because of the geometry of the sheared edge, the outer dimension of the part cut out of the sheet will be larger than the hole size. Thus, punch and die sizes for a round blank of diameter D_b are determined as

$$\text{Blanking punch diameter} = D_b - 2c \qquad (20.2a)$$

$$\text{Blanking die diameter} = D_b \qquad (20.2b)$$

Punch and die sizes for a round hole of diameter D_h are determined as:

$$\text{Hole punch diameter} = D_h \qquad (20.3a)$$

$$\text{Hole die diameter} = D_h + 2c \qquad (20.3b)$$

In order for the slug or blank to drop through the die, the die opening must have an **angular clearance** (see Figure 20.7) of 0.25° to 1.5° on each side.

Cutting Forces   Estimates of cutting force are important because this force determines the size (tonnage) of the press needed. Cutting force F in sheet metalworking can be determined by

$$F = StL \qquad (20.4)$$

where S = shear strength of the sheet metal, MPa (lb/in^2); t = stock thickness, mm (in), and L = length of the cut edge, mm (in). In blanking, punching, slotting, and similar operations, L is the perimeter length of the blank or hole being cut. The minor effect of clearance in determining the value of L can be neglected. If shear strength is unknown, an

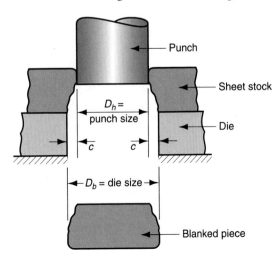

FIGURE 20.6   Die size determines blank size D_b; punch size determines hole size D_h.; c = clearance.

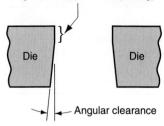

Straight portion (for resharpening)

Die

Die

Angular clearance

FIGURE 20.7 Angular clearance.

alternative way of estimating the cutting force is to use the tensile strength:

$$F = 0.7(TS)tL \qquad (20.5)$$

where TS = ultimate tensile strength MPa (lb/in^2).

These equations for estimating cutting force assume that the entire cut along the sheared edge length L is made at the same time. In this case the cutting force will be a maximum. It is possible to reduce the maximum force by using an angled cutting edge on the punch or die, as in Figure 20.3(b). The angle (called the **shear angle**), spreads the cut over time and reduces the force experienced at any one moment. However, the total energy required in the operation is the same, whether it is concentrated into a brief moment or distributed over a longer time period.

Example 20.1 Blanking Clearance and Force

A round disk of 150-mm diameter is to be blanked from a strip of 3.2-mm, half-hard cold-rolled steel whose shear strength = 310 MPa. Determine (a) the appropriate punch and die diameters, and (b) blanking force.

Solution: (a) From Table 20.1, the clearance allowance for half-hard cold-rolled steel is $A_c = 0.075$. Accordingly,

$$c = 0.075(3.2 \text{ mm}) = \mathbf{0.24 \text{ mm}}$$

The blank is to have a diameter = 150 mm, and die size determines blank size. Therefore,

$$\text{Die opening diameter} = \mathbf{150.00 \text{ mm}}$$
$$\text{Punch diameter} = 150 - 2(0.24) = \mathbf{149.52 \text{ mm}}$$

(b) To determine the blanking force, we assume that the entire perimeter of the part is blanked at one time. The length of the cut edge is

$$L = \pi D_b = 150\pi = 471.2 \text{ mm}$$

and the force is

$$F = 310(471.2)(3.2) = \mathbf{467,469 \text{ N}}[\sim 53 \text{ tons}]$$

20.1.3 OTHER SHEET-METAL-CUTTING OPERATIONS

In addition to shearing, blanking, and punching, there are several other cutting operations in presswork. The cutting mechanism in each case involves the same shearing action discussed above.

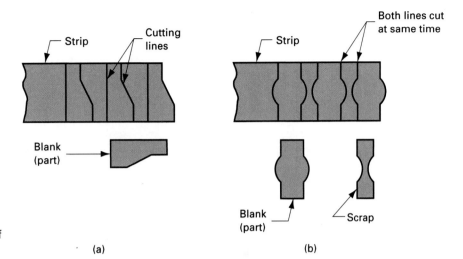

FIGURE 20.8   (a) Cutoff and (b) parting.

Cutoff and Parting   Cutoff is a shearing operation in which blanks are separated from a sheet-metal strip by cutting the opposite sides of the part in sequence, as shown in Figure 20.8(a). With each cut, a new part is produced. The features of a cutoff operation that distinguish it from a conventional shearing operation are (1) the cut edges are not necessarily straight, and (2) the blanks can be nested on the strip in such a way that scrap is avoided.

  Parting involves cutting a sheet-metal strip by a punch with two cutting edges that match the opposite sides of the blank, as shown in Figure 20.8(b). This might be required because the part outline has an irregular shape that precludes perfect nesting of the blanks on the strip. Parting is less efficient than cutoff in the sense that it results in some wasted material.

Slotting, Perforating, and Notching   Slotting is the term sometimes used for a punching operation that cuts out an elongated or rectangular hole, as pictured in Figure 20.9(a). *Perforating* involves the simultaneous punching of a pattern of holes in sheet metal, as in Figure 20.9(b). The hole pattern is usually for decorative purposes, or to allow passage of light, gas, or fluid.

  To obtain the desired outline of a blank, portions of the sheet metal are often removed by notching and seminotching. *Notching* involves cutting out a portion of metal from the side of the sheet or strip. *Seminotching* removes a portion of metal from the interior of the sheet. These operations are depicted in Figure 20.9(c). Seminotching might seem to the reader to be the same as a punching or slotting operation. The difference is

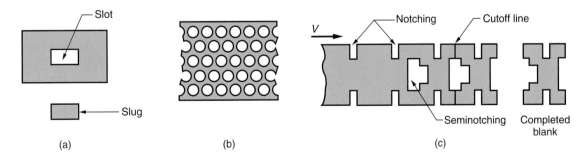

FIGURE 20.9   (a) Slotting, (b) perforating, (c) notching and seminotching. Symbol *v* indicates motion of strip.

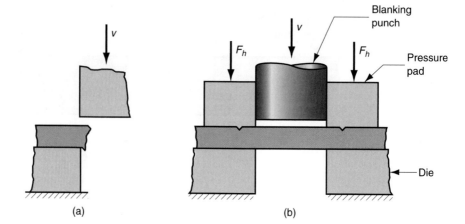

FIGURE 20.10
(a) Shaving and (b) fine blanking. Symbols: $v =$ motion of punch, $F_h =$ blank holding force.

(a)

(b)

that the metal removed by seminotching creates part of the blank outline, while punching and slotting create holes in the blank.

Trimming, Shaving, and Fine Blanking Trimming is a cutting operation performed on a formed part to remove excess metal and establish size. The term has the same basic meaning here as in forging (Section 19.4). A typical example in sheet metalwork is trimming the upper portion of a deep drawn cup to leave the desired dimensions on the cup.

Shaving is a shearing operation performed with very small clearance to obtain accurate dimensions and cut edges that are smooth and straight, as pictured in Figure 20.10(a). Shaving is typically performed as a secondary or finishing operation on parts that have been previously cut.

Fine blanking is a shearing operation used to blank sheet-metal parts with close tolerances and smooth, straight edges in one step, as illustrated in Figure 20.10(b). At the start of the cycle, a pressure pad with a V-shaped projection applies a holding force F_h against the work adjacent to the punch in order to compress the metal and prevent distortion. The punch then descends with a slower-than-normal velocity and smaller clearances to provide the desired dimensions and cut edges. The process is usually reserved for relatively small stock thicknesses.

20.2 BENDING OPERATIONS

Bending in sheet-metal work is defined as the straining of the metal around a straight axis, as in Figure 20.11. During the bending operation, the metal on the inside of the neutral plane is compressed, while the metal on the outside of the neutral plane is stretched. These strain conditions can be seen in Figure 20.11(b). The metal is plastically deformed so that the bend takes a permanent set upon removal of the stresses that caused it. Bending produces little or no change in the thickness of the sheet metal.

20.2.1 V-BENDING AND EDGE BENDING

Bending operations are performed using punch and die tooling. The two common bending methods and associated tooling are V-bending, performed with a V-die; and edge bending, performed with a wiping die. These methods are illustrated in Figure 20.12.

FIGURE 20.11
(a) Bending of sheet metal;
(b) both compression and tensile elongation of the metal occur in bending.

In *V-bending,* the sheet metal is bent between a V-shaped punch and die. Included angles ranging from very obtuse to very acute can be made with V-dies. V-bending is generally used for low-production operations. It is often performed on a press brake (Section 20.5.2), and the associated V-dies are relatively simple and inexpensive.

Edge bending involves cantilever loading of the sheet metal. A pressure pad is used to apply a force F_h to hold the base of the part against the die, while the punch forces the part to yield and bend over the edge of the die. In the setup shown in Figure 20.12(b), edge bending is limited to bends of 90° or less. More complicated wiping dies can be designed for bend angles greater than 90°. Because of the pressure pad, wiping dies are more complicated and costly than V-dies and are generally used for high-production work.

20.2.2 ENGINEERING ANALYSIS OF BENDING

Some of the important terms in sheet-metal bending are identified in Figure 20.11. The metal of thickness t is bent through an angle called the bend angle α. This results in a sheet-metal part with an included angle α', where $\alpha + \alpha' = 180°$. The bend radius R is normally specified on the inside of the part, rather than at the neutral axis, and is determined by the radius on the tooling used to perform the operation. The bend is made over the width of the workpiece w.

Bend Allowance If the bend radius is small relative to stock thickness, the metal tends to stretch during bending. It is important to be able to estimate the amount of stretching

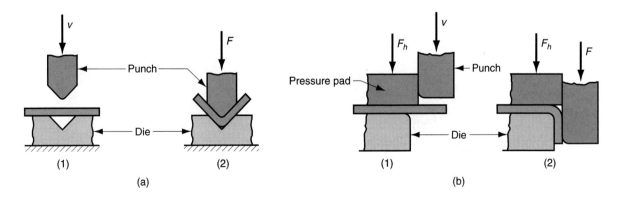

FIGURE 20.12 Two common bending methods: (a) V-bending and (b) edge bending; (1) before and (2) after bending. Symbols: v = motion, F = applied bending force, F_h = blank.

that occurs, if any, so that the final part length will match the specified dimension. The problem is to determine the length of the neutral axis before bending to account for stretching of the final bent section. This length is called the **bend allowance,** and it can be estimated as follows:

$$A_b = 2\pi \frac{\alpha}{360}(R + K_{ba}t) \tag{20.6}$$

where A_b = bend allowance, mm (in); α = bend angle, degrees; R = bend radius, mm (in); t = stock thickness, mm (in); and K_{ba} is factor to estimate stretching. The following design values are recommended for K_{ba} [3]: if $<2t$, $K_{ba} = 0.33$; and if $R \geq 2t$, $K_{ba} = 0.50$. The values of K_{ba} predict that stretching occurs only if bend radius is small relative to sheet thickness.

Springback When the bending pressure is removed at the end of the deformation operation, elastic energy remains in the bent part, causing it to recover partially toward its original shape. This elastic recovery is called **springback,** defined as the increase in included angle of the bent part relative to the included angle of the forming tool after the tool is removed. This is illustrated in Figure 20.13 and is expressed:

$$SB = \frac{\alpha' - \alpha'_t}{\alpha'_t} \tag{20.7}$$

where SB = springback; α' = included angle of the sheet-metal part, degrees; and α'_t = included angle of the bending tool, degrees. Although not as obvious, an increase in the bend radius also occurs due to elastic recovery. The amount of springback increases with modulus of elasticity E and yield strength Y of the work metal.

Compensation for springback can be accomplished by several methods. Two common methods are overbending and bottoming. In **overbending,** the punch angle and radius are fabricated slightly smaller than the specified angle on the final part so that the sheet metal springs back to the desired value. **Bottoming** involves squeezing the part at the end of the stroke, thus plastically deforming it in the bend region.

Bending Force The force required to perform bending depends on the geometry of the punch-and-die and the strength, thickness, and length of the sheet metal. The maximum

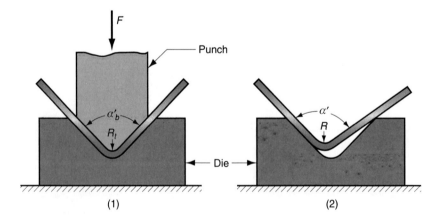

FIGURE 20.13 Springback in bending shows itself as a decrease in bend angle and an increase in bend radius: (1) during the operation, the work is forced to take the radius R_t and included angle α'_t = determined by the bending tool (punch in V-bending); (2) after the punch is removed, the work springs back to radius R and included angle α'. Symbol: F = applied bending force.

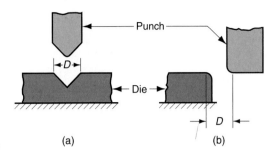

FIGURE 20.14 Die opening dimension D: (a) V-die, (b) wiping die.

bending force can be estimated by means of the following equation:

$$F = \frac{K_{bf}(TS)wt^2}{D} \qquad (20.8)$$

where F = bending force, N (lb); TS = tensile strength of the sheet metal, MPa (lb/in^2); w = width of part in the direction of the bend axis, mm (in); t = stock thickness, mm (in); and D = die opening dimension as defined in Figure 20.14, mm (in). Eq. (20.8) is based on bending of a simple beam in mechanics, and K_{bf} is a constant that accounts for differences encountered in an actual bending process. Its value depends on type of bending: for V-bending, $K_{bf} = 1.33$; and for edge bending, $K_{bf} = 0.33$.

**Example 20.2
Sheet-Metal
Bending**

A sheet-metal blank is to be bent as shown in Figure 20.15. The metal has a modulus of elasticity = 205 (10^3) MPa, yield strength = 275 MPa, and tensile strength = 450 MPa. Determine (a) the starting blank size and (b) the bending force if a V-die is used with a die opening dimension = 25 mm.

Solution: (a) The starting blank = 44.5 mm wide. Its length = $38 + A_b + 25$ (mm). For the included angle $\alpha' = 120°$, the bend angle $\alpha = 60°$. The value of K_{ba} in Eq. (20.6) = 0.33 since $R/t = 4.75/3.2 = 1.48$ (less than 2.0).

$$A_b = 2\pi \frac{60}{360}(4.75 + 0.33 \times 3.2) = 6.08 \text{ mm}$$

Length of the blank is therefore $38 + 6.08 + 25 = 69.08$ mm.
(b) Force is obtained from Eq. (20.8) using $K_{bf} = 1.33$.

$$F = \frac{1.33(450)(44.5)(3.2)^2}{2.5} = 10,909 \text{ N}$$

20.2.3 OTHER BENDING AND FORMING OPERATIONS

Some sheet-metal operations involve bending over a curved axis rather than a straight axis, or they have other features that differentiate them from the bending operations described above.

FIGURE 20.15 Sheet-metal part of Example 20.220.2 (dimensions in mm).

FIGURE 20.16
Flanging: (a) straight flanging, (b) stretch flanging, and (c) shrink flanging.

(a)   (b)   (c)

FIGURE 20.17
(a) Hemming, (b) seaming, and (c) curling.

(a)   (b)   (c)

Flanging, Hemming, Seaming, and Curling Flanging is a bending operation in which the edge of a sheet-metal part is bent at a 90° angle (usually) to form a rim or flange. It is often used to strengthen or stiffen sheet metal. The flange can be formed over a straight bend axis, as illustrated in Figure 20.16(a), or it can involve some stretching or shrinking of the metal, as in (b) and (c).

Hemming involves bending the edge of the sheet over on itself, in more than one bending step. This is often done to eliminate the sharp edge on the piece, to increase stiffness, and to improve appearance. *Seaming* is a related operation in which two sheet-metal edges are assembled. Hemming and seaming are illustrated in Figure 20.17(a) and (b).

Curling, also called *beading,* forms the edges of the part into a roll or curl, as in Figure 20.17(c). As in hemming, it is done for purposes of safety, strength, and aesthetics. Examples of products in which curling is used include hinges, pots and pans, and pocket-watch cases. These examples show that curling can be performed over straight or curved bend axes.

Miscellaneous Bending Operations Various other bending operations are depicted in Figure 20.18 to illustrate the variety of shapes that can be bent. Most of these operations are performed in relatively simple dies similar to V-dies.

20.3 DRAWING

Drawing is a sheet-metal-forming operation used to make cup-shaped, box-shaped, or other complex-curved and concave parts. It is performed by placing a piece of sheet metal over a die cavity and then pushing the metal into the opening with a punch, as in Figure 20.19. The blank must usually be held down flat against the die by a blankholder. Common parts made by drawing include beverage cans, ammunition shells, sinks, cooking pots, and automobile body panels.

20.3.1 MECHANICS OF DRAWING

Drawing of a cup-shaped part is the basic drawing operation, with dimensions and parameters as pictured in Figure 20.19. A blank of diameter D_b is drawn into a die cavity by means of a punch with diameter D_p. The punch and die must have corner radii, given by

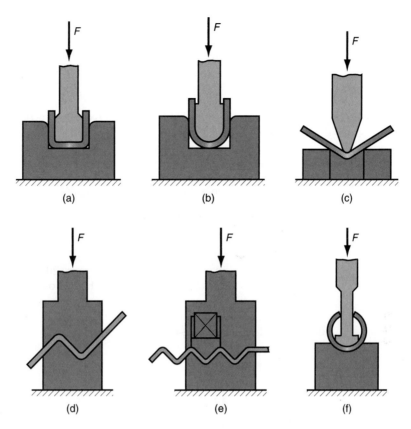

FIGURE 20.18
Miscellaneous bending operations: (a) channel bending, (b) U-bending, (c) air bending, (d) offset bending, (e) corrugating, and (f) tube forming. Symbol: F = applied force.

FIGURE 20.19
(a) Drawing of a cup-shaped part: (1) start of operation before punch contacts work, and (2) near end of stroke; and (b) corresponding workpart: (1) starting blank, and (2) drawn part. Symbols: c = clearance, D_b = blank diameter, D_p = punch diameter, R_d = die corner radius, R_p = punch corner radius, F = drawing force, F_h = holding force.

R_p and R_d. If the punch and die were to have sharp corners (R_p and $R_d = 0$), a hole-punching operation (and not a very good one) would be accomplished rather than a drawing operation. The sides of the punch and die are separated by a clearance c. This clearance in drawing is about 10% greater than the stock thickness:

$$c = 1.1\,t \tag{20.9}$$

The punch applies a downward force F to accomplish the deformation of the metal, and a downward holding force F_h is applied by the blankholder, as shown in the sketch.

As the punch proceeds downward toward its final bottom position, the work experiences a complex sequence of stresses and strains as it is gradually formed into the shape defined by the punch and die cavity. The stages in the deformation process are illustrated in Figure 20.20. As the punch first begins to push into the work, the metal is subjected to a **bending** operation. The sheet is simply bent over the corner of the punch and the corner of the die, as in Figure 20.20(2). The outside perimeter of the blank moves in toward the center in this first stage, but only slightly.

As the punch moves further down, a **straightening** action occurs in the metal that was previously bent over the die radius, as in Figure 20.20(3). The metal at the bottom of the cup, as well as along the punch radius, has been moved downward with the punch, but the metal that was bent over the die radius must now be straightened in order to be pulled into the clearance to form the wall of the cylinder. At the same time, more metal must be added to replace that being used in the cylinder wall. This new metal comes from the

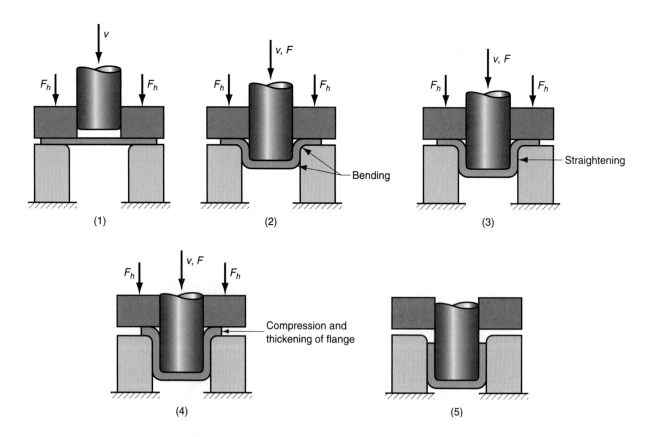

FIGURE 20.20   Stages in deformation of the work in deep drawing: (1) punch makes initial contact with work, (2) bending, (3) straightening, (4) friction and compression, and (5) final cup shape showing effects of thinning in the cup walls. Symbols: v = motion of punch, F = punch force, F_h = blankholder force.

outside edge of the blank. The metal in the outer portions of the blank is pulled or **drawn** toward the die opening to resupply the previously bent and straightened metal now forming the cylinder wall. This type of metal flow through a constricted space gives the drawing process its name.

During this stage of the process, friction and compression play important roles in the flange of the blank. In order for the material in the flange to move toward the die opening, **friction** between the sheet metal and the surfaces of the blankholder and the die must be overcome. Initially, static friction is involved until the metal starts to slide; then, after metal flow begins, dynamic friction governs the process. The magnitude of the holding force applied by the blankholder, as well as the friction conditions at the two interfaces, are determining factors in the success of this aspect of the drawing operation. Lubricants or drawing compounds are generally used to reduce friction forces. In addition to friction, **compression** is also occurring in the outer edge of the blank. As the metal in this portion of the blank is drawn toward the center, the outer perimeter becomes smaller. Because the volume of metal remains constant, the metal is squeezed and becomes thicker as the perimeter is reduced. This often results in wrinkling of the remaining flange of the blank, especially when thin sheet metal is drawn, or when the blankholder force is too low. It is a condition which cannot be corrected once it has occurred. The friction and compression effects are illustrated in Figure 20.20(4).

The holding force applied by the blankholder is now seen to be a critical factor in deep drawing. If it is too small, wrinkling occurs. If it is too large, it prevents the metal from flowing properly toward the die cavity, resulting in stretching and possible tearing of the sheet metal. Determining the proper holding force involves a delicate balance between these opposing factors.

Progressive downward motion of the punch results in a continuation of the metal flow caused by drawing and compression. In addition, some **thinning** of the cylinder wall occurs, as in Figure 20.20(5). The force being applied by the punch is opposed by the metal in the form of deformation and friction in the operation. A portion of the deformation involves stretching and thinning of the metal as it is pulled over the edge of the die opening. Up to 25% thinning of the side wall may occur in a successful drawing operation, mostly near the base of the cup.

20.3.2 ENGINEERING ANALYSIS OF DRAWING

It is important to assess the limitations on the amount of drawing that can be accomplished. This is often guided by simple measures that can be readily calculated for a given operation. In addition, drawing force and holding force are important process variables. Finally, the starting blank size must be determined.

Measures of Drawing   One of the measures of the severity of a deep drawing operation is the **drawing ratio** DR. This is most easily defined for a cylindrical shape as the ratio of blank diameter D_b to punch diameter D_p. In equation form,

$$DR = \frac{D_b}{D_p} \qquad (20.10)$$

The drawing ratio provides an indication, albeit a crude one, of the severity of a given drawing operation. The greater the ratio, the more severe the operation. An approximate upper limit on the drawing ratio is a value of 2.0. The actual limiting value for a given operation depends on punch and die corner radii (R_p and R_d), friction conditions, depth of draw, and characteristics of the sheet metal (e.g., ductility, degree of directionality of strength properties in the metal).

Another way to characterize a given drawing operation is by the **reduction** r, where

$$r = \frac{D_b - D_p}{D_b} \tag{20.11}$$

It is very closely related to drawing ratio. Consistent with the previous limit on DR ($DR \leq 2.0$), the value of reduction r should be less than 0.50.

A third measure in deep drawing is the **thickness-to-diameter ratio** t/D_b (thickness of the starting blank t divided by the blank diameter D_b). Often expressed as a percentage, it is desirable for the t/D_b ratio to be greater than 1%. As t/D_b decreases, tendency for wrinkling (Section 20.3.4) increases.

In cases where these limits on drawing ratio, reduction, and t/D_b ratio are exceeded by the design of the drawn part, the blank must be drawn in two or more steps, sometimes with annealing between the steps.

Example 20.3 Cup Drawing

A drawing operation is used to form a cylindrical cup with inside diameter = 75 mm and height = 50 mm. The starting blank size = 138 mm and the stock thickness = 2.4 mm. Based on these data, is the operation feasible?

Solution: To assess feasibility, we determine the drawing ratio, reduction, and thickness-to-diameter ratio.

$$DR = 138/75 = 1.84$$
$$r = (138 - 75)/138 = 0.4565 = 45.65\%$$
$$t/D_b = 2.4/138 = 0.017 = 1.7\%$$

According to these measures, the drawing operation is feasible. The drawing ratio is less than 2.0, the reduction is less than 50%, and the t/D_b ratio is greater than 1%. These are general guidelines frequently used to indicate technical feasibility.   ∎

Forces   The **drawing force** required to perform a given operation can be estimated roughly by the formula:

$$F = \pi D_p t (TS) \left(\frac{D_b}{D_p} - 0.7 \right) \tag{20.12}$$

where F = drawing force, N (lb); t = original blank thickness, mm (in); TS = tensile strength, MPa (lb/in^2); and D_b and D_p are the starting blank diameter and punch diameter, respectively, mm (in). The constant 0.7 is a correction factor to account for friction. Eq. (20.12) estimates the maximum force in the operation. The drawing force varies throughout the downward movement of the punch, usually reaching its maximum value at about one-third the length of the punch stroke.

The **holding force** is an important factor in a drawing operation. As a rough approximation, the holding pressure can be set at a value = 0.015 of the yield strength of the sheet metal [8]. This value is then multiplied by that portion of the starting area of the blank that is to be held by the blankholder. In equation form,

$$F_h = 0.015 Y \pi \left\{ D_b^2 - (D_p + 2.2t + 2R_d)^2 \right\} \tag{20.13}$$

where F_h = holding force in drawing, N (lb); Y = yield strength of the sheet metal, MPa (lb/in^2); t = starting stock thickness, mm (in); R_d = die corner radius, mm (in); and the other terms have been previously defined. The holding force is usually about one-third the drawing force [10].

**Example 20.4
Forces in
Drawing**

For the drawing operation of Example 20.3, determine (a) drawing force and (b) holding force, given that the tensile strength of the sheet metal (low-carbon steel) = 300 MPa and yield strength = 175 MPa. The die corner radius = 6 mm.

Solution: (a) Maximum drawing force is given by Eq. (20.12):

$$F = \pi(75)(2.4)(300)\left(\frac{138}{75} - 0.7\right) = 193,396 \text{ N}$$

(b) Holding force is estimated by Eq. (20.13):

$$F_h = 0.015(175)\,\pi(138^2 - (75 + 2.2 \times 2.4 + 2 \times 6)^2) = 86,824 \text{ N}$$

Blank Size Determination   For the final dimensions to be achieved on the cylindrical drawn shape, the correct starting blank diameter is needed. It must be large enough to supply sufficient metal to complete the cup. Yet if there is too much material, unnecessary waste will result. For drawn shapes other than cylindrical cups, the same problem of estimating the starting blank size exists, only the shape of the blank may be other than round.

The following is a reasonable method for estimating the starting blank diameter in a deep drawing operation that produces a round part (e.g., cylindrical cup and more complex shapes so long as they are axisymmetric). Because the volume of the final product is the same as that of the starting sheet-metal blank, then the blank diameter can be calculated by setting the initial blank volume equal to the final volume of the product and solving for diameter D_b. To facilitate the calculation, it is often assumed that negligible thinning of the part wall occurs.

20.3.3   OTHER DRAWING OPERATIONS

Our discussion has focused on a conventional cup-drawing operation that produces a simple cylindrical shape in a single step and uses a blankholder to facilitate the process. Let us consider some of the variations of this basic operation.

Redrawing   If the shape change required by the part design is too severe (drawing ratio is too high), complete forming of the part may require more than one drawing step. The second drawing step, and any further drawing steps if needed, are referred to as *redrawing.* A redrawing operation is illustrated in Figure 20.21.

When the part design indicates a drawing ratio that is too large to form the part in a single step, the following is a general guide to the amount of reduction that can be taken in

FIGURE 20.21
Redrawing of a cup:
(1) start of redraw, and
(2) end of stroke. Symbols:
v = punch velocity, F =
applied punch force, F_h =
blankholder force.

(1)   (2)

FIGURE 20.22 Reverse drawing: (1) start and (2) completion. Symbols: v = punch velocity, F = applied punch force, F_h = blankholder force.

(1)   (2)

each drawing operation [10]: For the first draw, the maximum reduction of the starting blank should be 40% to 45%; for the second draw (first redraw), the maximum reduction should be 30%; and for the third draw (second redraw), the maximum reduction should be 16%.

A related operation is **reverse drawing,** in which a drawn part is positioned face down on the die so that the second drawing operation produces a configuration such as that shown in Figure 20.22. Although it may seem that reverse drawing would produce a more severe deformation than redrawing, it is actually easier on the metal. The reason is that the sheet metal is bent in the same direction at the outside and inside corners of the die in reverse drawing; while in redrawing the metal is bent in the opposite directions at the two corners. Because of this difference, the metal experiences less strain hardening in reverse drawing and the drawing force is lower.

Drawing of Shapes Other than Cylindrical Cups Many products require drawing of shapes other than cylindrical cups. The variety of drawn shapes include square or rectangular boxes (as in sinks), stepped cups, cones, cups with spherical rather than flat bases, and irregular curved forms (as in automobile body panels). Each of these shapes presents unique technical problems in drawing. Eary and Reed [2] provide a detailed discussion of the drawing of these kinds of shapes.

Drawing Without a Blankholder One of the primary functions of the blankholder is to prevent wrinkling of the flange while the cup is being drawn. The tendency for wrinkling is reduced as the thickness-to-diameter ratio of the blank increases. If the t/D_b ratio is large enough, drawing can be accomplished without a blankholder, as in Figure 20.23.

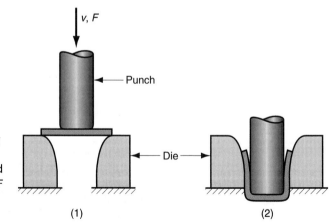

FIGURE 20.23 Drawing without a blankholder: (1) start of process, (2) end of stroke. Symbols v and F indicate motion and applied force, respectively.

(1)   (2)

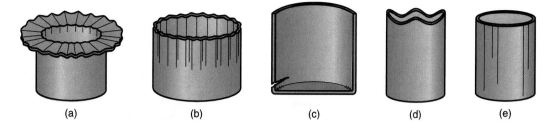

FIGURE 20.24   Common defects in drawn parts: (a) wrinkling can occur either in the flange or (b) in the wall, (c) tearing, (d) earing, and (e) surface scratches.

The limiting condition for drawing without a blankholder can be estimated from the following [5]:

$$D_b - D_p < 5t \tag{20.14}$$

The draw die must have the shape of a funnel or cone to permit the material to be drawn properly into the die cavity. When drawing without a blankholder is feasible, it has the advantages of lower cost tooling and a simpler press, because the need to separately control the movements of the blankholder and punch can be avoided.

20.3.4 DEFECTS IN DRAWING

Sheet-metal drawing is a more complex operation than cutting or bending, and more things can go wrong. A number of defects can occur in a drawn product, some of which we have already alluded to. Following is a list of common defects, with sketches in Figure 20.24:

(a) *Wrinkling in the flange.* Wrinkling in a drawn part consists of a series of ridges that form radially in the undrawn flange of the workpart due to compressive buckling.

(b) *Wrinkling in the wall.* If and when the wrinkled flange is drawn into the cup, these ridges appear in the vertical wall.

(c) *Tearing.* Tearing is an open crack in the vertical wall, usually near the base of the drawn cup, due to high tensile stresses that cause thinning and failure of the metal at this location. This type of failure can also occur as the metal is pulled over a sharp die corner.

(d) *Earing.* This is the formation of irregularities (called *ears*) in the upper edge of a deep drawn cup, caused by anisotropy in the sheet metal. If the material is perfectly isotropic, ears do not form.

(e) *Surface scratches.* Surface scratches can occur on the drawn part if the punch and die are not smooth or if lubrication is insufficient.

20.4 OTHER SHEET-METAL-FORMING OPERATIONS

In addition to bending and drawing, several other sheet-metal-forming operations can be accomplished on conventional presses. We classify these as (1) operations performed with metal tooling and (2) operations performed with flexible rubber tooling.

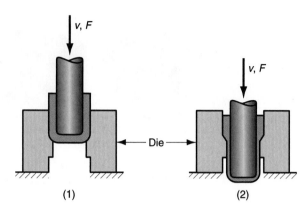

FIGURE 20.25 Ironing to achieve a more uniform wall thickness in a drawn cup: (1) start of process; (2) during process. Note thinning and elongation of walls. Symbols *v* and *F* indicate motion and applied force, respectively.

20.4.1 OPERATIONS PERFORMED WITH METAL TOOLING

Operations performed with metal tooling include (1) ironing, (2) coining and embossing, (3) lancing, and (4) twisting.

Ironing   In deep drawing the flange is compressed by the squeezing action of the blank perimeter seeking a smaller circumference as it is drawn toward the die opening. Because of this compression, the sheet metal near the outer edge of the blank becomes thicker as it moves inward. If the thickness of this stock is greater than the clearance between the punch and die, it will be squeezed to the size of the clearance, a process known as *ironing.*

Sometimes ironing is performed as a separate step that follows drawing. This case is illustrated in Figure 20.25. Ironing makes the cylindrical cup more uniform in wall thickness. The drawn part is therefore longer and more efficient in terms of material usage. Beverage cans and artillery shells, two very high-production items, include ironing among their processing steps to achieve economy in material usage.

Coining and Embossing   Coining is a bulk deformation operation discussed in the previous chapter. It is frequently used in sheet-metal work to form indentations and raised sections in the part. The indentations result in thinning of the sheet metal, and the raised sections result in thickening of the metal.

Embossing is a forming operation used to create indentations in the sheet, such as raised (or indented) lettering or strengthening ribs, as depicted in Figure 20.26. Some stretching and thinning of the metal are involved. This operation may seem similar to coining. However, embossing dies possess matching cavity contours, the punch containing the positive contour and the die containing the negative; whereas coining dies may have quite different cavities in the two die halves, thus causing more significant metal deformation than embossing.

FIGURE 20.26
Embossing: (a) cross section of punch and die configuration during pressing; (b) finished part with embossed ribs.

(a)

(b)

FIGURE 20.27 Lancing in several forms: (a) cutting and bending; (b) and (c) two types of cutting and forming.

(a) (b) (c)

Lancing Lancing is a combined cutting and bending or cutting and forming operation performed in one step to partially separate the metal from the sheet. Several examples are shown in Figure 20.27. Among other applications, lancing is used to make louvers in sheet-metal air vents for heating and air conditioning systems in buildings.

Twisting Twisting subjects the sheet metal to a torsion loading rather than a bending load, thus causing a twist in the sheet over its length. This type of operation has limited applications. It is used to make such products as fan and propeller blades. It can be performed in a conventional punch and die which has been designed to deform the part in the required twist shape.

20.4.2 RUBBER FORMING PROCESSES

The two operations discussed in this article are performed on conventional presses, but the tooling is unusual in that it uses a flexible element (made of rubber or similar material) to effect the forming operation. The operations are (1) the Guerin process, and (2) hydroforming.

Guerin Process The *Guerin process* uses a thick rubber pad (or other flexible material) to form sheet metal over a positive form block, as in Figure 20.28. The rubber pad is confined in a steel container. As the ram descends, the rubber gradually surrounds the sheet, applying pressure to deform it to the shape of the form block. It is limited to relatively

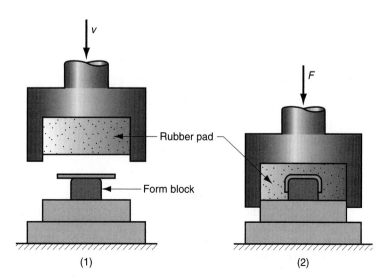

FIGURE 20.28 Guerin process: (1) before and (2) after. Symbols *v* and *F* indicate motion and applied force, respectively.

Rubber pad

Form block

(1) (2)

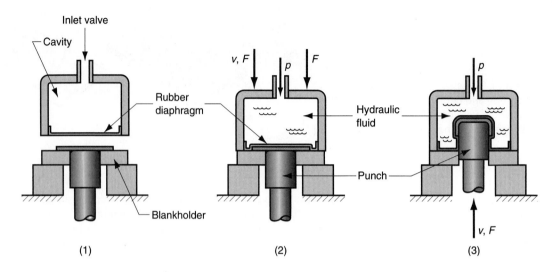

FIGURE 20.29   Hydroform process: (1) start-up, no fluid in cavity; (2) press closed, cavity pressurized with hydraulic fluid; (3) punch pressed into work to form part. Symbols: v = velocity, F = applied force, p = hydraulic pressure.

shallow forms, because the pressures developed by the rubber—up to about 10 MPa (1500 lb/in^2)—are not sufficient to prevent wrinkling in deeper formed parts.

The advantage of the Guerin process is the relatively low cost of the tooling. The form block can be made of wood, plastic, or other materials that are easy to shape, and the rubber pad can be used with different form blocks. These factors make rubber forming attractive in small-quantity production, such as the aircraft industry, where the process was developed.

Hydroforming   Hydroforming is similar to the Guerin process; the difference is that it substitutes a rubber diaphragm filled with hydraulic fluid in place of the thick rubber pad, as illustrated in Figure 20.29. This allows the pressure that forms the workpart to be increased—to around 100 MPa (15,000 lb/in^2)—thus preventing wrinkling in deep formed parts. In fact, deeper draws can be achieved with the hydroform process than with conventional deep drawing. This is because the uniform pressure in hydroforming forces the work to contact the punch throughout its length, thus increasing friction and reducing the tensile stresses that cause tearing at the base of the drawn cup.

20.5 DIES AND PRESSES FOR SHEET-METAL PROCESSES

In this section we examine the punch-and-die tooling and production equipment used in conventional sheet-metal processing.

20.5.1 DIES

Nearly all of the preceding pressworking operations are performed with conventional punch-and-die tooling. The tooling is referred to as a ***die.*** It is custom-designed for the particular part to be produced. The term ***stamping die*** is sometimes used for high-production dies. Typical materials for stamping dies are tool steel types D, A, O, and S (Table 6.5).

FIGURE 20.30
Components of a punch and die for a blanking operation.

Components of a Stamping Die   The components of a stamping die to perform a simple blanking operation are illustrated in Figure 20.30. The working components are the ***punch*** and ***die,*** which perform the cutting operation. They are attached to the upper and lower portions of the ***die set,*** respectively called the ***punch holder*** (or ***upper shoe***) and ***die holder*** (***lower shoe***). The die set also includes guide pins and bushings to ensure proper alignment between the punch and die during the stamping operation. The die holder is attached to the base of the press, and the punch holder is attached to the ram. Actuation of the ram accomplishes the pressworking operation.

In addition to these components, a die used for blanking or hole-punching must include a means of preventing the sheet metal from sticking to the punch when it is retracted upward after the operation. The newly created hole in the stock is the same size as the punch, and it tends to cling to the punch on its withdrawal. The device in the die that strips the sheet metal from the punch is called a ***stripper.*** It is often a simple plate attached to the die as in Figure 20.30, with a hole slightly larger than the punch diameter.

For dies that process strips or coils of sheet metal, a device is required to stop the sheet metal as it advances through the die between press cycles. That device is called (try to guess) a ***stop.*** Stops range from simple solid pins located in the path of the strip to block its forward motion, to more complex mechanisms synchronized to rise and retract with the actuation of the press. The simpler stop is shown in Figure 20.30.

There are other components in pressworking dies, but the preceding description provides an introduction to the terminology.

Types of Stamping Dies   Aside from differences in stamping dies related to the operations they perform (e.g., cutting, bending, drawing), other differences deal with the number of separate operations to be performed in each press actuation and how they are accomplished.

The type of die considered above performs a single blanking operation with each stroke of the press and is called a ***simple die.*** Other dies that perform a single operation include V-dies (Section 20.2.1). More complicated pressworking dies include compound dies, combination dies, and progressive dies. A ***compound die*** performs two operations at a single station, such as blanking and punching, or blanking and drawing [2]. A good example is a compound die that blanks and punches a washer. A ***combination die*** is less common; it performs two operations at two different stations in the die. Examples of applications include blanking two different parts (e.g., right-hand and left-hand parts), or blanking and then bending the same part [2].

A ***progressive die*** performs two or more operations on a sheet-metal coil at two or more stations with each press stroke. The part is fabricated progressively. The coil is fed

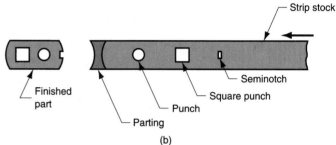

FIGURE 20.31
(a) Progressive die and
(b) associated strip
development.

from one station to the next and different operations (e.g., punching, notching, bending, and blanking) are performed at each station. When the part exits the final station it has been completed and separated (cut) from the remaining coil. Design of a progressive die begins with the layout of the part on the strip or coil and the determination of which operations are to be performed at each station. The result of this procedure is called the ***strip development.*** A progressive die and associated strip development are illustrated in Figure 20.31. Progressive dies can have a dozen or more stations. They are the most complicated and most costly stamping dies, economically justified only for complex parts requiring multiple operations at high-production rates.

20.5.2 PRESSES

A press used for sheet metalworking is a machine tool with a stationary ***bed*** and a powered ***ram*** (or ***slide***) that can be driven toward and away from the bed to perform various cutting and forming operations. A typical press, with principal components labeled, is diagrammed in Figure 20.32. The relative positions of the bed and ram are established by the ***frame,*** and the ram is driven by mechanical or hydraulic power. When a die is mounted in the press, the punch holder is attached to the ram, and the die holder is attached to a ***bolster plate*** of the press bed.

Presses are available in a variety of capacities, power systems, and frame types. The capacity of a press is its ability to deliver the required force and energy to accomplish the stamping operation. This is determined by the physical size of the press and by its power system. The power system refers to whether mechanical or hydraulic power is used and the type of drive used to transmit the power to the ram. Production rate is another important aspect of capacity. Type of frame refers to the physical construction of the press. There are two frame types in common use: gap frame and straight-sided frame.

Gap Frame Presses   The ***gap frame*** has the general configuration of the letter C and is often referred to as a ***C-frame.*** Gap frame presses provide good access to the die, and

FIGURE 20.32 Components of a typical (mechanical drive) stamping press.

they are usually open in the back to permit convenient ejection of stampings or scrap. The principal types of gap frame press are (a) solid gap frame, (b) adjustable bed, (c) open-back inclinable, (d) press brake, and (e) turret press.

The *solid gap frame* (sometimes called simply a *gap press*) has one-piece construction, as shown in Figure 20.32. Presses with this frame are rigid, yet the C-shape allows convenient access from the sides for feeding strip or coil stock. They are available in a range of sizes, with capacities up to around 9000 kN (1000 tons). The model shown in Figure 20.33 has a capacity of 1350 kN (150 tons). The *adjustable bed frame* press is a variation of the gap frame, in which an adjustable bed is added to accommodate various die sizes. The adjustment feature

FIGURE 20.33 Gap frame press for sheet metalworking. (Photo courtesy of E. W. Bliss Company, Hastings, Michigan.). Capacity = 1350 kN (150 tons).

FIGURE 20.34 Press brake with bed width of 9.15 m (30 ft) and capacity of 11,200 kN (1250 tons); two workers are shown positioning plate stock for bending. (Photo courtesy of Niagara Machine & Tool Works, Buffalo, New York.)

results in some sacrifice of tonnage capacity. The ***open-back inclinable*** press has a C-frame assembled to a base in such a way that the frame can be tilted back to various angles so that the stampings fall through the rear opening by gravity. Capacities of open-back inclinable presses range between 1 ton and around 2250 kN (250 tons). They can be operated at high speeds—up to around 1000 strokes per minute.

The ***press brake*** is a gap frame press with a very wide bed. The model in Figure 20.34 has a bed width of 9.15 m (30 ft). This allows a number of separate dies (simple V-bending dies are typical) to be set up in the bed, so that small quantities of stampings can be made economically. These low quantities of parts, sometimes requiring multiple bends at different angles, necessitate a manual operation. For a part requiring a series of bends, the operator moves the starting piece of sheet metal through the desired sequence of bending dies, actuating the press at each die, to complete the work needed.

Whereas press brakes are well adapted to bending operations, ***turret presses*** are suited to situations in which a sequence of punching, notching, and related cutting operations must be accomplished on sheet-metal parts, as in Figure 20.35. Turret presses have a C-frame, although this construction is not obvious in Figure 20.36. The conventional ram and punch is replaced by a turret containing many punches of different sizes and shapes. The turret works by indexing (rotating) to the position holding the punch to perform the required operation. Beneath the punch turret is a corresponding die turret that positions the die opening for each punch. Between the punch and die is the sheet-metal blank, held by an $x-y$ positioning system that operates by computer numerical control (Section 38.3). The blank is moved to the required coordinate position for each cutting operation.

Straight-sided Frame Presses For jobs requiring high tonnage, press frames with greater structural rigidity are needed. Straight-sided presses have full sides, giving it a

FIGURE 20.35 Several sheet-metal parts produced on a turret press, showing variety of possible hole shapes. (Photo courtesy of Strippet, Inc., Akron, New York.)

FIGURE 20.36 Computer numerical control turret press. (Photo courtesy of Strippet, Inc., Akron, New York.)

FIGURE 20.37 Straight-sided frame press. (Photo courtesy Greenerd Press & Machine Company, Inc., Nashua, New Hampshire.)

box-like appearance as in Figure 20.37. This construction increases the strength and stiffness of the frame. As a result, capacities up to 35,000 kN (4000 tons) are available in straight-sided presses for sheet metalwork. Large presses of this frame type are used for forging (Section 19.3).

In all of these presses, gap frame and straight-sided frame, the size is closely correlated to tonnage capacity. Larger presses are built to withstand higher forces in pressworking. Press size is also related to the speed at which it can operate. Smaller presses are generally capable of higher production rates than larger presses.

Power and Drive Systems Power systems on presses are either hydraulic or mechanical. *Hydraulic presses* use a large piston and cylinder to drive the ram. This power system typically provides longer ram strokes than mechanical drives and can develop the full tonnage force throughout the entire stroke. However, it is slower. Its application for sheet metal is normally limited to deep drawing and other forming operations where these load-stroke characteristics are advantageous. These presses are available with one or more independently operated slides, called single action (single slide), double action (two slides), and so on. Double-action presses are useful in deep drawing operations where it is required to separately control the punch force and the blank-holder force.

There are several types of drive mechanisms used on *mechanical presses.* These include eccentric, crankshaft, and knuckle joint, illustrated in Figure 20.38. They convert the rotational motion of a drive motor into the linear motion of the ram. A *flywheel* is used to store the energy of the drive motor for use in the stamping operation. Mechanical presses using these drives achieve very high forces at the bottom of their strokes, and are therefore quite suited to blanking and punching operations. The knuckle joint delivers very high force when it bottoms, and is therefore often used in coining operations.

FIGURE 20.38   Types of drives for sheet-metal presses: (a) eccentric, (b) crankshaft, and (c) knuckle joint.

20.6   SHEET-METAL OPERATIONS NOT PERFORMED ON PRESSES

A number of sheet-metal operations are not performed on conventional stamping presses. In this section we examine several of these processes: (1) stretch forming, (2) roll bending and forming, (3) spinning, and (4) high-energy-rate forming processes.

20.6.1   STRETCH FORMING

Stretch forming is a sheet-metal deformation process in which the sheet metal is intentionally stretched and simultaneously bent in order to achieve shape change. The process is illustrated in Figure 20.39 for a relatively simple and gradual bend. The workpart is gripped by one or more jaws on each end and then stretched and bent over a positive die containing the desired form. The metal is stressed in tension to a level above its yield point. When the tension loading is released, the metal has been plastically deformed. The combination of stretching and bending results in relatively little springback in the part. An estimate of the force required in stretch forming can be obtained by multiplying the cross-sectional area of the sheet in the direction of pulling by the flow stress of the metal. In equation form,

$$F = LtY_f \tag{20.15}$$

where F = stretching force, N (lb); L = length of the sheet in the direction perpendicular to stretching, mm (in); t = instantaneous stock thickness, mm (in); and Y_f = flow stress of the work metal, MPa (lb/in^2). The die force F_{die} shown in the figure can be determined by balancing vertical force components.

FIGURE 20.39   Stretch forming: (1) start of process; (2) form die is pressed into the work with force F_{die}, causing it to be stretched and bent over the form. F = stretching force.

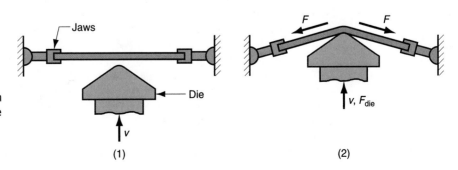

More complex contours than that shown in our figure are possible by stretch forming, but there are limitations on how sharp the curves in the sheet can be. Stretch forming is widely used in the aircraft and aerospace industries to economically produce large sheet-metal parts in the low quantities characteristic of those industries.

20.6.2 ROLL BENDING AND ROLL FORMING

FIGURE 20.40
Roll bending.

The operations described in this section use rolls to form sheet metal. ***Roll bending*** is an operation in which (usually) large sheet-metal or plate-metal parts are formed into curved sections by means of rolls. One possible arrangement of the rolls is pictured in Figure 20.40. As the sheet passes between the rolls, the rolls are brought toward each other to a configuration that achieves the desired radius of curvature on the work. Components for large storage tanks and pressure vessels are fabricated by roll bending. The operation can also be used to bend structural shapes, railroad rails, and tubes.

A related operation is ***roll straightening*** in which nonflat sheets (or other cross-sectional forms) are straightened by passing them between a series of rolls. The rolls subject the work to a sequence of decreasing small bends in opposite directions, thus causing it to be straight at the exit.

Roll forming (also called ***contour roll forming***) is a continuous bending process in which opposing rolls are used to produce long sections of formed shapes from coil or strip stock. Several pairs of rolls are usually required to progressively accomplish the bending of the stock into the desired shape. The process is illustrated in Figure 20.41 for a U-shaped section. Products made by roll forming include channels, gutters, metal siding sections (for homes), pipes and tubing with seams, and various structural sections. Although roll forming has the general appearance of a rolling operation (and the tooling certainly looks similar), the difference is that roll forming involves bending rather than compressing the work.

20.6.3 SPINNING

Spinning is a metal-forming process in which an axially symmetric part is gradually shaped over a mandrel or form by means of a rounded tool or roller. The tool or roller applies a very localized pressure (almost a point contact) to deform the work by axial and radial motions over the surface of the part. Basic geometric shapes typically produced by spinning include cups, cones, hemispheres, and tubes. There are three types of spinning operations: (1) conventional spinning, (2) shear spinning, and (3) tube spinning.

Conventional Spinning Conventional spinning is the basic spinning operation. As illustrated in Figure 20.42, a sheet-metal disk is held against the end of a rotating mandrel

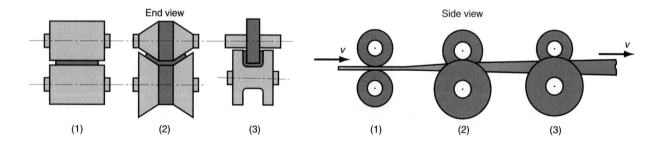

FIGURE 20.41 Roll forming of a continuous channel section: (1) straight rolls, (2) partial form, and (3) final form.

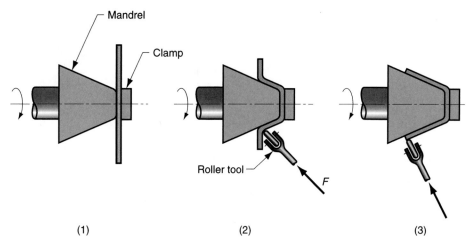

FIGURE 20.42
Conventional spinning:
(1) setup at start of process;
(2) during spinning; and
(3) completion of process.

of the desired inside shape of the final part, while the tool or roller deforms the metal against the mandrel. In some cases, the starting workpart is other than a flat disk. The process requires a series of steps, as indicated in the figure, to complete the shaping of the part. The tool position is controlled either by a human operator, using a fixed fulcrum to achieve the required leverage, or by an automatic method such as numerical control. These alternatives are **manual spinning** and **power spinning.** Power spinning has the capability to apply higher forces to the operation, resulting in faster cycle times and greater work size capacity. It also achieves better process control than manual spinning.

Conventional spinning bends the metal around a moving circular axis to conform to the outside surface of the axisymmetric mandrel. The thickness of the metal therefore remains unchanged (more or less) relative to the starting disk thickness. The diameter of the disk must therefore be somewhat larger than the diameter of the resulting part. The required starting diameter can be figured by assuming constant volume, before and after spinning.

Applications of conventional spinning include production of conical and curved shapes in low quantities. Very large diameter parts—up to 5 m (15 ft) or more—can be made by spinning. Alternative sheet-metal processes would require excessively high die costs. The form mandrel in spinning can be made of wood or other soft materials that are easy to shape. It is therefore a low-cost tool compared to the punch and die required for deep drawing, which might be a substitute process for some parts.

Shear Spinning   In shear spinning, the part is formed over the mandrel by a shear deformation process in which the outside diameter remains constant and the wall thickness is therefore reduced, as in Figure 20.43. This shear straining (and consequent thinning of the metal) distinguishes this process from the bending action in conventional spinning. Several other names have been used for shear spinning, including *flow turning,* *shear forming,* and *spin forging.* The process has been applied in the aerospace industry to form large parts such as rocket nose cones.

For the simple conical shape in our figure, the resulting thickness of the spun wall can be readily determined by the sine law relationship:

$$t_f = t \sin \alpha \tag{20.16}$$

where t_f = the final thickness of the wall after spinning, t = the starting thickness of the disk, and α = the mandrel angle (actually the half angle). Thinning is sometimes quantified by the spinning reduction r:

$$r = \frac{t - t_f}{t} \tag{20.17}$$

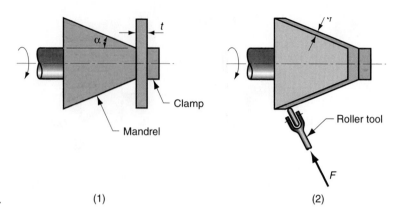

FIGURE 20.43 Shear spinning: (1) setup and (2) completion of process.

(1)   (2)

There are limits to the amount of thinning that the metal will endure in a spinning operation before fracture occurs. The maximum reduction correlates well with reduction of area in a tension test [8].

Tube Spinning Tube spinning is used to reduce the wall thickness and increase the length of a tube by means of a roller applied to the work over a cylindrical mandrel, as in Figure 20.44. Tube spinning is similar to shear spinning except that the starting workpiece is a tube rather than a flat disk. The operation can be performed by applying the roller against the work externally (using a cylindrical mandrel on the inside of the tube) or internally (using a die to surround the tube). It is also possible to form profiles in the walls of the cylinder, as in Figure 20.44(c), by controlling the path of the roller as it moves tangentially along the wall.

     Spinning reduction for a tube-spinning operation that produces a wall of uniform thickness can be determined as in shear spinning by Eq. (20.17).

20.6.4 HIGH-ENERGY-RATE FORMING

Several processes have been developed to form metals using large amounts of energy applied in a very short time. Owing to this feature, these operations are called ***high-***

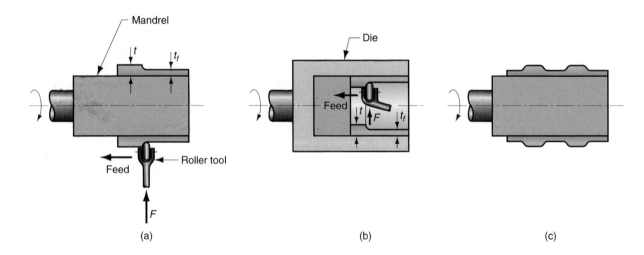

(a)   (b)   (c)

FIGURE 20.44 Tube spinning: (a) external; (b) internal; and (c) profiling.

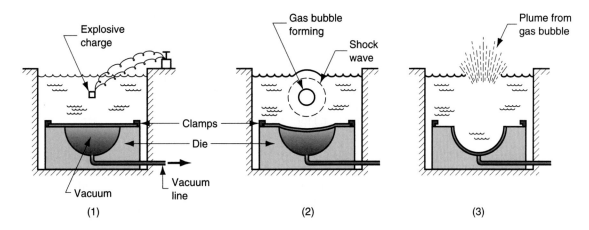

FIGURE 20.45   Explosive forming: (1) setup, (2) explosive is detonated, and (3) shock wave forms part and plume escapes water surface.

energy-rate forming (HERF) processes. They include explosive forming, electrohydraulic forming, and electromagnetic forming.

Explosive Forming   Explosive forming involves the use of an explosive charge to form sheet (or plate) metal into a die cavity. One method of implementing the process is illustrated in Figure 20.45. The workpart is clamped and sealed over the die, and a vacuum is created in the cavity beneath. The apparatus is then placed in a large vessel of water. An explosive charge is placed in the water at a certain distance above the work. Detonation of the charge results in a shock wave whose energy is transmitted by the water to cause rapid forming of the part into the cavity. The size of the explosive charge and the distance at which it is placed above the part are largely a matter of art and experience. Explosive forming is reserved for large parts, typical of the aerospace industry.

Electrohydraulic Forming   Electrohydraulic forming is a HERF process in which a shock wave to deform the work into a die cavity is generated by the discharge of electrical energy between two electrodes submerged in a transmission fluid (water). Owing to its principle of operation, this process is also called *electric discharge forming.* The setup for the process is illustrated in Figure 20.46. Electrical energy is accumulated in large capacitors and then released to the electrodes. Electrohydraulic forming is similar to explosive forming. The difference is in the method of generating the energy and the smaller amounts of energy that are released. This limits electrohydraulic forming to much smaller part sizes.

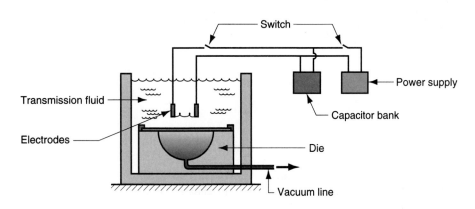

FIGURE 20.46
Electrohydraulic forming
setup.

FIGURE 20.47 Electromagnetic forming: (1) setup in which coil is inserted into tubular workpart surrounded by die; (2) formed part.

(1)　　　　　　　(2)

Electromagnetic Forming Electromagnetic forming, also called ***magnetic pulse forming,*** is a process in which sheet metal is deformed by the mechanical force of an electromagnetic field induced in the workpart by an energized coil. The coil, energized by a capacitor, produces a magnetic field. This generates eddy currents in the work that produce their own magnetic field. The induced field opposes the primary field, producing a mechanical force that deforms the part into the surrounding cavity. Developed in the 1960s, electromagnetic forming is the most widely used HERF process [10]. It is typically used to form tubular parts, as illustrated in Figure 20.47.

20.7 BENDING OF TUBE STOCK

Several methods of producing tubes and pipes are discussed in the previous chapter, and tube spinning is described in Section 20.6.3. In this section, we examine methods by which tubes are bent and otherwise formed. Bending of tube stock is more difficult than sheet stock because a tube tends to collapse and fold when attempts are made to bend it. Special flexible mandrels are usually inserted into the tube prior to bending to support the walls during the operation.

Some of the terms in tube bending are defined in Figure 20.48. The radius of the bend R is defined with respect to the centerline of the tube. When the tube is bent, the wall on the inside of the bend is in compression, and the wall at the outside is in tension. These stress conditions cause thinning and elongation of the outer wall and thickening and shortening of the inner wall. As a result, there is a tendency for the inner and outer walls to be forced toward each other to cause the cross section of the tube to flatten. Because of this flattening tendency, the minimum bend radius R that the tube can be bent is about 1.5 times the diameter D when a mandrel is used and 3.0 times D when no mandrel is used [10]. The exact value depends on the wall factor WF, which is the diameter D divided by wall thickness t. Higher values of WF increase the minimum bend

FIGURE 20.48 Dimensions and terms for a bent tube: $D =$ outside diameter of tube, $R =$ bend radius, $t =$ wall thickness.

FIGURE 20.49 Tube bending methods: (a) stretch bending, (b) draw bending, and (c) compression bending. For each method: (1) start of process, and (2) during bending. Symbols *v* and *F* indicate motion and applied force, respectively.

radius; that is, tube bending is more difficult for thin walls. Ductility of the work material is also an important factor in the process.

Several methods to bend tubes (and similar sections) are illustrated in Figure 20.49. ***Stretch bending*** is accomplished by pulling and bending the tube around a fixed form block, as in Figure 20.49(a). ***Draw bending*** is performed by clamping the tube against a form block, and then pulling the tube through the bend by rotating the block as in (b). A pressure bar is used to support the work as it is being bent. In ***compression bending,*** a wiper shoe is used to wrap the tube around the contour of a fixed form block, as in (c). ***Roll bending*** (Section 20.6.2), generally associated with the forming of sheet stock, is also used for bending tubes and other cross sections.

REFERENCES

[1] ***ASM Handbook,*** Vol. 14B, ***Metalworking: Sheet Forming.*** ASM International, Materials Park, Ohio, 2006.

[2] Eary, D. F., and Reed, E. A. ***Techniques of Press-working Sheet Metal,*** 2nd ed. Prentice-Hall, Inc., Englewood Cliffs, New Jersey, 1974.

[3] Hoffman, E. G. *Fundamentals of Tool Design,* 2nd ed. Society of Manufacturing Engineers, Dearborn, Michigan, 1984.

[4] Hosford, W. F., and Caddell, R. M. *Metal Forming: Mechanics and Metallurgy,* 3rd ed. Cambridge University Press, Cambridge, UK, 2007.

[5] Kalpakjian, S. *Manufacturing Processes for Engineering Materials,* 4th ed. Prentice Hall/Pearson, Upper Saddle River, New Jersey, 2003.

[6] Lange, K., et al. (eds.). *Handbook of Metal Forming.* Society of Manufacturing Engineers, Dearborn, Michigan, 1995.

[7] Mielnik, E. M. *Metalworking Science and Engineering.* McGraw-Hill, Inc., New York, 1991.

[8] Schey, J. A. *Introduction to Manufacturing Processes,* 3rd ed. McGraw-Hill Book Company, New York, 2000.

[9] Spitler, D., Lantrip, J., Nee, J., and Smith, D. A. *Fundamentals of Tool Design,* 5th ed. Society of Manufacturing Engineers, Dearborn, Michigan, 2003.

[10] Wick, C., et al. (eds.). *Tool and Manufacturing Engineers Handbook,* 4th ed. Vol. II, *Forming.* Society of Manufacturing Engineers, Dearborn, Michigan, 1984.

REVIEW QUESTIONS

20.1. Identify the three basic types of sheet metalworking operations.

20.2. In conventional sheet metalworking operations, (a) what is the name of the tooling and (b) what is the name of the machine tool used in the operations?

20.3. In blanking of a circular sheet-metal part, is the clearance applied to the punch diameter or the die diameter?

20.4. What is the difference between a cutoff operation and a parting operation?

20.5. What is the difference between a notching operation and a seminotching operation?

20.6. Describe each of the two types of sheet-metal-bending operations: V-bending and edge bending.

20.7. For what is the bend allowance intended to compensate?

20.8. What is springback in sheet-metal bending?

20.9. Define drawing in the context of sheet metalworking.

20.10. What are some of the simple measures used to assess the feasibility of a proposed cup-drawing operation?

20.11. Distinguish between redrawing and reverse drawing.

20.12. What are some of the possible defects in drawn sheet-metal parts?

20.13. What is an embossing operation?

20.14. What is stretch forming?

20.15. Identify the principal components of a stamping die that performs blanking.

20.16. What are the two basic categories of structural frames used in stamping presses?

20.17. What are the relative advantages and disadvantages of mechanical presses versus hydraulic presses in sheet metalworking?

20.18. What is the Guerin process?

20.19. Identify a major technical problem in tube bending.

20.20. Distinguish between roll bending and roll forming.

20.21. (**Video**) According to the video on sheet-metal shearing, what is the blade rake angle?

20.22. (**Video**) According to the video on sheet-metal bending, what are the principal terms used to describe bending on a press brake?

20.23. (**Video**) According to the video on sheet-metal stamping dies and processes, what are the factors that affect the formability of a metal?

20.24. (**Video**) Name the four forming processes listed in the video clip on sheet-metal stamping dies and processes.

20.25. (**Video**) List the factors that affect the hold down pressure in a drawing operation according to the video on sheet-metal stamping dies and processes.

MULTIPLE CHOICE QUIZ

There are 21 correct answers in the following multiple choice questions (some questions have multiple answers that are correct). To attain a perfect score on the quiz, all correct answers must be given. Each correct answer is worth 1 point. Each omitted answer or wrong answer reduces the score by 1 point, and each additional answer beyond the correct number of answers reduces the score by 1 point. Percentage score on the quiz is based on the total number of correct answers.

20.1. Most sheet metalworking operations are performed as which one of the following: (a) cold working, (b) hot working, or (c) warm working?

20.2. In a sheet-metal-cutting operation used to produce a flat part with a hole in the center, the part itself is called a blank, and the scrap piece that was cut out to make the hole is called a slug: (a) true or (b) false?

20.3. As sheet-metal stock hardness increases in a blanking operation, the clearance between punch and die should be (a) decreased, (b) increased, or (c) remain the same?

20.4. A circular sheet-metal slug produced in a hole punching operation will have the same diameter as (a) the die opening or (b) the punch?

20.5. The cutting force in a sheet-metal blanking operation depends on which mechanical property of the metal (one correct answer): (a) compressive strength, (b) modulus of elasticity, (c) shear strength, (d) strain rate, (e) tensile strength, or (f) yield strength?

20.6. Which of the following descriptions applies to a V-bending operation as compared to an edge-bending operation (two best answers): (a) costly tooling, (b) inexpensive tooling, (c) limited to 90° bends or less, (d) used for high production, (e) used for low production, and (f) uses a pressure pad to hold down the sheet metal?

20.7. Sheet-metal bending involves which of the following stresses and strains (two correct answers): (a) compressive, (b) shear, and (c) tensile?

20.8. Which one of the following is the best definition of bend allowance: (a) amount by which the die is larger than the punch, (b) amount of elastic recovery experienced by the metal after bending, (c) safety factor used in calculating bending force, or (d) length before bending of the straight sheet-metal section to be bent?

20.9. Springback in a sheet-metal-bending operation is the result of which one of the following: (a) elastic modulus of the metal, (b) elastic recovery of the metal, (c) overbending, (d) overstraining, or (e) yield strength of the metal?

20.10. Which of the following are variations of sheet metal-bending operations (two best answers): (a) coining, (b) flanging, (c) hemming, (d) ironing, (e) notching, (f) shear spinning, (g) trimming, and (h) tube bending?

20.11. The following are measures of feasibility for several proposed cup-drawing operations; which of the operations are likely to be feasible (three best answers): (a) $DR = 1.7$, (b) $DR = 2.7$, (c) $r = 0.35$, (d) $r = 0.65$, and (e) $t/D = 2\%$?

20.12. The holding force in drawing is most likely to be (a) greater than, (b) equal to, or (c) less than the maximum drawing force?

20.13. Which one of the following stamping dies is the most complicated: (a) blanking die, (b) combination die, (c) compound die, (d) edge-bending die, (e) progressive die, or (f) V-bending die?

20.14. Which one of the following press types is usually associated with the highest production rates in sheet-metal-stamping operations: (a) adjustable bed, (b) open-back inclinable, (c) press brake, (d) solid gap, or (e) straight-sided?

20.15. Which of the following processes are classified as high-energy-rate forming processes (two best answers): (a) electrochemical machining, (b) electromagnetic forming, (c) electron beam cutting, (d) explosive forming, (e) Guerin process, (f) hydroforming, (g) redrawing, and (h) shear spinning?

PROBLEMS

Cutting Operations

20.1. A power shears is used to cut soft cold-rolled steel that is 4.75 mm thick. At what clearance should the shears be set to yield an optimum cut?

20.2. A blanking operation is to be performed on 2.0-mm thick cold-rolled steel (half hard). The part is circular with diameter = 75.0 mm. Determine the appropriate punch and die sizes for this operation.

20.3. A compound die will be used to blank and punch a large washer out of 6061ST aluminum alloy sheet stock 3.50 mm thick. The outside diameter of the washer is 50.0 mm and the inside diameter is 15.0 mm. Determine (a) the punch and die sizes for the blanking operation, and (b) the punch and die sizes for the punching operation.

20.4. A blanking die is to be designed to blank the part outline shown in Figure P20.4. The material is 4-mm

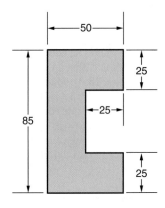

FIGURE P20.4 Blanked part for Problem 20.4 (dimensions in mm).

thick stainless steel (half hard). Determine the dimensions of the blanking punch and the die opening.

20.5. Determine the blanking force required in Problem 20.2, if the shear strength of the steel = 325 MPa and the tensile strength is 450 MPa.

20.6. Determine the minimum tonnage press to perform the blanking and punching operation in Problem 20.3. The aluminum sheet metal has a tensile strength = 310 MPa, a strength coefficient of 350 MPa, and a strain-hardening exponent of 0.12. (a) Assume that blanking and punching occur simultaneously. (b) Assume the punches are staggered so that punching occurs first, then blanking.

20.7. Determine the tonnage requirement for the blanking operation in Problem 20.4, given that the stainless steel has a yield strength = 500 MPa, a shear strength = 600 MPa, and a tensile strength = 700 MPa.

20.8. The foreman in the pressworking section comes to you with the problem of a blanking operation that is producing parts with excessive burrs. (a) What are the possible reasons for the burrs? (b) What can be done to correct the condition?

Bending

20.9. A bending operation is to be performed on 5.00-mm thick cold-rolled steel. The part drawing is given in Figure P20.9. Determine the blank size required.

FIGURE P20.9 Part in bending operation of Problem 20.9 (dimensions in mm).

20.10. Solve Problem 20.9 except that the bend radius R = 11.35 mm.

20.11. An L-shaped part is to be bent in a V-bending operation on a press brake from a flat blank 4.0 in by 1.5 in that is 5/32 in thick. The bend of 90° is to be made in the middle of the 4.0 in length. (a) Determine the dimensions of the two equal sides that will result after the bend, if the bend radius = 3/16 in. For convenience, these sides should be measured to the beginning of the bend radius. (b) Also, determine the length of the

part's neutral axis after the bend. (c) Where should the machine operator set the stop on the press brake relative to the starting length of the part?

20.12. A bending operation is to be performed on 4.0-mm thick cold-rolled steel sheet that is 25 mm wide and 100 mm long. The sheet is bent along the 25 mm direction, so that the bend is 25 mm long. The resulting sheet metal part has an acute angle of 30° and a bend radius of 6 mm. Determine (a) the bend allowance and (b) the length of the neutral axis of the part after the bend. (Hint: the length of the neutral axis before the bend = 100.0 mm).

20.13. Determine the bending force required in Problem 20.9 if the bend is to be performed in a V-die with a die opening dimension of 40 mm. The material has a tensile strength of 600 MPa and a shear strength of 430 MPa.

20.14. Solve Problem 20.13 except that the operation is performed using a wiping die with die opening dimension = 28 mm.

20.15. Determine the bending force required in Problem 20.11 if the bend is to be performed in a V-die with a die opening width dimension = 1.25 in. The material has a tensile strength = 70,000 lb/in².

20.16. Solve Problem 20.15 except that the operation is performed using a wiping die with die opening dimension = 0.75 in.

20.17. A sheet-metal part 3.0 mm thick and 20.0 mm long is bent to an included angle = 60° and a bend radius = 7.5 mm in a V-die. The metal has a yield strength = 220 MPa and a tensile strength = 340 MPa. Compute the required force to bend the part, given that the die opening dimension = 15 mm.

Drawing Operations

20.18. Derive an expression for the reduction r in drawing as a function of drawing ratio DR.

20.19. A cup is to be drawn in a deep drawing operation. The height of the cup is 75 mm and its inside diameter =

100 mm. The sheet-metal thickness = 2 mm. If the blank diameter = 225 mm, determine (a) drawing ratio, (b) reduction, and (c) thickness-to-diameter ratio. (d) Does the operation seem feasible?

20.20. Solve Problem 20.19 except that the starting blank size diameter = 175 mm.

20.21. A deep drawing operation is performed in which the inside of the cylindrical cup has a diameter of 4.25 in and a height = 2.65 in. The stock thickness = 3/16 in, and the starting blank diameter = 7.7 in. Punch and die radii = 5/32 in. The metal has a tensile strength = 65,000 lb/in^2, a yield strength = 32,000 lb/in^2, and a shear strength of 40,000 lb/in^2. Determine (a) drawing ratio, (b) reduction, (c) drawing force, and (d) blankholder force.

20.22. Solve Problem 20.21 except that the stock thickness t = 1/8 in.

20.23. A cup-drawing operation is performed in which the inside diameter = 80 mm and the height = 50 mm. The stock thickness = 3.0 mm, and the starting blank diameter = 150 mm. Punch and die radii = 4 mm. Tensile strength = 400 MPa and yield strength = 180 MPa for this sheet metal. Determine (a) drawing ratio, (b) reduction, (c) drawing force, and (d) blankholder force.

20.24. A deep drawing operation is to be performed on a sheet-metal blank that is 1/8 thick. The height (inside dimension) of the cup = 3.8 in and the diameter (inside dimension) = 5.0 in. Assuming the punch radius = 0, compute the starting diameter of the blank to complete the operation with no material left in the flange. Is the operation feasible (ignoring the fact that the punch radius is too small)?

20.25. Solve Problem 20.24 except use a punch radius = 0.375 in.

20.26. A drawing operation is performed on 3.0 mm stock. The part is a cylindrical cup with height = 50 mm and inside diameter = 70 mm. Assume the corner radius on the punch is zero. (a) Find the required starting blank size D_b. (b) Is the drawing operation feasible?

20.27. Solve Problem 20.26 except that the height = 60 mm.

20.28. Solve Problem 20.27 except that the corner radius on the punch = 10 mm.

20.29. The foreman in the drawing section of the shop brings to you several samples of parts that have been drawn in the shop. The samples have various defects. One has ears, another has wrinkles, and still a third has torn sections at its base. What are the causes of each of these defects and what remedies would you propose?

20.30. A cup-shaped part is to be drawn without a blankholder from sheet metal whose thickness = 0.25 in. The inside diameter of the cup = 2.5 in, its height = 1.5 in, and the corner radius at the base = 0.375 in. (a) What is the minimum starting blank diameter that can be used, according to Eq. (20.14)? (b) Does this blank diameter provide sufficient material to complete the cup?

Other Operations

20.31. A 20-in-long sheet-metal workpiece is stretched in a stretch forming operation to the dimensions shown in Figure P20.31. The thickness of the beginning stock is 3/16 in and the width is 8.5 in. The metal has a flow curve defined by a strength coefficient of 75,000 lb/in^2 and a strain hardening exponent of 0.20. The yield strength of the material is 30,000 lb/in^2. (a) Find the stretching force F required near the beginning of the operation when yielding first occurs. Determine (b) true strain experienced by the metal, (c) stretching force F, and (d) die force F_{die} at the very end when the part is formed as indicated in Figure P20.31(b).

20.32. Determine the starting disk diameter required to spin the part in Figure P20.32 using a conventional spinning operation. The starting thickness = 2.4 mm.

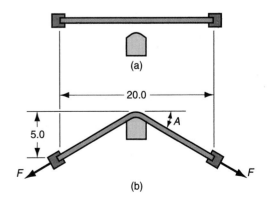

FIGURE P20.31   Stretch forming operation: (a) before, and (b) after (dimensions in inches).

FIGURE P20.32 Part (cross section) in conventional spinning (dimensions in mm).

20.33. If the part illustrated in Figure P20.32 were made by shear spinning, determine (a) the wall thickness along the cone-shaped portion, and (b) the spinning reduction r.

20.34. Determine the shear strain that is experienced by the material that is shear spun in Problem 20.33.

20.35. A 75-mm diameter tube is bent into a rather complex shape with a series of simple tube bending operations. The wall thickness on the tube = 4.75 mm. The tubes will be used to deliver fluids in a chemical plant. In one of the bends where the bend radius is 125 mm, the walls of the tube are flattening badly. What can be done to correct the condition?

Part VI   Material Removal Processes

THEORY OF METAL MACHINING

Chapter Contents

The *material removal processes* are a family of shaping operations (Figure 1.4) in which excess material is removed from a starting workpart so that what remains is the desired final geometry. The "family tree" is shown in Figure 21.1. The most important branch of the family is *conventional machining*, in which a sharp cutting tool is used to mechanically cut the material to achieve the desired geometry. The three principal machining processes are turning, drilling, and milling. The "other machining operations" in Figure 21.1 include shaping, planing, broaching, and sawing. This chapter begins our coverage of machining, which runs through Chapter 24.

Another group of material removal processes is the *abrasive processes,* which mechanically remove material by the action of hard, abrasive particles. This process group, which includes grinding, is covered in Chapter 25. The "other abrasive processes" in Figure 21.1 include honing, lapping, and superfinishing. Finally, there are the *nontraditional processes,* which use various energy forms other than a sharp cutting tool or abrasive particles to remove material. The energy forms include mechanical, electrochemical, thermal, and chemical.[1] The nontraditional processes are discussed in Chapter 26.

Machining is a manufacturing process in which a sharp cutting tool is used to cut away material to leave the

---

[1]Some of the mechanical energy forms in the nontraditional processes involve the use of abrasive particles, and so they overlap with the abrasive processes in Chapter 25.

FIGURE 21.1
Classification of material removal processes.

desired part shape. The predominant cutting action in machining involves shear deformation of the work material to form a chip; as the chip is removed, a new surface is exposed. Machining is most frequently applied to shape metals. The process is illustrated in the diagram of Figure 21.2.

Machining is one of the most important manufacturing processes. The Industrial Revolution and the growth of the manufacturing-based economies of the world can be traced largely to the development of the various machining operations (Historical Note 22.1). Machining is important commercially and technologically for several reasons:

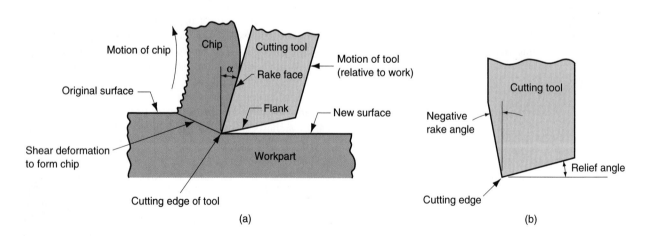

FIGURE 21.2 (a) A cross-sectional view of the machining process. (b) Tool with negative rake angle; compare with positive rake angle in (a).

> *Variety of work materials.* Machining can be applied to a wide variety of work materials. Virtually all solid metals can be machined. Plastics and plastic composites can also be cut by machining. Ceramics pose difficulties because of their high hardness and brittleness; however, most ceramics can be successfully cut by the abrasive machining processes discussed in Chapter 25.

> *Variety of part shapes and geometric features.* Machining can be used to create any regular geometries, such as flat planes, round holes, and cylinders. By introducing variations in tool shapes and tool paths, irregular geometries can be created, such as screw threads and T-slots. By combining several machining operations in sequence, shapes of almost unlimited complexity and variety can be produced.

> *Dimensional accuracy.* Machining can produce dimensions to very close tolerances. Some machining processes can achieve tolerances of ± 0.025 mm (± 0.001 in), much more accurate than most other processes.

> *Good surface finishes.* Machining is capable of creating very smooth surface finishes. Roughness values less than 0.4 microns (16 μ-in.) can be achieved in conventional machining operations. Some abrasive processes can achieve even better finishes.

On the other hand, certain disadvantages are associated with machining and other material removal processes:

> *Wasteful of material.* Machining is inherently wasteful of material. The chips generated in a machining operation are wasted material. Although these chips can usually be recycled, they represent waste in terms of the unit operation.

> *Time consuming.* A machining operation generally takes more time to shape a given part than alternative shaping processes such as casting or forging.

Machining is generally performed after other manufacturing processes such as casting or bulk deformation (e.g., forging, bar drawing). The other processes create the general shape of the starting workpart, and machining provides the final geometry, dimensions, and finish.

21.1 OVERVIEW OF MACHINING TECHNOLOGY

Machining is not just one process; it is a group of processes. The common feature is the use of a cutting tool to form a chip that is removed from the workpart. To perform the operation, relative motion is required between the tool and work. This relative motion is achieved in most machining operations by means of a primary motion, called the *cutting speed,* and a secondary motion, called the *feed.* The shape of the tool and its penetration into the work surface, combined with these motions, produces the desired geometry of the resulting work surface.

Types of Machining Operations There are many kinds of machining operations, each of which is capable of generating a certain part geometry and surface texture. We discuss these operations in considerable detail in Chapter 22, but for now it is appropriate to identify and define the three most common types: turning, drilling, and milling, illustrated in Figure 21.3.

In *turning,* a cutting tool with a single cutting edge is used to remove material from a rotating workpiece to generate a cylindrical shape, as in Figure 21.3(a). The speed motion in turning is provided by the rotating workpart, and the feed motion is achieved by the cutting tool moving slowly in a direction parallel to the axis of rotation of the workpiece. *Drilling* is used to create a round hole. It is accomplished by a rotating tool that typically has two

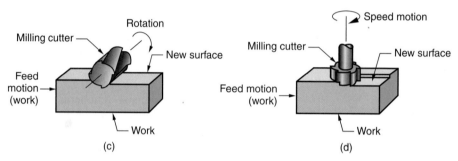

FIGURE 21.3   The three most common types of machining processes: (a) turning, (b) drilling, and two forms of milling: (c) peripheral milling, and (d) face milling.

cutting edges. The tool is fed in a direction parallel to its axis of rotation into the workpart to form the round hole, as in Figure 21.3(b). In **milling,** a rotating tool with multiple cutting edges is fed slowly across the work material to generate a plane or straight surface. The direction of the feed motion is perpendicular to the tool's axis of rotation. The speed motion is provided by the rotating milling cutter. The two basic forms of milling are peripheral milling and face milling, as in Figure 21.3(c) and (d).

Other conventional machining operations include shaping, planing, broaching, and sawing (Section 22.6). Also, grinding and similar abrasive operations are often included within the category of machining. These processes commonly follow the conventional machining operations and are used to achieve a superior surface finish on the workpart.

The Cutting Tool   A cutting tool has one or more sharp cutting edges and is made of a material that is harder than the work material. The cutting edge serves to separate a chip from the parent work material, as in Figure 21.2. Connected to the cutting edge are two surfaces of the tool: the rake face and the flank. The rake face, which directs the flow of the newly formed chip, is oriented at a certain angle called the **rake angle** α. It is measured relative to a plane perpendicular to the work surface. The rake angle can be positive, as in Figure 21.2(a), or negative as in (b). The flank of the tool provides a clearance between the tool and the newly generated work surface, thus protecting the surface from abrasion, which would degrade the finish. This flank surface is oriented at an angle called the **relief angle.**

Most cutting tools in practice have more complex geometries than those in Figure 21.2. There are two basic types, examples of which are illustrated in Figure 21.4: (a) single-point tools and (b) multiple-cutting-edge tools. A **single-point tool** has one cutting edge and is used for operations such as turning. In addition to the tool features shown in Figure 21.2, there is one tool point from which the name of this cutting tool is derived. During machining, the point of the tool penetrates below the original work surface of the part. The point is usually rounded to a certain radius, called the nose radius. **Multiple-cutting-edge tools** have more

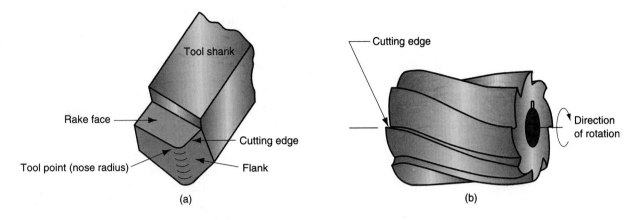

FIGURE 21.4   (a) A single-point tool showing rake face, flank, and tool point; and (b) a helical milling cutter, representative of tools with multiple cutting edges.

than one cutting edge and usually achieve their motion relative to the workpart by rotating. Drilling and milling use rotating multiple-cutting-edge tools. Figure 21.4(b) shows a helical milling cutter used in peripheral milling. Although the shape is quite different from a single-point tool, many elements of tool geometry are similar. Single-point and multiple-cutting-edge tools and the materials used in them are discussed in more detail in Chapter 23.

Cutting Conditions   Relative motion is required between the tool and work to perform a machining operation. The primary motion is accomplished at a certain ***cutting speed v***. In addition, the tool must be moved laterally across the work. This is a much slower motion, called the ***feed f***. The remaining dimension of the cut is the penetration of the cutting tool below the original work surface, called the ***depth of cut d***. Collectively, speed, feed, and depth of cut are called the ***cutting conditions***. They form the three dimensions of the machining process, and for certain operations (e.g., most single-point tool operations) they can be used to calculate the material removal rate for the process:

$$R_{MR} = vfd \qquad (21.1)$$

where R_{MR} = material removal rate, mm³/s (in³/min); v = cutting speed, m/s (ft/min), which must be converted to mm/s (in/min); f = feed, mm (in); and d = depth of cut, mm (in).

The cutting conditions for a turning operation are depicted in Figure 21.5. Typical units used for cutting speed are m/s (ft/min). Feed in turning is expressed in mm/rev

FIGURE 21.5   Cutting speed, feed, and depth of cut for a turning operation.

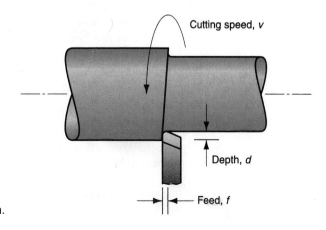

(in/rev), and depth of cut is expressed in mm (in). In other machining operations, interpretations of the cutting conditions may differ. For example, in a drilling operation, depth is interpreted as the depth of the drilled hole.

Machining operations usually divide into two categories, distinguished by purpose and cutting conditions: roughing cuts and finishing cuts. **Roughing** cuts are used to remove large amounts of material from the starting workpart as rapidly as possible, in order to produce a shape close to the desired form, but leaving some material on the piece for a subsequent finishing operation. **Finishing** cuts are used to complete the part and achieve the final dimensions, tolerances, and surface finish. In production machining jobs, one or more roughing cuts are usually performed on the work, followed by one or two finishing cuts. Roughing operations are performed at high feeds and depths—feeds of 0.4 to 1.25 mm/rev (0.015–0.050 in/rev) and depths of 2.5 to 20 mm (0.100–0.750 in) are typical. Finishing operations are carried out at low feeds and depths—feeds of 0.125 to 0.4 mm (0.005–0.015 in/rev) and depths of 0.75 to 2.0 mm (0.030–0.075 in) are typical. Cutting speeds are lower in roughing than in finishing.

A **cutting fluid** is often applied to the machining operation to cool and lubricate the cutting tool (cutting fluids are discussed in Section 23.4). Determining whether a cutting fluid should be used, and, if so, choosing the proper cutting fluid, is usually included within the scope of cutting conditions. Given the work material and tooling, the selection of these conditions is very influential in determining the success of a machining operation.

Machine Tools A machine tool is used to hold the workpart, position the tool relative to the work, and provide power for the machining process at the speed, feed, and depth that have been set. By controlling the tool, work, and cutting conditions, machine tools permit parts to be made with great accuracy and repeatability, to tolerances of 0.025 mm (0.001 in) and better. The term **machine tool** applies to any power-driven machine that performs a machining operation, including grinding. The term is also applied to machines that perform metal forming and pressworking operations (Chapters 19 and 20).

The traditional machine tools used to perform turning, drilling, and milling are lathes, drill presses, and milling machines, respectively. Conventional machine tools are usually tended by a human operator, who loads and unloads the workparts, changes cutting tools, and sets the cutting conditions. Many modern machine tools are designed to accomplish their operations with a form of automation called computer numerical control (Section 38.3).

21.2 THEORY OF CHIP FORMATION IN METAL MACHINING

The geometry of most practical machining operations is somewhat complex. A simplified model of machining is available that neglects many of the geometric complexities, yet describes the mechanics of the process quite well. It is called the **orthogonal** cutting model, Figure 21.6. Although an actual machining process is three-dimensional, the orthogonal model has only two dimensions that play active roles in the analysis.

21.2.1 THE ORTHOGONAL CUTTING MODEL

By definition, orthogonal cutting uses a wedge-shaped tool in which the cutting edge is perpendicular to the direction of cutting speed. As the tool is forced into the material, the chip is formed by shear deformation along a plane called the **shear plane,** which is oriented at an angle ϕ with the surface of the work. Only at the sharp cutting edge of the tool does failure of the material occur, resulting in separation of the chip from the parent

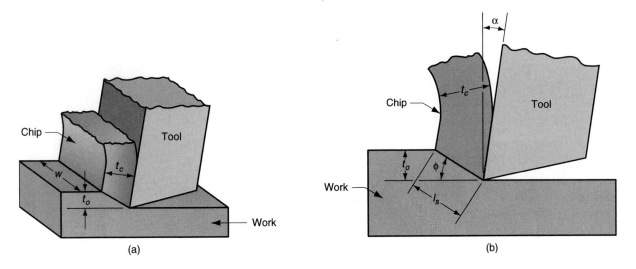

FIGURE 21.6   Orthogonal cutting: (a) as a three-dimensional process, and (b) how it reduces to two dimensions in the side view.

material. Along the shear plane, where the bulk of the mechanical energy is consumed in machining, the material is plastically deformed.

The tool in orthogonal cutting has only two elements of geometry: (1) rake angle and (2) clearance angle. As indicated previously, the rake angle α determines the direction that the chip flows as it is formed from the workpart; and the clearance angle provides a small clearance between the tool flank and the newly generated work surface.

During cutting, the cutting edge of the tool is positioned a certain distance below the original work surface. This corresponds to the thickness of the chip prior to chip formation, t_o. As the chip is formed along the shear plane, its thickness increases to t_c. The ratio of t_o to t_c is called the ***chip thickness ratio*** (or simply the ***chip ratio***) r:

$$r = \frac{t_o}{t_c} \tag{21.2}$$

Since the chip thickness after cutting is always greater than the corresponding thickness before cutting, the chip ratio will always be less than 1.0.

In addition to t_o, the orthogonal cut has a width dimension w, as shown in Figure 21.6(a), even though this dimension does not contribute much to the analysis in orthogonal cutting.

The geometry of the orthogonal cutting model allows us to establish an important relationship between the chip thickness ratio, the rake angle, and the shear plane angle. Let l_s be the length of the shear plane. We can make the substitutions: $t_o = l_s \sin\phi$, and $t_c = l_s \cos (\phi - \alpha)$. Thus,

$$r = \frac{l_s \sin \phi}{l_s \cos (\phi - \alpha)} = \frac{\sin \phi}{\cos (\phi - \alpha)}$$

This can be rearranged to determine ϕ as follows:

$$\tan \phi = \frac{r \cos \alpha}{1 - r \sin \alpha} \tag{21.3}$$

The shear strain that occurs along the shear plane can be estimated by examining Figure 21.7. Part (a) shows shear deformation approximated by a series of parallel plates sliding against one another to form the chip. Consistent with our definition of shear strain

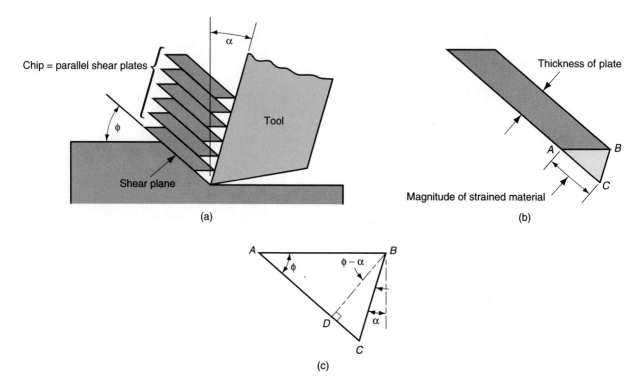

FIGURE 21.7   Shear strain during chip formation: (a) chip formation depicted as a series of parallel plates sliding relative to each other; (b) one of the plates isolated to illustrate the definition of shear strain based on this parallel plate model; and (c) shear strain triangle used to derive Eq. (21.4).

(Section 3.1.4), each plate experiences the shear strain shown in Figure 21.7(b). Referring to part (c), this can be expressed as

$$\gamma = \frac{AC}{BD} = \frac{AD + DC}{BD}$$

which can be reduced to the following definition of shear strain in metal cutting:

$$\gamma = \tan(\phi - \alpha) + \cot\phi \qquad (21.4)$$

**Example 21.1
Orthogonal
Cutting**

In a machining operation that approximates orthogonal cutting, the cutting tool has a rake angle $= 10°$. The chip thickness before the cut $t_o = 0.50$ mm and the chip thickness after the cut $t_c = 1.125$ in. Calculate the shear plane angle and the shear strain in the operation.

Solution: The chip thickness ratio can be determined from Eq. (21.2):

$$r = \frac{0.50}{1.125} = 0.444$$

The shear plane angle is given by Eq. (21.3):

$$\tan\phi = \frac{0.444\cos 10}{1 - 0.444\sin 10} = 0.4738$$

$$\phi = 25.4°$$

Finally, the shear strain is calculated from Eq. (21.4):

$$\gamma = \tan(25.4 - 10) + \cot 25.4$$
$$\gamma = 0.275 + 2.111 = 2.386$$

21.2.2 ACTUAL CHIP FORMATION

We should note that there are differences between the orthogonal model and an actual machining process. First, the shear deformation process does not occur along a plane, but within a zone. If shearing were to take place across a plane of zero thickness, it would imply that the shearing action must occur instantaneously as it passes through the plane, rather than over some finite (although brief) time period. For the material to behave in a realistic way, the shear deformation must occur within a thin shear zone. This more realistic model of the shear deformation process in machining is illustrated in Figure 21.8. Metal-cutting experiments have indicated that the thickness of the shear zone is only a few thousandths of an inch. Since the shear zone is so thin, there is not a great loss of accuracy in most cases by referring to it as a plane.

Second, in addition to shear deformation that occurs in the shear zone, another shearing action occurs in the chip after it has been formed. This additional shear is referred to as secondary shear to distinguish it from primary shear. Secondary shear results from friction between the chip and the tool as the chip slides along the rake face of the tool. Its effect increases with increased friction between the tool and chip. The primary and secondary shear zones can be seen in Figure 21.8.

Third, formation of the chip depends on the type of material being machined and the cutting conditions of the operation. Four basic types of chip can be distinguished, illustrated in Figure 21.9:

➤ **Discontinuous chip.** When relatively brittle materials (e.g., cast irons) are machined at low cutting speeds, the chips often form into separate segments (sometimes the segments are loosely attached). This tends to impart an irregular texture to the machined surface. High tool–chip friction and large feed and depth of cut promote the formation of this chip type.

➤ **Continuous chip.** When ductile work materials are cut at high speeds and relatively small feeds and depths, long continuous chips are formed. A good surface finish typically results when this chip type is formed. A sharp cutting edge on the tool and

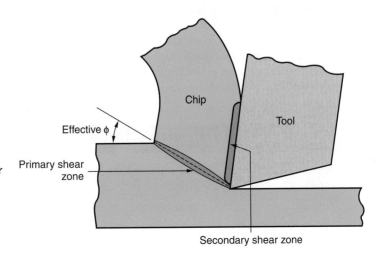

FIGURE 21.8 More realistic view of chip formation, showing shear zone rather than shear plane. Also shown is the secondary shear zone resulting from tool–chip friction.

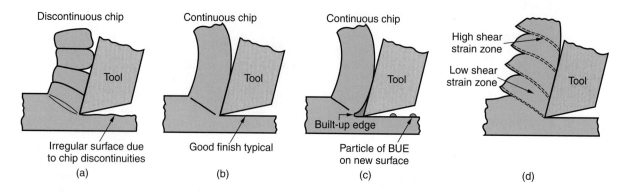

FIGURE 21.9 Four types of chip formation in metal cutting: (a) discontinuous, (b) continuous, (c) continuous with built-up edge, (d) serrated.

low tool–chip friction encourage the formation of continuous chips. Long, continuous chips (as in turning) can cause problems with regard to chip disposal and/or tangling about the tool. To solve these problems, turning tools are often equipped with chip breakers (Section 23.3.1).

➤ *Continuous chip with built-up edge.* When machining ductile materials at low-to-medium cutting speeds, friction between tool and chip tends to cause portions of the work material to adhere to the rake face of the tool near the cutting edge. This formation is called a built-up edge (BUE). The formation of a BUE is cyclical; it forms and grows, then becomes unstable and breaks off. Much of the detached BUE is carried away with the chip, sometimes taking portions of the tool rake face with it, which reduces the life of the cutting tool. Portions of the detached BUE that are not carried off with the chip become imbedded in the newly created work surface, causing the surface to become rough.

The preceding chip types were first classified by Ernst in the late 1930s [13]. Since then, the available metals used in machining, cutting tool materials, and cutting speeds have all increased, and a fourth chip type has been identified:

➤ *Serrated chips* (the term *shear-localized* is also used for this fourth chip type). These chips are semi-continuous in the sense that they possess a saw-tooth appearance that is produced by a cyclical chip formation of alternating high shear strain followed by low shear strain. This fourth type of chip is most closely associated with certain difficult-to-machine metals such as titanium alloys, nickel-base superalloys, and austenitic stainless steels when they are machined at higher cutting speeds. However, the phenomenon is also found with more common work metals (e.g., steels) when they are cut at high speeds [13].[2]

21.3 FORCE RELATIONSHIPS AND THE MERCHANT EQUATION

Several forces can be defined relative to the orthogonal cutting model. Based on these forces, shear stress, coefficient of friction, and certain other relationships can be defined.

---

[2]A more complete description of the serrated chip type can be found in Trent & Wright [12], pp. 348–367.

21.3.1 FORCES IN METAL CUTTING

Consider the forces acting on the chip during orthogonal cutting in Figure 21.10(a). The forces applied against the chip by the tool can be separated into two mutually perpendicular components: friction force and normal force to friction. The **_friction force F_** is the frictional force resisting the flow of the chip along the rake face of the tool. The **_normal force to friction N_** is perpendicular to the friction force. These two components can be used to define the coefficient of friction between the tool and the chip:

$$\mu = \frac{F}{N} \qquad (21.5)$$

The friction force and its normal force can be added vectorially to form a resultant force R, which is oriented at an angle β, called the friction angle. The friction angle is related to the coefficient of friction as

$$\mu = \tan \beta \qquad (21.6)$$

In addition to the tool forces acting on the chip, there are two force components applied by the workpiece on the chip: shear force and normal force to shear. The **_shear force F_s_** is the force that causes shear deformation to occur in the shear plane, and the **_normal force to shear F_n_** is perpendicular to the shear force. Based on the shear force, we can define the shear stress that acts along the shear plane between the work and the chip:

$$\tau = \frac{F_s}{A_s} \qquad (21.7)$$

where A_s = area of the shear plane. This shear plane area can be calculated as

$$A_s = \frac{t_o w}{\sin \phi} \qquad (21.8)$$

The shear stress in Eq. (21.7) represents the level of stress required to perform the machining operation. Therefore, this stress is equal to the shear strength of the work material ($\tau = S$) under the conditions at which cutting occurs.

Vector addition of the two force components F_s and F_n yields the resultant force R'. In order for the forces acting on the chip to be in balance, this resultant R' must be equal in magnitude, opposite in direction, and collinear with the resultant R.

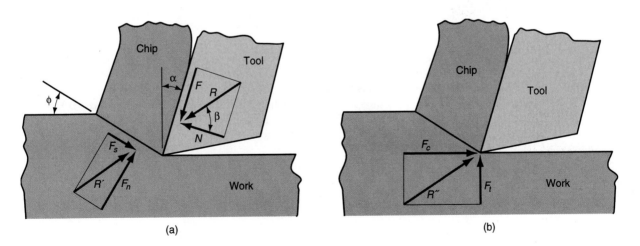

(a)   (b)

FIGURE 21.10   Forces in metal cutting: (a) forces acting on the chip in orthogonal cutting, and (b) forces acting on the tool that can be measured.

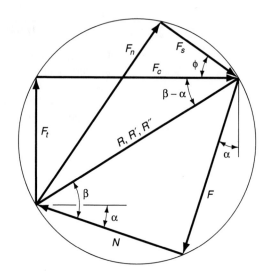

FIGURE 21.11 Force diagram showing geometric relationships between F, N, F_s, F_n, F_c, and F_t.

None of the four force components F, N, F_s, and F_n can be directly measured in a machining operation, because the directions in which they are applied vary with different tool geometries and cutting conditions. However, it is possible for the cutting tool to be instrumented using a force measuring device called a dynamometer, so that two additional force components acting against the tool can be directly measured: cutting force and thrust force. The **cutting force** F_c is in the direction of cutting, the same direction as the cutting speed v, and the **thrust force** F_t is perpendicular to the cutting force and is associated with the chip thickness before the cut t_o. The cutting force and thrust force are shown in Figure 21.10 (b) together with their resultant force R''. The respective directions of these forces are known, so the force transducers in the dynamometer can be aligned accordingly.

Equations can be derived to relate the four force components that cannot be measured to the two forces that can be measured. Using the force diagram in Figure 21.11, the following trigonometric relationships can be derived:

$$F = F_c \sin \alpha + F_t \cos \alpha \tag{21.9}$$

$$N = F_c \cos \alpha - F_t \sin \alpha \tag{21.10}$$

$$F_s = F_c \cos \phi - F_t \sin \phi \tag{21.11}$$

$$F_n = F_c \sin \phi + F_t \cos \phi \tag{21.12}$$

If cutting force and thrust force are known, these four equations can be used to calculate estimates of shear force, friction force, and normal force to friction. Based on these force estimates, shear stress and coefficient of friction can be determined.

Note that in the special case of orthogonal cutting when the rake angle $\alpha = 0$, Eqs. (21.9) and (21.10) reduce to $F = F_t$ and $N = F_c$, respectively. Thus, in this special case, friction force and its normal force could be directly measured by the dynamometer.

Example 21.2 Shear Stress in Machining

Suppose in Example 21.1 that cutting force and thrust force are measured during an orthogonal cutting operation: $F_c = 1559$ N and $F_t = 1271$ N. The width of the orthogonal cutting operation $w = 3.0$ mm. Based on these data, determine the shear strength of the work material.

Solution: From Example 21.1, rake angle $\alpha = 10°$, and shear plane angle $\phi = 25.4°$. Shear force can be computed from Eq. (21.11):

$$F_s = 1559 \cos 25.4 - 1271 \sin 25.4 = 863 \text{ N}$$

The shear plane area is given by Eq. (21.8):

$$A_s = \frac{(0.5)(3.0)}{\sin 25.4} = 3.497 \text{ mm}^2$$

Thus the shear stress, which equals the shear strength of the work material, is

$$\tau = S = \frac{863}{3.497} = 247 \text{ N/mm}^2 = 247 \text{ MPa}$$

This example demonstrates that cutting force and thrust force are related to the shear strength of the work material. The relationships can be established in a more direct way. Recalling from Eq. (21.7) that the shear force $F_s = S\,A_s$, the force diagram of Figure 21.11 can be used to derive the following equations:

$$F_c = \frac{St_o\,w\cos(\beta - \alpha)}{\sin\phi\cos(\phi + \beta - \alpha)} = \frac{F_s\cos(\beta - \alpha)}{\cos(\phi + \beta - \alpha)} \qquad (21.13)$$

and

$$F_t = \frac{S_t w\sin(\beta - \alpha)}{\sin\phi\cos(\phi + \beta - \alpha)} = \frac{F_s\sin(\beta - \alpha)}{\cos(\phi + \beta - \alpha)} \qquad (21.14)$$

These equations allow one to estimate cutting force and thrust force in an orthogonal cutting operation if the shear strength of the work material is known.

21.3.2 THE MERCHANT EQUATION

One of the important relationships in metal cutting was derived by Eugene Merchant [10]. Its derivation was based on the assumption of orthogonal cutting, but its general validity extends to three-dimensional machining operations. Merchant started with the definition of shear stress expressed in the form of the following relationship derived by combining Eqs. (21.7), (21.8), and (21.11):

$$\tau = \frac{F_c\cos\phi - F_t\sin\phi}{(t_o w/\sin\phi)} \qquad (21.15)$$

Merchant reasoned that, out of all the possible angles emanating from the cutting edge of the tool at which shear deformation could occur, there is one angle ϕ that predominates. This is the angle at which shear stress is just equal to the shear strength of the work material, and so shear deformation occurs at this angle. For all other possible shear angles, the shear stress is less than the shear strength, so chip formation cannot occur at these other angles. In effect, the work material will select a shear plane angle that minimizes energy. This angle can be determined by taking the derivative of the shear stress S in Eq. (21.15) with respect to ϕ and setting the derivative to zero. Solving for ϕ, we get the relationship named after Merchant:

$$\phi = 45 + \frac{\alpha}{2} - \frac{\beta}{2} \qquad (21.16)$$

Among the assumptions in the Merchant equation is that shear strength of the work material is a constant, unaffected by strain rate, temperature, and other factors. Because this assumption is violated in practical machining operations, Eq. (21.16) must be considered an approximate relationship rather than an accurate mathematical equation. Let us nevertheless consider its application in the following example.

Example 21.3 Estimating Friction Angle

Using the data and results from our previous examples, determine (a) the friction angle and (b) the coefficient of friction.

Solution: (a) From Example 21.1, $\alpha = 10°$, and $\phi = 25.4°$. Rearranging Eq. (21.16), the friction angle can be estimated:

$$\beta = 2\,(45) + 10 - 2\,(25.4) = 49.2°$$

(b) The coefficient of friction is given by Eq. (21.6):

$$\mu = \tan 49.2 = 1.16$$

Lessons Based on the Merchant Equation   The real value of the Merchant equation is that it defines the general relationship between rake angle, tool–chip friction, and shear plane angle. The shear plane angle can be increased by (1) increasing the rake angle and (2) decreasing the friction angle (and coefficient of friction) between the tool and the chip. Rake angle can be increased by proper tool design, and friction angle can be reduced by using a lubricant cutting fluid.

The importance of increasing the shear plane angle can be seen in Figure 21.12. If all other factors remain the same, a higher shear plane angle results in a smaller shear plane area. Since the shear strength is applied across this area, the shear force required to form the chip will decrease when the shear plane area is reduced. A greater shear plane angle results in lower cutting energy, lower power requirements, and lower cutting temperature. These are good reasons to try to make the shear plane angle as large as possible during machining.

Approximation of Turning by Orthogonal Cutting   The orthogonal model can be used to approximate turning and certain other single-point machining operations so long as the feed in these operations is small relative to depth of cut. Thus, most of the cutting will take place in the direction of the feed, and cutting on the point of the tool will be negligible. Figure 21.13 indicates the conversion from one cutting situation to the other.

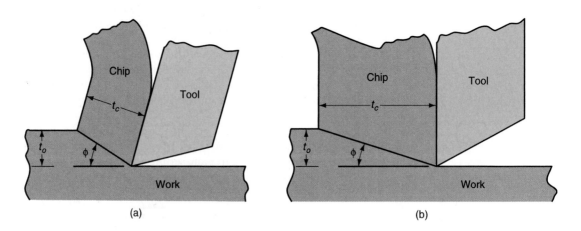

(a)   (b)

FIGURE 21.12   Effect of shear plane angle ϕ: (a) higher ϕ with a resulting lower shear plane area; (b) smaller ϕ with a corresponding larger shear plane area. Note that the rake angle is larger in (a), which tends to increase shear angle according to the Merchant equation.

FIGURE 21.13
Approximation of turning by the orthogonal model: (a) turning; and (b) the corresponding orthogonal cutting.

TABLE 21.1   Conversion key: turning operation vs. orthogonal cutting.

Turning Operation	Orthogonal Cutting Model
Feed $f =$	Chip thickness before cut t_o
Depth $d =$	Width of cut w
Cutting speed $v =$	Cutting speed v
Cutting force $F_c =$	Cutting force F_c
Feed force $F_f =$	Thrust force F_t

    The interpretation of cutting conditions is different in the two cases. The chip thickness before the cut t_o in orthogonal cutting corresponds to the feed f in turning, and the width of cut w in orthogonal cutting corresponds to the depth of cut d in turning. In addition, the thrust force F_t in the orthogonal model corresponds to the feed force F_f in turning. Cutting speed and cutting force have the same meanings in the two cases. Table 21.1 summarizes the conversions.

21.4   POWER AND ENERGY RELATIONSHIPS IN MACHINING

    A machining operation requires power. The cutting force in a production machining operation might exceed 1000 N (several hundred pounds), as suggested by Example 21.2. Typical cutting speeds are several hundred m/min. The product of cutting force and speed gives the power (energy per unit time) required to perform a machining operation:

$$P_c = F_c v \qquad (21.17)$$

where P_c = cutting power, N-m/s or W (ft-lb/min); F_c = cutting force, N (lb); and v = cutting speed, m/s (ft/min). In U.S. customary units, power is traditionally expressed as horsepower by dividing ft-lb/min by 33,000. Hence,

$$HP_c = \frac{F_c v}{33,000} \tag{21.18}$$

where HP_c = cutting horsepower, hp. The gross power required to operate the machine tool is greater than the power delivered to the cutting process because of mechanical losses in the motor and drive train in the machine. These losses can be accounted for by the mechanical efficiency of the machine tool:

$$P_g = \frac{P_c}{E} \quad \text{or} \quad HP_g = \frac{HP_c}{E} \tag{21.19}$$

where P_g = gross power of the machine tool motor, W; HP_g = gross horsepower; and E = mechanical efficiency of the machine tool. Typical values of E for machine tools are around 90%.

It is often useful to convert power into power per unit volume rate of metal cut. This is called the **unit power**, P_u (or **unit horsepower**, HP_u), defined:

$$P_u = \frac{P_c}{R_{MR}} \quad \text{or} \quad HP_u = \frac{HP_c}{R_{MR}} \tag{21.20}$$

where R_{MR} = material removal rate, mm³/s (in³/min). The material removal rate can be calculated as the product of $vt_o w$. This is Eq. (21.1) using the conversions from Table 21.1. Unit power is also known as the **specific energy** U.

$$U = P_u = \frac{P_c}{R_{MR}} = \frac{F_c v}{vt_o w} = \frac{F_c}{t_o w} \tag{21.21}$$

The units for specific energy are typically N-m/mm³ (in-lb/in³). However, the last expression in Eq. (21.21) suggests that the units might be reduced to N/mm² (lb/in²). It is more meaningful to retain the units as N-m/mm³ or J/mm³ (in-lb/in³).

**Example 21.4
Power
Relationships in
Machining**

Continuing with our previous examples, let us determine cutting power and specific energy in the machining operation if the cutting speed = 100 m/min. Summarizing the data and results from previous examples, t_o = 0.50 mm, w = 3.0 mm, F_c = 1557 N.

Solution: From Eq. (21.18), power in the operation is

$$P_c = (1557 \text{ N})(100 \text{ m/min}) = 155,700 \text{ N} - \text{m/min} = 155,700 \text{ J/min} = 2595 \text{ J/s} = 2595 \text{ W}$$

Specific energy is calculated from Eq. (21.21):

$$U = \frac{155,700}{100(10^3)(3.0)(0.5)} = \frac{155,700}{150,000} = 1.038 \text{ N-m/min}^3$$

Unit power and specific energy provide a useful measure of how much power (or energy) is required to remove a unit volume of metal during machining. Using this measure, different work materials can be compared in terms of their power and energy requirements. Table 21.2 presents a listing of unit horsepower and specific energy values for selected work materials.

The values in Table 21.2 are based on two assumptions: (1) the cutting tool is sharp, and (2) the chip thickness before the cut t_o = 0.25 mm (0.010 in). If these assumptions are not met, some adjustments must be made. For worn tools, the power required to perform the cut is greater, and this is reflected in higher specific energy and unit horsepower values. As an approximate guide, the values in the table should be multiplied by a factor between 1.00 and 1.25 depending on the degree of dullness of the tool. For sharp tools, the factor is

TABLE 21.2 Values of unit horsepower and specific energy for selected work materials using sharp cutting tools and chip thickness before the cut $t_o = 0.25$ mm (0.010 in).

Material	Brinell Hardness	Specific Energy U or Unit Power P_u		Unit Horsepower HP_u hp/(in^3/min)
		N-m/mm^3	in-lb/in^3	
Carbon steel	150–200	1.6	240,000	0.6
	201–250	2.2	320,000	0.8
	251–300	2.8	400,000	1.0
Alloy steels	200–250	2.2	320,000	0.8
	251–300	2.8	400,000	1.0
	301–350	3.6	520,000	1.3
	351–400	4.4	640,000	1.6
Cast irons	125–175	1.1	160,000	0.4
	175–250	1.6	240,000	0.6
Stainless steel	150–250	2.8	400,000	1.0
Aluminum	50–100	0.7	100,000	0.25
Aluminum alloys	100–150	0.8	120,000	0.3
Brass	100–150	2.2	320,000	0.8
Bronze	100–150	2.2	320,000	0.8
Magnesium alloys	50–100	0.4	60,000	0.15

Data compiled from [6], [8], [11], and other sources.

1.00. For tools in a finishing operation that are nearly worn out, the factor is around 1.10, and for tools in a roughing operation that are nearly worn out, the factor is 1.25.

Chip thickness before the cut t_o also affects the specific energy and unit horsepower values. As t_o is reduced, unit power requirements increase. This relationship is referred to as the *size effect.* For example, grinding, in which the chips are extremely small by comparison to most other machining operations, requires very high specific energy values. The U and HP_u values in Table 21.2 can still be used to estimate horsepower and energy for situations in which t_o is not equal to 0.25 mm (0.010 in) by applying a correction factor to account for any difference in chip thickness before the cut. Figure 21.14 provides values of this correction

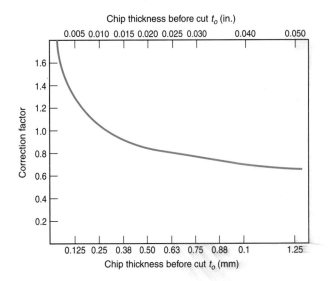

FIGURE 21.14 Correction factor for unit horsepower and specific energy when values of chip thickness before the cut t_o are different from 0.25 mm (0.010 in).

factor as a function of t_o. The unit horsepower and specific energy values in Table 21.2 should be multiplied by the appropriate correction factor when t_o is different from 0.25 mm (0.010 in).

In addition to tool sharpness and size effect, other factors also influence the values of specific energy and unit horsepower for a given operation. These other factors include rake angle, cutting speed, and cutting fluid. As rake angle or cutting speed are increased, or when cutting fluid is added, the U and HP_u values are reduced slightly. For our purposes in the end-of-chapter exercises, the effects of these additional factors can be ignored.

21.5 CUTTING TEMPERATURE

Of the total energy consumed in machining, nearly all of it ($\sim$98%) is converted into heat. This heat can cause temperatures to be very high at the tool–chip interface—over 600°C (1100°F) is not unusual. The remaining energy ($\sim$2%) is retained as elastic energy in the chip.

Cutting temperatures are important because high temperatures (1) reduce tool life, (2) produce hot chips that pose safety hazards to the machine operator, and (3) can cause inaccuracies in workpart dimensions due to thermal expansion of the work material. In this section, we discuss the methods of calculating and measuring temperatures in machining operations.

21.5.1 ANALYTICAL METHODS TO COMPUTE CUTTING TEMPERATURES

There are several analytical methods to calculate estimates of cutting temperature. References [3], [5], [9], and [15] present some of these approaches. We describe the method by Cook [5], which was derived using experimental data for a variety of work materials to establish parameter values for the resulting equation. The equation can be used to predict the increase in temperature at the tool–chip interface during machining:

$$\Delta T = \frac{0.4U}{\rho C} \left(\frac{vt_o}{K}\right)^{0.333}$$

(21.22)

where ΔT = mean temperature rise at the tool–chip interface, C° (F°); U = specific energy in the operation, N-m/mm^3 or J/mm^3 (in-lb/in^3); v = cutting speed, m/s (in/sec); t_o = chip thickness before the cut, m (in); ρC = volumetric specific heat of the work material, J/mm^3-C (in-lb/in^3-F); K = thermal diffusivity of the work material, m^2/s (in^2/sec).

Example 21.5
Cutting
Temperature

For the specific energy obtained in Example 21.4, calculate the increase in temperature above ambient temperature of 20°C. Use the given data from the previous examples in this chapter: v = 100 m/min, t_o = 0.50 mm. In addition, the volumetric specific heat for the work material = 3.0 (10^{-3}) J/mm^3-C, and thermal diffusivity = 50 (10^{-6}) m^2/s (or 50 mm^2/s).

Solution: Cutting speed must be converted to mm/s: v = (100 m/min)(10^3 mm/m)/(60 s/min) = 1667 mm/s. Eq. (21.22) can now be used to compute the mean temperature rise:

$$\Delta T = \frac{0.4(1.038)}{3.0(10^3)}{}^{\circ}C\left(\frac{1667(0.5)}{50}\right)^{0.333} = (138.4)(2.552) = 353{}^{\circ}C$$

∎

21.5.2 MEASUREMENT OF CUTTING TEMPERATURE

Experimental methods have been developed to measure temperatures in machining. The most frequently used measuring technique is the ***tool–chip thermocouple.*** This thermocouple consists of the tool and the chip as the two dissimilar metals forming the

FIGURE 21.15
Experimentally measured cutting temperatures plotted against speed for three work materials, indicating general agreement with Eq. (21.23). (Based on data in [9].)[3]

thermocouple junction. By properly connecting electrical leads to the tool and work-part (which is connected to the chip), the voltage generated at the tool–chip interface during cutting can be monitored using a recording potentiometer or other appropriate data-collection device. The voltage output of the tool–chip thermocouple (measured in mV) can be converted into the corresponding temperature value by means of calibration equations for the particular tool–work combination.

The tool–chip thermocouple has been utilized by researchers to investigate the relationship between temperature and cutting conditions such as speed and feed. Trigger [14] determined the speed–temperature relationship to be of the following general form:

$$T = K v^m \tag{21.23}$$

where T = measured tool–chip interface temperature and v = cutting speed. The parameters K and m depend on cutting conditions (other than v) and work material. Figure 21.15 plots temperature versus cutting speed for several work materials, with equations of the form of Eq. (21.23) determined for each material. A similar relationship exists between cutting temperature and feed; however, the effect of feed on temperature is not as strong as cutting speed. These empirical results tend to support the general validity of the Cook equation: Eq. (21.22).

REFERENCES

[1] *ASM Handbook,* Vol. 16, *Machining.* ASM International, Materials Park, Ohio, 1989.

[2] Black, J, and Kohser, R. *DeGarmo's Materials and Processes in Manufacturing,* 10th ed. John Wiley & Sons, Inc., Hoboken, New Jersey, 2008.

[3] Boothroyd, G., and Knight, W. A. *Fundamentals of Metal Machining and Machine Tools,* 3rd ed. CRC Taylor and Francis, Boca Raton, Florida, 2006.

[4] Chao, B. T., and Trigger, K. J. "Temperature Distribution at the Tool-Chip Interface in Metal

[3]The units reported in the Loewen and Shaw ASME paper [9] were °F for cutting temperature and ft/min for cutting speed. We have retained those units in the plots and equations of our figure.

Cutting," *ASME Transactions,* Vol. 77, October 1955, pp. 1107– 1121.

[5] Cook, N. "Tool Wear and Tool Life," *ASME Transactions, Journal of Engineering for Industry,* Vol. 95, November 1973, pp. 931–938.

[6] Drozda, T. J., and Wick, C. (eds.). *Tool and Manufacturing Engineers Handbook,* 4th ed., Vol. I, *Machining.* Society of Manufacturing Engineers, Dearborn, Michigan, 1983.

[7] Kalpakjian, S., and Schmid, R. *Manufacturing Processes for Engineering Materials,* 4th ed. Prentice Hall/Pearson, Upper Saddle River, New Jersey, 2003.

[8] Lindberg, R. A. *Processes and Materials of Manufacture,* 4th ed. Allyn and Bacon, Inc., Boston, 1990.

[9] Loewen, E. G., and Shaw, M. C. "On the Analysis of Cutting Tool Temperatures," *ASME Transactions,* Vol. 76, No. 2, February 1954, pp. 217–225.

[10] Merchant, M. E., "Mechanics of the Metal Cutting Process: II. Plasticity Conditions in Orthogonal Cutting," *Journal of Applied Physics,* Vol. 16, June 1945 pp. 318–324.

[11] Schey, J. A. *Introduction to Manufacturing Processes,* 3rd ed. McGraw-Hill Book Company, New York, 1999.

[12] Shaw, M. C. *Metal Cutting Principles,* 2nd ed. Oxford University Press, Oxford, UK, 2005.

[13] Trent, E. M., and Wright, P. K. *Metal Cutting,* 4th ed. Butterworth Heinemann, Boston, 2000.

[14] Trigger, K. J. "Progress Report No. 2 on Tool–Chip Interface Temperatures," *ASME Transactions,* Vol. 71, No. 2, February 1949, pp. 163–174.

[15] Trigger, K. J., and Chao, B. T. "An Analytical Evaluation of Metal Cutting Temperatures," *ASME Transactions,* Vol. 73, No. 1, January 1951, pp. 57–68.

REVIEW QUESTIONS

21.1. What are the three basic categories of material removal processes?

21.2. What distinguishes machining from other manufacturing processes?

21.3. Identify some of the reasons why machining is commercially and technologically important.

21.4. Name the three most common machining processes.

21.5. What are the two basic categories of cutting tools in machining? Give two examples of machining operations that use each of the tooling types.

21.6. What are the parameters of a machining operation that are included within the scope of cutting conditions?

21.7. Explain the difference between roughing and finishing operations in machining.

21.8. What is a machine tool?

21.9. What is an orthogonal cutting operation?

21.10. Why is the orthogonal cutting model useful in the analysis of metal machining?

21.11. Name and briefly describe the four types of chips that occur in metal cutting.

21.12. Identify the four forces that act upon the chip in the orthogonal metal cutting model but cannot be measured directly in an operation.

21.13. Identify the two forces that can be measured in the orthogonal metal cutting model.

21.14. What is the relationship between the coefficient of friction and the friction angle in the orthogonal cutting model?

21.15. Describe in words what the Merchant equation tells us.

21.16. How is the power required in a cutting operation related to the cutting force?

21.17. What is the specific energy in metal machining?

21.18. What does the term size effect mean in metal cutting?

21.19. What is a tool–chip thermocouple?

MULTIPLE CHOICE QUIZ

There are 17 correct answers in the following multiple choice questions (some questions have multiple answers that are correct). To attain a perfect score on the quiz, all correct answers must be given. Each correct answer is worth 1 point. Each omitted answer or wrong answer reduces the score by 1 point, and each additional answer beyond the correct number of answers reduces the score by 1 point. Percentage score on the quiz is based on the total number of correct answers.

21.1. Which of the following manufacturing processes are classified as material removal processes (two correct answers): (a) casting, (b) drawing, (c) extrusion, (d) forging, (e) grinding, (f) machining, (g) molding, (h) pressworking, and (i) spinning?

21.2. A lathe is used to perform which one of the following manufacturing operations: (a) broaching, (b) drilling, (c) lapping, (d) milling, or (e) turning?

21.3. With which one of the following geometric forms is the drilling operation most closely associated:

(a) external cylinder, (b) flat plane, (c) round hole, (d) screw threads, or (e) sphere?

21.4. If the cutting conditions in a turning operation are cutting speed $= 300$ ft/min, feed $= 0.010$ in/rev, and depth of cut $= 0.100$ in, which one of the following is the material removal rate: (a) 0.025 in³/min, (b) 0.3 in³/min, (c) 3.0 in³/min, or (d) 3.6 in³/min?

21.5. A roughing operation generally involves which one of the following combinations of cutting conditions: (a) high v, f, and d; (b) high v, low f and d; (c) low v, high f and d; or (d) low v, f, and d, where $v =$ cutting speed, $f =$ feed, and $d =$ depth?

21.6. Which of the following are characteristics of the orthogonal cutting model (three best answers): (a) a circular cutting edge is used, (b) a multiple-cutting-edge tool is used, (c) a single-point tool is used, (d) only two dimensions play an active role in the analysis, (e) the cutting edge is parallel to the direction of cutting speed, (f) the cutting edge is perpendicular to the direction of cutting speed, and (g) the two elements of tool geometry are rake and relief angle?

21.7. The chip thickness ratio is which one of the following: (a) t_c/t_o, (b) t_o/t_c, (c) f/d, or (d) t_o/w, where $t_c =$ chip thickness after the cut, $t_o =$ chip thickness before the cut, $f =$ feed, $d =$ depth, and $w =$ width of cut?

21.8. Which one of the four types of chip would be expected in a turning operation conducted at low cutting speed on a brittle work material: (a) continuous, (b) continuous with built-up edge, (c) discontinuous, or (d) serrated?

21.9. According to the Merchant equation, an increase in rake angle would have which of the following results, all other factors remaining the same (two best answers): (a) decrease in friction angle, (b) decrease in power requirements, (c) decrease in shear plane angle, (d) increase in cutting temperature, and (e) increase in shear plane angle?

21.10. In using the orthogonal cutting model to approximate a turning operation, the chip thickness before the cut t_o corresponds to which one of the following cutting conditions in turning: (a) depth of cut d, (b) feed f, or (c) speed v?

21.11. Which one of the following metals would usually have the lowest unit horsepower in a machining operation: (a) aluminum, (b) brass, (c) cast iron, or (d) steel?

21.12. For which one of the following values of chip thickness before the cut t_o would you expect the specific energy in machining to be the greatest: (a) 0.010 in, (b) 0.025 in, (c) 0.12 mm, or (d) 0.50 mm?

21.13. Which of the following cutting conditions has the strongest effect on cutting temperature: (a) feed or (b) speed?

PROBLEMS

Chip Formation and Forces in Machining

21.1. In an orthogonal cutting operation, the tool has a rake angle $= 15°$. The chip thickness before the cut $= 0.30$ mm and the cut yields a deformed chip thickness $= 0.65$ mm. Calculate (a) the shear plane angle and (b) the shear strain for the operation.

21.2. In Problem 21.1, suppose the rake angle were changed to $0°$. Assuming that the friction angle remains the same, determine (a) the shear plane angle, (b) the chip thickness, and (c) the shear strain for the operation.

21.3. In an orthogonal cutting operation, the 0.25-in wide tool has a rake angle of $5°$. The lathe is set so the chip thickness before the cut is 0.010 in. After the cut, the deformed chip thickness is measured to be 0.027 in. Calculate (a) the shear plane angle and (b) the shear strain for the operation.

21.4. In a turning operation, spindle speed is set to provide a cutting speed of 1.8 m/s. The feed and depth of cut of cut are 0.30 mm and 2.6 mm, respectively. The tool rake angle is $8°$. After the cut, the deformed chip thickness is measured to be 0.49 mm. Determine (a) shear plane angle, (b) shear strain, and (c) material removal rate. Use the orthogonal cutting model as an approximation of the turning process.

21.5. The cutting force and thrust force in an orthogonal cutting operation are 1470 N and 1589 N, respectively. The rake angle $= 5°$, the width of the cut $= 5.0$ mm, the chip thickness before the cut $= 0.6$, and the chip thickness ratio $= 0.38$. Determine (a) the shear strength of the work material and (b) the coefficient of friction in the operation.

21.6. The cutting force and thrust force have been measured in an orthogonal cutting operation to be 300 lb and 291 lb, respectively. The rake angle $= 10°$, width of cut $= 0.200$ in, chip thickness before the cut $= 0.015$, and chip thickness ratio $= 0.4$. Determine (a) the shear strength of the work material and (b) the coefficient of friction in the operation.

21.7. An orthogonal cutting operation is performed using a rake angle of 15°, chip thickness before the cut = 0.012 in and width of cut = 0.100 in. The chip thickness ratio is measured after the cut to be 0.55. Determine (a) the chip thickness after the cut, (b) shear angle, (c) friction angle, (d) coefficient of friction, and (e) shear strain.

21.8. The orthogonal cutting operation described in previous Problem 21.7 involves a work material whose shear strength is 40,000 lb/in². Based on your answers to the previous problem, compute (a) the shear force, (b) cutting force, (c) thrust force, and (d) friction force.

21.9. In an orthogonal cutting operation, the rake angle = −5°, chip thickness before the cut = 0.2 mm and width of cut = 4.0 mm. The chip ratio = 0.4. Determine (a) the chip thickness after the cut, (b) shear angle, (c) friction angle, (d) coefficient of friction, and (e) shear strain.

21.10. The shear strength of a certain work material = 50,000 lb/in². An orthogonal cutting operation is performed using a tool with a rake angle = 20° at the following cutting conditions: cutting speed = 100 ft/min, chip thickness before the cut = 0.015 in, and width of cut = 0.150 in. The resulting chip thickness ratio = 0.50. Determine (a) the shear plane angle, (b) shear force, (c) cutting force and thrust force, and (d) friction force.

21.11. Consider the data in Problem 21.10 except that rake angle is a variable, and its effect on the forces in parts (b), (c), and (d) is to be evaluated. (a) Using a spreadsheet calculator, compute the values of shear force, cutting force, thrust force, and friction force as a function of rake angle over a range of rake angles between the high value of 20° in Problem 21.10 and a low value of −10°. Use intervals of 5° between these limits. The chip thickness ratio decreases as rake angle is reduced and can be approximated by the following relationship: $r = 0.38 + 0.006\alpha$, where r = chip thickness and α =

rake angle. (b) What observations can be made from the computed results?

21.12. Solve previous Problem 21.10 except that the rake angle has been changed to −5° and the resulting chip thickness ratio = 0.35.

21.13. A carbon steel bar with 7.64 in diameter has a tensile strength of 65,000 lb/in² and a shear strength of 45,000 lb/in². The diameter is reduced using a turning operation at a cutting speed of 400 ft/min. The feed is 0.011 in/rev and the depth of cut is 0.120 in. The rake angle on the tool in the direction of chip flow is 13°. The cutting conditions result in a chip ratio of 0.52. Using the orthogonal model as an approximation of turning, determine (a) the shear plane angle, (b) shear force, (c) cutting force and feed force, and (d) coefficient of friction between the tool and chip.

21.14. Low carbon steel having a tensile strength of 300 MPa and a shear strength of 220 MPa is cut in a turning operation with a cutting speed of 3.0 m/s. The feed is 0.20 mm/rev and the depth of cut is 3.0 mm. The rake angle of the tool is 5° in the direction of chip flow. The resulting chip ratio is 0.45. Using the orthogonal model as an approximation of turning, determine (a) the shear plane angle, (b) shear force, (c) cutting force and feed force.

21.15. A turning operation is made with a rake angle of 10°, a feed of 0.010 in/rev and a depth of cut = 0.100 in. The shear strength of the work material is known to be 50,000 lb/in², and the chip thickness ratio is measured after the cut to be 0.40. Determine the cutting force and the feed force. Use the orthogonal cutting model as an approximation of the turning process.

21.16. Show how Eq. (21.3) is derived from the definition of chip ratio, Eq. (21.2), and Figure 21.5(b).

21.17. Show how Eq. (21.4) is derived from Figure 21.6.

21.18. Derive the force equations for F, N, F_s, and F_n (Eqs. (21.9) through (21.12) in the text) using the force diagram of Figure 21.11.

Power and Energy in Machining

21.19. In a turning operation on stainless steel with hardness = 200 HB, the cutting speed = 200 m/min, feed = 0.25 mm/rev, and depth of cut = 7.5 mm. How much power will the lathe draw in performing this operation if its mechanical efficiency = 90%. Use Table 21.2 to obtain the appropriate specific energy value.

21.20. In Problem 21.18, compute the lathe power requirements if feed = 0.50 mm/rev.

21.21. In a turning operation on aluminum, cutting speed = 900 ft/min, feed = 0.020 in/rev, and depth of cut = 0.250 in. What horsepower is required of

the drive motor, if the lathe has a mechanical efficiency = 87%? Use Table 21.2 to obtain the appropriate unit horsepower value.

21.22. In a turning operation on plain carbon steel whose Brinell hardness = 275 HB, the cutting speed is set at 200 m/min and depth of cut = 6.0 mm. The lathe motor is rated at 25 kW, and its mechanical efficiency = 90%. Using the appropriate specific energy value from Table 21.2, determine the maximum feed that can be set for this operation. Use of a spreadsheet calculator is recommended for the iterative calculations required in this problem.

21.23. A turning operation is to be performed on a 20 hp lathe that has an 87% efficiency rating. The roughing cut is made on alloy steel whose hardness is in the range 325 to 335 HB. The cutting speed is 375 ft/min, feed is 0.030 in/rev, and depth of cut is 0.150 in. Based on these values, can the job be performed on the 20 hp lathe? Use Table 21.2 to obtain the appropriate unit horsepower value.

21.24. Suppose the cutting speed in Problems 21.7 and 21.8 is 200 ft/min. From your answers to those problems, find (a) the horsepower consumed in the operation, (b) metal removal rate in in^3/min, (c) unit horsepower (hp-min/in^3), and (d) the specific energy (in-lb/in^3).

21.25. For Problem 21.12, the lathe has a mechanical efficiency = 0.83. Determine (a) the horsepower consumed by the turning operation; (b) horsepower that must be generated by the lathe; (c) unit horsepower and specific energy for the work material in this operation.

21.26. In a turning operation on low carbon steel (175 BHN), cutting speed = 400 ft/min, feed = 0.010 in/rev, and depth of cut = 0.075 in. The lathe has a mechanical efficiency = 0.85. Based on the unit horsepower values in Table 21.2, determine (a) the horsepower consumed by the turning operation and (b) the horsepower that must be generated by the lathe.

21.27. Solve Problem 21.25 except that the feed = 0.0075 in/rev and the work material is stainless steel (Brinell hardness = 240 HB).

21.28. A turning operation is carried out on aluminum (100 BHN). Cutting speed = 5.6 m/s, feed = 0.25 mm/rev, and depth of cut = 2.0 mm. The lathe has a mechanical efficiency = 0.85. Based on the specific energy values in Table 21.2, determine (a) the cutting power and (b) gross power in the turning operation, in Watts.

Cutting Temperature

21.33. Orthogonal cutting is performed on a metal whose mass specific heat = 1.0 J/g-C, density = 2.9 g/cm^3, and thermal diffusivity = 0.8 cm^2/s. The cutting speed is 4.5 m/s, uncut chip thickness is 0.25 mm, and width of cut is 2.2 mm. The cutting force is measured at 1170 N. Using Cook's equation, determine the cutting temperature if the ambient temperature = 22°C.

21.34. Consider a turning operation performed on steel whose hardness = 225 HB at a speed = 3.0 m/s, feed = 0.25 mm, and depth = 4.0 mm. Using values of thermal properties found in the tables and definitions of Section 4.1 and the appropriate

21.29. Solve Problem 21.27 but with the following changes: cutting speed = 1.3 m/s, feed = 0.75 mm/rev, and depth = 4.0 mm. Note that although the power used in this operation is only about 10% greater than in the previous problem, the metal removal rate is about 40% greater.

21.30. A turning operation is performed on an engine lathe using a tool with zero rake angle in the direction of chip flow. The work material is an alloy steel with hardness = 325 Brinell hardness. The feed is 0.015 in/rev, depth of cut is 0.125 in and cutting speed is 300 ft/min. After the cut, the chip thickness ratio is measured to be 0.45. (a) Using the appropriate value of specific energy from Table 21.2, compute the horsepower at the drive motor, if the lathe has an efficiency = 85%. (b) Based on horsepower, compute your best estimate of the cutting force for this turning operation. Use the orthogonal cutting model as an approximation of the turning process.

21.31. A lathe performs a turning operation on a workpiece of 6.0 in diameter. The shear strength of the work is 40,000 lb/in^2 and the tensile strength is 60,000 lb/in^2. The rake angle of the tool is 6°. The cutting speed = 700 ft/min, feed = 0.015 in/rev, and depth = 0.090 in. The chip thickness after the cut is 0.025 in. Determine (a) the horsepower required in the operation, (b) unit horsepower for this material under these conditions, and (c) unit horsepower as it would be listed in Table 21.2 for a t_o of 0.010 in. Use the orthogonal cutting model as an approximation of the turning process.

21.32. In a turning operation on an aluminum alloy workpiece, the feed = 0.020 in/rev, and depth of cut = 0.250 in. The motor horsepower of the lathe is 20 hp and it has a mechanical efficiency = 92%. The unit horsepower value = 0.25 hp/(in^3/min) for this aluminum grade. What is the maximum cutting speed that can be used on this job?

specific energy value from Table 21.2, compute an estimate of cutting temperature using the Cook equation. Assume ambient temperature = 20°C.

21.35. An orthogonal cutting operation is performed on a certain metal whose volumetric specific heat = 110 in-lb/in^3-F, and thermal diffusivity = 0.140 in^2/sec. The cutting speed = 350 ft/min, chip thickness before the cut = 0.008 in, and width of cut = 0.100 in. The cutting force is measured at 200 lb. Using Cook's equation, determine the cutting temperature if the ambient temperature = 70°F.

21.36. It is desired to estimate the cutting temperature for a certain alloy steel whose hardness = 240 Brinell. Use the appropriate value of specific energy from Table 21.2 and compute the cutting temperature by means of the Cook equation for a turning operation in which the cutting speed is 500 ft/min, feed is 0.005 in/rev, and depth of cut is 0.070 in. The work material has a volumetric specific heat of 210 in lb/in^3-F and a thermal diffusivity of 0.16 in^2/sec. Assume ambient temperature = 88°F.

21.37. An orthogonal machining operation removes metal at 1.8 in^3/min. The cutting force in the process = 300 lb. The work material has a thermal diffusivity = 0.18 in^2/sec and a volumetric specific heat = 124 in-lb/in^3-F. If the feed $f = t_o = 0.010$ in and width of cut = 0.100 in, use the Cook formula to compute the cutting temperature in the operation given that ambient temperature = 70°F.

21.38. A turning operation uses a cutting speed = 200 m/min, feed = 0.25 mm/rev, and depth of cut = 4.00 mm. The thermal diffusivity of the work material = 20 mm^2/s and the volumetric specific heat = 3.5 (10^{-3}) J/mm^3-C. If the temperature increase above ambient temperature (20°F) is measured by a tool–chip thermocouple to be 700°C, determine the specific energy for the work material in this operation.

21.39. During a turning operation, a tool–chip thermocouple was used to measure cutting temperature. The following temperature data were collected during the cuts at three different cutting speeds (feed and depth were held constant): (1) $v = 100$ m/min, $T = 505$°C, (2) $v = 130$ m/min, $T = 552$°C, (3) $v = 160$ m/min, $T = 592$°C. Determine an equation for temperature as a function of cutting speed that is in the form of the Trigger equation, Eq. (21.23).

22 MACHINING OPERATIONS AND MACHINE TOOLS

Chapter Contents

Machining is the most versatile and accurate of all manufacturing processes in its capability to produce a diversity of part geometries and geometric features. Casting can also produce a variety of shapes, but it lacks the precision and accuracy of machining. In this chapter, we describe the important machining operations and the machine tools used to perform them. Historical Note 22.1 provides a brief narrative of the development of machine tool technology.

22.1 MACHINING AND PART GEOMETRY

To introduce our topic in this chapter, let us provide an overview of the creation of part geometry by machining. Machined parts can be classified as rotational or nonrotational (Figure 22.1). A *rotational* workpart has a cylindrical or disk-like shape. The characteristic operation that produces this geometry is one in which a cutting tool removes material from a rotating workpart. Examples include turning and boring. Drilling is closely related except that an internal cylindrical shape is created and the tool rotates (rather than the work) in most drilling operations. A *nonrotational* (also called *prismatic*) workpart is block-like or plate-like, as in Figure 22.1(b). This geometry is achieved by linear motions of the workpart, combined with either rotating or linear tool motions. Operations in this category include milling, shaping, planing, and sawing.

Each machining operation produces a characteristic geometry due to two factors: (1) the relative motions between the tool and the workpart and (2) the shape of the cutting tool. We classify these operations by which part shape is created as generating and forming. In *generating,* the geometry of the workpart is determined by the feed trajectory of the cutting tool. The path followed by the tool during its feed motion is imparted to the work surface in order to create shape. Examples of generating the work

507

Historical Note 22.1 *Machine tool technology*

Material removal as a means of making things dates back to prehistoric times, when man learned to carve wood and chip stones to make hunting and farming implements. There is archaeological evidence that the ancient Egyptians used a rotating bowstring mechanism to drill holes.

Development of modern machine tools is closely related to the Industrial Revolution. When James Watt designed his steam engine in England around 1763, one of the technical problems he faced was to make the bore of the cylinder sufficiently accurate to prevent steam from escaping around the piston. John Wilkinson built a water-wheel powered *boring machine* around 1775, which permitted Watt to build his steam engine. This boring machine is often recognized as the first machine tool.

Another Englishman, Henry Maudsley, developed the first *screw-cutting lathe* around 1800. Although the turning of wood had been accomplished for many centuries, Maudsley's machine added a mechanized tool carriage with which feeding and threading operations could be performed with much greater precision than any means before.

Eli Whitney is credited with developing the first *milling machine* in the United States around 1818. Development of the *planer* and *shaper* occurred in England between 1800 and 1835, in response to the need to make components for the steam engine, textile equipment, and other machines associated with the Industrial Revolution. The powered *drill press* was developed by James Nasmyth around 1846, which permitted drilling of accurate holes in metal.

Most of the conventional boring machines, lathes, milling machines, planers, shapers, and drill presses used today have the same basic designs as the early versions developed during the last two centuries. Modern machining centers—machine tools capable of performing more than one type of cutting operation—were introduced in the late 1950s, after numerical control had been developed (Historical Note 38.1).

---

shape in machining include straight turning, taper turning, contour turning, peripheral milling, and profile milling, all illustrated in Figure 22.2. In each of these operations, material removal is accomplished by the speed motion in the operation, but part shape is determined by the feed motion. The feed trajectory may involve variations in depth or width of cut during the operation. For example, in the contour turning and profile milling operations shown in our figure, the feed motion results in changes in depth and width, respectively, as cutting proceeds.

In *forming,* the shape of the part is created by the geometry of the cutting tool. In effect, the cutting edge of the tool has the reverse of the shape to be produced on the part surface. Form turning, drilling, and broaching are examples of this case. In these operations, illustrated in Figure 22.3, the shape of the cutting tool is imparted to the work in order to create part geometry. The cutting conditions in forming usually include the primary speed motion combined with a feeding motion that is directed into the work.

(a)

(b)

FIGURE 22.1 Machined parts are classified as (a) rotational, or (b) nonrotational, shown here by block and flat parts.

FIGURE 22.2 Generating shape in machining: (a) straight turning, (b) taper turning, (c) contour turning, (d) plain milling, and (e) profile milling.

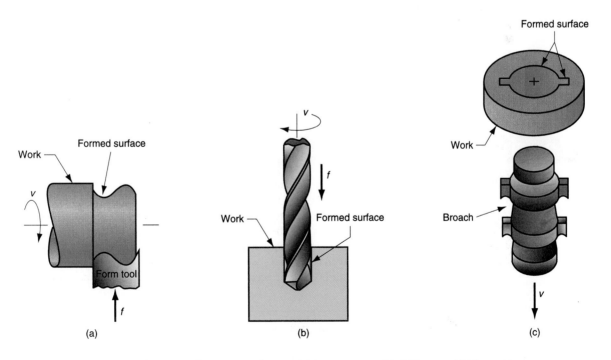

FIGURE 22.3 Forming to create shape in machining: (a) form turning, (b) drilling, and (c) broaching.

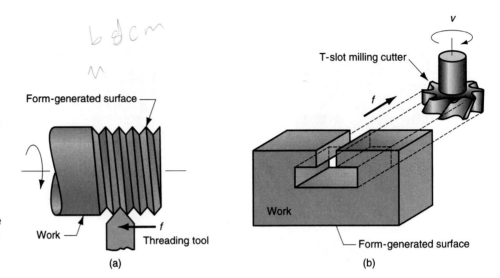

FIGURE 22.4
Combination of forming and generating to create shape: (a) thread cutting on a lathe, and (b) slot milling.

Depth of cut in this category of machining usually refers to the final penetration into the work after the feed motion has been completed.

Forming and generating are sometimes combined in one operation, as illustrated in Figure 22.4 for thread cutting on a lathe and slotting on a milling machine. In thread cutting, the pointed shape of the cutting tool determines the form of the threads, but the large feed rate generates the threads. In slotting (also called slot milling), the width of the cutter determines the width of the slot, but the feed motion creates the slot.

Machining is classified as a secondary process. In general, secondary processes follow basic processes, whose purpose is to establish the initial shape of a workpiece. Examples of basic processes include casting, forging, and bar rolling (to produce rod and bar stock). The shapes produced by these processes usually require refinement by secondary processes. Machining operations serve to transform the starting shapes into the final geometries specified by the part designer. For example, bar stock is the initial shape, but the final geometry after a series of machining operations is a shaft. We discuss basic and secondary processes in more detail and provide additional examples in Section 40.1.1 on process planning.

22.2 TURNING AND RELATED OPERATIONS

Turning is a machining process in which a single-point tool removes material from the surface of a rotating workpiece. The tool is fed linearly in a direction parallel to the axis of rotation to generate a cylindrical geometry, as illustrated in Figures 22.2(a) and 22.5. Single-point tools used in turning and other machining operations are discussed in Section 23.3.1. Turning is traditionally carried out on a machine tool called a *lathe,* which provides power to turn the part at a given rotational speed and to feed the tool at a specified rate and depth of cut. Included on the DVD that accompanies this text is a video clip on turning.

VIDEO CLIP _____

Turning and Lathe Basics. This clip contains four segments: (1) lathe types, (2) lathe turrets, (3) lathe workholding, and (4) turning operations.

FIGURE 22.5   Turning operation.

22.2.1 CUTTING CONDITIONS IN TURNING

The rotational speed in turning is related to the desired cutting speed at the surface of the cylindrical workpiece by the equation

$$N = \frac{v}{\pi D_o} \tag{22.1}$$

where N = rotational speed, rev/min; v = cutting speed, m/min (ft/min); and D_o = original diameter of the part, m (ft).

The turning operation reduces the diameter of the work from its original diameter D_o to a final diameter D_f, as determined by the depth of cut d:

$$D_f = D_o - 2d \tag{22.2}$$

The feed in turning is generally expressed in mm/rev (in/rev). This feed can be converted to a linear travel rate in mm/min (in/min) by the formula

$$f_r = Nf \tag{22.3}$$

where f_r = feed rate, mm/min (in/min); and f = feed, mm/rev (in/rev).

The time to machine from one end of a cylindrical workpart to the other is given by

$$T_m = \frac{L}{f_r} \tag{22.4}$$

where T_m = machining time, min; and L = length of the cylindrical workpart, mm (in). A more direct computation of the machining time is provided by the following equation:

$$T_m = \frac{\pi D_o L}{fv} \tag{22.5}$$

where D_o = work diameter, mm (in); L = workpart length, mm (in); f = feed, mm/rev (in/rev); and v = cutting speed, mm/min (in/min). As a practical matter, a small distance is usually added to the workpart length at the beginning and end of the piece to allow for approach and overtravel of the tool. Thus, the duration of the feed motion past the work will be longer than T_m.

The volumetric rate of material removal can be most conveniently determined by the following equation:

$$R_{MR} = vfd \qquad (22.6)$$

where R_{MR} = material removal rate, mm³/min (in³/min). In using this equation, the units for f are expressed simply as mm (in), in effect neglecting the rotational character of turning. Also, care must be exercised to ensure that the units for speed are consistent with those for f and d.

22.2.2 OPERATIONS RELATED TO TURNING

A variety of other machining operations can be performed on a lathe in addition to turning; these include the following, illustrated in Figure 22.6:

FIGURE 22.6 Machining operations other than turning that are performed on a lathe: (a) facing, (b) taper turning, (c) contour turning, (d) form turning, (e) chamfering, (f) cutoff, (g) threading, (h) boring, (i) drilling, and (j) knurling.

(a) ***Facing.*** The tool is fed radially into the rotating work on one end to create a flat surface on the end.

(b) ***Taper turning.*** Instead of feeding the tool parallel to the axis of rotation of the work, the tool is fed at an angle, thus creating a tapered cylinder or conical shape.

(c) ***Contour turning.*** Instead of feeding the tool along a straight line parallel to the axis of rotation as in turning, the tool follows a contour that is other than straight, thus creating a contoured form in the turned part.

(d) ***Form turning.*** In this operation, sometimes called ***forming,*** the tool has a shape that is imparted to the work by plunging the tool radially into the work.

(e) ***Chamfering.*** The cutting edge of the tool is used to cut an angle on the corner of the cylinder, forming what is called a "chamfer."

(f) ***Cutoff.*** The tool is fed radially into the rotating work at some location along its length to cut off the end of the part. This operation is sometimes referred to as ***parting.***

(g) ***Threading.*** A pointed tool is fed linearly across the outside surface of the rotating workpart in a direction parallel to the axis of rotation at a large effective feed rate, thus creating threads in the cylinder. Methods of machining screw threads are discussed in greater detail in Section 22.7.1.

(h) ***Boring.*** A single-point tool is fed linearly, parallel to the axis of rotation, on the inside diameter of an existing hole in the part.

(i) ***Drilling.*** Drilling can be performed on a lathe by feeding the drill into the rotating work along its axis. ***Reaming*** can be performed in a similar way.

(j) ***Knurling.*** This is not a machining operation because it does not involve cutting of material. Instead, it is a metal forming operation used to produce a regular cross-hatched pattern in the work surface.

Most lathe operations use single-point tools, which we discuss in Section 23.3.1. Turning, facing, taper turning, contour turning, chamfering, and boring are all performed with single-point tools. A threading operation is accomplished using a single-point tool designed with a geometry that shapes the thread. Certain operations require tools other than single-point. Form turning is performed with a specially designed tool called a form tool. The profile shape ground into the tool establishes the shape of the workpart. A cutoff tool is basically a form tool. Drilling is accomplished by a drill bit (Section 23.3.2). Knurling is performed by a knurling tool, consisting of two hardened forming rolls, each mounted between centers. The forming rolls have the desired knurling pattern on their surfaces. To perform knurling, the tool is pressed against the rotating workpart with sufficient pressure to impress the pattern onto the work surface.

22.2.3 THE ENGINE LATHE

The basic lathe used for turning and related operations is an ***engine lathe.*** It is a versatile machine tool, manually operated, and widely used in low and medium production. The term ***engine*** dates from the time when these machines were driven by steam engines.

Engine Lathe Technology   Figure 22.7 is a sketch of an engine lathe showing its principal components. The ***headstock*** contains the drive unit to rotate the spindle, which rotates the work. Opposite the headstock is the ***tailstock,*** in which a center is mounted to support the other end of the workpiece.

The cutting tool is held in a ***tool post*** fastened to the ***cross-slide,*** which is assembled to the ***carriage.*** The carriage is designed to slide along the ***ways*** of the lathe in order to feed the tool parallel to the axis of rotation. The ways are like tracks along which the carriage rides, and they

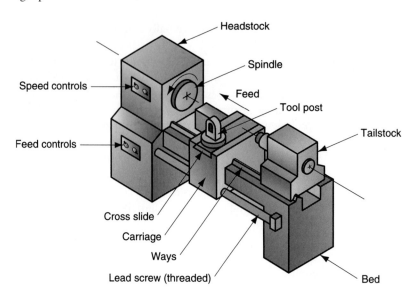

FIGURE 22.7 Diagram of an engine lathe, indicating its principal components.

are made with great precision to achieve a high degree of parallelism relative to the spindle axis. The ways are built into the **bed** of the lathe, providing a rigid frame for the machine tool.

The carriage is driven by a leadscrew that rotates at the proper speed to obtain the desired feed rate. The cross-slide is designed to feed in a direction perpendicular to the carriage movement. Thus, by moving the carriage, the tool can be fed parallel to the work axis to perform straight turning; or by moving the cross-slide, the tool can be fed radially into the work to perform facing, form turning, or cutoff operations.

The conventional engine lathe and most other machines described in this section are **horizontal turning machines;** that is, the spindle axis is horizontal. This is appropriate for the majority of turning jobs, in which the length is greater than the diameter. For jobs in which the diameter is large relative to length and the work is heavy, it is more convenient to orient the work so that it rotates about a vertical axis; these are **vertical turning machines.**

The size of a lathe is designated by swing and maximum distance between centers. The **swing** is the maximum workpart diameter that can be rotated in the spindle, determined as twice the distance between the centerline of the spindle and the ways of the machine. The actual maximum size of a cylindrical workpiece that can be accommodated on the lathe is smaller than the swing because the carriage and cross-slide assembly are in the way. The **maximum distance between centers** indicates the maximum length of a workpiece that can be mounted between headstock and tailstock centers. For example, a 350 mm × 1.2 m (14 in × 48 in) lathe designates that the swing is 350 mm (14 in) and the maximum distance between centers is 1.2 m (48 in).

Methods of Holding the Work in a Lathe   There are four common methods used to hold workparts in turning. These workholding methods consist of various mechanisms to grasp the work, center and support it in position along the spindle axis, and rotate it. The methods, illustrated in Figure 22.8, are (a) mounting the work between centers, (b) chuck, (c) collet, and (d) face plate. Our video clip on workholding illustrates the various aspects of fixturing for turning and other machining operations.

VIDEO CLIP _____

Introduction to Workholding. This clip contains four segments: (1) workholding of parts, (2) principles of workholding, (3) 3-2-1 locational workholding method, and (4) workpiece reclamping.

FIGURE 22.8   Four workholding methods used in lathes: (a) mounting the work between centers using a dog, (b) three-jaw chuck, (c) collet, and (d) faceplate for noncylindrical workparts.

Holding the work ***between centers*** refers to the use of two centers, one in the headstock and the other in the tailstock, as in Figure 22.8(a). This method is appropriate for parts with large length-to-diameter ratios. At the headstock center, a device called a ***dog*** is attached to the outside of the work and is used to drive the rotation from the spindle. The tailstock center has a cone-shaped point which is inserted into a tapered hole in the end of the work. The tailstock center is either a "live" center or a "dead" center. A ***live center*** rotates in a bearing in the tailstock, so that there is no relative rotation between the work and the live center, hence, no friction between the center and the workpiece. In contrast, a ***dead center*** is fixed to the tailstock, so that it does not rotate; instead, the workpiece rotates about it. Because of friction and the heat buildup that results, this setup is normally used at lower rotational speeds. The live center can be used at higher speeds.

The ***chuck,*** Figure 22.8(b), is available in several designs, with three or four jaws to grasp the cylindrical workpart on its outside diameter. The jaws are often designed so they can also grasp the inside diameter of a tubular part. A ***self-centering*** chuck has a mechanism to move the jaws in or out simultaneously, thus centering the work at the spindle axis. Other chucks allow independent operation of each jaw. Chucks can be used with or without a tailstock center. For parts with low length-to-diameter ratios, holding the part in the chuck in a cantilever fashion is usually sufficient to withstand the cutting forces. For long workbars, the tailstock center is needed for support.

A ***collet*** consists of a tubular bushing with longitudinal slits running over half its length and equally spaced around its circumference, as in Figure 22.8(c). The inside diameter of the collet is used to hold cylindrical work such as barstock. Owing to the slits, one end of the collet can be squeezed to reduce its diameter and provide a secure grasping pressure against the work. Because there is a limit to the reduction obtainable in a collet of any given diameter, these workholding devices must be made in various sizes to match the particular workpart size in the operation.

A *face plate,* Figure 22.8(d), is a workholding device that fastens to the lathe spindle and is used to grasp parts with irregular shapes. Because of their irregular shape, these parts cannot be held by other workholding methods. The faceplate is therefore equipped with the custom-designed clamps for the particular geometry of the part.

22.2.4 OTHER LATHES AND TURNING MACHINES

In addition to the engine lathe, other turning machines have been developed to satisfy particular functions or to automate the turning process. Among these machines are (1) toolroom lathe, (2) speed lathe, (3) turret lathe, (4) chucking machine, (5) automatic screw machine, and (6) numerically controlled lathe.

The toolroom lathe and speed lathe are closely related to the engine lathe. The *toolroom lathe* is smaller and has a wider available range of speeds and feeds. It is also built for higher accuracy, consistent with its purpose of fabricating components for tools, fixtures, and other high-precision devices.

The *speed lathe* is simpler in construction than the engine lathe. It has no carriage and cross-slide assembly, and therefore no leadscrew to drive the carriage. The cutting tool is held by the operator using a rest attached to the lathe for support. The speeds are higher on a speed lathe, but the number of speed settings is limited. Applications of this machine type include wood turning, metal spinning, and polishing operations.

A *turret lathe* is a manually operated lathe in which the tailstock is replaced by a turret that holds up to six cutting tools. These tools can be rapidly brought into action against the work one by one by indexing the turret. In addition, the conventional tool post used on an engine lathe is replaced by a four-sided turret that is capable of indexing up to four tools into position. Hence, because of the capacity to quickly change from one cutting tool to the next, the turret lathe is used for high-production work that requires a sequence of cuts to be made on the part.

As the name suggests, a *chucking machine* (nicknamed *chucker*) uses a chuck in its spindle to hold the workpart. The tailstock is absent on a chucker, so parts cannot be mounted between centers. This restricts the use of a chucking machine to short, light-weight parts. The setup and operation are similar to a turret lathe except that the feeding actions of the cutting tools are controlled automatically rather than by a human operator. The function of the operator is to load and unload the parts.

A *bar machine* is similar to a chucking machine except that a collet is used (instead of a chuck), which permits long bar stock to be fed through the headstock into position. At the end of each machining cycle, a cutoff operation separates the new part. The bar stock is then indexed forward to present stock for the next part. Feeding the stock as well as indexing and feeding the cutting tools is accomplished automatically. Owing to its high level of automatic operation, it is often called an *automatic bar machine.* One of its important applications is in the production of screws and similar small hardware items; the name *automatic screw machine* is frequently used for machines used in these applications.

Bar machines can be classified as single spindle or multiple spindle. A *single spindle bar machine* has one spindle that normally allows only one cutting tool to be used at a time on the single workpart being machined. Thus, while each tool is cutting the work, the other tools are idle. (Turret lathes and chucking machines are also limited by this sequential, rather than simultaneous, tool operation). To increase cutting tool utilization and production rate, *multiple spindle bar machines* are available. These machines have more than one spindle, so multiple parts are machined simultaneously by multiple tools. For example, a six-spindle automatic bar machine works on six parts at a time, as in Figure 22.9. At the end of each machining cycle, the spindles (including collets and workbars) are indexed (rotated) to the next position. In our illustration, each part is cut sequentially by five sets of cutting tools,

FIGURE 22.9   (a) Type of part produced on a six-spindle automatic bar machine; and (b) sequence of operations to produce the part: (1) feed stock to stop, (2) turn main diameter, (3) form second diameter and spotface, (4) drill, (5) chamfer, and (6) cutoff.

which takes six cycles (position 1 is for advancing the bar stock to a "stop"). With this arrangement, a part is completed at the end of each cycle. As a result, a six-spindle automatic screw machine has a very high production rate.

The sequencing and actuation of the motions on screw machines and chucking machines have traditionally been controlled by cams and other mechanical devices. The modern form of control is *computer numerical control* (CNC), in which the machine tool operations are controlled by a "program of instructions" consisting of alphanumeric code (Section 38.3). CNC provides a more sophisticated and versatile means of control than mechanical devices. This has led to the development of machine tools capable of more complex machining cycles and part geometries, and a higher level of automated operation than conventional screw machines and chucking machines. The CNC lathe is an example of these machines in turning. It is especially useful for contour turning operations and close tolerance work. Today, automatic chuckers and bar machines are implemented by CNC.

22.2.5 BORING MACHINES

Boring is similar to turning. It uses a single-point tool against a rotating workpart. The difference is that boring is performed on the inside diameter of an existing hole rather than the outside diameter of an existing cylinder. In effect, boring is an internal turning operation. Machine tools used to perform boring operations are called *boring machines* (also *boring mills*). One might expect that boring machines would have features in common with turning machines; indeed, as previously indicated, lathes are sometimes used to accomplish boring.

Boring mills can be horizontal or vertical. The designation refers to the orientation of the axis of rotation of the machine spindle or workpart. In a *horizontal boring* operation, the setup can be arranged in either of two ways. The first setup is one in which the work is fixtured to a rotating spindle, and the tool is attached to a cantilevered boring bar that feeds

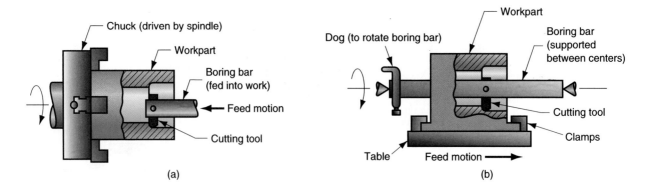

FIGURE 22.10 Two forms of horizontal boring: (a) boring bar is fed into a rotating workpart, and (b) work is fed past a rotating boring bar.

into the work, as illustrated in Figure 22.10(a). The boring bar in this setup must be very stiff to avoid deflection and vibration during cutting. To achieve high stiffness, boring bars are often made of cemented carbide, whose modulus of elasticity approaches 620×10^3 MPa $(90 \times 10^6$ lb/in^2). Figure 22.11 shows a carbide boring bar.

The second possible setup is one in which the tool is mounted to a boring bar, and the boring bar is supported and rotated between centers. The work is fastened to a feeding mechanism that feeds it past the tool. This setup, Figure 22.10(b), can be used to perform a boring operation on a conventional engine lathe.

A ***vertical boring machine*** is used for large, heavy workparts with large diameters; usually the workpart diameter is greater than its length. As in Figure 22.12, the part is clamped to a worktable that rotates relative to the machine base. Worktables up to 40 ft in diameter are available. The typical boring machine can position and feed several cutting

FIGURE 22.11 Boring bar made of cemented carbide (WC–Co) that uses indexable cemented carbide inserts. (Courtesy of Kennametal Inc., Latrobe, Pennsylvania.)

FIGURE 22.12
A vertical boring mill.

Cross-rail

Tool head

Workpart

Worktable

Side column

Side tool head

Base

tools simultaneously. The tools are mounted on tool heads that can be fed horizontally and vertically relative to the worktable. One or two heads are mounted on a horizontal cross-rail assembled to the machine tool housing above the worktable. The cutting tools mounted above the work can be used for facing and boring. In addition to the tools on the cross-rail, one or two additional tool heads can be mounted on the side columns of the housing to enable turning on the outside diameter of the work.

The tool heads used on a vertical boring machine often include turrets to accommodate several cutting tools. This results in a loss of distinction between this machine and a *vertical turret lathe.* Some machine tool builders make the distinction that the vertical turret lathe is used for work diameters up to 2.5 m (100 in), while the vertical boring machine is used for larger diameters [7]. Also, vertical boring mills are often applied to one-of-a-kind jobs, while vertical turret lathes are used for batch production.

22.3 DRILLING AND RELATED OPERATIONS

Drilling, Figure 22.3(b), is a machining operation used to create a round hole in a workpart. This contrasts with boring, which can only be used to enlarge an existing hole. Drilling is usually performed with a rotating cylindrical tool that has two cutting edges on its working end. The tool is called a *drill* or *drill bit* (described in Section 23.3.2). The most common drill bit is the twist drill, described in Section 23.3.2. The rotating drill feeds into the stationary workpart to form a hole whose diameter is equal to the drill diameter. Drilling is customarily performed on a *drill press,* although other machine tools also perform this operation. The video clip on hole making illustrates the drilling operation.

VIDEO CLIP _____

Basic Hole Making: Two segments are included in this clip: (1) the drill and (2) hole-making machines.

22.3.1 CUTTING CONDITIONS IN DRILLING

The cutting speed in a drilling operation is the surface speed at the outside diameter of the drill. It is specified in this way for convenience, even though nearly all of the cutting is actually performed at lower speeds closer to the axis of rotation. To set the desired cutting speed in drilling, it is necessary to determine the rotational speed of the drill. Letting N represent the spindle rev/min,

$$N = \frac{v}{\pi D} \qquad (22.7)$$

where $v =$ cutting speed, mm/min (in/min); and $D =$ the drill diameter, mm (in). In some drilling operations, the workpiece is rotated about a stationary tool, but the same formula applies.

Feed f in drilling is specified in mm/rev (in/rev). Recommended feeds are roughly proportional to drill diameter; higher feeds are used with larger diameter drills. Since there are (usually) two cutting edges at the drill point, the uncut chip thickness (chip load) taken by each cutting edge is half the feed. Feed can be converted to feed rate using the same equation as for turning:

$$f_r = Nf \qquad (22.8)$$

where $f_r =$ feed rate, mm/min (in/min).

Drilled holes are either through holes or blind holes, Figure 22.13. In **through holes,** the drill exits the opposite side of the work; in **blind holes,** it does not. The machining time required to drill a through hole can be determined by the following formula:

$$T_m = \frac{t + A}{f_r} \qquad (22.9)$$

where $T_m =$ machining (drilling) time, min; $t =$ work thickness, mm (in); $f_r =$ feed rate, mm/min (in/min); and $A =$ an approach allowance that accounts for the drill point angle, representing the distance the drill must feed into the work before reaching full diameter, Figure 22.10(a). This allowance is given by

$$A = 0.5\,D \tan\left(90 - \frac{\theta}{2}\right) \qquad (22.10)$$

where $A =$ approach allowance, mm (in); and $\theta =$ drill point angle. In drilling a through hole, the feed motion usually proceeds slightly beyond the opposite side of the work,

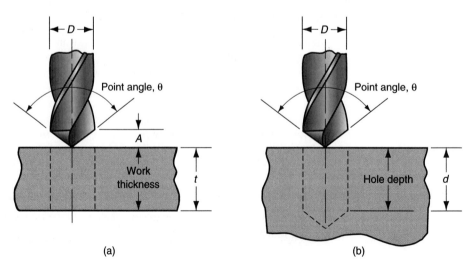

FIGURE 22.13 Two hole types: (a) through hole and (b) blind hole.

(a)

(b)

thus making the actual duration of the cut greater than T_m in Eq. (22.9) by a small amount.

In a blind-hole, hole depth d is defined as the distance from the work surface to the depth of the full diameter, Figure 22.13(b). Thus, for a blind hole, machining time is given by

$$T_m = \frac{d + A}{f_r} \qquad (22.11)$$

where $A =$ the approach allowance by Eq. (22.10).

The rate of metal removal in drilling is determined as the product of the drill cross-sectional area and the feed rate:

$$R_{MR} = \frac{\pi D^2 f_r}{4} \qquad (22.12)$$

This equation is valid only after the drill reaches full diameter and excludes the initial approach of the drill into the work.

22.3.2 OPERATIONS RELATED TO DRILLING

Several operations are related to drilling. These are illustrated in Figure 22.14 and described in this section. Most of the operations follow drilling; a hole must be made first by drilling, and then the hole is modified by one of the other operations. Centering and spot facing are exceptions to this rule. All of the operations use rotating tools.

(a) **Reaming.** Reaming is used to slightly enlarge a hole, to provide a better tolerance on its diameter, and to improve its surface finish. The tool is called a ***reamer***, and it usually has straight flutes.

(b) **Tapping.** This operation is performed by a ***tap*** and is used to provide internal screw threads on an existing hole. Tapping is discussed in more detail in Section 22.7.1.

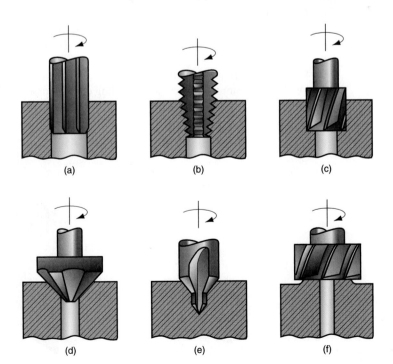

FIGURE 22.14
Machining operations related to drilling:
(a) reaming, (b) tapping,
(c) counterboring,
(d) countersinking,
(e) center drilling, and
(f) spot facing.

(c) ***Counterboring.*** Counterboring provides a stepped hole, in which a larger diameter follows a smaller diameter partially into the hole. A counterbored hole is used to seat bolt heads into a hole so the heads do not protrude above the surface.

(d) ***Countersinking.*** This is similar to counterboring, except that the step in the hole is cone-shaped for flat head screws and bolts.

(e) ***Centering.*** Also called center drilling, this operation drills a starting hole to accurately establish its location for subsequent drilling. The tool is called a ***center drill.***

(f) ***Spot facing.*** Spot facing is similar to milling. It is used to provide a flat machined surface on the workpart in a localized area.

22.3.3 DRILL PRESSES

The standard machine tool for drilling is the drill press. There are various types of drill press, the most basic of which is the upright drill, Figure 22.15. The ***upright drill*** stands on the floor and consists of a table for holding the workpart, a drilling head with powered spindle for the drill bit, and a base and column for support. A similar drill press, but smaller, is the ***bench drill,*** which is mounted on a table or bench rather than the floor.

The ***radial drill,*** Figure 22.16, is a large drill press designed to cut holes in large parts. It has a radial arm along which the drilling head can be moved and clamped. The head therefore can be positioned along the arm at locations that are a significant distance from the column to accommodate large work. The radial arm can also be swiveled about the column to drill parts on either side of the worktable.

The ***gang drill*** is a drill press consisting basically of two to six upright drills connected together in an in-line arrangement. Each spindle is powered and operated independently, and they share a common worktable, so that a series of drilling and related operations can be accomplished in sequence (e.g., centering, drilling, reaming, tapping) simply by sliding the workpart along the worktable from one spindle to the next. A related machine is the ***multiple-spindle drill,*** in which several drill spindles are connected together to drill multiple holes simultaneously into the workpart.

In addition, ***CNC drill presses*** are available to control the positioning of the holes in the workparts. These drill presses are often equipped with turrets to hold multiple tools that can be indexed under control of the CNC program. The term ***CNC turret drill*** is used for these machine tools.

Workholding on a drill press is accomplished by clamping the part in a vise, fixture, or jig. A ***vise*** is a general-purpose workholding device possessing two jaws that grasp the

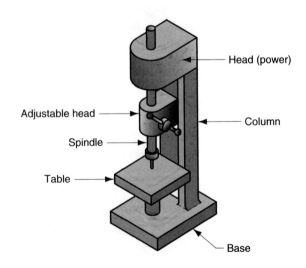

FIGURE 22.15   Upright drill press.

FIGURE 22.16 Radial drill press. (Courtesy of Willis Machinery and Tools Co., Toledo, Ohio.)

work in position. A *fixture* is a workholding device that is usually custom-designed for the particular workpart. The fixture can be designed to achieve higher accuracy in positioning the part relative to the machining operation, faster production rates, and greater operator convenience in use. A *jig* is a workholding device that is also specially designed for the workpart. The distinguishing feature between a jig and a fixture is that the jig provides a means of guiding the tool during the drilling operation. A fixture does not provide this tool guidance feature. A jig used for drilling is called a *drill jig.*

22.4 MILLING

Milling is a machining operation in which a workpart is fed past a rotating cylindrical tool with multiple cutting edges, as illustrated in Figure 22.2(d) and (e). (In rare cases, a tool with one cutting edge, called a *fly-cutter,* is used). The axis of rotation of the cutting tool is perpendicular to the direction of feed. This orientation between the tool axis and the feed direction is one of the features that distinguishes milling from drilling. In drilling, the cutting tool is fed in a direction parallel to its axis of rotation. The cutting tool in milling is called a *milling cutter* and the cutting edges are called teeth. Aspects of milling cutter geometry are discussed in Section 23.3.2. The conventional machine tool that performs this operation is a ***milling machine.*** The reader can view milling operations and the various milling machines in our video clip on milling and machining centers.

VIDEO CLIP _____

Milling and Machining Center Basics. View the segment titled Milling Cutters and Operations.

_____

The geometric form created by milling is a plane surface. Other work geometries can be created either by means of the cutter path or the cutter shape. Owing to the variety of shapes possible and its high production rates, milling is one of the most versatile and widely used machining operations.

Milling is an ***interrupted cutting*** operation; the teeth of the milling cutter enter and exit the work during each revolution. This interrupted cutting action subjects the teeth to a cycle of impact force and thermal shock on every rotation. The tool material and cutter geometry must be designed to withstand these conditions.

22.4.1 TYPES OF MILLING OPERATIONS

There are two basic types of milling operations, shown in Figure 22.17: (a) peripheral milling and (b) face milling. Most milling operations create geometry by generating the shape (Section 22.1).

Peripheral Milling   In peripheral milling, also called ***plain milling,*** the axis of the tool is parallel to the surface being machined, and the operation is performed by cutting edges on the outside periphery of the cutter. Several types of peripheral milling are shown in Figure 22.18: (a) ***slab milling,*** the basic form of peripheral milling in which the cutter width extends beyond the workpiece on both sides; (b) ***slotting,*** also called ***slot milling,*** in which the width of the cutter is less than the workpiece width, creating a slot in the work—when the cutter is very thin, this operation can be used to mill narrow slots or cut a workpart in two, called ***saw milling;*** (c) ***side milling,*** in which the cutter machines the side of the workpiece; (d) ***straddle milling,*** the same as side milling, only cutting takes place on both sides of the work; and ***form milling,*** in which the milling teeth have a

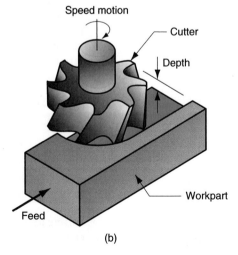

FIGURE 22.17   Two basic types of milling operations: (a) peripheral or plain milling and (b) face milling.

(a)

(b)

FIGURE 22.18
Peripheral milling: (a) slab milling, (b) slotting, (c) side milling, (d) straddle milling, and (e) form milling.

special profile that determines the shape of the slot that is cut in the work. Form milling is therefore classified as a forming operation (Section 22.1).

In peripheral milling, the direction of cutter rotation distinguishes two forms of milling: up milling and down milling, illustrated in Figure 22.19. In *up milling,* also called *conventional milling,* the direction of motion of the cutter teeth is opposite the feed direction when the teeth cut into the work. It is milling "against the feed." In *down milling,* also called *climb milling,* the direction of cutter motion is the same as the feed direction when the teeth cut the work. It is milling "with the feed."

The relative geometries of these two forms of milling result in differences in their cutting actions. In up milling, the chip formed by each cutter tooth starts out very thin and increases in thickness during the sweep of the cutter. In down milling, each chip starts out thick and reduces in thickness throughout the cut. The length of a chip in down milling is less than in up milling (the difference is exaggerated in our figure). This means that the cutter is engaged in the work for less time per volume of material cut, and this tends to increase tool life in down milling.

The cutting force direction is tangential to the periphery of the cutter for the teeth that are engaged in the work. In up milling, this has a tendency to lift the workpart as the cutter teeth exit the material. In down milling, this cutter force direction is downward, tending to hold the work against the milling machine table.

Face Milling   In face milling, the axis of the cutter is perpendicular to the surface being milled, and machining is performed by cutting edges on both the end and outside periphery of

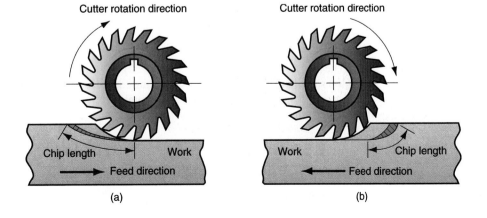

FIGURE 22.19   Two forms of peripheral milling operation with a 20-teeth cutter: (a) up milling, and (b) down milling.

FIGURE 22.20 Face milling: (a) conventional face milling, (b) partial face milling, (c) end milling, (d) profile milling, (e) pocket milling, and (f) surface contouring.

the cutter. As in peripheral milling, various forms of face milling exist, several of which are shown in Figure 22.20: (a) **conventional face milling,** in which the diameter of the cutter is greater than the workpart width, so the cutter overhangs the work on both sides; (b) **partial face milling,** where the cutter overhangs the work on only one side; (c) **end milling,** in which the cutter diameter is less than the work width, so a slot is cut into the part; (d) **profile milling,** a form of end milling in which the outside periphery of a flat part is cut; (e) **pocket milling,** another form of end milling used to mill shallow pockets into flat parts; and (f) **surface contouring,** in which a ball-nose cutter (rather than square-end cutter) is fed back and forth across the work along a curvilinear path at close intervals to create a three-dimensional surface form. The same basic cutter control is required to machine the contours of mold and die cavities, in which case the operation is called **die sinking.**

22.4.2 CUTTING CONDITIONS IN MILLING

The cutting speed is determined at the outside diameter of a milling cutter. This can be converted to spindle rotation speed using a formula that should now be familiar:

$$N = \frac{v}{\pi D} \tag{22.13}$$

The feed f in milling is usually given as a feed per cutter tooth; called the **chip load,** it represents the size of the chip formed by each cutting edge. This can be converted to feed rate by taking into account the spindle speed and the number of teeth on the cutter as follows:

$$f_r = N n_t f \tag{22.14}$$

where f_r = feed rate, mm/min (in/min); N = spindle speed, rev/min; n_t = number of teeth on the cutter; and f = chip load in mm/tooth (in/tooth).

Material removal rate in milling is determined using the product of the cross-sectional area of the cut and the feed rate. Accordingly, if a slab-milling operation is

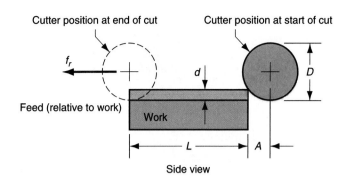

FIGURE 22.21 Slab (peripheral) milling showing entry of cutter into the workpiece.

cutting a workpiece with width w at a depth d, the material removal rate is

$$R_{MR} = wdf_r \qquad (22.15)$$

This neglects the initial entry of the cutter before full engagement. Eq. (22.15) can be applied to end milling, side milling, face milling, and other milling operations, making the proper adjustments in the computation of cross-sectional area of cut.

The time required to mill a workpiece of length L must account for the approach distance required to fully engage the cutter. First, consider the case of slab milling, Figure 22.21. To determine the time to perform a slab milling operation, the approach distance A to reach full cutter depth is given by

$$A = \sqrt{d(D - d)} \qquad (22.16)$$

where d = depth of cut, mm (in); and D = diameter of the milling cutter, mm (in). The time T_m in which the cutter is engaged milling the workpiece is therefore

$$T_m = \frac{L + A}{f_r} \qquad (22.17)$$

For face milling, let us consider the two possible cases pictured in Figure 22.22. The first case is when the cutter is centered over a rectangular workpiece as in Figure 22.22(a). The cutter feeds from right to left across the workpiece. In order for the cutter to reach the full width of the work, it must travel an approach distance given by the following:

$$A = 0.5\left(D - \sqrt{D^2 - w^2}\right) \qquad (22.18)$$

FIGURE 22.22 Face milling showing approach and overtravel distances for two cases: (a) when cutter is centered over the workpiece, and (b) when cutter is offset to one side over the work.

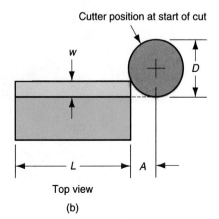

where D = cutter diameter, mm (in) and w = width of the workpiece, mm (in). If $D = w$, then Eq. (22.18) reduces to $A = 0.5D$. And if $D < w$, then a slot is cut into the work and $A = 0.5D$.

The second case is when the cutter is offset to one side of the work, as in Figure 22.22(b). In this case, the approach distance is given by

$$A = \sqrt{w(D - w)} \tag{22.19}$$

where w = width of the cut, mm (in). In either case, the machining time is given by

$$T_m = \frac{L + A}{f_r} \tag{22.20}$$

It should be emphasized in all of these milling scenarios that T_m represents the time the cutter teeth are engaged in the work, making chips. Approach and overtravel distances are usually added at the beginning and end of each cut to allow access to the work for loading and unloading. Thus the actual duration of the cutter feed motion is likely to be greater than T_m.

22.4.3 MILLING MACHINES

Milling machines must provide a rotating spindle for the cutter and a table for fastening, positioning, and feeding the workpart. Various machine tool designs satisfy these requirements. To begin with, milling machines can be classified as horizontal or vertical. A *horizontal milling machine* has a horizontal spindle, and this design is well suited for performing peripheral milling (e.g., slab milling, slotting, side and straddle milling) on workparts that are roughly cube shaped. A *vertical milling machine* has a vertical spindle, and this orientation is appropriate for face milling, end milling, surface contouring, and die-sinking on relatively flat workparts.

Other than spindle orientation, milling machines can be classified into the following types: (1) knee-and-column, (2) bed type, (3) planer type, (4) tracer mills, and (5) CNC milling machines.

The *knee-and-column milling machine* is the basic machine tool for milling. It derives its name from the fact that its two main components are a *column* that supports the spindle, and a *knee* (roughly resembling a human knee) that supports the worktable. It is available as either a horizontal or a vertical machine, as illustrated in Figure 22.23. In the horizontal version, an arbor usually supports the cutter. The *arbor* is basically a shaft that holds the milling cutter and is driven by the spindle. An overarm is provided on

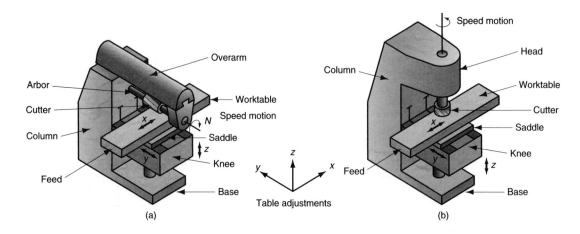

FIGURE 22.23 Two basic types of knee-and-column milling machine: (a) horizontal and (b) vertical.

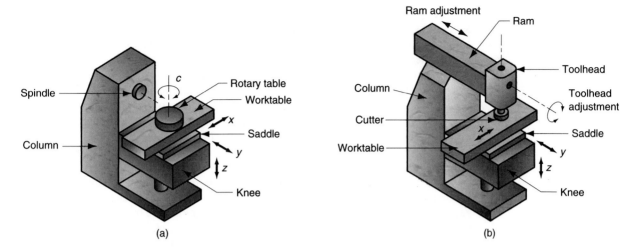

FIGURE 22.24   Special types of knee-and-column milling machine: (a) universal—overarm, arbor, and cutter omitted for clarity: and (b) ram type.

horizontal machines to support the arbor. On vertical knee-and-column machines, milling cutters can be mounted directly in the spindle without an arbor.

One of the features of the knee-and-column milling machine that makes it so versatile is its capability for worktable feed movement in any of the x–y–z axes. The worktable can be moved in the x-direction, the saddle can be moved in the y-direction, and the knee can be moved vertically to achieve the z-movement.

Two special knee-and-column machines should be identified. One is the ***universal*** milling machine, Figure 22.24(a), which has a table that can be swiveled in a horizontal plane (about a vertical axis) to any specified angle. This facilitates the cutting of angular shapes and helixes on workparts. Another special machine is the ***ram mill,*** Figure 22.24(b), in which the toolhead containing the spindle is located on the end of a horizontal ram; the ram can be adjusted in and out over the worktable to locate the cutter relative to the work. The toolhead can also be swiveled to achieve an angular orientation of the cutter with respect to the work. These features provide considerable versatility in machining a variety of work shapes.

Bed-type milling machines are designed for high production. They are constructed with greater rigidity than knee-and-column machines, thus permitting them to achieve heavier feed rates and depths of cut needed for high material removal rates. The characteristic construction of the bed-type milling machine is shown in Figure 22.25.

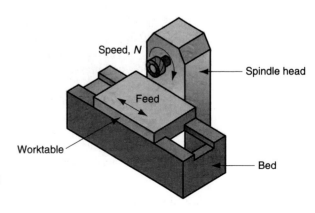

FIGURE 22.25   Simplex bed-type milling machine horizontal spindle.

The worktable is mounted directly to the bed of the machine tool, rather than using the less rigid knee-type design. This construction limits the possible motion of the table to longitudinal feeding of the work past the milling cutter. The cutter is mounted in a spindle head that can be adjusted vertically along the machine column. Single spindle bed machines are called *simplex* mills, as in Figure 22.25, and are available in either horizontal or vertical models. *Duplex* mills use two spindle heads. The heads are usually positioned horizontally on opposite sides of the bed to perform simultaneous operations during one feeding pass of the work. *Triplex* mills add a third spindle mounted vertically over the bed to further increase machining capability.

Planer type mills are the largest milling machines. Their general appearance and construction are those of a large planer (see Figure 22.31); the difference is that milling is performed instead of planing. Accordingly, one or more milling heads are substituted for the single-point cutting tools used on planers, and the motion of the work past the tool is a feed rate motion rather than a cutting speed motion. Planer mills are built to machine very large parts. The worktable and bed of the machine are heavy and relatively low to the ground, and the milling heads are supported by a bridge structure that spans across the table.

A *tracer mill,* also called a *profiling mill,* is designed to reproduce an irregular part geometry that has been created on a template. Using either manual feed by a human operator or automatic feed by the machine tool, a tracing probe is controlled to follow the template while a milling head duplicates the path taken by the probe to machine the desired shape. Tracer mills are of two types: (1) *x−y tracing,* in which the contour of a flat template is profile milled using two-axis control; and (2) *x–y–z tracing,* in which the probe follows a three-dimensional pattern using three-axis control. Tracer mills have been used for creating shapes that cannot easily be generated by a simple feeding action of the work against the milling cutter. Applications include molds and dies. In recent years, many of these applications have been taken over by CNC milling machines.

Computer numerical control milling machines are milling machines in which the cutter path is controlled by alphanumerical data rather than a physical template. They are especially suited to profile milling, pocket milling, surface contouring, and die sinking operations, in which two or three axes of the worktable must be simultaneously controlled to achieve the required cutter path. An operator is normally required to change cutters as well as load and unload workparts.

22.5 MACHINING CENTERS AND TURNING CENTERS

A *machining center,* illustrated in Figure 22.26, is a highly automated machine tool capable of performing multiple machining operations under computer numerical control in one setup with minimal human attention. Workers are needed to load and unload parts, which usually takes considerable less time than the machine cycle time, so one worker may be able to tend more than one machine. Typical operations performed on a machining center are milling and drilling, which use rotating cutting tools.

The typical features that distinguish a machining center from conventional machine tools and make it so productive include:

➢ *Multiple operations in one setup.* Most workparts require more than one operation to completely machine the specified geometry. Complex parts may require dozens of distinct machining operations, each requiring its own machine tool, setup, and cutting tool. Machining centers are capable of performing most or all of the operations at one location, thus minimizing setup time and production lead time.

➢ *Automatic tool changing.* To change from one machining operation to the next, the cutting tools must be changed. This is done on a machining center under CNC

FIGURE 22.26
A universal machining center. Capability to orient the workhead makes this a five-axis machine. (Courtesy of Cincinnati Milacron, Batavia, Ohio.)

program control by an automatic tool-changer designed to exchange cutters between the machine tool spindle and a ***tool storage carousels.*** Capacities of these carousels commonly range from 16 to 80 cutting tools. The machine in Figure 22.26 has two storage carousels on the left side of the column.

➤ ***Pallet shuttles.*** Some machining centers are equipped with pallet shuttles, which are automatically transferred between the spindle position and the loading station, as shown in Figure 22.26. Parts are fixtured on pallets that are attached to the shuttles. In this arrangement, the operator can be unloading the previous part and loading the next part while the machine tool is engaged in machining the current part. Nonproductive time on the machine is thereby reduced.

➤ ***Automatic workpart positioning.*** Many machining centers have more than three axes. One of the additional axes is often designed as a rotary table to position the part at some specified angle relative to the spindle. The rotary table permits the cutter to perform machining on four sides of the part in a single setup.

Machining centers are classified as horizontal, vertical, or universal. The designation refers to spindle orientation. Horizontal machining centers normally machine cube-shaped parts, in which the four vertical sides of the cube can be accessed by the cutter. Vertical machining centers are suited to flat parts on which the tool can machine the top surface. Universal machining centers have workheads that swivel their spindle axes to any angle between horizontal and vertical, as in Figure 22.26. Our video clip on machining centers shows several of these machines.

VIDEO CLIP ————————————————————————————————

Milling and Machining Center Basics. The relevant segments are: (1) vertical machining centers, (2) horizontal machining centers, and (3) machining center workholding.

FIGURE 22.27
Computer numerical control, four-axis turning center. (Courtesy of Cincinnati Milacron, Batavia, Ohio.).

FIGURE 22.28   Operation of a mill-turn center: (a) example part with turned, milled, and drilled surfaces; and (b) sequence of operations on a mill-turn center: (1) turn second diameter, (2) mill flat with part in programmed angular position, (3) drill hole with part in same programmed position, and (4) cutoff.

Success of CNC machining centers led to the development of CNC turning centers. A modern *CNC turning center,* Figure 22.27, is capable of performing various turning and related operations, contour turning, and automatic tool indexing, all under computer control. In addition, the most sophisticated turning centers can accomplish (1) workpart gaging (checking key dimensions after machining), (2) tool monitoring (sensors to indicate when the tools are worn), (3) automatic tool changing when tools become worn, and even (4) automatic workpart changing at the completion of the work cycle [14].

Another type of machine tool related to machining centers and turning centers is the *CNC mill-turn center.* This machine has the general configuration of a turning center; in addition, it can position a cylindrical workpart at a specified angle so that a rotating cutting tool (e.g., milling cutter) can machine features into the outside surface of the part, as illustrated in Figure 22.28. An ordinary turning center does not have the capability to stop the workpart at a defined angular position, and it does not possess rotating tool spindles.

Further progress in machine tool technology has taken the mill-turn center one step further by integrating additional capabilities into a single machine. The additional capabilities include (1) combining milling, drilling, and turning with grinding, welding, and inspection operations, all in one machine tool; (2) using multiple spindles simultaneously, either on a single workpiece or two different workpieces; and (3) automating the part handling function by adding industrial robots to the machine [2], [20]. The terms *multitasking machine* and *multifunction machine* are sometimes used for these products.

22.6   OTHER MACHINING OPERATIONS

In addition to turning, drilling, and milling, several other machining operations should be included in our survey: (1) shaping and planing, (2) broaching, and (3) sawing.

22.6.1   SHAPING AND PLANING

Shaping and planing are similar operations, both involving the use of a single-point cutting tool moved linearly relative to the workpart. In conventional shaping and planing, a straight, flat surface is created by this action. The difference between the two operations is illustrated in Figure 22.29. In shaping, the speed motion is accomplished by moving the cutting tool; while in planing, the speed motion is accomplished by moving the workpart.

Cutting tools used in shaping and planing are single-point tools (Section 23.3.1). Unlike turning, interrupted cutting occurs in shaping and planing, subjecting the tool to

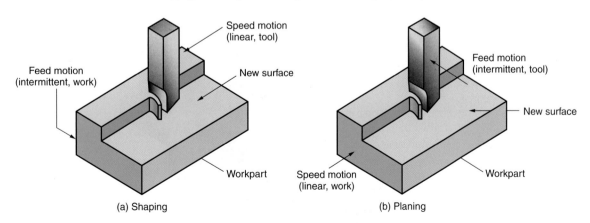

FIGURE 22.29   (a) Shaping, and (b) planing.

FIGURE 22.30
Components of a shaper.

an impact loading upon entry into the work. In addition, these machine tools are limited to low speeds due to their start-and-stop motion. The conditions normally dictate use of high-speed steel cutting tools.

Shaping Shaping is performed on a machine tool called a *shaper*, Figure 22.30. The components of the shaper include a *ram*, which moves relative to a *column* to provide the cutting motion, and a worktable that holds the part and accomplishes the feed motion. The motion of the ram consists of a forward stroke to achieve the cut, and a return stroke during which the tool is lifted slightly to clear the work and then reset for the next pass. On completion of each return stroke, the worktable is advanced laterally relative to the ram motion in order to feed the part. Feed is specified in mm/stroke (in/stroke). The drive mechanism for the ram can be either hydraulic or mechanical. Hydraulic drive has greater flexibility in adjusting the stroke length and a more uniform speed during the forward stroke, but it is more expensive than a mechanical drive unit. Both mechanical and hydraulic drives are designed to achieve higher speeds on the return (noncutting) stroke than on the forward (cutting) stroke, thereby increasing the proportion of time spent cutting.

Planing The machine tool for planing is a *planer.* Cutting speed is achieved by a reciprocating worktable that moves the part past the single-point cutting tool. The construction and motion capability of a planer permit much larger parts to be machined than on a shaper. Planers can be classified as open side planers or double-column planers. The *open-side planer,* also known as a *single-column planer,* Figure 22.31, has a single

FIGURE 22.31
Open-side planer.

FIGURE 22.32 Types of shapes that can cut by shaping and planing: (a) V-groove, (b) square groove, (c) T-slot, (d) dovetail slot, and (e) gear teeth.

(a)   (b)   (c)   (d)   (e)

column supporting the cross-rail on which a toolhead is mounted. Another toolhead can also be mounted and fed along the vertical column. Multiple toolheads permit more than one cut to be taken on each pass. At the completion of each stroke, each toolhead is moved relative to the cross-rail (or column) to achieve the intermittent feed motion. The configuration of the open-side planer permits very wide workparts to be machined.

A *double-column planer* has two columns, one on either side of the base and worktable. The columns support the cross-rail, on which one or more toolheads are mounted. The two columns provide a more rigid structure for the operation; however, the two columns limit the width of the work that can be handled on this machine.

Shaping and planing can be used to machine shapes other than flat surfaces. The restriction is that the cut surface must be straight. This allows the cutting of grooves, slots, gear teeth, and other shapes as illustrated in Figure 22.32. Special machines and tool geometries must be specified to cut some of these shapes. An important example is the *gear shaper,* a vertical shaper with a specially designed rotary feed table and synchronized tool head used to generate teeth on spur gears. Gear shaping and other methods of producing gears are discussed in Section 22.7.2.

22.6.2 BROACHING

Broaching is performed using a multiple-teeth cutting tool by moving the tool linearly relative to the work in the direction of the tool axis, as in Figure 22.33. The machine tool is called a *broaching machine,* and the cutting tool is called a *broach.* Aspects of broach geometry are discussed in Section 23.3.2. In certain jobs for which broaching can be used, it is a highly productive method of machining. Advantages include good surface finish, close tolerances, and a variety of work shapes. Owing to the complicated and often custom-shaped geometry of the broach, tooling is expensive.

There are two principal types of broaching: external (also called surface broaching) and internal. *External broaching* is performed on the outside surface of the work to create a certain cross-sectional shape on the surface. Figure 22.34(a) shows some possible cross sections that can be formed by external broaching. *Internal broaching* is accomplished on the internal surface of a hole in the part. Accordingly, a starting hole must be present in the

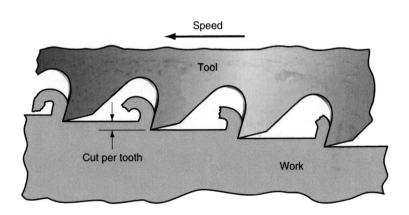

FIGURE 22.33 The broaching operation.

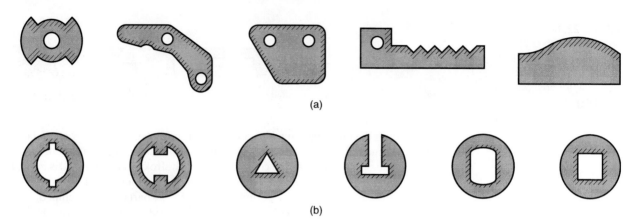

FIGURE 22.34 Work shapes that can be cut by: (a) external broaching, and (b) internal broaching. Cross-hatching indicates the surfaces broached.

part so as to insert the broach at the beginning of the broaching stroke. Figure 22.34(b) indicates some of the shapes that can be produced by internal broaching.

The basic function of a broaching machine is to provide a precise linear motion of the tool past a stationary work position, but there are various ways in which this can be done. Most broaching machines can be classified as either vertical or horizontal machines. The **vertical broaching machine** is designed to move the broach along a vertical path, while the **horizontal broaching machine** has a horizontal tool trajectory. Most broaching machines pull the broach past the work. However, there are exceptions to this pull action. One exception is a relatively simple type called a **broaching press,** used only for internal broaching, that pushes the tool through the workpart. Another exception is the **continuous broaching machine,** in which the workparts are fixtured to an endless belt loop and moved past a stationary broach. Because of its continuous operation, this machine can be used only for surface broaching.

22.6.3 SAWING

Sawing is a process in which a narrow slit is cut into the work by a tool consisting of a series of narrowly spaced teeth. Sawing is normally used to separate a workpart into two pieces, or to cut off an unwanted portion of a part. These operations are often referred to as **cutoff** operations. Since many factories require cutoff operations at some point in the production sequence, sawing is an important manufacturing process.

In most sawing operations, the work is held stationary and the **saw blade** is moved relative to it. Saw blade tooth geometry is discussed in Section 23.3.2. There are three basic types of sawing, as in Figure 22.35, according to the type of blade motion involved: (a) hacksawing, (b) bandsawing, and (c) circular sawing.

Hacksawing, Figure 22.35(a), involves a linear reciprocating motion of the saw against the work. This method of sawing is often used in cutoff operations. Cutting is accomplished only on the forward stroke of the saw blade. Because of this intermittent cutting action, hacksawing is inherently less efficient than the other sawing methods, both of which are continuous. The **hacksaw** blade is a thin straight tool with cutting teeth on one edge. Hacksawing can be done either manually or with a power hacksaw. A **power hacksaw** provides a drive mechanism to operate the saw blade at a desired speed; it also applies a given feed rate or sawing pressure.

Bandsawing involves a linear continuous motion, using a **bandsaw blade** made in the form of an endless flexible loop with teeth on one edge. The sawing machine is a **bandsaw,**

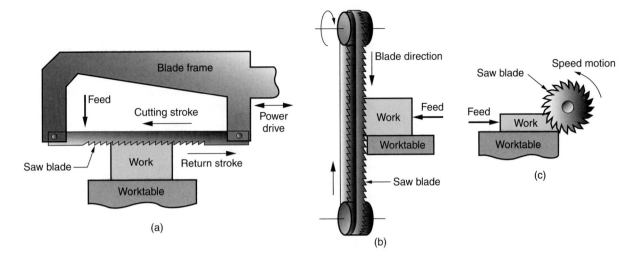

FIGURE 22.35   Three types of sawing operations: (a) power hacksaw, (b) bandsaw (vertical), and (c) circular saw.

which provides a pulley-like drive mechanism to continuously move and guide the bandsaw blade past the work. Bandsaws are classified as vertical or horizontal. The designation refers to the direction of saw blade motion during cutting. Vertical bandsaws are used for cutoff as well as other operations such as contouring and slotting. *Contouring* on a bandsaw involves cutting a part profile from flat stock. *Slotting* is the cutting of a thin slot into a part, an operation for which bandsawing is well suited. Contour sawing and slotting are operations in which the work is fed into the saw blade.

Vertical bandsaw machines can be operated either manually, in which the operator guides and feeds the work past the bandsaw blade, or automatically, in which the work is power fed past the blade. Recent innovations in bandsaw design have permitted the use of CNC to perform contouring of complex outlines. Some of the details of the vertical bandsawing operation are illustrated in Figure 22.35(b). Horizontal bandsaws are normally used for cutoff operations as alternatives to power hacksaws.

Circular sawing, Figure 22.35(c), uses a rotating saw blade to provide a continuous motion of the tool past the work. Circular sawing is often used to cut long bars, tubes, and similar shapes to specified length. The cutting action is similar to a slot milling operation, except that the saw blade is thinner and contains many more cutting teeth than a slot milling cutter. Circular sawing machines have powered spindles to rotate the saw blade and a feeding mechanism to drive the rotating blade into the work.

Two operations related to circular sawing are abrasive cutoff and friction sawing. In *abrasive cutoff,* an abrasive disk is used to perform cutoff operations on hard materials that would be difficult to saw with a conventional saw blade. In *friction sawing,* a steel disk is rotated against the work at very high speeds, resulting in friction heat that causes the material to soften sufficiently to permit penetration of the disk through the work. The cutting speeds in both of these operations are much faster than in circular sawing.

22.7   MACHINING OPERATIONS FOR SPECIAL GEOMETRIES

One of the reasons for the technological importance of machining is its capability to produce unique geometric features such as screw threads and gear teeth. In this section we discuss the cutting processes that are used to accomplish these shapes, most of which are adaptations of machining operations discussed earlier in the chapter.

22.7.1 SCREW THREADS

Threaded hardware components are widely used as fasteners in assembly (screws, bolts, and nuts, Section 32.1) and for transmission of motion in machinery (e.g., lead screws in positioning systems, Section 38.3.2). We can define threads as grooves that form a spiral around the outside of a cylinder (external threads) or the inside of a round hole (internal threads). We have previously considered the manufacture of threaded components in our coverage of thread rolling in Section 19.2. Thread rolling is by far the most common method for producing external threads, but the process is not economical for low production quantities and the work metal must be ductile. Metallic threaded components can also be made by casting, especially investment casting and die casting (Sections 11.2.4 and 11.3.3), and plastic parts with threads can be injection molded (Section 13.6). Finally, threaded components can be machined, and this is the topic we address here. The discussion is organized into external and internal thread machining.

External Threads   The simplest and most versatile method of cutting an external thread on a cylindrical workpart is **_single-point threading,_** which employs a single-point cutting tool on a lathe. This process is illustrated in Figure 22.6(g). The starting diameter of the workpiece is equal to the major diameter of the screw thread. The tool must have the profile of the thread groove, and the lathe must be capable of maintaining the same relationship between the tool and the workpiece on successive passes in order to cut a consistent spiral. This relationship is achieved by means of the lathe's lead screw (see Figure 22.7). More than one turning pass is usually required. The first pass takes a light cut; the tool is then retracted and rapidly traversed back to the starting point; and each ensuing pass traces the same spiral using ever greater depths of cut until the desired form of the thread groove has been established. Single-point threading is suitable for low or even medium production quantities, but less time-consuming methods are more economical for high production.

An alternative to using a single-point tool is a **_threading die,_** shown in Figure 22.36. To cut an external thread, the die is rotated around the starting cylindrical stock of the proper diameter, beginning at one end and proceeding to the other end. The cutting teeth at the opening of the die are tapered so that the starting depth of cut is less at the beginning of the operation, finally reaching full thread depth at the trailing side of the die. The pitch of the threading die teeth determines the pitch of the screw that is being cut. The die in Figure 22.36 has a slit that allows the size of the opening to be adjusted to compensate for tool wear on the teeth or to provide for minor differences in screw size. Threading dies cut the threads in a single pass rather than multiple passes as in single-point threading.

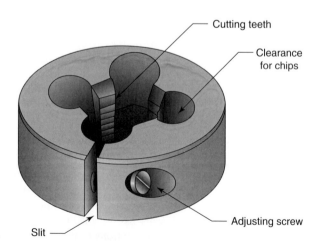

FIGURE 22.36   Threading die.

Threading dies are typically used in manual operations, in which the die is fixed in a holder that can be rotated by hand. If the workpiece has a head or other obstacle at the other end, the die must be unwound from the screw just created in order to remove it. This is not only time consuming, but it also risks possible damage to the thread surfaces. In mechanized threading operations, cycle times can be reduced by using *self-opening threading dies,* which are designed with an automatic device that opens the cutting teeth at the end of each cut. This eliminates the need to unwind the die from the work and avoids possible damage to the threads. Self-opening dies are equipped with four sets of cutting teeth, similar to the threading die in Figure 22.36, except that the teeth can be adjusted and removed for resharpening, and the toolholder mechanism possesses the self-opening feature. Different sets of cutting teeth are required for different thread sizes.

The term *thread chasing* is often applied to production operations that utilize self-opening dies. Two types of thread chasing equipment are available: (1) stationary self-opening dies, in which the workpiece rotates and the die does not, like a turning operation; and (2) revolving self-opening dies, in which the die rotates and the workpiece does not, like a drilling operation.

Two additional external threading operations should be mentioned: thread milling and thread grinding. *Thread milling* involves the use of a milling cutter to shape the threads of a screw. One possible setup is illustrated in Figure 22.37. In this operation a form-milling cutter, whose profile is that of the thread groove, is oriented at an angle equal to the helix angle of the thread and fed longitudinally as the workpiece is slowly rotated. In a variation of this operation, a multiple-form cutter is used, so that multiple screw threads can be cut simultaneously to increase production rates. Possible reasons for preferring thread milling over thread chasing include (1) the size of the thread is too large to be readily cut with a die and (2) thread milling is generally noted to produce more accurate and smoother threads.

Thread grinding is similar to thread milling except the cutter is a grinding wheel with the shape of the thread groove, and the rotational speed of the grinding wheel is much greater than in milling. The process can be used to completely form the threads or to finish

FIGURE 22.37   Thread milling using a form-milling cutter.

threads that have been formed by one of the previously discussed processes. Thread grinding is especially applicable for threads that have been hardened by heat treatment.

Internal Threads   The most common process for cutting internal threads is ***tapping,*** in which a cylindrical tool with cutting teeth arranged in a spiral whose pitch is equal to that of the screw threads, is simultaneously rotated and fed into a pre-existing hole. The operation is illustrated in Figure 22.14(b), and the cutting tool is called a ***tap.*** The end of the tool is slightly conical to facilitate entry into the hole. The initial hole size is approximately equal to the minor diameter of the screw thread. In the simplest version of the process, the tap is a solid piece and the tapping operation is performed on a drill press equipped with a tapping head, which allows penetration into the hole at a rate that corresponds to the screw pitch. At the end of the operation, the spindle rotation is reversed so the tap can be unscrewed from the hole.

In addition to solid taps, collapsible taps are available, just as self-opening dies are available for external threading. ***Collapsible taps*** have cutting teeth that automatically retract into the tool when the thread has been cut, allowing it to be quickly removed from the tapped hole without reversing spindle direction. Thus, shorter cycle times are possible.

Although production tapping can be accomplished on drill presses and other conventional machine tools (e.g., lathes, turret lathes), several types of specialized machines have been developed for higher production rates. Single-spindle tapping machines perform tapping one workpiece at a time, with manual or automatic loading and unloading of the starting blanks. Multiple-spindle tapping machines operate on multiple work parts simultaneously and provide for different hole sizes and screw pitches to be accomplished together. Finally gang drills (Section 22.3.3) can be set up to perform drilling, reaming, and tapping in rapid sequence on the same part.

22.7.2 GEARS

Gears are machinery components used to transmit motion and power between rotating shafts. As illustrated in Figure 22.38, the transmission of rotational motion is achieved

FIGURE 22.38
Two meshing spur gears.

between meshing gears by teeth located around their respective circumferences. The teeth have a special curved shape called an involute, which minimizes friction and wear between contacting teeth of meshing gears. Depending on the relative numbers of teeth of the two gears, the speed of rotation can be increased or decreased from one gear to the next, with a corresponding decrease or increase in torque. We examine these speed effects in our discussion of numerical control positioning systems in Section 38.3.2.

There are various gear types, the most basic and least complicated to produce is the ***spur gear*** represented in Figure 22.38. It has teeth that are parallel to the axis of the gear's rotation. A gear with teeth that form an angle relative to the axis of rotation is called a ***helical gear.*** The helical tooth design allows more than one tooth to be in contact for smoother operation. Spur and helical gears provide rotation between shafts whose axes are parallel. Other types, such as ***bevel gears,*** provide motion between shafts that are at an angle with each other, usually 90°. A ***rack*** is a straight gear (a gear of infinite radius), which allows rotational motion to be converted into linear motion (e.g., rack-and-pinion steering on automobiles). The variety of gear types is far too great for us to discuss them all, and the interested reader is referred to texts on machine design for coverage of gear design and mechanics. Our interest here is on the manufacture of gears.

Several of the shape processing operations discussed in previous chapters can be used to produce gears. These include investment casting, die casting, plastic injection molding, powder metallurgy, forging, and other bulk deformation operations (e.g., gear rolling, Section 19.2). The advantage of these operations over machining is material savings because no chips are produced. Sheet-metal stamping operations (Section 20.1) are used to produce thin gears used in watches and clocks. The gears produced by all of the preceding operations can often be used without further processing. In other cases, a basic shape processing operation such as casting or forging is used to produce a starting metal blank, and these parts are then machined to form the gear teeth. Finishing operations are often required to achieve the specified accuracies of the teeth dimensions.

The principal machining operations used to cut gear teeth are form milling, gear hobbing, gear shaping, and gear broaching. Form milling and gear broaching are considered to be forming operations in the sense of Section 22.1, while gear hobbing and gear shaping are classified as generating operations. Finishing processes for gear teeth include gear shaving, gear grinding, and burnishing. The video clip on gears and gear manufacturing illustrates the various aspects of gear technology. Many of the processes used to make gears are also used to produce splines, sprockets, and other special machinery components.

VIDEO CLIP _____

Gears and Gear Manufacturing. This clip contains two segments: (1) gear functions and (2) gear machining methods.

_____

Form Milling   In this process, illustrated in Figure 22.39, the teeth on a gear blank are machined individually by a form-milling cutter whose cutting edges have the shape of the spaces between the teeth on the gear. The machining operation is classified as forming (Section 22.1) because the shape of the cutter determines the geometry of the gear teeth. The disadvantage of form milling is that production rates are slow because each tooth space is created one at a time and the gear blank must be indexed between each pass to establish the correct size of the gear tooth, which also takes time. The advantage of form milling over gear hobbing (discussed next) is that the milling cutter is much less expensive. The slow production rates and relatively low-cost tooling make form milling appropriate for low-production quantities.

Gear Hobbing   Gear hobbing is also a milling operation, but the cutter, called a ***hob,*** is much more complex and therefore much more expensive than a form milling cutter. In

FIGURE 22.39 Form milling of gear teeth on a starting blank.

addition, special milling machines (called **hobbing machines**) are required to accomplish the relative speed and feed motions between the cutter and the gear blank. Gear hobbing is illustrated in Figure 22.40. As shown in the figure, the hob has a slight helix and its rotation must be coordinated with the much slower rotation of the gear blank in order for the hob's cutting teeth to mesh with the blank's teeth as they are being cut. This is accomplished for a spur gear by offsetting the axis of rotation of the hob by an amount equal to 90° less the helix angle relative to the axis of the gear blank. In addition to these rotary motions of the hob and the workpiece, a straight-line motion is also required to feed the hob relative to the gear blank throughout its thickness. Several teeth are cut simultaneously in hobbing, which allows for higher production rates than form milling. Accordingly, it is a widely used gear making process for medium and high production quantities.

Gear Shaping   In gear shaping, a reciprocating cutting tool motion is used rather than a rotational motion as in form milling and gear hobbing. Two quite different forms of shaping operation (Section 22.6.1) are used to produce gears. In the first type, a single-point tool takes multiple passes to gradually shape each tooth profile using computerized controls or a template. The gear blank is slowly rotated or indexed, with the same profile being imparted to each tooth. The procedure is slow and applied only in the fabrication of very large gears.

In the second type of gear shaping operation, the cutter has the general shape of a gear, with cutting teeth on one side. The axes of the cutter and the gear blank are parallel, as illustrated in Figure 22.41, and the action is similar to a pair of conjugate gears except

FIGURE 22.40   Gear hobbing.

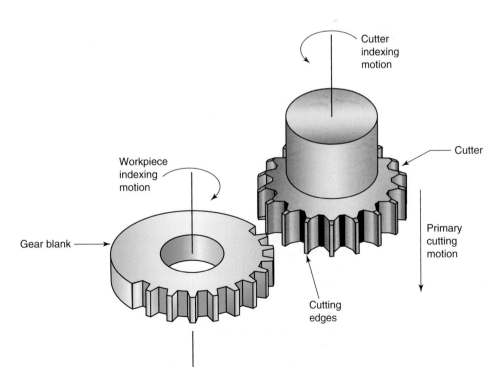

FIGURE 22.41   Gear shaping.

that the reciprocation of the cutter is gradually creating the form of the matching teeth in its mating component. At the beginning of the operation for a given gear blank, the cutter is fed into the blank after each stroke until the required depth has been reached. Then, after each successive pass of the tool, both the cutter and the blank are rotated a small amount (indexed) so as to maintain the same tooth spacing on each. Gear shaping by this second method is widely used in industry, and specialized machines (called *gear shapers*) are available to accomplish the process.

Gear Broaching   Broaching (Section 22.6.2) as a gear making process is noted for short production cycle times and high tooling cost. It is therefore economical only for high volumes. Good dimensional accuracy and fine surface finish are also features of gear broaching. The process can be applied for both external gears (the conventional gear) and internal gears (teeth on the inside of the gear). For making internal gears, the operation is similar to that shown in Figure 22.3(c), except the cross section of the tool consists of a series of gear-shaped cutting teeth of increasing size to form the gear teeth in successive steps as the broach is drawn through the work blank. To produce external gears, the broach is tubular with inward-facing teeth. As mentioned, the cost of tooling in both cases is high due to the complex geometry.

Finishing Operations   Some metal gears can be used without heat treatment, while those used in more demanding applications are usually heat treated to harden the teeth for maximum wear resistance. Unfortunately, heat treatment (Chapter 27) often results in warpage of the workpiece, and the proper gear-tooth shape must be restored. Whether heat treated or not, some type of finishing operation is generally required to improve dimensional accuracy and surface finish of the gear after machining. Finishing processes applied to gears that have not been heat treated include shaving and burnishing. Finishing processes applied to hardened gears include grinding, lapping, and honing (Chapter 25).

Gear shaving involves the use of a gear-shaped cutter that is meshed and rotated with the gear. Cutting action results from reciprocation of the cutter during rotation. Each tooth of the gear-shaped cutter has multiple cutting edges along its width, producing very small chips and removing very little metal from the surface of each gear tooth. Gear shaping is probably the most common industrial process for finishing gears. It is often applied to a gear prior to heat treatment, and then followed by grinding and/or lapping after heat treatment.

Gear burnishing is a plastic deformation process in which one or more hardened gear-shaped dies are rolled in contact with the gear, and pressure is applied by the dies to effect cold working of the gear teeth. Thus, the teeth are strengthened through strain hardening, and surface finish is improved.

Grinding, honing, and lapping are three finishing processes that can be used on hardened gears. *Gear grinding* can be based on either of two methods. The first is form grinding, in which the grinding wheel has the exact shape of the tooth spacing (similar to form milling), and a grinding pass or series of passes are made to finish form each tooth in the gear. The other method involves generating the tooth profile using a conventional straight-sided grinding wheel. Both of these grinding methods are very time consuming and expensive.

Honing and lapping, discussed in Section 25.2.1 and 25.2.2, respectively, are two finishing processes that can be adapted to gear finishing using very fine abrasives. The tools in both processes usually possess the geometry of a gear that meshes with the gear to be processed. Gear honing uses a tool that is made of either plastic impregnated with abrasives or steel coated with carbide. Gear lapping uses a cast iron tool (other metals are sometimes substituted), and the cutting action is accomplished by the lapping compound containing abrasives.

22.8 HIGH-SPEED MACHINING

One persistent trend throughout the history of metal machining has been the use of higher and higher cutting speeds. In recent years, there has been renewed interest in this area due to its potential for faster production rates, shorter lead times, reduced costs, and improved part quality. In its simplest definition, *high-speed machining* (HSM) means using cutting speeds that are significantly higher than those used in conventional machining operations. Some examples of cutting speed values for conventional and HSM are presented in Table 22.1, according to data compiled by Kennametal Inc.[1]

Other definitions of HSM have been developed to deal with the wide variety of work materials and tool materials used in machining. One popular HSM definition is the *DN ratio*—the bearing bore diameter (mm) multiplied by the maximum spindle speed (rev/min). For high-speed machining, the typical DN ratio is between 500,000 and 1,000,000. This definition allows larger diameter bearings to fall within the HSM range, even though they operate at lower rotational speeds than smaller bearings. Typical HSM spindle velocities range between 8000 and 35,000 rpm, although some spindles today are designed to rotate at 100,000 rpm.

Another HSM definition is based on the ratio of horsepower to maximum spindle speed, or *hp/rpm ratio.* Conventional machine tools usually have a higher hp/rpm ratio than machines equipped for high-speed machining. By this metric, the dividing line between conventional machining and HSM is around 0.005 hp/rpm. Thus, high-speed machining includes 50 hp spindles capable of 10,000 rpm (0.005 hp/rpm) and 15 hp spindles that can rotate at 30,000 rpm (0.0005 hp/rpm).

Other definitions emphasize higher production rates and shorter lead times, rather than functions of spindle speed. In this case, important noncutting factors come into play, such as high rapid traverse speeds and quick automatic tool changes ("chip-to-chip" times of 7 sec and less).

Requirements for high-speed machining include the following: (1) high-speed spindles using special bearings designed for high rpm operation; (2) high feed rate capability, typically around 50 m/min (2000 in/min); (3) CNC motion controls with "look-ahead"

TABLE 22.1   Comparison of cutting speeds used in conventional versus high-speed machining for selected work materials.

Work Material	Solid Tools (end mills, drills)[a]				Indexable Tools (face mills)[a]			
	Conventional Speed		High Cutting Speed		Conventional Speed		High Cutting Speed	
	m/min	ft/min	m/min	ft/min	m/min	ft/min	m/min	ft/min
Aluminum	300+	1000+	3000+	10,000+	600+	2000+	3600+	12,000+
Cast iron, soft	150	500	360	1200	360	1200	1200	4000
Cast iron, ductile	105	350	250	800	250	800	900	3000
Steel, free machining	105	350	360	1200	360	1200	600	2000
Steel, alloy	75	250	250	800	210	700	360	1200
Titanium	40	125	60	200	45	150	90	300

[a]Solid tools are made of one solid piece, indexable tools use indexable inserts. Appropriate tool materials include cemented carbide and coated carbide of various grades for all materials, ceramics for all materials, polycrystalline diamond tools for aluminum, and cubic boron nitride for steels (see Section 23.2 for discussion of these tool materials).
Source: Kennametal Inc., Latrobe, Pennsylvania [3].

---

[1]Kennametal Inc., Latrobe, Pennsylvania, is a leading cutting tool producer.

features that allow the controller to see upcoming directional changes and to make adjustments to avoid undershooting or overshooting the desired tool path; (4) balanced cutting tools, toolholders, and spindles to minimize vibration effects; (5) coolant delivery systems that provide pressures an order of magnitude greater than in conventional machining; and (6) chip control and removal systems to cope with the much larger metal removal rates in HSM. Also important are the cutting tool materials. As listed in Table 22.1, various tool materials are used for high-speed machining, and these materials are discussed in the following chapter.

Applications of HSM seem to divide into three categories [3]. One is in the aircraft industry, by companies such as Boeing, in which long airframe structural components are machined from large aluminum blocks. Much metal removal is required, mostly by milling. The resulting pieces are characterized by thin walls and large surface-to-volume ratios, but they can be produced more quickly and are more reliable than assemblies involving multiple components and riveted joints. A second category involves the machining of aluminum by multiple operations to produce a variety of components for industries such as automotive, computer, and medical. Multiple cutting operations mean many tool changes as well as many accelerations and decelerations of the tooling. Thus, quick tool changes and tool path control are important in these applications. The third application category for HSM is in the die and mold industry, which fabricates complex geometries from hard materials. In this case, high-speed machining involves much metal removal to create the mold or die cavity and finishing operations to achieve fine surface finishes.

REFERENCES

[1] Aronson, R. B. "Spindles are the Key to HSM," *Manufacturing Engineering,* October 2004, pp. 67–80.

[2] Aronson, R. B. "Multitalented Machine Tools," *Manufacturing Engineering,* January 2005, pp. 65–75.

[3] Ashley, S. "High-speed Machining Goes Mainstream," *Mechanical Engineering,* May 1995, pp. 56–61.

[4] *ASM Handbook,* Vol. 16, *Machining.* ASM International, Materials Park, Ohio, 1989.

[5] Black, J, and Kohser, R. *DeGarmo's Materials and Processes in Manufacturing,* 10th ed. John Wiley & Sons, Inc., Hoboken, New Jersey, 2008.

[6] Boston, O. W. *Metal Processing,* 2nd ed. John Wiley & Sons, Inc., New York, 1951.

[7] Drozda, T. J., and Wick, C. (eds.) *Tool and Manufacturing Engineers Handbook,* 4th ed. Vol. I, *Machining.* Society of Manufacturing Engineers, Dearborn, Michigan, 1983.

[ ] Eary, D. F., and Johnson, G. E. *Process Engineering for Manufacturing.* Prentice Hall, Inc., Englewood Cliffs, New Jersey, 1962.

[ ] Kalpakjian, S., and Schmid, S. R. *Manufacturing Engineering and Technology,* 4th ed. Prentice Hall, Upper Saddle River, New Jersey, 2003.

[ ] Kalpakjian, S., and Schmid S. R. *Manufacturing Processes for Engineering Materials,* 6th ed. Pearson Prentice Hall, Upper Saddle River, New Jersey, 2010.

[11] Krar, S. F., and Ratterman, E. *Superabrasives: Grinding and Machining with CBN and Diamond.* McGraw-Hill, Inc., New York, 1990.

[12] Lindberg, R. A. *Processes and Materials of Manufacture,* 4th ed. Allyn and Bacon, Inc., Boston, 1990.

[13] Marinac, D. "Smart Tool Paths for HSM," *Manufacturing Engineering,* November 2000, pp. 44–50.

[14] Mason, F., and Freeman, N. B. "Turning Centers Come of Age," Special Report 773, *American Machinist,* February 1985, pp. 97–116.

[15] *Modern Metal Cutting.* AB Sandvik Coromant, Sandvik, Sweden, 1994.

[16] Ostwald, P. F., and J. Munoz, *Manufacturing Processes and Systems,* 9th ed. John Wiley & Sons, Inc., New York, 1997.

[17] Rolt, L. T. C. *A Short History of Machine Tools.* The MIT Press, Cambridge, Massachusetts, 1965.

[18] Steeds, W. *A History of Machine Tools—1700–1910.* Oxford University Press, London, 1969.

[19] Trent, E. M., and Wright, P. K. *Metal Cutting,* 4th ed. Butterworth Heinemann, Boston, 2000.

[20] Witkorski, M., and Bingeman, A. "The Case for Multiple Spindle HMCs," *Manufacturing Engineering,* March 2004, pp. 139–148.

REVIEW QUESTIONS

22.1. What are the differences between rotational parts and prismatic parts in machining?

22.2. Distinguish between generating and forming when machining workpart geometries.

22.3. Give two examples of machining operations in which generating and forming are combined to create workpart geometry.

22.4. Describe the turning process.

22.5. What is the difference between threading and tapping?

22.6. How does a boring operation differ from a turning operation?

22.7. What is meant by the designation 12 in × 36 in lathe?

22.8. Name the various ways in which a workpart can be held in a lathe.

22.9. What is the difference between a live center and a dead center, when these terms are used in the context of workholding in a lathe?

22.10. How does a turret lathe differ from an engine lathe?

22.11. What is a blind hole?

22.12. What is the distinguishing feature of a radial drill press?

22.13. What is the difference between peripheral milling and face milling?

22.14. Describe profile milling.

22.15. What is pocket milling?

22.16. Describe the difference between up milling and down milling.

22.17. How does a universal milling machine differ from a conventional knee-and-column machine?

22.18. What is a machining center?

22.19. What is the difference between a machining center and a turning center?

22.20. What can a mill-turn center do that a conventional turning center cannot do?

22.21. How do shaping and planing differ?

22.22. What is the difference between internal broaching and external broaching?

22.23. Identify the three basic forms of sawing operation.

22.24. (**Video**) For what types of parts are vertical turret lathes used?

22.25. (**Video**) List the four axes for a vertical machining center with a rotational axis on the table.

22.26. (**Video**) What is the purpose of a tombstone that is used with a horizontal machining center?

22.27. (**Video**) List the three parts of a common twist drill.

22.28. (**Video**) What is a gang-drilling machine?

MULTIPLE CHOICE QUESTIONS

There are 23 correct answers in the following multiple choice questions (some questions have multiple answers that are correct). To attain a perfect score on the quiz, all correct answers must be given. Each correct answer is worth 1 point. Each omitted answer or wrong answer reduces the score by 1 point, and each additional answer beyond the correct number of answers reduces the score by 1 point. Percentage score on the quiz is based on the total number of correct answers.

22.1. Which of the following are examples of generating the workpart geometry in machining, as opposed to forming the geometry (two best answers): (a) broaching, (b) contour turning, (c) drilling, (d) profile milling, and (e) thread cutting?

22.2. In a turning operation, the change in diameter of the workpart is equal to which one one of the following: (a) 1 × depth of cut, (b) 2 × depth of cut, (c) 1 × feed, or (d) 2 × feed?

22.3. A lathe can be used to perform which of the following machining operations (three correct answers): (a) boring, (b) broaching, (c) drilling, (d) milling, (e) planing, and (f) turning?

22.4. A facing operation is normally performed on which one of the following machine tools: (a) drill press, (b) lathe, (c) milling machine, (d) planer, or (e) shaper?

22.5. Knurling is performed on a lathe, but it is not a metal cutting operation: (a) true or (b) false?

22.6. Which one of the following cutting tools cannot be used on a turret lathe: (a) broach, (b) cutoff tool, (c) drill bit, (d) single-point turning tool, or (e) threading tool?

22.7. Which one of the following turning machines permits very long bar stock to be used: (a) chucking machine, (b) engine lathe, (c) screw machine, (d) speed lathe, or (e) turret lathe?

22.8. The twist drill is the most common type of drill bit: (a) true or (b) false?

22.9. A tap is a cutting tool used to create which one of the following geometries: (a) external threads, (b) flat planar surfaces, (c) holes used in beer kegs, (d) internal threads, or (e) square holes?

22.10. Reaming is used for which of the following functions (three correct answers): (a) accurately locate a hole position, (b) enlarge a drilled hole, (c) improve surface finish on a hole, (d) improve tolerance on hole diameter, and (e) provide an internal thread?

22.11. End milling is most similar to which one of the following: (a) face milling, (b) peripheral milling, (c) plain milling, or (d) slab milling?

22.12. The basic milling machine is which one of the following: (a) bed type, (b) knee-and-column, (c) profiling mill, (d) ram mill, or (e) universal milling machine?

22.13. A planing operation is best described by which one of the following: (a) a single-point tool moves linearly past a stationary workpart, (b) a tool with multiple teeth moves linearly past a stationary workpart, (c) a workpart is fed linearly past a rotating cutting tool, or (d) a workpart moves linearly past a single-point tool?

22.14. A broaching operation is best described by which one of the following: (a) a rotating tool moves past a stationary workpart, (b) a tool with multiple teeth moves linearly past a stationary workpart, (c) a workpart is fed past a rotating cutting tool, or (d) a workpart moves linearly past a stationary single-point tool?

22.15. The three basic types of sawing, according to type of blade motion involved, are (a) abrasive cutoff, (b) bandsawing, (c) circular sawing, (d) contouring, (e) friction sawing, (f) hacksawing, and (g) slotting?

22.16. Gear hobbing is a special form of which one of the following machining operations: (a) grinding, (b) milling, (c) planing, (d) shaping, or (e) turning?

PROBLEMS

Turning and Related Operations

22.1. A cylindrical workpart 200 mm in diameter and 700 mm long is to be turned in an engine lathe. Cutting speed = 2.30 m/s, feed = 0.32 mm/rev, and depth of cut = 1.80 mm. Determine (a) cutting time, and (b) metal removal rate.

22.2. In a production turning operation, the foreman has decreed that a single pass must be completed on the cylindrical workpiece in 5.0 min. The piece is 400 mm long and 150 mm in diameter. Using a feed = 0.30 mm/rev and a depth of cut = 4.0 mm, what cutting speed must be used to meet this machining time requirement?

22.3. A facing operation is performed on an engine lathe. The diameter of the cylindrical part is 6 in and the length is 15 in. The spindle rotates at a speed of 180 rev/min. Depth of cut = 0.110 in, and feed = 0.008 in/rev. Assume the cutting tool moves from the outer diameter of the workpiece to exactly the center at a constant velocity. Determine (a) the velocity of the tool as it moves from the outer diameter towards the center and (b) the cutting time.

22.4. A tapered surface is to be turned on an automatic lathe. The workpiece is 750 mm long with minimum and maximum diameters of 100 mm and 200 mm at opposite ends. The automatic controls on the lathe permit the surface speed to be maintained at a constant value of 200 m/min by adjusting the rotational speed as a function of workpiece diameter. Feed = 0.25 mm/rev and depth of cut = 3.0 mm. The rough geometry of the piece has already been formed, and this operation will be the final cut. Determine (a) the time required to turn the taper and (b) the rotational speeds at the beginning and end of the cut.

22.5. In the taper turning job of Problem 22.4, suppose that the automatic lathe with surface speed control is not available and a conventional lathe must be used. Determine the rotational speed that would be required to complete the job in exactly the same time as your answer to part (a) of that problem.

22.6. A cylindrical work bar with 4.5 in diameter and 52 in length is chucked in an engine lathe and supported at the opposite end using a live center. A 46.0-in portion of the length is to be turned to a diameter of 4.25 in one pass at a speed of 450 ft/min. The metal removal rate should be 6.75 in³/min. Determine (a) the required depth of cut, (b) the required feed, and (c) the cutting time.

22.7. A 4.00-in-diameter workpiece that is 25 in long is to be turned down to a diameter of 3.50 in, using two passes on an engine lathe using a cutting speed = 300 ft/min, feed = 0.015 in/rev, and depth of cut = 0.125 in. The bar will be held in a chuck and supported on the opposite end in a live center. With this workholding setup, one end must be turned to diameter; then the bar must be reversed to turn the other end. Using an overhead crane available at the lathe, the time required to load and unload the bar is 5 min, and the time to reverse the bar is 3 min. For each turning cut an allowance must be added to the cut length for approach and overtravel. The total allowance (approach plus overtravel) = 0.50 in. Determine the total cycle time to complete this turning operation.

22.8. The end of a large tubular workpart is to be faced on a CNC vertical boring mill. The part has an outside diameter of 38.0 in and an inside diameter of 24.0 in. If the facing operation is performed at a rotational speed of 40.0 rev/min, feed of 0.015 in/rev, and depth of cut of 0.180 in, determine (a) the cutting time to complete the facing operation and the cutting speeds and metal removal rates at the beginning and end of the cut.

22.9. Solve Problem 22.8 except that the machine tool controls operate at a constant cutting speed by continuously adjusting rotational speed for the position of the tool relative to the axis of rotation. The rotational speed at the beginning of the cut = 40 rev/min, and is continuously increased thereafter to maintain a constant cutting speed.

Drilling

22.10. A drilling operation is to be performed with a 12.7-mm diameter twist drill in a steel workpart. The hole is a blind hole at a depth of 60 mm and the point angle is 118°. The cutting speed is 25 m/min and the feed is 0.30 mm/rev. Determine (a) the cutting time to complete the drilling operation, and (b) metal removal rate during the operation, after the drill bit reaches full diameter.

22.11. A two-spindle drill simultaneously drills a 1/2 in hole and a 3/4 in hole through a workpiece that is 1.0 in thick. Both drills are twist drills with point angles of 118°. Cutting speed for the material is 230 ft/min. The rotational speed of each spindle can be set individually. The feed rate for both holes must be set to the same value because the two spindles lower at the same rate. The feed rate is set so the total metal removal rate does not exceed 1.50 in³/min. Determine (a) the maximum feed rate (in/min) that can be used, (b) the individual feeds (in/rev) that result for each hole, and (c) the time required to drill the holes.

22.12. A CNC drill press is to perform a series of through-hole drilling operations on a 1.75-in thick aluminum plate that is a component in a heat exchanger. Each hole is 3/4 in diameter. There are 100 holes in all, arranged in a 10 × 10 matrix pattern, and the distance between adjacent hole centers (along the square) = 1.5 in. The cutting speed = 300 ft/min, the penetration feed (z-direction) = 0.015 in/rev, and the traverse rate between holes (x-y plane) = 15.0 in/min. Assume that x-y moves are made at a distance of 0.50 in above the work surface, and that this distance must be included in the penetration feed rate for each hole. Also, the rate at which the drill is retracted from each hole is twice the penetration feed rate. The drill has a point angle = 100°. Determine the time required from the beginning of the first hole to the completion of the last hole, assuming the most efficient drilling sequence will be used to accomplish the job.

22.13. A gun-drilling operation is used to drill a 9/64-in diameter hole to a certain depth. It takes 4.5 minutes to perform the drilling operation using high pressure fluid delivery of coolant to the drill point. The current spindle speed = 4000 rev/min, and feed = 0.0017 in/rev. In order to improve the surface finish in the hole, it has been decided to increase the speed by 20% and decrease the feed by 25%. How long will it take to perform the operation at the new cutting conditions?

Milling

22.14. A peripheral milling operation is performed on the top surface of a rectangular workpart which is 400 mm long × 60 mm wide. The milling cutter, which is 80 mm in diameter and has five teeth, overhangs the width of the part on both sides. Cutting speed = 70 m/min, chip load = 0.25 mm/tooth, and depth of cut = 5.0 mm. Determine (a) the actual machining time to make one pass across the surface and (b) the maximum material removal rate during the cut.

22.15. A face milling operation is used to machine 6.0 mm from the top surface of a rectangular piece of aluminum 300 mm long by 125 mm wide in a single pass. The cutter follows a path that is centered over the workpiece. It has four teeth and is 150 mm in diameter. Cutting speed = 2.8 m/s, and chip load = 0.27 mm/tooth. Determine (a) the actual machining time to make the pass across the surface and (b) the maximum metal removal rate during cutting.

22.16. A slab milling operation is performed on the top surface of a steel rectangular workpiece 12.0 in long by 2.5 in wide. The helical milling cutter, which has a 3.0 in diameter and ten teeth, is set up to overhang the width of the part on both sides. Cutting speed is 125 ft/min, feed is 0.006 in/tooth, and depth of cut = 0.300 in. Determine (a) the actual machining time to make one pass across the surface and (b) the maximum metal removal rate during the cut. (c) If an additional approach distance of 0.5 in is provided at the beginning of the pass (before cutting begins), and an overtravel distance is provided at the end of the pass equal to the cutter radius plus 0.5 in, what is the duration of the feed motion.

22.17. A face milling operation is performed on the top surface of a steel rectangular workpiece 12.0 in long by 2.5 in wide. The milling cutter follows a path that is centered over the workpiece. It has five teeth and a 3.0 in diameter. Cutting speed = 250 ft/min, feed = 0.006 in/tooth, and depth of cut = 0.150 in. Determine (a) the actual cutting time to make one pass across the surface and (b) the maximum metal removal rate during the cut. (c) If an additional approach distance of 0.5 in is provided at the beginning of the pass (before cutting begins), and an overtravel distance is provided at the end of the pass equal to the cutter radius plus 0.5 in, what is the duration of the feed motion.

22.18. Solve Problem 22.17 except that the workpiece is 5.0 in wide and the cutter is offset to one side so that the swath cut by the cutter = 1.0 in wide. This is called partial face milling, Figure 22.20(b).

22.19. A face milling operation removes 0.32 in depth of cut from the end of a cylinder that has a diameter of 3.90 in. The cutter has a 4-in diameter with 4 teeth, and its feed trajectory is centered over the circular face of the work. The cutting speed is 375 ft/min and the chip load is 0.006 in/tooth. Determine (a) the time to machine, (b) the average metal removal rate (considering the entire machining time), and (c) the maximum metal removal rate.

22.20. The top surface of a rectangular workpart is machined using a peripheral milling operation. The workpart is 735 mm long by 50 mm wide by 95 mm thick. The milling cutter, which is 60 mm in diameter and has five teeth, overhangs the width of the part equally on both sides. Cutting speed = 80 m/min, chip load = 0.30 mm/tooth, and depth of cut = 7.5 mm. (a) Determine the time required to make one pass across the surface, given that the setup and machine settings provide an approach distance of 5 mm before actual cutting begins and an overtravel distance of 25 mm after actual cutting has finished. (b) What is the maximum material removal rate during the cut?

Machining and Turning Centers

22.21. A three-axis CNC machining center is tended by a worker who loads and unloads parts between machining cycles. The machining cycle takes 5.75 min, and the worker takes 2.80 min using a hoist to unload the part just completed and load and fixture the next part onto the machine worktable. A proposal has been made to install a two-position pallet shuttle at the machine so that the worker and the machine tool can perform their respective tasks simultaneously rather than sequentially. The pallet shuttle would transfer the parts between the machine worktable and the load/unload station in 15 sec. Determine (a) the current cycle time for the operation and (b) the cycle time if the proposal is implemented. What is the percentage increase in hourly production rate that would result from using the pallet shuttle?

22.22. A part is produced using six conventional machine tools consisting of three milling machines and three drill presses. The machine cycle times on these machines are 4.7 min, 2.3 min, 0.8 min, 0.9 min, 3.4 min, and 0.5 min. The average load/unload time for each of these operations is 1.25 min. The corresponding setup times for the six machines are 1.55 hr, 2.82 hr, 57 min, 45 min, 3.15 hr, and 36 min, respectively. The total material handling time to carry one part between the machines is 20 min (consisting of five moves between six machines). A CNC machining center has been installed, and all six operations will be performed on it to produce the part. The setup time for the machining center for this job is 1.0 hr. In addition, the machine must be programmed for this part (called "part programming"), which takes 3.0 hr. The machine cycle time is the sum of the machine cycle times for the six machines. Load/unload time is 1.25 min. (a) What is the total time to produce one of these parts using the six conventional machines if the total consists of all setups, machine cycle times, load/unload times, and part transfer times between machines? (b) What is the total time to produce one of these parts using the CNC machining center if the total consists of the setup time, programming time, machine cycle time, and load/unload time, and what are the percent savings in total time compared to your answer in (a)? (c) If the same part is produced in a batch of 20 pieces, what is the total time to produce them under the same conditions as in (a) except that the total material handling time to carry the 20 parts in one unit load between the machines is 40 min? (d) If the part is produced in a batch of 20 pieces on the CNC machining center, what is the total time to produce them under the same conditions as in part (b), and what are the percent savings in total time compared to your answer in (c)? (e) In future orders of 20 pieces of the same part, the programming time will not be included in the total time because the part program has already been prepared and saved. In this case, how long does it take to produce the 20 parts using the machining center, and what are the percent savings in total time compared to your answer in (c)?

Other Operations

22.23. A shaper is used to reduce the thickness of a 50 mm part to 45 mm. The part is made of cast iron and has a tensile strength of 270 MPa and a Brinell hardness of 165 HB. The starting dimensions of the part are 750 mm × 450 mm × 50 mm. The cutting speed is 0.125 m/sec and the feed is 0.40 mm/pass. The shaper ram is hydraulically driven and has a return stroke time that is 50% of the cutting stroke time. An extra 150 mm must be added before and after the part for acceleration and deceleration to take place. Assuming the ram moves parallel to the long dimension of the part, how long will it take to machine?

22.24. An open side planer is to be used to plane the top surface of a rectangular workpart, 20.0 in × 45.0 in. The cutting speed is 30 ft/min, the feed is 0.015 in/pass, and the depth of cut is 0.250 in. The length of the stroke across the work must be set up so that 10 in are allowed at both the beginning and end of the stroke for approach and overtravel. The return stroke, including an allowance for acceleration and deceleration, takes 60% of the time for the forward stroke. The workpart is made of carbon steel with a tensile strength of 50,000 lb/in² and a Brinell hardness of 110 HB. How long will it take to complete the job, assuming that the part is oriented in such a way as to minimize the time?

22.25. High-speed machining is being considered to produce the aluminum part in Problem 22.15. All cutting conditions remain the same except for the cutting speed and the type of insert used in the cutter. Assume the cutting speed will be at the limit given in Table 22.1. Determine (a) the new time to machine the part and (b) the new metal removal rate. (c) Is this part a good candidate for high-speed machining? Explain.

23 CUTTING-TOOL TECHNOLOGY

Chapter Contents

Machining operations are accomplished using cutting tools. The high forces and temperatures during machining create a very harsh environment for the tool. If cutting force becomes too high, the tool fractures. If cutting temperature becomes too high, the tool material softens and fails. If neither of these conditions causes the tool to fail, continual wear of the cutting edge ultimately leads to failure.

Cutting tool technology has two principal aspects: tool material and tool geometry. The first is concerned with developing materials that can withstand the forces, temperatures, and wearing action in the machining process. The second deals with optimizing the geometry of the cutting tool for the tool material and for a given operation. These are the issues we address in the present chapter. It is appropriate to begin by considering tool life, because this is a prerequisite for much of our subsequent discussion on tool materials. It also seems appropriate to include a section on cutting fluids at the end of this chapter; cutting fluids are often used in machining operations to prolong the life of a cutting tool. In the DVD included with this book is a video clip on Cutting-Tool Materials.

VIDEO CLIP

Cutting-Tool Materials. This clip has three segments: (1) cutting-tool materials, which includes an overview of the different cutting-tool categories; (2) tool material quality trade-offs; and (3) tool failure modes.

23.1 TOOL LIFE

As suggested by our opening paragraph, there are three possible modes by which a cutting tool can fail in machining:

1. *Fracture failure.* This mode of failure occurs when the cutting force at the tool point becomes excessive, causing it to fail suddenly by brittle fracture.

2. ***Temperature failure.*** This failure occurs when the cutting temperature is too high for the tool material, causing the material at the tool point to soften, which leads to plastic deformation and loss of the sharp edge.

3. ***Gradual wear.*** Gradual wearing of the cutting edge causes loss of tool shape, reduction in cutting efficiency, an acceleration of wearing as the tool becomes heavily worn, and finally tool failure in a manner similar to a temperature failure.

Fracture and temperature failures result in premature loss of the cutting tool. These two modes of failure are therefore undesirable. Of the three possible tool failures, gradual wear is preferred because it leads to the longest possible use of the tool, with the associated economic advantage of that longer use.

Product quality must also be considered when attempting to control the mode of tool failure. When the tool point fails suddenly during a cut, it often causes damage to the work surface. This damage requires either rework of the surface or possible scrapping of the part. The damage can be avoided by selecting cutting conditions that favor gradual wearing of the tool rather than fracture or temperature failure, and by changing the tool before the final catastrophic loss of the cutting edge occurs.

23.1.1 TOOL WEAR

Gradual wear occurs at two principal locations on a cutting tool: the top rake face and the flank. Accordingly, two main types of tool wear can be distinguished: crater wear and flank wear, illustrated in Figures 23.1 and 23.2. We will use a single-point tool to explain tool wear and the mechanisms that cause it. ***Crater wear,*** Figure 23.2(a), consists of a cavity in the rake face of the tool that forms and grows from the action of the chip sliding against the surface. High stresses and temperatures characterize the tool–chip contact interface, contributing to the wearing action. The crater can be measured either by its depth or its area. ***Flank wear,*** Figure 23.2(b), occurs on the flank, or relief face, of the tool. It results from rubbing between the newly generated work surface and the flank face adjacent to the cutting edge. Flank wear is measured by the width of the wear band, FW. This wear band is sometimes called the flank wear ***land.***

Certain features of flank wear can be identified. First, an extreme condition of flank wear often appears on the cutting edge at the location corresponding to the original surface of the workpart. This is called ***notch wear.*** It occurs because the original work surface is harder and/or more abrasive than the internal material, which could be caused by work

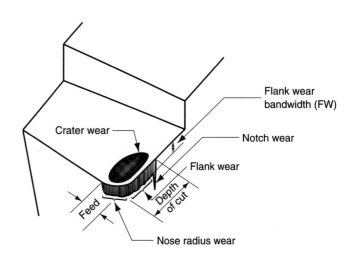

FIGURE 23.1 Diagram of worn cutting tool, showing the principal locations and types of wear that occur.

(a)

FIGURE 23.2 (a) Crater wear and (b) flank wear on a cemented carbide tool, as seen through a toolmaker's microscope. (Courtesy of Manufacturing Technology Laboratory, Lehigh University, photos by J. C. Keefe.)

(b)

hardening from cold drawing or previous machining, sand particles in the surface from casting, or other reasons. As a consequence of the harder surface, wear is accelerated at this location. A second region of flank wear that can be identified is ***nose radius wear;*** this occurs on the nose radius leading into the end cutting edge.

The mechanisms that cause wear at the tool–chip and tool–work interfaces in machining can be summarized as follows:

➤ ***Abrasion.*** This is a mechanical wearing action caused by hard particles in the work material gouging and removing small portions of the tool. This abrasive action occurs in both flank wear and crater wear; it is a significant cause of flank wear.

➤ ***Adhesion.*** When two metals are forced into contact under high pressure and temperature, adhesion or welding occur between them. These conditions are present between the chip and the rake face of the tool. As the chip flows across the tool, small particles of the tool are broken away from the surface, resulting in attrition of the surface.

➤ *Diffusion.* This is a process in which an exchange of atoms takes place across a close contact boundary between two materials (Section 4.3). In the case of tool wear, diffusion occurs at the tool–chip boundary, causing the tool surface to become depleted of the atoms responsible for its hardness. As this process continues, the tool surface becomes more susceptible to abrasion and adhesion. Diffusion is believed to be a principal mechanism of crater wear.

➤ *Chemical reactions.* The high temperatures and clean surfaces at the tool–chip interface in machining at high speeds can result in chemical reactions, in particular, oxidation, on the rake face of the tool. The oxidized layer, being softer than the parent tool material, is sheared away, exposing new material to sustain the reaction process.

➤ *Plastic deformation.* Another mechanism that contributes to tool wear is plastic deformation of the cutting edge. The cutting forces acting on the cutting edge at high temperature cause the edge to deform plastically, making it more vulnerable to abrasion of the tool surface. Plastic deformation contributes mainly to flank wear.

Most of these tool-wear mechanisms are accelerated at higher cutting speeds and temperatures. Diffusion and chemical reaction are especially sensitive to elevated temperature.

23.1.2 TOOL LIFE AND THE TAYLOR TOOL LIFE EQUATION

As cutting proceeds, the various wear mechanisms result in increasing levels of wear on the cutting tool. The general relationship of tool wear versus cutting time is shown in Figure 23.3. Although the relationship shown is for flank wear, a similar relationship occurs for crater wear. Three regions can usually be identified in the typical wear growth curve. The first is the **break-in period,** in which the sharp cutting edge wears rapidly at the beginning of its use. This first region occurs within the first few minutes of cutting. The break-in period is followed by wear that occurs at a fairly uniform rate. This is called the **steady-state wear** region. In our figure, this region is pictured as a linear function of time, although there are deviations from the straight line in actual machining. Finally, wear reaches a level at which the wear rate begins to accelerate. This marks the beginning of the **failure region,** in which cutting temperatures are higher, and the general efficiency of the machining process is reduced. If allowed to continue, the tool finally fails by temperature failure.

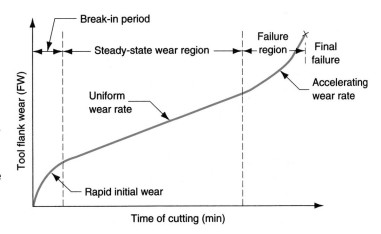

FIGURE 23.3 Tool wear as a function of cutting time. Flank wear (FW) is used here as the measure of tool wear. Crater wear follows a similar growth curve.

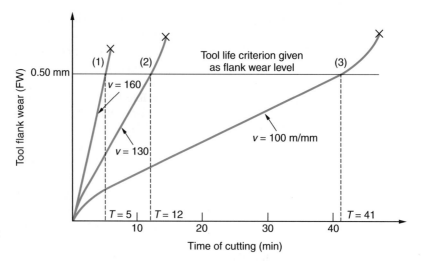

FIGURE 23.4 Effect of cutting speed on tool flank wear (FW) for three cutting speeds. Hypothetical values of speed and tool life are shown for a tool life criterion of 0.50-mm flank wear.

The slope of the tool wear curve in the steady-state region is affected by work material and cutting conditions. Harder work materials cause the wear rate (slope of the tool wear curve) to increase. Increased speed, feed, and depth of cut have a similar effect, with speed being the most important of the three. If the tool wear curves are plotted for several different cutting speeds, the results appear as in Figure 23.4. As cutting speed is increased, wear rate increases so the same level of wear is reached in less time.

Tool life is defined as the length of cutting time that the tool can be used. Operating the tool until final catastrophic failure is one way of defining tool life. This is indicated in Figure 23.4 by the end of each tool wear curve. However, in production, it is often a disadvantage to use the tool until this failure occurs because of difficulties in resharpening the tool and problems with work surface quality. As an alternative, a level of tool wear can be selected as a criterion of tool life, and the tool is replaced when wear reaches that level. A convenient tool life criterion is a certain flank wear value, such as 0.5 mm (0.020 in), illustrated as the horizontal line on the graph. When each of the three wear curves intersects that line, the life of the corresponding tool is defined as ended. If the intersection points are projected down to the time axis, the values of tool life can be identified, as we have done.

Taylor Tool Life Equation   If the tool life values for the three wear curves in Figure 23.4 are plotted on a natural log–log graph of cutting speed versus tool life, the resulting relationship is a straight line as shown in Figure 23.5.[1]

The discovery of this relationship around 1900 is credited to F. W. Taylor. It can be expressed in equation form and is called the Taylor tool life equation:

$$vT^n = C \tag{23.1}$$

where v = cutting speed, m/min (ft/min); T = tool life, min; and n and C are parameters whose values depend on feed, depth of cut, work material, tooling (material in particular), and the tool life criterion used.

The value of n is relative constant for a given tool material, whereas the value of C depends on tool material, work material, and cutting conditions. We will elaborate on these relationships when we discuss the various tool materials in Section 23.2.

---

[1]The reader may have noted in Figure 23.5 that we have plotted the dependent variable (tool life) on the horizontal axis and the independent variable (cutting speed) on the vertical axis. Although this is a reversal of the normal plotting convention, it nevertheless is the way the Taylor tool life relationship is usually presented.

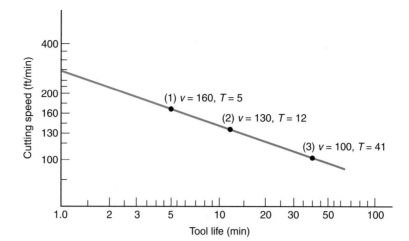

FIGURE 23.5 Natural log–log plot of cutting speed vs. tool life.

Basically, Eq. (23.1) states that higher cutting speeds result in shorter tool lives. Relating the parameters n and C to Figure 23.5, n is the slope of the plot (expressed in linear terms rather than in the scale of the axes), and C is the intercept on the speed axis. C represents the cutting speed that results in a 1-min tool life.

The problem with Eq. (23.1) is that the units on the right-hand side of the equation are not consistent with the units on the left-hand side. To make the units consistent, the equation should be expressed in the form

$$vT^n = C\left(T_{\text{ref}}^{\ n}\right) \tag{23.2}$$

where T_{ref} = a reference value for C. T_{ref} is simply 1 min when m/min (ft/min) and minutes are used for v and T, respectively.

The advantage of Eq. (23.2) is seen when it is desired to use the Taylor equation with units other than m/min (ft/min) and minutes—for example, if cutting speed were expressed as m/sec and tool life as sec. In this case, T_{ref} would be 60 sec and C would therefore be the same speed value as in Eq. (23.1), although converted to units of m/sec. The slope n would have the same numerical value as in Eq. (23.1).

Example 23.1 Taylor Tool Life Equation

Determine the values of C and n in the plot of Figure 23.5, using two of the three points on the curve and solving simultaneous equations of the form of Eq. (23.1).

Solution: Choosing the two extreme points: $v = 160$ m/min, $T = 5$ min; and $v = 100$ m/min, $T = 41$ min; we have

$$160(5)^n = C$$
$$100(41)^n = C$$

Setting the left-hand sides of each equation equal,

$$160(5)^n = 100(41)^n$$

Taking the natural logarithms of each term,

$$\ln(160) + n\ln(5) = \ln(100) + n\ln(41)$$
$$5.0752 + 1.6094\,n = 4.6052 + 3.7136\,n$$
$$0.4700 = 2.1042\,n$$
$$n = \frac{0.4700}{2.1042} = 0.223$$

Substituting this value of n into either starting equation, we obtain the value of C:

$$C = 160(5)^{0.223} = 229$$

or

$$C = 100(41)^{0.223} = 229$$

The Taylor tool life equation for the data of Figure 23.5 is therefore

$$vT^{0.223} = 229$$

■

An expanded version of Eq. (23.2) can be formulated to include the effects of feed, depth of cut, and even work material hardness:

$$vT^n f^m d^p H^p = KT_{ref}{}^n f_{ref}{}^m d_{ref}{}^p H_{ref}{}^q \qquad (23.3)$$

where f = feed, mm (in); d = depth of cut, mm (in); H = hardness, expressed in an appropriate hardness scale; m, p, and q are exponents whose values are experimentally determined for the conditions of the operation; K = a constant analogous to C in Eq. (23.2); and f_{ref}, d_{ref}, and H_{ref} are reference values for feed, depth of cut, and hardness.

The values of m and p, the exponents for feed and depth, are less than 1.0. This indicates the greater effect of cutting speed on tool life, because the exponent of v is 1.0. After speed, feed is next in importance, so m has a value greater than p. The exponent for work hardness, q, is also less than 1.0.

Perhaps the greatest difficulty in applying Eq. (23.3) in a practical machining operation is the tremendous amount of machining data that would be required to determine the parameters of the equation. Variations in work materials and testing conditions also cause difficulties by introducing statistical variations in the data. Equation (23.3) is valid in indicating general trends among its variables, but not in its ability to accurately predict tool life performance. To reduce these problems and make the scope of the equation more manageable, some of the terms are usually eliminated. For example, omitting depth and hardness reduces Eq. (23.3) to the following:

$$vT^n f^m = KT_{ref}{}^n f_{ref}{}^m \qquad (23.4)$$

where the terms have the same meaning as before, except that the constant K will have a slightly different interpretation.

Tool Life Criteria in Production   Although flank wear is the tool life criterion in our previous discussion of the Taylor equation, this criterion is not very practical in a factory environment because of the difficulties and time required to measure flank wear. Following are nine alternative tool life criteria that are more convenient to use in a production machining operation, some of which are admittedly subjective:

1. Complete failure of the cutting edge (fracture failure, temperature failure, or wearing until complete breakdown of the tool has occurred). This criterion has disadvantages, as discussed earlier.
2. Visual inspection of flank wear (or crater wear) by the machine operator (without a toolmaker's microscope). This criterion is limited by the operator's judgment and ability to observe tool wear with the naked eye.
3. Fingernail test across the cutting edge by the operator to test for irregularities.
4. Changes in the sound emitting from the operation, as judged by the operator.
5. Chips become ribbony, stringy, and difficult to dispose of.

6. Degradation of the surface finish on the work.

7. Increased power consumption in the operation, as measured by a wattmeter connected to the machine tool.

8. Workpiece count. The operator is instructed to change the tool after a certain specified number of parts have been machined.

9. Cumulative cutting time. This is similar to the previous workpiece count, except that the length of time the tool has been cutting is monitored. This is possible on machine tools controlled by computer; the computer is programmed to keep data on the total cutting time for each tool.

23.2   TOOL MATERIALS

The three modes of tool failure allow us to identify three important properties required in a tool material:

➤ **Toughness.** To avoid fracture failure, the tool material must possess high toughness. Toughness is the capacity of a material to absorb energy without failing. It is usually characterized by a combination of strength and ductility in the material.

➤ **Hot hardness.** Hot hardness is the ability of a material to retain its hardness at high temperatures. This is required because of the high-temperature environment in which the tool operates.

➤ **Wear resistance.** Hardness is the single most important property needed to resist abrasive wear. All cutting-tool materials must be hard. However, wear resistance in metal cutting depends on more than just tool hardness, because of the other tool-wear mechanisms. Other characteristics affecting wear resistance include surface finish on the tool (a smoother surface means a lower coefficient of friction), chemistry of tool and work materials, and whether a cutting fluid is used.

Cutting-tool materials achieve this combination of properties in varying degrees. In this section, the following cutting-tool materials are discussed: (1) high-speed steel and its predecessors, plain carbon and low alloy steels; (2) cast cobalt alloys; (3) cemented carbides, cermets, and coated carbides; (4) ceramics; (5) synthetic diamond and cubic boron nitride. Before examining these individual materials, a brief overview and technical comparison will be helpful. The historical development of these materials is described in Historical Note 23.1. Commercially, the most important tool materials are high-speed steel and cemented carbides, cermets, and coated carbides. These two categories account for more than 90% of the cutting tools used in machining operations.

Table 23.1 and Figure 23.6 present data on properties of various tool materials. The properties are those related to the requirements of a cutting tool: hardness, toughness, and hot hardness. Table 23.1 lists room temperature hardness and transverse rupture strength for selected materials. Transverse rupture strength (Section 3.1.3) is a property used to indicate toughness for hard materials. Figure 23.6 shows hardness as a function of temperature for several of the tool materials discussed in this section.

In addition to these property comparisons, it is useful to compare the materials in terms of the parameters n and C in the Taylor tool life equation. In general, the development of new cutting-tool materials has resulted in increases in the values of these two parameters. Table 23.2 provides a listing of representative values of n and C in the Taylor tool life equation for selected cutting-tool materials.

The chronological development of tool materials has generally followed a path in which new materials have permitted higher and higher cutting speeds to be achieved.

Historical Note 23.1   *Cutting-tool materials*

In 1800, England was leading the Industrial Revolution, and iron was the leading metal in the revolution. The best tools for cutting iron were made of cast steel by the crucible process, invented in 1742 by B. Huntsman. Cast steel, whose carbon content lies between wrought iron and cast iron, could be hardened by heat treatment to machine the other metals. In 1868, R. Mushet discovered that by alloying about 7% tungsten in crucible steel, a hardened tool steel was obtained by air quenching after heat treatment. Mushet's tool steel was far superior to its predecessor in machining.

Frederick W. Taylor stands as an important figure in the history of cutting tools. Starting around 1880 at Midvale Steel in Philadelphia and later at Bethlehem Steel in Bethlehem, Pennsylvania, he began a series of experiments that lasted a quarter century, yielding a much improved understanding of the metal-cutting process. Among the developments resulting from the work of Taylor and colleague Maunsel White at Bethlehem was *high-speed steel* (HSS), a class of highly alloyed tool steels that permitted substantially higher cutting speeds than previous cutting tools. The superiority of HSS resulted not only from greater alloying, but also from refinements in heat treatment. Tools of the new steel allowed cutting speeds more than twice those of Mushet's steel and almost four times those of plain carbon cast steels.

Tungsten carbide (WC) was first synthesized in the late 1890s. It took nearly three decades before a useful cutting tool material was developed by sintering the WC with a metallic binder to form *cemented carbides*. These were first used in metal cutting in the mid-1920s in Germany, and in the late 1920s in the United States (Historical Note 7.2). *Cermet* cutting tools based on titanium carbide were first introduced in the 1950s, but their commercial importance dates from the 1970s. The first *coated carbides*, consisting of one coating on a WC–Co substrate, were first used around 1970. Coating materials included TiC, TiN, and Al_2O_3. Modern coated carbides have three or more coatings of these and other hard materials.

Attempts to use *alumina ceramics* in machining date from the early 1900s in Europe. Their brittleness inhibited success in these early applications. Processing refinements over many decades have resulted in property improvements in these materials. U.S. commercial use of ceramic cutting tools dates from the mid-1950s.

The first industrial diamonds were produced by the General Electric Company in 1954. They were single crystal diamonds that were applied with some success in grinding operations starting around 1957. Greater acceptance of diamond cutting tools has resulted from the use of *sintered polycrystalline diamond* (SPD), dating from the early 1970s. A similar tool material, sintered *cubic boron nitride*, was first introduced in 1969 by GE under the trade name Borazon.

TABLE 23.1   Typical hardness values (at room temperature) and transverse rupture strengths for various tool materials.[a]

Material	Hardness	Transverse Rupture Strength	
		MPa	**lb/in^2**
Plain carbon steel	60 HRC	5200	750,000
High-speed steel	65 HRC	4100	600,000
Cast cobalt alloy	65 HRC	2250	325,000
Cemented carbide (WC)			
Low Co content	93 HRA, 1800 HK	1400	200,000
High Co content	90 HRA, 1700 HK	2400	350,000
Cermet (TiC)	2400 HK	1700	250,000
Alumina (Al_2O_3)	2100 HK	400	60,000
Cubic boron nitride	5000 HK	700	100,000
Polycrystalline diamond	6000 HK	1000	150,000
Natural diamond	8000 HK	1500	215,000

Compiled from [4], [9], [17], and other sources.
[a]*Note*: The values of hardness and TRS are intended to be comparative and typical. Variations in properties result from differences in composition and processing.

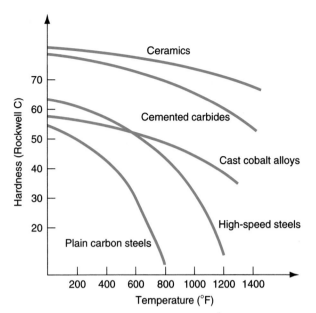

FIGURE 23.6 Typical hot hardness relationships for selected tool materials. Plain carbon steel shows a rapid loss of hardness as temperature increases. High-speed steel is substantially better, whereas cemented carbides and ceramics are significantly harder at elevated temperatures.

Table 23.3 identifies the cutting-tool materials, together with their approximate year of introduction and typical maximum allowable cutting speeds at which they can be used. Dramatic increases in machining productivity have been made possible because of advances in tool material technology, as indicated in our table. Machine tool practice has not always kept pace with cutting-tool technology. Limitations on horsepower, machine tool rigidity, spindle bearings, and the widespread use of older equipment in industry have acted to underutilize the possible upper speeds permitted by available cutting tools.

23.2.1 HIGH-SPEED STEEL AND ITS PREDECESSORS

Before the development of high-speed steel, plain carbon steel and Mushet's steel were the principal tool materials for metal cutting. Today, these steels are rarely used in

TABLE 23.2   Representative values of *n* and *C* in the Taylor tool life equation, Eq. (23.1), for selected tool materials.

Tool Material	*n*	Nonsteel Cutting m/min	(ft/min)	Steel Cutting m/min	ft/min
Plain carbon tool steel	0.1	70	(200)	20	60
High-speed steel	0.125	120	(350)	70	200
Cemented carbide	0.25	900	(2700)	500	1500
Cermet	0.25			600	2000
Coated carbide	0.25			700	2200
Ceramic	0.6			3000	10,000

Compiled from [4], [9], and other sources.
The parameter values are approximated for turning at feed = 0.25 mm/rev (0.010 in/rev) and depth = 2.5 mm (0.100 in). Nonsteel cutting refers to easy-to-machine metals such as aluminum, brass, and cast iron. Steel cutting refers to the machining of mild (unhardened) steel. It should be noted that significant variations in these values can be expected in practice.

TABLE 23.3   Cutting-tool materials with their approximate dates of initial use and allowable cutting speeds.

| | | Allowable Cutting Speed[a] | | | |
| | | Nonsteel Cutting | | Steel Cutting | |
Tool Material	Year of Initial Use	m/min	ft/min	m/min	ft/min
Plain carbon tool steel	1800s	Below 10	Below 30	Below 5	Below 15
High-speed steel	1900	25–65	75–200	17–33	50–100
Cast cobalt alloys	1915	50–200	150–600	33–100	100–300
Cemented carbides (WC)	1930	330–650	1000–2000	100–300	300–900
Cermets (TiC)	1950s			165–400	500–1200
Ceramics (Al_2O_3)	1955			330–650	1000–2000
Synthetic diamonds	1954, 1973	390–1300	1200–4000		
Cubic boron nitride	1969			500–800	1500–2500
Coated carbides	1970			165–400	500–1200

[a]Compiled from [9], [12], [16], [19], and other sources.

industrial machining applications. The plain carbon steels used as cutting tools could be heat-treated to achieve relatively high hardness (Rockwell C 60), because of their fairly high carbon content. However, because of low alloying levels, they possess poor hot hardness (Figure 23.6), which renders them unusable in metal cutting except at speeds too low to be practical by today's standards. Mushet's steel has been displaced by advances in tool steel metallurgy.

High-speed steel (HSS) is a highly alloyed tool steel capable of maintaining hardness at elevated temperatures better than high carbon and low alloy steels. Its good hot hardness permits tools made of HSS to be used at higher cutting speeds. Compared with the other tool materials at the time of its development, it was truly deserving of its name "high speed." A wide variety of high-speed steels are available, but they can be divided into two basic types: (1) tungsten-type, designated T-grades by the American Iron and Steel Institute (AISI); and (2) molybdenum-type, designated M-grades by AISI.

Tungsten-type HSS contains tungsten (W) as its principal alloying ingredient. Additional alloying elements are chromium (Cr), and vanadium (V). One of the original and best known HSS grades is T1, or 18-4-1 high-speed steel, containing 18% W, 4% Cr, and 1% V. *Molybdenum HSS* grades contain combinations of tungsten and molybdenum (Mo), plus the same additional alloying elements as in the T-grades. Cobalt (Co) is sometimes added to HSS to enhance hot hardness. Of course, high-speed steel contains carbon, the element common to all steels. Typical alloying contents and functions of each alloying element in HSS are listed in Table 23.4.

Commercially, high-speed steel is one of the most important cutting-tool materials in use today, despite the fact that it was introduced more than a century ago. HSS is especially suited to applications involving complicated tool geometries, such as drills, taps, milling cutters, and broaches. These complex shapes are generally easier and less expensive to produce from unhardened HSS than other tool materials. They can then be heat-treated so that cutting-edge hardness is very good (Rockwell C 65), whereas toughness of the internal portions of the tool is also good. HSS cutters possess better toughness than any of the harder nonsteel tool materials used for machining, such as cemented carbides and ceramics. Even for single-point tools, HSS is popular among machinists because of the ease with which desired tool geometry can be ground into the tool point. Over the years, improvements have been made in the metallurgical formulation and processing of HSS so that this class of tool material remains competitive in many applications. Also, HSS tools, drills in particular, are often coated with a thin film

TABLE 23.4 Typical contents and functions of alloying elements in high-speed steel.

Alloying Element	Typical Content in HSS, % by Weight	Functions in High-Speed Steel
Tungsten	T-type HSS: 12–20	Increases hot hardness
	M-type HSS: 1.5–6	Improves abrasion resistance through formation of hard carbides in HSS
Molybdenum	T-type HSS: none	Increases hot hardness
	M-type HSS: 5–10	Improves abrasion resistance through formation of hard carbides in HSS
Chromium	3.75–4.5	Depth hardenability during heat treatment
		Improves abrasion resistance through formation of hard carbides in HSS
		Corrosion resistance (minor effect)
Vanadium	1–5	Combines with carbon for wear resistance
		Retards grain growth for better toughness
Cobalt	0–12	Increases hot hardness
Carbon	0.75–1.5	Principal hardening element in steel
		Provides available carbon to form carbides with other alloying elements for wear resistance

of titanium nitride (TiN) to provide significant increases in cutting performance. Physical vapor deposition processes (Section 28.5.1) are commonly used to coat these HSS tools.

23.2.2 CAST COBALT ALLOYS

Cast cobalt alloy cutting tools consist of cobalt, around 40% to 50%; chromium, about 25% to 35%; and tungsten, usually 15% to 20%; with trace amounts of other elements. These tools are made into the desired shape by casting in graphite molds and then grinding to final size and cutting-edge sharpness. High hardness is achieved as cast, an advantage over HSS, which requires heat treatment to achieve its hardness. Wear resistance of the cast cobalts is better than high-speed steel, but not as good as cemented carbide. Toughness of cast cobalt tools is better than carbides but not as good as HSS. Hot hardness also lies between these two materials.

As might be expected from their properties, applications of cast cobalt tools are generally between those of high-speed steel and cemented carbides. They are capable of heavy roughing cuts at speeds greater than HSS and feeds greater than carbides. Work materials include both steels and nonsteels, as well as nonmetallic materials such as plastics and graphite. Today, cast cobalt alloy tools are not nearly as important commercially as either high-speed steel or cemented carbides. They were introduced around 1915 as a tool material that would allow higher cutting speeds than HSS. The carbides were subsequently developed and proved to be superior to the cast Co alloys in most cutting situations.

23.2.3 CEMENTED CARBIDES, CERMETS, AND COATED CARBIDES

Cermets are defined as composites of *cer*amic and *met*allic materials (Section 9.2.1). Technically speaking, cemented carbides are included within this definition; however, cermets based on WC–Co, including WC–TiC–TaC–Co, are known as carbides (cemented carbides) in common usage. In cutting-tool terminology, the term cermet is applied to ceramic-metal composites containing TiC, TiN, and certain other ceramics not including WC. One of the advances in cutting-tool materials involves the application of a very thin coating to a WC–Co substrate. These tools are called coated carbides. Thus, we have three important and closely related tool materials to discuss: (1) cemented carbides, (2) cermets, and (3) coated carbides.

Cemented Carbides   Cemented carbides (also called *sintered carbides*) are a class of hard tool material formulated from tungsten carbide (WC) using powder metallurgy techniques (Chapter 16) with cobalt (Co) as the binder (Sections 7.3.2, 9.2.1, and 17.3.1). There may be other carbide compounds in the mixture, such as titanium carbide (TiC) and/or tantalum carbide (TaC), in addition to WC.

The first cemented carbide cutting tools were made of WC–Co (Historical Note 7.2) and could be used to machine cast irons and nonsteel materials at cutting speeds faster than those possible with high-speed steel and cast cobalt alloys. However, when the straight WC–Co tools were used to cut steel, crater wear occurred rapidly, leading to early failure of the tools. A strong chemical affinity exists between steel and the carbon in WC, resulting in accelerated wear by diffusion and chemical reaction at the tool–chip interface for this work-tool combination. Consequently, straight WC–Co tools cannot be used effectively to machine steel. It was subsequently discovered that additions of titanium carbide and tantalum carbide to the WC–Co mix significantly retarded the rate of crater wear when cutting steel. These new WC–TiC–TaC–Co tools could be used for steel machining. The result is that cemented carbides are divided into two basic types: (1) nonsteel-cutting grades, consisting of only WC–Co; and (2) steel-cutting grades, with combinations of TiC and TaC added to the WC–Co.

The general properties of the two types of cemented carbides are similar: (1) high compressive strength but low-to-moderate tensile strength; (2) high hardness (90 to 95 HRA); (3) good hot hardness; (4) good wear resistance; (5) high thermal conductivity; (6) high modulus of elasticity—E values up to around 600×10^3 MPa (90×10^6 lb/in^2); and (7) toughness lower than high-speed steel.

Nonsteel-cutting grades refer to those cemented carbides that are suitable for machining aluminum, brass, copper, magnesium, titanium, and other nonferrous metals; anomalously, gray cast iron is included in this group of work materials. In the nonsteel-cutting grades, grain size and cobalt content are the factors that influence properties of the cemented carbide material. The typical grain size found in conventional cemented carbides ranges between 0.5 and 5 μm (20 and 200 μ-in). As grain size is increased, hardness and hot hardness decrease, but transverse rupture strength increases.[2] The typical cobalt content in cemented carbides used for cutting tools is 3% to 12%. The effect of cobalt content on hardness and transverse rupture strength is shown in Figure 9.9. As cobalt content increases, TRS improves at the expense of hardness and wear resistance. Cemented carbides with low percentages of cobalt content (3% to 6%) have high hardness and low TRS, whereas carbides with high Co (6% to 12%) have high TRS but lower hardness (Table 23.1). Accordingly, cemented carbides with higher cobalt are used for roughing operations and interrupted cuts (such as milling), while carbides with lower cobalt (therefore, higher hardness and wear resistance) are used in finishing cuts.

Steel-cutting grades are used for low carbon, stainless, and other alloy steels. For these carbide grades, titanium carbide and/or tantalum carbide is substituted for some of the tungsten carbide. TiC is the more popular additive in most applications. Typically, from 10% to 25% of the WC might be replaced by combinations of TiC and TaC. This composition increases the crater wear resistance for steel cutting, but tends to adversely

---

[2]The effect of grain size (GS) on transverse rupture strength (TRS) is more complicated than we are reporting. Published data indicate that the effect of GS on TRS is influenced by cobalt content. At lower Co contents (less than 10%), TRS does indeed increase as GS increases, but at higher Co contents (greater than 10%) TRS decreases as GS increases [4], [16].

TABLE 23.5 The ANSI C-grade classification system for cemented carbides.

Machining Application	Nonsteel-cutting Grades	Steel-cutting Grades	Cobalt and Properties
Roughing	C1	C5	High Co for max. toughness
General purpose	C2	C6	Medium to high Co
Finishing	C3	C7	Medium to low Co
Precision finishing	C4	C8	Low Co for max. hardness
Work materials	Al, brass, Ti, cast iron	Carbon and alloy steels	
Typical ingredients	WC–Co	WC–TiC–TaC–Co	

affect flank wear resistance for nonsteel-cutting applications. That is why two basic categories of cemented carbide are needed.

One of the important developments in cemented carbide technology in recent years is the use of very fine grain sizes (submicron sizes) of the various carbide ingredients (WC, TiC, and TaC). Although small grain size is usually associated with higher hardness but lower transverse rupture strength, the decrease in TRS is reduced or reversed at the submicron particle sizes. Therefore, these ultrafine grain carbides possess high hardness combined with good toughness.

Since the two basic types of cemented carbide were introduced in the 1920s and 1930s, the increasing number and variety of engineering materials have complicated the selection of the most appropriate cemented carbide for a given machining application. To address the problem of grade selection, two classification systems have been developed: (1) the ANSI (American National Standards Institute) C-grade system, developed in the United States starting around 1942; and (2) the ISO R513-1975(E) system, introduced by the International Organization for Standardization (ISO) around 1964. In the C-grade system, summarized in Table 23.5, machining grades of cemented carbide are divided into two basic groups, corresponding to nonsteel-cutting and steel-cutting categories. Within each group there are four levels, corresponding to roughing, general purpose, finishing, and precision finishing.

The ISO R513-1975(E) system, titled "Application of Carbides for Machining by Chip Removal," classifies all machining grades of cemented carbides into three basic groups, each with its own letter and color code, as summarized in Table 23.6. Within each group, the grades are numbered on a scale that ranges from maximum hardness to maximum toughness. Harder grades are used for finishing operations (high speeds, low feeds and depths), whereas tougher grades are used for roughing operations. The ISO classification system can also be used to recommend applications for cermets and coated carbides.

TABLE 23.6 ISO R513-1975(E) "Application of Carbides for Machining by Chip Removal."

Group	Carbide Type	Work Materials	Number Scheme (Cobalt and Properties)
P (blue)	Highly alloyed WC–TiC–TaC–Co	Steel, steel castings, ductile cast iron (ferrous metals with long chips)	P01 (low Co for maximum hardness) to P50 (high Co for maximum toughness)
M (yellow)	Alloyed WC–TiC–TaC–Co	Free-cutting steel, gray cast iron, austenitic stainless steel, superalloys	M10 (low Co for maximum hardness) to M40 (high Co for maximum toughness)
K (red)	Straight WC–Co	Nonferrous metals and alloys, gray cast iron (ferrous metals with short chips), nonmetallics	K01 (low Co for maximum hardness) to K40 (high Co for maximum toughness)

The two systems map into each other as follows: The ANSI C1 through C4-grades map into the ISO K-grades, but in reverse numerical order, and the ANSI C5 through C8 grades translate into the ISO P-grades, but again in reverse numerical order.

Cermets Although cemented carbides are technically classified as cermet composites, the term *cermet* in cutting-tool technology is generally reserved for combinations of TiC, TiN, and titanium carbonitride (TiCN), with nickel and/or molybdenum as binders. Some of the cermet chemistries are more complex (e.g., ceramics such as Ta_xNb_yC and binders such as Mo_2C). However, cermets exclude metallic composites that are primarily based on WC–Co. Applications of cermets include high-speed finishing and semifinishing of steels, stainless steels, and cast irons. Higher speeds are generally allowed with these tools compared with steel-cutting carbide grades. Lower feeds are typically used so that better surface finish is achieved, often eliminating the need for grinding.

Coated Carbides The development of coated carbides around 1970 represented a significant advance in cutting-tool technology. *Coated carbides* are a cemented carbide insert coated with one or more thin layers of wear-resistant material, such as titanium carbide, titanium nitride, and/or aluminum oxide (Al_2O_3). The coating is applied to the substrate by chemical vapor deposition or physical vapor deposition (Section 28.5). The coating thickness is only 2.5 to 13 μm (0.0001 to 0.0005 in). It has been found that thicker coatings tend to be brittle, resulting in cracking, chipping, and separation from the substrate.

The first generation of coated carbides had only a single layer coating (TiC, TiN, or Al_2O_3). More recently, coated inserts have been developed that consist of multiple layers. The first layer applied to the WC–Co base is usually TiN or TiCN because of good adhesion and similar coefficient of thermal expansion. Additional layers of various combinations of TiN, TiCN, Al_2O_3, and TiAlN are subsequently applied.

Coated carbides are used to machine cast irons and steels in turning and milling operations. They are best applied at high cutting speeds in situations in which dynamic force and thermal shock are minimal. If these conditions become too severe, as in some interrupted cut operations, chipping of the coating can occur, resulting in premature tool failure. In this situation, uncoated carbides formulated for toughness are preferred. When properly applied, coated carbide tools usually permit increases in allowable cutting speeds compared with uncoated cemented carbides.

Use of coated carbide tools is expanding to nonferrous metal and nonmetal applications for improved tool life and higher cutting speeds. Different coating materials are required, such as chromium carbide (CrC), zirconium nitride (ZrN), and diamond [11].

23.2.4 CERAMICS

Cutting tools made from ceramics were first used commercially in the United States in the mid-1950s, although their development and use in Europe dates back to the early 1900s. Today's ceramic cutting tools are composed primarily of fine-grained *aluminum oxide* (Al_2O_3), pressed and sintered at high pressures and temperatures with no binder into insert form (Section 17.2). The aluminum oxide is usually very pure (99% is typical), although some manufacturers add other oxides (such as zirconium oxide) in small amounts. In producing ceramic tools, it is important to use a very fine grain size in the alumina powder, and to maximize density of the mix through high-pressure compaction to improve the material's low toughness.

Aluminum oxide cutting tools are most successful in high-speed turning of cast iron and steel. Applications also include finish turning of hardened steels using high cutting speeds, low feeds and depths, and a rigid work setup. Many premature fracture failures of ceramic tools are because of non-rigid machine tool setups, which subject the tools to mechanical shock. When properly applied, ceramic cutting tools can be used to obtain very good surface finish. Ceramics are not recommended for heavy interrupted cut operations (e.g., rough milling) because of their low toughness. In addition to its use as inserts in conventional machining operations, Al_2O_3 is widely used as an abrasive in grinding and other abrasive processes (Chapter 25).

Other commercially available ceramic cutting-tool materials include silicon nitride (SiN), *sialon* (silicon nitride and aluminum oxide, $SiN–Al_2O_3$), aluminum oxide and titanium carbide ($Al_2O_3–TiC$), and aluminum oxide reinforced with single crystal-whiskers of silicon carbide. These tools are usually intended for special applications, a discussion of which is beyond our scope.

23.2.5 SYNTHETIC DIAMONDS AND CUBIC BORON NITRIDE

Diamond is the hardest material known (Section 7.5.1). By some measures of hardness, diamond is three to four times as hard as tungsten carbide or aluminum oxide. Since high hardness is one of the desirable properties of a cutting tool, it is natural to think of diamonds for machining and grinding applications. Synthetic diamond cutting tools are made of sintered polycrystalline diamond (SPD), which dates from the early 1970s. *Sintered polycrystalline diamond* is fabricated by sintering fine-grained diamond crystals under high temperatures and pressures into the desired shape. Little or no binder is used. The crystals have a random orientation and this adds considerable toughness to the SPD tools compared with single crystal diamonds. Tool inserts are typically made by depositing a layer of SPD about 0.5 mm (0.020 in) thick on the surface of a cemented carbide base. Very small inserts have also been made of 100% SPD.

Applications of diamond cutting tools include high-speed machining of nonferrous metals and abrasive nonmetals such as fiberglass, graphite, and wood. Machining of steel, other ferrous metals, and nickel-based alloys with SPD tools is not practical because of the chemical affinity that exists between these metals and carbon (a diamond, after all, is carbon).

Next to diamond, *cubic boron nitride* (Section 7.3.3) is the hardest material known, and its fabrication into cutting tool inserts is basically the same as SPD; that is, coatings on WC–Co inserts. Cubic boron nitride (symbolized cBN) does not react chemically with iron and nickel as SPD does; therefore, the applications of cBN-coated tools are for machining steel and nickel-based alloys. Both SPD and cBN tools are expensive, as one might expect, and the applications must justify the additional tooling cost.

23.3 TOOL GEOMETRY

A cutting tool must possess a shape that is suited to the machining operation. One important way to classify cutting tools is according to the machining process. Thus, we have turning tools, cutoff tools, milling cutters, drill bits, reamers, taps, and many other cutting tools that are named for the operation in which they are used, each with its own tool geometry—in some cases quite unique.

As indicated in Section 21.1, cutting tools can be divided into single-point tools and multiple-cutting-edge tools. Single-point tools are used in turning, boring, shaping, and planing. Multiple-cutting-edge tools are used in drilling, reaming, tapping, milling, broaching, and sawing. Many of the principles that apply to single-point tools also apply to the other cutting-tool types, simply because the mechanism of chip formation is basically the same for all machining operations.

23.3.1 SINGLE-POINT TOOL GEOMETRY

The general shape of a single-point cutting tool is illustrated in Figure 21.4(a). Figure 23.7 shows a more detailed drawing. The reader can observe single-point tools in action in our video clip on turning and lathe basics.

VIDEO CLIP _____

Turning and Lathe Basics. The relevant segment is titled "Turning Operations."

_____

We have previously treated the rake angle of a cutting tool as one parameter. In a single-point tool, the orientation of the rake face is defined by two angles, ***back rake angle*** (α_b) and ***side rake angle*** (α_s). Together, these angles are influential in determining the direction of chip flow across the rake face. The flank surface of the tool is defined by the ***end relief angle*** (ERA) and ***side relief angle*** (SRA). These angles determine the amount of clearance between the tool and the freshly cut work surface. The cutting edge of a single-point tool is divided into two sections, side cutting edge and end cutting edge. These two sections are separated by the tool point, which has a certain radius, called the nose radius. The ***side cutting edge angle*** (SCEA) determines the entry of the tool into the work and can be used to reduce the sudden force the tool experiences as it enters a workpart. ***Nose radius*** (NR) determines to a large degree the texture of the surface generated in the operation. A very pointed tool (small nose radius) results in very pronounced feed marks on the surface. We return to this issue of surface roughness in machining in Section 24.2.2. ***End cutting edge angle*** (ECEA) provides a clearance between the trailing edge of the tool and the newly generated work surface, thus reducing rubbing and friction against the surface.

In all, there are seven elements of tool geometry for a single-point tool. When specified in the following order, they are collectively called the ***tool geometry signature:*** back rake angle, side rake angle, end relief angle, side relief angle, end cutting edge angle, side cutting edge angle, and nose radius. For example, a single-point tool used in turning might have the following signature: 5, 5, 7, 7, 20, 15, 2/64 in.

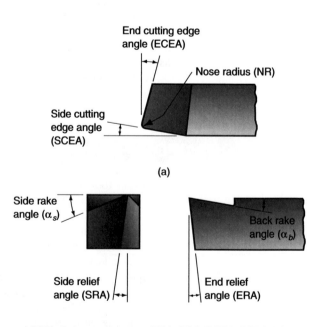

FIGURE 23.7 (a) Seven elements of single-point tool geometry, and (b) the tool signature convention that defines the seven elements.

(a)

(b) Tool signature: α_b, α_s, ERA, SRA, ECEA, SCEA, NR

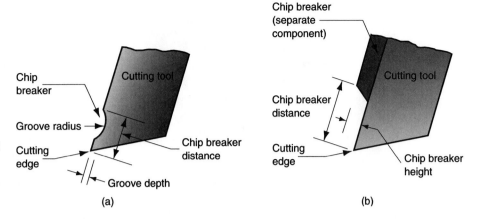

FIGURE 23.8 Two methods of chip breaking in single-point tools: (a) groove-type and (b) obstruction-type chip breakers.

Chip Breakers   Chip disposal is a problem that is often encountered in turning and other continuous operations. Long, stringy chips are often generated, especially when turning ductile materials at high speeds. These chips cause a hazard to the machine operator and the workpart finish, and they interfere with automatic operation of the turning process. ***Chip breakers*** are frequently used with single-point tools to force the chips to curl more tightly than they would naturally be inclined to do, thus causing them to fracture. There are two principal forms of chip breaker design commonly used on single-point turning tools, illustrated in Figure 23.8: (a) groove-type chip breaker designed into the cutting tool itself, and (b) obstruction-type chip breaker designed as an additional device on the rake face of the tool. The chip breaker distance can be adjusted in the obstruction-type device for different cutting conditions.

Effect of Tool Material on Tool Geometry   It was noted in our discussion of the Merchant equation (Section 21.3.2) that a positive rake angle is generally desirable because it reduces cutting forces, temperature, and power consumption. High-speed steel-cutting tools are almost always ground with positive rake angles, typically ranging from +5° to +20°. HSS has good strength and toughness, so that the thinner cross section of the tool created by high positive rake angles does not usually cause a problem with tool breakage. HSS tools are predominantly made of one piece. The heat treatment of high-speed steel can be controlled to provide a hard cutting edge while maintaining a tough inner core.

   With the development of the very hard tool materials (e.g., cemented carbides and ceramics), changes in tool geometry were required. As a group, these materials have higher hardness and lower toughness than HSS. Also, their shear and tensile strengths are low relative to their compressive strengths, and their properties cannot be manipulated through heat treatment like those of HSS. Finally, cost per unit weight for these very hard materials is higher than the cost of HSS. These factors have affected cutting-tool design for the very hard tool materials in several ways.

   First, the very hard materials must be designed with either negative rake or small positive angles. This change tends to load the tool more in compression and less in shear, thus favoring the high compressive strength of these harder materials. Cemented carbides, for example, are used with rake angles typically in the range from −5° to +10°. Ceramics have rake angles between −5° and −15°. Relief angles are made as small as possible (5° is typical) to provide as much support for the cutting edge as possible.

   Another difference is the way in which the cutting edge of the tool is held in position. The alternative ways of holding and presenting the cutting edge for a single-point tool are illustrated in Figure 23.9. The geometry of a HSS tool is ground from a solid shank, as shown in part (a) of the figure. The higher cost and differences in properties and processing of the harder tool materials have given rise to the use of inserts that are either brazed or mechanically clamped to a toolholder. Part (b) shows a brazed insert, in which a cemented

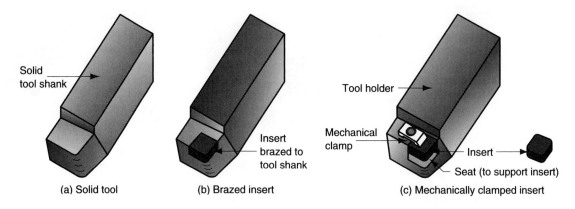

FIGURE 23.9 Three ways of holding and presenting the cutting edge for a single-point tool: (a) solid tool, typical of HSS; (b) brazed insert, one way of holding a cemented carbide insert; and (c) mechanically clamped insert, used for cemented carbides, ceramics, and other very hard tool materials.

carbide insert is brazed to a tool shank. The shank is made of tool steel for strength and toughness. Part (c) illustrates one possible design for mechanically clamping an insert in a toolholder. Mechanical clamping is used for cemented carbides, ceramics, and the other hard materials. The significant advantage of the mechanically clamped insert is that each insert contains multiple cutting edges. When an edge wears out, the insert is unclamped, indexed (rotated in the toolholder) to the next edge, and reclamped in the toolholder. When all of the cutting edges are worn, the insert is discarded and replaced.

Inserts Cutting-tool inserts are widely used in machining because they are economical and adaptable to many different types of machining operations: turning, boring, threading, milling, and even drilling. They are available in a variety of shapes and sizes for the variety of cutting situations encountered in practice. A square insert is shown in Figure 23.9(c). Other common shapes used in turning operations are displayed in Figure 23.10. In general,

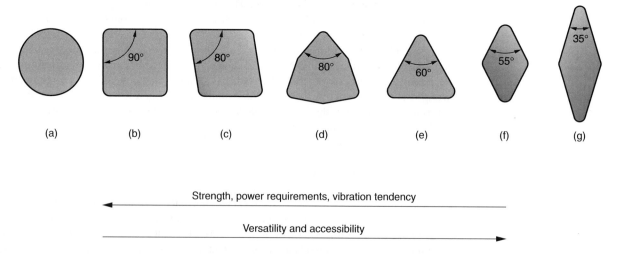

FIGURE 23.10 Common insert shapes: (a) round, (b) square, (c) rhombus with two 80° point angles, (d) hexagon with three 80° point angles, (e) triangle (equilateral), (f) rhombus with two 55° point angles, (g) rhombus with two 35° point angles. Also shown are typical features of the geometry. Strength, power requirements, and tendency for vibration increase as we move to the left; whereas versatility and accessibility tend to be better with the geometries at the right.

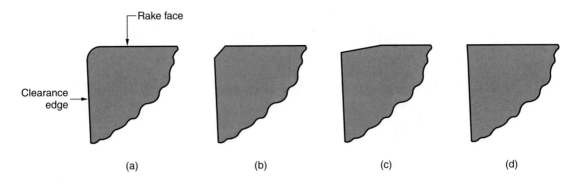

FIGURE 23.11   Three types of edge preparation that are applied to the cutting edge of an insert: (a) radius, (b) chamfer, (c) land, and (d) perfectly sharp edge (no edge preparation).

the largest point angle should be selected for strength and economy. Round inserts possess large point angles (and large nose radii) just because of their shape. Inserts with large point angles are inherently stronger and less likely to chip or break during cutting, but they require more power, and there is a greater likelihood of vibration. The economic advantage of round inserts is that they can be indexed multiple times for more cuts per insert. Square inserts present four cutting edges, triangular shapes have three edges, whereas rhombus shapes have only two. Fewer edges are a cost disadvantage. If both sides of the insert can be used (e.g., in most negative rake angle applications), then the number of cutting edges is doubled. Rhombus shapes are used (especially with acute point angles) because of their versatility and accessibility when a variety of operations are to be performed. These shapes can be more readily positioned in tight spaces and can be used not only for turning but also for facing (Figure 22.6(a)), and contour turning (Figure 22.6(c)).

Inserts are usually not made with perfectly sharp cutting edges, because a sharp edge is weaker and fractures more easily, especially for the very hard and brittle tool materials from which inserts are made (cemented carbides, coated carbides, cermets, ceramics, cBN, and diamond). Some kind of shape alteration is commonly performed on the cutting edge at an almost microscopic level. The effect of this ***edge preparation*** is to increase the strength of the cutting edge by providing a more gradual transition between the clearance edge and the rake face of the tool. Three common edge preparations are shown in Figure 23.11: (a) radius or edge rounding, also referred to as honed edge, (b) chamfer, and (c) land. For comparison, a perfectly sharp cutting edge is shown in (d). The radius in (a) is typically only about 0.025 mm (0.001 in), and the land in (c) is 15° or 20°. Combinations of these edge preparations are often applied to a single cutting edge to maximize the strengthening effect.

23.3.2 MULTIPLE-CUTTING-EDGE TOOLS

Most multiple-cutting-edge tools are used in machining operations in which the tool is rotated. Primary examples are drilling and milling. On the other hand, broaching and some sawing operations (hack sawing and band sawing) use multiple-cutting-edge tools that operate with a linear motion. Other sawing operations (circular sawing) use rotating saw blades.

Drills   Various cutting tools are available for hole making, but the ***twist drill*** is by far the most common. It comes in diameters ranging from about 0.15 mm (0.006 in) to as large as

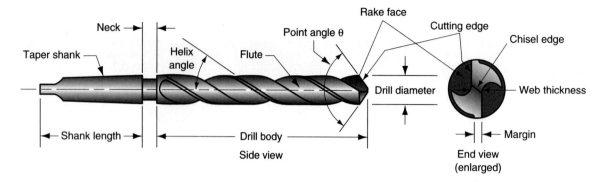

FIGURE 23.12 Standard geometry of a twist drill.

75 mm (3.0 in). Twist drills are widely used in industry to produce holes rapidly and economically. The video clip on hole making illustrates the twist drill.

VIDEO CLIP

Hole making. See the segment titled "The Drill."

The standard twist drill geometry is illustrated in Figure 23.12. The body of the drill has two spiral *flutes* (the spiral gives the twist drill its name). The angle of the spiral flutes is called the *helix angle,* a typical value of which is around 30°. While drilling, the flutes act as passageways for extraction of chips from the hole. Although it is desirable for the flute openings to be large to provide maximum clearance for the chips, the body of the drill must be supported over its length. This support is provided by the *web,* which is the thickness of the drill between the flutes.

The point of the twist drill has a conical shape. A typical value for the *point angle* is 118°. The point can be designed in various ways, but the most common design is a *chisel edge,* as in Figure 23.12. Connected to the chisel edge are two cutting edges (sometimes called lips) that lead into the flutes. The portion of each flute adjacent to the cutting edge acts as the rake face of the tool.

The cutting action of the twist drill is complex. The rotation and feeding of the drill bit result in relative motion between the cutting edges and the workpiece to form the chips. The cutting speed along each cutting edge varies as a function of the distance from the axis of rotation. Accordingly, the efficiency of the cutting action varies, being most efficient at the outer diameter of the drill and least efficient at the center. In fact, the relative velocity at the drill point is zero, so no cutting takes place. Instead, the chisel edge of the drill point pushes aside the material at the center as it penetrates into the hole; a large thrust force is required to drive the twist drill forward into the hole. Also, at the beginning of the operation, the rotating chisel edge tends to wander on the surface of the workpart, causing loss of positional accuracy. Various alternative drill point designs have been developed to address this problem.

Chip removal can be a problem in drilling. The cutting action takes place inside the hole, and the flutes must provide sufficient clearance throughout the length of the drill to allow the chips to be extracted from the hole. As the chip is formed it is forced through the flutes to the work surface. Friction makes matters worse in two ways. In addition to the usual friction in metal cutting between the chip and the rake face of the cutting edge, friction also results from rubbing between the outside diameter of the drill bit and the

FIGURE 23.13 Straight-flute drill that uses indexable inserts.

newly formed hole. This increases the temperature of the drill and work. Delivery of cutting fluid to the drill point to reduce the friction and heat is difficult because the chips are flowing in the opposite direction. Because of chip removal and heat, a twist drill is normally limited to a hole depth of about four times its diameter. Some twist drills are designed with internal holes running their lengths, through which cutting fluid can be pumped to the hole near the drill point, thus delivering the fluid directly to the cutting operation. An alternative approach with twist drills that do not have fluid holes is to use a "pecking" procedure during the drilling operation. In this procedure, the drill is periodically withdrawn from the hole to clear the chips before proceeding deeper.

Twist drills are normally made of high-speed steel. The geometry of the drill is fabricated before heat treatment, and then the outer shell of the drill (cutting edges and friction surfaces) is hardened while retaining an inner core that is relatively tough. Grinding is used to sharpen the cutting edges and shape the drill point.

Although twist drills are the most common hole-making tools, other drill types are also available. *Straight-flute drills* operate like twist drills except that the flutes for chip removal are straight along the length of the tool rather than spiraled. The simpler design of the straight-flute drill permits carbide tips to be used as the cutting edges, either as brazed or indexable inserts. Figure 23.13 illustrates the straight-flute indexable-insert drill. The cemented carbide inserts allow higher cutting speeds and greater production rates than HSS twist drills. However, the inserts limit how small the drills can be made. Thus, the diameter range of commercially available indexable-insert drills runs from about 16 mm (0.625 in) to about 127 mm (5 in) [9].

A straight-flute drill designed for deep-hole drilling is the *gun drill,* shown in Figure 23.14. Whereas the twist drill is usually limited to a depth-to-diameter ratio of 4:1, and the straight-flute drill to about 3:1, the gun drill can cut holes up to 125 times its diameter. As shown in our figure, the gun drill has a carbide cutting edge, a single flute for chip removal, and a coolant hole running its complete length. In the typical gun drilling operation, the work rotates around the stationary drill (opposite of most drilling operations), and the coolant flows into the cutting process and out of the hole along the flute, carrying the chips with it. Gun drills range in diameter from less than 2 mm (0.075 in) to about 50 mm (2 in).

It was previously mentioned that twist drills are available with diameters up to 75 mm (3 in). Twist drills that large are uncommon because so much metal is required in the drill bit. An alternative for large diameter holes is the *spade drill,* illustrated in Figure 23.15. Standard sizes range from 25 to 152 mm (1 to 6 in). The interchangeable drill bit is held in a

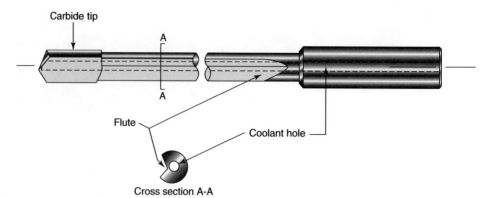

Carbide tip

Flute

Coolant hole

Cross section A-A

FIGURE 23.14 Gun drill.

toolholder, which provides rigidity during cutting. The mass of the spade drill is much less than a twist drill of the same diameter.

More information on hole-making tools can be found in several of our references [3] and [9].

Milling Cutters Classification of milling cutters is closely associated with the milling operations described in Section 22.4.1. The video clip on milling shows some of the tools in operation. The major types of milling cutters are the following:

➢ *Plain milling cutters.* These are used for peripheral or slab milling. As Figures 22.17 (a) and 22.18(a) indicate, they are cylinder shaped with several rows of teeth. The cutting edges are usually oriented at a helix angle (as in the figures) to reduce impact on entry into the work, and these cutters are called *helical milling cutters.* Tool geometry elements of a plain milling cutter are shown in Figure 23.16.

➢ *Form milling cutters.* These are peripheral milling cutters in which the cutting edges have a special profile that is to be imparted to the work. An important application is in gear making, in which the form milling cutter is shaped to cut the slots between adjacent gear teeth, thereby leaving the geometry of the gear teeth.

➢ *Face milling cutters.* These are designed with teeth that cut on both the periphery as well as the end of the cutter. Face milling cutters can be made of HSS, as in

Blade thickness

Chip splitters

Blade

Blade holder

Diameter

A

A

Chisel edge

Rake face

Cross-section A-A

FIGURE 23.15
Spade drill.

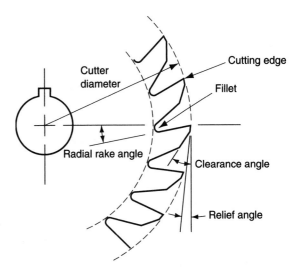

FIGURE 23.16   Tool geometry elements of an 18-tooth plain milling cutter.

Figure 22.17(b), or they can be designed to use cemented carbide inserts. Figure 23.17 shows a four-tooth face-milling cutter that uses inserts.

➢ **End milling cutters.** As shown in Figure 22.20(c), an end milling cutter looks like a drill bit, but close inspection indicates that it is designed for primary cutting with its peripheral teeth rather than its end. (A drill bit cuts only on its end as it penetrates into the work.) End mills are designed with square ends, ends with radii, and ball ends. End mills can be used for face milling, profile milling and pocketing, cutting slots, engraving, surface contouring, and die sinking.

VIDEO CLIP ——————————————————————————————

Milling and Machining Center Basics. See the segment on milling cutters and operations.

———

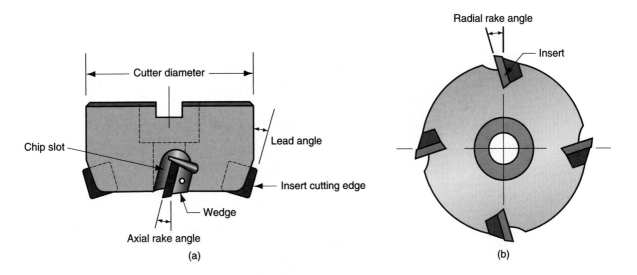

FIGURE 23.17   Tool geometry elements of a four-tooth face milling cutter: (a) side view and (b) bottom view.

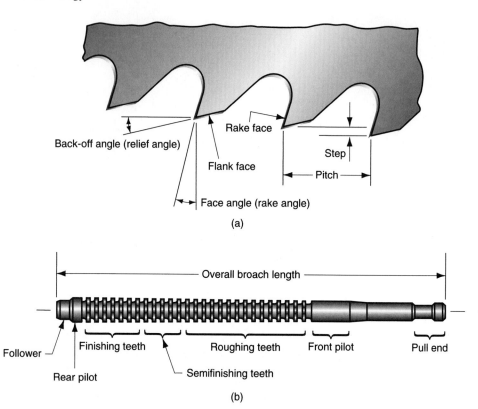

FIGURE 23.18 The broach: (a) terminology of the tooth geometry, and (b) a typical broach used for internal broaching.

Broaches The terminology and geometry of the broach are illustrated in Figure 23.18. The broach consists of a series of distinct cutting teeth along its length. Feed is accomplished by the increased step between successive teeth on the broach. This feeding action is unique among machining operations, because most operations accomplish feeding by a relative feed motion that is carried out by either the tool or the work. The total material removed in a single pass of the broach is the cumulative result of all the steps in the tool. The speed motion is accomplished by the linear travel of the tool past the work surface. The shape of the cut surface is determined by the contour of the cutting edges on the broach, particularly the final cutting edge. Owing to its complex geometry and the low speeds used in broaching, most broaches are made of HSS. In broaching of certain cast irons, the cutting edges are cemented carbide inserts either brazed or mechanically held in place on the broaching tool.

Saw Blades For each of the three sawing operations (Section 22.6.3), the saw blades possess certain common features, including tooth form, tooth spacing, and tooth set, as seen in Figure 23.19. **Tooth form** is concerned with the geometry of each cutting tooth. Rake angle, clearance angle, tooth spacing, and other features of geometry are shown in Figure 23.19(a). **Tooth spacing** is the distance between adjacent teeth on the saw blade. This parameter determines the size of the teeth and the size of the gullet between teeth. The gullet allows space for the formation of the chip by the adjacent cutting tooth. Different tooth forms are appropriate for different work materials and cutting situations. Two forms commonly used in hacksaw and bandsaw blades are shown in Figure 23.19(b). The **tooth set** permits the kerf cut by the saw blade to be wider than the width of the blade itself; otherwise the blade would bind against the walls of the slit made by the saw. Two common tooth sets are illustrated in Figure 23.19(c).

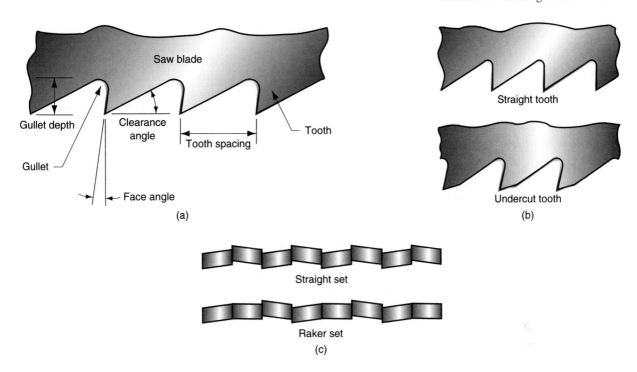

FIGURE 23.19   Features of saw blades: (a) nomenclature for saw blade geometries, (b) two common tooth forms, and (c) two types of tooth set.

23.4   CUTTING FLUIDS

A *cutting fluid* is any liquid or gas that is applied directly to the machining operation to improve cutting performance. Cutting fluids address two main problems: (1) heat generation at the shear zone and friction zone, and (2) friction at the tool–chip and tool–work interfaces. In addition to removing heat and reducing friction, cutting fluids provide additional benefits, such as washing away chips (especially in grinding and milling), reducing the temperature of the workpart for easier handling, reducing cutting forces and power requirements, improving dimensional stability of the workpart, and improving surface finish.

23.4.1   TYPES OF CUTTING FLUIDS

A variety of cutting fluids are commercially available. It is appropriate to discuss them first according to function and then to classify them according to chemical formulation.

Cutting Fluid Functions   There are two general categories of cutting fluids, corresponding to the two main problems they are designed to address: coolants and lubricants. *Coolants* are cutting fluids designed to reduce the effects of heat in the machining operation. They have a limited effect on the amount of heat energy generated in cutting; instead, they carry away the heat that is generated, thereby reducing the temperature of tool and workpiece. This helps to prolong the life of the cutting tool. The capacity of a cutting fluid to reduce temperatures in machining depends on its thermal properties.

Specific heat and thermal conductivity are the most important properties (Section 4.2.1). Water has high specific heat and thermal conductivity relative to other liquids, which is why water is used as the base in coolant-type cutting fluids. These properties allow the coolant to draw heat away from the operation, thereby reducing the temperature of the cutting tool.

Coolant-type cutting fluids seem to be most effective at relatively high cutting speeds, in which heat generation and high temperatures are problems. They are most effective on tool materials that are most susceptible to temperature failures, such as high-speed steels, and are used frequently in turning and milling operations, in which large amounts of heat are generated.

Lubricants are usually oil-based fluids (because oils possess good lubricating qualities) formulated to reduce friction at the tool–chip and tool–work interfaces. Lubricant cutting fluids operate by *extreme pressure lubrication,* a special form of lubrication that involves formation of thin solid salt layers on the hot, clean metal surfaces through chemical reaction with the lubricant. Compounds of sulfur, chlorine, and phosphorus in the lubricant cause the formation of these surface layers, which act to separate the two metal surfaces (i.e., chip and tool). These extreme pressure films are significantly more effective in reducing friction in metal cutting than conventional lubrication, which is based on the presence of liquid films between the two surfaces.

Lubricant-type cutting fluids are most effective at lower cutting speeds. They tend to lose their effectiveness at high speeds (above about 120 m/min [400 ft/min]) because the motion of the chip at these speeds prevents the cutting fluid from reaching the tool–chip interface. In addition, high cutting temperatures at these speeds cause the oils to vaporize before they can lubricate. Machining operations such as drilling and tapping usually benefit from lubricants. In these operations, built-up edge formation is retarded, and torque on the tool is reduced.

Although the principal purpose of a lubricant is to reduce friction, it also reduces the temperature in the operation through several mechanisms. First, the specific heat and thermal conductivity of the lubricant help to remove heat from the operation, thereby reducing temperatures. Second, because friction is reduced, the heat generated from friction is also reduced. Third, a lower coefficient of friction means a lower friction angle. According to Merchant's equation, Eq. (21.16), a lower friction angle causes the shear plane angle to increase, hence reducing the amount of heat energy generated in the shear zone.

There is typically an overlapping effect between the two types of cutting fluids. Coolants are formulated with ingredients that help reduce friction. And lubricants have thermal properties that, although not as good as those of water, act to remove heat from the cutting operation. Cutting fluids (both coolants and lubricants) manifest their effect on the Taylor tool life equation through higher C values. Increases of 10% to 40% are typical. The slope n is not significantly affected.

Chemical Formulation of Cutting Fluids   There are four categories of cutting fluids according to chemical formulation: (1) cutting oils, (2) emulsified oils, (3) semichemical fluids, and (4) chemical fluids. All of these cutting fluids provide both coolant and lubricating functions. The cutting oils are most effective as lubricants, whereas the other three categories are more effective as coolants because they are primarily water.

Cutting oils are based on oil derived from petroleum, animal, marine, or vegetable origin. Mineral oils (petroleum based) are the principal type because of their abundance and generally desirable lubricating characteristics. To achieve maximum lubricity, several types of oils are often combined in the same fluid. Chemical additives are also mixed with the oils to increase lubricating qualities. These additives contain compounds of sulfur, chlorine, and phosphorus, and are designed to react chemically with the chip and tool surfaces to form solid films (extreme pressure lubrication) that help to avoid metal-to-metal contact between the two.

Emulsified oils consist of oil droplets suspended in water. The fluid is made by blending oil (usually mineral oil) in water using an emulsifying agent to promote blending and stability of the emulsion. A typical ratio of water to oil is 30:1. Chemical additives based on sulfur, chlorine, and phosphorus are often used to promote extreme pressure lubrication. Because they contain both oil and water, the emulsified oils combine cooling and lubricating qualities in one cutting fluid.

Chemical fluids are chemicals in a water solution rather than oils in emulsion. The dissolved chemicals include compounds of sulfur, chlorine, and phosphorus, plus wetting agents. The chemicals are intended to provide some degree of lubrication to the solution. Chemical fluids provide good coolant qualities but their lubricating qualities are less than the other cutting fluid types. *Semichemical fluids* have small amounts of emulsified oil added to increase the lubricating characteristics of the cutting fluid. In effect, they are a hybrid class between chemical fluids and emulsified oils.

23.4.2 APPLICATION OF CUTTING FLUIDS

Cutting fluids are applied to machining operations in various ways. In this section we consider these application techniques. We also consider the problem of cutting-fluid contamination and what steps can be taken to address this problem.

Application Methods The most common method is *flooding,* sometimes called flood-cooling because it is generally used with coolant-type cutting fluids. In flooding, a steady stream of fluid is directed at the tool–work or tool–chip interface of the machining operation. A second method of delivery is *mist application,* primarily used for water-based cutting fluids. In this method the fluid is directed at the operation in the form of a high-speed mist carried by a pressurized air stream. Mist application is generally not as effective as flooding in cooling the tool. However, because of the high-velocity air stream, mist application may be more effective in delivering the cutting fluid to areas that are difficult to access by conventional flooding.

Manual application by means of a squirt can or paint brush is sometimes used for applying lubricants in tapping and other operations in which cutting speeds are low and friction is a problem. It is generally not preferred by most production machine shops because of its variability in application.

Cutting Fluid Filtration and Dry Machining Cutting fluids become contaminated over time with a variety of foreign substances, such as tramp oil (machine oil, hydraulic fluid, etc.), garbage (cigarette butts, food, etc.), small chips, molds, fungi, and bacteria. In addition to causing odors and health hazards, contaminated cutting fluids do not perform their lubricating function as well. Alternative ways of dealing with this problem are to: (1) replace the cutting fluid at regular and frequent intervals (perhaps twice per month); (2) use a filtration system to continuously or periodically clean the fluid; or (3) dry machining; that is, machine without cutting fluids. Because of growing concern about environmental pollution and associated legislation, disposing old fluids has become both costly and contrary to the general public welfare.

Filtration systems are being installed in numerous machine shops today to solve the contamination problem. Advantages of these systems include: (1) prolonged cutting fluid life between changes—instead of replacing the fluid once or twice per month, coolant lives of 1 year have been reported; (2) reduced fluid disposal cost, since disposal is much less frequent when a filter is used; (3) cleaner cutting fluid for better working environment and reduced health hazards; (4) lower machine tool maintenance; and (5) longer tool life. There are various types of filtration systems for filtering cutting fluids. For the interested reader, filtration systems and the benefits of using them are discussed in reference [19].

The third alternative is called *dry machining,* meaning that no cutting fluid is used. Dry machining avoids the problems of cutting fluid contamination, disposal, and filtration, but can lead to problems of its own: (1) overheating the tool, (2) operating at lower cutting speeds and production rates to prolong tool life, and (3) absence of chip removal benefits in grinding and milling. Cutting-tool producers have developed certain grades of carbides and coated carbides for use in dry machining.

REFERENCES

[1] Aronson, R. B. "Using High-Pressure Fluids," *Manufacturing Engineering,* June 2004, pp. 87–96.

[2] *ASM Handbook,* Vol. 16: *Machining,* ASM International, Materials Park, Ohio, 1989.

[3] Black, J, and Kohser, R. *DeGarmo's Materials and Processes in Manufacturing,* 10th ed., John Wiley & Sons, Hoboken, New Jersey, 2008.

[4] Brierley, R. G., and Siekman, H. J. *Machining Principles and Cost Control.* McGraw-Hill, New York, 1964.

[5] Carnes, R., and Maddock, G. "Tool Steel Selection," *Advanced Materials & Processes,* June 2004, pp. 37–40.

[6] Cook, N. H. "Tool Wear and Tool Life," *ASME Transactions, Journal of Engineering for Industry,* Vol. *95,* November 1973, pp. 931–938.

[7] Davis, J. R. (ed.), *ASM Specialty Handbook Tool Materials,* ASM International, Materials Park, Ohio, 1995.

[8] Destephani, J. "The Science of pCBN," *Manufacturing Engineering,* January 2005, pp. 53–62.

[9] Drozda, T. J., and Wick, C. (eds.). *Tool and Manufacturing Engineers Handbook,* 4th ed., Vol. I. Machining, Society of Manufacturing Engineers, Dearborn, Michigan, 1983.

[10] Esford, D. "Ceramics Take a Turn," *Cutting Tool Engineering,* Vol. *52,* No. 7, July 2000, pp. 40–46.

[11] Koelsch, J. R. "Beyond TiN," *Manufacturing Engineering,* October 1992, pp. 27–32.

[12] Krar, S. F., and Ratterman, E. *Superabrasives: Grinding and Machining with CBN and Diamond.* McGraw-Hill, New York, 1990.

[13] Liebhold, P. "The History of Tools," *Cutting Tool Engineer,* June 1989, pp. 137–138.

[14] *Machining Data Handbook,* 3rd ed., Vols. I. and II. Metcut Research Associates, Inc., Cincinnati, Ohio, 1980.

[15] *Modern Metal Cutting,* AB Sandvik Coromant, Sandvik, Sweden, 1994.

[16] Owen, J. V. "Are Cermets for Real?" *Manufacturing Engineering,* October 1991, pp. 28–31.

[17] Pfouts, W. R. "Cutting Edge Coatings," *Manufacturing Engineering,* July 2000, pp. 98–107.

[18] Schey, J. A. *Introduction to Manufacturing Processes,* 3rd ed. McGraw-Hill, New York, 1999.

[19] Shaw, M. C. *Metal Cutting Principles,* 2nd ed. Oxford University Press, Oxford, England, 2005.

[20] Spitler, D., Lantrip, J., Nee, J., and Smith, D. A. *Fundamentals of Tool Design,* 5th ed., Society of Manufacturing Engineers, Dearborn, Michigan, 2003.

[21] Tlusty, J. *Manufacturing Processes and Equipment,* Prentice Hall, Upper Saddle River, New Jersey, 2000.

REVIEW QUESTIONS

23.1. What are the two principal aspects of cutting-tool technology?

23.2. Name the three modes of tool failure in machining.

23.3. What are the two principal locations on a cutting tool where tool wear occurs?

23.4. Identify the mechanisms by which cutting tools wear during machining.

23.5. What is the physical interpretation of the parameter C in the Taylor tool life equation?

23.6. In addition to cutting speed, what other cutting variables are included in the expanded version of the Taylor tool life equation?

23.7. What are some of the tool life criteria used in production machining operations?

23.8. Identify three desirable properties of a cutting-tool material.

23.9. What are the principal alloying ingredients in high-speed steel?

23.10. What is the difference in ingredients between steel cutting grades and nonsteel-cutting grades of cemented carbides?

23.11. Identify some of the common compounds that form the thin coatings on the surface of coated carbide inserts.

23.12. Name the seven elements of tool geometry for a single point cutting tool.

23.13. Why are ceramic cutting tools generally designed with negative rake angles?

23.14. Identify the alternative ways by which a cutting tool is held in place during machining.

23.15. Name the two main categories of cutting fluid according to function.

23.16. Name the four categories of cutting fluid according to chemistry.

23.17. What are the principal lubricating mechanisms by which cutting fluids work?

23.18. What are the methods by which cutting fluids are applied in a machining operation?

23.19. Why are cutting fluid filter systems becoming more common and what are their advantages?

23.20. Dry machining is being considered by machine shops because of certain problems inherent in the use of cutting fluids. What are those problems associated with the use of cutting fluids?

23.21. What are some of the new problems introduced by machining dry?

23.22. (**Video**) List the two principal categories of cutting tools.

23.23. (**Video**) According to the video clip, what is the objective in selection of cutting tools for a given operation?

23.24. (**Video**) What are the factors a machinist should know to select the proper tooling? List at least five.

23.25. (**Video**) List five characteristics of a good tool material.

MULTIPLE CHOICE QUIZ

There are 19 correct answers in the following multiple-choice questions (some questions have multiple answers that are correct). To attain a perfect score on the quiz, all correct answers must be given. Each correct answer is worth 1 point. Each omitted answer or wrong answer reduces the score by 1 point, and each additional answer beyond the correct number of answers reduces the score by 1 point. Percentage score on the quiz is based on the total number of correct answers.

23.1. Of the following cutting conditions, which one has the greatest effect on tool wear: (a) cutting speed, (b) depth of cut, or (c) feed?

23.2. As an alloying ingredient in high-speed steel, tungsten serves which of the following functions (two best answers): (a) forms hard carbides to resist abrasion, (b) improves strength and hardness, (c) increases corrosion resistance, (d) increases hot hardness, and (e) increases toughness?

23.3. Cast cobalt alloys typically contain which of the following main ingredients (three best answers): (a) aluminum, (b) cobalt, (c) chromium, (d) iron, (e) nickel, (f) steel, and (g) tungsten?

23.4. Which of the following is not a common ingredient in cemented carbide cutting tools (two correct answers): (a) Al_2O_3, (b) Co, (c) CrC, (d) TiC, and (e) WC?

23.5. An increase in cobalt content has which of the following effects on WC-Co cemented carbides (two best answers): (a) decreases hardness, (b) decreases transverse rupture strength, (c) increases hardness, (d) increases toughness, and (e) increases wear resistance?

23.6. Steel-cutting grades of cemented carbide are typically characterized by which of the following ingredients (three correct answers): (a) Co, (b) Fe, (c) Mo, (d) Ni, (e) TiC, and (f) WC?

23.7. If you had to select a cemented carbide for an application involving finish turning of steel, which C-grade would you select (one best answer): (a) C1, (b) C3, (c) C5, or (d) C7?

23.8. Which of the following processes are used to provide the thin coatings on the surface of coated carbide inserts (two best answers): (a) chemical vapor deposition, (b) electroplating, (c) physical vapor deposition, (d) pressing and sintering, and (e) spray painting?

23.9. Which one of the following materials has the highest hardness: (a) aluminum oxide, (b) cubic boron nitride, (c) high-speed steel, (d) titanium carbide, or (e) tungsten carbide?

23.10. Which of the following are the two main functions of a cutting fluid in machining (two best answers): (a) improve surface finish on the workpiece, (b) reduce forces and power, (c) reduce friction at the tool–chip interface, (d) remove heat from the process, and (e) wash away chips?

PROBLEMS

Tool Life and the Taylor Equation

23.1. Flank wear data were collected in a series of turning tests using a coated carbide tool on hardened alloy steel at a feed of 0.30 mm/rev and a depth of 4.0 mm. At a speed of 125 m/min, flank wear = 0.12 mm at 1 min, 0.27 mm at 5 min, 0.45 mm at 11 min, 0.58 mm at 15 min, 0.73 at 20 min, and 0.97 mm at

25 min. At a speed of 165 m/min, flank wear = 0.22 mm at 1 min, 0.47 mm at 5 min, 0.70 mm at 9 min, 0.80 mm at 11 min, and 0.99 mm at 13 min. The last value in each case is when final tool failure occurred. (a) On a single piece of linear graph paper, plot flank wear as a function of time for both speeds. Using 0.75 mm of flank wear as the criterion of tool failure, determine the tool lives for the two cutting speeds. (b) On a piece of natural log-log paper, plot your results determined in the previous part. From the plot, determine the values of n and C in the Taylor Tool Life Equation. (c) As a comparison, calculate the values of n and C in the Taylor equation solving simultaneous equations. Are the resulting n and C values the same?

23.2. Solve Problem 23.1 except that the tool life criterion is 0.50 mm of flank land wear rather than 0.75 mm.

23.3. A series of turning tests were conducted using a cemented carbide tool, and flank wear data were collected. The feed was 0.010 in/rev and the depth was 0.125 in. At a speed of 350 ft/min, flank wear = 0.005 in at 1 min, 0.008 in at 5 min, 0.012 in at 11 min, 0.0.015 in at 15 min, 0.021 in at 20 min, and 0.040 in at 25 min. At a speed of 450 ft/min, flank wear = 0.007 in at 1 min, 0.017 in at 5 min, 0.027 in at 9 min, 0.033 in at 11 min, and 0.040 in at 13 min. The last value in each case is when final tool failure occurred. (a) On a single piece of linear graph paper, plot flank wear as a function of time. Using 0.020 in of flank wear as the criterion of tool failure, determine the tool lives for the two cutting speeds. (b) On a piece of natural log–log paper, plot your results determined in the previous part. From the plot, determine the values of n and C in the Taylor Tool Life Equation. (c) As a comparison, calculate the values of n and C in the Taylor equation solving simultaneous equations. Are the resulting n and C values the same?

23.4. Solve Problem 23.3 except the tool life wear criterion is 0.015 in of flank wear. What cutting speed should be used to get 20 minutes of tool life?

23.5. Tool life tests on a lathe have resulted in the following data: (1) at a cutting speed of 375 ft/min, the tool life was 5.5 min; (2) at a cutting speed of 275 ft/min, the tool life was 53 min. (a) Determine the parameters n and C in the Taylor tool life equation. (b) Based on the n and C values, what is the likely tool material used in this operation? (c) Using your equation, compute the tool life that corresponds to a cutting speed of 300 ft/min. (d) Compute the cutting speed that corresponds to a tool life $T = 10$ min.

23.6. Tool life tests in turning yield the following data: (1) when cutting speed is 100 m/min, tool life is 10 min; (2) when cutting speed is 75 m/min, tool life is 30 min. (a) Determine the n and C values in the Taylor tool life equation. Based on your equation, compute (b) the tool life for a speed of 110 m/min, and (c) the speed corresponding to a tool life of 15 min.

23.7. Turning tests have resulted in 1-min tool life at a cutting speed = 4.0 m/s and a 20-min tool life at a speed = 2.0 m/s. (a) Find the n and C values in the Taylor tool life equation. (b) Project how long the tool would last at a speed of 1.0 m/s.

23.8. A 15.0-in × 2.0-in-workpart is machined in a face milling operation using a 2.5-in diameter fly cutter with a single carbide insert. The machine is set for a feed of 0.010 in/tooth and a depth of 0.20 in. If a cutting speed of 400 ft/min is used, the tool lasts for three pieces. If a cutting speed of 200 ft/min is used, the tool lasts for 12 parts. Determine the Taylor tool life equation.

23.9. In a production turning operation, the workpart is 125 mm in diameter and 300 mm long. A feed of 0.225 mm/rev is used in the operation. If cutting speed = 3.0 m/s, the tool must be changed every five workparts; but if cutting speed = 2.0 m/s, the tool can be used to produce 25 pieces between tool changes. Determine the Taylor tool life equation for this job.

23.10. For the tool life plot of Figure 23.5, show that the middle data point ($v = 130$ m/min, $T = 12$ min) is consistent with the Taylor equation determined in Example Problem 23.1.

23.11. In the tool wear plots of Figure 23.4, complete failure of the cutting tool is indicated by the end of each wear curve. Using complete failure as the criterion of tool life instead of 0.50 mm flank wear, the resulting data are: (1) $v = 160$ m/min, $T = 5.75$ min; (2) $v = 130$ m/min, $T = 14.25$ min; and (3) $v = 100$ m/min, $T = 47$ min. Determine the parameters n and C in the Taylor tool life equation for this data.

23.12. The Taylor equation for a certain set of test conditions is $vT^{.25} = 1000$, where the U.S. customary units are used: ft/min for v and min for T. Convert this equation to the equivalent Taylor equation in the International System of units (metric), where v is in m/sec and T is in seconds. Validate the metric equation using a tool life = 16 min. That is, compute the corresponding cutting speeds in ft/min and m/sec using the two equations.

23.13. A series of turning tests are performed to determine the parameters n, m, and K in the expanded version of the Taylor equation, Eq. (23.4). The following data were obtained during the tests: (1) cutting speed = 1.9 m/s, feed = 0.22 mm/rev, tool life = 10 min; (2) cutting speed = 1.3 m/s, feed = 0.22 mm/

rev, tool life $= 47$ min; and (3) cutting speed $= 1.9$ m/s, feed $= 0.32$ mm/rev, tool life $= 8$ min. (a) Determine n, m, and K. (b) Using your equation, compute the tool life when the cutting speed is 1.5 m/s and the feed is 0.28 mm/rev.

23.14. Eq. (23.4) in the text relates tool life to speed and feed. In a series of turning tests conducted to determine the parameters n, m, and K, the following data were collected: (1) $v = 400$ ft/min, $f = 0.010$ in/rev, $T = 10$ min; (2) $v = 300$ ft/min, $f = 0.010$ in/rev, $T = 35$ min; and (3) $v = 400$ ft/min, $f = 0.015$ in/rev, $T = 8$ min. Determine n, m, and K. What is the physical interpretation of the constant K?

23.15. The n and C values in Table 23.2 are based on a feed rate of 0.25 mm/rev and a depth of cut $= 2.5$ mm. Determine how many cubic mm of steel would be removed for each of the following tool materials, if a 10-min tool life were required in each case: (a) plain carbon steel, (b) high speed steel, (c) cemented carbide, and (d) ceramic. Use of a spreadsheet calculator is recommended.

23.16. A drilling operation is performed in which 0.5 in diameter holes are drilled through cast iron plates that are 1.0 in thick. Sample holes have been drilled to determine the tool life at two cutting speeds. At 80 surface ft/min, the tool lasted for exactly 50 holes. At 120 surface ft/min, the tool lasted for exactly five holes. The feed of the drill was 0.003 in/rev. (Ignore effects of drill entrance and exit from the hole. Consider the depth of cut to be exactly 1.00 in, corresponding to the plate thickness.) Determine the values of n and C in the Taylor tool life equation for the above sample data, where cutting speed v is expressed in ft/min, and tool life T is expressed in min.

23.17. The outside diameter of a cylinder made of titanium alloy is to be turned. The starting diameter is 400 mm and the length is 1100 mm. The feed is 0.35 mm/rev and the depth of cut is 2.5 mm. The cut will be made with a cemented carbide cutting tool whose Taylor tool life parameters are: $n = 0.24$ and $C = 450$. Units for the Taylor equation are min for tool life and m/min for cutting speed. Compute the cutting speed that will allow the tool life to be just equal to the cutting time for this part.

23.18. The outside diameter of a roll for a steel rolling mill is to be turned. In the final pass, the starting diameter $= 26.25$ in and the length $= 48.0$ in. The cutting conditions will be: feed $= 0.0125$ in/rev, and depth of cut $= 0.125$ in. A cemented carbide cutting tool is to be used and the parameters of the Taylor tool life equation for this setup are: $n = 0.25$ and $C = 1300$. Units for the Taylor equation are min for tool life and ft/min for cutting speed. It is desirable to operate at a cutting speed so that the tool will not need to be changed during the cut. Determine the cutting speed that will make the tool life equal to the time required to complete the turning operation.

23.19. The workpart in a turning operation is 88 mm in diameter and 400 mm long. A feed of 0.25 mm/rev is used in the operation. If cutting speed $= 3.5$ m/s, the tool must be changed every three workparts; but if cutting speed $= 2.5$ m/s, the tool can be used to produce 20 pieces between tool changes. Determine the cutting speed that will allow the tool to be used for 50 parts between tool changes.

23.20. In a production turning operation, the steel workpart has a 4.5 in diameter and is 17.5 in long. A feed of 0.012 in/rev is used in the operation. If cutting speed $= 400$ ft/min, the tool must be changed every four workparts; but if cutting speed $= 275$ ft/min, the tool can be used to produce 15 pieces between tool changes. A new order for 25 pieces has been received but the dimensions of the workpart have been changed. The new diameter is 3.5 in, and the new length is 15.0 in. The work material and tooling remain the same, and the feed and depth are also unchanged, so the Taylor tool life equation determined for the previous workparts is valid for the new parts. Determine the cutting speed that will allow one cutting tool to be used for the new order.

23.21. The outside diameter of a cylinder made of a steel alloy is to be turned. The starting diameter is 300 mm and the length is 625 mm. The feed is 0.35 mm/rev and the depth of cut is 2.5 mm. The cut will be made with a cemented carbide cutting tool whose Taylor tool life parameters are: $n = 0.24$ and $C = 450$. Units for the Taylor equation are min for tool life and m/min for cutting speed. Compute the cutting speed that will allow the tool life to be just equal to the cutting time for three of these parts.

Tooling Applications

23.22. Specify the ANSI C-grade or grades (C1 through C8 in Table 23.5) of cemented carbide for each of the following situations: (a) turning the diameter of a high carbon steel shaft from 4.2 in to 3.5 in, (b) making a final face milling pass using a shallow depth of cut and feed on a titanium part, (c) boring out the cylinders of an alloy steel automobile engine block before honing, and (d) cutting the threads on the inlet and outlet of a large brass valve.

23.23. A certain machine shop uses four cemented carbide grades in its operations. The chemical composition of these grades are as follows: Grade 1 contains 95% WC and 5% Co; Grade 2 contains 82% WC, 4% Co, and 14% TiC; Grade 3 contains 80% WC, 10% Co, and 10% TiC; and Grade 4 contains 89% WC and 11% Co. (a) Which grade should be used for finish turning of unhardened steel? (b) Which grade should be used for rough milling of aluminum? (c) Which grade should be used for finish turning of brass? (d) Which of the grades listed would be suitable for machining cast iron? For each case, explain your recommendation.

23.24. List the ISO R513-1975(E) group (letter and color in Table 23.6) and whether the number would be toward the lower or higher end of the ranges for each of the following situations: (a) milling the head gasket surface of an aluminum cylinder head of an automobile (cylinder head has a hole for each cylinder and must be very flat and smooth to mate up with the block), (b) rough turning a hardened steel shaft, (c) milling a fiber-reinforced polymer composite that requires a precise finish, and (d) milling the rough shape in a die made of steel before it is hardened.

23.25. A turning operation is performed on a steel shaft with diameter = 5.0 in and length = 32 in. A slot or keyway has been milled along its entire length. The turning operation reduces the shaft diameter. For each of the following tool materials, indicate whether it is a reasonable candidate to use in the operation: (a) plain carbon steel, (b) high-speed steel, (c) cemented carbide, (d) ceramic, and (e) sintered polycrystalline diamond. For each material that is not a good candidate, give the reason why it is not.

Cutting Fluids

23.26. In a milling operation with no coolant, a cutting speed of 500 ft/min is used. The current cutting conditions (dry) yield Taylor tool life equation parameters of $n = 0.25$ and $C = 1300$ (ft/min). When a coolant is used in the operation, the cutting speed can be increased by 20% and still maintain the same tool life. Assuming n does not change with the addition of coolant, what is the resulting change in the value of C?

23.27. In a turning operation using high-speed steel tooling, cutting speed = 110 m/min. The Taylor tool life equation has parameters $n = 0.140$ and $C = 150$ (m/min) when the operation is conducted dry. When a coolant is used in the operation, the value of C is increased by 15%. Determine the percent increase in tool life that results if the cutting speed is maintained at 110 m/min.

23.28. A production turning operation on a steel workpiece normally operates at a cutting speed of 125 ft/min using high-speed steel tooling with no cutting fluid. The appropriate n and C values in the Taylor equation are given in Table 23.2 in the text. It has been found that the use of a coolant type cutting fluid will allow an increase of 25 ft/min in the speed without any effect on tool life. If it can be assumed that the effect of the cutting fluid is simply to increase the constant C by 25, what would be the increase in tool life if the original cutting speed of 125 ft/min were used in the operation?

23.29. A high speed steel 6.0 mm twist drill is being used in a production drilling operation on mild steel. A cutting oil is applied by the operator by brushing the lubricant onto the drill point and flutes prior to each hole. The cutting conditions are: speed = 25 m/min, and feed = 0.10 mm/rev, and hole depth = 40 mm. The foreman says that the "speed and feed are right out of the handbook" for this work material. Nevertheless, he says, "the chips are clogging in the flutes, resulting in friction heat, and the drill bit is failing prematurely because of overheating." What's the problem? What do you recommend to solve it?

24 ECONOMIC AND PRODUCT DESIGN CONSIDERATIONS IN MACHINING

Chapter Contents

In this chapter, we conclude our coverage of traditional machining technology by discussing several remaining topics. The first topic is machinability, which is concerned with how work material properties affect machining performance. The second topic is concerned with the tolerances and surface finishes (Chapter 5) that can be expected in machining processes. Third, we consider how to select cutting conditions (speed, feed, and depth of cut) in a machining operation. This selection determines to a large extent the economic success of a given operation. Finally, we provide some guidelines for product designers to consider when they design parts that are to be produced by machining.

24.1 MACHINABILITY

Properties of the work material have a significant influence on the success of the machining operation. These properties and other characteristics of the work are often summarized in the term "machinability." *Machinability* denotes the relative ease with which a material (usually a metal) can be machined using appropriate tooling and cutting conditions.

There are various criteria used to evaluate machinability, the most important of which are: (1) tool life, (2) forces and power, (3) surface finish, and (4) ease of chip disposal. Although machinability generally refers to the work material, it should be recognized that machining performance depends on more than just material. The type of machining operation, tooling, and cutting conditions are also important factors. In addition, the machinability criterion is a source of variation. One material may yield a longer tool life, whereas another material provides a better surface finish. All of these factors make evaluation of machinability difficult.

Machinability testing usually involves a comparison of work materials. The machining performance of a test material is measured relative to that of a base (standard) material. Possible measures of performance in machinability testing include: (1) tool life, (2) tool wear, (3) cutting force, (4) power in the operation, (5) cutting temperature, and (6) material removal rate under standard test conditions. The relative performance is expressed as an index number, called the machinability rating (MR). The base material used as the standard is given a machinability rating of 1.00. B1112 steel is often used as the base material in machinability comparisons. Materials that are easier to machine than the base have ratings greater than 1.00, and materials that are more difficult to machine have ratings less than 1.00. Machinability ratings are often expressed as percentages rather than index numbers. Let us illustrate how a machinability rating might be determined using a tool life test as the basis of comparison.

Example 24.1
Machinability

A series of tool life tests are conducted on two work materials under identical cutting conditions, varying only speed in the test procedure. The first material, defined as the base material, yields a Taylor tool life equation $vT^{0.28} = 350$, and the other material (test material) yields a Taylor equation $vT^{0.27} = 440$, where speed is in m/min and tool life is in min. Determine the machinability rating of the test material using the cutting speed that provides a 60-min tool life as the basis of comparison. This speed is denoted by v_{60}.

Solution: The base material has a machinability rating = 1.0. Its v_{60} value can be determined from the Taylor tool life equation as follows:

$$v_{60} = \left(350/60^{0.28}\right) = 111 \text{ m/min}$$

The cutting speed at a 60-min tool life for the test material is determined similarly:

$$v_{60} = \left(440/60^{0.27}\right) = 146 \text{ m/min}$$

Accordingly, the machinability rating can be calculated as

$$MR(\text{for the test material}) = \frac{146}{111} = 1.31 \ (131\%)$$

Many work material factors affect machining performance. Important mechanical properties include hardness and strength. As hardness increases, abrasive wear of the tool increases so that tool life is reduced. Strength is usually indicated as tensile strength, even though machining involves shear stresses. Of course, shear strength and tensile strength are correlated. As work material strength increases, cutting forces, specific energy, and cutting temperature increase, making the material more difficult to machine. On the other hand, very low hardness can be detrimental to machining performance. For example, low carbon steel, which has relatively low hardness, is often too ductile to machine well. High ductility causes tearing of the metal as the chip is formed, resulting in poor finish, and problems with chip disposal. Cold drawing is often used on low carbon bars to increase surface hardness and promote chip-breaking during cutting.

A metal's chemistry has an important effect on properties; and in some cases, chemistry affects the wear mechanisms that act on the tool material. Through these relationships, chemistry affects machinability. Carbon content has a significant effect on the properties of steel. As carbon is increased, the strength and hardness of the steel increase; this reduces machining performance. Many alloying elements added to steel to enhance properties are detrimental to machinability. Chromium, molybdenum, and tungsten form carbides in steel, which increase tool wear and reduce machinability. Manganese and nickel add strength and toughness to steel, which reduce machinability. Certain elements can be added to steel to improve machining performance, such as

TABLE 24.1   Approximate values of Brinell hardness and typical machinability ratings for selected work materials.

Work Material	Brinell Hardness	Machinability Rating[a]	Work Material	Brinell Hardness	Machinability Rating[a]
Base steel: B1112	180–220	1.00	Tool steel (unhardened)	200–250	0.30
Low carbon steel:	130–170	0.50	Cast iron		
C1008, C1010, C1015			Soft	60	0.70
Medium carbon steel:	140–210	0.65	Medium hardness	200	0.55
C1020, C1025, C1030			Hard	230	0.40
High carbon steel:	180–230	0.55	Super alloys		
C1040, C1045, C1050			Inconel	240–260	0.30
Alloy steels24[b]			Inconel X	350–370	0.15
1320, 1330, 3130, 3140	170–230	0.55	Waspalloy	250–280	0.12
4130	180–200	0.65	Titanium		
4140	190–210	0.55	Plain	160	0.30
4340	200–230	0.45	Alloys	220–280	0.20
4340 (casting)	250–300	0.25	Aluminum		
6120, 6130, 6140	180–230	0.50	2-S, 11-S, 17-S	Soft	5.00[c]
8620, 8630	190–200	0.60	Aluminum alloys (soft)	Soft	2.00[d]
B1113	170–220	1.35	Aluminum alloys (hard)	Hard	1.25[d]
Free machining steels	160–220	1.50	Copper	Soft	0.60
Stainless steel			Brass	Soft	2.00[d]
301, 302	170–190	0.50	Bronze	Soft	0.65[d]
304	160–170	0.40			
316, 317	190–200	0.35			
403	190–210	0.55			
416	190–210	0.90			

Values are estimated average values based on [1], [4], [5], [7], and other sources. Ratings represent relative cutting speeds for a given tool life (see Example 24.1).
[a]Machinability ratings are often expressed as percents (index number × 100%).
[b]Our list of alloy steels is by no means complete. We have attempted to include some of the more common alloys and to indicate the range of machinability ratings among these steels.
[c]The machinability of aluminum varies widely. It is expressed here as MR = 5.00, but the range is probably from 3.00 to 10.00 or more.
[d]Aluminum alloys, brasses, and bronzes also vary significantly in machining performance. Different grades have different machinability ratings. For each case, we have attempted to reduce the variation to a single average value to indicate relative performance with other work materials.

lead, sulfur, and phosphorus. The additives have the effect of reducing the coefficient of friction between the tool and chip, thereby reducing forces, temperature, and built-up edge formation. Better tool life and surface finish result from these effects. Steel alloys formulated to improve machinability are referred to as *free machining steels* (Section 6.2.3).

Similar relationships exist for other work materials. Table 24.1 lists selected metals and their approximate machinability ratings. These ratings are intended to summarize the machining performance of the materials.

24.2   TOLERANCES AND SURFACE FINISH

Machining operations are used to produce parts with defined geometries to tolerances and surface finishes specified by the product designer. In this section we examine these issues of tolerance and surface finish in machining.

TABLE 24.2 Typical tolerances and surface roughness values (arithmetic average) achievable in machining operations.

Machining Operation	Tolerance Capability —Typical		Surface Roughness AA—Typical		Machining Operation	Tolerance Capability —Typical		Surface Roughness AA—Typical	
	mm	in	μm	μ-in		mm	in	μm	μ-in
Turning, boring			0.8	32	Reaming			0.4	16
Diameter $D < 25$ mm	±0.025	±0.001			Diameter $D < 12$ mm	±0.025	±0.001		
25 mm $< D < 50$ mm	±0.05	±0.002			12 mm $< D < 25$ mm	±0.05	±0.002		
Diameter $D > 50$ mm	±0.075	±0.003			Diameter $D > 25$ mm	±0.075	±0.003		
Drilling*			0.8	32	Milling			0.4	16
Diameter $D < 2.5$ mm	±0.05	±0.002			Peripheral	±0.025	±0.001		
2.5 mm $< D < 6$ mm	±0.075	±0.003			Face	±0.025	±0.001		
6 mm $< D < 12$ mm	±0.10	±0.004			End	±0.05	±0.002		
12 mm $< D < 25$ mm	±0.125	±0.005			Shaping, slotting	±0.025	±0.001	1.6	63
Diameter $D > 25$ mm	±0.20	±0.008			Planing	±0.075	±0.003	1.6	63
Broaching	±0.025	±0.001	0.2	8	Sawing	±0.50	±0.02	6.0	250

*Drilling tolerances are typically expressed as biased bilateral tolerances (e.g., + 0.010/–0.002).
Values in this table are expressed as closest bilateral tolerance (e.g., ±0.006).
Compiled from various sources, including [2], [5], [7], [8], [12], and [15].

24.2.1 TOLERANCES IN MACHINING

There is variability in any manufacturing process, and tolerances are used to set permissible limits on this variability (Section 5.1.1). Machining is often selected when tolerances are close, because it is more accurate than most other shape-making processes. Table 24.2 indicates typical tolerances that can be achieved for most machining operations examined in Chapter 22. It should be mentioned that the values in this tabulation represent ideal conditions, yet conditions that are readily achievable in a modern factory. If the machine tool is old and worn, process variability will likely be greater than the ideal, and these tolerances would be difficult to maintain. On the other hand, newer machine tools can achieve closer tolerances than those listed.

Tighter tolerances usually mean higher costs. For example, if the product designer specifies a tolerance of ±0.10 mm on a hole diameter of 6.0 mm, this tolerance could be achieved by a drilling operation, according to Table 24.2. However, if the designer specifies a tolerance of ±0.025 mm, then an additional reaming operation is needed to satisfy this tighter requirement. This is not to suggest that looser tolerances are always good. It often happens that closer tolerances and lower variability in the machining of the individual components will lead to fewer problems in assembly, final product testing, field service, and customer acceptance. Although these costs are not always as easy to quantify as direct manufacturing costs, they can nevertheless be significant. Tighter tolerances that push a factory to achieve better control over its manufacturing processes may lead to lower total operating costs for the company over the long run.

24.2.2 SURFACE FINISH IN MACHINING

Because machining is often the manufacturing process that determines the final geometry and dimensions of the part, it is also the process that determines the part's surface texture (Section 5.3.2). Table 24.2 lists typical surface roughness values that can be achieved in various machining operations. These finishes should be readily achievable by modern, well-maintained machine tools.

Let us examine how surface finish is determined in a machining operation. The roughness of a machined surface depends on many factors that can be grouped as follows: (1) geometric factors, (2) work material factors, and (3) vibration and machine tool factors. Our discussion of surface finish in this section examines these factors and their effects.

Geometric Factors   These are the machining parameters that determine the surface geometry of a machined part. They include: (1) type of machining operation; (2) cutting tool geometry, most importantly nose radius; and (3) feed. The surface geometry that would result from these factors is referred to as the "ideal" or "theoretical" surface roughness, which is the finish that would be obtained in the absence of work material, vibration, and machine tool factors.

Type of operation refers to the machining process used to generate the surface. For example, peripheral milling, facing milling, and shaping all produce a flat surface; however, the surface geometry is different for each operation because of differences in tool shape and the way the tool interacts with the surface. A sense of the differences can be seen in Figure 5.14 showing various possible lays of a surface.

Tool geometry and feed combine to form the surface geometry. The shape of the tool point is the important tool geometry factor. The effects can be seen for a single-point tool in Figure 24.1. With the same feed, a larger nose radius causes the feed marks to be less pronounced, thus leading to a better finish. If two feeds are compared with the same nose radius, the larger feed increases the separation between feed marks, leading to an increase in the value of ideal surface roughness. If feed rate is large enough and the nose radius is small enough so that the end cutting edge participates in creating the new surface, then the end cutting-edge angle will affect surface geometry. In this case, a higher ECEA will result in a higher surface roughness value. In theory, a zero ECEA would yield a perfectly smooth surface; however, imperfections in the tool, work material, and machining process preclude achieving such an ideal finish.

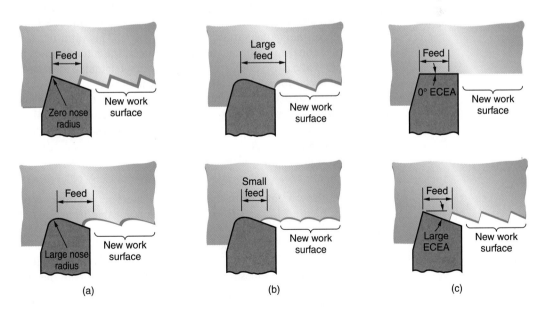

FIGURE 24.1   Effect of geometric factors in determining the theoretical finish on a work surface for single-point tools: (a) effect of nose radius, (b) effect of feed, and (c) effect of end cutting-edge angle.

The effects of nose radius and feed can be combined in an equation to predict the ideal average roughness for a surface produced by a single-point tool. The equation applies to operations such as turning, shaping, and planing

$$R_i = \frac{f^2}{32NR} \qquad (24.1)$$

where R_i = theoretical arithmetic average surface roughness, mm (in); f = feed, mm (in); and NR = nose radius on the tool point, mm (in).

The equation assumes that the nose radius is not zero and that feed and nose radius will be the principal factors that determine the geometry of the surface. The values for R_i will be in units of mm (in), which can be converted to μm (μ-in). Eq. (24.1) can also be used to estimate the ideal surface roughness in face milling with insert tooling, using f to represent the chip load (feed per tooth).

Equation (24.1) assumes a sharp cutting tool. As the tool wears, the shape of the cutting point changes, which is reflected in the geometry of the work surface. For slight amounts of tool wear, the effect is not noticeable. However, when tool wear becomes significant, especially nose radius wear, surface roughness deteriorates compared with the ideal values given by the preceding equations.

Work Material Factors Achieving the ideal surface finish is not possible in most machining operations because of factors related to the work material and its interaction with the tool. Work material factors that affect finish include: (1) built-up edge effects—as the BUE cyclically forms and breaks away, particles are deposited on the newly created work surface, causing it to have a rough "sandpaper" texture; (2) damage to the surface caused by the chip curling back into the work; (3) tearing of the work surface during chip formation when machining ductile materials; (4) cracks in the surface caused by discontinuous chip formation when machining brittle materials; and (5) friction between the tool flank and the newly generated work surface. These work material factors are influenced by cutting speed and rake angle, such that an increase in cutting speed or rake angle generally improves surface finish.

The work material factors usually cause the actual surface finish to be worse than the ideal. An empirical ratio can be developed to convert the ideal roughness value into an estimate of the actual surface roughness value. This ratio takes into account BUE formation, tearing, and other factors. The value of the ratio depends on cutting speed as well as work material. Figure 24.2 shows the ratio of actual to ideal surface roughness as a function of speed for several classes of work material.

The procedure for predicting the actual surface roughness in a machining operation is to compute the ideal surface roughness value and then multiply this value by the ratio of actual to ideal roughness for the appropriate class of work material. This can be summarized as

$$R_a = r_{ai}R_i \qquad (24.2)$$

where R_a = the estimated value of actual roughness; r_{ai} = ratio of actual to ideal surface finish from Figure 24.2, and R_i = ideal roughness value from Eq. (24.1).

**Example 24.2
Surface
Roughness**

A turning operation is performed on C1008 steel (a relatively ductile material) using a tool with a nose radius = 1.2 mm. The cutting conditions are speed = 100 m/min, and feed = 0.25 mm/rev. Compute an estimate of the surface roughness in this operation.

Solution: The ideal surface roughness can be calculated from Eq. (24.1):

$$R_i = (0.25)^2/(32 \times 1.2) = 0.0016 \text{ mm} = 1.6 \; \mu\text{m}$$

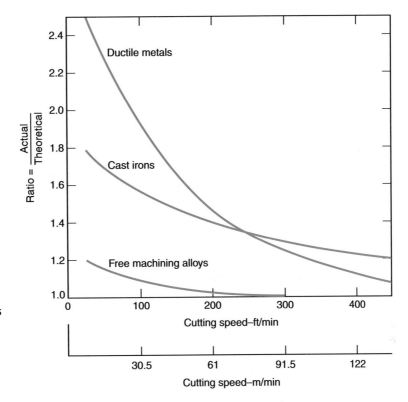

FIGURE 24.2 Ratio of actual surface roughness to ideal surface roughness for several classes of materials. (*Source*: General Electric Co. data [14].)

From the chart in Figure 24.2, the ratio of actual to ideal roughness for ductile metals at 100 m/min is approximately 1.25. Accordingly, the actual surface roughness for the operation would be (approximately)

$$R_a = 1.25 \times 1.6 = 2.0 \ \mu m$$

Vibration and Machine Tool Factors These factors are related to the machine tool, tooling, and setup in the operation. They include chatter or vibration in the machine tool or cutting tool; deflections in the fixturing, often resulting in vibration; and backlash in the feed mechanism, particularly on older machine tools. If these machine tool factors can be minimized or eliminated, the surface roughness in machining will be determined primarily by geometric and work material factors described in the preceding.

Chatter or vibration in a machining operation can result in pronounced waviness in the work surface. When chatter occurs, a distinctive noise occurs that can be recognized by any experienced machinist. Possible steps to reduce or eliminate vibration include: (1) adding stiffness and/or damping to the setup, (2) operating at speeds that do not cause cyclical forces whose frequency approaches the natural frequency of the machine tool system, (3) reducing feeds and depths to reduce forces in cutting, and (4) changing the cutter design to reduce forces. Workpiece geometry can sometimes play a role in chatter. Thin cross sections tend to increase the likelihood of chatter, requiring additional supports to alleviate the condition.

24.3 SELECTION OF CUTTING CONDITIONS

One of the practical problems in machining is selecting the proper cutting conditions for a given operation. This is one of the tasks in process planning (Section 40.1). For each operation, decisions must be made about machine tool, cutting tool(s), and cutting conditions. These decisions must give due consideration to workpart machinability, part geometry, surface finish, and so forth.

24.3.1   SELECTING FEED AND DEPTH OF CUT

Cutting conditions in a machining operation consist of speed, feed, depth of cut, and cutting fluid (whether a cutting fluid is to be used and, if so, type of cutting fluid). Tooling considerations are usually the dominant factor in decisions about cutting fluids (Section 23.4). Depth of cut is often predetermined by workpiece geometry and operation sequence. Many jobs require a series of roughing operations followed by a final finishing operation. In the roughing operations, depth is made as large as possible within the limitations of available horsepower, machine tool and setup rigidity, strength of the cutting tool, and so on. In the finishing cut, depth is set to achieve the final dimensions for the part.

The problem then reduces to selection of feed and speed. In general, values of these parameters should be decided in the order: *feed first, speed second.* Determining the appropriate feed rate for a given machining operation depends on the following factors:

➢ *Tooling.* What type of tooling will be used? Harder tool materials (e.g., cemented carbides, ceramics, etc.) tend to fracture more readily than high-speed steel. These tools are normally used at lower feed rates. HSS can tolerate higher feeds because of its greater toughness.

➢ *Roughing or finishing.* Roughing operations involve high feeds, typically 0.5 to 1.25 mm/rev (0.020 to 0.050 in/rev) for turning; finishing operations involve low feeds, typically 0.125 to 0.4 mm/rev (0.005 to 0.015 in/rev) for turning.

➢ *Constraints on feed in roughing.* If the operation is roughing, how high can the feed rate be set? To maximize metal removal rate, feed should be set as high as possible. Upper limits on feed are imposed by cutting forces, setup rigidity, and sometimes horsepower.

➢ *Surface finish requirements in finishing.* If the operation is finishing, what is the desired surface finish? Feed is an important factor in surface finish, and computations like those in Example 24.2 can be used to estimate the feed that will produce a desired surface finish.

24.3.2   OPTIMIZING CUTTING SPEED

Selection of cutting speed is based on making the best use of the cutting tool, which normally means choosing a speed that provides a high metal removal rate yet suitably long tool life. Mathematical formulas have been derived to determine optimal cutting speed for a machining operation, given that the various time and cost components of the operation are known. The original derivation of these *machining economics* equations is credited to W. Gilbert [10]. The formulas allow the optimal cutting speed to be calculated for either of two objectives: (1) maximum production rate, or (2) minimum unit cost. Both objectives seek to achieve a balance between material removal rate and tool life. The formulas are based on a known Taylor tool life equation for the tool used in the operation. Accordingly, feed, depth of cut, and work material have already been set. The derivation will be illustrated for a turning operation. Similar derivations can be developed for other types of machining operations [3].

Maximizing Production Rate   For maximum production rate, the speed that minimizes machining time per workpiece is determined. Minimizing cutting time per unit is equivalent to maximizing production rate. This objective is important in cases when the production order must be completed as quickly as possible.

In turning, there are three time elements that contribute to the total production cycle time for one part:

1. **Part handling time T_h.** This is the time the operator spends loading the part into the machine tool at the beginning of the production cycle and unloading the part after machining is completed. Any additional time required to reposition the tool for the start of the next cycle should also be included here.

2. **Machining time T_m.** This is the time the tool is actually engaged in machining during the cycle.

3. **Tool change time T_t.** At the end of the tool life, the tool must be changed, which takes time. This time must be apportioned over the number of parts cut during the tool life. Let n_p = the number of pieces cut in one tool life (the number of pieces cut with one cutting edge until the tool is changed); thus, the tool change time per part = T_t/n_p.

The sum of these three time elements gives the total time per unit product for the operation cycle

$$T_c = T_h + T_m + \frac{T_t}{n_p} \qquad (24.3)$$

where T_c = production cycle time per piece, min; and the other terms are defined in the preceding.

The cycle time T_c is a function of cutting speed. As cutting speed is increased, T_m decreases and T_t/n_p increases; T_h is unaffected by speed. These relationships are shown in Figure 24.3.

The cycle time per part is minimized at a certain value of cutting speed. This optimal speed can be identified by recasting Eq. (24.3) as a function of speed. Machining time in a straight turning operation is given by previous Eq. (22.5)

$$T_m = \frac{\pi DL}{vf}$$

where T_m = machining time, min; D = workpart diameter, mm (in); L = workpart length, mm (in); f = feed, mm/rev (in/rev); and v = cutting speed, mm/min for consistency of units (in/min for consistency of units).

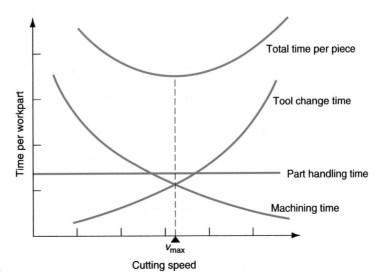

FIGURE 24.3   Time elements in a machining cycle plotted as a function of cutting speed. Total cycle time per piece is minimized at a certain value of cutting speed. This is the speed for maximum production rate.

The number of pieces per tool n_p is also a function of speed. It can be shown that

$$n_p = \frac{T}{T_m} \tag{24.4}$$

where T = tool life, min/tool; and T_m = machining time per part, min/pc. Both T and T_m are functions of speed; hence, the ratio is a function of speed

$$n_p = \frac{fC^{1/n}}{\pi DL v^{1/n-1}} \tag{24.5}$$

The effect of this relation is to cause the T_t/n_p term in Eq. (24.3) to increase as cutting speed increases. Substituting Eqs. (22.5) and (24.5) into Eq. (24.3) for T_c, we have

$$T_c = T_h + \frac{\pi DL}{fv} + \frac{T_t\left(\pi DL v^{1/n-1}\right)}{fC^{1/n}} \tag{24.6}$$

The cycle time per piece is a minimum at the cutting speed at which the derivative of Eq. (24.6) is zero

$$\frac{dT_c}{dv} = 0$$

Solving this equation yields the cutting speed for maximum production rate in the operation

$$v_{\max} = \frac{C}{\left[\left(\frac{1}{n} - 1\right)T_t\right]^n} \tag{24.7}$$

where $v_{\max}$ is expressed in m/min (ft/min). The corresponding tool life for maximum production rate is

$$T_{\max} = \left(\frac{1}{n} - 1\right)T_t \tag{24.8}$$

Minimizing Cost per Unit   For minimum cost per unit, the speed that minimizes production cost per piece for the operation is determined. To derive the equations for this case, we begin with the four cost components that determine total cost of producing one part during a turning operation:

1. *Cost of part handling time.* This is the cost of the time the operator spends loading and unloading the part. Let C_o = the cost rate (e.g., \$/min) for the operator and machine. Thus the cost of part handling time = $C_o T_h$.
2. *Cost of machining time.* This is the cost of the time the tool is engaged in machining. Using C_o again to represent the cost per minute of the operator and machine tool, the cutting time cost = $C_o T_m$.
3. *Cost of tool change time.* The cost of tool change time = $C_o T_t/n_p$.
4. *Tooling cost.* In addition to the tool change time, the tool itself has a cost that must be added to the total operation cost. This cost is the cost per cutting edge C_t, divided by the number of pieces machined with that cutting edge n_p. Thus, tool cost per workpiece is given by C_t/n_p.

Tooling cost requires an explanation, because it is affected by different tooling situations. For disposable inserts (e.g., cemented carbide inserts), tool cost is determined as

$$C_t = \frac{P_t}{n_e} \tag{24.9}$$

where C_t = cost per cutting edge, \$/tool life; P_t = price of the insert, \$/insert; and n_e = number of cutting edges per insert.

This depends on the insert type; for example, triangular inserts that can be used only one side (positive rake tooling) have three edges/insert; if both sides of the insert can be used (negative rake tooling), there are six edges/insert; and so forth.

For regrindable tooling (e.g., high-speed steel solid shank tools, brazed carbide tools), the tool cost includes purchase price plus cost to regrind:

$$C_t = \frac{P_t}{n_e} + T_g C_g \qquad (24.10)$$

where C_t = cost per tool life, \$/tool life; P_t = purchase price of the solid shank tool or brazed insert, \$/tool; n_g = number of tool lives per tool, which is the number of times the tool can be ground before it can no longer be used (5 to 10 times for roughing tools and 10 to 20 times for finishing tools); T_g = time to grind or regrind the tool, min/tool life; and C_g = grinder's rate, \$/min.

The sum of the four cost components gives the total cost per unit product C_c for the machining cycle:

$$C_c = C_o T_h + C_o T_m + \frac{C_o T_t}{n_p} + \frac{C_t}{n_p} \qquad (24.11)$$

C_c is a function of cutting speed, just as T_c is a function of v. The relationships for the individual terms and total cost as a function of cutting speed are shown in Figure 24.4. Eq. (24.11) can be rewritten in terms of v to yield:

$$C_c = C_o T_h + \frac{C_o \pi DL}{fv} + \frac{(C_o T_t + C_t)(\pi DL v^{1/n-1})}{fC^{1/n}} \qquad (24.12)$$

The cutting speed that obtains minimum cost per piece for the operation can be determined by taking the derivative of Eq. (24.12) with respect to v, setting it to zero, and solving for v_{min}

$$v_{min} = C\left(\frac{n}{1-n} \cdot \frac{C_o}{C_o T_t + C_t}\right)^n \qquad (24.13)$$

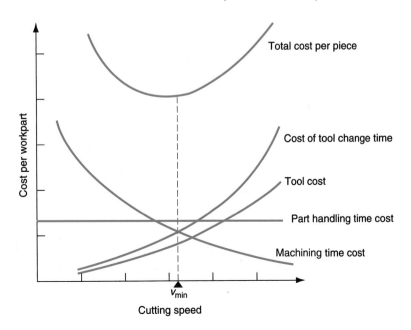

FIGURE 24.4 Cost components in a machining operation plotted as a function of cutting speed. Total cost per piece is minimized at a certain value of cutting speed. This is the speed for minimum cost per piece.

The corresponding tool life is given by

$$T_{min} = \left(\frac{1}{n} - 1\right)\left(\frac{C_o T_t + C_t}{C_o}\right) \tag{24.14}$$

**Example 24.3
Determining
Cutting Speeds in
Machining
Economics**

Suppose a turning operation is to be performed with HSS tooling on mild steel, with Taylor tool life parameters $n = 0.125$, $C = 70$ m/min (Table 23.2). Workpart length = 500 mm and diameter = 100 mm. Feed = 0.25 mm/rev. Handling time per piece = 5.0 min, and tool change time = 2.0 min. Cost of machine and operator = \$30/hr, and tooling cost = \$3 per cutting edge. Find: (a) cutting speed for maximum production rate, and (b) cutting speed for minimum cost.

Solution: (a) Cutting speed for maximum production rate is given by Eq. (24.7)

$$v_{max} = 70\left(\frac{0.125}{0.875} \cdot \frac{1}{2}\right)^{0.125} = 50 \text{ m/min}$$

(b) Converting $C_o = \$30$/hr to \$0.5/min, minimum cost cutting speed is given by Eq. (24.13)

$$v_{min} = 70\left(\frac{0.125}{0.875} \cdot \frac{0.5}{0.5(2) + 3.00}\right)^{0.125} = 42 \text{ m/min}$$

**Example 24.4
Production Rate
and Cost in
Machining
Economics**

Determine the hourly production rate and cost per piece for the two cutting speeds computed in Example 24.3.

Solution: (a) For the cutting speed for maximum production, $v_{max} = 50$ m/min, let us calculate machining time per piece and tool life.

$$\text{Machining time } T_m = \frac{\pi(0.5)(0.1)}{(0.25)(10^{-3})(50)} = 12.57 \text{ min/pc}$$

$$\text{Tool life } T = \left(\frac{70}{50}\right)^8 = 14.76 \text{ min/cutting edge}$$

From this we see that the number of pieces per tool $n_p = 14.76/12.57 = 1.17$. Use $n_p = 1$. From Eq. (24.3), average production cycle time for the operation is

$$T_c = 5.0 + 12.57 + 2.0/1 = 19.57 \text{ min/pc}$$

Corresponding hourly production rate $R_p = 60/19.57 = 3.1$ pc/hr. From Eq. (24.11), average cost per piece for the operation is

$$C_c = 0.5(5.0) + 0.5(12.57) + 0.5(2.0)/1 + 3.00/1 = \$12.79/\text{pc}$$

(b) For the cutting speed for minimum production cost per piece, $v_{min} = 42$ m/min, the machining time per piece and tool life are calculated as follows

$$\text{Machining time } T_m = \frac{\pi(0.5)(0.1)}{(0.25)(10^{-3})(42)} = 14.96 \text{ min/pc}$$

$$\text{Tool life } T = \left(\frac{70}{42}\right)^8 = 59.54 \text{ min/cutting edge}$$

The number of pieces per tool $n_p = 59.54/14.96 = 3.98 \rightarrow$ Use $n_p = 3$ to avoid failure during the fourth workpiece. Average production cycle time for the operation is

$$T_c = 5.0 + 14.96 + 2.0/3 = 20.63 \text{ min/pc.}$$

Corresponding hourly production rate $R_p = 60/20.63 = 2.9$ pc/hr. Average cost per piece for the operation is

$$C_c = 0.5(5.0) + 0.5(14.96) + 0.5(2.0)/3 + 3.00/3 = \$11.32/\text{pc}$$

Note that production rate is greater for v_{max} and cost per piece is minimum for v_{min}. ∎

Some Comments on Machining Economics Some practical observations can be made relative to these optimum cutting speed equations. First, as the values of C and n increase in the Taylor tool life equation, the optimum cutting speed increases by either Eq. (24.7) or Eq. (24.13). Cemented carbides and ceramic cutting tools should be used at speeds that are significantly higher than for high-speed steel tools.

Second, as the tool change time and/or tooling cost (T_{tc} and C_t) increase, the cutting speed equations yield lower values. Lower speeds allow the tools to last longer, and it is wasteful to change tools too frequently if either the cost of tools or the time to change them is high. An important effect of this tool cost factor is that disposable inserts usually possess a substantial economic advantage over regrindable tooling. Even though the cost per insert is significant, the number of edges per insert is large enough and the time required to change the cutting edge is low enough that disposable tooling generally achieves higher production rates and lower costs per unit product.

Third, v_{max} is always greater than v_{min}. The C_t/n_p term in Eq. (24.13) has the effect of pushing the optimum speed value to the left in Figure 24.4, resulting in a lower value than in Figure 24.3. Rather than taking the risk of cutting at a speed above v_{max} or below v_{min}, some machine shops strive to operate in the interval between v_{min} and v_{max}—an interval sometimes referred to as the "high-efficiency range."

The procedures outlined for selecting feeds and speeds in machining are often difficult to apply in practice. The best feed rate is difficult to determine because the relationships between feed and surface finish, force, horsepower, and other constraints are not readily available for each machine tool. Experience, judgment, and experimentation are required to select the proper feed. The optimum cutting speed is difficult to calculate because the Taylor equation parameters C and n are not usually known without prior testing. Testing of this kind in a production environment is expensive.

24.4 PRODUCT DESIGN CONSIDERATIONS IN MACHINING

Several important aspects of product design have already been considered in our discussion of tolerance and surface finish (Section 24.2). In this section, we present some design guidelines for machining, compiled from sources [1], [5], and [15]:

➢ If possible, parts should be designed that do not need machining. If this is not possible, then minimize the amount of machining required on the parts. In general, a lower-cost product is achieved through the use of net shape processes such as precision casting, closed die forging, or (plastic) molding; or near net shape processes such as impression die forging. Reasons why machining may be required include close tolerances; good surface finish; and special geometric features such as threads, precision holes, cylindrical sections with high degree of roundness, and similar shapes that cannot be achieved except by machining.

➤ Tolerances should be specified to satisfy functional requirements, but process capabilities should also be considered. See Table 24.2 for tolerance capabilities in machining. Excessively close tolerances add cost but may not add value to the part. As tolerances become tighter (smaller), product costs generally increase because of additional processing, fixturing, inspection, sortation, rework, and scrap.

➤ Surface finish should be specified to meet functional and/or aesthetic requirements, but better finishes generally increase processing costs by requiring additional operations such as grinding or lapping.

➤ Machined features such as sharp corners, edges, and points should be avoided; they are often difficult to accomplish by machining. Sharp internal corners require pointed cutting tools that tend to break during machining. Sharp external corners and edges tend to create burrs and are dangerous to handle.

➤ Deep holes that must be bored should be avoided. Deep hole boring requires a long boring bar. Boring bars must be stiff, and this often requires use of high modulus materials such as cemented carbide, which is expensive.

➤ Machined parts should be designed so they can be produced from standard available stock. Choose exterior dimensions equal to or close to the standard stock size to minimize machining; for example, rotational parts with outside diameters that are equal to standard bar stock diameters.

➤ Parts should be designed to be rigid enough to withstand forces of cutting and workholder clamping. Machining of long narrow parts, large flat parts, parts with thin walls, and similar shapes should be avoided if possible.

➤ Undercuts as in Figure 24.5 should be avoided because they often require additional setups and operations and/or special tooling; they can also lead to stress concentrations in service.

➤ Materials with good machinability should be selected by the designer (Section 24.1). As a rough guide, the machinability rating of a material correlates with the allowable cutting speed and production rate that can be used. Thus, parts made of materials with low machinability cost more to produce. Parts that are hardened by heat treatment must usually be finish ground or machined with higher cost tools after hardening to achieve final size and tolerance.

➤ Machined parts should be designed with features that can be produced in a minimum number of setups—one setup if possible. This usually means geometric features that can be accessed from one side of the part (see Figure 24.6).

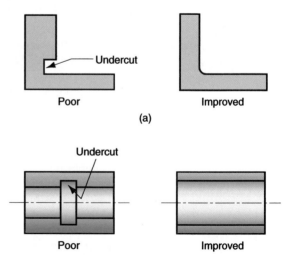

FIGURE 24.5 Two machined parts with undercuts: cross sections of (a) bracket and (b) rotational part. Also shown is how the part design might be improved.

FIGURE 24.6 Two parts with similar hole features: (a) holes that must be machined from two sides, requiring two setups, and (b) holes that can all be machined from one side.

(a)

(b)

➤ Machined parts should be designed with features that can be achieved with standard cutting tools. This means avoiding unusual hole sizes, threads, and features with unusual shapes requiring special form tools. In addition, it is helpful to design parts such that the number of individual cutting tools needed in machining is minimized; this often allows the part to be completed in one setup on a machine such as a machining center (Section 22.5).

REFERENCES

[1] Bakerjian, R. (ed.). *Tool and Manufacturing Engineers Handbook.* 4th ed. Vol VI, *Design for Manufacturability.* Society of Manufacturing Engineers, Dearborn, Michigan, 1992.

[2] Black, J, and Kohser, R. *DeGarmo's Materials and Processes in Manufacturing,* 10th ed., John Wiley & Sons, Hoboken, New Jersey, 2008.

[3] Boothroyd, G., and Knight, W. A. *Fundamentals of Metal Machining and Machine Tools.* 3rd ed. CRC Taylor & Francis, Boca Raton, Florida, 2006.

[4] Boston, O. W. *Metal Processing.* 2nd ed. John Wiley & Sons, New York, 1951.

[5] Bralla, J. G. (ed.). *Design for Manufacturability Handbook.* 2nd ed. McGraw-Hill, New York, 1998.

[6] Brierley, R. G., and Siekman, H. J. *Machining Principles and Cost Control.* McGraw-Hill, New York, 1964.

[7] Drozda, T. J., and Wick, C. (eds.). *Tool and Manufacturing Engineers Handbook.* 4th ed. Vol I, *Machining.* Society of Manufacturing Engineers, Dearborn, Michigan, 1983.

[8] Eary, D. F., and Johnson, G. E. *Process Engineering: for Manufacturing.* Prentice-Hall, Englewood Cliffs, New Jersey, 1962.

[9] Ewell, J. R. "Thermal Coefficients—A Proposed Machinability Index." *Technical Paper MR67-200.* Society of Manufacturing Engineers, Dearborn, Michigan, 1967.

[10] Gilbert, W. W. "Economics of Machining." *Machining—Theory and Practice.* American Society for Metals, Metals Park, Ohio, 1950, pp. 465–485.

[11] Groover, M. P. "A Survey on the Machinability of Metals." *Technical Paper MR76-269.* Society of Manufacturing Engineers, Dearborn, Michigan, 1976.

[12] *Machining Data Handbook.* 3rd ed. Vols. I. and II, Metcut Research Associates, Cincinnati, Ohio, 1980.

[13] Schaffer, G. H. "The Many Faces of Surface Texture." Special Report 801, *American Machinist & Automated Manufacturing.* June 1988 pp. 61–68.

[14] *Surface Finish.* Machining Development Service, Publication A-5, General Electric Company, Schenectady, New York (no date).

[15] Trucks, H. E., and Lewis, G. *Designing for Economical Production.* 2nd ed. Society of Manufacturing Engineers, Dearborn, Michigan, 1987.

[16] Van Voast, J. *United States Air Force Machinability Report.* Vol. 3. Curtis-Wright Corporation, 1954.

REVIEW QUESTIONS

24.1. Define machinability.

24.2. What are the criteria by which machinability is commonly assessed in a production machining operation?

24.3. Name some of the important mechanical and physical properties that affect the machinability of a work material.

24.4. Why do costs tend to increase when better surface finish is required on a machined part?

24.5. What are the basic factors that affect surface finish in machining?

24.6. What are the parameters that have the greatest influence in determining the ideal surface roughness R_i in a turning operation?

24.7. Name some of the steps that can be taken to reduce or eliminate vibrations in machining.

24.8. What are the factors on which the selection of feed in a machining operation should be based?

24.9. The unit cost in a machining operation is the sum of four cost terms. The first three terms are: (1) part load/unload cost, (2) cost of time the tool is actually cutting the work, and (3) cost of the time to change the tool. What is the fourth term?

24.10. Which cutting speed is always lower for a given machining operation, cutting speed for minimum cost or cutting speed for maximum production rate? Why?

MULTIPLE CHOICE QUIZ

There are 14 correct answers in the following multiple choice questions (some questions have multiple answers that are correct). To attain a perfect score on the quiz, all correct answers must be given. Each correct answer is worth 1 point. Each omitted answer or wrong answer reduces the score by 1 point, and each additional answer beyond the correct number of answers reduces the score by 1 point. Percentage score on the quiz is based on the total number of correct answers.

24.1. Which of the following criteria are generally recognized to indicate good machinability (four best answers): (a) ease of chip disposal, (b) high cutting temperatures, (c) high power requirements, (d) high value of R_a, (e) long tool life, (f) low cutting forces, and (g) zero shear plane angle?

24.2. Of the various methods for testing machinability, which one of the following is the most important: (a) cutting forces, (b) cutting temperature, (c) horsepower consumed in the operation, (d) surface roughness, (e) tool life, or (f) tool wear?

24.3. A machinability rating greater than 1.0 indicates that the work material is (a) easier to machine than the base metal or (b) more difficult to machine than the base metal, where the base metal has a rating = 1.0?

24.4. In general, which one of the following materials has the highest machinability: (a) aluminum, (b) cast iron, (c) copper, (d) low carbon steel, (e) stainless steel, (f) titanium alloys, or (g) unhardened tool steel?

24.5. Which one of the following operations is generally capable of the closest tolerances: (a) broaching, (b) drilling, (c) end milling, (d) planing, or (e) sawing?

24.6. When cutting a ductile work material, an increase in cutting speed will generally (a) degrade surface finish, which means a higher value of R_a or (b) improve surface finish, which means a lower value of R_a?

24.7. Which one of the following operations is generally capable of the best surface finishes (lowest value of R_a): (a) broaching, (b) drilling, (c) end milling, (d) planing, or (e) turning?

24.8. Which of the following time components in the average production machining cycle is affected by cutting speed (two correct answers): (a) part loading and unloading time, and (b) setup time for the machine tool, (c) time the tool is engaged in cutting, and (d) average tool change time per piece?

24.9. Which cutting speed is always lower for a given machining operation: (a) cutting speed for maximum production rate, or (b) cutting speed for minimum cost?

24.10. A high tooling cost and/or tool change time will tend to (a) decrease, (b) have no effect on, or (c) increase the cutting speed for minimum cost?

PROBLEMS

Machinability

24.1. A machinability rating is to be determined for a new work material using the cutting speed for a 60-min tool life as the basis of comparison. For the base material (B1112 steel), test data resulted in Taylor equation parameter values of $n = 0.29$ and $C = 500$, where speed is in m/min and tool life is in min. For the new material, the parameter values were $n = 0.21$ and $C = 400$. These results were obtained using cemented carbide tooling. (a) Compute a machinability rating for the new material.

(b) Suppose the machinability criterion were the cutting speed for a 10-min tool life rather than the present criterion. Compute the machinability rating for this case. (c) What do the results of the two calculations show about the difficulties in machinability measurement?

24.2. A small company uses a band saw to cut through 2-inch metal bar stock. A material supplier is pushing a new material that is supposed to be more machinable while providing similar mechanical properties. The company does not have access to sophisticated measuring devices, but they do have a stopwatch. They have acquired a sample of the new material and cut both the present material and the new material with the same band saw settings. In the process, they measured how long it took to cut through each material. To cut through the present material, it took an average of 2 minutes, 20 seconds. To cut through the new material, it took an average of 2 minutes, 6 seconds. (a) Develop a machinability rating system based on time to cut through the 2.0-inch bar stock, using the present material as the base material. (b) Using your rating system, determine the machinability rating for the new material.

24.3. A machinability rating is to be determined for a new work material. For the base material (B1112),

test data resulted in a Taylor equation with parameters $n = 0.29$ and $C = 490$. For the new material, the Taylor parameters were $n = 0.23$ and $C = 430$. Units in both cases are: speed in m/min and tool life in min. These results were obtained using cemented carbide tooling. (a) Compute a machinability rating for the new material using cutting speed for a 30-min tool life as the basis of comparison. (b) If the machinability criterion were tool life for a cutting speed of 150 m/min, what is the machinability rating for the new material?

24.4. Tool life turning tests have been conducted on B1112 steel with high-speed steel tooling, and the resulting parameters of the Taylor equation are: $n = 0.13$ and $C = 225$. B1112 is the base metal and has a machinability rating = 1.00 (100%). During the tests, feed = 0.010 in/rev, and depth of cut = 0.100 in. Based on this information, and machinability data given in Table 24.1, determine the cutting speed you would recommend for the following work materials, if the tool life desired in operation is 30 min (the same feed and depth of cut are to be used): (a) C1008 low carbon steel with 150 Brinell hardness, (b) 4130 alloy steel with 190 Brinell hardness, and (c) B1113 steel with 170 Brinell hardness.

Surface Roughness

24.5. In a turning operation on cast iron, the nose radius on the tool = 1.5 mm, feed = 0.22 mm/rev, and speed = 1.8 m/s. Compute an estimate of the surface roughness for this cut.

24.6. A turning operation uses a 2/64 in nose radius cutting tool on a free machining steel with a feed rate = 0.010 in/rev and a cutting speed = 300 ft/min. Determine the surface roughness for this cut.

24.7. A single-point HSS tool with a 3/64 in nose radius is used in a shaping operation on a ductile steel workpart. The cutting speed is 120 ft/min. The feed is 0.014 in/pass and depth of cut is 0.135 in. Determine the surface roughness for this operation.

24.8. A part to be turned in an engine lathe must have a surface finish of 1.6 μm. The part is made of a free-machining aluminum alloy. Cutting speed = 150 m/min, and depth of cut = 4.0 mm. The nose radius on the tool = 0.75 mm. Determine the feed that will achieve the specified surface finish.

24.9. Solve previous Problem 24.8 except that the part is made of cast iron instead of aluminum and the cutting speed is reduced to 100 m/min.

24.10. A part to be turned in an engine lathe must have a surface finish of 1.5 μm. The part is made of aluminum. The cutting speed is 1.5 m/s and the depth is 3.0

mm. The nose radius on the tool = 1.0 mm. Determine the feed that will achieve the specified surface finish.

24.11. The surface finish specification in a turning job is 0.8 μm. The work material is cast iron. Cutting speed = 75 m/min, feed = 0.3 mm/rev, and depth of cut = 4.0 mm. The nose radius of the cutting tool must be selected. Determine the minimum nose radius that will obtain the specified finish in this operation.

24.12. A face milling operation is to be performed on a cast iron part to finish the surface to 36 μ-in. The cutter uses four inserts and its diameter is 3.0 in. The cutter rotates at 475 rev/min. To obtain the best possible finish, a type of carbide insert with 4/64 in nose radius is to be used. Determine the required feed rate (in/min) that will achieve the 36 μ-in finish.

24.13. A face milling operation is not yielding the required surface finish on the work. The cutter is a four-tooth insert type face milling cutter. The machine shop foreman thinks the problem is that the work material is too ductile for the job, but this property tests well within the ductility range for the material specified by the designer. Without knowing any more about the job, what changes in (a) cutting conditions and (b) tooling would you suggest to improve the surface finish?

24.14. A turning operation is to be performed on C1010 steel, which is a ductile grade. It is desired to achieve a surface finish of 64 μ-in, while at the same time maximizing the metal removal rate. It has been decided that the speed should be in the range 200 ft/min to 400 ft/min, and that the depth of cut will be 0.080 in. The tool nose radius = 3/64 in. Determine the speed and feed combination that meets these criteria.

Machining Economics

24.15. A high-speed steel tool is used to turn a steel workpart that is 300 mm long and 80 mm in diameter. The parameters in the Taylor equation are: $n = 0.13$ and $C = 75$ (m/min) for a feed of 0.4 mm/rev. The operator and machine tool rate = $30/hr, and the tooling cost per cutting edge = $4. It takes 2.0 min to load and unload the workpart and 3.50 min to change tools. Determine (a) cutting speed for maximum production rate, (b) tool life in min of cutting, and (c) cycle time and cost per unit of product.

24.16. Solve Problem 24.15 except that in part (a) determine cutting speed for minimum cost.

24.17. A cemented carbide tool is used to turn a part with a length of 14.0 in and diameter = 4.0 in. The parameters in the Taylor equation are: $n = 0.25$ and $C = 1000$ (ft/min). The rate for the operator and machine tool = $45/hr, and the tooling cost per cutting edge = $2.50. It takes 2.5 min to load and unload the workpart and 1.50 min to change tools. The feed = 0.015 in/rev. Determine (a) cutting speed for maximum production rate, (b) tool life in min of cutting, and (c) cycle time and cost per unit of product.

24.18. Solve Problem 24.17 except that in part (a) determine cutting speed for minimum cost.

24.19. Compare disposable and regrindable tooling. The same grade of cemented carbide tooling is available in two forms for turning operations in a certain machine shop: disposable inserts and brazed inserts. The parameters in the Taylor equation for this grade are: $n = 0.25$ and $C = 300$ (m/min) under the cutting conditions considered here. For the disposable inserts, price of each insert = $6, there are four cutting edges per insert, and the tool change time = 1.0 min (this is an average of the time to index the insert and the time to replace it when all edges have been used). For the brazed insert, the price of the tool = $30 and it is estimated that it can be used a total of 15 times before it must be scrapped. The tool change time for the regrindable tooling = 3.0 min. The standard time to grind or regrind the cutting edge is 5.0 min, and the grinder is paid at a rate = $20/hr. Machine time on the lathe costs $24/hr. The workpart to be used in the comparison is 375 mm long and 62.5 mm in diameter, and it takes 2.0 min to load and unload the work. The feed = 0.30 mm/rev. For the two tooling cases, compare (a) cutting speeds for minimum cost, (b) tool lives, (c) cycle time and cost per unit of production. Which tool would you recommend?

24.20. Solve Problem 24.19 except that in part (a) determine the cutting speeds for maximum production rate.

24.21. Three tool materials are to be compared for the same finish turning operation on a batch of 150 steel parts: high-speed steel, cemented carbide, and ceramic. For the high-speed steel tool, the Taylor equation parameters are: $n = 0.130$ and $C = 80$ (m/min). The price of the HSS tool is $20 and it is estimated that it can be ground and reground 15 times at a cost of $2 per grind. Tool change time is 3 min. Both carbide and ceramic tools are in insert form and can be held in the same mechanical toolholder. The Taylor equation parameters for the cemented carbide are: $n = 0.30$ and $C = 650$ (m/min); and for the ceramic: $n = 0.6$ and $C = 3,500$ (m/min). The cost per insert for the carbide is $8 and for the ceramic is $10. There are six cutting edges per insert in both cases. Tool change time is 1.0 min for both tools. The time to change a part is 2.5 min. The feed is 0.30 mm/rev, and depth of cut is 3.5 mm. The cost of machine time is $40/hr. The part is 73.0 mm in diameter and 250 mm in length. Setup time for the batch is 2.0 hr. For the three tooling cases, compare: (a) cutting speeds for minimum cost, (b) tool lives, (c) cycle time, (d) cost per production unit, (e) total time to complete the batch and production rate. (f) What is the proportion of time spent actually cutting metal for each tooling? Use of a spreadsheet calculator is recommended.

24.22. Solve Problem 24.21 except that in parts (a) and (b) determine the cutting speeds and tool lives for maximum production rate. Use of a spreadsheet calculator is recommended.

24.23. A vertical boring mill is used to bore the inside diameter of a large batch of tube-shaped parts. The diameter = 28.0 in and the length of the bore = 14.0 in. Current cutting conditions are: speed = 200 ft/min, feed = 0.015 in/rev, and depth = 0.125 in. The parameters of the Taylor equation for the cutting tool in the operation are: $n = 0.23$ and $C = 850$ (ft/

min). Tool change time = 3.0 min, and tooling cost = $3.50 per cutting edge. The time required to load and unload the parts = 12.0 min, and the cost of machine time on this boring mill = $42/hr. Management has decreed that the production rate must be increased by 25%. Is that possible? Assume that feed must remain unchanged to achieve the required surface finish. What is the current production rate and the maximum possible production rate for this job?

24.24. An NC lathe cuts two passes across a cylindrical workpiece under automatic cycle. The operator loads and unloads the machine. The starting diameter of the work is 3.00 in and its length = 10 in. The work cycle consists of the following steps (with element times given in parentheses where applicable): (1) Operator loads part into machine, starts cycle (1.00 min); (2) NC lathe positions tool for first pass (0.10 min); (3) NC lathe turns first pass (time depends on cutting speed); (4) NC lathe repositions tool for second pass (0.4 min); (5) NC lathe turns second pass (time depends on cutting speed); and (6) Operator unloads part and places in tote pan (1.00 min). In addition, the cutting tool must be periodically changed. This tool change time takes 1.00 min. The feed rate = 0.007 in/rev and the depth of cut for each pass = 0.100 in. The cost of the operator and machine = $39/hr and the tool cost = $2/cutting edge. The applicable Taylor tool life equation has parameters: $n = 0.26$ and $C = 900$ (ft/min). Determine (a) the cutting speed for minimum cost per piece, (b) the average time required to complete one production cycle, (c) cost of the production cycle. (d) If the setup time for this job is 3.0 hours and the batch size = 300 parts, how long will it take to complete the batch?

24.25. As indicated in Section 23.4, the effect of a cutting fluid is to increase the value of C in the Taylor tool life equation. In a certain machining situation using HSS tooling, the C value is increased from $C = 200$ to $C = 225$ owing to the use of the cutting fluid. The n value is the same with or without fluid at $n = 0.125$. Cutting speed used in the operation is $v = 125$ ft/min. Feed = 0.010 in/rev and depth = 0.100 in. The effect of the cutting fluid can be to either increase cutting speed (at the same tool life) or increase tool life (at the same cutting speed). (a) What is the cutting speed that would result from using the cutting fluid if tool life remains the same as with no fluid? (b) What is the tool life that would result if the cutting speed remained at 125 ft/min? (c) Economically, which effect is better, given that tooling cost = $2 per cutting edge, tool change time = 2.5 min, and operator and machine rate = $30/hr? Justify you answer with calculations, using cost per cubic in of metal machined as the criterion of comparison. Ignore effects of workpart handling time.

24.26. In a turning operation on ductile steel, it is desired to obtain an actual surface roughness of 63 μ-in with a 2/64 in nose radius tool. The ideal roughness is given by Eq. (24.1) and an adjustment will have to be made using Figure 24.2 to convert the 63 μ-in actual roughness to an ideal roughness, taking into account the material and cutting speed. Disposable inserts are used at a cost of $1.75 per cutting edge (each insert costs $7 and there are four edges per insert). Average time to change each insert = 1.0 min. The workpiece length = 30.0 in and its diameter = 3.5 in. The machine and operator's rate = $39 per hour including applicable overheads. The Taylor tool life equation for this tool and work combination is given by: $vT^{0.23}f^{0.55} = 40.75$, where $T =$ tool life, min; $v =$ cutting speed, ft/min; and $f =$ feed, in/rev. Solve for (a) the feed in in/rev that will achieve the desired actual finish, (b) cutting speed for minimum cost per piece at the feed determined in (a). *Hint:* To solve (a) and (b) requires an iterative computational procedure. Use of a spreadsheet calculator is recommended for this iterative procedure.

24.27. Solve Problem 24.26 only using maximum production rate as the objective rather than minimum piece cost. Use of a spreadsheet calculator is recommended.

24.28. Verify that the derivative of Eq. (24.6) results in Eq. (24.7).

24.29. Verify that the derivative of Eq. (24.12) results in Eq. (24.13).

25 GRINDING AND OTHER ABRASIVE PROCESSES

Chapter Contents

Abrasive machining involves material removal by the action of hard, abrasive particles that are usually in the form of a bonded wheel. Grinding is the most important abrasive process. In terms of number of machine tools in use, grinding is the most common of all metalworking operations [11]. Other traditional abrasive processes include honing, lapping, superfinishing, polishing, and buffing. The abrasive machining processes are generally used as finishing operations, although some abrasive processes are capable of high material removal rates rivaling those of conventional machining operations.

The use of abrasives to shape parts is probably the oldest material removal process (Historical Note 25.1). Abrasive processes are important commercially and technologically for the following reasons:

➢ They can be used on all types of materials ranging from soft metals to hardened steels and hard nonmetallic materials such as ceramics and silicon.
➢ Some of these processes can produce extremely fine surface finishes, to 0.025 μm (1 μ-in).
➢ For certain abrasive processes, dimensions can be held to extremely close tolerances.

Abrasive water jet cutting and ultrasonic machining are also abrasive processes, because material removal is accomplished by means of abrasives. However, they are commonly classified as nontraditional processes and are covered in the following chapter.

25.1 GRINDING

Grinding is a material removal process accomplished by abrasive particles that are contained in a bonded grinding wheel rotating at very high surface speeds. The grinding wheel is usually disk-shaped, and is precisely balanced for

Historical Note 25.1   *Development of abrasive processes*

Use of abrasives predates any of the other machining operations. There is archaeological evidence that ancient people used abrasive stones such as sandstone found in nature to sharpen tools and weapons and scrape away unwanted portions of softer materials to make domestic implements.

Grinding became an important technical trade in ancient Egypt. The large stones used to build the Egyptian pyramids were cut to size by a rudimentary grinding process. The grinding of metals dates to around 2000 BCE and was a highly valued skill at that time.

Early abrasive materials were those found in nature, such as sandstone, which consists primarily of quartz (SiO_2); emery, consisting of corundum (Al_2O_3) plus equal or lesser amounts of the iron minerals hematite (Fe_2O_3) and magnetite (Fe_3O_4); and diamond. The first grinding wheels were likely cut out of sandstone and were no doubt rotated under manual power. However, grinding wheels made in this way were not consistent in quality.

In the early 1800s, the first solid bonded grinding wheels were produced in India. They were used to grind gems, an important trade in India at the time. The abrasives were corundum, emery, or diamond. The bonding material was natural gum-resin shellac. The technology was exported to Europe and the United States, and other bonding materials were subsequently introduced: rubber bond in the mid-1800s, vitrified bond around 1870, shellac bond around 1880, and resinoid bond in the 1920s with the development of the first thermosetting plastics (phenol-formaldehyde).

In the late 1800s, synthetic abrasives were first produced: silicon carbide (SiC) and aluminum oxide (Al_2O_3). By manufacturing the abrasives, chemistry and size of the individual abrasive grains could be controlled more closely, resulting in higher quality grinding wheels.

The first real grinding machines were made by the U.S. firm Brown & Sharpe in the 1860s for grinding parts for sewing machines, an important industry during the period. Grinding machines also contributed to the development of the bicycle industry in the 1890s and later the U.S. automobile industry. The grinding process was used to size and finish heat-treated (hardened) parts in these products.

The superabrasives diamond and cubic boron nitride are products of the twentieth century. Synthetic diamonds were first produced by the General Electric Company in 1955. These abrasives were used to grind cemented carbide cutting tools, and today this remains one of the important applications of diamond abrasives. Cubic boron nitride (cBN), second only to diamond in hardness, was first synthesized in 1957 by GE using a similar process to that for making artificial diamonds. Cubic BN has become an important abrasive for grinding hardened steels.

high rotational speeds. The reader can see grinding in action in our video clip titled Basics of Grinding.

VIDEO CLIP

Basics of Grinding. This clip contains four segments: (1) CNC grinding, (2) grinding wheel ring testing, (3) wheel dressing, and (4) grinding fluids.

Grinding can be likened to the milling process. Cutting occurs on either the periphery or the face of the grinding wheel, similar to peripheral and face milling. Peripheral grinding is much more common than face grinding. The rotating grinding wheel consists of many cutting teeth (the abrasive particles), and the work is fed relative to the wheel to accomplish material removal. Despite these similarities, there are significant differences between grinding and milling: (1) the abrasive grains in the wheel are much smaller and more numerous than the teeth on a milling cutter; (2) cutting speeds in grinding are much higher than in milling; (3) the abrasive grits in a grinding wheel are randomly oriented and possess on average a very high negative rake angle; and (4) a grinding wheel is self-sharpening—as the wheel wears, the abrasive particles become dull and either fracture to create fresh cutting edges or are pulled out of the surface of the wheel to expose new grains.

25.1.1 THE GRINDING WHEEL

A grinding wheel consists of abrasive particles and bonding material. The bonding material holds the particles in place and establishes the shape and structure of the wheel. These two ingredients and the way they are fabricated determine the five basic parameters of a grinding wheel: (1) abrasive material, (2) grain size, (3) bonding material, (4) wheel grade, and (5) wheel structure. To achieve the desired performance in a given application, each of the parameters must be carefully selected.

Abrasive Material   Different abrasive materials are appropriate for grinding different work materials. General properties of an abrasive material used in grinding wheels include high hardness, wear resistance, toughness, and friability. Hardness, wear resistance, and toughness are desirable properties of any cutting-tool material. *Friability* refers to the capacity of the abrasive material to fracture when the cutting edge of the grain becomes dull, thereby exposing a new sharp edge.

The development of grinding abrasives is described in our historical note. Today, the abrasive materials of greatest commercial importance are aluminum oxide, silicon carbide, cubic boron nitride, and diamond. They are briefly described in Table 25.1, together with their relative hardness values.

Grain Size   The grain size of the abrasive particle is important in determining surface finish and material removal rate. Small grit sizes produce better finishes, whereas larger grain sizes permit larger material removal rates. Thus, a choice must be made between these two objectives when selecting abrasive grain size. The selection of grit size also depends to some extent on the hardness of the work material. Harder work materials require smaller grain sizes to cut effectively, whereas softer materials require larger grit sizes.

The grit size is measured using a screen mesh procedure, as explained in Section 16.1. In this procedure, smaller grit sizes have larger numbers and vice versa. Grain sizes used in grinding wheels typically range between 8 and 250. Grit size 8 is very coarse and size 250 is very fine. Even finer grit sizes are used for lapping and superfinishing (Section 25.2).

Bonding Materials   The bonding material holds the abrasive grains and establishes the shape and structural integrity of the grinding wheel. Desirable properties of the bond

TABLE 25.1   Abrasives of greatest importance in grinding.

Abrasive	Description	Knoop Hardness
Aluminum oxide (Al_2O_3)	Most common abrasive material (Section 7.3.1), used to grind steel and other ferrous, high-strength alloys.	2100
Silicon carbide (SiC)	Harder than Al_2O_3, but not as tough (Section 7.2). Applications include ductile metals such as aluminum, brass, and stainless steel, as well as brittle materials such as some cast irons and certain ceramics. Cannot be used effectively for grinding steel because of the strong chemical affinity between the carbon in SiC and the iron in steel.	2500
Cubic boron nitride (cBN)	When used as an abrasive, cBN (Section 7.3.3) is produced under the trade name Borazon by the General Electric Company. cBN grinding wheels are used for hard materials such as hardened tool steels and aerospace alloys.	5000
Diamond	Diamond abrasives occur naturally and are also made synthetically (Section 7.5.1). Diamond wheels are generally used in grinding applications on hard, abrasive materials such as ceramics, cemented carbides, and glass.	7000

TABLE 25.2   Bonding materials used in grinding wheels.

Bonding Material	Description
Vitrified bond	Consists chiefly of baked clay and ceramic materials. Most grinding wheels in common use are vitrified bonded wheels. They are strong and rigid, resistant to elevated temperatures, and relatively unaffected by water and oil that might be used in grinding fluids.
Silicate bond	Consists of sodium silicate (Na_2SO_3). Applications are generally limited to situations in which heat generation must be minimized, such as grinding cutting tools.
Rubber bond	Most flexible of the bonding materials and used as a bonding material in cutoff wheels.
Resinoid bond	Consists of various thermosetting resin materials, such as phenol-formaldehyde. It has very high strength and is used for rough grinding and cutoff operations.
Shellac bond	Relatively strong but not rigid; often used in applications requiring a good finish.
Metallic bond	Metal, usually bronze, is the common bond material for diamond and cBN grinding wheels. Particulate processing (Chapters 16 and 17) is used to bond the metal matrix and abrasive grains to the outside periphery of the wheel, thus conserving the costly abrasive materials.

material include strength, toughness, hardness, and temperature resistance. The bonding material must be able to withstand the centrifugal forces and high temperatures experienced by the grinding wheel, resist shattering in shock loading of the wheel, and hold the abrasive grains rigidly in place to accomplish the cutting action while allowing those grains that are worn to be dislodged so that new grains can be exposed. Bonding materials commonly used in grinding wheels are identified and briefly described in Table 25.2.

Wheel Structure and Wheel Grade   *Wheel structure* refers to the relative spacing of the abrasive grains in the wheel. In addition to the abrasive grains and bond material, grinding wheels contain air gaps or pores, as illustrated in Figure 25.1. The volumetric proportions of grains, bond material, and pores can be expressed as

$$P_g + P_b + P_p = 1.0 \qquad (25.1)$$

where P_g = proportion of abrasive grains in the total wheel volume, P_b = proportion of bond material, and P_p = proportion of pores (air gaps).

Wheel structure is measured on a scale that ranges between "open" and "dense." An open structure is one in which P_p is relatively large, and P_g is relatively small. That is, there are more pores and fewer grains per unit volume in a wheel of open structure. By contrast, a

FIGURE 25.1   Typical structure of a grinding wheel.

TABLE 25.3 Marking system for conventional grinding wheels as defined by ANSI Standard B74.13-1977 [3].

30　A　46　H　6　V　XX

Manufacturer's private marking for wheel (optional).

Bond type: B = Resinoid, BF = resinoid reinforced, E = Shellac, R = Rubber, RF = rubber reinforced, S = Silicate, V = Vitrified.

Structure: Scale ranges from 1 to 15: 1 = very dense structure, 15 = very open structure.

Grade: Scale ranges from A to Z: A· = soft, M = medium, Z = hard.

Grain size: Coarse = grit sizes 8 to 24, Medium = grit sizes 30 to 60, Fine = grit sizes 70 to 180, Very fine = grit sizes 220 to 600.

Abrasive type: A = aluminum oxide, C = silicon carbide.

Prefix: Manufacturer's symbol for abrasive (optional).

dense structure is one in which P_p is relatively small, and P_g is larger. Generally, open structures are recommended in situations in which clearance for chips must be provided. Dense structures are used to obtain better surface finish and dimensional control.

Wheel grade indicates the grinding wheel's bond strength in retaining the abrasive grits during cutting. This is largely dependent on the amount of bonding material present in the wheel structure—P_b in Eq. (25.1). Grade is measured on a scale that ranges between soft and hard. "Soft" wheels lose grains readily, whereas "hard" wheels retain their abrasive grains. Soft wheels are generally used for applications requiring low material removal rates and grinding of hard work materials. Hard wheels are typically used to achieve high stock removal rates and for grinding of relative soft work materials.

Grinding Wheel Specification The preceding parameters can be concisely designated in a standard grinding wheel marking system defined by the American National Standards Institute (ANSI) [3]. This marking system uses numbers and letters to specify abrasive type, grit size, grade, structure, and bond material. Table 25.3 presents an abbreviated version of the ANSI Standard, indicating how the numbers and letters are interpreted. The standard also provides for additional identifications that might be used by the grinding wheel manufacturers. The ANSI Standard for diamond and cubic boron nitride grinding wheels is slightly different than for conventional wheels. The marking system for these newer grinding wheels is presented in Table 25.4.

TABLE 25.4 Marking system for diamond and cubic boron nitride grinding wheels as defined by ANSI Standard B74.13-1977 [3].

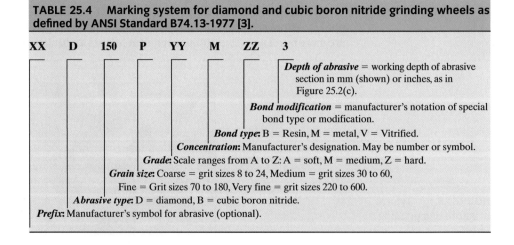

XX　D　150　P　YY　M　ZZ　3

Depth of abrasive = working depth of abrasive section in mm (shown) or inches, as in Figure 25.2(c).

Bond modification = manufacturer's notation of special bond type or modification.

Bond type: B = Resin, M = metal, V = Vitrified.

Concentration: Manufacturer's designation. May be number or symbol.

Grade: Scale ranges from A to Z: A = soft, M = medium, Z = hard.

Grain size: Coarse = grit sizes 8 to 24, Medium = grit sizes 30 to 60, Fine = Grit sizes 70 to 180, Very fine = grit sizes 220 to 600.

Abrasive type: D = diamond, B = cubic boron nitride.

Prefix: Manufacturer's symbol for abrasive (optional).

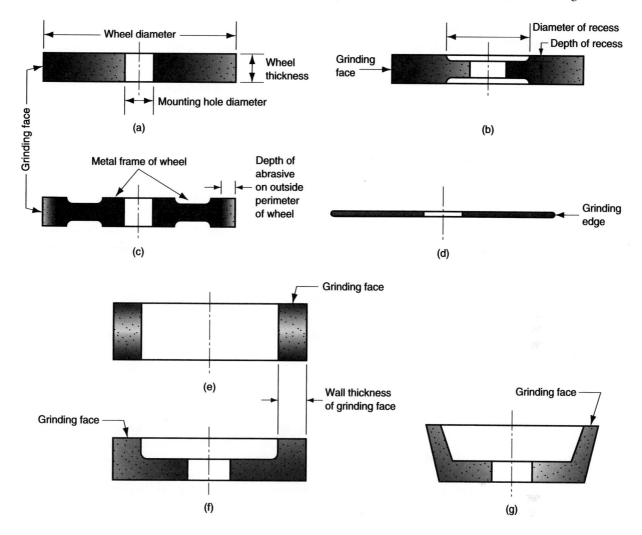

FIGURE 25.2   Some of the standard grinding wheel shapes: (a) straight, (b) recessed two sides, (c) metal wheel frame with abrasive bonded to outside circumference, (d) abrasive cutoff wheel, (e) cylinder wheel, (f) straight cup wheel, and (g) flaring cup wheel.

Grinding wheels come in a variety of shapes and sizes, as shown in Figure 25.2. Configurations (a), (b), and (c) are peripheral grinding wheels, in which material removal is accomplished by the outside circumference of the wheel. A typical abrasive cutoff wheel is shown in (d), which also involves peripheral cutting. Wheels (e), (f), and (g) are face grinding wheels, in which the flat face of the wheel removes material from the work surface.

25.1.2   ANALYSIS OF THE GRINDING PROCESS

The cutting conditions in grinding are characterized by very high speeds and very small cut size, compared to milling and other traditional machining operations. Using surface grinding to illustrate, Figure 25.3(a) shows the principal features of the process. The peripheral speed of the grinding wheel is determined by the rotational speed of the wheel:

$$v = \pi D N \tag{25.2}$$

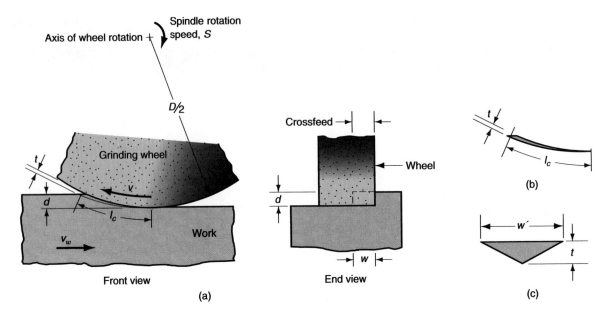

FIGURE 25.3 (a) The geometry of surface grinding, showing the cutting conditions; (b) assumed longitudinal shape and (c) cross section of a single chip.

where v = surface speed of wheel, m/min (ft/min); N = spindle speed, rev/min; and D = wheel diameter, m (ft).

Depth of cut d, called the **infeed**, is the penetration of the wheel below the original work surface. As the operation proceeds, the grinding wheel is fed laterally across the surface on each pass by the work. This is called the **crossfeed**, and it determines the width of the grinding path w in Figure 25.3(a). This width, multiplied by depth d determines the cross-sectional area of the cut. In most grinding operations, the work moves past the wheel at a certain speed v_w, so that the material removal rate is

$$R_{MR} = v_w w d \qquad (25.3)$$

Each grain in the grinding wheel cuts an individual chip whose longitudinal shape before cutting is shown in Figure 25.3(b) and whose assumed cross-sectional shape is triangular, as in Figure 25.3(c). At the exit point of the grit from the work, where the chip cross section is largest, this triangle has height t and width w'.

In a grinding operation, we are interested in how the cutting conditions combine with the grinding wheel parameters to affect (1) surface finish, (2) forces and energy, (3) temperature of the work surface, and (4) wheel wear.

Surface Finish Most commercial grinding is performed to achieve a surface finish that is superior to that which can be accomplished with conventional machining. The surface finish of the workpart is affected by the size of the individual chips formed during grinding. One obvious factor in determining chip size is grit size—smaller grit sizes yield better finishes.

Let us examine the dimensions of an individual chip. From the geometry of the grinding process in Figure 25.3, it can be shown that the average length of a chip is given by

$$l_c = \sqrt{Dd} \qquad (25.4)$$

where l_c is the length of the chip, mm (in); D = wheel diameter, mm (in); and d = depth of cut, or infeed, mm (in).

This assumes the chip is formed by a grit that acts throughout the entire sweep arc shown in the diagram.

Figure 25.3(c) shows the assumed cross section of a chip in grinding. The cross-sectional shape is triangular with width w' being greater than the thickness t by a factor called the grain aspect ratio r_g, defined by

$$r_g = \frac{w'}{t}$$ (25.5)

Typical values of grain aspect ratio are between 10 and 20.

The number of active grits (cutting teeth) per square inch on the outside periphery of the grinding wheel is denoted by C. In general, smaller grain sizes give larger C values. C is also related to the wheel structure. A denser structure means more grits per area. Based on the value of C, the number of chips formed per time n_c is given by

$$n_c = vwC$$ (25.6)

where v = wheel speed, mm/min (in/min); w = crossfeed, mm (in); and C = grits per area on the grinding wheel surface, grits/mm^2 (grits/in^2).

It stands to reason that surface finish will be improved by increasing the number of chips formed per unit time on the work surface for a given width w. Therefore, according to Eq. (25.6), increasing v and/or C will improve finish.

Forces and Energy   If the force required to drive the work past the grinding wheel were known, the specific energy in grinding could be determined as

$$U = \frac{F_c v}{v_w w d}$$ (25.7)

where U = specific energy, J/mm^3 (in-lb/in^3); F_c = cutting force, which is the force to drive the work past the wheel, N (lb); v = wheel speed, m/min (ft/min); v_w = work speed, mm/min (in/min); w = width of cut, mm (in); and d = depth of cut, mm (in).

In grinding, the specific energy is much greater than in conventional machining. There are several reasons for this. First is the *size effect* in machining. As discussed, the chip thickness in grinding is much smaller than for other machining operations, such as milling. According to the size effect (Section 21.4), the small chip sizes in grinding cause the energy required to remove each unit volume of material to be significantly higher than in conventional machining—roughly 10 times higher.

Second, the individual grains in a grinding wheel possess extremely negative rake angles. The average rake angle is about –30°, with values on some individual grains believed to be as low as –60°. These very low rake angles result in low values of shear plane angle and high shear strains, both of which mean higher energy levels in grinding.

Third, specific energy is higher in grinding because not all of the individual grits are engaged in actual cutting. Because of the random positions and orientations of the grains in the wheel, some grains do not project far enough into the work surface to accomplish cutting. Three types of grain actions can be recognized, as illustrated in Figure 25.4: (a) *cutting,* in which the grit projects far enough into the work surface to form a chip and remove material; (b) *plowing,* in which the grit projects into the work, but not far enough to cause cutting; instead, the work surface is deformed and energy is consumed without any material removal; and (c) *rubbing,* in which the grit contacts the surface during its sweep, but only rubbing friction occurs, thus consuming energy without removing any material.

The size effect, negative rake angles, and ineffective grain actions combine to make the grinding process inefficient in terms of energy consumption per volume of material removed.

Using the specific energy relationship in Eq. (25.7), and assuming that the cutting force acting on a single grain in the grinding wheel is proportional to $r_g t$, it can be

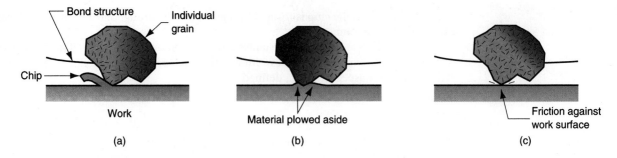

FIGURE 25.4 Three types of grain action in grinding: (a) cutting, (b) plowing, and (c) rubbing.

shown [10] that

$$F'_c = K_1 \left(\frac{r_g v_w}{vC}\right)^{0.5} \left(\frac{d}{D}\right)^{0.25}$$ (25.8)

where F'_c is the cutting force acting on an individual grain, K_1 is a constant of proportionality that depends on the strength of the material being cut and the sharpness of the individual grain, and the other terms have been previously defined.

The practical significance of this relationship is that F'_c affects whether an individual grain will be pulled out of the grinding wheel, an important factor in the wheel's capacity to "resharpen" itself. Referring back to our discussion on wheel grade, a hard wheel can be made to appear softer by increasing the cutting force acting on an individual grain through appropriate adjustments in v_w, v, and d, according to Eq. (25.8).

Temperatures at the Work Surface Because of the size effect, high negative rake angles, and plowing and rubbing of the abrasive grits against the work surface, the grinding process is characterized by high temperatures. Unlike conventional machining operations in which most of the heat energy generated in the process is carried off in the chip, much of the energy in grinding remains in the ground surface [11], resulting in high work surface temperatures. The high surface temperatures have several possible damaging effects, primarily surface burns and cracks. The burn marks show themselves as discolorations on the surface caused by oxidation. Grinding burns are often a sign of metallurgical damage immediately beneath the surface. The surface cracks are perpendicular to the wheel speed direction. They indicate an extreme case of thermal damage to the work surface.

A second harmful thermal effect is softening of the work surface. Many grinding operations are carried out on parts that have been heat-treated to obtain high hardness. High grinding temperatures can cause the surface to lose some of its hardness. Third, thermal effects in grinding can cause residual stresses in the work surface, possibly decreasing the fatigue strength of the part.

It is important to understand what factors influence work surface temperatures in grinding. Experimentally, it has been observed that surface temperature is dependent on energy per surface area ground (closely related to specific energy U). Because this varies inversely with chip thickness, it can be shown that surface temperature T_s is related to grinding parameters as follows [10]:

$$T_s = K_2 d^{0.75} \left(\frac{r_g C v}{v_w}\right)^{0.5} D^{0.25}$$ (25.9)

where K_2 = a constant of proportionality.

The practical implication of this relationship is that surface damage owing to high work temperatures can be mitigated by decreasing depth of cut d, wheel speed v, and number of active grits per square inch on the grinding wheel C, or by increasing work speed v_w. In addition, dull grinding wheels and wheels that have a hard grade and dense structure tend to cause thermal problems. Of course, using a cutting fluid can also reduce grinding temperatures.

Wheel Wear   Grinding wheels wear, just as conventional cutting tools wear. Three mechanisms are recognized as the principal causes of wear in grinding wheels: (1) grain fracture, (2) attritious wear, and (3) bond fracture. ***Grain fracture*** occurs when a portion of the grain breaks off, but the rest of the grain remains bonded in the wheel. The edges of the fractured area become new cutting edges on the grinding wheel. The tendency of the grain to fracture is called ***friability.*** High friability means that the grains fracture more readily because of the cutting forces on the grains F_c'.

   Attritious wear involves dulling of the individual grains, resulting in flat spots and rounded edges. Attritious wear is analogous to tool wear in a conventional cutting tool. It is caused by similar physical mechanisms including friction and diffusion, as well as chemical reactions between the abrasive material and the work material in the presence of very high temperatures.

   Bond fracture occurs when the individual grains are pulled out of the bonding material. The tendency toward this mechanism depends on wheel grade, among other factors. Bond fracture usually occurs because the grain has become dull because of attritious wear, and the resulting cutting force is excessive. Sharp grains cut more efficiently with lower cutting forces; hence, they remain attached in the bond structure.

   The three mechanisms combine to cause the grinding wheel to wear as depicted in Figure 25.5. Three wear regions can be identified. In the first region, the grains are initially sharp, and wear is accelerated because of grain fracture. This corresponds to the "break-in" period in conventional tool wear. In the second region, the wear rate is fairly constant, resulting in a linear relationship between wheel wear and volume of metal removed. This region is characterized by attritious wear, with some grain and bond fracture. In the third region of the wheel wear curve, the grains become dull, and the amount of plowing and rubbing increases relative to cutting. In addition, some of the chips become clogged in the pores of the wheel. This is called ***wheel loading,*** and it impairs the cutting action and leads to higher heat and work surface temperatures. As a consequence, grinding efficiency decreases, and the volume of wheel removed increases relative to the volume of metal removed.

   The ***grinding ratio*** is a term used to indicate the slope of the wheel wear curve. Specifically

$$GR = \frac{V_w}{V_g} \tag{25.10}$$

where GR = the grinding ratio, V_w = the volume of work material removed, and V_g = the corresponding volume of the grinding wheel that is worn in the process.

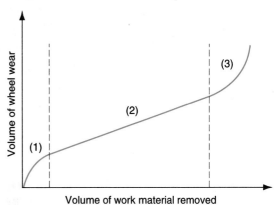

FIGURE 25.5   Typical wear curve of a grinding wheel. Wear is conveniently plotted as a function of volume of material removed, rather than as a function of time. (Based on [16].)

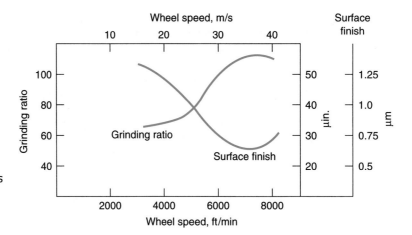

FIGURE 25.6 Grinding ratio and surface finish as a function of wheel speed. (Based on data in Krabacher [14].)

The grinding ratio has the most significance in the linear wear region of Figure 25.5. Typical values of *GR* range between 95 and 125 [5], which is about five orders of magnitude less than the analogous ratio in conventional machining. Grinding ratio is generally increased by increasing wheel speed *v*. The reason for this is that the size of the chip formed by each grit is smaller with higher speeds, so the amount of grain fracture is reduced. Because higher wheel speeds also improve surface finish, there is a general advantage in operating at high grinding speeds. However, when speeds become too high, attritious wear and surface temperatures increase. As a result, the grinding ratio is reduced and the surface finish is impaired. This effect was originally reported by Krabacher [14], as in Figure 25.6.

When the wheel is in the third region of the wear curve, it must be resharpened by a procedure called ***dressing,*** which consists of (1) breaking off the dulled grits on the outside periphery of the grinding wheel in order to expose fresh sharp grains and (2) removing chips that have become clogged in the wheel. It is accomplished by a rotating disk, an abrasive stick, or another grinding wheel operating at high speed, held against the wheel being dressed as it rotates. Although dressing sharpens the wheel, it does not guarantee the shape of the wheel. ***Truing*** is an alternative procedure that not only sharpens the wheel, but also restores its cylindrical shape and ensures that it is straight across its outside perimeter. The procedure uses a diamond-pointed tool (other types of truing tools are also used) that is fed slowly and precisely across the wheel as it rotates. A very light depth is taken (0.025 mm or less) against the wheel.

25.1.3 APPLICATION CONSIDERATIONS IN GRINDING

In this section, we attempt to bring together the previous discussion of wheel parameters and theoretical analysis of grinding and consider their practical application. We also consider grinding fluids, which are commonly used in grinding operations.

Application Guidelines There are many variables in grinding that affect the performance and success of the operation. The guidelines listed in Table 25.5 are helpful in sorting out the many complexities and selecting the proper wheel parameters and grinding conditions.

Grinding Fluids The proper application of cutting fluids has been found to be effective in reducing the thermal effects and high work surface temperatures described previously. When used in grinding operations, cutting fluids are called grinding fluids. The functions

TABLE 25.5   Application guidelines for grinding.

Application Problem or Objective	Recommendation or Guideline
Grinding steel and most cast irons	Select aluminum oxide as the abrasive.
Grinding most nonferrous metals	Select silicon carbide as the abrasive.
Grinding hardened tool steels and certain aerospace alloys	Select cubic boron nitride as the abrasive.
Grinding hard abrasive materials such as ceramics, cemented carbides, and glass	Select diamond as the abrasive.
Grinding soft metals	Select a large grit size and harder grade wheel.
Grinding hard metals	Select a small grit size and softer grade wheel.
Optimize surface finish	Select a small grit size and dense wheel structure. Use high wheel speeds (v), lower work speeds (v_w).
Maximize material removal rate	Select a large grit size, more open wheel structure, and vitrified bond.
To minimize heat damage, cracking, and warping of the work surface	Maintain sharpness of the wheel. Dress the wheel frequently. Use lighter depths of cut (d), lower wheel speeds (v), and faster work speeds (v_w).
If the grinding wheel glazes and burns	Select wheel with a soft grade and open structure.
If the grinding wheel breaks down too rapidly	Select wheel with a hard grade and dense structure.

Compiled from [8], [11], and [16].

performed by grinding fluids are similar to those performed by cutting fluids (Section 23.4). Reducing friction and removing heat from the process are the two common functions. In addition, washing away chips and reducing temperature of the work surface are very important in grinding.

Types of grinding fluids by chemistry include grinding oils and emulsified oils. The grinding oils are derived from petroleum and other sources. These products are attractive because friction is such an important factor in grinding. However, they pose hazards in terms of fire and operator health, and their cost is high relative to emulsified oils. In addition, their capacity to carry away heat is less than fluids based on water. Accordingly, mixtures of oil in water are most commonly recommended as grinding fluids. These are usually mixed with higher concentrations than emulsified oils used as conventional cutting fluids. In this way, the friction reduction mechanism is emphasized.

25.1.4   GRINDING OPERATIONS AND GRINDING MACHINES

Grinding is traditionally used to finish parts whose geometries have already been created by other operations. Accordingly, grinding machines have been developed to grind plain flat surfaces, external and internal cylinders, and contour shapes such as threads. The contour shapes are often created by special formed wheels that have the opposite of the desired contour to be imparted to the work. Grinding is also used in tool rooms to form the geometries on cutting tools. In addition to these traditional uses, applications of grinding are expanding to include more high speed, high material removal operations. Our discussion of operations and machines in this section includes the following types:

FIGURE 25.7 Four types of surface grinding: (a) horizontal spindle with reciprocating worktable, (b) horizontal spindle with rotating worktable, (c) vertical spindle with reciprocating worktable, and (d) vertical spindle with rotating worktable.

(1) surface grinding, (2) cylindrical grinding, (3) centerless grinding, (4) creep feed grinding, and (5) other grinding operations.

Surface Grinding   Surface grinding is normally used to grind plain flat surfaces. It is performed using either the periphery of the grinding wheel or the flat face of the wheel. Because the work is normally held in a horizontal orientation, peripheral grinding is performed by rotating the wheel about a horizontal axis, and face grinding is performed by rotating the wheel about a vertical axis. In either case, the relative motion of the workpart is achieved by reciprocating the work past the wheel or by rotating it. These possible combinations of wheel orientations and workpart motions provide the four types of surface grinding machines illustrated in Figure 25.7.

Of the four types, the horizontal spindle machine with reciprocating worktable is the most common, shown in Figure 25.8. Grinding is accomplished by reciprocating the work longitudinally under the wheel at a very small depth (infeed) and by feeding the wheel transversely into the work a certain distance between strokes. In these operations, the width of the wheel is usually less than that of the workpiece.

In addition to its conventional application, a grinding machine with horizontal spindle and reciprocating table can be used to form special contoured surfaces by employing a formed grinding wheel. Instead of feeding the wheel transversely across the work as it reciprocates, the wheel is *plunge-fed* vertically into the work. The shape of the formed wheel is therefore imparted to the work surface.

Grinding machines with vertical spindles and reciprocating tables are set up so that the wheel diameter is greater than the work width. Accordingly, these operations can be performed without using a transverse feed motion. Instead, grinding is accomplished by reciprocating the work past the wheel, and feeding the wheel vertically into the work to the desired dimension. This configuration is capable of achieving a very flat surface on the work.

FIGURE 25.8 Surface grinder with horizontal spindle and reciprocating worktable.

Of the two types of rotary table grinding in Figure 25.7(b) and (d), the vertical spindle machines are more common. Owing to the relatively large surface contact area between wheel and workpart, vertical spindle-rotary table grinding machines are capable of high metal removal rates when equipped with appropriate grinding wheels.

Cylindrical Grinding   As its name suggests, cylindrical grinding is used for rotational parts. These grinding operations divide into two basic types (Figure 25.9): (a) external cylindrical grinding and (b) internal cylindrical grinding.

External cylindrical grinding (also called *center-type grinding* to distinguish it from centerless grinding) is performed much like a turning operation. The grinding machines used for these operations closely resemble a lathe in which the tool post has been replaced by a high-speed motor to rotate the grinding wheel. The cylindrical workpiece is rotated between centers to provide a surface speed of 18 to 30 m/min (60 to 100 ft/min) [16], and the grinding wheel, rotating at 1200 to 2000 m/min (4000 to 6500 ft/min), is engaged to perform the cut. There are two types of feed motion possible, traverse feed and plunge-cut, shown in Figure 25.10. In traverse feed, the grinding wheel is fed in a direction parallel to the axis of rotation of the workpart. The infeed is set within a range typically from 0.0075 to 0.075 mm (0.0003 to 0.003 in). A longitudinal reciprocating motion is sometimes given to either the

FIGURE 25.9 Two types of cylindrical grinding: (a) external, and (b) internal.

(a)   (b)

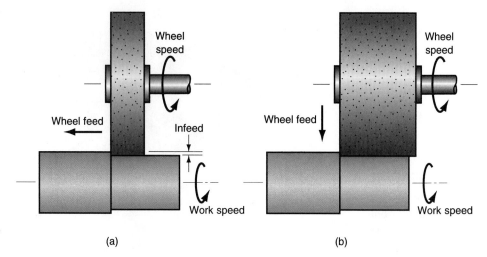

FIGURE 25.10 Two types of feed motion in external cylindrical grinding: (a) traverse feed, and (b) plunge-cut.

(a) (b)

work or the wheel to improve surface finish. In plunge-cut, the grinding wheel is fed radially into the work. Formed grinding wheels use this type of feed motion.

External cylindrical grinding is used to finish parts that have been machined to approximate size and heat treated to desired hardness. The parts include axles, crankshafts, spindles, bearings and bushings, and rolls for rolling mills. The grinding operation produces the final size and required surface finish on these hardened parts.

Internal cylindrical grinding operates somewhat like a boring operation. The workpiece is usually held in a chuck and rotated to provide surface speeds of 20 to 60 m/min (75 to 200 ft/min) [16]. Wheel surface speeds similar to external cylindrical grinding are used. The wheel is fed in either of two ways: traverse feed, Figure 25.9(b), or plunge feed. Obviously, the wheel diameter in internal cylindrical grinding must be smaller than the original bore hole. This often means that the wheel diameter is quite small, necessitating very high rotational speeds in order to achieve the desired surface speed. Internal cylindrical grinding is used to finish the hardened inside surfaces of bearing races and bushing surfaces.

Centerless Grinding Centerless grinding is an alternative process for grinding external and internal cylindrical surfaces. As its name suggests, the workpiece is not held between centers. This results in a reduction in work handling time; hence, centerless grinding is often used for high-production work. The setup for *external centerless grinding* (Figure 25.11), consists of two wheels: the grinding wheel and a regulating wheel. The workparts, which may be many individual short pieces or long rods (e.g., 3 to 4 m long), are supported by a rest blade and fed through between the two wheels. The grinding wheel does the cutting,

FIGURE 25.11 External centerless grinding.

(End view) (Side view)

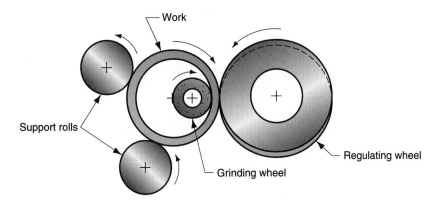

FIGURE 25.12 Internal centerless grinding.

rotating at surface speeds of 1200 to 1800 m/min (4000 to 6000 ft/min). The regulating wheel rotates at much lower speeds and is inclined at a slight angle I to control throughfeed of the work. The following equation can be used to predict throughfeed rate, based on inclination angle and other parameters of the process [16]:

$$f_r = \pi D_r N_r \sin I \tag{25.11}$$

where f_r = throughfeed rate, mm/min (in/min); D_r = diameter of the regulating wheel, mm (in); N_r = rotational speed of the regulating wheel, rev/min; and I = inclination angle of the regulating wheel.

The typical setup in ***internal centerless grinding*** is shown in Figure 25.12. In place of the rest blade, two support rolls are used to maintain the position of the work. The regulating wheel is tilted at a small inclination angle to control the feed of the work past the grinding wheel. Because of the need to support the grinding wheel, throughfeed of the work as in external centerless grinding is not possible. Therefore this grinding operation cannot achieve the same high-production rates as in the external centerless process. Its advantage is that it is capable of providing very close concentricity between internal and external diameters on a tubular part such as a roller bearing race.

Creep Feed Grinding   A relatively new form of grinding is creep feed grinding, developed around 1958. Creep feed grinding is performed at very high depths of cut and very low feed rates; hence, the name creep feed. The comparison with conventional surface grinding is illustrated in Figure 25.13.

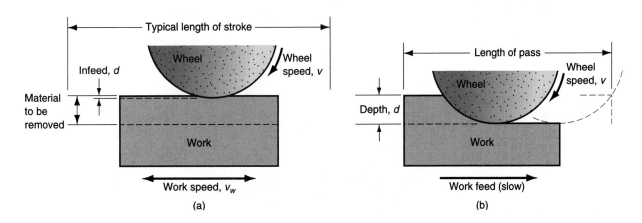

FIGURE 25.13   Comparison of (a) conventional surface grinding and (b) creep feed grinding.

Depths of cut in creep feed grinding are 1000 to 10,000 times greater than in conventional surface grinding, and the feed rates are reduced by about the same proportion. However, material removal rate and productivity are increased in creep feed grinding because the wheel is continuously cutting. This contrasts with conventional surface grinding in which the reciprocating motion of the work results in significant lost time during each stroke.

Creep feed grinding can be applied in both surface grinding and external cylindrical grinding. Surface grinding applications include grinding of slots and profiles. The process seems especially suited to those cases in which depth-to-width ratios are relatively large. The cylindrical applications include threads, formed gear shapes, and other cylindrical components. The term **deep grinding** is used in Europe to describe these external cylindrical creep feed grinding applications.

The introduction of grinding machines designed with special features for creep feed grinding has spurred interest in the process. The features include [11] high static and dynamic stability, highly accurate slides, two to three times the spindle power of conventional grinding machines, consistent table speeds for low feeds, high-pressure grinding fluid delivery systems, and dressing systems capable of dressing the grinding wheels during the process. Typical advantages of creep feed grinding include: (1) high material removal rates, (2) improved accuracy for formed surfaces, and (3) reduced temperatures at the work surface.

Other Grinding Operations   Several other grinding operations should be briefly mentioned to complete our review. These include tool grinding, jig grinding, disk grinding, snag grinding, and abrasive belt grinding.

Cutting tools are made of hardened tool steel and other hard materials. **Tool grinders** are special grinding machines of various designs to sharpen and recondition cutting tools. They have devices for positioning and orienting the tools to grind the desired surfaces at specified angles and radii. Some tool grinders are general purpose while others cut the unique geometries of specific tool types. General-purpose tool and cutter grinders use special attachments and adjustments to accommodate a variety of tool geometries. Single-purpose tool grinders include gear cutter sharpeners, milling cutter grinders of various types, broach sharpeners, and drill point grinders.

Jig grinders are grinding machines traditionally used to grind holes in hardened steel parts to high accuracies. The original applications included pressworking dies and tools. Although these applications are still important, jig grinders are used today in a broader range of applications in which high accuracy and good finish are required on hardened components. Numerical control is available on modern jig grinding machines to achieve automated operation.

Disk grinders are grinding machines with large abrasive disks mounted on either end of a horizontal spindle as in Figure 25.14. The work is held (usually manually) against the flat

FIGURE 25.14   Typical configuration of a disk grinder.

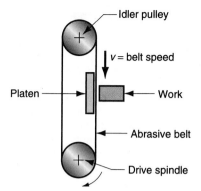

FIGURE 25.15   Abrasive belt grinding.

surface of the wheel to accomplish the grinding operation. Some disk grinding machines have double opposing spindles. By setting the disks at the desired separation, the workpart can be fed automatically between the two disks and ground simultaneously on opposite sides. Advantages of the disk grinder are good flatness and parallelism at high production rates.

The *snag grinder* is similar in configuration to a disk grinder. The difference is that the grinding is done on the outside periphery of the wheel rather than on the side flat surface. The grinding wheels are therefore different in design than those in disk grinding. Snag grinding is generally a manual operation, used for rough grinding operations such as removing the flash from castings and forgings, and smoothing weld joints.

Abrasive belt grinding uses abrasive particles bonded to a flexible (cloth) belt. A typical setup is illustrated in Figure 25.15. Support of the belt is required when the work is pressed against it, and this support is provided by a roll or platen located behind the belt. A flat platen is used for work that will have a flat surface. A soft platen can be used if it is desirable for the abrasive belt to conform to the general contour of the part during grinding. Belt speed depends on the material being ground; a range of 750 to 1700 m/min (2500 to 5500 ft/min) is typical [16]. Owing to improvements in abrasives and bonding materials, abrasive belt grinding is being used increasingly for heavy stock removal rates, rather than light grinding, which was its traditional application. The term *belt sanding* refers to the light grinding applications in which the workpart is pressed against the belt to remove burrs and high spots, and produce an improved finish quickly by hand.

25.2   RELATED ABRASIVE PROCESSES

Other abrasive processes include honing, lapping, superfinishing, polishing, and buffing. They are used exclusively as finishing operations. The initial part shape is created by some other process; then the part is finished by one of these operations to achieve superior surface finish. The usual part geometries and typical surface roughness values for these processes are indicated in Table 25.6. For comparison, we also present corresponding data for grinding.

Another class of finishing operations, called mass finishing (Section 28.1.2), is used to finish parts in bulk rather than individually. These mass finishing methods are also used for cleaning and deburring.

25.2.1   HONING

Honing is an abrasive process performed by a set of bonded abrasive sticks. A common application is to finish the bores of internal combustion engines. Other applications include bearings, hydraulic cylinders, and gun barrels. Surface finishes of around 0.12 μm

TABLE 25.6   Usual part geometries for honing, lapping, superfinishing, polishing, and buffing.

Process	Usual Part Geometry	Surface Roughness μm	Surface Roughness μ-in
Grinding, medium grit size	Flat, external cylinders, round holes	0.4–1.6	16–63
Grinding, fine grit size	Flat, external cylinders, round holes	0.2–0.4	8–16
Honing	Round hole (e.g., engine bore)	0.1–0.8	4–32
Lapping	Flat or slightly spherical (e.g., lens)	0.025–0.4	1–16
Superfinishing	Flat surface, external cylinder	0.013–0.2	0.5–8
Polishing	Miscellaneous shapes	0.025–0.8	1–32
Buffing	Miscellaneous shapes	0.013–0.4	0.5–16

(5 μ-in) or slightly better are typically achieved in these applications. In addition, honing produces a characteristic cross-hatched surface that tends to retain lubrication during operation of the component, thus contributing to its function and service life.

The honing process for an internal cylindrical surface is illustrated in Figure 25.16. The honing tool consists of a set of bonded abrasive sticks. Four sticks are used on the tool shown in the figure, but the number depends on hole size. Two to four sticks would be used for small holes (e.g., gun barrels), and a dozen or more would be used for larger diameter holes. The motion of the honing tool is a combination of rotation and linear reciprocation, regulated in such a way that a given point on the abrasive stick does not trace the same path repeatedly. This rather complex motion accounts for the cross-hatched pattern on the bore surface. Honing speeds are 15 to 150 m/min (50 to 500 ft/min) [4]. During the process, the sticks are pressed outward against the hole surface to produce the desired abrasive cutting action. Hone pressures of 1 to 3 MPa (150 to 450 lb/in^2) are typical. The honing tool is supported in the hole by two universal joints, thus causing the tool to follow the previously defined hole axis. Honing enlarges and finishes the hole but cannot change its location.

Grit sizes in honing range between 30 and 600. The same trade-off between better finish and faster material removal rates exists in honing as in grinding. The amount of material removed from the work surface during a honing operation may be as much as

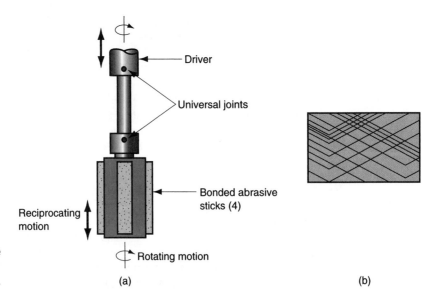

FIGURE 25.16   The honing process: (a) the honing tool used for internal bore surface, and (b) cross-hatched surface pattern created by the action of the honing tool.

(a)   (b)

FIGURE 25.17
The lapping process in lens-making.

0.5 mm (0.020 in), but is usually much less than this. A cutting fluid must be used in honing to cool and lubricate the tool and to help remove the chips.

25.2.2 LAPPING

Lapping is an abrasive process used to produce surface finishes of extreme accuracy and smoothness. It is used in the production of optical lenses, metallic bearing surfaces, gages, and other parts requiring very good finishes. Metal parts that are subject to fatigue loading or surfaces that must be used to establish a seal with a mating part are often lapped.

Instead of a bonded abrasive tool, lapping uses a fluid suspension of very small abrasive particles between the workpiece and the lapping tool. The process is illustrated in Figure 25.17 as applied in lens-making. The fluid with abrasives is referred to as the *lapping compound* and has the general appearance of a chalky paste. The fluids used to make the compound include oils and kerosene. Common abrasives are aluminum oxide and silicon carbide with typical grit sizes between 300 and 600. The lapping tool is called a *lap,* and it has the reverse of the desired shape of the workpart. To accomplish the process, the lap is pressed against the work and moved back and forth over the surface in a figure-eight or other motion pattern, subjecting all portions of the surface to the same action. Lapping is sometimes performed by hand, but lapping machines accomplish the process with greater consistency and efficiency.

Materials used to make the lap range from steel and cast iron to copper and lead. Wood laps have also been made. Because a lapping compound is used rather than a bonded abrasive tool, the mechanism by which this process works is somewhat different than grinding and honing. It is hypothesized that two alternative cutting mechanisms are at work in lapping [4]. The first mechanism is that the abrasive particles roll and slide between the lap and the work, with very small cuts occurring in both surfaces. The second mechanism is that the abrasives become embedded in the lap surface and the cutting action is very similar to grinding. It is likely that lapping is a combination of these two mechanisms, depending on the relative hardnesses of the work and the lap. For laps made of soft materials, the embedded grit mechanism is emphasized; and for hard laps, the rolling and sliding mechanism dominates.

25.2.3 SUPERFINISHING

Superfinishing is an abrasive process similar to honing. Both processes use a bonded abrasive stick moved with a reciprocating motion and pressed against the surface to be finished. Superfinishing differs from honing in the following respects [4]: (1) the strokes are shorter, 5 mm (3/16 in); (2) higher frequencies are used, up to 1500 strokes per minute; (3) lower pressures are applied between the tool and the surface, below 0.28 MPa (40 lb/in^2); (4) workpiece speeds are lower, 15 m/min (50 ft/min) or less; and (5) grit sizes are generally smaller. The relative motion between the abrasive stick and the work surface is varied so

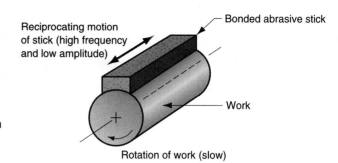

Reciprocating motion of stick (high frequency and low amplitude)

Bonded abrasive stick

Work

Rotation of work (slow)

FIGURE 25.18 Superfinishing on an external cylindrical surface.

that individual grains do not retrace the same path. A cutting fluid is used to cool the work surface and wash away chips. In addition, the fluid tends to separate the abrasive stick from the work surface after a certain level of smoothness is achieved, thus preventing further cutting action. The result of these operating conditions is mirror-like finishes with surface roughness values around 0.025 μm (1 μ-in). Superfinishing can be used to finish flat and external cylindrical surfaces. The process is illustrated in Figure 25.18 for the latter geometry.

25.2.4 POLISHING AND BUFFING

Polishing is used to remove scratches and burrs and to smooth rough surfaces by means of abrasive grains attached to a polishing wheel rotating at high speed—around 2300 m/min (7500 ft/min). The wheels are made of canvas, leather, felt, and even paper; thus, the wheels are somewhat flexible. The abrasive grains are glued to the outside periphery of the wheel. After the abrasives have been worn down and used up, the wheel is replenished with new grits. Grit sizes of 20 to 80 are used for rough polishing, 90 to 120 for finish polishing, and above 120 for fine finishing. Polishing operations are often accomplished manually.

Buffing is similar to polishing in appearance, but its function is different. Buffing is used to provide attractive surfaces with high luster. Buffing wheels are made of materials similar to those used for polishing wheels—leather, felt, cotton, etc.—but buffing wheels are generally softer. The abrasives are very fine and are contained in a buffing compound that is pressed into the outside surface of the wheel while it rotates. This contrasts with polishing in which the abrasive grits are glued to the wheel surface. As in polishing, the abrasive particles must be periodically replenished. Buffing is usually done manually, although machines have been designed to perform the process automatically. Speeds are generally 2400 to 5200 m/min (8000 to 17,000 ft/min).

REFERENCES

[1] Aronson, R. B. "More Than a Pretty Finish," *Manufacturing Engineering,* February 2005, pp. 57–69.

[2] Andrew, C., Howes, T. D., and Pearce, T. R. A. *Creep Feed Grinding*. Holt, Rinehart and Winston, London, 1985.

[3] *ANSI Standard B74. 13-1977*, "Markings for Identifying Grinding Wheels and Other Bonded Abrasives." American National Standards Institute, New York, 1977.

[4] Armarego, E. J. A., and Brown, R. H. *The Machining of Metals*. Prentice-Hall, Englewood Cliffs, New Jersey, 1969.

[5] Bacher, W. R., and Merchant, M. E. "On the Basic Mechanics of the Grinding Process," *Transactions ASME,* Series B, Vol. *80* No. 1, 1958, pp. 141.

[6] Black, J, and Kohser, R. *DeGarmo's Materials and Processes in Manufacturing*, 10th ed. John Wiley & Sons, Hoboken, New Jersey, 2008.

[7]   Black, P. H. *Theory of Metal Cutting*. McGraw-Hill, New York, 1961.

[8]   Boothroyd, G., and Knight, W. A. *Fundamentals of Metal Machining and Machine Tools*. 3rd ed. CRC Taylor and Francis, Boca Raton, Florida, 2006.

[9]   Boston, O. W. *Metal Processing*. 2nd ed. John Wiley & Sons, New York, 1951.

[10] Cook, N. H. *Manufacturing Analysis*. Addison-Wesley, Inc., Reading, Massachusetts, 1966.

[11] Drozda, T. J., and Wick, C. (eds.). *Tool and Manufacturing Engineers Handbook*. 4th ed. Vol. I, *Machining,* Society of Manufacturing Engineers, Dearborn, Michigan, 1983.

[12] Eary, D. F., and Johnson, G. E. *Process Engineering: for Manufacturing*. Prentice-Hall, Englewood Cliffs, New Jersey, 1962.

[13] Kaiser, R. "The Facts about Grinding." *Manufacturing Engineering*. Vol. 125, No. 3, September 2000, pp. 78–85.

[14] Krabacher, E. J. "Factors Influencing the Performance of Grinding Wheels." *Transactions ASME,* Series B, Vol. 81, No. 3, 1959, pp. 187–199.

[15] Krar, S. F. *Grinding Technology*. 2nd ed. Delmar Publishers, Florence, Kentucky, 1995.

[16] *Machining Data Handbook*. 3rd ed. Vol. I. and II. Metcut Research Associates, Cincinnati, Ohio, 1980.

[17] Malkin, S. *Grinding Technology: Theory and Applications of Machining with Abrasives*. 2nd ed. Industrial Press, New York, 2008.

[18] Phillips, D. "Creeping Up." *Cutting Tool Engineering.* Vol. 52, No. 3, March 2000, pp. 32–43.

[19] Rowe, W. *Principles of Modern Grinding Technology*, William Andrew, Elsevier Applied Science Publishers, New York, 2009.

[20] Salmon, S. "Creep-Feed Grinding Is Surprisingly Versatile." *Manufacturing Engineering,* November 2004, pp. 59–64.

REVIEW QUESTIONS

25.1.   Why are abrasive processes technologically and commercially important?

25.2.   What are the five principal parameters of a grinding wheel?

25.3.   What are some of the principal abrasive materials used in grinding wheels?

25.4.   Name some of the principal bonding materials used in grinding wheels.

25.5.   What is wheel structure?

25.6.   What is wheel grade?

25.7.   Why are specific energy values so much higher in grinding than in traditional machining processes such as milling?

25.8.   Grinding creates high temperatures. How is temperature harmful in grinding?

25.9.   What are the three mechanisms of grinding wheel wear?

25.10. What is dressing, in reference to grinding wheels?

25.11. What is truing, in reference to grinding wheels?

25.12. What abrasive material would one select for grinding a cemented carbide cutting tool?

25.13. What are the functions of a grinding fluid?

25.14. What is centerless grinding?

25.15. How does creep feed grinding differ from conventional grinding?

25.16. How does abrasive belt grinding differ from a conventional surface grinding operation?

25.17. Name some of the abrasive operations available to achieve very good surface finishes.

25.18. (**Video**) Describe a wheel ring test.

25.19. (**Video**) List two purposes of dressing a grinding wheel.

25.20. (**Video**) What is the purpose of using a coolant in the grinding process?

MULTIPLE CHOICE QUIZ

There are 16 correct answers in the following multiple choice questions (some questions have multiple answers that are correct). To attain a perfect score on the quiz, all correct answers must be given. Each correct answer is worth 1 point. Each omitted answer or wrong answer reduces the score by 1 point, and each additional answer beyond the correct number of answers reduces the score by 1 point. Percentage score on the quiz is based on the total number of correct answers.

25.1.   Which one of the following conventional machining processes is closest to grinding: (a) drilling, (b) milling, (c) shaping, or (d) turning?

25.2.   Of the following abrasive materials, which one has the highest hardness: (a) aluminum oxide, (b) cubic boron nitride, or (c) silicon carbide?

25.3.   Smaller grain size in a grinding wheel tends to (a) degrade surface finish, (b) have no effect on surface finish, or (c) improve surface finish?

25.4.   Which of the following would tend to give higher material removal rates: (a) larger grain size, or (b) smaller grain size?

25.5. Which of the following will improve surface finish in grinding (three best answers): (a) denser wheel structure, (b) higher wheel speed, (c) higher workspeeds, (d) larger infeed, (e) lower infeed, (f) lower wheel speed, (g) lower workspeed, and (h) more open wheel structure?

25.6. Which one of the following abrasive materials is most appropriate for grinding steel and cast iron: (a) aluminum oxide, (b) cubic boron nitride, (c) diamond, or (d) silicon carbide?

25.7. Which one of the following abrasive materials is most appropriate for grinding hardened tool steel: (a) aluminum oxide, (b) cubic boron nitride, (c) diamond, or (d) silicon carbide?

25.8. Which one of the following abrasive materials is most appropriate for grinding nonferrous metals: (a) aluminum oxide, (b) cubic boron nitride, (c) diamond, or (d) silicon carbide?

25.9. Which of the following will help to reduce the incidence of heat damage to the work surface in grinding (four correct answers): (a) frequent dressing or truing of the wheel, (b) higher infeeds, (c) higher wheel speeds, (d) higher workspeeds, (e) lower infeeds, (f) lower wheel speeds, and (g) lower workspeeds?

25.10. Which one of the following abrasive processes achieves the best surface finish: (a) centerless grinding, (b) honing, (c) lapping, or (d) superfinishing?

25.11. The term deep grinding refers to which one of the following: (a) alternative name for any creep feed grinding operation, (b) external cylindrical creep feed grinding, (c) grinding operation performed at the bottom of a hole, (d) surface grinding that uses a large crossfeed, or (e) surface grinding that uses a large infeed?

PROBLEMS

25.1. In a surface grinding operation wheel diameter = 150 mm and infeed = 0.07 mm. Wheel speed = 1450 m/min, workspeed = 0.25 m/s, and crossfeed = 5 mm. The number of active grits per area of wheel surface = 0.75 grits/mm^2. Determine (a) average length per chip, (b) metal removal rate, and (c) number of chips formed per unit time for the portion of the operation when the wheel is engaged in the work.

25.2. The following conditions and settings are used in a certain surface grinding operation: wheel diameter = 6.0 in, infeed = 0.003 in, wheel speed = 4750 ft/min, workspeed = 50 ft/min, and crossfeed = 0.20 in. The number of active grits per square inch of wheel surface = 500. Determine (a) average length per chip, (b) metal removal rate, and (c) number of chips formed per unit time for the portion of the operation when the wheel is engaged in the work.

25.3. An internal cylindrical grinding operation is used to finish an internal bore from an initial diameter of 250 mm to a final diameter of 252.5 mm. The bore is 125 mm long. A grinding wheel with an initial diameter of 150 mm and a width of 20 mm is used. After the operation, the diameter of the grinding wheel has been reduced to 149.75 mm. Determine the grinding ratio in this operation.

25.4. In a surface grinding operation performed on hardened plain carbon steel, the grinding wheel has a diameter = 200 mm and width = 25 mm. The wheel rotates at 2400 rev/min, with a depth of cut (infeed) = 0.05 mm/pass and a crossfeed = 3.50 mm. The reciprocating speed of the work is 6 m/min, and the operation is performed dry. Determine (a) length of contact between the wheel and the work and (b) volume rate of metal removed. (c) If there are 64 active grits/cm^2 of wheel surface, estimate the number of chips formed per unit time. (d) What is the average volume per chip? (e) If the tangential cutting force on the work = 25 N, compute the specific energy in this operation?

25.5. An 8-in diameter grinding wheel, 1.0 in wide, is used in a surface grinding job performed on a flat piece of heat-treated 4340 steel. The wheel rotates to achieve a surface speed of 5000 ft/min, with a depth of cut (infeed) = 0.002 in per pass and a crossfeed = 0.15 in. The reciprocating speed of the work is 20 ft/min, and the operation is performed dry. (a) What is the length of contact between the wheel and the work? (b) What is the volume rate of metal removed? (c) If there are 300 active grits/in^2 of wheel surface, estimate the number of chips formed per unit time. (d) What is the average volume per chip? (e) If the tangential cutting force on the workpiece = 7.3 lb, what is the specific energy calculated for this job?

25.6. A surface grinding operation is being performed on a 6150 steel workpart (annealed, approximately 200 BHN). The designation on the grinding wheel is C-24-D-5-V. The wheel diameter = 7.0 in and its width = 1.00 in. Rotational speed = 3000 rev/min. The depth (infeed) = 0.002 in per pass, and the crossfeed = 0.5 in. Workspeed = 20 ft/min. This operation has been a source of trouble right from the beginning. The surface finish is not as good as the 16 μ-in specified on the part print, and there are signs of metallurgical damage on the surface. In addition, the wheel seems to become clogged almost as soon as the operation begins. In short, nearly everything that can go wrong with the job has gone wrong. (a) Determine the rate of metal removal when the wheel is engaged in the work. (b) If the number of active grits per square inch = 200, determine the average chip length and the number of chips formed per time. (c) What changes would you recommend in the grinding wheel to help solve the problems encountered? Explain why you made each recommendation.

25.7. The grinding wheel in a centerless grinding operation has a diameter = 200 mm, and the regulating wheel diameter = 125 mm. The grinding wheel rotates at 3000 rev/min and the regulating wheel rotates at 200 rev/min. The inclination angle of the regulating wheel = 2.5°. Determine the through-feed rate of cylindrical workparts that are 25.0 mm in diameter and 175 mm long.

25.8. A centerless grinding operation uses a regulating wheel that is 150 mm in diameter and rotates at 500 rev/min. At what inclination angle should the regulating wheel be set, if it is desired to feed a workpiece with length = 3.5 m and diameter = 18 mm through the operation in exactly 30 sec?

25.9. In a certain centerless grinding operation, the grinding wheel diameter = 8.5 in, and the regulating wheel diameter = 5 in. The grinding wheel rotates at 3500 rev/min and the regulating wheel rotates at 150 rev/min. The inclination angle of the regulating wheel = 3°. Determine the throughfeed rate of cylindrical workparts that have the following dimensions: diameter = 1.25 in and length = 8 in.

25.10. It is desired to compare the cycle times required to grind a particular workpiece using traditional surface grinding and using creep feed grinding. The workpiece is 200 mm long, 30 mm wide, and 75 mm thick. To make a fair comparison, the grinding wheel in both cases is 250 mm in diameter, 35 mm in width, and rotates at 1500 rev/min. It is desired to remove 25 mm of material from the surface. When traditional grinding is used, the infeed is set at 0.025 mm, and the wheel traverses twice (forward and back) across the work surface during each pass before resetting the infeed. There is no crossfeed since the wheel width is greater than the work width. Each pass is made at a workspeed of 12 m/min, but the wheel overshoots the part on both sides. With acceleration and deceleration, the wheel is engaged in the work for 50% of the time on each pass. When creep feed grinding is used, the depth is increased by 1000 and the forward feed is decreased by 1000. How long will it take to complete the grinding operation (a) with traditional grinding and (b) with creep feed grinding?

25.11. In a certain grinding operation, the grade of the grinding wheel should be "M" (medium), but the only available wheel is grade "T" (hard). It is desired to make the wheel appear softer by making changes in cutting conditions. What changes would you recommend?

25.12. An aluminum alloy is to be ground in an external cylindrical grinding operation to obtain a good surface finish. Specify the appropriate grinding wheel parameters and the grinding conditions for this job.

25.13. A high-speed steel broach (hardened) is to be resharpened to achieve a good finish. Specify the appropriate parameters of the grinding wheel for this job.

25.14. Based on equations in the text, derive an equation to compute the average volume per chip formed in the grinding process.

26 NONTRADITIONAL MACHINING AND THERMAL CUTTING PROCESSES

Chapter Contents

Conventional machining processes (i.e., turning, drilling, milling) use a sharp cutting tool to form a chip from the work by shear deformation. In addition to these conventional methods, there is a group of processes that uses other mechanisms to remove material. The term *nontraditional machining* refers to this group that removes excess material by various techniques involving mechanical, thermal, electrical, or chemical energy (or combinations of these energies). They do not use a sharp cutting tool in the conventional sense.

The nontraditional processes have been developed since World War II largely in response to new and unusual machining requirements that could not be satisfied by conventional methods. These requirements, and the resulting commercial and technological importance of the nontraditional processes, include:

➤ The need to machine newly developed metals and nonmetals. These new materials often have special properties (e.g., high strength, high hardness, high toughness) that make them difficult or impossible to machine by conventional methods.

➤ The need for unusual and/or complex part geometries that cannot easily be accomplished and in some cases are impossible to achieve by conventional machining.

➤ The need to avoid surface damage that often accompanies the stresses created by conventional machining.

Many of these requirements are associated with the aerospace and electronics industries, which have become increasingly important in recent decades.

There are literally dozens of nontraditional machining processes, most of which are unique in their range of applications. In the present chapter, we discuss those that are most important commercially. More detailed discussions of these nontraditional methods are presented in several of the references.

The nontraditional processes are often classified according to principal form of energy used to effect material removal. By this classification, there are four types:

1. *Mechanical.* Mechanical energy in some form other than the action of a conventional cutting tool is used in these nontraditional processes. Erosion of the work material by a high velocity stream of abrasives or fluid (or both) is a typical form of mechanical action in these processes.
2. *Electrical.* These nontraditional processes use electrochemical energy to remove material; the mechanism is the reverse of electroplating.
3. *Thermal.* These processes use thermal energy to cut or shape the workpart. The thermal energy is generally applied to a very small portion of the work surface, causing that portion to be removed by fusion and/or vaporization. The thermal energy is generated by the conversion of electrical energy.
4. *Chemical.* Most materials (metals particularly) are susceptible to chemical attack by certain acids or other etchants. In chemical machining, chemicals selectively remove material from portions of the workpart, whereas other portions of the surface are protected by a mask.

26.1 MECHANICAL ENERGY PROCESSES

In this section we examine several of the nontraditional processes that use mechanical energy other than a sharp cutting tool: (1) ultrasonic machining, (2) water jet processes, and (3) other abrasive processes.

26.1.1 ULTRASONIC MACHINING

Ultrasonic machining (USM) is a nontraditional machining process in which abrasives contained in a slurry are driven at high velocity against the work by a tool vibrating at low amplitude and high frequency. The amplitudes are around 0.075 mm (0.003 in), and the frequencies are approximately 20,000 Hz. The tool oscillates in a direction perpendicular to the work surface, and is fed slowly into the work, so that the shape of the tool is formed in the part. However, it is the action of the abrasives, impinging against the work surface, that performs the cutting. The general arrangement of the USM process is depicted in Figure 26.1.

Common tool materials used in USM include soft steel and stainless steel. Abrasive materials in USM include boron nitride, boron carbide, aluminum oxide, silicon carbide,

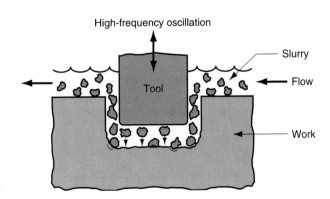

FIGURE 26.1
Ultrasonic machining.

and diamond. Grit size (Section 16.1.1) ranges between 100 and 2000. The vibration amplitude should be set approximately equal to the grit size, and the gap size should be maintained at about two times grit size. To a significant degree, grit size determines the surface finish on the new work surface. In addition to surface finish, material removal rate is an important performance variable in ultrasonic machining. For a given work material, the removal rate in USM increases with increasing frequency and amplitude of vibration.

The cutting action in USM operates on the tool as well as the work. As the abrasive particles erode the work surface, they also erode the tool, thus affecting its shape. It is therefore important to know the relative volumes of work material and tool material removed during the process—similar to the grinding ratio (Section 25.1.2). This ratio of stock removed to tool wear varies for different work materials, ranging from around 100:1 for cutting glass down to about 1:1 for cutting tool steel.

The slurry in USM consists of a mixture of water and abrasive particles. Concentration of abrasives in water ranges from 20% to 60% [5]. The slurry must be continuously circulated to bring fresh grains into action at the tool–work gap. It also washes away chips and worn grits created by the cutting process.

The development of ultrasonic machining was motivated by the need to machine hard, brittle work materials, such as ceramics, glass, and carbides. It is also successfully used on certain metals, such as stainless steel and titanium. Shapes obtained by USM include non-round holes, holes along a curved axis, and coining operations, in which an image pattern on the tool is imparted to a flat work surface.

26.1.2 PROCESSES USING WATER JETS

The two processes described in this section remove material by means of high-velocity streams of water or a combination of water and abrasives.

Water Jet Cutting Water jet cutting (WJC) uses a fine, high-pressure, high-velocity stream of water directed at the work surface to cause cutting of the work, as illustrated in Figure 26.2. To obtain the fine stream of water a small nozzle opening of diameter 0.1 to 0.4 mm (0.004 to 0.016 in) is used. To provide the stream with sufficient energy for cutting, pressures up to 400 MPa (60,000 lb/in^2) are used, and the jet reaches velocities up to 900 m/s (3000 ft/sec). The fluid is pressurized to the desired level by a hydraulic pump. The nozzle unit consists of a holder made of stainless steel, and a jewel nozzle made of sapphire, ruby, or

FIGURE 26.2 Water jet cutting.

diamond. Diamond lasts the longest but costs the most. Filtration systems must be used in WJC to separate the swarf produced during cutting.

Cutting fluids in WJC are polymer solutions, preferred because of their tendency to produce a coherent stream. We have discussed cutting fluids before in the context of conventional machining (Section 23.4), but never has the term been more appropriately applied than in WJC.

Important process parameters include standoff distance, nozzle opening diameter, water pressure, and cutting feed rate. As in Figure 26.2, the ***standoff distance*** is the separation between the nozzle opening and the work surface. It is generally desirable for this distance to be small to minimize dispersion of the fluid stream before it strikes the surface. A typical standoff distance is 3.2 mm (0.125 in). Size of the nozzle orifice affects the precision of the cut; smaller openings are used for finer cuts on thinner materials. To cut thicker stock, thicker jet streams and higher pressures are required. The cutting feed rate refers to the velocity at which the WJC nozzle is traversed along the cutting path. Typical feed rates range from 5 mm/s (12 in/min) to more than 500 mm/s (1200 in/min), depending on work material and its thickness [5]. The WJC process is usually automated using computer numerical control or industrial robots to manipulate the nozzle unit along the desired trajectory.

Water jet cutting can be used effectively to cut narrow slits in flat stock such as plastic, textiles, composites, floor tile, carpet, leather, and cardboard. Robotic cells have been installed with WJC nozzles mounted as the robot's tool to follow cutting patterns that are irregular in three dimensions, such as cutting and trimming of automobile dashboards before assembly [9]. In these applications, advantages of WJC include: (1) no crushing or burning of the work surface typical in other mechanical or thermal processes, (2) minimum material loss because of the narrow cut slit, (3) no environmental pollution, and (4) ease of automating the process. A limitation of WJC is that the process is not suitable for cutting brittle materials (e.g., glass) because of their tendency to crack during cutting.

Abrasive Water Jet Cutting   When WJC is used on metallic workparts, abrasive particles must usually be added to the jet stream to facilitate cutting. This process is therefore called ***abrasive water jet cutting*** (AWJC). Introduction of abrasive particles into the stream complicates the process by adding to the number of parameters that must be controlled. Among the additional parameters are abrasive type, grit size, and flow rate. Aluminum oxide, silicon dioxide, and garnet (a silicate mineral) are typical abrasive materials, at grit sizes ranging between 60 and 120. The abrasive particles are added to the water stream at approximately 0.25 kg/min (0.5 lb/min) after it has exited the WJC nozzle.

The remaining process parameters include those that are common to WJC: nozzle opening diameter, water pressure, and standoff distance. Nozzle orifice diameters are 0.25 to 0.63 mm (0.010 to 0.025 in)—somewhat larger than in water jet cutting to permit higher flow rates and more energy to be contained in the stream before injection of abrasives. Water pressures are about the same as in WJC. Standoff distances are somewhat less to minimize the effect of dispersion of the cutting fluid that now contains abrasive particles. Typical standoff distances are between 1/4 and 1/2 of those in WJC.

26.1.3   OTHER NONTRADITIONAL ABRASIVE PROCESSES

Two additional mechanical energy processes use abrasives to accomplish deburring, polishing, or other operations in which very little material is removed.

Abrasive Jet Machining   Not to be confused with AWJC is the process called abrasive jet machining (AJM), a material removal process caused by the action of a high-velocity stream of gas containing small abrasive particles, as in Figure 26.3. The gas is dry, and pressures of 0.2 to 1.4 MPa (25 to 200 lb/in^2) are used to propel it through nozzle orifices

FIGURE 26.3 Abrasive jet machining (AJM).

of diameter 0.075 to 1.0 mm (0.003 to 0.040 in) at velocities of 2.5 to 5.0 m/s (500 to 1000 ft/min). Gases include dry air, nitrogen, carbon dioxide, and helium.

The process is usually performed manually by an operator who directs the nozzle at the work. Typical distances between nozzle tip and work surface range between 3 mm and 75 mm (0.125 in and 3 in). The workstation must be set up to provide proper ventilation for the operator.

AJM is normally used as a finishing process rather than a production cutting process. Applications include deburring, trimming and deflashing, cleaning, and polishing. Cutting is accomplished successfully on hard, brittle materials (e.g., glass, silicon, mica, and ceramics) that are in the form of thin flat stock. Typical abrasives used in AJM include aluminum oxide (for aluminum and brass), silicon carbide (for stainless steel and ceramics), and glass beads (for polishing). Grit sizes are small, 15 to 40 μm (0.0006 to 0.0016 in) in diameter, and must be uniform in size for a given application. It is important not to recycle the abrasives because used grains become fractured (and therefore smaller in size), worn, and contaminated.

Abrasive Flow Machining   This process was developed in the 1960s to deburr and polish difficult-to-reach areas using abrasive particles mixed in a viscoelastic polymer that is forced to flow through or around the part surfaces and edges. The polymer has the consistency of putty. Silicon carbide is a typical abrasive. Abrasive flow machining (AFM) is particularly well-suited for internal passageways that are often inaccessible by conventional methods. The abrasive-polymer mixture, called the media, flows past the target regions of the part under pressures ranging between 0.7 and 20 MPa (100 and 3000 lb/in^2). In addition to deburring and polishing, other AFM applications include forming radii on sharp edges, removing rough surfaces on castings, and other finishing operations. These applications are found in industries such as aerospace, automotive, and die-making. The process can be automated to economically finish hundreds of parts per hour.

A common setup is to position the workpart between two opposing cylinders, one containing media and the other empty. The media is forced to flow through the part from the first cylinder to the other, and then back again, as many times as necessary to achieve the desired material removal and finish.

26.2   ELECTROCHEMICAL MACHINING PROCESSES

An important group of nontraditional processes use electrical energy to remove material. This group is identified by the term *electrochemical processes,* because electrical energy is used in combination with chemical reactions to accomplish material removal. In effect, these processes are the reverse of electroplating (Section 28.3.1). The work material must be a conductor in the electrochemical machining processes.

26.2.1 ELECTROCHEMICAL MACHINING

The basic process in this group is electrochemical machining (ECM). Electrochemical machining removes metal from an electrically conductive workpiece by anodic dissolution, in which the shape of the workpiece is obtained by a formed electrode tool in close proximity to, but separated from, the work by a rapidly flowing electrolyte. ECM is basically a deplating operation. As illustrated in Figure 26.4, the workpiece is the anode, and the tool is the cathode. The principle underlying the process is that material is deplated from the anode (the positive pole) and deposited onto the cathode (the negative pole) in the presence of an electrolyte bath (Section 4.5). The difference in ECM is that the electrolyte bath flows rapidly between the two poles to carry off the deplated material, so that it does not become plated onto the tool.

The electrode tool, usually made of copper, brass, or stainless steel, is designed to possess approximately the inverse of the desired final shape of the part. An allowance in the tool size must be provided for the gap that exists between the tool and the work. To accomplish metal removal, the electrode is fed into the work at a rate equal to the rate of metal removal from the work. Metal removal rate is determined by Faraday's First Law, which states that the amount of chemical change produced by an electric current (i.e., the amount of metal dissolved) is proportional to the quantity of electricity passed (current $\times$ time):

$$V = CIt \tag{26.1}$$

where V = volume of metal removed, mm^3 (in^3); C = a constant called the specific removal rate that depends on atomic weight, valence, and density of the work material, mm^3/amp-s (in^3/amp-min); I = current, amps; and t = time, s (min).

Based on Ohm's law, current $I = E/R$, where E = voltage and R = resistance. Under the conditions of the ECM operation, resistance is given by

$$R = \frac{gr}{A} \tag{26.2}$$

where g = gap between electrode and work, mm (in); r = resistivity of electrolyte, ohm-mm (ohm-in); and A = surface area between work and tool in the working frontal gap, mm^2 (in^2).

Substituting this expression for R into Ohm's law, we have

$$I = \frac{EA}{gr} \tag{26.3}$$

FIGURE 26.4
Electrochemical
machining (ECM).

TABLE 26.1   Typical values of specific removal rate C for selected work materials in electrochemical machining.

Work Material[a]	Specific Removal Rate C		Work Material[a]	Specific Removal Rate C	
	mm³/amp-sec	in³/amp-min		mm³/amp-sec	in³/amp-min
Aluminum (3)	3.44×10^{-2}	1.26×10^{-4}	Steels:		
Copper (1)	7.35×10^{-2}	2.69×10^{-4}	Low alloy	3.0×10^{-2}	1.1×10^{-4}
Iron (2)	3.69×10^{-2}	1.35×10^{-4}	High alloy	2.73×10^{-2}	1.0×10^{-4}
Nickel (2)	3.42×10^{-2}	1.25×10^{-4}	Stainless	2.46×10^{-2}	0.9×10^{-4}
			Titanium (4)	2.73×10^{-2}	1.0×10^{-4}

Compiled from data in [8].
[a]Most common valence given in parentheses () is assumed in determining specific removal rate C. For different valence, multiply C by most common valence and divide by actual valence.

And substituting this equation back into the equation defining Faraday's law

$$V = \frac{C(EAt)}{gr} \qquad (26.4)$$

It is convenient to convert this equation into an expression for feed rate, the rate at which the electrode (tool) can be advanced into the work. This conversion can be accomplished in two steps. First, let us divide Eq. (26.4) by At (area × time) to convert volume of metal removed into a linear travel rate

$$\frac{V}{At} = f_r = \frac{CE}{gr} \qquad (26.5)$$

where f_r = feed rate, mm/s (in/min). Second, let us substitute I/A in place of $E/(gr)$, as provided by Eq. (26.3).
    Thus, feed rate in ECM is

$$f_r = \frac{CI}{A} \qquad (26.6)$$

where A = the frontal area of the electrode, mm² (in²).
    This is the projected area of the tool in the direction of the feed into the work. Values of specific removal rate C are presented in Table 26.1 for various work materials. We should note that this equation assumes 100% efficiency of metal removal. The actual efficiency is in the range 90% to 100% and depends on tool shape, voltage and current density, and other factors.

Example 26.1
Electrochemical
Machining

An ECM operation is to be used to cut a hole into a plate of aluminum that is 12 mm thick. The hole has a rectangular cross section, 10 mm × 30 mm. The ECM operation will be accomplished at a current = 1200 amps. Efficiency is expected to be 95%. Determine feed rate and time required to cut through the plate.

Solution: From Table 26.1, specific removal rate C for aluminum = 3.44×10^{-2} mm³/A-s. The frontal area of the electrode A = 10 mm × 30 mm = 300 mm². At a current level of 1200 amps, feed rate is

$$f_r = 0.0344 \text{ mm}^3/\text{A-s} \left(\frac{1200}{300} \text{ A/mm}^2 \right) = 0.1376 \text{ mm/s}$$

At an efficiency of 95%, the actual feed rate is

$$f_r = 0.1376 \text{ mm/s} (0.95) = 0.1307 \text{ mm/s}$$

Time to machine through the 12-mm plate is

$$T_m = \frac{12.0}{0.1307} = 91.8\,\text{s} = 1.53\,\text{min}$$

The preceding equations indicate the important process parameters for determining metal removal rate and feed rate in electrochemical machining: gap distance g, electrolyte resistivity r, current I, and electrode frontal area A. Gap distance needs to be controlled closely. If g becomes too large, the electrochemical process slows down. However, if the electrode touches the work, a short circuit occurs, which stops the process altogether. As a practical matter, gap distance is usually maintained within a range 0.075 to 0.75 mm (0.003 to 0.030 in).

Water is used as the base for the electrolyte in ECM. To reduce electrolyte resistivity, salts such as NaCl or $NaNO_3$ are added in solution. In addition to carrying off the material that has been removed from the workpiece, the flowing electrolyte also serves the function of removing heat and hydrogen bubbles created in the chemical reactions of the process. The removed work material is in the form of microscopic particles that must be separated from the electrolyte through centrifuge, sedimentation, or other means. The separated particles form a thick sludge whose disposal is an environmental problem associated with ECM.

Large amounts of electrical power are required to perform ECM. As the equations indicate, rate of metal removal is determined by electrical power, specifically the current density that can be supplied to the operation. The voltage in ECM is kept relatively low to minimize arcing across the gap.

Electrochemical machining is generally used in applications in which the work metal is very hard or difficult to machine, or the workpart geometry is difficult (or impossible) to accomplish by conventional machining methods. Work hardness makes no difference in ECM, because the metal removal is not mechanical. Typical ECM applications include: (1) *die sinking,* which involves the machining of irregular shapes and contours into forging dies, plastic molds, and other shaping tools; (2) multiple hole drilling, in which many holes can be drilled simultaneously with ECM and conventional drilling would probably require the holes to be made sequentially; (3) holes that are not round, because ECM does not use a rotating drill; and (4) deburring (Section 26.2.2).

Advantages of ECM include: (1) little surface damage to the workpart, (2) no burrs as in conventional machining, (3) low tool wear (the only tool wear results from the flowing electrolyte), and (4) relatively high metal removal rates for hard and difficult-to-machine metals. Disadvantages of ECM are: (1) significant cost of electrical power to drive the operation and (2) problems of disposing of the electrolyte sludge.

26.2.2 ELECTROCHEMICAL DEBURRING AND GRINDING

Electrochemical deburring (ECD) is an adaptation of ECM designed to remove burrs or to round sharp corners on metal workparts by anodic dissolution. One possible setup for ECD is shown in Figure 26.5. The hole in the workpart has a sharp burr of the type that is

FIGURE 26.5
Electrochemical
deburring (ECD).

FIGURE 26.6
Electrochemical grinding
(ECG).

produced in a conventional through-hole drilling operation. The electrode tool is designed to focus the metal removal action on the burr. Portions of the tool not being used for machining are insulated. The electrolyte flows through the hole to carry away the burr particles. The same ECM principles of operation also apply to ECD. However, since much less material is removed in electrochemical deburring, cycle times are much shorter. A typical cycle time in ECD is less than a minute. The time can be increased if it is desired to round the corner in addition to removing the burr.

Electrochemical grinding (ECG) is a special form of ECM in which a rotating grinding wheel with a conductive bond material is used to augment the anodic dissolution of the metal workpart surface, as illustrated in Figure 26.6. Abrasives used in ECG include aluminum oxide and diamond. The bond material is either metallic (for diamond abrasives) or resin bond impregnated with metal particles to make it electrically conductive (for aluminum oxide). The abrasive grits protruding from the grinding wheel at the contact with the workpart establish the gap distance in ECG. The electrolyte flows through the gap between the grains to play its role in electrolysis.

Deplating is responsible for 95% or more of the metal removal in ECG, and the abrasive action of the grinding wheel removes the remaining 5% or less, mostly in the form of salt films that have been formed during the electrochemical reactions at the work surface. Because most of the machining is accomplished by electrochemical action, the grinding wheel in ECG lasts much longer than a wheel in conventional grinding. The result is a much higher grinding ratio. In addition, dressing of the grinding wheel is required much less frequently. These are the significant advantages of the process. Applications of ECG include sharpening of cemented carbide tools and grinding of surgical needles, other thin wall tubes, and fragile parts.

26.3 THERMAL ENERGY PROCESSES

Material removal processes based on thermal energy are characterized by very high local temperatures—hot enough to remove material by fusion or vaporization. Because of the high temperatures, these processes cause physical and metallurgical damage to the new work surface. In some cases, the resulting finish is so poor that subsequent processing is required to smooth the surface. In this section we examine several thermal energy processes that have commercial importance: (1) electric discharge machining and electric discharge wire cutting, (2) electron beam machining, (3) laser beam machining, (4) arc cutting processes, and (5) oxyfuel cutting processes.

26.3.1 ELECTRIC DISCHARGE PROCESSES

Electric discharge processes remove metal by a series of discrete electrical discharges (sparks) that cause localized temperatures high enough to melt or vaporize the metal in the immediate vicinity of the discharge. The two main processes in this category are (1) electric discharge machining and (2) wire electric discharge machining. These processes can be used only on electrically conducting work materials. The video clip on electric discharge machining illustrates the various types of EDM.

VIDEO CLIP

Electric Discharge Machining. This clip contains three segments: (1) the EDM process, (2) ram EDM, and (3) wire EDM.

Electric Discharge Machining   Electric discharge machining (EDM) is one of the most widely used nontraditional processes. An EDM setup is illustrated in Figure 26.7. The shape of the finished work surface is produced by a formed electrode tool. The sparks occur across a small gap between tool and work surface. The EDM process must take place in the presence of a dielectric fluid, which creates a path for each discharge as the fluid becomes ionized in the gap. The discharges are generated by a pulsating direct current power supply connected to the work and the tool.

Figure 26.7(b) shows a close-up view of the gap between the tool and the work. The discharge occurs at the location where the two surfaces are closest. The dielectric fluid ionizes at this location to create a path for the discharge. The region in which discharge occurs is heated to extremely high temperatures, so that a small portion of the work surface is suddenly melted and removed. The flowing dielectric then flushes away the small particle (call it a "chip"). Because the surface of the work at the location of the previous discharge is now separated from the tool by a greater distance, this location is less likely to be the site of another spark until the surrounding regions have been reduced to the same level or below. Although the individual discharges remove metal at very

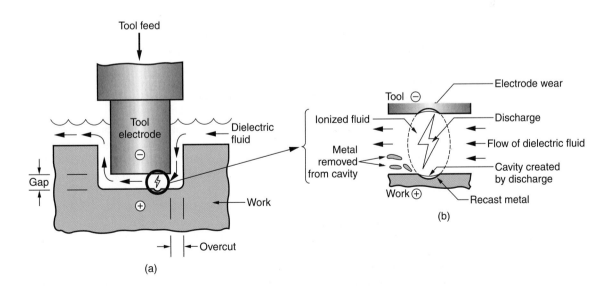

FIGURE 26.7   Electric discharge machining (EDM): (a) overall setup, and (b) close-up view of gap, showing discharge and metal removal.

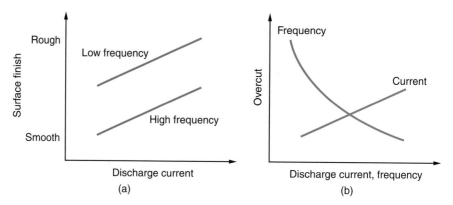

FIGURE 26.8
(a) Surface finish in EDM as a function of discharge current and frequency of discharges. (b) Overcut in EDM as a function of discharge current and frequency of discharges.

localized points, they occur hundreds or thousands of times per second so that a gradual erosion of the entire surface occurs in the area of the gap.

Two important process parameters in EDM are discharge current and frequency of discharges. As either of these parameters is increased, metal removal rate increases. Surface roughness is also affected by current and frequency, as shown in Figure 26.8(a). The best surface finish is obtained in EDM by operating at high frequencies and low discharge currents. As the electrode tool penetrates into the work, overcutting occurs. *Overcut* in EDM is the distance by which the machined cavity in the workpart exceeds the size of the tool on each side of the tool, as illustrated in Figure 26.7(a). It is produced because the electrical discharges occur at the sides of the tool as well as its frontal area. Overcut is a function of current and frequency, as seen in Figure 26.8(b), and can amount to several hundredths of a millimeter.

The high spark temperatures that melt the work also melt the tool, creating a small cavity in the surface opposite the cavity produced in the work. Tool wear is usually measured as the ratio of work material removed to tool material removed (similar to the grinding ratio). This wear ratio ranges between 1.0 and 100 or slightly above, depending on the combination of work and electrode materials. Electrodes are made of graphite, copper, brass, copper tungsten, silver tungsten, and other materials. The selection depends on the type of power supply circuit available on the EDM machine, the type of work material that is to be machined, and whether roughing or finishing is to be done. Graphite is preferred for many applications because of its melting characteristics. In fact, graphite does not melt. It vaporizes at very high temperatures, and the cavity created by the spark is generally smaller than for most other EDM electrode materials. Consequently, a high ratio of work material removed to tool wear is usually obtained with graphite tools.

The hardness and strength of the work material are not factors in EDM, because the process is not a contest of hardness between tool and work. The melting point of the work material is an important property, and metal removal rate can be related to melting point approximately by the following empirical formula, based on an equation described in Weller [17]:

$$R_{MR} = \frac{KI}{T_m^{1.23}} \tag{26.7}$$

where R_{MR} = metal removal rate, mm^3/s (in^3/min); K = constant of proportionality whose value = 664 in SI units (5.08 in U.S. customary units); I = discharge current, amps; and T_m = melting temperature of work metal, °C (°F).

Melting points of selected metals are listed in Table 4.1.

**Example 26.2
Electric Discharge
Machining**

Copper is to be machined in an EDM operation. If discharge current = 25 amps, what is the expected metal removal rate?

Solution: From Table 4.1, the melting point of copper is found to be 1083°C. Using Eq. (26.7), the anticipated metal removal rate is

$$R_{MR} = \frac{664(25)}{1083^{1.23}} = 3.07 \text{ mm}^3/\text{s}$$

Dielectric fluids used in EDM include hydrocarbon oils, kerosene, and distilled or deionized water. The dielectric fluid serves as an insulator in the gap except when ionization occurs in the presence of a spark. Its other functions are to flush debris out of the gap and remove heat from tool and workpart.

Applications of electric discharge machining include both tool fabrication and parts production. The tooling for many of the mechanical processes discussed in this book are often made by EDM, including molds for plastic injection molding, extrusion dies, wire drawing dies, forging and heading dies, and sheet metal stamping dies. As in ECM, the term *die sinking* is used for operations in which a mold cavity is produced, and the EDM process is sometimes referred to as *ram EDM.* For many of the applications, the materials used to fabricate the tooling are difficult (or impossible) to machine by conventional methods. Certain production parts also call for application of EDM. Examples include delicate parts that are not rigid enough to withstand conventional cutting forces, hole drilling where the axis of the hole is at an acute angle to the surface so that a conventional drill would be unable to start the hole, and production machining of hard and exotic metals.

Electric Discharge Wire Cutting   Electric discharge wire cutting (EDWC), commonly called *wire EDM,* is a special form of electric discharge machining that uses a small diameter wire as the electrode to cut a narrow kerf in the work. The cutting action in wire EDM is achieved by thermal energy from electric discharges between the electrode wire and the workpiece. Wire EDM is illustrated in Figure 26.9. The workpiece is fed past the wire to achieve the desired cutting path, somewhat in the manner of a bandsaw operation. Numerical control is used to control the workpart motions during cutting. As it cuts, the wire is slowly and continuously advanced between a supply spool and a take-up spool to present a fresh electrode of constant diameter to the work. This helps to maintain a constant kerf width during cutting. As in EDM, wire EDM must be carried out in the presence of a dielectric. This is applied by nozzles directed at the tool–work interface as in our figure, or the workpart is submerged in a dielectric bath.

Wire diameters range from 0.076 to 0.30 mm (0.003 to 0.012 in), depending on required kerf width. Materials used for the wire include brass, copper, tungsten, and

FIGURE 26.9   Electric discharge wire cutting (EDWC), also called wire EDM.

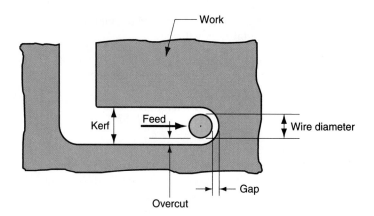

FIGURE 26.10
Definition of kerf and overcut in electric discharge wire cutting.

molybdenum. Dielectric fluids include deionized water or oil. As in EDM, an overcut exists in wire EDM that makes the kerf larger than the wire diameter, as shown in Figure 26.10. This overcut is in the range 0.020 to 0.050 mm (0.0008 to 0.002 in). Once cutting conditions have been established for a given cut, the overcut remains fairly constant and predictable.

Although EDWC seems similar to a bandsaw operation, its precision far exceeds that of a bandsaw. The kerf is much narrower, corners can be made much sharper, and the cutting forces against the work are nil. In addition, hardness and toughness of the work material do not affect cutting performance. The only requirement is that the work material must be electrically conductive.

The special features of wire EDM make it ideal for making components for stamping dies. Because the kerf is so narrow, it is often possible to fabricate punch and die in a single cut, as suggested by Figure 26.11. Other tools and parts with intricate outline shapes, such as lathe form tools, extrusion dies, and flat templates, are made with electric discharge wire cutting.

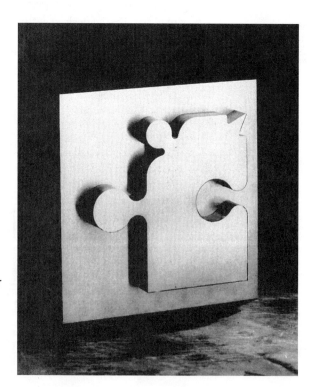

FIGURE 26.11 Irregular outline cut from a solid metal slab by wire EDM. (Photo courtesy of LeBlond Makino Machine Tool Company, Amelia, Ohio.)

High-voltage cable

Vacuum chamber

Electron beam gun

Cathode grid (−)

Anode (+)

Valve

Electron beam

Magnetic lens

Magnetic deflection coil

Workpiece

FIGURE 26.12 Electron beam machining (EBM).

26.3.2 ELECTRON BEAM MACHINING

Electron beam machining (EBM) is one of several industrial processes that use electron beams. Besides machining, other applications of the technology include heat treating (Section 27.5.2) and welding (Section 30.4). *Electron beam machining* uses a high velocity stream of electrons focused on the workpiece surface to remove material by melting and vaporization. A schematic of the EBM process is illustrated in Figure 26.12. An electron beam gun generates a continuous stream of electrons that is accelerated to approximately 75% of the speed of light and focused through an electromagnetic lens on the work surface. The lens is capable of reducing the area of the beam to a diameter as small as 0.025 mm (0.001 in). On impinging the surface, the kinetic energy of the electrons is converted into thermal energy of extremely high density that melts or vaporizes the material in a very localized area.

Electron beam machining is used for a variety of high-precision cutting applications on any known material. Applications include drilling of extremely small diameter holes—down to 0.05 mm (0.002 in) diameter, drilling of holes with very high depth-to-diameter ratios—more than 100:1, and cutting of slots that are only about 0.001 in (0.025 mm) wide. These cuts can be made to very close tolerances with no cutting forces or tool wear. The process is ideal for micromachining and is generally limited to cutting operations in thin parts—in the range 0.25 to 6.3 mm (0.010 to 0.250 in) thick. EBM must be carried out in a vacuum chamber to eliminate collision of the electrons with gas molecules. Other limitations include the high energy required and expensive equipment.

26.3.3 LASER BEAM MACHINING

Lasers are being used for a variety of industrial applications, including heat treatment (Section 27.5.2), welding (Section 30.4), measurement (Section 42.6.2), as well as scribing, cutting, and drilling (described here). The term *laser* stands for *l*ight *a*mplification by *s*timulated *e*mission of *r*adiation. A laser is an optical transducer that converts electrical energy into a highly coherent light beam. A laser light beam has several properties that distinguish it from other forms of light. It is monochromatic (theoretically,

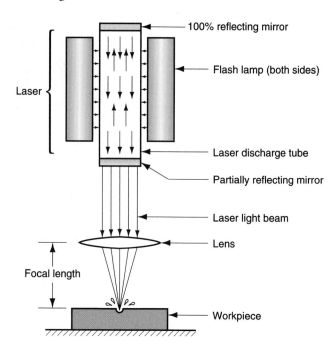

FIGURE 26.13   Laser beam machining (LBM).

the light has a single wave length) and highly collimated (the light rays in the beam are almost perfectly parallel). These properties allow the light generated by a laser to be focused, using conventional optical lenses, onto a very small spot with resulting high power density. Depending on the amount of energy contained in the light beam, and its degree of concentration at the spot, the various laser processes identified in the preceding can be accomplished.

Laser beam machining (LBM) uses the light energy from a laser to remove material by vaporization and ablation. The setup for LBM is illustrated in Figure 26.13. The types of lasers used in LBM are carbon dioxide gas lasers and solid-state lasers (of which there are several types). In laser beam machining, the energy of the coherent light beam is concentrated not only optically but also in terms of time. The light beam is pulsed so that the released energy results in an impulse against the work surface that produces a combination of evaporation and melting, with the melted material evacuating the surface at high velocity.

LBM is used to perform various types of drilling, slitting, slotting, scribing, and marking operations. Drilling small diameter holes is possible—down to 0.025 mm (0.001 in). For larger holes, above 0.50-mm (0.020-in) diameter, the laser beam is controlled to cut the outline of the hole. LBM is not considered a mass production process, and it is generally used on thin stock. The range of work materials that can be machined by LBM is virtually unlimited. Ideal properties of a material for LBM include high light energy absorption, poor reflectivity, good thermal conductivity, low specific heat, low heat of fusion, and low heat of vaporization. Of course, no material has this ideal combination of properties. The actual list of work materials processed by LBM includes metals with high hardness and strength, soft metals, ceramics, glass and glass epoxy, plastics, rubber, cloth, and wood.

26.3.4 ARC-CUTTING PROCESSES

The intense heat from an electric arc can be used to melt virtually any metal for the purpose of welding or cutting. Most arc-cutting processes use the heat generated by an arc between an electrode and a metallic workpart (usually a flat plate or sheet) to melt a

Tungsten electrode

Torch body

Outer shell

Primary gas

Plasma arc

Secondary gas

Work

Molten metal removed

Kerf

Plasma arc torch

Standoff distance

FIGURE 26.14   Plasma arc cutting (PAC).

kerf that separates the part. The most common arc-cutting processes are (1) plasma arc cutting and (2) air carbon arc cutting [11].

Plasma Arc Cutting   A *plasma* is defined as a superheated, electrically ionized gas. Plasma arc cutting (PAC) uses a plasma stream operating at temperatures in the range 10,000°C to 14,000°C (18,000°F to 25,000°F) to cut metal by melting, as shown in Figure 26.14. The cutting action operates by directing the high-velocity plasma stream at the work, thus melting it and blowing the molten metal through the kerf. The plasma arc is generated between an electrode inside the torch and the anode workpiece. The plasma flows through a water-cooled nozzle that constricts and directs the stream to the desired location on the work. The resulting plasma jet is a high-velocity, well-collimated stream with extremely high temperatures at its center, hot enough to cut through metal in some cases 150 mm (6 in) thick.

Gases used to create the plasma in PAC include nitrogen, argon, hydrogen, or mixtures of these gases. These are referred to as the primary gases in the process. Secondary gases or water are often directed to surround the plasma jet to help confine the arc and clean the kerf of molten metal as it forms.

Most applications of PAC involve cutting of flat metal sheets and plates. Operations include hole piercing and cutting along a defined path. The desired path can be cut either by use of a hand-held torch manipulated by a human operator, or by directing the cutting path of the torch under numerical control (NC). For faster production and higher accuracy, NC is preferred because of better control over the important process variables such as standoff distance and feed rate. Plasma arc cutting can be used to cut nearly any electrically conductive metal. Metals frequently cut by PAC include plain carbon steel, stainless steel, and aluminum. The advantage of NC in these applications is high productivity. Feed rates along the cutting path can be as high as 200 mm/s (450 in/min) for 6-mm (0.25-in) aluminum plate and 85 mm/s (200 in/min) for 6-mm (0.25-in) steel plate [8]. Feed rates must be reduced for thicker stock. For example, the maximum feed rate for cutting 100-mm (4-in) thick aluminum stock is around 8 mm/s (20 in/min) [8]. Disadvantages of PAC are (1) the cut surface is rough, and (2) metallurgical damage at the surface is the most severe among the nontraditional metalworking processes.

Air Carbon Arc Cutting   In this process, the arc is generated between a carbon electrode and the metallic work, and a high-velocity air jet is used to blow away the melted portion of the metal. This procedure can be used to form a kerf for severing the piece, or to gouge a cavity in the part. Gouging is used to prepare the edges of plates for welding, for example to create a U-groove in a butt joint (Section 29.2.1). Air carbon arc cutting is used on a variety of metals, including cast iron, carbon steel, low alloy, and stainless steels, and various nonferrous alloys. Spattering of the molten metal is a hazard and a disadvantage of the process.

Other Arc-Cutting Processes   Various other electric arc processes are used for cutting applications, although not as widely as plasma arc and air carbon arc cutting. These other processes include: (1) gas metal arc cutting, (2) shielded metal arc cutting, (3) gas tungsten arc cutting, and (4) carbon arc cutting. The technologies are the same as those used in arc welding (Section 30.1), except that the heat of the electric arc is used for cutting.

26.3.5 OXYFUEL-CUTTING PROCESSES

A widely used family of thermal cutting processes, popularly known as *flame cutting,* use the heat of combustion of certain fuel gases combined with the exothermic reaction of the metal with oxygen. The cutting torch used in these processes is designed to deliver a mixture of fuel gas and oxygen in the proper amounts, and to direct a stream of oxygen to the cutting region. The primary mechanism of material removal in oxyfuel cutting (OFC) is the chemical reaction of oxygen with the base metal. The purpose of the oxyfuel combustion is to raise the temperature in the region of cutting to support the reaction. These processes are commonly used to cut ferrous metal plates, in which the rapid oxidation of iron occurs according to the following reactions [11]:

$$Fe + O \rightarrow FeO + heat \tag{26.8a}$$

$$3Fe + 2O_2 \rightarrow Fe_3O_4 + heat \tag{26.8b}$$

$$2Fe + 1.5O_2 \rightarrow Fe_2O_3 + heat \tag{26.8c}$$

The second of these reactions, Eq. (26.8b), is the most significant in terms of heat generation.

The cutting mechanism for nonferrous metals is somewhat different. These metals are generally characterized by lower melting temperatures than the ferrous metals, and they are more oxidation resistant. In these cases, the heat of combustion of the oxyfuel mixture plays a more important role in creating the kerf. Also, to promote the metal oxidation reaction, chemical fluxes or metallic powders are often added to the oxygen stream.

Fuels used in OFC include acetylene (C_2H_2), MAPP (methylacetylene-propadiene—C_3H_4), propylene (C_3H_6), and propane (C_3H_8). Flame temperatures and heats of combustion for these fuels are listed in Table 30.2. Acetylene burns at the highest flame temperature and is the most widely used fuel for welding and cutting. However, there are certain hazards with the storage and handling of acetylene that must be considered (Section 30.3.1).

OFC processes are performed either manually or by machine. Manually operated torches are used for repair work, cutting of scrap metal, trimming of risers from sand castings, and similar operations that generally require minimal accuracy. For production work, machine flame cutting allows faster speeds and greater accuracies. This equipment is often numerically controlled to allow profiled shapes to be cut.

26.4 CHEMICAL MACHINING

Chemical machining (CHM) is a nontraditional process in which material is removed by means of a strong chemical etchant. Applications as an industrial process began shortly after World War II in the aircraft industry. The use of chemicals to remove unwanted material from a workpart can be accomplished in several ways, and different terms have been developed to distinguish the applications. These terms include chemical milling, chemical blanking, chemical engraving, and photochemical machining (PCM). They all use the same mechanism of material removal, and it is appropriate to discuss the general characteristics of chemical machining before defining the individual processes.

26.4.1 MECHANICS AND CHEMISTRY OF CHEMICAL MACHINING

The chemical machining process consists of several steps. Differences in applications and the ways in which the steps are implemented account for the different forms of CHM. The steps are:

1. ***Cleaning.*** The first step is a cleaning operation to ensure that material will be removed uniformly from the surfaces to be etched.
2. ***Masking.*** A protective coating called a maskant is applied to certain portions of the part surface. This maskant is made of a material that is chemically resistant to the etchant (the term ***resist*** is used for this masking material). It is therefore applied to those portions of the work surface that are not to be etched.
3. ***Etching.*** This is the material removal step. The part is immersed in an etchant that chemically attacks those portions of the part surface that are not masked. The usual method of attack is to convert the work material (e.g., a metal) into a salt that dissolves in the etchant and is thereby removed from the surface. When the desired amount of material has been removed, the part is withdrawn from the etchant and washed to stop the process.
4. ***Demasking.*** The maskant is removed from the part.

The two steps in chemical machining that involve significant variations in methods, materials, and process parameters are masking and etching—steps 2 and 3.

Maskant materials include neoprene, polyvinylchloride, polyethylene, and other polymers. Masking can be accomplished by any of three methods: (1) cut and peel, (2) photographic resist, and (3) screen resist. The ***cut and peel*** method applies the maskant over the entire part by dipping, painting, or spraying. The resulting thickness of the maskant is 0.025 to 0.125 mm (0.001 to 0.005 in). After the maskant has hardened, it is cut using a scribing knife and peeled away in the areas of the work surface that are to be etched. The maskant cutting operation is performed by hand, usually guiding the knife with a template. The cut and peel method is generally used for large workparts, low production quantities, and where accuracy is not a critical factor. This method cannot hold tolerances tighter than ± 0.125 mm (± 0.005 in) except with extreme care.

As the name suggests, the ***photographic resist*** method (called the ***photoresist*** method for short) uses photographic techniques to perform the masking step. The masking materials contain photosensitive chemicals. They are applied to the work surface and exposed to light through a negative image of the desired areas to be etched. These areas of the maskant can then be removed from the surface using photographic developing techniques. This procedure leaves the desired surfaces of the part protected by the maskant and the remaining areas unprotected, vulnerable to chemical etching. Photoresist masking techniques are normally applied where small parts are produced in high quantities, and close tolerances are required. Tolerances closer than ± 0.0125 mm (± 0.0005 in) can be held [17].

The ***screen resist*** method applies the maskant by means of silk screening methods. In these methods, the maskant is painted onto the workpart surface through a silk or stainless steel mesh. Embedded in the mesh is a stencil that protects those areas to be etched from being painted. The maskant is thus painted onto the work areas that are not to be etched. The screen resist method is generally used in applications that are between

TABLE 26.2   Common work materials and etchants in CHM, with typical penetration rates and etch factors.

Work Material	Etchant	Penetration Rates		Etch Factor
		mm/min	in/min	
Aluminum	$FeCl_3$	0.020	0.0008	1.75
and alloys	NaOH	0.025	0.001	1.75
Copper and alloys	$FeCl_3$	0.050	0.002	2.75
Magnesium and alloys	H_2SO_4	0.038	0.0015	1.0
Silicon	$HNO_3 : HF : H_2O$	very slow		NA
Mild steel	$HCl : HNO_3$	0.025	0.001	2.0
	$FeCl_3$	0.025	0.001	2.0
Titanium	HF	0.025	0.001	1.0
and alloys	$HF : HNO_3$	0.025	0.001	1.0

Compiled from [5], [8], and [17].
NA, Data not available.

the other two masking methods in terms of accuracy, part size, and production quantities. Tolerances of ±0.075 mm (±0.003 in) can be achieved with this masking method.

Selection of the **etchant** depends on work material to be etched, desired depth and rate of material removal, and surface finish requirements. The etchant must also be matched with the type of maskant that is used to ensure that the maskant material is not chemically attacked by the etchant. Table 26.2 lists some of the work materials machined by CHM together with the etchants that are generally used on these materials. Also included in the table are penetration rates and etch factors. These parameters are explained next.

Material removal rates in CHM are generally indicated as penetration rates, mm/min (in/min), because rate of chemical attack of the work material by the etchant is directed into the surface. The penetration rate is unaffected by surface area. Penetration rates listed in Table 26.2 are typical values for the given material and etchant.

Depths of cut in chemical machining are as much as 12.5 mm (0.5 in) for aircraft panels made out of metal plates. However, many applications require depths that are only several hundredths of a millimeter. Along with the penetration into the work, etching also occurs sideways under the maskant, as illustrated in Figure 26.15. The effect is referred to as the **undercut,** and it must be accounted for in the design of the mask for the resulting cut to have the specified dimensions. For a given work material, the undercut is directly related to the depth of cut. The constant of proportionality for the material is called the etch factor, defined as

$$F_e = \frac{d}{u} \qquad (26.9)$$

where F_e = etch factor; d = depth of cut, mm (in); and u = undercut, mm (in).

The dimensions u and d are defined in Figure 26.15. Different work materials have different etch factors in chemical machining. Some typical values are presented in Table 26.2.

FIGURE 26.15
Undercut in chemical machining.

FIGURE 26.16   Sequence of processing steps in chemical milling: (1) clean raw part, (2) apply maskant, (3) scribe, cut, and peel the maskant from areas to be etched, (4) etch, and (5) remove maskant and clean to yield finished part.

The etch factor can be used to determine the dimensions of the cutaway areas in the maskant, so that the specified dimensions of the etched areas on the part can be achieved.

26.4.2 CHM PROCESSES

In this section, we describe the principle chemical machining processes: (1) chemical milling, (2) chemical blanking, (3) chemical engraving, and (4) photochemical machining.

Chemical Milling   Chemical milling was the first CHM process to be commercialized. During World War II, an aircraft company in the United States began to use chemical milling to remove metal from aircraft components. They referred to their process as the "chem-mill" process. Today, chemical milling is still used largely in the aircraft industry, to remove material from aircraft wing and fuselage panels for weight reduction. It is applicable to large parts where substantial amounts of metal are removed during the process. The cut and peel maskant method is employed. A template is generally used that takes into account the undercut that will result during etching. The sequence of processing steps is illustrated in Figure 26.16.

Chemical milling produces a surface finish that varies with different work materials. Table 26.3 provides a sampling of the values. Surface finish depends on depth of penetration. As depth increases, finish becomes worse, approaching the upper side of the ranges given in the table. Metallurgical damage from chemical milling is very small, perhaps around 0.005 mm (0.0002 in) into the work surface.

Chemical Blanking   Chemical blanking uses chemical erosion to cut very thin sheetmetal parts—down to 0.025 mm (0.001 in) thick and/or for intricate cutting patterns. In both

TABLE 26.3 Surface finishes expected in chemical milling.		
	Surface Finishes Range	
Work Material	**μm**	**μ-in**
Aluminum and alloys	1.8–4.1	70–160
Magnesium	0.8–1.8	30–70
Mild steel	0.8–6.4	30–250
Titanium and alloys	0.4–2.5	15–100

Compiled from [8] and [17].

instances, conventional punch-and-die methods do not work because the stamping forces damage the sheet metal, or the tooling cost would be prohibitive, or both. Chemical blanking produces parts that are burr free, an advantage over conventional shearing operations.

Methods used for applying the maskant in chemical blanking are either the photoresist method or the screen resist method. For small and/or intricate cutting patterns and close tolerances, the photoresist method is used. Tolerances as close as ±0.0025 mm (±0.0001 in) can be held on 0.025 mm (0.001 in) thick stock using the photoresist method of masking. As stock thickness increases, more generous tolerances must be allowed. Screen resist masking methods are not nearly so accurate as photoresist. The small work size in chemical blanking excludes the cut and peel maskant method.

Using the screen resist method to illustrate, the steps in chemical blanking are shown in Figure 26.17. Because chemical etching takes place on both sides of the part in chemical blanking, it is important that the masking procedure provides accurate registration between the two sides. Otherwise, the erosion into the part from opposite directions will not line up. This is especially critical with small part sizes and intricate patterns.

Application of chemical blanking is generally limited to thin materials and/or intricate patterns for reasons given in the preceding. Maximum stock thickness is around 0.75 mm (0.030 in). Also, hardened and brittle materials can be processed by chemical blanking where mechanical methods would surely fracture the work. Figure 26.18 presents a sampling of parts produced by the chemical blanking process.

Chemical Engraving   Chemical engraving is a chemical machining process for making name plates and other flat panels that have lettering and/or artwork on one side. These plates and panels would otherwise be made using a conventional engraving machine or similar process. Chemical engraving can be used to make panels with either recessed lettering or raised lettering, simply by reversing the portions of the panel to be etched. Masking is done by either the photoresist or screen resist methods. The sequence in chemical engraving is similar to the other CHM processes, except that a filling operation follows etching. The purpose of filling is to apply paint or other coating into the recessed areas that have been created by etching. Then, the panel is immersed in a solution that dissolves the resist but does not attack the coating material. Thus, when the resist is removed, the coating remains in the etched areas but not in the areas that were masked. The effect is to highlight the pattern.

Photochemical Machining   Photochemical machining (PCM) is chemical machining in which the photoresist method of masking is used. The term can therefore be applied correctly to chemical blanking and chemical engraving when these methods use the photographic resist method. PCM is employed in metalworking when close tolerances

FIGURE 26.18 Parts made by chemical blanking. (Courtesy of Buckbee-Mears, St. Paul.)

and/or intricate patterns are required on flat parts. Photochemical processes are also used extensively in the electronics industry to produce intricate circuit designs on semi-conductor wafers (Section 34.3).

Figure 26.19 shows the sequence of steps in photochemical machining as it is applied to chemical blanking. There are various ways to photographically expose the desired image onto the resist. The figure shows the negative in contact with the surface of the resist during exposure. This is contact printing, but other photographic printing methods are available that expose the negative through a lens system to enlarge or reduce

FIGURE 26.19
Sequence of processing steps in photochemical machining: (1) clean raw part; (2) apply resist (maskant) by dipping, spraying, or painting; (3) place negative on resist; (4) expose to ultraviolet light; (5) develop to remove resist from areas to be etched; (6) etch (shown partially etched); (7) etch (completed); (8) remove resist and clean to yield finished part.

the size of the pattern printed on the resist surface. Photoresist materials in current use are sensitive to ultraviolet light but not to light of other wavelengths. Therefore, with proper lighting in the factory, there is no need to carry out the processing steps in a dark room environment. Once the masking operation is accomplished, the remaining steps in the procedure are similar to the other chemical machining methods.

In photochemical machining, the term corresponding to etch factor is ***anisotropy,*** which is defined as the depth of cut d divided by the undercut u (see Figure 26.17). This is the same definition as in Eq. (26.9).

26.5   APPLICATION CONSIDERATIONS

Typical applications of nontraditional processes include special geometric features and work materials that cannot be readily processed by conventional techniques. In this section, we examine these issues. We also summarize the general performance characteristics of nontraditional processes.

Workpart Geometry and Work Materials   Some of the special workpart shapes for which nontraditional processes are well suited are listed in Table 26.4 along with the nontraditional processes that are likely to be appropriate.

As a group, the nontraditional processes can be applied to nearly all work materials, metals and nonmetals. However, certain processes are not suited to certain work materials. Table 26.5 relates applicability of the nontraditional processes to various types of materials. Several of the processes can be used on metals but not nonmetals. For example, ECM, EDM, and PAM require work materials that are electrical conductors. This generally limits their applicability to metal parts. Chemical machining depends on the availability of an appropriate etchant for the given work material. Because metals are more susceptible to chemical attack by various etchants, CHM is commonly used to process metals. With some exceptions, USM, AJM, EBM, and LBM can be used on both

TABLE 26.4   Workpart geometric features and appropriate nontraditional processes.

Geometric Feature	Likely Process
Very small holes. Diameters less than 0.125 mm (0.005 in), in some cases down to 0.025 mm (0.001 in), generally smaller than the diameter range of conventional drill bits.	EBM, LBM
Holes with large depth-to-diameter ratios, e.g., $d/D > 20$. Except for gun drilling, these holes cannot be machined in conventional drilling operations.	ECM, EDM
Holes that are not round. Non-round holes cannot be drilled with a rotating drill bit.	EDM, ECM
Narrow slots in slabs and plates of various materials. The slots are not necessarily straight. In some cases, the slots have extremely intricate shapes.	EBM, LBM, WJC, wire EDM, AWJC
Micromachining. In addition to cutting small holes and narrow slits, there are other material removal applications in which the workpart and/or areas to be cut are very small.	PCM, LBM, EBM
Shallow pockets and surface details in flat parts. There is a significant range in the sizes of the parts in this category, from microscopic integrated circuit chips to large aircraft panels.	CHM
Special contoured shapes for mold and die applications. These applications are sometimes referred to as die-sinking.	EDM, ECM

TABLE 26.5 Applicability of selected nontraditional machining processes to various work materials. For comparison, conventional milling and grinding are included in the compilation.

Work Material	Nontraditional Processes								Conventional Processes	
	Mech		Elec		Thermal			Chem		
	USM	WJC	ECM	EDM	EBM	LBM	PAC	CHM	Milling	Grinding
Aluminum	C	C	B	B	B	B	A	A	A	A
Steel	B	D	A	A	B	B	A	A	A	A
Super alloys	C	D	A	A	B	B	A	B	B	B
Ceramic	A	D	D	D	A	A	D	C	D	C
Glass	A	D	D	D	B	B	D	B	D	C
Silicon[a]			D	D	B	B	D	B	D	B
Plastics	B	B	D	D	B	B	D	C	B	C
Cardboard[b]	D	A	D	D				D	D	D
Textiles[c]	D	A	D	D				D	D	D

Compiled from [17] and other sources.

A, Good application; B, fair application, C, poor application; D, not applicable; and blank entries indicate no data available during compilation.

[a]Refers to silicon used in fabricating integrated circuit chips.

[b]Includes other paper products.

[c]Includes felt, leather, and similar materials.

metals and nonmetals. WJC is generally limited to the cutting of plastics, cardboards, textiles, and other materials that do not possess the strength of metals.

Performance of Nontraditional Processes The nontraditional processes are generally characterized by low material removal rates and high specific energies relative to conventional machining operations. The capabilities for dimensional control and surface finish of the nontraditional processes vary widely, with some of the processes providing high accuracies and good finishes, and others yielding poor accuracies and finishes. Surface damage is also a consideration. Some of these processes produce very little metallurgical damage at and immediately below the work surface, whereas others (mostly the thermal-based processes) do considerable damage to the surface. Table 26.6 compares these features

TABLE 26.6 Machining characteristics of the nontraditional machining processes

Work Material	Nontraditional Processes								Conventional Processes	
	Mech		Elec		Thermal			Chem		
	USM	WJC	ECM	EDM	EBM	LBM	PAC	CHM	Milling	Grinding
Material removal rates	C	C	B	C	D	D	A	B–D[a]	A	B
Dimensional control	A	B	B	A–D[b]	A	A	D	A–B[b]	B	A
Surface finish	A	A	B	B–D[b]	B	B	D	B	B–C[b]	A
Surface damage[c]	B	B	A	D	D	D	D	A	B	B–C[b]

Compiled from [17].

A, Excellent; B, good, C, fair, D, poor.

[a]Rating depends on size of work and masking method.

[b]Rating depends on cutting conditions.

[c]In surface damage a good rating means low surface damage and poor rating means deep penetration of surface damage; thermal processes can cause damage up to 0.020 in (0.50 mm) below the new work surface.

of the prominent nontraditional methods, using conventional milling and surface grinding for comparison. Inspection of the data reveals wide differences in machining characteristics. In comparing the characteristics of nontraditional and conventional machining, it must be remembered that nontraditional processes are generally used where conventional methods are not practical or economical.

REFERENCES

[1] Aronson, R. B. "Spindles are the Key to HSM." "Waterjets Move into the Mainstream." *Manufacturing Engineering*, April 2005, pp. 69–74.

[2] Bellows, G., and Kohls, J. B. "Drilling without Drills," Special Report 743, *American Machinist*, March 1982, pp. 173–188.

[3] Benedict, G. F. *Nontraditional Manufacturing Processes.* Marcel Dekker, New York, 1987.

[4] Dini, J. W. "Fundamentals of Chemical Milling." Special Report 768, *American Machinist*, July 1984, pp. 99–114.

[5] Drozda, T. J., and C. Wick (eds.). *Tool and Manufacturing Engineers Handbook.* 4th ed. Vol. I, *Machining.* Society of Manufacturing Engineers, Dearborn, Michigan, 1983.

[6] El-Hofy, H. *Advanced Machining Processes: Nontraditional and Hybrid Machining Processes,* McGraw-Hill Professional, New York, 2005.

[7] Guitrau, E. "Sparking Innovations." *Cutting Tool Engineering.* Vol. 52, No. 10, October 2000, pp. 36–43.

[8] *Machining Data Handbook.* 3rd ed., Vol. 2. Machinability Data Center, Metcut Research Associates Inc., Cincinnati, Ohio, 1980.

[9] Mason, F. "Water Jet Cuts Instrument Panels." *American Machinist & Automated Manufacturing*, July 1988, pp. 126–127.

[10] McGeough, J. A. *Advanced Methods of Machining.* Chapman and Hall, London, England, 1988.

[11] O'Brien, R. L. *Welding Handbook.* 8th ed. Vol. 2, *Welding Processes.* American Welding Society, Miami, Florida, 1991.

[12] Pandey, P. C., and Shan, H. S. *Modern Machining Processes.* Tata McGraw-Hill, New Delhi, India, 1980.

[13] Vaccari, J. A. "The Laser's Edge in Metalworking." Special Report 768, *American Machinist.* August 1984, pp. 99–114.

[14] Vaccari, J. A. "Thermal Cutting." Special Report 778, *American Machinist*, July 1988, pp. 111–126.

[15] Vaccari, J. A. "Advances in Laser Cutting." *American Machinist & Automated Manufacturing*, March 1988, pp. 59–61.

[16] Waurzyniak, P. "EDM's Cutting Edge." *Manufacturing Engineering*, Vol. *123*, No. 5, November 1999, pp. 38–44.

[17] Weller, E. J. (ed.). *Nontraditional Machining Processes.* 2nd ed. Society of Manufacturing Engineers, Dearborn, Michigan, 1984.

[18] www.engineershandbook.com/MfgMethods.

REVIEW QUESTIONS

26.1. Why are the nontraditional material removal processes important?

26.2. There are four categories of nontraditional machining processes, based on principal energy form. Name the four categories.

26.3. How does the ultrasonic machining process work?

26.4. Describe the water jet cutting process.

26.5. What is the difference between water jet cutting, abrasive water jet cutting, and abrasive jet cutting?

26.6. Name the three main types of electrochemical machining.

26.7. Identify the two significant disadvantages of electrochemical machining.

26.8. How does increasing discharge current affect metal removal rate and surface finish in electric discharge machining?

26.9. What is meant by the term overcut in electric discharge machining?

26.10. Identify two major disadvantages of plasma arc cutting.

26.11. What are some of the fuels used in oxyfuel cutting?

26.12. Name the four principal steps in chemical machining.

26.13. What are the three methods of performing the masking step in chemical machining?

26.14. What is a photoresist in chemical machining?

26.15. (**Video**) What are the three layers of a part's surface after undergoing EDM?

26.16. (**Video**) What are two other names for ram type EDMs?

26.17. (**Video**) Name the four subsystems in a RAM EDM process.

26.18. (**Video**) Name the four subsystems in a wire EDM process.

MULTIPLE CHOICE QUIZ

There are 17 correct answers in the following multiple choice questions (some questions have multiple answers that are correct). To attain a perfect score on the quiz, all correct answers must be given. Each correct answer is worth 1 point. Each omitted answer or wrong answer reduces the score by 1 point, and each additional answer beyond the correct number of answers reduces the score by 1 point. Percentage score on the quiz is based on the total number of correct answers.

26.1. Which of the following processes use mechanical energy as the principal energy source (three correct answers): (a) electrochemical grinding, (b) laser beam machining, (c) conventional milling, (d) ultrasonic machining, (e) water jet cutting, and (f) wire EDM?

26.2. Ultrasonic machining can be used to machine both metallic and nonmetallic materials: (a) true or (b) false?

26.3. Applications of electron beam machining are limited to metallic work materials because of the need for the work to be electrically conductive: (a) true or (b) false?

26.4. Which one of the following is closest to the temperatures used in plasma arc cutting: (a) 2750°C (5000°F), (b) 5500°C (10,000°F), (c) 8300°C (15,000°F), (d) 11,000°C (20,000°F), or (e) 16,500°C (30,000°F)?

26.5. Chemical milling is used in which of the following applications (two best answers): (a) drilling holes with high depth-to-diameter ratio, (b) making intricate patterns in thin sheet metal, (c) removing material to make shallow pockets in metal, (d) removing metal from aircraft wing panels, and (e) cutting of plastic sheets?

26.6. Etch factor is equal to which of the following in chemical machining (more than one): (a) anisotropy, (b) CIt, (c) d/u, and (d) u/d; where C = specific removal rate, d = depth of cut, I = current, t = time, and u = undercut?

26.7. Of the following processes, which one is noted for the highest material removal rates: (a) electric discharge machining, (b) electrochemical machining, (c) laser beam machining, (d) oxyfuel cutting, (e) plasma arc cutting, (f) ultrasonic machining, or (g) water jet cutting?

26.8. Which one of the following processes would be appropriate to drill a hole with a square cross section, 0.25 inch on a side and 1-inch deep in a steel workpiece: (a) abrasive jet machining, (b) chemical milling, (c) EDM, (d) laser beam machining, (e) oxyfuel cutting, (f) water jet cutting, or (g) wire EDM?

26.9. Which of the following processes would be appropriate for cutting a narrow slot, less than 0.015 inch wide, in a 3/8-in-thick sheet of fiber-reinforced plastic (two best answers): (a) abrasive jet machining, (b) chemical milling, (c) EDM, (d) laser beam machining, (e) oxyfuel cutting, (f) water jet cutting, and (g) wire EDM?

26.10. Which one of the following processes would be appropriate for cutting a hole of 0.003 inch diameter through a plate of aluminum that is 1/16 in thick: (a) abrasive jet machining, (b) chemical milling, (c) EDM, (d) laser beam machining, (e) oxyfuel cutting, (f) water jet cutting, and (g) wire EDM?

26.11. Which of the following processes could be used to cut a large piece of 1/2-inch plate steel into two sections (two best answers): (a) abrasive jet machining, (b) chemical milling, (c) EDM, (d) laser beam machining, (e) oxyfuel cutting, and (f) water jet cutting, and (g) wire EDM?

PROBLEMS

Application Problems

26.1. For the following application, identify one or more nontraditional machining processes that might be used, and present arguments to support your selection. Assume that either the part geometry or the work material (or both) preclude the use of conventional machining. The application is a matrix of 0.1 mm (0.004 in) diameter holes in a plate of 3.2 mm (0.125 in) thick hardened tool steel. The matrix is rectangular, 75 by 125 mm (3.0 by 5.0 in) with the separation between holes in each direction = 1.6 mm (0.0625 in).

26.2. For the following application, identify one or more nontraditional machining processes that might be used, and present arguments to support your selection. Assume that either the part geometry or the work material (or both) preclude the use of conventional machining. The application is an engraved aluminum printing plate to be used in an offset printing press to make 275×350 mm (11×14 in) posters of Lincoln's Gettysburg address.

26.3. For the following application, identify one or more nontraditional machining processes that might be used, and present arguments to support your selection. Assume that either the part geometry or the work material (or both) preclude the use of conventional machining. The application is a through-hole in the shape of the letter L in a 12.5 mm (0.5 in) thick plate of glass. The size of the "L" is 25×15 mm (1.0×0.6 in) and the width of the hole is 3 mm (1/8 in).

26.4. For the following application, identify one or more nontraditional machining processes that might be used, and present arguments to support your selection. Assume that either the part geometry or the work material (or both) preclude the use of conventional machining. The application is a blind-hole in the shape of the letter G in a 50 mm (2.0 in) cube of steel. The overall size of the "G" is 25×19 mm (1.0×0.75 in), the depth of the hole is 3.8 mm (0.15 in), and its width is 3 mm (1/8 in).

26.5. Much of the work at the Cut-Anything Company involves cutting and forming of flat sheets of fiberglass for the pleasure boat industry. Manual methods based on portable saws are currently used to perform the cutting operation, but production is slow and scrap rates are high. The foreman says the company should invest in a plasma arc cutting machine, but the plant manager thinks it would be too expensive. What do you think? Justify your answer by indicating the characteristics of the process that make PAC attractive or unattractive in this application.

26.6. A furniture company that makes upholstered chairs and sofas must cut large quantities of fabrics. Many of these fabrics are strong and wear-resistant, which properties make them difficult to cut. What nontraditional process(es) would you recommend to the company for this application? Justify your answer by indicating the characteristics of the process that make it attractive.

Electrochemical Machining

26.7. The frontal working area of the electrode in an ECM operation is 2000 mm^2. The applied current = 1800 amps and the voltage = 12 volts. The material being cut is nickel (valence = 2), whose specific removal rate is given in Table 26.1. (a) If the process is 90% efficient, determine the rate of metal removal in mm^3/min. (b) If the resistivity of the electrolyte = 140 ohm-mm, determine the working gap.

26.8. In an electrochemical machining operation, the frontal working area of the electrode is 2.5 in^2. The applied current = 1500 amps, and the voltage = 12 volts. The material being cut is pure aluminum, whose specific removal rate is given in Table 26.1. (a) If the ECM process is 90% efficient, determine the rate of metal removal in in^3/hr. (b) If the resistivity of the electrolyte = 6.2 ohm-in, determine the working gap.

26.9. A square hole is to be cut using ECM through a plate of pure copper (valence = 1) that is 20 mm thick. The hole is 25 mm on each side, but the electrode used to cut the hole is slightly less that 25 mm on its sides to allow for overcut, and its shape includes a hole in its center to permit the flow of electrolyte and reduce the area of the cut. This tool design results in a frontal area of 200 mm^2. The applied current = 1000 amps. Using an efficiency of 95%, determine how long it will take to cut the hole.

26.10. A 3.5 in diameter through hole is to be cut in a block of pure iron (Valence = 2) by electrochemical machining. The block is 2.0 in thick. To speed the cutting process, the electrode tool will have a center hole of 3.0 in which will produce a center core that can be removed after the tool breaks through. The outside diameter of the electrode is undersized to allow for overcut. The overcut is expected to be 0.005 in on a side. If the efficiency of the ECM operation is 90%, what current will be required to complete the cutting operation in 20 minutes?

Electric Discharge Machining

26.11. An electric discharge machining operation is being performed on two work materials: tungsten and tin. Determine the amount of metal removed in the operation after 1 hour at a discharge current of 20 amps for each of these metals. Use metric units and express the answers in mm^3/hr. From Table 4.1, the melting temperatures of tungsten and tin are 3410°C and 232°C, respectively.

26.12. An electric discharge machining operation is being performed on two work materials: tungsten and zinc. Determine the amount of metal removed in the operation after 1 hour at a discharge amperage = 20 amps for each of these metals. Use U.S. Customary units and express the answer in in^3/hr. From Table 4.1, the melting temperatures of tungsten and zinc are 6170°F and 420°F, respectively.

26.13. Suppose the hole in Problem 26.10 were to be cut using EDM rather than ECM. Using a discharge current = 20 amps (which would be typical for EDM), how long would it take to cut the hole? From Table 4.1, the melting temperature of iron is 2802°F.

26.14. A metal removal rate of 0.01 in³/min is achieved in a certain EDM operation on a pure iron workpart. What metal removal rate would be achieved on nickel in this EDM operation if the same discharge current were used? The melting temperatures of iron and nickel are 2802°F and 2651°F, respectively.

26.15. In a wire EDM operation performed on 7-mm-thick C1080 steel using a tungsten wire electrode whose diameter = 0.125 mm, past experience suggests that the overcut will be 0.02 mm, so that the kerf width will be 0.165 mm. Using a discharge current = 10 amps, what is the allowable feed rate that can be used in the operation? Estimate the melting temperature of 0.80% carbon steel from the phase diagram in Figure 6.4.

26.16. A wire EDM operation is to be performed on a slab of 3/4-in-thick aluminum using a brass wire electrode whose diameter = 0.005 in. It is anticipated that the overcut will be 0.001 in, so that the kerf width will be 0.007 in. Using a discharge current = 7 amps, what is the expected allowable feed rate that can be used in the operation? The melting temperature of aluminum is 1220°F.

26.17. A wire EDM operation is used to cut out punch-and-die components from 25-mm-thick tool steel plates. However, in preliminary cuts, the surface finish on the cut edge is poor. What changes in discharge current and frequency of discharges should be made to improve the finish?

Chemical Machining

26.18. Chemical milling is used in an aircraft plant to create pockets in wing sections made of an aluminum alloy. The starting thickness of one workpart of interest is 20 mm. A series of rectangular-shaped pockets 12 mm deep are to be etched with dimensions 200 mm by 400 mm. The corners of each rectangle are radiused to 15 mm. The part is an aluminum alloy and the etchant is NaOH. The penetration rate for this combination is 0.024 mm/min and the etch factor is 1.75. Determine (a) metal removal rate in mm³/min, (b) time required to etch to the specified depth, and (c) required dimensions of the opening in the cut and peel maskant to achieve the desired pocket size on the part.

26.19. In a chemical milling operation on a flat mild steel plate, it is desired to cut an ellipse-shaped pocket to a depth of 0.4 in. The semiaxes of the ellipse are $a = 9.0$ in and $b = 6.0$ in. A solution of hydrochloric and nitric acids will be used as the etchant. Determine (a) metal removal rate in in³/hr, (b) time required to etch to depth, and (c) required dimensions of the opening in the cut and peel maskant required to achieve the desired pocket size on the part.

26.20. In a certain chemical blanking operation, a sulfuric acid etchant is used to remove material from a sheet of magnesium alloy. The sheet is 0.25 mm thick. The screen resist method of masking was used to permit high production rates to be achieved. As it turns out, the process is producing a large proportion of scrap. Specified tolerances of ±0.025 mm are not being achieved. The foreman in the CHM department complains that there must be something wrong with the sulfuric acid. "Perhaps the concentration is incorrect," he suggests. Analyze the problem and recommend a solution.

26.21. In a chemical blanking operation, stock thickness of the aluminum sheet is 0.015 in. The pattern to be cut out of the sheet is a hole pattern, consisting of a matrix of 0.100-in diameter holes. If photochemical machining is used to cut these holes, and contact printing is used to make the resist (maskant) pattern, determine the diameter of the holes that should be used in the pattern.

Part VII   Property Enhancing and Surface Processing Operations

27

HEAT TREATMENT OF METALS

Chapter Contents

The manufacturing processes covered in the preceding chapters involve the creation of part geometry. We now consider processes that either enhance the properties of the workpart (Chapter 27) or apply some surface treatment to it, such as cleaning or coating (Chapter 28). Property-enhancing operations are performed to improve mechanical or physical properties of the work material. They do not alter part geometry, at least not intentionally. The most important property-enhancing operations are heat treatments. *Heat treatment* involves various heating and cooling procedures performed to effect microstructural changes in a material, which in turn affect its mechanical properties. Its most common applications are on metals, discussed in this chapter. Similar treatments are performed on glass-ceramics (Section 7.4.3), tempered glass (Section 12.3.1), and powder metals and ceramics (Sections 16.3.3 and 17.2.3).

Heat treatment operations can be performed on a metallic workpart at various times during its manufacturing sequence. In some cases, the treatment is applied before shaping (e.g., to soften the metal so that it can be more easily formed while hot). In other cases, heat treatment is used to relieve the effects of strain hardening that occur during forming, so that the material can be subjected to further deformation. Heat treatment can also be accomplished at or near the end of the sequence to achieve the final strength and hardness required in the finished product. The principal heat treatments are annealing, martensite formation in steel, precipitation hardening, and surface hardening.

27.1 ANNEALING

Annealing consists of heating the metal to a suitable temperature, holding at that temperature for a certain time (called *soaking*), and slowly cooling. It is performed on a metal for any of the following reasons: (1) to reduce hardness and brittleness, (2) to alter microstructure so that desirable mechanical properties can be obtained, (3) to soften metals for improved machinability or formability, (4) to recrystallize cold-worked (strain-hardened) metals, and (5) to relieve residual stresses induced by prior processes. Different terms are used in annealing, depending on the details of the process and the temperature used relative to the recrystallization temperature of the metal being treated.

Full annealing is associated with ferrous metals (usually low and medium carbon steels); it involves heating the alloy into the austenite region, followed by slow cooling in the furnace to produce coarse pearlite. *Normalizing* involves similar heating and soaking cycles, but the cooling rates are faster. The steel is allowed to cool in air to room temperature. This results in fine pearlite, higher strength and hardness, but lower ductility than the full anneal treatment.

Cold-worked parts are often annealed to reduce effects of strain hardening and increase ductility. The treatment allows the strain-hardened metal to recrystallize partially or completely, depending on temperatures, soaking periods, and cooling rates. When annealing is performed to allow for further cold working of the part, it is called a *process anneal.* When performed on the completed (cold-worked) part to remove the effects of strain hardening and where no subsequent deformation will be accomplished, it is simply called an *anneal.* The process itself is pretty much the same, but different terms are used to indicate the purpose of the treatment.

If annealing conditions permit full recovery of the cold-worked metal to its original grain structure, then *recrystallization* has occurred. After this type of anneal, the metal has the new geometry created by the forming operation, but its grain structure and associated properties are essentially the same as before cold working. The conditions that tend to favor recrystallization are higher temperature, longer holding time, and slower cooling rate. If the annealing process only permits partial return of the grain structure toward its original state, it is termed a *recovery anneal.* Recovery allows the metal to retain most of the strain hardening obtained in cold working, but the toughness of the part is improved.

The preceding annealing operations are performed primarily to accomplish functions other than stress relief. However, annealing is sometimes performed solely to relieve residual stresses in the workpiece. Called *stress-relief annealing,* it helps to reduce distortion and dimensional variations that might otherwise occur in the stressed parts.

27.2 MARTENSITE FORMATION IN STEEL

The iron–carbon phase diagram in Figure 6.4 indicates the phases of iron and iron carbide (cementite) present under equilibrium conditions. It assumes that cooling from high temperature is slow enough to permit austenite to decompose into a mixture of ferrite and cementite (Fe_3C) at room temperature. This decomposition reaction requires diffusion and other processes that depend on time and temperature to transform the metal into its preferred final form. However, under conditions of rapid cooling, so that the equilibrium reaction is inhibited, austenite transforms into a nonequilibrium phase called martensite. **Martensite** is a hard, brittle phase that gives steel its unique ability to be strengthened to very high levels. Our video clip on heat treatment gives an overview of the heat treatment of steel.

VIDEO CLIP ————————————————————————————————

Heat Treatment: View the segment on the iron–carbon phase diagram.

27.2.1 THE TIME-TEMPERATURE-TRANSFORMATION CURVE

The nature of the martensite transformation can best be understood using the time-temperature-transformation curve (TTT curve) for eutectoid steel, illustrated in Figure 27.1. The TTT curve shows how cooling rate affects the transformation of austenite into various possible phases. The phases can be divided between (1) alternative forms of ferrite and cementite and (2) martensite. Time is displayed (logarithmically for convenience) along the horizontal axis, and temperature is scaled on the vertical axis. The curve is interpreted by starting at time zero in the austenite region (somewhere above the A_1 temperature line for the given composition) and proceeding downward and to the right along a trajectory representing how the metal is cooled as a function of time. The TTT curve shown in the figure is for a specific composition of steel (0.80% carbon). The shape of the curve is different for other compositions.

At slow cooling rates, the trajectory proceeds through the region indicating transformation into pearlite or bainite, which are alternative forms of ferrite–carbide mixtures. Because these transformations take time, the TTT diagram shows two lines — the start and finish of the transformation as time passes, indicated for the different phase regions by the subscripts s and f, respectively. **Pearlite** is a mixture of ferrite and carbide

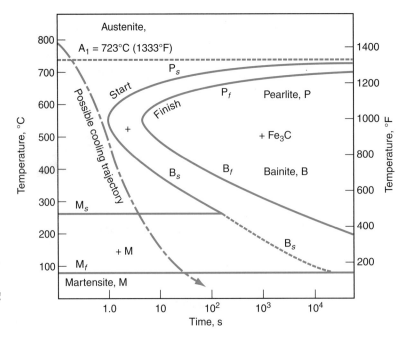

FIGURE 27.1 The TTT curve, showing the transformation of austenite into other phases as a function of time and temperature for a composition of about 0.80% C steel. The cooling trajectory shown here yields martensite.

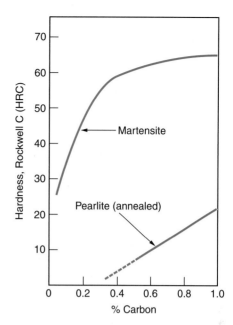

FIGURE 27.2 Hardness of plain carbon steel as a function of carbon content in (hardened) martensite and pearlite (annealed).

phases in the form of thin parallel plates. It is obtained by slow cooling from austenite, so that the cooling trajectory passes through P_s above the "nose" of the TTT curve. **Bainite** is an alternative mixture of the same phases that can be produced by initial rapid cooling to a temperature somewhat above M_s, so that the nose of the TTT curve is avoided; this is followed by much slower cooling to pass through B_s and into the ferrite–carbide region. Bainite has a needle-like or feather-like structure consisting of fine carbide regions.

If cooling occurs at a sufficiently rapid rate (indicated by the dashed line in Figure 27.1), austenite is transformed into martensite. **Martensite** is a unique phase consisting of an iron–carbon solution whose composition is the same as the austenite from which it was derived. The face-centered cubic structure of austenite is transformed into the body-centered tetragonal (BCT) structure of martensite almost instantly—without the time-dependent diffusion process needed to separate ferrite and iron carbide in the preceding transformations.

During cooling, the martensite transformation begins at a certain temperature M_s, and finishes at a lower temperature M_f, as shown in our TTT diagram. At points between these two levels, the steel is a mixture of austenite and martensite. If cooling is stopped at a temperature between the M_s and M_f lines, the austenite will transform to bainite as the time-temperature trajectory crosses the B_s threshold. The level of the M_s line is influenced by alloying elements, including carbon. In some cases, the M_s line is depressed below room temperature, making it impossible for these steels to form martensite by traditional heat-treating methods.

The extreme hardness of martensite results from the lattice strain created by carbon atoms trapped in the BCT structure, thus providing a barrier to slip. Figure 27.2 shows the significant effect that the martensite transformation has on the hardness of steel for increasing carbon contents.

27.2.2 THE HEAT TREATMENT PROCESS

The heat treatment to form martensite consists of two steps: austenitizing and quenching. These steps are often followed by tempering to produce tempered martensite. **Austenitizing** involves heating the steel to a sufficiently high temperature that it is converted

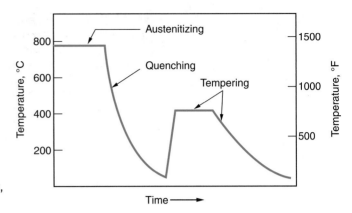

FIGURE 27.3 Typical heat treatment of steel: austenitizing, quenching, and tempering.

entirely or partially to austenite. This temperature can be determined from the phase diagram for the particular alloy composition. The transformation to austenite involves a phase change, which requires time as well as heat. Accordingly, the steel must be held at the elevated temperature for a sufficient period of time to allow the new phase to form and the required homogeneity of composition to be achieved.

The *quenching* step involves cooling the austenite rapidly enough to avoid passing through the nose of the TTT curve, as indicated in the cooling trajectory shown in Figure 27.1. The cooling rate depends on the quenching medium and the rate of heat transfer within the steel workpiece. Various quenching media are used in commercial heat treatment practice: (1) brine—salt water, usually agitated; (2) fresh water—still, not agitated; (3) still oil; and (4) air. Quenching in agitated brine provides the fastest cooling of the heated part surface, whereas air quench is the slowest. Trouble is, the more effective the quenching media is at cooling, the more likely it is to cause internal stresses, distortion, and cracks in the product.

The rate of heat transfer within the part depends largely on its mass and geometry. A large cubic shape will cool much more slowly than a small, thin sheet. The coefficient of thermal conductivity k of the particular composition is also a factor in the flow of heat in the metal. There is considerable variation in k for different grades of steel; for example, plain low carbon steel has a typical k value equal to 0.046 J/sec-mm-C (2.2 Btu/hr-in-F), whereas a highly alloyed steel might have one-third that value.

Martensite is hard and brittle. *Tempering* is a heat treatment applied to hardened steels to reduce brittleness, increase ductility and toughness, and relieve stresses in the martensite structure. It involves heating and soaking at a temperature below the austenitizing level for about 1 hour, followed by slow cooling. This results in precipitation of very fine carbide particles from the martensitic iron–carbon solution, and gradually transforms the crystal structure from BCT to BCC. This new structure is called *tempered martensite.* A slight reduction in strength and hardness accompanies the improvement in ductility and toughness. The temperature and time of the tempering treatment control the degree of softening in the hardened steel, because the change from untempered to tempered martensite involves diffusion.

Taken together, the three steps in the heat treatment of steel to form tempered martensite can be pictured as in Figure 27.3. There are two heating and cooling cycles, the first to produce martensite and the second to temper the martensite.

27.2.3 HARDENABILITY

Hardenability refers to the relative capacity of a steel to be hardened by transformation to martensite. It is a property that determines the depth below the quenched surface to

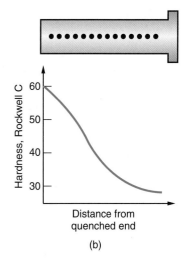

FIGURE 27.4 The Jominy end-quench test: (a) setup of the test, showing end quench of the test specimen; and (b) typical pattern of hardness readings as a function of distance from quenched end.

which the steel is hardened, or the severity of the quench required to achieve a certain hardness penetration. Steels with good hardenability can be hardened more deeply below the surface and do not require high cooling rates. Hardenability does not refer to the maximum hardness that can be attained in the steel; that depends on the carbon content.

The hardenability of a steel is increased through alloying. Alloying elements having the greatest effect are chromium, manganese, molybdenum (and nickel, to a lesser extent). The mechanism by which these alloying ingredients operate is to extend the time before the start of the austenite-to-pearlite transformation in the TTT diagram. In effect, the TTT curve is moved to the right, thus permitting slower quenching rates during quenching. Therefore, the cooling trajectory is able to follow a less hastened path to the M_s line, more easily avoiding the nose of the TTT curve.

The most common method for measuring hardenability is the ***Jominy end-quench test.*** The test involves heating a standard specimen of diameter = 25.4 mm (1.0 in) and length = 102 mm (4.0 in) into the austenite range, and then quenching one end with a stream of cold water while the specimen is supported vertically as shown in Figure 27.4 (a). The cooling rate in the test specimen decreases with increased distance from the quenched end. Hardenability is indicated by the hardness of the specimen as a function of distance from quenched end, as in Figure 27.4(b).

27.3 PRECIPITATION HARDENING

Precipitation hardening involves the formation of fine particles (precipitates) that act to block the movement of dislocations and thus strengthen and harden the metal. It is the principal heat treatment for strengthening alloys of aluminum, copper, magnesium, nickel, and other nonferrous metals. Precipitation hardening can also be used to strengthen certain steel alloys. When applied to steels, the process is called ***maraging*** (an abbreviation of martensite and aging), and the steels are called maraging steels (Section 6.2.3).

The necessary condition that determines whether an alloy system can be strengthened by precipitation hardening is the presence of a sloping solvus line, as shown in the phase diagram of Figure 27.5(a). A composition that can be precipitation hardened is one

FIGURE 27.5
Precipitation hardening:
(a) phase diagram of an
alloy system consisting of
metals A and B that can be
precipitation hardened;
and (b) heat treatment:
(1) solution treatment,
(2) quenching, and (3)
precipitation treatment.

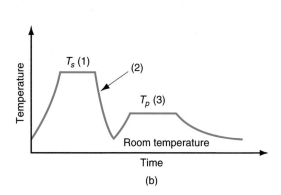

that contains two phases at room temperature, but which can be heated to a temperature that dissolves the second phase. Composition C satisfies this requirement. The heat treatment process consists of three steps, illustrated in Figure 27.5(b): (1) *solution treatment,* in which the alloy is heated to a temperature T_s above the solvus line into the alpha phase region and held for a period sufficient to dissolve the beta phase; (2) *quenching* to room temperature to create a supersaturated solid solution; and (3) *precipitation treatment,* in which the alloy is heated to a temperature T_p, below T_s, to cause precipitation of fine particles of the beta phase. This third step is called *aging,* and for this reason the whole heat treatment is sometimes called *age hardening.* However, aging can occur in some alloys at room temperature, and so the term *precipitation hardening* seems more precise for the three-step heat treatment process under discussion here. When the aging step is performed at room temperature, it is called *natural aging.* When it is accomplished at an elevated temperature, as in our figure, the term *artificial aging* is often used.

It is during the aging step that high strength and hardness are achieved in the alloy. The combination of temperature and time during the precipitation treatment (aging) is critical in bringing out the desired properties in the alloy. At higher precipitation treatment temperatures, as in Figure 27.6(a), the hardness peaks in a relatively short time; whereas at lower temperatures, as in Figure 27.6(b), more time is required to harden the alloy but its maximum hardness is likely to be greater than in the first case. As seen in the plot, continuation of the aging process results in a reduction in hardness and strength properties, called *overaging.* Its overall effect is similar to annealing.

FIGURE 27.6   Effect of
temperature and time
during precipitation
treatment (aging): (a) high
precipitation tempera-
ture; and (b) lower pre-
cipitation temperature.

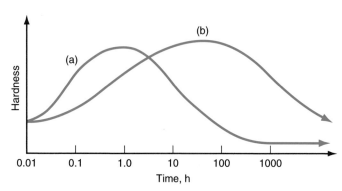

27.4 SURFACE HARDENING

Surface hardening refers to any of several thermochemical treatments applied to steels in which the composition of the part surface is altered by addition of carbon, nitrogen, or other elements. The most common treatments are carburizing, nitriding, and carbonitriding. These processes are commonly applied to low carbon steel parts to achieve a hard, wear-resistant outer shell while retaining a tough inner core. The term *case hardening* is often used for these treatments.

Carburizing is the most common surface-hardening treatment. It involves heating a part of low carbon steel in the presence of a carbon-rich environment so that C is diffused into the surface. In effect the surface is converted to high carbon steel, capable of higher hardness than the low-C core. The carbon-rich environment can be created in several ways. One method involves the use of carbonaceous materials such as charcoal or coke packed in a closed container with the parts. This process, called *pack carburizing,* produces a relatively thick layer on the part surface, ranging from around 0.6 to 4 mm (0.025 to 0.150 in). Another method, called *gas carburizing,* uses hydrocarbon fuels such as propane (C_3H_8) inside a sealed furnace to diffuse carbon into the parts. The case thickness in this treatment is thin, 0.13 to 0.75 mm (0.005 to 0.030 in). Another process is *liquid carburizing,* which employs a molten salt bath containing sodium cyanide (NaCN), barium chloride ($BaCl_2$), and other compounds to diffuse carbon into the steel. This process produces surface layer thicknesses generally between those of the other two treatments. Typical carburizing temperatures are 875° to 925°C (1600° to 1700°F), well into the austenite range.

Carburizing followed by quenching produces a case hardness of around HRC=60. However, because the internal regions of the part consist of low carbon steel, and its hardenability is low, it is unaffected by the quench and remains relatively tough and ductile to withstand impact and fatigue stresses.

Nitriding is a treatment in which nitrogen is diffused into the surfaces of special alloy steels to produce a thin hard casing without quenching. To be most effective, the steel must contain certain alloying ingredients such as aluminum (0.85% to 1.5%) or chromium (5% or more). These elements form nitride compounds that precipitate as very fine particles in the casing to harden the steel. Nitriding methods include: *gas nitriding,* in which the steel parts are heated in an atmosphere of ammonia (NH_3) or other nitrogen-rich gas mixture; and *liquid nitriding,* in which the parts are dipped in molten cyanide salt baths. Both processes are carried out at around 500°C (950°F). Case thicknesses range as low as 0.025 mm (0.001 in) and up to around 0.5 mm (0.020 in), with hardnesses up to HRC 70.

As its name suggests, *carbonitriding* is a treatment in which both carbon and nitrogen are absorbed into the steel surface, usually by heating in a furnace containing carbon and ammonia. Case thicknesses are usually 0.07 to 0.5 mm (0.003 to 0.020 in), with hardnesses comparable with those of the other two treatments.

Two additional surface-hardening treatments diffuse chromium and boron, respectively, into the steel to produce casings that are typically only 0.025 to 0.05 mm (0.001 to 0.002 in) thick. *Chromizing* requires higher temperatures and longer treatment times than the preceding surface-hardening treatments, but the resulting casing is not only hard and wear resistant, it is also heat and corrosion resistant. The process is usually applied to low carbon steels. Techniques for diffusing chromium into the surface include: packing the steel parts in chromium-rich powders or granules, dipping in a molten salt bath containing Cr and Cr salts, and chemical vapor deposition (Section 28.5.2).

Boronizing is performed on tool steels, nickel- and cobalt-based alloys, and cast irons, in addition to plain carbon steels, using powders, salts, or gas atmospheres containing boron. The process results in a thin casing with high abrasion resistance and low coefficient of friction. Casing hardnesses reach 70 HRC. When boronizing is used on low carbon and low alloy steels, corrosion resistance is also improved.

27.5 HEAT TREATMENT METHODS AND FACILITIES

Most heat treatment operations are performed in furnaces. In addition, other techniques can be used to selectively heat only the work surface or a portion of the work surface. Thus, we divide this section into two categories of methods and facilities for heat treatment [11]: (1) furnaces and (2) selective surface-hardening methods.

It should be mentioned that some of the equipment described here is used for other processes in addition to heat treatment; these include melting metals for casting (Section 11.4.1); heating before warm and hot working (Section 18.3); brazing, soldering, and adhesive curing (Chapter 31); and semiconductor processing (Chapter 34).

27.5.1 FURNACES FOR HEAT TREATMENT

Furnaces vary greatly in heating technology, size and capacity, construction, and atmosphere control. They usually heat the workparts by a combination of radiation, convection, and conduction. Heating technologies divide between fuel-fired and electric heating. *Fuel-fired furnaces* are normally *direct-fired,* which means that the work is exposed directly to the combustion products. Fuels include gases (such as natural gas or propane) and oils that can be atomized (such as diesel fuel and fuel oil). The chemistry of the combustion products can be controlled by adjusting the fuel-air or fuel-oxygen mixture to minimize scaling (oxide formation) on the work surface. *Electric furnaces* use electric resistance for heating; they are cleaner, quieter, and provide more uniform heating, but they are more expensive to purchase and operate.

A conventional furnace is an enclosure designed to resist heat loss and accommodate the size of the work to be processed. Furnaces are classified as batch or continuous. *Batch furnaces* are simpler, basically consisting of a heating system in an insulated chamber, with a door for loading and unloading the work. *Continuous furnaces* are generally used for higher production rates and provide a means of moving the work through the interior of the heating chamber.

Special atmospheres are required in certain heat treatment operations, such as some of the surface hardening treatments we have discussed. These atmospheres include carbon- and nitrogen-rich environments for diffusion of these elements into the surface of the work. Atmosphere control is desirable in conventional heat treatment operations to avoid excessive oxidation or decarburization.

Other furnace types include salt bath and fluidized bed. *Salt bath furnaces* consist of vessels containing molten salts of chlorides and/or nitrates. Parts to be treated are immersed in the molten media. *Fluidized bed furnaces* have a container in which small inert particles are suspended by a high-velocity stream of hot gas. Under proper conditions, the aggregate behavior of the particles is fluid-like; thus, rapid heating of parts immersed in the particle bed occurs.

27.5.2 SELECTIVE SURFACE-HARDENING METHODS

These methods heat only the surface of the work, or local areas of the work surface. They differ from surface-hardening methods (Section 27.4) in that no chemical changes occur. Here the treatments are only thermal. The selective surface hardening methods include

FIGURE 27.7 Typical induction heating setup. High-frequency alternating current in a coil induces current in the workpart to effect heating.

flame hardening, induction hardening, high-frequency resistance heating, electron beam heating, and laser beam heating.

Flame hardening involves heating the work surface by means of one or more torches followed by rapid quenching. As a hardening process, it is applied to carbon and alloy steels, tool steels, and cast irons. Fuels include acetylene (C_2H_2), propane (C_3H_8), and other gases. The name flame hardening invokes images of a highly manual operation with general lack of control over the results; however, the process can be set up to include temperature control, fixtures for positioning the work relative to the flame, and indexing devices that operate on a precise cycle time, all of which provide close control over the resulting heat treatment. It is fast and versatile, lending itself to high production as well as big components such as large gears that exceed the capacity of furnaces.

Induction heating involves application of electromagnetically induced energy supplied by an induction coil to an electrically conductive workpart. Induction heating is widely used in industry for processes such as brazing, soldering, adhesive curing, and various heat treatments. When used for hardening steel, quenching follows heating. A typical setup is illustrated in Figure 27.7. The induction heating coil carries a high-frequency alternating current that induces a current in the encircled workpart to effect heating. The surface, a portion of the surface, or the entire mass of the part can be heated by the process. Induction heating provides a fast and efficient method of heating any electrically conductive material. Heating cycle times are short, so the process lends itself to high production as well as midrange production.

High-frequency (HF) resistance heating is used to harden specific areas of steel work surfaces by application of localized resistance heating at high frequency (400 kHz typical). A typical setup is shown in Figure 27.8. The apparatus consists of a water-cooled proximity

FIGURE 27.8 Typical setup for high-frequency resistance heating.

conductor located over the area to be heated. Contacts are attached to the workpart at the outer edges of the area. When the HF current is applied, the region beneath the proximity conductor is heated rapidly to high temperature—heating to the austenite range typically requires less than a second. When the power is turned off, the area, usually a narrow line as in our figure, is quenched by heat transfer to the surrounding metal. Depth of the treated area is around 0.63 mm (0.025 in); hardness depends on carbon content of the steel and can range up to 60 HRC [11].

Electron beam (EB) heating involves localized surface hardening of steel in which the electron beam is focused onto a small area, resulting in rapid heat buildup. Austenitizing temperatures can often be achieved in less than a second. When the directed beam is removed, the heated area is immediately quenched and hardened by heat transfer to the surrounding cold metal. A disadvantage of EB heating is that best results are achieved when the process is performed in a vacuum. A special vacuum chamber is needed, and time is required to draw the vacuum, thus slowing production rates.

Laser beam (LB) heating uses a high-intensity beam of coherent light focused on a small area. The beam is usually moved along a defined path on the work surface, causing heating of the steel into the austenite region. When the beam is moved, the area is immediately quenched by heat conduction to the surrounding metal. ***Laser*** is an acronym for *l*ight *a*mplification by *s*timulated *e*mission of *r*adiation. The advantage of LB over EB heating is that laser beams do not require a vacuum to achieve best results. Energy density levels in EB and LB heating are lower than in cutting or welding.

REFERENCES

[1] ***ASM Handbook.*** Vol. 4, ***Heat Treating.*** ASM International, Materials Park, Ohio, 1991.

[2] Babu, S. S., and Totten, G. E. ***Steel Heat Treatment Handbook,*** 2nd ed. CRC Taylor & Francis, Boca Raton, Florida, 2006.

[3] Brick, R. M., Pense, A. W., and Gordon, R. B. ***Structure and Properties of Engineering Materials.*** 4th ed. McGraw-Hill, New York, 1977.

[4] Chandler, H. (ed.). ***Heat Treater's Guide: Practices and Procedures for Irons and Steels.*** ASM International, Materials Park, Ohio, 1995.

[5] Chandler, H. (ed.). ***Heat Treater's Guide: Practices and Procedures for Nonferrous Alloys.*** ASM International, Materials Park, Ohio, 1996.

[6] Dossett, J. L., and Boyer, H. E. ***Practical Heat Treating,*** 2nd ed. 2006.

[7] Flinn, R. A., and Trojan, P. K. ***Engineering Materials and Their Applications.*** 5th ed. John Wiley & Sons, New York, 1995.

[8] Guy, A. G., and Hren, J. J. ***Elements of Physical Metallurgy.*** 3rd ed. Addison-Wesley, Reading, Massachusetts, 1974.

[9] Ostwald, P. F., and Munoz, J. ***Manufacturing Processes and Systems.*** 9th ed. John Wiley & Sons, New York, 1997.

[10] Vaccari, J. A."Fundamentals of heat treating." Special Report 737, ***American Machinist.*** September 1981, pp. 185–200.

[11] Wick, C. and Veilleux, R. F. (eds.). ***Tool and Manufacturing Engineers Handbook.*** 4th ed. Vol. 3, ***Materials, Finishing, and Coating.*** Section 2: Heat Treatment. Society of Manufacturing Engineers, Dearborn, Michigan, 1985.

REVIEW QUESTIONS

27.1. Why are metals heat treated?

27.2. Identify the important reasons why metals are annealed.

27.3. What is the most important heat treatment for hardening steels?

27.4. What is the mechanism by which carbon strengthens steel during heat treatment?

27.5. What information is conveyed by the TTT curve?

27.6. What function is served by tempering?

27.7. Define hardenability.

27.8. Name some of the elements that have the greatest effect on the hardenability of steel.

27.9. Indicate how the hardenability alloying elements in steel affect the TTT curve.

27.10. Define precipitation hardening.

27.11. How does carburizing work?

27.12. Identify the selective surface-hardening methods.

27.13. (**Video**) List three properties of ferrite at room temperature.

27.14. (**Video**) How does austenite differ from ferrite?

MULTIPLE CHOICE QUIZ

There are 12 correct answers in the following multiple choice questions (some questions have multiple answers that are correct). To attain a perfect score on the quiz, all correct answers must be given. Each correct answer is worth 1 point. Each omitted answer or wrong answer reduces the score by 1 point, and each additional answer beyond the correct number of answers reduces the score by 1 point. Percentage score on the quiz is based on the total number of correct answers.

27.1. Which of the following are the usual objectives of heat treatment (three best answers): (a) increase hardness, (b) increase melting temperature, (c) increase recrystallization temperature, (d) reduce brittleness, (e) reduce density, and (f) relieve stresses?

27.2. Of the following quenching media, which one produces the most rapid cooling rate: (a) air, (b) brine, (c) oil, or (d) pure water?

27.3. On which one of the following metals is the treatment called austenitizing be performed: (a) aluminum alloys, (b) brass, (c) copper alloys, or (d) steel?

27.4. The treatment in which the brittleness of martensite is reduced is called which one of the following: (a) aging, (b) annealing, (c) austenitizing, (d) normalizing, (e) quenching, or (f) tempering?

27.5. The Jominy end-quench test is designed to indicate which one of the following: (a) cooling rate, (b) ductility, (c) hardenability, (d) hardness, or (e) strength?

27.6. In precipitation hardening, the hardening and strengthening of the metal occurs in which one of the following steps: (a) aging, (b) quenching, or (c) solution treatment?

27.7. Which one of the following surface-hardening treatments is the most common: (a) boronizing, (b) carbonitriding, (c) carburizing, (d) chromizing, or (e) nitriding?

27.8. Which of the following are selective surface-hardening methods (three correct answers): (a) austenitizing, (b) electron beam heating, (c) fluidized bed furnaces, (d) induction heating, (e) laser beam heating, and (f) vacuum furnaces?

28 SURFACE PROCESSING OPERATIONS

Chapter Contents

The processes discussed in this chapter operate on the surfaces of parts and/or products. The major categories of surface processing operations are (1) cleaning, (2) surface treatments, and (3) coating and thin film deposition. Cleaning refers to industrial cleaning processes that remove soils and contaminants that result from previous processing or the factory environment. They include both chemical and mechanical cleaning methods. Surface treatments are mechanical and physical operations that alter the part surface in some way, such as improving its finish or impregnating it with atoms of a foreign material to change its chemistry and physical properties.

Coating and thin film deposition include various processes that apply a layer of material to a surface. Products made of metal are almost always coated by electroplating (e.g., chrome plating), painting, or other process. Principal reasons for coating a metal are to (1) provide corrosion protection, (2) enhance product appearance (e.g., providing a specified color or texture), (3) increase wear resistance and/or reduce friction of the surface, (4) increase electrical conductivity, (5) increase electrical resistance, (6) prepare a metallic surface for subsequent processing, and (7) rebuild surfaces worn or eroded during service. Nonmetallic materials are also sometimes coated. Examples include (1) plastic parts coated to give them a metallic appearance; (2) antireflection coatings on optical glass lenses; and (3) certain coating and deposition processes used in the fabrication of semiconductor chips (Chapter 34) and printed circuit boards (Chapter 35). In all cases, good adhesion must be achieved between coating and substrate, and for this to occur the substrate surface must be very clean.

28.1 INDUSTRIAL CLEANING PROCESSES

Most workparts must be cleaned one or more times during their manufacturing sequence. Chemical and/or mechanical processes are used to accomplish this cleaning. Chemical cleaning methods use chemicals to remove unwanted oils and soils from the workpiece surface. Mechanical cleaning involves removal of substances from a surface by mechanical operations of various kinds. These operations often serve other functions such as removing burrs, improving smoothness, adding luster, and enhancing surface properties.

28.1.1 CHEMICAL CLEANING

A typical surface is covered with various films, oils, dirt, and other contaminants (Section 5.3.1). Although some of these substances may operate in a beneficial way (such as the oxide film on aluminum), it is usually desirable to remove contaminants from the surface. In this section, we discuss some general considerations related to cleaning, and we survey the principal chemical cleaning processes used in industry.

Some of the important reasons why manufactured parts (and products) must be cleaned are (1) to prepare the surface for subsequent industrial processing, such as a coating application or adhesive bonding; (2) to improve hygiene conditions for workers and customers; (3) to remove contaminants that might chemically react with the surface; and (4) to enhance appearance and performance of the product.

General Considerations in Cleaning There is no single cleaning method that can be used for all cleaning tasks. Just as various soaps and detergents are required for different household jobs (laundry, dishwashing, pot scrubbing, bathtub cleaning, and so forth), various cleaning methods are also needed to solve different cleaning problems in industry. Important factors in selecting a cleaning method are (1) the contaminant to be removed, (2) degree of cleanliness required, (3) substrate material to be cleaned, (4) purpose of the cleaning, (5) environmental and safety factors, (6) size and geometry of the part, and (7) production and cost requirements.

Various kinds of contaminants build up on part surfaces, either due to previous processing or the factory environment. To select the best cleaning method, one must first identify what must be cleaned. Surface contaminants found in the factory usually divide into one of the following categories: (1) oil and grease, which includes lubricants used in metalworking; (2) solid particles such as metal chips, abrasive grits, shop dirt, dust, and similar materials; (3) buffing and polishing compounds; and (4) oxide films, rust, and scale.

Degree of cleanliness refers to the amount of contaminant remaining after a given cleaning operation. Parts being prepared to accept a coating (e.g., paint, metallic film) or adhesive must be very clean; otherwise, adhesion of the coated material is jeopardized. In other cases, it may be desirable for the cleaning operation to leave a residue on the part surface for corrosion protection during storage, in effect replacing one contaminant on the surface by another that is beneficial. Degree of cleanliness is often difficult to measure in a quantifiable way. A simple test is a *wiping method,* in which the surface is wiped with a clean white cloth, and the amount of soil absorbed by the cloth is observed. It is a nonquantitative but easy test to use.

The substrate material must be considered in selecting a cleaning method, so that damaging reactions are not caused by the cleaning chemicals. To cite several examples: aluminum is dissolved by most acids and alkalis; magnesium is attacked by many acids; copper is attacked by oxidizing acids (e.g., nitric acid); steels are resistant to alkalis but react with virtually all acids.

Some cleaning methods are appropriate to prepare the surface for painting, while others are better for plating. Environmental protection and worker safety are becoming increasingly important in industrial processes. Cleaning methods and the associated chemicals should be selected to avoid pollution and health hazards.

Chemical Cleaning Processes   Chemical cleaning uses various types of chemicals to effect contaminant removal from the surface. The major chemical cleaning methods are (1) alkaline cleaning, (2) emulsion cleaning, (3) solvent cleaning, (4) acid cleaning, and (5) ultrasonic cleaning. In some cases, chemical action is augmented by other energy forms; for example, ultrasonic cleaning uses high-frequency mechanical vibrations combined with chemical cleaning. In the following paragraphs, we review these chemical methods.

Alkaline cleaning is the most widely used industrial cleaning method. As its name indicates, it employs an alkali to remove oils, grease, wax, and various types of particles (metal chips, silica, carbon, and light scale) from a metallic surface. Alkaline cleaning solutions consist of low-cost, water-soluble salts such as sodium and potassium hydroxide (NaOH, KOH), sodium carbonate (Na_2CO_3), borax ($Na_2B_4O_7$), phosphates and silicates of sodium and potassium, combined with dispersants and surfactants in water. The cleaning method is commonly by immersion or spraying, usually at temperatures of 50°C to 95°C (120°F–200°F). Following application of the alkaline solution, a water rinse is used to remove the alkali residue. Metal surfaces cleaned by alkaline solutions are typically electroplated or conversion coated.

Electrolytic cleaning, also called *electrocleaning,* is a related process in which a 3-V to 12-V direct current is applied to an alkaline cleaning solution. The electrolytic action results in the generation of gas bubbles at the part surface, causing a scrubbing action that aids in removal of tenacious dirt films.

Emulsion cleaning uses organic solvents (oils) dispersed in an aqueous solution. The use of suitable emulsifiers (soaps) results in a two-phase cleaning fluid (oil-in-water), which functions by dissolving or emulsifying the soils on the part surface. The process can be used on either metal or nonmetallic parts. Emulsion cleaning must be followed by alkaline cleaning to eliminate all residues of the organic solvent prior to plating.

In *solvent cleaning,* organic soils such as oil and grease are removed from a metallic surface by means of chemicals that dissolve the soils. Common application techniques include hand-wiping, immersion, spraying, and vapor degreasing. *Vapor degreasing* uses hot vapors of solvents to dissolve and remove oil and grease on part surfaces. The common solvents include trichlorethylene (C_2HCl_3), methylene chloride (CH_2Cl_2), and perchlorethylene (C_2Cl_4), all of which have relatively low boiling points.[1] The vapor degreasing process consists of heating the liquid solvent to its boiling point in a container to produce hot vapors. Parts to be cleaned are then introduced into the vapor, which condenses on the relatively cold part surfaces, dissolving the contaminants and dripping to the bottom of the container. Condensing coils near the top of the container prevent any vapors from escaping the container into the surrounding atmosphere. This is important because these solvents are classified as hazardous air pollutants under the 1992 Clean Air Act [10].

Acid cleaning removes oils and light oxides from metal surfaces by soaking, spraying, or manual brushing or wiping. The process is carried out at ambient or elevated temperatures. Common cleaning fluids are acid solutions combined with water-miscible solvents, wetting and emulsifying agents. Cleaning acids include hydrochloric (HCl), nitric (HNO_3), phosphoric (H_3PO_4), and sulfuric (H_2SO_4), the selection depending on the base metal and purpose of the cleaning. For example, phosphoric acid produces a light phosphate film on the metallic surface, which can be a useful preparation for painting. A closely related cleaning process is *acid pickling,* which involves a more severe treatment to remove thicker oxides, rusts, and scales; it generally results in some etching of the metallic surface, which serves to improve organic paint adhesion.

Ultrasonic cleaning combines chemical cleaning and mechanical agitation of the cleaning fluid to provide a highly effective method for removing surface contaminants. The cleaning fluid is generally an aqueous solution containing alkaline detergents. The

---

[1]The highest boiling point of the three solvents is 121°C (250°F) for C_2Cl_4.

mechanical agitation is produced by high-frequency vibrations of sufficient amplitude to cause cavitation—formation of low-pressure vapor bubbles or cavities. As the vibration wave passes a given point in the liquid, the low-pressure region is followed by a high-pressure front that implodes the cavity, thereby producing a shock wave capable of penetrating contaminant particles adhering to the work surface. This rapid cycle of cavitation and implosion occurs throughout the liquid medium, thus making ultrasonic cleaning effective even on complex and intricate internal shapes. The cleaning process is performed at frequencies between 20 and 45 kHz, and the cleaning solution is usually at an elevated temperature, typically 65°C to 85°C (150°F–190°F).

28.1.2 MECHANICAL CLEANING AND SURFACE TREATMENTS

Mechanical cleaning involves the physical removal of soils, scales, or films from the work surface of the workpart by means of abrasives or similar mechanical action. The processes used for mechanical cleaning often serve other functions in addition to cleaning, such as deburring and improving surface finish.

Blast Finishing and Shot Peening

Blast finishing uses the high-velocity impact of particulate media to clean and finish a surface. The most well known of these methods is *sand blasting*, which uses grits of sand (SiO_2) as the blasting media. Various other media are also used in blast finishing, including hard abrasives such as aluminum oxide (Al_2O_3) and silicon carbide (SiC), and soft media such as nylon beads and crushed nut shells. The media is propelled at the target surface by pressurized air or centrifugal force. In some applications, the process is performed wet, in which fine particles in a water slurry are directed under hydraulic pressure at the surface.

In *shot peening,* a high-velocity stream of small cast steel pellets (called *shot*) is directed at a metallic surface with the effect of cold working and inducing compressive stresses into the surface layers. Shot peening is used primarily to improve fatigue strength of metal parts. Its purpose is therefore different from blast finishing, although surface cleaning is accomplished as a by-product of the operation.

Tumbling and Other Mass Finishing

Tumbling, vibratory finishing, and similar operations comprise a group of finishing processes known as mass finishing methods. *Mass finishing* involves the finishing of parts in bulk by a mixing action inside a container, usually in the presence of an abrasive media. The mixing causes the parts to rub against the media and each other to achieve the desired finishing action. Mass finishing methods are used for deburring, descaling, deflashing, polishing, radiusing, burnishing, and cleaning. The parts include stampings, castings, forgings, extrusions, and machined parts. Even plastic and ceramic parts are sometimes subjected to these mass finishing operations to achieve desired finishing results. The parts processed by these methods are usually small and are therefore uneconomical to finish individually.

Mass finishing methods include tumbling, vibratory finishing, and several techniques that utilize centrifugal force. *Tumbling* (also called *barrel finishing* and *tumbling barrel finishing*) involves the use of a horizontally oriented barrel of hexagonal or octagonal cross-section in which parts are mixed by rotating the barrel at speeds of 10 to 50 rev/min. Finishing is performed by a "landslide" action of the media and parts as the barrel revolves. As pictured in Figure 28.1, the contents rise in the barrel due to rotation, followed by a tumbling down of the top layer due to gravity. This cycle of rising and tumbling occurs continuously and, over time, subjects all of the parts to the same desired finishing action. However, because only the top layer of parts is being finished at any moment, barrel finishing is a relatively slow process compared to other mass finishing methods. It often takes several hours of tumbling to complete the processing.

FIGURE 28.1 Diagram of tumbling (barrel finishing) operation showing "landslide" action of parts and abrasive media to finish the parts.

Sliding layer of media and parts

Barrel

Media and parts

Side view

Rotation of barrel

Other drawbacks of barrel finishing include high noise levels and large floor space requirements.

Vibratory finishing was introduced in the late 1950s as an alternative to tumbling. The vibrating vessel subjects all parts to agitation with the abrasive media, as opposed to only the top layer as in barrel finishing. Consequently, processing times for vibratory finishing are significantly reduced. The open tubs used in this method permit inspection of the parts during processing, and noise is reduced.

Most of the *media* in these operations are abrasive; however, some media perform nonabrasive finishing operations such as burnishing and surface hardening. The media may be natural or synthetic materials. Natural media include corundum, granite, limestone, and even hardwood. The problem with these materials is that they are generally softer (and therefore wear more rapidly) and nonuniform in size (and sometimes clog in the workparts). Synthetic media can be made with greater consistency, both in size and hardness. These materials include Al_2O_3 and SiC, compacted into a desired shape and size using a bonding material such as a polyester resin. The shapes for these media include spheres, cones, angle-cut cylinders, and other regular geometric forms, as in Figure 28.2(a). Steel is also used as a mass finishing medium in shapes such as those shown in Figure 28.2(b) for burnishing, surface hardening, and light deburring operations. The shapes shown in Figure 28.2 come in various sizes. Selection of media is based on part size and shape, as well as finishing requirements.

In most mass finishing processes, a compound is used with the media. The mass finishing *compound* is a combination of chemicals for specific functions such as cleaning, cooling, rust inhibiting (of steel parts and steel media), and enhancing brightness and color of the parts (especially in burnishing).

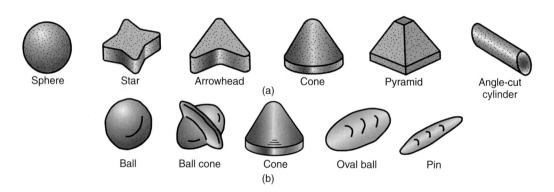

Sphere   Star   Arrowhead   Cone   Pyramid   Angle-cut cylinder

(a)

Ball   Ball cone   Cone   Oval ball   Pin

(b)

FIGURE 28.2 Typical preformed media shapes used in mass finishing operations: (a) abrasive media for finishing, and (b) steel media for burnishing.

28.2   DIFFUSION AND ION IMPLANTATION

In this section we discuss two processes in which the surface of a substrate is impregnated with foreign atoms that alter its chemistry and properties.

28.2.1   DIFFUSION

Diffusion involves the alteration of surface layers of a material by diffusing atoms of a different material (usually an element) into the surface (Section 4.3). The diffusion process impregnates the surface layers of the substrate with the foreign element, but the surface still contains a high proportion of substrate material. A typical profile of composition as a function of depth below the surface for a diffusion coated metal part is illustrated in Figure 28.3. The characteristic of a diffusion impregnated surface is that the diffused element has a maximum percentage at the surface and rapidly declines with distance below the surface. The diffusion process has important applications in metallurgy and semiconductor manufacture.

In metallurgical applications, diffusion is used to alter the surface chemistry of metals in a number of processes and treatments. One important example is surface hardening, typified by *carburizing, nitriding, carbonitriding, chromizing,* and *boronizing* (Section 27.4). In these treatments, one or more elements (C and/or Ni, Cr, or Bo) are diffused into the surface of iron or steel.

There are other diffusion processes in which corrosion resistance and/or high-temperature oxidation resistance are main objectives. Aluminizing and siliconizing are important examples. *Aluminizing,* also known as *calorizing,* involves diffusion of aluminum into carbon steel, alloy steels, and alloys of nickel and cobalt. The treatment is accomplished by either (1) *pack diffusion,* in which workparts are packed with Al powders and baked at high temperature to create the diffusion layer; or (2) a *slurry method,* in which the workparts are dipped or sprayed with a mixture of Al powders and binders, then dried and baked.

Siliconizing is a treatment of steel in which silicon is diffused into the part surface to create a layer with good corrosion and wear resistance and moderate heat resistance. The treatment is carried out by heating the work in powders of silicon carbide (SiC) in an atmosphere containing vapors of silicon tetrachloride ($SiCl_4$). Siliconizing is less common than aluminizing.

FIGURE 28.3   Characteristic profile of diffused element as a function of distance below surface in diffusion. The plot given here is for carbon diffused into iron. (Source: [6].)

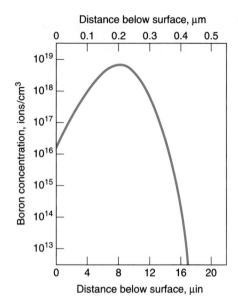

FIGURE 28.4 Profile of surface chemistry as treated by ion implantation. (Source: [17].) Shown here is a typical plot for boron implanted in silicon. Note the difference in profile shape and depth of altered layer compared to diffusion in Figure 28.3.

Semiconductor Applications In semiconductor processing, diffusion of an impurity element into the surface of a silicon chip is used to change the electrical properties at the surface to create devices such as transistors and diodes. We examine how diffusion is used to accomplish this *doping*, as it is called, and other semiconductor processes in Chapter 34.

28.2.2 ION IMPLANTATION

Ion implantation is an alternative to diffusion when the latter method is not feasible because of the high temperatures required. The ion implantation process involves embedding atoms of one (or more) foreign element(s) into a substrate surface using a high-energy beam of ionized particles. The result is an alteration of the chemical and physical properties of the layers near the substrate surface. Penetration of atoms produces a much thinner altered layer than diffusion, as indicated by a comparison of Figures 28.3 and 28.4. Also, the concentration profile of the impregnated element is quite different from the characteristic diffusion profile.

Advantages of ion implantation include (1) low-temperature processing, (2) good control and reproducibility of penetration depth of impurities, and (3) solubility limits can be exceeded without precipitation of excess atoms. Ion implantation finds some of its applications as a substitute for certain coating processes, where its advantages include (4) no problems with waste disposal as in electroplating and many coating processes, and (5) no discontinuity between coating and substrate. Principal applications of ion implantation are in modifying metal surfaces to improve properties and fabrication of semiconductor devices.

28.3 PLATING AND RELATED PROCESSES

Plating involves the coating of a thin metallic layer onto the surface of a substrate material. The substrate is usually metallic, although methods are available to plate plastic and ceramic parts. The most familiar and widely used plating technology is electroplating.

28.3.1 ELECTROPLATING

Electroplating, also known as *electrochemical plating,* is an electrolytic process (Section 4.5) in which metal ions in an electrolyte solution are deposited onto a cathode workpart. The setup is shown in Figure 28.5. The anode is generally made of the metal being plated and thus serves as the source of the plate metal. Direct current from an external power supply is passed between the anode and the cathode. The electrolyte is an aqueous solution of acids, bases, or salts; it conducts electric current by the movement of plate metal ions in solution. For optimum results, parts must be chemically cleaned just prior to electroplating.

Principles of Electroplating   Electrochemical plating is based on Faraday's two physical laws. Briefly for our purposes, the laws state: (1) the mass of a substance liberated in electrolysis is proportional to the quantity of electricity passed through the cell; and (2) the mass of the material liberated is proportional to its electrochemical equivalent (ratio of atomic weight to valence). The effects can be summarized in the equation

$$V = CIt \qquad (28.1)$$

where V = volume of metal plated, mm^3 (in^3); C = plating constant, which depends on electrochemical equivalent and density, mm^3/amp-s (in^3/amp-min); I = current, amps; and t = time during which current is applied, s (min). The product It (current × time) is the electrical charge passed in the cell, and the value of C indicates the amount of plating material deposited onto the cathodic workpart per electrical charge.

For most plating metals, not all of the electrical energy in the process is used for deposition; some energy may be consumed in other reactions, such as the liberation of hydrogen at the cathode. This reduces the amount of metal plated. The actual amount of metal deposited on the cathode (workpart) divided by the theoretical amount given by Eq. (28.1) is called the *cathode efficiency.* Taking the cathode efficiency into account, a more realistic equation for determining the volume of metal plated is

$$V = ECIt \qquad (28.2)$$

where E = cathode efficiency, and the other terms are defined as before. Typical values of cathode efficiency E and plating constant C for different metals are presented in Table 28.1. The average plating thickness can be determined from the following:

$$d = \frac{V}{A} \qquad (28.3)$$

FIGURE 28.5   Setup for electroplating.

TABLE 28.1   Typical cathode efficiencies in electroplating and values of plating constant C.

Plate Metal[a]	Electrolyte	Cathode Efficiency (%)	Plating Constant C[a] mm³/amp-s	in³/amp-min
Cadmium (2)	Cyanide	90	6.73×10^{-2}	2.47×10^{-4}
Chromium (3)	Chromium-acid-sulfate	15	2.50×10^{-2}	0.92×10^{-4}
Copper (1)	Cyanide	98	7.35×10^{-2}	2.69×10^{-4}
Gold (1)	Cyanide	80	10.6×10^{-2}	3.87×10^{-4}
Nickel (2)	Acid sulfate	95	3.42×10^{-2}	1.25×10^{-4}
Silver (1)	Cyanide	100	10.7×10^{-2}	3.90×10^{-4}
Tin (4)	Acid sulfate	90	4.21×10^{-2}	1.54×10^{-4}
Zinc (2)	Chloride	95	4.75×10^{-2}	1.74×10^{-4}

Compiled from [17].

[a]Most common valence given in parenthesis ( ); this is the value assumed in determining the plating constant C. For a different valence, compute the new C by multiplying C value in the table by the most common valence and then dividing by the new valence.

where d = plating depth or thickness, mm (in); V = volume of plate metal from Eq. (28.2); and A = surface area of plated part, mm^2 (in^2).

**Example 28.1
Electroplating**

A steel part with surface area $A = 125 \text{ cm}^2$ is to be nickel plated. What average plating thickness will result if 12 amps are applied for 15 min in an acid sulfate electrolyte bath?

Solution: From Table 28.1, the cathode efficiency for nickel is $E = 0.95$ and the plating constant $C = 3.42(10^{-2})$ mm³/amp-s. Using Eq. (28.2), the total amount of plating metal deposited onto the part surface in 15 min is given by

$$V = 0.95(3.42 \times 10^{-2})(12)(15)(60) = 350.9 \text{ mm}^3$$

This is spread across an area $A = 125 \text{ cm}^2 = 12{,}500 \text{ mm}^2$, so the average plate thickness is

$$d = \frac{350.9}{12500} = \mathbf{0.028 \text{ mm}}$$

Methods and Applications   A variety of equipment are available for electroplating, the choice depending on part size and geometry, throughput requirements, and plating metal. The principal methods are (1) barrel plating, (2) rack plating, and (3) strip plating. ***Barrel plating*** is performed in rotating barrels that are oriented either horizontally or at an oblique angle (35°). The method is suited to the plating of many small parts in a batch. Electrical contact is maintained through the tumbling action of the parts themselves and by means of an externally connected conductor that projects into the barrel. There are limitations to barrel plating; the tumbling action inherent in the process may damage soft metal parts, threaded components, parts requiring good finishes, and heavy parts with sharp edges.

   Rack plating is used for parts that are too large, heavy, or complex for barrel plating. The racks are made of heavy-gauge copper wire, formed into suitable shapes for holding the parts and conducting current to them. The racks are fabricated so that workparts can be hung on hooks, or held by clips, or loaded into baskets. To avoid plating of the copper itself, the racks are covered with insulation except in locations where part contact occurs. The racks containing the parts are moved through a sequence of tanks that perform the electroplating operation. ***Strip plating*** is a high-production method in which the work consists of a continuous strip that is pulled through the plating solution by means of a take-up reel. Plated wire is an example of a suitable application. Small sheet-metal parts held in a long strip can also be plated by this method. The process can be set up so that only specific regions of the parts are plated, for example, contact points plated with gold on electrical connectors.

Common coating metals in electroplating include zinc, nickel, tin, copper, and chromium. Steel is the most common substrate metal. Precious metals (gold, silver, platinum) are plated on jewelry. Gold is also used for electrical contacts.

Zinc-plated steel products include fasteners, wire goods, electric switch boxes, and various sheet-metal parts. The zinc coating serves as a sacrificial barrier to the corrosion of the steel beneath. An alternative process for coating zinc onto steel is galvanizing (Section 28.3.4). *Nickel plating* is used for corrosion resistance and decorative purposes over steel, brass, zinc die castings, and other metals. Applications include automotive trim and other consumer goods. Nickel is also used as a base coat under a much thinner chrome plate. *Tin plate* is still widely used for corrosion protection in "tin cans" and other food containers. Tin plate is also used to improve solderability of electrical components.

Copper has several important applications as a plating metal. It is widely used as a decorative coating on steel and zinc, either alone or alloyed with zinc as brass plate. It also has important plating applications in printed circuit boards (Section 35.2). Finally, copper is often plated on steel as a base beneath nickel and/or chrome plate. *Chromium plate* (popularly known as *chrome plate*) is valued for its decorative appearance and is widely used in automotive products, office furniture, and kitchen appliances. It also produces one of the hardest of all electroplated coatings, and so it is widely used for parts requiring wear resistance (e.g., hydraulic pistons and cylinders, piston rings, aircraft engine components, and thread guides in textile machinery).

28.3.2 ELECTROFORMING

This process is virtually the same as electroplating but its purpose is quite different. Electroforming involves electrolytic deposition of metal onto a pattern until the required thickness is achieved; the pattern is then removed to leave the formed part. Whereas typical plating thickness is only about 0.05 mm (0.002 in) or less, electroformed parts are often substantially thicker, so the production cycle is proportionally longer.

Patterns used in electroforming are either solid or expendable. Solid patterns have a taper or other geometry that permits removal of the electroplated part. Expendable patterns are destroyed during part removal; they are used when part shape precludes a solid pattern. Expendable patterns are either fusible or soluble. The fusible type is made of low-melting alloys, plastic, wax, or other material that can be removed by melting. When nonconductive materials are used, the pattern must be metallized to accept the electro-deposited coating. Soluble patterns are made of a material that can be readily dissolved by chemicals; for example, aluminum can be dissolved in sodium hydroxide (NaOH).

Electroformed parts are commonly fabricated of copper, nickel, and nickel cobalt alloys. Applications include fine molds for lenses, compact discs (CDs), and videodiscs (DVDs); copper foil used to produce blank printed circuit boards; and plates for embossing and printing. Molds for compact discs and videodiscs represent a demanding application because the surface details that must be imprinted on the disc are measured in μm ($1\ \mu$m = 10^{-6} m). These details are readily obtained in the mold by electroforming.

28.3.3 ELECTROLESS PLATING

Electroless plating is a plating process driven entirely by chemical reactions—no external source of electric current is required. Deposition of metal onto a part surface occurs in an aqueous solution containing ions of the desired plating metal. The process uses a reducing agent, and the workpart surface acts as a catalyst for the reaction.

The metals that can be electroless plated are limited; and for those that can be processed by this technique, the cost is generally greater than electrochemical plating. The most common electroless plating metal is nickel and certain of its alloys (Ni–Co, Ni–P, and Ni–B). Copper and, to a lesser degree, gold are also used as plating metals. Nickel plating by this process is used for applications requiring high resistance to corrosion and wear. Electroless copper plating is used to plate through holes of printed circuit boards (Section 35.2.4). Cu can also be plated onto plastic parts for decorative purposes. Advantages sometimes cited for electroless plating include (1) uniform plate thickness on complex part geometries (a problem with electroplating); (2) the process can be used on both metallic and nonmetallic substrates; and (3) no need for a DC power supply to drive the process.

28.3.4 HOT DIPPING

Hot dipping is a process in which a metal substrate is immersed in a molten bath of a second metal; upon removal, the second metal is coated onto the first. Of course, the first metal must possess a higher melting temperature than the second. The most common substrate metals are steel and iron. Zinc, aluminum, tin, and lead are the common coating metals. Hot dipping works by forming transition layers of varying alloy compositions. Next to the substrate are normally intermetallic compounds of the two metals; at the exterior are solid solution alloys consisting predominantly of the coating metal. The transition layers provide excellent adhesion of the coating.

The primary purpose of hot dipping is corrosion protection. Two mechanisms normally operate to provide this protection: (1) barrier protection—the coating simply serves as a shield for the metal beneath; and (2) sacrificial protection—the coating corrodes by a slow electrochemical process to preserve the substrate.

Hot dipping goes by different names, depending on coating metal: *galvanizing* is when zinc (Zn) is coated onto steel or iron; *aluminizing* refers to coating of aluminum (Al) onto a substrate; *tinning* is coating of tin (Sn); and *terneplate* describes the plating of lead–tin alloy onto steel. Galvanizing is by far the most important hot dipping process, dating back about 200 years. It is applied to finished steel and iron parts in a batch process; and to sheet, strip, piping, tubing, and wire in an automated continuous process. Coating thickness is typically 0.04 to 0.09 mm (0.0016–0.0035 in). Thickness is controlled largely by immersion time. Bath temperature is maintained at around 450°C (850°F).

Commercial use of aluminizing is on the rise, gradually increasing in market share relative to galvanizing. Hot-dipped aluminum coatings provide excellent corrosion protection, in some cases five times more effective than galvanizing [17]. Tin plating by hot dipping provides a nontoxic corrosion protection for steel in applications for food containers, dairy equipment, and soldering applications. Hot dipping has gradually been overtaken by electroplating as the preferred commercial method for plating of tin onto steel. Terneplating involves hot dipping of a lead–tin alloy onto steel. The alloy is predominantly lead (only 2%–15% Sn); however, tin is required to obtain satisfactory adhesion of the coating. Terneplate is the lowest cost of the coating methods for steel, but its corrosion protection is limited.

28.4 CONVERSION COATING

Conversion coating refers to a family of processes in which a thin film of oxide, phosphate, or chromate is formed on a metallic surface by chemical or electrochemical reaction. Immersion and spraying are the two common methods of exposing the metal surface to the reacting chemicals. The common metals treated by conversion coating are steel (including galvanized steel), zinc, and aluminum. However, nearly any metal product can benefit from the treatment. The important reasons for using a conversion coating process are (1) to provide corrosion protection, (2) to prepare the surface for painting, (3) to increase wear resistance, (4) to permit the surface to better hold lubricants for metal forming processes, (5) to increase electrical resistance of surface, (6) to provide a decorative finish, and (7) for part identification [17].

Conversion coating processes divide into two categories: (1) chemical treatments, which involve a chemical reaction only, and (2) anodizing, which consists of an electro-chemical reaction to produce an oxide coating (anodize is a contraction of ***anodic oxidize***).

28.4.1 CHEMICAL CONVERSION COATINGS

These processes expose the base metal to certain chemicals that form thin, nonmetallic surface films. Similar reactions occur in nature; the oxidation of iron and aluminum are examples. Whereas rusting is progressively destructive of iron, formation of a thin Al_2O_3 coating on aluminum protects the base metal. It is the purpose of these chemical conversion treatments to accomplish the latter effect. The two main processes are phosphate and chromate coating.

Phosphate coating transforms the base metal surface into a protective phosphate film by exposure to solutions of certain phosphate salts (e.g., Zn, Mg, and Ca) together with dilute phosphoric acid (H_3PO_4). The coatings range in thickness from 0.0025 to 0.05 mm (0.0001–0.002 in). The most common base metals are zinc and steel, including galvanized steel. The phosphate coating serves as a useful preparation for painting in the automotive and heavy appliance industries.

Chromate coating converts the base metal into various forms of chromate films using aqueous solutions of chromic acid, chromate salts, and other chemicals. Metals treated by this method include aluminum, cadmium, copper, magnesium, and zinc (and their alloys). Immersion of the base part is the common method of application. Chromate conversion coatings are somewhat thinner than phosphate, typically less than 0.0025 mm (0.0001 in). Usual reasons for chromate coating are (1) corrosion protection, (2) base for painting, and (3) decorative purposes. Chromate coatings can be clear or colorful; available colors include olive drab, bronze, yellow, or bright blue.

28.4.2 ANODIZING

Although the previous processes are normally performed without electrolysis, anodizing is an electrolytic treatment that produces a stable oxide layer on a metallic surface. Its most common applications are with aluminum and magnesium, but it is also applied to zinc, titanium, and other less common metals. Anodized coatings are used primarily for decorative purposes; they also provide corrosion protection.

It is instructive to compare anodizing to electroplating, since they are both electro-lytic processes. Two differences stand out. (1) In electrochemical plating, the workpart to be coated is the cathode in the reaction. By contrast, in anodizing, the work is the anode, whereas the processing tank is cathodic. (2) In electroplating, the coating is grown by adhesion of ions of a second metal to the base metal surface. In anodizing, the surface coating is formed through chemical reaction of the substrate metal into an oxide layer.

Anodized coatings usually range in thickness between 0.0025 and 0.075 mm (0.0001 and 0.003 in). Dyes can be incorporated into the anodizing process to create a wide variety of colors; this is especially common in aluminum anodizing. Very thick coatings up to 0.25 mm (0.010 in) can also be formed on aluminum by a special process called **hard anodizing;** these coatings are noted for high resistance to wear and corrosion.

28.5 VAPOR DEPOSITION PROCESSES

The vapor deposition processes form a thin coating on a substrate by either condensation or chemical reaction of a gas onto the surface of the substrate. The two categories of processes that fall under this heading are physical vapor deposition and chemical vapor deposition.

28.5.1 PHYSICAL VAPOR DEPOSITION

Physical vapor deposition (PVD) is a group of thin film processes in which a material is converted into its vapor phase in a vacuum chamber and condensed onto a substrate surface as a very thin layer. PVD can be used to apply a wide variety of coating materials: metals, alloys, ceramics and other inorganic compounds, and even certain polymers. Possible substrates include metals, glass, and plastics. Thus, PVD represents a versatile coating technology, applicable to an almost unlimited combination of coating substances and substrate materials.

Applications of PVD include thin decorative coatings on plastic and metal parts such as trophies, toys, pens and pencils, watchcases, and interior trim in automobiles. The coatings are thin films of aluminum (around 150 nm) coated with clear lacquer to give a high gloss silver or chrome appearance. Another use of PVD is to apply antireflection coatings of magnesium fluoride (MgF_2) onto optical lenses. PVD is applied in the fabrication of electronic devices, principally for depositing metal to form electrical connections in integrated circuits. Finally, PVD is widely used to coat titanium nitride (TiN) onto cutting tools and plastic injection molds for wear resistance.

All physical vapor deposition processes consist of the following steps: (1) synthesis of the coating vapor, (2) vapor transport to the substrate, and (3) condensation of vapors onto the substrate surface. These steps are generally carried out inside a vacuum chamber, so evacuation of the chamber must precede the actual PVD process.

Synthesis of the coating vapor can be accomplished by any of several methods, such as electric resistance heating or ion bombardment to vaporize an existing solid (or liquid). These and other variations result in several PVD processes. They are grouped into three principal types: (1) vacuum evaporation, (2) sputtering, and (3) ion plating. Table 28.2 presents a summary of these processes.

TABLE 28.2   Summary of physical vapor deposition (PVD) processes.

PVD Process	Features and Comparisons	Coating Materials
Vacuum evaporation	Equipment is relatively low-cost and simple; deposition of compounds is difficult; coating adhesion not as good as other PVD processes	Ag, Al, Au, Cr, Cu, Mo, W
Sputtering	Better throwing power and coating adhesion than vacuum evaporation, can coat compounds, slower deposition rates and more difficult process control than vacuum evaporation	Al_2O_3, Au, Cr, Mo, SiO_2, Si_3N_4, TiC, TiN
Ion plating	Best coverage and coating adhesion of PVD processes, most complex process control, higher deposition rates than sputtering	Ag, Au, Cr, Mo, Si_3N_4, TiC, TiN

Compiled from [2].

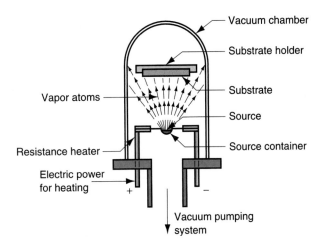

FIGURE 28.6 Setup for vacuum evaporation physical vapor deposition.

Vacuum Evaporation   Certain materials (mostly pure metals) can be deposited onto a substrate by first transforming them from solid to vapor state in a vacuum and then letting them condense on the substrate surface. The setup for the vacuum evaporation process is shown in Figure 28.6. The material to be deposited, called the source, is heated to a sufficiently high temperature that it evaporates (or sublimes). Since heating is accomplished in a vacuum, the temperature required for vaporization is significantly below the corresponding temperature required at atmospheric pressure. Also, the absence of air in the chamber prevents oxidation of the source material at the heating temperatures.

Various methods can be used to heat and vaporize the material. A container must be provided to hold the source material before vaporization. Among the important vaporization methods are resistance heating and electron beam bombardment. ***Resistance heating*** is the simplest technology. A refractory metal (e.g., W, Mo) is formed into a suitable container to hold the source material. Current is applied to heat the container, which then heats the material in contact with it. One problem with this heating method is possible alloying between the holder and its contents, so that the deposited film becomes contaminated with the metal of the resistance heating container. In ***electron beam evaporation,*** a stream of electrons at high velocity is directed to bombard the surface of the source material to cause vaporization. By contrast with resistance heating, very little energy acts to heat the container, thus minimizing contamination of the container material with the coating.

Whatever the vaporization technique, evaporated atoms leave the source and follow straight-line paths until they collide with other gas molecules or strike a solid surface. The vacuum inside the chamber virtually eliminates other gas molecules, thus reducing the probability of collisions with source vapor atoms. The substrate surface to be coated is usually positioned relative to the source so that it is the likely solid surface on which the vapor atoms will be deposited. A mechanical manipulator is sometimes used to rotate the substrate so that all surfaces are coated. Upon contact with the relative cool substrate surface, the energy level of the impinging atoms is suddenly reduced to the point where they cannot remain in a vapor state; they condense and become attached to the solid surface, forming a deposited thin film.

Sputtering   If the surface of a solid (or liquid) is bombarded by atomic particles of sufficiently high energy, individual atoms of the surface may acquire enough energy due to the collision that they are ejected from the surface by transfer of momentum. This is the process known as sputtering. The most convenient form of high energy particle is an ionized gas, such as argon, energized by means of an electric field to form a plasma. As a PVD process, ***sputtering*** involves bombardment of the cathodic coating material with argon ions (Ar^+), causing surface atoms to escape and then be deposited onto a substrate, forming a thin film on

FIGURE 28.7 One possible setup for sputtering, a form of physical vapor deposition.

the substrate surface. The substrate must be placed close to the cathode and is usually heated to improve bonding of the coating atoms. A typical arrangement is shown in Figure 28.7.

Whereas vacuum evaporation is generally limited to metals, sputtering can be applied to nearly any material—metallic and nonmetallic elements; alloys, ceramics, and polymers. Films of alloys and compounds can be sputtered without changing their chemical compositions. Films of chemical compounds can also be deposited by employing reactive gases that form oxides, carbides, or nitrides with the sputtered metal.

Drawbacks of sputtering PVD include (1) slow deposition rates and (2) since the ions bombarding the surface are a gas, traces of the gas can usually be found in the coated films, and the entrapped gases sometimes affect mechanical properties adversely.

Ion Plating Ion plating uses a combination of sputtering and vacuum evaporation to deposit a thin film onto a substrate. The process works as follows. The substrate is set up to be the cathode in the upper part of the chamber, and the source material is placed below it. A vacuum is then established in the chamber. Argon gas is admitted and an electric field is applied to ionize the gas (Ar^+) and establish a plasma. This results in ion bombardment (sputtering) of the substrate so that its surface is scrubbed to a condition of atomic cleanliness (interpret this as "very clean"). Next, the source material is heated sufficiently to generate coating vapors. The heating methods used here are similar to those used in vacuum evaporation: resistance heating, electron beam bombardment, and so on. The vapor molecules pass through the plasma and coat the substrate. Sputtering is continued during deposition, so that the ion bombardment consists not only of the original argon ions but also source material ions that have been energized while being subjected to the same energy field as the argon. The effect of these processing conditions is to produce films of uniform thickness and excellent adherence to the substrate.

Ion plating is applicable to parts having irregular geometries, due to the scattering effects that exist in the plasma field. An example of interest here is TiN coating of high-speed steel cutting tools (e.g., drill bits). In addition to coating uniformity and good adherence, other advantages of the process include high deposition rates, high film densities, and the capability to coat the inside walls of holes and other hollow shapes.

28.5.2 CHEMICAL VAPOR DEPOSITION

Physical vapor deposition involves deposition of a coating by condensation onto a substrate from the vapor phase; it is strictly a physical process. By comparison, *chemical vapor deposition* (CVD) involves the interaction between a mixture of gases and the surface of a heated substrate, causing chemical decomposition of some of the gas constituents and formation of a solid film on the substrate. The reactions take place in an enclosed reaction chamber. The reaction product (either a metal or a compound) nucleates and grows on the substrate surface to form the coating. Most CVD reactions require heat. However, depending on the chemicals involved, the reactions can be driven by other possible energy sources, such as ultraviolet light or plasma. CVD includes a wide range of pressures and temperatures; and it can be applied to a great variety of coating and substrate materials.

Industrial metallurgical processes based on chemical vapor deposition date back to the 1800s (e.g., the Mond process in Table 28.3). Modern interest in CVD is focused on its coating applications such as coated cemented carbide tools, solar cells, depositing refractory metals on jet engine turbine blades, and other applications where resistance to wear, corrosion, erosion, and thermal shock are important. In addition, CVD is an important technology in integrated circuit fabrication.

Advantages typically cited for CVD include (1) capability to deposit refractory materials at temperatures below their melting or sintering temperatures; (2) control of grain size is possible; (3) the process is carried out at atmospheric pressure—it does not

---

TABLE 28.3   Some examples of reactions in chemical vapor deposition (CVD).

1. The **Mond process** includes a CVD process for decomposition of nickel from nickel carbonyl ($Ni(CO)_4$), which is an intermediate compound formed in reducing nickel ore:

$$Ni(CO)_4 \xrightarrow{200°C\,(400°F)} Ni + 4CO \qquad (28.4)$$

2. Coating of titanium carbide (TiC) onto a substrate of cemented tungsten carbide (WC–Co) to produce a high-performance cutting tool:

$$TiCl_4 + CH_4 \xrightarrow[\text{excess } H_2]{1000°C\,(1800°F)} TiC + 4HCl \qquad (28.5)$$

3. Coating of titanium nitride (TiN) onto a substrate of cemented tungsten carbide (WC–Co) to produce a high-performance cutting tool:

$$TiCl_4 + 0.5N_2 + 2H_2 \xrightarrow{900°C\,(1650°F)} TiN + 4HCl \qquad (28.6)$$

4. Coating of aluminum oxide (Al_2O_3) onto a substrate of cemented tungsten carbide (WC–Co) to produce a high-performance cutting tool:

$$2AlCl_3 + 3CO_2 + 3H_2 \xrightarrow{500°C\,(900°F)} Al_2O_3 + 3CO + 6HCl \qquad (28.7)$$

5. Coating of silicon nitride (Si_3N_4) onto silicon (Si), a process in semiconductor manufacturing:

$$3SiF_4 + 4NH_3 \xrightarrow{1000°C\,(1800°F)} Si_3N_4 + 12\,HF \qquad (28.8)$$

6. Coating of silicon dioxide (SiO_2) onto silicon (Si), a process in semiconductor manufacturing:

$$2SiCl_3 + 3H_2O + 0.5O_2 \xrightarrow{900°C\,(1600°F)} 2SiO_2 + 6\,HCl \qquad (28.9)$$

7. Coating of the refractory metal tungsten (W) onto a substrate, such as a jet engine turbine blade:

$$WF_6 + 3H_2 \xrightarrow{600°C\,(1100°F)} W + 6HF \qquad (28.10)$$

Compiled from [6], [13], and [17].

FIGURE 28.8
Photomicrograph of the cross section of a coated carbide cutting tool (Kennametal Grade KC792M); chemical vapor deposition was used to coat TiN and TiCN onto the surface of a WC–Co substrate, followed by a TiN coating applied by physical vapor deposition. (Photo courtesy of Kennametal Inc., Latrobe, Pennsylvania.)

require vacuum equipment; and (4) good bonding of coating to substrate surface [1]. Disadvantages include (1) corrosive and/or toxic nature of chemicals generally necessitates a closed chamber as well as special pumping and disposal equipment; (2) certain reaction ingredients are relatively expensive; and (3) material utilization is low.

CVD Materials and Reactions   In general, metals that are readily electroplated are not good candidates for CVD, owing to the hazardous chemicals that must be used and the costs of safeguarding against them. Metals suitable for coating by CVD include tungsten, molybdenum, titanium, vanadium, and tantalum. Chemical vapor deposition is especially suited to the deposition of compounds, such as aluminum oxide (Al_2O_3), silicon dioxide (SiO_2), silicon nitride (Si_3N_4), titanium carbide (TiC), and titanium nitride (TiN). Figure 28.8 illustrates the application of both CVD and PVD to provide multiple wear-resistant coatings on a cemented carbide cutting tool.

The commonly used reacting gases or vapors are metallic hydrides (MH_x), chlorides (MCl_x), fluorides (MF_x), and carbonyls ($M(CO)_x$), where M = the metal to be deposited and x is used to balance the valences in the compound. Other gases, such as hydrogen (H_2), nitrogen (N_2), methane (CH_4), carbon dioxide (CO_2), and ammonia (NH_3) are used in some of the reactions. Table 28.3 presents some examples of CVD reactions that result in deposition of a metal or ceramic coating onto a suitable substrate. Typical temperatures at which these reactions are carried out are also given.

Processing Equipment   Chemical vapor deposition processes are carried out in a reactor, which consists of (1) reactant supply system, (2) deposition chamber, and (3) recycle/disposal system. Although reactor configurations differ depending on the application, one possible CVD reactor is illustrated in Figure 28.9. The purpose of the reactant supply system is to deliver reactants to the deposition chamber in the proper proportions.

FIGURE 28.9 A typical reactor used in chemical vapor deposition.

Different types of supply system are required, depending on whether the reactants are delivered as gas, liquid, or solid (e.g., pellets, powders).

The deposition chamber contains the substrates and chemical reactions that lead to deposition of reaction products onto the substrate surfaces. Deposition occurs at elevated temperatures, and the substrate must be heated by induction heating, radiant heat, or other means. Deposition temperatures for different CVD reactions range from 250°C to 1950°C (500°F–3500°F), so the chamber must be designed to meet these temperature demands.

The third component of the reactor is the recycle/disposal system, whose function is to render harmless the byproducts of the CVD reaction. This includes collection of materials that are toxic, corrosive, and/or flammable, followed by proper processing and disposition.

Alternative Forms of CVD   What we have described is ***atmospheric pressure chemical vapor deposition,*** in which the reactions are carried out at or near atmospheric pressure. For many reactions, there are advantages in performing the process at pressures well below atmospheric. This is called ***low-pressure chemical vapor deposition*** (LPCVD), in which the reactions occur in a partial vacuum. Advantages of LPCVD include (1) uniform thickness, (2) good control over composition and structure, (3) low-temperature processing, (4) fast deposition rates, and (5) high throughput and lower processing costs [13]. The technical problem in LPCVD is designing the vacuum pumps to create the partial vacuum when the reaction products are not only hot but may also be corrosive. These pumps must often include systems to cool and trap the corrosive gases before they reach the actual pumping unit.

Another variation of CVD is ***plasma-assisted chemical vapor deposition*** (PACVD), in which deposition onto a substrate is accomplished by reacting the ingredients in a gas that has been ionized by means of an electric discharge (i.e., a plasma). In effect, the energy contained in the plasma rather than thermal energy is used to activate the chemical reactions. Advantages of PACVD include (1) lower substrate temperatures, (2) better covering power, (3) better adhesion, and (4) faster deposition rates [6]. Applications include deposition of silicon nitride (Si_3N_4) in semiconductor processing, TiN and TiC coatings for tools, and polymer coatings. The process is also known as plasma-enhanced chemical vapor deposition, plasma chemical vapor deposition, or just simply plasma deposition.

28.6 ORGANIC COATINGS

Organic coatings are polymers and resins, produced either naturally or synthetically, usually formulated to be applied as liquids that dry or harden as thin surface films on substrate materials. These coatings are valued for the variety of colors and textures possible, their capacity to protect the substrate surface, low cost, and ease with which they can be applied. In this section, we consider the compositions of organic coatings and the methods to apply them. Although most organic coatings are applied in liquid form, some are applied as powders; we consider this alternative in Section 28.6.2.

Organic coatings are formulated to contain the following: (1) binders, which give the coating its properties; (2) dyes or pigments, which lend color to the coating; (3) solvents, to dissolve the polymers and resins and add proper fluidity to the liquid; and (4) additives.

Binders in organic coatings are polymers and resins that determine the solid-state properties of the coating, such as strength, physical properties, and adhesion to the substrate surface. The binder holds the pigments and other ingredients in the coating during and after application to the surface. The most common binders in organic coatings are natural oils (used to produce oil-based paints), and resins of polyesters, polyurethanes, epoxies, acrylics, and cellulosics.

Dyes and pigments provide color to the coating. *Dyes* are soluble chemicals that color the coating liquid but do not conceal the surface beneath. Thus, dye-colored coatings are generally transparent or translucent. *Pigments* are solid particles of uniform, microscopic size that are dispersed in the coating liquid but insoluble in it. They not only color the coating; they also hide the surface below. Since pigments are particulate matter, they also tend to strengthen the coating.

Solvents are used to dissolve the binder and certain other ingredients in the liquid coating composition. Common solvents used in organic coatings are aliphatic and aromatic hydrocarbons, alcohols, esters, ketones, and chlorinated solvents. Different solvents are required for different binders. *Additives* in organic coatings include surfactants (to facilitate spreading on the surface), biocides and fungicides, thickeners, freeze/thaw stabilizers, heat and light stabilizers, coalescing agents, plasticizers, defoamers, and catalysts to promote cross-linking. These ingredients are formulated to obtain a wide variety of coatings, such as paints, lacquers, and varnishes.

28.6.1 APPLICATION METHODS

The method of applying an organic coating to a surface depends on factors such as composition of the coating liquid, required thickness of the coating, production rate and cost considerations, part size, and environmental requirements. For any of the application methods, it is of utmost importance that the surface be properly prepared. This includes cleaning and possible treatment of the surface such as phosphate coating. In some cases, metallic surfaces are plated prior to organic coating for maximum corrosion protection.

With any coating method, transfer efficiency is a critical measure. *Transfer efficiency* is the proportion of paint supplied to the process that is actually deposited onto the work surface. Some methods yield as low as a 30% transfer efficiency (meaning that 70% of the paint is wasted and cannot be recovered).

Available methods of applying liquid organic coatings include brushing and rolling, spray coating, immersion, and flow coating. In some cases, several successive coatings are applied to the substrate surface to achieve the desired result. An automobile car body is an important example; the following is a typical sequence applied to the sheet-metal car body in a mass-production automobile: (1) phosphate coat applied by dipping, (2) primer coat applied by dipping, (3) color paint coat applied by spray coating, and (4) clear coat (for high gloss and added protection) applied by spraying.

Brushing and *rolling* are the two most familiar application methods to most people. They have a high transfer efficiency—approaching 100%. Manual brushing and rolling methods are suited to low production but not mass production. While brushing is quite versatile, rolling is limited to flat surfaces.

Spray coating is a widely used production method for applying organic coatings. The process forces the coating liquid to atomize into a fine mist immediately prior to deposition onto the part surface. When the droplets hit the surface, they spread and flow together to form a uniform coating within the localized region of the spray. If done properly, spray coating provides a uniform coating over the entire work surface.

Spray coating can be performed manually in spray painting booths, or it can be set up as an automated process. Transfer efficiency is relatively low (as low as 30%) with these methods. Efficiency can be improved by *electrostatic spraying,* in which the workpart is grounded electrically and the atomized droplets are electrostatically charged. This causes the droplets to be drawn to the part surfaces, increasing transfer efficiencies to values up to 90% [17]. Spraying is utilized extensively in the automotive industry for applying external paint coats to car bodies. It is also used for coating appliances and other consumer products.

Immersion applies large amounts of liquid coating to the workpart and allows the excess to drain off and be recycled. The simplest method is *dip coating,* in which a part is immersed in an open tank of liquid coating material; when the part is withdrawn, the excess liquid drains back into the tank. A variation of dip coating is *electrocoating,* in which the part is electrically charged and then dipped into a paint bath that has been given an opposite charge. This improves adhesion and permits use of water-based paints (which reduce fire and pollution hazards).

In *flow coating,* workparts are moved through an enclosed paint booth, where a series of nozzles shower the coating liquid onto the part surfaces. Excess liquid drains back into a sump, which allows it to be reused.

Once applied, the organic coating must convert from liquid to solid. The term *drying* is often used to describe this conversion process. Many organic coatings dry by evaporation of their solvents. However, in order to form a durable film on the substrate surface, a further conversion is necessary, called curing. *Curing* involves a chemical change in the organic resin in which polymerization or cross-linking occurs to harden the coating.

The type of resin determines the type of chemical reaction that takes place in curing. The principal methods by which curing is effected in organic coatings are [17] (1) *ambient temperature curing,* which involves evaporation of the solvent and oxidation of the resin (most lacquers cure by this method); (2) *elevated temperature curing,* in which elevated temperatures are used to accelerate solvent evaporation, as well as polymerization and cross-linking of the resin; (3) *catalytic curing,* in which the starting resins require reactive agents mixed immediately prior to application to bring about polymerization and cross-linking (epoxy and polyurethane paints are examples); and (4) *radiation curing,* in which various forms of radiation, such as microwaves, ultraviolet light, and electron beams, are required to cure the resin.

28.6.2 POWDER COATING

The organic coatings discussed above are liquid systems consisting of resins that are soluble (or at least miscible) in a suitable solvent. Powder coatings are different. They are applied as dry, finely pulverized, solid particles that are melted on the surface to form a uniform liquid film, after which they resolidify into a dry coating. Powder coating systems have grown significantly in commercial importance among organic coatings since the mid-1970s.

Powder coatings are classified as thermoplastic or thermosetting. Common thermoplastic powders include polyvinylchloride, nylon, polyester, polyethylene, and polypropylene. They are generally applied as relatively thick coatings, 0.08 to 0.30 mm (0.003–0.012 in). Common thermosetting coating powders are epoxy, polyester, and acrylic. They are applied as uncured resins that polymerize and cross-link on heating or reaction with other ingredients. Coating thicknesses are typically 0.025 to 0.075 mm (0.001–0.003 in).

There are two principal application methods for powder coatings: spraying and fluidized bed. In the **spraying** method, an electrostatic charge is given to each particle in order to attract it to an electrically grounded part surface. Several spray gun designs are available to impart the charge to the powders. The spray guns can be operated manually or by industrial robots. Compressed air is used to propel the powders to the nozzle. The powders are dry when sprayed, and any excess particles that do not attach to the surface can be recycled (unless multiple paint colors are mixed in the same spray booth). Powders can be sprayed onto a part at room temperature, followed by heating of the part to melt the powders; or they can be sprayed onto a part that has been heated to above the melting point of the powder, which usually provides a thicker coating.

The **fluidized bed** is a less commonly used alternative to electrostatic spraying. In this method, the workpart to be coated is preheated and passed through a fluidized bed, in which powders are suspended (fluidized) by an airstream. The powders attach themselves to the part surface to form the coating. In some implementations of this coating method, the powders are electrostatically charged to increase attraction to the grounded part surface.

28.7 PORCELAIN ENAMELING AND OTHER CERAMIC COATINGS

Porcelain is a ceramic made from kaolin, feldspar, and quartz (Chapter 7). It can be applied to substrate metals such as steel, cast iron, and aluminum as a vitreous porcelain enamel. Porcelain coatings are valued for their beauty, color, smoothness, ease of cleaning, chemical inertness, and general durability. **Porcelain enameling** is the name given to the technology of these ceramic coating materials and the processes by which they are applied.

Porcelain enameling is used in a wide variety of products, including bathroom fixtures (e.g., sinks, bathtubs, lavatories), household appliances (e.g., ranges, water heaters, washing machines, dishwashers), kitchen ware, hospital utensils, jet engine components, automotive mufflers, and electronic circuit boards. Compositions of the porcelains vary, depending on product requirements. Some porcelains are formulated for color and beauty, while others are designed for functions such as resistance to chemicals and weather, ability to withstand high service temperatures, hardness and abrasion resistance, and electrical resistance.

As a process, porcelain enameling consists of (1) preparing the coating material, (2) applying to the surface, (3) drying, if needed, and (4) firing. Preparation involves converting the glassy porcelain into fine particles, called *frit*, that are milled to proper and consistent size. The methods for applying the frit are similar to methods used for applying organic coatings, even though the starting material is entirely different. Some application methods involve mixing frit with water as a carrier (the mixture is called a **slip**), while other methods apply the porcelain as dry powder. The techniques include spraying, electrostatic spraying, flow coating, dipping, and electrodeposition. Firing is accomplished at temperatures around 800°C (1500°F). Firing is a **sintering** process (Section 17.1.4) in which the frit is transformed into nonporous vitreous porcelain. Coating thickness ranges from around 0.075 to 2 mm (0.08–0.003 in). The processing sequence may be repeated several times to obtain the desired thickness.

In addition to porcelain, other ceramics are used as coatings for special purposes. These coatings usually contain a high content of alumina, which makes them more suited to refractory applications. Techniques for applying the coatings are similar to the preceding, except firing temperatures are higher.

28.8 THERMAL AND MECHANICAL COATING PROCESSES

These processes apply discrete coatings that are generally thicker than coatings deposited by other processes considered in this chapter. They are based on either thermal or mechanical energy.

28.8.1 THERMAL SURFACING PROCESSES

These methods use thermal energy in various forms to apply a coating whose function is to provide resistance to corrosion, erosion, wear, and high temperature oxidation. The processes include (1) thermal spraying, (2) hard facing, and (3) the flexible overlay process.

In ***thermal spraying***, molten and semimolten coating materials are sprayed onto a substrate, where they solidify and adhere to the surface. A wide variety of coating materials can be applied; the categories are pure metals and metal alloys; ceramics (oxides, carbides, and certain glasses); other metallic compounds (sulfides, silicides); cermet composites; and certain plastics (epoxy, nylon, Teflon, and others). The substrates include metals, ceramics, glass, some plastics, wood, and paper. Not all coatings can be applied to all substrates. When the process is used to apply a metallic coating, the terms ***metallizing*** or ***metal spraying*** are used.

Technologies used to heat the coating material are oxyfuel flame, electric arc, and plasma arc. The starting coating material is in the form of wire or rod, or powders. When wire (or rod) is used, the heating source melts the leading end of the wire, thereby separating it from the solid stock. The molten material is then atomized by a high-velocity gas stream (compressed air or other source), and the droplets are spattered against the work surface. When powder stock is used, a powder feeder dispenses the fine particles into a gas stream, which transports them into the flame, where they are melted. The expanding gases in the flame propel the molten (or semimolten) powders against the workpiece. Coating thickness in thermal spraying is generally greater than in other deposition processes; the typical range is 0.05 to 2.5 mm (0.002–0.100 in).

The first applications of thermal spray coating were to rebuild worn areas on used machinery components and to salvage workparts that had been machined undersize. Success of the technique has led to its use in manufacturing as a coating process for corrosion resistance, high temperature protection, wear resistance, electrical conductivity, electrical resistance, electromagnetic interference shielding, and other functions.

Hard facing is a surfacing technique in which alloys are applied as welded deposits to substrate metals. What distinguishes hard facing is that fusion occurs between the coating and the substrate, as in fusion welding (Chapter 29), whereas the bond in thermal spraying is typically mechanical interlocking that does not stand up as well to abrasive wear. Thus, hard facing is especially suited to components requiring good wear resistance. Applications include coating new parts and repairing used part surfaces that are heavily worn, eroded, or corroded. An advantage of hard facing that should be mentioned is that it is readily accomplished outside of the relatively controlled factory environment by many of the common welding processes, such as oxyacetylene gas welding and arc welding. Some of the common surfacing materials include steel and iron alloys, cobalt-based alloys, and nickel-based alloys. Coating thickness is usually 0.75 to 2.5 mm (0.030–0.125 in), although thicknesses as great as 9 mm (3/8 in) are possible.

The ***flexible overlay process*** is capable of depositing a very hard coating material, such as tungsten carbide (WC), onto a substrate surface. This is an important advantage of the process compared to other methods, permitting coating hardness up to about 70 Rockwell C. The process can also be used to apply coatings only to selected regions of a workpart. In the flexible overlay process, a cloth impregnated with hard ceramic or metal powders and another cloth impregnated with brazing alloy are laid onto a substrate and heated to fuse the powders to the surface. Thickness of overlay coatings is usually 0.25 to 2.5 mm (0.010–0.100 in). In addition to coatings of WC and WC–Co, cobalt-based and nickel-based alloys are also applied. Applications include chain saw teeth, rock drill bits, oil drill collars, extrusion dies, and similar parts requiring good wear resistance.

28.8.2 MECHANICAL PLATING

In this coating process, mechanical energy is used to build a metallic coating onto the surface. In mechanical plating, the parts to be coated, together with plating metal powders, glass beads, and special chemicals to promote the plating action, are tumbled in a barrel. The metallic powders are microscopic in size—5 μm (0.0002 in) in diameter; while the glass beads are much larger—2.5 mm (0.10 in) in diameter. As the mixture is tumbled, the mechanical energy from the rotating barrel is transmitted through the glass beads to pound the metal powders against the part surface, causing a mechanical or metallurgical bond to result. The deposited metals must be malleable in order to achieve a satisfactory bond with the substrate. Plating metals include zinc, cadmium, tin, and lead. The term *mechanical galvanizing* is used for parts that are zinc coated. Ferrous metals are most commonly coated; other metals include brass and bronze. Typical applications include fasteners such as screws, bolts, nuts, and nails. Plating thickness in mechanical plating is usually 0.005 to 0.025 mm (0.0002–0.001 in). Zinc is mechanically plated to a thickness of around 0.075 mm (0.003 in).

REFERENCES

[1] *ASM Handbook,* Vol. 5, *Surface Engineering.* ASM International, Materials Park, Ohio, 1993.

[2] Budinski, K. G. *Surface Engineering for Wear Resistance.* Prentice Hall, Inc., Englewood Cliffs, New Jersey, 1988.

[3] Durney, L. J. (ed.). *The Graham's Electroplating Engineering Handbook,* 4th ed. Chapman & Hall, London, 1996.

[4] Freeman, N. B. "A New Look at Mass Finishing," Special Report 757, *American Machinist,* August 1983, pp. 93–104.

[5] George, J. *Preparation of Thin Films.* Marcel Dekker, Inc., New York, 1992.

[6] Hocking, M. G., Vasantasree, V., and Sidky, P. S. *Metallic and Ceramic Coatings.* Addison-Wesley Longman, Ltd., Reading, Massachusetts, 1989.

[7] *Metal Finishing;* Guidebook and Directory Issue. Metals and Plastics Publications, Inc., Hackensack, New Jersey, 2000.

[8] Morosanu, C. E. *Thin Films by Chemical Vapour Deposition.* Elsevier, Amsterdam, The Netherlands, 1990.

[9] Murphy, J. A. (ed.). *Surface Preparation and Finishes for Metals.* McGraw-Hill Book Company, New York, 1971.

[10] Sabatka, W. "Vapor Degreasing." Available at: www.pfonline.com.

[11] Satas, D. (ed.). *Coatings Technology Handbook,* 2nd ed. Marcel Dekker, Inc., New York, 2000.

[12] Stuart, R. V. *Vacuum Technology, Thin Films, and Sputtering.* Academic Press, New York, 1983.

[13] Sze, S. M. *VLSI Technology,* 2nd ed. McGraw-Hill Book Company, New York, 1988.

[14] Tracton, A. A. (ed.) *Coatings Technology Handbook,* 3rd ed. CRC Taylor & Francis, Boca Raton, Florida, 2006.

[15] Tucker, Jr., R. C. "Surface Engineering Technologies," *Advanced Materials & Processes,* April 2002, pp. 36–38.

[16] Tucker, Jr., R. C. "Considerations in the Selection of Coatings," *Advanced Materials & Processes,* March 2004, pp. 25–28.

[17] Wick, C., and Veilleux, R. (eds.). *Tool and Manufacturing Engineers Handbook,* 4th ed., Vol III, *Materials, Finishes, and Coating.* Society of Manufacturing Engineers, Dearborn, Michigan, 1985.

REVIEW QUESTIONS

28.1. What are some of the important reasons why manufactured parts must be cleaned?

28.2. Mechanical surface treatments are often performed for reasons other than or in addition to cleaning. What are the reasons?

28.3. What are the basic types of contaminants that must be cleaned from metallic surfaces in manufacturing?

28.4. Name some of the important chemical cleaning methods.

28.5. In addition to surface cleaning, what is the main function performed by shot peening?

28.6. What is meant by the term mass finishing?

28.7. What is the difference between diffusion and ion implantation?

28.8. What is calorizing?

28.9. Why are metals coated?

28.10. Identify the most common types of coating processes.

28.11. What are the two basic mechanisms of corrosion protection?

28.12. What is the most commonly plated substrate metal?

28.13. One of the mandrel types in electroforming is a solid mandrel. How is the part removed from a solid mandrel?

28.14. How does electroless plating differ from electrochemical plating?

28.15. What is a conversion coating?

28.16. How does anodizing differ from other conversion coatings?

28.17. What is physical vapor deposition?

28.18. What is the difference between physical vapor deposition and chemical vapor deposition?

28.19. What are some of the applications of PVD?

28.20. Name the commonly used coating materials deposited by PVD onto cutting tools?

28.21. What are some of the advantages of chemical vapor deposition?

28.22. What are the two most common titanium compounds that are coated onto cutting tools by chemical vapor deposition?

28.23. Identify the four major types of ingredients in organic coatings.

28.24. What is meant by the term transfer efficiency in organic coating technology?

28.25. Describe the principal methods by which organic coatings are applied to a surface.

28.26. The terms drying and curing have different meanings; indicate the distinction.

28.27. In porcelain enameling, what is frit?

MULTIPLE CHOICE QUIZ

There are 20 correct answers in the following multiple choice questions (some questions have multiple answers that are correct). To attain a perfect score on the quiz, all correct answers must be given. Each correct answer is worth 1 point. Each omitted answer or wrong answer reduces the score by 1 point, and each additional answer beyond the correct number of answers reduces the score by 1 point. Percentage score on the quiz is based on the total number of correct answers.

28.1. Which of the following are reasons why workparts must be cleaned in industry (four best answers): (a) to avoid air pollution, (b) to avoid water pollution, (c) to enhance appearance, (d) to enhance mechanical properties of the surface, (e) to improve hygiene conditions for workers, (f) to improve surface finish, (g) to prepare the surface for subsequent processing, and (h) to remove contaminants that might chemically attack the surface?

28.2. Which of the following chemicals are associated with alkaline cleaning (two correct answers): (a) borax, (b) hydrochloric acid, (c) propane, (d) sodium hydroxide, (e) sulfuric acid, and (f) trichlorethylene?

28.3. In sand blasting, which one of the following blast media is used: (a) Al_2O_3, (b) crushed nut shells, (c) nylon beads, (d) SiC, or (e) SiO_2?

28.4. Which of the following processes generally produces a deeper penetration of atoms in the impregnated surface: (a) diffusion or (b) ion implantation?

28.5. Calorizing is the same as which one of the following surface processes: (a) aluminizing, (b) doping, (c) hot-sand blasting, or (d) siliconizing?

28.6. Which one of the following plate metals produces the hardest surface on a metallic substrate: (a) cadmium, (b) chromium, (c) copper, (d) nickel, or (e) tin?

28.7. Which one of the following plating metals is associated with the term galvanizing: (a) iron, (b) lead, (c) steel, (d) tin, or (e) zinc?

28.8. Which of the following processes involves electrochemical reactions (two correct answers): (a) anodizing, (b) chromate coatings, (c) electroless plating, (d) electroplating, and (e) phosphate coatings?

28.9. With which one of the following metals is anodizing most commonly associated (one answer): (a) aluminum, (b) magnesium, (c) steel, (d) titanium, or (e) zinc?

28.10. Sputtering is a form of which one of the following: (a) chemical vapor deposition, (b) defect in arc welding, (c) diffusion, (d) ion implantation, or (e) physical vapor deposition?

28.11. Which one of the following gases is the most commonly used in sputtering and ion plating: (a) argon, (b) chlorine, (c) neon, (d) nitrogen, or (e) oxygen?

28.12. The principal methods of applying powder coatings are which of the following (two best answers): (a) brushing, (b) electrostatic spraying, (c) fluidized bed, (d) immersion, and (e) roller coating?

28.13. Porcelain enamel is applied to a surface in which one of the following forms: (a) liquid emulsion, (b) liquid solution, (c) molten liquid, or (d) powders?

28.14. Hard facing utilizes which one of the following basic processes: (a) arc welding, (b) brazing, (c) dip coating, (d) electroplating, or (e) mechanical deformation to work harden the surface?

PROBLEMS

Electroplating

28.1. What volume (cm^3) and weight (g) of zinc will be deposited onto a cathodic workpart if 10 amps of current are applied for 1 hour?

28.2. A sheet metal steel part with surface area $= 100$ cm^2 is to be zinc plated. What average plating thickness will result if 15 amps are applied for 12 minutes in a chloride electrolyte solution?

28.3. A sheet metal steel part with surface area $= 15.0$ in^2 is to be chrome plated. What average plating thickness will result if 15 amps are applied for 10 minutes in a chromic acid-sulfate bath?

28.4. Twenty-five jewelry pieces, each with a surface area $= 0.5$ in^2 are to be gold plated in a batch plating operation. (a) What average plating thickness will result if 8 amps are applied for 10 min in a cyanide bath? (b) What is the value of the gold that will be plated onto each piece if one ounce of gold is valued at $900? The density of gold $= 0.698$ lb/in^3.

28.5. A part made of sheet steel is to be nickel plated. The part is a rectangular flat plate that is 0.075 cm thick and whose face dimensions are 14 cm by 19 cm. The plating operation is carried out in an acid sulfate electrolyte, using a current $= 20$ amps for a duration $= 30$ min. Determine the average thickness of the plated metal resulting from this operation.

28.6. A steel sheet metal part has total surface area $= 36$ in^2. How long will it take to deposit a copper plating (assume valence $= +1$) of thickness $= 0.001$ in onto the surface if 15 amps of current are applied?

28.7. Increasing current is applied to a workpart surface in an electroplating process according to the relation $I = 12.0 + 0.2t$, where $I =$ current, amps; and $t =$ time, min. The plating metal is chromium, and the part is submersed in the plating solution for a duration of 20 min. What volume of coating will be applied in the process?

28.8. A batch of 100 parts is to be nickel plated in a barrel plating operation. The parts are identical, each with a surface area $A = 7.8$ in^2. The plating process applies a current $I = 120$ amps, and the batch takes 40 minutes to complete. Determine the average plating thickness on the parts.

28.9. A batch of 40 identical parts is to be chrome plated using racks. Each part has a surface are $= 22.7$ cm^2. If it is desired to plate an average thickness $= 0.010$ mm on the surface of each part, how long should the plating operation be allowed to run at a current $= 80$ amps?

Part VIII   Joining and Assembly Processes

<table>
<tr><td>29</td><td># FUNDAMENTALS OF WELDING</td></tr>
</table>

Chapter Contents

In this part of the book, we consider the processes that are used to join two or more parts into an assembled entity. These processes are labeled in the lower stem of Figure 1.4. The term *joining* is generally used for welding, brazing, soldering, and adhesive bonding, which form a permanent joint between the parts—a joint that cannot easily be separated. The term *assembly* usually refers to mechanical methods of fastening parts together. Some of these methods allow for easy disassembly, while others do not. Mechanical assembly is covered in Chapter 32. Brazing, soldering, and adhesive bonding are discussed in Chapter 31. We begin our coverage of the joining and assembly processes with welding, covered in this chapter and the following.

Welding is a materials joining process in which two or more parts are coalesced at their contacting surfaces by a suitable application of heat and/or pressure. Many welding processes are accomplished by heat alone, with no pressure applied; others by a combination of heat and pressure; and still others by pressure alone, with no external heat supplied. In some welding processes a *filler* material is added to facilitate coalescence. The assemblage of parts that are joined by welding is called a *weldment.* Welding is most commonly associated with metal parts, but the process is also used for joining plastics. Our discussion of welding will focus on metals.

Welding is a relatively new process (Historical Note 29.1). Its commercial and technological importance derives from the following:

> ➤ Welding provides a permanent joint. The welded parts become a single entity.
> ➤ The welded joint can be stronger than the parent materials if a filler metal is used that has strength properties superior to those of the parents, and if proper welding techniques are used.
> ➤ Welding is usually the most economical way to join components in terms of material usage and fabrication costs. Alternative mechanical methods of assembly require more complex shape alterations (e.g., drilling of holes) and addition of fasteners (e.g., rivets or bolts). The resulting mechanical assembly is usually heavier than a corresponding weldment.
> ➤ Welding is not restricted to the factory environment. It can be accomplished "in the field."

Although welding has the advantages indicated above, it also has certain limitations and drawbacks (or potential drawbacks):

> ➤ Most welding operations are performed manually and are expensive in terms of labor cost. Many welding operations are considered "skilled trades," and the labor to perform these operations may be scarce.
> ➤ Most welding processes are inherently dangerous because they involve the use of high energy.
> ➤ Since welding accomplishes a permanent bond between the components, it does not allow for convenient disassembly. If the product must occasionally be disassembled (e.g., for repair or maintenance), then welding should not be used as the assembly method.
> ➤ The welded joint can suffer from certain quality defects that are difficult to detect. The defects can reduce the strength of the joint.

Historical Note 29.1   *Origins of welding*

Although welding is considered a relatively new process as practiced today, its origins can be traced to ancient times. Around 1000 BCE, the Egyptians and others in the eastern Mediterranean area learned to accomplish forge welding (Section 30.5.2). It was a natural extension of hot forging, which they used to make weapons, tools, and other implements. Forge-welded articles of bronze have been recovered by archeologists from the pyramids of Egypt. From these early beginnings through the Middle Ages, the blacksmith trade developed the art of welding by hammering to a high level of maturity. Welded objects of iron and other metals dating from these times have been found in India and Europe.

It was not until the 1800s that the technological foundations of modern welding were established. Two important discoveries were made, both attributed to English scientist Sir Humphrey Davy: (1) the electric arc, and (2) acetylene gas.

Around 1801, Davy observed that an electric arc could be struck between two carbon electrodes. However, not until the mid-1800s, when the electric

generator was invented, did electrical power become available in amounts sufficient to sustain *arc welding*. It was a Russian, Nikolai Benardos, working out of a laboratory in France, who was granted a series of patents for the carbon arc–welding process (one in England in 1885, and another in the United States in 1887). By the turn of the century, carbon arc welding had become a popular commercial process for joining metals.

Benardos' inventions seem to have been limited to carbon arc welding. In 1892, an American named Charles Coffin was awarded a U.S. patent for developing an arc–welding process utilizing a metal electrode. The unique feature was that the electrode added filler metal to the weld joint (the carbon arc process does not deposit filler). The idea of coating the metal electrode (to shield the welding process from the atmosphere) was developed later, with enhancements to the metal arc–welding process being made in England and Sweden starting around 1900.

Between 1885 and 1900, several forms of *resistance welding* were developed by Elihu Thompson. These included spot welding and seam welding, two joining methods widely used today in sheet metalworking.

Although Davy discovered acetylene gas early in the 1800s, *oxyfuel gas welding* required the subsequent development of torches for combining acetylene and oxygen around 1900. During the 1890s, hydrogen and natural gas were mixed with oxygen for welding, but the oxyacetylene flame achieved significantly higher temperatures.

These three welding processes—arc welding, resistance welding, and oxyfuel gas welding—constitute by far the majority of welding operations performed today.

29.1 OVERVIEW OF WELDING TECHNOLOGY

Welding involves localized coalescence or joining together of two metallic parts at their faying surfaces. The *faying surfaces* are the part surfaces in contact or close proximity that are to be joined. Welding is usually performed on parts made of the same metal, but some welding operations can be used to join dissimilar metals.

29.1.1 TYPES OF WELDING PROCESSES

Some 50 different types of welding operations have been cataloged by the American Welding Society. They use various types or combinations of energy to provide the required power. We can divide the welding processes into two major groups: (1) fusion welding and (2) solid-state welding.

Fusion Welding   Fusion-welding processes use heat to melt the base metals. In many fusion welding operations, a filler metal is added to the molten pool to facilitate the process and provide bulk and strength to the welded joint. A fusion-welding operation in which no filler metal is added is referred to as an *autogenous* weld. The fusion category includes the most widely used welding processes, which can be organized into the following general groups (initials in parentheses are designations of the American Welding Society):

➢ *Arc welding* (AW). Arc welding refers to a group of welding processes in which heating of the metals is accomplished by an electric arc, as shown in Figure 29.1. Some arc-welding operations also apply pressure during the process and most utilize a filler metal.

➢ *Resistance welding* (RW). Resistance welding achieves coalescence using heat from electrical resistance to the flow of a current passing between the faying surfaces of two parts held together under pressure.

➢ *Oxyfuel gas welding* (OFW). These joining processes use an oxyfuel gas, such as a mixture of oxygen and acetylene, to produce a hot flame for melting the base metal and filler metal, if one is used.

FIGURE 29.1 Basics of arc welding: (1) before the weld; (2) during the weld (the base metal is melted and filler metal is added to the molten pool); and (3) the completed weldment. There are many variations of the arc-welding process.

➤ Other fusion-welding processes. Other welding processes that produce fusion of the metals joined include **electron beam welding** and **laser beam welding.**

Certain arc and oxyfuel processes are also used for cutting metals (Sections 26.3.4 and 26.3.5).

Solid-State Welding   Solid-state welding refers to joining processes in which coalescence results from application of pressure alone or a combination of heat and pressure. If heat is used, the temperature in the process is below the melting point of the metals being welded. No filler metal is utilized. Representative welding processes in this group include:

➤ **Diffusion welding** (DFW). Two surfaces are held together under pressure at an elevated temperature and the parts coalesce by solid-state diffusion.
➤ **Friction welding** (FRW). Coalescence is achieved by the heat of friction between two surfaces.
➤ **Ultrasonic welding** (USW). Moderate pressure is applied between the two parts and an oscillating motion at ultrasonic frequencies is used in a direction parallel to the contacting surfaces. The combination of normal and vibratory forces results in shear stresses that remove surface films and achieve atomic bonding of the surfaces.

In Chapter 30, we describe the various welding processes in greater detail. The preceding survey should provide a sufficient framework for our discussion of welding terminology and principles in the present chapter.

29.1.2 WELDING AS A COMMERCIAL OPERATION

The principal applications of welding are (1) construction, such as buildings and bridges; (2) piping, pressure vessels, boilers, and storage tanks; (3) shipbuilding; (4) aircraft and aerospace; and (5) automotive and railroad [1]. Welding is performed in a variety of locations and in a variety of industries. Owing to its versatility as an assembly technique for commercial products, many welding operations are performed in factories. However, several of the traditional processes, such as arc welding and oxyfuel gas welding, use equipment that can be readily moved, so these operations are not limited to the factory. They can be performed at construction sites, in shipyards, at customers' plants, and in automotive repair shops.

Most welding operations are labor intensive. For example, arc welding is usually performed by a skilled worker, called a **welder,** who manually controls the path or placement of the weld to join individual parts into a larger unit. In factory operations in which arc welding is manually performed, the welder often works with a second worker, called a **fitter.** It is the fitter's job to arrange the individual components for the welder prior to making the weld. Welding fixtures and positioners are used for this purpose. A **welding fixture** is a device for clamping and holding the components in fixed position for welding. It is custom-fabricated for the particular geometry of the weldment and therefore must be economically justified on the basis of the quantities of assemblies to be produced. A **welding positioner** is a device that holds the parts and also moves the assemblage to the desired position for welding. This differs from a welding fixture that only holds the parts in a single fixed position. The desired position is usually one in which the weld path is flat and horizontal.

The Safety Issue   Welding is inherently dangerous to human workers. Strict safety precautions must be practiced by those who perform these operations. The high temperatures of the molten metals in welding are an obvious danger. In gas welding, the fuels (e.g., acetylene) are a fire hazard. Most of the processes use high energy to cause melting of the part surfaces to be joined. In many welding processes, electrical power is the source of thermal energy, so there is the hazard of electrical shock to the worker. Certain welding processes have their own particular perils. In arc welding, for example, ultraviolet radiation is emitted that is injurious to human vision. A special helmet that includes a dark viewing window must be worn by the welder. This window filters out the dangerous radiation but is so dark that it renders the welder virtually blind, except when the arc is struck. Sparks, spatters of molten metal, smoke, and fumes add to the risks associated with welding operations. Ventilation facilities must be used to exhaust the dangerous fumes generated by some of the fluxes and molten metals used in welding. If the operation is performed in an enclosed area, special ventilation suits or hoods are required.

Automation in Welding   Because of the hazards of manual welding, and in efforts to increase productivity and improve product quality, various forms of mechanization and automation have been developed. The categories include machine welding, automatic welding, and robotic welding.

   Machine welding can be defined as mechanized welding with equipment that performs the operation under the continuous supervision of an operator. It is normally accomplished by a welding head that is moved by mechanical means relative to a stationary work, or by moving the work relative to a stationary welding head. The human worker must continually observe and interact with the equipment to control the operation.

   If the equipment is capable of performing the operation without control by a human operator, it is referred to as *automatic welding.* A human worker is usually present to oversee the process and detect variations from normal conditions. What distinguishes automatic welding from machine welding is a weld cycle controller to regulate the arc movement and workpiece positioning without continuous human attention. Automatic welding requires a welding fixture and/or positioner to position the work relative to the welding head. It also requires a higher degree of consistency and accuracy in the component parts used in the weldment. For these reasons, automatic welding can be justified only for large quantity production.

   In *robotic welding,* an industrial robot or programmable manipulator is used to automatically control the movement of the welding head relative to the work (Section 38.4.3). The versatile reach of the robot arm permits the use of relatively simple fixtures, and the robot's capacity to be reprogrammed for new part configurations allows this form of automation to be justified for relatively low production quantities. A typical robotic arc-welding cell consists of two welding fixtures and a human fitter to load and unload parts while the robot welds. In addition to arc welding, industrial robots are also used in automobile final assembly plants to perform resistance welding on car bodies (Figure 39.11).

29.2   THE WELD JOINT

Welding produces a solid connection between two pieces, called a weld joint. A *weld joint* is the junction of the edges or surfaces of parts that have been joined by welding. This section covers two classifications related to weld joints: (1) types of joints and (2) the types of welds used to join the pieces that form the joints.

29.2.1   TYPES OF JOINTS

There are five basic types of joints for bringing two parts together for joining. The five joint types are not limited to welding; they apply to other joining and fastening

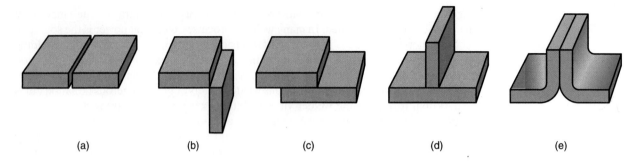

(a)   (b)   (c)   (d)   (e)

FIGURE 29.2 Five basic types of joints: (a) butt, (b) corner, (c) lap, (d) tee, and (e) edge.

techniques as well. With reference to Figure 29.2, the five joint types can be defined as follows:

(a) **Butt joint.** In this joint type, the parts lie in the same plane and are joined at their edges.

(b) **Corner joint.** The parts in a corner joint form a right angle and are joined at the corner of the angle.

(c) **Lap joint.** This joint consists of two overlapping parts.

(d) **Tee joint.** In a tee joint, one part is perpendicular to the other in the approximate shape of the letter "T."

(e) **Edge joint.** The parts in an edge joint are parallel with at least one of their edges in common, and the joint is made at the common edge(s).

29.2.2 TYPES OF WELDS

Each of the preceding joints can be made by welding. It is appropriate to distinguish between the joint type and the way in which it is welded—the weld type. Differences among weld types are in geometry (joint type) and welding process.

A *fillet weld* is used to fill in the edges of plates created by corner, lap, and tee joints, as in Figure 29.3. Filler metal is used to provide a cross section approximately the shape of a right triangle. It is the most common weld type in arc and oxyfuel welding because it requires minimum edge preparation—the basic square edges of the parts are used. Fillet welds can be single or double (i.e., welded on one side or both) and can be continuous or intermittent (i.e., welded along the entire length of the joint or with unwelded spaces along the length).

Groove welds usually require that the edges of the parts be shaped into a groove to facilitate weld penetration. The grooved shapes include square, bevel, V, U, and J, in

FIGURE 29.3 Various forms of fillet welds: (a) inside single fillet corner joint; (b) outside single fillet corner joint; (c) double fillet lap joint; and (d) double fillet tee joint. Dashed lines show the original part edges.

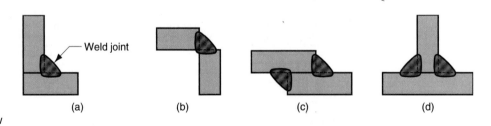

Weld joint

(a)   (b)   (c)   (d)

FIGURE 29.4 Some typical groove welds: (a) square groove weld, one side; (b) single bevel groove weld; (c) single V-groove weld; (d) single U-groove weld; (e) single J-groove weld; (f) double V-groove weld for thicker sections. Dashed lines show the original part edges.

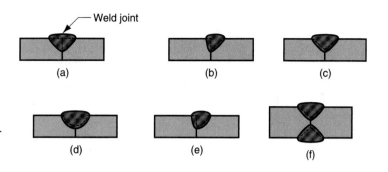

single or double sides, as shown in Figure 29.4. Filler metal is used to fill in the joint, usually by arc or oxyfuel welding. Preparation of the part edges beyond the basic square edge, although requiring additional processing, is often done to increase the strength of the welded joint or where thicker parts are to be welded. Although most closely associated with a butt joint, groove welds are used on all joint types except lap.

Plug welds and *slot welds* are used for attaching flat plates, as shown in Figure 29.5, using one or more holes or slots in the top part and then filling with filler metal to fuse the two parts together.

Spot welds and seam welds, used for lap joints, are diagrammed in Figure 29.6. A *spot weld* is a small fused section between the surfaces of two sheets or plates. Multiple spot welds are typically required to join the parts. It is most closely associated with resistance welding. A *seam weld* is similar to a spot weld except it consists of a more or less continuously fused section between the two sheets or plates.

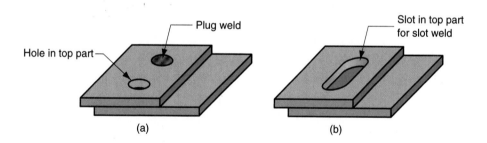

FIGURE 29.5 (a) Plug weld; and (b) slot weld.

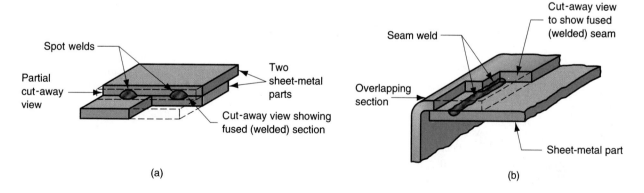

FIGURE 29.6 (a) Spot weld; and (b) seam weld.

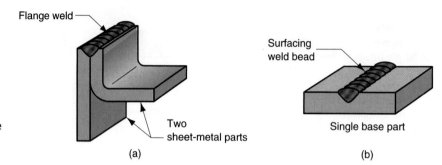

FIGURE 29.7 (a) Flange weld; and (b) surfacing weld.

Flange welds and surfacing welds are shown in Figure 29.7. A *flange weld* is made on the edges of two (or more) parts, usually sheet metal or thin plate, at least one of the parts being flanged as in Figure 29.7(a). A *surfacing weld* is not used to join parts, but rather to deposit filler metal onto the surface of a base part in one or more weld beads. The weld beads can be made in a series of overlapping parallel passes, thereby covering large areas of the base part. The purpose is to increase the thickness of the plate or to provide a protective coating on the surface.

29.3 PHYSICS OF WELDING

Although several coalescing mechanisms are available for welding, fusion is by far the most common means. In this section, we consider the physical relationships that allow fusion welding to be performed. We first examine the issue of power density and its importance, and then we define the heat and power equations that describe a welding process.

29.3.1 POWER DENSITY

To accomplish fusion, a source of high-density heat energy is supplied to the faying surfaces, and the resulting temperatures are sufficient to cause localized melting of the base metals. If a filler metal is added, the heat density must be high enough to melt it also. Heat density can be defined as the power transferred to the work per unit surface area, W/mm^2 (Btu/sec-in^2). The time to melt the metal is inversely proportional to the power density. At low power densities, a significant amount of time is required to cause melting. If power density is too low, the heat is conducted into the work as rapidly as it is added at the surface, and melting never occurs. It has been found that the minimum power density required to melt most metals in welding is about 10 W/mm^2 (6 Btu/sec-in^2). As heat density increases, melting time is reduced. If power density is too high—above around 10^5 W/mm^2 (60,000 Btu/sec-in^2)—the localized temperatures vaporize the metal in the affected region. Thus, there is a practical range of values for power density within which welding can be performed. Differences among welding processes in this range are (1) the rate at which welding can be performed and/or (2) the size of the region that can be welded. Table 29.1 provides a comparison of power densities for the major fusion welding processes. Oxyfuel gas welding is capable of developing large amounts of heat, but the heat density is relatively low because it is spread over a large area. Oxyacetylene gas, the hottest of the OFW fuels, burns at a top temperature of around 3500°C (6300°F). By comparison, arc welding produces high energy over a smaller area, resulting in local temperatures of 5500°C to 6600°C (10,000°F–12,000°F). For metallurgical reasons, it is desirable to melt the metal with minimum energy, and high power densities are generally preferable.

TABLE 29.1   Comparison of several fusion welding processes on the basis of their power densities.

Welding Process	Approximate Power Density	
	W/mm^2	Btu/sec-in^2
Oxyfuel welding	10	6
Arc welding	50	30
Resistanc welding	1000	600
Laser beam welding	9000	5000
Electron beam welding	10,000	6000

Power density can be computed as the power entering the surface divided by the corresponding surface area:

$$PD = \frac{P}{A} \tag{29.1}$$

where PD = power density, W/mm^2 (Btu/sec-in^2); P = power entering the surface, W (Btu/sec); and A = surface area over which the energy is entering, mm^2 (in^2). The issue is more complicated than indicated by Eq. (29.1). One complication is that the power source (e.g., the arc) is moving in many welding processes, which results in preheating ahead of the operation and postheating behind it. Another complication is that power density is not uniform throughout the affected surface; it is distributed as a function of area, as demonstrated by the following example.

Example 29.1 Power Density in Welding

A heat source transfers 3000 W to the surface of a metal part. The heat impinges the surface in a circular area, with intensities varying inside the circle. The distribution is as follows: 70% of the power is transferred within a circle of diameter = 5 mm, and 90% is transferred within a concentric circle of diameter = 12 mm. What are the power densities in (a) the 5-mm diameter inner circle and (b) the 12-mm-diameter ring that lies around the inner circle?

Solution: (a) The inner circle has an area $A = \frac{\pi(5)^2}{4} = 19.63$ mm^2.

The power inside this area $P = 0.70 \times 3000 = 2100$ W.

Thus the power density $PD = \frac{2100}{19.63} = 107$ W/mm^2.

(b) The area of the ring outside the inner circle is $A = \frac{\pi(12^2 - 5^2)}{4} = 93.4$ mm^2.

The power in this region $P = 0.9\,(3000) - 2100 = 600$ W.

The power density is therefore $PD\frac{600}{93.4} = 6.4$ W/mm^2.

Observation: The power density seems high enough for melting in the inner circle, but probably not sufficient in the ring that lies outside this inner circle.   ■

29.3.2 HEAT BALANCE IN FUSION WELDING

The quantity of heat required to melt a given volume of metal depends on (1) the heat to raise the temperature of the solid metal to its melting point, which depends on the metal's volumetric specific heat, (2) the melting point of the metal, and (3) the heat to transform the metal from solid to liquid phase at the melting point, which depends on the metal's heat of fusion. To a reasonable approximation, this quantity of heat can be estimated by

TABLE 29.2   Melting temperatures on the absolute temperature scale for selected metals.

Metal	Melting Temperature		Metal	Melting Temperature	
	°Kª	°Rᵇ		°Kª	°Rᵇ
Aluminum alloys	930	1680	Steels		
Cast iron	1530	2760	Low carbon	1760	3160
Copper and alloys			Medium carbon	1700	3060
Pure	1350	2440	High carbon	1650	2960
Brass, navy	1160	2090	Low alloy	1700	3060
Bronze (90 Cu–10 Sn)	1120	2010	Stainless steels		
Inconel	1660	3000	Austenitic	1670	3010
Magnesium	940	1700	Martensitic	1700	3060
Nickel	1720	3110	Titanium	2070	3730

Based on values in [2].
ªKelvin scale = Centigrade (Celsius) temperature + 273.
ᵇRankine scale = Fahrenheit temperature + 460.

the following equation [5]:

$$U_m = K T_m^2 \tag{29.2}$$

where U_m = the unit energy for melting (i.e., the quantity of heat required to melt a unit volume of metal starting from room temperature), J/mm³ (Btu/in³); T_m = melting point of the metal on an absolute temperature scale, °K (°R); and K = constant whose value is 3.33×10^{-6} when the Kelvin scale is used (and $K = 1.467 \times 10^{-5}$ for the Rankine temperature scale). Absolute melting temperatures for selected metals are presented in Table 29.2.

Not all of the energy generated at the heat source is used to melt the weld metal. There are two heat transfer mechanisms at work, both of which reduce the amount of generated heat that is used by the welding process. The situation is depicted in Figure 29.8. The first mechanism involves the transfer of heat between the heat source and the surface of the work. This process has a certain **heat transfer factor** f_1, defined as the ratio of the actual heat received by the workpiece divided by the total heat generated at the source. The second mechanism involves the conduction of heat away from the weld area to be dissipated throughout the work metal, so that only a portion of the heat transferred to the surface is available for melting. This **melting factor** f_2 is the proportion of heat received at the work surface that can be used for melting. The combined effect of these two factors is

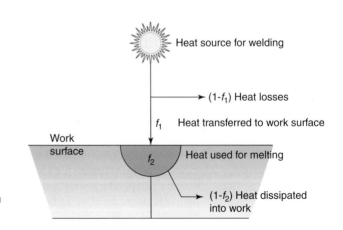

FIGURE 29.8   Heat transfer mechanisms in fusion welding.

to reduce the heat energy available for welding as follows:

$$H_w = f_1 f_2 H \tag{29.3}$$

where H_w = net heat available for welding, J (Btu), f_1 = heat transfer factor, f_2 = the melting factor, and H = the total heat generated by the welding process, J (Btu).

The factors f_1 and f_2 range in value between zero and one. It is appropriate to separate f_1 and f_2 in concept, even though they act in concert during the welding process. The heat transfer factor f_1 is determined largely by the welding process and the capacity to convert the power source (e.g., electrical energy) into usable heat at the work surface. Arc-welding processes are relatively efficient in this regard, while oxyfuel gas-welding processes are relatively inefficient.

The melting factor f_2 depends on the welding process, but it is also influenced by the thermal properties of the metal, joint configuration, and work thickness. Metals with high thermal conductivity, such as aluminum and copper, present a problem in welding because of the rapid dissipation of heat away from the heat contact area. The problem is exacerbated by welding heat sources with low energy densities (e.g., oxyfuel welding) because the heat input is spread over a larger area, thus facilitating conduction into the work. In general, a high power density combined with a low conductivity work material results in a high melting factor.

We can now write a balance equation between the energy input and the energy needed for welding:

$$H_w = U_m V \tag{29.4}$$

where H_w = net heat energy used by the welding operation, J (Btu); U_m = unit energy required to melt the metal, J/mm³ (Btu/in³); and V = the volume of metal melted, mm³ (in³).

Most welding operations are rate processes; that is, the net heat energy H_w is delivered at a given rate, and the weld bead is made at a certain travel velocity. This is characteristic for example of most arc-welding, many oxyfuel gas-welding operations, and even some resistance welding operations. It is therefore appropriate to express Eq. (30) as a rate balance equation:

$$R_{Hw} = U_m R_{WV} \tag{29.5}$$

where R_{Hw} = rate of heat energy delivered to the operation for welding, J/s = W (Btu/min); and R_{WV} = volume rate of metal welded, mm³/s (in³/min). In the welding of a continuous bead, the volume rate of metal welded is the product of weld area A_w and travel velocity v. Substituting these terms into the above equation, the rate balance equation can now be expressed as

$$R_{Hw} = f_1 f_2 R_H = U_m A_w v \tag{29.6}$$

where f_1 and f_2 are the heat transfer and melting factors; R_H = rate of input energy generated by the welding power source, W (Btu/min); A_w = weld cross-sectional area, mm² (in²); and v = the travel velocity of the welding operation, mm/s (in/min). In Chapter 30, we examine how the power density in Eq. (29.1) and the input energy rate for Eq. (29.6) are generated for some of the individual welding processes.

**Example 29.2
Welding Travel
Speed**

The power source in a particular welding setup generates 3500 W that can be transferred to the work surface with a heat transfer factor = 0.7. The metal to be welded is low carbon steel, whose melting temperature, from Table 29.2, is 1760°K. The melting factor in the operation is 0.5. A continuous fillet weld is to be made with a cross-sectional area = 20 mm². Determine the travel speed at which the welding operation can be accomplished.

Solution: Let us first find the unit energy required to melt the metal U_m from Eq. (29.2).

$$U_m = 3.33(10^{-6}) \times 1760^2 = 10.3 \text{ J/mm}^3$$

Rearranging Eq. (29.6) to solve for travel velocity, we have $v = \dfrac{f_1 f_2 R_H}{U_m A_w}$, and solving for the conditions of the problem, $v = \dfrac{0.7\,(0.5)\,(3500)}{10.3\,(20)} = 5.95$ mm/s. ∎

29.4 FEATURES OF A FUSION-WELDED JOINT

Most weld joints are fusion welded. As illustrated in the cross-sectional view of Figure 29.9 (a), a typical fusion-weld joint in which filler metal has been added consists of several zones: (1) fusion zone, (2) weld interface, (3) heat-affected zone, and (4) unaffected base metal zone.

The **fusion zone** consists of a mixture of filler metal and base metal that have completely melted. This zone is characterized by a high degree of homogeneity among the component metals that have been melted during welding. The mixing of these components is motivated largely by convection in the molten weld pool. Solidification in the fusion zone has similarities to a casting process. In welding the mold is formed by the unmelted edges or surfaces of the components being welded. The significant difference between solidification in casting and in welding is that epitaxial grain growth occurs in welding. The reader may recall that in casting, the metallic grains are formed from the melt by nucleation of solid particles at the mold wall, followed by grain growth. In welding, by contrast, the nucleation stage of solidification is avoided by the mechanism of **epitaxial grain growth,** in which atoms from the molten pool solidify on preexisting lattice sites of the adjacent solid base metal. Consequently, the grain structure in the fusion zone near the heat-affected zone tends to mimic the crystallographic orientation of the surrounding heat-affected zone. Further into the fusion zone, a preferred orientation develops in which the grains are roughly perpendicular to the boundaries of the weld interface. The resulting structure in the solidified fusion zone tends to feature coarse columnar grains, as depicted in Figure 29.9(b). The grain structure depends on various factors, including welding process, metals being welded (e.g., identical metals vs. dissimilar metals welded), whether a filler metal is used, and the feed rate at which welding is accomplished. A detailed discussion of welding metallurgy is beyond the scope of this text, and interested readers can consult any of several references [1], [4], [5].

The second zone in the weld joint is the **weld interface,** a narrow boundary that separates the fusion zone from the heat-affected zone. The interface consists of a thin band of base metal that was melted or partially melted (localized melting within the grains) during the welding process but then immediately solidified before any mixing with the metal in the fusion zone. Its chemical composition is therefore identical to that of the base metal.

The third zone in the typical fusion weld is the **heat-affected zone** (HAZ). The metal in this zone has experienced temperatures that are below its melting point, yet high

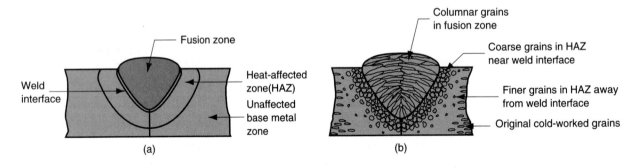

FIGURE 29.9   Cross section of a typical fusion-welded joint: (a) principal zones in the joint and (b) typical grain structure.

enough to cause microstructural changes in the solid metal. The chemical composition in the heat-affected zone is the same as the base metal, but this region has been heat treated due to the welding temperatures so that its properties and structure have been altered. The amount of metallurgical damage in the HAZ depends on factors such as the amount of heat input and peak temperatures reached, distance from the fusion zone, length of time the metal has been subjected to the high temperatures, cooling rate, and the metal's thermal properties. The effect on mechanical properties in the heat-affected zone is usually negative, and it is in this region of the weld joint that welding failures often occur.

As the distance from the fusion zone increases, the **unaffected base metal zone** is finally reached, in which no metallurgical change has occurred. Nevertheless, the base metal surrounding the HAZ is likely to be in a state of high residual stress, the result of shrinkage in the fusion zone.

REFERENCES

[1] *ASM Handbook,* Vol. 6, *Welding, Brazing, and Soldering.* ASM International, Materials Park, Ohio, 1993.

[2] Cary, H. B., and Helzer, S. C. *Modern Welding Technology,* 6th ed. Pearson/Prentice-Hall, Upper Saddle River, New Jersey, 2005.

[3] Datsko, J. *Material Properties and Manufacturing Processes.* John Wiley & Sons, Inc., New York, 1966.

[4] Messler, R. W., Jr. *Principles of Welding: Processes, Physics, Chemistry, and Metallurgy.* John Wiley & Sons, Inc., New York, 1999.

[5] *Welding Handbook,* 9th ed., Vol. 1, *Welding Science and Technology.* American Welding Society, Miami, Florida, 2007.

[6] Wick, C., and Veilleux, R. F. *Tool and Manufacturing Engineers Handbook,* 4th ed., Vol. IV, *Quality Control and Assembly.* Society of Manufacturing Engineers, Dearborn, Michigan, 1987.

REVIEW QUESTIONS

29.1. What are the advantages and disadvantages of welding compared to other types of assembly operations?

29.2. What were the two discoveries of Sir Humphrey Davy that led to the development of modern welding technology?

29.3. What is meant by the term faying surface?

29.4. Define the term fusion weld.

29.5. What is the fundamental difference between a fusion weld and a solid state weld?

29.6. What is an autogenous weld?

29.7. Discuss the reasons why most welding operations are inherently dangerous.

29.8. What is the difference between machine welding and automatic welding?

29.9. Name and sketch the five joint types.

29.10. Define and sketch a fillet weld.

29.11. Define and sketch a groove weld.

29.12. Why is a surfacing weld different from the other weld types?

29.13. Why is it desirable to use energy sources for welding that have high heat densities?

29.14. What is the unit melting energy in welding, and what are the factors on which it depends?

29.15. Define and distinguish the two terms heat transfer factor and melting factor in welding.

29.16. What is the heat-affected zone in a fusion weld?

MULTIPLE CHOICE QUIZ

There are 14 correct answers in the following multiple choice questions (some questions have multiple answers that are correct). To attain a perfect score on the quiz, all correct answers must be given. Each correct answer is worth 1 point. Each

omitted answer or wrong answer reduces the score by 1 point, and each additional answer beyond the correct number of answers reduces the score by 1 point. Percentage score on the quiz is based on the total number of correct answers.

29.1. Welding can only be performed on metals that have the same melting point; otherwise, the metal with the lower melting temperature always melts while the other metal remains solid: (a) true, (b) false?

29.2. A fillet weld can be used to join which of the following joint types (three correct answers): (a) butt, (b) corner, (c) edge, (d) lap, and (e) tee?

29.3. A fillet weld has a cross-sectional shape that is approximately which one of the following: (a) rectangular, (b) round, (c) square, or (d) triangular?

29.4. Groove welds are most closely associated with which one of the following joint types: (a) butt, (b) corner, (c) edge, (d) lap, or (e) tee?

29.5. A flange weld is most closely associated with which one of the following joint types: (a) butt, (b) corner, (c) edge, (d) lap, or (e) tee?

29.6. For metallurgical reasons, it is desirable to melt the weld metal with minimum energy input. Which one of the following heat sources is most consistent with this objective: (a) high power, (b) high power density, (c) low power, or (d) low power density?

29.7. The amount of heat required to melt a given volume of metal depends strongly on which of the following properties (three best answers):

(a) coefficient of thermal expansion, (b) heat of fusion, (c) melting temperature, (d) modulus of elasticity, (e) specific heat, (f) thermal conductivity, and (g) thermal diffusivity?

29.8. The heat transfer factor in welding is correctly defined by which one of the following descriptions: (a) the proportion of the heat received at the work surface that is used for melting, (b) the proportion of the total heat generated at the source that is received at the work surface, (c) the proportion of the total heat generated at the source that is used for melting, or (d) the proportion of the total heat generated at the source that is used for welding?

29.9. The melting factor in welding is correctly defined by which one of the following descriptions: (a) the proportion of the heat received at the work surface that is used for melting, (b) the proportion of the total heat generated at the source that is received at the work surface, (c) the proportion of the total heat generated at the source that is used for melting, or (d) the proportion of the total heat generated at the source that is used for welding?

29.10. Weld failures always occur in the fusion zone of the weld joint, since this is the part of the joint that has been melted: (a) true, (b) false?

PROBLEMS

Power Density

29.1. A heat source can transfer 3500 J/sec to a metal part surface. The heated area is circular, and the heat intensity decreases as the radius increases, as follows: 70% of the heat is concentrated in a circular area that is 3.75 mm in diameter. Is the resulting power density enough to melt metal?

29.2. In a laser beam welding process, what is the quantity of heat per unit time (J/sec) that is transferred to the material if the heat is concentrated in circle with a diameter of 0.2 mm? Assume the power density provided in Table 29.1.

29.3. A welding heat source is capable of transferring 150 Btu/min to the surface of a metal part. The heated area is approximately circular, and the heat intensity decreases with increasing radius as follows: 50% of the power is transferred within a circle of diameter = 0.1 in and 75% is transferred within a concentric circle of diameter = 0.25 in. What are the power densities in (a) the 0.1-in diameter inner circle and (b) the 0.25-in diameter ring that lies around the inner circle? (c) Are these power densities sufficient for melting metal?

Unit Melting Energy

29.4. Compute the unit energy for melting for the following metals: (a) aluminum and (b) plain low carbon steel.

29.5. Compute the unit energy for melting for the following metals: (a) copper and (b) titanium.

29.6. Make the calculations and plot on linearly scaled axes the relationship for unit melting energy as a

function of temperature. Use temperatures as follows to construct the plot: 200°C, 400°C, 600°C, 800°C, 1000°C, 1200°C, 1400°C, 1600°C, 1800°C, and 2000°C. On the plot, mark the positions of some of the welding metals in Table 29.2. Use of a spreadsheet program is recommended for the calculations.

29.7. Make the calculations and plot on linearly scaled axes the relationship for unit melting energy as a function of temperature. Use temperatures as follows to construct the plot: 500°F, 1000°F, 1500°F, 2000°F, 2500°F, 3000°F, and 3500°F. On the plot, mark the positions of some of the welding metals in Table 29.2. Use of a spreadsheet program is recommended for the calculations.

29.8. A fillet weld has a cross-sectional area of 25.0 mm^2 and is 300 mm long. (a) What quantity of heat (in J) is required to accomplish the weld, if the metal to be welded is low carbon steel? (b) How much heat must be generated at the welding source, if the heat transfer factor is 0.75 and the melting factor = 0.63?

29.9. A U-groove weld is used to butt weld 2 pieces of 7.0-mm-thick titanium plate. The U-groove is prepared using a milling cutter so the radius of the groove is 3.0 mm. During welding, the penetration of the weld causes an additional 1.5 mm of material to be melted. The final cross-sectional area of the weld can be approximated by a semicircle with a radius of 4.5 mm. The length of the weld is 200 mm. The melting factor of the setup is 0.57 and the heat transfer factor is 0.86. (a) What is the quantity of heat (in J) required to melt the volume of metal in this weld (filler metal plus base metal)? Assume the resulting top surface of the weld bead is flush with the top surface of the plates. (b) What is the required heat generated at the welding source?

Energy Balance in Welding

29.14. The welding power generated in a particular arc-welding operation = 3000 W. This is transferred to the work surface with a heat transfer factor = 0.9. The metal to be welded is copper whose melting point is given in Table 29.2. Assume that the melting factor = 0.25. A continuous fillet weld is to be made with a cross-sectional area = 15.0 mm^2. Determine the travel speed at which the welding operation can be accomplished.

29.15. Solve the previous problem except that the metal to be welded is high carbon steel, the cross-sectional area of the weld = 25.0 mm^2, and the melting factor = 0.6.

29.16. A welding operation on an aluminum alloy makes a groove weld. The cross-sectional area of the weld is 30.0 mm^2. The welding velocity is 4.0 mm/sec. The heat transfer factor is 0.92 and the melting factor is 0.48. The melting temperature of the aluminum alloy is 650°C. Determine the rate of heat generation required at the welding source to accomplish this weld.

29.10. A groove weld has a cross-sectional area = 0.045 in^2 and is 10 in long. (a) What quantity of heat (in Btu) is required to accomplish the weld, if the metal to be welded is medium carbon steel? (b) How much heat must be generated at the welding source, if the heat transfer factor = 0.9 and the melting factor = 0.7?

29.11. Solve the previous problem, except that the metal to be welded is aluminum, and the corresponding melting factor is half the value for steel.

29.12. In a controlled experiment, it takes 3700 J to melt the amount of metal that is in a weld bead with a cross-sectional area of 6.0 mm^2 that is 150.0 mm long. (a) Using Table 29.2, what is the most likely metal? (b) If the heat transfer factor is 0.85 and the melting factor is 0.55 for a welding process, how much heat must be generated at the welding source to accomplish the weld?

29.13. Compute the unit melting energy for (a) aluminum and (b) steel as the sum of: (1) the heat required to raise the temperature of the metal from room temperature to its melting point, which is the volumetric specific heat multiplied by the temperature rise; and (2) the heat of fusion, so that this value can be compared to the unit melting energy calculated by Eq. (29.2). Use either the SI units or U.S. customary units. Find the values of the properties needed in these calculations either in this text or in other references. Are the values close enough to validate Eq. (29.2)?

29.17. The power source in a particular welding operation generates 125 Btu/min, which is transferred to the work surface with heat transfer factor = 0.8. The melting point for the metal to be welded = 1800°F and its melting factor = 0.5. A continuous fillet weld is to be made with a cross-sectional area = 0.04 in^2. Determine the travel speed at which the welding operation can be accomplished.

29.18. In a certain welding operation to make a fillet weld, the cross-sectional area = 0.025 in^2 and the travel speed = 15 in/min. If the heat transfer factor = 0.95 and melting factor = 0.5, and the melting point = 2000°F for the metal to be welded, determine the rate of heat generation required at the heat source to accomplish this weld.

29.19. A fillet weld is used to join 2 medium carbon steel plates each having a thickness of 5.0 mm. The plates are joined at a 90° angle using an inside fillet corner joint. The velocity of the welding head is 6 mm/sec. Assume the cross section of the weld bead approximates a right isosceles triangle with a leg length of 4.5 mm, the heat transfer factor is 0.80, and the melting factor is 0.58. Determine the rate of heat generation required at the welding source to accomplish the weld.

29.20. A spot weld was made using an arc-welding process. In a spot-welding operation, two 1/16-in thick aluminum plates were joined. The melted metal formed a nugget that had a diameter of 1/4 in. The operation required the power to be on for 4 sec. Assume the final nugget had the same thickness as the two aluminum plates (1/8 in thick), the heat transfer factor was 0.80 and the melting factor was 0.50. Determine the rate of heat generation that was required at the source to accomplish this weld.

29.21. A surfacing weld is to be applied to a rectangular low carbon steel plate that is 200 mm by 350 mm. The filler metal to be added is a harder (alloy) grade of steel, whose melting point is assumed to be the same. A thickness of 2.0 mm will be added to the plate, but with penetration into the base metal, the total thickness melted during welding = 6.0 mm, on average. The surface will be applied by making a series of parallel, overlapped welding beads running lengthwise on the plate. The operation will be carried out automatically with the beads laid down in one long continuous operation at a travel speed = 7.0 mm/s, using welding passes separated by 5 mm. Assume the welding bead is rectangular in cross section: 5 mm by 6 mm. Ignore the minor complications of the turnarounds at the ends of the plate. Assuming the heat transfer factor = 0.8 and the melting factor = 0.6, determine (a) the rate of heat that must be generated at the welding source, and (b) how long will it take to complete the surfacing operation.

29.22. An axle-bearing surface made of high carbon steel has worn beyond its useful life. When it was new, the diameter was 4.00 in. In order to restore it, the diameter was turned to 3.90 in to provide a uniform surface. Next the axle was built up so that it was oversized by the deposition of a surface weld bead, which was deposited in a spiral pattern using a single pass on a lathe. After the weld buildup, the axle was turned again to achieve the original diameter of 4.00 in. The weld metal deposited was a similar composition to the steel in the axle. The length of the bearing surface was 7.0 in. During the welding operation, the welding apparatus was attached to the tool holder, which was fed toward the head of the lathe as the axle rotated. The axle rotated at a speed of 4.0 rev/min. The weld bead height was 3/32 in above the original surface. In addition, the weld bead penetrated 1/16 in into the surface of the axle. The width of the weld bead was 0.25 in, thus the feed on the lathe was set to 0.25 in/rev. Assuming the heat transfer factor was 0.80 and the melting factor was 0.65, determine (a) the relative velocity between the workpiece and the welding head, (b) the rate of heat generated at the welding source, and (c) how long it took to complete the welding portion of this operation.

30 WELDING PROCESSES

Chapter Contents

Welding processes divide into two major categories: (1) *fusion welding,* in which coalescence is accomplished by melting the two parts to be joined, in some cases adding filler metal to the joint; and (2) *solid-state welding,* in which heat and/or pressure are used to achieve coalescence, but no melting of the base metals occurs and no filler metal is added.

Fusion welding is by far the more important category. It includes (1) arc welding, (2) resistance welding, (3) oxyfuel gas welding, and (4) other fusion welding processes— ones that cannot be classified as any of the first three types. Fusion welding processes are discussed in the first four sections of this chapter. Section 30.5 covers solid-state welding. And in the final three sections of the chapter, we examine issues common to all welding operations: weld quality, weldability, and design for welding.

30.1 ARC WELDING

Arc welding (AW) is a fusion-welding process in which coalescence of the metals is achieved by the heat of an electric arc between an electrode and the work. The same basic process is also used in arc cutting (Section 26.3.4). A generic AW process is shown in Figure 30.1. An electric arc is a discharge of electric current across a gap in a circuit. It is sustained by the presence of a thermally ionized column of gas (called a plasma) through which current flows. To initiate the arc in an AW process, the electrode is brought into contact with the work and then quickly separated from it by a short distance. The electric energy from the arc thus formed produces temperatures of 5500°C (10,000°F) or higher, sufficiently hot to melt any metal. A pool of molten metal, consisting of base metal(s) and filler metal (if one is used) is formed near the tip of the electrode. In most arc-welding processes, filler metal is added during the operation to increase the volume and strength of the weld joint. As the

FIGURE 30.1 The basic configuration and electrical circuit of an arc-welding process.

electrode is moved along the joint, the molten weld pool solidifies in its wake. Our Video Clip on welding illustrates the various forms of arc welding described in this section.

VIDEO CLIP ─────────────────────────────

Welding: View the segment on arc welding.

Movement of the electrode relative to the work is accomplished by either a human welder (manual welding) or by mechanical means (i.e., machine welding, automatic welding, or robotic welding). One of the troublesome aspects of manual arc welding is that the quality of the weld joint depends on the skill and work ethic of the human welder. Productivity is also an issue. It is often measured as ***arc time*** (also called ***arc-on time***)— the proportion of hours worked that arc welding is being accomplished:

$$\text{Arc time} = (\text{time arc is on})/(\text{hours worked}) \qquad (30.1)$$

This definition can be applied to an individual welder or to a mechanized workstation. For manual welding, arc time is usually around 20%. Frequent rest periods are needed by the welder to overcome fatigue in manual arc welding, which requires hand-eye coordination under stressful conditions. Arc time increases to about 50% (more or less, depending on the operation) for machine, automatic, and robotic welding.

30.1.1 GENERAL TECHNOLOGY OF ARC WELDING

Before describing the individual AW processes, it is instructional to examine some of the general technical issues that apply to these processes.

Electrodes Electrodes used in AW processes are classified as consumable or nonconsumable. ***Consumable electrodes*** provide the source of the filler metal in arc welding. These electrodes are available in two principal forms: rods (also called sticks) and wire. Welding rods are typically 225 to 450 mm (9–18 in) long and 9.5 mm (3/8 in) or less in diameter. The problem with consumable welding rods, at least in production welding operations, is that they must be changed periodically, reducing arc time of the welder. Consumable weld wire has the advantage that it can be continuously fed into the weld pool from spools containing long lengths of wire, thus avoiding the frequent interruptions that occur when using welding sticks. In both rod and wire forms, the electrode is consumed by the arc during the welding process and added to the weld joint as filler metal.

Nonconsumable electrodes are made of tungsten (or carbon, rarely), which resists melting by the arc. Despite its name, a nonconsumable electrode is gradually depleted during the welding process (vaporization is the principal mechanism), analogous to the gradual wearing of a cutting tool in a machining operation. For AW processes that utilize nonconsumable electrodes, any filler metal used in the operation must be supplied by means of a separate wire that is fed into the weld pool.

Arc Shielding   At the high temperatures in arc welding, the metals being joined are chemically reactive to oxygen, nitrogen, and hydrogen in the air. The mechanical properties of the weld joint can be seriously degraded by these reactions. Thus, some means to shield the arc from the surrounding air is provided in nearly all AW processes. Arc shielding is accomplished by covering the electrode tip, arc, and molten weld pool with a blanket of gas or flux, or both, which inhibit exposure of the weld metal to air.

Common shielding gases include argon and helium, both of which are inert. In the welding of ferrous metals with certain AW processes, oxygen and carbon dioxide are used, usually in combination with Ar and/or He, to produce an oxidizing atmosphere or to control weld shape.

A *flux* is a substance used to prevent the formation of oxides and other unwanted contaminants, or to dissolve them and facilitate removal. During welding, the flux melts and becomes a liquid slag, covering the operation and protecting the molten weld metal. The slag hardens upon cooling and must be removed later by chipping or brushing. Flux is usually formulated to serve several additional functions: (1) provide a protective atmosphere for welding, (2) stabilize the arc, and (3) reduce spattering.

The method of flux application differs for each process. The delivery techniques include (1) pouring granular flux onto the welding operation, (2) using a stick electrode coated with flux material in which the coating melts during welding to cover the operation, and (3) using tubular electrodes in which flux is contained in the core and released as the electrode is consumed. These techniques are discussed further in our descriptions of the individual AW processes.

Power Source in Arc Welding   Both direct current (DC) and alternating current (AC) are used in arc welding. AC machines are less expensive to purchase and operate, but are generally restricted to welding of ferrous metals. DC equipment can be used on all metals with good results and is generally noted for better arc control.

In all arc-welding processes, power to drive the operation is the product of the current I passing through the arc and the voltage E across it. This power is converted into heat, but not all of the heat is transferred to the surface of the work. Convection, conduction, radiation, and spatter account for losses that reduce the amount of usable heat. The effect of the losses is expressed by the heat transfer factor f_1 (Section 29.3). Some representative values of f_1 for several AW processes are given in Table 30.1. Heat transfer factors are

TABLE 30.1 Heat transfer factors for several arc-welding processes.	
Arc-Welding Process[a]	**Typical Heat Transfer Factor f_1**
Shielded metal arc welding	0.9
Gas metal arc welding	0.9
Flux-cored arc welding	0.9
Submerged arc welding	0.95
Gas tungsten arc welding	0.7

Compiled from [5].
[a]The arc-welding processes are described in Sections 30.1.2 and 30.1.3.

greater for AW processes that use consumable electrodes because most of the heat consumed in melting the electrode is subsequently transferred to the work as molten metal. The process with the lowest f_1 value in Table 30.1 is gas tungsten arc welding, which uses a nonconsumable electrode. Melting factor f_2 (Section 29.3) further reduces the available heat for welding. The resulting power balance in arc welding is defined by

$$R_{Hw} = f_1 f_2 IE = U_m A_w v \tag{30.2}$$

where E = voltage, V; I = current, A; and the other terms were defined in Section 29.3. The units of R_{Hw} are watts (current multiplied by voltage), which equal J/sec. This can be converted to Btu/sec by recalling that 1 Btu = 1055 J, and thus 1 Btu/sec = 1055 watts.

Example 30.1
Power in Arc
Welding

A gas tungsten arc-welding operation is performed at a current of 300 A and voltage of 20 V. The melting factor $f_2 = 0.5$, and the unit melting energy for the metal $U_m = 10$ J/mm^3. Determine (a) power in the operation, (b) rate of heat generation at the weld, and (c) volume rate of metal welded.

Solution: (a) The power in this arc-welding operation is

$$P = IE = (300\ \text{A})(20\ \text{V}) = 6000\ \text{W}$$

(b) From Table 30.1, the heat transfer factor $f_1 = 0.7$. The rate of heat used for welding is given by

$$R_{Hw} = f_1 f_2 IE = (0.7)(0.5)(6000) = 2100\ \text{W} = 2100\ \text{J/s}$$

(c) The volume rate of metal welded is

$$R_{VW} = (2100\ \text{J/s})/(10\ \text{J/mm}^3) = 210\ \text{mm}^3/\text{s}$$

30.1.2   AW PROCESSES—CONSUMABLE ELECTRODES

A number of important arc-welding processes use consumable electrodes. These are discussed in this section. Symbols for the welding processes are those used by the American Welding Society.

Shielded Metal Arc Welding   *Shielded metal arc welding* (SMAW) is an AW process that uses a consumable electrode consisting of a filler metal rod coated with chemicals that provide flux and shielding. The process is illustrated in Figures 30.2 and 30.3. The welding stick (SMAW is sometimes called *stick welding*) is typically 225 to 450 mm (9–18 in) long and 2.5 to 9.5 mm (3/32–3/8 in) in diameter. The filler metal used in the rod must be compatible with the metal to be welded, the composition usually being very close to that of the base metal. The coating consists of powdered cellulose (i.e., cotton and wood powders) mixed with oxides, carbonates, and other ingredients, held together by a silicate binder. Metal powders are also sometimes included in the coating to increase the amount of filler metal and to add alloying elements. The heat of the welding process melts the coating to provide a protective atmosphere and slag for the welding operation. It also helps to stabilize the arc and regulate the rate at which the electrode melts.

During operation the bare metal end of the welding stick (opposite the welding tip) is clamped in an electrode holder that is connected to the power source. The holder has an insulated handle so that it can be held and manipulated by a human welder. Currents typically used in SMAW range between 30 and 300 A at voltages from 15 to 45 V. Selection of the proper power parameters depends on the metals being welded, electrode type and length, and depth of weld penetration required. Power supply, connecting cables, and electrode holder can be bought for a few thousand dollars.

FIGURE 30.2 Shielded metal arc welding (stick welding) performed by a (human) welder. (Photo courtesy of Hobart Brothers, Troy, Ohio.)

Shielded metal arc welding is usually performed manually. Common applications include construction, pipelines, machinery structures, shipbuilding, job shop fabrication, and repair work. It is preferred over oxyfuel welding for thicker sections—above 5 mm (3/16 in)—because of its higher power density. The equipment is portable and low cost, making SMAW highly versatile and probably the most widely used of the AW processes. Base metals include steels, stainless steels, cast irons, and certain nonferrous alloys. It is not used or seldom used for aluminum and its alloys, copper alloys, and titanium.

A disadvantage of shielded metal arc welding as a production operation is the use of the consumable electrode stick. As the sticks are used up, they must periodically be changed. This reduces the arc time with this welding process. Another limitation is the current level that can be used. Because the electrode length varies during the operation and this length affects the resistance heating of the electrode, current levels must be maintained within a safe range or the coating will overheat and melt prematurely when starting a new welding stick. Some of the other AW processes overcome the limitations of welding stick length in SMAW by using a continuously fed wire electrode.

Gas Metal Arc Welding *Gas metal arc welding* (GMAW) is an AW process in which the electrode is a consumable bare metal wire, and shielding is accomplished by flooding the arc

FIGURE 30.3 Shielded metal arc welding (SMAW).

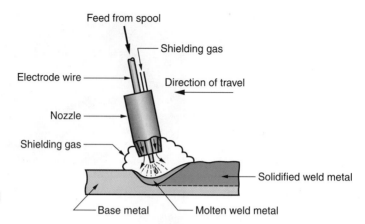

FIGURE 30.4 Gas metal arc welding (GMAW).

with a gas. The bare wire is fed continuously and automatically from a spool through the welding gun, as illustrated in Figure 30.4. A welding gun is shown in Figure 30.5. Wire diameters ranging from 0.8 to 6.5 mm (1/32–1/4 in) are used in GMAW, the size depending on the thickness of the parts being joined and the desired deposition rate. Gases used for shielding include inert gases such as argon and helium, and active gases such as carbon dioxide. Selection of gases (and mixtures of gases) depends on the metal being welded, as well as other factors. Inert gases are used for welding aluminum alloys and stainless steels,

FIGURE 30.5 Welding gun for gas metal arc welding. (Courtesy of Lincoln Electric Company, Cleveland, Ohio.)

while CO_2 is commonly used for welding low and medium carbon steels. The combination of bare electrode wire and shielding gases eliminates the slag covering on the weld bead and thus precludes the need for manual grinding and cleaning of the slag. The GMAW process is therefore ideal for making multiple welding passes on the same joint.

The various metals on which GMAW is used and the variations of the process itself have given rise to a variety of names for gas metal arc welding. When the process was first introduced in the late 1940s, it was applied to the welding of aluminum using inert gas (argon) for arc shielding. The name applied to this process was **_MIG welding_** (for **_m_**etal **_i_**nert **_g_**as welding). When the same welding process was applied to steel, it was found that inert gases were expensive and CO_2 was used as a substitute. Hence the term **_CO$_2$ welding_** was applied. Refinements in GMAW for steel welding have led to the use of gas mixtures, including CO_2 and argon, and even oxygen and argon.

GMAW is widely used in fabrication operations in factories for welding a variety of ferrous and nonferrous metals. Because it uses continuous weld wire rather than welding sticks, it has a significant advantage over SMAW in terms of arc time when performed manually. For the same reason, it also lends itself to automation of arc welding. The electrode stubs remaining after stick welding also wastes filler metal, so the utilization of electrode material is higher with GMAW. Other features of GMAW include elimination of slag removal (since no flux is used), higher deposition rates than SMAW, and good versatility.

Flux-Cored Arc Welding   This arc-welding process was developed in the early 1950s as an adaptation of shielded metal arc welding to overcome the limitations imposed by the use of stick electrodes. **_Flux-cored arc welding_** (FCAW) is an arc-welding process in which the electrode is a continuous consumable tubing that contains flux and other ingredients in its core. Other ingredients may include deoxidizers and alloying elements. The tubular flux-cored "wire" is flexible and can therefore be supplied in the form of coils to be continuously fed through the arc-welding gun. There are two versions of FCAW: (1) self-shielded and (2) gas shielded. In the first version of FCAW to be developed, arc shielding was provided by a flux core, thus leading to the name **_self-shielded flux-cored arc welding._** The core in this form of FCAW includes not only fluxes but also ingredients that generate shielding gases for protecting the arc. The second version of FCAW, developed primarily for welding steels, obtains arc shielding from externally supplied gases, similar to gas metal arc welding. This version is called **_gas-shielded flux-cored arc welding._** Because it utilizes an electrode containing its own flux together with separate shielding gases, it might be considered a hybrid of SMAW and GMAW. Shielding gases typically employed are carbon dioxide for mild steels or mixtures of argon and carbon dioxide for stainless steels. Figure 30.6 illustrates the FCAW process, with the gas (optional) distinguishing between the two types.

FIGURE 30.6   Flux-cored arc welding. The presence or absence of externally supplied shielding gas distinguishes the two types: (1) self-shielded, in which the core provides the ingredients for shield-ing; and (2) gas shielded, in which external shielding gases are supplied.

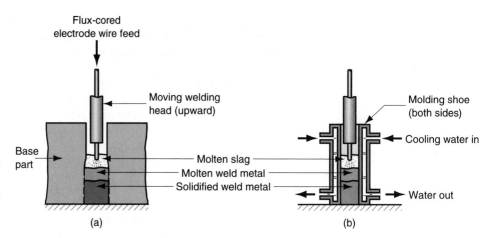

FIGURE 30.7
Electrogas welding using flux-cored electrode wire: (a) front view with molding shoe removed for clarity, and (b) side view showing molding shoes on both sides.

FCAW has advantages similar to GMAW, due to continuous feeding of the electrode. It is used primarily for welding steels and stainless steels over a wide stock thickness range. It is noted for its capability to produce very-high-quality weld joints that are smooth and uniform.

Electrogas Welding   Electrogas welding (EGW) is an AW process that uses a continuous consumable electrode (either flux-cored wire or bare wire with externally supplied shielding gases) and molding shoes to contain the molten metal. The process is primarily applied to vertical butt welding, as pictured in Figure 30.7. When the flux-cored electrode wire is employed, no external gases are supplied, and the process can be considered a special application of self-shielded FCAW. When a bare electrode wire is used with shielding gases from an external source, it is considered a special case of GMAW. The molding shoes are water cooled to prevent their being added to the weld pool. Together with the edges of the parts being welded, the shoes form a container, almost like a mold cavity, into which the molten metal from the electrode and base parts is gradually added. The process is performed automatically, with a moving weld head traveling vertically upward to fill the cavity in a single pass.

Principal applications of electrogas welding are steels (low- and medium-carbon, low-alloy, and certain stainless steels) in the construction of large storage tanks and in shipbuilding. Stock thicknesses from 12 to 75 mm (0.5–3.0 in) are within the capacity of EGW. In addition to butt welding, it can also be used for fillet and groove welds, always in a vertical orientation. Specially designed molding shoes must sometimes be fabricated for the joint shapes involved.

Submerged Arc Welding   This process, developed during the 1930s, was one of the first AW processes to be automated. **Submerged arc welding** (SAW) is an arc-welding process that uses a continuous, consumable bare wire electrode, and arc shielding is provided by a cover of granular flux. The electrode wire is fed automatically from a coil into the arc. The flux is introduced into the joint slightly ahead of the weld arc by gravity from a hopper, as shown in Figure 30.8. The blanket of granular flux completely submerges the welding operation, preventing sparks, spatter, and radiation that are so hazardous in other AW processes. Thus, the welding operator in SAW need not wear the somewhat cumbersome face shield required in the other operations (safety glasses and protective gloves, of course, are required). The portion of the flux closest to the arc is melted, mixing with the molten weld metal to remove impurities and then solidifying on top of the weld joint to form a glass-like slag. The slag and unfused flux granules on top provide good protection from the atmosphere and good thermal insulation for the weld area, resulting in relatively slow cooling and a high-quality weld joint, noted for toughness and ductility. As depicted in our

FIGURE 30.8 Submerged arc welding (SAW).

sketch, the unfused flux remaining after welding can be recovered and reused. The solid slag covering the weld must be chipped away, usually by manual means.

Submerged arc welding is widely used in steel fabrication for structural shapes (e.g., welded I-beams); longitudinal and circumferential seams for large diameter pipes, tanks, and pressure vessels; and welded components for heavy machinery. In these kinds of applications, steel plates of 25-mm (1.0-in) thickness and heavier are routinely welded by this process. Low-carbon, low-alloy, and stainless steels can be readily welded by SAW; but not high-carbon steels, tool steels, and most nonferrous metals. Because of the gravity feed of the granular flux, the parts must always be in a horizontal orientation, and a backup plate is often required beneath the joint during the welding operation.

30.1.3   AW PROCESSES—NONCONSUMABLE ELECTRODES

The AW processes discussed above use consumable electrodes. Gas tungsten arc welding, plasma arc welding, and several other processes use nonconsumable electrodes.

Gas Tungsten Arc Welding   Gas tungsten arc welding (GTAW) is an AW process that uses a nonconsumable tungsten electrode and an inert gas for arc shielding. The term ***TIG welding*** (*t*ungsten *i*nert *g*as welding) is often applied to this process (in Europe, ***WIG welding*** is the term—the chemical symbol for tungsten is W, for Wolfram). GTAW can be implemented with or without a filler metal. Figure 30.9 illustrates the latter case. When a filler metal is used, it is added to the weld pool from a separate rod or wire, being melted by the heat of the arc rather than transferred across the arc as in the consumable electrode AW processes. Tungsten is a good electrode material due to its high melting point of 3410°C (6170°F). Typical shielding gases include argon, helium, or a mixture of these gas elements.

GTAW is applicable to nearly all metals in a wide range of stock thicknesses. It can also be used for joining various combinations of dissimilar metals. Its most common applications

FIGURE 30.9   Gas tungsten arc welding (GTAW).

FIGURE 30.10 Plasma arc welding (PAW).

are for aluminum and stainless steel. Cast irons, wrought irons, and of course tungsten are difficult to weld by GTAW. In steel welding applications, GTAW is generally slower and more costly than the consumable electrode AW processes, except when thin sections are involved and very-high-quality welds are required. When thin sheets are TIG welded to close tolerances, filler metal is usually not added. The process can be performed manually or by machine and automated methods for all joint types. Advantages of GTAW in the applications to which it is suited include high-quality welds, no weld spatter because no filler metal is transferred across the arc, and little or no postweld cleaning because no flux is used.

Plasma Arc Welding   *Plasma arc welding* (PAW) is a special form of gas tungsten arc welding in which a constricted plasma arc is directed at the weld area. In PAW, a tungsten electrode is contained in a specially designed nozzle that focuses a high-velocity stream of inert gas (e.g., argon or argon–hydrogen mixtures) into the region of the arc to form a high-velocity, intensely hot plasma arc stream, as in Figure 30.10. Argon, argon–hydrogen, and helium are also used as the arc-shielding gases.

Temperatures in plasma arc welding reach 17,000°C (30,000°F) or greater, hot enough to melt any known metal. The reason why temperatures are so high in PAW (significantly higher than those in GTAW) derives from the constriction of the arc. Although the typical power levels used in PAW are below those used in GTAW, the power is highly concentrated to produce a plasma jet of small diameter and very high power density.

Plasma arc welding was introduced around 1960 but was slow to catch on. In recent years its use is increasing as a substitute for GTAW in applications such as automobile subassemblies, metal cabinets, door and window frames, and home appliances. Owing to the special features of PAW, its advantages in these applications include good arc stability, better penetration control than most other AW processes, high travel speeds, and excellent weld quality. The process can be used to weld almost any metal, including tungsten. Difficult-to-weld metals with PAW include bronze, cast irons, lead, and magnesium. Other limitations include high equipment cost and larger torch size than other AW operations, which tends to restrict access in some joint configurations.

Other Arc-Welding and Related Processes   The preceding AW processes are the most important commercially. There are several others that should be mentioned, which are special cases or variations of the principal AW processes.

Carbon arc welding (CAW) is an arc-welding process in which a nonconsumable carbon (graphite) electrode is used. It has historical importance because it was the first arc-welding process to be developed, but its commercial importance today is practically nil. The carbon arc process is used as a heat source for brazing and for repairing iron castings. It can also be used in some applications for depositing wear-resistant materials on surfaces. Graphite electrodes for welding have been largely superseded by tungsten (in GTAW and PAW).

FIGURE 30.11   Stud arc welding (SW): (1) stud is positioned; (2) current flows from the gun, and stud is pulled from base to establish arc and create a molten pool; (3) stud is plunged into molten pool; and (4) ceramic ferrule is removed after solidification.

Stud welding (SW) is a specialized AW process for joining studs or similar components to base parts. A typical SW operation is illustrated in Figure 30.11, in which shielding is obtained by the use of a ceramic ferrule. To begin with, the stud is chucked in a special weld gun that automatically controls the timing and power parameters of the steps shown in the sequence. The worker must only position the gun at the proper location against the base workpart to which the stud will be attached and pull the trigger. SW applications include threaded fasteners for attaching handles to cookware, heat radiation fins on machinery, and similar assembly situations. In high-production operations, stud welding usually has advantages over rivets, manually arc-welded attachments, and drilled and tapped holes.

30.2 RESISTANCE WELDING

Resistance welding (RW) is a group of fusion-welding processes that uses a combination of heat and pressure to accomplish coalescence, the heat being generated by electrical resistance to current flow at the junction to be welded. The principal components in resistance welding are shown in Figure 30.12 for a resistance spot-welding operation, the most widely used process in the group. The components include workparts to be welded

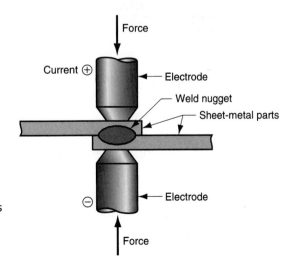

FIGURE 30.12
Resistance welding (RW), showing the components in spot welding, the predominant process in the RW group.

(usually sheet metal parts), two opposing electrodes, a means of applying pressure to squeeze the parts between the electrodes, and an AC power supply from which a controlled current can be applied. The operation results in a fused zone between the two parts, called a *weld nugget* in spot welding.

By comparison to arc welding, resistance welding uses no shielding gases, flux, or filler metal; and the electrodes that conduct electrical power to the process are non-consumable. RW is classified as fusion welding because the applied heat almost always causes melting of the faying surfaces. However, there are exceptions. Some welding operations based on resistance heating use temperatures below the melting points of the base metals, so fusion does not occur.

30.2.1   POWER SOURCE IN RESISTANCE WELDING

The heat energy supplied to the welding operation depends on current flow, resistance of the circuit, and length of time the current is applied. This can be expressed by the equation

$$H = I^2 R t \tag{30.3}$$

where $H =$ heat generated, J (to convert to Btu divide by 1055); $I =$ current, A; $R =$ electrical resistance, Ω; and $t =$ time, s.

The current used in resistance welding operations is very high (5000 to 20,000 A, typically), although voltage is relatively low (usually below 10 V). The duration t of the current is short in most processes, perhaps lasting 0.1 to 0.4 s in a typical spot-welding operation.

The reason why such a high current is used in RW is because (1) the squared term in Eq. (30.3) amplifies the effect of current, and (2) the resistance is very low (around 0.0001 Ω). Resistance in the welding circuit is the sum of (1) resistance of the electrodes, (2) resistances of the workparts, (3) contact resistances between electrodes and workparts, and (4) contact resistance of the faying surfaces. Thus, heat is generated in all of these regions of electrical resistance. The ideal situation is for the faying surfaces to be the largest resistance in the sum, since this is the desired location of the weld. The resistance of the electrodes is minimized by using metals with very low resistivities, such as copper. Also, the electrodes are often water cooled to dissipate the heat that is generated there. The workpart resistances are a function of the resistivities of the base metals and the part thicknesses. The contact resistances between the electrodes and the parts are determined by the contact areas (i.e., size and shape of the electrode) and the condition of the surfaces (e.g., cleanliness of the work surfaces and scale on the electrode). Finally, the resistance at the faying surfaces depends on surface finish, cleanliness, contact area, and pressure. No paint, oil, dirt, or other contaminants should be present to separate the contacting surfaces.

**Example 30.2
Resistance
Welding**

A resistance spot-welding operation is performed on two pieces of 1.5-mm-thick sheet steel using 12,000 A for a 0.20 s duration. The electrodes are 6 mm in diameter at the contacting surfaces. Resistance is assumed to be 0.0001 Ω, and the resulting weld nugget is 6 mm in diameter and 2.5 mm thick. The unit melting energy for the metal $U_m = 12.0$ J/ mm^3. What portion of the heat generated was used to form the weld nugget, and what portion was dissipated into the work metal, electrodes, and surrounding air?

Solution: The heat generated in the operation is given by Eq. (30.3) as.

$$H = (12,000)^2 (0.0001)(0.2) = 2880 \text{ J}$$

The volume of the weld nugget (assumed disc-shaped) is

$$v = 2.5 \frac{\pi (6)^2}{4} 70.7 \text{ mm}^3.$$

The heat required to melt this volume of metal is $H_w = 70.7(12.0) = 848$ J. The remaining heat, $2880 - 848 = 2032$ J (70.6% of the total), is lost into the work metal, electrodes, and surrounding air. In effect, this loss represents the combined effect of the heat transfer factor f_1 and the melting factor f_2 (Section 29.3).   ∎

Success in resistance welding depends on pressure as well as heat. The principal functions of pressure in RW are to (1) force contact between the electrodes and the workparts and between the two work surfaces prior to applying current, and (2) press the faying surfaces together to accomplish coalescence when the proper welding temperature has been reached.

General advantages of resistance welding include (1) no filler metal is required, (2) high production rates are possible, (3) lends itself to mechanization and automation, (4) operator skill level is lower than that required for arc welding, and (5) good repeatability and reliability. Drawbacks are (1) equipment cost is high—usually much higher than most arc-welding operations, and (2) types of joints that can be welded are limited to lap joints for most RW processes.

30.2.2 RESISTANCE-WELDING PROCESSES

The resistance-welding processes of most commercial importance are spot, seam, and projection welding. These processes are illustrated in our Video Clip on welding.

VIDEO CLIP

Welding: View the segment titled Resistance Welding.

Resistance Spot Welding   Resistance spot welding is by far the predominant process in this group. It is widely used in mass production of automobiles, appliances, metal furniture, and other products made of sheet metal. If one considers that a typical car body has approximately 10,000 individual spot welds, and that the annual production of automobiles throughout the world is measured in tens of millions of units, the economic importance of resistance spot welding can be appreciated.

Resistance spot welding (RSW) is an RW process in which fusion of the faying surfaces of a lap joint is achieved at one location by opposing electrodes. The process is used to join sheet-metal parts of thickness 3 mm (0.125 in) or less, using a series of spot welds, in situations where an airtight assembly is not required. The size and shape of the weld spot is determined by the electrode tip, the most common electrode shape being round, but hexagonal, square, and other shapes are also used. The resulting weld nugget is typically 5 to 10 mm (0.2–0.4 in) in diameter, with a heat-affected zone extending slightly beyond the nugget into the base metals. If the weld is made properly, its strength will be comparable to that of the surrounding metal. The steps in a spot welding cycle are depicted in Figure 30.13.

Materials used for RSW electrodes consist of two main groups: (1) copper-based alloys and (2) refractory metal compositions such as copper and tungsten combinations. The second group is noted for superior wear resistance. As in most manufacturing processes, the tooling in spot welding gradually wears out as it is used. Whenever practical, the electrodes are designed with internal passageways for water cooling.

Because of its widespread industrial use, various machines and methods are available to perform spot-welding operations. The equipment includes rocker-arm and press-type spot-welding machines, and portable spot-welding guns. ***Rocker-arm spot welders,*** shown in Figure 30.14, have a stationary lower electrode and a movable upper electrode that can be raised and lowered for loading and unloading the work. The upper electrode is mounted

FIGURE 30.13 (a) Steps in a spot-welding cycle, and (b) plot of squeezing force and current during cycle. The sequence is: (1) parts inserted between open electrodes, (2) electrodes close and force is applied, (3) weld time—current is switched on, (4) current is turned off but force is maintained or increased (a reduced current is sometimes applied near the end of this step for stress relief in the weld region), and (5) electrodes are opened, and the welded assembly is removed.

on a rocker arm (hence the name) whose movement is controlled by a foot pedal operated by the worker. Modern machines can be programmed to control force and current during the weld cycle.

Press-type spot welders are intended for larger work. The upper electrode has a straight-line motion provided by a vertical press that is pneumatically or hydraulically powered. The press action permits larger forces to be applied, and the controls usually permit programming of complex weld cycles.

The previous two machine types are both stationary spot welders, in which the work is brought to the machine. For large, heavy work it is difficult to move and position the part into stationary machines. For these cases, *portable spot-welding guns* are available in

FIGURE 30.14 Rocker-arm spot-welding machine.

FIGURE 30.15
Resistance seam
welding (RSEW).

various sizes and configurations. These devices consist of two opposing electrodes contained in a pincer mechanism. Each unit is lightweight so that it can be held and manipulated by a human worker or an industrial robot. The gun is connected to its own power and control source by means of flexible electrical cables and air hoses. Water cooling for the electrodes, if needed, can also be provided through a water hose. Portable spot-welding guns are widely used in automobile final assembly plants to spot weld car bodies. Some of these guns are operated by people, but industrial robots have become the preferred technology, illustrated in Figure 38.16.

Resistance Seam Welding   In resistance seam welding (RSEW), the stick-shaped electrodes in spot welding are replaced by rotating wheels, as shown in Figure 30.15, and a series of overlapping spot welds are made along the lap joint. The process is capable of producing air-tight joints, and its industrial applications include the production of gasoline tanks, automobile mufflers, and various other fabricated sheet metal containers. Technically, RSEW is the same as spot welding, except that the wheel electrodes introduce certain complexities. Since the operation is usually carried out continuously, rather than discretely, the seams should be along a straight or uniformly curved line. Sharp corners and similar discontinuities are difficult to deal with. Also, warping of the parts becomes more of a factor in resistance seam welding, and fixtures are required to hold the work in position and minimize distortion.

The spacing between the weld nuggets in resistance seam welding depends on the motion of the electrode wheels relative to the application of the weld current. In the usual method of operation, called ***continuous motion welding,*** the wheel is rotated continuously at a constant velocity, and current is turned on at timing intervals consistent with the desired spacing between spot welds along the seam. Frequency of the current discharges is normally set so that overlapping weld spots are produced. But if the frequency is reduced sufficiently, then there will be spaces between the weld spots, and this method is termed ***roll spot welding.*** In another variation, the welding current remains on at a constant level (rather than being pulsed) so that a truly continuous welding seam is produced. These variations are depicted in Figure 30.16.

An alternative to continuous motion welding is ***intermittent motion welding,*** in which the electrode wheel is periodically stopped to make the spot weld. The amount of wheel rotation between stops determines the distance between weld spots along the seam, yielding patterns similar to (a) and (b) in Figure 30.16.

Seam-welding machines are similar to press-type spot welders except that electrode wheels are used rather than the usual stick-shaped electrodes. Cooling of the work and wheels is often necessary in RSEW, and this is accomplished by directing water at the top and underside of the workpart surfaces near the electrode wheels.

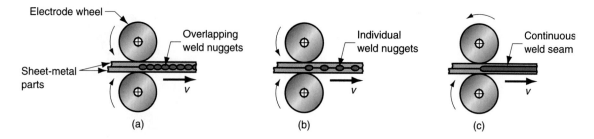

FIGURE 30.16 Different types of seams produced by electrode wheels: (a) conventional resistance seam welding, in which overlapping spots are produced; (b) roll spot welding; and (c) continuous resistance seam.

Resistance Projection Welding Resistance projection welding (RPW) is an RW process in which coalescence occurs at one or more relatively small contact points on the parts. These contact points are determined by the design of the parts to be joined, and may consist of projections, embossments, or localized intersections of the parts. A typical case in which two sheet-metal parts are welded together is described in Figure 30.17. The part on top has been fabricated with two embossed points to contact the other part at the start of the process. It might be argued that the embossing operation increases the cost of the part, but this increase may be more than offset by savings in welding cost.

There are variations of resistance projection welding, two of which are shown in Figure 30.18. In one variation, fasteners with machined or formed projections can be permanently joined to sheet or plate by RPW, facilitating subsequent assembly operations. Another variation, called **cross-wire welding,** is used to fabricate welded wire products such as wire fence, shopping carts, and stove grills. In this process, the contacting surfaces of the round wires serve as the projections to localize the resistance heat for welding.

Other Resistance-Welding Operations In addition to the principal RW processes described above, several additional processes in this group should be identified: flash, upset, percussion, and high-frequency resistance welding.

In *flash welding* (FW), normally used for butt joints, the two surfaces to be joined are brought into contact or near contact and electric current is applied to heat the surfaces to the melting point, after which the surfaces are forced together to form the weld. The two steps are outlined in Figure 30.19. In addition to resistance heating, some arcing occurs (called *flashing,* hence the name of the welding process), depending on the extent of

FIGURE 30.17
Resistance projection welding (RPW): (1) at start of operation, contact between parts is at projections; and (2) when current is applied, weld nuggets similar to those in spot welding are formed at the projections.

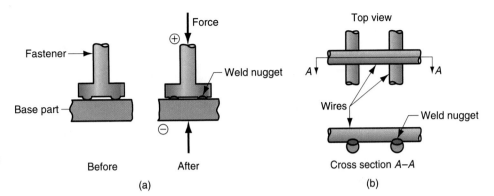

FIGURE 30.18
Variations of resistance projection welding: (a) welding of a machined or formed fastener onto a sheet-metal part; and (b) cross-wire welding.

contact between the faying surfaces, so flash welding is sometimes classified in the arc-welding group. Current is usually stopped during upsetting. Some metal, as well as contaminants on the surfaces, is squeezed out of the joint and must be subsequently machined to provide a joint of uniform size.

Applications of flash welding include butt welding of steel strips in rolling-mill operations, joining ends of wire in wire drawing, and welding of tubular parts. The ends to be joined must have the same cross sections. For these kinds of high-production applications, flash welding is fast and economical, but the equipment is expensive.

Upset welding (UW) is similar to flash welding except that in UW the faying surfaces are pressed together during heating and upsetting. In flash welding, the heating and pressing steps are separated during the cycle. Heating in UW is accomplished entirely by electrical resistance at the contacting surfaces; no arcing occurs. When the faying surfaces have been heated to a suitable temperature below the melting point, the force pressing the parts together is increased to cause upsetting and coalescence in the contact region. Thus, upset welding is not a fusion-welding process in the same sense as the other welding processes we have discussed. Applications of UW are similar to those of flash welding: joining ends of wire, pipes, tubes, and so on.

Percussion welding (PEW) is also similar to flash welding, except that the duration of the weld cycle is extremely short, typically lasting only 1 to 10 ms. Fast heating is accomplished by rapid discharge of electrical energy between the two surfaces to be joined, followed immediately by percussion of one part against the other to form the weld. The heating is very localized, making this process attractive for electronic applications in which the dimensions are very small and nearby components may be sensitive to heat.

High-frequency resistance welding (HFRW) is a resistance-welding process in which a high-frequency alternating current is used for heating, followed by the rapid application of an upsetting force to cause coalescence, as in Figure 30.20(a). The frequencies are 10 to 500 kHz, and the electrodes make contact with the work in the immediate vicinity of the weld joint. In a variation of the process, called *high-frequency*

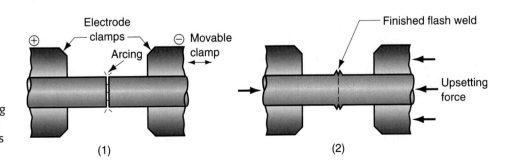

FIGURE 30.19 Flash welding (FW): (1) heating by electrical resistance; and (2) upsetting—parts are forced together.

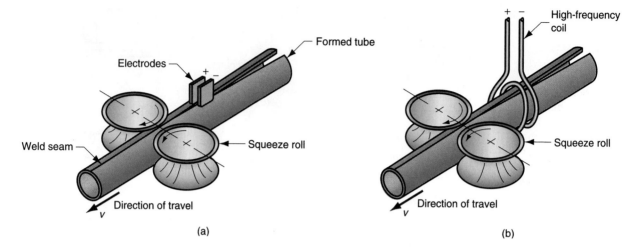

FIGURE 30.20 Welding of tube seams by: (a) high-frequency resistance welding, and (b) high-frequency induction welding.

induction welding (HFIW), the heating current is induced in the parts by a high-frequency induction coil, as in Figure 30.20(b). The coil does not make physical contact with the work. The principal applications of both HFRW and HFIW are continuous butt welding of the longitudinal seams of metal pipes and tubes.

30.3 OXYFUEL GAS WELDING

Oxyfuel gas welding (OFW) is the term used to describe the group of FW operations that burn various fuels mixed with oxygen to perform welding. The OFW processes employ several types of gases, which is the primary distinction among the members of this group. Oxyfuel gas is also commonly used in cutting torches to cut and separate metal plates and other parts (Section 26.3.5). The most important OFW process is oxyacetylene welding.

30.3.1 OXYACETYLENE WELDING

Oxyacetylene welding (OAW) is a fusion-welding process performed by a high-temperature flame from combustion of acetylene and oxygen. The flame is directed by a welding torch. A filler metal is sometimes added, and pressure is occasionally applied in OAW between the contacting part surfaces. A typical OAW operation is sketched in Figure 30.21.

FIGURE 30.21 A typical oxyacetylene welding operation (OAW).

FIGURE 30.22 The neutral flame from an oxyacetylene torch, indicating temperatures achieved.

When filler metal is used, it is typically in the form of a rod with diameters ranging from 1.6 to 9.5 mm (1/16–3/8 in). Composition of the filler must be similar to that of the base metals. The filler is often coated with a **flux** that helps to clean the surfaces and prevent oxidation, thus creating a better weld joint.

Acetylene (C_2H_2) is the most popular fuel among the OFW group because it is capable of higher temperatures than any of the others—up to 3480°C (6300°F). The flame in OAW is produced by the chemical reaction of acetylene and oxygen in two stages. The first stage is defined by the reaction

$$C_2H_2 + O_2 \rightarrow 2CO + H_2 + \text{heat} \qquad (30.4a)$$

the products of which are both combustible, which leads to the second-stage reaction

$$2CO + H_2 + 1.5O_2 \rightarrow 2CO_2 + H_2O + \text{heat} \qquad (30.4b)$$

The two stages of combustion are visible in the oxyacetylene flame emitted from the torch. When the mixture of acetylene and oxygen is in the ratio 1:1, as described in Eq. (30.4), the resulting **neutral flame** is shown in Figure 30.22. The first-stage reaction is seen as the inner cone of the flame (which is bright white), while the second-stage reaction is exhibited by the outer envelope (which is nearly colorless but with tinges ranging from blue to orange). The maximum temperature of the flame is reached at the tip of the inner cone; the second-stage temperatures are somewhat below those of the inner cone. During welding, the outer envelope spreads out and covers the work surfaces being joined, thus shielding them from the surrounding atmosphere.

Total heat liberated during the two stages of combustion is 55×10^6 J/m³ (1470 Btu/ft³) of acetylene. However, because of the temperature distribution in the flame, the way in which the flame spreads over the work surface, and losses to the air, power densities and heat transfer factors in oxyacetylene welding are relatively low; $f_1 = 0.10$ to 0.30.

Example 30.3
Heat Generation in Oxyacetylene Welding

An oxyacetylene torch supplies 0.3 m³ of acetylene per hour and an equal volume rate of oxygen for an OAW operation on 4.5-mm-thick steel. Heat generated by combustion is transferred to the work surface with a heat transfer factor $f_1 = 0.20$. If 75% of the heat from the flame is concentrated in a circular area on the work surface that is 9.0 mm in diameter, find (a) rate of heat liberated during combustion, (b) rate of heat transferred to the work surface, and (c) average power density in the circular area.

Solution: (a) The rate of heat generated by the torch is the product of the volume rate of acetylene times the heat of combustion:

$$R_H = (0.3 \text{ m}^3/\text{hr})(55 \times 10^6 \text{ J/m}^3) = 16.5 \times 10^6 \text{ J/hr or } 4583 \text{ J/s}$$

(b) With a heat transfer factor $f_1 = 0.20$, the rate of heat received at the work surface is

$$f_1 R_H = 0.20(4583) = 917 \text{ J/s}$$

(c) The area of the circle in which 75% of the heat of the flame is concentrated is

$$A = \frac{\pi(9)^2}{4} = 63.6 \text{ mm}^2$$

The power density in the circle is found by dividing the available heat by the area of the circle:

$$PD = \frac{0.75(917)}{63.6} = 10.8 \text{ W/mm}^2$$

The combination of acetylene and oxygen is highly flammable, and the environment in which OAW is performed is therefore hazardous. Some of the dangers relate specifically to the acetylene. Pure C_2H_2 is a colorless, odorless gas. For safety reasons, commercial acetylene is processed to have a characteristic garlic odor. One of the physical limitations of the gas is that it is unstable at pressures much above 1 atm (0.1 MPa or 15 lb/in^2). Accordingly, acetylene storage cylinders are packed with a porous filler material (such as asbestos, balsa wood, and other materials) saturated with acetone (CH_3COCH_3). Acetylene dissolves in liquid acetone; in fact, acetone dissolves about 25 times its own volume of acetylene, thus providing a relatively safe means of storing this welding gas. The welder wears eye and skin protection (goggles, gloves, and protective clothing) as an additional safety precaution, and different screw threads are standard on the acetylene and oxygen cylinders and hoses to avoid accidental connection of the wrong gases. Proper maintenance of the equipment is imperative. OAW equipment is relatively inexpensive and portable. It is therefore an economical, versatile process that is well suited to low-quantity production and repair jobs. It is rarely used to weld sheet and plate stock thicker than 6.4 mm (1/4 in) because of the advantages of arc welding in such applications. Although OAW can be mechanized, it is usually performed manually and is hence dependent on the skill of the welder to produce a high-quality weld joint.

30.3.2 ALTERNATIVE GASES FOR OXYFUEL WELDING

Several members of the OFW group are based on gases other than acetylene. Most of the alternative fuels are listed in Table 30.2, together with their burning temperatures and combustion heats. For comparison, acetylene is included in the list. Although oxy-acetylene is the most common OFW fuel, each of the other gases can be used in certain applications—typically limited to welding of sheet metal and metals with low melting

TABLE 30.2 Gases used in oxyfuel welding and/or cutting, with flame temperatures and heats of combustion.

Fuel	Temperature[a]		Heat of Combustion	
	°C	°F	MJ/m³	Btu/ft³
Acetylene (C_2H_2)	3087	5589	54.8	1470
MAPP[b] (C_3H_4)	2927	5301	91.7	2460
Hydrogen (H_2)	2660	4820	12.1	325
Propylene[c] (C_3H_6)	2900	5250	89.4	2400
Propane (C_3H_8)	2526	4579	93.1	2498
Natural gas[d]	2538	4600	37.3	1000

Compiled from [10].

[a]Neutral flame temperatures are compared since this is the flame that would most commonly be used for welding.

[b]MAPP is the commercial abbreviation for methylacetylene-propadiene.

[c]Propylene is used primarily in flame cutting.

[d]Data are based on methane gas (CH_4); natural gas consists of ethane (C_2H_6) as well as methane; flame temperature and heat of combustion vary with composition.

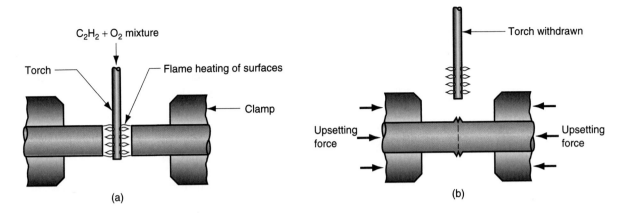

FIGURE 30.23   An application of pressure gas welding: (a) heating of the two parts, and (b) applying pressure to form the weld.

temperatures, and brazing (Section 31.1). In addition, some users prefer these alternative gases for safety reasons.

The fuel that competes most closely with acetylene in burning temperature and heating value is methylacetylene-propadiene. It is a fuel developed by the Dow Chemical Company sold under the trade name ***MAPP*** (we are grateful to Dow for the abbreviation). MAPP (C_3H_4) has heating characteristics similar to acetylene and can be stored under pressure as a liquid, thus avoiding the special storage problems associated with C_2H_2.

When hydrogen is burned with oxygen as the fuel, the process is called ***oxy-hydrogen welding*** (OHW). As shown in Table 30.2, the welding temperature in OHW is below that possible in oxyacetylene welding. In addition, the color of the flame is not affected by differences in the mixture of hydrogen and oxygen, and therefore it is more difficult for the welder to adjust the torch.

Other fuels used in OFW include propane and natural gas. Propane (C_3H_8) is more closely associated with brazing, soldering, and cutting operations than with welding. Natural gas consists mostly of ethane (C_2H_6) and methane (CH_4). When mixed with oxygen it achieves a high temperature flame and is becoming more common in small welding shops.

Pressure Gas Welding   This is a special OFW process, distinguished by type of application rather than fuel gas. ***Pressure gas welding*** (PGW) is a fusion-welding process in which coalescence is obtained over the entire contact surfaces of the two parts by heating them with an appropriate fuel mixture (usually oxyacetylene gas) and then applying pressure to bond the surfaces. A typical application is illustrated in Figure 30.23. Parts are heated until melting begins on the surfaces. The heating torch is then withdrawn, and the parts are pressed together and held at high pressure while solidification occurs. No filler metal is used in PGW.

30.4   OTHER FUSION-WELDING PROCESSES

Some fusion-welding processes cannot be classified as arc, resistance, or oxyfuel welding. Each of these other processes uses a unique technology to develop heat for melting; and typically, the applications are unique.

Electron-Beam Welding   Electron-beam welding (EBW) is a fusion-welding process in which the heat for welding is produced by a highly focused, high-intensity stream of electrons impinging against the work surface. The equipment is similar to that used for electron-beam machining (Section 26.3.2). The electron beam gun operates at high voltage to accelerate the electrons (e.g., 10–150 kV typical), and beam currents are low (measured in milliamps). The power in EBW is not exceptional, but power density is. High power density is achieved by focusing the electron beam on a very small area of the work surface, so that the power density PD is based on

$$PD = \frac{f_1 EI}{A} \tag{30.5}$$

where PD = power density, W/mm^2 (W/in^2, which can be converted to Btu/sec-in^2 by dividing by 1055.); f_1 = heat transfer factor (typical values for EBW range from 0.8–0.95 [9]); E = accelerating voltage, V; I = beam current, A; and A = the work surface area on which the electron beam is focused, mm^2 (in^2). Typical weld areas for EBW range from 13×10^{-3} to 2000×10^{-3} mm^2 (20×10^{-6} to 3000×10^{-6} in^2).

The process had its beginnings in the 1950s in the atomic power field. When first developed, welding had to be carried out in a vacuum chamber to minimize the disruption of the electron beam by air molecules. This requirement was, and still is, a serious inconvenience in production, due to the time required to evacuate the chamber prior to welding. The pump-down time, as it is called, can take as long as an hour, depending on the size of the chamber and the level of vacuum required. Today, EBW technology has progressed to where some operations are performed without a vacuum. Three categories can be distinguished: (1) **high-vacuum welding** (EBW-HV), in which welding is carried out in the same vacuum as beam generation; (2) **medium-vacuum welding** (EBW-MV), in which the operation is performed in a separate chamber where only a partial vacuum is achieved; and (3) **nonvacuum welding** (EBW-NV), in which welding is accomplished at or near atmospheric pressure. The pump-down time during workpart loading and unloading is reduced in medium-vacuum EBW and minimized in nonvacuum EBW, but there is a price paid for this advantage. In the latter two operations, the equipment must include one or more vacuum dividers (very small orifices that impede air flow but permit passage of the electron beam) to separate the beam generator (which requires a high vacuum) from the work chamber. Also, in nonvacuum EBW, the work must be located close to the orifice of the electron beam gun, approximately 13 mm (0.5 in) or less. Finally, the lower vacuum processes cannot achieve the high weld qualities and depth-to-width ratios accomplished by EBW-HV.

Any metals that can be arc welded can be welded by EBW, as well as certain refractory and difficult-to-weld metals that are not suited to AW. Work sizes range from thin foil to thick plate. EBW is applied mostly in the automotive, aerospace, and nuclear industries. In the automotive industry, EBW assembly includes aluminum manifolds, steel torque converters, catalytic converters, and transmission components. In these and other applications, electron-beam welding is noted for high-quality welds with deep and/ or narrow profiles, limited heat-affected zone, and low thermal distortion. Welding speeds are high compared to other continuous welding operations. No filler metal is used, and no flux or shielding gases are needed. Disadvantages of EBW include high equipment cost, need for precise joint preparation and alignment, and the limitations associated with performing the process in a vacuum, as we have already discussed. In addition, there are safety concerns because EBW generates X-rays from which humans must be shielded.

Laser-Beam Welding   Laser-beam welding (LBW) is a fusion-welding process in which coalescence is achieved by the energy of a highly concentrated, coherent light beam focused on the joint to be welded. The term **laser** is an acronym for *l*ight *a*mplification by *s*timulated *e*mission of *r*adiation. This same technology is used for laser-beam machining (Section 26.3.3). LBW is normally performed with shielding gases (e.g.,

FIGURE 30.24
Electroslag welding (ESW): (a) front view with molding shoe removed for clarity; (b) side view showing schematic of molding shoe. Setup is similar to electrogas welding (Figure 30.7) except that resistance heating of molten slag is used to melt the base and filler metals.

helium, argon, nitrogen, and carbon dioxide) to prevent oxidation. Filler metal is not usually added.

LBW produces welds of high quality, deep penetration, and narrow heat-affected zone. These features are similar to those achieved in electron-beam welding, and the two processes are often compared. There are several advantages of LBW over EBW: no vacuum chamber is required, no X-rays are emitted, and laser beams can be focused and directed by optical lenses and mirrors. On the other hand, LBW does not possess the capability for the deep welds and high depth-to-width ratios of EBW. Maximum depth in laser welding is about 19 mm (0.75 in), whereas EBW can be used for weld depths of 50 mm (2 in) or more; and the depth-to-width ratios in LBW are typically limited to around 5:1. Because of the highly concentrated energy in the small area of the laser beam, the process is often used to join small parts.

Electroslag Welding   This process uses the same basic equipment as in some arc-welding operations, and it utilizes an arc to initiate welding. However, it is not an AW process because an arc is not used during welding. *Electroslag welding* (ESW) is a fusion-welding process in which coalescence is achieved by hot, electrically conductive molten slag acting on the base parts and filler metal. As shown in Figure 30.24, the general configuration of ESW is similar to electrogas welding. It is performed in a vertical orientation (shown here for butt welding), using water-cooled molding shoes to contain the molten slag and weld metal. At the start of the process, granulated conductive flux is put into the cavity. The consumable electrode tip is positioned near the bottom of the cavity, and an arc is generated for a short while to start melting the flux. Once a pool of slag has been created, the arc is extinguished and the current passes from the electrode to the base metal through the conductive slag, so that its electrical resistance generates heat to maintain the welding process. Since the density of the slag is less than that of the molten metal, it remains on top to protect the weld pool. Solidification occurs from the bottom, while additional molten metal is supplied from above by the electrode and the edges of the base parts. The process gradually continues until it reaches the top of the joint.

Thermit Welding   *Thermit* is a trademark name for *thermite*, a mixture of aluminum powder and iron oxide that produces an exothermic reaction when ignited. It is used in incendiary bombs and for welding. As a welding process, the use of Thermit dates from around 1900. *Thermit welding* (TW) is a fusion-welding process in which the heat for coalescence is produced by superheated molten metal from the chemical reaction of Thermit. Filler metal is obtained from the liquid metal; and although the process is used for joining, it has more in common with casting than it does with welding.

Finely mixed powders of aluminum and iron oxide (in a 1:3 mixture), when ignited at a temperature of around 1300°C (2300°F), produce the following chemical

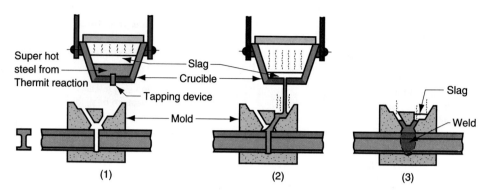

FIGURE 30.25 Thermit welding: (1) Thermit ignited; (2) crucible tapped, superheated metal flows into mold; (3) metal solidifies to produce weld joint.

reaction:

$$8Al + 3Fe_3O_4 \rightarrow 9Fe + 4Al_2O_3 + heat \qquad (30.6)$$

The temperature from the reaction is around 2500°C (4500°F), resulting in superheated molten iron plus aluminum oxide that floats to the top as a slag and protects the iron from the atmosphere. In Thermit welding, the superheated iron (or steel if the mixture of powders is formulated accordingly) is contained in a crucible located above the joint to be welded, as indicated by our diagram of the TW process in Figure 30.25. After the reaction is complete (about 30 s, irrespective of the amount of Thermit involved), the crucible is tapped and the liquid metal flows into a mold built specially to surround the weld joint. Because the entering metal is so hot, it melts the edges of the base parts, causing coalescence upon solidification. After cooling, the mold is broken away, and the gates and risers are removed by oxyacetylene torch or other method.

Thermit welding has applications in joining of railroad rails (as pictured in our figure), and repair of cracks in large steel castings and forgings such as ingot molds, large diameter shafts, frames for machinery, and ship rudders. The surface of the weld in these applications is often sufficiently smooth so that no subsequent finishing is required.

30.5 SOLID-STATE WELDING

In solid state welding, coalescence of the part surfaces is achieved by (1) pressure alone, or (2) heat and pressure. For some solid-state processes, time is also a factor. If both heat and pressure are used, the amount of heat by itself is not sufficient to cause melting of the work surfaces. In other words, fusion of the parts would not occur using only the heat that is externally applied in these processes. In some cases, the combination of heat and pressure, or the particular manner in which pressure alone is applied, generates sufficient energy to cause localized melting of the faying surfaces. Filler metal is not added in solid-state welding.

30.5.1 GENERAL CONSIDERATIONS IN SOLID-STATE WELDING

In most of the solid-state processes, a metallurgical bond is created with little or no melting of the base metals. To metallurgically bond two similar or dissimilar metals, the two metals must be brought into intimate contact so that their cohesive atomic forces attract each other. In normal physical contact between two surfaces, such intimate contact is prohibited by the presence of chemical films, gases, oils, and so on. In order for atomic bonding to succeed, these films and other substances must be removed. In fusion welding (as well as other joining processes such as brazing and soldering), the films are dissolved or burned away by high temperatures, and atomic bonding is established by the melting and solidification of the metals in these processes. But in solid-state welding, the films and other contaminants must be removed by other means to allow metallurgical bonding to take place. In some cases, a thorough cleaning of the surfaces is done just before the welding process; while in other cases, the cleaning action is accomplished as an integral part of bringing the part surfaces together. To summarize, the essential ingredients for a successful solid-state weld are that the two surfaces must be very clean, and they must be brought into very close physical contact with each other to permit atomic bonding.

Welding processes that do not involve melting have several advantages over fusion-welding processes. If no melting occurs, then there is no heat-affected zone, and so the metal surrounding the joint retains its original properties. Many of these processes produce welded joints that comprise the entire contact interface between the two parts, rather than at distinct spots or seams, as in most fusion-welding operations. Also, some of these processes are quite applicable to bonding dissimilar metals, without concerns about relative thermal expansions, conductivities, and other problems that usually arise when dissimilar metals are melted and then solidified during joining.

30.5.2 SOLID STATE-WELDING PROCESSES

The solid-state welding group includes the oldest joining process as well as some of the most modern. Each process in this group has its own unique way of creating the bond at the faying surfaces. We begin our coverage with forge welding, the first welding process.

Forge Welding Forge welding is of historic significance in the development of manufacturing technology. The process dates from about 1000 BCE, when blacksmiths of the ancient world learned to join two pieces of metal (Historical Note 30.1). *Forge welding* is a welding process in which the components to be joined are heated to hot working temperatures and then forged together by hammer or other means. Considerable skill was required by the craftsmen who practiced it in order to achieve a good weld by present-day standards. The process may be of historic interest; however, it is of minor commercial importance today except for its variants that are discussed below.

Cold Welding Cold welding (CW) is a solid-state welding process accomplished by applying high pressure between clean contacting surfaces at room temperature. The faying surfaces must be exceptionally clean for CW to work, and cleaning is usually done by degreasing and wire brushing immediately before joining. Also, at least one of the metals to be welded, and preferably both, must be very ductile and free of work hardening. Metals such as soft aluminum and copper can be readily cold welded. The applied compression forces in the process result in cold working of the metal parts, reducing thickness by as much as 50%; but they also cause localized plastic deformation at the contacting surfaces, resulting in coalescence. For small parts, the forces may be applied by simple hand-operated tools. For heavier work, powered presses are required to exert the necessary force. No heat is applied from external sources in CW, but the deformation process raises the temperature of the work somewhat. Applications of CW include making electrical connections.

Roll Welding Roll welding is a variation of either forge welding or cold welding, depending on whether external heating of the workparts is accomplished prior to the process. *Roll welding* (ROW) is a solid-state welding process in which pressure sufficient to cause coalescence is applied by means of rolls, either with or without external application of heat. The process is illustrated in Figure 30.26. If no external heat is supplied, the process is called *cold-roll welding;* if heat is supplied, the term *hot-roll welding* is used. Applications of roll welding include cladding stainless steel to mild or

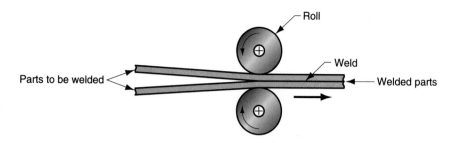

FIGURE 30.26 Roll welding (ROW).

low alloy steel for corrosion resistance, making bimetallic strips for measuring tempera-ture, and producing "sandwich" coins for the U.S. mint.

Hot Pressure Welding Hot pressure welding (HPW) is another variation of forge welding in which coalescence occurs from the application of heat and pressure sufficient to cause considerable deformation of the base metals. The deformation disrupts the surface oxide film, thus leaving clean metal to establish a good bond between the two parts. Time must be allowed for diffusion to occur across the faying surfaces. The operation is usually carried out in a vacuum chamber or in the presence of a shielding medium. Principal applications of HPW are in the aerospace industry.

Diffusion Welding Diffusion welding (DFW) is a solid-state welding process that results from the application of heat and pressure, usually in a controlled atmosphere, with sufficient time allowed for diffusion and coalescence to occur. Temperatures are well below the melting points of the metals (about $0.5\ T_m$ is the maximum), and plastic deformation at the surfaces is minimal. The primary mechanism of coalescence is solid-state diffusion, which involves migration of atoms across the interface between contacting surfaces. Applications of DFW include the joining of high-strength and refractory metals in the aerospace and nuclear industries. The process is used to join both similar and dissimilar metals, and in the latter case a filler layer of a different metal is often sandwiched between the two base metals to promote diffusion. The time for diffusion to occur between the faying surfaces can be significant, requiring more than an hour in some applications [10].

Explosion Welding Explosion welding (EXW) is a solid-state welding process in which rapid coalescence of two metallic surfaces is caused by the energy of a detonated explosive. It is commonly used to bond two dissimilar metals, in particular to clad one metal on top of a base metal over large areas. Applications include production of corrosion-resistant sheet and plate stock for making processing equipment in the chemical and petroleum industries. The term *explosion cladding* is used in this context. No filler metal is used in EXW, and no external heat is applied. Also, no diffusion occurs during the process (the time is too short). The nature of the bond is metallurgical, in many cases combined with a mechanical interlocking that results from a rippled or wavy interface between the metals.

The process for cladding one metal plate on another can be described with reference to Figure 30.27. In this setup, the two plates are in a parallel configuration, separated by a certain gap distance, with the explosive charge above the upper plate, called the *flyer plate.* A buffer layer (e.g., rubber, plastic) is often used between the explosive and the flyer plate to protect its surface. The lower plate, called the *backer* metal, rests on an anvil for support. When detonation is initiated, the explosive charge propagates from one end of the flyer plate to the other, caught in the stop-action view shown in Figure 30.27(2). One of the difficulties in comprehending what happens in EXW is the common misconception that an explosion occurs instantaneously; it is actually a progressive reaction, although admittedly very rapid—propagating at rates as high as

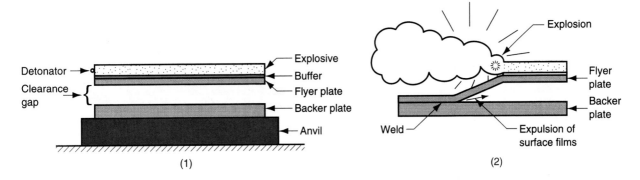

FIGURE 30.27   Explosive welding (EXW): (1) setup in the parallel configuration, and (2) during detonation of the explosive charge.

8500 m/s (28,000 ft/sec). The resulting high-pressure zone propels the flyer plate to collide with the backer metal progressively at high velocity, so that it takes on an angular shape as the explosion advances, as illustrated in our sketch. The upper plate remains in position in the region where the explosive has not yet detonated. The high-speed collision, occurring in a progressive and angular fashion as it does, causes the surfaces at the point of contact to become fluid, and any surface films are expelled forward from the apex of the angle. The colliding surfaces are thus chemically clean, and the fluid behavior of the metal, which involves some interfacial melting, provides intimate contact between the surfaces, leading to metallurgical bonding. Variations in collision velocity and impact angle during the process can result in a wavy or rippled interface between the two metals. This kind of interface strengthens the bond because it increases the contact area and tends to mechanically interlock the two surfaces.

Friction Welding   Friction welding is a widely used commercial process, amenable to automated production methods. The process was developed in the (former) Soviet Union and introduced into the United States around 1960. ***Friction welding*** (FRW) is a solid-state welding process in which coalescence is achieved by frictional heat combined with pressure. The friction is induced by mechanical rubbing between the two surfaces, usually by rotation of one part relative to the other, to raise the temperature at the joint interface to the hot working range for the metals involved. Then the parts are driven toward each other with sufficient force to form a metallurgical bond. The sequence is portrayed in Figure 30.28 for welding two cylindrical parts, the typical application. The axial compression force upsets the parts, and a flash is produced by the material displaced. Any surface films that may have been on the contacting surfaces are expunged during the process. The flash must be subsequently trimmed (e.g., by turning) to provide a smooth surface in the weld region. When properly carried out, no melting occurs at the faying surfaces. No filler metal, flux, or shielding gases are normally used.

Nearly all FRW operations use rotation to develop the frictional heat for welding. There are two principal drive systems, distinguishing two types of FRW: (1) continuous-drive friction welding, and (2) inertia friction welding. In ***continuous-drive friction welding,*** one part is driven at a constant rotational speed and forced into contact with the stationary part at a certain force level so that friction heat is generated at the interface. When the proper hot working temperature has been reached, braking is applied to stop the rotation abruptly, and simultaneously the pieces are forced together at forging pressures. In ***inertia friction welding,*** the rotating part is connected to a flywheel, which is brought up to a predetermined speed. Then the flywheel is disengaged from the drive motor, and the parts are forced together. The kinetic energy stored in the flywheel is

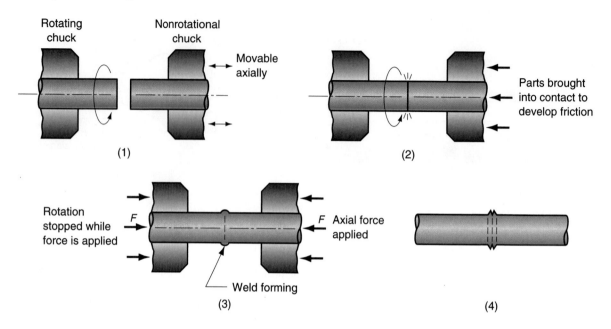

FIGURE 30.28   Friction welding (FRW): (1) rotating part, no contact; (2) parts brought into contact to generate friction heat; (3) rotation stopped and axial pressure applied; and (4) weld created.

dissipated in the form of friction heat to cause coalescence at the abutting surfaces. The total cycle for these operations is about 20 seconds.

Machines used for friction welding have the appearance of an engine lathe. They require a powered spindle to turn one part at high speed, and a means of applying an axial force between the rotating part and the nonrotating part. With its short cycle times, the process lends itself to mass production. It is applied in the welding of various shafts and tubular parts in industries such as automotive, aircraft, farm equipment, petroleum, and natural gas. The process yields a narrow heat-affected zone and can be used to join dissimilar metals. However, at least one of the parts must be rotational, flash must usually be removed, and upsetting reduces the part lengths (which must be taken into consideration in product design).

The conventional friction welding operations discussed above utilize a rotary motion to develop the required friction between faying surfaces. A more recent version of the process is ***linear friction welding,*** in which a linear reciprocating motion is used to generate friction heat between the parts. This eliminates the requirement for at least one of the parts to be rotational (e.g., cylindrical, tubular).

Friction Stir Welding   Friction stir welding (FSW), illustrated in Figure 30.29, is a solid state welding process in which a rotating tool is fed along the joint line between two workpieces, generating friction heat and mechanically stirring the metal to form the weld seam. The process derives its name from this stirring or mixing action. FSW is distinguished from conventional FRW by the fact that friction heat is generated by a separate wear-resistant tool rather than by the parts themselves. FSW was developed in 1991 at The Welding Institute in Cambridge, UK.

The rotating tool is stepped, consisting of a cylindrical shoulder and a smaller probe projecting beneath it. During welding, the shoulder rubs against the top surfaces of the two parts, developing much of the friction heat, while the probe generates additional heat by mechanically mixing the metal along the butt surfaces. The probe has a geometry designed to facilitate the mixing action. The heat produced by the combination of friction

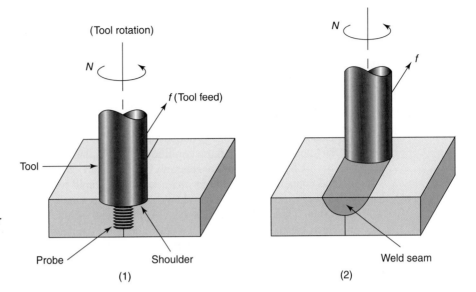

FIGURE 30.29 Friction stir welding (FSW): (1) rotating tool just prior to feeding into joint and (2) partially completed weld seam. N = tool rotation, f = tool feed.

and mixing does not melt the metal but softens it to a highly plastic condition. As the tool is fed forward along the joint, the leading surface of the rotating probe forces the metal around it and into its wake, developing forces that forge the metal into a weld seam. The shoulder serves to constrain the plasticized metal flowing around the probe.

The FSW process is used in the aerospace, automotive, railway, and shipbuilding industries. Typical applications are butt joints on large aluminum parts. Other metals, including steel, copper, and titanium, as well as polymers and composites have also been joined using FSW. Advantages in these applications include (1) good mechanical properties of the weld joint, (2) avoidance of toxic fumes, warping, shielding issues, and other problems associated with arc welding, (3) little distortion or shrinkage, and (4) good weld appearance. Disadvantages include (1) an exit hole is produced when the tool is withdrawn from the work, and (2) heavy-duty clamping of the parts is required.

Ultrasonic Welding   Ultrasonic welding (USW) is a solid-state welding process in which two components are held together under modest clamping force, and oscillatory shear stresses of ultrasonic frequency are applied to the interface to cause coalescence. The operation is illustrated in Figure 30.30 for lap welding, the typical application. The oscillatory motion between the two parts breaks down any surface films to allow

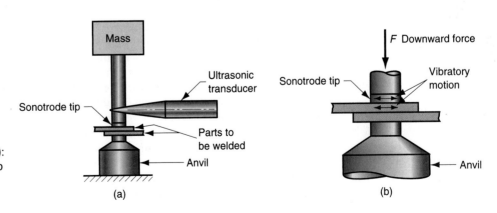

FIGURE 30.30 Ultrasonic welding (USW): (a) general setup for a lap joint; and (b) close-up of weld area.

intimate contact and strong metallurgical bonding between the surfaces. Although heating of the contacting surfaces occurs due to interfacial rubbing and plastic deformation, the resulting temperatures are well below the melting point. No filler metals, fluxes, or shielding gases are required in USW.

The oscillatory motion is transmitted to the upper workpart by means of a ***sonotrode,*** which is coupled to an ultrasonic transducer. This device converts electrical power into high-frequency vibratory motion. Typical frequencies used in USW are 15 to 75 kHz, with amplitudes of 0.018 to 0.13 mm (0.0007–0.005 in). Clamping pressures are well below those used in cold welding and produce no significant plastic deformation between the surfaces. Welding times under these conditions are less than 1 sec.

USW operations are generally limited to lap joints on soft materials such as aluminum and copper. Welding harder materials causes rapid wear of the sonotrode contacting the upper workpart. Workparts should be relatively small, and welding thicknesses less than 3 mm (1/8 in) is the typical case. Applications include wire terminations and splicing in electrical and electronics industries (eliminates the need for soldering), assembly of aluminum sheet-metal panels, welding of tubes to sheets in solar panels, and other tasks in small parts assembly.

30.6 WELD QUALITY

The purpose of any welding process is to join two or more components into a single structure. The physical integrity of the structure thus formed depends on the quality of the weld. Our discussion of weld quality deals primarily with arc welding, the most widely used welding process and the one for which the quality issue is the most critical and complex.

Residual Stresses and Distortion   The rapid heating and cooling in localized regions of the work during fusion welding, especially arc welding, result in thermal expansion and contraction that cause residual stresses in the weldment. These stresses, in turn, can cause distortion and warping of the welded assembly.

The situation in welding is complicated because (1) heating is very localized, (2) melting of the base metals occurs in these local regions, and (3) the location of heating and melting is in motion (at least in arc welding). Consider, for example, butt welding of two plates by arc-welding as shown in Figure 30.31(a). The operation begins at one end and travels to the opposite end. As it proceeds, a molten pool is formed from the base metal (and filler metal, if used) that quickly solidifies behind the moving arc. The portions of the work immediately adjacent to the weld bead become extremely hot and expand, while portions removed from the weld remain relatively cool. The weld pool quickly solidifies in the cavity between the two parts, and as it and the surrounding metal cool and contract, shrinkage occurs across the width of the weldment, as seen in Figure 30.31(b). The weld seam is left in residual tension, and reactionary compressive stresses are set up in regions of the parts away from the weld. Residual stresses and shrinkage also occurs along the length of the weld bead. Since the outer regions of the base parts have remained relatively cool and dimensionally unchanged, while the weld bead has solidified from very high temperatures and then contracted, residual tensile stresses remain longitudinally in the weld bead. These transverse and longitudinal stress patterns are depicted in Figure 30.31(c). The net result of these residual stresses, transversely and longitudinally, is likely to cause warping in the welded assembly as shown in Figure 30.31(d).

The arc-welded butt joint in our example is only one of a variety of joint types and welding operations. Thermally induced residual stresses and the accompanying distortion are a potential problem in nearly all fusion-welding processes and in certain solid-state welding operations in which significant heating takes place. Following are some

FIGURE 30.31 (a) Butt welding two plates; (b) shrinkage across the width of the welded assembly; (c) transverse and longitudinal residual stress pattern; and (d) likely warping in the welded assembly.

techniques to minimize warping in a weldment: (1) *Welding fixtures* can be used to physically restrain movement of the parts during welding. (2) *Heat sinks* can be used to rapidly remove heat from sections of the welded parts to reduce distortion. (3) *Tack welding* at multiple points along the joint can create a rigid structure prior to continuous seam welding. (4) *Welding conditions* (speed, amount of filler metal used, etc.) can be selected to reduce warping. (5) The base parts can be *preheated* to reduce the level of thermal stresses experienced by the parts. (6) *Stress relief* heat treatment can be performed on the welded assembly, either in a furnace for small weldments, or using methods that can be used in the field for large structures. (7) *Proper design* of the weldment itself can reduce the degree of warping.

Welding Defects In addition to residual stresses and distortion in the final assembly, other defects can occur in welding. Following is a brief description of each of the major categories, based on a classification in Cary [3]:

➢ *Cracks.* Cracks are fracture-type interruptions either in the weld itself or in the base metal adjacent to the weld. This is perhaps the most serious welding defect because it constitutes a discontinuity in the metal that significant reduces weld strength. Several forms are defined in Figure 30.32. Welding cracks are caused by embrittlement or low ductility of the weld and/or base metal combined with high restraint during contraction. Generally, this defect must be repaired.

➢ *Cavities.* These include various porosity and shrinkage voids. *Porosity* consists of small voids in the weld metal formed by gases entrapped during solidification. The shapes of the voids vary between spherical (blow holes) to elongated (worm holes). Porosity usually results from inclusion of atmospheric gases, sulfur in the weld metal, or contaminants on the surfaces. *Shrinkage voids* are cavities formed by shrinkage during solidification. Both of these cavity-type defects are similar to defects found in castings and emphasize the close kinship between casting and welding.

➢ *Solid inclusions.* These are nonmetallic solid materials trapped inside the weld metal. The most common form is slag inclusions generated during arc-welding

FIGURE 30.32 Various forms of welding cracks.

processes that use flux. Instead of floating to the top of the weld pool, globules of slag become encased during solidification of the metal. Another form of inclusion is metallic oxides that form during the welding of metals such as aluminum, which normally has a surface coating of Al_2O_3.

➢ *Incomplete fusion.* Several forms of this defect are illustrated in Figure 30.33. Also known as *lack of fusion*, it is simply a weld bead in which fusion has not occurred throughout the entire cross section of the joint. A related defect is *lack of penetration* which means that fusion has not penetrated deeply enough into the root of the joint.

➢ *Imperfect shape or unacceptable contour.* The weld should have a certain desired profile for maximum strength, as indicated in Figure 30.34(a) for a single V-groove weld. This weld profile maximizes the strength of the welded joint and avoids

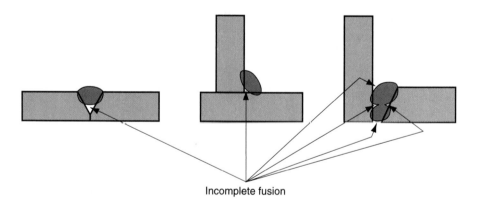

FIGURE 30.33 Several forms of incomplete fusion.

Incomplete fusion

Good profile   Undercut   Underfill   Overlap

(a)   (b)   (c)   (d)

FIGURE 30.34 (a) Desired weld profile for single V-groove weld joint. Same joint but with several weld defects: (b) *undercut*, in which a portion of the base metal part is melted away; (c) *underfill*, a depression in the weld below the level of the adjacent base metal surface; and (d) *overlap*, in which the weld metal spills beyond the joint onto the surface of the base part but no fusion occurs.

incomplete fusion and lack of penetration. Some of the common defects in weld shape and contour are illustrated in Figure 30.34.

➤ *Miscellaneous defects.* This category includes **arc strikes,** in which the welder accidentally allows the electrode to touch the base metal next to the joint, leaving a scar on the surface; and **excessive spatter,** in which drops of molten weld metal splash onto the surface of the base parts.

Inspection and Testing Methods   A variety of inspection and testing methods are available to check the quality of the welded joint. Standardized procedures have been developed and specified over the years by engineering and trade societies such as the American Welding Society (AWS). For purposes of discussion, these inspection and testing procedures can be divided into three categories: (1) visual, (2) nondestructive, and (3) destructive.

Visual inspection is no doubt the most widely used welding inspection method. An inspector visually examines the weldment for (1) conformance to dimensional specifications on the part drawing, (2) warping, and (3) cracks, cavities, incomplete fusion, and other visible defects. The welding inspector also determines if additional tests are warranted, usually in the nondestructive category. The limitation of visual inspection is that only surface defects are detectable; internal defects cannot be discovered by visual methods.

Nondestructive evaluation (NDE) includes various methods that do not damage the specimen being inspected. *Dye-penetrant* and *fluorescent-penetrant tests* are methods for detecting small defects such as cracks and cavities that are open to the surface. Fluorescent penetrants are highly visible when exposed to ultraviolet light, and their use is therefore more sensitive than dyes.

Several other NDE methods should be mentioned. **Magnetic particle testing** is limited to ferromagnetic materials. A magnetic field is established in the subject part, and magnetic particles (e.g., iron filings) are sprinkled on the surface. Subsurface defects such as cracks and inclusions reveal themselves by distorting the magnetic field, causing the particles to be concentrated in certain regions on the surface. **Ultrasonic testing** involves the use of high-frequency sound waves (>20 kHz) directed through the specimen. Discontinuities (e.g., cracks, inclusions, porosity) are detected by losses in sound transmission. **Radiographic testing** uses X-rays or gamma radiation to detect flaws internal to the weld metal. It provides a photographic film record of any defects.

Destructive testing methods in which the weld is destroyed either during the test or to prepare the test specimen. They include mechanical and metallurgical tests. **Mechanical tests** are similar in purpose to conventional testing methods such as tensile tests and shear tests (Chapter 3). The difference is that the test specimen is a weld joint. Figure 30.35 presents a sampling of the mechanical tests used in welding. **Metallurgical tests** involve the preparation of metallurgical specimens of the weldment to examine such features as

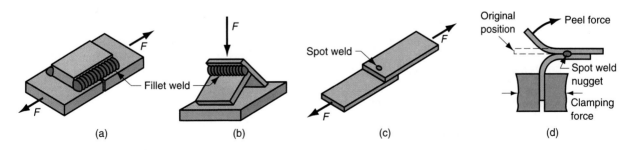

FIGURE 30.35   Mechanical tests used in welding: (a) tension–shear test of arc weldment, (b) fillet break test, (c) tension–shear test of spot weld, (d) peel test for spot weld.

metallic structure, defects, extent and condition of heat-affected zone, presence of other elements, and similar phenomena.

30.7 WELDABILITY

Weldability is the capacity of a metal or combination of metals to be welded into a suitably designed structure, and for the resulting weld joint(s) to possess the required metallurgical properties to perform satisfactorily in the intended service. Good weldability is characterized by the ease with which the welding process is accomplished, absence of weld defects, and acceptable strength, ductility, and toughness in the welded joint.

Factors that affect weldability include (1) welding process, (2) base metal properties, (3) filler metal, and (4) surface conditions. The welding process is significant. Some metals or metal combinations that can be readily welded by one process are difficult to weld by others. For example, stainless steel can be readily welded by most AW processes, but is considered a difficult metal for oxyfuel welding.

Properties of the base metal affect welding performance. Important properties include melting point, thermal conductivity, and coefficient of thermal expansion. One might think that a lower melting point would mean easier welding. However, some metals melt too easily for good welding (e.g., aluminum). Metals with high thermal conductivity tend to transfer heat away from the weld zone, which can make them hard to weld (e.g., copper). High thermal expansion and contraction in the metal causes distortion problems in the welded assembly.

Dissimilar metals pose special problems in welding when their physical and/or mechanical properties are substantially different. Differences in melting temperature are an obvious problem. Differences in strength or coefficient of thermal expansion may result in high residual stresses that can lead to cracking. If a filler metal is used, it must be compatible with the base metal(s). In general, elements mixed in the liquid state that form a solid solution upon solidification will not cause a problem. Embrittlement in the weld joint may occur if the solubility limits are exceeded.

Surface conditions of the base metals can adversely affect the operation. For example, moisture can result in porosity in the fusion zone. Oxides and other solid films on the metal surfaces can prevent adequate contact and fusion from occurring.

30.8 DESIGN CONSIDERATIONS IN WELDING

If an assembly is to be permanently welded, the designer should follow certain guidelines (compiled from [2], [3], and other sources):

> *Design for welding.* The most basic guideline is that the product should be designed from the start as a welded assembly, and not as a casting or forging or other formed shape.
> *Minimum parts.* Welded assemblies should consist of the fewest number of parts possible. For example, it is usually more cost efficient to perform simple bending operations on a part than to weld an assembly from flat plates and sheets.

The following guidelines apply to arc welding:

> *Good fit-up of parts* to be welded is important to maintain dimensional control and minimize distortion. Machining is sometimes required to achieve satisfactory fit-up.
> The assembly must provide access room to allow the welding gun to reach the welding area.

FIGURE 30.36 Welding positions (defined here for groove welds): (a) flat, (b) horizontal, (c) vertical, and (d) overhead.

(a)   (b)   (c)   (d)

> Whenever possible, design of the assembly should allow *flat welding* to be performed, since this is the fastest and most convenient welding position. The possible welding positions are defined in Figure 30.36. The overhead position is the most difficult.

The following design guidelines apply to resistance spot welding:

> Low-carbon sheet steel up to 3.2 mm (0.125 in) is the ideal metal for resistance spot welding.
> Additional strength and stiffness can be obtained in large flat sheet metal components by: (1) spot welding reinforcing parts into them, or (2) forming flanges and embossments into them.
> The spot-welded assembly must provide access for the electrodes to reach the welding area.
> Sufficient overlap of the sheet-metal parts is required for the electrode tip to make proper contact in spot welding. For example, for low-carbon sheet steel, the overlap distance should range from about six times stock thickness for thick sheets of 3.2 mm (0.125 in) to about 20 times thickness for thin sheets, such as 0.5 mm (0.020 in).

REFERENCES

[1] *ASM Handbook,* Vol. 6, *Welding, Brazing, and Soldering.* ASM International, Materials Park, Ohio, 1993.

[2] Bralla, J. G. (Editor in Chief). *Design for Manufacturability Handbook,* 2nd ed. McGraw-Hill Book Company, New York, 1998.

[3] Cary, H. B., and Helzer S. C. *Modern Welding Technology,* 6th ed. Pearson/Prentice-Hall, Upper Saddle River, New Jersey, 2005.

[4] Galyen, J., Sear, G., and Tuttle, C. A. *Welding, Fundamentals and Procedures,* 2nd ed. Prentice-Hall, Inc., Upper Saddle River, New Jersey, 1991.

[5] Jeffus, L. F. *Welding: Principles and Applications,* 6th ed. Delmar Cengage Learning, Clifton Park, New York, 2007.

[6] Messler, R. W., Jr. *Principles of Welding: Processes, Physics, Chemistry, and Metallurgy.* John Wiley & Sons, Inc., New York, 1999.

[7] Stotler, T., and Bernath, J. "Friction Stir Welding Advances," *Advanced Materials and Processes,* March 2009, pp 35–37.

[8] Stout, R. D., and Ott, C. D. *Weldability of Steels,* 4th ed. Welding Research Council, New York, 1987.

[9] *Welding Handbook,* 9th ed., Vol. 1, *Welding Science and Technology.* American Welding Society, Miami, Florida, 2007.

[10] *Welding Handbook,* 9th ed., Vol. 2, *Welding Processes.* American Welding Society, Miami, Florida, 2007.

[11] Wick, C., and Veilleux, R. F. (eds.). *Tool and Manufacturing Engineers Handbook,* 4th ed. Vol. IV, *Quality Control and Assembly.* Society of Manufacturing Engineers, Dearborn, Michigan, 1987.

REVIEW QUESTIONS

30.1. Name the principal groups of processes included in fusion welding.

30.2. What is the fundamental feature that distinguishes fusion welding from solid-state welding?

30.3. Define what an electrical arc is.

30.4. What do the terms arc-on time and arc time mean?

30.5. Electrodes in arc welding are divided into two categories. Name and define the two types.

30.6. What are the two basic methods of arc shielding?

30.7. Why is the heat transfer factor in arc-welding processes that utilize consumable electrodes greater than in those that use nonconsumable electrodes?

30.8. Describe the shielded metal arc-welding process.

30.9. Why is the shielded metal arc-welding process difficult to automate?

30.10. Describe submerged arc welding.

30.11. Why are the temperatures much higher in plasma arc welding than in other arc-welding processes?

30.12. Define resistance welding.

30.13. What are the desirable properties of a metal that would provide good weldability for resistance welding?

30.14. Describe the sequence of steps in the cycle of a resistance spot-welding operation.

30.15. What is resistance-projection welding?

30.16. Describe cross-wire welding.

30.17. Why is the oxyacetylene welding process favored over the other oxyfuel welding processes?

30.18. Define pressure gas welding.

30.19. Electron-beam welding has a significant disadvantage in high-production applications. What is that disadvantage?

30.20. Laser-beam welding and electron-beam welding are often compared because they both produce very high power densities. LBW has certain advantages over EBW. What are they?

30.21. There are several modern-day variations of forge welding, the original welding process. Name them.

30.22. There are two basic types of friction welding. Describe and distinguish the two types.

30.23. What is friction stir welding, and how is it different from friction welding?

30.24. What is a sonotrode in ultrasonic welding?

30.25. Distortion (warping) is a serious problem in fusion welding, particularly arc welding. What are some of the techniques that can be taken to reduce the incidence and extent of distortion?

30.26. What are some of the important welding defects?

30.27. What are the three basic categories of inspection and testing techniques used for weldments? Name some typical inspections and/or tests in each category.

30.28. What are the factors that affect weldability?

30.29. What are some of the design guidelines for weldments that are fabricated by arc welding?

30.30. (**Video**) According to the video, what are four possible functions of the electrodes in resistance spot welding?

MULTIPLE CHOICE QUIZ

There are 23 correct answers in the following multiple choice questions (some questions have multiple answers that are correct). To attain a perfect score on the quiz, all correct answers must be given. Each correct answer is worth 1 point. Each omitted answer or wrong answer reduces the score by 1 point, and each additional answer beyond the correct number of answers reduces the score by 1 point. Percentage score on the quiz is based on the total number of correct answers.

30.1. The feature that distinguishes fusion welding from solid-state welding is that melting of the faying surfaces occurs during fusion welding but not in solid-state welding: (a) true or (b) false?

30.2. Which of the following processes are classified as fusion welding (three correct answers): (a) electrogas welding, (b) electron-beam welding, (c) explosion welding, (d) forge welding, (e) laser-beam welding, and (f) ultrasonic welding?

30.3. Which of the following processes are classified as fusion welding (two correct answers): (a) diffusion welding, (b) friction welding, (c) pressure gas welding, (d) resistance welding, and (e) roll welding?

30.4. Which of the following processes are classified as solid-state welding (three correct answers): (a) diffusion welding, (b) friction stir welding, (c) resistance spot welding, (d) roll welding, (e) Thermit welding, and (f) upset welding?

30.5. An electric arc is a discharge of current across a gap in an electrical circuit. The arc is sustained in arc-welding processes by the transfer of molten metal across the gap between the electrode and the work: (a) true or (b) false?

30.6. Which one of the following arc-welding processes uses a nonconsumable electrode: (a) FCAW, (b) GMAW, (c) GTAW, or (d) SMAW?

30.7. MIG welding is a term sometimes applied when referring to which one of the following processes: (a) FCAW, (b) GMAW, (c) GTAW, or (d) SMAW?

30.8. "Stick" welding is a term sometimes applied when referring to which one of the following processes: (a) FCAW, (b) GMAW, (c) GTAW, or (d) SMAW?

30.9. Which one of the following arc-welding processes uses an electrode consisting of continuous consumable tubing containing flux and other ingredients in its core: (a) FCAW, (b) GMAW, (c) GTAW, or (d) SMAW?

30.10. Which one of the following arc-welding processes produces the highest temperatures: (a) CAW, (b) PAW, (c) SAW, or (d) TIG welding?

30.11. Resistance-welding processes make use of the heat generated by electrical resistance to achieve fusion of the two parts to be joined; no pressure is used in these processes, and no filler metal is added: (a) true or (b) false?

30.12. Metals that are easiest to weld in resistance welding are ones that have low resistivities since low resistivity assists in the flow of electrical current: (a) true or (b) false?

30.13. Oxyacetylene welding is the most widely used oxy-fuel welding process because acetylene mixed with an equal volume of oxygen burns hotter than any other commercially available fuel: (a) true or (b) false?

30.14. The term "laser" stands for "light actuated system for effective reflection": (a) true or (b) false?

30.15. Which of the following solid-state welding processes applies heat from an external source (two best answers): (a) diffusion welding, (b) forge welding, (c) friction welding, and (d) ultrasonic welding?

30.16. The term weldability takes into account not only the ease with which a welding operation can be performed, but also the quality of the resulting weld: (a) true or (b) false?

30.17. Copper is a relatively easy metal to weld because its thermal conductivity is high: (a) true or (b) false?

PROBLEMS

Arc Welding

30.1. A SMAW operation is accomplished in a work cell using a fitter and a welder. The fitter takes 5.5 min to place the unwelded components into the welding fixture at the beginning of the work cycle, and 2.5 min to unload the completed weldment at the end of the cycle. The total length of the several weld seams to be made is 2000 mm, and the travel speed used by the welder averages 400 mm/min. Every 750 mm of weld length, the welding stick must be changed, which takes 0.8 min. While the fitter is working, the welder is idle (resting); and while the welder is working, the fitter is idle. (a) Determine the average arc time in this welding cycle. (b) How much improvement in arc time would result if the welder used FCAW (manually operated), given that the spool of flux-cored weld wire must be changed every five weldments, and it takes the welder 5.0 min to accomplish the change? (c) What are the production rates for these two cases (weldments completed per hour)?

30.2. In the previous problem, suppose an industrial robot cell were installed to replace the welder. The cell consists of the robot (using GMAW instead of SMAW or FCAW), two welding fixtures, and the fitter who loads and unloads the parts. With two fixtures, fitter and robot work simultaneously, the robot welding at one fixture while the fitter unloads and loads at the other. At the end of each work cycle, they switch places. The electrode wire spool must be changed every five workparts, which task requires 5.0 minutes and is accomplished by the fitter. Determine (a) arc time and (b) production rate for this work cell.

30.3. A shielded metal arc-welding operation is performed on steel at a voltage = 30 V and a current = 225 A. The heat transfer factor = 0.85 and melting factor = 0.75. The unit melting energy for steel = 10.2 J/mm^3. Determine (a) the rate of heat generation at the weld and (b) the volume rate of metal welded.

30.4. A GTAW operation is performed on low carbon steel, whose unit melting energy is 10.3 J/mm^3. The welding voltage is 22 V and the current is 135 A. The heat transfer factor is 0.7 and the melting factor is 0.65. If filler metal wire of 3.5 mm diameter is added to the operation, the final weld bead is composed of 60% volume of filler and 40% volume base metal. If the travel speed in the operation is 5 mm/s, determine (a) cross-sectional area of the weld bead, and (b) the feed rate (mm/s) at which the filler wire must be supplied.

30.5. A flux-cored arc-welding operation is performed to butt weld two austenitic stainless steel plates together. The welding voltage is 21 V and the current is 185 A. The cross-sectional area of the weld seam = 75 mm^2 and the melting factor of the stainless steel is assumed to be 0.60. Using tabular data and equations given in this and the preceding chapter, determine the likely value for travel speed v in the operation.

30.6. A flux-cored arc-welding process is used to join two low alloy steel plates at a 90° angle using an outside fillet weld. The steel plates are 1/2 in thick. The weld bead consists of 55% metal from the electrode and the remaining 45% from the steel plates. The melting factor of the steel is 0.65 and the heat transfer factor is 0.90. The welding current is 75 A and the voltage is 16 V. The velocity of the welding head is 40 in/min. The diameter of the electrode is 0.10 in. There is a core of flux running through the center of the electrode that has a diameter of 0.05 in and contains flux (compounds that do not become part of the weld bead). (a) What is the cross-sectional area of the weld bead? (b) How fast must the electrode be fed into the workpiece?

30.7. A gas metal arc-welding test is performed to determine the value of melting factor f_2 for a certain metal and operation. The welding voltage = 25 V, current = 125 A, and heat transfer factor is assumed to be = 0.90, a typical value for GMAW. The rate at which the filler metal is added to the weld is 0.50 in³ per min, and measurements indicate that the final weld bead consists of 57% filler metal and 43% base metal. The unit melting energy for the metal is known to be 75 Btu/in³. (a) Find the melting factor. (b) What is the travel speed if the cross-sectional area of the weld bead = 0.05 in²?

30.8. A continuous weld is to be made around the circumference of a round steel tube of diameter = 6.0 ft, using a submerged arc-welding operation under automatic control at a voltage of 25 V and current of 300 A. The tube is slowly rotated under a stationary welding head. The heat transfer factor for SAW is = 0.95 and the assumed melting factor = 0.7. The cross-sectional area of the weld bead is 0.12 in². If the unit melting energy for the steel = 150 Btu/in³, determine (a) the rotational speed of the tube and (b) the time required to complete the weld.

Resistance Welding

30.9. An RSW operation is used to make a series of spot welds between two pieces of aluminum, each 2.0 mm thick. The unit melting energy for aluminum = 2.90 J/mm³. Welding current = 6000 A, and time duration = 0.15 sec. Assume that the resistance = 75 micro-Ω. The resulting weld nugget measures 5.0 mm in diameter by 2.5 mm thick. How much of the total energy generated is used to form the weld nugget?

30.10. An RSW operation is used to join two pieces of sheet steel having a unit melting energy of 130 Btu/in³. The sheet steel has a thickness of 1/8 in. The weld duration will be set at 0.25 sec with a current of 11,000 A. Based on the electrode diameter, the weld nugget will have a diameter of 0.30 in. Experience has shown that 40% of the supplied heat melts the nugget and the rest is dissipated by the metal. If the electrical resistance between the surfaces is 130 micro-Ω, what is the thickness of the weld nugget assuming it has a uniform thickness?

30.11. The unit melting energy for a certain sheet metal is 9.5 J/mm³. The thickness of each of the two sheets to be spot welded is 3.5 mm. To achieve required strength, it is desired to form a weld nugget that is 5.5 mm in diameter and 5.0 mm thick. The weld duration will be set at 0.3 sec. If it is assumed that the electrical resistance between the surfaces is 140 micro-Ω, and that only one-third of the electrical energy generated will be used to form the weld nugget (the rest being dissipated), determine the minimum current level required in this operation.

30.12. A resistance spot-welding operation is performed on two pieces of 0.040-in thick sheet steel (low carbon). The unit melting energy for steel = 150 Btu/in³. Process parameters are: current = 9500 A and time duration = 0.17 sec. This results in a weld nugget of diameter = 0.19 in and thickness = 0.060 in. Assume the resistance = 100 micro-Ω. Determine (a) the average power density in the interface area defined by the weld nugget, and (b) the proportion of energy generated that went into formation of the weld nugget.

30.13. A resistance seam-welding operation is performed on two pieces of 2.5-mm-thick austenitic stainless steel to fabricate a container. The weld current in the operation is 10,000 A, the weld duration = 0.3 sec, and the resistance at the interface is 75 micro-Ω. Continuous motion welding is used, with 200-mm-diameter electrode wheels. The individual weld nuggets formed in this RSEW operation have diameter = 6 mm and thickness = 3 mm (assume the weld nuggets are disc-shaped). These weld nuggets must be contiguous to form a sealed seam. The power unit driving the process requires an off-time between spot welds of 1.0 s. Given these conditions, determine (a) the unit melting energy of stainless steel using the methods of the previous chapter, (b) the proportion of energy generated that goes into the formation of each weld nugget, and (c) the rotational speed of the electrode wheels.

30.14. Suppose in the previous problem that a roll spot-welding operation is performed instead of seam welding. The interface resistance increases to 100 micro-Ω, and the center-to-center separation between weld nuggets is 25 mm. Given the conditions from the previous problem, with the changes noted here, determine (a) the proportion of energy generated that goes into the formation of each weld nugget, and (b) the rotational speed of the electrode wheels. (c) At this higher rotational speed, how much does the wheel move during the current on-time, and might this have the effect of elongating the weld nugget (making it elliptical rather than round)?

30.15. Resistance projection welding is used to simultaneously weld two thin, steel plates together at four locations. One of the pieces of steel plate is preformed with projections that have a diameter of 0.25 in and a height of 0.20 in. The duration of current flow during the weld is 0.30 sec and all four projections are welded simultaneously. The plate steel has a unit melting energy of 140 Btu/in^3 and a resistance between plates of 90.0 micro-Ω. Experience has shown that 55% of the heat is dissipated by the metal and 45% melts the weld nugget. Assume the volume of the nuggets will be twice the volume of the projections because metal from

both plates is melted. How much current is required for the process?

30.16. An experimental power source for spot welding is designed to deliver current as a ramp function of time: $I = 100,000\,t$, where I = amp and t = sec. At the end of the power-on time, the current is stopped abruptly. The sheet metal being spot welded is low carbon steel whose unit melting energy = 10 J/mm^3. The resistance R = 85 micro-Ω. The desired weld nugget diameter = 4 mm and thickness = 2 mm (assume a disc-shaped nugget). It is assumed that 1/4 of the energy generated from the power source will be used to form the weld nugget. Determine the power-on time the current must be applied in order to perform this spot-welding operation.

Oxyfuel Welding

30.17. Suppose in Example 30.3 in the text that the fuel used in the welding operation is MAPP instead of acetylene, and the proportion of heat concentrated in the 9 mm circle is 60% instead of 75 %. Compute (a) rate of heat liberated during combustion, (b) rate of heat transferred to the work surface, and (c) average power density in the circular area.

30.18. An oxyacetylene torch supplies 8.5 ft^3 of acetylene per hour and an equal volume rate of oxygen

for an OAW operation on 1/4 in steel. Heat generated by combustion is transferred to the work surface with a heat transfer factor of 0.3. If 80% of the heat from the flame is concentrated in a circular area on the work surface whose diameter = 0.40 in, find: (a) rate of heat liberated during combustion, (b) rate of heat transferred to the work surface, and (c) average power density in the circular area.

Electron Beam Welding

30.19. The voltage in an EBW operation is 45 kV. The beam current is 60 milliamp. The electron beam is focused on a circular area that is 0.25 mm in diameter. The heat transfer factor is 0.87. Calculate the average power density in the area in watt/mm^2.

30.20. An electron-beam welding operation is to be accomplished to butt weld two sheet-metal parts that are 3.0 mm thick. The unit melting energy = 5.0 J/mm^3. The weld joint is to be 0.35 mm wide, so that the cross section of the fused metal is 0.35 mm by 3.0 mm. If accelerating voltage = 25 kV, beam current = 30 milliamp, heat transfer factor f_1 = 0.85, and melting factor f_2 = 0.75, determine the travel speed at which this weld can be made along the seam.

30.21. An electron-beam welding operation will join two pieces of steel plate together. The plates are 1.00 in thick. The unit melting energy is 125 Btu/in^3. The diameter of the work area focus of the beam is 0.060 in, hence the width of the weld will be 0.060 in. The accelerating voltage is 30 kV and the beam current is 35 milliamp. The heat transfer factor is 0.70 and the melting factor is 0.55. If the beam moves at a speed of 50 in/min, will the beam penetrate the full thickness of the plates?

30.22. An electron-beam welding operation uses the following process parameters: accelerating voltage = 25 kV, beam current = 100 milliamp, and the circular area on which the beam is focused has a diameter = 0.020 in. If the heat transfer factor = 90%, determine the average power density in the area in Btu/sec in^2.

31 BRAZING, SOLDERING, AND ADHESIVE BONDING

Chapter Contents

In this chapter, we consider three joining processes that are similar to welding in certain respects: brazing, soldering, and adhesive bonding. Brazing and soldering both use filler metals to join and bond two (or more) metal parts to provide a permanent joint. It is difficult, although not impossible, to disassemble the parts after a brazed or soldered joint has been made. In the spectrum of joining processes, brazing and soldering lie between fusion welding and solid-state welding. A filler metal is added in brazing and soldering as in most fusion-welding operations; however, no melting of the base metals occurs, which is similar to solid-state welding. Despite these anomalies, brazing and soldering are generally considered to be distinct from welding. Brazing and soldering are attractive compared to welding under circumstances where (1) the metals have poor weldability, (2) dissimilar metals are to be joined, (3) the intense heat of welding may damage the components being joined, (4) the geometry of the joint does not lend itself to any of the welding methods, and/or (5) high strength is not a requirement.

Adhesive bonding shares certain features in common with brazing and soldering. It utilizes the forces of attachment between a filler material and two closely spaced surfaces to bond the parts. The differences are that the filler material in adhesive bonding is not metallic, and the joining process is carried out at room temperature or only modestly above.

31.1 BRAZING

Brazing is a joining process in which a filler metal is melted and distributed by capillary action between the faying surfaces of the metal parts being joined. No melting of the base metals occurs in brazing; only the filler melts. In brazing the filler metal (also called the *brazing metal*), has a melting temperature (liquidus) that is above 450°C (840°F) but below the melting point (solidus) of the base metal(s) to be joined. If the joint is properly designed and the brazing operation has been properly performed, the brazed joint will be stronger than the filler metal out of which it has been formed upon solidification. This rather remarkable result is due to the small part clearances used in brazing, the metallurgical bonding that occurs between base and filler metals, and the geometric constrictions that are imposed on the joint by the base parts.

Brazing has several advantages compared to welding: (1) any metals can be joined, including dissimilar metals; (2) certain brazing methods can be performed quickly and consistently, thus permitting high cycle rates and automated production; (3) some methods allow multiple joints to be brazed simultaneously; (4) brazing can be applied to join thin-walled parts that cannot be welded; (5) in general, less heat and power are required than in fusion welding; (6) problems with the heat-affected zone in the base metal near the joint are reduced; and (7) joint areas that are inaccessible by many welding processes can be brazed, since capillary action draws the molten filler metal into the joint.

Disadvantages and limitations of brazing include (1) joint strength is generally less than that of a welded joint; (2) although strength of a good brazed joint is greater than that of the filler metal, it is likely to be less than that of the base metals; (3) high service temperatures may weaken a brazed joint; and (4) the color of the metal in the brazed joint may not match the color of the base metal parts, a possible aesthetic disadvantage.

Brazing as a production process is widely used in a variety of industries, including automotive (e.g., joining tubes and pipes), electrical equipment (e.g., joining wires and cables), cutting tools (e.g., brazing cemented carbide inserts to shanks), and jewelry making. In addition, the chemical processing industry and plumbing and heating contractors join metal pipes and tubes by brazing. The process is used extensively for repair and maintenance work in nearly all industries.

31.1.1 BRAZED JOINTS

Brazed joints are commonly of two types: butt and lap (Section 29.2.1). However, the two types have been adapted for the brazing process in several ways. The conventional butt joint provides a limited area for brazing, thus jeopardizing the strength of the joint. To increase the faying areas in brazed joints, the mating parts are often scarfed or stepped or otherwise altered, as shown in Figure 31.1. Of course, additional processing is usually required in the making of the parts for these special joints. One of the particular difficulties associated with a scarfed joint is the problem of maintaining the alignment of the parts before and during brazing.

Lap joints are more widely used in brazing, since they can provide a relatively large interface area between the parts. An overlap of at least three times the thickness of the thinner part is generally considered good design practice. Some adaptations of the lap joint for brazing are illustrated in Figure 31.2. An advantage of brazing over welding in

FIGURE 31.1
(a) Conventional butt joint, and adaptations of the butt joint for brazing: (b) scarf joint, (c) stepped butt joint, (d) increased cross section of the part at the joint.

FIGURE 31.2 (a) Conventional lap joint, and adaptations of the lap joint for brazing: (b) cylindrical parts, (c) sandwiched parts, and (d) use of sleeve to convert butt joint into lap joint.

lap joints is that the filler metal is bonded to the base parts throughout the entire interface area between the parts, rather than only at the edges (as in fillet welds made by arc welding) or at discrete spots (as in resistance spot welding).

Clearance between mating surfaces of the base parts is important in brazing. The clearance must be large enough so as not to restrict molten filler metal from flowing throughout the entire interface. Yet if the joint clearance is too great, capillary action will be reduced and there will be areas between the parts where no filler metal is present. Joint strength is affected by clearance, as depicted in Figure 31.3. There is an optimum clearance value at which joint strength is maximized. The issue is complicated by the fact that the optimum depends on base and filler metals, joint configuration, and processing conditions. Typical brazing clearances in practice are 0.025 to 0.25 mm (0.001 to 0.010 in). These values represent the joint clearance at the brazing temperature, which may be different from room temperature clearance, depending on thermal expansion of the base metal(s).

Cleanliness of the joint surfaces prior to brazing is also important. Surfaces must be free of oxides, oils, and other contaminants in order to promote wetting and capillary attraction during the process, as well as bonding across the entire interface. Chemical

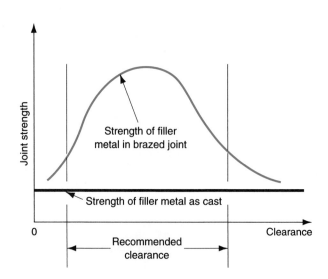

FIGURE 31.3 Joint strength as a function of joint clearance.

TABLE 31.1   Common filler metals used in brazing and the base metals on which they are used.

Filler Metal	Typical Composition	Approximate Brazing Temperature		Base Metals
		°C	°F	
Aluminum and silicon	90 Al, 10 Si	600	1100	Aluminum
Copper	99.9 Cu	1120	2050	Nickel copper
Copper and phosphorous	95 Cu, 5 P	850	1550	Copper
Copper and zinc	60 Cu, 40 Zn	925	1700	Steels, cast irons, nickel
Gold and silver	80 Au, 20 Ag	950	1750	Stainless steel, nickel alloys
Nickel alloys	Ni, Cr, others	1120	2050	Stainless steel, nickel alloys
Silver alloys	Ag, Cu, Zn, Cd	730	1350	Titanium, Monel, Inconel, tool steel, nickel

Compiled from [5] and [7].

treatments such as solvent cleaning (Section 28.1.1) and mechanical treatments such as wire brushing and sand blasting (Section 28.1.2) are used to clean the surfaces. After cleaning and during the brazing operation, fluxes are used to maintain surface cleanliness and promote wetting for capillary action in the clearance between faying surfaces.

31.1.2   FILLER METALS AND FLUXES

Common filler metals used in brazing are listed in Table 31.1 along with the principal base metals on which they are typically used. To qualify as a brazing metal, the following characteristics are needed: (1) melting temperature must be compatible with the base metal, (2) surface tension in the liquid phase must be low for good wettability, (3) fluidity of the molten metal must be high for penetration into the interface, (4) the metal must be capable of being brazed into a joint of adequate strength for the application, and (5) chemical and physical interactions with base metal (e.g., galvanic reaction) must be avoided. Filler metals are applied to the brazing operation in various ways, including wire, rod, sheets and strips, powders, pastes, preformed parts made of braze metal designed to fit a particular joint configuration, and cladding on one of the surfaces to be brazed. Several of these techniques are illustrated in Figures 31.4 and 31.5. Braze metal pastes, shown in Figure 31.5, consist of filler metal powders mixed with fluid fluxes and binders.

Brazing fluxes serve a similar purpose as in welding; they dissolve, combine with, and otherwise inhibit the formation of oxides and other unwanted byproducts in the brazing process. Use of a flux does not substitute for the cleaning steps described above. Characteristics of a good flux include (1) low melting temperature, (2) low viscosity so that it can be displaced by the filler metal, (3) facilitates wetting, and (4) protects the joint until solidification of the filler metal. The flux should also be easy to remove after brazing. Common ingredients for brazing fluxes include borax, borates, fluorides, and chlorides. Wetting agents are also included in the mix to reduce surface tension of the molten filler metal and to improve wettability. Forms of flux include powders, pastes, and slurries. Alternatives to using a flux are to perform the operation in a vacuum or a reducing atmosphere that inhibits oxide formation.

31.1.3   BRAZING METHODS

There are various methods used in brazing. Referred to as brazing processes, they are differentiated by their heating sources.

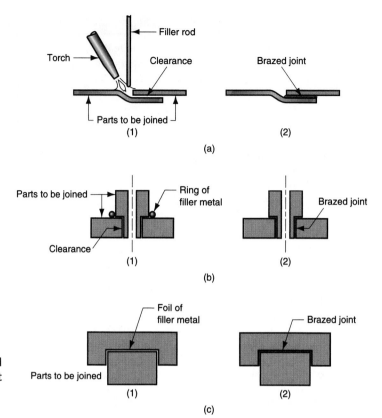

FIGURE 31.4 Several techniques for applying filler metal in brazing: (a) torch and filler rod; (b) ring of filler metal at entrance of gap; and (c) foil of filler metal between flat part surfaces. Sequence: (1) before, and (2) after.

FIGURE 31.5 Application of brazing paste to joint by dispenser. (Courtesy of Fusion, Inc., Willoughby, Ohio.)

Torch Brazing   In torch brazing, flux is applied to the part surfaces and a torch is used to direct a flame against the work in the vicinity of the joint. A reducing flame is typically used to inhibit oxidation. After the workpart joint areas have been heated to a suitable temperature, filler wire is added to the joint, usually in wire or rod form. Fuels used in torch brazing include acetylene, propane, and other gases, with air or oxygen. The selection of the mixture depends on heating requirements of the job. Torch brazing is often performed manually, and skilled workers must be employed to control the flame, manipulate the hand-held torches, and properly judge the temperatures; repair work is a common application. The method can also be used in mechanized production operations, in which parts and brazing metal are loaded onto a conveyor or indexing table and passed under one or more torches.

Furnace Brazing   Furnace brazing uses a furnace to supply heat for brazing and is best suited to medium and high production. In medium production, usually in batches, the component parts and brazing metal are loaded into the furnace, heated to brazing temperature, and then cooled and removed. High-production operations use flow-through furnaces, in which parts are placed on a conveyor and are transported through the various heating and cooling sections. Temperature and atmosphere control are important in furnace brazing; the atmosphere must be neutral or reducing. Vacuum furnaces are sometimes used. Depending on the atmosphere and metals being brazed, the need for a flux may be eliminated.

Induction Brazing   Induction brazing utilizes heat from electrical resistance to a high-frequency current induced in the work. The parts are preloaded with filler metal and placed in a high-frequency AC field—the parts do not directly contact the induction coil. Frequencies range from 5 kHz to 5 MHz. High-frequency power sources tend to provide surface heating, while lower frequencies cause deeper heat penetration into the work and are appropriate for heavier sections. The process can be used to meet low- to high-production requirements.

Resistance Brazing   Heat to melt the filler metal in this process is obtained by resistance to flow of electrical current through the parts. As distinguished from induction brazing, the parts are directly connected to the electrical circuit in resistance brazing. The equipment is similar to that used in resistance welding, except that a lower power level is required for brazing. The parts with filler metal preplaced are held between electrodes while pressure and current are applied. Both induction and resistance brazing achieve rapid heating cycles and are used for relatively small parts. Induction brazing seems to be the more widely used of the two processes.

Dip Brazing   In dip brazing, either a molten salt bath or a molten metal bath accomplishes heating. In both methods, assembled parts are immersed in the baths contained in a heating pot. Solidification occurs when the parts are removed from the bath. In the *salt bath method,* the molten mixture contains fluxing ingredients and the filler metal is preloaded onto the assembly. In the *metal bath method,* the molten filler metal is the heating medium; it is drawn by capillary action into the joint during submersion. A flux cover is maintained on the surface of the molten metal bath. Dip brazing achieves fast heating cycles and can be used to braze many joints on a single part or on multiple parts simultaneously.

Infrared Brazing   This method uses heat from a high-intensity infrared lamp. Some IR lamps are capable of generating up to 5000 W of radiant heat energy, which can be directed at the workparts for brazing. The process is slower than most of the other processes reviewed above, and is generally limited to thin sections.

Braze Welding   This process differs from the other brazing processes in the type of joint to which it is applied. As pictured in Figure 31.6, braze welding is used for filling a more

FIGURE 31.6 Braze welding. The joint consists of braze (filler) metal; no base metal is fused in the joint.

conventional weld joint, such as the V-joint shown. A greater quantity of filler metal is deposited than in brazing, and no capillary action occurs. In braze welding, the joint consists entirely of filler metal; the base metal does not melt and is therefore not fused into the joint as in a conventional fusion welding process. The principal application of braze welding is repair work.

31.2 SOLDERING

Soldering is similar to brazing and can be defined as a joining process in which a filler metal with melting point (liquidus) not exceeding 450°C (840°F) is melted and distributed by capillary action between the faying surfaces of the metal parts being joined. As in brazing, no melting of the base metals occurs, but the filler metal wets and combines with the base metal to form a metallurgical bond. Details of soldering are similar to those of brazing, and many of the heating methods are the same. Surfaces to be soldered must be precleaned so they are free of oxides, oils, and so on. An appropriate flux must be applied to the faying surfaces, and the surfaces are heated. Filler metal, called **solder,** is added to the joint, which distributes itself between the closely fitting parts.

In some applications, the solder is precoated onto one or both of the surfaces—a process called **tinning,** irrespective of whether the solder contains any tin. Typical clearances in soldering range from 0.075 to 0.125 mm (0.003–0.005 in), except when the surfaces are tinned, in which case a clearance of about 0.025 mm (0.001 in) is used. After solidification, the flux residue must be removed.

As an industrial process, soldering is most closely associated with electronics assembly (Chapter 35). It is also used for mechanical joints, but not for joints subjected to elevated stresses or temperatures. Advantages attributed to soldering include (1) low energy input relative to brazing and fusion welding, (2) variety of heating methods available, (3) good electrical and thermal conductivity in the joint, (4) capability to make air-tight and liquid-tight seams for containers, and (5) easy to repair and rework.

The biggest disadvantages of soldering are (1) low joint strength unless reinforced by mechanically means and (2) possible weakening or melting of the joint in elevated temperature service.

31.2.1 JOINT DESIGNS IN SOLDERING

As in brazing, soldered joints are limited to lap and butt types, although butt joints should not be used in load-bearing applications. Some of the brazing adaptations of these joints also apply to soldering, and soldering technology has added a few more variations of its own to deal with the special part geometries that occur in electrical connections. In soldered mechanical joints of sheet-metal parts, the edges of the sheets are often bent over and interlocked before soldering, as shown in Figure 31.7, to increase joint strength.

For electronics applications, the principal function of the soldered joint is to provide an electrically conductive path between two parts being joined. Other design considerations in these types of soldered joints include heat generation (from the electrical resistance of the joint) and vibration. Mechanical strength in a soldered electrical connection is often

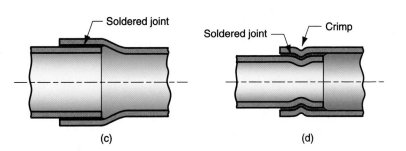

FIGURE 31.7
Mechanical interlocking in soldered joints for increased strength: (a) flat lock seam; (b) bolted or riveted joint; (c) copper pipe fittings—lap cylindrical joint; and (d) crimping (forming) of cylindrical lap joint.

achieved by deforming one or both of the metal parts to accomplish a mechanical joint between them, or by making the surface area larger to provide maximum support by the solder. Several possibilities are sketched in Figure 31.8.

31.2.2 SOLDERS AND FLUXES

Solders and fluxes are the materials used in soldering. Both are critically important in the joining process.

Solders   Most solders are alloys of tin and lead, since both metals have low melting points (see Figure 6.3). Their alloys possess a range of liquidus and solidus temperatures to achieve good control of the soldering process for a variety of applications. Lead is poisonous and its percentage is minimized in most solder compositions. Tin is chemically active at soldering temperatures and promotes the wetting action required for successful joining. In soldering copper, common in electrical connections, intermetallic compounds of copper and tin are formed that strengthen the bond. Silver and antimony are also sometimes used in soldering

FIGURE 31.8
Techniques for securing the joint by mechanical means prior to soldering in electrical connections: (a) crimped lead wire on printed circuit board (PCB); (b) plated through hole on PCB to maximize solder contact surface; (c) hooked wire on flat terminal; and (d) twisted wires.

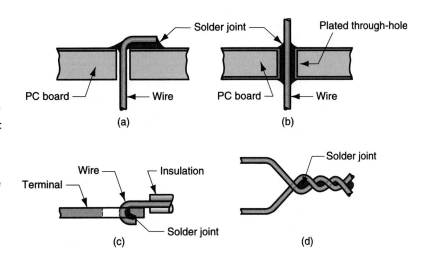

TABLE 31.2   Some common solder alloy compositions with their melting temperatures and applications.

Filler Metal	Approximate Composition	Approximate Melting Temperature		Principal Applications
		°C	°F	
Lead–silver	96 Pb, 4 Ag	305	580	Elevated temperature joints
Tin–antimony	95 Sn, 5 Sb	238	460	Plumbing and heating
Tin–lead	63 Sn, 37 Pb	183[a]	361[a]	Electrical/electronics
	60 Sn, 40 Pb	188	370	Electrical/electronics
	50 Sn, 50 Pb	199	390	General purpose
	40 Sn, 60 Pb	207	405	Automobile radiators
Tin–silver	96 Sn, 4 Ag	221	430	Food containers
Tin–zinc	91 Sn, 9 Zn	199	390	Aluminum joining
Tin–silver–copper	95.5 Sn, 3.9 Ag, 0.6 Cu	217	423	Electronics: surface mount technology

Compiled from [2], [3], [4], and [13].
[a]Eutectic composition—lowest melting point of tin–lead compositions.

alloys. Table 31.2 lists various solder alloy compositions, indicating their approximate soldering temperatures and principal applications. Lead-free solders are becoming increasingly important as legislation to eliminate lead from solders is enacted.

Soldering Fluxes   Soldering fluxes should do the following: (1) be molten at soldering temperatures, (2) remove oxide films and tarnish from the base part surfaces, (3) prevent oxidation during heating, (4) promote wetting of the faying surfaces, (5) be readily displaced by the molten solder during the process, and (6) leave a residue that is noncorrosive and nonconductive. Unfortunately, there is no single flux that serves all of these functions perfectly for all combinations of solder and base metals. The flux formulation must be selected for a given application.

Soldering fluxes can be classified as organic or inorganic. *Organic fluxes* are made of either rosin (i.e., natural rosin such as gum wood, which is not water-soluble) or water-soluble ingredients (e.g., alcohols, organic acids, and halogenated salts). The water-soluble type facilitates cleanup after soldering. Organic fluxes are most commonly used for electrical and electronics connections. They tend to be chemically reactive at elevated soldering temperatures but relatively noncorrosive at room temperatures. *Inorganic fluxes* consist of inorganic acids (e.g., muriatic acid) and salts (e.g., combinations of zinc and ammonium chlorides) and are used to achieve rapid and active fluxing where oxide films are a problem. The salts become active when melted, but are less corrosive than the acids. When solder wire is purchased with an *acid core* it is in this category.

Both organic and inorganic fluxes should be removed after soldering, but it is especially important in the case of inorganic acids to prevent continued corrosion of the metal surfaces. Flux removal is usually accomplished using water solutions except in the case of rosins, which require chemical solvents. Recent trends in industry favor water-soluble fluxes over rosins because chemical solvents used with rosins are harmful to the environment and to humans.

31.2.3   SOLDERING METHODS

Many of the methods used in soldering are the same as those used in brazing, except that less heat and lower temperatures are required for soldering. These methods include torch soldering, furnace soldering, induction soldering, resistance soldering, dip soldering, and infrared soldering. There are other soldering methods, not used in brazing, that should be described here. These methods are hand soldering, wave soldering, and reflow soldering.

Hand Soldering   Hand soldering is performed manually using a hot soldering iron. A *bit,* made of copper, is the working end of a soldering iron. Its functions are (1) to deliver heat to the parts being soldered, (2) to melt the solder, (3) to convey molten solder to the joint, and (4) to withdraw excess solder. Most modern soldering irons are heated by electrical resistance. Some are designed as fast-heating *soldering guns,* which are popular in electronics assembly for intermittent (on/off) operation actuated by a trigger. They are capable of making a solder joint in about a second.

Wave Soldering   Wave soldering is a mechanized technique that allows multiple lead wires to be soldered to a printed circuit board (PCB) as it passes over a wave of molten solder. The typical setup is one in which a PCB, on which electronic components have been placed with their lead wires extending through the holes in the board, is loaded onto a conveyor for transport through the wave-soldering equipment. The conveyor supports the PCB on its sides, so that its underside is exposed to the processing steps, which consist of the following: (1) flux is applied using any of several methods, including foaming, spraying, or brushing; (2) preheating (using light bulbs, heating coils, and infrared devices) to evaporate solvents, activate the flux, and raise the temperature of the assembly; and (3) wave soldering, in which the liquid solder is pumped from a molten bath through a slit onto the bottom of the board to make the soldering connections between the lead wires and the metal circuit on the board. This third step is illustrated in Figure 31.9. The board is often inclined slightly, as depicted in the sketch, and a special tinning oil is mixed with the molten solder to lower its surface tension. Both of these measures help to inhibit buildup of excess solder and formation of "icicles" on the bottom of the board. Wave soldering is widely applied in electronics to produce printed circuit board assemblies (Section 35.3.2).

Reflow Soldering   This process is also widely used in electronics to assemble surface mount components to printed circuit boards (Section 35.4.2). In the process, a solder paste consisting of solder powders in a flux binder is applied to spots on the board where electrical contacts are to be made between surface mount components and the copper circuit. The components are then placed on the paste spots, and the board is heated to melt the solder, forming mechanical and electrical bonds between the component leads and the copper on the circuit board.

Heating methods for reflow soldering include vapor phase reflow and infrared reflow. In *vapor phase reflow soldering,* an inert fluorinated hydrocarbon liquid is vaporized by heating in an oven; it subsequently condenses on the board surface where it transfers its heat of vaporization to melt the solder paste and form solder joints on the

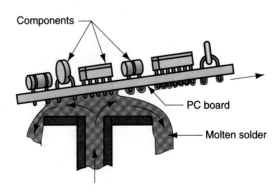

FIGURE 31.9   Wave soldering, in which molten solder is delivered up through a narrow slot onto the underside of a printed circuit board to connect the component lead wires.

printed circuit boards. In *infrared reflow soldering,* heat from an infrared lamp is used to melt the solder paste and form joints between component leads and circuit areas on the board. Additional heating methods to reflow the solder paste include use of hot plates, hot air, and lasers.

31.3   ADHESIVE BONDING

Use of adhesives dates back to ancient times (Historical Note 31.1) and adhesive bonding was probably the first of the permanent joining methods. Today, adhesives are used in a wide range of bonding and sealing applications for joining similar and dissimilar materials such as metals, plastics, ceramics, wood, paper, and cardboard. Although well-established as a joining technique, adhesive bonding is considered a growth area among assembly technologies because of the tremendous opportunities for increased applications.

Historical Note 31.1   *Adhesive bonding*

Adhesives date from ancient times. Carvings 3300 years old show a glue pot and brush for gluing veneer to wood planks. The ancient Egyptians used gum from the Acacia tree for various assembly and sealing purposes. Bitumen, an asphalt adhesive, was used in ancient times as a cement and mortar for construction in Asia Minor. The Romans used pine wood tar and beeswax to caulk their ships. Glues derived from fish, stag horns, and cheese were used in the early centuries after Christ for assembling components of wood.

In more modern times, adhesives have become an important joining process. Plywood, which relies on the use of adhesives to bond multiple layers of wood, was developed around 1900. Phenol formaldehyde was the first synthetic adhesive developed, around 1910, and its primary use was in bonding of wood products such as plywood. During World War II, phenolic resins were developed for adhesive bonding of certain aircraft components. In the 1950s, epoxies were first formulated. And since the 1950s a variety of additional adhesives have been developed, including anaerobics, various new polymers, and second-generation acrylics.

Adhesive bonding is a joining process in which a filler material is used to hold two (or more) closely spaced parts together by surface attachment. The filler material that binds the parts together is the *adhesive.* It is a nonmetallic substance—usually a polymer. The parts being joined are called *adherends.* Adhesives of greatest interest in engineering are *structural adhesives,* which are capable of forming strong, permanent joints between strong, rigid adherends. A large number of commercially available adhesives are cured by various mechanisms and suited to the bonding of various materials. *Curing* refers to the process by which the adhesive's physical properties are changed from a liquid to a solid, usually by chemical reaction, to accomplish the surface attachment of the parts. The chemical reaction may involve polymerization, condensation, or vulcanization. Curing is often motivated by heat and/or a catalyst, and pressure is sometimes applied between the two parts to activate the bonding process. If heat is required, the curing temperatures are relatively low, and so the materials being joined are usually unaffected—an advantage for adhesive bonding. The curing or hardening of the adhesive takes time, called *curing time* or *setting time.* In some cases this time is significant—generally a disadvantage in manufacturing.

Joint strength in adhesive bonding is determined by the strength of the adhesive itself and the strength of attachment between adhesive and each of the adherends. One of the criteria often used to define a satisfactory adhesive joint is that if a failure should occur due to excessive stresses, it occurs in one of the adherends rather than at an interface or within the adhesive itself. The strength of the attachment results from several mechanisms, all depending on the particular adhesive and adherends: (1) chemical bonding, in which the adhesive unites with the adherends and forms a primary chemical bond upon hardening; (2) physical interactions, in which secondary bonding forces result between the atoms of the opposing surfaces; and (3) mechanical interlocking, in which the surface roughness of the adherend causes the hardened adhesive to become entangled or trapped in its microscopic surface asperities.

For these adhesion mechanisms to operate with best results, the following conditions must prevail: (1) surfaces of the adherend must be clean—free of dirt, oil, and oxide films that would interfere with achieving intimate contact between adhesive and adherend; special preparation of the surfaces is often required; (2) the adhesive in its initial liquid form must achieve thorough wetting of the adherend surface; and (3) it is usually helpful for the surfaces to be other than perfectly smooth—a slightly roughened surface increases the effective contact area and promotes mechanical interlocking. In addition, the joint must be designed to exploit the particular strengths of adhesive bonding and avoid its limitations.

31.3.1 JOINT DESIGN

Adhesive joints are not generally as strong as those by welding, brazing, or soldering. Accordingly, consideration must be given to the design of joints that are adhesively bonded. The following design principles are applicable: (1) Joint contact area should be maximized. (2) Adhesive joints are strongest in shear and tension as in Figure 31.10(a) and (b), and joints should be designed so that the applied stresses are of these types. (3) Adhesive bonded joints are weakest in cleavage or peeling as in Figure 31.10(c) and (d), and adhesive bonded joints should be designed to avoid these types of stresses.

Typical joint designs for adhesive bonding that illustrate these design principles are presented in Figure 31.11. Some joint designs combine adhesive bonding with other joining methods to increase strength and/or provide sealing between the two components. Some of the possibilities are shown in Figure 31.12. For example, the combination of adhesive bonding and spot welding is called **weldbonding.**

In addition to the mechanical configuration of the joint, the application must be selected so that the physical and chemical properties of adhesive and adherends are compatible under the service conditions to which the assembly will be subjected. Adherend materials include metals, ceramics, glass, plastics, wood, rubber, leather, cloth, paper, and cardboard. Note that the list includes materials that are rigid and flexible, porous and

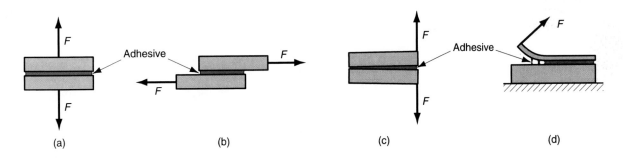

FIGURE 31.10   Types of stresses that must be considered in adhesive bonded joints: (a) tension, (b) shear, (c) cleavage, and (d) peeling.

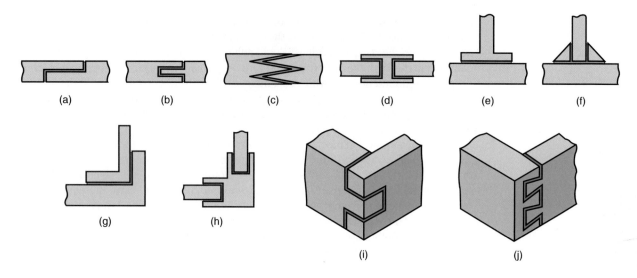

FIGURE 31.11 Some joint designs for adhesive bonding: (a) through (b) butt joints; (c) and (d) T-joints; and (e) through (f) corner joints.

nonporous, metallic and nonmetallic, and that similar or dissimilar substances can be bonded together.

31.3.2 ADHESIVE TYPES

A large number of commercial adhesives are available. They can be classified into three categories: (1) natural, (2) inorganic, and (3) synthetic.

Natural adhesives are derived from natural sources (e.g., plants and animals), including gums, starch, dextrin, soy flour, and collagen. This category of adhesive is generally limited to low-stress applications, such as cardboard cartons, furniture, and bookbinding; or where large surface areas are involved (e.g., plywood). *Inorganic adhesives* are based principally on sodium silicate and magnesium oxychloride. Although relatively low in cost, they are also low in strength—a serious limitation in a structural adhesive.

Synthetic adhesives constitute the most important category in manufacturing. They include a variety of thermoplastic and thermosetting polymers, many of which are listed and briefly described in Table 31.3. They are cured by various mechanisms, such as

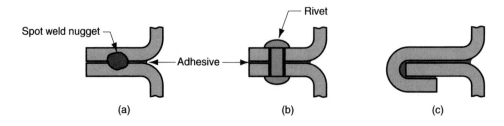

FIGURE 31.12 Adhesive bonding combined with other joining methods: (a) weldbonding—spot welded and adhesive bonded; (b) riveted (or bolted) and adhesive bonded; and (c) formed plus adhesive bonded.

TABLE 31.3   Important synthetic adhesives.

Adhesive	Description and Applications
Anaerobic	Single-component, thermosetting, acrylic-based. Cures by free radical mechanism at room temperature. Applications: sealant, structural assembly.
Modified acrylics	Two-component thermoset, consisting of acrylic-based resin and initiator/hardener. Cures at room temperature after mixing. Applications: fiberglass in boats, sheet metal in cars and aircraft.
Cyanoacrylate	Single-component, thermosetting, acrylic-based that cures at room temperature on alkaline surfaces. Applications: rubber to plastic, electronic components on circuit boards, plastic and metal cosmetic cases.
Epoxy	Includes a variety of widely used adhesives formulated from epoxy resins, curing agents, and filler/modifiers that harden upon mixing. Some are cured when heated. Applications: aluminum bonding applications and honeycomb panels for aircraft, sheet-metal reinforcements for cars, lamination of wooden beams, seals in electronics.
Hot melt	Single-component, thermoplastic adhesive hardens from molten state after cooling from elevated temperatures. Formulated from thermoplastic polymers including ethylene vinyl acetate, polyethylene, styrene block copolymer, butyl rubber, polyamide, polyurethane, and polyester. Applications: packaging (e.g., cartons, labels), furniture, footwear, bookbinding, carpeting, and assemblies in appliances and cars.
Pressure-sensitive tapes and films	Usually one component in solid form that possesses high tackiness resulting in bonding when pressure is applied. Formed from various polymers of high-molecular weight. Can be single-sided or double-sided. Applications: solar panels, electronic assemblies, plastics to wood and metals.
Silicone	One or two components, thermosetting liquid, based on silicon polymers. Curing by room-temperature vulcanization to rubbery solid. Applications: seals in cars (e.g., windshields), electronic seals and insulation, gaskets, bonding of plastics.
Urethane	One or two components, thermosetting, based on urethane polymers. Applications: bonding of fiberglass and plastics.

Compiled from [8], [10], and [14].

(1) mixing a catalyst or reactive ingredient with the polymer immediately prior to applying, (2) heating to initiate the chemical reaction, (3) radiation curing, such as ultraviolet light, and (4) curing by evaporation of water from the liquid or paste adhesive. In addition, some synthetic adhesives are applied as films or as pressure-sensitive coatings on the surface of one of the adherends.

31.3.3 ADHESIVE APPLICATION TECHNOLOGY

Industrial applications of adhesive bonding are widespread and growing. Major users are automotive, aircraft, building products, and packaging industries; other industries include footwear, furniture, bookbinding, electrical, and shipbuilding. Table 31.3 indicates some of the specific applications for which synthetic adhesives are used. In this section we consider several issues relating to adhesives application technology.

Surface Preparation   In order for adhesive bonding to succeed, part surfaces must be extremely clean. The strength of the bond depends on the degree of adhesion between adhesive and adherend, and this depends on the cleanliness of the surface. In most cases, additional processing steps are required for cleaning and surface preparation, the methods varying with different adherend materials. For metals, solvent wiping is often used for cleaning, and abrading the surface by sand blasting or other process usually improves adhesion. For nonmetallic parts, solvent cleaning is generally used, and the surfaces are sometimes mechanically abraded or chemically etched to increase roughness. It is desirable to accomplish the adhesive bonding process as soon as possible after these treatments, since surface oxidation and dirt accumulation increase with time.

Application Methods   The actual application of the adhesive to one or both part surfaces is accomplished in a number of ways. The following list, though incomplete, provides a sampling of the techniques used in industry:

> *Brushing,* performed manually, uses a stiff-bristled brush. Coatings are often uneven.
> *Flowing,* using manually operated pressure-fed flow guns, has more consistent control than brushing.
> *Manual rollers,* similar to paint rollers, are used to apply adhesive from a flat container.
> *Silk screening* involves brushing the adhesive through the open areas of the screen onto the part surface, so that only selected areas are coated.
> *Spraying* uses an air-driven (or airless) spray gun for fast application over large or difficult-to-reach areas.
> *Automatic applicators* include various automatic dispensers and nozzles for use on medium- and high-speed production applications. Figure 31.13 illustrates the use of a dispenser for assembly.
> *Roll coating* is a mechanized technique in which a rotating roller is partially submersed in a pan of liquid adhesive and picks up a coating of the adhesive, which is then transferred to the work surface. Figure 31.14 shows one possible application, in which the work is a thin, flexible material (e.g., paper, cloth, leather, plastic). Variations of the method are used for coating adhesive onto wood, wood composite, cardboard, and similar materials with large surface areas.

FIGURE 31.13   Adhesive is dispensed by a manually controlled dispenser to bond parts during assembly. (Courtesy of EFD, Inc., East Providence, Rhode Island.)

FIGURE 31.14   Roll coating of adhesive onto thin, flexible material such as paper, cloth, or flexible polymer.

Advantages and Limitations   Advantages of adhesive bonding are (1) the process is applicable to a wide variety of materials; (2) parts of different sizes and cross sections can be joined—fragile parts can be joined by adhesive bonding; (3) bonding occurs over the entire surface area of the joint, rather than in discrete spots or along seams as in fusion welding, thereby distributing stresses over the entire area; (4) some adhesives are flexible after bonding and are thus tolerant of cyclical loading and differences in thermal expansion of adherends; (5) low temperature curing avoids damage to parts being joined; (6) sealing as well as bonding can be achieved; and (7) joint design is often simplified (e.g., two flat surfaces can be joined without providing special part features such as screw holes).

Principal limitations of this technology include (1) joints are generally not as strong as other joining methods; (2) adhesive must be compatible with materials being joined; (3) service temperatures are limited; (4) cleanliness and surface preparation prior to application of adhesive are important; (5) curing times can impose a limit on production rates; and (6) inspection of the bonded joint is difficult.

REFERENCES

[1] Adams, R. S. (ed.). *Adhesive Bonding: Science, Technology, and Applications.* CRC Taylor & Francis, Boca Raton, Florida, 2005.

[2] Bastow, E. "Five Solder Families and How They Work," *Advanced Materials & Processes,* December 2003, pp. 26–29.

[3] Bilotta, A. J. *Connections in Electronic Assemblies.* Marcel Dekker, Inc., New York, 1985.

[4] Bralla, J. G. (Editor in Chief). *Design for Manufacturability Handbook,* 2nd ed. McGraw-Hill Book Company, New York, 1998.

[5] *Brazing Manual,* 3rd ed. American Welding Society, Miami, Florida, 1976.

[6] Brockman, W., Geiss, P. L., Klingen, J., and Schroeder, K. B. *Adhesive Bonding: Materials, Applications, and Technology.* John Wiley & Sons, Hoboken, New Jersey, 2009.

[7] Cary, H. B., and Helzer, S. C. *Modern Welding Technology,* 6th ed. Pearson/Prentice Hall, Upper Saddle River, New Jersey, 2005.

[8] Doyle, D. J. "The Sticky Six—Steps for Selecting Adhesives," *Manufacturing Engineering,* June 1991, pp. 39–43.

[9] Driscoll, B., and Campagna, J. "Epoxy, Acrylic, and Urethane Adhesives," *Advanced Materials & Processes,* August 2003, pp. 73–75.

[10] Hartshorn, S. R. (ed.). *Structural Adhesives, Chemistry and Technology.* Plenum Press, New York, 1986.

[11] Humpston, G., and Jacobson, D. M. *Principles of Brazing.* ASM International, Materials Park, Ohio, 2005.

[12] Humpston, G., and Jacobson, D. M. *Principles of Soldering.* ASM International, Materials Park, Ohio, 2004.

[13] Lambert, L. P. *Soldering for Electronic Assemblies.* Marcel Dekker, Inc., New York, 1988.

[14] Lincoln, B., Gomes, K. J., and Braden, J. F. *Mechanical Fastening of Plastics.* Marcel Dekker, Inc., New York, 1984.

[15] Petrie, E. M. *Handbook of Adhesives and Sealants,* 2nd ed. McGraw-Hill, New York, 2006.

[16] Schneberger, G. L. (ed.). *Adhesives in Manufacturing.* CRC Taylor & Francis, Boca Raton, Florida, 1983.

[17] Shields, J. *Adhesives Handbook,* 3rd ed. Butterworths Heinemann, Woburn, UK, 1984.

[18] Skeist, I. (ed.). *Handbook of Adhesives,* 3rd ed. Chapman & Hall, New York, 1990.

[19] *Soldering Manual,* 2nd ed. American Welding Society, Miami, Florida, 1978.

[20] *Welding Handbook,* 9th ed., Vol. 2, *Welding Processes.* American Welding Society, Miami, Florida, 2007.

[21] Wick, C., and Veilleux, R. F. (eds.). *Tool and Manufacturing Engineers Handbook,* 4th ed., Vol. 4, *Quality Control and Assembly.* Society of Manufacturing Engineers, Dearborn, Michigan, 1987.

REVIEW QUESTIONS

31.1. How do brazing and soldering differ from the fusion-welding processes?

31.2. How do brazing and soldering differ from the solid-state welding processes?

31.3. What is the technical difference between brazing and soldering?

31.4. Under what circumstances would brazing or soldering be preferred over welding?

31.5. What are the two joint types most commonly used in brazing?

31.6. Certain changes in joint configuration are usually made to improve the strength of brazed joints. What are some of these changes?

31.7. The molten filler metal in brazing is distributed throughout the joint by capillary action. What is capillary action?

31.8. What are the desirable characteristics of a brazing flux?

31.9. What is dip brazing?

31.10. Define braze welding.

31.11. What are some of the disadvantages and limitations of brazing?

31.12. What are the two most common alloying metals used in solders?

31.13. What are the functions served by the bit of a soldering iron in hand soldering?

31.14. What is wave soldering?

31.15. List the advantages often attributed to soldering as an industrial joining process?

31.16. What are the disadvantages and drawbacks of soldering?

31.17. What is meant by the term structural adhesive?

31.18. An adhesive must cure in order to bond. What is meant by the term curing?

31.19. What are some of the methods used to cure adhesives?

31.20. Name the three basic categories of commercial adhesives.

31.21. What is an important precondition for the success of an adhesive bonding operation?

31.22. What are some of the methods used to apply adhesives in industrial production operations?

31.23. Identify some of the advantages of adhesive bonding compared to alternative joining methods.

31.24. What are some of the limitations of adhesive bonding?

MULTIPLE CHOICE QUIZ

There are 20 correct answers in the following multiple choice questions (some questions have multiple answers that are correct). To attain a perfect score on the quiz, all correct answers must be given. Each correct answer is worth 1 point. Each omitted answer or wrong answer reduces the score by 1 point, and each additional answer beyond the correct number of answers reduces the score by 1 point. Percentage score on the quiz is based on the total number of correct answers.

31.1. In brazing, the base metals melt at temperatures above 840°F (450°C) while in soldering they melt at 840°F (450°C) or below: (a) true or (b) false?

31.2. The strength of a brazed joint is typically (a) equal to, (b) stronger than, or (c) weaker than the filler metal out of which it is made?

31.3. Scarfing in the brazing of a butt joint involves the wrapping of a sheath around the two parts to be joined to contain the molten filler metal during the heating process: (a) true or (b) false?

31.4. Best clearances between surfaces in brazing are which one of the following: (a) 0.0025 to 0.025 mm (0.0001–0.001 in.), (b) 0.025 to 0.250 mm (0.001–0.010 in.), (c) 0.250 to 2.50 mm (0.010–0.100 in.), or (d) 2.5 to 5.0 mm (0.10–0.20 in.)?

31.5. Which of the following is an advantage of brazing (three best answers): (a) annealing of the base parts is a by-product of the process, (b) dissimilar metals can be joined, (c) less heat and energy required than fusion welding, (d) metallurgical improvements in the base metals, (e) multiple joints can be brazed simultaneously, (f) parts can be readily disassembled, and (g) stronger joint than welding?

31.6. Which of the following soldering methods are not used for brazing (two correct answers): (a) dip soldering, (b) infrared soldering, (c) soldering iron, (d) torch soldering, and (e) wave soldering?

31.7. Which one of the following is not a function of a flux in brazing or soldering: (a) chemically etch the surfaces to increase roughness for better adhesion of the filler metal, (b) promote wetting of the surfaces, (c) protect the faying surfaces during the process, or (d) remove or inhibit formation of oxide films?

31.8. Which of the following metals are used in solder alloys (four correct answers): (a) aluminum, (b) antimony, (c) gold, (d) iron, (e) lead, (f) nickel, (g) silver, (h) tin, and (i) titanium?

31.9. A soldering gun is capable of injecting molten solder metal into the joint area: (a) true, or (b) false?

31.10. In adhesive bonding, which one of the following is the term used for the parts that are joined: (a) adherend, (b) adherent, (c) adhesive, (d) adhibit, or (e) ad infinitum?

31.11. Weldbonding is an adhesive joining method in which heat is used to melt the adhesive: (a) true or (b) false?

31.12. Adhesively bonded joints are strongest under which type of stresses (two best answers): (a) cleavage, (b) peeling, (c) shear, and (d) tension?

31.13. Roughening of the faying surfaces tends to (a) have no effect on, (b) increase, or (c) reduce the strength of an adhesively bonded joint because it increases the effective area of the joint and promotes mechanical interlocking?

32 MECHANICAL ASSEMBLY

Chapter Contents

Mechanical assembly uses various methods to mechanically attach two (or more) parts together. In most cases, the method involves the use of discrete hardware components, called *fasteners,* that are added to the parts during the assembly operation. In other cases, the method involves the shaping or reshaping of one of the components being assembled, and no separate fasteners are required. Many consumer products are produced using mechanical assembly: automobiles, large and small appliances, telephones, furniture, utensils—even wearing apparel is "assembled" by mechanical means. In addition, industrial products such as airplanes, machine tools, and construction equipment almost always involve mechanical assembly.

Mechanical fastening methods can be divided into two major classes: (1) those that allow for disassembly, and (2) those that create a permanent joint. Threaded fasteners (e.g., screws, bolts, and nuts) are examples of the first class, and rivets illustrate the second. There are good reasons why mechanical assembly is often preferred over other joining processes discussed in previous chapters. The main reasons are (1) ease of assembly and (2) ease of disassembly (for the fastening methods that permit disassembly).

Mechanical assembly is usually accomplished by unskilled workers with a minimum of special tooling and in a relatively short time. The technology is simple, and the results are easily inspected. These factors are advantageous not only in the factory, but also during field installation. Large products that are too big and heavy to be transported completely assembled can be shipped in smaller subassemblies and then put together at the customer's site.

Ease of disassembly applies, of course, only to the mechanical fastening methods that permit disassembly. Periodic disassembly is required for many products so that maintenance and repair can be performed; for example, to replace worn-out components, make adjustments, and so forth. Permanent joining techniques such as welding do not allow for disassembly.

For purposes of organization, we divide mechanical assembly methods into the following categories: (1) threaded fasteners, (2) rivets, (3) interference fits, (4) other mechanical fastening methods, and (5) molded-in inserts and integral fasteners. These categories are described in Sections 32.1 through 32.5. In Section 32.6, we discuss an important issue in assembly: design for assembly. Assembly of electronic products includes mechanical techniques. However, electronics assembly represents a unique and specialized field, which is covered in Chapter 35.

32.1 THREADED FASTENERS

Threaded fasteners are discrete hardware components that have external or internal threads for assembly of parts. In nearly all cases, they permit disassembly. Threaded fasteners are the most important category of mechanical assembly; the common threaded fastener types are screws, bolts, and nuts.

32.1.1 SCREWS, BOLTS, AND NUTS

Screws and bolts are threaded fasteners that have external threads. There is a technical distinction between a screw and a bolt that is often blurred in popular usage. A *screw* is an externally threaded fastener that is generally assembled into a blind threaded hole. Some types, called *self-tapping screws,* possess geometries that permit them to form or cut the matching threads in the hole. A *bolt* is an externally threaded fastener that is inserted through holes in the parts and "screwed" into a nut on the opposite side. A *nut* is an internally threaded fastener having standard threads that match those on bolts of the same diameter, pitch, and thread form. The typical assemblies that result from the use of screws and bolts are illustrated in Figure 32.1.

Screws and bolts come in a variety of standard sizes, threads, and shapes. Table 32.1 provides a selection of common threaded fastener sizes in metric units (ISO standard) and U.S. customary units (ANSI standard). (ISO is the abbreviation for the International Standards Organization. ANSI is the abbreviation for the American National Standards Institute.)

The metric specification consists of the nominal major diameter, mm, followed by the pitch, mm. For example, a specification of 4-0.7 means a 4.0-mm major diameter and a pitch of 0.7 mm. The U.S. standard specifies either a number designating the major diameter (up to 0.2160 in) or the nominal major diameter, in, followed by the number of threads per inch. For example, the specification 1/4-20 indicates a major diameter of 0.25 in and 20 threads per inch. Both coarse pitch and fine pitch standards are given in our table.

Additional technical data on these and other standard threaded fastener sizes can be found in design texts and handbooks. The United States has been gradually converting

FIGURE 32.1 Typical assemblies using: (a) bolt and nut, and (b) screw.

TABLE 32.1 Selected standard threaded fastener sizes in metric and U.S. customary units.

ISO (Metric) Standard			ANSI (U.S.C.S) Standard			
Nominal Diameter, mm	Coarse Pitch, mm	Fine Pitch, mm	Nominal Size	Major Diameter, in	Threads/in, Coarse (UNC)[a]	Threads/in, Fine (UNF)[a]
2	0.4		2	0.086	56	64
3	0.5		4	0.112	40	48
4	0.7		6	0.138	32	40
5	0.8		8	0.164	32	36
6	1.0		10	0.190	24	32
8	1.25		12	0.216	24	28
10	1.5	1.25	1/4	0.250	20	28
12	1.75	1.25	3/8	0.375	16	24
16	2.0	1.5	1/2	0.500	13	20
20	2.5	1.5	5/8	0.625	11	18
24	3.0	2.0	3/4	0.750	10	16
30	3.5	2.0	1	1.000	8	12

[a]UNC, unified coarse; UNF, unified fine (in the ANSI standard).

to metric thread sizes, which will reduce proliferation of specifications. It should be noted that differences among threaded fasteners have tooling implications in manufacturing. To use a particular type of screw or bolt, the assembly worker must have tools that are designed for that fastener type. For example, there are numerous head styles available on bolts and screws, the most common of which are shown in Figure 32.2. The geometries of these heads, as well as the variety of sizes available, require different hand tools (e.g., screwdrivers) for the worker. One cannot turn a hex-head bolt with a conventional flat-blade screwdriver.

Screws come in a greater variety of configurations than bolts, since their functions vary more. The types include machine screws, capscrews, setscrews, and self-tapping screws. *Machine screws* are the generic type, designed for assembly into tapped holes. They are sometimes assembled to nuts, and in this usage they overlap with bolts. *Capscrews* have the same geometry as machine screws but are made of higher strength metals and to closer tolerances. *Setscrews* are hardened and designed for assembly functions such as fastening collars, gears, and pulleys to shafts as shown in Figure 32.3(a). They come in various geometries, some of which are illustrated in Figure 32.3(b). A *self-tapping screw* (also called a *tapping screw*) is designed to form or cut threads in a preexisting hole into which it is being turned. Figure 32.4 shows two of the typical thread geometries for self-tapping screws.

Flat head   Fillester head   Truss head   Hexagon head

FIGURE 32.2 Various head styles available on screws and bolts. There are additional head styles not shown.

Phillips head   Hex (internal) head   Square (internal) head

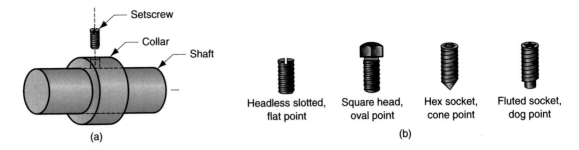

(a)

(b)

FIGURE 32.3   (a) Assembly of collar to shaft using a setscrew; (b) various setscrew geometries (head types and points).

FIGURE 32.4   Self-tapping screws:
(a) thread-forming and (b) thread-cutting.   (a)   (b)

Most threaded fasteners are produced by cold forming (Section 19.2). Some are machined (Sections 22.2.2, 22.3.2, and 22.7.1), but this is usually a more expensive thread-making process. A variety of materials are used to make threaded fasteners, steels being the most common because of their good strength and low cost. These include low and medium carbon as well as alloy steels. Fasteners made of steel are usually plated or coated for superficial resistance to corrosion. Nickel, chromium, zinc, black oxide, and similar coatings are used for this purpose. When corrosion or other factors deny the use of steel fasteners, other materials must be used, including stainless steels, aluminum alloys, nickel alloys, and plastics (however, plastics are suited to low stress applications only).

32.1.2   OTHER THREADED FASTENERS AND RELATED HARDWARE

Additional threaded fasteners and related hardware include studs, screw thread inserts, captive threaded fasteners, and washers. A *stud* (in the context of fasteners) is an externally threaded fastener, but without the usual head possessed by a bolt. Studs can be used to assemble two parts using two nuts as shown in Figure 32.5(a). They are available with threads on one end or both as in Figure 32.5(b) and (c).

FIGURE 32.5   (a) Stud and nuts used for assembly. Other stud types: (b) threads on one end only and (c) double-end stud.

(a)   (b)   (c)

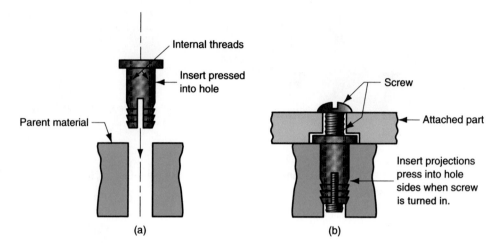

FIGURE 32.6 Screw thread inserts: (a) before insertion, and (b) after insertion into hole and screw is turned into the insert.

Screw thread inserts are internally threaded plugs or wire coils made to be inserted into an unthreaded hole and to accept an externally threaded fastener. They are assembled into weaker materials (e.g., plastic, wood, and light-weight metals such as magnesium) to provide strong threads. There are many designs of screw thread inserts, one example of which is illustrated in Figure 32.6. Upon subsequent assembly of the screw into the insert, the insert barrel expands into the sides of the hole, securing the assembly.

Captive threaded fasteners are threaded fasteners that have been permanently preassembled to one of the parts to be joined. Possible preassembly processes include welding, brazing, press fitting, or cold forming. Two types of captive threaded fasteners are illustrated in Figure 32.7.

A *washer* is a hardware component often used with threaded fasteners to ensure tightness of the mechanical joint; in its simplest form, it is a flat thin ring of sheet metal. Washers serve various functions. They (1) distribute stresses that might otherwise be concentrated at the bolt or screw head and nut, (2) provide support for large clearance

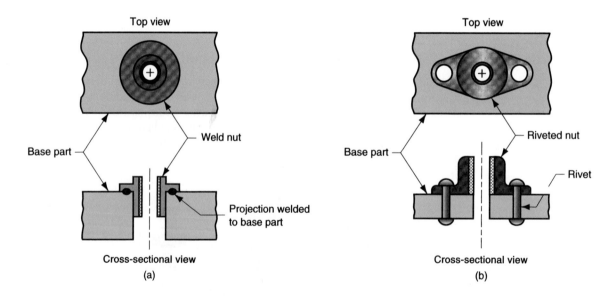

FIGURE 32.7 Captive threaded fasteners: (a) weld nut and (b) riveted nut.

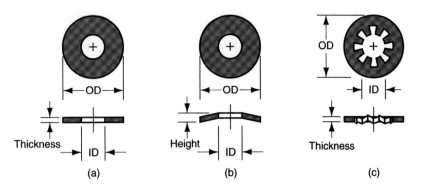

FIGURE 32.8 Types of washers: (a) plain (flat) washers; (b) spring washers, used to dampen vibration or compensate for wear; and (c) lockwasher designed to resist loosening of the bolt or screw.

holes in the assembled parts, (3) increase spring tension, (4) protect part surfaces, (5) seal the joint, and (6) resist inadvertent unfastening [13]. Three washer types are illustrated in Figure 32.8.

32.1.3 STRESSES AND STRENGTHS IN BOLTED JOINTS

Typical stresses acting on a bolted or screwed joint include both tensile and shear, as depicted in Figure 32.9. Shown in the figure is a bolt-and-nut assembly. Once tightened, the bolt is loaded in tension, and the parts are loaded in compression. In addition, forces may be acting in opposite directions on the parts, which results in a shear stress on the bolt cross section. Finally, there are stresses applied on the threads throughout their engagement length with the nut in a direction parallel to the axis of the bolt. These shear stresses can cause *stripping* of the threads. (This failure can also occur on the internal threads of the nut.)

The strength of a threaded fastener is generally specified by two measures: (1) tensile strength, which has the traditional definition (Section 3.1.1), and (2) proof strength. *Proof strength* is roughly equivalent to yield strength; specifically, it is the maximum tensile stress to which an externally threaded fastener can be subjected without permanent deformation. Typical values of tensile and proof strength for steel bolts are given in Table 32.2.

The problem that can arise during assembly is that the threaded fasteners are overtightened, causing stresses that exceed the strength of the fastener material. Assuming a bolt-and-nut assembly as shown in Figure 32.9, failure can occur in one of the following ways: (1) external threads (e.g., bolt or screw) can strip, (2) internal threads (e.g., nut) can strip, or (3) the bolt can break because of excessive tensile stresses on its cross-sectional

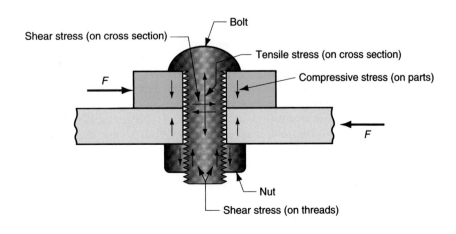

FIGURE 32.9 Typical stresses acting on a bolted joint.

TABLE 32.2   Typical values of tensile and proof strengths for steel bolts and screws, diameters range from 6.4 mm (0.25 in) to 38 mm (1.50 in).

Material	Proof Stress		Tensile Stress	
	MPa	lb/in^2	MPa	lb/in^2
Low/medium C steel	228	33,000	414	60,000
Alloy steel	830	120,000	1030	150,000

Source: [13].

area. Thread stripping, failures (1) and (2), is a shear failure and occurs when the length of engagement is too short (less than about 60% of the nominal bolt diameter). This can be avoided by providing adequate thread engagement in the fastener design. Tensile failure (3) is the most common problem. The bolt breaks at about 85% of its rated tensile strength because of combined tensile and torsion stresses during tightening [2].

The tensile stress to which a bolt is subjected can be calculated as the tensile load applied to the joint divided by the applicable area:

$$\sigma = \frac{F}{A_s} \tag{32.1}$$

where σ = stress, MPa (lb/in^2); F = load, N (lb); and A_s = tensile stress area, mm^2 (in^2).

This stress is compared to the bolt strength values listed in Table 32.2. The tensile stress area for a threaded fastener is the cross-sectional area of the minor diameter. This area can be calculated directly from one of the following equations [2], depending on whether the bolt is metric standard or American standard. For the metric standard (ISO), the formula is

$$A_s = \frac{\pi}{4}(D - 0.9382p)^2 \tag{32.2}$$

where D = nominal size (basic major diameter) of the bolt or screw, mm; and p = thread pitch, mm.

For the American standard (ANSI), the formula is

$$A_s = \frac{\pi}{4}\left(D - \frac{0.9743}{n}\right)^2 \tag{32.3}$$

where D = nominal size (basic major diameter) of the bolt or screw, in; and n = the number of threads per inch.

32.1.4 TOOLS AND METHODS FOR THREADED FASTENERS

The basic function of the tools and methods for assembling threaded fasteners is to provide relative rotation between the external and internal threads, and to apply sufficient torque to secure the assembly. Available tools range from simple hand-held screwdrivers or wrenches to powered tools with sophisticated electronic sensors to ensure proper tightening. It is important that the tool match the screw or bolt and/or the nut in style and size, since there are so many heads available. Hand tools are usually made with a single point or blade, but powered tools are generally designed to use interchangeable bits. The powered tools operate by pneumatic, hydraulic, or electric power.

Whether a threaded fastener serves its intended purpose depends to a large degree on the amount of torque applied to tighten it. Once the bolt or screw (or nut) has been rotated until it is seated against the part surface, additional tightening will increase the tension in the fastener (and simultaneously the compression in the parts being held together); and the tightening will be resisted by an increasing torque. Thus, there is a correlation between the torque required to tighten the fastener and the tensile stress experienced by it. To achieve the desired function in the assembled joint (e.g., to improve fatigue resistance) and to lock the threaded fasteners, the product designer will often specify the tension force that should be applied. This force is called the ***preload***. The following relationship can be used to determine the required torque to obtain a specified preload [13]:

$$T = C_t DF \qquad (32.4)$$

where T = torque, N-mm (lb-in); C_t = the torque coefficient whose value typically ranges between 0.15 and 0.25, depending on the thread surface conditions; D = nominal bolt or screw diameter, mm (in); and F = specified preload tension force, N (lb).

Various methods are employed to apply the required torque, including (1) operator feel—not very accurate, but adequate for most assemblies; (2) torque wrenches, which measure the torque as the fastener is being turned; (3) stall-motors, which are motorized wrenches designed to stall when the required torque is reached, and (4) torque-turn tightening, in which the fastener is initially tightened to a low torque level and then rotated a specified additional amount (e.g., a quarter turn).

32.2   RIVETS AND EYELETS

Rivets are widely used for achieving a permanent mechanically fastened joint. Riveting is a fastening method that offers high production rates, simplicity, dependability, and low cost. Despite these apparent advantages, its applications have declined in recent decades in favor of threaded fasteners, welding, and adhesive bonding. Riveting is one of the primary fastening processes in the aircraft and aerospace industries for joining skins to channels and other structural members.

A ***rivet*** is an unthreaded, headed pin used to join two (or more) parts by passing the pin through holes in the parts and then forming (upsetting) a second head in the pin on the opposite side. The deforming operation can be performed hot or cold (hot working or cold working), and by hammering or steady pressing. Once the rivet has been deformed, it cannot be removed except by breaking one of the heads. Rivets are specified by their length, diameter, head, and type. Rivet type refers to five basic geometries that affect how the rivet will be upset to form the second head. The five types are defined in Figure 32.10. In addition, there are special rivets for special applications.

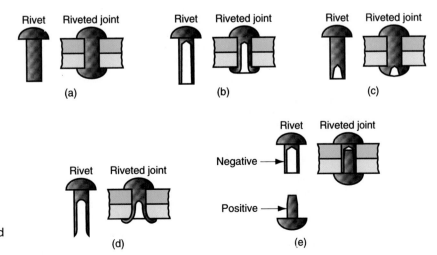

FIGURE 32.10   Five basic rivet types, also shown in assembled configuration: (a) solid, (b) tubular, (c) semi-tubular, (d) bifurcated, and (e) compression.

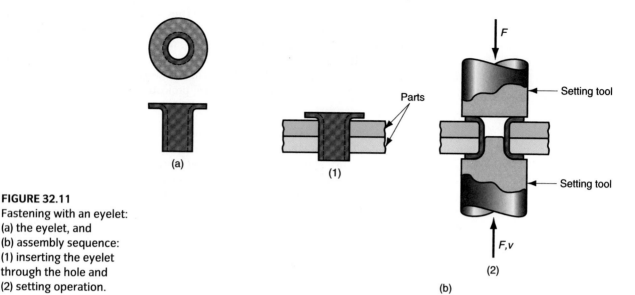

FIGURE 32.11
Fastening with an eyelet:
(a) the eyelet, and
(b) assembly sequence:
(1) inserting the eyelet
through the hole and
(2) setting operation.

Rivets are used primarily for lap joints. The clearance hole into which the rivet is inserted must be close to the diameter of the rivet. If the hole is too small, rivet insertion will be difficult, thus reducing production rate. If the hole is too large, the rivet will not fill the hole and may bend or compress during formation of the opposite head. Rivet design tables are available to specify the optimum hole sizes.

The tooling and methods used in riveting can be divided into the following categories: (1) impact, in which a pneumatic hammer delivers a succession of blows to upset the rivet; (2) steady compression, in which the riveting tool applies a continuous squeezing pressure to upset the rivet; and (3) a combination of impact and compression. Much of the equipment used in riveting is portable and manually operated. Automatic drilling-and-riveting machines are available for drilling the holes and then inserting and upsetting the rivets.

Eyelets are thin-walled tubular fasteners with a flange on one end, usually made from sheet metal, as in Figure 32.11(a). They are used to produce a permanent lap joint between two (or more) flat parts. Eyelets are substituted for rivets in low-stress applications to save material, weight, and cost. During fastening, the eyelet is inserted through the part holes, and the straight end is formed over to secure the assembly. The forming operation is called *setting* and is performed by opposing tools that hold the eyelet in position and curl the extended portion of its barrel. Figure 32.11(b) illustrates the sequence for a typical eyelet design. Applications of this fastening method include automotive subassemblies, electrical components, toys, and apparel.

32.3   ASSEMBLY METHODS BASED ON INTERFERENCE FITS

Several assembly methods are based on mechanical interference between the two mating parts being joined. This interference, which occurs either during assembly or after the parts are joined, holds the parts together. The methods include press fitting, shrink and expansion fits, snap fits, and retaining rings.

Press Fitting   A press fit assembly is one in which the two components have an interference fit between them. The typical case is where a pin (e.g., a straight cylindrical

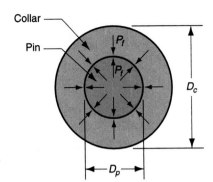

FIGURE 32.12   Cross section of a solid pin or shaft assembled to a collar by interference fit.

pin) of a certain diameter is pressed into a hole of a slightly smaller diameter. Standard pin sizes are commercially available to accomplish a variety of functions, such as (1) locating and locking the components—used to augment threaded fasteners by holding two (or more) parts in fixed alignment with each other; (2) pivot points, to permit rotation of one component about the other; and (3) shear pins. Except for (3), the pins are normally hardened. Shear pins are made of softer metals so as to break under a sudden or severe shearing load to save the rest of the assembly. Other applications of press fitting include assembly of collars, gears, pulleys, and similar components onto shafts.

The pressures and stresses in an interference fit can be estimated using several applicable formulas. If the fit consists of a round solid pin or shaft inside a collar (or similar component), as depicted in Figure 32.12, and the components are made of the same material, the radial pressure between the pin and the collar can be determined by [13]:

$$p_f = \frac{Ei\left(D_c^2 - D_p^2\right)}{D_p D_c^2} \tag{32.5}$$

where p_f = radial or interference fit pressure, MPa (lb/in²); E = modulus of elasticity for the material; i = interference between the pin (or shaft) and the collar; that is, the starting difference between the inside diameter of the collar hole and the outside diameter of the pin, mm (in); D_c = outside diameter of the collar, mm (in); and D_p = pin or shaft diameter, mm (in).

The maximum effective stress occurs in the collar at its inside diameter and can be calculated as

$$\text{Max } \sigma_e = \frac{2p_f D_c^2}{D_c^2 - D_p^2} \tag{32.6}$$

where Max σ_e = the maximum effective stress, MPa (lb/in²), and p_f is the interference fit pressure computed from Eq. (32.5).

In situations in which a straight pin or shaft is pressed into the hole of a large part with geometry other than that of a collar, we can alter the previous equations by taking the outside diameter D_c to be infinite, thus reducing the equation for interference pressure to

$$p_f = \frac{Ei}{D_p} \tag{32.7}$$

and the corresponding maximum effective stress becomes

$$\text{Max } \sigma_e = 2p_f \tag{32.8}$$

In most cases, particularly for ductile metals, the maximum effective stress should be compared with the yield strength of the material, applying an appropriate safety factor, as in the following:

$$\text{Max } \sigma_e \leq \frac{Y}{SF} \tag{32.9}$$

where Y = yield strength of the material, and SF is the applicable safety factor.

Various pin geometries are available for interference fits. The basic type is a **straight pin,** usually made from cold-drawn carbon steel wire or bar stock, ranging in diameter from 1.6 to 25 mm (1/16 to 1.0 in). They are unground, with chamfered or square ends (chamfered ends facilitate press fitting). **Dowel pins** are manufactured to more precise specifications than straight pins, and can be ground and hardened. They are used to fix the alignment of assembled components in dies, fixtures, and machinery. **Taper pins** possess a taper of 6.4 mm (0.25 in) per foot and are driven into the hole to establish a fixed relative position between the parts. Their advantage is that they can readily be driven back out of the hole.

Additional pin geometries are commercially available, including **grooved pins**—solid straight pins with three longitudinal grooves in which the metal is raised on either side of each groove to cause interference when the pin is pressed into a hole; **knurled pins,** pins with a knurled pattern that causes interference in the mating hole; and **coiled pins,** also called **spiral pins,** which are made by rolling strip stock into a coiled spring.

Shrink and Expansion Fits   These terms refer to the assembly of two parts that have an interference fit at room temperature. The typical case is a cylindrical pin or shaft assembled into a collar. To assemble by **shrink fitting,** the external part is heated to enlarge it by thermal expansion, and the internal part either remains at room temperature or is cooled to contract its size. The parts are then assembled and brought back to room temperature, so that the external part shrinks, and if previously cooled the internal part expands, to form a strong interference fit. An **expansion fit** is when only the internal part is cooled to contract it for assembly; once inserted into the mating component, it warms to room temperature, expanding to create the interference assembly. These assembly methods are used to fit gears, pulleys, sleeves, and other components onto solid and hollow shafts.

Various methods are used to heat and/or cool the workparts. Heating equipment includes torches, furnaces, electric resistance heaters, and electric induction heaters. Cooling methods include conventional refrigeration, packing in dry ice, and immersion in cold liquids, including liquid nitrogen. The resulting change in diameter depends on the coefficient of thermal expansion and the temperature difference that is applied to the part. If we assume that the heating or cooling has produced a uniform temperature throughout the work, then the change in diameter is given by

$$D_2 - D_1 = \alpha \, D_1 (T_2 - T_1) \tag{32.10}$$

where α = the coefficient of linear thermal expansion, mm/mm-°C (in/in-°F) for the material (see Table 4.1); T_2 = the temperature to which the parts have been heated or cooled, °C (°F); T_1 = starting ambient temperature; D_2 = diameter of the part at T_2, mm (in); and D_1 = diameter of the part at T_1.

Equations (32.5) through (32.9) for computing interference pressures and effective stresses can be used to determine the corresponding values for shrink and expansion fits.

Snap Fits and Retaining Rings   Snap fits are a variation of interference fits. A **snap fit** involves joining two parts in which the mating elements possess a temporary interference while being pressed together, but once assembled they interlock to maintain the assembly. A typical example is shown in Figure 32.13: as the parts are pressed together, the mating elements elastically deform to accommodate the interference, subsequently allowing the

FIGURE 32.13 Snap fit assembly, showing cross sections of two mating parts: (1) before assembly and (2) parts snapped together.

(1)

(2)

FIGURE 32.14 Retaining ring assembled into a groove on a shaft.

Groove in shaft

Shaft

Retaining ring

parts to snap together; once in position, the elements become connected mechanically so that they cannot easily be disassembled. The parts are usually designed so that a slight interference exists after assembly.

Advantages of snap fit assembly include (1) the parts can be designed with self-aligning features, (2) no special tooling is required, and (3) assembly can be accomplished very quickly. Snap fitting was originally conceived as a method that would be ideally suited to industrial robotics applications; however, it is no surprise that assembly techniques that are easier for robots are also easier for human assembly workers.

A *retaining ring,* also known as a *snap ring,* is a fastener that snaps into a circumferential groove on a shaft or tube to form a shoulder, as in Figure 32.14. The assembly can be used to locate or restrict the movement of parts mounted on the shaft. Retaining rings are available for both external (shaft) and internal (bore) applications. They are made from either sheet metal or wire stock, heat treated for hardness and stiffness. To assemble a retaining ring, a special pliers tool is used to elastically deform the ring so that it fits over the shaft (or into the bore) and then is released into the groove.

32.4 OTHER MECHANICAL FASTENING METHODS

In addition to the mechanical assembly techniques discussed in the preceding, there are several additional methods that involve the use of fasteners. These include stitching, stapling, sewing, and cotter pins.

Stitching, Stapling, and Sewing   Industrial stitching and stapling are similar operations involving the use of U-shaped metal fasteners. *Stitching* is a fastening operation in which a stitching machine is used to form the U-shaped stitches one at a time from steel wire and immediately drive them through the two parts to be joined. Figure 32.15 illustrates several

FIGURE 32.15
Common types of wire
stitches: (a) unclinched,
(b) standard loop,
(c) bypass loop, and
(d) flat clinch.

FIGURE 32.16   Cotter
pins: (a) offset head,
standard point;
(b) symmetric head,
hammerlock point;
(c) square point;
(d) mitered point; and
(e) chisel point.

types of wire stitches. The parts to be joined must be relatively thin, consistent with the stitch size, and the assembly can involve various combinations of metal and nonmetal materials. Applications of industrial stitching include light sheetmetal assembly, metal hinges, electrical connections, magazine binding, corrugated boxes, and final product packaging. Conditions that favor stitching in these applications are (1) high-speed operation, (2) elimination of the need for prefabricated holes in the parts, and (3) desirability of using fasteners that encircle the parts.

In *stapling,* preformed U-shaped staples are punched through the two parts to be attached. The staples are supplied in convenient strips. The individual staples are lightly stuck together to form the strip, but they can be separated by the stapling tool for driving. The staples come with various point styles to facilitate their entry into the work. Staples are usually applied by means of portable pneumatic guns, into which strips containing several hundred staples can be loaded. Applications of industrial stapling include: furniture and upholstery, assembly of car seats, and various light-gage sheetmetal and plastic assembly jobs.

Sewing is a common joining method for soft, flexible parts such as cloth and leather. The method involves the use of a long thread or cord interwoven with the parts so as to produce a continuous seam between them. The process is widely used in the needle trades industry for assembling garments.

Cotter Pins   Cotter pins are fasteners formed from half-round wire into a single two-stem pin, as in Figure 32.16. They vary in diameter, ranging between 0.8 mm (0.031 in) and 19 mm (3/4 in), and in point style, several of which are shown in the figure. Cotter pins are inserted into holes in the mating parts and their legs are split to lock the assembly. They are used to secure parts onto shafts and similar applications.

32.5   MOLDING INSERTS AND INTEGRAL FASTENERS

These assembly methods form a permanent joint between parts by shaping or reshaping one of the components through a manufacturing process such as casting, molding, or sheet-metal forming.

FIGURE 32.17
Examples of molded-in inserts: (a) threaded bushing and (b) threaded stud.

Inserts in Moldings and Castings   This method involves the placement of a component into a mold before plastic molding or metal casting, so that it becomes a permanent and integral part of the molding or casting. Inserting a separate component is preferable to molding or casting its shape if the superior properties (e.g., strength) of the insert material are required, or the geometry achieved through the use of the insert is too complex or intricate to incorporate into the mold. Examples of inserts in molded or cast parts include internally threaded bushings and nuts, externally threaded studs, bearings, and electrical contacts. Some of these are illustrated in Figure 32.17. Internally threaded inserts must be placed into the mold with threaded pins to prevent the molding material from flowing into the threaded hole.

Placing inserts into a mold has certain disadvantages in production: (1) design of the mold becomes more complicated; (2) handling and placing the insert into the cavity takes time that reduces production rate; and (3) inserts introduce a foreign material into the casting or molding, and in the event of a defect, the cast metal or plastic cannot be easily reclaimed and recycled. Despite these disadvantages, use of inserts is often the most functional design and least-cost production method.

Integral Fasteners   Integral fasteners involve deformation of component parts so they interlock and create a mechanically fastened joint. This assembly method is most common for sheetmetal parts. The possibilities, Figure 32.18, include (a) *lanced tabs* to attach wires or shafts to sheet-metal parts; (b) *embossed protrusions,* in which bosses are formed in one part and flattened over the mating assembled part; (c) *seaming,* where the edges of two separate sheet-metal parts or the opposite edges of the same part are bent over to form the fastening seam—the metal must be ductile in order for the bending to be feasible; (d) *beading,* in which a tube-shaped part is attached to a smaller shaft (or other round part) by deforming the outer diameter inward to cause an interference around the entire circumference; and (e) *dimpling*—forming of simple round indentations in an outer part to retain an inner part.

Crimping, in which the edges of one part are deformed over a mating component, is another example of integral assembly. A common example involves squeezing the barrel of an electrical terminal onto a wire (Section 35.5.1).

32.6 DESIGN FOR ASSEMBLY

Design for assembly (DFA) has received much attention in recent years because assembly operations constitute a high labor cost for many manufacturing companies. The key to successful design for assembly can be simply stated [3]: (1) design the product with as few parts as possible, and (2) design the remaining parts so they are easy to assemble. The cost of assembly is determined largely during product design, because that is when the number

FIGURE 32.18 Integral fasteners: (a) lanced tabs to attach wires or shafts to sheetmetal, (b) embossed protrusions, similar to riveting, (c) single-lock seaming, (d) beading, and (e) dimpling. Numbers in parentheses indicate sequence in (b), (c), and (d).

of separate components in the product is determined, and decisions are made about how these components will be assembled. Once these decisions have been made, there is little that can be done in manufacturing to influence assembly costs (except, of course, to manage the operations well).

In this section we consider some of the principles that can be applied during product design to facilitate assembly. Most of the principles have been developed in the context of mechanical assembly, although some of them apply to the other assembly and joining processes. Much of the research in design for assembly has been motivated by the increasing use of automated assembly systems in industry. Accordingly, our discussion is divided into two sections, the first dealing with general principles of DFA, and the second concerned specifically with design for automated assembly.

FIGURE 32.19
Symmetrical parts are generally easier to insert and assemble: (a) only one rotational orientation possible for insertion; (b) two possible orientations; (c) four possible orientations; and (d) infinite rotational orientations.

(a)   (b)   (c)   (d)

32.6.1 GENERAL PRINCIPLES OF DFA

Most of the general principles apply to both manual and automated assembly. Their goal is to achieve the required design function by the simplest and lowest cost means. The following recommendations have been compiled from [1], [3], [4], and [6]:

> *Use the fewest number of parts possible to reduce the amount of assembly required.* This principle is implemented by combining functions within the same part that might otherwise be accomplished by separate components (e.g., using a plastic molded part instead of an assembly of sheet metal parts).

> *Reduce the number of threaded fasteners required.* Instead of using separate threaded fasteners, design the component to utilize snap fits, retaining rings, integral fasteners, and similar fastening mechanisms that can be accomplished more rapidly. Use threaded fasteners only where justified (e.g., where disassembly or adjustment is required).

> *Standardize fasteners.* This is intended to reduce the number of sizes and styles of fasteners required in the product. Ordering and inventory problems are reduced, the assembly worker does not have to distinguish between so many separate fasteners, the workstation is simplified, and the variety of separate fastening tools is reduced.

> *Reduce parts orientation difficulties.* Orientation problems are generally reduced by designing a part to be symmetrical and minimizing the number of asymmetric features. This allows easier handling and insertion during assembly. This principle is illustrated in Figure 32.19.

> *Avoid parts that tangle.* Certain part configurations are more likely to become entangled in parts bins, frustrating assembly workers or jamming automatic feeders. Parts with hooks, holes, slots, and curls exhibit more of this tendency than parts without these features. See Figure 32.20.

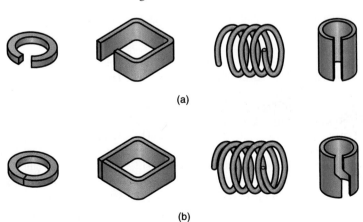

(a)

(b)

FIGURE 32.20   (a) Parts that tend to tangle and (b) parts designed to avoid tangling.

32.6.2 DESIGN FOR AUTOMATED ASSEMBLY

Methods suitable for manual assembly are not necessarily the best methods for automated assembly. Some assembly operations readily performed by a human worker are quite difficult to automate (e.g., assembly using bolts and nuts). To automate the assembly process, parts fastening methods must be specified during product design that lend themselves to machine insertion and joining techniques and do not require the senses, dexterity, and intelligence of human assembly workers. Following are some recommendations and principles that can be applied in product design to facilitate automated assembly [6], [10]:

> *Use modularity in product design.* Increasing the number of separate tasks that are accomplished by an automated assembly system will reduce the reliability of the system. To alleviate the reliability problem, Riley [10] suggests that the design of the product be modular in which each module or subassembly has a maximum of 12 or 13 parts to be produced on a single assembly system. Also, the subassembly should be designed around a base part to which other components are added.

> *Reduce the need for multiple components to be handled at once.* The preferred practice for automated assembly is to separate the operations at different stations rather than to simultaneously handle and fasten multiple components at the same workstation.

> *Limit the required directions of access.* This means that the number of directions in which new components are added to the existing subassembly should be minimized. Ideally, all components should be added vertically from above, if possible.

> *High-quality components.* High performance of an automated assembly system requires that consistently good-quality components are added at each workstation. Poor quality components cause jams in feeding and assembly mechanisms that result in downtime.

> *Use of snap fit assembly.* This eliminates the need for threaded fasteners; assembly is by simple insertion, usually from above. It requires that the parts be designed with special positive and negative features to facilitate insertion and fastening.

REFERENCES

[1] Andreasen, M., Kahler, S., and Lund, T. *Design for Assembly.* Springer-Verlag, New York, 1988.

[2] Blake, A. *What Every Engineer Should Know About Threaded Fasteners.* Marcel Dekker, New York, 1986.

[3] Boothroyd, G., Dewhurst, P., and Knight, W. *Product Design for Manufacture and Assembly.* 2nd ed. CRC Taylor & Francis, Boca Raton, Florida, 2001.

[4] Bralla, J. G. (Editor-in-Chief). *Design for Manufacturability Handbook,* 2nd ed. McGraw-Hill, New York, 1998.

[5] Dewhurst, P., and Boothroyd, G. "Design for Assembly in Action," *Assembly Engineering,* January 1987, pp. 64–68.

[6] Groover, M. P. *Automation, Production Systems, and Computer Integrated Manufacturing,* 3rd ed. Pearson Prentice-Hall, Upper Saddle River, New Jersey, 2008.

[7] Groover, M. P., Weiss, M., Nagel, R. N., and Odrey, N. G. *Industrial Robotics: Technology, Programming, and Applications.* McGraw-Hill, New York, 1986.

[8] Nof, S. Y., Wilhelm, W. E., and Warnecke, H-J. *Industrial Assembly.* Chapman & Hall, New York, 1997.

[9] Parmley, R. O. (ed.). *Standard Handbook of Fastening and Joining,* 3rd ed. McGraw-Hill, New York, 1997.

[10] Riley, F. J. *Assembly Automation, A. Management Handbook,* 2nd ed. Industrial Press, New York, 1999.

[11] Speck, J. A. *Mechanical Fastening, Joining, and Assembly.* Marcel Dekker, New York, 1997.

[12] Whitney, D. E. *Mechanical Assemblies.* Oxford University Press, New York, 2004.

[13] Wick, C., and Veilleux, R. F. (eds.). *Tool and Manufacturing Engineers Handbook,* 4th ed., Vol. IV, *Quality Control and Assembly.* Society of Manufacturing Engineers, Dearborn, Michigan, 1987.

REVIEW QUESTIONS

32.1. How does mechanical assembly differ from the other methods of assembly discussed in previous chapters (e.g., welding, brazing, etc.)?

32.2. What are some of the reasons why assemblies must be sometimes disassembled?

32.3. What is the technical difference between a screw and a bolt?

32.4. What is a stud (in the context of threaded fasteners)?

32.5. What is torque-turn tightening?

32.6. Define proof strength as the term applies in threaded fasteners.

32.7. What are the three ways in which a threaded fastener can fail during tightening?

32.8. What is a rivet?

32.9. What is the difference between a shrink fit and expansion fit in assembly?

32.10. What are the advantages of snap fitting?

32.11. What is the difference between industrial stitching and stapling?

32.12. What are integral fasteners?

32.13. Identify some of the general principles and guidelines for design for assembly.

32.14. Identify some of the general principles and guidelines that apply specifically to automated assembly.

MULTIPLE CHOICE QUIZ

There are 16 correct answers in the following multiple choice questions (some questions have multiple answers that are correct). To attain a perfect score on the quiz, all correct answers must be given. Each correct answer is worth 1 point. Each omitted answer or wrong answer reduces the score by 1 point, and each additional answer beyond the correct number of answers reduces the score by 1 point. Percentage score on the quiz is based on the total number of correct answers.

32.1. Which of the following are reasons why mechanical assembly is often preferred over other forming processes (two best answers): (a) ease of assembly, (b) ease of disassembly, (c) economies of scale, (d) involves melting of the base parts, (e) no heat affected zone in the base parts, and (f) specialization of labor?

32.2. Most externally threaded fasteners are produced by which one of the following processes: (a) cutting the threads, (b) milling the threads, (c) tapping, (d) thread rolling, or (e) turning the threads?

32.3. Which of the following methods and tools are used for applying the required torque to achieve a desired preload of a threaded fastener (three best answers): (a) arbor press, (b) preload method, (c) sense of feel by a human operator, (d) snap fit, (e) stall-motor wrenches, (f) torque wrench, and (g) use of lockwashers?

32.4. Which of the following are the common ways in which threaded fasteners fail during tightening (two best answers): (a) excessive compressive stresses on the head of the fastener because of force applied by the tightening tool, (b) excessive compressive stresses on the shank of the fastener, (c) excessive shear stresses on the shank of the fastener, (d) excessive tensile stresses on the head of the fastener because of force applied by the tightening tool, (e) excessive tensile stresses on the shank of the fastener, and (f) stripping of the internal or external threads?

32.5. The difference between a shrink fit and an expansion fit is that in a shrink fit the internal part is cooled to a sufficiently low temperature to reduce its size for assembly, whereas in an expansion fit, the external part is heated sufficiently to increase its size for assembly: (a) true or (b) false?

32.6. Advantages of snap fit assembly include which of the following (three best answers): (a) components can be designed with features to facilitate part mating, (b) ease of disassembly, (c) no heat affected zone, (d) no special tools are required, (e) parts can be assembled quickly, and (f) stronger joint than with most other assembly methods?

32.7. The difference between industrial stitching and stapling is that the U-shaped fasteners are formed during the stitching process while in stapling the fasteners are preformed: (a) true or (b) false?

32.8. From the standpoint of assembly cost, it is more desirable to use many small threaded fasteners rather than few large ones to distribute the stresses more uniformly: (a) true or (b) false?

32.9. Which of the following are considered good product design rules for automated assembly (two best answers): (a) design the assembly with the fewest number of components possible, (b) design the product using bolts and nuts to allow for disassembly, (c) design with many different fastener types to maximize design flexibility, (d) design parts with asymmetric features to mate with other parts having corresponding (but reverse) features, and (e) limit the required directions of access when adding components to a base part?

PROBLEMS

Threaded Fasteners

32.1. A 5-mm-diameter bolt is to be tightened to produce a preload = 250 N. If the torque coefficient = 0.23, determine the torque that should be applied.

32.2. A 3/8-24 UNF nut and bolt (3/8 in nominal diameter, 24 threads/in) are inserted through a hole in two stacked steel plates. They are tightened so the plates are clamped together with a force of 1000 lb. The torque coefficient is 0.20. (a) What is the torque required to tighten them? (b) What is the resulting stress in the bolt?

32.3. An alloy steel Metric 10×1.5 screw (10-mm diameter, pitch $p = 1.5$ mm) is to be turned into a threaded hole and tightened to one/half of its proof strength. According to Table 32.2, the proof strength = 830 MPa. Determine the maximum torque that should be used if the torque coefficient = 0.18.

32.4. A Metric 16×2 bolt (16-mm diameter, 2-mm pitch) is subjected to a torque of 15 N-m during tightening. If the torque coefficient is 0.24, determine the tensile stress on the bolt.

32.5. A 1/2-13 screw is to be preloaded to a tension force = 1000 lb. Torque coefficient = 0.22. Determine the torque that should be used to tighten the bolt.

32.6. Threaded metric fasteners are available in several systems, two of which are coarse and fine (Table 32.1). Finer threads are not cut as deep and as a result have a larger tensile stress area for the same nominal diameter. (a) Determine the maximum preload that can be safely achieved for coarse pitch and fine pitch threads for a 12-mm bolt. (b) Determine the percent increase in preload of fine threads compared with course threads. Coarse pitch = 1.75 mm and fine pitch = 1.25 mm. Assume the proof strength for both bolts is 600 MPa.

32.7. A torque wrench is used on a 3/4-10 UNC bolt in an automobile final assembly plant. A torque of 70 ft-lb is generated by the wrench. If the torque coefficient = 0.17, determine the tensile stress in the bolt.

32.8. The designer has specified that a 3/8-16 UNC low-carbon bolt (3/8 in nominal diameter, 16 threads/in) in a certain application should be stressed to its proof stress of 33,000 lb/in^2 (see Table 32.2). Determine the maximum torque that should be used if $C = 0.25$.

32.9. A 300-mm-long wrench is used to tighten a Metric 20×2.5 bolt. The proof strength of the bolt for the particular alloy is 380 MPa. The torque coefficient is 0.21. Determine the maximum force that can be applied to the end of the wrench so that the bolt does not permanently deform.

32.10. A 1-8 UNC low carbon steel bolt (diameter = 1.0 in, 8 threads/in) is currently planned for a certain application. It is to be preloaded to 75% of its proof strength, which is 33,000 lb/in^2 (Table 32.2). However, this bolt is too large for the size of the components involved, and a higher strength but smaller bolt would be preferable. Determine (a) the smallest nominal size of an alloy steel bolt (proof strength = 120,000 lb/in^2) that could be used to achieve the same preload from the following standard UNC sizes used by the company: 1/4-20, 5/16-18, 3/8-16, 1/2-13, 5/8-11, or 3/4-10; and (b) compare the torque required to obtain the preload for the original 1-in bolt and the alloy steel bolt selected in part (a) if the torque coefficient in both cases = 0.20.

Interference Fits

32.1. A dowel pin made of steel (elastic modulus = 209,000 MPa) is to be press fitted into a steel collar. The pin has a nominal diameter of 16 mm, and the collar has an outside diameter of 27 mm. (a) Compute the radial pressure and the maximum effective stress if the interference between the shaft OD and the collar ID is 0.03 mm. (b) Determine the effect of increasing the outside diameter of the collar to 39 mm on the radial pressure and the maximum effective stress.

32.2. A pin made of alloy steel is press-fitted into a hole machined in the base of a large machine. The hole has a diameter of 2.497 in. The pin has a diameter of 2.500 in. The base of the machine is 4 ft $\times$ 8 ft. The base and pin have a modulus of elasticity of 30×10^6 lb/in^2, a yield strength of 85,000 lb/in^2, and a tensile strength of 120,000 lb/in^2. Determine (a) the radial pressure between the pin and the base and (b) the maximum effective stress in the interface.

32.3. A gear made of aluminum (modulus of elasticity = 69,000 MPa) is press fitted onto an aluminum shaft. The gear has a diameter of 55 mm at the base of its teeth. The nominal internal diameter of the gear = 30 mm and the interference = 0.10 mm. Compute: (a) the radial pressure between the shaft and the gear, and (b) the maximum effective stress in the gear at its inside diameter.

32.4. A steel collar is press fitted onto a steel shaft. The modulus of elasticity of steel is 30×10^6 lb/in^2. The collar has an internal diameter of 2.498 in and the shaft has an outside diameter = 2.500 in. The outside diameter of the collar is 4.000 in. Determine the radial (interference) pressure on the assembly, and (b) the maximum effective stress in the collar at its inside diameter.

32.5. The yield strength of a certain metal = 50,000 lb/in^2 and its modulus of elasticity = 22×10^6 lb/in^2. It is to be used for the outer ring of a press-fit assembly with a mating shaft made of the same metal. The nominal inside diameter of the ring is 1.000 in and its outside diameter = 2.500 in. Using a safety factor = 2.0, determine the maximum interference that should be used with this assembly.

32.6. A shaft made of aluminum is 40.0 mm in diameter at room temperature (21°C). Its coefficient of thermal expansion = 24.8×10^{-6} mm/mm per °C. If it must be reduced in size by 0.20 mm in order to be expansion fitted into a hole, determine the temperature to which the shaft must be cooled.

32.7. A steel ring has an inside diameter = 30 mm and an outside diameter = 50 mm at room temperature (21°C). If the coefficient of thermal expansion of steel = 12.1×10^{-6} mm/mm per °C, determine the inside diameter of the ring when heated to 500°C.

32.8. A steel collar is to be heated from room temperature (70°F) to 700°F. Its inside diameter = 1.000 in, and its outside diameter = 1.625 in. If the coefficient of thermal expansion of the steel is = 6.7×10^{-6} in/in per °F, determine the increase in the inside diameter of the collar.

32.9. A bearing for the output shaft of a 200 hp motor is to be heated to expand it enough to press on the shaft. At 70°F the bearing has an inside diameter of 4.000 in and an outside diameter of 7.000 in. The shaft has an outside diameter of 4.004 in. The modulus of elasticity for the shaft and bearing is 30×10^6 lb/in^2 and the coefficient of thermal expansion is 6.7×10^{-6} in/in per °F. (a) At what temperature will the bearing have 0.005 of clearance to fit over the shaft? (b) After it is assembled and cooled, what is the radial pressure between the bearing and shaft? (c) Determine the maximum effective stress in the bearing.

32.10. A steel collar whose outside diameter = 3.000 in at room temperature is to be shrink fitted onto a steel shaft by heating it to an elevated temperature while the shaft remains at room temperature. The shaft diameter = 1.500 in. For ease of assembly when the collar is heated to an elevated temperature of 1000°F, the clearance between the shaft and the collar is to be 0.007 in. Determine (a) the initial inside diameter of the collar at room temperature so that this clearance is satisfied, (b) the radial pressure and (c) maximum effective stress on the resulting interference fit at room temperature (70°F). For steel, the elastic modulus = 30,000,000 lb/in^2 and coefficient of thermal expansion = 6.7×10^{-6} in/in per °F.

32.11. A pin is to be inserted into a collar using an expansion fit. Properties of the pin and collar metal are: coefficient of thermal expansion is 12.3×10^{-6} m/m/°C, yield strength is 400 MPa, and modulus of elasticity is 209 GPa. At room temperature (20°C), the outer and inner diameters of the collar = 95.00 mm and 60.00 mm, respectively, and the pin has a diameter = 60.03 mm. The pin is to be reduced in size for assembly into the collar by cooling to a sufficiently low temperature that there is a clearance of 0.06 mm. (a) What is the temperature to which the pin must be cooled for assembly? (b) What is the radial pressure at room temperature after assembly? (c) What is the safety factor in the resulting assembly?

Part IX   Special Processing and Assembly Technologies

33   RAPID PROTOTYPING

Chapter Contents

In this part of the book, we discuss a collection of processing and assembly technologies that do not fit neatly into our classification scheme in Figure 1.4. They are technologies that have been adapted from the conventional manufacturing and assembly operations or developed from scratch to serve the special functions or needs of designers and manufacturers. Rapid prototyping, covered in the present chapter, is a collection of processes used to fabricate a model, part, or tool in minimum possible time. Chapters 34 and 35 discuss technologies used in electronics manufacturing, an activity of significant economic importance. Chapter 34 covers integrated circuit processing, and Chapter 35 covers electronics assembly and packaging. Chapters 36 and 37 survey some of the technologies used to produce very small parts and products. Chapter 36 describes microfabrication technologies used to produce items measured in microns (10^{-6} m), whereas Chapter 37 discusses nanofabrication technologies for producing items measured in nanometers (10^{-9} m). The processes covered in these five chapters are relatively new. Rapid prototyping dates from about 1988. Modern electronics production techniques date from around 1960 (Historical Note 34.1), although dramatic advances have been made in electronics processing since that time. The microfabrication technologies discussed in Chapter 36 followed soon after electronics processing. Finally, the nanofabrication technologies represent an emerging field today that dates from the 1990s.

Rapid prototyping (RP) is a family of fabrication methods to make engineering prototypes in minimum possible lead times based on a computer-aided design (CAD) model of the item. The traditional method of fabricating a prototype part is machining, which can require significant lead times—up to several weeks, sometimes longer, depending on part complexity, difficulty in ordering materials, and scheduling production equipment. A number of rapid prototyping techniques are now available that allow a part to be produced in hours or days rather than weeks, given that a computer model of the part has been generated on a CAD system.

33.1   FUNDAMENTALS OF RAPID PROTOTYPING

The special need that motivates the variety of rapid prototyping technologies arises because product designers would like to have a physical model of a new part or product design rather than a computer model or line drawing. The creation of a prototype is an integral step in the design procedure. A *virtual prototype*, which is a computer model of the part design on a CAD system, may not be adequate for the designer to visualize the part. It certainly is not sufficient to conduct real physical tests on the part, although it is possible to perform simulated tests by finite element analysis or other methods. Using one of the available RP technologies, a solid physical part can be created in a relatively short time (hours if the company possesses the RP equipment or days if the part fabrication must be contracted to an outside firm specializing in RP). The designer can therefore visually examine and physically feel the part and begin to perform tests and experiments to assess its merits and shortcomings.

Available rapid prototyping technologies can be divided into two basic categories: (1) material removal processes and (2) material addition processes. The *material removal RP* alternative involves machining (Chapter 22), primarily milling and drilling, using a dedicated Computer Numerical Control (CNC) machine that is available to the design department on short notice. To use CNC, a part program must be prepared from the CAD model (Section 38.3.3). The starting material is often a solid block of wax, which is very easy to machine, and the part and chips can be melted and resolidified for reuse when the current prototype is no longer needed. Other starting materials can also be used, such as wood, plastics, or metals (e.g., a machinable grade of aluminum or brass). The CNC machines used for rapid prototyping are often small, and the terms *desktop milling* or *desktop machining* are sometimes used for this technology. Maximum starting block sizes in desktop machining are typically 180 mm (7 in) in the x-direction, 150 mm (6 in) in the y-direction, and 150 mm (6 in) in the z-direction [2].

The principal emphasis in this chapter is on *material-addition RP* technologies, all of which work by adding layers of material one at a time to build the solid part from bottom to top. Starting materials include (1) liquid monomers that are cured layer by layer into solid polymers, (2) powders that are aggregated and bonded layer by layer, and (3) solid sheets that are laminated to create the solid part. In addition to starting material, what distinguishes the various material addition RP technologies is the method of building and adding the layers to create the solid part. Some techniques use lasers to solidify the starting material, another deposits a soft plastic filament in the outline of each layer, while others bond solid layers together. There is a correlation between the starting material and the part-building techniques, as we shall see in our discussion of RP technologies.

The common approach to prepare the control instructions (part program) in all of the current material addition RP techniques involves the following steps [6]:

1. *Geometric modeling.* This consists of modeling the component on a CAD system to define its enclosed volume. Solid modeling is the preferred technique because it

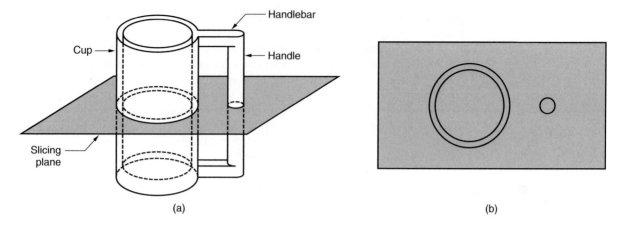

FIGURE 33.1   Conversion of a solid model of an object into layers (only one layer is shown).

provides a complete and unambiguous mathematical representation of the geometry. For rapid prototyping, the important issue is to distinguish the interior (mass) of the part from its exterior, and solid modeling provides for this distinction.

2. ***Tessellation of the geometric model***[1]***.*** In this step, the CAD model is converted into a format that approximates its surfaces by triangles or polygons, with their vertices arranged to distinguish the object's interior from its exterior. The common tessellation format used in rapid prototyping is STL, which has become the de facto standard input format for nearly all RP systems.

3. ***Slicing of the model into layers.*** In this step, the model in STL[2] file format is sliced into closely spaced parallel horizontal layers. Conversion of a solid model into layers is illustrated in Figure 33.1. These layers are subsequently used by the RP system to construct the physical model. By convention, the layers are formed in the x-y plane orientation, and the layering procedure occurs in the z-axis direction. For each layer, a curing path is generated, called the STI file, which is the path that will be followed by the RP system to cure (or otherwise solidify) the layer.

As our brief overview indicates, there are several different technologies used for material addition rapid prototyping. This heterogeneity has spawned several alternative names for rapid prototyping, including ***layer manufacturing***, ***direct CAD manufacturing***, and ***solid freeform fabrication***. The term ***rapid prototyping and manufacturing*** (RPM) is also being used more frequently to indicate that the RP technologies can be applied to make production parts and production tooling, not just prototypes.

33.2 RAPID PROTOTYPING TECHNOLOGIES

The 25 or so RP techniques currently developed can be classified in various ways. Let us adopt a classification system recommended in [6], which is consistent with the way we classify part-shaping processes in this book (after all, rapid prototyping is a part-shaping

---

[1]More generally, the term ***tessellation*** refers to the laying out or creation of a mosaic, such as one consisting of small colored tiles affixed to a surface for decoration.

[2]STL stands for STereoLithography, one of the primary technologies used for rapid prototyping, developed by 3D Systems Inc.

process). The classification method is based on the form of the starting material in the RP process: (1) liquid-based, (2) solid-based, and (3) powder-based. We discuss examples of each class in the following three sections.

33.2.1 LIQUID-BASED RAPID PROTOTYPING SYSTEMS

The starting material in these technologies is a liquid. About a dozen RP technologies are in this category, of which we have selected the following to describe: (1) stereolithography, (2) solid ground curing, and (3) droplet deposition manufacturing.

Stereolithography   This was the first material addition RP technology, dating from about 1988 and introduced by 3D Systems, Inc. based on the work of inventor Charles Hull. There are more installations of stereolithography than any other RP technology. Stereolithography (STL) is a process for fabricating a solid plastic part out of a photosensitive liquid polymer using a directed laser beam to solidify the polymer. The general setup for the process is illustrated in Figure 33.2. Part fabrication is accomplished as a series of layers, in which one layer is added onto the previous layer to gradually build the desired three-dimensional geometry. A part fabricated by STL is illustrated in Figure 33.3.

The stereolithography apparatus consists of (1) a platform that can be moved vertically inside a vessel containing the photosensitive polymer, and (2) a laser whose beam can be controlled in the *x-y* direction. At the start of the process, the platform is positioned vertically near the surface of the liquid photopolymer, and a laser beam is directed through a curing path that comprises an area corresponding to the base (bottom layer) of the part. This and subsequent curing paths are defined by the STI file (step 3 in preparing the control instructions described in the preceding). The action of the laser is to harden (cure) the photosensitive polymer where the beam strikes the liquid, forming a solid layer of plastic that adheres to the platform. When the initial layer is completed, the platform is lowered by a distance equal to the layer thickness, and a second layer is formed on top of the first by the laser, and so on. Before each new layer is cured, a wiper blade is passed over the viscous liquid resin to ensure that its level is the same throughout the surface. Each layer consists of its own area shape, so that the succession of layers, one on top of the previous, creates the solid part shape. Each layer is 0.076 to 0.50 mm (0.003 to 0.020 in) thick. Thinner layers provide better resolution and allow more intricate part shapes; but processing time is

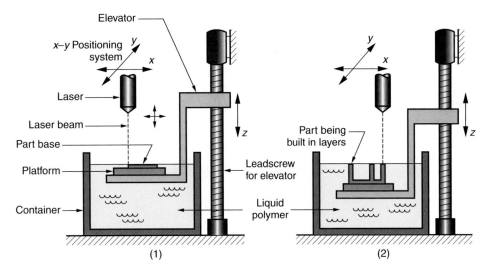

FIGURE 33.2
Stereolithography: (1) at the start of the process, in which the initial layer is added to the platform; and (2) after several layers have been added so that the part geometry gradually takes form.

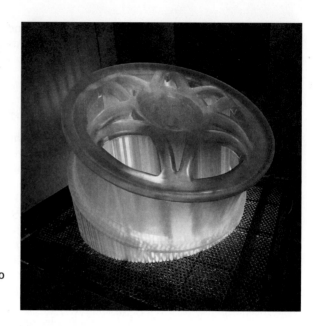

FIGURE 33.3 A part produced by stereolithography. (Photo courtesy of 3D Systems, Inc.)

greater. Photopolymers are typically acrylic [13], although use of epoxy for STL has also been reported [10]. The starting materials are liquid monomers. Polymerization occurs upon exposure to ultraviolet light produced by helium-cadmium or argon ion lasers. Scan speeds of STL lasers typically range between 500 and 2500 mm/s.

The time required to build the part by this layering process ranges from 1 hour for small parts of simple geometry up to several dozen hours for complex parts. Other factors that affect cycle time are scan speed and layer thickness. The part build time in stereolithography can be estimated by determining the time to complete each layer and then summing the times for all layers. First, the time to complete a single layer is given by the following equation:

$$T_i = \frac{A_i}{vD} + T_r \tag{33.1}$$

where T_i = time to complete layer i, seconds, where the subscript i is used to identify the layer; A_i = area of layer i, mm^2 (in^2); v = average scanning speed of the laser beam at the surface, mm/s (in/sec); D = diameter of the laser beam at the surface (called the "spot size," assumed circular), mm (in); and T_r = repositioning time between layers, s.

In the case of stereolithography, the repositioning time involves lowering the worktable in preparation for the next layer to be fabricated. Other RP techniques require analogous repositioning steps between layers. The average scanning speed v must include any effects of interruptions in the scanning path (e.g., because of gaps between areas of the part in a given layer). Once the T_i values have been determined for all layers, then the build cycle time can be determined:

$$T_c = \sum_{i=1}^{n_l} T_i \tag{33.2}$$

where T_c = the STL build cycle time, s; and n_l = the number of layers used to approximate the part.[3]

---

[3]Although these equations have been developed here for stereolithography, similar formulas can be developed for the other RP material addition technologies discussed in this chapter, because they all use the same layer-by-layer fabrication method.

After all of the layers have been formed, the photopolymer is about 95% cured. The piece is therefore "baked" in a fluorescent oven to completely solidify the polymer. Excess polymer is removed with alcohol, and light sanding is sometimes used to improve smoothness and appearance.

Depending on its design and orientation, a part may contain overhanging features that have no means of support during the bottom-up approach used in stereolithography. For example, in the part of Figure 33.1, if the lower half of the handle and the lower handlebar were eliminated, the upper portion of the handle would be unsupported during fabrication. In these cases, extra pillars or webs may need to be added to the part simply for support purposes. Otherwise, the overhangs may float away or otherwise distort the desired part geometry. These extra features must be trimmed away after the process is completed.

Solid Ground Curing   Like stereolithography, solid ground curing (SGC) works by curing a photosensitive polymer layer by layer to create a solid model based on CAD geometric data. Instead of using a scanning laser beam to accomplish the curing of a given layer, the entire layer is exposed to an ultraviolet light source through a mask that is positioned above the surface of the liquid polymer. The hardening process takes 2 to 3 seconds for each layer. SGC systems are sold under the name ***Solider system*** by Cubital Ltd.

The starting data in SGC is similar to that used in stereolithography: a CAD geometric model of the part that has been sliced into layers. For each layer, the step-by-step procedure in SGC is illustrated in Figure 33.4 and described here: (1) A mask is

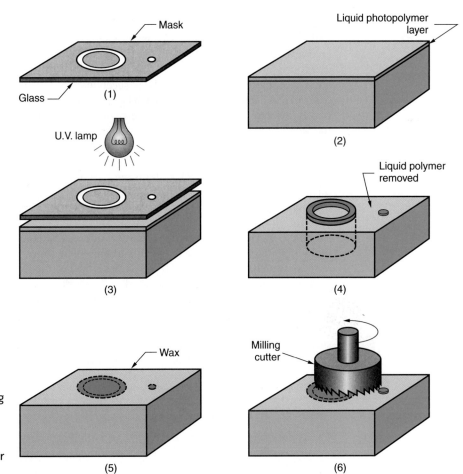

FIGURE 33.4   Solid ground curing process for each layer: (1) mask preparation, (2) applying liquid photopolymer layer, (3) mask positioning and exposure of layer, (4) uncured polymer removed from surface, (5) wax filling, (6) milling for flatness and thickness.

created on a glass plate by electrostatically charging a negative image of the layer onto the surface. The imaging technology is basically the same as that used in photocopiers. (2) A thin flat layer of liquid photopolymer is distributed over the surface of the work platform. (3) The mask is positioned above the liquid polymer surface and exposed by a high-powered (e.g., 2000 W) ultraviolet lamp. The portions of the liquid polymer layer that are unprotected by the mask are solidified in about 2 seconds. The shaded areas of the layer remain in the liquid state. (4) The mask is removed, the glass plate is cleaned and made ready for a subsequent layer in step 1. Meanwhile, the liquid polymer remaining on the surface is removed in a wiping and vacuuming procedure. (5) The now-open areas of the layer are filled in with hot wax. When hardened, the wax acts to support overhanging sections of the part. (6) When the wax has cooled and solidified, the polymer-wax surface is milled to form a flat layer of specified thickness, ready to receive the next application of liquid photopolymer in step 2. Although we have described SGC as a sequential process, certain steps are accomplished in parallel. Specifically, the mask preparation step 1 for the next layer is performed simultaneously with the layer fabrication steps 2 through 6, using two glass plates during alternating layers.

The sequence for each layer takes about 90 seconds. Throughput time to produce a part by SGC is claimed to be about eight times faster than competing RP systems [6]. The solid cubic form created in SGC consists of solid polymer and wax. The wax provides support for fragile and overhanging features of the part during fabrication, but can be melted away later to leave the free-standing part. No post curing of the completed prototype model is required, as in stereolithography.

Droplet Deposition Manufacturing   These systems operate by melting the starting material and shooting small droplets onto a previously formed layer. The liquid droplets cold weld to the surface to form a new layer. The deposition of droplets for each new layer is controlled by a moving x-y spray nozzle workhead whose path is based on a cross section of a CAD geometric model that has been sliced into layers (similar to the other RP systems described in the preceding). After each layer has been applied, the platform supporting the part is lowered a certain distance corresponding to the layer thickness, in preparation for the next layer. The term droplet deposition manufacturing (DDM) refers to the fact that small particles of work material are deposited as projectile droplets from the workhead nozzle.

Several commercial RP systems are based on this general operating principle, the differences being in the type of material that is deposited and the corresponding technique by which the workhead operates to melt and apply the material. An important criterion that must be satisfied by the starting material is that it be readily melted and solidified. Work materials used in DDM include wax and thermoplastics. Metals with low melting point, such as tin, zinc, lead, and aluminum, have also been tested.

One of the more popular BPM systems is the Personal Modeler, available from BPM Technology, Inc. Wax is commonly used as the work material. The ejector head operates using a piezoelectric oscillator that shoots droplets of wax at a rate of 10,000 to 15,000 per second. The droplets are of uniform size at about 0.076 mm (0.003 in) diameter, which flatten to about 0.05-mm (0.002-in) solidified thickness on impact against the existing part surface. After each layer has been deposited, the surface is milled or thermally smoothed to achieve accuracy in the z-direction. Layer thickness is about 0.09 mm (0.0035 in).

33.2.2   SOLID-BASED RAPID PROTOTYPING SYSTEMS

The common feature in these RP systems is that the starting material is solid. In this section we discuss two solid-based RP systems: (1) laminated-object manufacturing and (2) fused-deposition modeling.

FIGURE 33.5
Laminated-object
manufacturing.

Laminated-Object Manufacturing   The principal company offering laminated-object manufacturing (LOM) systems is Helisys, Inc. Of interest is that much of the early research and development work on LOM was funded by National Science Foundation. The first commercial LOM unit was shipped in 1991.

Laminated-object manufacturing produces a solid physical model by stacking layers of sheet stock that are each cut to an outline corresponding to the cross-sectional shape of a CAD model that has been sliced into layers. The layers are bonded one on top of the previous one before cutting. After cutting, the excess material in the layer remains in place to support the part during building. Starting material in LOM can be virtually any material in sheet stock form, such as paper, plastic, cellulose, metals, or fiber-reinforced materials. Stock thickness is 0.05 to 0.50 mm (0.002 to 0.020 in). In LOM, the sheet material is usually supplied with adhesive backing as rolls that are spooled between two reels, as in Figure 33.5. Otherwise, the LOM process must include an adhesive coating step for each layer.

The data preparation phase in LOM consists of slicing the geometric model using the STL file for the given part. The slicing function is accomplished by LOMSlice™, the special software used in laminated-object manufacturing. Slicing the STL model in LOM is performed after each layer has been physically completed and the vertical height of the part has been measured. This provides a feedback correction to account for the actual thickness of the sheet stock being used, a feature unavailable on most other RP systems. With reference to Figure 33.5, the LOM process for each layer can be described as follows, picking up the action with a sheet of stock in place and bonded to the previous stack: (1) LOMSlice™ computes the cross-sectional perimeter of the STL model based on the measured height of the physical part at the current layer of completion. (2) A laser beam is used to cut along the perimeter, as well as to crosshatch the exterior portions of the sheet for subsequent removal. The laser is typically a 25 or 50 W CO_2 laser. The cutting trajectory is controlled by means of an x-y positioning system. The cutting depth is controlled so that only the top layer is cut. (3) The platform holding the stack is lowered, and the sheet stock is advanced between supply roll and take-up spool for the next layer. The platform is then raised to a height consistent with the stock thickness and a heated roller moves across the new layer to bond it to the previous layer. The height of the physical stack is measured in preparation for the next slicing computation by LOMSlice™.

When all of the layers are completed, the new part is separated from the excess external material using a hammer, putty knife, and wood carving tools. The part can then be sanded to smooth and blend the layer edges. A sealing application is recommended, using a urethane, epoxy, or other polymer spray to prevent moisture absorption and damage. LOM part sizes can be relatively large among RP processes, with work volumes up to 800 mm × 500 mm by 550 mm (32 in × 20 in × 22 in). More common work volumes are 380 mm × 250 mm × 350 mm (15 in × 10 in × 14 in).

Several low-cost systems based on the LOM build method are available. For example, the JP System 5, available from Schroff Development Corporation, uses a mechanical knife rather than a laser to cut the sheet stock for each layer. This system is intended as a teaching tool and requires manual assembly of the layers.

Fused-Deposition Modeling   Fused-deposition modeling (FDM) is an RP process in which a filament of wax or polymer is extruded onto the existing part surface from a workhead to complete each new layer. The workhead is controlled in the *x-y* plane during each layer and then moves up by a distance equal to one layer in the *z*-direction. The starting material is a solid filament with typical diameter = 1.25 mm (0.050 in) fed from a spool into the workhead that heats the material to about 0.5°C (1°F) above its melting point before extruding it onto the part surface. The extrudate is solidified and cold welded to the cooler part surface in about 0.1 second. The part is fabricated from the base up, using a layer-by-layer procedure similar to other RP systems.

FDM was developed by Stratasys Inc., which sold its first machine in 1990. The starting data is a CAD geometric model that is processed by Stratasys's software modules QuickSlice; and SupportWork[TM]. QuickSlice; is used to slice the model into layers, and SupportWork[TM] is used to generate any support structures that are required during the build process. If supports are needed, a dual extrusion head and a different material is used to create the supports. The second material is designed to readily be separated from the primary modeling material. The slice (layer) thickness can be set anywhere from 0.05 to 0.75 mm (0.002 to 0.030 in). About 400 mm of filament material can be deposited per second by the extrusion workhead in widths (called the ***road width***) that can be set between 0.25 and 2.5 mm (0.010 to 0.100 in). Starting materials are wax and several polymers, including ABS, polyamide, polyethylene, and polypropylene. These materials are nontoxic, allowing the FDM machine to be set up in an office environment.

33.2.3 POWDER-BASED RAPID PROTOTYPING SYSTEMS

The common feature of the RP technologies described in this section is that the starting material is powder.[4] We discuss two RP systems in this category: (1) selective laser sintering and (2) three-dimensional printing.

Selective Laser Sintering   Selective laser sintering (SLS) uses a moving laser beam to sinter heat-fusible powders in areas corresponding to the CAD geometric model one layer at a time to build the solid part. After each layer is completed, a new layer of loose powders is spread across the surface using a counter-rotating roller. The powders are preheated to just below their melting point to facilitate bonding and reduce distortion. Layer by layer, the powders are gradually bonded into a solid mass that forms the three-dimensional part geometry. In areas not sintered by the laser beam, the powders remain loose so they can be poured out of the completed part. Meanwhile, they serve to support the solid regions of the part as fabrication proceeds. Layer thickness is 0.075 to 0.50 mm (0.003 to 0.020 in).

SLS was developed at the University of Texas (Austin) as an alternative to stereolithography, and SLS machines are currently marketed by DTM Corp. It is a more versatile process than stereolithography in terms of possible work materials. Current materials used

---

[4]The definition, characteristics, and production of powders are described in Chapters 16 and 17.

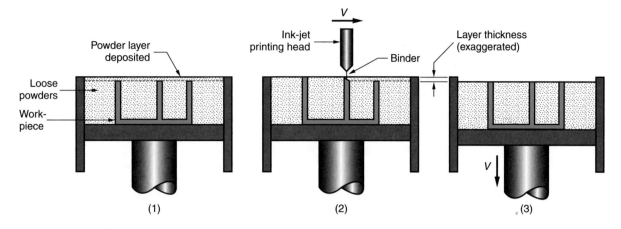

FIGURE 33.6   Three-dimensional printing: (1) powder layer is deposited, (2) ink-jet printing of areas that will become the part, and (3) piston is lowered for next layer (key: v = motion).

in selective laser sintering include polyvinylchloride, polycarbonate, polyester, poly-urethane, ABS, nylon, and investment casting wax. These materials are less expensive than the photosensitive resins used in stereolithography. They are also nontoxic and can be sintered using low power (25 to 50 W) CO_2 lasers. Metal and ceramic powders are also being used in SLS.

Three-Dimensional Printing   This RP technology was developed at Massachusetts Institute of Technology. Three-dimensional printing (3DP) builds the part in the usual layer-by-layer fashion using an ink-jet printer to eject an adhesive bonding material onto successive layers of powders. The binder is deposited in areas corresponding to the cross sections of the solid part, as determined by slicing the CAD geometric model into layers. The binder holds the powders together to form the solid part, while the unbonded powders remain loose to be removed later. While the loose powders are in place during the build process, they provide support for overhanging and fragile features of the part. When the build process is completed, the part is heat treated to strengthen the bonding, followed by removal of the loose powders. To further strengthen the part, a sintering step can be applied to bond the individual powders.

The part is built on a platform whose level is controlled by a piston. Let us describe the process for one cross section with reference to Figure 33.6: (1) A layer of powder is spread on the existing part-in-process. (2) An ink-jet printing head moves across the surface, ejecting droplets of binder on those regions that are to become the solid part. (3) When the printing of the current layer is completed, the piston lowers the platform for the next layer.

Starting materials in 3DP are powders of ceramic, metal, or cermet, and binders that are polymeric or colloidal silica or silicon carbide [10], [13]. Typical layer thickness ranges from 0.10 to 0.18 mm (0.004 to 0.007 in). The ink-jet printing head moves across the layer at a speed of about 1.5 m/s (59 in/sec), with ejection of liquid binder determined during the sweep by raster scanning. The sweep time, together with the spreading of the powders, permits a cycle time per layer of about 2 seconds [13].

33.3   APPLICATION ISSUES IN RAPID PROTOTYPING

Applications of rapid prototyping can be classified into three categories: (1) design, (2) engineering analysis and planning, and (3) tooling and manufacturing.

Design   This was the initial application area for RP systems. Designers are able to confirm their design by building a real physical model in minimum time using rapid prototyping. The features and functions of the part can be communicated to others more easily using a physical model than by a paper drawing or displaying it on a CAD system monitor. Benefits to design attributed to rapid prototyping include [2]: (1) reduced lead times to produce prototype components, (2) improved ability to visualize the part geometry because of its physical existence, (3) earlier detection and reduction of design errors, and (4) increased capability to compute mass properties of components and assemblies.

Engineering Analysis and Planning   The existence of an RP-fabricated part allows for certain types of engineering analysis and planning activities to be accomplished that would be more difficult without the physical entity. Some of the possibilities are (1) comparison of different shapes and styles to optimize aesthetic appeal of the part; (2) analysis of fluid flow through different orifice designs in valves fabricated by RP; (3) wind tunnel testing of different streamline shapes using physical models created by RP; (4) stress analysis of a physical model; (5) fabrication of preproduction parts by RP as an aid in process planning and tool design; and (6) combining medical imaging technologies, such as magnetic resonance imaging (MRI), with RP to create models for doctors in planning surgical procedures or fabricating prostheses or implants.

Tooling and Manufacturing   The trend in RP applications is toward its greater use in the fabrication of production tooling and for actual manufacture of parts. When RP is adopted to fabricate production tooling, the term ***rapid tool making*** (RTM) is often used. RTM applications divide into two approaches [4]: ***indirect*** RTM method, in which a pattern is created by RP and the pattern is used to fabricate the tool, and ***direct*** RTM method, in which RP is used to make the tool itself. Examples of indirect RTM include (1) use of an RP-fabricated part as the master in making a silicon rubber mold that is subsequently used as a production mold, (2) RP patterns to make the sand molds in sand casting (Section 11.1), (3) fabrication of patterns of low-melting point materials (e.g., wax) in limited quantities for investment casting (Section 11.2.4), and (4) making electrodes for EDM (Section 26.3.1) [6], [10]. Examples of direct RTM include: (1) RP-fabricated mold cavity inserts that can be sprayed with metal to produce injection molds for a limited quantity of production plastic parts (Section 13.6) and (2) 3-D printing to create a die geometry in metallic powders followed by sintering and infiltration to complete the fabrication of the die [4], [6], [10].

Examples of actual part production include [10]: (1) small batch sizes of plastic parts that could not be economically injection molded because of the high cost of the mold, (2) parts with intricate internal geometries that could not be made using conventional technologies without assembly, and (3) one-of-a-kind parts such as bone replacements that must be made to correct size for each user.

Not all RP technologies can be used for all of these tooling and manufacturing examples. Interested readers should consult more complete treatments of the RP technologies for specific details on these and other examples.

Problems with Rapid Prototyping   The principal problems with current RP technologies include (1) part accuracy, (2) limited variety of materials, and (3) mechanical performance of the fabricated parts.

Several sources of error limit part accuracy in RP systems: (1) mathematical, (2) process related, or (3) material related [13]. Mathematical errors include approximations of part surfaces used in RP data preparation and differences between the slicing thicknesses and actual layer thicknesses in the physical part. The latter differences result in z-axis dimensional errors. An inherent limitation in the physical part is the steps between layers, especially as layer thickness is increased, resulting in a staircase appearance for sloping part surfaces. Process-related errors are those that result from the particular part building technology used in the RP system. These errors degrade the shape of each layer as well as the registration between adjacent layers. Process errors can also affect the z-axis dimension. Finally, material-related errors include shrinkage and distortion. An allowance for shrinkage can be made by enlarging the CAD model of the part based on previous experience with the process and materials.

Current rapid prototyping systems are limited in the variety of materials they can process. For example, the most common RP technology, stereolithography, is limited to photosensitive polymers. In general, the materials used in RP systems are not as strong as the production part materials that will be used in the actual product. This limits the mechanical performance of the prototypes and the amount of realistic testing that can be done to verify the design during product development.

REFERENCES

[1] Ashley, S. "Rapid Prototyping Is Coming of Age," *Mechanical Engineering,* July 1995, pp. 62–68.

[2] Bakerjian, R., and Mitchell, P. (eds.). *Tool and Manufacturing Engineers Handbook,* 4th ed., Vol. VI, *Design for Manufacturability.* Society of Manufacturing Engineers, Dearborn, Michigan, 1992, Chapter 7.

[3] Destefani, J. "Plus or Minus," *Manufacturing Engineering,* April 2005, pp. 93–97.

[4] Hilton, P. "Making the Leap to Rapid Tool Making," *Mechanical Engineering,* July 1995, pp. 75–76.

[5] Kai, C. C., and Fai, L. K. "Rapid Prototyping and Manufacturing: The Essential Link between Design and Manufacturing," Chapter 6 in *Integrated Product and Process Development: Methods, Tools, and Technologies,* J. M. Usher, U. Roy, and H. R. Parsaei (eds.). John Wiley & Sons, New York, 1998, pp. 151–183.

[6] Kai, C. C., Fai, L. K., and Chu-Sing, L. *Rapid Prototyping: Principles and Applications.* 2nd ed. World Scientific Publishing Co., Singapore, 2003.

[7] Kochan, D., Kai, C. C. and Zhaohui, D. "Rapid Prototyping Issues in the 21st Century," *Computers in Industry,* Vol. 39, pp. 3–10, 1999.

[8] Noorani, R. I., *Rapid Prototyping: Principles and Applications,* John Wiley & Sons, Hoboken, New Jersey, 2006.

[9] Pacheco, J. M. *Rapid Prototyping,* Report MTIAC SOAR-93-01. Manufacturing Technology Information Analysis Center, IIT Research Institute, Chicago, 1993.

[10] Pham, D. T., and Gault, R. S. "A Comparison of Rapid Prototyping Technologies," *International Journal of Machine Tools and Manufacture,* Vol. 38, pp. 1257–1287, 1998.

[11] Tseng, A. A., Lee, M. H., and Zhao, B. "Design and Operation of a Droplet Deposition System for Freeform Fabrication of Metal Parts," *ASME Journal of Eng. Mat. Tech.,* Vol. 123, No. 1, 2001.

[12] Wohlers, T., "Direct Digital Manufacturing," *Manufacturing Engineering,* January 2009, pp. 73–81.

[13] Yan, X., and Gu, P. "A Review of Rapid Prototyping Technologies and Systems," *Computer-Aided Design,* Vol. 28, No. 4, pp. 307–318, 1996.

REVIEW QUESTIONS

33.1. What is rapid prototyping? Provide a definition of the term.

33.2. What are the three types of starting materials in rapid prototyping?

33.3. Besides the starting material, what other feature distinguishes the rapid prototyping technologies?

33.4. What is the common approach used in all of the material addition technologies to prepare the control instructions for the RP system?

33.5. Of all of the current rapid prototyping technologies, which one is the most widely used?

33.6. Describe the RP technology called solid ground curing.

33.7. Describe the RP technology called laminated-object manufacturing.

33.8. What is the starting material in fused-deposition modeling?

MULTIPLE CHOICE QUIZ

There are 11 correct answers in the following multiple choice questions (some questions have multiple answers that are correct). To attain a perfect score on the quiz, all correct answers must be given. Each correct answer is worth 1 point. Each omitted answer or wrong answer reduces the score by 1 point, and each additional answer beyond the correct number of answers reduces the score by 1 point. Percentage score on the quiz is based on the total number of correct answers.

33.1. Machining is never used for rapid prototyping because it takes too long: (a) true or (b) false?

33.2. Which of the following rapid prototyping processes starts with a photosensitive liquid polymer to fabricate a component (two correct answers): (a) ballistic particle manufacturing, (b) fused-deposition modeling, (c) selective laser sintering, (d) solid ground curing, and (e) stereolithography?

33.3. Of all of the current material addition rapid prototyping technologies, which one is the most widely used: (a) ballistic particle manufacturing, (b) fused deposition modeling, (c) selective laser sintering, (d) solid ground curing, and (e) stereolithography?

33.4. Which one of the following RP technologies uses solid sheet stock as the starting material: (a) ballistic particle manufacturing, (b) fused-deposition

modeling, (c) laminated-object manufacturing, (d) solid ground curing, or (e) stereolithography?

33.5. Which of the following RP technologies uses powders as the starting material (two correct answers): (a) ballistic particle manufacturing, (b) fused-deposition modeling, (c) selective laser sintering, (d) solid ground curing, and (e) three-dimensional printing?

33.6. Rapid prototyping technologies are never used to make production parts: (a) true or (b) false?

33.7. Which of the following are problems with the current material addition rapid prototyping technologies (three best answers): (a) inability of the designer to design the part, (b) inability to convert a solid part into layers, (c) limited material variety, (d) part accuracy, (e) part shrinkage, and (f) poor machinability of the starting material?

PROBLEMS

33.1. A prototype of a tube with a square cross section is to be fabricated using stereolithography. The outside dimension of the square = 100 mm and the inside dimension = 90 mm (wall thickness = 5 mm except at corners). The height of the tube (z-direction) = 80 mm. Layer thickness = 0.10 mm. The diameter of the laser beam ("spot size") = 0.25 mm, and the beam is moved across the surface of the photopolymer at a velocity of 500 mm/s. Compute an estimate for the time required to build the part, if 10 s are lost each layer to lower the height of the platform that holds the part. Neglect the time for postcuring.

33.2. Solve Problem 33.1 except that the layer thickness = 0.40 mm.

33.3. The part in Problem 33.1 is to be fabricated using fused deposition modeling instead of stereolithography. Layer thickness is to be 0.20 mm and the width of the extrudate deposited on the surface of the part = 1.25 mm. The extruder workhead moves in the x-y plane at a speed of 150 mm/s. A delay of 10 s is experienced between each layer to reposition the workhead. Compute an estimate for the time required to build the part.

33.4. Solve Problem 33.3, except using the following additional information. It is known that the diameter of

the filament fed into the extruder workhead is 1.25 mm, and the filament is fed into the workhead from its spool at a rate of 30.6 mm of length per second while the workhead is depositing material. Between layers, the feed rate from the spool is zero.

33.5. A cone-shaped part is to be fabricated using stereolithography. The radius of the cone at its base = 35 mm and its height = 40 mm. The layer thickness = 0.20 mm. The diameter of the laser beam = 0.22 mm, and the beam is moved across the surface of the photopolymer at a velocity of 500 mm/s. Compute an estimate for the time required to build the part, if 10 seconds are lost each layer to lower the height of the platform that holds the part. Neglect post-curing time.

33.6. The cone-shaped part in Problem 33.5 is to be built using laminated-object manufacturing. Layer thickness = 0.20 mm. The laser beam can cut the sheet stock at a velocity of 500 mm/s. Compute an estimate for the time required to build the part, if 10 seconds are lost each layer to lower the height of the platform that holds the part and advance the sheet stock in preparation for the next layer. Ignore cutting of the cross-hatched areas outside of the part since the cone should readily drop out of the stack owing to its geometry.

33.7. Stereolithography is to be used to build the part in Figure 33.1 in the text. Dimensions of the part are: height = 125 mm, outside diameter = 75 mm, inside diameter = 65 mm, handle diameter = 12 mm, handle distance from cup = 70 mm measured from center (axis) of cup to center of handle. The handle bars connecting the cup and handle at the top and bottom of the part have a rectangular cross section and are 10 mm thick and 12 mm wide. The thickness at the base of the cup is 10 mm. The laser beam diameter = 0.25 mm, and the beam can be moved across the surface of the photopolymer at = 500 mm/s. Layer thickness = 0.20 mm. Compute an estimate of the time required to build the part, if 10 seconds are lost each layer to lower the height of the platform that holds the part. Neglect post-curing time.

33.8. A prototype of a part is to be fabricated using stereolithography. The part is shaped like a right triangle whose base = 36 mm, height = 48 mm, and thickness = 25 mm. In application, the part will stand on its base, which is 36 mm by 25 mm. In the stereolithography process, the layer thickness = 0.20 mm. The diameter of the laser beam ("spot size") = 0.15 mm, and the beam is moved across the surface of the photopolymer at a velocity of 400 mm/s. Compute the minimum possible time required to build the part, if 8 seconds are lost each layer to lower the height of the platform that holds the part. Neglect the time for postcuring.

34 PROCESSING OF INTEGRATED CIRCUITS

Chapter Contents

An *integrated circuit* (IC) is a collection of electronic devices such as transistors, diodes, and resistors that have been fabricated and electrically intraconnected onto a small flat chip of semiconductor material. The IC was invented in 1959 and has been the subject of continual development ever since (Historical Note 34.1). Silicon (Si) is the most widely used semiconductor material for ICs, because of its combination of properties and low cost. Less common semiconductor chips are made of germanium (Ge) and gallium arsenide (GaAs). Because the circuits are fabricated into one solid piece of material, the term *solid-state* electronics is used to denote these devices.

The most fascinating aspect of microelectronics technology is the huge number of devices that can be packed onto a single small chip. Various terms have been developed to define the level of integration and density of packing, such as large-scale integration (LSI) and very-large-scale integration (VLSI). Table 34.1 lists these terms, their definitions (although there is not complete agreement over the dividing lines between levels), and the period during which the technology was or is being introduced.

34.1 OVERVIEW OF IC PROCESSING

Structurally, an integrated circuit consists of hundreds, thousands, or millions of microscopic electronic devices that have been fabricated and electrically intraconnected within the surface of a silicon chip. A *chip*, also called a *die*, is a square or rectangular flat plate that is about 0.5 mm (0.020 in) thick and typically 5 to 25 mm (0.200 to 1.0 in) on a side. Each electronic device (i.e., transistor, diode, etc.) on the chip surface consists of separate layers and regions with different electrical properties combined to perform the particular

Historical Note 34.1   *Integrated circuit technology*

The history of integrated circuits includes inventions of electronic devices and the processes for making these devices. The development of radar immediately before World War II (1939 to 1945) identified germanium and silicon as important semiconductor elements for the diodes used in radar circuitry. Owing to the importance of radar in the war, commercial sources of germanium and silicon were developed.

In 1947, the transistor was developed at the Bell Telephone Laboratories by J. Bardeen and W. Brattain. An improved version was subsequently invented by W. Shockley of Bell Labs in 1952. These three inventors shared the 1956 Nobel Prize in Physics for their research on semiconductors and the discovery of the transistor. The interest of the Bell Labs was to develop electronic switching systems that were more reliable than the electromechanical relays and vacuum tubes used at that time.

In February 1959, J. Kilby of Texas Instruments Inc. filed a patent application for the fabrication of multiple electronic devices and their intraconnection to form a circuit on a single piece of semiconductor material. Kilby was describing an integrated circuit (IC). In May 1959, J. Hoerni of Fairchild Semiconductor Corp. applied for a patent describing the planar process for fabricating transistors. In July of the same year, R. Noyce also of Fairchild filed a patent application similar to the Kilby invention but specifying the use of planar technology and adherent leads.

Although filed later than Kilby's, Noyce's patent was issued first, in 1961 (the Kilby patent was awarded in 1964). This discrepancy in dates and similarity in inventions have resulted in considerable controversy over who was really the inventor of the IC. The issue was argued in legal suits stretching all the way to the U.S. Supreme Court. The high court refused to hear the case, leaving stand a lower court ruling that favored several of Noyce's claims. The result (at the risk of oversimplifying) is that Kilby is generally credited with the concept of the monolithic integrated circuit, whereas Noyce is credited with the method for fabricating it.

The first commercial ICs were introduced by Texas Instruments in March 1960. Early integrated circuits contained about 10 devices on a small silicon chip—about 3 mm (0.12 in) square. By 1966, silicon had overtaken germanium as the preferred semiconductor material. Since that year, Si has been the predominant material in IC fabrication. Since the 1960s, a continual trend toward miniaturization and increased integration of multiple devices in a single chip has occurred in the electronics industry (the progress can be seen in Table 34.1), leading to the components described in this chapter.

electronic function of the device. A typical cross section of such a MOSFET[1] device is illustrated in Figure 34.1. The devices are electrically connected to one another by very fine lines of conducting material, usually aluminum, so that the intraconnected devices (that is, the integrated circuit) function in the specified way. Conducting lines and pads are also

TABLE 34.1   Levels of integration in microelectronics.

Integration Level	Number of Devices on a Chip	Approx. Year Introduced
Small-scale integration (SSI)	10–50	1959
Medium-scale integration (MSI)	$50-10^3$	1960s
Large-scale integration (LSI)	10^3-10^4	1970s
Very-large-scale integration (VLSI)	10^4-10^6	1980s
Ultra-large-scale integration (ULSI)	10^6-10^8	1990s
Giga-scale integration	10^9-10^{10}	2000s

---

[1]MOSFET stands for Metal-Oxide-Semiconductor Field-Effect Transistor. A transistor is a semiconductor device capable of performing various functions such as amplifying, controlling, or generating electrical signals. A field-effect transistor is one in which current flows between source and drain regions through a channel, the flow depending on the application of voltage to the channel gate. A metal-oxide-semiconductor FET uses silicon dioxide to separate the channel and gate metallization.

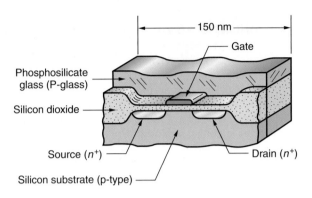

FIGURE 34.1 Cross section of a transistor (specifically, a MOSFET) in an integrated circuit. Approximate size of the device is shown; feature sizes within the device can be as small as 40 nm.

provided to electrically connect the IC to leads, which in turn permit the IC to be connected to external circuits.

To allow the IC to be connected to the outside world, and to protect it from damage, the chip is attached to a lead frame and encapsulated inside a suitable package, as in Figure 34.2. The package is an enclosure, usually made of plastic or ceramic, which provides mechanical and environmental protection for the chip and includes leads by which the IC can be electrically connected to external circuits. The leads are attached to conducting pads on the chip that access the IC.

34.1.1 PROCESSING SEQUENCE

The sequence to fabricate a silicon-based IC chip begins with the processing of silicon (Section 7.5.2). Briefly, silicon of very high purity is reduced in several steps from sand (silicon dioxide, SiO_2). The silicon is grown from a melt into a large solid single crystal log, with typical length of 1 to 3 m (3 to 10 ft) and diameter up to 300 mm (12 in). The log, called a *boule*, is then sliced into thin wafers, which are disks of thickness equal to about 0.5 mm (0.020 in).

After suitable finishing and cleaning, the wafers are ready for the sequence of processes by which microscopic features of various chemistries will be created in their surface to form the electronic devices and their intraconnections. The sequence consists of several types of processes, most of them repeated many times. A total of 200 or more processing steps may be required to produce a modern IC. Basically, the objective in the sequence is to add, alter, or remove a layer of material in selected regions of the wafer surface. The layering steps in IC fabrication are sometimes referred to as the *planar process*, because the processing relies on the geometric form of the silicon wafer being a plane. The processes by which the layers are added include thin film deposition techniques such as physical vapor deposition

FIGURE 34.2
Packaging of an integrated circuit chip: (a) cutaway view showing the chip attached to a lead frame and encapsulated in a plastic enclosure, and (b) the package as it would appear to a user. This type of package is called a dual in-line package (DIP).

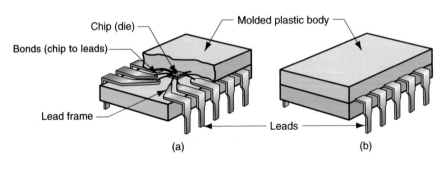

and chemical vapor deposition (Section 28.5), and existing layers are altered by diffusion and ion implantation (Section 28.2). Additional layer-forming techniques, such as thermal oxidation, are also employed. Layers are removed in selected regions by etching, using chemical etchants (usually acid solutions) and other more advanced technologies such as plasma etching.

The addition, alteration, and removal of layers must be done selectively; that is, only in certain extremely small regions of the wafer surface to create the device details such as in Figure 34.1. To distinguish which regions will be affected in each processing step, a procedure involving *lithography* is used. In this technique, masks are formed on the surface to protect certain areas and allow other areas to be exposed to the particular process (e.g., film deposition, etching). By repeating the steps many times, exposing different areas in each step, the starting silicon wafer is gradually transformed into many integrated circuits.

Processing of the wafer is organized in such a way that many individual chip surfaces are formed on a single wafer. Because the wafer is round with diameters ranging from 150 to 300 mm (6 to 12 in), whereas the final chip may only be 12 mm (0.5 in) square, it is possible to produce hundreds of chips on a single wafer. At the conclusion of planar processing, each IC on the wafer is visually and functionally tested, the wafer is diced into individual chips, and each chip that passes the quality test is packaged as in Figure 34.2.

Summarizing the preceding discussion, the production of silicon-based integrated circuits consists of the following stages, portrayed in Figure 34.3: (1) *Silicon processing*, in which sand is reduced to very pure silicon and then shaped into wafers; (2) *IC fabrication*, consisting of multiple processing steps that add, alter, and remove thin layers in selected regions to form the electronic devices; lithography is used to define the regions to be processed on the surface of the wafer; and (3) *IC packaging*, in which the wafer is tested, cut into individual dies (IC chips), and the dies are encapsulated in an appropriate package.

The presentation in subsequent sections of our chapter is concerned with the details of these processing stages. Section 34.2 deals with silicon processing. Section 34.3 discusses lithography and Section 34.3 examines the processes used in conjunction with lithography to add, alter, or remove layers. We consider an example of IC fabrication in Section 34.5. Section 34.6 describes die cutting and packaging of the chips. Finally, Section 34.7 covers yield analysis in IC fabrication.

Before beginning our coverage of the processing details, it is important to note that the microscopic dimensions of the devices in integrated circuits impose special requirements on the environment in which IC fabrication is accomplished.

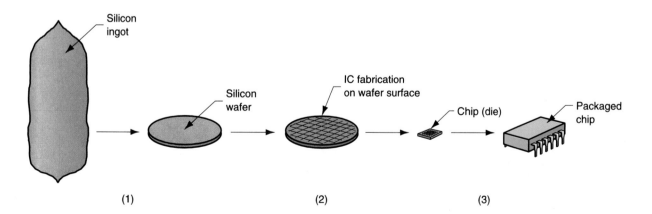

FIGURE 34.3   Sequence of processing steps in the production of integrated circuits: (1) pure silicon is formed from the molten state into an ingot and then sliced into wafers; (2) fabrication of integrated circuits on the wafer surface; and (3) wafer is cut into chips and packaged.

34.1.2 CLEAN ROOMS

Much of the processing sequence for integrated circuits must be carried out in a clean room, the ambiance of which is more like a hospital operating room than a production factory. Cleanliness is dictated by the microscopic feature sizes in an IC, the scale of which continues to decrease with each passing year. Figure 34.4 shows the trend in IC device feature size; also displayed in the same figure are common airborne particles that are potential contaminants in IC processing. These particles can cause defects in the integrated circuits, reducing yields and increasing costs.

A clean room provides protection from these contaminants. The air is purified to remove most of the particles from the processing environment; temperature and humidity are also controlled. A standard classification system is used to specify the cleanliness of a clean room. In the system, a number (in increments of ten) is used to indicate the quantity of particles of size 0.5 μm or greater in one cubic foot of air.[2] Thus, a ***class 100*** clean room must maintain a count of particles of size 0.5 μm or greater at less than 100/ft³. Modern VLSI processing requires ***class 10*** clean rooms, which means that the number of particles of size equal to or greater than 0.5 μm is less than 10/ft³. The air in the clean room is air conditioned to a temperature of 21°C (70°F) and 45% relative humidity. The air is passed through a high-efficiency particulate air (HEPA) filter to capture particle contaminants.

Humans are the biggest source of contaminants in IC processing; emanating from humans are bacteria, tobacco smoke, viruses, hair, and other particles. Human workers in IC processing areas are required to wear special clothing, generally consisting of white cloaks, gloves, and hair nets. Where extreme cleanliness is required, workers are completely encased in bunny suits. Processing equipment is a second major source of contaminants; machinery produces wear particles, oil, dirt, and similar contaminants. IC processing is usually accomplished in laminar-flow hooded work areas, which can be purified to greater levels of cleanliness than the general environment of the clean room.

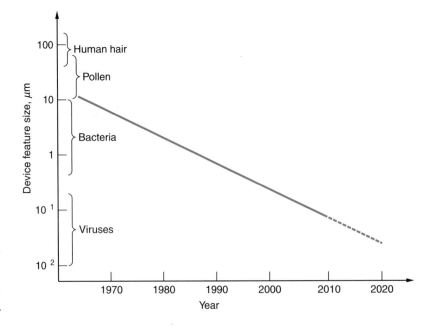

FIGURE 34.4 Trend in device feature size in IC fabrication; also shown is the size of common airborne particles that can contaminate the processing environment. Minimum feature sizes for logic type ICs are expected to be about 13 nm in the year 2016 [10].

---

[2]Only in the United States would we mix metric units (0.5 μm) with U.S. customary units (ft³).

In addition to the very pure atmosphere provided by the clean room, the chemicals and water used in IC processing must be very clean and free of particles. Modern practice requires that chemicals and water be filtered before use.

34.2 SILICON PROCESSING

Microelectronic chips are fabricated on a substrate of semiconductor material. Silicon is the leading semiconductor material today, constituting more than 95% of all semiconductor devices produced in the world. Our discussion in this introductory treatment will be limited to Si. The preparation of the silicon substrate can be divided into three steps: (1) production of electronic grade silicon, (2) crystal growing, and (3) shaping of Si into wafers.

34.2.1 PRODUCTION OF ELECTRONIC GRADE SILICON

Silicon is one of the most abundant materials in the Earth's crust (Table 7.1), occurring naturally as silica (e.g., sand) and silicates (e.g., clay). Electronic grade silicon (EGS) is polycrystalline silicon of ultra high purity—so pure that the impurities are in the range of parts per billion (ppb). They cannot be measured by conventional chemical laboratory techniques but must be inferred from measurements of resistivity on test ingots. The reduction of the naturally occurring Si compound to EGS involves the following processing steps.

The first step is carried out in a submerged-electrode arc furnace. The principal raw material for silicon is *quartzite*, which is very pure SiO_2. The charge also includes coal, coke, and wood chips as sources of carbon for the various chemical reactions that occur in the furnace. The net product consists of metallurgical grade silicon (MGS), and the gases SiO and CO. MGS is only about 98% Si, which is adequate for metallurgical alloying but not for electronics components. The major impurities (making up the remaining 2% of MGS) include aluminum, calcium, carbon, iron, and titanium.

The second step involves grinding the brittle MGS and reacting the Si powders with anhydrous HCl to form trichlorsilane:

$$Si + 3HCl \, (gas) \rightarrow SiHCl_3 \, (gas) + H_2 \, (gas) \qquad (34.1)$$

The reaction is performed in a fluidized-bed reactor at temperatures around 300°C (550°F). Trichlorsilane ($SiHCl_3$), although shown as a gas in Eq. (34.1), is a liquid at room temperature. Its low boiling point of 32°C (90°F) permits it to be separated from the leftover impurities of MGS by fractional distillation.

The final step in the process is reduction of the purified trichlorsilane by means of hydrogen gas. The process is carried out at temperatures up to 1000°C (1800°F), and a simplified equation of the reaction can be written as follows:

$$SiHCl_3 \, (gas) + H_2 \, (gas) \rightarrow Si + 3HCl \, (gas) \qquad (34.2)$$

The product of this reaction is electronic grade silicon—nearly 100% pure Si.

34.2.2 CRYSTAL GROWING

The silicon substrate for microelectronic chips must be made of a single crystal whose unit cell is oriented in a certain direction. The properties of the substrate and the way it is processed are both influenced by these requirements. Accordingly, silicon used as the raw material in semiconductor device fabrication must not only be of ultra high purity, as in

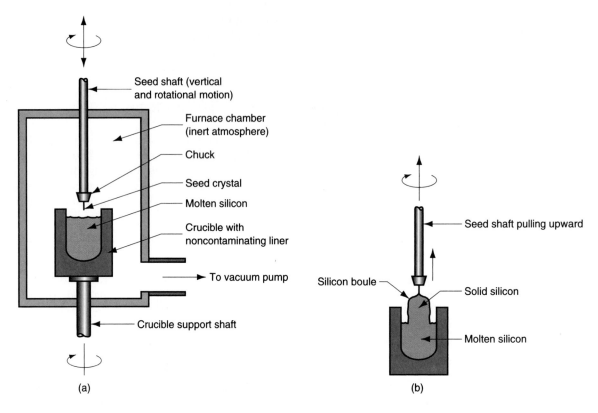

FIGURE 34.5 The Czochralski process for growing single-crystal ingots of silicon: (a) initial setup prior to start of crystal pulling, and (b) during crystal pulling to form the boule.

electronic grade silicon; it must also be prepared in the form of a single crystal and then cut in a direction that achieves the desired planar orientation. The crystal-growing process is covered here, whereas the next section details the cutting operation.

The most widely used crystal-growing method in the semiconductor industry is the **Czochralski process**, illustrated in Figure 34.5, in which a single crystal ingot, called a **boule**, is pulled upward from a pool of molten silicon. The setup includes a furnace, a mechanical apparatus for pulling the boule, a vacuum system, and supporting controls. The furnace consists of a crucible and heating system contained in a vacuum chamber. The crucible is supported by a mechanism that permits rotation during the crystal-pulling procedure. Chunks of EGS are placed in the crucible and heated to a temperature slightly above the melting point of silicon: $1410°C$ ($2570°F$). Heating is by induction or resistance, the latter being used for large melt sizes. The molten silicon is doped[3] before boule pulling to make the crystal either p-type or n-type.

To initiate crystal growing, a seed crystal of silicon is dipped into the molten pool and then withdrawn upward under carefully controlled conditions. At first the pulling rate (vertical velocity of the pulling apparatus) is relatively rapid, which causes a single crystal of silicon to solidify against the seed, forming a thin neck. The velocity is then reduced, causing the neck to grow into the desired larger diameter of the boule while maintaining its single crystal structure. In addition to pulling rate, rotation of the crucible and other process parameters are used to

---

[3]The term **dope** (doped, doping) refers to the introduction of impurities into the semiconductor material to alter its electrical properties, making the semiconductor either an n-type (excess electrons in its structure) or a p-type (missing electrons in its structure).

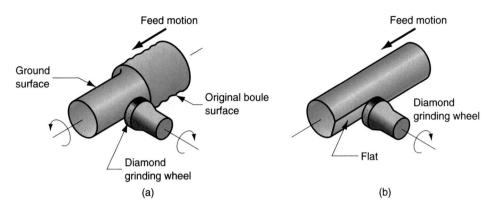

FIGURE 34.6 Grinding operations used in shaping the silicon ingot: (a) a form of cylindrical grinding provides diameter and roundness control, and (b) a flat ground on the cylinder.

control boule size. Single-crystal ingots of diameter $= 300\,mm$ (12 in) and up to 3 m (10 ft) long are commonly produced for subsequent fabrication of microelectronic chips.

It is important to avoid contamination of the silicon during crystal growing, because contaminants, even in small amounts, can dramatically alter the electrical properties of Si. To minimize unwanted reactions with silicon and the introduction of contaminants at the elevated temperatures of crystal growing, the procedure is carried out either in an inert gas (argon or helium) or a vacuum. Choice of crucible material is also important; fused silica (SiO_2), although not perfect for the application, represents the best available material and is used almost exclusively. Gradual dissolution of the crucible introduces oxygen as an unintentional impurity in the silicon boule. Unfortunately, the level of oxygen in the melt increases during the process, leading to a variation in concentration of the impurity throughout the length and diameter of the ingot.

34.2.3 SHAPING OF SILICON INTO WAFERS

A series of processing steps are used to reduce the boule into thin, disc-shaped wafers. The steps can be grouped as follows: (1) ingot preparation, (2) wafer slicing, and (3) wafer preparation. In ingot preparation, the seed and tang ends of the ingot are first cut off, as well as portions of the ingot that do not meet the strict resistivity and crystallographic requirements for subsequent IC processing. Next, a form of cylindrical grinding, as shown in Figure 34.6(a), is used to shape the ingot into a more perfect cylinder, because the crystal-growing process cannot achieve sufficient control over diameter and roundness. One or more flats are then ground along the length of the ingot, as in Figure 34.6(b). After the wafers have been cut from the ingot, these flats serve several functions: (1) identification, (2) orientation of the ICs relative to crystal structure, and (3) mechanical location during processing.

The ingot is now ready to be sliced into wafers, using the abrasive cutoff process illustrated in Figure 34.7. A very thin saw blade with diamond grit bonded to the internal diameter serves as the cutting edge. Use of the ID for slicing rather than the OD of the saw blade provides better control over flatness, thickness, parallelism, and surface characteristics of the wafer. The wafers are cut to a thickness of around 0.5 to 0.7 mm (0.020 to 0.028 in),

FIGURE 34.7 Wafer slicing using a diamond abrasive cutoff saw.

FIGURE 34.8   Two of the steps in wafer preparation: (a) contour grinding to round the wafer rim, and (b) surface polishing.

depending on diameter (greater thicknesses for larger wafer diameters). For every wafer cut, a certain amount of silicon is wasted because of the kerf width of the saw blade. To minimize kerf loss, the blades are made as thin as possible—around 0.33 mm (0.013 in).

Next the wafer must be prepared for the subsequent processes and handling in IC fabrication. After slicing, the rims of the wafers are rounded using a contour-grinding operation, such as in Figure 34.8(a). This reduces chipping of the wafer edges during handling and minimizes accumulation of photoresist solutions at the wafer rims. The wafers are then chemically etched to remove surface damage that occurred during slicing. This is followed by a flat polishing operation to provide very smooth surfaces that will accept the subsequent photolithography processes. The polishing step, seen in Figure 34.8(b), uses a slurry of very fine silica (SiO_2) particles in an aqueous solution of sodium hydroxide (NaOH). The NaOH oxidizes the Si wafer surface, and the abrasive particles remove the oxidized surface layers—about 0.025 mm (0.001 in) is removed from each side during polishing. Finally, the wafer is chemically cleaned to remove residues and organic films.

It is of interest to know how many IC chips can be fabricated on a wafer of a given size. The number depends on the chip size relative to the wafer size. Assuming that the chips are square, the following equation can be used to estimate the number of chips on the wafer:

$$n_c = 0.34 \left(\frac{D_w}{L_c} \right)^{2.25} \tag{34.3}$$

where n_c = estimated number of chips on the wafer; D_w = diameter of the processable area of the wafer, assumed circular, mm (in); and L_c = side dimension of the chip, assumed square, mm (in).

The diameter of the processable area of the wafer will be slightly less than the outside diameter of the wafer. The actual number of chips on the wafer may be different from the value given by Eq. (34.3), depending on the way the chips are laid out on the wafer.

34.3 LITHOGRAPHY

An IC consists of many microscopic regions on the wafer surface that make up the transistors, other devices, and intraconnections in the circuit design. In the planar process, the regions are fabricated by a sequence of steps to add layers to selected areas of the surface. The form of each layer is determined by a geometric pattern representing circuit design information that is transferred to the wafer surface by a procedure known as lithography—basically the same procedure used by artists and printers for centuries.

Several lithographic technologies are used in semiconductor processing: (1) photolithography, (2) electron lithography, (3) x-ray lithography, and (4) ion lithography. As their names indicate, the differences are in the type of radiation used to transfer the mask pattern to the surface by exposing the photoresist. The traditional technique is photolithography, and most of our discussion will be directed at this topic. The reader may recall that photolithography is used in some chemical machining processes (Section 26.4).

34.3.1 PHOTOLITHOGRAPHY

Photolithography, also known as *optical lithography*, uses light radiation to expose a coating of photoresist on the surface of the silicon wafer; a mask containing the required geometric pattern for each layer separates the light source from the wafer, so that only the portions of the photoresist not blocked by the mask are exposed. The *mask* consists of a flat plate of transparent glass onto which a thin film of an opaque substance has been deposited in certain areas to form the desired pattern. Thickness of the glass plate is around 2 mm (0.080 in), whereas the deposited film is only a few μm thick—for some film materials, less than 1 μm. The mask itself is fabricated by lithography, the pattern being based on circuit design data, usually in the form of digital output from the CAD system used by the circuit designer.

Photoresists   A photoresist is an organic polymer that is sensitive to light radiation in a certain wavelength range; the sensitivity causes either an increase or decrease in solubility of the polymer to certain chemicals. Typical practice in semiconductor processing is to use photoresists that are sensitive to ultraviolet light. UV light has a short wavelength compared with visible light, permitting sharper imaging of microscopic circuit details on the wafer surface. It also permits the fabrication and photoresist areas in the plant to be illuminated at low light levels outside the UV band.

Two types of photoresists are available: positive and negative. A *positive resist* becomes more soluble in developing solutions after exposure to light. A *negative resist* becomes less soluble (the polymer cross–links and hardens) when exposed to light. Figure 34.9 illustrates the operation of both resist types. Negative resists have better adhesion to SiO_2 and metal surfaces and good etch resistance. However, positive resists achieve better resolution, which has made it the more widely used technique as IC feature sizes have become smaller and smaller.

Exposure Techniques   The resists are exposed through the mask by one of three exposure techniques: (a) contact printing, (b) proximity printing, and (c) projection printing, illustrated in Figure 34.10. In *contact printing*, the mask is pressed against the resist coating during exposure. This results in high resolution of the pattern onto the wafer

FIGURE 34.9 Application of (a) positive resist and (b) negative resist in photolithography; for both types, the sequence shows: (1) exposure through the mask and (2) remaining resist after developing.

surface; an important disadvantage is that physical contact with the wafers gradually wears out the mask. In ***proximity printing***, the mask is separated from the resist coating by a distance of 10 to 25 μm (0.0004 to 0.001 in). This eliminates mask wear, but resolution of the image is slightly reduced. ***Projection printing*** involves the use of a high-quality lens (or mirror) system to project the image through the mask onto the wafer. This has become the preferred technique because it is noncontact (thus, no mask wear), and the mask pattern can be reduced through optical projection to obtain high resolution.

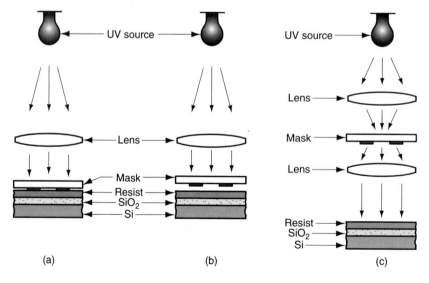

FIGURE 34.10
Photolithography
exposure techniques:
(a) contact printing,
(b) proximity printing, and
(c) projection printing.

FIGURE 34.11
Photolithography process applied to a silicon wafer: (1) prepare surface; (2) apply photo-resist; (3) soft-bake; (4) align mask and expose; (5) develop resist; (6) hard-bake; (7) etch; (8) strip resist.

Processing Sequence in Photolithography Let us examine a typical processing sequence for a silicon wafer in which photolithography is employed. The surface of the silicon has been oxidized to form a thin film of SiO_2 on the wafer. It is desired to remove the SiO_2 film in certain regions as defined by the mask pattern. The sequence for a positive resist proceeds as follows, illustrated in Figure 34.11. (1) *Prepare surface*. The wafer is properly cleaned to promote wetting and adhesion of the resist. (2) *Apply photoresist*. In semiconductor processing, photoresists are applied by feeding a metered amount of liquid resist onto the center of the wafer and then spinning the wafer to spread the liquid and achieve a uniform coating thickness. Desired thickness is around 1 μm (0.00004 in), which gives good resolution yet minimizes pinhole defects. (3) *Soft-bake*. Purpose of this pre-exposure bake is to remove solvents, promote adhesion, and harden the resist. Typical soft-bake temperatures are around 90°C (190°F) for 10 to 20 min. (4) *Align mask and expose*. The pattern mask is aligned relative to the wafer and the resist is exposed through the mask by one of the methods described in the preceding. Alignment must be accomplished with very high precision, using optical-mechanical equipment designed specifically for the purpose. If the wafer has been previously processed by lithography so that a pattern has already been formed in the wafer, then subsequent masks must be accurately registered relative to the existing pattern. Exposure of the resist depends on the same basic rule as in photography—the exposure is a function of light intensity × time. A mercury arc lamp or other source of UV light is used. (5) *Develop resist*. The exposed wafer is next immersed in a developing solution, or the solution is sprayed onto the wafer surface. For the positive resist in our example, the exposed areas are dissolved in the developer, thus leaving the SiO_2 surface uncovered in these areas. Development is usually followed by a rinse to stop development and remove residual chemicals. (6) *Hard-bake*. This baking step expels volatiles remaining from the developing solution and increases adhesion of the resist, especially at the newly created edges of the resist film. (7) *Etch*. Etching removes the SiO_2 layer at selected regions where the resist has been removed. (8) *Strip resist*. After etching, the resist coating that remains on the surface must be removed. Stripping is accomplished using either wet or dry techniques. Wet stripping uses liquid chemicals; a mixture of sulfuric acid and hydrogen peroxide (H_2SO_4–H_2O_2) is common. Dry stripping uses plasma etching with oxygen as the reactive gas.

Although our example describes the use of photolithography to remove a thin film of SiO_2 from a silicon substrate, the same basic procedure is followed for other processing steps. The purpose of photolithography in all of these steps is to expose specific regions beneath the photoresist layer so that the process can be performed on these exposed regions. In the processing of a given wafer, photolithography is repeated as many times as needed to produce the desired integrated circuit, each time using a different mask to define the appropriate pattern.

34.3.2 OTHER LITHOGRAPHY TECHNIQUES

As feature size in integrated circuits continues to decrease and conventional UV photolithography becomes increasingly inadequate, other lithography techniques that offer higher resolution are growing in importance. These techniques are extreme ultraviolet lithography, electron beam lithography, x-ray lithography, and ion lithography. In the following paragraphs, we provide brief descriptions of these alternatives. For each technique, special resist materials are required that react to the particular type of radiation.

Extreme ultraviolet lithography (EUV) represents a refinement of conventional UV lithography through the use of shorter wavelengths during exposure. The ultraviolet wavelength spectrum ranges from about 10 nm to 380 nm (nm = nanometer = 10^{-9} m), the upper end of which is close to the visible light range (approximately 400 to 700 nm wavelengths). EUV technology permits the feature size of an integrated circuit to be reduced to at about 0.04 μm, compared with about 0.1 μm with conventional UV exposure.

Compared with UV and EUV lithography, *electron-beam (E-Beam) lithography* virtually eliminates diffraction during exposure of the resist, thus permitting higher resolution of the image. Another potential advantage is that a scanning E-beam can be directed to expose only certain regions of the wafer surface, thus eliminating the need for a mask. Unfortunately, high-quality electron-beam systems are expensive. Also, because of the time-consuming sequential nature of the exposure method, production rates are low compared with the mask techniques of optical lithography. Accordingly, use of E-beam lithography tends to be limited to small production quantities. E-beam techniques are widely used for making the masks in UV lithography.

X-ray lithography has been under development since around 1972. As in E-beam lithography, the wavelengths of X-rays are much shorter than UV light (x-ray wavelength ranges from 0.005 nm to several dozen nm, overlapping the lower end of the UV range). Thus, they hold the promise of sharper imaging during exposure of the resist. X-rays are difficult to focus during lithography. Consequently, contact or proximity printing must be used, and a small x-ray source must be used at a relatively large distance from the wafer surface to achieve good image resolution through the mask.

Ion lithography systems divide into two categories: (1) focused ion beam systems, whose operation is similar to a scanning E-beam system and avoids the need for a mask; and (2) masked ion beam systems, which expose the resist through a mask by proximity printing. As with E-beam and x-ray systems, ion lithography produces higher image resolution than conventional UV photolithography.

34.4 LAYER PROCESSES USED IN IC FABRICATION

The steps required to produce an integrated circuit consist of chemical and physical processes that add, alter, or remove regions on the silicon wafer that have been defined by photolithography. These regions constitute the insulating, semiconducting, and conducting areas that form the devices and their intraconnections in the integrated circuits. The layers are fabricated one at a time, step by step, each layer having a different configuration and each requiring a separate photolithography mask, until all of the microscopic details of the electronic devices and conducting paths have been constructed on the wafer surface.

In this section we consider the wafer processes used to add, alter, and subtract layers. Processes that add or alter layers to the surface include (1) thermal oxidation, used to grow a layer of silicon dioxide onto the silicon substrate; (2) chemical vapor deposition, a versatile process used to apply various types of layers in IC fabrication; (3) diffusion and ion implantation, used to alter the chemistry of an existing layer or substrate; and (4) various metallization processes that add metal layers to provide regions of electrical conduction on the wafer. Finally (5), several etching processes are used to remove portions of the layers that have been added to achieve the desired details of the integrated circuit.

34.4.1   THERMAL OXIDATION

Oxidation of the silicon wafer may be performed multiple times during fabrication of an integrated circuit. Silicon dioxide (SiO_2) is an insulator, contrasted with the semiconducting properties of Si. The ease with which a thin film of SiO_2 can be produced on the surface of a silicon wafer is one of the attractive features of silicon as a semiconductor material.

Silicon dioxide serves a number of important functions in IC fabrication: (1) It is used as a mask to prevent diffusion or ion implantation of dopants into silicon. (2) It can be used to isolate devices in the circuit. (3) It provides electrical insulation between levels in multilevel metallization systems.

Several processes are used to form SiO_2 in semiconductor manufacturing, depending on when during chip fabrication the oxide must be added. The most common process is thermal oxidation, appropriate for growing SiO_2 films on silicon substrates. In ***thermal oxidation***, the wafer is exposed to an oxidizing atmosphere at elevated temperature; either oxygen or steam atmospheres are used, with the following reactions, respectively:

$$Si + O_2 \rightarrow SiO_2 \tag{34.4}$$

or

$$Si + 2H_2O \rightarrow SiO_2 + 2H_2 \tag{34.5}$$

Typical temperatures used in thermal oxidation of silicon range from 900°C to 1300°C (1650°F to 2350°F). By controlling temperature and time, oxide films of predictable thicknesses can be obtained. The equations show that silicon at the surface of the wafer is consumed in the reaction, as seen in Figure 34.12. To grow a SiO_2 film of thickness d requires a layer of silicon that is $0.44d$ thick.

When a silicon dioxide film must be applied to surfaces other than silicon, then direct thermal oxidation is not appropriate. An alternative process must be used, such as chemical vapor deposition.

34.4.2   CHEMICAL VAPOR DEPOSITION

Chemical vapor deposition (CVD) involves growth of a thin film on the surface of a heated substrate by chemical reactions or decomposition of gases (Section 28.5.2). CVD is widely

FIGURE 34.12   Growth of SiO_2 film on a silicon substrate by thermal oxidation, showing changes in thickness that occur: (1) before oxidation and (2) after thermal oxidation.

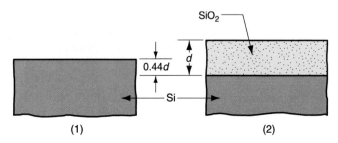

used in the processing of integrated circuit wafers to add layers of silicon dioxide, silicon nitride (Si_3N_4), and silicon. Plasma-enhanced CVD is often used because it permits the reactions to take place at lower temperatures.

Typical CVD Reactions in IC Fabrication   In the case of silicon dioxide, if the surface of the wafer is only silicon (e.g., at the start of IC fabrication), then thermal oxidation is the appropriate process by which to form a layer of SiO_2. If the oxide layer must be grown over materials other than silicon, such as aluminum or silicon nitride, then some alternative technique must be used, such as CVD. Chemical vapor deposition of SiO_2 is accomplished by reacting a silicon compound such as silane (SiH_4) with oxygen onto a heated substrate. The reaction is carried out at around 425°C (800°F) and can be summarized by

$$SiH_4 + O_2 \rightarrow SiO_2 + 2H_2 \tag{34.6}$$

The density of the silicon dioxide film and its bonding to the substrate is generally poorer than that achieved by thermal oxidation. Consequently, CVD is used only when the preferred process is not feasible; that is, when the substrate surface is not silicon, or when the high temperatures used in thermal oxidation cannot be tolerated. CVD can be used to deposit layers of doped SiO_2, such as phosphorus-doped silicon dioxide (called P-glass).

Silicon nitride is used as a masking layer during oxidation of silicon. Si_3N_4 has a low oxidation rate compared with Si, so a nitride mask can be used to prevent oxidation in coated areas on the silicon surface. Silicon nitride is also used as a passivation layer (protecting against sodium diffusion and moisture). A conventional CVD process for coating Si_3N_4 onto a silicon wafer involves reaction of silane and ammonia (NH_3) at around 800°C (1500°F) as follows:

$$3SiH_4 + 4NH_3 \rightarrow Si_3N_4 + 12H_2 \tag{34.7}$$

Plasma-enhanced CVD is also used for basically the same coating reaction, the advantage being that it can be performed at much lower temperatures—around 300°C (600°F).

Polycrystalline silicon (called ***polysilicon*** to distinguish it from silicon having a single crystal structure such as the boule and wafer) has a number of uses in IC fabrication, including [14]: conducting material for leads, gate electrodes in MOS devices, and contact material in shallow junction devices. Chemical vapor deposition to coat polysilicon onto a wafer involves reduction of silane at temperatures around 600°C (1100°F), as expressed by the following:

$$SiH_4 \rightarrow Si + 2H_2 \tag{34.8}$$

Epitaxial Deposition   A related process for growing a film onto a substrate is epitaxial deposition, distinguished by the feature that the film has a crystalline structure that is an extension of the substrate's structure. If the film material is the same as the substrate (e.g., silicon on silicon), then its crystal lattice will be identical to and a continuation of the wafer crystal. Two primary techniques to perform epitaxial deposition are vapor-phase epitaxy and molecular-beam epitaxy.

Vapor-phase epitaxy is the most important in semiconductor processing and is based on chemical vapor deposition. In growing silicon on silicon, the process is accomplished under closely controlled conditions at higher temperatures than conventional CVD of Si, using diluted reacting gases to slow the process so that an epitaxial layer can be successfully formed. Various reactions are possible, including Eq. (34.8), but the most widely used industrial process involves hydrogen reduction of silicon tetrachloride gas ($SiCl_4$) at around 1100°C (2000°F) as follows:

$$SiCl_4 + 2H_2 \rightarrow Si + 4HCl \tag{34.9}$$

The melting point of silicon is 1410°C (2570°F), so the preceding reaction is carried out at temperatures below T_m for Si, considered an advantage for vapor-phase epitaxy.

Molecular-beam epitaxy uses a vacuum evaporation process (Section 28.5.1), in which silicon together with one or more dopants are vaporized and transported to the substrate in a vacuum chamber. Its advantage is that it can be carried out at lower temperatures than CVD; processing temperatures are 400°C to 800°C (750°F to 1450°F). However, throughput is relatively low and equipment is expensive.

34.4.3 INTRODUCTION OF IMPURITIES INTO SILICON

IC technology relies on the ability to alter the electrical properties of silicon by introducing impurities into selected regions at the surface. Adding impurities into the silicon surface is called *doping*. The doped regions are used to create p-n junctions that form the transistors, diodes, and other devices in the circuit. A silicon-dioxide mask produced by thermal oxidation and photolithography is used to isolate the silicon regions that are to be doped. Common elements used as impurities are boron (B), which forms electron acceptor regions in the silicon substrate (p-type regions); and phosphorus (P), arsenic (As), and antimony (Sb), which form electron donor regions (n-type regions). The predominant technique by which silicon is doped with these elements is ion implantation.

In ion implantation, vaporized ions of the impurity element are accelerated by an electric field and directed at the silicon substrate surface (Section 28.2.2). The atoms penetrate into the surface, losing energy and finally stopping at some depth in the crystal structure, the average depth being determined by the mass of the ion and the acceleration voltage. Higher voltages produce greater depths of penetration, typically several hundred Angstroms (1 Angstrom $= 10^{-4}$ mm $= 10^{-1}$ nm). Advantages of ion implantation are that it can be accomplished at room temperature and provides exact doping density.

The problem with ion implantation is that the ion collisions disrupt and damage the crystal lattice structure. Very-high-energy collisions can transform the starting crystalline material into an amorphous structure. This problem is solved by annealing at temperatures between 500°C and 900°C (1000°F and 1800°F), which allows the lattice structure to repair itself and return to its crystal state.

34.4.4 METALLIZATION

Conductive materials must be deposited onto the wafer during processing to serve several functions: (1) form certain components (e.g., gates) of devices in the IC; (2) provide intraconnecting conduction paths between devices on the chip; and (3) connect the chip to external circuits. To satisfy these functions the conducting materials must be formed into very fine patterns. The process of fabricating these patterns is known as *metallization*, and it combines various thin film deposition technologies with photolithography. In this section we consider the materials and processes used in metallization. Connecting the chip to external circuitry also involves IC packaging, which is explored in Section 34.6.

Metallization Materials   Materials used in the metallization of silicon-based integrated circuits must have certain desirable properties, some of which relate to electrical function, whereas others relate to manufacturing processing. The desirable properties of a metallization material are [5], [14]: (1) low resistivity; (2) low-contact resistance with silicon, (3) good adherence to the underlying material, usually Si or SiO_2; (4) ease of deposition, compatible with photolithography; (5) chemical stability–noncorroding, nonreactive, and noncontaminating; (6) physical stability during temperatures encountered in processing; and (7) good lifetime stability.

Although no material meets all of these requirements perfectly, aluminum satisfies most of them either well or adequately, and it is the most widely used metallization material. Aluminum is usually alloyed with small amounts of (1) silicon to reduce reactivity with silicon in the substrate, and (2) copper to inhibit electromigration of Al atoms caused by current flow when the IC is in service. Other materials used for metallization in integrated circuits include polysilicon (Si); gold (Au); refractory metals (e.g., W, Mo); silicides (e.g., WSi_2, $MoSi_2$, $TaSi_2$); and nitrides (e.g., TaN, TiN, and ZrN). These other materials are generally used in applications such as gates and contacts. Aluminum is generally favored for device intraconnections and top level connections to external circuitry.

Metallization Processes   A number of processes are available to accomplish metallization in IC fabrication: physical vapor deposition, chemical vapor deposition, and electroplating. Among PVD processes, vacuum evaporation and sputtering are applicable (Section 28.5.1). *Vacuum evaporation* can be applied for aluminum metallization. Vaporization is usually accomplished by resistance heating or electron beam evaporation. Evaporation is difficult or impossible for depositing refractory metals and compounds. *Sputtering* can be used for depositing aluminum as well as refractory metals and certain metallizing compounds. It achieves better step coverage than evaporation, often important after many processing cycles when the surface contour has become irregular. However, deposition rates are lower and equipment is more expensive.

Chemical vapor deposition is also applicable as a metallization technique. Its processing advantages include excellent step coverage and good deposition rates. Materials suited to CVD include tungsten, molybdenum, and most of the silicides used in semiconductor metallization. CVD for metallization in semiconductor processing is less common than PVD. Finally, *electroplating* (Section 28.3.1) is occasionally used in IC fabrication to increase the thickness of thin films.

34.4.5   ETCHING

All of the preceding processes in this section involve addition of material to the wafer surface, either in the form of a thin film or the doping of the surface with an impurity element. Certain steps in IC manufacturing require material removal from the surface; this is accomplished by etching away the unwanted material. Etching is usually done selectively, by coating surface areas that are to be protected and leaving other areas exposed for etching. The coating may be an etch-resistant photoresist, or it may be a previously applied layer of material such as silicon dioxide. We briefly encountered etching in our description of photolithography. This section gives some of the technical details of this step in IC fabrication.

There are two main categories of etching process in semiconductor processing: wet chemical etching and dry plasma etching. Wet chemical etching is the older of the two processes and is easier to use. However, there are certain disadvantages that have resulted in growing use of dry plasma etching.

TABLE 34.2   Some common chemical etchants used in semiconductor processing.

Material to be Removed	Etchant (usually in aqueous solution)
Aluminum (Al)	Mixture of phosphoric acid (H_3PO_4), nitric acid (HNO_3), and acetic acid (CH_3COOH).
Silicon (Si)	Mixture of nitric acid (HNO_3) and hydrofluoric acid (HF)
Silicon dioxide (SiO_2)	Hydrofluoric acid (HF)
Silicon nitride (Si_3N_4)	Hot phosphoric acid (H_3PO_4)

FIGURE 34.13   Profile of a properly etched layer.

Wet Chemical Etching   Wet chemical etching involves the use of an aqueous solution, usually an acid, to etch away a target material. The etching solution is selected because it chemically attacks the specific material to be removed and not the protective layer used as a mask. Some of the common etchants used to remove materials in wafer processing are listed in Table 34.2.

In its simplest form, the process can be accomplished by immersing the masked wafers in an appropriate etchant for a specified time and then immediately transferring them to a thorough rinsing procedure to stop the etching. Process variables such as immersion time, etchant concentration, and temperature are important in determining the amount of material removed. A properly etched layer will have a profile as shown in Figure 34.13. Note that the etching reaction is *isotropic* (it proceeds equally in all directions), resulting in an undercut below the protective mask. In general, wet chemical etching is isotropic, and so the mask pattern must be sized to compensate for this effect.

Note also that the etchant does not attack the layer below the target material in our illustration. In the ideal case, an etching solution can be formulated that will react only with the target material and not with other materials in contact with it. In practical cases, the other materials exposed to the etchant may be attacked but to a lesser degree than the target material. The *etch selectivity* of the etchant is the ratio of etching rates between the target material and some other material, such as the mask or substrate material. For example, etch selectivity of hydrofluoric acid for SiO_2 over Si is infinite.

If process control is inadequate, either under-etching or over-etching can occur, as in Figure 34.14. Underetching, in which the target layer is not completely removed, results when the etching time is too short and/or the etching solution is too weak. Over-etching involves too much of the target material being removed, resulting in loss of pattern definition and possible damage to the layer beneath the target layer. Over-etching is caused by overexposure to the etchant.

Dry Plasma Etching   This etching process uses an ionized gas to etch a target material. The ionized gas is created by introducing an appropriate gas mixture into a vacuum chamber

FIGURE 34.14   Two problems in etching: (a) under-etching and (b) over-etching.

FIGURE 34.15 (a) A fully anisotropic etch, with $A = \infty$; and (b) a partially anisotropic etch, with A = approximately 1.3.

and using radio frequency (RF) electrical energy to ionize a portion of the gas, thus creating a plasma. The high-energy plasma reacts with the target surface, vaporizing the material to remove it. There are several ways in which a plasma can be used to etch a material; the two principal processes in IC fabrication are plasma etching and reactive ion etching.

In *plasma etching*, the function of the ionized gas is to generate atoms or molecules that are chemically very reactive, so that the target surface is chemically etched upon exposure. The plasma etchants are usually based on fluorine or chlorine gases. Etch selectivity is generally more of a problem in plasma etching than in wet chemical etching. For example, etch selectivity for SiO_2 over Si in a typical plasma etching process is 15 at best [4], compared with infinity with HF chemical etching.

An alternative function of the ionized gas can be to physically bombard the target material, causing atoms to be ejected from the surface. This is the process of sputtering, one of the techniques in physical vapor deposition. When used for etching, the process is called *sputter etching*. Although this form of etching has been applied in semiconductor processing, it is much more common to combine sputtering with plasma etching as described in the preceding, which results in the process known as *reactive ion etching*. This produces both chemical and physical etching of the target surface.

The advantage of the plasma etching processes over wet chemical etching is that they are much more *anisotropic*. This property can be readily defined with reference to Figure 34.15. In (a), a fully anisotropic etch is shown; the undercut is zero. The degree to which an etching process is anisotropic is defined as the ratio:

$$A = \frac{d}{u} \tag{34.10}$$

where A = degree of anisotropy; d = depth of etch, which in most cases will be the thickness of the etched layer; and u = the undercut dimension, as illustrated in Figure 34.15(b).

Wet chemical etching usually yields A values around 1.0, indicating isotropic etching. In sputter etching, ion bombardment of the surface is nearly perpendicular, resulting in A values approaching infinity—almost fully anisotropic. Plasma etching and reactive ion etching have high degrees of anisotropy, but below those achieved in sputter etching. As IC feature sizes continue to shrink, anisotropy becomes increasingly important for achieving the required dimensional tolerances.

34.5   INTEGRATING THE FABRICATION STEPS

In Sections 34.3 and 34.4, we examined the individual processing technologies used in IC fabrication. In this section, we show how these technologies are combined into the sequence of steps to produce an integrated circuit.

The planar processing sequence consists of fabricating a series of layers of various materials in selected areas on a silicon substrate. The layers form insulating, semiconducting,

TABLE 34.3   Layer materials added or altered in IC fabrication and associated processes.

Layer Material (function)	Typical Fabrication Processes
Si, polysilicon (semiconductor)	CVD
Si, epitaxial (semiconductor)	Vapor phase epitaxy
Si doping (n-type or p-type)	Ion implantation, diffusion
SiO_2 (insulator, mask)	Thermal oxidation, CVD
Si_3N_4 (mask)	CVD
Al (conductor)	PVD, CVD
P-glass (protection)	CVD

or conducting regions on the substrate to create the particular electronic devices required in the integrated circuit. The layers might also serve the temporary function of masking certain areas so that a particular process is only applied to desired portions of the surface. The masks are subsequently removed.

The layers are formed by thermal oxidation, epitaxial growth, deposition techniques (CVD and PVD), diffusion, and ion implantation. In Table 34.3, we summarize the processes typically used to add or alter a layer of a given material type. The use of lithography to apply a particular process only to selected regions of the surface is illustrated in Figure 34.16.

An example will be useful here to show the process integration in IC fabrication. We will use an n-channel metal oxide semiconductor (NMOS) logic device to illustrate the processing sequence. The sequence for NMOS integrated circuits is less complex than for CMOS or bipolar technologies, although the processes for these IC categories are basically similar. The device to be fabricated is illustrated in Figure 34.1.

The starting substrate is a lightly doped p-type silicon wafer, which will form the base of the n-channel transistor. The processing steps are illustrated in Figure 34.17 and described here (some details have been simplified, and the metallization process for intraconnecting devices has been omitted): (1) A layer of Si_3N_4 is deposited by CVD onto the Si substrate using photolithography to define the regions. This layer of Si_3N_4 will serve as a mask for the thermal oxidation process in the next step. (2) SiO_2 is grown in the exposed regions of the surface by thermal oxidation. The SiO_2 regions are insulating and will become the means by which this device is isolated from other devices in the circuit. (3) The Si_3N_4 mask is stripped by etching. (4) Another thermal oxidation is done to add a thin gate oxide layer to previously uncoated surfaces and increase the thickness of the previous SiO_2 layer. (5) Polysilicon is deposited by CVD onto the surface and then doped n-type using ion implantation. (6) The polysilicon is selectively etched using photolithography to leave the gate electrode of the transistor. (7) The source and drain regions (n+) are formed by ion implantation of arsenic (As) into the substrate. An implantation energy level is selected that will penetrate the thin SiO_2 layer but not the polysilicon gate

FIGURE 34.16   Formation of layers selectively through the use of masks: (a) thermal oxidation of silicon, (b) selective doping, and (c) deposition of a material onto a substrate.

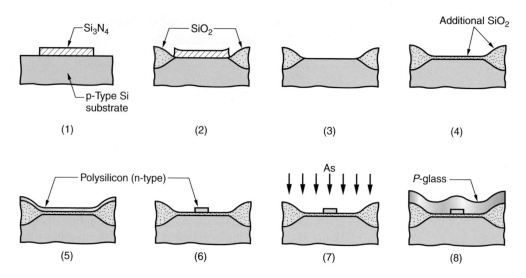

FIGURE 34.17   IC fabrication sequence: (1) Si$_3$N$_4$ mask is deposited by CVD on Si substrate; (2) SiO$_2$ is grown by thermal oxidation in unmasked regions; (3) the Si$_3$N$_4$ mask is stripped; (4) a thin layer of SiO$_2$ is grown by thermal oxidation; (5) polysilicon is deposited by CVD and doped n$^+$ using ion implantation; (6) the poly-Si is selectively etched using photolithography to define the gate electrode; (7) source and drain regions are formed by doping n$^+$ in the substrate; (8) P-glass is deposited onto the surface for protection.

or the thicker SiO$_2$ isolation layer. (8) Phosphosilicate glass (P-glass) is deposited onto the surface by CVD to protect the circuitry beneath.

34.6   IC PACKAGING

After all of the processing steps on the wafer have been completed, a final series of operations must be accomplished to transform the wafer into individual chips, ready to connect to external circuits and prepared to withstand the harsh environment of the world outside the clean room. These final steps are referred to as IC packaging. (As we shall see in the following chapter, packaging extends beyond the preparation of individual IC chips.)

Packaging of integrated circuits is concerned with design issues such as (1) electrical connections to external circuits; (2) materials to encase the chip and protect it from the environment (humidity, corrosion, temperature, vibration, mechanical shock); (3) heat dissipation; (4) performance, reliability, and service life; and (5) cost.

There are also manufacturing issues in packaging, including: (1) chip separation— cutting the wafer into individual chips, (2) connecting it to the package, (3) encapsulating the chip, and (4) circuit testing. The manufacturing issues are the ones of greatest interest in this section. Although most of the design issues are properly left to other texts [8], [11], and [13], let us examine some of the engineering aspects of IC packages and the types of IC packages available, before describing the package processing steps to make them.

34.6.1   IC PACKAGE DESIGN

In this section we consider three topics related to the design of an integrated circuit package: (1) the number of input/output terminals required for an IC of a given size, (2) the materials used in IC packages, and (3) package styles.

Determining the Number of Input/Output Terminals   The basic engineering problem in IC packaging is to connect the many internal circuits to input/output (I/O) terminals so that the appropriate electrical signals can be communicated between the IC and the outside world. As the number of devices in an IC increases, the required number of I/O terminals (leads) also increases. The problem is of course aggravated by trends in semiconductor technology that have led to decreases in device size and increases in the number of devices that can be packed into an IC. Fortunately, the number of I/O terminals does not have to equal the number of devices in the IC. The dependency between the two values is given by Rent's rule, named after the IBM engineer (E. F. Rent) who defined the following relationship around 1960:

$$n_{io} = C\, n_c^{\,m} \tag{34.11}$$

where n_{io} = the number of input/output terminals required; n_c = the number of circuits in the IC, usually taken to be the number of logic gates; and C and m are parameters in the equation.

Commonly accepted C and m values are 4.5 and 0.5 for a modern VLSI microprocessor circuit. However, the parameters in Rent's rule depend on the type of circuit. Memory devices require far fewer I/O terminals than microprocessors because of the column and row structure of memory units. Typical values for a static memory device are $C = 6.0$ and $m = 0.12$.

IC Package Materials   Package sealing involves encapsulating the IC chip in an appropriate packaging material. Two material types dominate current packaging technology: ceramic and plastic. Metal was used in early packaging designs but is today no longer important, except for lead frames.

The common ceramic packaging material is alumina (Al_2O_3). Advantages of ceramic packaging include hermetic sealing of the IC chip and the fact that highly complex packages can be produced. Disadvantages include poor dimensional control because of shrinkage during firing and the high dielectric constant of alumina.

Plastic IC packages are not hermetically sealed, but their cost is lower than ceramic. They are generally used for mass produced ICs, where very high reliability is not required. Plastics used in IC packaging include epoxies, polyimides, and silicones.

IC Package Styles   A wide variety of IC package styles is available to meet the input/output requirements indicated in the preceding. In nearly all applications, the IC is a component in a larger electronic system and must be attached to a printed circuit board (PCB). There are two broad categories of component mounting to a PCB, shown in Figure 34.18: through-hole and surface mount. In ***through-hole mounting***, also known as ***pin-in-hole*** (PIH) technology, the IC package and other electronic components (e.g., discrete resistors, capacitors) have leads that are inserted through holes in the board and are soldered on the underside. In ***surface-mount technology*** (SMT), the components are attached to the surface of the board (or in some cases, both top and bottom surfaces). Several lead configurations are available in SMT, as illustrated in (b), (c), and (d) of the figure.

FIGURE 34.18   Types of component lead attachment on a printed circuit board: (a) through-hole, and several styles of surface-mount technology; (b) butt lead; (c) "J" lead; and (d) gull-wing.

FIGURE 34.19   Dual in-line package with 16 terminals, shown as a through-hole configuration.

The major styles of IC packages include (1) dual in-line package, (2) square package, and (3) pin grid array. Some of these are available in both through-hole and surface-mount styles, whereas others are designed for only one mounting method.

The **_dual in-line package_** (DIP) is currently the most common form of IC package, available in both through-hole and surface-mount configurations. It has two rows of leads (terminals) on either side of a rectangular body, as in Figure 34.19. Spacing between leads (center-to-center distance) in the conventional through-hole DIP is 2.54 mm (0.1 in), and the number of leads ranges between 8 and 64. Hole spacing in the through-hole DIP style is limited by the ability to drill holes closely together in a printed circuit board. This limitation can be relaxed with surface-mount technology because the leads are not inserted into the board; standard lead spacing on surface-mount DIPs is 1.27 mm (0.05 in).

The number of terminals in a DIP is limited by its rectangular shape in which leads project only from two sides; that means that the number of leads on either side is $n_{io}/2$. For high values of n_{io} (between 48 and 64), differences in conducting lengths between leads in the middle of the DIP and those on the ends cause problems in high-speed electrical characteristics. Some of these problems are addressed with a square package, in which the leads are arranged around the periphery so that the number of terminals on a side is $n_{io}/4$. A surface-mount square package is illustrated in Figure 34.20.

Even with a square chip package, there is still a practical upper limit on terminal count dictated by the manner in which the leads in the package are linearly allocated. The number of leads on a package can be maximized by using a square matrix of pins. A **_pin grid array_** (PGA) consists of a two-dimensional array of pin terminals on the underside of a square chip enclosure. In the ideal, the entire bottom surface of the package is fully occupied by pins, so that the pin count in each direction is square root of n_{io}. However, as a practical matter, the center area of the package has no pins because this region contains the IC chip.

34.6.2 PROCESSING STEPS IN IC PACKAGING

The packaging of an IC chip in manufacturing can be divided into the following steps: (1) wafer testing, (2) chip separation, (3) die bonding, (4) wire bonding, and (5) package sealing. After packaging, a final functional test is performed on each packaged IC.

FIGURE 34.20   Square package for surface mounting with gull wing leads.

Wafer Testing   Current semiconductor processing techniques provide several hundred individual ICs per wafer. It is convenient to perform certain functional tests on the ICs while they are still together on the wafer—before chip separation. Testing is accomplished by computer-controlled test equipment that uses a set of needle probes configured to match the connecting pads on the surface of the chip; *multiprobe* is the term used for this testing procedure. When the probes contact the pads, a series of DC tests are carried out to indicate short circuits and other faults; this is followed by a functional test of the IC. Chips that fail the test are marked with an ink dot; these defects are not packaged. Each IC is positioned in its turn beneath the probes for testing, using a high precision x-y table to index the wafer from one chip site to the next.

Chip Separation   The next step after testing is to cut the wafer into individual chips (dice). A thin diamond-impregnated saw blade is used to perform the cutting operation. The sawing machine is highly automatic and its alignment with the "streets" between circuits is very accurate. The wafer is attached to a piece of adhesive tape that is mounted in a frame. The adhesive tape holds the individual chips in place during and after sawing; the frame is a convenience in subsequent handling of the chips. Chips with ink dots are now discarded.

Die Bonding   The individual chips must next be attached to their individual packages, a procedure called die bonding. Owing to the miniature size of the chips, automated handling systems are used to pick the separated chips from the tape frame and place them for die bonding. Various techniques have been developed to bond the chip to the packaging substrate; we describe two methods here: eutectic die bonding and epoxy die bonding. *Eutectic die bonding*, used for ceramic packages, consists of (1) depositing a thin film of gold on the bottom surface of the chip; (2) heating the base of the ceramic package to a temperature above 370°C (698°F), the eutectic temperature of the Au–Si system; and (3) bonding the chip to the metallization pattern on the heated base. In *epoxy die bonding*, used for plastic VLSI packaging, a small amount of epoxy is dispensed on the package base (the lead frame), and the chip is positioned on the epoxy; the epoxy is then cured, bonding the chip to the surface.

Wire Bonding   After the die is bonded to the package, electrical connections are made between the contact pads on the chip surface and the package leads. The connections are generally made using small-diameter wires of aluminum or gold, as illustrated in Figure 34.21. Typical wire diameters for aluminum are 0.05 mm (0.002 in), and gold wire diameters are about half that diameter. (Au has higher electrical conductivity than Al, but is more expensive.) Aluminum wires are bonded by ultrasonic bonding, whereas gold wires are bonded by thermocompression, thermosonic, or ultrasonic means. *Ultrasonic bonding* uses ultrasonic energy to weld the wire to the pad surface. *Thermocompression bonding* involves heating the end of the wire to form a molten ball, and then the ball is pressed into the pad to form the bond. *Thermosonic bonding* combines ultrasonic and thermal energies to form the

FIGURE 34.21   Typical wire connection between chip contact pad and lead.

bond. Automatic wire bonding machines are used to perform these operations at rates up to 200 bonds per minute.

Package Sealing   As mentioned, the two common packaging materials are ceramic and plastic. The processing methods are different for the two materials. *Ceramic packages* are made from a dispersion of ceramic powder (Al_2O_3 is most common) in a liquid binder (e.g., polymer and solvent). The mix is first formed into thin sheets and dried, and then cut to size. Holes are punched for interconnections. The required wiring paths are then fabricated onto each sheet, and metal is filled into the holes. The sheets are then laminated by pressing and sintering to form a monolithic (single stone) body.

Two types of *plastic package* are available, postmolded and premolded. In *postmolded packages*, an epoxy thermosetting plastic is transfer molded around the assembled chip and lead frame (after wire bonding), in effect transforming the pieces into one solid body. However, the molding process can be harsh on the delicate bond wires, and premolded packages are an alternative. In *premolded packaging*, an enclosure base is molded before encapsulation and then the chip and lead frame are connected to it, adding a solid lid or other material to provide protection.

Final Testing   Upon completion of the packaging sequence, each IC must undergo a final test, the purpose of which is to (1) determine which units, if any, have been damaged during packaging; and (2) measure performance characteristics of each device.

Burn-in test procedures sometimes include elevated temperature testing, in which the packaged IC is placed in an oven at temperatures around 125°C (250°F) for 24 hours and then tested. A device that fails such a test would have been likely to have failed early during service. If the device is intended for environments in which wide temperature variations occur, a temperature cycle test is appropriate. This test subjects each device to a series of temperature reversals, between values around –50°C (–60°F) on the lower side and 125°C (250°F) on the upper side. Additional tests for devices requiring high reliability might include mechanical vibration tests and hermetic (leak) tests.

34.7   YIELDS IN IC PROCESSING

The fabrication of integrated circuits consists of many processing steps performed in sequence. In wafer processing in particular, there may be hundreds of distinct operations through which the wafer passes. At each step, there is a chance that something may go wrong, resulting in the loss of the wafer or portions of it corresponding to individual chips. A simple probability model to predict the final yield of good product is

$$Y = Y_1 Y_2 \ldots Y_n$$

where Y = final yield; Y_1, Y_2, Y_n are the yields of each processing step; and n = total number of steps in the processing sequence.

As a practical matter, this model, although perfectly valid, is difficult to use because of the large number of steps involved and the variability of yields for each step. It is more convenient to divide the processing sequence into major phases, as we have organized our discussion in this chapter (see Figure 34.3), and to define the yields for each phase. The first phase involves growth of the single-crystal boule. The term *crystal yield* Y_c refers to the amount of single-crystal material in the boule compared with the starting amount of electronic grade silicon. The typical crystal yield is about 50%. After crystal growing, the boule is sliced into wafers, the yield for which is described as the *crystal-to-slice yield* Y_s. This depends on the amount of material lost during grinding of the boule, the width of the saw blade relative to the wafer thickness during slicing, and other losses. A typical value might be 50%, although much of the lost silicon during grinding and slicing is recyclable.

The next phase is wafer processing to fabricate the individual ICs. From a yield viewpoint, this can be divided into wafer yield and multiprobe yield. *Wafer yield* Y_w refers to the number of wafers that survive processing compared to the starting quantity. Certain wafers are designated as test pieces or similar uses and therefore result in losses and a reduction in yield; in other cases, wafers are broken or processing conditions go awry. Typical values of wafer yield are around 70% if testing losses are included. For wafers that come through processing and are multiprobe tested, only a certain proportion pass the test, called the *multiprobe yield* Y_m. Multiprobe yield is highly variable and can range from very low values (less than 10%) to relatively high values (more than 90%), depending on IC complexity and worker skill in the processing areas.

Following packaging, final testing of the IC is performed. This invariably produces additional losses, resulting in a *final test yield* Y_t in the range 90% to 95%. If the five phase yields are combined, the final yield can be estimated by

$$Y = Y_c Y_s Y_w Y_m Y_t \tag{34.12}$$

Given the typical values at each step, the final yield compared with the starting amount of silicon is quite low.

The heart of IC fabrication is wafer processing, the yield from which is measured in multiprobe testing Y_m. Yields in the other areas are fairly predictable, but not in wafer-fab. Two types of processing defects can be distinguished in wafer processing: (1) area defects and (2) point defects. *Area defects* are those that affect major areas of the wafer, possibly the entire surface. These are caused by variations or incorrect settings in process parameters. Examples include added layers that are too thin or too thick, insufficient diffusion depths in doping, and over- or under-etching. In general these defects are correctable by improved process control or development of alternative processes that are superior. For example, doping by ion implantation has largely replaced diffusion, and dry plasma etching has been substituted for wet chemical etching to better control feature dimensions.

Point defects occur at very localized areas on the wafer surface, affecting only one or a limited number of ICs in a particular area. They are commonly caused by dust particles either on the wafer surface or the lithographic masks. Point defects also include dislocations in the crystal lattice structure (Section 2.3.2). These point defects are distributed in some way over the surface of the wafer, resulting in a yield that is a function of the density of the defects, their distribution over the surface, and the processed area of the wafer. If the area defects are assumed negligible, and the point defects are assumed uniform over the surface area of the wafer, the resulting yield can be modeled by the equation

$$Y_m = \frac{1}{1 + AD} \tag{34.13}$$

where Y_m = the yield of good chips as determined in multiprobe; A = the area processed, cm^2 (in^2); and D = density of point defects, defects/cm^2 (defects/in^2). This equation is based on *Bose-Einstein* statistics and has been found to be a good predictor of wafer processing performance, especially for highly integrated chips (VLSI and beyond).

Wafer processing is the key to successful fabrication of integrated circuits. For an IC producer to be profitable, high yields must be achieved during this phase of manufacturing. This is accomplished using the purest possible starting materials, the latest equipment technologies, good process control over the individual processing steps, maintenance of clean room conditions, and efficient and effective inspection and testing procedures.

REFERENCES

[1] Bakoglu, H. B. *Circuits, Interconnections, and Packaging for VLSI.* Addison-Wesley Longman, Reading, Massachusetts, 1990.

[2] Coombs, C. F., Jr. (ed.) *Printed Circuits Handbook,* 6th ed. McGraw-Hill, New York, 2006.

[3] Edwards, P. R. *Manufacturing Technology in the Electronics Industry.* Chapman & Hall, London, 1991.

[4] *Encyclopedia of Chemical Technology.* 4th ed. John Wiley & Sons, New York, 2000.

[5] Gise, P., and Blanchard, R. *Modern Semiconductor Fabrication Technology.* Prentice-Hall, Upper Saddle River, New Jersey, 1986.

[6] Harper, C. *Electronic Materials and Processes Handbook,* 3rd ed., McGraw-Hill, New York, 2009.

[7] Jackson, K. A., and Schroter, W. (eds.). *Handbook of Semiconductor Technology.* Vol. 2, *Processing of Semiconductors.* John Wiley & Sons, New York, 2000.

[8] Manzione, L. T. *Plastic Packaging of Microelectronic Devices.* AT&T Bell Laboratories, published by Van Nostrand Reinhold, New York, 1990.

[9] May, G. S., and Spanos, C. J. *Fundamentals of Semiconductor Manufacturing and Process Control.* John Wiley & Sons, Hoboken, New Jersey, 2006.

[10] National Research Council (NRC). *Implications of Emerging Micro- and Nanotechnologies.* Committee on Implications of Emerging Micro- and Nanotechnologies, The National Academies Press, Washington, D.C., 2002.

[11] Pecht, M. (ed.). *Handbook of Electronic Package Design.* Marcel Dekker, New York, 1991.

[12] Runyan, W. R., and Bean, K. E. *Semiconductor Integrated Circuit Processing Technology.* Addison-Wesley Longman, Reading, Massachusetts, 1990.

[13] Seraphim, D. P., Lasky, R., and Li, C-Y. (eds.). *Principles of Electronic Packaging.* McGraw-Hill, New York, 1989.

[14] Sze, S. M. (ed.). *VLSI Technology.* McGraw-Hill, New York, 2004.

[15] Ulrich, R. K., and Brown, W. D. *Advanced Electronic Packaging.* 2nd ed. IEEE Press and John Wiley & Sons, Hoboken, New Jersey, 2006.

[16] Van Zant, P. *Microchip Fabrication.* 5th ed. McGraw-Hill, New York, 2005.

REVIEW QUESTIONS

34.1. What is an integrated circuit?

34.2. Name some of the important semiconductor materials.

34.3. Describe the planar process.

34.4. What are the three major stages in the production of silicon-based integrated circuits?

34.5. What is a clean room and explain the classification system by which clean rooms are rated?

34.6. What are some of the significant sources of contaminants in IC processing?

34.7. What is the name of the process most commonly used to grow single crystal ingots of silicon for semiconductor processing?

34.8. What are the alternatives to photolithography in IC processing?

34.9. What is a photoresist?

34.10. Why is ultraviolet light favored over visible light in photolithography?

34.11. Name the three exposure techniques in photolithography.

34.12. What layer material is produced by thermal oxidation in IC fabrication?

34.13. Define epitaxial deposition.

34.14. What are some of the important design functions of IC packaging?

34.15. What is Rent's rule?

34.16. Name the two categories of component mounting to a printed circuit board.

34.17. What is a DIP?

34.18. What is the difference between postmolding and premolding in plastic IC chip packaging?

MULTIPLE CHOICE QUIZ

There are 16 correct answers in the following multiple choice questions (some questions have multiple answers that are correct). To attain a perfect score on the quiz, all correct answers must be given. Each correct answer is worth 1 point. Each omitted answer or wrong answer reduces the score by 1 point, and each additional answer beyond the correct number of answers reduces the score by 1 point. Percentage score on the quiz is based on the total number of correct answers.

34.1. How many electronic devices would be contained in an IC chip for it to be classified in the VLSI category: (a) 1000, (b) 10,000, (c) 1 million, or (d) 100 million?

34.2. An alternative name for chip in semiconductor processing is which one of the following (one best answer): (a) component, (b) device, (c) die, (d) package, or (e) wafer?

34.3. Which one of the following is the source of silicon for semiconductor processing: (a) pure Si in nature, (b) SiC, (c) Si_3N_4, or (d) SiO_2?

34.4. Which one of the following is the most common form of radiation used in photolithography: (a) electronic beam radiation, (b) incandescent light, (c) infrared light, (d) ultraviolet light, or (e) X-ray?

34.5. After exposure to light, a positive resist becomes (a) less soluble or (b) more soluble to the chemical developing fluid?

34.6. Which of the following processes are used to add layers of various materials in IC fabrication (three best answers): (a) chemical vapor deposition, (b) diffusion, (c) ion implantation, (d) physical vapor deposition, (e) plasma etching, (f) thermal oxidation, and (g) wet etching?

34.7. Which of the following are doping processes in IC fabrication (two best answers): (a) chemical vapor deposition, (b) diffusion, (c) ion implantation, (d) physical vapor deposition, (e) plasma etching, (f) thermal oxidation, and (g) wet etching?

34.8. Which one of the following is the most common metal for intraconnection of devices in a silicon integrated circuit: (a) aluminum, (b) copper, (c) gold, (d) nickel, (e) silicon, or (f) silver?

34.9. Which etching process produces the more anisotropic etch in IC fabrication: (a) plasma etching or (b) wet chemical etching?

34.10. Which of the following are the two principal packaging materials used in IC packaging: (a) aluminum, (b) aluminum oxide, (c) copper, (d) epoxies, and (e) silicon dioxide?

34.11. Which of the following metals are commonly used for wire bonding of chip pads to the lead frame (two best answers): (a) aluminum, (b) copper, (c) gold, (d) nickel, (e) silicon, and (f) silver?

PROBLEMS

Silicon Processing and IC Fabrication

34.1. A single crystal boule of silicon is grown by the Czochralski process to an average diameter of 320 mm with length = 1500 mm. The seed and tang ends are removed, which reduces the length to 1150 mm. The diameter is ground to 300 mm. A 90-mm-wide flat is ground on the surface that extends from one end to the other. The ingot is then sliced into wafers of thickness = 0.50 mm, using an abrasive saw blade whose thickness = 0.33 mm. Assuming that the seed and tang portions cut off the ends of the starting boule were conical in shape, determine (a) the original volume of the boule, mm^3; (b) how many wafers are cut from it, assuming the entire 1150 mm length can be sliced; and (c) the volumetric proportion of silicon in the starting boule that is wasted during processing.

34.2. A silicon boule is grown by the Czochralski process to a diameter of 5.25 in and a length of 5 ft. The seed and tang ends are cut off, reducing the effective length to 48.00 in. Assume that the seed and tang portions are conical in shape. The diameter is ground to 4.921 in (125 mm). A primary flat of width 1.625 in is ground on the surface the entire

length of the ingot. The ingot is then sliced into wafers 0.025 in thick, using an abrasive saw blade whose thickness = 0.0128 in. Determine (a) the original volume of the boule, in^3; (b) how many wafers are cut from it, assuming the entire 4 ft length can be sliced, and (c) what is the volumetric proportion of silicon in the starting boule that is wasted during processing?

34.3. The processable area on a 156-mm-diameter wafer is a 150-mm-diameter circle. How many square IC chips can be processed within this area, if each chip is 7.5 mm on a side? Assume the cut lines (streets) between chips are of negligible width.

34.4. Solve Problem 34.3, only use a wafer size of 257 mm whose processable area has a diameter = 250 mm. What is the percent increase in (a) wafer diameter, (b) processable wafer area, and (c) number of chips, compared to the values in the previous problem?

34.5. A 6.0-in wafer has a processable area with a 5.85-in diameter. How many square IC chips can be fabricated within this area, if each chip is 0.50 in on a side? Assume the cut lines (streets) between chips are of negligible width.

34.6. Solve Problem 34.5, only use a wafer size of 12.0 in whose processable area has a diameter = 11.75 in. What is the percent increase in (a) processable area on the wafer and (b) number of chips on the wafer compared with the 200% increase in wafer diameter?

34.7. A 250 mm diameter silicon wafer has a processable area that is circular with a diameter = 225 mm. The IC chips that will be fabricated on the wafer surface are square with 20 mm on a side. However, the processable area on each chip is only 18 mm by 18 mm. The density of circuits within each chip's processable area is 465 circuits per mm^2. (a) How many IC chips can be placed onto the wafer? (b) Using Rent's Rule with $C = 3.8$ and $m = 0.43$, how many input/output terminals (pins) will be needed for each chip package?

34.8. A 12-inch diameter silicon wafer has a processable area that is circular with a diameter = 11.4 in. The IC chips that will be fabricated on the wafer surface are square with 0.75 in on a side, including an allowance for subsequent chip separation. However, the processable area on each chip is only 0.60 in by 0.60 in. The density of circuits within each chip's processable area is 100,000 circuits per square inch. (a) How many IC chips can be placed onto the wafer? (b) Using Rent's Rule with $C = 3.8$ and $m = 0.43$, how many input/output terminals (pins) will be needed for each chip package?

IC Packaging

34.12. An integrated circuit used in a microprocessor will contain 1000 logic gates. Use Rent's rule with $C = 3.8$ and $m = 0.6$ to determine the approximate number of input/output pins required in the package.

34.13. A dual-in-line package has a total of 48 leads. Use Rent's rule with $C = 4.5$ and $m = 0.5$ to determine the approximate number of logic gates that could be fabricated in the IC chip for this package.

34.14. It is desired to determine the effect of package style on the number of circuits (logic gates) that can be fabricated onto an IC chip to which the package is assembled. Using Rent's rule with $C = 4.5$ and $m = 0.5$, compute the estimated number of devices (logic gates) that could be placed on the chip in the following cases: (a) a DIP with 16 I/O pins on a side—a total of 32 pins; (b) a square chip carrier with 16 pins on a side—a total of 64 I/O pins; and (c) a pin grid array with 16 by 16 pins—a total of 256 pins.

34.15. An integrated circuit used in a memory module contains 2^{24} memory circuits. Sixteen of these integrated circuits are packaged onto a board to

34.9. A silicon boule has been processed through grinding to provide a cylinder whose diameter = 285 mm and whose length = 900 mm. Next, it will be sliced into wafers 0.7 mm thick using a cut-off saw with a kerf = 0.5 mm. The wafers thus produced will be used to fabricate as many IC chips as possible for the personal computer market. Each IC has a market value to the company of $98. Each chip is square with 15 mm on a side. The processable area of each wafer is defined by a diameter = 270 mm. Estimate the value of all of the IC chips that could be produced, assuming an overall yield of 80% good product.

34.10. The surface of a silicon wafer is thermally oxidized, resulting in a SiO_2 film that is 100 nm thick. If the starting thickness of the wafer was exactly 0.400 mm, what is the final wafer thickness after thermal oxidation?

34.11. It is desired to etch out a region of a silicon dioxide film on the surface of a silicon wafer. The SiO_2 film is 100 nm thick. The width of the etched-out area is specified to be 650 nm. (a) If the degree of anisotropy for the etchant in the process is known to be 1.25, what should be the size of the opening in the mask through which the etchant will operate? (b) If plasma etching is used instead of wet etching, and the degree of anisotropy for plasma etching is infinity, what should be the size of the mask opening?

provide a 256 Mbyte memory module. Use Rent's rule, Eq. (34.11), with $C = 6.0$ and $m = 0.12$ to determine the approximate number of input/output pins required in each of the integrated circuits.

34.16. In the equation for Rent's rule with $C = 4.5$ and $m = 0.5$, determine the value of n_{io} and n_c at which the number of logic gates equals the number of I/O terminals in the package.

34.17. A static memory device will have a two-dimensional array with 64×64 cells. Determine the number of input/output pins required using Rent's rule with $C = 6.0$ and $m = 0.12$.

34.18. To produce a 10 megabit memory chip, how many I/O pins are predicted by Rent's rule ($C = 6.0$ and $m = 0.12$)?

34.19. The first IBM personal computer was based on the Intel 8088 CPU, which was released in 1979. The 8088 had 29,000 transistors and 40 I/O pins. The final version of the Pentium III (1 GHz) was released in 2000. It contained 28,000,000 transistors and had 370 I/O pins. (a) Determine the Rent's rule coefficient values m and C assuming that a transistor can be considered a circuit. (b) Use the value of m and C to predict the number of I/O pins required for the first Pentium 4 assuming that it is manufactured with 42,000,000 transistors. (c) The first Pentium 4, released in 2001, used 423 I/O pins. Comment on the accuracy of your prediction.

34.20. Suppose it is desired to produce a memory device that will be contained in a dual-in-line package with 32 I/O leads. How many memory cells can be contained in the device, as estimated by (a) Rent's rule with $C = 6.0$ and $m = 0.12$?

34.21. A 12-inch diameter silicon wafer has a processable area that is circular with a diameter = 11.4 in. The IC chips that will be fabricated on the wafer surface are square with 0.75 in on a side, including an allowance for subsequent chip separation. However, the processable area on each chip is only 0.60 in by 0.60 in. The density of circuits within

each chip's processable area is 100,000 circuits per square inch. (a) How many IC chips can be placed onto the wafer? (b) Using Rent's Rule with $C = 3.8$ and $m = 0.43$, how many input/output terminals (pins) will be needed for each chip package?

34.22. A 250 mm diameter silicon wafer has a processable area that is circular with a diameter = 225 mm. The IC chips that will be fabricated on the wafer surface are square with 20 mm on a side. However, the processable area on each chip is only 18 mm by 18 mm. The density of circuits within each chip's processable area is 465 circuits per mm^2. (a) How many IC chips can be placed onto the wafer? (b) Using Rent's Rule with $C = 4.5$ and $m = 0.35$, how many input/output terminals (pins) will be needed for each chip package?

Yields in IC Processing

34.23. Given that crystal yield = 55%, crystal-to-slice yield = 60%, wafer yield = 75%, multiprobe yield = 65%, and final test yield = 95%, if a starting boule weighs 125 kg, what is the final weight of silicon that is represented by the non-defective chips after final test?

34.24. On a particular production line in a wafer fabrication facility, the crystal yield is 60%, the crystal-to-slice yield is 60%, wafer yield is 90%, multiprobe is 70%, and final test yield is 80%. (a) What is the overall yield for the production line? (b) If wafer yield and multiprobe yield are combined into the same reporting category, what overall yield for the two operations would be expected?

34.25. A silicon wafer with a diameter of 200 mm is processed over a circular area whose diameter = 190 mm. The chips to be fabricated are square with 10 mm on a side. The density of point defects in the surface area is 0.0047 defects/cm^2. Determine an

estimate of the number of good chips using the Bose-Einstein yield computation.

34.26. A 12-in wafer is processed over a circular area of diameter = 11.75 in. The density of point defects in the surface area is 0.018 defects/in^2. The chips to be fabricated are square with an area of 0.16 in^2 each. Determine an estimate of the number of good chips using the Bose-Einstein yield computation.

34.27. The yield of good chips in multiprobe for a certain batch of wafers is 83%. The wafers have a diameter of 150 mm with a processable area that is 140 mm in diameter. If the defects are all assumed to be point defects, determine the density of point defects using the Bose-Einstein method of estimating yield.

34.28. A silicon wafer has a processable area of 35.0 in^2. The yield of good chips on this wafer is $Y_m = 75\%$. If the defects are all assumed to be point defects, determine the density of point defects using the Bose-Einstein method of estimating yield.

35 ELECTRONICS ASSEMBLY AND PACKAGING

Chapter Contents

Integrated circuits constitute the heart of an electronic system, but the complete system consists of much more than packaged ICs. The ICs and other components are mounted and interconnected on printed circuit boards, which in turn are interconnected and contained in a chassis or cabinet. Chip packaging (Section 34.6) is only part of the total electronic package. In this chapter we consider the remaining levels of the package and how they are manufactured and assembled.

35.1 ELECTRONICS PACKAGING

The electronics package is the physical means by which the components in a system are electrically interconnected and interfaced to external devices; it includes the mechanical structure that holds and protects the circuitry. A well-designed electronics package serves the following functions: (1) power distribution and signal interconnection, (2) structural support, (3) circuit protection from physical and chemical hazards in the environment, (4) dissipation of heat generated by the circuits, and (5) minimum delays in signal transmission within the system.

For complex systems containing many components and interconnections, the electronics package is organized into levels that comprise a *packaging hierarchy,* illustrated in Figure 35.1 and summarized in Table 35.1. The lowest level is the *zero level,* which refers to the intraconnections on the semiconductor chip. The packaged chip, consisting of the IC in a plastic or ceramic enclosure and connected to the package leads, constitutes the *first level of packaging*.

Packaged chips and other components are assembled to a printed circuit board (PCB) using two technologies (Section 35.6.1): (1) *pin-in-hole* (PIH) technology and (2) *surface-mount technology* (SMT). The chip package styles and assembly techniques are different for PIH and SMT. In many cases, both assembly technologies are

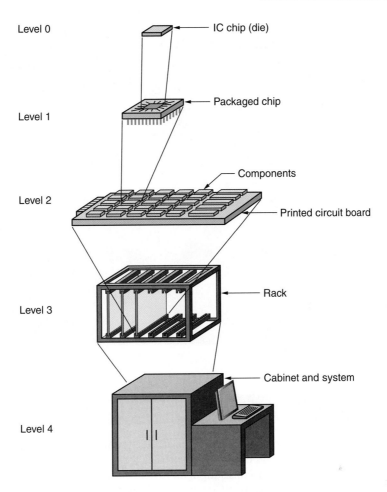

Level 0 — IC chip (die)

Level 1 — Packaged chip

Level 2 — Components
— Printed circuit board

Level 3 — Rack

Level 4 — Cabinet and system

FIGURE 35.1
Packaging hierarchy in a
large electronic system.

employed in the same board. Printed circuit board assembly represents the ***second level of packaging***. Figure 35.2 shows a variety of PCB assemblies of both PIH and SMT types.

The assembled PCBs are, in turn, connected to a chassis or other framework; this is the ***third level of packaging***. This third level may consist of a ***rack*** that holds the boards, using wiring cables to make the interconnections. In major electronic systems, such as large computers, the PCBs are typically mounted onto a larger printed circuit board called a ***back plane,*** which has conduction paths to permit interconnection between the smaller boards attached to it. This latter configuration is known as ***card-on-board*** (COB) packaging, in which the smaller PCBs are called cards and the back plane is the board.

The ***fourth level of packaging*** consists of wiring and cabling inside the cabinet that contains the electronic system. For systems of relatively low complexity, the packaging may not include all of the possible levels in the hierarchy.

TABLE 35.1	Packaging hierarchy.
Level	**Description of Interconnection**
0	Intraconnections on the chip
1	Chip-to-package interconnections to form IC package
2	IC package to circuit board interconnections
3	Circuit board to rack; card-on-board packaging
4	Wiring and cabling connections in cabinet

FIGURE 35.2 A collection of printed circuit board assemblies showing both pin-in-hole and surface-mount technologies. (Photo courtesy of Phoenix Technologies, Inc.)

35.2 PRINTED CIRCUIT BOARDS

A printed circuit board consists of one or more thin sheets of insulating material, with thin copper lines on one or both surfaces that interconnect the components attached to the board. In boards consisting of more than one layer, copper conducting paths are interleaved between the layers. PCBs are used in packaged electronic systems to hold components, provide electrical interconnections among them, and make connections to external circuits. They have become standard building blocks in virtually all electronic systems that contain packaged ICs and other components (Historical Note 35.1). PCBs are so important and widely used because (1) they provide a convenient structural platform for the components; (2) a board with correctly routed interconnections can be mass produced consistently, without the variability usually associated with hand wiring; (3) nearly all of the soldering connections between components and the PCB can be accomplished in a one-step mechanized operation, (4) an assembled PCB gives reliable performance; and (5) in complex electronic systems, each assembled PCB can be detached from the system for service, repair, or replacement.

Historical Note 35.1   *Printed circuit boards*

Before printed circuit boards, electrical and electronic components were manually fastened to a sheet-metal chassis and then hand wired and soldered to form the desired circuit. The usual sheet metal was aluminum. In the late 1950s, various plastic boards became commercially available. These boards, which provided electrical insulation, gradually replaced the aluminum chassis. The first plastics were phenolics, followed by glass-fiber–reinforced epoxies. The boards came with predrilled holes spaced at standard intervals in both directions. This inspired the use of electronic components that matched these hole spacings. The dual-in-line package evolved during this period.

The components in these circuit boards were hand-wired, which proved increasingly difficult and prone to human error as component densities increased and circuits became more complex. The printed circuit board, with etched copper foil on its surface to form the wiring interconnections, was developed to solve these problems with manual wiring.

Initial techniques to design the circuit masks involved a manual inking procedure, in which the designer attempted to route the conducting tracks to provide the required connections and avoid short circuits on a large sheet of paper or vellum. This became more difficult as the number of components on the board increased and the conducting lines interconnecting the components became finer. Computer programs were developed to aid the designer in solving the routing problem. However, in many cases, it was impossible to find a solution with no intersecting tracks (short circuits). To solve the problem, jumper wires were hand-soldered to the board to make these connections. As the number of jumper wires increased, the problem of human error again appeared. Multilayer boards were introduced to deal with this routing issue.

The initial technique for "printing" the circuit pattern onto the copper-clad board was screen printing. As track widths became finer and finer, photolithography was substituted.

35.2.1 STRUCTURES, TYPES, AND MATERIALS FOR PCBs

A *printed circuit board* (PCB), also called a *printed wiring board* (PWB), is a laminated flat panel of insulating material designed to provide electrical interconnections between electronic components attached to it. Interconnections are made by thin conducting paths on the surface of the board or in alternating layers sandwiched between layers of insulating material. The conducting paths are made of copper and are called *tracks*. Other copper areas, called *lands,* are also available on the board surface for attaching and electrically connecting components.

Insulation materials in PCBs are usually polymer composites reinforced with glass fabrics or paper. Polymers include epoxy (most widely used), phenolic, and polyimide. E-glass is the usual fiber in glass-reinforcing fabrics, especially in epoxy PCBs; paper is a common reinforcing layer for phenolic boards. The usual thickness of the substrate layer is 0.8 to 3.2 mm (0.031 to 0.125 in), and copper foil thickness is around 0.04 mm (0.0015 in). The materials forming the PCB structure must be electrically insulating, strong and rigid, resistant to warpage, dimensionally stable, heat resistant, and flame retardant. Chemicals are often added to the polymer composite to obtain the last two characteristics.

There are three principal types of printed circuit board, shown in Figure 35.3: (a) *single-sided* board, in which copper foil is only on one side of the insulation substrate; (b) *double-sided* board, in which the copper foil is on both sides of the substrate; and (c) *multilayer* board, consisting of alternating layers of conducting foil and insulation. In all three structures, the insulation layers are constructed of multiple laminates of epoxy-glass sheets (or other composite) bonded together to form a strong and rigid structure. Multilayer boards are used for complex circuit assemblies in which a large number of components must be interconnected with many track routings, thus requiring more conducting paths than can be accommodated in one or two copper layers. Four layers is the most common multilayer configuration, but boards with up to 24 conducting layers are produced.

35.2.2 PRODUCTION OF THE STARTING BOARDS

Single- and double-sided boards can be purchased from suppliers that specialize in mass producing them in standard sizes. The boards are then custom-processed by a circuit fabricator to create the specified circuit pattern and board size for a given application. Multilayer boards

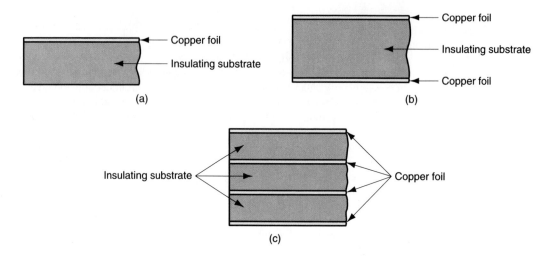

FIGURE 35.3   Three types of printed circuit board structure: (a) single-sided, (b) double-sided, and (c) multilayer.

are fabricated from standard single- and double-sided boards. The circuit fabricator processes the boards separately to form the required circuit pattern for each layer in the final structure, and then the individual boards are bonded together with additional layers of epoxy-fabric. Processing of multilayer boards is more involved and more expensive than the other types; the reason for using them is that they provide better performance for large systems than using a much greater number of lower-density boards of simpler construction.

The copper foil used to clad the starting boards is produced by a continuous electroforming process (Section 28.3.2), in which a rotating smooth metal drum is partially submersed in an electrolytic bath containing copper ions. The drum is the cathode in the circuit, causing the copper to plate onto its surface. As the drum rotates out of the bath, the thin copper foil is peeled from its surface. The process is ideal for producing the very thin copper foil needed for PCBs.

Production of the starting boards consists of pressing multiple sheets of woven glass fiber that have been impregnated with partially cured epoxy (or other thermosetting polymer). The number of sheets used in the starting sandwich determines the thickness of the final board. Copper foil is placed on one or both sides of the epoxy-glass laminated stack, depending on whether single- or double-sided boards are to be made. For single-sided boards, a thin release film is used on one side in place of the copper foil to prevent sticking of the epoxy in the press. Pressing is accomplished between two steam-heated platens of a hydraulic press. The combination of heat and pressure compacts and cures the epoxy-glass layers to bond and harden the laminates into a single-piece board. The board is then cooled and trimmed to remove excess epoxy that has been squeezed out around the edges.

The completed board consists of a glass-fabric–reinforced epoxy panel, clad with copper over its surface area on one or both sides. It is now ready for the circuit fabricator. Panels are usually produced in large standard widths designed to match the board handling systems on wave-soldering equipment, automatic insertion machines, and other PCB processing and assembly facilities. If the electronic design calls for a smaller size, several units can be processed together on the same larger board and then separated later.

35.2.3 PROCESSES USED IN PCB FABRICATION

The circuit fabricator employs a variety of processing operations to produce a finished PCB, ready for assembly of components. The operations include cleaning, shearing, hole drilling or punching, pattern imaging, etching, and electroless and electrolytic plating. Most of these processes have been discussed. In this section we focus on the details that are relevant to PCB fabrication. Our discussion follows approximately the order in which the processes are performed on a board. However, there are differences in processing sequence between different board types, and we examine these differences in Section 35.2.4. Some of the operations in PCB fabrication must be performed under clean room conditions to avoid defects in the printed circuits, especially for boards with fine tracks and details.

Board Preparation   Initial preparation of the board consists of shearing, hole-making, and other shaping operations to create tabs, slots, and similar features in the board. If necessary, the starting panel may have to be sheared to size for compatibility with the circuit fabricator's equipment. The holes, called tooling holes, are made by drilling or punching and are used for positioning the board during subsequent processing. The sequence of fabrication steps requires close alignment from one process to the next, and these holes are used with locating pins at each operation to achieve accurate registration. Three tooling holes per board are usually sufficient for this purpose; hole size is about 3.2 mm (0.125 in), larger than the circuit holes to be drilled later.

   The board is typically bar coded for identification purposes in this preparation phase. Finally, a cleaning process is used to remove dirt and grease from the board surface. Although cleanliness requirements are not as stringent as in IC fabrication, small particles of dirt and dust can cause defects in the circuit pattern of a printed circuit board; and surface films of grease can inhibit etching and other chemical processes. Cleanliness is essential for producing reliable PCBs.

Hole Drilling   In addition to tooling holes, functional circuit holes are required in PCBs as (1) *insertion holes* for inserting component leads in through-hole boards, (2) *via holes,* which are later copper-plated and used as conducting paths from one side of the board to the other, and (3) holes to fasten certain components such as heat sinks and connectors to the board. These holes are either drilled or punched, using the tooling holes for location. Cleaner holes can be produced by drilling, so most holes in PCB fabrication are drilled. A stack of three or four panels may be drilled in the same operation, using a computer numerically controlled (CNC) drill press that receives its programming instructions from the design database. For high-production jobs, multiple-spindle drills are sometimes used, permitting all of the holes in the board to be drilled in one feed motion.

   Standard twist drills (Section 23.3.2) are used to drill the holes, but the application makes a number of unusual demands on the drill and drilling equipment. Perhaps the biggest single problem is the small hole size in printed circuit boards; drill diameter is generally less than 1.27 mm (0.050 in), but some high-density boards require hole sizes of 0.15 mm (0.006 in) or even less [8]. Such small drill bits lack strength, and their capacity to dissipate heat is low.

   Another difficulty is the unique work material. The drill bit must first pass through a thin copper foil and then proceed through an abrasive epoxy-glass composite. Different drills would normally be specified for these materials, but a single drill must suffice in PCB drilling. The small hole size, combined with the stacking of several boards or multilayer boards, result in a high depth-to-diameter ratio, aggravating the problem of chip extraction from the hole. Other requirements placed on the operation include high accuracy in hole location, smooth hole walls, and absence of burrs on the holes. Burrs are usually formed when the drill enters or exits a hole; thin sheets of material are often placed on top of and beneath the stack of boards to inhibit burr formation on the boards themselves.

   Finally, any cutting tool must be used at a certain cutting speed to operate at best efficiency. For a drill bit, cutting speed is measured at the diameter. For very small drill sizes, this means extremely high rotational speeds—up to 100,000 rev/min in some cases. Special spindle bearings and motors are required to achieve these speeds.

Circuit Pattern Imaging and Etching   There are two basic methods by which the circuit pattern is transferred to the copper surface of the board: screen printing and photolithography. Both methods involve the use of a resist coating on the board surface that determines where etching of the copper will occur to create the tracks and lands of the circuit.

Screen printing was the first method used for PCBs. It is indeed a printing technique, and the term *"printed circuit board"* can be traced to this method. In ***screen printing,*** a stencil screen containing the circuit pattern is placed on the board, and liquid resist is squeezed through the mesh to the surface beneath. This method is simple and inexpensive, but its resolution is limited. It is normally used only for applications in which track widths are greater than about 0.25 mm (0.010 in).

The second method of transferring the circuit pattern is ***photolithography,*** in which a light-sensitive resist material is exposed through a mask to transfer the circuit pattern. The procedure is similar to the corresponding process in IC fabrication (Section 34.3.1); some of the details in PCB processing will be described here.

Photoresists used by circuit fabricators are available in two forms: liquid or dry film. Liquid resists can be applied by roller or spraying. Dry film resists are more commonly used in PCB fabrication. They consist of three layers: a film of photosensitive polymer sandwiched between a polyester support sheet on one side and a removable plastic cover sheet on the other side. The cover sheet prevents the photosensitive material from sticking during storage and handling. Although more expensive than liquid resists, they can be applied in coatings of uniform thickness, and their processing in photolithography is simpler. To apply, the cover sheet is removed, and the resist film is placed on the copper surface to which it readily adheres. Hot rollers are used to press and smooth the resist onto the surface.

Alignment of the masks relative to the board relies on the use of registration holes that are aligned with the tooling holes on the board. Contact printing is used to expose the resist beneath the mask. The resist is then developed, which involves removal of the unexposed regions of the negative resist from the surface.

After resist development, certain areas of the copper surface remain covered by resist while other areas are now unprotected. The covered areas correspond to circuit tracks and lands, while uncovered areas correspond to open regions between. ***Etching*** removes the copper cladding in the unprotected regions from the board surface, usually by means of a chemical etchant. Etching is the step that transforms the solid copper film into the interconnections for an electrical circuit.

Etching is done in an etching chamber in which the etchant is sprayed onto the surface of the board that is now partially coated with resist. Various etchants are used to remove copper, including ammonium persulfate ($(NH_4)_2S_2O_4$), ammonium hydroxide (NH_4OH), cupric chloride ($CuCl_2$), and ferric chloride ($FeCl_3$). Each has its relative advantages and disadvantages. Process parameters (e.g., temperature, etchant concentration, and duration) must be closely controlled to avoid over- or under-etching, as in IC fabrication. After etching, the board must be rinsed and the remaining resist chemically stripped from the surface.

Plating   In printed circuit boards, plating is needed on the hole surfaces to provide conductive paths from one side to the other in double-sided boards, or between layers in multilayer boards. Two types of plating process are used in PCB fabrication: electroplating and electroless plating (Section 28.3.3). Electroplating has a higher deposition rate than electroless plating but requires that the coated surface be metallic (conductive); electroless plating is slower but does not require a conductive surface.

After drilling of the via holes and insertion holes, the walls of the holes consist of epoxy-glass insulation material, which is nonconductive. Accordingly, electroless plating must be used initially to provide a thin coating of copper on the hole walls. Once the thin film of copper has been applied, electrolytic plating is then used to increase coating thickness on the hole surfaces to between 0.025 and 0.05 mm (0.001 and 0.002 in).

FIGURE 35.4   A section of a double-sided PCB, showing various features accomplished during fabrication: tracks and lands, and copper-plated insertion and via holes.

Gold is another metal sometimes plated onto printed circuit boards. It is used as a very thin coating on PCB edge connectors to provide superior electrical contact. Coating thickness is only about 2.5 μm (0.0001 in).

35.2.4   PCB FABRICATION SEQUENCE

In this section we describe the processing sequence for various board types. The sequence is concerned with transforming a copper-clad board of reinforced polymer into a printed circuit board, a procedure called **_circuitization_**. The desired result is illustrated in Figure 35.4 for a double-sided board.

Circuitization   Three methods of circuitization can be used to determine which regions of the board will be coated with copper [12]: (1) subtractive, (2) additive, and (3) semiadditive.

In the **_subtractive method,_** open portions of the copper cladding on the starting board are etched away from the surface, so that the tracks and lands of the desired circuit remain. The process is called "subtractive" because copper is removed from the board surface. The steps in the subtractive method are described in Figure 35.5.

The **_additive method_** starts with a board surface that is not copper clad, such as the uncoated surface of a single-sided board. However, the uncoated surface is treated with a chemical, called a **_buttercoat,_** which acts as the catalyst for electroless plating. The steps in the method are outlined in Figure 35.6.

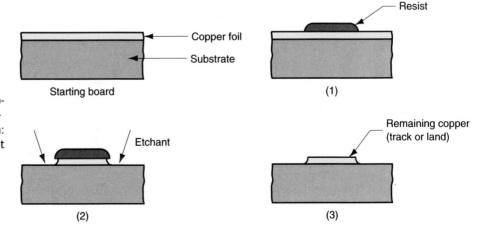

FIGURE 35.5   The subtractive method of circuitization in PCB fabrication: (1) apply resist to areas not to be etched, using photolithography to expose the areas that are to be etched, (2) etch, and (3) strip resist.

FIGURE 35.6   The additive method of circuitization in PCB fabrication: (1) a resist film is applied to the surface using photolithography to expose the areas to be copper plated; (2) the exposed surface is chemically activated to serve as a catalyst for electroless plating; (3) copper is plated on exposed areas; and (4) resist is stripped.

The *semiadditive method* uses a combination of additive and subtractive steps. The starting board has a very thin copper film on its surface—5 μm (0.0002 in) or less. The method proceeds as described in Figure 35.7.

Processing of Different Board Types   Processing methods differ for the three PCB types: single-sided, double-sided, and multilayer. A *single-sided board* begins fabrication

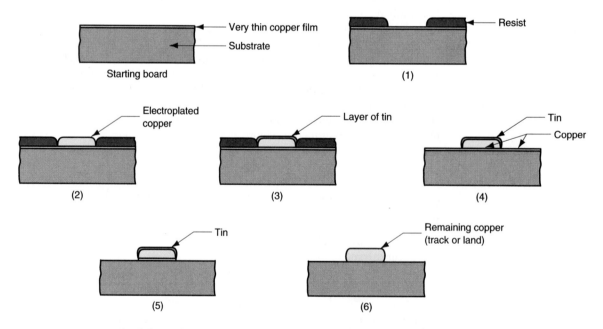

FIGURE 35.7   The semiadditive method of circuitization in PCB fabrication: (1) Apply resist to areas that will not be plated; (2) electroplate copper, using the thin copper film for conduction; (3) apply tin on top of plated copper; (4) strip resist; (5) etch remaining thin film of copper on the surface, while the tin serves as a resist for the electroplated copper; and (6) strip tin from copper.

FIGURE 35.8 Typical cross section of a multilayer printed circuit board.

as a flat sheet of insulating material clad on one side with copper film. The subtractive method is used to produce the circuit pattern in the copper cladding.

A *double-sided board* involves a somewhat more complex processing sequence because it has circuit tracks on both sides that must be electrically connected. The interconnection is accomplished by means of copper-plated via holes that run from lands on one surface of the board to lands on the opposite surface, as shown in Figure 35.4. A typical fabrication sequence for a double-sided board (copper-clad on both sides) uses the semiadditive method. After hole drilling, electroless plating is used to initially plate the holes, followed by electroplating to increase plating thickness.

A *multilayer board* is structurally the most complex of the three types, and this complexity is reflected in its manufacturing sequence. The laminated construction can be seen in Figure 35.8, which highlights some of the features of a multilayer PCB. The fabrication steps for the individual layers are basically the same as those used for single- and double-sided boards. What makes multilayer board fabrication more complicated is that (1) all of the layers, each with its own circuit design, must first be processed; then (2) the layers must be joined together to form one integral board; and finally (3) the assembled board must itself be put through its own processing sequence.

A multilayer board consists of *logic layers,* which carry electrical signals between components on the board, and *voltage layers,* which are used to distribute power. Logic layers are generally fabricated from double-sided boards, whereas voltage layers are usually made from single-sided boards. Thinner insulating substrates are used for multi-layer boards than for their standalone single- and double-sided counterparts, so that a suitable thickness of the final board can be achieved.

In the second stage, the individual layers are assembled together. The procedure starts with copper foil on the bottom outside, and then adds the individual layers, separating one from the next by one or more sheets of glass fabric impregnated with partially cured epoxy. After all layers have been sandwiched together, a final copper foil is placed on the stack to form the top outer layer. Layers are then bonded into a single board by heating the assembly under pressure to cure the epoxy. After curing, any excess resin squeezed out of the sandwich around the edges is trimmed away.

At the start of the third stage of fabrication, the board consists of multiple layers bonded together, with copper foil cladded on its outer surfaces. Its construction can therefore be likened to that of a double-sided board; and its processing is likewise similar. The sequence consists of drilling additional through-holes, plating the holes to establish conduction paths between the two exterior copper films as well as certain internal copper layers, and the use of photolithography and etching to form the circuit pattern on the outer copper surfaces.

Testing and Finishing Operations   After a circuit has been fabricated on the board surface, it must be inspected and tested to ensure that it functions according to design specifications and contains no quality defects. Two procedures are common: (1) visual inspection and (2) continuity testing. In *visual inspection,* the board is examined visually to detect open and short circuits, errors in drilled hole locations, and other faults that can be observed without applying electrical power to the board. Visual inspections, performed not only after fabrication but also at various critical stages during production, are accomplished by human eye or machine vision (Section 42.6.3).

Continuity testing involves the use of contact probes brought simultaneously into contact with track and land areas on the board surface. The setup consists of an array of probes that are forced under light pressure to make contact with specified points on the board surface. Electrical connections between contact points can be quickly checked in this procedure.

Several additional processing steps must be performed on the bare board to prepare it for assembly. The first of these finishing operations is the application of a thin solder layer on the track and land surfaces. This layer serves to protect the copper from oxidation and contamination. It is carried out either by electroplating or by bringing the copper side into contact with rotating rollers that are partially submersed in molten solder.

A second operation involves application of a coating of solder resist to all areas of the board surface except the lands that are to be subsequently soldered in assembly. The solder resist coating is chemically formulated to resist adhesion of solder; thus, in the subsequent soldering processes, solder adheres only to land areas. Solder resist is usually applied by screen printing.

Finally, an identification legend is printed onto the surface, again by screen printing. The legend indicates where the different components are to be placed on the board in final assembly. In modern industrial practice, a bar code is also printed on the board for production control purposes.

35.3   PRINTED CIRCUIT BOARD ASSEMBLY

A printed circuit board assembly consists of electronic components (e.g., IC packages, resistors, capacitors) as well as mechanical components (e.g., fasteners, heat sinks) mounted on a printed circuit board. This is level 2 in electronic packaging (Table 35.1). As indicated, PCB assembly is based on either pin-in-hole (PIH) or surface-mount technologies (SMT). Some PCB assemblies include both PIH and SMT components. Our discussion here deals exclusively with PIH assemblies. In Section 35.4, we consider surface-mount technology and combinations of the two types. The scope of electronic assembly also includes higher packaging levels such as assemblies of multiple PCBs electrically connected and mechanically contained in a chassis or cabinet. In Section 35.5, we explore the technologies by which electrical connections are made at these higher levels.

In printed circuit assemblies using PIH technology, the lead pins must be inserted into through-holes in the circuit board. Once inserted, the leads are soldered into place in the holes in the board. In double-sided and multilayer boards, the hole surfaces into which the leads are inserted are generally copper plated, giving rise to the name *plated through-hole* (PTH) for these cases. After soldering, the boards are cleaned and tested, and those boards not passing the test are reworked if possible. Thus, we can divide the processing of PIH assemblies into the following steps: (1) component insertion, (2) soldering, (3) cleaning, (4) testing, and (5) rework.

35.3.1   COMPONENT INSERTION

In component insertion, the leads of components are inserted into their proper through-holes in the PCB. A single board may be populated with hundreds of separate components (DIPs, resistors, etc.), all of which need to be inserted into the board. In modern electronic assembly plants, most component insertions are accomplished by automatic insertion machines. A small proportion is done by hand for nonstandard components that cannot be accommodated on automatic machines. These cases include switches and connectors as well as resistors, capacitors, and certain other components. Although the proportion of component insertions accomplished manually in industry is low, their cost is high because of much lower production rates than automatic insertions. Industrial robots (Section 38.4) are sometimes used to substitute for human labor in these component insertion tasks.

Automatic insertion machines are either semiautomatic or fully automatic. The semiautomatic type involves insertion of the component by a mechanical insertion device whose position relative to the board is controlled by a human operator. Fully automatic insertion machines comprise the preferred category because they are faster and their need for human attention is limited to loading components and fixing jams when they occur. Automatic insertion machines are controlled by a program that is usually prepared directly from circuit design data. Components are loaded into these machines in the form of reels, magazines, or other carriers that maintain proper orientation of the components until insertion.

The insertion operation involves (1) preforming the leads, (2) insertion of leads into the board holes, and then (3) cropping and clinching the leads on the other side of the board. Preforming is needed only for some component types and involves bending of leads that are initially straight into a U-shape for insertion. Many components come with properly shaped leads and require little or no preforming.

Insertion is accomplished by a workhead designed for the component type. Components inserted by automatic machines are grouped into three basic categories: (a) axial lead, (b) radial lead, and (c) dual-in-line package. The dual-in-line package (Section 34.6.1) is a very common package for integrated circuits. Typical axial and radial lead components are pictured in Figure 35.9. Axial components are shaped like a cylinder, with leads projecting from each end. Typical components of this type include resistors, capacitors, and diodes. Their leads must be bent, as suggested in our figure, to be inserted. Radial components have parallel leads and have various body shapes, one of which is shown in Figure 35.9(b). This type of component is exemplified by light-emitting diodes, potentiometers, resistor networks, and fuse holders. These configurations are sufficiently different that separate insertion machines with the appropriate workhead designs must be used to handle each category. Accurate positioning of the board beneath the workhead before each insertion is performed by a high-speed x-y positioning table.

Once the leads have been inserted through the holes in the board, they are clinched and cropped. Clinching involves bending the leads, as in Figure 35.10, to mechanically secure the component to the board until soldering. If this were not done, the component is at risk of being knocked out of its holes during handling. In cropping, the leads are cut to proper length; otherwise, there is a possibility that they might become bent and cause a short circuit with nearby circuit tracks or components. These operations are performed automatically on the underside of the board by the insertion machine.

The three types of insertion machines, corresponding to the three basic component configurations, can be joined to form an integrated circuit board assembly line. The

FIGURE 35.9   Two of the three basic component types used with automatic insertion machines: (a) axial lead and (b) radial lead. The third type, dual-in-line package (DIP), is illustrated in Figure 34.19.

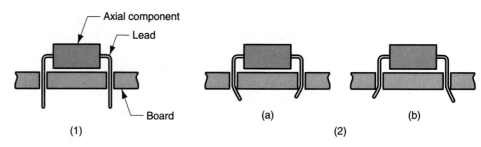

FIGURE 35.10
Clinching and cropping of component leads: (1) as inserted, and (2) after bending and cutting; leads can be bent either (a) inward or (b) outward.

integration is accomplished by means of a conveyor system that transfers boards from one machine type to the next. A computer control system is used to track the progress of each board as it moves through the cell and download the correct programs to each workstation.

35.3.2 SOLDERING

The second basic step in PCB assembly is soldering. For inserted components, the most important soldering techniques are wave soldering and hand soldering. These methods as well as other aspects of soldering are discussed in Section 31.2.

Wave Soldering Wave soldering is a mechanized technique in which printed circuit boards containing inserted components are moved by conveyor over a standing wave of molten solder (Figure 31.9). The position of the conveyor is such that only the underside of the board, with component leads projecting through the holes, is in contact with the solder. The combination of capillary action and the upward force of the wave cause the liquid solder to flow into the clearances between leads and through-holes to obtain a good solder joint. The tremendous advantage of wave soldering is that all of the solder joints on a board are made in a single pass through the process.

Hand Soldering Hand soldering involves a skilled operator using a soldering iron to make circuit connections. Compared with wave soldering, hand soldering is slow because solder joints are made one at a time. As a production method, it is generally used only for small lot production and rework. As with other manual tasks, human error can result in quality problems. Hand soldering is sometimes used after wave soldering to add delicate components that would be damaged in the harsh environment of the wave-soldering chamber. Manual methods have certain advantages in PCB assembly that should be noted: (1) Heat is localized and can be directed at a small target area; (2) equipment is inexpensive compared with wave soldering; and (3) energy consumption is considerably less.

35.3.3 CLEANING, TESTING, AND REWORK

The final processing steps in PCB assembly are cleaning, testing, and rework. Visual inspections are also performed on the board to detect obvious flaws.

Cleaning After soldering, contaminants are present on the printed circuit assembly. These foreign substances include flux, oil and grease, salts, and dirt, some of which can cause chemical degradation of the assembly or interfere with its electronic functions. One or more chemical cleaning operations (Section 28.1.1) must be carried out to remove these undesirable materials. Traditional cleaning methods for PCB assemblies include hand cleaning with appropriate solvents and vapor degreasing with chlorinated solvents. Concern over environmental hazards in recent years has motivated the search for effective water-based solvents to replace the chlorinated and fluorinated chemicals traditionally used in vapor degreasing.

Testing   Visual inspection is used to detect for board substrate damage, missing or damaged components, soldering faults, and similar quality defects that can be observed by eye. Machine vision systems are being used to perform these inspections automatically in a growing number of installations.

Test procedures must be performed on the completed assembly to verify its functionality. The board design must allow for this testing by including test points in the circuit layout. These test points are convenient locations in the circuit where probes can make contact for testing. Individual components in the circuit are tested by contacting the component leads, applying input test signals, and measuring the output signals. More sophisticated procedures include digital function tests, in which the entire circuit or major subcircuits are tested using a programmed sequence of input signals and measuring the corresponding output signals to simulate operating conditions.

Another test used for printed circuit board assemblies is the substitution test, in which a production unit is plugged into a mock-up of the working system and energized to perform its functions. If the assembly performs in a satisfactory way, it is deemed as passing the test. It is then unplugged and the next production unit is substituted in the mock-up.

Finally, a burn-in test is performed on certain types of PCB assemblies that may be subject to "infant mortality." Some boards contain defects that are not revealed in normal functional tests but which are likely to cause failure of the circuit during early service. Burn-in tests operate the assemblies under power for a certain period of time, such as 24 or 72 hours, sometimes at elevated temperatures, such as 40°C (100°F), to force these defects to manifest their failures during the testing period. Boards not subject to infant mortality will survive this test and provide long service life.

Rework   When inspection and testing indicate that one or more components on the board are defective or certain solder joints are faulty, it usually makes sense to try to repair the assembly rather than discard it together with all of the remaining good components. This repair step is an integral part of electronic assembly plant operations. Common rework tasks include touchup (repair of solder faults), replacement of defective or missing components, and repair of copper film that has lifted from the substrate surface. These tasks are manual operations, requiring skilled workers using soldering irons.

35.4   SURFACE-MOUNT TECHNOLOGY

One effect of the growing complexity of electronic systems has been the need for greater packing densities in printed circuit assemblies. Conventional PCB assemblies that use leaded components inserted into through-holes have the following inherent limitations in terms of packing density: (1) components can be mounted on only one side of the board, and (2) center-to-center distance between lead pins in leaded components must be a minimum of 1.0 mm (0.04 in) and is usually 2.5 mm (0.10 in).

Surface-mount technology uses an assembly method in which component leads are soldered to lands on the surface of the board rather than into holes running through the board (Historical Note 35.2). By eliminating the need for leads inserted into through holes in the board, several advantages accrue [6]: (1) smaller components can be made, with leads closer together; (2) packing densities can be increased; (3) components can be mounted on both sides of the board; (4) smaller PCBs can be used for the same electronic system; and (5) drilling of the many through holes during board fabrication is eliminated—via holes to interconnect layers are still required. Typical areas on the board surface taken by SMT components range between 20% and 60% compared with through-hole components.

Historical Note 35.2   *Surface-mount technology*

Surface-mount technology (SMT) traces its origins to the electronic systems in the aerospace and military industries of the 1960s. The first components were small, flat ceramic packages with gull-wing leads. The initial reason why these packages were attractive, compared with through-hole technology, was the fact that they could be placed on both sides of a printed circuit board—in effect, doubling the component density. In addition, the SMT package could be made smaller than a comparable through-hole package, further increasing component densities on the printed circuit board.

In the early 1970s, further advances in SMT were made in the form of leadless components—components with ceramic packages that had no discrete leads. This permitted even greater circuit densities in military and aerospace electronics. In the late 1970s, plastic SMT packages became available, motivating the widespread use of surface-mount technology. The computer and automotive industries have become important users of SMT, and their demand for SMT components has contributed to the significant growth in this technology.

Despite these advantages, the electronics industry has not fully adopted SMT to the exclusion of PIH technology. There are several reasons: (1) Owing to their smaller size, surface mount components are more difficult to handle and assemble by humans; (2) SMT components are generally more expensive than leaded components, although this disadvantage may change as SMT production techniques are perfected; (3) inspection, testing, and rework of the circuit assemblies is generally more difficult in SMT because of the smaller scale involved; and (4) certain types of components are not available in surface mount form. This final limitation results in some electronic assemblies that contain both surface-mount and leaded components.

The same basic steps are required to assemble surface-mount components to PCBs as in pin-in-hole technology. The components must be placed on the board and soldered, followed by cleaning, testing, and rework. The methods of placement and soldering the components, as well as certain of the testing and rework procedures, are different in surface mount technology. Component placement in SMT means correctly locating the component on the PCB and affixing it sufficiently to the surface until soldering provides a permanent mechanical and electrical connection. Two alternative placement and soldering methods are available: (1) adhesive bonding of components and wave soldering, and (2) solder paste and reflow soldering. It turns out that certain types of SMT components are more suited to one method, whereas other types are more suited to the other.

35.4.1 ADHESIVE BONDING AND WAVE SOLDERING

The steps in this method are described in Figure 35.11. Various adhesives (Section 31.3) are used for affixing components to the board surface. Most common are epoxies and acrylics. The adhesive is applied by one of three methods: (1) brushing liquid adhesive through a screen stencil; (2) using an automatic dispensing machine with a programmable *x-y* positioning system; or (3) using a pin transfer method, in which a fixture consisting of pins arranged according to where adhesive must be applied is dipped into the liquid adhesive and then positioned onto the board surface to deposit adhesive in the required spots.

The components are then placed onto the board surface by automatic placement machines operating under computer control. The term "*on*sertion" machines is used for these units, to distinguish them from *in*sertion machines used in PIH technology. Onsertion machines operate at cycle rates of up to four components placed per second.

After component placement, the adhesive is cured. Depending on adhesive type, curing is by heat, ultraviolet (UV) light, or a combination of UV and infrared (IR) radiation.

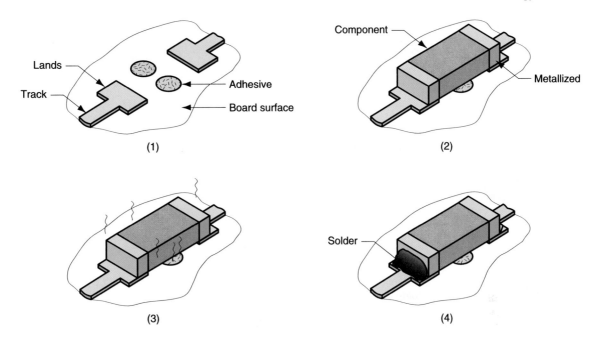

FIGURE 35.11   Adhesive bonding and wave soldering, shown here for a discrete capacitor or resistor component: (1) adhesive is applied to areas on the board where components are to be placed; (2) components are placed onto adhesive-coated areas; (3) adhesive is cured; and (4) solder joints are made by wave soldering.

With the surface-mount components now bonded to the PCB surface, the board is put through wave soldering. The operation differs from its PIH counterpart in that the components themselves pass through the molten solder wave. Technical problems sometimes encountered in SMT wave soldering include components uprooted from the board, components shifting position, and larger components creating shadows that inhibit proper soldering of neighboring components.

35.4.2 SOLDER PASTE AND REFLOW SOLDERING

In this method, a solder paste is used to affix components to the surface of the circuit board. The sequence of steps is depicted in Figure 35.12.

A *solder paste* is a suspension of solder powders in a flux binder. It has three functions: (1) it is the solder—typically 80% to 90% of total paste volume, (2) it is the flux, and (3) it is the adhesive that secures the components to the surface of the board. Methods of applying the solder paste to the board surface include screen printing and syringe dispensing. Properties of the paste must be compatible with these application methods; the paste must flow yet not be so liquid that it spreads beyond the localized area where it is applied.

After solder paste application, components are placed on the board by the same type of onsertion machines used with the adhesive bonding assembly method. A low-temperature baking operation is performed to dry the flux binder; this reduces gas escape during soldering. Finally, the solder reflow process (Section 31.2.3) heats the solder paste sufficiently that the solder particles melt to form a high-quality mechanical and electrical joint between the component leads and the circuit lands on the board.

As in PIH technology, the various operations required to assemble SMT printed circuit boards are accomplished using integrated production lines, as shown in Figure 35.13.

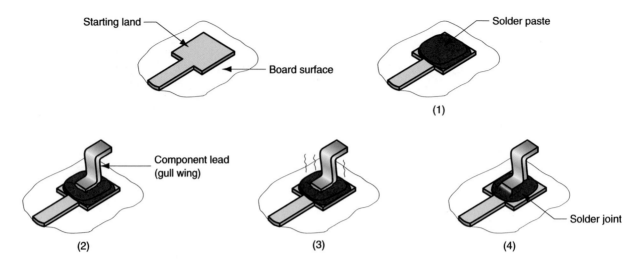

FIGURE 35.12 Solder paste and reflow method: (1) apply solder paste to desired land areas, (2) place components onto board, (3) bake paste, and (4) solder reflow.

35.4.3 COMBINED SMT-PIH ASSEMBLY

Our discussion of the SMT assembly methods has assumed a relatively simple circuit board with SMT components on only one side. These cases are unusual because most SMT circuit assemblies combine surface-mounted and pin-in-hole components on the same board. In addition, SMT assemblies can be populated on both sides of the board, whereas PIH components are normally limited to one side only. The assembly sequence must be altered to allow for these additional possibilities, although the basic processing steps described in the two preceding sections are the same.

One possibility is for the SMT and PIH components to be on the same side of the board. For this case, a typical sequence would consist of the steps described in Figure 35.14. More complex PCB assemblies consist of SMT-PIH components as in our figure, but with SMT components on both sides of the board.

FIGURE 35.13 SMT production line. Stations include board launching, screen printing of solder paste, several component placement operations, and solder reflow oven. (Photo courtesy of Universal Instruments Corp.)

FIGURE 35.14 Typical process sequence for combined SMT-PIH assemblies with components on same side of board: (1) apply solder paste on lands for SMT components, (2) place SMT components on the board, (3) bake, (4) reflow solder, (5) insert PIH components, and (6) wave solder PIH components. This would be followed by cleaning, testing, and rework.

35.4.4 CLEANING, INSPECTION, TESTING, AND REWORK

After the components have been connected to the board, the assembly must be cleaned, inspected for solder faults, circuit tested, and reworked if necessary.

Inspection of soldering quality is somewhat more difficult for surface-mounted circuits (SMCs) because these assemblies are generally more densely packed, the solder joints are smaller, and their geometries are different from joints in through-hole assemblies. One of the problems is the way SMCs are held in place during soldering. In PIH assembly, the components are mechanically fastened in place by clinched leads. In SMT assembly, components are held by adhesive or paste. At soldering temperatures this method of attachment is not as secure, and component shifting sometimes occurs. Another problem with the smaller sizes in SMT is a greater likelihood of solder bridges forming between adjacent leads, resulting in short circuits.

The smaller scale also poses problems in SMT circuit testing because less space is available around each component. Contact probes must be physically smaller and more probes are required because SMT assemblies are more densely populated. One way of dealing with this issue is to design the circuit layout with extra lands whose only purpose is to provide a test probe contact site. Unfortunately, including these test lands runs counter to the goal of achieving higher packing densities on the board.

Manual rework in surface-mount assemblies is more difficult than in conventional PIH assemblies, again due to the smaller component sizes. Special tools are required, such as small-bit soldering irons, magnifying devices, and instruments for grasping and manipulating the small parts.

35.5 ELECTRICAL CONNECTOR TECHNOLOGY

PCB assemblies must be connected to back planes, and into racks and cabinets, and these cabinets must be connected to other cabinets and systems by means of cables. The growing use of electronics in so many types of products has made electrical connections an important technology. The performance of any electronic system depends on the reliability of the individual connections linking the elements of the system together. In this section we examine connector technology that is usually applied at the third and higher levels of electronics packaging.

To begin, there are two basic methods of making electrical connections: (1) soldering and (2) pressure connections. Soldering was discussed in Section 31.2 and throughout the current chapter. It is the most widely used technology in electronics. ***Pressure connections*** are electrical connections in which mechanical forces are used to establish electrical continuity between components. They can be divided into two types: permanent and separable.

35.5.1 PERMANENT CONNECTIONS

A permanent connection involves high-pressure contact between two metal surfaces, in which one or both of the parts is mechanically deformed during the assembly process. Permanent connection methods include crimping, press fit technology, and insulation displacement.

Crimping of Connector Terminals   This connection method is used to assemble wire to electrical terminals. Although assembly of the wire to the terminal forms a permanent joint, the terminal itself is designed to be connected and disconnected to its mating component. There are a variety of terminal styles, some of which are shown in Figure 35.15, and they are available in various sizes. They all must be connected to conductor wire, and crimping is the operation for doing this. ***Crimping*** involves the mechanical deformation of the terminal barrel to form a permanent connection with the stripped end of a wire inserted into it. The crimping operation squeezes and closes the barrel around the bare wire. Crimping is performed by hand tools or crimping machines. The terminals are supplied either as individual pieces or on long strips that can be fed into a crimping machine. Properly accomplished, the crimped joint will have low electrical resistance and high mechanical strength.

Press Fit Technology   Press fit in electrical connections is similar to that in mechanical assembly, but the part configurations are different. Press fit technology is widely used in the electronics industry to assemble terminal pins into metal-plated through-holes in large PCBs. In that context, a ***press fit*** involves an interference fit between the terminal pin and the plated hole into which it has been inserted. There are two categories of terminal pins: (a) solid and (b) compliant, as in Figure 35.16. Within these categories, pin designs vary among suppliers. The solid pin is rectangular in cross section and is designed so that its corners press and even cut into the metal of the plated hole to form a good electrical connection. The compliant pin is designed as a spring-loaded device that conforms to the hole contour but presses against the walls of the hole to achieve electrical contact.

Insulation Displacement   Insulation displacement is a method of making a permanent electrical connection in which a sharp, prong-shaped contact pierces the insulation and squeezes against the wire conductor to form an electrical connection. The method is illustrated in Figure 35.17 and is commonly used to make simultaneous connections between multiple contacts and flat cable. The flat cable, called ***ribbon cable,*** consists of a number of parallel wires

FIGURE 35.15   Some of the terminal styles available for making separable electrical connections: (a) slotted tongue, (b) ring tongue, and (c) flanged spade.

(a)   (b)   (c)

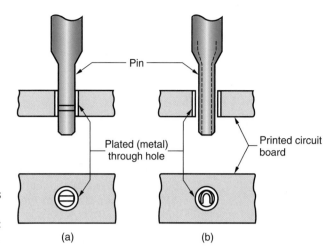

FIGURE 35.16 Two types of terminal pins in electronics press fit technology: (a) solid, and (b) compliant.

(a)   (b)

held in a fixed arrangement by the insulation surrounding them. It is often terminated in multiple pin connectors that are widely used in electronics to make electrical connections between major subassemblies. In these applications, the insulation displacement method reduces wiring errors and speeds harness assembly. To make the assembly, the cable is placed in a nest and a press is used to drive the connector contacts through the insulation and against the metal wires.

35.5.2 SEPARABLE CONNECTORS

Separable connections are designed to permit disassembly and reassembly; they are meant to be connected and disconnected multiple times. When connected they must provide metal-to-metal contact between mating components with high reliability and low electrical resistance. Separable connection devices typically consist of multiple contacts, contained in a plastic molded housing, designed to mate with a compatible connector or individual wires or terminals. They are used for making electrical connections between various combinations of cables, printed circuit boards, components, and individual wires.

A wide selection of connectors is available to serve many different applications. The design issues in choosing among them include (1) power level (e.g., whether the connector is used for power or signal transmission), (2) cost, (3) number of individual conductors involved, (4) types of devices and circuits to be connected, (5) space limitations, (6) ease of joining the connector to its leads, (7) ease of connecting with the mating terminal or connector, and (8) frequency of connection and disconnection. Some of the principal connector types are cable connectors, terminal blocks, sockets, and connectors with low or zero insertion force.

FIGURE 35.17
Insulation displacement method of joining a connector contact to flat wire cable: (1) starting position, (2) contacts pierce insulation, and (3) after connection.

(1)   (2)   (3)

FIGURE 35.18 Multiple pin connector and mating receptacle, both attached to cables. (Courtesy of AMP Inc., now a division of Tyco Industries.)

Cable connectors are devices that are permanently connected to cables (one or both ends) and are designed to be plugged into and unplugged from a mating connector. A power cord connector that plugs into a wall receptacle is a familiar example. Other styles include the type of multiple pin connector and mating receptacle shown in Figure 35.18, used to provide signal transmission between electronic subassemblies. Other multiple pin connector styles are used to attach printed circuit boards to other subassemblies in the electronic system.

Terminal blocks consist of a series of evenly spaced receptacles that allow connections between individual terminals or wires. The terminals or wires are often attached to the block by means of screws or other mechanical fastening mechanisms to permit disassembly. A conventional terminal block is illustrated in Figure 35.19.

A *socket* in electronics refers to a connection device mounted to a PCB, into which IC packages and other components can be inserted. Sockets are permanently attached to the PCB by soldering and/or press fitting, but they provide a separable connection method for the components, which can be conveniently added, removed, or replaced in the PCB assembly. Sockets are therefore an alternative to soldering in electronics packaging.

Insertion and withdrawal forces can be a problem in the use of pin connectors and PCB sockets. These forces increase in proportion to the number of pins involved. Possible damage can result when components with many contacts are assembled. This problem has motivated the development of connectors with *low insertion force* (LIF) or *zero insertion force* (ZIF), in which special mechanisms have been devised to reduce or eliminate the forces required to push the positive and negative connectors together and disconnect them.

FIGURE 35.19 Terminal block that uses screws to attach terminals. (Photo courtesy of AMP, Inc., now a division of Tyco Industries.)

REFERENCES

[1] Arabian, J. *Computer Integrated Electronics Manufacturing and Testing.* Marcel Dekker, New York, 1989.

[2] Bakoglu, H. B. *Circuits, Interconnections, and Packaging for VLSI.* Addison-Wesley, Reading, Massachusetts, 1990.

[3] Bilotta, A. J. *Connections in Electronic Assemblies.* Marcel Dekker, New York, 1985.

[4] Capillo, C. *Surface Mount Technology.* McGraw-Hill, New York, 1990.

[5] Coombs, C. F. Jr. (ed.). *Printed Circuits Handbook,* 6th ed. McGraw-Hill, New York, 2007.

[6] Edwards, P. R. *Manufacturing Technology in the Electronics Industry.* Chapman & Hall, London, 1991.

[7] Harper, C. *Electronic Materials and Processes Handbook,* 3rd ed. McGraw-Hill, New York, 2009.

[8] Kear, F. W. *Printed Circuit Assembly Manufacturing,* Marcel Dekker, New York, 1987.

[9] Lambert, L. P. *Soldering for Electronic Assemblies.* Marcel Dekker, New York, 1988.

[10] Marks, L. and Caterina, J. *Printed Circuit Assembly Design.* McGraw-Hill, New York, 2000.

[11] Prasad, R. P. *Surface Mount Technology: Principles and Practice,* 2nd ed. New York, Springer, 1997.

[12] Seraphim, D. P., Lasky, R., and Li, C-Y. (eds.). *Principles of Electronic Packaging.* McGraw-Hill, New York, 1989.

[13] Ulrich, R. K., and Brown, W. D. *Advanced Electronic Packaging.* 2nd ed. IEEE Press and John Wiley & Sons, Hoboken, New Jersey, 2006.

REVIEW QUESTIONS

35.1. What are the functions of a well-designed electronics package?

35.2. Identify the levels of packaging hierarchy in electronics.

35.3. What is the difference between a track and a land on a printed circuit board?

35.4. Define a printed circuit board (PCB).

35.5. Name the three principal types of printed circuit board.

35.6. What is a via hole in a printed circuit board?

35.7. What are the two basic methods by which the circuit pattern is transferred to the copper surface of the boards?

35.8. What is etching used for in PCB fabrication?

35.9. What is continuity testing, and when is it performed in the PCB fabrication sequence?

35.10. What are the two main categories of printed circuit board assemblies, as distinguished by the method of attaching components to the board?

35.11. What are some of the reasons and defects that make rework an integral step in the PCB fabrication sequence?

35.12. Identify some of the advantages of surface-mount technology over conventional through-hole technology.

35.13. Identify some of the limitations and disadvantages of surface-mount technology?

35.14. What are the two methods of component placement and soldering in surface-mount technology?

35.15. What is a solder paste?

35.16. Identify the two basic methods of making electrical connections.

35.17. Define crimping in the context of electrical connections.

35.18. What is press fit technology in electrical connections?

35.19. What is a terminal block?

35.20. What is a pin connector?

MULTIPLE CHOICE QUIZ

There are 14 correct answers in the following multiple choice questions (some questions have multiple answers that are correct). To attain a perfect score on the quiz, all correct answers must be given. Each correct answer is worth 1 point. Each omitted answer or wrong answer reduces the score by 1 point, and each additional answer beyond the correct number of answers reduces the score by 1 point. Percentage score on the quiz is based on the total number of correct answers.

35.1. The second level of packaging refers to which one of the following: (a) component to printed circuit board, (b) IC chip to package, (c) intraconnections on the IC chip, or (d) wiring and cabling connections?

35.2. Surface-mount technology is included within which one of the following levels of packaging: (a) zeroth, (b) first, (c) second, (d) third, or (e) fourth?

35.3. Card-on-board (COB) packaging refers to which one of the following levels in the electronics packaging hierarchy: (a) zeroth, (b) first, (c) second, (d) third, or (e) fourth?

35.4. Which of the following polymeric materials is commonly used as an ingredient in the insulation layer of a printed circuit board (two correct answers): (a) copper, (b) E-glass, (c) epoxy, (d) phenolic, (e) polyethylene, and (f) polypropylene?

35.5. Typical thickness of the copper layer in a printed circuit board is which one of the following: (a) 0.100 inch, (b) 0.010 inch, (c) 0.001 inch, or (d) 0.0001 inch?

35.6. Photolithography is widely used in PCB fabrication. Which of the following is the most common resist type used in the processing of PCBs: (a) negative resists or (b) positive resists?

35.7. Which of the following plating processes has the higher deposition rate in PCB fabrication: (a) electroless plating or (b) electroplating?

35.8. In addition to copper, which one of the following is another common metal plated onto a PCB: (a) aluminum, (b) gold, (c) nickel, or (d) tin?

35.9. Which of the following are the soldering processes used to attach components to printed circuit boards in through-hole technology (two best answers): (a) hand soldering, (b) infrared soldering, (c) reflow soldering, (d) torch soldering, and (e) wave soldering?

35.10. In general, which of the following technologies results in greater problems during rework: (a) surface-mount technology, or (b) through-hole technology?

35.11. Which of the following electrical connection methods produce a separable connection (two correct answers): (a) crimping of terminals, (b) press fitting, (c) soldering, (d) terminal blocks, and (e) sockets?

36 MICROFABRICATION TECHNOLOGIES

Chapter Contents

An important trend in engineering design and manufacturing is the growth in the number of products and/or components of products whose features sizes are measured in microns (1 μm $= 10^{-3}$ mm $= 10^{-6}$ m). Several terms have been applied to these miniaturized items. The term *microelectromechanical systems* (MEMS) emphasizes the miniaturization of systems consisting of both electronic and mechanical components. The word *micromachines* is sometimes used for these devices. *Microsystem technology* (MST) is a more general term that refers to the products (not necessarily limited to electromechanical products) as well as the fabrication technologies to produce them. A related term is *nanotechnology,* which refers to even smaller products whose dimensions are measured in nanometers (1 nm $= 10^{-3}$ μm $= 10^{-9}$ m). Figure 36.1 indicates the relative sizes and other factors associated with these terms. We discuss microfabrication techniques in the current chapter and nanofabrication in Chapter 37.

36.1 MICROSYSTEM PRODUCTS

Designing products that are smaller and comprised of even smaller parts and subassemblies means less material usage, lower power requirements, greater functionality per unit space, and accessibility to regions that are forbidden to larger products. In most cases, smaller products should mean lower prices because less material is used; however, the price of a given product is influenced by the costs of research, development, and production, and how these costs can be spread over the number of units sold. The economies of scale that result in lower-priced products have not yet fully been realized in microsystems technology, except for a limited number of cases that we shall examine in this section.

Key: nm = nanometer, μm = micron or micrometer, mm = millimeter, m = meter

FIGURE 36.1   Terminology and relative sizes for microsystems and related technologies.

36.1.1 TYPES OF MICROSYSTEM DEVICES

Microsystem products can be classified by type of device (e.g., sensor, actuator) or by application area (e.g., medical, automotive). The device categories are as follows [1]:

➢ **Microsensors.** A sensor is a device that detects or measures some physical phenomenon such as heat or pressure. It includes a transducer that converts one form of physical variable into another form (e.g., a piezoelectric device converts mechanical force into electrical current) plus the physical packaging and external connections. Most microsensors are fabricated on a silicon substrate using the same processing technologies as those used for integrated circuits (Chapter 34). Microscopic-sized sensors have been developed for measuring force, pressure, position, speed, acceleration, temperature, flow, and a variety of optical, chemical, environmental, and biological variables. The term **hybrid microsensor** is often used when the sensing element (transducer) is combined with electronic components in the same device. Figure 36.2 shows a micrograph of a micro-accelerometer developed at Motorola Co.

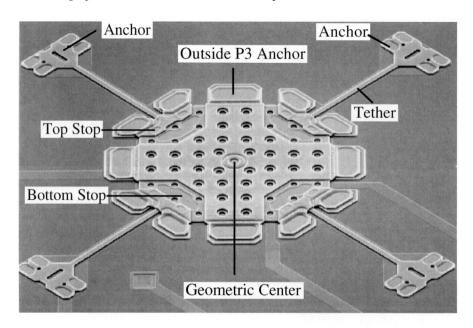

FIGURE 36.2
Micrograph of a micro-accelerometer. (Photo courtesy of A. A. Tseng, Arizona State University [4].)

FIGURE 36.3
A microscopic gear and a human hair. The image was made using a scanning electron microscope. The gear is high-density polyethylene molded by a process similar to the LIGA process (Section 36.3.3) except that the mold cavity was fabricated using a focused ion beam. (Photo courtesy of W. Hung, Texas A&M University, and M. Ali, Nanyang Technological University.)

➤ *Microactuators.* Like a sensor, an actuator converts a physical variable of one type into another type, but the converted variable usually involves some mechanical action (e.g., a piezoelectric device oscillating in response to an alternating electrical field). An actuator causes a change in position or the application of force. Examples of microactuators include valves, positioners, switches, pumps, and rotational and linear motors [1].

➤ *Microstructures and microcomponents.* These terms are used to denote a microsized part that is not a sensor or actuator. Examples of microstructures and microcomponents include microscopic gears, lenses, mirrors, nozzles, and beams. These items must be combined with other components (microscopic or otherwise) to provide a useful function. Figure 36.3 shows a microscopic gear alongside a human hair for comparison.

➤ *Microsystems and micro-instruments.* These terms denote the integration of several of the preceding components together with the appropriate electronics package into a miniature system or instrument. Microsystems and micro-instruments tend to be very application specific; for example, microlasers, optical chemical analyzers, and microspectrometers. The economics of manufacturing these kinds of systems have tended to make commercialization difficult.

36.1.2 MICROSYSTEM APPLICATIONS

The preceding microdevices and systems have been applied in a wide variety of fields. There are many problem areas that can be approached best using very small devices. Some important examples are the following:

Ink-Jet Printing Heads This is currently one of the largest applications of MST, because a typical ink-jet printer uses up several cartridges each year. The operation of an ink-jet printing head is depicted in Figure 36.4. An array of resistance heating elements is located above a corresponding array of nozzles. Ink is supplied by a reservoir and flows between the heaters and nozzles. Each heating element can be independently activated under microprocessor control in microseconds. When activated by a pulse of current, the liquid ink immediately beneath the heater boils to form a vapor bubble, forcing ink to be expelled through the nozzle opening. The ink hits the paper and dries almost immediately to form a dot that is part of an alphanumeric character or other image. Meanwhile, the vapor bubble collapses, drawing more ink from the reservoir to replenish the supply. Today's ink-jet printers possess resolutions of 1200 dots per inch (dpi), which converts to a nozzle separation of only about 21 μm, certainly in the microsystem range.

FIGURE 36.4 Diagram of an ink-jet printing head.

Thin-Film Magnetic Heads   Read-write heads are key components in magnetic storage devices. These heads were previously manufactured from horseshoe magnets that were manually wound with insulated copper wire. Because the reading and writing of magnetic media with higher-bit densities are limited by the size of the read-write head, hand-wound horseshoe magnets were a limitation on the technological trend toward greater storage densities. Development of thin-film magnetic heads at IBM Corporation was an important breakthrough in digital storage technology as well as a significant success story for microfabrication technologies. Thin-film read-write heads are produced annually in hundreds of millions of units, with a market of several billions of dollars per year.

A simplified sketch of the read-write head is presented in Figure 36.5, showing its MST parts. The copper conductor coils are fabricated by electroplating copper through a resist mold. The cross section of the coil is about 2 to 3 μm on a side. The thin-film cover, only a few μm thick, is made of nickel–iron alloy. The miniature size of the read-write head has permitted the significant increases in bit densities of magnetic storage media. The small sizes are made possible by microfabrication technologies.

Compact Discs   Compact discs (CDs) and digital versatile discs (DVDs)[1] are important commercial products today, as storage media for audio, video, games, and computer software

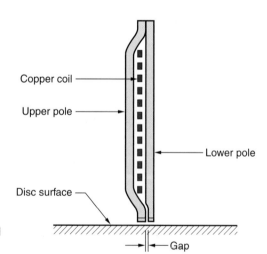

FIGURE 36.5   Thin-film magnetic read-write head (simplified).

---

[1]The DVD was originally called a digital video disc because its primary applications were motion picture videos. However, DVDs of various formats are now used for data storage and other computer applications, games, and high-quality audio.

and data applications. A CD disk is molded of polycarbonate (Section 8.2), which has ideal optical and mechanical properties for the application. The disk is 120 mm in diameter and 1.2 mm thick. The data consist of small pits (depressions) in a helical track that begins at a diameter of 46 mm and ends at about 117 mm. The tracks in the spiral are separated by about 1.6 μm. Each pit in the track is about 0.5 μm wide and about 0.8 μm to 3.5 μm long. These dimensions certainly qualify CDs as products of microsystem technology. The corresponding dimensions of DVDs are even smaller, permitting much higher data storage capacities.

Although most of the microfabrication processes are discussed in Section 36.2, let us briefly describe the production sequence for CDs here, because it is rather unique and uses several processes that are quite conventional. As consumer products, music CDs are mass-produced by plastic injection molding (Section 13.6). To make the mold, a master is created from a smooth, thin layer of positive photoresist coated onto a 300-mm diameter glass plate. A modulated laser beam writes the data onto the photoresist by exposing microscopic regions on the surface as the plate is rotated and moved slowly and precisely to create the spiral track. When the photoresist is developed, the exposed regions are removed. These regions in the master will correspond to the pits in the CD. A thin layer of nickel is then deposited onto the surface of the master by sputtering (Section 28.5.1). Electroforming (Section 28.3.2) is then used to build up the thickness of the nickel (to several mm), thus creating a negative impression of the master. This is called the "*father*". Several impressions are made of the father by the same electroforming process, in effect creating a negative impression of the father, whose surface geometry is identical to the original glass plate master. These impressions are called "*mothers*". Finally, the mothers are used to create the actual mold impressions (called "*stampers*"), again by electroforming, and these are used to mass-produce the CDs.[2] The process sequence is similar for DVDs but more involved because of the smaller scale and different data format requirements.

Once molded, the pitted side of the polycarbonate disk is coated with aluminum by sputtering to create a mirror surface. To protect this layer, a thin polymer coating (e.g., acrylic) is deposited onto the metal. Thus, the final compact disk is a sandwich with a relatively thick polycarbonate substrate on one side, a thin polymer layer on the other side, and in between a very thin layer of aluminum. In subsequent operation, the laser beam of a CD player (or other data reader) is directed through the polycarbonate substrate onto the reflective surface, and the reflected beam is interpreted as a sequence of binary digits.

Automotive   Microsensors and other microdevices are widely used in modern automotive products. Use of these microsystems is consistent with the increased application of on-board electronics to accomplish control and safety functions for the vehicle. The functions include electronic engine control, cruise control, anti-lock braking systems, air-bag deployment, automatic transmission control, power steering, all-wheel drive, automatic stability control, on-board navigation systems, and remote locking and unlocking, not to mention air conditioning and radio. These control systems and safety features require sensors and actuators, and a growing number of these are microscopic in size. There are currently 20 to 100 sensors installed in a modern automobile, depending on make and model. In 1970 there were virtually no on-board sensors. Some specific on-board microsensors are listed in Table 36.1.

Medical   Opportunities for using microsystems technology in this area are tremendous. Indeed, significant strides have already been made, and many of the traditional medical and

---

[2]The reason for the rather involved mold-making sequence is because the pitted surfaces of the impressions degrade after multiple uses. A father can be used to make three to six mothers, and each mother can be used to make three to six stampers, before their respective surfaces become degraded. A stamper (mold) can be used to produce only a few thousand disks, so if the production run is for several hundred thousand CDs, more than one stamper must be used during the run to produce all high-quality CDs.

TABLE 36.1   Microsensors installed in a modern automobile.

Microdevice	Application(s)
Accelerometer	Air-bag release, anti-lock brakes, active suspension system
Angular speed sensor	Intelligent navigation systems
Level sensors	Sense oil and gasoline levels
Optical sensor	Automatic headlight control
Position sensor	Transmission, engine timing,
Pressure sensors	Optimize fuel consumption, sense oil pressure, fluid pressures of hydraulic systems (e.g., suspension systems), lumbar seat support pressure, climate control, tire pressure
Proximity and range sensors	Sense distances from front and rear bumpers for parking control and collision prevention
Temperature sensors	Cabin climate control, engine management system
Torque sensor	Drive train

Compiled from [1] and [5].

surgical methods have already been transformed by MST. One of the driving forces behind the use of microscopic devices is the principle of minimal-invasive therapy, which involves the use of very small incisions or even available body orifices to access the medical problem of concern. Advantages of this approach over the use of relatively large surgical incisions include less patient discomfort, quicker recovery, fewer and smaller scars, shorter hospital stays, and lower health insurance costs.

Among the techniques based on miniaturization of medical instrumentation is the field of endoscopy,[3] now routinely used for diagnostic purposes and with growing applications in surgery. It is standard medical practice today to use endoscopic examination accompanied by laparoscopic surgery for hernia repair and removal of organs such as gall bladder and appendix. Growing use of similar procedures is expected in brain surgery, operating through one or more small holes drilled through the skull.

Other applications of MST in the medical field now include or are expected to include (1) angioplasty, in which damaged blood vessels and arteries are repaired using surgery, lasers, or miniaturized inflatable balloons at the end of a catheter that is inserted into the vein; (2) telemicrosurgery, in which a surgical operation is performed remotely using a stereo microscope and microscopic surgical tools; (3) artificial prostheses, such as heart pacemakers and hearing aids; (4) implantable sensor systems to monitor physical variables in the human body such as blood pressure and temperature; (5) drug delivery devices that can be swallowed by a patient and then activated by remote control at the exact location intended for treatment, such as the intestine, and (6) artificial eyes.

Chemical and Environmental   A principal role of microsystem technology in chemical and environmental applications is the analysis of substances to measure trace amounts of chemicals or detect harmful contaminants. A variety of chemical microsensors have been developed. They are capable of analyzing very small samples of the substance of interest. Micropumps are sometimes integrated into these systems so that the proper amounts of the substance can be delivered to the sensor component.

Other Applications   There are many other applications of microsystem technology beyond those described in the preceding. We list some examples in the following:

---

[3]The use of a small instrument (i.e., an endoscope) to visually examine the inside of a hollow body organ such as the rectum or colon.

> *Scanning probe microscope.* This is a technology for measuring microscopic details of surfaces, allowing surface structures to be examined at the nanometer level. In order to operate in this dimensional range, the instruments require probes that are only a few μm in length and that scan the surface at a distance measured in nm. These probes are produced using microfabrication techniques.[4]

> *Biotechnology.* In biotechnology, the specimens of interest are often microscopic in size. To study these specimens, manipulators and other tools are needed that are of the same size scale. Microdevices are being developed for holding, moving, sorting, dissecting, and injecting the small samples of biomaterials under a microscope.

> *Electronics.* Printed circuit board (PCB) and connector technologies are discussed in Chapter 35, but they should also be cited here in the context of MST. Miniaturization trends in electronics have forced PCBs, contacts, and connectors to be fabricated with smaller and more complex physical details, and with mechanical structures that are more consistent with the microdevices discussed in this chapter than with the integrated circuits discussed in Chapter 34.

36.2 MICROFABRICATION PROCESSES

Many of the products in microsystem technology are based on silicon, and most of the processing techniques used in the fabrication of microsystems are borrowed from the microelectronics industry. There are several important reasons why silicon is a desirable material in MST: (1) The microdevices in MST often include electronic circuits, so both the circuit and the microdevice can be fabricated in combination on the same substrate. (2) In addition to its desirable electronic properties, silicon also possesses useful mechanical properties, such as high strength and elasticity, good hardness, and relatively low density.[5] (3) The technologies for processing silicon are well-established, owing to their widespread use in microelectronics. (4) Use of single-crystal silicon permits the production of physical features to very close tolerances.

Microsystem technology often requires silicon to be fabricated along with other materials to obtain a particular microdevice. For example, microactuators often consist of several components made of different materials. Accordingly, microfabrication techniques consist of more than just silicon processing. Our coverage of the microfabrication processes is organized into three sections: (1) silicon layering processes, (2) the LIGA process, and (3) other processes accomplished on a microscopic scale.

36.2.1 SILICON LAYER PROCESSES

The first application of silicon in microsystems technology was in the fabrication of Si piezoresistive sensors for the measurement of stress, strain, and pressure in the early 1960s [5]. Silicon is now widely used in MST to produce sensors, actuators, and other microdevices. The basic processing technologies are those used to produce integrated circuits (Chapter 34). However, it should be noted that certain differences exist between the processing of ICs and the fabrication of the microdevices covered in this chapter:

1. The aspect ratios in microfabrication are generally much greater than in IC fabrication. *Aspect ratio* is defined as the height-to-width ratio of the features produced, as illustrated in Figure 36.6. Typical aspect ratios in semiconductor processing are about

---

[4]Scanning probe microscopes are discussed in Section 37.2.2.
[5]Silicon is discussed in Section 7.5.2.

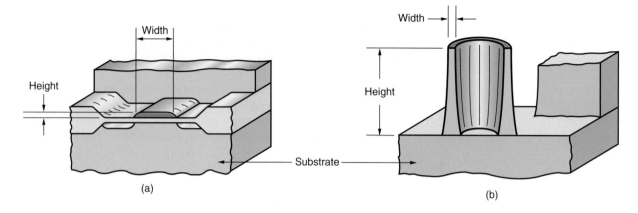

FIGURE 36.6 Aspect ratio (height-to-width ratio) typical in (a) fabrication of integrated circuits and (b) microfabricated components.

1.0 or less, whereas in microfabrication the corresponding ratio might be as high as 400 [5].

2. The sizes of the devices made in microfabrication are often much larger than in IC processing, where the prevailing trend in microelectronics is inexorably toward greater circuit densities and miniaturization.

3. The structures produced in microfabrication often include cantilevers and bridges and other shapes requiring gaps between layers. These kinds of structures are uncommon in IC fabrication.

4. The silicon processing techniques are sometimes supplemented to obtain a three-dimensional structure or other physical feature in the microsystem.

Notwithstanding these differences, let us nevertheless recognize that most of the silicon-processing steps used in microfabrication are the same or very similar to those used to produce ICs. After all, silicon is the same material whether it is used for integrated circuits or microdevices. The processing steps are listed in Table 36.2, together with brief descriptions and text references in which the reader can obtain more detailed descriptions. All of these process steps are discussed in previous chapters. As in IC fabrication, the various processes in Table 36.2 add, alter, or remove layers of material from a substrate according to geometric data contained in lithographic masks. Lithography is the fundamental technology that determines the shape of the microdevice being fabricated.

Regarding our preceding list of differences between IC fabrication and microdevice fabrication, the issue of aspect ratio should be addressed in more detail. The structures in IC processing are basically planar, whereas three-dimensional structures are more likely to be required in microsystems. The features of microdevices are likely to possess large height-to-width ratios. These 3-D features can be produced in single-crystal silicon by wet etching, provided the crystal structure is oriented to allow the etching process to proceed anisotropically. Chemical wet etching of polycrystalline silicon is isotropic, with the formation of cavities under the edges of the resist, as illustrated in Figure 34.13. However, in single-crystal Si, the etching rate depends on the orientation of the lattice structure. In Figure 36.7, the three crystal faces of silicon's cubic lattice structure are illustrated. Certain etching solutions, such as potassium hydroxide (KOH) and sodium hydroxide (NaOH), have a very low etching rate in the direction of the (111) crystal face. This permits the formation of distinct geometric structures with sharp edges in a single-crystal Si substrate whose lattice is oriented to favor etch penetration vertically or at sharp angles into the substrate. Structures

TABLE 36.2   Silicon layering processes used in microfabrication.

Process	Brief Description	Text Reference
Lithography	Printing process used to transfer copies of a mask pattern onto the surface of silicon or other solid material (e.g., silicon dioxide). The usual technique in microfabrication is photolithography.	Section 34.3
Thermal oxidation	(Layer addition) Oxidation of silicon surface to form silicon dioxide layer.	Section 34.4.1
Chemical vapor deposition	(Layer addition) Formation of a thin film on the surface of a substrate by chemical reactions or decomposition of gases.	Sections 28.5.2 and 34.4.2
Physical vapor deposition	(Layer addition) Family of deposition processes in which a material is converted to vapor phase and condensed onto a substrate surface as a thin film. PVD processes include vacuum evaporation and sputtering.	Section 28.5.1
Electroplating and electroforming	(Layer addition) Electrolytic process in which metal ions in solution are deposited onto a cathode work material.	Sections 28.3.1 and 28.3.2
Electroless plating	(Layer addition) Deposition in an aqueous solution containing ions of the plating metal with no external electric current. Work surface acts as catalyst for the reaction.	Section 28.3.3
Thermal diffusion (doping)	(Layer alteration) Physical process in which atoms migrate from regions of high concentration into regions of low concentration.	Sections 28.2.1 and 34.4.3
Ion implantation (doping)	(Layer alteration) Embedding atoms of one or more elements in a substrate using a high-energy beam of ionized particles.	Sections 28.2.2 and 34.4.3
Wet etching	(Layer removal) Application of a chemical etchant in aqueous solution to etch away a target material, usually in conjunction with a mask pattern.	Section 34.4.5
Dry etching	(Layer removal) Dry plasma etching using an ionized gas to etch a target material.	Section 34.4.5

such as those in Figure 36.8 can be created using this procedure. It should be noted that anisotropic wet etching is also desirable in IC fabrication (Section 34.4.5), but its consequence is greater in microfabrication because of the larger aspect ratios. The term **bulk micromachining** is used for the relatively deep wet etching process into single-crystal silicon substrate (Si wafer); whereas the term **surface micromachining** refers to the planar structuring of the substrate surface, using much more shallow layering processes.

Bulk micromachining can be used to create thin membranes in a microstructure. However, a method is needed to control the etching penetration into the silicon, so as to leave the membrane layer. A common method used for this purpose is to dope the silicon substrate with boron atoms, which significantly reduce the etching rate of the silicon. The

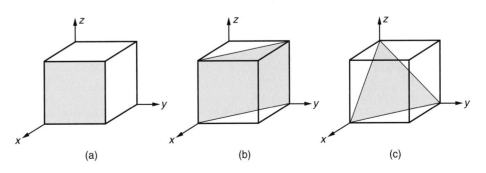

FIGURE 36.7 Three crystal faces in the silicon cubic lattice structure: (a) (100) crystal face, (b) (110) crystal face, and (c) (111) crystal face.

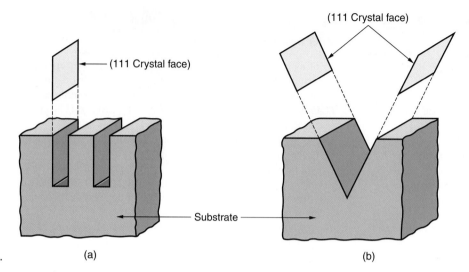

FIGURE 36.8 Several structures that can be formed in single-crystal silicon substrate by bulk micromachining: (a) (110) silicon and (b) (100) silicon.

processing sequence is shown in Figure 36.9. In step (2), epitaxial deposition is used to apply the upper layer of silicon so that it will possess the same single-crystal structure and lattice orientation as the substrate (Section 34.4.2). This is a requirement of bulk micromachining that will be used to provide the deeply etched region in subsequent processing. The use of boron doping to establish the etch resistant layer of silicon is called the ***p$^+$etch-stop technique.***

Surface micromachining can be used to construct cantilevers, overhangs, and similar structures on a silicon substrate, as shown in part (5) of Figure 36.10. The cantilevered beams in the figure are parallel to but separated by a gap from the silicon surface. Gap size and beam thickness are in the micron range. The process sequence to fabricate this type of structure is depicted in the earlier parts of Figure 36.10.

Dry etching, which involves material removal through the physical and/or chemical interaction between the ions in an ionized gas (plasma) and the atoms of a surface that has been exposed to the ionized gas (Section 34.4.5), provides anisotropic etching in almost any material. Its anisotropic penetration characteristic is not limited to a single-crystal silicon substrate. On the other hand, etch selectivity is more of a problem in dry etching; that is, any surfaces exposed to the plasma are attacked.

A procedure called the ***lift-off technique*** is used in microfabrication to pattern metals such as platinum on a substrate. These structures are used in certain chemical

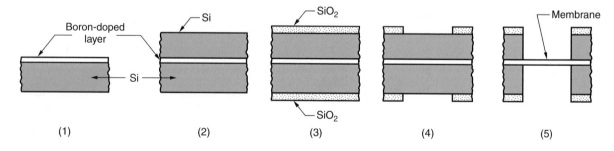

FIGURE 36.9 Formation of a thin membrane in a silicon substrate: (1) silicon substrate is doped with boron, (2) a thick layer of silicon is applied on top of the doped layer by epitaxial deposition, (3) both sides are thermally oxidized to form a SiO$_2$ resist on the surfaces, (4) the resist is patterned by lithography, and (5) anisotropic etching is used to remove the silicon except in the boron-doped layer.

FIGURE 36.10 Surface micromachining to form cantilevers: (1) on the silicon substrate is formed a silicon dioxide layer, whose thickness will determine the gap size for the cantilevered member; (2) portions of the SiO_2 layer are etched using lithography; (3) a polysilicon layer is applied; (4) portions of the polysilicon layer are etched using lithography; and (5) the SiO_2 layer beneath the cantilevers is selectively etched.

sensors, but are difficult to produce by wet etching. The processing sequence in the lift-off technique is illustrated in Figure 36.11.

36.2.2 LIGA PROCESS

LIGA is an important process in MST. It was developed in Germany in the early 1980s. The letters ***LIGA*** stand for the German words ***LI***thographie (in particular, x-ray lithography, although other lithographic exposure methods are also used, such as ion beams in Figure 36.3), ***G***alvanoformung (translated electrodeposition or electroforming), and ***A***bformtechnik (plastic molding). The letters also indicate the LIGA processing sequence. These processing steps have each been described in previous sections of our book: x-ray lithography in Section 35.3.2; electrodeposition and electroforming in Sections 28.3.1 and 28.3.2, respectively; and plastic molding processes in Sections 13.6 and 13.7. Let us examine how they are integrated in LIGA technology.

The LIGA processing steps are illustrated in Figure 36.12. Let us elaborate on the brief description provided in the figure's caption: (1) A thick layer of (x-ray) radiation-sensitive resist is applied to a substrate. Layer thickness can range between several microns to centimeters, depending on the size of the part(s) to be produced. The common resist material used in LIGA is polymethylmethacrylate (PMMA, Section 8.2.2 under "Acrylics"). The substrate must be a conductive material for the subsequent electro-deposition processes performed. The resist is exposed through a mask to high-energy x-ray radiation. (2) The irradiated areas of the positive resist are chemically removed from the substrate surface, leaving the unexposed portions standing as a three-dimensional plastic structure. (3) The regions where the resist has been removed are filled with metal using electrodeposition. Nickel is the common plating metal used in LIGA. (4) The remaining resist structure is stripped (removed), yielding a three-dimensional metal structure.

FIGURE 36.11 The lift-off technique: (1) resist is applied to substrate and structured by lithography; (2) platinum is deposited onto surfaces; and (3) resist is removed, taking with it the platinum on its surface but leaving the desired platinum microstructure.

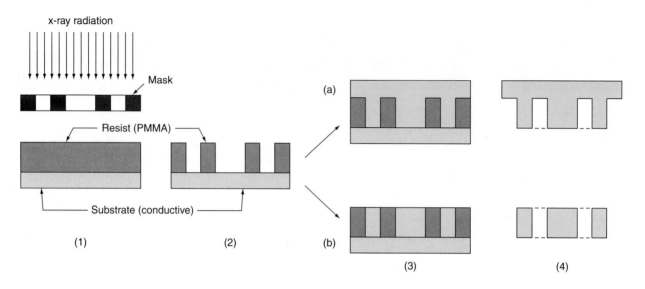

FIGURE 36.12   LIGA processing steps: (1) thick layer of resist applied and x-ray exposure through mask, (2) exposed portions of resist removed, (3) electrodeposition to fill openings in resist, (4) resist stripped to provide (a) a mold or (b) a metal part.

Depending on the geometry created, this metallic structure may be (a) the mold used for producing plastic parts by injection molding, reaction injection molding, or compression molding. In the case of injection molding, in which thermoplastic parts are produced, these parts may be used as ''lost molds'' in investment casting (Section 11.2.4). Alternatively, (b) the metal part may be a pattern for fabricating plastic molds that will be used to produce more metallic parts by electrodeposition.

As our description indicates, LIGA can produce parts by several different methods. This is one of the greatest advantages of this microfabrication process: (1) LIGA is a versatile process. Other advantages include (2) high aspect ratios are possible—large height-to-width ratios in the fabricated part; (3) wide range of part sizes is feasible, with heights ranging from micrometers to centimeters; and (4) close tolerances can be achieved. A significant disadvantage of LIGA is that it is a very expensive process, so large quantities of parts are usually required to justify its application. Also, the required use of x-ray radiation is a disadvantage.

36.2.3   OTHER MICROFABRICATION PROCESSES

MST research is providing several additional fabrication techniques, most of which are variations of lithography or adaptations of macro-scale processes. In this section we discuss several of these additional techniques.

Soft Lithography   This term is used for processes that use an elastomeric flat mold (similar to a rubber ink stamp) to create a pattern on a substrate surface. The sequence for creating the mold is illustrated in Figure 36.13. A master pattern is fabricated on a silicon surface using one of the lithography processes such as UV photolithography or electron beam lithography. This master pattern is then used to produce the flat mold for the soft lithography process. The common mold material is polydimethylsiloxane (PDMS, a silicon rubber, Section 8.4.3). After the PDMS has cured, it is peeled away from the pattern and attached to a substrate for support and handling.

FIGURE 36.13   Steps in mold-making for soft lithography: (1) master pattern fabricated by traditional lithography, (2) polydimethylsiloxane flat mold is cast from the master pattern, and (3) cured flat mold is peeled off pattern for use.

Two of the soft lithography processes are micro-imprint lithography and micro-contact printing. In ***micro-imprint lithography,*** the mold is pressed into the surface of a soft resist to displace the resist away from certain regions of the substrate for subsequent etching. The process sequence is illustrated in Figure 36.14. The flat mold consists of raised and depressed regions, and the raised regions correspond to areas on the resist surface that will be displaced to expose the substrate. The resist material is a thermoplastic polymer that has been softened by heating before pressing. The alteration of the resist layer is by mechanical deformation rather than electromagnetic radiation, as in the more traditional lithography methods. The compressed regions of the resist layer are subsequently removed by anisotropic etching (Section 34.4.5). The etching process also reduces the thickness of the remaining resist layer, but enough remains to protect the substrate for subsequent processing. Micro-imprint lithography can be set up for high production rates at modest cost. A mask is not required in the imprint procedure, although the mold requires an analogous preparation.

The same type of flat stamp can be used in a printing mode, in which case the process is called ***micro-contact printing.*** In this form of soft lithography, the mold is used to transfer a pattern of a substance to a substrate surface, much like ink can be transferred to a paper surface. This process allows very thin layers to be fabricated onto the substrate.

Nontraditional and Traditional Processes in Microfabrication   A number of nontraditional machining processes (Chapter 26), as well as conventional manufacturing processes, are important in microfabrication. ***Photochemical machining*** (PCM, Section 26.4.2) is an essential process in IC processing and microfabrication, but we have referred to it in our descriptions here and in Chapter 34 as wet chemical etching (combined with photolithography). PCM is often used with conventional processes of ***electroplating, electroforming,*** and/or ***electroless plating*** (Section 28.3) to add layers of metallic materials according to microscopic pattern masks.

FIGURE 36.14   Steps in micro-imprint lithography: (1) mold positioned above and (2) pressed into resist, (3) mold is lifted, and (4) remaining resist is removed from substrate surface in defined regions.

FIGURE 36.15 Ultra-high-precision milling of grooves in aluminum foil.

Other nontraditional processes capable of micro-level processing include [5]: (1) *electric discharge machining,* used to cut holes as small as 0.3 mm in diameter with aspect ratios (depth-to-diameter) as high as 100; (2) *electron-beam machining,* for cutting holes of diameter smaller than 100 μm in hard-to-machine materials; (3) *laser-beam machining,* which can produce complex profiles and holes as small as 10 μm in diameter with aspect ratios (depth-to-width or depth-to-diameter) approaching 50; (4) *ultrasonic machining,* capable of drilling holes in hard and brittle materials as small as 50 μm in diameter; and (5) *wire electric discharge cutting,* or *wire-EDM,* which can cut very narrow swaths with aspect ratios (depth-to-width) greater than 100.

Trends in conventional machining have included its capabilities for taking smaller and smaller cut sizes and associated tolerances. Referred to as *ultra-high-precision machining,* the enabling technologies have included single-crystal diamond cutting tools and position control systems with resolutions as fine as 0.01 μm [5]. Figure 36.15 depicts one reported application, the milling of grooves in aluminum foil using a single-point diamond fly-cutter. The aluminum foil is 100 μm thick, and the grooves are 85 μm wide and 70 μm deep. Similar ultra-high-precision machining is being applied today to produce products such as computer hard discs, photocopier drums, mold inserts for compact disk reader heads, and high-definition TV projection lenses.

Rapid Prototyping Technologies Several rapid prototyping (RP) methods (Chapter 33) have been adapted to produce micro-sized parts [7]. RP methods use a layer additive approach to build three-dimensional components, based on a CAD (computer-aided design) geometric model of the component. Each layer is very thin, typically as low as 0.05 mm thick, which approaches the scale of microfabrication technologies. By making the layers even thinner, microcomponents can be fabricated.

One approach is called *electrochemical fabrication* (EFAB), which involves the electrochemical deposition of metallic layers in specific areas that are determined by pattern masks created by "slicing" a CAD model of the object to be made (Section 33.1). The deposited layers are generally 5 to 10 μm thick, with feature sizes as small as 20 μm in width. EFAB is carried out at temperatures below 60°C (140°F) and does not require a clean room environment. However, the process is slow, requiring about 40 minutes to apply each layer, or about 36 layers (a height between 180 and 360 μm) per 24-hour period. To overcome this disadvantage, the mask for each layer can contain multiple copies of the part slice pattern, permitting many parts to be produced simultaneously in a batch process.

Another RP approach, called ***microstereolithography,*** is based on stereolithography (STL, Section 33.2.1), but the scale of the processing steps is reduced in size. Whereas the layer thickness in conventional stereolithography ranges between 75 μm and 500 μm, microstereolithography (MSTL) uses layer thicknesses between 10 and 20 μm typically, with even thinner layers possible. The laser spot size in STL is typically around 250 μm in diameter, whereas MSTL uses a spot size as small as 1 or 2 μm. Another difference in MSTL is that the work material is not limited to a photosensitive polymer. Researchers report success in fabricating 3-D microstructures from ceramic and metallic materials. The difference is that the starting material is a powder rather than a liquid.

Photofabrication   This term applies to an industrial process in which ultraviolet exposure through a pattern mask causes a significant modification in the chemical solubility of an optically clear material. The change is manifested in the form of an increase in solubility to certain etchants. For example, hydrofluoric acid etches the UV-exposed photosensitive glass between 15 and 30 times faster than the same glass that has not been exposed. Masking is not required during etching, the difference in solubility being the determining factor in which portions of the glass are removed.

Origination of photofabrication actually preceded the microprocessing of silicon. Now, with the growing interest in microfabrication technologies, there is a renewed interest in the older technology. Examples of modern materials used in photofabrication include Corning Glass Works' Fotoform™ glasses and Fotoceram™ ceramics, and DuPont's Dycril and Templex photosensitive solid polymers. When processing these materials, aspect ratios of around 3:1 can be obtained with the polymers and 20:1 with the glasses and ceramics.

REFERENCES

[1] Fatikow, S., and Rembold, U. ***Microsystem Technology and Microrobotics.*** Springer-Verlag, Berlin, 1997.

[2] Hornyak, G. L., Moore, J. J., Tibbals, H. F., and Dutta, J. ***Fundamentals of Nanotechnology,*** CRC Taylor & Francis, Boca Raton, Florida, 2009.

[3] Jackson, M. J. ***Micro and Nanomanufacturing,*** Springer, New York, 2007.

[4] Li, G., and Tseng, A. A. "Low Stress Packaging of a Micromachined Accelerometer," ***IEEE Transactions on Electronics Packaging Manufacturing,*** Vol. 24, No. 1, January 2001, pp. 18–25.

[5] Madou, M. ***Fundamentals of Microfabrication.*** CRC Press, Boca Raton, Florida, 1997.

[6] Madou, M. ***Manufacturing Techniques for Microfabrication and Nanotechnology.*** CRC Taylor & Francis, Boca Raton, Florida, 2009.

[7] O'Connor, L., and Hutchinson, H. "Skyscrapers in a Microworld," ***Mechanical Engineering,*** Vol. 122, No. 3, March 2000, pp. 64–67.

[8] National Research Council (NRC). ***Implications of Emerging Micro- and Nanotechnologies.*** Committee on Implications of Emerging Micro- and Nanotechnologies, The National Academies Press, Washington, D.C., 2002.

[9] Paula, G. "An Explosion in Microsystems Technology," ***Mechanical Engineering,*** Vol. 119, No. 9, September 1997, pp. 71–74.

[10] Tseng, A. A., and Mon, J-I. "NSF 2001 Workshop on Manufacturing of Micro-Electro Mechanical Systems," in ***Proceedings of the 2001 NSF Design, Service, and Manufacturing Grantees and Research Conference,*** National Science Foundation, 2001.

REVIEW QUESTIONS

36.1. Define microelectromechanical system.

36.2. What is the approximate size scale in microsystem technology?

36.3. Why is it reasonable to believe that microsystem products would be available at lower costs than products of larger, more conventional size?

36.4. What is a hybrid microsensor?

36.5. What are some of the basic types of microsystem devices?

36.6. Name some products that represent microsystem technology.

36.7. Why is silicon a desirable work material in micro-system technology?

36.8. What is meant by the term *aspect ratio* in micro-system technology?

36.9. What is the difference between bulk micromachining and surface micromachining?

36.10. What are the three steps in the LIGA process?

MULTIPLE CHOICE QUIZ

There are 14 correct answers in the following multiple choice questions (some questions have multiple answers that are correct). To attain a perfect score on the quiz, all correct answers must be given. Each correct answer is worth 1 point. Each omitted answer or wrong answer reduces the score by 1 point, and each additional answer beyond the correct number of answers reduces the score by 1 point. Percentage score on the quiz is based on the total number of correct answers.

36.1. Microsystem technology includes which of the following (three best answers): (a) LIGA technology, (b) microelectromechanical systems, (c) micromachines, (d) nanotechnology, (e) and precision engineering?

36.2. Which of the following are current applications of microsystem technology in modern automobiles (three best answers): (a) air-bag release sensors, (b) alcohol blood level sensors, (c) driver identification sensors for theft prevention, (d) oil pressure sensors, and (e) temperature sensors for cabin climate control?

36.3. The polymer used to make compact discs (CDs) and digital versatile discs (DVDs) is which one of the following: (a) amino resin, (b) epoxy resin, (c) polyamides, (d) polycarbonate, (e) polyethylene, or (f) polypropylene?

36.4. The most common work material used in microsystem technology is which one of the following: (a) boron, (b) gold, (c) nickel, (d) potassium hydroxide, or (e) silicon?

36.5. The aspect ratio in microsystem technology is best defined by which one of the following: (a) degree of anisotropy in etched features, (b) height-to-width ratio of the fabricated features, (c) height-to-width ratio of the MST device, (d) length-to-width ratio of the fabricated features, or (e) thickness-to-length ratio of the MST device?

36.6. Which of the following forms of radiation have wavelengths shorter than the wavelength of ultraviolet light used in photolithography (two correct answers): (a) electron beam radiation, (b) natural light, and (c) x-ray radiation?

36.7. Bulk micromachining refers to a relatively deep wet etching process into a single-crystal silicon substrate: (a) true or (b) false?

36.8. In the LIGA process, the letters LIGA stand for which one of the following: (a) let it go already, (b) little grinding apparatus, (c) lithographic applications, (d) lithography, electrodeposition, and plastic molding, or (e) lithography, grinding, and alteration?

36.9. Photofabrication is the same process as photolithography. (a) true or (b) false?

37 NANOFABRICATION TECHNOLOGIES

Chapter Contents

The trend in miniaturization is continuing beyond the micrometer range into the nanometer (nm) scale. **Nanotechnology** refers to the fabrication and application of entities whose feature sizes range from less than 1 nm to 100 nm (1 nm $= 10^{-3}$ μm $= 10^{-6}$ mm $= 10^{-9}$ m).[1] The entities include films, coatings, dots, lines, wires, tubes, structures, and systems. The prefix *nano* is used for these items; thus, we have new words such as nanotube, nanostructure, nanoscale, and nanoscience entering our vocabulary. **Nanoscience** is the field of scientific study that is concerned with objects in the same size range. **Nanoscale** refers to dimensions within this range and slightly below, which overlaps on the lower end with the sizes of atoms and molecules. For example, the smallest atom is Helium, with a diameter close to 0.1 nm. Uranium has a diameter of about 0.22 nm and is the largest of the naturally occurring atoms. Molecules tend to be larger because they consist of multiple atoms. Molecules made up of about 30 atoms are roughly 1 nm in size, depending on the elements involved. Thus, nanoscience involves the behavior of individual molecules and the principles that explain this behavior, and nanotechnology involves the application of these principles to create useful products.

In the previous chapter, we provided an overview of the products and devices in microsystem technology. Let us do the same for nanotechnology. What are the currently available and potential future products and materials? Nanotechnology involves not just a reduction in scale by three orders of magnitude. The science is different when the sizes of the entities approach the molecular and atomic levels. We discuss some of these differences in Section 37.2. Finally, in Section 37.3, we describe the major categories of fabrication processes used in nanotechnology.

---

[1]The dividing line between nanotechnology and microsystem technology (Chapter 36) is considered to be 100 nm $= 0.1$ μm [7]. This is illustrated approximately in Figure 36.1.

37.1   NANOTECHNOLOGY PRODUCTS

Most products in nanotechnology are not just smaller versions of microsystem technology (MST) products; they also include new materials, coatings, and unique entities that are not included within the scope of MST. Nanoscale products and processes that have been around for a while include the following:

➤ The colorful stained-glass windows of churches built during the Middle Ages were based on gold particles of nanometer scale embedded in the glass. Depending on their size, the particles can take on a variety of different colors.

➤ Modern photography has roots dating back more than 150 years and depends on the formation of silver nanoparticles to create the image in the photograph.

➤ Nanoscale particles of carbon are used as reinforcing fillers in automobile tires.

➤ Catalytic converters required in the exhaust systems of modern automobiles make use of nanoscale coatings of platinum and palladium on a ceramic honeycomb structure. The metal coatings act as catalysts to convert harmful emission gases into harmless gases.

We should also mention that the fabrication technology for integrated circuits now includes feature sizes that are in the nanotech range. Of course, integrated circuits have been produced since the 1960s, but only in recent years have nanoscale features been achieved.

Other more recent products exploiting applications of nanotechnology include cosmetics, sun lotions, car polishes and waxes, coatings for eyeglass lenses, and scratch-resistant paints. All of these categories contain nanoscale particles (nanoparticles), which qualifies them as products of nanotechnology. A more complete list of examples of present and future products and materials based on nanotechnology is presented in Table 37.1. For

TABLE 37.1   Examples of present and future products and materials that are based on nanotechnology.

Computers. Carbon nanotubes (Section 37.1.3) are strong candidates to substitute for silicon-based electronics as the limits of size reduction are approached in the lithography-based processes used to make integrated circuits on silicon wafers. These limits are expected to be reached around the year 2015.

Materials. Nanoscale particles (nanodots) and fibers (nanowires) may prove to be useful reinforcing agents for composite materials. For example, the truck bed for one of General Motors' Hummer vehicles is made with nanocomposites. Entirely new material systems, not known today, may be possible with nanotechnology.

Nanoparticle catalysts. Metal nanoparticles and coatings of noble metals (e.g., gold, platinum) on ceramic substrates act as catalysts for certain chemical reactions. Catalytic converters in automobiles are an important example.

Cancer drugs. Nanoscale drugs are being developed that will be designed to match the specific genetic profile of the cancer cells of a patient and to attack and destroy the cells. For example, Abraxine is a nanoscale protein-based medicine produced by American Pharmaceutical that is used to treat metastatic breast cancer.

Solar energy. Nanoscale surface films have the potential to absorb more of the sun's electromagnetic energy than existing photovoltaic receptacles. Developments in this area may reduce our reliance on fossil fuels for power generation.

Coatings. Nanoscale coatings and ultra-thin films are being developed that will increase scratch resistance of surfaces (eyeglass lenses with such coatings are already available), stain resistance of fabrics, and self-cleaning capabilities for windows and other surfaces (the "lotus effect").

Flat screen displays for television and computer monitors. TV screens based on carbon nanotubes are being introduced. They are expected to be brighter, less expensive, and more energy efficient than current displays. They will be produced by Samsung Electronics of South Korea.

Portable medical laboratories. Instruments based on nanotechnology will provide fast analysis of a variety of ailments such as diabetes and HIV.

Batteries. Carbon nanotubes may be future components in high-powered batteries and storage devices for hydrogen. Hydrogen storage will no doubt play a role in converting from fossil-fuel motors to hydrogen-based engines.

Light sources. Lamps are being developed based on nanotechnology that use a fraction of the energy of an incandescent light bulb and never burn out.

Based mostly on [1] and [24].

an extensive list of nanotech products, the interested reader can consult www.nano-techproject.org/inventories/consumer [27].

An important product category in microsystem technology is microelectrome-chanical systems (MEMS), which have found quite a few applications in the computer, medical, and automotive industries (Section 36.1.2). With the advent of nanotechnology, there has been growing interest in the notion of extending the development of these kinds of devices into the nanoscale range. *Nanoelectromechanical systems* (NEMS) are the sub-micron sized counterparts of MEMS devices, only their smaller sizes would result in even greater potential advantages. An important NEMS structural product currently produced is the probe used in atomic force microscopes (Section 37.2.2). The sharp point on the probe is of nanoscale size. Nanosensors are another developing application. Nanosensors would be more accurate, faster responding, and operate with lower power requirements than larger sensors. Current NEMS sensor applications include accelerometers and chemical sensors. It has been suggested that multiple nanosensors could be distributed throughout the subject area to collect data, thus providing the benefit of multiple readings of the variable of interest, rather than using a single larger sensor at one location.

Formidable technical problems arise in applications of *nanomachines,* defined as nanosystems consisting of movable parts and at least two different materials [7]. The problems result from the fact that the part surfaces cannot be made smooth at the atomic and molecular sizes. Other surface characteristics also come into play, as discussed in Section 37.2.1.

37.1.1 CARBON NANOSTRUCTURES

Two structures of significant scientific and commercial interest in nanotechnology are carbon buckyballs and nanotubes. They are basically graphite layers that have been formed into spheres and tubes, respectively.

The name *buckyball* refers to the molecule C_{60}, which contains exactly 60 carbon atoms and is shaped like a soccer ball, as in Figure 37.1. The original name of the molecule was *buckministerfullerene,* after the architect/inventor R. Buckminister Fuller, who designed the geodesic dome that resembles the C_{60} structure. Today, C_{60} is simply called a *fullerene,* which refers to any closed hollow carbon molecules that consist of 12 pentagonal and various numbers of hexagonal faces. In the case of

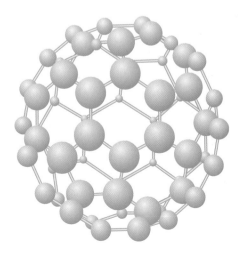

FIGURE 37.1 Fullerine structure of the C_{60} molecule. (Reprinted by permission from [17].)

C_{60}, the 60 atoms are arranged symmetrically into 12 pentagonal faces and 20 hexagonal faces to form a ball. These molecular balls can be bonded together by van der Waals forces (Section 2.2) to form crystals whose lattice structure is face-centered cubic (Figure 2.8(b), Section 2.3.1). The separation between any molecule and its closest neighbor in the C_{60} lattice structure is 1 nm.

Fullerenes are of interest for a number of reasons. One is their electrical properties and the capability to alter these properties. A C_{60} crystal has the properties of an insulator. However, when doped with an alkaline metal such as potassium (forming K_3C_{60}), it is transformed into an electrical conductor. Moreover, it exhibits properties of a super-conductor at temperatures of around 18°K. Another potential application area for the C_{60} fullerenes is in the medical field. The C_{60} molecule has many possible attachment points for focused drug treatments. Other possible medical applications for buckyballs include antioxidants, burn creams, and diagnostic imaging.

Carbon nanotubes (CNTs) are another molecular structure consisting of carbon atoms bonded together in the shape of a long tube. The atoms can be arranged into a number of alternative configurations, three of which are illustrated in Figure 37.2. The nanotubes shown in the figure are all single-walled nanotubes (SWNT), but multi-walled structures (MWNT) can also be fabricated, which are tubes within a tube. A SWNT has a typical diameter of a few nanometers (down to 1 nm) and a length of around 100 nm, and it is closed at both ends.

The electrical properties of nanotubes are unusual. Depending on the structure and diameter, nanotubes can have metallic (conducting) or semiconducting properties. Conductivity of metallic nanotubes can be superior to that of copper by six orders of magnitude [7]. The explanation for this is that nanotubes contain few of the defects existing in metals that tend to scatter electrons, thus increasing electrical resistance. Because nanotubes have such low resistance, high currents do not increase their

FIGURE 37.2 Several possible structures of carbon nanotubes: (a) arm-chair, (b) zigzag, and (c) chiral. (Reprinted by permission from [17].)

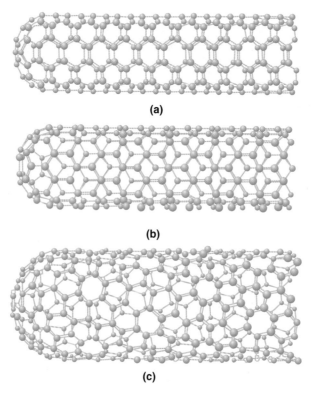

(a)

(b)

(c)

temperature the way metals heat up under the same electrical loads. Thermal conductivity of metallic nanotubes is also very high. These electrical and thermal properties are of significant interest to manufacturers of computers and integrated circuits because they may allow higher clock speeds of processors without the heat buildup problems currently encountered as the density of components on a silicon chip increases. Clock speeds 10^4 times faster than current-day processors may be possible [17], along with much higher densities.

Another electrical property of carbon nanotubes is field emission, in which electrons are emitted from the ends of the tubes at very high rates when an electrical field is applied parallel to the axis of a nanotube. The possible commercial applications of field emission properties of nanotubes include flat panel displays for televisions and computer monitors.

Mechanical properties are another reason for the interest in single-walled nanotubes. Compared with steel, density is only 1/6, modulus of elasticity is five times higher, and tensile strength is 100 times greater [7]. Yet, when SWNTs are bent, they exhibit great resilience to return to their previous shape without damage. These mechanical properties present opportunities for using them in applications ranging from reinforcing materials in polymer matrix composites (Section 9.4) to fiber cloths in bulletproof vests. Ironically, multi-walled nanotubes are not as strong.

37.1.2 THE NATIONAL NANOTECHNOLOGY INITIATIVE

In the year 2000, a national initiative on nanotechnology was enacted by the U.S. Congress at a funding level of $400 million starting in 2001. Funding levels have increased in what is now called the National Nanotechnology Initiative (NNI). A total of $3.7 billion was allocated over the 4-year period starting in 2005, making it the largest federally funded R&D program since the Apollo Space Program. The NNI Act mandated the coordination of nanotechnology research and development activities in the various federal agencies that are involved in this technology, including the Departments of Defense and Energy, the National Science Foundation, National Institutes of Health, National Institute of Standards and Technology, and the National Aeronautics and Space Administration. In addition, the Act defined nine areas of nanotechnology development (referred to as the NNI Grand Challenges) that will affect the lives of virtually all U.S. citizens. Table 37.2 briefly describes the nine areas of nanotechnology development to provide an overview of the future opportunities envisioned for this technology.

37.2 INTRODUCTION TO NANOSCIENCE

The fields of nanoscience and nanotechnology are interdisciplinary. They rely on the synergistic contributions of chemistry, physics, various engineering disciplines, and computer science. The fields of biology and medical science are also involved. Biology operates in the nanoscale range. Proteins, basic substances in living organisms, are large molecules ranging in size between about 4 nm and 50 nm. Proteins are made up of amino acids (organic acids containing the amino group NH_2), whose molecular size is about 0.5 nm. Each protein molecule consists of combinations of various amino acid molecules[2] connected together to form a long chain (a nanowire). This long macromolecule twists and turns to compact itself into a mass with a cross section in the 4- to 50-nm range. Other

---

[2]There are more than 100 different amino acids that occur naturally, but most of the proteins found in living organisms consist of only 20 of these amino acid types.

TABLE 37.2   Nine areas of nanotechnology development identified in the National Nanotechnology Initiative (NNI).
Nanostructured materials by design. The objective is to develop materials that are stronger, harder, lighter, safer, and smarter; and to also devise materials that possess self-repairing characteristics. The research will focus on (1) understanding the relationships between a material's nanostructure and its macroscopic properties and (2) development of new methods of fabrication and measurement.
Nanoelectronics, optoelectronics, and magnetics. The objectives include developing new devices and fabrication technologies in these areas for integration into existing systems and new architectures (e.g., new circuit architectures to address the limits of present trends in silicon-based integrated circuit fabrication technologies).
Advanced health care, therapeutics, and diagnosis. The objectives are to (1) improve health of humans by the development of new biosensors and medical imaging technologies, (2) develop nano-based devices that can be used to direct the delivery of medications to targeted sites in the human body, (3) improve biological implants by means of nanoscale processing of the implant interface with the bone, (4) develop nanoscale-based devices to enable sight and hearing, and (5) devise improved diagnostic techniques using gene sequencing methods.
Nanoscaled processes for environmental improvement. The objectives are to (1) find new methods to measure pollutants based on nanotechnology, (2) develop new ways of removing submicroscopic pollutants from the air and water, and (3) extend scientific knowledge about nanoscale phenomena that are important to maintaining environmental quality and reducing undesirable emissions.
Efficient energy conversion and storage. The objectives include developing (1) more efficient energy sources using nanocrystal catalysts, (2) more efficient solar cells, (3) efficient photoactive materials for solar conversion of materials into fuels, and (4) high-efficiency light sources. Additional activities include exploring the use of carbon nanotubes for high-density storage of hydrogen and improving the efficiency of heat exchangers using fluids with suspended nanocrystalline particles.
Microcraft space exploration and industrialization. The objectives are to (1) reduce the size of spacecraft by an order of magnitude, (2) use the light weight and high strength of nanostructured materials to reduce fuel consumption, (3) enable autonomous decision making and increased data storage by means of nanoelectronics and nanomagnetics, and (4) use self-repairing materials to extend the reach of space exploration.
Bionanosensor devices for communicable disease and biological threat detection. The objectives include (1) improving detection of and response to threats from chemical and biological warfare and from human disease, (2) increasing human capabilities and improving health by means of nanoscale devices, and (3) performing research on the compatibility between nanoscale materials and living tissue.
Application to economical and safe transportation. The objectives include developing (1) more efficient transportation modes using nanomaterials that are lighter and have lower failure rates, (2) more durable materials for roads and bridges, (3) smart materials and devices capable of detecting imminent failure and performing self-repair processes, (4) nanoscale coatings with low friction and low corrosion properties, and (5) nanoscale performance sensors.
National security. The general objective is to achieve military dominance at lower cost and manpower, and to reduce the risks of personnel engaged in combat. Proposed research and development activities include (1) improving knowledge superiority by increasing processor speed, storage capacity, access speed, display technology, and communications capability, (2) use of materials with better properties for military systems, and (3) sensor technologies to protect combat personnel and enhance their fighting capabilities.

Compiled from [14].

biological entities of nanoscale size include chlorophyll molecules in plants (about 1 nm), hemoglobin in blood (7 nm), and flu viruses (60 nm). Biological cells are orders of magnitude larger. For example, a red blood cell is disk-shaped with diameter of about 8000 nm (8 μm) and thickness of about 1500 nm (1.5 μm). The diameter of the human hair shown in Figure 36.3 is approximately 100,000 nm (0.1 mm).

Our focus in this chapter is on nanoscale entities that are nonbiological. As in biology, nanotechnology deals with objects that are not much bigger than the atoms and molecules that comprise them. In Section 37.2.1, we discuss these "size effects" and how material properties are affected when the dimensions of an entity are measured in nanometers. The inability to "see" nanoscale objects has inhibited developments in nanotechnology until recently. The advent of scanning probe microscopes in the 1980s has allowed objects at the molecular level to be visualized and measured. These types of microscopes are described in Section 37.2.2.

37.2.1 SIZE MATTERS

One of the physical effects that occurs with very small objects is that their surface properties become much more important relative to their bulk properties. Consider the surface-to-volume ratio of a given amount of material as its dimensions are changed. Let us start with a cubic block of material that is 1 m on each side. Its total surface area is 6 m^2, and its volume is 1 m^3, giving it a surface-to-volume ratio of 6-to-1. If that same volume of material were now compressed into a flat square plate that is 1 μm thick (0.00004 in, or about 1/100 the diameter of a human hair), its dimensions would be 1000 m on each side, and its total surface area (top, bottom, and edges) would be 2,000,000.004 m^2 (1000 × 1000 m^2 on each of two sides, plus 0.001 m^2 on each of the four edges). This would give it a surface-to-volume ratio of slightly greater than 2,000,000-to-1.

Next, suppose the flat plate were sliced in two directions to create cubes that are 1 μm × 1 μm × 1 μm. The total number of cubes would be 10^{18}, and the surface area of each cube would be 6 $μm^2$ or $6(10^{-12})$ m^2. Multiplying the surface area of each cube by the number of cubes gives a total surface area of 6,000,000 m^2, or a surface-to-volume ratio of 6,000,000-to-1 for the original amount of material.

A cube that is 1 μm on each side is surely small, but in nanometers, it is 1000 nm on each edge. Suppose the molecules of this material are cube-shaped, and from our earlier discussion, each molecule measures 1-nm on a side (admittedly, the molecular cube shape is a stretch, but the 1-nm size is plausible). This means that the 1-μm cube contains 10^9 molecules, of which $6(10^6)$ are on the surface of the cube. This leaves $10^9 - 6(10^6) = 994(10^6)$ molecules that are internal (beneath the surface). The ratio of internal to surface molecules is 994-to-6 or 165.667-to-1. By comparison, the same ratio for a cube with 1 m on a side is about 10^{27}-to-1. As the size of the cube decreases, the ratio of internal-to-surface molecules continues to get smaller and smaller, until finally, we have a cube that is 1 nm on a side (the size of the molecule itself), and there are no internal molecules. What this numerical exercise demonstrates is that as the size of an object decreases, approaching nanometer dimensions, the surface molecules become increasingly important relative to the internal molecules simply because of their increasing numerical proportion. Thus, the surface properties of the materials out of which nanometer-sized objects are made become more influential in determining the behavior of the objects, and the relative influence of the bulk properties of the material is reduced.

Recall from Section 2.2 that there are two types of atomic bonding: (1) primary bonds that are generally associated with combining atoms into molecules, and (2) secondary bonds that attract molecules together to form bulk materials. One of the implications of the large surface-to-volume ratio of nanoscale objects is that the secondary bonds that exist between molecules assume greater importance because the shape and properties of an object not much bigger than the molecules comprising it tend to depend on these secondary bonding forces. Accordingly, the material properties and behaviors of nanoscale structures are different from those of structures with dimensions in the macroscale or even microscale. These differences can sometimes be exploited to create materials and products with improved electronic, magnetic, and/or optical properties. Two examples of recently developed materials in this category are (1) carbon nanotubes (Section 37.1.1), and (2) magnetoresistive materials for use in high-density magnetic memories. Nanotechnology will enable the development of entirely new classes of materials.

Another difference that arises between nanoscale objects and their macroscopic counterparts is that material behavior tends to be influenced by quantum mechanics rather than bulk properties. *Quantum mechanics* is a branch of physics that is concerned with the notion that all forms of energy (e.g., electricity, light) occur in discrete units or packets when observed on a small enough scale. The discrete units or packets are called quanta (plural of quantum), which cannot be further subdivided. For example, electricity is conducted in units of electrons. An electrical charge of less than one electron is not possible. In light energy, the quanta are photons. In magnetic energy, they are called magnons. For every type of energy there are comparable units. All physical phenomena exhibit quantum behavior at the submicroscopic level. On a macroscopic level, the energy appears to be continuous because it is being released in very large quantities of quanta.

The movement of electrons in microelectronics is of particular interest because of the significant reductions in size that continue to be achieved in the fabrication of integrated circuits. The feature sizes of the devices in integrated circuits produced in 2009 are on the order of 50 nm. They are projected to decrease in size to about 20 nm by around 2015. At a feature size of around 10 nm, the effects of quantum mechanics become significant, changing the way a device operates. As feature size continues to be reduced toward just a few nanometers, the proportion of surface atoms in the device increases relative to those beneath the surface, which means that the electrical characteristics are no longer determined exclusively by the bulk properties of the material. As device size continues to decrease and density of components on a chip continues to increase, the electronics industry is approaching the limits of technological feasibility of the current fabrication processes discussed in Chapter 34.

37.2.2 SCANNING PROBE MICROSCOPES

Conventional optical microscopes use visible light focused through optical lenses to provide enlarged images of very small objects. However, the wavelength of visible light is 400 to 700 nm, which is greater than the dimensions of nanosized objects. Thus, these objects cannot be seen with conventional optical microscopes. The most powerful optical microscopes provide magnifications of about 1000 times, allowing resolutions of about 0.0002 mm (200 nm). Electron microscopes, which allow specimens to be visualized using a beam of electrons instead of light, were developed in the 1930s. The electron beam can be considered as a form of wave motion, but one that has a much shorter effective wavelength. (Today's electron microscopes permit magnifications of about 1,000,000 times and resolutions of about 1 nm). To obtain an image of a surface, the electron beam is scanned across the surface of an object in a raster pattern, similar to the way a cathode ray scans the surface of a television screen.

For making observations on the nanoscale level, an improvement over the electron microscope is the family of scanning probe instruments that date from the 1980s. They possess magnification capabilities approximately 10 times greater than an electron microscope. In a scanning probe microscope (SPM), the probe consists of a needle with a very sharp tip. The point size approaches the size of a single atom. In operation, the probe is moved along the surface of the specimen at a distance of only one nanometer or so, and any of several properties of the surface are measured, depending on the type of scanning probe device. The two scanning probe microscopes of greatest interest in nanotechnology are the scanning tunneling microscope and the atomic force microscope.

The *scanning tunneling microscope* (STM) was the first scanning probe instrument to be developed. It is called a tunneling microscope because its operation is based on the quantum mechanics phenomenon known as *tunneling,* in which individual electrons in a solid material can jump beyond the surface of the solid into space. The probability of electrons being in this space beyond the surface decreases exponentially in proportion to the distance from the surface. This sensitivity to distance is exploited in the STM by positioning the probe tip very close to the surface (i.e., 1 nm) and applying a small voltage between the two. This causes electrons of surface atoms to be attracted to the small positive charge of the tip, and they tunnel across the gap to the probe. As the probe is moved along the surface, variations in the resulting current occur because of the positions of individual

FIGURE 37.3   An atomic force microscope image of silicon dioxide letters on a silicon substrate. The oxide lines of the letters are about 20 nm wide. (Image courtesy of IBM Corporation.)

atoms on the surface. Alternatively, if the elevation of the tip above the surface is allowed to float by maintaining a constant current, then the vertical deflection of the tip can be measured as it traverses the surface. These variations in current or deflection can be used to create images or topographical maps of the surface on an atomic or molecular scale.

A limitation of the scanning tunneling microscope is that it can only be used on surfaces of conducting materials. By comparison, the ***atomic force microscope*** (AFM) can be used on any material; it uses a probe attached to a delicate cantilever that deflects because of the force exerted by the surface on the probe as it traverses the specimen surface. The AFM responds to various types of forces, depending on the application. The forces include mechanical owing to physical contact of the probe with the specimen surface, and non-contact, such as van der Waals forces (Section 2.2), capillary forces, magnetic forces,[3] and others. The vertical deflection of the probe is measured optically, based on the interference pattern of a light beam or the reflection of a laser beam from the cantilever. Figure 37.3 shows an image generated by an AFM.

Our discussion here has focused on the use of scanning probe microscopes for observing surfaces. In Section 37.3.2, we describe applications of these instruments for manipulating individual atoms, molecules, and other nanoscale clusters of atoms or molecules.

37.3 NANOFABRICATION PROCESSES

Creating products at least some of whose feature sizes are in the nanometer range requires fabrication techniques that are often quite different from those used to process bulk materials and macro-sized products. The fabrication processes for nanometer-scale materials and structures can be divided into two basic categories:

1. ***Top-down approaches,*** which adapt some of the lithography-based microfabrication techniques discussed in Chapters 34 and 36 to nanoscale object sizes. They involve mostly subtractive processes (material removal) to achieve the desired geometry.

---

[3]The term *magnetic force microscope* (MFM) is used when the forces are magnetic. The principle of operation is similar to that of the reading head on a hard disk drive.

2. ***Bottom-up approaches,*** in which atoms and molecules are manipulated and combined into larger structures. These might be described as additive processes because they construct the nanoscale entity from smaller components.

Our organization in this section is based on these two approaches. Because the processing methods associated with the top-down approaches have been discussed in two previous chapters, our coverage in Section 37.3.1 will emphasize how these processes must be modified for the nanoscale. Section 37.3.2 discusses the bottom-up approaches, which are perhaps of greater interest here because of their uniqueness and special relevance to nanotechnology.

37.3.1 TOP-DOWN PROCESSING APPROACHES

The top-down approaches for fabricating nanoscale objects involve the processing of bulk materials (e.g., silicon wafers) and thin films using lithographic techniques like those used in the fabrication of integrated circuits and microsystems. The top-down approaches also include other precision machining techniques (Section 36.2.3) that have been adapted for making nanostructures. The term ***nanomachining*** is used for these processes that involve material removal when applied in the sub-micron scale. Nanostructures have been machined out of materials such as silicon, silicon carbide, diamond, and silicon nitride [23]. Nano-machining must often be coupled with thin-film deposition processes such as physical vapor deposition and chemical vapor deposition (Section 28.5) to achieve the desired structure and combination of materials.

As the feature sizes of the components in an integrated circuit (IC) become smaller and smaller, fabrication techniques based on optical lithography become limited because of the wavelengths of visible light. Ultraviolet light is currently used to fabricate ICs because its shorter wavelengths permit smaller features to be fabricated, thus allowing higher densities of components in the IC. The current technology being refined for IC fabrication is called extreme ultraviolet (EUV) lithography (Section 34.3.2). It uses UV light with a wavelength as short as 13 nm, which is certainly within the nanotechnology range. However, certain technical problems must be addressed when EUV lithography is used at these very short UV wavelengths. The problems include (1) new photoresists that are sensitive to this wavelength must be used, (2) focusing systems must be based on all reflective optics, and (3) plasma sources based on laser irradiation of the element xenon [14] must be used.

Other lithography techniques are available for use in fabricating nanoscale structures. These include electron-beam lithography, x-ray lithography, and micro- or nano-imprint lithography. Electron-beam and x-ray lithography are discussed in the context of integrated circuit processing in Section 34.3.2. ***Electron-beam lithography*** (EBL) operates by directing a highly focused beam of electrons along the desired pattern across the surface of a material, thus exposing the surface areas using a sequential process without the need for a mask. Although EBL is capable of resolutions on the order of 10 nm, its sequential operation makes it relatively slow compared with masking techniques and thus it is unsuited to mass production. ***X-ray lithography*** can produce patterns with resolutions around 20 nm, and it uses masking techniques, which makes high production possible. However, x-rays are difficult to focus and require contact or proximity printing (Section 34.3.1). In addition, the equipment is expensive for production applications, and x-rays are hazardous to humans.

Two of the processes known as ***soft lithography*** are described in our previous chapter on microfabrication (Section 36.2.3). The processes are ***micro-imprint lithography,*** in which a patterned flat mold (similar to a rubber stamp) is used to mechanically deform a thermoplastic resist on the surface of a substrate in preparation for etching, and ***micro-contact printing,*** in which the stamp is dipped into a substance and then pressed against a substrate. This transfers a very thin layer of the substance onto the substrate surface in the pattern defined by the stamp. These same processes can be applied to nanofabrication, in which case they are called ***nano-imprint lithography*** and ***nano-contact printing***. Nano-imprint lithography can produce pattern resolutions of approximately 5 nm [23]. One of the original applications of nano-contact printing was to transfer a thin film of thiols (a family of organic compounds derived from hydrogen sulfide) onto a gold surface. The uniqueness of the application was that the film was only one molecule thick (called a monolayer, Section 37.3.2), which certainly qualifies as nanoscale.

37.3.2 BOTTOM-UP PROCESSING APPROACHES

In the bottom-up approaches, the starting materials are atoms, molecules, and ions. The processes bring these basic building blocks together, in some cases one at a time, to fabricate the desired nanoscale entity. Our coverage consists of three approaches that are of considerable interest in nanotechnology: (1) production of carbon nanotubes, (2) nano-fabrication by scanning probe techniques, and (3) self-assembly.

Production of Carbon Nanotubes   The remarkable properties and potential applications of carbon nanotubes are discussed in Section 37.1.1. Carbon nanotubes can be produced by several techniques. In the following paragraphs we discuss three: (1) laser evaporation, (2) carbon arc techniques, and (3) chemical vapor deposition.

In the ***laser evaporation method,*** the starting raw material is a graphite workpiece containing small amounts of cobalt and nickel. These metal traces perform the role of catalyst, acting as nucleation sites for the subsequent formation of the nanotubes. The graphite is placed in a quartz tube filled with argon gas and heated to 1200°C (2200°F). A pulsed laser beam is focused on the workpiece, causing the carbon atoms to evaporate from the bulk graphite. The argon moves the carbon atoms out of the high-temperature region of the tube and into an area in which a water-cooled copper apparatus is located. The carbon atoms condense on the cold copper, and as they do, they form nanotubes with diameters of 10 to 20 nm and lengths of about 100 μm.

The ***carbon arc technique*** uses two carbon electrodes that are 5 to 20 μm in diameter and separated by 1 mm. The electrodes are located in a partially evacuated container (about 2/3 of 1 atmospheric pressure) with helium flowing in it. To start the process, a voltage of about 25 V is applied across the two electrodes, causing carbon atoms to be ejected from the positive electrode and carried to the negative electrode where they form nanotubes. The structure of the nanotubes depends on whether a catalyst is used. If no catalyst is used, then multi-walled nanotubes are produced. If trace amounts of cobalt, iron, or nickel are placed in the interior of the positive electrode, then the process creates single-walled nanotubes that are 1 to 5 nm in diameter and about 1 μm long.

Chemical vapor deposition (Section 28.5.2) can be used to produce carbon nano-tubes. In one variation of CVD, the starting work material is a hydrocarbon gas such as methane (CH_4). The gas is heated to 1100°C (2000°F), causing it to decompose and release carbon atoms. The atoms then condense on a cool substrate to form nanotubes with open ends rather than the closed ends characteristic of the other fabrication techniques. The substrate may contain iron or other metals that act as catalysts for the process. The metal catalyst acts as a nucleation site for creation of the nanotube, and it also controls the orientation of the structure. An alternative CVD process called HiPCO (high-pressure carbon monoxide decomposition process) starts with carbon monoxide (CO) and uses carbon pentacarbonyl ($Fe(CO)_5$) as the catalyst to produce high-purity single-walled nanotubes at 900°C to 1100°C (1700 to 2000°F) and 30 to 50 atm [7].

Production of nanotubes by CVD has the advantage that it can be operated continuously, which makes it economically attractive for mass production.

Nanofabrication by Scanning Probe Techniques   Scanning probe microscopy (SPM) techniques are described in Section 37.2.2 in the context of measuring and observing nanometer-scale features and objects. In addition to viewing a surface, the scanning tunneling microscope (STM) and atomic force microscope (AFM) can also be used to manipulate individual atoms, molecules, or clusters of atoms or molecules that adhere to a substrate surface by the forces of adsorption (weak chemical bonds). Clusters of atoms or molecules are called *nanoclusters,* and their size is just a few nanometers [23]. Figure 37.4(a) illustrates the variation in either current or deflection of the STM probe tip as it is moved across a surface upon which is located an adsorbed atom. As the tip moves over the surface immediately above the adsorbed atom, there is an increase in the signal. Although the bonding force that attracts the atom to the surface is weak, it is significantly greater than the force of attraction created by the tip, simply because the distance is greater. However, if the probe tip is moved close enough to the adsorbed atom so that its force of attraction is greater than the adsorption force, the atom will be dragged along the surface, as suggested in Figure 37.4(b). In this way, individual atoms or molecules can be manipulated to create various nanoscale structures. A notable STM example accomplished at the IBM Research Labs was the fabrication of the company logo out of xenon atoms adsorbed onto a nickel surface in an area 5×16 nm. This scale is considerably smaller than the lettering in Figure 37.3 (which is also nanoscale, as noted in the caption).

The manipulation of individual atoms or molecules by scanning tunneling microscopy techniques can be classified as lateral manipulation and vertical manipulation. In lateral manipulation, atoms or molecules are transferred horizontally along the surface by the attractive or repulsive forces exerted by the STM tip, as in Figure 37.5(b). In vertical manipulation, the atoms or molecules are lifted from the surface and deposited at a different location to form a structure. Although this kind of STM manipulation of atoms and molecules is of scientific interest, there are technological limitations that inhibit its commercial

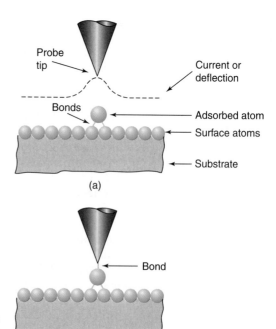

FIGURE 37.4   Manipulation of individual atoms by means of scanning tunneling microscopy techniques: (a) probe tip is maintained a distance from the surface that is sufficient to avoid disturbance of adsorbed atom and (b) probe tip is moved closer to the surface so that the adsorbed atom is attracted to the tip.

FIGURE 37.5 Dip-pen nanolithography, in which the tip of an atomic force microscope is used to deposit molecules through the water meniscus that forms naturally between the tip and the substrate.

application, at least in high production of nanotech products. One of the limitations is that it must be carried out in a very high vacuum environment to prevent stray atoms or molecules from interfering with the process. Another limitation is that the surface of the substrate must be cooled to temperatures approaching absolute zero ($-273°$C or $-460°$F) to reduce thermal diffusion that would gradually distort the atomic structure being formed. These limitations make it a very slow and expensive process.

The atomic force microscope is also used for similar nanoscale manipulations. In comparing the AFM and STM applications, the AFM is more versatile because it is not restricted to conductive surfaces as is the STM and it can be used under normal room conditions. On the other hand, the AFM has a lower resolution than the STM. Consequently, the STM can be used to manipulate single atoms, whereas the AFM is better suited to the manipulation of larger molecules and nanoclusters [23].

Another scanning probe technique, one that shows promise for practical applications, is called dip-pen nanolithography. In **dip-pen nanolithography** (DPN), the tip of an atomic force microscope is used to transfer molecules to a substrate surface by means of a solvent meniscus, as shown in Figure 37.5. The process is somewhat analogous to using an old-fashioned quill pen to transfer ink to a paper surface via capillary forces. In DPN, the AFM tip serves as the nib of the pen, and the substrate becomes the surface onto which the dissolved molecules (i.e., the ink) are deposited. The deposited molecules must have a chemical affinity for the substrate material, just as wet ink adheres to paper. DPN can be used to "write" patterns of molecules onto a surface, where the patterns are of submicron dimension. In addition, DPN can be used to deposit different types of molecules at different locations on the substrate surface.

Self-Assembly   Self-assembly is a fundamental process in nature. The natural formation of a crystalline structure during the slow cooling of molten minerals is an example of nonliving self-assembly. The growth of living organisms is an example of biological self-assembly. In both instances, entities at the atomic and molecular level combine on their own into larger entities, proceeding in a constructive manner toward the creation of some deliberate thing. If the thing is a living organism, the intermediate entities are biological cells, and the organism is grown through an additive process that exhibits massive replication of individual cell formations, yet the final result is often remarkably intricate and complex (e.g., a human being).

One of the promising bottom-up approaches in nanotechnology involves the emulation of nature's self-assembly process to produce materials and systems that have nanometer-scale features or building blocks, but the final product may be larger than nanoscale. It may be of micro- or macro-scale size, at least in some of its dimensions. The term **biomimetics** describes this process of building artificial, non-biological entities by imitating nature's methods. Desirable attributes of atomic or molecular self-assembly processes in nanotechnology include the following: (1) they can be carried out rapidly; (2) they occur automatically and do not require any central control; (3) they exhibit massive replication; and (4) they can be performed under mild environmental conditions (at or near atmospheric pressure and room temperature). Self-assembly is likely to be the most important of the nanofabrication processes because of its low cost, capacity for producing structures over a range of sizes (from nanoscale to macroscale), and general applicability to a wide variety of products [18].

An underlying principle behind self-assembly is that of minimum energy. Physical entities such as atoms and molecules seek out a state that minimizes the total energy of the system of which they are components. This principle has the following implications for self-assembly:

1. There must be some mechanism for the movement of the entities (e.g., atoms, molecules, ions) in the system, thus causing the entities to come into close proximity with one another. Possible mechanisms for this movement include diffusion, convection in a fluid, and electric fields.

2. There must be some form of molecular recognition among the entities. Molecular recognition refers to the tendency of one molecule (or atom or ion) to be attracted to and bind with another molecule (or atom or ion), for example, the way sodium and chlorine are attracted to each other to form table salt.

3. The molecular recognition among the entities causes them to join in such a way that the resulting physical arrangement of the entities achieves a state of minimum energy. The joining process involves chemical bonding, usually the weaker secondary types (e.g., van der Waals bonds).

We have previously encountered several instances of molecular-level self-assembly in this book. Let us cite two examples here: (1) crystal formation and (2) polymerization. Crystal formation in metals, ceramics, and certain polymers is a form of self-assembly. Growing silicon boules in the Czochralski process (Section 34.2.2) for fabrication of integrated circuits is a good illustration. Using a starting seed crystal, very pure molten silicon is formed into a large cylindrical solid whose repeating lattice structure matches that of the seed throughout its volume. The lattice spacing in the crystal structure is of nanometer proportions, but the replication exhibits long-range order.

It can be argued that polymers are products of nanometer-scale self-assembly. The process of polymerization (Section 8.1.1) involves the joining of individual monomers (individual molecules such as ethylene C_2H_4) to form very large molecules (macromolecules such as polyethylene), often in the form of a long chain with thousands of repeating units. Copolymers (Section 8.1.2) represent a more complex self-assembly process, in which two different types of starting monomers are joined in a regular repeating structure. An example is the copolymer synthesized from ethylene and propylene (C_3H_6). In these polymer examples, the repeating units are of nanometer size, and they form by a massive self-assembly process into bulk materials that have significant commercial value.

The technologies for producing silicon boules and polymers precede the current scientific interest in nanotechnology. Of greater relevance in this chapter are self-assembly fabrication techniques that have been developed under the nanotechnology banner. These self-assembly processes, most of which are still in the research stage, include the following categories: (1) fabrication of nanoscale objects, including molecules, macromolecules, clusters of molecules, nanotubes, and crystals; and (2) formation of two-dimensional arrays such as self-assembled monolayers (surface films that are one molecule thick) and three-dimensional networks of molecules.

We have already discussed some of the processes in category 1. Let us consider the self-assembly of surface films as an important example of category 2. Surface films are two-dimensional coatings formed on a solid (three-dimensional) substrate. Most surface films are inherently thin, yet the thickness is typically measured in micrometers or even millimeters (or fractions thereof), well above the nanometer scale. Of interest here are surface films whose thicknesses are measured in nanometers. Of particular interest in nanotechnology are surface films that self-assemble, are only one molecule thick, and in which the molecules are organized in some orderly fashion. These types of films are

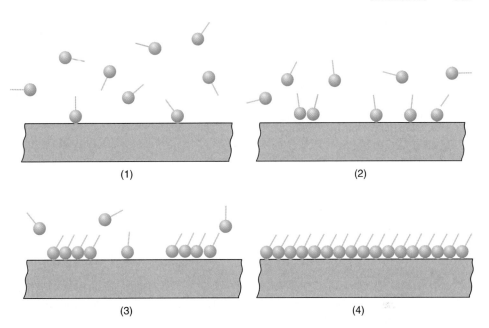

FIGURE 37.6
Typical sequence in the formation of a monolayer of a thiol onto a gold substrate: (1) some of the layering molecules in motion above the substrate become attracted to the surface, (2) they are adsorbed on the surface, (3) they form islands, (4) the islands grow until the surface is covered. (Based on a figure in [9].)

called self-assembled monolayers (SAMs). Multilayered structures are also possible that possess order and are two or more molecules thick.

The substrate materials for self-assembled monolayers and multilayers include a variety of metallic and other inorganic materials. The list includes gold, silver, copper, silicon, and silicon dioxide. Noble metals have the advantage of not forming an oxide surface film that would interfere with the reactions that generate the desired layer of interest. Layering materials include thiols, sulfides, and disulfides. The layering material must be capable of being adsorbed onto the surface material. The typical process sequence in the formation of the monolayer of a thiol on gold is illustrated in Figure 37.6. (We mentioned this combination of thiol on a gold surface in Section 37.3.1 in the context of nano-contact printing.) Layering molecules move freely above the substrate surface and are adsorbed onto the surface. Contact occurs between adsorbed molecules on the surface, and they form stable islands. The islands become larger and gradually join together through the addition of more molecules laterally on the surface, until the substrate is completely covered. Bonding to the gold surface is provided by the sulfur atom in the thiol, sulfide, or disulfide layer. In some applications, self-assembled monolayers can be formed into desired patterns or regions on the substrate surface using techniques such as nano-contact printing and dip-pen nanolithography.

REFERENCES

[1] Baker, S., and Aston, A. "The Business of Nano-tech," *Business Week,* February 14, 2005, pp. 64–71.

[2] Balzani, V., Credi, A., and Venturi, M. *Molecular Devices and Machines—A Journey into the Nano World.* Wiley-VCH Verlag GmbH & Co. KGaA, Weinheim, Germany, 2003.

[3] Bashir, R. "Biologically Mediated Assembly of Artificial Nanostructures and Microstructures," Chapter 5 in *Handbook of Nanoscience, Engineering, and Technology*, W. A. Goddard, III, D. W. Brenner, S. E. Lyshevski and G. J. Iafrate (eds.). CRC Press, Boca Raton, Florida, 2003.

[4] Chaiko, D. J. "Nanocomposite Manufacturing," *Advanced Materials & Processes,* June 2003, pp. 44–46.

[5] Drexler, K. E. *Nanosystems: Molecular Machinery, Manufacturing, and Computation.* Wiley-Interscience, John Wiley & Sons, New York, 1992.

[6] Fujita, H. (ed.). *Micromachines as Tools for Nano-technology.* Springer-Verlag, Berlin, 2003.

[7] Hornyak, G. L., Moore, J. J., Tibbals, H. F., and Dutta, J. *Fundamentals of Nanotechnology*, CRC Taylor & Francis, Boca Raton, Florida, 2009.

[8] Jackson, M. L., *Micro and Nanomanufacturing*, Springer, New York, 2007.

[9] Kohler, M., and Fritsche, W. *Nanotechnology: An Introduction to Nanostructuring Techniques*. Wiley-VCH Verlag GmbH & Co. KGaA, Weinheim, Germany, 2004.

[10] Lyshevski, S. E."Nano- and Micromachines in NEMS and MEMS," Chapter 23 in *Handbook of Nanoscience, Engineering, and Technology*, W. A. Goddard, III, D. W. Brenner, S. E. Lyshevski, and G. J. Iafrate (eds.). CRC Press, Boca Raton, Florida, 2003, pp. 23–27.

[11] Maynor, B. W., and Liu, J. *"Dip-Pen Lithography,"Encyclopedia of Nanoscience and Nanotechnology*, American Scientific Publishers, 2004, pp. 429–441.

[12] Meyyappan, M., and Srivastava, D."Carbon Nanotubes," Chapter 18 in *Handbook of Nanoscience, Engineering, and Technology*, W. A. Goddard, III, D. W. Brenner, S. E. Lyshevski, and G. J. Iafrate (eds.). CRC Press, Boca Raton, Florida, 2003. pp. 18–1 to 18–26.

[13] Morita, S., Wiesendanger, R., and Meyer, E. (eds.). *Noncontact Atomic Force Microscopy*. Springer-Verlag, Berlin, 2002.

[14] National Research Council (NRC). *Implications of Emerging Micro- and Nanotechnologies*. Committee on Implications of Emerging Micro- and Nanotechnologies, The National Academies Press, Washington, D.C., 2002.

[15] Nazarov, A. A., and Mulyukov, R. R."Nanostructured Materials," Chapter 22 in *Handbook of Nanoscience,*

Engineering, and Technology, W. A. Goddard, III, D. W. Brenner, S. E. Lyshevski, and G. J. Iafrate (eds.). CRC Press, Boca Raton, Florida, 2003. 22–1 to 22–41.

[16] Piner, R. D., Zhu, J., Xu, F., Hong, S., and Mirkin, C. A. "Dip-Pen Nanolithography," *Science,* Vol. 283, January 29, 1999, pp. 661–663.

[17] Poole, Jr., C. P., and Owens, F. J. *Introduction to Nanotechnology*. Wiley-Interscience, John Wiley & Sons, Hoboken, New Jersey, 2003.

[18] Ratner, M., and Ratner, D. *Nanotechnology: A Gentle Introduction to the Next Big Idea*. Prentice Hall PTR, Pearson Education, Inc., Upper Saddle River, New Jersey, 2003.

[19] Rietman, E. A. *Molecular Engineering of Nanosystems*. Springer-Verlag, Berlin, 2000.

[20] Rubahn, H.-G. *Basics of Nanotechnology*, 3rd ed., Wiley-VCH, Weinheim, Germany, 2008.

[21] Schmid, G. (ed.). *Nanoparticles: From Theory to Application*. Wiley-VCH Verlag GmbH & Co. KGaA, Weinheim, Germany, 2004.

[22] Torres, C. M. S. (ed.). *Alternative Lithography: Unleashing the Potentials of Nanotechnology*. Kluwer Academic/Plenum Publishers, New York, 2003.

[23] Tseng, A. A. (ed.), *Nanofabrication Fundamentals and Applications*, World Scientific, Singapore, 2008.

[24] Weber, A. "Nanotech: Small Products, Big Potential," *Assembly,* February 2004, pp. 54–59.

[25] Website: en.wikipedia.org/wiki/nanotechnology

[26] Website: www.nanotechproject.org/inventories/consumer

[27] Website: www.research.ibm.com/nanoscience

[28] Website: www.zurich.ibm.com/st/atomic_manipulation

REVIEW QUESTIONS

37.1. What is the range of feature sizes of entities associated with nanotechnology?

37.2. Identify some of the present and future products associated with nanotechnology.

37.3. What is a buckyball?

37.4. What is a carbon nanotube?

37.5. What are the scientific and technical disciplines associated with nanoscience and nanotechnology?

37.6. Why is biology so closely associated with nanoscience and nanotechnology?

37.7. The behavior of nanoscale structures is different from macroscale and even microscale structures

because of two factors mentioned in the text. What are those two factors?

37.8. What is a scanning probe instrument, and why is it so important in nanoscience and nanotechnology?

37.9. What is tunneling, as referred to in the scanning tunneling microscope?

37.10. What are the two basic categories of approaches used in nanofabrication?

37.11. Why is photolithography based on visible light not used in nanotechnology?

37.12. What are the lithography techniques used in nanofabrication?

37.13. How is nano-imprint lithography different from micro-imprint lithography?

37.14. What are the limitations of scanning tunneling microscope in nanofabrication that inhibit its commercial application?

37.15. What is self-assembly in nanofabrication?

37.16. What are the desirable features of atomic or molecular self-assembly processes in nanotechnology?

MULTIPLE CHOICE QUIZ

There are 18 correct answers in the following multiple choice questions (some questions have more than one correct answer). To achieve a perfect score on the quiz, all correct answers must be given. Each correct answer is worth 1 point. Each omitted answer or wrong answer reduces the total score by 1 point, and each additional answer beyond the correct number of answers reduces the score by 1 point. The percentage score on the quiz is based on the total number of correct answers.

37.1. Nanotechnology refers to the fabrication and application of entities whose feature sizes are in which of the following ranges (one best answer): (a) 0.1 nm to 10 nm, (b) 1 nm to 100 nm, or (c) 100 nm to 1000 nm?

37.2. One nanometer is equivalent to which of the following (two correct answers): (a) 1×10^{-3} μm, (b) 1×10^{-6} m, (c) 1×10^{-9} m, and (d) 1×10^{6} mm.

37.3. NNI stands for which one of the following: (a) Nanoscience Naval Institute, (b) Nanoscience Nonsense and Ignorance, (c) National Nanotechnology Initiative, or (d) Nanotechnology News Identification?

37.4. The surface-to-volume ratio of a cube that is 1×10^{-6} m on each edge is significantly greater than the surface-to-volume ratio of a cube that is 1 m on each edge: (a) true or (b) false?

37.5. The proportion of surface molecules relative to internal molecules is significantly greater for a cube that is 1×10^{-6} m on each edge than for a cube that is 1 m on each edge: (a) true or (b) false?

37.6. Which one of the following microscopes can achieve the greatest magnification: (a) electron microscope, (b) optical microscope, or (c) scanning tunneling microscope?

37.7. Which of the following are correct statements about a buckyball (three best answers): (a) it contains 60 atoms, (b) it contains 100 atoms, (c) it contains 600 atoms, (d) it is a carbon atom, (e) it is a carbon molecule, (f) it is shaped like a basketball, (g) it is shaped like a tube, and (h) it is shaped like a volleyball?

37.8. Which of the following are considered techniques that fall within the category called top-down approaches to nanofabrication (three best answers): (a) biological evolution, (b) electron-beam lithography, (c) micro-imprint lithography, (d) scanning probe techniques, (e) self-assembly, and (f) x-ray lithography?

37.9. Which of the following are considered techniques that fall within the category called bottom-up approaches to nanofabrication (three best answers): (a) electron beam lithography, (b) extreme ultraviolet lithography, (c) chemical vapor deposition to produce carbon nanotubes, (d) nano-imprint lithography, (e) scanning probe techniques, (f) self-assembly, and (g) x-ray lithography?

37.10. Dip-pen nanolithography uses which one of the following techniques and/or devices: (a) atomic force microscope, (b) chemical vapor deposition, (c) electron beam lithography, (d) nano-imprint lithography, or (e) self-assembly?

37.11. A self-assembled monolayer has a thickness that is which one of the following: (a) one micrometer, (b) one millimeter, (c) one molecule, or (d) one nanometer?

Part X Manufacturing Systems

38 AUTOMATION TECHNOLOGIES FOR MANUFACTURING SYSTEMS

Chapter Contents

In this part of the book, we consider the manufacturing systems that are commonly associated with the production and assembly processes discussed in preceding chapters. A *manufacturing system* can be defined as a collection of integrated equipment and human resources that performs one or more processing and/or assembly operations on a starting work material, part, or set of parts. The integrated equipment consists of production machines, material handling and positioning devices, and computer systems. Human resources are required either full-time or part-time to keep the equipment operating. The position of the manufacturing systems in the larger production system is shown in Figure 38.1. As the diagram indicates, the manufacturing systems are located in the factory. They accomplish the value-added work on the part or product.

Manufacturing systems include both automated and manually operated systems. The distinction between the two categories is not always clear, because many manufacturing systems consist of both automated and manual work elements (e.g., a machine tool that operates on a semiautomatic processing cycle but which must be loaded and unloaded each cycle by a human worker). Our coverage includes both categories and is organized into two chapters: Chapter 38 on automation technologies and Chapter 39 on integrated manufacturing systems. Chapter 38 provides an introductory

FIGURE 38.1 The position of the manufacturing systems in the larger production system.

treatment of automation technology and the components that make up an automated system. We also discuss two important automation technologies used in manufacturing: numerical control and industrial robotics. In Chapter 39, we examine how these automation technologies are integrated into more sophisticated manufacturing systems. Topics include production lines, cellular manufacturing, flexible manufacturing systems, and computer integrated manufacturing. A more detailed discussion of the topics in these two chapters can be found in [5].

38.1 AUTOMATION FUNDAMENTALS

Automation can be defined as the technology by which a process or procedure is performed without human assistance. Humans may be present as observers or even participants, but the process itself operates under its own self-direction. Automation is implemented by means of a control system that executes a program of instructions. To automate a process, power is required to operate the control system and to drive the process itself.

38.1.1 THREE COMPONENTS OF AN AUTOMATED SYSTEM

As indicated above, an automated system consists of three basic components: (1) power, (2) a program of instructions, and (3) a control system to carry out the instructions. The relationship among these components is shown in Figure 38.2.

The form of power used in most automated systems is electrical. The advantages of electrical power include (1) it is widely available, (2) it can be readily converted to other forms of power such as mechanical, thermal, or hydraulic, (3) it can be used at very low power levels for functions such as signal processing, communication, data storage, and data processing, and (4) it can be stored in long-life batteries [5].

FIGURE 38.2 Elements of an automated system: (1) power, (2) program of instructions, and (3) control system.

In a manufacturing process, power is required to accomplish the activities associated with the particular process. Examples of these activities include (1) melting a metal in a casting operation, (2) driving the motions of a cutting tool relative to a workpiece in a machining operation, and (3) pressing and sintering parts in a powder metallurgy process. Power is also used to accomplish any material handling activities needed in the process, such as loading and unloading parts, if these activities are not performed manually. Finally, power is used to operate the control system.

The activities in an automated process are determined by a program of instructions. In the simplest automated processes, the only instruction may be to maintain a certain controlled variable at a specified level, such as regulating the temperature in a heat treatment furnace. In more complex processes, a sequence of activities is required during the work cycle, and the order and details of each activity are defined by the program of instructions. Each activity involves changes in one or more process parameters, such as changing the x-coordinate position of a machine tool worktable, opening or closing a valve in a fluid flow system, or turning a motor on or off. Process parameters are inputs to the process. They may be continuous (continuously variable over a given range, such as the x-position of a worktable) or discrete (On or Off). Their values affect the outputs of the process, which are called process variables. Like process parameters, process variables can be continuous or discrete. Examples include the actual position of the machine worktable, the rotational speed of a motor shaft, or whether a warning light is on or off. The program of instructions specifies the changes in process parameters and when they should occur during the work cycle, and these changes determine the resulting values of the process variables. For example, in computer numerical control, the program of instructions is called a part program. The numerical control (NC) part program specifies the individual sequence of steps required to machine a given part, including worktable and cutter positions, cutting speeds, feeds, and other details of the operation.

In some automated processes, the work cycle program must contain instructions for making decisions or reacting to unexpected events during the work cycle. Examples of situations requiring this kind of capability include (1) variations in raw materials that require adjusting certain process parameters to compensate, (2) interactions and communications with human such as responding to requests for system status information, (3) safety monitoring requirements, and (4) equipment malfunctions.

The program of instructions is executed by a control system, the third basic component of an automated system. Two types of control system can be distinguished: closed loop and open loop. A *closed loop system*, also known as a *feedback control system*, is one in which the process variable of interest (output of the process) is compared with the corresponding process parameter (input to the process), and any difference between them is used to drive the output value into agreement with the input. Figure 38.3(a) shows the six elements of a closed loop system: (1) input parameter, (2) process, (3) output variable, (4) feedback sensor, (5) controller, and (6) actuator. The input parameter represents the desired value of the output variable. The process is the operation or activity being controlled; more specifically, the output variable is being controlled by the system. A sensor is used to measure the output variable and feed back its value to the controller, which compares output with input and makes the required adjustment to reduce any difference. The adjustment is made by means of one or more actuators, which are hardware devices that physically accomplish the control actions.

The other type of control system is an open loop system, presented in Figure 38.3(b). As shown in the diagram, an *open loop system* executes the program of instructions without a feedback loop. No measurement of the output variable is made, so there is no comparison between output and input in an open loop system. In effect, the controller relies on the expectation that the actuator will have the intended effect on the output variable. Thus, there is always a risk in an open loop system that the actuator will not function properly or that its actuation will not have the expected effect on the output. On

FIGURE 38.3 Two basic types of control systems: (a) closed loop and (b) open loop.

the other hand, the advantage of an open loop system is that its cost is less than a comparable closed loop system.

38.1.2 TYPES OF AUTOMATION

Automated systems used in manufacturing can be classified into three basic types: (1) fixed automation, (2) programmable automation, and (3) flexible automation.

Fixed Automation In fixed automation, the processing or assembly steps and their sequence are fixed by the equipment configuration. The program of instructions is determined by the equipment design and cannot be easily changed. Each step in the sequence usually involves a simple action, such as feeding a rotating spindle along a linear trajectory. Although the work cycle consists of simple operations, integrating and coordinating the actions can result in the need for a rather sophisticated control system, and computer control is often required.

Typical features of fixed automation include (1) high initial investment for specialized equipment, (2) high production rates, and (3) little or no flexibility to accommodate product variety. Automated systems with these features can be justified for parts and products that are produced in very large quantities. The high investment cost can be spread over many units, thus making the cost per unit relatively low compared to alternative production methods. The automated production lines discussed in the following chapter are examples of fixed automation.

Programmable Automation As its name suggests, the equipment in programmable automation is designed with the capability to change the program of instructions to allow production of different parts or products. New programs can be prepared for new parts, and the equipment can read each program and execute the encoded instructions. Thus the features that characterize programmable automation are (1) high investment in general purpose equipment that can be reprogrammed, (2) lower production rates than fixed automation, (3) ability to cope with product variety by reprogramming the equipment, and (4) suitability for batch production of various part or product styles. Examples of programmable automation include computer numerical control and industrial robotics, discussed in Sections 38.3 and 38.4, respectively.

Flexible Automation Suitability for batch production is mentioned as one of the features of programmable automation. As discussed in Chapter 1, the disadvantage of batch production is that lost production time occurs between batches due to equipment and/or tooling changeovers that are required to accommodate the next batch. Thus, programmable automation usually suffers from this disadvantage. Flexible automation is an extension of programmable automation in which there is virtually no lost production time for setup changes and/or reprogramming. Any required changes in the program of instructions and/or setup can be accomplished quickly; that is, within the time needed to move the next work unit into position at the machine. A flexible system is therefore capable of producing a mixture of different parts or products one right after the other instead of in batches. Features usually associated with flexible automation include (1) high investment cost for custom-engineered equipment, (2) medium production rates, and (3) continuous production of different part or product styles.

Using some terminology developed in Chapter 1, we might say that fixed automation is applicable in situations of hard product variety, programmable automation is applicable to medium product variety, and flexible automation can be used for soft product variety.

38.2 HARDWARE COMPONENTS FOR AUTOMATION

Automation and process control are implemented using various hardware devices that interact with the production operation and associated processing equipment. Sensors are required to measure the process variables. Actuators are used to drive the process parameters. And various additional devices are needed to interface the sensors and actuators with the process controller, which is usually a digital computer.

38.2.1 SENSORS

A sensor is a device that converts a physical stimulus or variable of interest (e.g., temperature, force, pressure, or other characteristic of the process) into a more convenient physical form (e.g., electrical voltage) for the purpose of measuring the variable. The conversion allows the variable to be interpreted as a quantitative value.

Sensors of various types are available to collect data for feedback control in manufacturing automation. They are often classified according to type of stimulus; thus, we have mechanical, electrical, thermal, radiation, magnetic, and chemical variables. Within each category, there are multiple variables that can be measured. Within the mechanical category, the physical variables include position, velocity, force, torque, and many others. Electrical variables include voltage, current, and resistance. And so on for the other major categories.

In addition to type of stimulus, sensors are also classified as analog or discrete. An *analog sensor* measures a continuous analog variable and converts it into a continuous signal such as electrical voltage. Thermocouples, strain gages, and ammeters are examples of analog sensors. A *discrete sensor* produces a signal that can have only a limited number of values. Within this category, we have binary sensors and digital sensors. A *binary sensor* can take on only two possible values, such as Off and On, or 0 and 1. Limit switches operate this way. A *digital sensor* produces a digital output signal, either in the form of parallel status bits, such as a photoelectric sensor array) or a series of pulses that can be counted, such as an optical encoder. Digital sensors have an advantage that they can be readily interfaced to a digital computer, whereas the signals from analog sensors must be converted to digital in order to be read by the computer.

For a given sensor, there is a relationship between the value of the physical stimulus and the value of the signal produced by the sensor. This input/output relationship is called the sensor's **transfer function**, which can be expressed as:

$$S = f(s) \tag{38.1}$$

where S = the output signal of the sensor (typically voltage), s = the stimulus or input, and $f(s)$ is the functional relationship between them. The ideal form for an analog sensor is a proportional relationship:

$$S = C + ms \tag{38.2}$$

where C = the value of the sensor output when the stimulus value is zero, and m = the constant of proportionality between s and S. The constant m indicates how much the output S is affected by the input s. This is referred to as the **sensitivity** of the measuring device. For example, a standard Chromel/Alumel thermocouple produces 40.6 microvolts per °C change in temperature.

A binary sensor (e.g., limit switch, photoelectric switch) exhibits a binary relationship between stimulus and sensor output:

$$S = 1 \text{ if } s > 0 \text{ and } S = 0 \text{ if } s \leq 0 \tag{38.3}$$

Before a measuring device can be used, it must be calibrated, which basically means determining the transfer function of the sensor; specifically, how is the value of the stimulus s determined from the value of the output signal S? Ease of calibration is one criterion by which a measuring device can be selected. Other criteria include accuracy, precision, operating range, speed of response, reliability and cost.

38.2.2 ACTUATORS

In automated systems, an actuator is a device that converts a control signal into a physical action, which usually refers to a change in a process input parameter. The action is typically mechanical, such as a change in position of a worktable or rotational speed of a motor. The control signal is generally a low level signal, and an amplifier may be required to increase the power of the signal to drive the actuator.

Actuators can be classified according to type of amplifier as (1) electrical, (2) hydraulic, or (3) pneumatic. Electrical actuators include AC and DC electric motors, stepper motors, and solenoids. The operations of two types of electric motors (servomotors and stepper motors) are described in Section 38.3.2, which deals with the analysis of positioning systems. Hydraulic actuators utilize hydraulic fluid to amplify the control signal and are often specified when large forces are required in the application. Pneumatic actuators are driven by compressed air, which is commonly used in factories. All three actuator types are available as linear or rotational devices. This designation distinguishes whether the output action is a linear motion or a rotational motion. Electric motors and stepper motors are more common as rotational actuators, whereas most hydraulic and pneumatic actuators provide a linear output.

38.2.3 INTERFACE DEVICES

Interface devices allow the process to be connected to the computer controller and vice versa. Sensor signals from the manufacturing process are fed into the computer, and

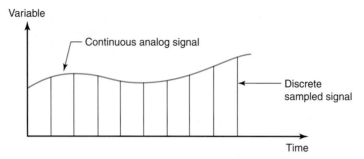

FIGURE 38.4
An analog-to-digital converter works by converting a continuous analog signal into a series of discrete sampled data.

command signals are sent to actuators that operate the process. In this section, we discuss the hardware devices that enable this communication between the process and the controller. The devices include analog-to-digital converters, digital-to-analog converters, contact input/output interfaces, and pulse counters and generators.

Continuous analog signals from sensors attached to the process must be transformed into digital values that can be used by the control computer, a function that is accomplished by an ***analog-to-digital converter*** (ADC). As illustrated in Figure 38.4, an ADC (1) samples the continuous signal at periodic intervals, (2) converts the sampled data into one of a finite number of defined amplitude levels, and (3) encodes each amplitude level into a sequence of binary digits that can be interpreted by the control computer. Important characteristics of an analog-to-digital converter include sampling rate and resolution. Sampling rate is the frequency with which the continuous signal is sampled. A faster sampling rate means that the actual form of the continuous signal can be more closely approximated. Resolution refers to the precision with which the analog value can be converted into binary code. This depends on the number of bits used in the encoding procedure, the more bits, the higher the resolution. Unfortunately, using more bits requires more time to make the conversion, which can impose a practical limit on the sampling rate.

A ***digital-to-analog converter*** (DAC) accomplishes the reverse process of the ADC. It converts the digital output of the control computer into a quasi-continuous signal capable of driving an analog actuator or other analog device. The DAC performs its function in two steps: (1) decoding, in which the sequence of digital output values is transformed into a corresponding series of analog values at discrete time intervals, and (2) data holding, in which each analog value is changed into a continuous signal during the duration of the time interval. In the simplest case, the continuous signal consists of a series of step functions, as in Figure 38.5, which are used to drive the analog actuator.

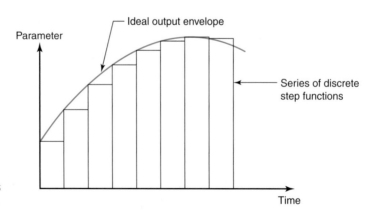

FIGURE 38.5
An analog-to-digital converter works by converting a continuous analog signal into a series of discrete sampled data.

Many automated systems operate by turning on and off motors, switches, and other devices to respond to conditions and as a function of time. These control devices use binary variables. They can have either of two possible values, 1 or 0, interpreted as On or Off, object present or not present, high or low voltage level, and so on. Binary sensors commonly used in process control systems include limit switches and photocells. Common binary actuators solenoids, valves, clutches, lights, control relays, and certain motors.

Contact input/output interfaces are components used to communicate binary data back and forth between the process and the control computer. A *contact input interface* is a device that reads binary data into the computer from an external source. It consists of a series of binary electrical contacts that indicate the status of a binary device such as a limit switch attached to the process. The status of each contact is periodically scanned by the computer to update values used by the control program. A *contact output interface* is a device used to communicate on/off signals from the computer to external binary components such as solenoids, alarms, and indicator lights. It can also be used to turn on and off constant speed motors.

As mentioned earlier, discrete data sometimes exist in the form of a series of pulses. For example, an optical encoder (discussed in Section 38.3.2) emits its measurement of position and velocity as a series of pulses. A *pulse counter* is a device that converts a series of pulses from an external source into a digital value, which is entered into the control computer. In addition to reading the output of an optical encoder, applications of pulse counters include counting the number of parts flowing along a conveyor past a photo-electric sensor. The opposite of a pulse counter is a *pulse generator*, a device that produces a series of electrical pulses based on digital values generated by a control computer. Both the number and frequency of the pulses are controlled. An important pulse generator application is to drive stepper motors, which respond to each step by rotating through a small incremental angle, called a step angle.

38.2.4 PROCESS CONTROLLERS

Most process control systems use some type of digital computer as the controller. Whether control involves continuous or discrete parameters and variables, or a combination of continuous and discrete, a digital computer can be connected to the process to communicate and interact with it using the interface devices discussed in Section 38.2.3. Requirements generally associated with real-time computer control include the following:

➢ The capability of the computer to respond to incoming signals from the process and if necessary, to interrupt execution of a current program to service the incoming signal.

➢ The capability to transmit commands to the process that are implemented by means of actuators connected to the process. These commands may be the response to incoming signals from the process.

➢ The capability to execute certain actions at specific points in time during process operation.

➢ The capability to communicate and interact with other computers that may be connected to the process. The term *distributed process control* is used to describe a control system in which multiple microcomputers are used to share the process control workload.

➢ The capability to accept input from operating personnel for purposes such as entering new programs or data, editing existing programs, and stopping the process in an emergency.

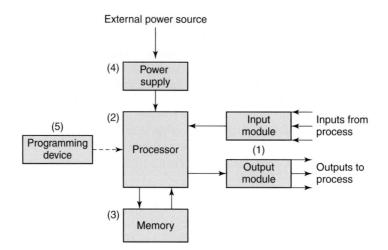

FIGURE 38.6 Major components of a programmable logic controller.

A widely used process controller that satisfies these requirements is a programmable logic controller. A **programmable logic controller** (PLC) is a microcomputer-based controller that uses stored instructions in programmable memory to implement logic, sequencing, timing, counting, and arithmetic control functions, through digital or analog input/output modules, for controlling various machines and processes. The major components of a PLC, shown in Figure 38.6, are (1) **input and output modules**, which connect the PLC to the industrial equipment to be controlled; (2) **processor**—the central processing unit (CPU), which executes the logic and sequencing functions to control the process by operating on the input signals and determining the proper output signals specified by the control program; (3) **PLC memory**, which is connected to the processor and contains the logic and sequencing instructions; (4) **power supply**—115 V AC is typically used to drive the PLC. In addition, (5) a **programming device** (usually detachable) is used to enter the program into the PLC.

Programming involves entry of the control instructions to the PLC using the programming device. The most common control instructions include logical operations, sequencing, counting, and timing. Many control applications require additional instructions for analog control, data processing, and computations. A variety of PLC programming languages have been developed, ranging from ladder logic diagrams to structured text. A discussion of these languages is beyond the scope of this text, and the reader is referred to our references.

Advantages associated with programmable logic controllers include (1) programming a PLC is easier than wiring a relay control panel; (2) PLCs can be reprogrammed, whereas conventional hard-wired controls must be rewired and are often scrapped instead because of the difficulty in rewiring; (3) a PLC can be interfaced with the plant computer system more readily than conventional controls; (4) PLCs require less floor space than relay controls, and (5) PLCs offer greater reliability and easier maintenance.

38.3 COMPUTER NUMERICAL CONTROL

Numerical control (NC) is a form of programmable automation in which the mechanical actions of a piece of equipment are controlled by a program containing coded alphanumeric data. The data represent relative positions between a workhead and a workpart. The workhead is a tool or other processing element, and the workpart is the object being processed. The operating principle of NC is to control the motion of the workhead relative to the workpart and to control the sequence in which the motions are carried out. The first application of numerical control was in machining (Historical Note 38.1), and this is still an important application area. NC machine tools are shown in Figures 22.26 and 22.27. Our video clip on computer numerical control shows the various types of CNC machines and operations.

VIDEO CLIP _____

Computer Numerical Control. The clip contains two segments: (1) computer numerical controls and (2) CNC principles.

Historical Note 38.1   *Numerical control* [3], [5]

The initial development work on numerical control is credited to John Parsons and Frank Stulen at the Parsons Corporation in Michigan in the late 1940s. Parsons was a machining contractor for the U.S. Air Force and had devised a means of using numerical coordinate data to move the worktable of a milling machine for producing complex parts for aircraft. On the basis of Parson's work, the Air Force awarded a contract to the company in 1949 to study the feasibility of the new control concept for machine tools. The project was subcontracted to the Massachusetts Institute of Technology to develop a prototype machine tool that utilized the new numerical data principle. The M.I.T. study confirmed that the concept was feasible and proceeded to adapt a three-axis vertical milling machine using combined analog-digital controls. The name *numerical control* (NC) was given to the system by which the machine tool motions were accomplished. The prototype machine was demonstrated in 1952.

The accuracy and repeatability of the NC system was far better than the manual machining methods then available. The potential for reducing nonproductive time in the machining cycle was also apparent. In 1956, the Air Force sponsored the development of NC machine tools at several different companies. These machines were placed in operation at various aircraft plants between 1958 and 1960. The advantages of NC soon became clear, and aerospace companies began placing orders for new NC machines.

The importance of part programming was clear from the start. The Air Force continued to encourage the development and application of NC by sponsoring research at M.I.T. for a part programming language to control NC machines. This research resulted in the development of *APT* in 1958 (APT stands for Automatically Programmed Tooling). APT is a part programming language by which a user could write the machining instructions in simple English-like statements, and the statements were coded to be interpreted by the NC system.

38.3.1 THE TECHNOLOGY OF NUMERICAL CONTROL

In this section we define the components of a numerical control system, and then proceed to describe the coordinate axis system and motion controls.

Components of an NC System   A numerical control system consists of three basic components: (1) part program, (2) machine control unit, and (3) processing equipment. The **part program** (the term commonly used in machine tool technology) is the detailed set of commands to be followed by the processing equipment. It is the program of instructions in the NC control system. Each command specifies a position or motion that is to be accomplished by the work head relative to the workpart. A position is defined by its *x-y-z* coordinates. In machine tool applications, additional details in the NC program include spindle rotation speed, spindle direction, feed rate, tool change instructions, and other commands related to the operation. The part program is prepared by a **part programmer**, a person who is familiar with the details of the programming language and also understands the technology of the processing equipment.

The **machine control unit** (MCU) in modern NC technology is a microcomputer that stores and executes the program by converting each command into actions by the processing equipment, one command at a time. The MCU consists of both hardware and software. The hardware includes the microcomputer, components to interface with the processing equipment, and certain feedback control elements. The software in the MCU includes control system software, calculation algorithms, and translation software to convert the NC part program into a usable format for the MCU. The MCU also permits the part program to be edited in case the program contains errors, or changes in cutting conditions are required. Because the MCU is a computer, the term **computer numerical control** (CNC) is often used to distinguish this type of NC from its technological predecessors that were based entirely on hard-wired electronics.

The **processing equipment** accomplishes the sequence of processing steps to transform the starting workpart into a completed part. It operates under the control of the MCU according to the instructions in the part program. We survey the variety of applications and processing equipment in Section 38.3.4.

Coordinate System and Motion Control in NC   A standard coordinate axis system is used to specify positions in numerical control. The system consists of the three linear axes (x, y, z) of the Cartesian coordinate system, plus three rotational axes (a, b, c), as shown in Figure 38.7(a). The rotational axes are used to rotate the workpart to present different surfaces for machining, or to orient the tool or workhead at some angle relative to the part. Most NC systems do not require all six axes. The simplest NC systems (e.g., plotters, pressworking machines for flat sheet-metal stock, and component insertion machines) are positioning systems whose locations can be defined in an x-y plane. Programming of these machines involves specifying a sequence of x-y coordinates. By contrast, some machine tools have five-axis control to shape complex workpart geometries. These systems typically include three linear axes plus two rotational axes.

The coordinates for a rotational NC system are illustrated in Figure 38.7(b). These systems are associated with turning operations on NC lathes. Although the work rotates, this is not one of the controlled axes in a conventional NC turning system. The cutting path of the tool relative to the rotating workpiece is defined in the x-z plane, as shown in our figure.

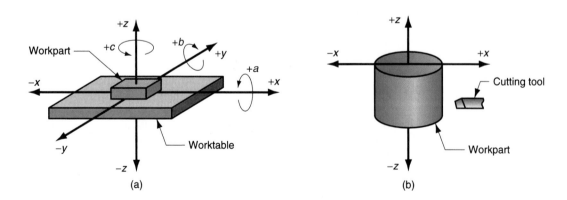

(a)   (b)

FIGURE 38.7   Coordinate systems used in numerical control: (a) for flat and prismatic work, and (b) for rotational work.

In many NC systems, the relative movements between the processing tool and the workpart are accomplished by fixing the part to a worktable and then controlling the positions and motions of the table relative to a stationary or semistationary workhead. Most machine tools and component insertion machines are based on this method of operation. In other systems, the workpart is held stationary and the work head is moved along two or three axes. Flame cutters, *x-y* plotters, and coordinate measuring machines operate in this mode.

Motion control systems based on NC can be divided into two types: (1) point-to-point and (2) continuous path. ***Point-to-point systems***, also called ***positioning systems***, move the workhead (or workpiece) to a programmed location with no regard for the path taken to get to that location. Once the move is completed, some processing action is accomplished by the workhead at the location, such as drilling or punching a hole. Thus, the program consists of a series of point locations at which operations are performed.

Continuous path systems provide continuous simultaneous control of more than one axis, thus controlling the path followed by the tool relative to the part. This permits the tool to perform a process while the axes are moving, enabling the system to generate angular surfaces, two-dimensional curves, or three-dimensional contours in the workpart. This operating scheme is required in drafting machines, certain milling and turning operations, and flame cutting. In machining, continuous path control also goes by the name ***contouring***.

An important aspect of continuous path motion is ***interpolation***, which is concerned with calculating the intermediate points along a path to be followed by the workhead relative to the part. Two common forms of interpolation are linear and circular. ***Linear interpolation*** is used for straight line paths, in which the part programmer specifies the coordinates of the beginning point and end point of the straight line as well as the feed rate to be used. The interpolator then computes the travel speeds of the two or three axes that will accomplish the specified trajectory. ***Circular interpolation*** allows the workhead to follow a circular arc by specifying the coordinates of its beginning and end points together with either the center or radius of the arc. The interpolator computes a series of small straight line segments that will approximate the arc within a defined tolerance.

Another aspect of motion control is concerned with whether the positions in the coordinate system are defined absolutely or incrementally. In ***absolute positioning***, the workhead locations are always defined with respect to the origin of the axis system. In ***incremental positioning***, the next workhead position is defined relative to the present location. The difference is illustrated in Figure 38.8.

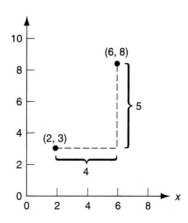

FIGURE 38.8   Absolute vs. incremental positioning. The workhead is at point (2,3) and is to be moved to point (6,8). In absolute positioning, the move is specified by $x = 6$, $y = 8$; while in incremental positioning, the move is specified by $x = 4$, $y = 5$.

FIGURE 38.9 Motor and leadscrew arrangement in an NC positioning system.

38.3.2 ANALYSIS OF NC POSITIONING SYSTEMS

The function of the positioning system is to convert the coordinates specified in the NC part program into relative positions between the tool and workpart during processing. Let us consider how a simple positioning system, shown in Figure 38.9, might operate. The system consists of a worktable on which a workpart is fixtured. The purpose of the table is to move the part relative to a tool or workhead. To accomplish this purpose, the worktable is moved linearly by means of a rotating leadscrew that is driven by a motor. For simplicity, only one axis is shown in our sketch. To provide x-y capability, the system shown would be piggybacked on top of a second axis perpendicular to the first. The leadscrew has a certain pitch p, mm/thread (in/thread) or mm/rev (in/rev). Thus, the table is moved a distance equal to the leadscrew pitch for each revolution. The velocity at which the worktable moves is determined by the rotational speed of the leadscrew.

Two basic types of motion control are used in NC: (a) open loop and (b) closed loop, as shown in Figure 38.10. The difference is that an open-loop system operates

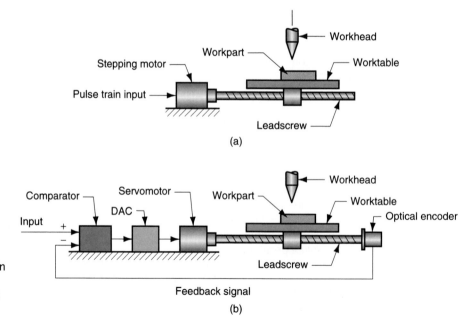

FIGURE 38.10 Two types of motion control in numerical control: (a) open loop and (b) closed loop.

without verifying that the desired position of the worktable has been achieved. A closed-loop control system uses feedback measurement to verify that the position of the worktable is indeed the location specified in the program. Open-loop systems are less expensive than closed-loop systems and are appropriate when the force resisting the actuating motion is minimal, as in point-to-point drilling, for example. Closed-loop systems are normally specified for machine tools that perform continuous path operations such as milling or turning, in which the resisting forces can be significant.

Open-Loop Positioning Systems To turn the leadscrew, an open-loop positioning system typically uses a stepping motor (a.k.a. stepper motor). In NC, the stepping motor is driven by a series of electrical pulses generated by the machine control unit. Each pulse causes the motor to rotate a fraction of one revolution, called the step angle. The allowable step angles must conform to the relationship

$$\alpha = \frac{360}{n_s} \tag{38.1}$$

where α = step angle, degrees; and n_s = the number of step angles for the motor, which must be an integer. The angle through which the motor shaft rotates is given by

$$A_m = \alpha n_p \tag{38.2}$$

where A_m = angle of motor shaft rotation, degrees; n_p = number of pulses received by the motor; and α = step angle, here defined as degrees/pulse. Finally, the rotational speed of the motor shaft is determined by the frequency of pulses sent to the motor:

$$N_m = \frac{60\alpha f_p}{360} \tag{38.3}$$

where N_m = speed of motor shaft rotation, rev/min; f_p = frequency of pulses driving the stepper motor, Hz (pulses/sec), the constant 60 converts pulses/sec to pulses/min; the constant 360 converts degrees of rotation to full revolutions; and α = step angle of the motor, as before.

The motor shaft drives the leadscrew that determines the position and velocity of the worktable. The connection is often designed using a gear reduction to increase the precision of table movement. However, the angle of rotation and rotational speed of the leadscrew are reduced by this gear ratio. The relationships are as follows:

$$A_m = r_g A_{ls} \tag{38.4a}$$

and

$$N_m = r_g N_{ls} \tag{38.4b}$$

where A_m and N_m are the angle of rotation, degrees, and rotational speed, rev/min, of the motor, respectively; A_{ls} and N_{ls} are the angle of rotation, degrees, and rotational speed, rev/min, of the leadscrew, respectively; and r_g = gear reduction between the motor shaft and the leadscrew; for example, a gear reduction of 2 means that the motor shaft rotates through two revolutions for each rotation of the leadscrew.

The linear position of the table in response to the rotation of the leadscrew depends on the leadscrew pitch p, and can be determined as follows:

$$x = \frac{pA_{ls}}{360} \tag{38.5}$$

where x = x-axis position relative to the starting position, mm (in); p = pitch of the leadscrew, mm/rev (in/rev); and $A_{ls}/360$ = the number of revolutions (and partial revolutions) of the leadscrew. By combining Eqs. (38.2), (38.4a), and (38.5) and rearranging, the number of pulses required to achieve a specified x-position increment in a point-to-point system can be found:

$$n_p = \frac{360 r_g x}{p\alpha} = \frac{r_g n_s A_{ls}}{360} \tag{38.6}$$

The velocity of the worktable in the direction of the leadscrew axis can be determined as follows:

$$v_t = f_r = N_{ls} p \tag{38.7}$$

where v_t = table travel speed, mm/min (in/min); f_r = table feed rate, mm/min (in/min); N_{ls} = rotational speed of the leadscrew, rev/min; and p = leadscrew pitch, mm/rev (in/rev). The rotational speed of the leadscrew depends on the frequency of pulses driving the stepping motor:

$$N_{ls} = \frac{60 f_p}{n_s r_g} \tag{38.8}$$

where N_{ls} = leadscrew rotational speed, rev/min; f_p = pulse train frequency, Hz (pulses/sec); n_s = steps/rev, or pulses/rev, and r_g = gear reduction between the motor and the leadscrew. For a two-axis table with continuous path control, the relative velocities of the axes are coordinated to achieve the desired travel direction. Finally, the required pulse frequency to drive the table at a specified feed rate can be obtained by combining Eqs. (38.7) and (38.8) and rearranging to solve for f_p:

$$f_p = \frac{v_t n_s r_g}{60 p} = \frac{f_r n_s r_g}{60 p} = \frac{N_{ls} n_s r_g}{60} = \frac{N_m n_s}{60} \tag{38.9}$$

Example 38.1 Open-Loop Positioning

A stepping motor has 48 step angles. Its output shaft is coupled to a leadscrew with a 4:1 gear reduction (four turns of the motor shaft for each turn of the leadscrew). The leadscrew pitch = 5.0 mm. The worktable of a positioning system is driven by the leadscrew. The table must move a distance of 75.0 mm from its current position at a travel speed of 400 mm/min. Determine (a) how many pulses are required to move the table the specified distance and (b) the motor speed and (c) pulse frequency required to achieve the desired table speed.

Solution: (a) To move a distance $x = 75$ mm, the leadscrew must rotate through an angle calculated as follows:

$$A_{ls} = \frac{360 x}{p} = \frac{360(75)}{5} = 5400°$$

With 48 step angles and a gear reduction of 4, the number of pulses to move the table 75 mm is

$$n_p = \frac{4(48)(5400)}{360} = 2880 \text{ pulses}$$

(b) Equation (38.7) can be used to find the leadscrew speed corresponding to the table speed of 400 mm/min,

$$N_{ls} = \frac{v_t}{p} = \frac{400}{5.0} = 80.0 \text{ rev/min}$$

The motor speed will be 4 times as fast:

$$N_m = r_g N_{ls} = 4(80) = 320 \text{ rev/min}$$

(c) Finally, the pulse rate is given by Eq. (38.13):

$$f_p = \frac{320(48)}{60} = 256 \text{ Hz}$$

∎

Closed-Loop Positioning Systems Closed-loop NC systems, Figure 38.10(b), use servomotors and feedback measurements to ensure that the desired position is achieved. A common feedback sensor used in NC (and also industrial robots) is the optical rotary encoder, illustrated in Figure 38.11. It consists of a light source, a photocell, and a disk containing a series of slots through which the light source can shine to energize the photocell. The disk is connected to a rotating shaft, which in turn is connected directly to the leadscrew. As the leadscrew rotates, the slots cause the light source to be seen by the photocell as a series of flashes, which are converted into an equivalent series of electrical pulses. By counting the pulses and computing the frequency of the pulse train, the leadscrew angle and rotational speed can be determined, and thus worktable position and speed can be calculated using the pitch of the leadscrew.

The equations describing the operation of a closed-loop positioning system are analogous to those for an open-loop system. In the basic optical encoder, the angle between slots in the disk must satisfy the following requirement:

$$\alpha = \frac{360}{n_s} \tag{38.10}$$

where α = angle between slots, degrees/slot; and n_s = the number of slots in the disk, slots/rev; and 360 = degrees/rev. For a certain angular rotation of the leadscrew, the encoder generates a number of pulses given by

$$n_p = \frac{A_{ls}}{\alpha} = \frac{A_{ls} n_s}{360} \tag{38.11}$$

where n_p = pulse count; A_{ls} = angle of rotation of the leadscrew, degrees; and α = angle between slots in the encoder, degrees/pulse. The pulse count can be used to determine the linear x-axis position of the worktable by factoring in the leadscrew pitch. Thus,

$$x = \frac{p n_p}{n_s} = \frac{p A_{ls}}{360} \tag{38.12}$$

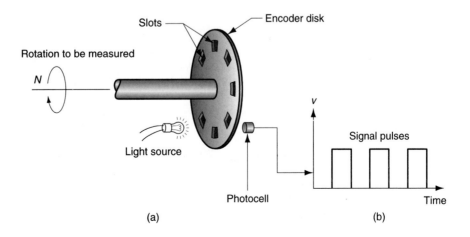

FIGURE 38.11 Optical encoder: (a) apparatus, and (b) series of pulses emitted to measure rotation of disk.

Slots — Encoder disk

Rotation to be measured

N

Light source

Photocell

v

Signal pulses

Time

(a)

(b)

Similarly, the feed rate at which the worktable moves is obtained from the frequency of the pulse train:

$$v_t = f_r = \frac{60 \, p f_p}{n_s} \tag{38.13}$$

where v_t = table travel speed, mm/min (in/min); f_r = feed rate, mm/min (in/min); p = pitch, mm/rev (in/rev); f_p = frequency of the pulse train, Hz (pulses/sec); n_s = number of slots in the encoder disk, pulses/rev; and 60 converts seconds to minutes. The speed relationship given by Eq. (38.7) is also valid for a closed-loop positioning system.

The series of pulses generated by the encoder is compared with the coordinate position and feed rate specified in the part program, and the difference is used by the machine control unit to drive a servomotor that in turn drives the leadscrew and worktable. As with the open-loop system, a gear reduction between the servomotor and the leadscrew can also be used, so Eqs. (38.4) are applicable. A digital-to-analog converter is used to convert the digital signals used by the MCU into a continuous analog signal to operate the drive motor. Closed-loop NC systems of the type described here are appropriate when there is force resisting the movement of the table. Most metal-machining operations fall into this category, particularly those involving continuous path control such as milling and turning.

Example 38.2 NC Closed-Loop Positioning

An NC worktable is driven by a closed-loop positioning system consisting of a servomotor, leadscrew, and optical encoder. The leadscrew has a pitch = 5.0 mm and is coupled to the motor shaft with a gear ratio of 4:1 (four turns of the motor for each turn of the leadscrew). The optical encoder generates 100 pulses/rev of the leadscrew. The table has been programmed to move a distance of 75.0 mm at a feed rate = 400 mm/min. Determine (a) how many pulses are received by the control system to verify that the table has moved exactly 75.0 mm; and (b) the pulse rate and (c) motor speed that correspond to the specified feed rate.

Solution: (a) Rearranging Eq. (38.12) to find n_p,

$$n_p = \frac{x n_s}{p} = \frac{75(100)}{5} = 1500 \text{ pulses}$$

(b) The pulse rate corresponding to 400 mm/min can be obtained by rearranging Eq. (38.13):

$$f_p = \frac{f_r n_s}{60 \, p} = \frac{400(100)}{60(5)} = 133.33 \text{ Hz}$$

(c) Leadscrew rotational speed is the table velocity divided by the pitch:

$$N_{ls} = \frac{f_r}{p} = 80 \text{ rev/min}$$

With a gear ratio r_g = 4.0, the motor speed N = 4(80) = 320 rev/min ∎

Precision in Positioning Three critical measures of precision in positioning are control resolution, accuracy, and repeatability. These terms are most easily explained by considering a single axis of the position system.

Control resolution refers to the system's ability to divide the total range of the axis movement into closely spaced points that can be distinguished by the control unit. **Control resolution** is defined as the distance separating two adjacent control points in the axis movement. Control points are sometimes called **addressable points** because they are locations along the axis to which the worktable can be directed to go. It is desirable for the control resolution to be as small as possible. This depends on limitations imposed by (1) the electromechanical components of the positioning system, and/or (2) the number of bits used by the controller to define the axis coordinate location.

The electromechanical factors that limit resolution include leadscrew pitch, gear ratio in the drive system, and the step angle in a stepping motor (for an open-loop system) or the angle between slots in an encoder disk (for a closed-loop system). Together, these factors determine a control resolution, or minimum distance that the worktable can be moved. For example, the control resolution for an open-loop system driven by a stepper motor with a gear reduction between the motor shaft and the leadscrew is given by

$$CR_1 = \frac{p}{n_s r_g} \tag{38.14a}$$

where CR_1 = control resolution of the electromechanical components, mm (in); p = leadscrew pitch, mm/rev (in/rev); n_s = number of steps/rev; and r_g = gear reduction.

The corresponding expression for a closed-loop positioning system is similar but does not include the gear reduction because the encoder is connected directly to the leadscrew. There is no gear reduction. Thus, control resolution for a closed-loop system is defined as follows:

$$CR_1 = \frac{p}{n_s} \tag{38.14b}$$

where n_s in this case refers to the number of slots in the optical encoder.

Although unusual in modern computer technology, the second possible factor that could limit control resolution is the number of bits defining the axis coordinate value. For example, this limitation may be imposed by the bit storage capacity of the controller. If B = the number of bits in the storage register for the axis, then the number of control points into which the axis range can be divided $= 2^B$. Assuming that the control points are separated equally within the range, then

$$CR_2 = \frac{L}{2^B - 1} \tag{38.15}$$

where CR_2 = control resolution of the computer control system, mm (in); and L = axis range, mm (in). The control resolution of the positioning system is the maximum of the two values; that is,

$$CR = \text{Max}\{CR_1, CR_2\} \tag{38.16}$$

It is generally desirable for $CR_2 \leq CR_1$, meaning that the electromechanical system is the limiting factor in control resolution.

When a positioning system is directed to move the worktable to a given control point, the capability of the system to move to that point will be limited by mechanical errors. These errors are due to a variety of inaccuracies and imperfections in the mechanical system, such as play between the leadscrew and the worktable, backlash in the gears, and deflection of machine components. It is convenient to assume that the errors form a statistical distribution about the control point that is an unbiased normal distribution with mean = 0. If we further assume that the standard deviation of the distribution is constant over the range of the axis under consideration, then nearly all of the mechanical errors (99.73%) are contained within ±3 standard deviations of the

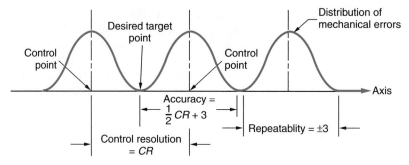

FIGURE 38.12
A portion of a linear positioning system axis, with definition of control resolution, accuracy, and repeatability.

control point. This is pictured in Figure 38.12 for a portion of the axis range, which includes three control points.

Given these definitions of control resolution and mechanical error distribution, let us now consider accuracy and repeatability. Accuracy is defined in a worst-case scenario in which the desired target point lies exactly between two adjacent control points. Since the system can only move to one or the other of the control points, there will be an error in the final position of the worktable. If the target were closer to one of the control points, then the table would be moved to the closer control point and the error would be smaller. It is appropriate to define accuracy in the worst case. The ***accuracy*** of any given axis of a positioning system is the maximum possible error that can occur between the desired target point and the actual position taken by the system; in equation form,

$$\text{Accuracy} = 0.5\, CR + 3\sigma \tag{38.17}$$

where CR = control resolution, mm (in); and σ = standard deviation of the error distribution, mm (in).

Repeatability refers to the capability of a positioning system to return to a given control point that has been previously programmed. This capability can be measured in terms of the location errors encountered when the system attempts to position itself at the control point. Location errors are a manifestation of the mechanical errors of the positioning system, which are defined by an assumed normal distribution, as described above. Thus, the ***repeatability*** of any given axis of a positioning system can be defined as the range of mechanical errors associated with the axis; this reduces to

$$\text{Repeatability} = \pm 3\sigma \tag{38.18}$$

Example 38.3 Control Resolution, Accuracy, and Repeatability

Referring back to Example 38.1, the mechanical inaccuracies in the open-loop positioning system can be described by a normal distribution whose standard deviation = 0.005 mm. The range of the worktable axis is 550 mm, and there are 16 bits in the binary register used by the digital controller to store the programmed position. Determine (a) control resolution, (b) accuracy, and (c) repeatability for the positioning system.

Solution: (a) Control resolution is the greater of CR_1 and CR_2 as defined by Eqs. (38.14a) and (38.15):

$$CR_1 = \frac{p}{n_s r_g} = \frac{5.0}{48(4)} = 0.0260 \text{ mm}$$

$$CR_2 = \frac{L}{2^B - 1} = \frac{550}{2^{16} - 1} = \frac{550}{65,535} = 0.0084 \text{ mm}$$

$$CR = \text{Max}\{0.0260, 0.0084\} = 0.0260 \text{ mm}$$

(b) Accuracy is given by Eq. (38.17):

$$\text{Accuracy} = 0.5(0.0260) + 3(0.005) = 0.0280 \text{ mm}$$

(c) Repeatability $= \pm 3(0.005) = \pm 0.015$ mm.

38.3.3 NC PART PROGRAMMING

In machine tool applications, the task of programming the system is called NC part programming because the program is prepared for a given part. It is usually accomplished by someone familiar with the metalworking process who has learned the programming procedure for the particular equipment in the plant. For other processes, other terms may be used for programming, but the principles are similar and a trained individual is needed to prepare the program. Computer systems are used extensively to prepare NC programs.

Part programming requires the programmer to define the points, lines, and surfaces of the workpart in the axis system, and to control the movement of the cutting tool relative to these defined part features. Several part programming techniques are available, the most important of which are (1) manual part programming, (2) computer-assisted part programming, (3) CAD/CAM-assisted part programming, and (4) manual data input.

Manual Part Programming For simple point-to-point machining jobs, such as drilling operations, manual programming is often the easiest and most economical method. Manual part programming uses basic numerical data and special alphanumeric codes to define the steps in the process. For example, to perform a drilling operation, a command of the following type is entered:

$$\text{n010x70.0 \quad y85.5 \quad f175 \quad s500}$$

Each "word" in the statement specifies a detail in the drilling operation. The n-word (n010) is simply a sequence number for the statement. The x- and y-words indicate the x and y coordinate positions ($x = 70.0$ mm and $y = 85.5$ mm). The f-word and s-word specify the feed rate and spindle speed to be used in the drilling operation (feed rate $= 175$ mm/min and spindle speed $= 500$ rev/min). The complete NC part program consists of a sequence of statements similar to the above command.

Computer-Assisted Part Programming Computer-assisted part programming involves the use of a high-level programming language. It is suited to the programming of more complex jobs than manual programming. The first part programming language was APT (Automatically Programmed Tooling), developed as an extension of the original NC machine tool research and first used in production around 1960.

In APT, the part programming task is divided into two steps: (1) definition of part geometry and (2) specification of tool path and operation sequence. In step 1, the part programmer defines the geometry of the workpart by means of basic geometric elements such as points, lines, planes, circles, and cylinders. These elements are defined using APT geometry statements, such as

$$\text{P1} = \text{POINT}/25.0, \ 150.0$$

$$\text{L1} = \text{LINE}/\text{P1}, \ \text{P2}$$

P1 is a point defined in the x-y plane located at $x = 25$ mm and $y = 150$ mm. L1 is a line that goes through points P1 and P2. Similar statements can be used to define circles, cylinders, and other geometry elements. Most workpart shapes can be described using statements like these to define their surfaces, corners, edges, and hole locations.

Specification of the tool path is accomplished with APT motion statements. A typical statement for point-to-point operation is

GOTO/P1

This directs the tool to move from its current location to a position defined by P1, where P1 has been defined by a previous APT geometry statement. Continuous path commands use geometry elements such as lines, circles, and planes. For example, the command

GORGT/L3, PAST, L4

directs the tool to go right (GORGT) along line L3 until it is positioned just past line L4 (of course, L4 must be a line that intersects L3).

Additional APT statements are used to define operating parameters such as feed rates, spindle speeds, tool sizes, and tolerances. When completed, the part programmer enters the APT program into the computer, where it is processed to generate low-level statements (similar to statements prepared in manual part programming) that can be used by a particular machine tool.

CAD/CAM-Assisted Part Programming   The use of CAD/CAM takes computer-assisted part programming a step further by using a computer graphics system (CAD/CAM system) to interact with the programmer as the part program is being prepared. In the conventional use of APT, a complete program is written and then entered into the computer for processing. Many programming errors are not detected until computer processing. When a CAD/CAM system is used, the programmer receives immediate visual verification when each statement is entered, to determine whether the statement is correct. When part geometry is entered by the programmer, the element is graphically displayed on the monitor. When the tool path is constructed, the programmer can see exactly how the motion commands will move the tool relative to the part. Errors can be corrected immediately rather than after the entire program has been written.

Interaction between programmer and programming system is a significant benefit of CAD/CAM-assisted programming. There are other important benefits of using CAD/CAM in NC part programming. First, the design of the product and its components may have been accomplished on a CAD/CAM system. The resulting design database, including the geometric definition of each part, can be retrieved by the NC programmer to use as the starting geometry for part programming. This retrieval saves valuable time compared to reconstructing the part from scratch using the APT geometry statements.

Second, special software routines are available in CAD/CAM-assisted part programming to automate portions of the tool path generation, such as profile milling around the outside periphery of a part, milling a pocket into the surface of a part, surface contouring, and certain point-to-point operations. These routines are called by the part programmer as special *macro* commands. Their use results in significant savings in programming time and effort.

Manual Data Input   Manual data input (MDI) is a method in which a machine operator enters the part program in the factory. The method involves use of a CRT display with graphics capability at the machine tool controls. NC part programming statements are entered using a menu-driven procedure that requires minimum training of the machine tool operator. Because part programming is simplified and does not require a special staff of NC part programmers, MDI is a way for small machine shops to economically implement numerical control into their operations.

38.3.4 APPLICATIONS OF NUMERICAL CONTROL

Machining is an important application area for numerical control, but the operating principle of NC can be applied to other operations as well. There are many industrial processes in which the position of a workhead must be controlled relative to the part or product being worked on. We divide the applications into two categories: (1) machine tool applications, and (2) nonmachine tool applications. It should be noted that the applications are not all identified by the name numerical control in their respective industries.

In the machine tool category, NC is widely used for *machining operations* such as turning, drilling, and milling (Sections 22.2, 22.3, and 22.4, respectively). The use of NC in these processes has motivated the development of highly automated machine tools called *machining centers*, which change their own cutting tools to perform a variety of machining operations under NC program control (Section 22.5). In addition to machining, other numerically controlled machine tools include (1) grinding machines (Section 25.1); (2) sheet metal pressworking machines (Section 20.5.2); (3) tube-bending machines (Section 20.7); and (4) thermal cutting processes (Section 26.3).

In the nonmachine tool category, NC applications include (1) tape-laying machines and filament-winding machines for composites (Section 15.2.3 and Section 15.4); (2) welding machines, both arc welding (Section 31.1) and resistance welding (Section 31.2); (3) component-insertion machines in electronics assembly (Sections 35.3 and 35.4); (4) drafting machines; and (5) coordinate measuring machines for inspection (Section 42.6.1).

Benefits of NC relative to manually operated equipment in these applications include (1) reduced nonproductive time, which results in shorter cycle times, (2) lower manufacturing lead times, (3) simpler fixturing, (4) greater manufacturing flexibility, (5) improved accuracy, and (6) reduced human error.

38.4 INDUSTRIAL ROBOTICS

An industrial robot is a general-purpose programmable machine possessing certain anthropomorphic features. The most obvious anthropomorphic, or human-like, feature is the robot's mechanical arm, or manipulator. The control unit for a modern industrial robot is a computer that can be programmed to execute rather sophisticated subroutines, thus providing the robot with an intelligence that sometimes seems almost human. The robot's manipulator, combined with a high-level controller, allows an industrial robot to perform a variety of tasks such as loading and unloading production machine, spot welding, and spray painting. Robots are typically used as substitutes for human workers in these tasks. The first industrial robot was installed in a die-casting operation at Ford Motor Company. The robot's job was to unload die castings from the die-casting machine.

In this section, we consider various aspects of robot technology and applications, including how industrial robots are programmed to perform their tasks.

38.4.1 ROBOT ANATOMY

An industrial robot consists of a mechanical manipulator and a controller to move it and perform other related functions. The mechanical manipulator consists of joints and links that can position and orient the end of the manipulator relative to its base. The controller unit consists of electronic hardware and software to operate the joints in a coordinated fashion to execute the programmed work cycle. *Robot anatomy* is concerned with the

FIGURE 38.13   The manipulator of a modern industrial robot. (Photo courtesy of Adept Technology, Inc., Pleasanton, California.)

mechanical manipulator and its construction. Figure 38.13 shows one of the common industrial robot configurations.

Manipulator Joints and Links   A joint in a robot is similar to a joint in a human body. It provides relative movement between two parts of the body. Connected to each joint are an input link and an output link. Each joint moves its output link relative to its input link. The robot manipulator consists of a series of link–joint–link combinations. The output link of one joint is the input link for the next joint. Typical industrial robots have five or six joints. The coordinated movement of these joints gives the robot its ability to move, position, and orient objects to perform useful work. Manipulator joints can be classified as linear or rotating, indicating the motion of the output link relative to the input link.

Manipulator Design   Using joints of the two basic types, each joint separated from the previous by a link, the manipulator is constructed. Most industrial robots are mounted to the floor. We can identify the base as link 0; this is the input link to joint 1 whose output is link 1, which is the input to joint 2 whose output link is link 2; and so forth, for the number of joints in the manipulator.

   Robot manipulators can usually be divided into two sections: arm-and-body assembly and wrist assembly. There are typically three joints associated with the arm-and-body assembly, and two or three joints associated with the wrist. The function of the arm-and-body is to position an object or tool, and the wrist function is to properly orient the object or tool. Positioning is concerned with moving the part or tool from one location to another. Orientation is concerned with precisely aligning the object relative to some stationary location in the work area.

To accomplish these functions, arm-and-body designs differ from those of the wrist. Positioning requires large spatial movements, while orientation requires twisting and rotating motions to align the part or tool relative to a fixed position in the workplace. The arm-and-body consists of large links and joints, whereas the wrist consists of short links. The arm-and-body joints often consist of both linear and rotating types, while the wrist joints are almost always rotating types.

There are five basic arm-and-body configurations available in commercial robots, identified in Figure 38.14. The design shown in part (e) of the figure and in Figure 38.13 is called a SCARA robot, which stands for "selectively compliant assembly robot arm." It is similar to a jointed arm anatomy, except that the shoulder and elbow joints have vertical axes of rotation, thus providing rigidity in the vertical direction but relative compliance in the horizontal direction.

FIGURE 38.14   Five common anatomies of commercial industrial robots: (a) polar, (b) cylindrical, (c) Cartesian coordinate, (d) jointed-arm, and (e) SCARA, or selectively compliant assembly robot arm.

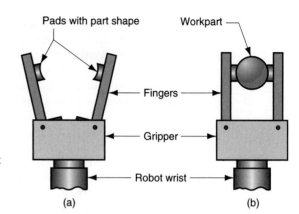

FIGURE 38.15   A robot gripper: (a) open and (b) closed to grasp a workpart.

The wrist is assembled to the last link in any of these arm-and-body configurations. The SCARA is sometimes an exception because it is almost always used for simple handling and assembly tasks involving vertical motions. Therefore, a wrist is not usually present at the end of its manipulator. Substituting for the wrist on the SCARA is usually a gripper to grasp components for movement and/or assembly.

Work Volume and Precision of Motion   One of the important technical considerations of an industrial robot is the size of its work volume. **Work volume** is defined as the envelope within which a robot manipulator can position and orient the end of its wrist. This envelope is determined by the number of joints, as well as their types and ranges, and the sizes of the links. Work volume is important because it plays a significant role in determining which applications a robot can perform.

The definitions of control resolution, accuracy, and repeatability developed in Section 38.3.2 for NC positioning systems apply to industrial robots. A robot manipulator is, after all, a positioning system. In general, the links and joints of robots are not nearly as rigid as their machine tool counterparts, and so the accuracy and repeatability of their movements are not as good.

End Effectors   An industrial robot is a general-purpose machine. For a robot to be useful in a particular application, it must be equipped with special tooling designed for the application. An **end effector** is the special tooling that connects to the robot's wrist-end to perform the specific task. There are two general types of end effector: tools and grippers. A **tool** is used when the robot must perform a processing operation. The special tools include spot-welding guns, arc-welding tools, spray-painting nozzles, rotating spindles, heating torches, and assembly tools (e.g., automatic screwdriver). The robot is programmed to manipulate the tool relative to the workpart being processed.

Grippers are designed to grasp and move objects during the work cycle. The objects are usually workparts, and the end effector must be designed specifically for the part. Grippers are used for part placement applications, machine loading and unloading, and palletizing. Figure 38.15 shows a typical gripper configuration.

38.4.2 CONTROL SYSTEMS AND ROBOT PROGRAMMING

The robot's controller consists of the electronic hardware and software to control the joints during execution of a programmed work cycle. Most robot control units today are based on a microcomputer system. The control systems in robotics can be classified as follows:

1. ***Playback with point-to-point (PTP) control.*** As in numerical control, robot motion systems can be divided into point-to-point and continuous path. The program for a point-to-point playback robot consists of a series of point locations and the sequence in which these points must be visited during the work cycle. During programming, these points are recorded into memory, and then subsequently played back during execution of the program. In a point-to-point motion, the path taken to get to the final position is not controlled.

2. ***Playback with continuous path (CP) control.*** Continuous path control is similar to PTP, except motion paths rather than individual points are stored in memory. In certain types of regular CP motions, such as a straight line path between two point locations, the trajectory required by the manipulator is computed by the controller unit for each move. For irregular continuous motions, such as a path followed in spray painting, the path is defined by a series of closely spaced points that approximate the irregular smooth path. Robots capable of continuous path motions can also execute point-to-point movements.

3. ***Intelligent control.*** Modern industrial robots exhibit characteristics that often make them appear to be acting intelligently. These characteristics include the ability to respond to sophisticated sensors such as machine vision, make decisions when things go wrong during the work cycle, make computations, and communicate with humans. Robot intelligence is implemented using powerful microprocessors and advanced programming techniques.

Robots execute a stored program of instructions that define the sequence of motions and positions in the work cycle, much like a part program in NC. In addition to motion instructions, the program may include instructions for other functions such as interacting with external equipment, responding to sensors, and processing data.

There are two basic methods used to teach modern robots their programs: lead-through programming and computer programming languages. ***Leadthrough programming*** involves a "teach-by-showing" method in which the manipulator is moved by the programmer through the sequence of positions in the work cycle. The controller records each position in memory for subsequent playback. Two procedures for leading the robot through the motion sequence are available: powered leadthrough and manual leadthrough. In ***powered leadthrough***, the manipulator is driven by a control box that has toggle switches or press buttons to control the movements of the joints. Using the control box, the programmer moves the manipulator to each location, recording the corresponding joint positions into memory. Powered leadthrough is the common method for programming playback robots with point-to-point control. ***Manual leadthrough*** is typically used for playback robots with continuous path control. In this method, the programmer physically moves the manipulator wrist through the motion cycle. For spray painting and certain other jobs, this is a more convenient means of programming the robot.

Computer programming languages for programming robots have evolved from the use of microcomputer controllers. The first commercial language was introduced around 1979. Computer languages provide a convenient way to integrate certain nonmotion functions into the work cycle, such as decision logic, interlocking with other equipment, and interfacing with sensors. A more thorough discussion of robot programming is presented in reference [6].

38.4.3 APPLICATIONS OF INDUSTRIAL ROBOTS

Some industrial work lends itself to robot applications. The following are the important characteristics of a work situation that tend to promote the substitution of a robot in place of a human worker: (1) the work environment is hazardous for humans, (2) the work cycle is repetitive, (3) the work is performed at a stationary location, (4) part or tool handling would be difficult for humans, (5) it is a multishift operation, (6) there are long production runs and infrequent changeovers, and (7) part positioning and orientation are established at the beginning of the work cycle, since most robots cannot see.

Applications of industrial robots that tend to match these characteristics can be divided into three basic categories: (1) material handling, (2) processing operations, and (3) assembly and inspection.

Material handling applications involve the movement of materials or parts from one location and orientation to another. To accomplish this relocation task, the robot is equipped with a gripper. As noted earlier, the gripper must be custom-designed to grasp the particular part in the application. Material handling applications include material transfer (part placement, palletizing, depalletizing) and machine loading and/or unloading (e.g., machine tools, presses, and plastic molding).

Processing operations require the robot to manipulate a tool as its end effector. The applications include spot welding, continuous arc welding, spray coating, and certain metal cutting and deburring operations in which the robot manipulates a special tool. In each of these operations, the tool is used as the robot's end effector. An application of spot welding is illustrated in Figure 38.16. Spot welding is a common application of industrial robots in the automotive industry.

Assembly and inspection applications cannot be classified neatly in either of the previous categories; they sometimes involve part handling and other times manipulation of a tool. *Assembly* applications often involve the stacking of one part onto another part—basically a part handling task. In other assembly operations a tool is manipulated, such as an automatic screwdriver. Similarly, *inspection* operations sometimes require the robot to position a workpart relative to an inspection device, or to load a part into an inspection machine; other applications involve the manipulation of a sensor to perform an inspection.

FIGURE 38.16
A portion of an automobile assembly line in which robots perform spot-welding operations. (Photo courtesy of Ford Motor Company, Dearborn, Michigan.)

REFERENCES

[1] Asfahl, C. R. *Robots and Manufacturing Automation.* John Wiley & Sons, Inc., New York, 1992.

[2] Bollinger, J. G., and Duffie N. A. *Computer Control of Machines and Processes.* Addison-Wesley Longman, Inc., New York, 1989.

[3] Chang, C-H., and Melkanoff, M. A. *NC Machine Programming and Software Design,* 3rd ed. Prentice Hall, Inc., Upper Saddle River, New Jersey, 2005.

[4] Engelberger, J. F. *Robotics in Practice: Management and Applications of Robotics in Industry.* AMACOM, New York, 1985.

[5] Groover, M. P. *Automation, Production Systems, and Computer Integrated Manufacturing,* 3rd ed. Pearson/Prentice Hall, Upper Saddle River, New Jersey, 2008.

[6] Groover, M. P., Weiss, M., Nagel, R. N., and Odrey, N. G. *Industrial Robotics: Technology, Programming, and Applications.* McGraw-Hill, New York, 1986.

[7] Hughes, T. A., *Programmable Controllers,* 4th ed. Instrumentation, Systems, and Automation Society, Research Triangle Park, North Carolina, 2005.

[8] Pessen, D. W. *Industrial Automation.* John Wiley & Sons, Inc., New York, 1989.

[9] Seames W. *Computer Numerical Control, Concepts and Programming.* Delmar-Thomson Learning, Albany, New York, 2002.

[10] Webb, J. W., and Reis, R. A. *Programmable Logic Controllers: Principles and Applications,* 5th ed. Pearson/Prentice Hall, Upper Saddle River, New Jersey, 2003.

[11] Weber, A. "Robot Dos and Don'ts," *Assembly,* February 2005, pp. 50–57.

REVIEW QUESTIONS

38.1. Define the term *manufacturing system*.

38.2. What are the three basic components of an automated system?

38.3. What are some of the advantages of using electrical power in an automated system?

38.4. What is the difference between a closed-loop control system and an open-loop control system?

38.5. What is the difference between fixed automation and programmable automation?

38.6. What is a sensor?

38.7. What is an actuator in an automated system?

38.8. What is a contact input interface?

38.9. What is a programmable logic controller?

38.10. Identify and briefly describe the three basic components of a numerical control (NC) system.

38.11. What is the difference between point-to-point and continuous path in a motion control system?

38.12. What is the difference between absolute positioning and incremental positioning?

38.13. What is the difference between an open-loop positioning system and a closed-loop positioning system?

38.14. Under what circumstances is a closed-loop positioning system preferable to an open-loop system?

38.15. Explain the operation of an optical encoder.

38.16. Why should the electromechanical system rather than the controller storage register be the limiting factor in control resolution?

38.17. What is manual data input in NC part programming?

38.18. Identify some of the non-machine tool applications of numerical control.

38.19. What are some of the benefits usually cited for NC compared to using manual alternative methods?

38.20. What is an industrial robot?

38.21. How is an industrial robot similar to numerical control?

38.22. What is an end effector?

38.23. In robot programming, what is the difference between powered leadthrough and manual leadthrough?

MULTIPLE CHOICE QUIZ

There are 21 correct answers in the following multiple choice questions (some questions have multiple answers that are correct). To attain a perfect score on the quiz, all correct answers must be given. Each correct answer is worth 1 point. Each omitted answer or wrong answer reduces the score by 1 point, and each additional answer beyond the correct number of answers reduces the score by 1 point. Percentage score on the quiz is based on the total number of correct answers.

38.1. The three components of an automated system are which of the following: (a) actuators, (b) communication system, (c) control system, (d) feedback loop, (e) humans, (f) power, (g) program of instructions, and (h) sensors?

38.2. The three basic types of automated systems used in manufacturing are fixed automation, programmable automation, and flexible automation. Flexible automation is an extension of programmable automation in which there is virtually no lost production time for setup changes or reprogramming: (a) true or (b) false?

38.3. The input/output relationship of a sensor is called which one of the following: (a) analog, (b) converter, (c) sensitivity, or (d) transfer function?

38.4. A stepper motor is which one of the following types of devices: (a) actuator, (b) interface device, (c) pulse counter, or (d) sensor?

38.5. A contact input interface is a device that reads analog data into the computer from an external source: (a) true of (b) false?

38.6. A programmable logic controller normally replaces which one of the following in control applications: (a) computer numerical control, (b) distributed process control, (c) humans, (d) industrial robots, or (e) relay control panel?

38.7. The standard coordinate system for numerical control machine tools is based on which one of the following: (a) Cartesian coordinates, (b) cylindrical coordinates, or (c) polar coordinates?

38.8. Identify which of the following applications are point-to-point and not continuous path operations (three correct answers): (a) arc welding, (b) drilling, (c) hole punching in sheet metal, (d) milling, (e) spot welding, and (f) turning?

38.9. The ability of a positioning system to return to a previously defined location is measured by which one of the following terms: (a) accuracy, (b) control resolution, or (c) repeatability?

38.10. The APT command GORGT is which of the following (two best answers): (a) continuous path command, (b) geometry statement involving a volume of revolution about a central axis, (c) name of the humanoid in the latest Star Wars movie, (d) point-to-point command, and (e) tool path command in which the tool must go right in the next move?

38.11. The arm-and-body of a robot manipulator generally performs which one of the following functions in an application: (a) holds the end effector, (b) orients the end effector within the work volume, or (c) positions the wrist within the work volume?

38.12. A SCARA robot is normally associated with which one of the following applications: (a) arc welding, (b) assembly, (c) inspection, (d) machine loading and unloading, or (e) resistance welding?

38.13. In robotics, spray-painting applications are classified as which of the following: (a) continuous path operation or (b) point-to-point operation?

38.14. Which of the following are characteristics of work situations that tend to promote the substitution of a robot in place of a human worker (three best answers): (a) frequent job changeovers, (b) hazardous work environment, (c) repetitive work cycle, (d) multiple work shifts, and (e) task requires mobility?

PROBLEMS

Open-Loop Positioning Systems

38.1. A leadscrew with a 7.5 mm pitch drives a worktable in a numerical control positioning system. The leadscrew is powered by a stepping motor which has 200 step angles. The worktable is programmed to move a distance of 120 mm from its present position at a travel speed of 300 mm/min. Determine (a) the number of pulses required to move the table the specified distance and (b) the required motor speed and pulse rate to achieve the desired table speed.

38.2. Referring to Problem 38.1, the mechanical inaccuracies in the open-loop positioning system can be described by a normal distribution whose standard deviation = 0.005 mm. The range of the worktable axis is 500 mm, and there are 12 bits in the binary register used by the digital controller to store the programmed position. For the positioning system, determine (a) control resolution, (b) accuracy, and (c) repeatability. (d) What is the minimum number of bits that the binary register should have so that the mechanical drive system becomes the limiting component on control resolution?

38.3. A stepping motor has 200 step angles. Its output shaft is directly coupled to leadscrew with pitch = 0.250 in. A worktable is driven by the leadscrew. The table must move a distance of 5.00 in from its present position at a travel speed of 20.0 in/min. Determine (a) the number of pulses required to move the table the specified distance and (b) the required motor speed and pulse rate to achieve the specified table speed.

38.4. A stepping motor with 100 step angles is coupled to a leadscrew through a gear reduction of 9:1 (9 rotations of the motor for each rotation of the leadscrew). The leadscrew has 5 threads/in. The worktable driven by the leadscrew must move a distance = 10.00 in at a feed rate of 30.0 in/min. Determine (a) number of pulses required to move the table, and (b) the required motor speed and pulse rate to achieve the desired table speed.

38.5. The drive unit for a positioning table is driven by a leadscrew directly coupled to the output shaft of a stepping motor. The pitch of the leadscrew = 0.18 in. The table must have a linear speed = 35 in/min, and a positioning accuracy = 0.001 in. Mechanical errors in the motor, leadscrew, and table connection are characterized by a normal distribution with standard deviation = 0.0002 in. Determine (a) the minimum number of step angles in the stepping motor to achieve the accuracy, (b) the associated step angle, and (c) the frequency of the pulse train required to drive the table at the desired speed.

38.6. The positioning table for a component insertion machine uses a stepping motor and leadscrew mechanism. The design specifications require a table speed of 40 in/min and an accuracy = 0.0008 in. The pitch of the leadscrew = 0.2 in, and the gear ratio = 2:1 (two turns of the motor for each turn of the leadscrew). The mechanical errors in the motor, gear box, leadscrew, and table connection are characterized by a normal distribution with standard deviation = 0.0001 in. Determine (a) the minimum number of step angles in the stepping motor, and (b) the frequency of the pulse train required to drive the table at the desired maximum speed.

38.7. The drive unit of a positioning table for a component insertion machine is based on a stepping motor and leadscrew mechanism. The specifications are for the table speed to be 25 mm/s over a 600 mm range and for the accuracy to be 0.025 mm. The pitch of the leadscrew = 4.5 mm, and the gear ratio = 5:1 (five turns of the motor for each turn of the leadscrew). The mechanical errors in the motor, gear box, leadscrew, and table connection are characterized by a normal distribution with standard deviation = 0.005 mm. Determine (a) the

minimum number of step angles in the stepping motor, and (b) the frequency of the pulse train required to drive the table at the desired maximum speed for the stepping motor in part (a).

38.8. The two axes of an x-y positioning table are each driven by a stepping motor connected to a leadscrew with a 10:1 gear reduction. The step angle on each stepping motor is 7.5°. Each leadscrew has a pitch = 5.0 mm and provides an axis range = 300.0 mm. There are 16 bits in each binary register used by the controller to store position data for the two axes. (a) What is the control resolution of each axis? (b) What are the required rotational speeds and corresponding pulse train frequencies of each stepping motor in order to drive the table at 600 mm/min in a straight line from point (25,25) to point (100,150)? Ignore acceleration.

38.9. The y-axis of an x-y positioning table is driven by a stepping motor that is connected to a leadscrew with a 3:1 gear reduction (three turns of the motor for each turn of the leadscrew). The stepping motor has 72 step angles. The leadscrew has 5 threads/in and provides an axis range = 30.0 in. There are 16 bits in each binary register used by the controller to store position data for the axis. (a) What is the control resolution of the y-axis? Determine (b) the required rotational speed of the y-axis stepping motor and (c) the corresponding pulse train frequency to drive the table in a straight line from point ($x = 20$ in, $y = 25$ in) to point ($x = 4.5$ in, $y = 7.5$ in) in exactly 30 sec. Ignore acceleration.

38.10. The two axes of an x-y positioning table are each driven by a stepping motor connected to a leadscrew with a 4:1 gear reduction. The number of step angles on each stepping motor is 200. Each leadscrew has a pitch = 5.0 mm and provides an axis range = 400.0 mm. There are 16 bits in each binary register used by the controller to store position data for the two axes. (a) What is the control resolution of each axis? (b) What are the required rotational speeds and corresponding pulse train frequencies of each stepping motor in order to drive the table at 600 mm/min in a straight line from point (25,25) to point (300,150)? Ignore acceleration.

Closed-Loop Positioning Systems

38.11. A numerical control (NC) machine tool table is powered by a servomotor, leadscrew, and optical encoder. The leadscrew has a pitch = 5.0 mm and is connected to the motor shaft with a gear ratio of 16:1 (16 turns of the motor for each turn of the leadscrew). The optical encoder is connected

directly to the leadscrew and generates 200 pulses/rev of the leadscrew. The table must move a distance = 100 mm at a feed rate = 500 mm/min. Determine (a) the pulse count received by the control system to verify that the table has moved exactly 100 mm; and (b) the pulse rate and (c) motor speed that correspond to the feed rate of 500 mm/min.

38.12. The worktable of a numerical control machine tool is driven by a closed-loop positioning system which consists of a servomotor, leadscrew, and optical encoder. The leadscrew has 4 threads/in and is coupled directly to the motor shaft (gear ratio = 1:1). The optical encoder generates 200 pulses per motor revolution. The table has been programmed to move a distance of 7.5 in at a feed rate = 20.0 in/min. (a) How many pulses are received by the control system to verify that the table has moved the programmed distance? What are (b) the pulse rate and (c) motor speed that correspond to the specified feed rate?

38.13. A leadscrew coupled directly to a dc servomotor is used to drive one of the table axes of an NC milling machine. The leadscrew has 5 threads/in. The optical encoder attached to the leadscrew emits 100 pulses/rev of the leadscrew. The motor rotates at a maximum speed of 800 rev/min. Determine (a) the control resolution of the system, expressed in linear travel distance of the table axis; (b) the frequency of the pulse train emitted by the optical encoder when the servomotor operates at maximum speed; and (c) the travel speed of the table at the maximum rpm of the motor.

38.14. Solve the previous problem only the servomotor is connected to the leadscrew through a gear box whose reduction ratio = 12:1 (12 revolutions of the motor for each revolution of the leadscrew).

38.15. A leadscrew connected directly to a DC servomotor is the drive system for a positioning table. The leadscrew pitch = 4 mm. The optical encoder attached to the leadscrew emits 250 pulses/rev of the leadscrew. Determine (a) the control resolution of the system, expressed in linear travel distance of the table axis, (b) the frequency of the pulse train emitted by the optical encoder when the servomotor operates at 14 rev/s, and (c) the travel speed of the table at the operating speed of the motor.

38.16. A milling operation is performed on an NC machining center. Total travel distance = 300 mm in a direction parallel to one of the axes of the worktable. Cutting speed = 1.25 m/s and chip load = 0.05 mm. The end milling cutter has four teeth and its diameter = 20.0 mm. The axis uses a DC servomotor whose output shaft is coupled to a leadscrew with pitch = 6.0 mm. The feedback sensing device connected to

the leadscrew is an optical encoder that emits 250 pulses per revolution. Determine (a) feed rate and time to complete the cut, and (b) rotational speed of the motor and the pulse rate of the encoder at the feed rate indicated.

38.17. An end milling operation is carried out along a straight line path that is 325 mm long. The cut is in a direction parallel to the x-axis on an NC machining center. Cutting speed = 30 m/min and chip load = 0.06 mm. The end milling cutter has two teeth and its diameter = 16.0 mm. The x-axis uses a DC servomotor connected directly to a leadscrew whose pitch = 6.0 mm. The feedback sensing device is an optical encoder that emits 400 pulses per revolution. Determine (a) feed rate and time to complete the cut, and (b) rotational speed of the motor and the pulse rate of the encoder at the feed rate indicated.

38.18. A DC servomotor drives the x-axis of a NC milling machine table. The motor is coupled to the table lead screw using a 4:1 gear reduction (four turns of the motor for each turn of the lead screw). The lead screw pitch = 6.25 mm. An optical encoder is connected to the lead screw. The optical encoder emits 500 pulses per revolution. To execute a certain programmed instruction, the table must move from point (x = 87.5 mm, y = 35.0) to point (x = 25.0 mm, y = 180.0 mm) in a straight-line trajectory at a feed rate = 200 mm/min. Determine (a) the control resolution of the system for the x-axis only, (b) the corresponding rotational speed of the motor, and (c) frequency of the pulse train emitted by the optical encoder at the desired feed rate.

38.19. A DC servomotor drives the y-axis of a NC milling machine table. The motor is coupled to the table lead screw with a gear reduction of 2:1 (two turns of the motor shaft for each single rotation of the lead screw). There are 2 threads/cm in the lead screw. An optical encoder is directly connected to the lead screw (1:1 gear ratio). The optical encoder emits 100 pulses per revolution. To execute a certain programmed instruction, the table must move from point (x = 25.0 mm, y = 28.0) to point (x = 155.0 mm, y = 275.0 mm) in a straight-line trajectory at a feed rate = 200 mm/min. For the y-axis only, determine: (a) the control resolution of the mechanical system, (b) rotational speed of the motor, and (c) frequency of the pulse train emitted by the optical encoder at the desired feed rate.

Industrial Robotics

38.20. The largest axis of a Cartesian coordinate robot has a total range of 750 mm. It is driven by pulley system capable of a mechanical accuracy = 0.25

mm and repeatability = ±0.15 mm. Determine the minimum number of bits required in the binary register for the axis in the robot's control memory.

38.21. A stepper motor serves as the drive unit for the linear joint of an industrial robot. The joint must have an accuracy of 0.25 mm. The motor is attached to a leadscrew through a 2:1 gear reduction (two turns of the motor for one turn of the leadscrew). The pitch of the leadscrew is 5.0 mm. The mechanical errors in the system (due to backlash of the leadscrew and the gear reducer) can be represented by a normal distribution with standard deviation = ±0.05 mm. Specify the number of step angles that the motor must have in order to meet the accuracy requirement.

38.22. The designer of a polar configuration robot is considering a portion of the manipulator consisting of a rotational joint connected to its output link. The output link is 25 in long and the rotational joint has a range of 75°. The accuracy of the joint–link combination, expressed as a linear measure at the end of the link which results from rotating the joint, is specified as 0.030 in. The mechanical inaccuracies of the joint result in a repeatability error = ±0.030° of rotation. It is assumed that the link is perfectly rigid, so there are no additional errors due to deflection. (a) Show that the specified accuracy can be achieved, given the repeatability error. (b) Determine the minimum number of bits required in the binary register of the robot's control memory to achieve the specified accuracy.

39 INTEGRATED MANUFACTURING SYSTEMS

Chapter Contents

The manufacturing systems discussed in this chapter consist of multiple workstations and/or machines whose operations are integrated by means of a material handling subsystem that moves parts or products between stations. In addition, most of these systems use computer control to coordinate the actions of the stations and material handling equipment and to collect data on overall system performance. Thus, the components of an integrated manufacturing system are (1) workstations and/or machines, (2) material handling equipment, and (3) computer control. In addition, human workers are required to manage the system, and workers may be used to operate the individual workstations and machines.

Integrated manufacturing systems include manual and automated production lines, manufacturing cells (from which the term "cellular manufacturing" is derived), and flexible manufacturing systems, all of which are described in this chapter. In the final section we define computer integrated manufacturing (CIM), the ultimate integrated manufacturing system. Let us begin by providing a concise overview of material handling, the physical integrator in integrated manufacturing systems.

39.1 MATERIAL HANDLING

Material handling is defined as "the movement, storage, protection and control of materials throughout the manufacturing and distribution process"[1] The term is usually associated with activities that occur inside a facility, as contrasted with transportation between facilities that involves rail, truck, air, or waterway delivery of goods.

Materials must be moved during the sequence of manufacturing operations that convert them into final product.

---

[1] This definition is published each year in the *Annual Report of the Material Handling Industry of America* (MHIA), the trade association for material handling companies doing business in North America.

Material handling functions in manufacturing include (1) loading and positioning work units at each workstation, (2) unloading work units from the station, and (3) transporting work units between workstations. Loading involves moving the work units into the production machine from a location in close proximity to or within the workstation. Positioning means locating the work units in a fixed orientation relative to the processing or assembly operation. At the end of the operation, the work units are unloaded or removed from the station. Loading and unloading are accomplished manually or by automated devices such as industrial robots. If the manufacturing operations require multiple workstations, then the units must be transported from one station to the next in the sequence. In many cases, a temporary storage function must also be provided by the material handling system, as work units await their turn at each workstation. The purpose of storage in this instance is to make sure that work is always present at each station, so that idle time of workers and equipment is avoided.

Material handling equipment and methods used in manufacturing can be divided into the following general categories: (1) material transport, (2) storage, and (3) unitizing.

Material transport equipment is used to move parts and materials between workstations in the factory. This movement may include intermediate stops for temporary storage of work-in-process. There are five main types of material transport equipment: (1) industrial trucks, the most important of which are fork lift trucks, (2) automated guided vehicles, (3) rail-guided vehicles, (4) conveyors, and (5) hoists and cranes. This equipment is briefly described in Table 39.1.

Two general categories of material transport equipment can be distinguished, according to the type of routing between workstations: fixed and variable. In *fixed routing,* all of the work units are moved through the same sequence of stations. This implies that the processing sequence required on all work units is either identical or very similar. Fixed routing is used on manual assembly lines and automated production lines. Typical material handling equipment used in fixed routing includes conveyors and rail-guided vehicles. In *variable routing,* different work units are moved through different workstation sequences, meaning that the manufacturing system processes or assembles different types of parts or products. Manufacturing cells and flexible manufacturing systems usually operate this way. Typical handling equipment found in variable routing includes industrial trucks, automated guided vehicles, and hoists and cranes.

Storage systems in factories are used for temporary storage of raw materials, work-in-process, and finished products. Storage systems can be classified into two general categories:

TABLE 39.1   Five types of material transport equipment.

Type	Description	Typical Production Applications
Industrial trucks	Powered trucks include fork lift trucks as in Figure 39.1(a). Hand trucks include wheeled platforms and dollies	Movement of pallet and container loads in factories and warehouses. Hand trucks used for small loads over short distances
Automated guided vehicles	Independently operated, self-propelled vehicles guided along defined pathways, as in Figure 39.1(b). Powered by on-board batteries	Movement of parts and products in assembly lines and flexible manufacturing systems
Rail-guided vehicles	Motorized vehicles guided by a fixed rail system. Powered by electrified rail	Monorails used for overhead delivery of large components and subassemblies
Conveyors	Apparatus to move items along fixed path using chain, moving belt, rollers (Figure 39.1(c), or other mechanical drive	Movement of large quantities of items between specific locations. Movement of product on production lines
Hoists and cranes	Apparatus used for vertical lifting (hoists) and horizontal movement (cranes)	Lifting and transporting heavy materials and loads

FIGURE 39.1 Several types of material handling equipment: (a) fork lift truck, (b) automated guided vehicle, and (c) roller conveyor.

TABLE 39.2 Types of material handling methods and systems generally associated with the four types of plant layout.

Layout Type	Features	Typical Methods and Equipment
Fixed-position	Product is large and heavy, low production rates	Cranes, hoists, fork lift trucks
Process	Medium and hard product variety, low and medium production rates	Fork lift trucks, automated guided vehicles, manual loading at workstations
Cellular	Soft product variety, medium production rates	Conveyors, manual handling for loading and moving between stations
Product	No product variety or soft product variety, high production rates	Conveyors for product flow, fork lift trucks or automated guided vehicles to deliver parts to stations

(1) conventional storage methods and equipment, which include bulk storage in an open area, rack systems, and shelves; and (2) automated storage systems, which include rack systems served by automatic cranes that store and retrieve pallet loads.

Finally, unitizing refers to containers used to hold individual items during transport and storage, as well as equipment used to make up such unit loads. Containers include pallets, tote pans, boxes, and baskets that hold parts during handling. Unitizing equipment includes palletizers that are used to load and stack cartons onto pallets and depalletizers that are used to accomplish the unloading operation. Palletizers and depalletizers are generally associated with cartons of finished product leaving a facility and boxes of raw materials coming into the facility, respectively.

In Section 1.4.1, we described four types of plant layout: (1) fixed position layout, (2) process layout, (3) cellular layout, and (4) product layout. In general, different types of material handling methods and equipment are associated with these four types, as summarized in Table 39.2.

39.2 FUNDAMENTALS OF PRODUCTION LINES

Production lines are an important class of manufacturing system when large quantities of identical or similar products are to be made. They are suited to situations where the total work to be performed on the product or part consists of many separate steps. Examples include

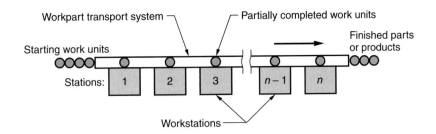

FIGURE 39.2 General configuration of a production line.

assembled products (e.g., automobiles and appliances) and mass-produced machined parts on which multiple machining operations are required (e.g., engine blocks and transmission housings). In a production line, the total work is divided into small tasks, and workers or machines perform these tasks with great efficiency. For purposes of organization we divide production lines into two basic types: manual assembly lines and automated production lines. However, hybrid lines consisting of both manual and automated operations are not uncommon. Before examining these particular systems, let us consider some of the general issues involved in production line design and operation.

A ***production line*** consists of a series of workstations arranged so that the product moves from one station to the next, and at each location a portion of the total work is performed on it, as depicted in Figure 39.2. The production rate of the line is limited by its slowest station. Workstations whose pace is faster than the slowest will ultimately be limited by that bottleneck station. Transfer of the product along the line is usually accomplished by a conveyor system or mechanical transfer device, although some manual lines simply pass the product from worker to worker by hand. Production lines are associated with mass production. If the product quantities are high and the work can be divided into separate tasks that can be assigned to individual workstations, then a production line is the most appropriate manufacturing system.

39.2.1 METHODS OF WORK TRANSPORT

There are various ways of moving work units from one workstation to the next. The two basic categories are manual and mechanized.

Manual Methods of Work Transport   Manual methods involve passing the work units between stations by hand. These methods are associated with manual assembly lines. In some cases, the output of each station is collected in a box or tote pan; when the box is full it is moved to the next station. This can result in a significant amount of in-process inventory, which is undesirable. In other cases, work units are moved individually along a flat table or unpowered conveyor (e.g., a roller conveyor). When the task is finished at each station, the worker simply pushes the unit toward the downstream station. Space is usually allowed for one or more units to collect between stations, thereby relaxing the requirement for all workers to perform their respective tasks in sync. One problem associated with manual methods of work transport is the difficulty in controlling the production rate on the line. Workers tend to work at a slower pace unless some mechanical means of pacing them is provided.

Mechanized Methods of Work Transport   Powered mechanical systems are commonly used to move work units along a production line. These systems include lift-and-carry devices, pick-and-place mechanisms, powered conveyors (e.g., overhead chain conveyors, belt conveyors, and chain-in-floor conveyors), and other material handling equipment, sometimes combining several types on the same line. Three major types of work transfer systems are used on production lines: (1) continuous transfer, (2) synchronous transfer, and (3) asynchronous transfer.

Continuous transfer systems consist of a continuously moving conveyor that operates at a constant velocity. The continuous transfer system is most common on manual assembly lines. Two cases are distinguished: (1) parts are fixed to the conveyor and (2) parts can be removed from the conveyor. In the first case, the product is usually large and heavy (e.g., automobile, washing machine) and cannot be removed from the line. The worker must therefore walk along with the moving conveyor to complete the assigned task for that unit while it is in the station. In the second case, the product is small enough that it can be removed from the conveyor to facilitate the work at each station. Some of the pacing benefits are lost in this arrangement, since each worker is not required to finish the assigned tasks within a fixed time period. On the other hand, this case allows greater flexibility to each worker to deal with any technical problems that may be encountered on a particular work unit.

In ***synchronous transfer systems,*** work units are simultaneously moved between stations with a quick, discontinuous motion. These systems are also known by the name ***intermittent transfer,*** which characterizes the type of motion experienced by the work units. Synchronous transfer includes positioning of the work at the stations, which is a requirement for automated lines that use this mode of transfer. Synchronous transfer is not common for manual lines, because the task at each and every station must be finished within the cycle time or the product will leave the station as an incomplete unit. This rigid pacing discipline is stressful to human workers. By contrast, this type of pacing lends itself to automated operation.

Asynchronous transfer allows each work unit to depart its current station when processing has been completed. Each unit moves independently, rather than synchronously. Thus, at any given moment, some units on the line are moving between stations, while others are positioned at stations. Associated with the operation of an asynchronous transfer system is the tactical use of queues between stations. Small queues of work units are permitted to form in front of each station, so that variations in worker task times will be averaged and stations will always have work waiting for them. Asynchronous transfer is used for both manual and automated production systems.

39.2.2 PRODUCT VARIATIONS

Production lines can be designed to cope with variations in product models. Three types of line can be distinguished: (1) single model line, (2) batch model line, and (3) mixed model line. A ***single model line*** is one that produces only one model, and there is no variation in the model. Thus, the tasks performed at each station are the same on all product units.

Batch model and mixed model lines are designed to produce two or more different product models on the same line, but they use different approaches for dealing with the model variations. As its name suggests, a ***batch model line*** produces each model in batches. The workstations are set up to produce the desired quantity of the first model; then the stations are reconfigured to produce the desired quantity of the next model; and so on. Production time is lost between batches due to the setup changes. Assembled products are often made using this approach when the demand for each product is medium and the product variety is also medium. The economics in this case favor the use of one production line for several products rather than using many separate lines for each model.

A ***mixed model line*** also produces multiple models; however, the models are intermixed on the same line rather than being produced in batches. While a particular model is being worked on at one station, a different model is being processed at the next station. Each station is equipped with the necessary tools and is sufficiently versatile to perform the variety of tasks needed to produce any model that moves through it. Many consumer products are assembled on mixed model lines when the level of product variety is soft. Prime examples are automobiles and major appliances, which are characterized by variations in models and options.

39.3   MANUAL ASSEMBLY LINES

The manual assembly line was an important development in integrated manufacturing systems. It is of global importance today in the manufacture of assembled products including automobiles and trucks, consumer electronic products, appliances, power tools, and other products made in large quantities.

A *manual assembly line* consists of multiple workstations arranged sequentially, at which assembly operations are performed by human workers, as in Figure 39.3. The usual procedure on a manual line begins with "launching" a base part onto the front end of the line. A work carrier is often required to hold the part during its movement along the line. The base part travels through each of the stations where workers perform tasks that progressively build the product. Components are added to the base part at each station, so that all tasks have been completed when the product exits the final station. Processes accomplished on manual assembly lines include mechanical fastening operations (Chapter 32), spot welding (Section 30.2), hand soldering (Section 31.2), and adhesive joining (Section 31.3).

39.3.1   CYCLE TIME ANALYSIS

Equations can be developed to determine the required number of workers and work-stations on a manual assembly line to meet a given annual demand. Suppose the problem is to design a single model line to satisfy annual demand for a certain product. Management must decide how many shifts per week the line will operate and the number of hours per shift. If we assume 50 weeks per year, then the required hourly production rate of the line will be given by

$$R_p = \frac{D_a}{50 S_w H_{sh}} \tag{39.1}$$

FIGURE 39.3   A portion of a manual assembly line. Each worker performs a task at his/her workstation. A conveyor moves parts on work carriers from one station to the next.

Components to be added

Conveyor

Base part

Work carrier

v

where R_p = the actual average production rate, units/hr; D_a = annual demand for the product, units/yr; S_w = number of shifts/wk; and H_{sh} = hr/shift. If the line operates 52 weeks rather than 50, then $R_p = D_a/52S_wH_{sh}$. The corresponding average production time per unit is the reciprocal of R_p

$$T_p = \frac{60}{R_p} \tag{39.2}$$

where T_p = actual average production time, converted to minutes.

Unfortunately, the line may not be able to operate for the entire time given by 50 S_wH_{sh}, because of lost time due to reliability problems. These reliability problems include mechanical and electrical failures, tools wearing out, power outages, and similar malfunctions. Accordingly, the line must operate at a faster time than T_p to compensate for these problems. If E = line efficiency, which is the proportion of uptime on the line, then the cycle time of the line T_c is given by

$$T_c = ET_p = \frac{60E}{R_p} \tag{39.3}$$

Any product contains a certain work content that represents all of the tasks that are to be accomplished on the line. This work content requires an amount of time called the **work content time** T_{wc}. This is the total time required to make the product on the line. If we assume that the work content time can be divided evenly among the workers, so that every worker has an equal workload whose time to perform equals T_c, then the minimum possible number of workers w_{min} in the line can be determined as

$$w_{min} = \text{Minimum Integer} \geq \frac{T_{wc}}{T_c} \tag{39.4}$$

If each worker is assigned to a separate workstation, then the number of workstations is equal to the number of workers; that is $n_{min} = w_{min}$.

There are two practical reasons why this minimum number of workers cannot be achieved: (1) **imperfect balancing,** in which some workers are assigned an amount of work that requires less time than T_c, and this inefficiency increases the total number of workers needed on the line; and (2) **repositioning losses,** in which some time is lost at each station to reposition the work or the worker, so that the service time actually available at each station is less than T_c, and this will also increase the number of workers on the line.

39.3.2 LINE BALANCING AND REPOSITIONING LOSSES

One of the biggest technical problems in designing and operating a manual assembly line is line balancing. This is the problem of assigning tasks to individual workers so that all workers have an equal amount of work. Recall that the entirety of work to be accomplished on the line is given by the work content. This total work content can be divided into **minimum rational work elements,** each element concerned with adding a component or joining them or performing some other small portion of the total work content. The notion of a minimum rational work element is that it is the smallest practical amount of work into which the total job can be divided. Different work elements require different times, and when they are grouped into logical tasks and assigned to workers, the task times will not be equal. Thus, simply due to the variable nature of element times, some workers will end up with more work, while other workers will have less. The cycle time of the assembly line is determined by the station with the longest task time.

One might think that although the work element times are different, it should be possible to find groups of elements whose sums (task times) are nearly equal, if not perfectly equal. What makes it difficult to find suitable groups is that there are several constraints on this combinatorial problem. First, the line must be designed to achieve some desired production rate, which establishes the cycle time T_c at which the line must operate, as provided by Eq. (39.3). Therefore, the sum of the work element times assigned to each station must be less than or equal to T_c.

Second, there are restrictions on the order in which the work elements can be performed. Some elements must be done before others. For example, a hole must be drilled before it can be tapped. A screw that will use the tapped hole to attach a mating component cannot be fastened before the hole has been drilled and tapped. These kinds of requirements on the work sequence are called **precedence constraints.** They complicate the line balancing problem. A certain element that might be allocated to a worker to obtain a task time $= T_c$ cannot be added because it violates a precedence constraint.

These and other limitations make it virtually impossible to achieve perfect balancing of the line, which means that some workers will require more time to complete their tasks than others. Methods of solving the line balancing problem, that is, allocating work elements to stations, are discussed in other references—excellent references indeed, such as [10]. The inability to achieve perfect balancing results in some idle time at most stations. Because of this idle time, the actual number of workers required on the line will be greater than the number of workstations given by Eq. (39.4).

A measure of the total idle time on a manual assembly line is given by the **balancing efficiency** E_b, defined as the total work content time divided by the total available service time on the line. The total work content time is equal to the sum of the times of all work elements that are to be accomplished on the line. The total available service time on the line $= wT_s$, where $w =$ number of workers on the line; and $T_s =$ the longest service time on the line; that is, $T_s = \text{Max}\{T_{si}\}$ for $i = 1, 2, \ldots n$, where $T_{si} =$ the service time (task time) at station i, min.

The reader may wonder why we are using a new term T_s rather than the previously defined cycle time T_c. The reason is that there is another time loss in the operation of a production line in addition to idle time from imperfect balancing. Let us call it the **repositioning time** T_r. It is the time required in each cycle to reposition the worker, or the work unit, or both. On a continuous transfer line where work units are attached to the line and move at a constant speed, T_r is the time taken by the worker to walk from the unit just completed to the next unit coming into the station. In all manual assembly lines, there will be some lost time due to repositioning. We assume that T_r is the same for all workers, although in fact repositioning may require different times at different stations. We can relate T_s, T_c, and T_r as follows:

$$T_c = T_s + T_r \tag{39.5}$$

The definition of balancing efficiency E_b can now be written in equation form as follows:

$$E_b = \frac{T_{wc}}{wT_s} \tag{39.6}$$

A perfect line balance yields a value of $E_b = 1.00$. Typical line balancing efficiencies in industry range between 0.90 and 0.95.

Equation (39.6) can be rearranged to obtain the actual number of workers required on a manual assembly line:

$$w = \text{Minimum Integer} \geq \frac{T_{wc}}{T_s E_b} \tag{39.7}$$

The utility of this relationship suffers from the fact that the balancing efficiency E_b depends on w in Eq. (39.6). Unfortunately, we have an equation where the thing to be determined depends on a parameter that, in turn, depends on the thing itself. Notwithstanding this drawback, Eq. (39.7) defines the relationship among the parameters in a manual assembly line. Using a typical value of E_b based on similar previous lines, it can be used to estimate the number of workers required to produce a given assembly.

Example 39.1
Manual Assembly
Line

A manual assembly line is being planned for a product whose annual demand = 90,000 units. A continuously moving conveyor will be used with work units attached. Work content time = 55 min. The line will run 50 wk/yr, 5 shifts/wk, and 8 hr/day. Each worker will be assigned to a separate workstation. Based on previous experience, assume line efficiency = 0.95, balancing efficiency = 0.93, and repositioning time = 9 sec. Determine (a) hourly production rate to meet demand, (b) number of workers and workstations required, and (c) for comparison, the ideal minimum value as given by w_{min} as given by Eq. (39.4).

Solution: (a) Hourly production rate required to meet annual demand is given by Eq. (39.1):

$$R_p = \frac{90,000}{50(5)(8)} = \textbf{45 units/hr}$$

(b) With a line efficiency of 0.95, the ideal cycle time is

$$T_c = \frac{60(0.95)}{45} = \textbf{1.2667 min}$$

Given that repositioning time $T_r = 9$ sec $= 0.15$ min, the service time is

$$T_s = 1.2667 - 0.150 = 1.1167 \, min$$

Workers required to operate the line, by Eq. (39.7) equals

$$w = \text{Minimum Integer} \geq \frac{55}{1.1167(0.93)} = 52.96 \rightarrow \textbf{53 workers}$$

With one worker per station, $n = \textbf{53 workstations}$.
(c) This compares with the ideal minimum number of workers given by Eq. (39.4):

$$w_{min} = \text{Minimum Integer} \geq \frac{55}{1.2667} = 43.42 \rightarrow \textbf{44 workers}$$

It is clear that lost time due to repositioning and imperfect line balancing take their toll in the overall efficiency of a manual assembly line. ∎

The number of workstations on a manual assembly line does not necessarily equal the number of workers. For large products, it may be possible to assign more than one worker to a station. This practice is common in final assembly plants that build cars and trucks. For example, two workers in a station might perform assembly tasks on opposite sides of the vehicle. The number of workers in a given station is called the station *manning level* M_i. Averaging the manning levels over the entire line,

$$M = \frac{w}{n} \tag{39.8}$$

where M = average manning level for the assembly line; w = number of workers on the line; and n = number of stations. Naturally, w and n must be integers. Multiple manning conserves valuable floor space in the factory because it reduces the number of stations required.

Another factor that affects manning level on an assembly line is the number of *automated stations* on the line, including stations that employ industrial robots (Section 38.4). Automation reduces the required labor force on the line, although it increases the need for technically trained personnel to service and maintain the automated stations. The automobile industry makes extensive use of robotic workstations to perform spot welding and spray painting on sheet-metal car bodies. The robots accomplish these operations with greater repeatability than human workers can, which translates into higher product quality.

39.4   AUTOMATED PRODUCTION LINES

Manual assembly lines generally use a mechanized transfer system to move parts between workstations, but the stations themselves are operated by human workers. An ***automated production line*** consists of automated workstations connected by a parts transfer system that is coordinated with the stations. In the ideal, no human workers are on the line, except to perform auxiliary functions such as tool changing, loading and unloading parts at the beginning and end of the line, and repair and maintenance activities. Modern automated lines are highly integrated systems, operating under computer control.

Operations performed by automated stations tend to be simpler than those performed by humans on manual lines. The reason is that simpler tasks are easier to automate. Operations that are difficult to automate are those requiring multiple steps, judgment, or human sensory capability. Tasks that are easy to automate consist of single work elements, quick actuating motions, and straight-line feed motions as in machining.

39.4.1   TYPES OF AUTOMATED LINES

Automated production lines can be divided into two basic categories: (1) those that perform processing operations such as machining, and (2) those that perform assembly operations. An important type in the processing category is the transfer line.

Transfer Lines and Similar Processing Systems   A ***transfer line*** consists of a sequence of workstations that perform production operations, with automatic transfer of work units between stations. Machining is the most common processing operation, as depicted in Figure 39.4. Automatic transfer systems for sheet metalworking and assembly are also available. In the case of machining, the workpiece typically starts as a metal casting or forging, and a series of machining operations are performed to accomplish the high-precision details (e.g., holes, threads, and finished flat surfaces).

FIGURE 39.4
A machining transfer line, an important type of automated production line.

FIGURE 39.5
Configuration of a
dial-indexing machine.

Transfer lines are usually expensive pieces of equipment, sometimes costing millions of dollars; they are designed for high part quantities. The amount of machining accomplished on the workpart is often significant, but since the work is divided among many stations, production rates are high and unit costs are low compared to alternative production methods. Synchronous transfer of work units between stations is commonly used on automated machining lines.

A variation of the automated transfer line is the *dial indexing machine,* Figure 39.5, in which workstations are arranged around a circular worktable, called a dial. The worktable is actuated by a mechanism that provides partial rotation of the table on each work cycle. The number of rotational positions is designed to match the number of workstations around the periphery of the table. Although the configuration of a dial-indexing machine is quite different from a transfer line, its operation and application are quite similar.

Automated Assembly Systems   Automated assembly systems consist of one or more workstations that perform assembly operations, such as adding components and/or affixing them to the work unit. Automated assembly systems can be divided into single station cells and multiple station systems. *Single station assembly cells* are often organized around an industrial robot that has been programmed to perform a sequence of assembly steps. The robot cannot work as fast as a series of specialized automatic stations, so single station cells are used for jobs in the medium production range.

Multiple station assembly systems are appropriate for high production. They are widely used for mass production of small products such as ball-point pens, cigarette lighters, flashlights, and similar items consisting of a limited number of components. The number of components and assembly steps is limited because system reliability decreases rapidly with increasing complexity.

Multiple station assembly systems are available in several configurations, pictured in Figure 39.6: (a) in-line, (b) rotary, and (c) carousel. The in-line configuration is the conventional

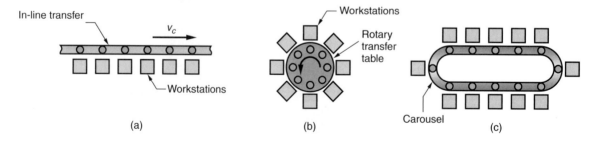

FIGURE 39.6   Three common configurations of multiple station assembly systems: (a) in-line, (b) rotary, and (c) carousel.

transfer line adapted to perform assembly work. These systems are not as massive as their machining counterparts. Rotary systems are usually implemented as dial indexing machines. Carousel assembly systems are arranged as a loop. They can be designed with a greater number of workstations than a rotary system. Owing to the loop configuration, the carousel allows the work carriers to be automatically returned to the starting point for reuse, an advantage shared with rotary systems but not with transfer lines unless provision for their return is made in the design.

39.4.2 ANALYSIS OF AUTOMATED PRODUCTION LINES

Line balancing is a problem on an automated line, just as it is on a manual assembly line. The total work content must be allocated to individual workstations. However, since the tasks assigned to automated stations are generally simpler, and the line often contains fewer stations, the problem of defining what work should be done at each station is not as difficult for an automated line as for a manual line.

A more significant problem in automated lines is reliability. The line consists of multiple stations, interconnected by a work transfer system. It operates as an integrated system, and when one station malfunctions, the entire system is adversely affected. To analyze the operation of an automated production line, let us assume a system that performs processing operations and uses synchronous transfer. This model includes transfer lines as well as dial indexing machines. It does not include automated assembly systems, which require an adaptation of the model [10]. Our terminology will borrow symbols from the first two sections: n = number of workstations on the line; T_c = ideal cycle time on the line; T_r = repositioning time, called the transfer time in a transfer line; and T_{si} = the service time at station i. The ideal cycle time T_c is the service time (processing time) for the slowest station on the line plus the transfer time; that is,

$$T_c = T_r + \text{Max}\{T_{si}\} \tag{39.9}$$

In the operation of a transfer line, periodic breakdowns cause downtime on the entire line. Let F = frequency with which breakdowns occur, causing a line stoppage; and T_d = average time the line is down when a breakdown occurs. The downtime includes the time for the repair crew to swing into action, diagnose the cause of the failure, fix it, and restart the line.

Based on these definitions, we can formulate the following expression for the actual average production time T_p:

$$T_p = T_c + FT_d \tag{39.10}$$

where F = downtime frequency, line stops/cycle; and T_d = downtime in minutes per line stop. Thus, FT_d = average downtime per cycle. The actual average production rate $R_p = 60/T_p$, as previously given in Eq. (39.2). It is of interest to compare this rate with the ideal production rate given by

$$R_c = \frac{60}{T_c} \tag{39.11}$$

where R_p and R_c are expressed in pc/hr, given that T_p and T_c are expressed in minutes.

Based on these definitions, we can define the line efficiency E for a transfer line. In the context of automated production systems, E refers to the proportion of uptime on the line and is more a measure of reliability than efficiency:

$$E = \frac{T_c}{T_c + FT_d} \tag{39.12}$$

This is the same relationship as earlier Eq. (39.3), since $T_p = T_c + FT_d$. It should be noted that the same definition of line efficiency applies to manual assembly lines, except that technological breakdowns are not as much of a problem on manual lines (human workers are more reliable than electromechanical equipment, at least in the sense we are discussing here).

Line downtime is usually associated with failures at individual workstations. Reasons for downtime include scheduled and unscheduled tool changes, mechanical and electrical malfunctions, hydraulic failures, and normal equipment wear. Let p_i = probability or frequency of a failure at station i, then

$$F = \sum_{i-1}^{n} p_i \qquad (39.13)$$

If all p_i are assumed equal, or an average value of p_i is computed, in either case calling it p, then

$$F = np \qquad (39.14)$$

Both of these equations clearly indicate that the frequency of line stops increases with the number of stations on the line. Stated another way, reliability of the line decreases as we add more stations.

Example 39.2
Automated
Transfer Line

An automated transfer line has 20 stations and an ideal cycle time of 1.0 min. Probability of a station failure is $p = 0.01$, and the average downtime when a breakdown occurs is 10 min. Determine (a) average production rate R_p and (b) line efficiency E.

Solution: The frequency of breakdowns on the line is given by $F = pn = 0.01(20) = 0.20$. The actual average production time is therefore

$$T_p = 1.0 + 0.20(10) = 3.0 \text{ min}$$

(a) Production rate is therefore

$$R_p = \frac{60}{T_p} = \frac{60}{3.0} = 20 \text{ pc/hr}$$

Note that this is far lower than the ideal production rate:

$$R_c = \frac{60}{T_c} = \frac{60}{1.0} = 60 \text{ pc/hr}$$

(b) Line efficiency is computed as

$$E = \frac{T_c}{T_p} = \frac{1.0}{3.0} = 0.333 \text{(or 33.3\%)}$$

From this example we see that if a production line operates like this, it spends more time down than up. Achieving high efficiencies is a real problem in automated production lines. ∎

The cost of operating an automated production line is the investment cost of the equipment and installation, plus the cost of maintenance, utilities, and labor assigned to the line. These costs are converted to an equivalent uniform annual cost and divided by the number of hours of operation per year to provide an hourly rate. This hourly cost rate can be used to figure the unit cost of processing a workpart on the line

$$C_p = \frac{C_o T_p}{60} \qquad (39.15)$$

where C_p = unit processing cost, \$/part; C_o = hourly rate of operating the line, as defined above, \$/hr; T_p = actual average production time per workpart, min/part; and the constant 60 converts the hourly cost rate to \$/min for consistency of units.

39.5   CELLULAR MANUFACTURING

Cellular manufacturing refers to the use of work cells that specialize in the production of families of parts or products made in medium quantities. Parts (and products) in this quantity range are traditionally made in batches, and batch production requires downtime for setup changeovers and has high inventory carrying costs. Cellular manufacturing is based on an approach called group technology (GT), which minimizes the disadvantages of batch production by recognizing that although the parts are different, they also possess similarities. When these similarities are exploited in production, operating efficiencies are improved. The improvement is typically achieved by organizing the production around manufacturing cells. Each cell is designed to produce one part family (or a limited number of part families), thereby following the principle of specialization of operations. The cell includes special production equipment and custom-designed tools and fixtures, so that the production of the part families can be optimized. In effect, each cell becomes a factory within the factory.

39.5.1   PART FAMILIES

A central feature of cellular manufacturing and group technology is the part family. A *part family* is a group of parts that possess similarities in geometric shape and size, or in the processing steps used in their manufacture. It is not unusual for a factory that produces 10,000 different parts to be able to group most of those parts into 20 to 30 part families. In each part family the processing steps are similar. There are always differences among parts in a family, but the similarities are close enough that the parts can be grouped into the same family. Figures 39.7 and 39.8 show two different part families. The parts shown in Figure 39.7 have the same size and shape; however, their processing requirements are quite different because of differences in work material, production quantities, and design tolerances. Figure 39.8 shows several parts with geometries that differ, but their manufacturing requirements are quite similar.

There are several ways by which part families are identified in industry. One method involves visual inspection of all the parts made in the factory (or photos of the parts) and using best judgment to group them into appropriate families. Another approach, called *production flow analysis,* uses information contained on route sheets (Section 40.1.1) to classify parts. In effect, parts with similar manufacturing steps are grouped into the same family.

A third method, usually the most expensive but most useful, is parts classification and coding. *Parts classification and coding* involve the identification of similarities and differences among parts and relating these parts by means of a numerical coding scheme. Most classification and coding systems are one of the following: (1) systems based on part design

FIGURE 39.7   Two parts that are identical in shape and size but quite different in manufacturing: (a) 1,000,000 units/yr, tolerance = ± 0.010 in, 1015 CR steel, nickel plate; and (b) 100/yr, tolerance = ± 0.001 in, 18-8 stainless steel.

(a)   (b)

FIGURE 39.8   Ten parts that are different in size and shape, but quite similar in terms of manufacturing. All parts are machined from cylindrical stock by turning; some parts require drilling and/or milling.

attributes, (2) systems based on part manufacturing attributes, and (3) systems based on both design and manufacturing attributes. Common part design and manufacturing attributes used in GT systems are presented in Table 39.3. Because each company produces a unique set of parts and products, a classification and coding system that may be satisfactory for one company is not necessarily appropriate for another company. Each company must design its own coding scheme. Parts classification and coding systems are described more thoroughly in several of our references [8], [10], [11].

Benefits often cited for a well-designed classification and coding system include (1) facilitates formation of part families, (2) permits quick retrieval of part design drawings, (3) reduces design duplication because similar or identical part designs can be retrieved and reused rather than designed from scratch, (4) promotes design standardization, (5) improves cost estimating and cost accounting, (6) facilitates numerical control (NC) part programming by allowing new parts to use the same basic part program as existing parts in the same family, (7) allows sharing of tools and fixtures, and (8) aids computer-aided process planning (Section 40.1.3) because standard process plans can be correlated to part family code numbers, so that existing process plans can be reused or edited for new parts in the same family.

TABLE 39.3   Design and manufacturing attributes typically included in a parts classification and coding system.

Part Design Attributes		Part Manufacturing Attributes	
Major dimensions	Material type	Major process	Major dimensions
Basic external shape	Part function	Operation sequence	Basic external shape
Basic internal shape	Tolerances	Batch size	Length/diameter ratio
Length/diameter ratio	Surface finish	Annual production	Material type
		Machine tools	Tolerances
		Cutting tools	Surface finish

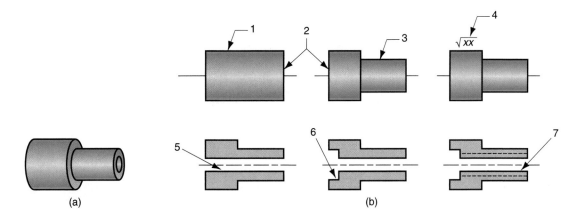

FIGURE 39.9 Composite part concept: (a) the composite part for a family of machined rotational parts, and (b) the individual features of the composite part.

39.5.2 MACHINE CELLS

To fully exploit the similarities among parts in a family, production should be organized using machine cells designed to specialize in making those particular parts. One of the principles in designing a group technology machine cell is the composite part concept.

Composite Part Concept   Members of a part family possess similar design and/or manufacturing features. There is usually a correlation between part design features and the manufacturing operations that produce those features. Round holes are made by drilling; cylindrical shapes are made by turning; and so on.

The ***composite part*** for a given family (not to be confused with a part made of composite material) is a hypothetical part that includes all of the design and manufacturing attributes of the family. In general, an individual part in the family will have some of the features that characterize the family, but not all of them. A production cell designed for the part family would include those machines required to make the composite part. Such a cell would be capable of producing any member of the family, simply by omitting those operations corresponding to features not possessed by the particular part. The cell would also be designed to allow for size variations within the family as well as feature variations.

To illustrate, consider the composite part in Figure 39.9(a). It represents a family of rotational parts with features defined in part (b) of the figure. Associated with each feature is a certain machining operation, as summarized in Table 39.4. A machine cell to produce

TABLE 39.4   Design features of the composite part in Figure 39.3 and the manufacturing operations required to shape those features.

Label	Design Feature	Corresponding Manufacturing Operation
1	External cylinder	Turning
2	Face of cylinder	Facing
3	Cylindrical step	Turning
4	Smooth surface	External cylindrical grinding
5	Axial hole	Drilling
6	Counterbore	Bore, counterbore
7	Internal threads	Tapping

FIGURE 39.10   Types of group technology machine cells: (a) single machine, (b) multiple machines with manual handling, (c) multiple machines with mechanized handling, (d) flexible manufacturing cell, and (e) flexible manufacturing system. Key: Man = manual operation; Aut = automated station.

this part family would be designed with the capability to accomplish all of the operations in the last column of the table.

Machine Cell Designs   Machine cells can be classified according to number of machines and level of automation. The possibilities are (a) single machine, (b) multiple machines with manual handling, (c) multiple machines with mechanized handling, (d) flexible manufacturing cell, or (e) flexible manufacturing system. These production cells are depicted in Figure 39.10.

The *single machine cell* has one machine that is manually operated. The cell would also include fixtures and tools to allow for feature and size variations within the part family produced by the cell. The machine cell required for the part family of Figure 39.9 would probably be of this type.

Multiple machine cells have two or more manually operated machines. These cells are distinguished by the method of workpart handling in the cell, manual or mechanized. Manual handling means that parts are moved within the cell by workers, usually the machine operators. Mechanized handling refers to conveyorized transfer of parts from one machine to the next. This may be required by the size and weight of the parts made in the cell, or simply to increase production rate. Our sketch depicts the work flow as being a line; other layouts are also possible, such as U-shaped or loop.

Flexible manufacturing cells and *flexible manufacturing systems* consist of automated machines with automated handling. Given the special nature of these integrated manufacturing systems and their importance, we devote Section 39.6 to their discussion.

Benefits and Problems in Group Technology   The use of machine cells and group technology provide substantial benefits to companies that have the discipline and perseverance to implement it. The potential benefits include the following: (1) GT promotes standardization of tooling, fixturing, and setups; (2) material handling is reduced because parts are moved within a machine cell rather than the entire factory; (3) production scheduling is simplified; (4) manufacturing lead time is reduced; (5) work-in-process is reduced; (6) process planning is simpler; (7) worker satisfaction usually improves working in a cell; and (8) higher quality work is accomplished.

There are several problems in implementing machine cells, however. One obvious problem is rearranging production machines in the plant into the appropriate machine cells. It takes time to plan and accomplish this rearrangement, and the machines are not producing during the changeover. The biggest problem in starting a GT program is identifying the part families. If the plant makes 10,000 different parts, reviewing all of the part drawings and grouping the parts into families are substantial tasks that consume a significant amount of time.

39.6   FLEXIBLE MANUFACTURING SYSTEMS AND CELLS

A flexible manufacturing system (FMS) is a highly automated GT machine cell, consisting of a group of processing stations (usually computer numerical control [CNC] machine tools), interconnected by an automated material handling and storage system, and controlled by an integrated computer system. An FMS is capable of processing a variety of different part styles simultaneously under NC program control at the different workstations.

An FMS relies on the principles of group technology. No manufacturing system can be completely flexible. It cannot produce an infinite range of parts or products. There are limits to how much flexibility can be incorporated into an FMS. Accordingly, a flexible manufacturing system is designed to produce parts (or products) within a range of styles, sizes, and processes. In other words, an FMS is capable of producing a single part family or a limited range of part families.

Flexible manufacturing systems vary in terms of number of machine tools and level of flexibility. When the system has only a few machines, the term *flexible manufacturing cell* (FMC) is sometimes used. Both cell and system are highly automated and computer controlled. The difference between an FMS and an FMC is not always clear, but it is sometimes based on the number of machines (workstations) included. The flexible manufacturing system consists of four or more machines, while a flexible manufacturing cell consists of three or fewer machines [10].

To qualify as being flexible, a manufacturing system should satisfy several criteria. The tests of flexibility in an automated production system are the capability to (1) process different part styles in a nonbatch mode, (2) accept changes in production schedule, (3) respond gracefully to equipment malfunctions and breakdowns in the system, and (4) accommodate the introduction of new part designs. These capabilities are made possible by the use of a central computer that controls and coordinates the components of the system. The most important criteria are (1) and (2); criteria (3) and (4) are softer and can be implemented at various levels of sophistication.

39.6.1   INTEGRATING THE FMS COMPONENTS

An FMS consists of hardware and software that must be integrated into an efficient and reliable unit. It also includes human personnel. In this section we examine these components and how they are integrated.

Hardware Components   FMS hardware includes workstations, material handling system, and central control computer. The workstations are CNC machines in a machining type system, plus inspection stations, parts cleaning and other stations, as required. A central chip conveyor system is often installed below floor level.

The material handling system is the means by which parts are moved between stations. The material handling system usually includes a limited capability to store parts. Handling systems suitable for automated manufacturing include roller conveyors, automated guided vehicles, and industrial robots. The most appropriate type depends on part size and geometry, as well as factors relating to economics and compatibility with other FMS components. Non-rotational parts are often moved in a FMS on pallet fixtures, so the pallets are designed for the particular handling system, and the fixtures are designed to accommodate the various part geometries in the family. Rotational parts are often handled by robots, if weight is not a limiting factor.

The handling system establishes the basic layout of the FMS. Five layout types can be distinguished: (1) in-line, (2) loop, (3) ladder, (4) open field, and (5) robot-centered cell. Types 1, 3, 4, and 5 are shown in Figure 39.11. Type 2 is shown in Figure 39.10(e). The *in-line layout* uses a linear transfer system to move parts between processing stations and load/unload station(s). The in-line transfer system is usually capable of two-directional movement; if not, then the FMS operates much like a transfer line, and the different part styles made on the system must follow the same basic processing sequence due to the one-direction flow. The *loop layout* consists of a conveyor loop with workstations located around its periphery. This configuration permits any processing sequence, because any station is accessible from any other station. This is also true for the *ladder layout,* in which workstations are located on the rungs of the ladder. The *open field layout* is the most complex FMS configuration, and consists of several loops tied together. Finally, the robot-centered cell consists of a robot whose work volume includes the load/unload positions of the machines in the cell.

The FMS also includes a central computer that is interfaced to the other hardware components. In addition to the central computer, the individual machines and other components generally have microcomputers as their individual control units. The function of the central computer is to coordinate the activities of the components so as to achieve a smooth overall operation of the system. It accomplishes this function by means of software.

FMS Software and Control Functions   FMS software consists of modules associated with the various functions performed by the manufacturing system. For example, one function involves downloading NC part programs to the individual machine tools; another function is concerned with controlling the material handling system; another is concerned with tool management; and so on. Table 39.5 lists the functions included in the operation of a typical FMS. Associated with each function is one or more software modules. Terms other than those in our table may be used in a given installation. The functions and modules are largely application specific.

Human Labor   An additional component in the operation of a flexible manufacturing system or cell is human labor. Duties performed by human workers include (1) loading and unloading parts from the system, (2) changing and setting cutting tools, (3) maintenance and repair of equipment, (4) NC part programming, (5) programming and operating the computer system, and (6) overall management of the system.

39.6.2   APPLICATIONS OF FLEXIBLE MANUFACTURING SYSTEMS

Flexible manufacturing systems are typically used for midvolume, midvariety production. If the part or product is made in high quantities with no style variations, then a transfer line

(a)

(b)

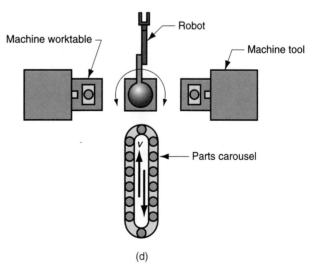

(c)

(d)

FIGURE 39.11 Four of the five FMS layout types: (a) in-line, (b) ladder, (c) open field, and (d) robot-centered cell. Key: Aut = automated station; L/UL = load/unload station; Insp = inspection station; AGV = automated guided vehicle; AGVS = automated guided vehicle system.

TABLE 39.5 Typical computer functions implemented by application software modules in a flexible manufacturing system.

Function	Description
NC part programming	Development of NC programs for new parts introduced into the system. This includes a language package such as APT
Production control	Product mix, machine scheduling, and other planning functions
NC program download	Part program commands must be downloaded to individual stations from the central computer
Machine control	Individual workstations require controls, usually computer numerical control
Workpart control	Monitor status of each workpart in the system, status of pallet fixtures, orders on loading/unloading pallet fixtures
Tool management	Functions include tool inventory control, tool status relative to expected tool life, tool changing and resharpening, and transport to and from tool grinding
Transport control	Scheduling and control of workpart handling system
System management	Compiles management reports on performance (utilization, piece counts, production rates, etc.). FMS simulation sometimes included

NC, numerical control; APT, automatically programmed tool; FMS, flexible manufacturing system.

or similar dedicated manufacturing system is most appropriate. If the parts are low volume with high variety, then a stand-alone NC machine or even manual methods would be more appropriate. These application characteristics are summarized in Figure 39.12.

Flexible machining systems comprise the most common application of FMS technology. Owing to the inherent flexibilities and capabilities of computer numerical control, it is possible to connect several CNC machine tools to a small central computer, and to devise automated material handling methods for transferring parts between machines. Figure 39.13 shows a flexible machining system consisting of five CNC machining centers and an in-line transfer system to pick up parts from a central load/unload station and move them to the appropriate machining stations.

In addition to machining systems, other types of flexible manufacturing systems have also been developed, although the state of technology in these other processes has not

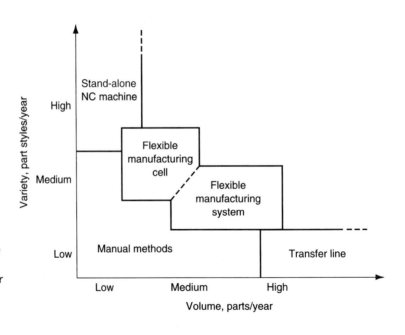

FIGURE 39.12
Application characteristics of flexible manufacturing systems and cells relative to other types of manufacturing systems.

FIGURE 39.13 A five-station flexible manufacturing system. (Photo courtesy of Cincinnati Milacron, Batavia, Ohio.)

permitted the rapid implementation that has occurred in machining. The other types of systems include assembly, inspection, sheet-metal processing (punching, shearing, bending, and forming), and forging.

Most of the experience in flexible manufacturing systems has been gained in machining applications. For flexible machining systems, the benefits usually given are (1) higher machine utilization than a conventional machine shop—relative utilizations are 40% to 50% for conventional batch-type operations and about 75% for a FMS due to better work handling, off-line setups, and improved scheduling; (2) reduced work-in-process due to continuous production rather than batch production; (3) lower manufacturing lead times; and (4) greater flexibility in production scheduling.

39.7 COMPUTER INTEGRATED MANUFACTURING

Distributed computer networks are widely used in modern manufacturing plants to monitor and/or control the integrated systems described in this chapter. Even though some of the operations are manually accomplished (e.g., manual assembly lines and manned cells), computer systems are utilized for production scheduling, data collection, record keeping, performance tracking, and other information-related functions. In the more automated systems (e.g., transfer lines and flexible manufacturing cells), computers directly control the operations. The term *computer integrated manufacturing* refers to the pervasive use of computer systems throughout the organization, not only to monitor and control the operations, but also to design the product, plan the manufacturing processes, and accomplish the business functions related to production. One might say that CIM is the ultimate integrated manufacturing system. In this final section of Part X, we outline the scope of CIM and provide a bridge to Part XI on manufacturing support systems.

To begin, let us identify four general functions that have to be accomplished in most manufacturing enterprises: (1) product design, (2) manufacturing planning, (3) manufacturing control, and (4) business functions. Product design is usually an iterative process that includes recognition of a need for a product, problem definition, creative synthesis of a solution, analysis and optimization, evaluation, and documentation. The overall quality of the resulting design is likely to be the most important factor upon which the commercial success of a product depends. In addition, a very significant portion of the final cost of the product is determined by decisions made during product design. Manufacturing planning is concerned with converting the engineering drawings and specifications that define the product design into a plan for producing the product. Manufacturing planning includes decisions on which parts will be purchased (the "make-or-buy decision"), how each "make" part will be produced, the equipment that will be used, how the work will be scheduled, and so on. Most of these decisions are discussed in Chapter 40 on manufacturing engineering and Chapter 41 on production planning. Manufacturing control includes not only control of the individual processes and equipment in the plant, but also the supporting functions such as shop floor control and quality control, discussed in Chapters 41 and 42, respectively. Finally, the business functions include order entry, cost accounting, payroll, customer billing, and other business-oriented information activities related to manufacturing.

Computer systems play an important role in these four general functions, and their integration within the organization is a distinguishing feature of computer integrated manufacturing, as depicted in Figure 39.14. Computer systems associated with product design are called CAD systems (for computer-aided design). Design systems and software include geometric modeling, engineering analysis packages such as finite element modeling, design review and evaluation, and automated drafting. Computer systems that support manufacturing planning are called CAM systems (for computer-aided manufacturing) and include computer-aided process planning, NC part programming, production scheduling, and planning packages such as manufacturing resource planning (discussed in Chapter 41). Manufacturing control systems include those used in process control, shop floor control, inventory control, and computer-aided inspection for quality control. And computerized business systems are used for order entry, customer billing, and other business functions. Customer

FIGURE 39.14   Four general functions in a manufacturing organization and how computer integrated manufacturing systems support these functions.

orders are entered by the company's sales force or by the customers themselves into the computerized order entry system. The orders include product specifications that provide the inputs to the design department. Based on these inputs, new products are designed on the company's CAD system. The design details serve as inputs to the manufacturing engineering group, where computer-aided process planning, computer-aided tool design, and related activities are performed in advance of actual production. The output from manufacturing engineering provides much of the input data required for manufacturing resource planning and production scheduling. Thus, computer integrated manufacturing provides the information flows required to accomplish the actual production of the product.

Today, computer integrated manufacturing is implemented in many companies using *enterprise resource planning* (ERP), an extension of manufacturing resource planning that organizes and integrates the information flows in a company through a single, central data base. The functions covered by ERP spread well beyond manufacturing operations; they include sales, marketing, purchasing, logistics, distribution, inventory control, finance, and human resources. ERP users within a company access and interact with the system using personal computers at their own workplaces, whether they are located in offices or in the factory.

REFERENCES

[1] Black, J. T. *The Design of the Factory with a Future.* McGraw-Hill, New York, 1990.

[2] Black, J. T. "An Overview of Cellular Manufacturing Systems and Comparison to Conventional Systems," *Industrial Engineering,* November 1983, pp. 36–84.

[3] Boothroyd, G., Poli, C., and Murch, L. E. *Automatic Assembly.* Marcel Dekker, New York, 1982.

[4] Buzacott, J. A. "Prediction of the Efficiency of Production Systems without Internal Storage," *International Journal of Production Research,* Vol. 6, No. 3, 1968, pp. 173–188.

[5] Buzacott, J. A., and Shanthikumar, J. G. *Stochastic Models of Manufacturing Systems.* Prentice-Hall, Upper Saddle River, New Jersey, 1993.

[6] Chang, T-C., Wysk, R. A., and Wang, H-P. *Computer-Aided Manufacturing,* 3rd ed. Prentice-Hall, Upper Saddle River, New Jersey, 2005.

[7] Chow, W-M. *Assembly Line Design.* Marcel Dekker, New York, 1990.

[8] Gallagher, C. C., and Knight, W. A. *Group Technology.* Butterworth & Co., Ltd., London, 1973.

[9] Groover, M. P. "Analyzing Automatic Transfer Lines," *Industrial Engineering,* Vol. 7, No. 11, 1975, pp. 26–31.

[10] Groover, M. P. *Automation, Production Systems, and Computer Integrated Manufacturing,* 3rd ed. Pearson Prentice-Hall, Upper Saddle River, New Jersey, 2008.

[11] Ham, I., Hitomi, K., and Yoshida, T. *Group Technology.* Kluwer Nijhoff Publishers, Hingham, Massachusetts, 1985.

[12] Houtzeel, A. "The Many Faces of Group Technology," *American Machinist,* January 1979, pp. 115–120.

[13] Luggen, W. W. *Flexible Manufacturing Cells and Systems.* Prentice Hall, Inc., Englewood Cliffs, New Jersey, 1991.

[14] Maleki, R. A. *Flexible Manufacturing Systems: The Technology and Management.* Prentice Hall, Inc., Englewood Cliffs, New Jersey, 1991.

[15] Moodie, C., Uzsoy, R., and Yih, Y. *Manufacturing Cells: A Systems Engineering View.* Taylor & Francis, Ltd., London, 1995.

[16] Parsai, H., Leep, H., and Jeon, G. *The Principles of Group Technology and Cellular Manufacturing.* John Wiley & Sons, Hoboken, New Jersey, 2006.

[17] Riley, F. J. *Assembly Automation, A. Management Handbook,* 2nd ed. Industrial Press, New York, 1999.

[18] Weber, A. "Is Flexibility a Myth?" *Assembly,* May 2004, pp. 50–59.

REVIEW QUESTIONS

39.1. What are the main components of an integrated manufacturing system?

39.2. What are the principal material handling functions in manufacturing?

39.3. Name the five main types of material transport equipment.

39.4. What is the difference between fixed routing and variable routing in material transport systems?

39.5.   What is a production line?

39.6.   What are the advantages of a mixed model line over a batch model line for producing different product styles?

39.7.   What are some of the limitations of a mixed model line compared to a batch model line?

39.8.   Describe how manual methods are used to move parts between workstations on a production line.

39.9.   Briefly define the three types of mechanized workpart transfer systems used in production lines.

39.10. Why are parts sometimes fixed to the conveyor in a continuous transfer system in manual assembly?

39.11. Why must a production line be paced at a rate higher than that required to satisfy the demand for the product?

39.12. Repositioning time on a synchronous transfer line is known by a different name; what is that name?

39.13. Why are single station assembly cells generally not suited to high-production jobs?

39.14. What are some of the reasons for downtime on a machining transfer line?

39.15. Define group technology.

39.16. What is a part family?

39.17. Define cellular manufacturing.

39.18. What is the composite part concept in group technology?

39.19. What is a flexible manufacturing system?

39.20. What are the criteria that should be satisfied to make an automated manufacturing system flexible?

39.21. Name some of the flexible manufacturing system software and control functions.

39.22. What are the advantages of flexible manufacturing system technology, compared to conventional batch operations?

39.23. Define computer integrated manufacturing.

MULTIPLE CHOICE QUIZ

There are 21 correct answers in the following multiple choice questions (some questions have multiple answers that are correct). To attain a perfect score on the quiz, all correct answers must be given. Each correct answer is worth 1 point. Each omitted answer or wrong answer reduces the score by 1 point, and each additional answer beyond the correct number of answers reduces the score by 1 point. Percentage score on the quiz is based on the total number of correct answers.

39.1.   Material handling is usually not associated with transportation between facilities that involves rail, truck, air, or waterway delivery of goods: (a) true or (b) false?

39.2.   Fixed routing is associated with which of the following types of manufacturing systems (two best answers): (a) automated production lines, (b) automated storage systems, (c) cellular manufacturing systems, (d) flexible manufacturing systems, (e) job shops, and (f) manual assembly lines?

39.3.   Which of the following types of material handling equipment are typically used in a process type layout (two best answers): (a) conveyors, (b) cranes and hoists, (c) fork lift trucks, and (d) rail-guided vehicles?

39.4.   Batch model production lines are most suited to which one of the following production situations: (a) job shop, (b) mass production, or (c) medium production?

39.5.   Precedence constraints are best described by which one of the following: (a) launching sequence in a mixed model line, (b) limiting value of the sum of element times that can be assigned to a worker or station, (c) order of workstations along the line, or (d) sequence in which the work elements must be done?

39.6.   Which of the following phrases are most appropriate to describe the characteristics of tasks that are performed at automated workstations (three best answers): (a) complex, (b) consists of multiple work elements, (c) involves a single work element, (d) involves straight-line motions, (e) requires sensory capability, and (f) simple?

39.7.   The transfer line is most closely associated with which one of the following types of production operations: (a) assembly, (b) automotive chassis fabrication, (c) machining, (d) pressworking, or (e) spot welding?

39.8.   A dial indexing machine uses which one of the following types of workpart transfer: (a) asynchronous, (b) continuous, (c) parts passed by hand, or (d) synchronous?

39.9.   Production flow analysis is a method of identifying part families that uses data from which one of the following sources: (a) bill of materials, (b) engineering drawings, (c) master schedule, (d) production schedule, or (e) route sheets?

39.10. Most parts classification and coding systems are based on which of the following types of part attributes (two best answers): (a) annual production rate, (b) date of design, (c) design, (d) manufacturing, and (e) weight?

39.11. What is the dividing line between a manufacturing cell and a flexible manufacturing system: (a) two machines, (b) four machines, or (c) six machines?

39.12. A machine capable of producing different part styles in a batch mode of operation qualifies as a flexible manufacturing system: (a) true or (b) false?

39.13. The physical layout of a flexible manufacturing system is determined principally by which one of the following: (a) computer system, (b) material handling system, (c) part family, (d) processing equipment, or (e) weight of parts processed?

39.14. Industrial robots can, in general, most easily handle which one of the following part types in a flexible machining system: (a) heavy parts, (b) metal parts, (c) nonrotational parts, (d) plastic parts, or (e) rotational parts?

39.15. Flexible manufacturing systems and cells are generally applied in which one of the following areas: (a) high-variety, low-volume production, (b) low variety, (c) low volume, (d) mass production, (e) medium-volume, medium-variety production?

39.16. Which one of the following technologies is most closely associated with flexible machining systems: (a) lasers, (b) machine vision, (c) manual assembly lines, (d) numerical control, or (e) transfer lines?

PROBLEMS

Manual Assembly Lines

39.1. A manual assembly line is being designed for a product with annual demand = 100,000 units. The line will operate 50 wk/yr, 5 shifts/wk, and 7.5 hr/shift. Work units will be attached to a continuously moving conveyor. Work content time = 42.0 min. Assume line efficiency = 0.97, balancing efficiency = 0.92, and repositioning time = 6 sec. Determine (a) hourly production rate to meet demand, (b) number of workers required, and (c) the number of workstations required if the estimated manning level is 1.4.

39.2. A manual assembly line produces a small appliance whose work content time = 25.9 min. Desired production rate = 50 units/hr. Repositioning time = 6 sec, line efficiency = 95%, and balancing efficiency is 93%. How many workers are on the line?

39.3. A single model manual assembly line produces a product whose work content time = 47.8 min. The line has 24 workstations with a manning level = 1.25. Available shift time per day = 8 hr, but downtime during the shift reduces actual production time to 7.6 hr on average. This results in an average daily production of 256 units/day. Repositioning time per worker is 8% of cycle time. Determine (a) line efficiency, (b) balancing efficiency, and (c) repositioning time.

39.4. A final assembly plant for a certain automobile model is to have a capacity of 240,000 units annually. The plant will operate 50 wk/yr, 2 shifts/day, 5 days/wk, and 8.0 hr/shift. It will be divided into three departments: (1) body shop, (2) paint shop, (3) trim-chassis-final department. The body shop welds the car bodies using robots, and the paint shop coats the bodies. Both of these departments are highly automated. Trim-chassis-final has no automation. There are 15.5 hr of direct labor content on each car in this department, where cars are moved by a continuous conveyor. Determine (a) hourly production rate of the plant, (b) number of workers and workstations required in trim-chassis-final if no automated stations are used, the average manning level is 2.5, balancing efficiency = 93%, proportion uptime = 95%, and a repositioning time of 0.15 min is allowed for each worker.

39.5. A product whose total work content time = 50 min is to be assembled on a manual production line. The required production rate is 30 units/hr. From previous experience with similar products, it is estimated that the manning level will be close to 1.5. Assume that the uptime proportion and line balancing efficiency are both = 1.0. If 9 sec will be lost from the cycle time for repositioning, determine (a) the cycle time and (b) the numbers of workers and stations that will be needed on the line.

39.6. A manual assembly line has 17 workstations with one operator per station. Total work content time to assemble the product = 22.2 min. The production rate of the line = 36 units/hr. A synchronous transfer system is used to advance the products from one station to the next, and the transfer time = 6 sec. The workers remain seated along the line. Proportion uptime = 0.90. Determine the balance efficiency.

39.7. A production line with four automatic workstations (the other stations are manual) produces a certain product whose total assembly work content time = 55.0 min of direct manual labor. The production rate on the line is 45 units/hr. Because of the automated stations, uptime efficiency = 89%. The manual stations each have one worker. It is known that 10% of the cycle time is lost due to repositioning. If the balancing efficiency = 0.92 on the manual stations, find (a) cycle time, (b) number of workers and (c) workstations on the line. (d) What is the average manning level on the line, where the average includes the automatic stations?

39.8. Production rate for a certain assembled product is 47.5 units/hr. The total assembly work content time = 32 min of direct manual labor. The line operates at 95% uptime. Ten workstations have two workers on opposite sides of the line so that both sides of the product can be worked on simultaneously. The remaining stations have one worker. Repositioning time lost by each worker is 0.2 min/cycle. It is known that the number of workers on the line is two more than the number required for perfect balance. Determine (a) number of workers, (b) number of workstations, (c) the balancing efficiency, and (d) average manning level.

39.9. The total work content for a product assembled on a manual production line is 48 min. The work is transported using a continuous overhead conveyor that operates at a speed of 3 ft/min. There are 24 workstations on the line, one-third of which have two workers; the remaining stations each have one worker. Repositioning time per worker is 9 sec, and uptime efficiency of the line is 95%. (a) What is the maximum possible hourly production rate if line is assumed to be perfectly balanced? (b) If the actual production rate is only 92% of the maximum possible rate determined in part (a), what is the balance efficiency on the line?

Automated Production Lines

39.10. An automated transfer line has 20 stations and operates with an ideal cycle time of 1.50 min. Probability of a station failure = 0.008 and average downtime when a breakdown occurs is 10.0 min. Determine (a) the average production rate and (b) the line efficiency.

39.11. A dial-indexing table has six stations. One station is used for loading and unloading, which is accomplished by a human worker. The other five perform processing operations. The longest process takes 25 sec and the indexing time = 5 sec. Each station has a frequency of failure = 0.015. When a failure occurs it takes an average of 3.0 min to make repairs and restart. Determine (a) hourly production rate and (b) line efficiency.

39.12. A seven-station transfer line has been observed over a 40-hour period. The process times at each station are as follows: station 1, 0.80 min; station 2, 1.10 min; station 3, 1.15 min; station 4, 0.95 min; station 5, 1.06 min; station 6, 0.92 min; and station 7, 0.80 min. The transfer time between stations = 6 sec. The number of downtime occurrences = 110, and hours of downtime = 14.5 hr. Determine (a) the number of parts produced during the week, (b) the average actual production rate in parts/hr, and (c) the line efficiency. (d) If the

balancing efficiency were computed for this line, what would its value be?

39.13. A 12-station transfer line was designed to operate with an ideal production rate = 50 parts/hr. However, the line does not achieve this rate, since the line efficiency = 0.60. It costs $75/hr to operate the line, exclusive of materials. The line operates 4000 hr/yr. A computer monitoring system has been proposed that will cost $25,000 (installed) and will reduce downtime on the line by 25%. If the value added per unit produced = $4.00, will the computer system pay for itself within 1 year of operation? Use expected increase in revenues resulting from the computer system as the criterion. Ignore material costs in your calculations.

39.14. An automated transfer line is to be designed. Based on previous experience, the average downtime per occurrence = 5.0 min, and the probability of a station failure that leads to a downtime occurrence $p = 0.01$. The total work content time = 9.8 min and is to be divided evenly amongst the workstations, so that the ideal cycle time for each station = 9.8/n. Determine (a) the optimum number of stations on the line n that will maximize production rate, and (b) the production rate and proportion uptime for your answer to part (a).

40   MANUFACTURING ENGINEERING

Chapter Contents

This final part of the book is concerned with ***manufacturing support systems***, which are the set of procedures and systems used by a company to solve the technical and logistics problems encountered in planning the processes, ordering materials, controlling production, and ensuring that the company's products meet required quality specifications. The position of the manufacturing support systems in the overall operations of the company is portrayed in Figure 40.1. As with the manufacturing systems in the factory, the manufacturing support systems include people. People make the systems work. Unlike the manufacturing systems in the factory, most of the support systems do not directly contact the product during its processing and assembly. Instead, they plan and control the activities in the factory to ensure that the products are completed and delivered to the customer on time, in the right quantities, and to the highest quality standards.

The quality control system is one of the manufacturing support systems, but it also consists of facilities located in the factory—inspection equipment used to measure and gage the materials being processed and products being assembled. We cover the quality control system in Chapter 42 on quality control and inspection. Many of the traditional measurement and gaging techniques used in inspection are described in Chapter 5. Other manufacturing support systems covered in this part of the book are production planning and control, Chapter 41, and manufacturing engineering in the present chapter.

Manufacturing engineering is a technical staff function that is concerned with planning the manufacturing processes for the economic production of high-quality

FIGURE 40.1
The position of the manufacturing support systems in the production system.

products. Its principal role is to engineer the transition of the product from design specification to physical product. Its overall goal is to optimize production in a particular organization. The scope of manufacturing engineering includes many activities and responsibilities that depend on the type of production operations accomplished by the organization. The usual activities include the following:

➤ *Process planning.* As our definition suggests, this is the principal activity of manufacturing engineering. Process planning includes (a) deciding what processes and methods should be used and in what sequence, (b) determining tooling requirements, (c) selecting production equipment and systems, and (d) estimating costs of production for the selected processes, tooling, and equipment.

➤ *Problem solving and continuous improvement.* Manufacturing engineering provides staff support to the operating departments (parts fabrication and product assembly) to solve technical production problems. It should also be engaged in continuous efforts to reduce production costs, increase productivity, and improve product quality.

➤ *Design for manufacturability.* In this function, which chronologically precedes the other two, manufacturing engineers serve as manufacturability advisors to product designers. The objective is to develop product designs that not only meet functional and performance requirements, but that also can be produced at reasonable cost with minimum technical problems at highest possible quality in the shortest possible time.

Manufacturing engineering must be performed in any industrial organization that is engaged in production. The manufacturing engineering department usually reports to the manager of manufacturing in a company. In some companies the department is known by other names, such as process engineering or production engineering. Often included under manufacturing engineering are tool design, tool fabrication, and various technical support groups.

40.1 PROCESS PLANNING

Process planning involves determining the most appropriate manufacturing processes and the order in which they should be performed to produce a given part or product specified by design engineering. If it is an assembled product, process planning includes deciding the appropriate sequence of assembly steps. The process plan must be developed within the limitations imposed by available processing equipment and productive capacity of the factory. Parts or subassemblies that cannot be made internally must be

TABLE 40.1   Decisions and details required in process planning.

Processes and sequence. The process plan should briefly describe all processing steps used on the work unit (e.g., part, assembly) in the order in which they are performed.

Equipment selection. In general, manufacturing engineers try to develop process plans that utilize existing equipment. When this is not possible, the component in question must be purchased (Section 40.1.2), or new equipment must be installed in the plant.

Tools, dies, molds, fixtures, and ***gages.*** The process planner must decide what tooling is needed for each process. Actual design is usually delegated to the tool design department, and fabrication is accomplished by the tool room.

Cutting tools and ***cutting conditions*** for machining operations. These are specified by the process planner, industrial engineer, shop foreman, or machine operator, often with reference to standard handbook recommendations.

Methods. Methods include hand and body motions, workplace layout, small tools, hoists for lifting heavy parts, and so forth. Methods must be specified for manual operations (e.g., assembly) and manual portions of machine cycles (e.g., loading and unloading a production machine). Methods planning is traditionally done by industrial engineers.

Work standards. Work measurement techniques are used to establish time standards for each operation.

Estimating production costs. This is often accomplished by cost estimators with help from the process planner.

purchased from external suppliers. In some cases, items that can be produced internally may be purchased from outside vendors for economic or other reasons.

40.1.1 TRADITIONAL PROCESS PLANNING

Traditionally, process planning has been accomplished by manufacturing engineers who are knowledgeable in the particular processes used in the factory and are able to read engineering drawings. Based on their knowledge, skill, and experience, they develop the processing steps in the most logical sequence required to make each part. Table 40.1 lists the many details and decisions usually included within the scope of process planning. Some of these details are often delegated to specialists, such as tool designers; but manufacturing engineering is responsible for them.

Process Planning for Parts   The processes needed to manufacture a given part are determined largely by the material out of which it is to be made. The material is selected by the product designer based on functional requirements. Once the material has been selected, the choice of possible processes is narrowed considerably. In our coverage of engineering materials, we provided guides to the processing of the four material groups: metals (Section 6.5), ceramics (Section 7.6), polymers (Section 8.5), and composite materials (Section 9.5)

A typical processing sequence to fabricate a discrete part consists of (1) a basic process, (2) one or more secondary processes, (3) operations to enhance physical properties, and (4) finishing operations, illustrated in Figure 40.2. Basic and secondary processes are shaping processes (Section 1.3.1) which alter the geometry of a workpart. A ***basic process*** establishes

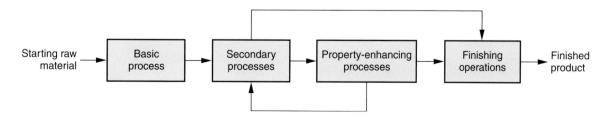

FIGURE 40.2   Typical sequence of processes required in part fabrication.

the initial geometry of the part. Examples include metal casting, forging, and sheet-metal rolling. In most cases, the starting geometry must be refined by a series of *secondary processes*. These operations transform the basic shape into the final geometry. There is a correlation between the secondary processes that might be used and the basic process that provides the initial form. For example, when sand casting or forging are the basic processes, machining operations are generally the secondary processes. When a rolling mill produces strips or coils of sheet metal, the secondary processes are stamping operations such as blanking, punching, and bending. Selection of certain basic processes minimizes the need for secondary processes. For example, if plastic injection molding is the basic process, secondary operations are usually not required because molding is capable of providing the detailed geometric features with good dimensional accuracy.

Shaping operations are generally followed by operations to enhance physical properties and/or finish the product. *Operations to enhance properties* include heat treating operations on metal components and glassware. In many cases, parts do not require these property-enhancing steps in their processing sequence. This is indicated by the alternate arrow path in our figure. *Finishing operations* are the final operations in the sequence; they usually provide a coating on the workpart (or assembly) surface. Examples of these processes are electroplating and painting.

In some cases, property-enhancing processes are followed by additional secondary operations before proceeding to finishing, as suggested by the return loop in Figure 40.2. An example is a machined part that is hardened by heat treatment. Prior to heat treatment, the part is left slightly oversized to allow for distortion. After hardening, it is reduced to final size and tolerance by finish grinding. Another example, again in metal parts fabrication, is when annealing is used to restore ductility to the metal after cold working to permit further deformation of the workpiece.

Table 40.2 presents some of the typical processing sequences for various materials and basic processes. The task of the process planner usually begins after the basic process has provided the initial shape of the part. Machined parts begin as bar stock or castings or forgings, and the basic processes for these starting shapes are often external to the fabricating plant. Stampings begin as sheet metal coils or strips purchased from the mill. These are the raw materials supplied from external suppliers for the secondary processes and subsequent operations to be performed in the factory. Determining the most appropriate processes and the order in which they must be accomplished relies on the skill, experience, and judgment of the process planner. Some of the basic guidelines and considerations used by process planners to make these decisions are outlined in Table 40.3.

The Route Sheet   The process plan is prepared on a form called a *route sheet*, a typical example of which is shown in Figure 40.3 (some companies use other names for this form). It

---

TABLE 40.2   Some typical process sequences.

Basic Process	Secondary Process(es)	Property-Enhancing Processes	Finishing Operations
Sand casting	Machining	(none)	Painting
Die casting	(none, net shape)	(none)	Painting
Casting of glass	Pressing, blow molding	(none)	(none)
Injection molding	(none, net shape)	(none)	(none)
Rolling of bar stock	Machining	Heat treatment (optional)	Electroplating
Rolling of sheet metal	Blanking, bending, drawing	(none)	Electroplating
Forging	Machining (near net shape)	(none)	Painting
Extrusion of aluminum	Cut to length	(none)	Anodize
Atomize metal powders	Pressing of powder metal part	Sintering	Painting

Compiled from [5].

TABLE 40.3   Guidelines and considerations in deciding processes and their sequence in process planning.

Design requirements. The sequence of processes must satisfy the dimensions, tolerances, surface finish, and other specifications established by product design.

Quality requirements. Processes must be selected that satisfy quality requirements in terms of tolerances, surface integrity, consistency and repeatability, and other quality measures.

Production volume and *rate.* Is the product in the category of low, medium, or high production? The selection of processes and systems is strongly influenced by volume and production rate.

Available processes. If the product and its components are to be made in-house, the process planner must select processes and equipment already available in the factory.

Material utilization. It is desirable for the process sequence to make efficient use of materials and minimize waste. When possible, net shape or near net shape processes should be selected.

Precedence constraints. These are technological sequencing requirements that determine or restrict the order in which the processing steps can be performed. A hole must be drilled before it can be tapped; a powder-metal part must be pressed before sintering; a surface must be cleaned before painting; and so on.

Reference surfaces. Certain surfaces of the part must be formed (usually by machining) near the beginning of the sequence so they can serve as locating surfaces for other dimensions that are formed subsequently. For example, if a hole is to be drilled a certain distance from the edge of a given part, that edge must first be machined.

Minimize setups. The number of separate machine setups should be minimized. Wherever possible, operations should be combined at the same workstation. This saves time and reduces material handling.

Eliminate unnecessary steps. The process sequence should be planned with the minimum number of processing steps. Unnecessary operations should be avoided. Design changes should be requested to eliminate features not absolutely needed, thereby eliminating the processing steps associated with those features.

Flexibility. Where feasible, the process should be sufficiently flexible to accommodate engineering design changes. This is often a problem when special tooling must be designed to produce the part; if the part design is changed, the special tooling may be rendered obsolete.

Safety. Worker safety must be considered in process selection. This makes good economic sense, and it is the law (Occupational Safety and Health Act).

Minimum cost. The process sequence should be the production method that satisfies all of the above requirements and also achieves the lowest possible product cost.

Part No: 031393	Part Name: Housing, valve		Rev. 2	Page 1 of 2
Matl: 416 Stainless	Size: 2.0 dia × 5. long		Planner: MPG	Date: 3/13/XX

No.	Operation	Dept.	Machine	Tooling, gages	Setup time	Cycle time
10	Face; rough & finish turn to 1.473 ± 0.003 dia. × 1.250 ± 0.003 length; face shoulder to 0.313 ± 0.002; finish turn to 1.875 ± 0.002 dia.; form 3 grooves at 0.125 width × 0.063 deep.	L	325	G857	1.0 h	8.22 m
20	Reverse; face to 4.750 ± 0.005 length; finish turn to 1.875 ± 0.002 dia.; drill 1.000 + 0.006, −0.002 dia. axial hole.	L	325		0.5 h	3.10 m
30	Drill & ream 3 radial holes at 0.375 ± 0.002 dia.	D	114	F511	0.3 h	2.50 m
40	Mill 0.500 ± 0.004 wide × 0.375 ± 0.003 deep slot.	M	240	F332	0.3 h	1.75 m
50	Mill 0.750 ± 0.004 wide × 0.375 ± 0.003 deep flat.	M	240	F333	0.3 h	1.60 m

FIGURE 40.3 Typical route sheet for specifying the process plan.

is called a route sheet because it specifies the sequence of operations and equipment that will be visited by the part during its production. The route sheet is to the process planner what the engineering drawing is to the product designer. It is the official document that specifies the details of the process plan. The route sheet should include all manufacturing operations to be performed on the workpart, listed in the proper order in which they are to be accomplished. For each operation, the following should be listed: (1) a brief description of the operation indicating the work to be done, surfaces to be processed with references to the part drawing, and dimensions (and tolerances, if not specified on part drawing) to be achieved; (2) the equipment on which the work is to be performed; and (3) any special tooling required, such as dies, molds, cutting tools, jigs or fixtures, and gages. In addition, some companies include cycle time standards, setup times, and other data on the route sheet.

Sometimes a more detailed *operation sheet* is also prepared for each operation listed in the routing. This is retained in the particular department where the operation is performed. It indicates the specific details of the operation, such as cutting speeds, feeds, and tools, and other instructions useful to the machine operator. Setup sketches are sometimes also included.

Process Planning for Assemblies   For low production, assembly is generally done at individual workstations and a worker or team of workers performs the assembly work elements to complete the product. In medium and high production, assembly is usually performed on production lines (Section 39.4). In either case, there is a precedence order in which the work must be accomplished.

Process planning for assembly involves preparation of the assembly instructions that must be performed. For single stations, the documentation is similar to the processing route sheet in Figure 40.3. It contains a list of the assembly steps in the order in which they must be accomplished. For assembly line production, process planning consists of allocating work elements to particular stations along the line, a procedure called *line balancing* (Section 39.3.2). In effect, the assembly line routes the work units to individual stations, and the line balancing solution determines what assembly steps must be performed at each station. As with process planning for parts, any tools and fixtures needed to accomplish a given assembly work element must be decided, and the workplace layout must be designed.

40.1.2 MAKE OR BUY DECISION

Inevitably, the question arises as to whether a given part should be purchased from an outside vendor or made internally. First of all, it should be recognized that virtually all manufacturers purchase their starting materials from suppliers. A machine shop buys bar stock from a metals distributor and castings from a foundry. A plastic molder obtains molding compound from a chemical company. A pressworking company purchases sheet metal from a rolling mill. Very few companies are vertically integrated all the way from raw materials to finished product.

Given that a company purchases at least some of its starting materials, it is reasonable to ask whether the company should purchase at least some of the parts that would otherwise be made in its own factory. The answer to the question is the *make or buy decision*. The make versus buy question is probably appropriate to ask for every component used by the company.

Cost is the most important factor in deciding whether a part should be made in-house or purchased. If the vendor is significantly more proficient in the processes required to make the component, it is likely that the internal production cost will be greater than the purchase price even when a profit is included for the vendor. On the other hand, if purchasing the part results in idle equipment in the factory, then an apparent cost advantage for the vendor may be a disadvantage for the home factory. Consider the following example.

Example 40.1
Make or Buy Cost
Comparison

Suppose that the quoted price for a certain component from a vendor is $8.00 per unit for 1000 units. The same part made in the home factory would cost $9.00. The cost breakdown on the make alternative is as follows:

$$
\begin{aligned}
\text{Unit material cost} &= \$2.25 \text{ per unit} \\
\text{Direct labor} &= \$2.00 \text{ per unit} \\
\text{Labor overhead at } 150\% &= \$3.00 \text{ per unit} \\
\text{Equipment fixed cost} &= \underline{\$1.75 \text{ per unit}} \\
\text{Total} &= \$9.00 \text{ per unit}
\end{aligned}
$$

Should the component by bought or made in-house?

Solution: Although the vendor's quote seems to favor the buy decision, let us consider the possible effect on the factory if we decide to accept the quote. The equipment fixed cost is an allocated cost based on an investment that has already been made. If it turns out that the equipment is rendered idle by the decision to buy the part, then one might argue that the fixed cost of $1.75 continues even if the equipment is not in use. Similarly, the overhead cost of $3.00 consists of factory floor space, indirect labor, and other costs that will also continue even if the part is bought. By this reasoning, the decision to purchase might cost the company as much as $8.00 + $1.75 + $3.00 = $12.75 per unit if it results in idle time in the factory on the machine that would have been used to make the part.

On the other hand, if the equipment can be used to produce other components for which the internal prices are less than the corresponding external quotes, then a buy decision makes good economic sense.   ■

Make or buy decisions are rarely as clear as in Example 40.1. Some of the other factors that enter the decision are listed in Table 40.4. Although these factors appear to be subjective, they all have cost implications, either directly or indirectly. In recent years, major companies have placed strong emphasis on building close relationships with parts suppliers. This trend has been especially prevalent in the automobile industry, where long-term agreements have been reached between each carmaker and a limited number of vendors who are able to deliver high-quality components reliably on schedule.

TABLE 40.4   Key factors in the make or buy decision.

Factor	Explanation and Effect on Make/Buy Decision
Process available in-house	If a given process is not available internally, then the obvious decision is to purchase. Vendors often develop proficiency in a limited set of processes that makes them cost competitive in external-internal comparisons. There are exceptions to this guideline, in which a company decides that, in its long-term strategy, it must develop a proficiency in a manufacturing process technology that it does not currently possess.
Production quantity	Number of units required. High volume tends to favor make decisions. Low quantities tend to favor buy decisions.
Product life	Long product life favors internal production.
Standard items	Standard catalog items, such as bolts, screws, nuts, and many other types of components are produced economically by suppliers specializing in those products. It is almost always better to purchase these standard items.
Supplier reliability	The reliable supplier gets the business.
Alternative source	In some cases, factories buy parts from vendors as an alternative source to their own production plants. This is an attempt to ensure uninterrupted supply of parts, or to smooth production in peak demand periods.

40.1.3 COMPUTER-AIDED PROCESS PLANNING

During the last several decades, there has been considerable interest in ***computer-aided process planning*** (CAPP)—automating the process planning function by means of computer systems. Shop people knowledgeable in manufacturing processes are gradually retiring. An alternative approach to process planning is needed, and CAPP systems provide this alternative. Computer-aided process planning systems are designed around either of two approaches: retrieval systems and generative systems.

Retrieval CAPP Systems   Retrieval CAPP systems, also known as ***variant CAPP systems***, are based on group technology and parts classification and coding (Section 39.5). In these systems, a standard process plan is stored in computer files for each part code number. The standard plans are based on current part routings in use in the factory, or on an ideal plan that is prepared for each family. Retrieval CAPP systems operate as indicated in Figure 40.4. The user begins by identifying the GT code of the part for which the process plan is to be determined. A search is made of the part family file to determine if a standard route sheet exists for the given part code. If the file contains a process plan for the part, it is retrieved and displayed for the user. The standard process plan is examined to determine whether modifications are necessary. Although the new part has the same code number, minor differences in the processes might be required to make the part. The standard plan is edited accordingly. The capacity to alter an existing process plan is why retrieval CAPP systems are also called variant systems.

If the file does not contain a standard process plan for the given code number, the user may search the file for a similar code number for which a standard routing exists. By editing the existing process plan, or by starting from scratch, the user develops the process plan for the new part. This becomes the standard process plan for the new part code number.

The final step is the process plan formatter, which prints the route sheet in the proper format. The formatter may call other application programs: determining cutting conditions for machine tool operations, calculating standard times for machining operations, or computing cost estimates.

Generative CAPP Systems   Generative CAPP systems are an alternative to retrieval systems. Rather than retrieving and editing existing plans from a database, a generative system creates the process plan using systematic procedures that might be applied by a human planner. In a fully generative CAPP system, the process sequence is planned without human assistance and without predefined standard plans.

Designing a generative CAPP system is a problem in the field of expert systems, a branch of artificial intelligence. ***Expert systems*** are computer programs capable of solving complex problems that normally require a human who has years of education and experience.

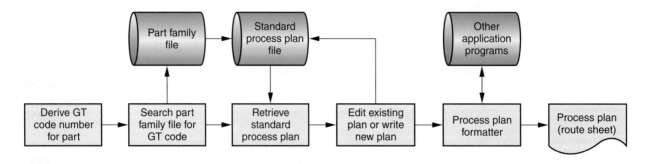

FIGURE 40.4   Operation of a retrieval computer-aided process planning system. (Source: [5].)

Process planning fits that definition. Several ingredients are required in a fully generative CAPP system:

1. ***Knowledge base.*** The technical knowledge of manufacturing and the logic used by successful process planners must be captured and coded into a computer program. An expert system applied to process planning requires the knowledge and logic of human process planners to be incorporated into a knowledge base. Generative CAPP systems then use the knowledge base to solve process planning problems; that is, to create route sheets.

2. ***Computer-compatible part description.*** Generative process planning requires a computer-compatible description of the part. The description contains all the pertinent data needed to plan the process sequence. Two possible descriptions are (1) the geometric model of the part developed on a CAD system during product design, or (2) a group technology code number of the part defining its features in significant detail.

3. ***Inference engine.*** A generative CAPP system requires the capability to apply the planning logic and process knowledge contained in the knowledge base to a given part description. The CAPP system applies its knowledge base to solve a specific problem of planning the process for a new part. This problem-solving procedure is referred to as the inference engine in the terminology of expert systems. By using its knowledge base and inference engine, the CAPP system synthesizes a new process plan for each new part presented to it.

Benefits of CAPP   Benefits of computer-automated process planning include the following: (1) process rationalization and standardization—automated process planning leads to more logical and consistent process plans than when traditional process planning is used; (2) increased productivity of process planners—the systematic approach and availability of standard process plans in the data files permit a greater number of process plans to be developed by the user; (3) reduced lead time to prepare process plans; (4) improved legibility compared to manually prepared route sheets; and (5) ability to interface CAPP programs with other application programs, such as cost estimating, work standards, and others.

40.2   PROBLEM SOLVING AND CONTINUOUS IMPROVEMENT

Problems arise in manufacturing that require technical staff support beyond what is normally available in the line organization of the production departments. Providing this technical support is one of the responsibilities of manufacturing engineering. The problems are usually specific to the particular technologies of the processes performed in the operating department. In machining, the problems may relate to selection of cutting tools, fixtures that do not work properly, parts with out-of-tolerance conditions, or non-optimal cutting conditions. In plastic molding, the problems may be excessive flash, parts sticking in the mold, or any of several defects that can occur in a molded part. These problems are technical, and engineering expertise is often required to solve them.

In some cases, the solution may require a design change; for example, changing the tolerance on a part dimension to eliminate a finish grinding operation while still achieving the functionality of the part. The manufacturing engineer is responsible for developing the proper solution to the problem and proposing the engineering change to the design department.

One of the areas that is ripe for improvement is setup time. The procedures involved in changing over from one production setup to the next (i.e., in batch production) are time consuming and costly. Manufacturing engineers are responsible for analyzing changeover procedures and finding ways to reduce the time required to perform them. Some of the approaches used in setup reduction are described in Section 41.4.

In addition to solving current technical problems ("fire fighting," as it might be called), the manufacturing engineering department is also responsible for continuous improvement projects. Continuous improvement means constantly searching for and implementing ways to reduce cost, improve quality, and increase productivity in manufacturing. It is accomplished one project at a time. Depending on the type of problem area, it may involve a project team whose membership includes not only manufacturing engineers, but also other personnel such as product designers, quality engineers, and production workers.

40.3 CONCURRENT ENGINEERING AND DESIGN FOR MANUFACTURABILITY

Much of the process planning function described in Section 40.1 is preempted by decisions made in product design. Decisions on material, part geometry, tolerances, surface finish, grouping of parts into subassemblies, and assembly techniques limit the available manufacturing processes that can be used to make a given part. If the product engineer designs an aluminum sand casting with features that can be achieved only by machining (e.g., flat surfaces with good finishes, close tolerances, and threaded holes), then the process planner has no choice but to plan for sand casting followed by the required machining operations. If the product designer specifies a collection of sheet-metal stampings to be assembled by threaded fasteners, then the process planner must lay out the series of blanking, punching, and forming steps to fabricate the stampings and then assemble them. In both of these examples, a plastic molded part may be a superior design, both functionally and economically. It is important for the manufacturing engineer to act as an advisor to the design engineer in matters of manufacturability because manufacturability matters, not only to the production departments but to the design engineer. A product design that is functionally superior and at the same time can be produced at minimum cost holds the greatest promise of success in the marketplace. Successful careers in design engineering are built on successful products.

Terms often associated with this attempt to favorably influence the manufacturability of a product are *design for manufacturing* (DFM) and *design for assembly* (DFA). Of course, DFM and DFA are inextricably coupled, so let us refer to them as DFM/A. The scope of DFM/A is expanded in some companies to include not only manufacturability issues but also marketability, testability, serviceability, maintainability, and so forth. This broader view calls for inputs from many departments in addition to design and manufacturing engineering. The approach is called *concurrent engineering*. Our discussion is organized into two sections: DFM/A and concurrent engineering.

40.3.1 DESIGN FOR MANUFACTURING AND ASSEMBLY

Design for manufacturing and assembly is an approach to product design that systematically includes considerations of manufacturability and assemblability in the design. DFM/A includes organizational changes and design principles and guidelines.

To implement DFM/A, a company must change its organizational structure, either formally or informally, to provide closer interaction and better communication between design and manufacturing personnel. This is often accomplished by forming project teams consisting of product designers, manufacturing engineers, and other specialties (e.g., quality engineers, material scientists) to design the product. In some companies, design engineers are required to spend some career time in manufacturing to learn about the problems encountered in making things. Another possibility is to assign manufacturing engineers to the product design department as full-time consultants.

DFM/A also includes principles and guidelines that indicate how to design a given product for maximum manufacturability. Many of these are universal design guidelines, such as those presented in Table 40.5. They are rules of thumb that can be applied to nearly

TABLE 40.5   General principles and guidelines in design for manufacturing and assembly.

Minimize number of components. Assembly costs are reduced. The final product is more reliable because there are fewer connections. Disassembly for maintenance and field service is easier. Reduced part count usually means automation is easier to implement. Work-in-process is reduced, and there are fewer inventory control problems. Fewer parts need to be purchased, which reduces ordering costs.

Use standard commercially available components. Design time and effort are reduced. Design of custom-engineered components is avoided. There are fewer part numbers. Inventory control is facilitated. Quantity discounts may be possible.

Use common parts across product lines. There is an opportunity to apply group technology (Section 39.5). Implementation of manufacturing cells may be possible. Quantity discounts may be possible.

Design for ease of part fabrication. Net shape and near net shape processes may be feasible. Part geometry is simplified, and unnecessary features are avoided. Unnecessary surface finish requirements should be avoided; otherwise, additional processing may be needed.

Design parts with tolerances that are within process capability. Tolerances tighter than the process capability (Section 42.2) should be avoided; otherwise, additional processing or sortation will be required. Bilateral tolerances should be specified.

Design the product to be foolproof during assembly. Assembly should be unambiguous. Components should be designed so they can be assembled only one way. Special geometric features must sometimes be added to components to achieve foolproof assembly.

Minimize use of flexible components. Flexible components include parts made of rubber, belts, gaskets, cables, etc. Flexible components are generally more difficult to handle and assemble.

Design for ease of assembly. Part features such as chamfers and tapers should be designed on mating parts. Design the assembly using base parts to which other components are added. The assembly should be designed so that components are added from one direction, usually vertically. Threaded fasteners (screws, bolts, nuts) should be avoided where possible, especially when automated assembly is used; instead, fast assembly techniques such as snap fits and adhesive bonding should be employed. The number of distinct fasteners should be minimized.

Use modular design. Each subassembly should consist of five to fifteen parts. Maintenance and repair are facilitated. Automated and manual assembly are implemented more readily. Inventory requirements are reduced. Final assembly time is minimized.

Shape parts and products for ease of packaging. The product should be designed so that standard packaging cartons can be used, which are compatible with automated packaging equipment. Shipment to customer is facilitated.

Eliminate or reduce adjustment required. Adjustments are time-consuming in assembly. Designing adjustments into the product means more opportunities for out-of-adjustment conditions to arise.

Compiled from [1], [2], [9].

any product design situation. In addition, several of our chapters on manufacturing processes include DFM/A principles that are specific to those processes.

The guidelines are sometimes in conflict. For example, one guideline for part design is to make the geometry as simple as possible. Yet, in design for assembly, it is often desirable to combine features of several assembled parts into a single component to reduce part count and assembly time. In these instances, design for manufacture conflicts with design for assembly, and a compromise must be found that achieves the best balance between opposing sides of the conflict.

Benefits typically cited for DFM/A include (1) shorter time to bring the product to market, (2) smoother transition into production, (3) fewer components in the final product, (4) easier assembly, (5) lower costs of production, (6) higher product quality, and (7) greater customer satisfaction [1], [2].

40.3.2 CONCURRENT ENGINEERING

Concurrent engineering refers to an approach to product design in which companies attempt to reduce the elapsed time required to bring a new product to market by integrating design engineering, manufacturing engineering, and other functions in the company. The traditional approach to launch a new product tends to separate the two functions, as illustrated in Figure 40.5(a). Product design develops the new design, sometimes with small regard for the manufacturing capabilities possessed by the company. There is little interaction between design engineers and manufacturing engineers who might provide advice on these capabilities and how the product design might be altered to accommodate them. It is as if a wall exists between the two functions; when design engineering completes the design, it tosses the drawings and specifications over the wall so that process planning can commence.

In a company that practices concurrent engineering (also known as *simultaneous engineering*), manufacturing planning begins while the product design is being developed, as pictured in Figure 40.5(b). Manufacturing engineering becomes involved early in the product development cycle. In addition, other functions are also involved, such as field service, quality engineering, the manufacturing departments, vendors supplying critical components, and in some cases customers who will use the product. All of these functions can contribute to a product design that not only performs well functionally, but is also manufacturable, assemblable, inspectable, testable, serviceable, maintainable, free of defects, and safe. All viewpoints have been combined to design a product of high quality that will deliver customer satisfaction. And through early involvement, rather than a procedure of reviewing the final design and suggesting changes after it is too late to conveniently make them, the total product development cycle is substantially reduced.

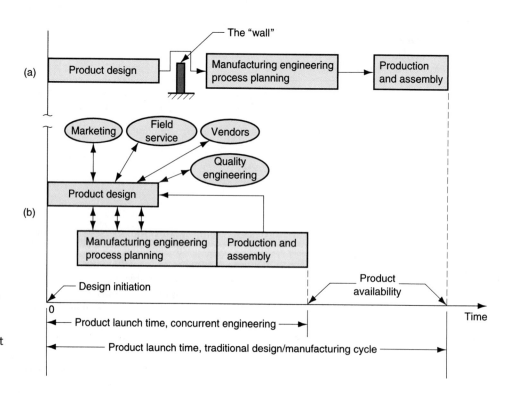

FIGURE 40.5
Comparison of: (a) traditional product development cycle, and (b) product development using concurrent engineering.

In addition to design for manufacturing and assembly, other objectives include design for quality, design for life cycle, and design for cost.

With the importance of quality in international competition, and the demonstrated success of those companies that have been able to produce products of high quality, one must conclude that *design for quality* is very important. Chapter 42 is devoted to the topic of quality control and includes a discussion of several quality approaches related to product design.

Design for life cycle refers to the product after it has been manufactured. In many cases, a product can involve a significant cost to the customer beyond the purchase price. These costs include installation, maintenance and repair, spare parts, future upgrading of the product, safety during operation, and disposition of the product at the end of its useful life. The price paid for the product may be a small portion of its total cost when life cycle costs are included. Some customers (e.g., federal government) consider life cycle costs in their purchasing decisions. The manufacturer must often include service contracts that limit customer vulnerability to excessive maintenance and service costs. In these cases, accurate estimates of these life cycle costs must be included in the total product cost.

A product's cost is a major factor in determining its commercial success. Cost affects the price charged for the product and the profit made on it. *Design for product cost* refers to the efforts of a company to identify the impact of design decisions on overall product costs and to control those costs through optimal design. Many of the DFM/A guidelines are directed at reducing product cost.

REFERENCES

[1] Bakerjian, R., and Mitchell, P. *Tool and Manufacturing Engineers Handbook,* 4th ed., Vol. VI, *Design for Manufacturability.* Society of Manufacturing Engineers, Dearborn, Michigan, 1992.

[2] Chang, C-H., and Melkanoff, M. A. *NC Machine Programming and Software Design,* 3rd ed. Prentice Hall, Inc., Upper Saddle River, New Jersey, 2005.

[3] Eary, D. F., and Johnson, G. E. *Process Engineering for Manufacturing.* Prentice Hall, Inc., Englewood Cliffs, New Jersey, 1962.

[4] Groover, M. P., and Zimmers, E. W., Jr. *CAD/CAM: Computer-Aided Design and Manufacturing.* Prentice Hall, Englewood Cliffs, New Jersey, 1984.

[5] Groover, M. P. *Automation, Production Systems, and Computer Integrated Manufacturing,* 3rd ed. Pearson Prentice Hall, Upper Saddle River, New Jersey, 2008.

[6] Kane, G. E. "The Role of the Manufacturing Engineer," Technical paper MM70-222. Society of Manufacturing Engineers, Dearborn, Michigan, 1970.

[7] Koenig, D. T. *Manufacturing Engineering.* Hemisphere Publishing Corporation (Harper & Row, Publishers, Inc.), Washington, DC, 1987.

[8] Kusiak, A. (ed.). *Concurrent Engineering: Automation, Tools, and Techniques.* John Wiley & Sons, Inc., New York, 1993.

[9] Martin, J. M. "The Final Piece of the Puzzle," *Manufacturing Engineering,* September 1988, pp. 46–51.

[10] Nevins, J. L., and Whitney, D. E. (eds.). *Concurrent Design of Products and Processes.* McGraw-Hill, New York, 1989.

[11] Tanner, J. P. *Manufacturing Engineering,* 2nd ed. CRC Taylor & Francis, Boca Raton, Florida, 1990.

[12] Usher, J. M., Roy, U., and Parsaei, H. R. (eds.). *Integrated Product and Process Development.* John Wiley & Sons, Inc., New York, 1998.

[13] Veilleux, R. F., and Petro, L. W. *Tool and Manufacturing Engineers Handbook,* 4th ed., Vol. V, *Manufacturing Management.* Society of Manufacturing Engineers, Dearborn, Michigan, 1988.

REVIEW QUESTIONS

40.1. Define manufacturing engineering.

40.2. What are the principal activities in manufacturing engineering?

40.3. Identify some of the details and decisions that are included within the scope of process planning.

40.4. What is a route sheet?

40.5. What is the difference between a basic process and a secondary process?

40.6. What is a precedence constraint in process planning?

40.7. In the make or buy decision, why is it that purchasing a component from a vendor may cost more than producing the component internally, even though the quoted price from the vendor is lower than the internal price?

40.8. Identify some of the important factors that should enter into the make or buy decision.

40.9. Name three of the general principles and guidelines in design for manufacturability.

40.10. What is concurrent engineering and what are its important components?

40.11. What is meant by the term design for life cycle?

MULTIPLE CHOICE QUIZ

There are 19 correct answers in the following multiple choice questions (some questions have multiple answers that are correct). To attain a perfect score on the quiz, all correct answers must be given. Each correct answer is worth 1 point. Each omitted answer or wrong answer reduces the score by 1 point, and each additional answer beyond the correct number of answers reduces the score by 1 point. Percentage score on the quiz is based on the total number of correct answers.

40.1. The manufacturing engineering department in an organization is best described as which one of the following: (a) branch of the sales department, (b) concurrent engineers, (c) management, (d) product designers, (e) production supervisors, or (f) technical staff function?

40.2. Which of the following are the usual responsibilities of the manufacturing engineering department (four best answers): (a) advising on design for manufacturability, (b) facilities planning, (c) marketing the product, (d) plant management, (e) process improvement, (f) process planning, (g) product design, (h) solving technical problems in the production departments, and (i) supervision of production workers?

40.3. Which of the following are considered basic processes, as opposed to secondary processes (four correct answers): (a) annealing, (b) anodizing, (c) drilling, (d) electroplating, (e) forward hot extrusion to produce aluminum bar stock, (f) impression die forging, (g) rolling of sheet steel, (h) sand casting, (i) sheet-metal stamping, (j) spot welding, (k) surface grinding of hardened steel, (l) tempering of martensitic steel, and (m) turning?

40.4. Which of the following would be considered secondary processes, as opposed to basic processes (four correct answers): (a) annealing, (b) arc welding, (c) drilling, (d) electroplating, (e) extrusion to produce steel automotive components, (f) impression die forging, (g) painting, (h) plastic injection molding, (i) rolling of sheet steel, (j) sand casting, (k) sheet-metal stamping, (l) sintering of pressed ceramic powders, and (m) ultrasonic machining?

40.5. Which of the following are operations to enhance physical properties (three correct answers): (a) annealing, (b) anodizing, (c) die casting, (d) drilling, (e) electroplating, (f) rolling of nickel alloys, (g) sheet metal drawing, (h) sintering of pressed ceramic powders, (i) surface grinding of hardened steel, (j) tempering of martensitic steel, (k) turning, and (l) ultrasonic cleaning?

40.6. A route sheet is a document whose principal function is which one of the following: (a) continuous improvement, (b) design for manufacturability, (c) provides authorization for material handlers to move the part, (d) quality inspection procedure, (e) specifies the process plan, or (f) specifies the detailed method for a given operation?

40.7. In a make or buy situation, the decision should always be to purchase the component if the vendor's quoted price is less than the in-house estimated cost of the component: (a) true or (b) false?

40.8. Which one of the following types of computer-aided process planning relies on parts classification and coding in group technology: (a) generative CAPP, (b) retrieval CAPP, (c) traditional process planning, or (d) none of the preceding?

41 PRODUCTION PLANNING AND CONTROL

Chapter Contents

Production planning and control are the manufacturing support systems concerned with logistics problems in the production function. *Production planning* is concerned with planning what products are to be produced, in what quantities, and when. It also considers the resources required to accomplish the plan. *Production control* determines whether the resources to execute the plan have been provided and, if not, takes the necessary action to correct the deficiency. The scope of production planning and control includes *inventory control*, which is concerned with having appropriate stock levels available of raw materials, work-in-process, and finished goods.

Problems in production planning and control differ for different types of manufacturing. One of the important factors is the relationship between product variety and production quantity (Section 1.1.2). At one extreme is *job shop production*, in which a number of different product types are each produced in low quantities. The products are often complex, consisting of many components, each of which must be processed through multiple operations. Solving the logistics problems in such a plant requires detailed planning—scheduling and coordinating the large numbers of different components and processing steps for the different products.

At the other extreme is *mass production*, in which a single product (with perhaps some limited model variations) is produced in very large quantities (millions of units). The logistics problems in mass production are simple if the product and process are simple. In more complex cases, the product is an assembly consisting of many components (e.g., automobiles and household appliances) and the facility is organized as a production line (Section 39.2). The logistics problem in operating such a plant is to get each component to the right workstation at the right time so that it can be assembled to the product as it passes through that station. Failure to solve this problem can result in stoppage of the entire production line for lack of a critical part.

To distinguish between these two extremes in terms of the issues in production planning and control, we can say that the planning function is emphasized in a job shop, whereas the

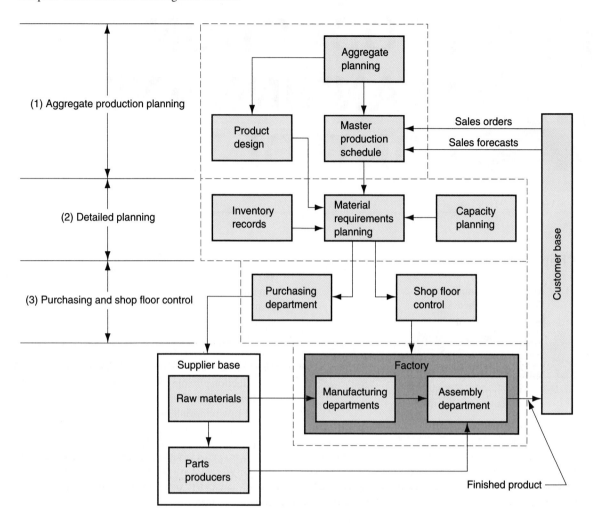

FIGURE 41.1   Activities in a production planning and control system.

control function is emphasized in the mass production of assembled products. There are many variations between these extremes, with accompanying differences in the way production planning and control are implemented.

Figure 41.1 presents a block diagram depicting the activities of a modern production planning and control system and their interrelationships. The activities can be divided into three phases: (1) aggregate production planning, (2) detailed planning of material and capacity requirements, and (3) purchasing and shop floor control. Our discussion of production planning and control in this chapter is organized around this framework.

41.1   AGGREGATE PLANNING AND THE MASTER PRODUCTION SCHEDULE

Any manufacturing firm must have a business plan, and the plan must include what products will be produced, how many, and when. The manufacturing plan should take into account current orders and sales forecasts, inventory levels, and plant capacity considerations. There are several types of manufacturing plans in terms of planning horizon: (1) *long-range plans*

TABLE 41.1   (a) Aggregate production plan, and (b) corresponding master production schedule for a hypothetical product line.

(a)

Product line	Week									
	1	2	3	4	5	6	7	8	9	10
P models	—	—	—	—	—	—	—	50	150	250
Q models	400	400	400	300	300	300	300	250	250	250
R models	100	100	150	150	200	200	200	250	300	350

(b)

Product	Week									
	1	2	3	4	5	6	7	8	9	10
Model P1								50	75	100
Model P2									50	50
Model P3									25	50
Model P4										50
Model Q1	200	200	200	100	100	100	100	50	50	50
Model Q2 (etc.)	200	200	200	200	200	200	200	200	200	200

that deal with a time horizon that is more than one year in the future; (2) *medium-range plans* that are concerned with the period 6 months to 1 year in the future; and (3) *short-range plans* that consider near future horizons such as days or weeks.

Long-range planning is the responsibility of the highest level executives of the company. It is concerned with corporate goals and strategies, future product lines, financial planning for the future, and obtaining the resources (personnel, facilities, and equipment) necessary so that the company will have a future. As the planning horizon is reduced, the company's long-range plan must be translated into medium-range and short-range plans that become increasingly specific. At the medium-range level are the aggregate production plan and the master production schedule, examined in this section. In the short range are material and capacity requirements planning, and detailed scheduling of the orders.

The *aggregate production plan* indicates production output levels for major product lines rather than specific products. It must be coordinated with the sales and marketing plan of the company and must consider current inventory levels. Aggregate planning is therefore a high-level corporate planning activity, although details of the planning process are delegated to staff. The aggregate plan must reconcile the marketing plans for current products and new products under development against the capacity resources available to make those products.

The planned output levels of the major product lines listed in the aggregate schedule must be converted into a very specific schedule of individual products. This is called the *master production schedule*, and it lists the products to be manufactured, when they should be completed, and in what quantities. A hypothetical master schedule is presented in Table 41.1(b) for a limited product set, with the corresponding aggregate plan for the product line in Table 41.1(a).

Products listed in the master schedule generally are divided into three categories: (1) firm customer orders, (2) forecasted demand, and (3) spare parts. Customer orders for specific products usually obligate the company to a delivery date that has been promised to a customer by the sales department. The second category consists of production output levels based on forecasted demand, in which statistical forecasting techniques are applied to previous demand patterns, estimates by the sales staff, and other sources. The forecast often dominates the master schedule. The third category is demand for individual component parts—repair parts to be stocked in the firm's service department. Some companies exclude this third category from the master schedule because it does not represent end products.

The master production schedule is a medium-range plan because it must consider the lead times required to order raw materials and components, fabricate parts in the factory, and then assemble and test the final products. Depending on type of product, these lead times can run from several months to more than a year. However, although it deals with the midterm horizon, it is a dynamic plan. It is usually considered to be fixed in the near term, meaning that changes are disallowed within about a 6-week horizon. However, adjustments in the schedule are possible beyond six weeks to deal with shifts in demand or new product opportunities. It should therefore be noted that the aggregate production plan is not the only input to the master schedule. Other drivers that may cause it to deviate from the aggregate plan include new customer orders and changes in sales forecast over the near term.

41.2 INVENTORY CONTROL

Inventory control is concerned with achieving a balance between two competing objectives: (1) minimizing the cost of maintaining inventory and (2) maximizing service to customers. Inventory costs include investment costs, storage costs, and cost of possible obsolescence or spoilage. Investment cost is often the dominant factor; a typical case is when the company invests borrowed money at a certain interest rate in materials that have not yet been delivered to the customer. All of these costs are referred to as *carrying costs*. The company can minimize carrying cost by maintaining zero inventories. However, customer service is likely to suffer, and customers may decide to take their business elsewhere. This has a cost, referred to as *stock-out cost*. A reasonable company wants to minimize stock-out cost and provide a high level of customer service. Customers are of two kinds: (1) external customers, the kind we usually associate with the word, and (2) internal customers, which are the operating departments, final assembly, and other units in the company that depend on the ready availability of materials and parts.

41.2.1 TYPES OF INVENTORY

Various types of inventory are encountered in manufacturing. The categories of greatest interest in production planning and control are raw materials, purchased components, in-process inventory (work-in-process), and finished products.

Different inventory control procedures are appropriate, depending on which type we are attempting to manage. An important distinction is between items subject to independent versus dependent demand. *Independent demand* means that the demand or consumption of the item is unrelated to demand for other items. End products and spare parts experience independent demand. Customers purchase end products and spare parts, and their decisions to do so are unrelated to the purchase of other items.

Dependent demand refers to the fact that demand for the item is directly related to demand for something else, usually because the item is a component of an end product subject to independent demand. Consider an automobile—an end product, the demand for which is independent. Each car has four tires (five, if we include the spare), whose demand depends on the demand for the automobile. Thus, the tires used on new automobiles are examples of dependent demand. For every car made in the final assembly plant, four tires must be ordered. The same is true of the thousands of other components used on an automobile. Once the decision is made to produce a car, all of these components must be supplied to build it.

Tires represent an interesting example because they not only experience dependent demand in the new car business, but also independent demand in the replacement tire market.

Different production and inventory control procedures must be used for independent and dependent demand. Forecasting procedures are commonly used to determine future production levels of independent demand products. Production of the components used on these products is then determined directly from the product quantities to be made. Two different inventory control systems are required for the two cases: (1) order point systems and (2) material requirements planning. Order point systems are covered in the following section. Material requirements planning is covered in Section 41.3.1.

41.2.2 ORDER POINT SYSTEMS

Order point systems address two related problems encountered when controlling inventories of independent demand items: how much to order and when to order. The first problem can be solved using economic order quantity formulas. The second problem can be solved using reorder points.

Economic Order Quantity   The problem of determining the appropriate quantity that should be ordered or produced arises in cases of independent demand products, in which demand for the item is relatively constant during the period under consideration and the production rate is significantly greater than demand rate. This is the typical ***make-to-stock*** situation. A similar problem is encountered in some dependent demand situations, when usage of the components in the final product is fairly steady over time and it makes sense to pay some inventory carrying costs in order to reduce the frequency of setups. In both of these situations, the inventory level is gradually depleted over time and then quickly replenished to some maximum level determined by the order quantity, as depicted in Figure 41.2.

One can derive a total cost equation for the sum of carrying cost and setup cost for the inventory model in Figure 41.2. The model has a sawtooth appearance, representing the gradual consumption of product down to a zero, followed by immediate replenishment up to a maximum level Q. Based on this behavior, the average inventory level is one-half the maximum level Q. The total annual inventory cost equation is

$$TIC = \frac{C_h Q}{2} + \frac{C_{su} D_a}{Q} \qquad (41.1)$$

where TIC = total annual inventory cost (carrying plus ordering costs), \$/yr; Q = order quantity, pc/order; C_h = holding cost (cost of carrying the inventory), \$/pc/yr; C_{su} = cost of setting up for an order, \$/setup or \$/order; and D_a = annual demand for the item, pc/yr. In the equation, the ratio $D_a Q$ = the number of orders (batches of parts produced) per year; it therefore gives the number of setups per year.

The carrying cost C_h is generally taken to be directly proportional to the value of the item; that is,

$$C_h = h C_p \qquad (41.2)$$

where C_p = cost per piece, \$/unit; and h = annual holding cost rate, which includes interest and storage charges, $(\text{yr})^{-1}$.

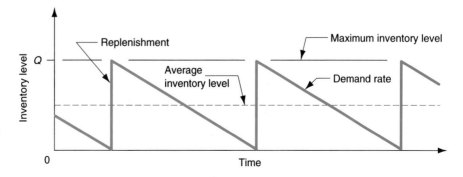

FIGURE 41.2   Model of inventory level over time in the typical make-to-stock situation.

The setup cost C_{su} includes the cost of idle production equipment during the changeover time between batches, as well as whatever labor costs are involved in the setup changes. Thus,

$$C_{su} = T_{su}C_{dt} \tag{41.3}$$

where T_{su} = setup or changeover time between batches, hr; and C_{dt} = cost rate of machine downtime, $/hr. In cases where parts are ordered from an outside vendor, the price quoted by the vendor usually includes a setup cost, either directly or in the form of quantity discounts. C_{su} should also include the internal costs of placing the order to the vendor.

It should be noted that Eq. (41.1) excludes the actual annual cost of part production, which is D_aC_p. If this cost is included, then annual total cost is given by

$$TC = D_aC_p + \frac{C_hQ}{2} + \frac{C_{su}D_a}{Q} \tag{41.4}$$

By taking the derivative of either Eq. (41.1) or Eq. (41.4) and setting it to zero, we obtain the economic order quantity formula that minimizes the sum of carrying costs and setup costs:

$$EOQ = \sqrt{\frac{2D_aC_{su}}{C_h}} \tag{41.5}$$

where EOQ = economic order quantity (number of parts that should be produced in the batch), pc; and the other terms are previously defined.

Example 41.1
Economic Order
Quantity

A product is made to stock. Annual demand rate is 12,000 units. One unit of product costs $10 and the holding cost rate = 24%/yr. Setting up to produce a batch of products requires changeover of equipment, which takes 4 hr. The cost of equipment downtime plus labor = $100/hr. Determine the economic order quantity and the total inventory costs for this case.

Solution: Setup cost $C_{su} = 4 \times \$100 = \$400$. Holding cost per unit $= 0.24 \times \$10 = \$2.40$. Using these values and the annual demand rate in the EOQ formula, we have

$$EOQ = \sqrt{\frac{2(12,000)(400)}{2.40}} = 2,000 \text{ units}$$

Total inventory costs are given by the TIC equation:

$$TIC = 0.5(2.40)(2,000) + 400(12,000/2,000) = \$4800$$

Including the actual production costs in the annual total, by Eq. (41.4) we have

$$TC = 12,000(10) + 4,800 = \$124,800$$ ∎

The EOQ formula has been a widely used model for deciding "optimum production runs." There are variations of Eqs. (41.1) and (41.4) that take into account additional factors such as rate of production. While the mathematical accuracy of the formula cannot be disputed, it is instructive to note some of the difficulties encountered in its application. One difficulty is concerned with the values of the parameters in the equation, namely setup or ordering cost and inventory carrying costs. These costs are often difficult to evaluate; yet they have an important effect on the calculated EOQ value.

A second difficulty is concerned with an erroneous tenet of manufacturing philosophy that has been promulgated by the use of the EOQ formula in the United States. It is that long production runs represent an optimum strategy in batch manufacturing. No matter how much it costs to change the setup, the formula gives the optimum production batch size. The higher

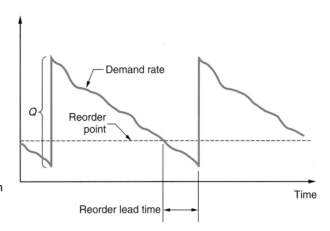

FIGURE 41.3   Operation of a reorder point inventory system.

the setup cost, the longer the production run. A preferable approach is to develop ways to reduce the setup cost by reducing the time required to accomplish a changeover. Setup reduction is an important component of just-in-time production, and we consider some of the ways to reduce setup time in Section 41.4.

When to Reorder   Determining when to reorder can be accomplished in several ways. Here we describe the reorder point system that is widely used in industry. Refer to Figure 41.3, which provides a more realistic view of the likely variations in demand rate than Figure 41.2. In a *reorder point system*, when the inventory level for a given stock item declines to some point defined as the reorder point, this is the signal to place an order to restock the item. The reorder point is set at a high enough level so as to minimize the probability that a stock-out will occur during the period between when the reorder point is reached and a new batch is received.

Reorder point policies can be implemented using computerized inventory control systems. These systems are programmed to continuously monitor the inventory level as transactions are made, and to automatically generate an order for a new batch when the level falls below the reorder point. A noncomputerized system, called the *two-bin approach*, starts with two equally sized bins both filled with parts of a certain type, but parts are only withdrawn from one of the bins to satisfy demand. When the supply in that bin has been exhausted, an order is placed to replenish it, and the other bin is used to satisfy demand. Switching back and forth between bins in this way provides a very simple method of inventory control. In effect, the reorder point is signaled when one of the bins becomes empty.

41.3   MATERIAL AND CAPACITY REQUIREMENTS PLANNING

We present two alternative techniques for planning and controlling production and inventory. In this section we cover procedures used for job shop and midrange production of assembled products. In Section 41.4, we examine procedures more appropriate for high production.

41.3.1   MATERIAL REQUIREMENTS PLANNING

Material requirements planning (MRP) is a computational procedure used to convert the master production schedule for end products into a detailed schedule for raw materials and components used in the end products. The detailed schedule indicates the quantities of each item, when it must be ordered, and when it must be delivered to achieve the master schedule. ***Capacity requirements planning*** (Section 41.3.2) coordinates labor and equipment resources with material requirements.

Material requirements planning is most appropriate for job shop and batch production of products that have multiple components, each of which must be purchased and/or fabricated. It is the proper technique for determining quantities of dependent demand items that constitute the inventories of manufacturing: raw materials, purchased parts, work-in-process, and so forth.

Material requirements planning is relatively straightforward in concept. Its application is complicated by the sheer magnitude of the data that must be processed. The master schedule specifies the production of final products in terms of month-by-month deliveries. Each product may contain hundreds of components. These components are produced from raw materials, some of which are common among the components (e.g., sheet steel for stampings). Some of the components themselves may be common to several different products (these are called ***common use items*** in MRP). For each product, the components are assembled into simple subassemblies, which are added to form other subassemblies, and so on, until the final products are completed. Each step in the sequence consumes time. All of these factors must be accounted for in material requirements planning. Although each calculation is simple, the large number of calculations and massive amounts of data require that MRP be implemented by computer.

The lead time for a job is the time that must be allowed to complete the job from start to finish. There are two kinds of lead times in MRP: ordering lead times and manufacturing lead times. ***Ordering lead time*** is the time required from initiation of the purchase requisition to receipt of the item from the vendor. If the item is a raw material stocked by the vendor, the ordering lead time should be relatively short, perhaps a few weeks. If the item is fabricated, the lead time may be substantial, perhaps several months. ***Manufacturing lead time*** is the time required to produce the item in the company's own plant, from order release to completion.

Inputs to the MRP System   For the MRP processor to function properly, it must receive inputs from several files: (1) master production schedule, (2) product design data, in the form of a bill of materials file, (3) inventory records, and (4) capacity requirements planning. Figure 41.1 shows the data flow into the MRP processor and the recipients of its output reports.

The master production schedule was discussed in Section 41.1. The ***bill-of-materials file*** lists the component parts and subassemblies that make up each product. It is used to compute the requirements for raw materials and components used in the end products listed in the master schedule. Figure 41.4 shows a (simplified) structure of an assembled product. The product consists of two subassemblies, each consisting of three parts. The number of each item in the next level above in the product structure is indicated in parentheses.

The ***inventory record file*** identifies each item (by part number) and provides a time-phased record of its inventory status. This means that not only is the current quantity of the item listed, but also any future changes in inventory status that will occur and when they will occur.

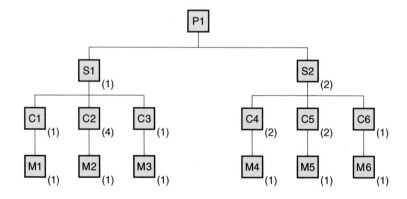

FIGURE 41.4   Product structure for assembled product P1. (Based on data in [3].)

These data include gross requirements for the item (how many units will be needed to build products in the master schedule), scheduled receipts, on-hand status, and planned order releases. Each of these data sets indicates the changes by time period in the schedule (e.g., month, week).

How MRP Works   Based on inputs from the master schedule, bill-of-materials file, and inventory record file, the MRP processor computes how many of each component and raw material will be needed in future time periods by "exploding" the end product schedule into successively lower levels in the product structure. The MRP computations must deal with several complicating factors. First, component and subassembly quantities must be adjusted for any inventories on hand or on order. Second, quantities of common use items must be combined during parts explosion to obtain a total requirement for each component and raw material in the schedule. Third, the time-phased delivery of end products must be converted into time-phased requirements for components and materials by factoring in the appropriate lead times. For every unit of final product listed in the MPS, the required number of components of each type must be ordered or fabricated, taking into account its ordering or manufacturing lead times. For each component, the raw material must be ordered, accounting for its ordering lead time. And assembly lead times must be considered in the scheduling of subassemblies and final products.

Example 41.2 Material Requirements Planning

Consider the requirements planning procedure for one of the components in product P1: C4. The required deliveries for P1 are indicated in the master production schedule in Table 41.1 (b). According to the product structure in Figure 41.4, two units of C4 are required to make subassembly S2, and two S2 units are required to make the final product P1. One unit of raw material M4 is used to make each unit of C4. Ordering, manufacturing, and assembly lead times for these items are known. For P1 and S2, the lead time is 1 week; for C4, the lead time is 2 weeks; and for M4, the lead time is 3 weeks. The inventory status of M4 is 50 units currently on hand, and zero units of component C4 and S2. There are no scheduled requirements, receipts, or order releases indicated in the inventory record for these items. Neither material M4 nor component C4 is used on any other product—they are not common use items. Determine the time-phased requirements for M4, C4, and S2 to meet the master schedule for product P1. Orders for P1 beyond period 10 are ignored in this problem.

Solution: Table 41.2 presents the solution to this MRP problem. Delivery requirements for P1 must be offset by 1 week to obtain the planned order releases. S2 must be exploded by 2 units per P1 unit and offset by 1 week to obtain its order release. C4 is exploded by 2 units per S2 unit and offset by 2 weeks to obtain its requirement. And M4 is offset by its 3-week ordering time to obtain its release date, taking into account current stock of M4 on hand. ∎

Output Reports and Benefits of MRP   MRP generates various output reports that can be used in planning and managing plant operations. The reports include (1) order releases, which authorize the placement of orders planned by the MRP system; (2) planned order releases in future periods; (3) rescheduling notices, indicating changes in due dates for open orders; (4) cancellation notices, which indicate that certain open orders have been canceled due to changes in the master schedule; (5) inventory status reports; (6) performance reports; (7) exception reports, showing deviations from schedule, overdue orders, scrap, and so forth; and (8) inventory forecasts, which project inventory levels in future periods.

Many benefits are claimed for a well-designed MRP system, including (1) inventory reductions, (2) faster response to changes in demand, (3) reduced setup and changeover costs, (4) better machine utilization, (5) improved ability to respond to changes in the master schedule, and (6) helpful in developing the master schedule. Despite these claims, MRP systems have been implemented in industry with varying degrees of success. Some of the reasons for unsuccessful MRP implementations include (1) inappropriate application, (2) MRP calculations based on inaccurate data, and (3) absence of capacity planning.

TABLE 41.2 Material requirements solution to Example 41.2

Period		1	2	3	4	5	6	7	8	9	10
Item: **Product P1**											
Gross requirements									50	75	100
Scheduled receipts											
On hand	0										
Net requirements									50	75	100
Planned order releases								50	75	100	
Item: **Subassembly S2**											
Gross requirements								100	150	200	
Scheduled receipts											
On hand	0										
Net requirements								100	150	200	
Planned order releases							100	150	200		
Item: **Component C4**											
Gross requirements							200	300	400		
Scheduled receipts											
On hand	0										
Net requirements							200	300	400		
Planned order releases					200	300	400				
Item: **Raw material M4**											
Gross requirements					200	300	400				
Scheduled receipts											
On hand	50	50	50	50	50						
Net requirements					150	300	400				
Planned order releases		150	300	400							

41.3.2 CAPACITY REQUIREMENTS PLANNING

Capacity requirements planning is concerned with determining the labor and equipment requirements needed to meet the master production schedule. It is also concerned with identifying the firm's long-term future capacity needs. Capacity planning also serves to identify production resource limitations so that a realistic master production schedule can be planned.

A realistic master schedule must take into account the manufacturing capabilities of the plant that is to make the products. The firm must be aware of its production capacity and must plan for changes in capacity to meet changing production requirements specified in the master schedule. The relationship between capacity planning and other functions in production planning and control is shown in Figure 41.1. The master schedule is reduced to material and component requirements using MRP. These requirements provide estimates of the required labor hours and other resources needed to produce the components. The required resources are then compared to plant capacity over the planning horizon. If the master schedule is not compatible with plant capacity, adjustments must be made either in the schedule or in plant capacity.

Plant capacity can be adjusted in the short term and in the long term. Short-term capacity adjustments include (1) *employment levels*—increasing or decreasing direct labor in the plant based on changes in capacity requirements; (2) *shift hours*—increasing or decreasing the number of labor hours per shift through the use of overtime or reduced hours; (3) *number of work shifts*—increasing or decreasing the number of shifts worked per production period by authorizing evening and night shifts and/or weekend shifts; (4) *inventory stockpiling*—this tactic is used to maintain steady employment levels during slow demand periods; (5) *order backlogs*—delaying deliveries to the customer during busy periods when production resources are insufficient to keep up with demand; and (6) *subcontracting*—contracting work to outside shops during busy periods or taking in extra work during slack periods.

Long-term capacity adjustments include possible changes in production capacity that generally require long lead times, including the following types of decisions: (1) *new equipment*—investments in additional machines, more productive machines, or new types of machines to match future changes in product design; (2) *new plants*—construction of new plants or purchase of existing plants from other companies; and (3) *plant closings*—closing plants not needed in the future.

41.4   JUST-IN-TIME AND LEAN PRODUCTION

Just-in-time (JIT) is an approach to production that was developed by Toyota Motors in Japan to minimize inventories. Work-in-process and other inventories are viewed as waste that should be eliminated. Inventory ties up investment funds and takes up space. To reduce this form of waste, the JIT approach includes a number of principles and procedures aimed at reducing inventories, either directly or indirectly. Indeed, the scope of JIT is so broad that it is often referred to as a philosophy. JIT is an important component of "lean production," a principal goal of which is to reduce waste in production operations (Section 1.5.1). *Lean production* can be defined as "an adaptation of mass production in which workers and work cells are made more flexible and efficient by adopting methods that reduce waste in all forms.[1]

Just-in-time procedures have proven most effective in high-volume repetitive manufacturing, such as the automobile industry [4]. The potential for in-process inventory accumulation in this type of manufacturing is significant because both the quantities of products and the number of components per product are large. A just-in-time system produces exactly the right number of each component required to satisfy the next operation in the manufacturing sequence just when that component is needed—"just in time." The ideal batch size is one part. As a practical matter, more than one part are produced at a time, but the batch size is kept small. Under JIT, producing too many units is to be avoided as much as producing too few units. This is a production discipline that contrasts with traditional U.S. practice, which has promoted use of large in-process inventories to deal with problems such as machine breakdowns, defective components, and other obstacles to smooth production. The U.S. approach might be described as a "just-in-case" philosophy.

Although the principal theme in JIT is inventory reduction, this cannot simply be mandated. Several requisites must be pursued to make it possible: (1) stable production schedules; (2) small batch sizes and short setup times; (3) on-time delivery; (4) defect-free components and materials; (5) reliable production equipment; (6) pull system of production control; (7) a work force that is capable, committed, and cooperative; and (8) a dependable supplier base.

Stable Schedule   For JIT to be successful, work must flow smoothly with minimal perturbations from normal operations. Perturbations require changes in operating

---

[1]M. P. Groover, *Automation, Production Systems, and Computer-Integrated Manufacturing* [APSCIM], p. 834.

procedures—increases and decreases in production rate, unscheduled setups, variations from the regular work routine, and other exceptions. Perturbations in downstream operations (i.e., final assembly) tend to be amplified in upstream operations (i.e., parts feeding). A master production schedule that remains relatively constant over time is one way of achieving smooth work flow and minimizing disturbances and changes in production.

Small Batch Sizes and Setup Reduction   Two requirements for minimizing inventories are small batch sizes and short setup times. We examined the relationship between batch size and setup time in the *EOQ* formula, Eq. (41.5). Instead of using the EOQ formula just to compute batch quantities, efforts should be focused on reducing setup time, thereby permitting smaller batches and lower work-in-process levels. Some of the approaches used to reduce setup time include (1) performing as much of the setup as possible while the previous job is still running; (2) using quick-acting clamping devices instead of bolts and nuts; (3) eliminating or minimizing adjustments in the setup; and (4) using group technology and cellular manufacturing so that similar part styles are produced on the same equipment.

On-Time Delivery, Zero Defects, and Reliable Equipment   Success of JIT production requires near perfection in on-time delivery, parts quality, and equipment reliability. The small lot sizes and parts buffers used in JIT require parts to be delivered before stock-outs occur at downstream stations. Otherwise, production must be suspended at these stations for lack of parts. If the delivered parts are defective, they cannot be used in assembly. This tends to promote zero defects in parts fabrication. Workers inspect their own output to make sure it is right before it proceeds to the next operation. Low work-in-process also requires reliable production equipment. Machines that break down cannot be tolerated in a JIT production system. This emphasizes the need for reliable equipment designs and preventive maintenance.

Pull System of Production Control   Just-in-time requires a ***pull system*** of production control, in which the order to produce parts at a given workstation comes from the downstream station that uses those parts. As the supply of parts becomes exhausted at a given station, it "places an order" at the upstream workstation to replenish the supply. This order provides the authorization for the upstream station to produce the needed parts. This procedure, repeated at each workstation throughout the plant, has the effect of pulling parts through the production system. By contrast, a ***push system*** of production operates by supplying parts to each station in the plant, in effect driving the work from upstream stations to downstream stations. MRP is a push system. The risk in a push system is to overload the factory by scheduling more work than it can handle. This results in large queues of parts in front of machines that cannot keep up with arriving work. A poorly implemented MRP system, one that does not include capacity planning, manifests this risk.

One famous pull system is the ***kanban*** system used by Toyota Motors. Kanban (pronounced kahn-bahn) is a Japanese word meaning ***card***. The kanban system of production control is based on the use of cards to authorize production and work flow in the plant. There are two types of kanbans: (1) production kanbans, and (2) transport kanbans. A ***production kanban*** authorizes production of a batch of parts. The parts are placed in containers, so the batch must consist of just enough parts to fill the container. Production of additional parts is not permitted. The ***transport kanban*** authorizes movement of the container of parts to the next station in the sequence.

Refer to Figure 41.5 as we explain how two workstations, one that feeds the other, operate in a kanban system. The figure shows four stations, but B and C are the stations we want to focus on here. Station B is the supplier in this pair, and station C is the consumer. Station C supplies downstream station D. B is supplied by upstream station A. When station C starts work on a full container, a worker removes the transport kanban from that container and takes it back to B. The worker finds a full container of parts at B that have just been produced, removes the production kanban from that container, and places it on a rack at B.

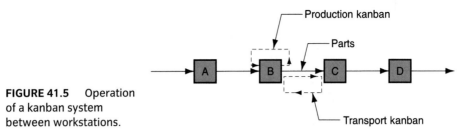

FIGURE 41.5 Operation of a kanban system between workstations.

The worker then places the transport kanban in the full container, which authorizes its movement to station C. The production kanban on the rack at station B authorizes production of a new batch of parts. Station B produces more than one part style, perhaps for several other downstream stations in addition to C. The scheduling of work is determined by the order in which the production kanbans are placed on the rack.

The kanban pull system between stations A and B and between stations C and D operates the same as it does between stations B and C, described here. This system of production control avoids unnecessary paperwork. The cards are used over and over rather than generating new production and transport orders every cycle. An apparent disadvantage is the considerable labor involvement in material handling (moving the cards and containers between stations); however, it is claimed that this promotes teamwork and cooperation among workers.

Workforce and Supplier Base Another requirement of a JIT production system is workers who are cooperative, committed, and capable of performing multiple tasks. The workers must be flexible to produce a variety of part styles at their respective stations, to inspect the quality of their work, and to deal with minor technical problems with the production equipment so that major breakdowns do not occur.

Just-in-time extends to the material and component suppliers of the company. Suppliers must be held to the same standards of on-time delivery, zero defects, and other JIT requisites as the company itself. Some of the vendor policies used by companies to implement JIT include (1) reducing the total number of suppliers, (2) selecting suppliers with proven records for meeting quality and delivery standards, (3) establishing long-term partnerships with suppliers, and (4) selecting suppliers that are located near the company's manufacturing plant.

41.5 SHOP FLOOR CONTROL

The third phase in production planning and control (Figure 41.1) is concerned with releasing production orders, monitoring and controlling progress of the orders, and acquiring up-to-date information on order status. The purchasing department is responsible for these functions among suppliers. The term *shop floor control* is used to describe these functions when accomplished in the company's own factories. In basic terms, shop floor control is concerned with managing work-in-progress in the factory. It is most relevant in job shop and batch production, where there are a variety of different orders in the shop that must be scheduled and tracked according to their relative priorities.

A typical shop floor control system consists of three modules: (1) order release, (2) order scheduling, and (3) order progress. The three modules and how they relate to other functions in the factory are depicted in Figure 41.6. They are implemented by a combination of computer systems and human resources.

Order Release Order release in shop floor control generates the documents needed to process a production order through the factory. The documents are sometimes called the

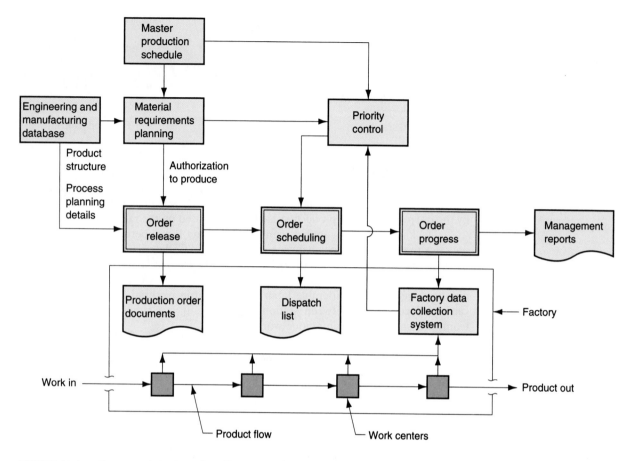

FIGURE 41.6   Three modules in a shop floor control system, and interconnections with other production planning and control functions.

shop packet; it typically consists of (1) route sheet, (2) material requisitions to draw the starting materials from stores, (3) job cards to report direct labor time used on the order, (4) move tickets to authorize transport of parts to subsequent work centers in the routing, and (5) parts list—required for assembly jobs. In a traditional factory, these documents move with the production order and are used to track its progress through the shop. In modern factories, automated methods such as bar code technology are used to monitor order status, making some or all of these paper documents unnecessary.

Order release is driven by two principal inputs, as indicated in Figure 41.6: (1) material requirements planning, which provides the authorization to produce; and (2) engineering and manufacturing database that indicates product structure and process planning details required to generate the documents that accompany the order through the shop.

Order Scheduling   Order scheduling assigns the production orders to work centers in the factory. It serves as the dispatching function in production planning and control. In order scheduling, a dispatch list is prepared indicating which orders should be accomplished at each work center. It also provides relative priorities for the different jobs (e.g., by showing due dates for each job). The dispatch list helps the department foreman assign work and allocate resources to achieve the master schedule.

Order scheduling addresses two problems in production planning and control: machine loading and job sequencing. To schedule production orders through the factory, they must first be assigned to work centers. Assigning orders to work centers is called **machine loading**. Loading all of the work centers in the plant is called **shop loading**. Since the number of production orders is likely to exceed the number of work centers, each work center will have a queue of orders waiting to be processed. A given production machine may have 10 to 20 jobs waiting to be processed.

Job sequencing is the problem of deciding the order in which to process jobs through a given machine. The processing sequence is decided by means of priorities among the jobs in the queue. The relative priorities are determined by a function called **priority control**. Some of the rules used to establish priorities for production orders in a plant include (1) **first-come-first-serve**—orders are processed in the sequence in which they arrive at the work center; (2) **earliest due date**—orders with earlier due dates are given higher priorities; (3) **shortest processing time**—orders with shorter processing times are given higher priorities; (4) **least slack time**—orders with the least slack in their schedule are given higher priorities (slack time is defined as the difference between the time remaining until due date and the process time remaining); and (5) **critical ratio**—orders with the lowest critical ratio are given higher priorities (the critical ratio is defined as the ratio of the time remaining until due date divided by the process time remaining).

The relative priorities of the orders may change over time. Expected demand could be higher or lower for certain products, equipment breakdowns could occur, orders could be cancelled, or there could be defects in raw materials, among other reasons. Priority control reviews the relative priorities of the production orders and adjusts the dispatch list accordingly. When an order is completed at one work center, it moves to the next machine in its routing. The order becomes part of the machine loading for the next work center, and priority control is again used to determine the sequence among jobs to be processed at that machine.

Order Progress   Order progress in shop floor control monitors the status of the orders, work-in-process, and other parameters in the plant that indicate progress and production performance. The objective in order progress is to provide information to manage production based on data collected from the factory.

Various techniques are available to collect data from factory operations. The techniques range from clerical procedures requiring workers to submit paper forms that are later compiled, to fully automated techniques requiring no human participation. The term **factory data collection system** is sometimes used to identify these techniques. More complete coverage of this topic is presented in [3].

Information presented to management is often summarized in the form of reports. The reports include (1) **work order status reports**, which indicate status of production orders, including the work center where each order is located, processing hours remaining before each order will be completed, whether the job is on time, and priority level; (2) **progress reports** that report shop performance during a certain time period such as a week or month—how many orders were completed during the period, how many orders should have been completed during the period but were not completed, and so forth; and (3) **exception reports** that indicate deviations from the production schedule, such as overdue jobs. These reports are helpful to management in deciding resource allocation issues, authorizing overtime, and identifying problem areas that adversely affect achievement of the master production schedule.

REFERENCES

[1] Bedworth, D. D., and Bailey, J. E. **Integrated Production Control Systems,** 2nd ed. John Wiley & Sons, New York, 1987.

[2] Chase, R. B., and Aquilano, N. J., et al. **Production and Operations Management,** 10th ed. McGraw-Hill-Irwin, Boston, 2001.

[3] Groover, M. P. **Automation, Production Systems, and Computer Integrated Manufacturing,** 3rd ed. Pearson Prentice-Hall, Upper Saddle River, New Jersey, 2008.

[4] Monden, Y. **Toyota Production System,** 3rd ed. Engineering and Management Press, Norcross, Georgia, 1998.

[5] Orlicky, J. *Material Requirements Planning.* McGraw-Hill, New York, 1975.

[6] Silver, E. A., Pyke, D. F., and Peterson, R. *Inventory Management and Production Planning and Control,* 3rd ed. John Wiley & Sons, New York, 1998.

[7] Veilleux, R. F., and Petro, L. W. (eds.). *Tool and Manufacturing Engineers Handbook,* 4th ed., Vol. V, *Manufacturing Management.* Society of Manufacturing Engineers, Dearborn, Michigan, 1988.

[8] Vollman, T. E., Berry, W. L., Whybark, D. C., and Jacobs, F. R. *Manufacturing Planning and Control Systems for Supply Chain Management,* 5th ed. McGraw-Hill, New York, 2005.

REVIEW QUESTIONS

41.1. What is meant by the term make-to-stock production?

41.2. How does aggregate planning differ from the master production scheduling?

41.3. What are the product categories usually listed in the master production schedule?

41.4. What is the difference between dependent and independent demand for products?

41.5. Define reorder point inventory system.

41.6. In material requirements planning, what are common use items?

41.7. Identify the inputs to the material requirements planning processor in material requirements planning.

41.8. What are some of the resource changes that can be made to increase plant capacity in the short run?

41.9. Identify the principal objective in just-in-time production.

41.10. How is a pull system distinguished from a push system in production and inventory control?

41.11. What are the three phases in shop floor control?

MULTIPLE CHOICE QUIZ

There are 15 correct answers in the following multiple choice questions (some questions have multiple answers that are correct). To attain a perfect score on the quiz, all correct answers must be given. Each correct answer is worth 1 point. Each omitted answer or wrong answer reduces the score by 1 point, and each additional answer beyond the correct number of answers reduces the score by 1 point. Percentage score on the quiz is based on the total number of correct answers.

41.1. Which one of the following terms best describes the overall function of production planning and control: (a) inventory control, (b) manufacturing logistics, (c) manufacturing engineering, (d) mass production, or (e) product design?

41.2. Which of the following are the three categories of items usually listed in the master production schedule: (a) components used to build the final products, (b) firm customer orders, (c) general product lines, (d) orders for maintenance and spare parts, (e) sales forecasts, and (f) spare tires?

41.3. Inventory carrying costs include which of the following (two best answers): (a) equipment downtime, (b) interest, (c) production, (d) setup, (e) spoilage, (f) stock-out, and (g) storage?

41.4. Which of the following are the three terms in the economic order quantity formula: (a) annual demand rate, (b) batch size, (c) cost per piece, (d) holding cost, (e) interest rate, and (f) setup cost?

41.5. Order point inventory systems are intended for which of the following (two best answers): (a) dependent demand items, (b) independent demand items, (c) low production quantities, (d) mass production quantities, and (e) mid-range production quantities?

41.6. With which of the following manufacturing resources is capacity requirements planning primarily concerned (two best answers): (a) component parts, (b) direct labor, (c) inventory storage space, (d) production equipment, and (e) raw materials?

41.7. The word kanban is most closely associated with which one of the following: (a) capacity planning, (b) economic order quantity, (c) just-in-time production, (d) master production schedule, or (e) material requirements planning?

41.8. Machine loading refers most closely to which one of the following: (a) assigning jobs to a work center, (b) floor foundation in the factory, (c) managing work-in-process in the factory, (d) releasing orders to the shop, or (e) sequencing jobs through a machine?

PROBLEMS

Inventory Control

41.1. A product is made to stock. Annual demand is 86,000 units. Each unit costs $9.50 and the annual holding cost rate is 22%. Setup cost to produce this product is $800. Determine (a) economic order quantity and (b) total inventory costs for this situation.

41.2. Given that annual demand for a product is 20,000 units, cost per unit = $6.00, holding cost rate = 2.5%/month, changeover (setup) time between products averages 2.0 hr, and downtime cost during changeover = $200/hr, determine (a) economic order quantity and (b) total inventory costs for this situation.

41.3. A product is produced in batches. Batch size = 2000 units. Annual demand = 50,000 units, and unit cost of the product = $4.00. Setup time to run a batch = 2.5 hr, cost of downtime on the affected equipment is figured at $250/hr, and annual holding cost rate = 30%. What would the annual savings be if the product were produced in the economic order quantity?

41.4. Assembly of a product requires that a component part be ordered and stocked. Demand for the product is constant throughout the year at 7800 units annually. The cost to place an order is $95. The cost of the part is $56 and the holding cost rate is 22%. When units are ordered, they take two weeks to arrive. Determine (a) the economic order quantity and (b) the reorder point. (c) The parts are prepackaged in multiples of 100. It saves the supplier unpacking and repackaging time if they can ship in multiples of 100. The supplier has offered to reduce the price by $1 per unit if even multiples of

100 are purchased. How much would be saved (if anything) by taking this offer?

41.5. A certain piece of production equipment is used to produce various components for an assembled product. To keep in-process inventories low, it is desired to produce the components in batch sizes of 150 units. Demand for each product is 2500 units per year. Production downtime costs an estimated $200/hr. All of the components made on the equipment are of approximately equal unit cost, which is $9.00. Holding cost rate = 30%/yr. In how many minutes must the changeover between batches be completed in order for 150 units to be the economic order quantity?

41.6. Current setup time on a certain machine is 3.0 hr. Cost of downtime on this machine is estimated at $200/hr. Annual holding cost per part made on the equipment, $C_h = \$1.00$. Annual demand for the part is 15,000 units. Determine (a) EOQ and (b) total inventory costs for this data. Also, determine (c) EOQ and (d) total inventory costs, if the changeover time could be reduced to 6 minutes.

41.7. The two-bin approach is used to control inventory for a particular low-cost component. Each bin holds 1200 units. The annual usage of the component is 45,000 units. Cost to order the component is around $70. (a) What is the imputed holding cost per unit for this data? (b) If the actual annual holding cost per unit is only 7 cents, what lot size should be ordered? (c) How much more is the current two-bin approach costing the company annually, compared to the economic order quantity?

Material Requirements Planning

41.8. Quantity requirements are to be planned for component C2 in product P1. Required deliveries for P1 are given in Table 41.1. Ordering, manufacturing, and assembly lead times are as follows: for P1 and C2, the lead time is one week; and for S1 and M2, the lead time is two weeks. Given the product structure in Figure 41.4, determine the time-phased requirements for M2, C2, and S1 to meet the master schedule for P1. Assume no common use items and all on-hand inventories and scheduled receipts are zero. Use a format similar to Table 41.2 and develop a spreadsheet calculator to solve. Ignore demand for P1 beyond period 10.

41.9. Requirements are to be planned for component C5 in product P1. Required deliveries for P1 are given in Table 41.1. Ordering, manufacturing, and assembly lead times are as follows: for P1 and S2, the lead time is one week; for C5, the lead time is three weeks; and for M5, the lead time is 2 weeks. Given the product structure in Figure 41.4, determine the time-phased requirements for M5, C5, and S2 to meet the master schedule for P1. Assume no common use items. On-hand inventories are 200 units for M5 and 100 units for C5, zero for S2. Use a format similar to Table 41.2 and develop a spreadsheet calculator to solve. Ignore demand for P1 beyond period 10.

41.10. Solve Problem 41.9 except that the following is known in addition to the information given:

scheduled receipts of M5 are 250 units in period (week) 3 and 50 units in period (week) 4.

Order Scheduling

41.11. Four products are to be manufactured in Department A, and it is desired to determine how to allocate resources in that department to meet the required demand for these products for a certain week. For product 1, demand = 750/wk, setup time = 6 hr, and operation time = 4.0 min. For product 2, demand = 900/wk, setup time = 5 hr, and operation time = 3.0 min. For product 3, demand = 400/wk, setup time = 7 hr, and operation time = 2.0 min. For product 4, demand = 400/wk, setup time = 6 hr, and operation time = 3.0 min. The plant normally operates one shift (7.0 hours per shift), 5 days per week and there are currently three work centers in the department. Propose a way of scheduling the machines to meet the weekly demand.

41.12. In the previous problem, propose a way of scheduling to meet the weekly demand if there were four machines instead of three.

41.13. The current date in the production calendar is day 14. There are three orders (A, B, and C) to be processed at a particular work center. The orders

arrived in the sequence A-B-C at the work center. For order A, the remaining process time = 8 days, and the due date is day 24. For order B, the remaining process time = 14 days, and the due date is day 33. For order C, the remaining process time = 6 days, and the due date is day 26. Determine the sequence of the orders that would be scheduled using (a) first-come-first-serve, (b) earliest due date, (c) shortest processing time, (d) least slack time, and (e) critical ratio.

41.14. Five jobs are waiting to be scheduled on a machine. For order A, the remaining process time = 5 days, and the due date is day 8. For order B, the remaining process time = 7 days, and the due date is day 16. For order C, the remaining process time = 11 days, and the due date is day 22. For order D, the remaining process time = 9 days, and the due date is day 31. For order E, the remaining process time = 10 days, and the due date is day 26. Determine a production schedule based on (a) shortest processing time, (b) earliest due date, (c) critical ratio, and (d) least slack time. All times are listed in days.

42 QUALITY CONTROL AND INSPECTION

Chapter Contents

Traditionally, *quality control* (QC) has been concerned with detecting poor quality in manufactured products and taking corrective action to eliminate it. QC has often been limited to inspecting the product and its components, and deciding whether the dimensions and other features conformed to design specifications. If they did, the product was shipped. The modern view of quality control encompasses a broader scope of activities, including various quality programs such as statistical process control and Six Sigma as well as modern inspection technologies such as coordinate measuring machines and machine vision. In this chapter, we discuss these and other quality and inspection topics that are relevant today in modern manufacturing operations. Let us begin our coverage by defining product quality.

42.1 PRODUCT QUALITY

The dictionary defines quality as "the degree of excellence which a thing possesses," or "the features that make something what it is"—its characteristic elements and attributes. The American Society for Quality (ASQ) defines quality as "the totality of features and characteristics of a product or service that bear on its ability to satisfy given needs" [2].

In a manufactured product, quality has two aspects [4]: (1) product features and (2) freedom from deficiencies. *Product features* are the characteristics of the product that result from design. They are the functional and aesthetic features of the item intended to appeal to and provide satisfaction to the customer. In an automobile, these features include the size of the car, its styling, the finish of the body, gas mileage, reliability, reputation of the manufacturer, and similar aspects. They also include the available options for the customer to choose. The sum of a product's features usually defines its *grade*, which relates to the level in the market at which the product is aimed. Cars (and most other products) come in different grades. Some cars provide basic transportation because that is what some customers want, while others are upscale for consumers willing to spend more to own a "better product." The features of a product are decided in design, and they generally determine the inherent cost of the product. Superior features and more of them mean higher cost.

Freedom from deficiencies means that the product does what it is supposed to do (within the limitations of its design features), that it is absent of defects and out-of-tolerance conditions, and that no parts are missing. This aspect of quality includes components and subassemblies of the product as well as the product itself. Freedom from deficiencies means conforming to design specifications, which is accomplished in manufacturing. Although the inherent cost to make a product is a function of its design, minimizing the product's cost to the lowest possible level within the limits set by its design is largely a matter of avoiding defects, tolerance deviations, and other errors during production. Costs of these deficiencies make a long list indeed: scrapped parts, larger lot sizes for scrap allowances, rework, reinspection, sortation, customer complaints and returns, warranty costs and customer allowances, lost sales, and lost good will in the marketplace.

Thus, product features are the aspect of quality for which the design department is responsible. Product features determine to a large degree the price that a company can charge for its products. Freedom from deficiencies is the quality aspect for which the manufacturing departments are responsible. The ability to minimize these deficiencies has an important influence on the cost of the product. These generalities oversimplify the way things work, because the responsibility for high quality extends well beyond the design and manufacturing functions in an organization.

42.2 PROCESS CAPABILITY AND TOLERANCES

In any manufacturing operation, variability exists in the process output. In a machining operation, which is one of the most accurate processes, the machined parts may appear to be identical, but close inspection reveals dimensional differences from one part to the next. Manufacturing variations can be divided into two types: random and assignable.

Random variations are caused by many factors: human variability within each operation cycle, variations in raw materials, machine vibration, and so on. Individually, these factors may not amount to much, but collectively the errors can be significant enough to cause trouble unless they are within the tolerances for the part. Random variations typically form a normal statistical distribution. The output of the process tends to cluster about the mean value, in terms of the product's quality characteristic of interest (e.g., length, diameter). A large proportion of the population of parts is centered around the mean, with fewer parts away from the mean. When the only variations in the process are of this type, the process is said to be in *statistical control*. This kind of variability will continue so long as the process is operating normally. It is when the process deviates from this normal operating condition that variations of the second type appear.

Assignable variations indicate an exception from normal operating conditions. Something has occurred in the process that is not accounted for by random variations. Reasons for assignable variations include operator mistakes, defective raw materials, tool failures, machine malfunctions, and so on. Assignable variations in manufacturing usually betray themselves by causing the output to deviate from the normal distribution. The process is no longer in statistical control.

Process capability relates to the normal variations inherent in the output when the process is in statistical control. By definition, *process capability* equals ± 3 standard deviations about the mean output value (a total of 6 standard deviations),

$$PC = \mu \pm 3\sigma \qquad (42.1)$$

where PC = process capability; μ = process mean, which is set at the nominal value of the product characteristic when bilateral tolerancing is used (Section 5.1.1); and σ = standard deviation of the process. Assumptions underlying this definition are (1) steady state operation has been achieved and the process is in statistical control, and (2) the output is normally distributed. Under these assumptions, 99.73% of the parts produced will have output values that fall within $\pm 3.0\sigma$ of the mean.

The process capability of a given manufacturing operation is not always known, and experiments must be conducted to assess it. Methods are available to estimate the natural tolerance limits based on a sampling of the process.

The issue of tolerances is critical to product quality. Design engineers tend to assign dimensional tolerances to components and assemblies based on their judgment of how size variations will affect function and performance. Conventional wisdom is that closer tolerances beget better performance. Small regard is given to the cost resulting from tolerances that are unduly tight relative to process capability. As tolerance is reduced, the cost of achieving the tolerance tends to increase because additional processing steps may be needed and/or more accurate and expensive production machines may be required. The design engineer should be aware of this relationship. Although primary consideration must be given to function in assigning tolerances, cost is also a factor, and any relief that can be given to the manufacturing departments in the form of wider tolerances without sacrificing product function is worthwhile.

Design tolerances must be compatible with process capability. It serves no useful purpose to specify a tolerance of ± 0.025 mm (± 0.001 in) on a dimension if the process capability is significantly wider than ± 0.025 mm (± 0.001 in). Either the tolerance should be opened further (if design functionality permits), or a different manufacturing process should be selected. Ideally, the specified tolerance should be greater than the process capability. If function and available processes prevent this, then sorting must be included in the manufacturing sequence to inspect every unit and separate those that meet specification from those that do not.

Design tolerances can be specified as being equal to process capability as defined in Eq. (42.1). The upper and lower boundaries of this range are known as the ***natural tolerance limits***. When design tolerances are set equal to the natural tolerance limits, then 99.73% of the parts will be within tolerance and 0.27% will be outside the limits. Any increase in the tolerance range will reduce the percentage of defective parts.

Tolerances are not usually set at their natural limits by product design engineers; tolerances are specified based on the allowable variability that will achieve required function and performance. It is useful to know the ratio of the specified tolerance relative to the process capability. This is indicated by the ***process capability index***

$$PCI = \frac{T}{6\sigma} \qquad (42.2)$$

where PCI = process capability index; T = tolerance range — the difference between upper and lower limits of the specified tolerance; and 6σ = natural tolerance limits. The underlying assumption in this definition is that the process mean is set equal to the nominal design specification, so that the numerator and denominator in Eq. (42.2) are centered about the same value.

Table 42.1 shows the effect of various multiples of standard deviation on defect rate (i.e., proportion of out-of-tolerance parts). The desire to achieve very-low-fraction defect rates has led to the popular notion of "six sigma" limits in quality control. Achieving Six Sigma limits virtually eliminates defects in a manufactured product, assuming the process is maintained within statistical control. As we shall see later in the chapter, Six Sigma quality programs do not quite live up to their names. Before addressing that issue, let us discuss a widely used quality control technique: statistical process control.

TABLE 42.1   Defect rate when tolerance is defined in terms of number of standard deviations of the process, given that the process is operating in statistical control.

No. of Standard Deviations	Process Capability Index	Defect Rate, %	Defective Parts per Million
±1.0	0.333	31.74%	317,400
±2.0	0.667	4.56%	45,600
±3.0	1.00	0.27%	2,700
±4.0	1.333	0.0063%	63
±5.0	1.667	0.000057%	0.57
±6.0	2.00	0.0000002%	0.002

42.3   STATISTICAL PROCESS CONTROL

Statistical process control (SPC) involves the use of various statistical methods to assess and analyze variations in a process. SPC methods include simply keeping records of the production data, histograms, process capability analysis, and control charts. Control charts are the most widely used SPC method, and this section will focus on them.

The underlying principle in control charts is that the variations in any process divide into two types (Section 42.2): (1) random variations, which are the only variations present if the process is in statistical control, and (2) assignable variations that indicate a departure from statistical control. It is the objective of a control chart to identify when the process has gone out of statistical control, thus signaling that some corrective action should be taken.

A ***control chart*** is a graphical technique in which statistics computed from measured values of a certain process characteristic are plotted over time to determine if the process remains in statistical control. The general form of the control chart is illustrated in Figure 42.1. The chart consists of three horizontal lines that remain constant over time: a center, a lower control limit (LCL), and an upper control limit (UCL). The center is usually set at the nominal design value. The upper and lower control limits are generally set at ±3 standard deviations of the sample means.

It is highly unlikely that a random sample drawn from the process will lie outside the upper or lower control limits while the process is in statistical control. Thus, if it happens that a sample value does fall outside these limits, it is interpreted to mean that the process is out of control. Therefore, an investigation is undertaken to determine the

FIGURE 42.1   Control chart.

reason for the out-of-control condition, with appropriate corrective action to eliminate the condition. By similar reasoning, if the process is found to be in statistical control, and there is no evidence of undesirable trends in the data, then no adjustments should be made since they would introduce an assignable variation to the process. The philosophy, "If it ain't broke, don't fix it," is applicable to control charts.

There are two basic types of control charts: (1) control charts for variables and (2) control charts for attributes. Control charts for variables require a measurement of the quality characteristic of interest. Control charts for attributes simply require a determination of whether a part is defective or how many defects there are in the sample.

42.3.1 CONTROL CHARTS FOR VARIABLES

A process that is out of statistical control manifests this condition in the form of significant changes in process mean and/or process variability. Corresponding to these possibilities, there are two principal types of control charts for variables: $\bar{x}$ chart and R chart. The $\bar{x}$ **chart** (call it "x bar chart") is used to plot the average measured value of a certain quality characteristic for each of a series of samples taken from the production process. It indicates how the process mean changes over time. The **R chart** plots the range of each sample, thus monitoring the variability of the process and indicating whether it changes over time.

A suitable quality characteristic of the process must be selected as the variable to be monitored on the $\bar{x}$ and R charts. In a mechanical process, this might be a shaft diameter or other critical dimension. Measurements of the process itself must be used to construct the two control charts.

With the process operating smoothly and absent of assignable variations, a series of samples (e.g., $m = 20$ or more is generally recommended) of small size (e.g., $n = 4, 5,$ or 6 parts per sample) are collected and the characteristic of interest is measured for each part. The following procedure is used to construct the center, LCL, and UCL for each chart:

1. Compute the mean $\bar{x}$ and range R for each of the m samples.
2. Compute the grand mean $\bar{\bar{x}}$, which is the mean of the $\bar{x}$ values for the m samples; this will be the center for the $\bar{x}$ chart.
3. Compute $\bar{R}$, which is the mean of the R values for the m samples; this will be the center for the R chart.
4. Determine the upper and lower control limits, UCL and LCL, for the $\bar{x}$ and R charts. The approach is based on statistical factors tabulated in Table 42.2 that have been derived specifically for these control charts. Values of the factors depend on sample size n. For the $\bar{x}$

TABLE 42.2   Constants for the $\bar{x}$ and R charts.

Sample Size n	$\bar{x}$ Chart A_2	R Chart D_3	R Chart D_4
3	1.023	0	2.574
4	0.729	0	2.282
5	0.577	0	2.114
6	0.483	0	2.004
7	0.419	0.076	1.924
8	0.373	0.136	1.864
9	0.337	0.184	1.816
10	0.308	0.223	1.777

chart:

$$LCL = \bar{\bar{x}} - A_2\overline{R} \quad \text{and} \quad UCL = \bar{\bar{x}} - A_2\overline{R} \qquad (42.3)$$

and for the R chart:

$$LCL = D_3\overline{R} \quad \text{and} \quad UCL = D_4\overline{R} \qquad (42.4)$$

Example 42.1 $\bar{x}$ **and R Charts**

Eight samples ($m = 8$) of size 4 ($n = 4$) have been collected from a manufacturing process that is in statistical control, and the dimension of interest has been measured for each part. It is desired to determine the values of the center, LCL, and UCL for $\bar{x}$ and R charts. The calculated $\bar{x}$ values (units are cm) for the eight samples are 2.008, 1.998, 1.993, 2.002, 2.001, 1.995, 2.004, and 1.999. The calculated R values (cm) are, respectively, 0.027, 0.011, 0.017, 0.009, 0.014, 0.020, 0.024, and 0.018.

Solution: The calculation of $\bar{x}$ and R values above comprise step 1 in our procedure. In step 2, we compute the grand mean of the sample averages.

$$\bar{\bar{x}} = (2.008 + 1.998 + \cdots + 1.999)/8 = 2.000$$

In step 3, the mean value of R is computed.

$$\overline{R} = (0.027 + 0.011 + \cdots + 0.018)/8 = 0.0175$$

In step 4, the values of LCL and UCL are determined based on factors in Table 42.2. First, using Eq. (42.3) for the $\bar{x}$ chart,

$$LCL = 2.000 - 0.729(0.0175) = 1.9872$$
$$UCL = 2.000 + 0.729(0.0175) = 2.0128$$

and for the R chart using Eq. (42.4),

$$LCL = 0(0.0175) = 0$$
$$UCL = 2.282(0.0175) = 0.0399$$

The two control charts are constructed in Figure 42.2 with the sample data plotted in the charts.

42.3.2 CONTROL CHARTS FOR ATTRIBUTES

Control charts for attributes do not use a measured quality variable; instead, they monitor the number of defects present in the sample or the fraction defect rate as the plotted statistic. Examples of these kinds of attributes include number of defects per automobile, fraction of bad parts in a sample, existence or absence of flash in plastic moldings, and number of flaws in a roll of sheet steel. The two principal types of control charts for attributes are the *p chart*, which plots the fraction defect rate in successive samples; and the *c chart*, which plots the number of defects, flaws, or other nonconformities per sample.

p Chart In the *p* chart, the quality characteristic of interest is the proportion (*p* for proportion) of nonconforming or defective units. For each sample, this proportion p_i is the ratio of the number of nonconforming or defective items d_i over the number of units in the

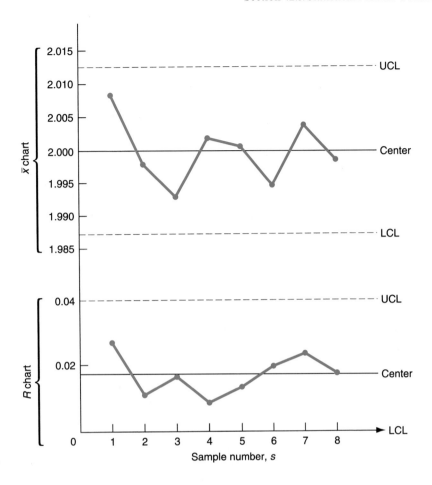

FIGURE 42.2   Control charts for Example 42.2.

sample n (we assume samples of equal size in constructing and using the control chart)

$$p_i = \frac{d_i}{n} \qquad (42.5)$$

where i is used to identify the sample. If the p_i values for a sufficient number of samples are averaged, the mean value $\bar{p}$ is a reasonable estimate of the true value of p for the process. The p chart is based on the binomial distribution, where p is the probability of a nonconforming unit. The center in the p chart is the computed value of $\bar{p}$ for m samples of equal size n collected while the process is operating in statistical control.

$$\bar{p} = \frac{\sum\limits_{i=1}^{m} p_i}{m} \qquad (42.6)$$

The control limits are computed as three standard deviations on either side of the center. Thus,

$$\text{LCL} = \bar{p} - 3\sqrt{\frac{\bar{p}(1-\bar{p})}{n}} \quad \text{and} \quad \text{UCL} = \bar{p} + 3\sqrt{\frac{\bar{p}(1-\bar{p})}{n}} \qquad (42.7)$$

where the standard deviation of $\bar{p}$ in the binomial distribution is given by

$$\sigma_p = \sqrt{\frac{\bar{p}(1-\bar{p})}{n}}$$

If the value of $\bar{p}$ is relatively low and the sample size n is small, then the lower control limit computed by the first of these equations is likely to be a negative value. In this case, let LCL = 0 (the fraction defect rate cannot be less than zero).

c Chart   In the c chart (c for count), the number of defects in the sample are plotted over time. The sample may be a single product such as an automobile, and c = number of quality defects found during final inspection. Or the sample may be a length of carpeting at the factory prior to cutting, and c = number of imperfections discovered in that strip. The c chart is based on the Poisson distribution, where c = parameter representing the number of events occurring within a defined sample space (defects per car, imperfections per specified length of carpet). Our best estimate of the true value of c is the mean value over a large number of samples drawn while the process is in statistical control:

$$\bar{c} = \frac{\sum_{i=1}^{m} c_i}{m} \tag{42.8}$$

This value of $\bar{c}$ is used as the center for the control chart. In the Poisson distribution, the standard deviation is the square root of parameter c. Thus, the control limits are:

$$\text{LCL} = \bar{c} - 3\sqrt{\bar{c}} \quad \text{and} \quad \text{UCL} = \bar{c} + 3\sqrt{\bar{c}} \tag{42.9}$$

42.3.3   INTERPRETING THE CHARTS

When control charts are used to monitor production quality, random samples are drawn from the process of the same size n used to construct the charts. For $\bar{x}$ and R charts, the $\bar{x}$ and R values of the measured characteristic are plotted on the control chart. By convention, the points are usually connected, as in our figures. To interpret the data, one looks for signs that indicate the process is not in statistical control. The most obvious sign is when $\bar{x}$ or R (or both) lie outside the LCL or UCL limits. This indicates an assignable cause such as bad starting materials, new operator, broken tooling, or similar factors. An out-of-limit $\bar{x}$ indicates a shift in the process mean. An out-of-limit R shows that the variability of the process has changed. The usual effect is that R increases, indicating variability has risen. Less obvious conditions may reveal process problems, even though the sample points lie within the $\pm 3\sigma$ limits. These conditions include (1) trends or cyclical patterns in the data, which may mean wear or other factors that occur as a function of time; (2) sudden changes in average level of the data; and (3) points consistently near the upper or lower limits.

The same kinds of interpretations that apply to the $\bar{x}$ chart and R chart are also applicable to the p chart and c chart.

42.4   QUALITY PROGRAMS IN MANUFACTURING

Statistical process control is widely used for monitoring the quality of manufactured parts and products. Several additional quality programs are also used in industry, and in this section we briefly describe four of them: (1) total quality management, (2) Six Sigma, (3) Taguchi methods, and (4) ISO 9000. These programs are not alternatives to statistical process control; in fact, the tools used in SPC are included within the methodologies of total quality management and Six Sigma.

42.4.1   TOTAL QUALITY MANAGEMENT

Total quality management (TQM) is a management approach to quality that pursues three main goals: (1) achieving customer satisfaction, (2) encouraging the involvement of the entire workforce, and (3) continuous improvement.

The customer and customer satisfaction are a central focus of TQM, and products are designed and manufactured with this focus in mind. The product must be designed with the features that customers want, and it must be manufactured free of deficiencies. Within the scope of customer satisfaction is the recognition that there are two categories of customers: (1) external customers and (2) internal customers. External customers are those who purchase the company's products and services. Internal customers are inside the company, such as the company's final assembly department which is the customer of the parts production departments. For the total organization to be effective and efficient, satisfaction must be achieved in both categories of customers.

In TQM, worker involvement in the quality efforts of the organization extends from the top executives through all levels beneath. There is recognition of the important influence that product design has on product quality and how decisions made during design affect the quality that can be achieved in manufacturing. In addition, production workers are made responsible for the quality of their own output, rather than rely on inspectors to uncover defects after the parts are already produced. TQM training, including use of the tools of statistical process control, is provided to all workers. The pursuit of high quality is embraced by every member of the organization.

The third goal of TQM is continuous improvement; that is, adopting the attitude that it is always possible to make something better, whether it is a product or a process. Continuous improvement in an organization is generally implemented using worker teams that have been organized to solve specific problems that are identified in production. The problems are not limited to quality issues. They may include productivity, cost, safety, or any other area of interest to the organization. Team members are selected on the basis of their knowledge and expertise in the problem area. They are drawn from various departments and serve part-time on the team, meeting several times per month until they are able to make recommendations and/or solve the problem. Then the team is disbanded.

42.4.2 SIX SIGMA

The Six Sigma quality program originated and was first used at Motorola Corporation in the 1980s. It has been adopted by many other companies in the United States and was briefly discussed in Section 1.5 as one of the trends in manufacturing. Six Sigma is quite similar to total quality management in its emphasis on management involvement, worker teams to solve specific problems, and the use of SPC tools such as control charts. The major difference between Six Sigma and TQM is that Six Sigma establishes measurable targets for quality based on the number of standard deviations (sigma σ) away from the mean in the Normal distribution. Six sigma implies near perfection in the process in the normal distribution. A process operating at the 6σ level in a Six Sigma program produces no more than 3.4 defects per million, where a defect is anything that might result in lack of customer satisfaction.

As in TQM, worker teams participate in problem-solving projects. A project requires the Six Sigma team to (1) define the problem, (2) measure the process and assess current performance, (3) analyze the process, (4) recommend improvements, and (5) develop a control plan to implement and sustain the improvements. The responsibility of management in Six Sigma is to identify important problems in their operations and sponsor the teams to address those problems.

Statistical Basis of Six Sigma   An underlying assumption in Six Sigma is that the defects in any process can be measured and quantified. Once quantified, the causes of the defects can be identified, and improvements can be made to eliminate or reduce the defects. The effects of any improvements can be assessed using the same measurements in a before-and-after comparison. The comparison is often summarized as a sigma level; for example, the process is now operating at the 4.8-sigma level whereas before it was only operating at the

TABLE 42.3   Sigma Levels and Corresponding Defects Per Million in a Six Sigma Program.

Sigma level	Defects per million	Sigma level	Defects per million	Sigma level	Defects per million	Sigma level	Defects per million
6.0σ	3.4						
5.8σ	8.5	4.8σ	483	3.8σ	10,724	2.8σ	96,801
5.6σ	21	4.6σ	968	3.6σ	17,864	2.6σ	135,666
5.4σ	48	4.4σ	1,866	3.4σ	28,716	2.4σ	184,060
5.2σ	108	4.2σ	3,467	3.2σ	44,565	2.2σ	241,964
5.0σ	233	4.0σ	6,210	3.0σ	66,807	2.0σ	308,538

2.6-sigma level. The relationship between sigma level and defects per million (DPM) is listed in Table 42.3 for a Six Sigma program. We see that the DPM was previously at 135,666 defects per 1,000,000 in our example, whereas it has now been reduced to 108 DPM.

A traditional measure for good process quality is $\pm 3\sigma$ (three sigma level). It implies that the process is stable and in statistical control, and the variable representing the output of the process is normally distributed. Under these conditions, 99.73% of the output will be within the $\pm 3\sigma$ range, and 0.27% or 2700 parts per million will lie outside these limits (0.135% or 1350 parts per million beyond the upper limit and the same number beyond the lower limit). But wait a minute, if we look up 3.0 sigma in Table 42.3, we find that there are 66,807 defects per million. Why is there a difference between the standard normal distribution value (2700 DPM and the value given in Table 42.3 (66,807 DPM)? There are two reasons for this discrepancy. First, the values in Table 42.3 refer to only one tail of the distribution, so that an appropriate comparison with the standard normal tables would only use one tail of the distribution (1350 DPM). Second, and much more significant, is that when Motorola devised the Six Sigma program, they considered the operation of processes over long periods of time, and processes over long periods tend to experience shifts from their original process means. To compensate for these shifts, Motorola decided to adjust the standard normal values by 1.5σ. To summarize, Table 42.3 includes only one tail of the normal distribution, and it shifts the distribution by 1.5 sigma relative to the standard normal distribution. These effects can be seen in Figure 42.3.

Measuring the Sigma Level   In a Six Sigma project, the performance level of the process of interest is reduced to a sigma level. This is done at two points during the project: (1) after measurements have been taken of the process as it is currently operating and (2) after process improvements have been made to assess the effect of the improvements. This provides a before-and-after comparison. High sigma values represent good performance; low sigma values mean poor performance.

To find the sigma level, the number of defects per million must first be determined. There are three measures of defects per million used in Six Sigma. The first and most important is the defects per million opportunities (DPMO), which considers that there may be more than one type of defect that can occur in each unit (product or service). More complex products are likely to have more opportunities for defects, while simple products have fewer opportunities. Thus, DPMO accounts for the complexity of the product and allows entirely different kinds of products or services to be compared. Defects per million opportunities is calculated using the following equation:

$$DPMO = 1,000,000\,\frac{N_d}{N_u N_o} \tag{42.10}$$

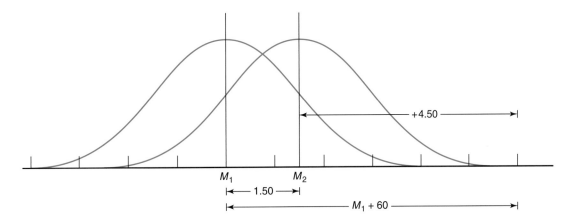

FIGURE 42.3 Normal distribution shift by 1.5σ from original mean and consideration of only one tail of the distribution (at right). Key: μ_1 = mean of original distribution, μ_2 = mean of shifted distribution, σ = standard distribution.

where N_d = total number of defects found, N_u = number of units in the population of interest, and N_o = number of opportunities for a defect per unit. The constant 1,000,000 converts the ratio into defects per million.

Other measures besides DPMO are defects per million (DPM), which measures all of the defects found in the population, and defective units per million (DUPM), which counts the number of defective units in the population and recognizes that there may be more than one type of defect in any defective unit. The following two equations can be used to compute DPM and DUPM:

$$DPM = 1,000,000\,\frac{N_d}{N_u} \tag{42.11}$$

$$DUPM = 1,000,000\,\frac{N_{du}}{N_u} \tag{42.12}$$

where N_{du} = number of defective units in the population, and the other terms are the same as for Eq. (42.10). Once the values of DPMO, DPM, and DUPM have been determined, Table 42.3 can be used to convert these values to their corresponding sigma levels.

Example 42.2 Determining the Sigma Level of a Process

A final assembly plant that makes dishwashers inspects for 23 features that are considered important for overall quality. During the previous month, 9056 dishwashers were produced. During inspection, 479 defects among the 23 features were found, and 226 dishwashers had one or more defect. Determine DPMO, DPM, and DUPM for these data and convert each to its corresponding sigma level.

Solution: Summarizing the data, N_u = 9056, N_o = 23, N_d = 479, and N_{du} = 226. Thus,

$$DPMO = 1,000,000\,\frac{479}{9056(23)} = 2300$$

The corresponding sigma level is about 4.3 from Table 42.3.

$$DPM = 1,000,000\,\frac{479}{9056} = 52,893$$

The corresponding sigma level is about 3.1.

$$DUPM = 1,000,000 \frac{226}{9056} = 24,956$$

The corresponding sigma level is about 3.4.

42.4.3 TAGUCHI METHODS

Genichi Taguchi has had an important influence on the development of quality engineering, especially in the design area—both product design and process design. In this section we review two of the Taguchi methods: (1) the loss function and (2) robust design. More complete coverage can be found among our references [5], [10].

The Loss Function   Taguchi defines quality as "the loss a product costs society from the time the product is released for shipment" [10]. Loss includes costs to operate, failure to function, maintenance and repair costs, customer dissatisfaction, injuries caused by poor design, and similar costs. Some of these losses are difficult to quantify in monetary terms, but they are nevertheless real. Defective products (or their components) that are exposed before shipment are not considered part of this loss. Instead, any expense to the company resulting from scrap or rework of defective product is a manufacturing cost rather than a quality loss.

Loss occurs when a product's functional characteristic differs from its nominal or target value. Although functional characteristics do not translate directly into dimensional features, the loss relationship is most readily understood in terms of dimensions. When the dimension of a component deviates from its nominal value, the component's function is adversely affected. No matter how small the deviation, there is some loss in function. The loss increases at an accelerating rate as the deviation grows, according to Taguchi. If we let $x =$ the quality characteristic of interest and $N =$ its nominal value, then the loss function will be a U-shaped curve as in Figure 42.4(a). A quadratic equation can be used to describe this curve:

$$L(x) = k(x - N)^2 \tag{42.13}$$

where $L(x) =$ loss function; $k =$ constant of proportionality; and x and N are defined above. At some level of deviation $(x_2 - N) = -(x_1 - N)$, the loss will be prohibitive, and it is necessary to scrap or rework the product. This level identifies one possible way of specifying the tolerance limit for the dimension.

In the traditional approach to quality control, tolerance limits are defined and any product within those limits is acceptable. Whether the quality characteristic (e.g., the dimension) is close to the nominal value or close to one of the tolerance limits, it is acceptable. Trying to visualize this approach in terms analogous to the preceding relation, we obtain the discontinuous loss function in Figure 42.4(b). The reality is that products closer to the nominal

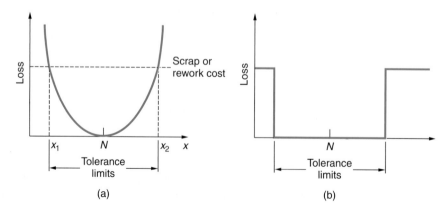

FIGURE 42.4   (a) The quadratic quality loss function. (b) Loss function implicit in traditional tolerance specification.

specification are better quality and will provide greater customer satisfaction. In order to improve quality and customer satisfaction, one must attempt to reduce the loss by designing the product and process to be as close as possible to the target value.

Example 42.3 Taguchi Loss Function

A certain product has a critical dimension that is specified as 20.00 ± 0.04 cm. Repair records indicate that if the tolerance is exceeded, there is a 75% probability that the product will be returned to the manufacturer at a cost of $80 for replacement and shipping. (a) Estimate the constant k in the Taguchi loss function, Eq. (42.13). (b) Using the loss function constant determined in (a), what would be the value of the loss function if the company could maintain a tolerance of ± 0.01 cm instead of ± 0.04 cm?

Solution: In Eq. (42.13), the value of $(x - N)$ is the tolerance 0.04 cm. The loss is the expected cost of replacement and shipping, which is calculated as follows:

$$E\{L(x)\} = 0.75(\$80) + 0.25(0) = \$60$$

Using this expected cost in the loss function, the value of k can be determined as follows:

$$60 = k(0.04)^2 = 0.0016k$$

$$k = 60/0.0016 = \$37,500$$

Accordingly, the Taguchi loss function is $L(x) = 37,500(x - N)$.
(b) For a tolerance of ± 0.01 cm, the loss function is determined as follows:

$$L(x) = 37,500(0.01)^2 = 37,500(0.0001) = \$3.75$$

This is a significant reduction from the $60.00 using a tolerance of ± 0.04 cm. ■

Robust Design   A basic purpose of quality control is to minimize variations. Taguchi calls the variations noise factors. A *noise factor* is a source of variation that is impossible or difficult to control and that affects the functional characteristics of the product. Three types of noise factors can be distinguished: (1) unit-to-unit, (2) internal, and (3) external.

Unit-to-unit noise factors consist of inherent random variations in the process or product caused by variability in raw materials, machinery, and human participation. These are noise factors we have previously called random variations in the process. They are associated with a production process that is in statistical control.

Internal noise factors are sources of variation that are internal to the product or process. They include time-dependent factors such as wear of mechanical components, spoilage of raw materials, and fatigue of metal parts; and operational errors, such as improper settings on the product or machine tool. An *external noise factor* is a source of variation that is external to the product or process, such as outside temperature, humidity, raw material supply, and input voltage. Internal and external noise factors constitute what we have previously called assignable variations.

A *robust design* is one in which the product's function and performance are relatively insensitive to variations in design and manufacturing parameters. It involves the design of both the product and process so that the manufactured product will be relatively unaffected by all noise factors. An example of a robust product design is an automobile whose ignition starter works as well in Minneapolis, Minnesota in winter as in Meridian, Mississippi in summer. An example of robust process design is a metal extrusion operation that produces good product despite temperature variations in the starting billet.

42.4.4 ISO 9000

ISO 9000 is a set of international standards that relate to the quality of the products (and services, if applicable) delivered by a given facility. The standards were developed by the

International Organization for Standardization (ISO), which is based in Geneva, Switzerland. ISO 9000 establishes standards for the systems and procedures used by the facility that determine the quality of its products. ISO 9000 is not a standard for the products themselves. Its focus is on systems and procedures, which include the facility's organizational structure, responsibilities, methods, and resources needed to manage quality. ISO 9000 is concerned with the activities used by the facility to ensure that its products achieve customer satisfaction.

ISO 9000 can be implemented in two ways, formally and informally. Formal implementation means that the facility becomes registered, which certifies that the facility meets the requirements of the standard. Registration is obtained through a third-party agency that conducts on-site inspections and reviews the facility's quality systems and procedures. A benefit of registration is that it qualifies the facility to do business with companies that require ISO 9000 registration, which is common in the European Economic Community where certain products are regulated and ISO 9000 registration is required for companies making these products.

Informal implementation of ISO 9000 means that the facility practices the standards or portions thereof simply to improve its quality systems. Such improvements are worthwhile, even without formal certification, for companies desiring to deliver high quality products.

42.5 INSPECTION PRINCIPLES

Inspection involves the use of measurement and gaging techniques to determine whether a product, its components, subassemblies, or starting materials conform to design specifications. The design specifications are established by the product designer, and for mechanical products they refer to dimensions, tolerances, surface finish, and similar features. Dimensions, tolerances, and surface finish were defined in Chapter 5, and many of the measuring instruments and gages for assessing these specifications were described in that chapter.

Inspection is performed before, during, and after manufacturing. The incoming materials and starting parts are inspected upon receipt from suppliers; work units are inspected at various stages during their production; and the final product should be inspected prior to shipment to the customer.

We should clarify the distinction between inspection and testing, which is a closely related topic. Whereas inspection determines the quality of the product relative to design specifications, testing generally refers to the functional aspects of the product. Does the product operate the way it is supposed to operate? Will it continue to operate for a reasonable period of time? Will it operate in environments of extreme temperature and humidity? In quality control, *testing* is a procedure in which the product, subassembly, part, or material is observed under conditions that might be encountered during service. For example, a product might be tested by operating it for a certain period of time to determine whether it functions properly. If it passes the test, it is approved for shipment to the customer.

Testing of a component or material is sometimes damaging or destructive. In these cases, the items must be tested on a sampling basis. The expense of destructive testing is significant, and great efforts are made to develop methods that do not destroy the item. These methods are referred to as *nondestructive testing* or *nondestructive evaluation*.

Inspections divide into two types: (1) *inspection by variables*, in which the product or part dimensions of interest are measured by the appropriate measuring instruments; and (2) *inspection by attributes*, in which the parts are gauged to determine whether they are within tolerance limits. The advantage of measuring a part dimension is that data are obtained about its actual value. The data might be recorded over time and used to analyze trends in the manufacturing process. Adjustments in the process can be made based on the data so that future parts are produced closer to the nominal design value.

When a part dimension is simply gaged, all that is known is whether it is within tolerance or too big or too small. On the positive side, gaging can be done quickly and at low cost.

42.5.1 MANUAL AND AUTOMATED INSPECTION

Inspection procedures are often performed manually. The work is usually boring and monotonous, yet the need for precision and accuracy is high. Hours are sometimes required to measure the important dimensions of only one part. Because of the time and cost of manual inspection, statistical sampling procedures are generally used to reduce the need to inspect every part.

Sampling versus 100% Inspection   When sampling inspection is used, the number of parts in the sample is generally small compared to the quantity of parts produced. The sample size may be only 1% of the production run. Because not all of the items in the population are measured, there is a risk in any sampling procedure that defective parts will slip through. One of the goals in statistical sampling is to define the expected risk; that is, to determine the average defect rate that will pass through the sampling procedure. The risk can be reduced by increasing the sample size and the frequency with which samples are collected. But the fact remains that 100% good quality cannot be guaranteed in a sampling inspection procedure.

Theoretically, the only way to achieve 100% good quality is by 100% inspection; thus, all defects are screened and only good parts pass through the inspection procedure. However, when 100% inspection is done manually, two problems are encountered. The first is the expense involved. Instead of dividing the cost of inspecting the sample over the number of parts in the production run, the unit inspection cost is applied to every part in the batch. Inspection cost sometimes exceeds the cost of making the part. Second, in 100% manual inspection, there are almost always errors associated with the procedure. The error rate depends on the complexity and difficulty of the inspection task and how much judgment is required by the human inspector. These factors are compounded by operator fatigue. Errors mean that a certain number of poor quality parts will be accepted and a certain number of good quality parts will be rejected. Therefore, 100% inspection using manual methods is no guarantee of 100% good quality product.

Automated 100% Inspection   Automation of the inspection process offers a possible way to overcome the problems associated with 100% manual inspection. *Automated inspection* is defined as automation of one or more steps in the inspection procedure, such as (1) automated presentation of parts by an automated handling system, with a human operator still performing the actual inspection process (e.g., visually examining parts for flaws); (2) manual loading of parts into an automatic inspection machine; and (3) fully automated inspection cell in which parts are both presented and inspected automatically. Inspection automation can also include (4) computerized data collection from electronic measuring instruments.

Automated 100% inspection can be integrated with the manufacturing process to accomplish some action relative to the process. The actions can be one or both of the following: (1) parts sortation, and/or (2) feedback of data to the process. *Parts sortation* means separating parts into two or more quality levels. The basic sortation includes two levels: acceptable and unacceptable. Some situations require more than two levels, such as acceptable, reworkable, and scrap. Sortation and inspection may be combined in the same station. An alternative approach is to locate one or more inspections along the processing line, and instructions are sent to a sortation station at the end of the line indicating what action is required for each part.

Feedback of inspection data to the upstream manufacturing operation allows compensating adjustments to be made in the process to reduce variability and improve quality. If inspection measurements indicate that the output is drifting toward one of the tolerance limits (e.g., due to tool wear), corrections can be made to process parameters to move the output toward the nominal value. The output is thereby maintained within a smaller variability range than possible with sampling inspection methods.

42.5.2 CONTACT VERSUS NONCONTACT INSPECTION

There are a variety of measurement and gaging technologies available for inspection. The possibilities can be divided into contact and noncontact inspection methods. *Contact inspection* involves the use of a mechanical probe or other device that makes contact with the object being inspected. By its nature, contact inspection is usually concerned with measuring or gaging some physical dimension of the part. It is accomplished manually or automatically. Most of the traditional measuring and gaging devices described in Chapter 5 relate to contact inspection. An example of an automated contact measuring system is the coordinate measuring machine (Section 42.6.1).

Noncontact inspection methods utilize a sensor located a certain distance from the object to measure or gage the desired feature(s). Typical advantages of noncontact inspection are (1) faster inspection cycles, and (2) avoidance of damage to the part that might result from contact. Noncontact methods can often be accomplished on the production line without any special handling. By contrast, contact inspection usually requires special positioning of the part, necessitating its removal from the production line. Also, noncontact inspection methods are inherently faster because they employ a stationary probe that does not require positioning for every part. By contrast, contact inspection requires positioning of the contact probe against the part, which takes time.

Noncontact inspection technologies can be classified as optical or nonoptical. Prominent among the optical methods are lasers (Section 42.6.2) and machine vision (Section 42.6.3). Nonoptical inspection sensors include electrical field techniques, radiation techniques, and ultrasonics (Section 42.6.4).

42.6 MODERN INSPECTION TECHNOLOGIES

Advanced technologies are substituting for manual measuring and gaging techniques in modern manufacturing plants. They include contact and noncontact sensing methods. We begin our coverage with an important contact inspection technology: coordinate measuring machines.

42.6.1 COORDINATE MEASURING MACHINES

A coordinate measuring machine (CMM) consists of a contact probe and a mechanism to position the probe in three dimensions relative to surfaces and features of a workpart. See Figure 42.5. The location coordinates of the probe can be accurately recorded as it contacts the part surface to obtain part geometry data.

In a CMM, the probe is fastened to a structure that allows movement of the probe relative to the part, which is fixtured on a worktable connected to the structure. The structure must be rigid to minimize deflections that contribute to measurement errors. The machine in Figure 42.5 has a bridge structure, one of the most common designs. Special features are used in CMM structures to build high accuracy and precision into the measuring machine, including use of low-friction air-bearings and mechanical isolation of the CMM to reduce vibrations. An important aspect in a CMM is the contact probe and

FIGURE 42.5
Coordinate measuring machine. (Courtesy of Brown & Sharpe Manufacturing Company, North Kingstown, Rhode Island.)

its operation. Modern "touch-trigger" probes have a sensitive electrical contact that emits a signal when the probe is deflected from its neutral position in the slightest amount. On contact, the coordinate positions are recorded by the CMM controller, adjusting for overtravel and probe size.

Positioning of the probe relative to the part can be accomplished either manually or under computer control. Methods of operating a CMM can be classified as (1) manual control, (2) manual computer-assisted, (3) motorized computer-assisted, and (4) direct computer control.

In **manual control**, a human operator physically moves the probe along the axes to contact the part and record the measurements. The probe is free-floating for easy movement. Measurements are indicated by digital read-out, and the operator can record the measurement manually or automatically (paper print-out). Any trigonometric calculations must be made by the operator. The **manual computer-assisted** CMM is capable of computer data processing to perform these calculations. Types of computations include simple conversions from U.S customary units to SI, determining the angle between two planes, and determining hole-center locations. The probe is still free-floating to permit the operator to bring it into contact with part surfaces.

Motorized computer-assisted CMMs power drive the probe along the machine axes under operator guidance. A joystick or similar device is used to control the motion. Low-power stepping motors and friction clutches are used to reduce the effects of collisions between probe and part. The *direct computer-control* CMM operates like a CNC machine tool. It is a computerized inspection machine that operates under program control. The basic capability of a CMM is to determine coordinate values where its probe contacts the surface of a part. Computer control permits the CMM to accomplish more sophisticated measurements and inspections, such as (1) determining center location of a hole or cylinder, (2) definition of a plane, (3) measurement of flatness of a surface or parallelism between two surfaces, and (4) measurement of an angle between two planes.

Advantages of using coordinate measuring machines over manual inspection methods include (1) higher productivity—a CMM can perform complex inspection procedures in much less time than traditional manual methods; (2) greater inherent accuracy and precision than conventional methods; and (3) reduced human error through automation of the inspection procedure and associated computations [8]. A CMM is a general-purpose machine that can be used to inspect a variety of part configurations.

42.6.2 MEASUREMENTS WITH LASERS

Recall that laser stands for light amplification by stimulated emission of radiation. Applications of lasers include cutting (Section 26.3.3) and welding (Section 30.4). These applications involve the use of solid-state lasers capable of focusing sufficient power to melt or sublimate the work material. Lasers for measurement applications are low-power gas lasers such as helium-neon, which emits light in the visible range. The light beam from a laser is (1) highly monochromatic, which means the light has a single wave length, and (2) highly collimated, which means the light rays are parallel. These properties have motivated a growing list of laser applications in measurement and inspection. We describe two here.

Scanning Laser Systems   The scanning laser uses a laser beam deflected by a rotating mirror to produce a beam of light that sweeps past an object, as in Figure 42.6. A photodetector on the far side of the object senses the light beam during its sweep except for the short time when it is interrupted by the object. This time period can be measured

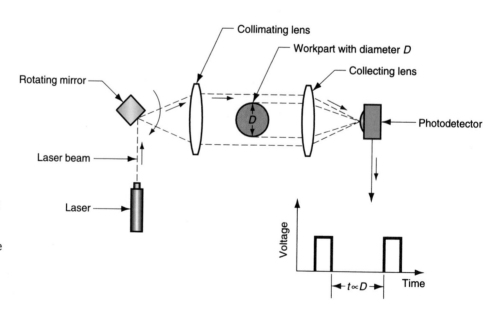

FIGURE 42.6 Scanning laser system for measuring diameter of cylindrical workpart; time of interruption of light beam is proportional to diameter *D*.

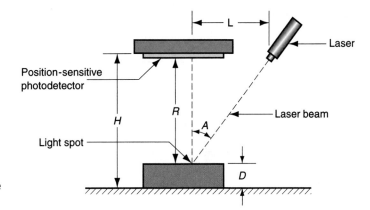

FIGURE 42.7 Laser triangulation to measure part dimension *D*.

quickly with great accuracy. A microprocessor system measures the time interruption that is related to the size of the object in the path of the laser beam, and converts the time to a linear dimension. Scanning laser beams can be applied in high production on-line inspection and gaging. Signals can be sent to production equipment to make adjustments in the process and/or activate a sortation device on the production line. Applications of scanning laser systems include rolling-mill operations, wire extrusion, machining, and grinding.

Laser Triangulation   Triangulation is used to determine the distance of an object from two known locations by means of trigonometric relationships of a right triangle. The principle can be applied in dimensional measurements using a laser system, as in Figure 42.7. The laser beam is focused on an object to form a light spot on the surface. A position-sensitive optical detector is used to determine the location of the spot. The angle *A* of the beam directed at the object and the distance *H* are fixed and known. Given that the photodetector is located a fixed distance above the worktable, the part depth *D* in the setup of Figure 42.7 is determined from

$$D = H - R = H - L \cot A \qquad (42.14)$$

where L is determined by the position of the light spot on the workpart.

42.6.3 MACHINE VISION

Machine vision involves the acquisition, processing, and interpretation of image data by computer for some useful application. Vision systems can be classified as two dimensional or three dimensional. Two-dimensional systems view the scene as a 2-D image, which is quite adequate for applications involving a planar object. Examples include dimensional measuring and gaging, verifying the presence of components, and checking for features on a flat (or almost flat) surface. Three-dimensional vision systems are required for applications requiring a 3-D analysis of the scene, where contours or shapes are involved. The majority of current applications are 2-D, and our discussion will focus (excuse the pun) on this technology.

Operation of Machine Vision Systems   Operation of a machine vision system consists of three steps, depicted in Figure 42.8: (1) image acquisition and digitization, (2) image processing and analysis, and (3) interpretation.

Image acquisition and digitizing is accomplished by a video camera connected to a digitizing system to store the image data for subsequent processing. With the camera focused on the subject, an image is obtained by dividing the viewing area into a matrix of discrete picture elements (called *pixels*), in which each element assumes a value proportional to the

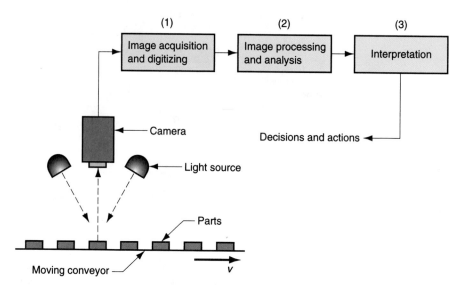

FIGURE 42.8 Operation of a machine vision system.

light intensity of that portion of the scene. The intensity value for each pixel is converted to its equivalent digital value by analog-to-digital conversion. Image acquisition and digitizing is depicted in Figure 42.9 for a ***binary vision*** system, in which the light intensity is reduced to either of two values (black or white = 0 or 1), as in Table 42.4. The pixel matrix in our illustration is only 12 × 12; a real vision system would have many more pixels for better resolution. Each set of pixel values is a ***frame***, which consists of the set of digitized pixel values. The frame is stored in computer memory. The process of reading all the pixel values in a frame is performed 30 times per second in United States, 25 cycle/s in European systems.

The ***resolution*** of a vision system is its ability to sense fine details and features in the image. This depends on the number of pixels used. Common pixel arrays include 640 (horizontal) × 480 (vertical), 1024 × 768, or 1040 × 1392 picture elements. The more pixels in the vision system, the higher its resolution. However, system cost increases as pixel count increases. Also, time required to read the picture elements and process the data increases with number of pixels. In addition to binary vision systems, more sophisticated vision systems distinguish various gray levels in the image that permit them to determine surface characteristics such as texture. Called ***gray-scale vision***, these systems typically use 4, 6, or 8 bits of memory. Other vision systems can recognize colors.

FIGURE 42.9 Image acquisition and digitizing: (a) the scene consists of a dark-colored part against a light background; (b) a 12 × 12 matrix of pixels imposed on the scene.

(a)

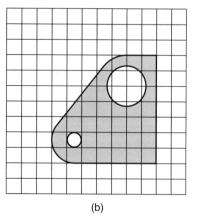

(b)

TABLE 42.4		Pixel values in a binary vision system for the image in Figure 42.8.									
1	1	1	1	1	1	1	1	1	1	1	1
1	1	1	1	1	1	1	1	1	1	1	1
1	1	1	1	1	1	1	1	1	1	1	1
1	1	1	1	1	1	1	0	0	0	1	1
1	1	1	1	1	1	0	1	1	0	1	1
1	1	1	1	1	0	0	1	1	0	1	1
1	1	1	1	0	0	0	0	0	0	1	1
1	1	1	0	0	0	0	0	0	0	1	1
1	1	1	0	1	0	0	0	0	0	1	1
1	1	1	1	0	0	0	0	0	0	1	1
1	1	1	1	1	1	1	1	1	1	1	1
1	1	1	1	1	1	1	1	1	1	1	1

The second function in machine vision is *image processing and analysis*. The data for each frame must be analyzed within the time required to complete one scan (1/30 s or 1/25 s). Several techniques have been developed to analyze image data, including edge detection and feature extraction. *Edge detection* involves determining the locations of boundaries between an object and its surroundings. This is accomplished by identifying contrast in light intensity between adjacent pixels at the borders of the object. *Feature extraction* is concerned with determining feature values of an image. Many machine vision systems identify an object in the image by means of its features. Features of an object include area, length, width, or diameter of the object, perimeter, center of gravity, and aspect ratio. Feature extraction algorithms are designed to determine these features based on the object's area and boundaries. Area of an object can be determined by counting the number of pixels that make up the object. Length can be found by measuring the distance (in pixels) between two opposite edges of the part.

Interpretation of the image is the third function. It is accomplished by extracted features. Interpretation is usually concerned with recognizing the object—identifying the object in the image by comparing it to predefined models or standard values. One common interpretation technique is *template matching*, which refers to methods that compare one or more features of an image with corresponding features of a model (template) stored in computer memory.

Machine Vision Applications   The interpretation function in machine vision is generally related to applications, which divide into four categories: (1) inspection, (2) part identification, (3) visual guidance and control, and (4) safety monitoring.

Inspection is the most important category, accounting for about 90% of all industrial applications. The applications are in mass production, where the time to program and install the system can be divided by many thousands of units. Typical inspection tasks include: (1) *dimensional measurement or gaging*, which involves measuring or gaging certain dimensions of parts or products moving along a conveyor; (2) *verification functions*, which include verifying presence of components in an assembled product, presence of a hole in a workpart, and similar tasks; and (3) *identification of flaws and defects*, such as identifying flaws in a printed label in the form of mislocation, poorly printed text, numbering, or graphics on the label.

Part identification applications include counting different parts flowing past on a conveyor, part sorting, and character recognition. *Visual guidance and control* involves a vision system interfaced with a robot or similar machine to control the movement of the machine. Examples include seam tracking in continuous arc welding, part positioning, part reorientation, and picking parts from a bin. In *safety monitoring* applications, the vision system monitors the production operation to detect irregularities that might indicate a hazardous condition to equipment or humans.

42.6.4 OTHER NONCONTACT INSPECTION TECHNIQUES

In addition to optical inspection methods, there are various nonoptical techniques used in inspection. These include sensor techniques based on electrical fields, radiation, and ultrasonics.

Under certain conditions, *electrical fields* created by an electrical probe can be used for inspection. The fields include reluctance, capacitance, and inductance; they are affected by an object in the vicinity of the probe. In a typical application, the workpart is positioned in a fixed relationship to the probe. By measuring the effect of the object on the electrical field, an indirect measurement of certain part characteristics can be made, such as dimensional features, thickness of sheet material, and flaws (cracks and voids below the surface) in the material.

Radiation techniques use X-ray radiation to inspect metals and weldments. The amount of radiation absorbed by the metal object indicates thickness and presence of flaws in the part or welded section. For example, X-ray inspection is used to measure thickness of sheet metal in rolling (Section 19.1). Data from the inspection is used to adjust the gap between rolls in the rolling mill.

Ultrasonic techniques use high-frequency sound (>20,000 Hz) to perform various inspection tasks. One of the techniques analyses the ultrasonic waves emitted by a probe and reflected off the object. During the setup for the inspection procedure, an ideal test part is positioned in front of the probe to obtain a reflected sound pattern. This sound pattern is used as the standard against which production parts are subsequently compared. If the reflected pattern from a given part matches the standard, the part is accepted. If a match is not obtained, the part is rejected.

REFERENCES

[1] DeFeo, J. A., Gryna, F. M., and Chua, R. C. H. *Juran's Quality Planning and Analysis for Enterprise Quality,* 5th ed., McGraw-Hill, New York, 2006.

[2] Evans, J. R., and Lindsay, W. M. *The Management and Control of Quality,* 6th ed. Thomson/South-Western College Publishing Company, Mason, Ohio, 2005.

[3] Groover, M. P. *Automation, Production Systems, and Computer Integrated Manufacturing,* 3rd ed. Prentice Hall, Upper Saddle River, New Jersey, 2008.

[4] Juran, J. M., and Gryna, F. M. *Quality Planning and Analysis,* 3rd ed. McGraw-Hill, New York, 1993.

[5] Lochner, R. H., and Matar, J. E. *Designing for Quality.* ASQC Quality Press, Milwaukee, Wisconsin, 1990.

[6] Montgomery, D. C. *Introduction to Statistical Quality Control,* 6th ed. John Wiley & Sons, Inc., Hoboken, New Jersey, 2008.

[7] Pyzdek, T., and Keller, P. *Quality Engineering Handbook.* 2nd ed. CRC Taylor & Francis, Boca Raton, Florida, 2003.

[8] Schaffer, G. H. "Taking the Measure of CMMs." Special Report 749, *American Machinist,* October 1982, pp. 145–160.

[9] Schaffer, G. H. "Machine Vision: A Sense for CIM." Special Report 767, *American Machinist,* June 1984, pp. 101–120.

[10] Taguchi, G., Elsayed, E. A., and Hsiang, T. C. *Quality Engineering in Production Systems.* McGraw-Hill, New York, 1989.

[11] Wick, C., and Veilleux, R. F. *Tool and Manufacturing Engineers Handbook,* 4th ed. Vol. IV, *Quality Control and Assembly.* Society of Manufacturing Engineers, Dearborn, Michigan, 1987.

REVIEW QUESTIONS

42.1. What are the two principal aspects of product quality?

42.2. How is a process operating in statistical control distinguished from one that is not?

42.3. Define process capability.

42.4. What are the natural tolerance limits?

42.5. What is the difference between control charts for variables and control charts for attributes?

42.6. Identify the two types of control charts for variables.

42.7. What are the two basic types of control charts for attributes?

42.8. When interpreting a control chart, what does one look for to identify problems?

42.9. What are the three main goals in total quality management (TQM)?

42.10. What is the difference between external customers and internal customers in TQM?

42.11. At what company was the Six Sigma quality program first used?

42.12. Why is the normal statistical table used in a Six Sigma program different from the standard normal tables found in textbooks on probability and statistics?

42.13. A Six Sigma program uses three measures of defects per million (DPM) to assess the performance of a given process. Name the three measures of DPM.

42.14. What is meant by robust design, as defined by Taguchi?

42.15. Automated inspection can be integrated with the manufacturing process to accomplish certain actions. What are these possible actions?

42.16. Give an example of a noncontact inspection technique.

42.17. What is a coordinate measuring machine?

42.18. Describe a scanning laser system.

42.19. What is a binary vision system?

42.20. Name some of the nonoptical noncontact sensor technologies available for inspection.

MULTIPLE CHOICE QUIZ

There are 23 correct answers in the following multiple choice questions (some questions have multiple answers that are correct). To attain a perfect score on the quiz, all correct answers must be given. Each correct answer is worth 1 point. Each omitted answer or wrong answer reduces the score by 1 point, and each additional answer beyond the correct number of answers reduces the score by 1 point. Percentage score on the quiz is based on the total number of correct answers.

42.1. Which of the following quality aspects would be classified as examples of freedom from deficiencies rather than product features (two correct answers): (a) components within tolerance, (b) location of on/off switch, (c) no missing parts, (d) product weight, (e) reliability, and (f) reputation of the company?

42.2. If the product tolerance is set so that the process capability index = 1.0, then the percentage of parts that are within tolerance will be closest to which one of the following when the process is operating in statistical control: (a) 35%, (b) 65%, (c) 95%, (d) 99%, or (e) 100%?

42.3. In a control chart, the upper control limit is set equal to which one of the following: (a) process mean, (b) process mean plus three standard deviations, (c) upper design tolerance limit, or (d) upper value of the maximum range R?

42.4. The R chart is used for which one of the following product or part characteristics: (a) number of rejects in the sample, (b) number of reworked parts in a sample, (c) radius of a cylindrical part, or (d) range of sample values?

42.5. Which one of the following best describes the situations for which the c chart is most suited: (a) control of defective parts, (b) mean value of part characteristic of interest, (c) number of defects in a sample, or (d) proportion of defects in a sample?

42.6. Which of the following identify a likely out-of-control condition in a control chart (three correct answers): (a) consistently increasing value of $\bar{x}$, (b) points near the central line, (c) points oscillating back and forth across the central line, (d) R outside the control limits of the R chart, (e) sample points consistently slightly above the central line, and (f) $\bar{x}$ outside the control limits of the $\bar{x}$ chart?

42.7. Which of the following are the three main goals in a total quality management program: (a) achieving customer satisfaction, (b) computing defects per million, (c) continuous improvement, (d) developing robust product and process designs, (e) encouraging the involvement of the entire workforce, (f) forming worker teams, (g) statistical process control, and (h) zero defects?

42.8. Which one of the following measures in a Six Sigma program allows products of different complexity to be directly compared: (a) defects per million units, (b) defects per million opportunities, or (c) defective units per million units?

42.9. Which of the following principles and/or approaches are generally credited to G. Taguchi (two correct answers): (a) acceptance sampling, (b) control charts, (c) loss function, (d) Pareto priority index, and (e) robust design?

42.10. Which of the following phrases relating to ISO 9000 are correct (three correct answers): (a) certified by the International Standards Office located in Geneva, Switzerland, (b) developed by the International Organization for Standardization located somewhere in Europe, (c) establishes standards for the quality systems and procedures used by a facility, (d) establishes standards for the products and services delivered by a facility, and (e) registration in ISO 9000 obtained through a third-party agency that certifies the facility's quality systems?

42.11. The two basic types of inspection are inspection by variables and inspection by attributes. The second of these inspections uses which one of the following: (a) destructive testing, (b) gaging, (c) measuring, or (d) nondestructive testing?

42.12. Automated 100% inspection can be integrated with the manufacturing process to accomplish which of the following (two best answers): (a)

better design of products, (b) feedback of data to adjust the process, (c) 100% perfect quality, and (d) sortation of good parts from defects?

42.13. Which one of the following is an example of contact inspection: (a) coordinate measuring systems, (b) machine vision, (c) radiation techniques, (d) scanning laser systems, and (e) ultrasonic techniques?

42.14. Which one of the following is the most important application of vision systems: (a) inspection, (b) object identification, (c) safety monitoring, or (d) visual guidance and control of a robotic manipulator?

PROBLEMS

Note: Problems identified with an asterisk (*) in this set require the use of statistical tables not included in this text.

Process Capability and Tolerances

42.1. An automatic turning process is set up to produce parts with a mean diameter = 6.255 cm. The process is in statistical control and the output is normally distributed with a standard deviation = 0.004 cm. Determine the process capability.

42.2. *In Problem 42.1, the design specification on the part is: diameter = 6.250 ± 0.013 cm. (a) What proportion of parts fall outside the tolerance limits? (b) If the process were adjusted so that its mean diameter = 6.250 cm and the standard deviation remained the same, what proportion of parts would fall outside the tolerance limits?

42.3. A sheet-metal bending operation produces bent parts with an included angle = 92.1°. The process is in statistical control and the values of included angle are normally distributed with a standard deviation = 0.23°. The design specification on

the angle = 90 ± 2°. (a) Determine the process capability. (b) If the process could be adjusted so that its mean = 90.0°, determine the value of the process capability index.

42.4. A plastic extrusion process produces round extrudate with a mean diameter = 28.6 mm. The process is in statistical control and the output is normally distributed with standard deviation = 0.53 mm. Determine the process capability.

42.5. *In Problem 42.4, the design specification on the diameter is 28.0 ± 2.0 mm. (a) What proportion of parts fall outside the tolerance limits? (b) If the process were adjusted so that its mean diameter = 28.0 mm and the standard deviation remained the same, what proportion of parts would fall outside the tolerance limits? (c) With the adjusted mean at 28.0 mm, determine the value of the process capability index.

Control Charts

42.6. In 12 samples of size $n = 7$, the average value of the sample means is $\bar{\bar{x}} = 6.860$ cm for the dimension of interest, and the mean of the ranges of the samples is $\bar{R} = 0.027$ cm. Determine (a) lower and upper control limits for the $\bar{x}$ chart and (b) lower and upper control limits for the R chart. (c) What is your best estimate of the standard deviation of the process?

42.7. In nine samples of size $n = 10$, the grand mean of the samples is $\bar{\bar{x}} = 100$ for the characteristic of interest, and the mean of the ranges of the samples is $\bar{R} = 8.5$. Determine (a) lower and upper control limits for the $\bar{x}$ chart and (b) lower and upper control limits for the R chart. (c) Based on the data given, estimate the standard deviation of the process?

42.8. Ten samples of size $n = 8$ have been collected from a process in statistical control, and the dimension of interest has been measured for each part. The calculated values of $\bar{x}$ for each sample are (mm) 9.22, 9.15, 9.20, 9.28, 9.19, 9.12, 9.20, 9.24, 9.17, and 9.23. The values of R are (mm) 0.24, 0.17, 0.30, 0.26, 0.26, 0.19, 0.21, 0.32, 0.21, and 0.23, respectively. (a) Determine the values of the center, LCL, and UCL for the $\bar{x}$ and R charts. (b) Construct the control charts and plot the sample data on the charts.

42.9. Seven samples of 5 parts each have been collected from an extrusion process that is in statistical control, and the diameter of the extrudate has been measured for each part. The calculated values of $\bar{x}$ for each sample are (in) 1.002, 0.999, 0.995, 1.004, 0.996,

0.998, and 1.006. The values of R are (in) 0.010, 0.011, 0.014, 0.020, 0.008, 0.013, and 0.017, respectively. (a) Determine the values of the center, LCL, and UCL for $\bar{x}$ and R charts. (b) Construct the control charts and plot the sample data on the charts.

42.10. A p chart is to be constructed. Six samples of 25 parts each have been collected, and the average number of defects per sample was 2.75. Determine the center, LCL and UCL for the p chart.

42.11. Ten samples of equal size are taken to prepare a p chart. The total number of parts in these ten samples was 900 and the total number of defects counted was 117. Determine the center, LCL and UCL for the p chart.

42.12. The yield of good chips during a certain step in silicon processing of integrated circuits averages

91%. The number of chips per wafer is 200. Determine the center, LCL, and UCL for the p chart that might be used for this process.

42.13. The upper and lower control limits for a p chart are: LCL = 0.19 and UCL = 0.24. Determine the sample size n that is used with this control chart.

42.14. The upper and lower control limits for a p chart are: LCL = 0 and UCL = 0.20. Determine the minimum possible sample size n that is compatible with this control chart.

42.15. Twelve cars were inspected after final assembly. The number of defects found ranged between 87 and 139 defect per car with an average of 116. Determine the center and upper and lower control limits for the c chart that might be used in this situation.

Quality Programs

42.16. A foundry that casts turbine blades inspects for eight features that are considered critical-to-quality. During the previous month, 1236 castings were produced. During inspection, 47 defects among the eight features were found, and 29 castings had one or more defects. Determine DPMO, DPM, and DUPM in a Six Sigma program for these data and convert each to its corresponding sigma level.

42.17. In the previous problem, if the foundry desired to improve its quality performance to the 5.0 sigma level in all three measures of DPM, how many defects and defective units would they produce in an annual production quantity of 15,000 castings? Assume the same eight features are used to assess quality.

42.18. The inspection department in an automobile final assembly plant inspects cars coming off the production line against 55 quality features considered important to customer satisfaction. The department counts the number of defects found per 100 cars, which is the same type of metric used by a national consumer advocate agency. During a 1-month period, a total of 16,582 cars rolled off the assembly line. These cars included a total of 6045 defects of the 55 features, which translates to 36.5 defects per 100 cars. In addition, a total of 1955 cars

had one or more of the defects during this month. Determine DPMO, DPM, and DUPM in a Six Sigma program for these data and convert each to its corresponding sigma level.

42.19. A company produces a certain part whose most important dimension is 37.50 ± 0.025 in. If the tolerance is exceeded, the customer will return the part to the manufacturer at a cost of $200 in rework and replacement expenses. (a) Determine the constant k in the Taguchi loss function, Eq. (42.13). (b) The company can add a finish grinding operation that will allow the tolerance to be reduced to ±0.010 in. Using the loss function from part (a) what is the value of the loss associated with this new tolerance?

42.20. The additional operation in the preceding problem will add $2.00 to the current cost of the part, which is $13.50. If the rate of returns from the customer at the tolerance of ±0.025 in is 2.1%, and it is expected to drop to zero returns using the new tolerance, should the company add the finish grinding operation to the manufacturing sequence for the part? Answer this question using the basic cost and return rate data without consideration of the Taguchi loss function.

Laser Measurement Technologies

42.21. A laser triangulation system has the laser mounted at a 35° angle from the vertical. The distance between the worktable and the photodetector is 24.0000 in. Determine (a) the distance between the laser and the photodetector when no part is present and (b) the height of a part when the distance between the laser and photo-detector is 12.0250 in.

42.22. A laser triangulation system is used to determine

the height of a steel block. The system has a photosensitive detector that is located 750.000 mm above the working surface and the laser is mounted at a 30.00° angle from the vertical. With no part on the worktable, the position of the laser reflection on the photo sensor is recorded. After a part is placed on the worktable, the laser reflection shifts 70.000 mm toward the laser. Determine the height of the object.

INDEX